ATOMIC MASSES OF THE ELEMENTS

Name	Symbol	Atomic Number	Atomic Mass[a]	Name	Symbol	Atomic Number	Atomic Mass[a]
Actinium	Ac	89	(227)[b]	Mendelevium	Md	101	(258)
Aluminum	Al	13	26.98	Mercury	Hg	80	200.6
Americium	Am	95	(243)	Molybdenum	Mo	42	95.94
Antimony	Sb	51	121.8	Neodymium	Nd	60	144.2
Argon	Ar	18	39.95	Neon	Ne	10	20.18
Arsenic	As	33	74.92	Neptunium	Np	93	(237)
Astatine	At	85	(210)	Nickel	Ni	28	58.69
Barium	Ba	56	137.3	Niobium	Nb	41	92.91
Berkelium	Bk	97	(247)	Nitrogen	N	7	14.01
Beryllium	Be	4	9.012	Nobelium	No	102	(259)
Bismuth	Bi	83	209.0	Osmium	Os	76	190.2
Bohrium	Bh	107	(264)	Oxygen	O	8	16.00
Boron	B	5	10.81	Palladium	Pd	46	106.4
Bromine	Br	35	79.90	Phosphorus	P	15	30.97
Cadmium	Cd	48	112.4	Platinum	Pt	78	195.1
Calcium	Ca	20	40.08	Plutonium	Pu	94	(244)
Californium	Cf	98	(251)	Polonium	Po	84	(209)
Carbon	C	6	12.01	Potassium	K	19	39.10
Cerium	Ce	58	140.1	Praseodymium	Pr	59	140.9
Cesium	Cs	55	132.9	Promethium	Pm	61	(145)
Chlorine	Cl	17	35.45	Protactinium	Pa	91	231.0
Chromium	Cr	24	52.00	Radium	Ra	88	(226)
Cobalt	Co	27	58.93	Radon	Rn	86	(222)
Copernicium	Cn	112	(285)	Rhenium	Re	75	186.2
Copper	Cu	29	63.55	Rhodium	Rh	45	102.9
Curium	Cm	96	(247)	Roentgenium	Rg	111	(272)
Darmstadtium	Ds	110	(271)	Rubidium	Rb	37	85.47
Dubnium	Db	105	(262)	Ruthenium	Ru	44	101.1
Dysprosium	Dy	66	162.5	Rutherfordium	Rf	104	(261)
Einsteinium	Es	99	(252)	Samarium	Sm	62	150.4
Erbium	Er	68	167.3	Scandium	Sc	21	44.96
Europium	Eu	63	152.0	Seaborgium	Sg	106	(266)
Fermium	Fm	100	(257)	Selenium	Se	34	78.96
Flerovium	Fl	114	(289)	Silicon	Si	14	28.09
Fluorine	F	9	19.00	Silver	Ag	47	107.9
Francium	Fr	87	(223)	Sodium	Na	11	22.99
Gadolinium	Gd	64	157.3	Strontium	Sr	38	87.62
Gallium	Ga	31	69.72	Sulfur	S	16	32.07
Germanium	Ge	32	72.64	Tantalum	Ta	73	180.9
Gold	Au	79	197.0	Technetium	Tc	43	(99)
Hafnium	Hf	72	178.5	Tellurium	Te	52	127.6
Hassium	Hs	108	(265)	Terbium	Tb	65	158.9
Helium	He	2	4.003	Thallium	Tl	81	204.4
Holmium	Ho	67	164.9	Thorium	Th	90	232.0
Hydrogen	H	1	1.003	Thulium	Tm	69	168.9
Indium	In	49	114.8	Tin	Sn	50	118.7
Iodine	I	53	126.9	Titanium	Ti	22	47.87
Iridium	Ir	77	192.2	Tungsten	W	74	183.8
Iron	Fe	26	55.85	Uranium	U	92	238.0
Krypton	Kr	36	83.80	Vanadium	V	23	50.94
Lanthanum	La	57	138.9	Xenon	Xe	54	131.3
Lawrencium	Lr	103	(262)	Ytterbium	Yb	70	173.0
Lead	Pb	82	207.2	Yttrium	Y	39	88.91
Lithium	Li	3	6.941	Zinc	Zn	30	65.41
Livermorium	Lv	116	(293)	Zirconium	Zr	40	91.22
Lutetium	Lu	71	175.0	—	—	113	(284)
Magnesium	Mg	12	24.31	—	—	115	(288)
Manganese	Mn	25	54.94	—	—	117	(293)
Meitnerium	Mt	109	(268)	—	—	118	(294)

[a]Values for atomic masses are given to four significant figures.
[b]Values in parentheses are the mass number of an important radioactive isotope.

CHEMISTRY

KAREN C. TIMBERLAKE

CHEMISTRY

An Introduction to General, Organic, and Biological Chemistry

ELEVENTH EDITION

Prentice Hall

Upper Saddle River Boston Columbus Indianapolis New York San Francisco
Amsterdam Cape Town Dubai London Madrid Milan Munich Paris Montréal Toronto
Delhi Mexico City São Paulo Sydney Hong Kong Seoul Singapore Taipei Tokyo

Editor in Chief: *Adam Jaworski*
Marketing Manager: *Erin Gardner*
Associate Editor: *Jessica Neumann*
Editorial Assistant: *Lisa Tarabokjia*
Marketing Assistant: *Nicola Houston*
Managing Editor, Chemistry and Geosciences: *Gina M. Cheselka*
Senior Project Manager, Production: *Beth Sweeten*
Senior Media Producer: *Angela Bernhardt*
Mastering Media Producer: *Kate Brayton*
Senior Media Production Supervisor: *Liz Winer*
Media Editor: *David Chavez*
Senior Technical Art Specialist: *Connie Long*
Art Studio: *Precision Graphics*
Art Director: *Derek Bacchus*
Interior and Cover Design: *Tamara Newnam*
Senior Manufacturing and Operations Manager: *Nick Sklitsis*
Operations Specialist: *Maura Zaldivar*
Image Permissions Coordinator: *Elaine Soares*
Photo Researcher: *Eric Shrader*
Production Supervision/Composition: *Prepare, Inc.*
Cover Photograph: David Alayeto/Getty Images

Credits and acknowledgments borrowed from other sources and reproduced, with permission, in this textbook appear on p. C-1.

Library of Congress Cataloging-in-Publication Data

Timberlake, Karen C.
Chemistry : an introduction to general, organic, and biological chemistry / Karen C. Timberlake.—11th ed.
p. cm.
Includes bibliographical references and index.
ISBN 978-0-321-69345-7 (alk. paper)
1. Chemistry—Textbooks. I. Title.
QD31.3.T55 2012
540—dc22

2010029053

ISBN-10: 0-321-69345-0
ISBN-13: 978-0-321-69345-7

Printed in the United States

10 9 8 7 6 5 4 3

Prentice Hall
is an imprint of

www.pearsonhighered.com

BRIEF CONTENTS

CONTENTS

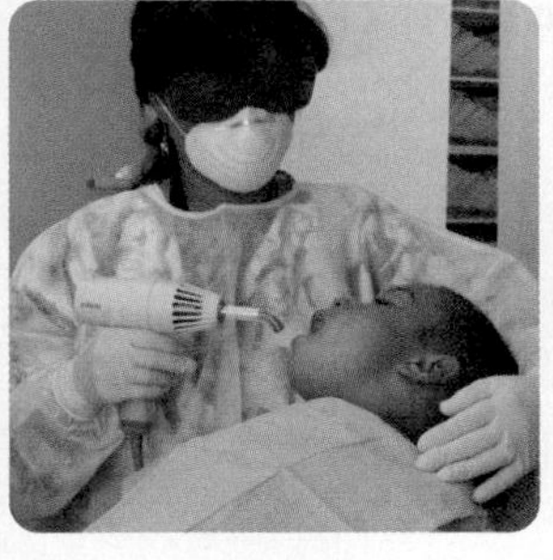

4 Compounds and Their Bonds 125

5 Chemical Quantities and Reactions 167

6 Gases 213

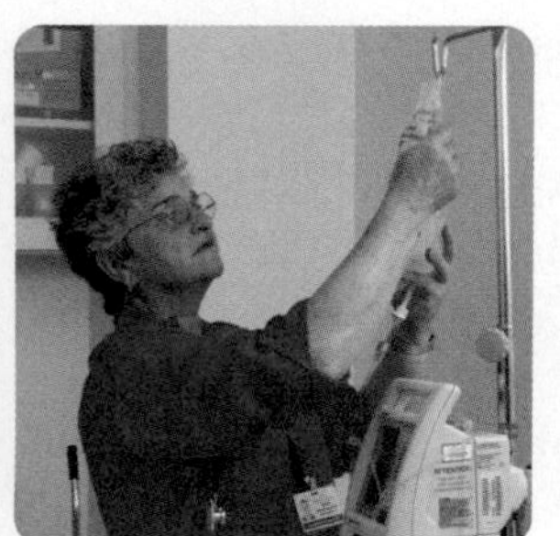

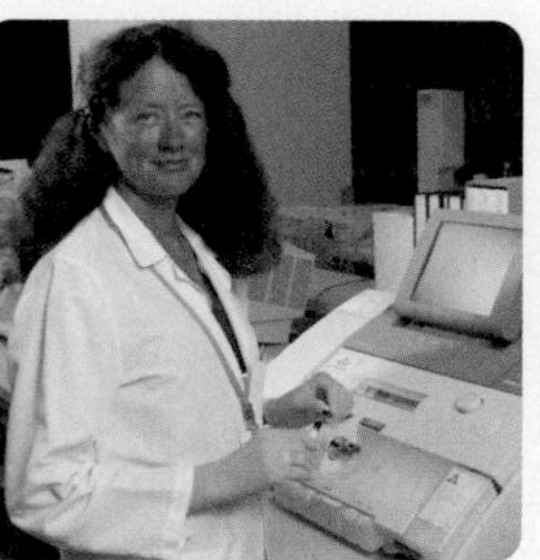

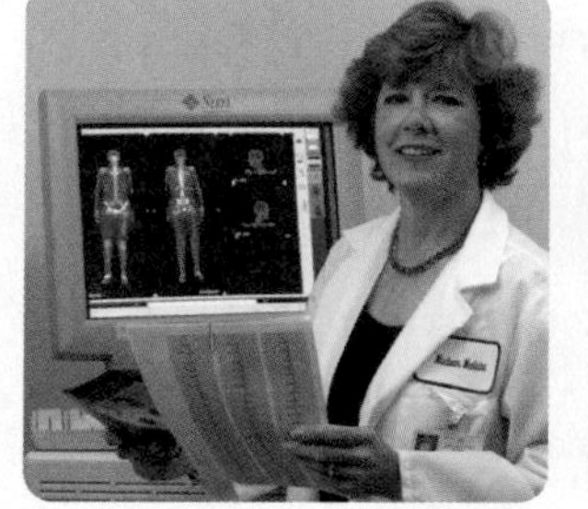

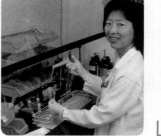

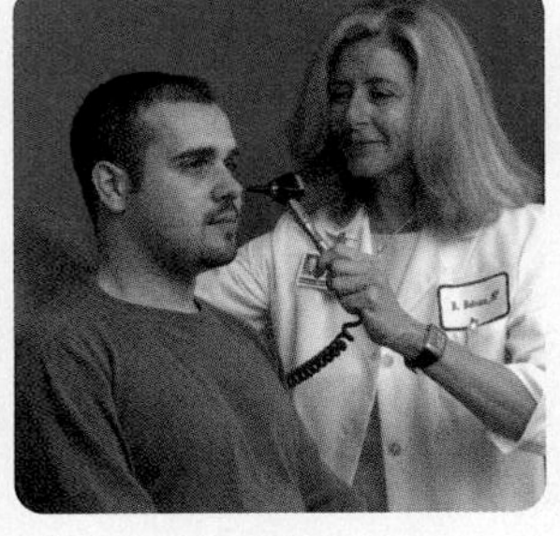

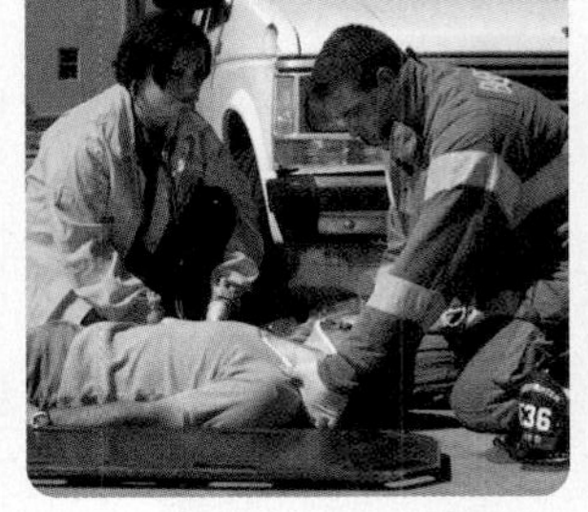

APPLICATIONS AND ACTIVITIES

Explore Your World

Chemistry Link to Health

Career Focus

Chemistry Link to the Environment

Chemistry Link to Industry

Chemistry Link to History

Guide to Problem Solving

About the Author

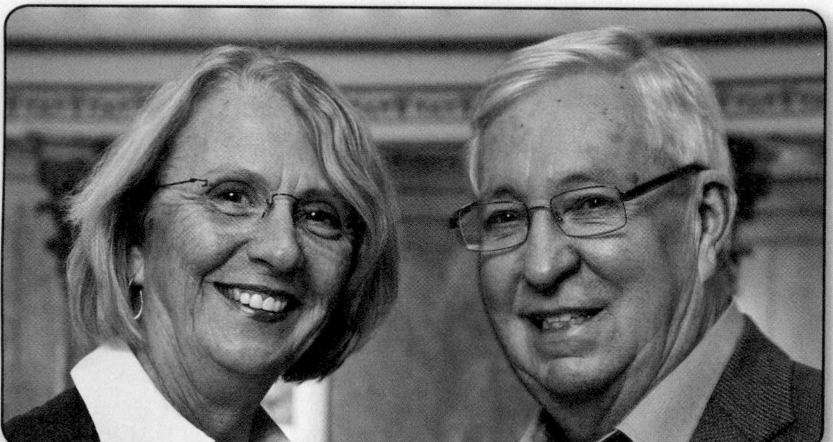

Karen and Bill Timberlake at the Natural History Museum where they are supporters of children's environmental programs.

KAREN TIMBERLAKE is Professor Emerita of Chemistry at Los Angeles Valley College, where she taught chemistry for allied health and preparatory chemistry for 36 years. She received her bachelor's degree in chemistry from the University of Washington and her master's degree in biochemistry from the University of California at Los Angeles.

Professor Timberlake has been writing chemistry textbooks for 34 years. During that time, her name has become associated with the strategic use of pedagogical tools that promote student success in chemistry and the application of chemistry to real-life situations. More than one million students have learned chemistry using texts, laboratory manuals, and study guides written by Karen Timberlake. In addition to *Chemistry: An Introduction to General, Organic, and Biological Chemistry*, Eleventh Edition, she is also the author of *General, Organic, and Biological Chemistry, Structures of Life*, Third Edition, and *Basic Chemistry*, Third Edition.

Professor Timberlake belongs to numerous scientific and educational organizations including the American Chemical Society (ACS) and the National Science Teachers Association (NSTA). She was the Western Regional Winner of Excellence in College Chemistry Teaching Award given by the Chemical Manufacturers Association. She received the McGuffey Award in Physical Sciences from the Textbook Authors Association for her textbook *Chemistry: An Introduction to General, Organic, and Biological Chemistry*, Eighth Edition. She received the "Texty" Textbook Excellence Award from the Textbook Authors Association for the first edition of *Basic Chemistry*. She has participated in education grants for science teaching including the Los Angeles Collaborative for Teaching Excellence (LACTE) and a Title III grant at her college. She speaks at conferences and educational meetings on the use of student-centered teaching methods in chemistry to promote the learning success of students.

Her husband, William Timberlake, who has contributed to writing this text, is Professor Emeritus of Chemistry at Los Angeles Harbor College, where he taught preparatory and organic chemistry for 36 years. He received his bachelor's degree in chemistry from Carnegie Mellon University and his master's degree in organic chemistry from the University of California at Los Angeles. When the Professors Timberlake are not writing textbooks, they relax by hiking, traveling, trying new restaurants, cooking, playing tennis, and taking care of their grandchildren, Daniel and Emily.

DEDICATION

I dedicate this book to

- My husband for his patience, loving support, and preparation of late meals
- My son, John, daughter-in-law, Cindy, grandson, Daniel, and granddaughter, Emily, for the precious things in life
- The wonderful students over many years whose hard work and commitment always motivated me and put purpose in my writing

Students learn chemistry using real-world examples

"Discovery consists of seeing what everybody has seen and thinking what nobody has thought."

—Albert Szent-Gyorgi

Feature	Description	Benefit	Page
Chapter-Opening Interview	Chapters begin with **interviews** with a professional in a career such as nursing, physical therapy, dentistry, agriculture, and food science.	Show you how health professionals use chemistry every day	319
Explore Your World	**Explore Your World** features are hands-on activities that use everyday materials to encourage you to explore selected chemistry topics.	Provide interactions with chemistry and support critical thinking	304
UPDATED! Chemistry Link to Health	**Chemistry Links to Health** apply chemical concepts to relevant topics of health and medicine such as weight loss and weight gain, trans fats, anabolic steroids, alcohol abuse, genetic diseases, viruses, and cancer.	Provide you with connections that illustrate the importance of understanding chemistry in real life health and medical situations	256
UPDATED! Chemistry Link to the Environment	**Chemistry Links to the Environment** relates chemistry to environmental topics such as global warming, radon in our homes, acid rain, and biodiesel.	Help you extend your understanding of the impact of chemistry on the environment	227
NEW! Chemistry Link to Industry	**Chemistry Links to Industry** describes industrial and commercial applications in the oil industry and the commercial production of margarine and solid shortening.	Show you how chemistry is applied to industry and manufacturing	370
NEW! Chemistry Link to History	**Chemistry Links to History** describes the historical development of chemical ideas.	Help you understand the role of chemistry in a historical setting	3
Career Focus	The **Career Focus** features within the chapters are additional examples of allied health professionals using chemistry.	Illustrate how chemistry is important in various fields within allied health	59

Engage students in the world of chemistry

"I never teach my pupils; I only attempt to provide the conditions in which they can learn."
—Albert Einstein

Feature	Description	Benefit	Page
Learning Goals LEARNING GOAL Give the IUPAC and common names for alcohols and phenols; identify common names for thiols and ethers.	**Learning Goals** at the beginning of each section identify the key concepts for that section and provide a roadmap for your study.	Help you focus your studying by emphasizing what is most important in each section	411
Writing Style The periodic table is an arrangement of the elements by increasing atomic number. A vertical column or group on the periodic table contains elements with similar properties. A horizontal row is called a period.	Timberlake's accessible **writing style** is based on careful development of chemical concepts suited to the skills and backgrounds of allied health students.	Helps you understand new terms and chemical concepts	119
UPDATED! Concept Maps Gases; are described by; Kinetic molecular theory; states that gas particles; Are far apart; Move fast; and; Exert pressure; behave according to; Gas laws; that relate; P and V Boyle's law; V and T Charles's law; P and T Gay-Lussac's law; P, V, and T Combined gas law; volume and moles are related by; V and n Avogadro's law; at STP gives; Molar volume; for a gas or; Gas mixture; where gases exert; Partial pressures; used to find; Moles or volume of gases; in; Reactions of gases	**Concept Maps** at the end of each chapter show how all the key concepts fit together.	Encourage learning by providing a **visual** guide to the interrelationship among all the concepts in each chapter	240
UPDATED! Macro-to-Micro Art 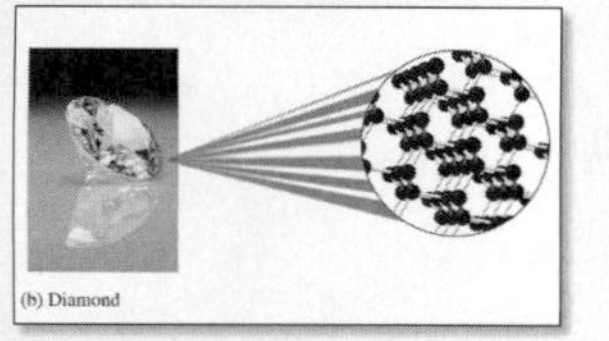 (b) Diamond	**Macro-to-Micro Art** utilizes photographs and drawings to illustrate recognizable objects and their atomic structure.	Helps you connect the world of atoms and molecules to the macroscopic world	89
UPDATED! Art Program 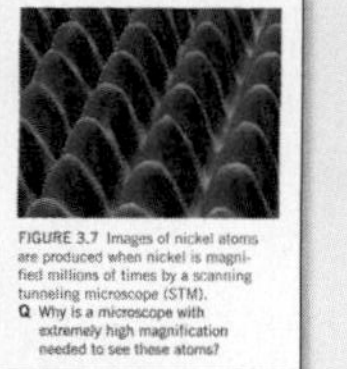 FIGURE 3.7 Images of nickel atoms are produced when nickel is magnified millions of times by a scanning tunneling microscope (STM). Q Why is a microscope with extremely high magnification needed to see these atoms?	The **art program** is beautifully rendered, pedagogically effective, and includes questions with all the figures.	Helps you think critically using photos and illustrations	96
NEW! Chapter Review 1.6 Significant Figures in Calculations Learning Goal: Adjust calculated answers to give the correct number of significant figures. In multiplication or division, the final answer is written so it has the same number of significant figures as the measurement with the fewest significant figures. In addition or subtraction, the final answer is written so it has the same number of decimal places as the measurement with the fewest decimal places.	The **Chapter Review** at the end of each chapter includes Learning Goals and new visual thumbnails to summarize the key points in each section.	Helps you determine your mastery of the chapter concepts and study for your tests	39

Many tools show students how to solve problems

"The whole art of teaching is only the art of awakening the natural curiosity of young minds."

—Anatole France

Feature	Description	Benefit	Page
UPDATED! Guide to Problem Solving (GPS)	**Guides to Problem Solving (GPS)**	Visually guide you step-by-step through each problem solving strategy	29
End-of-Section Questions and Problems	**Questions** and **Problems** are placed at the end of each section. Problems are paired and the Answers to the odd-numbered problems are given at the end of each chapter.	Encourage you to become involved immediately in the process of problem solving	226
NEW! Concept Checks	**Concept Checks** that transition from conceptual ideas to problem solving strategies are placed throughout each chapter.	Allow you to check your understanding of new chemical terms and ideas as they are introduced in the chapter	564
UPDATED! Sample Problems with Study Checks	Numerous **Sample Problems** in each chapter demonstrate the application of each new concept to problem solving. The worked-out solutions give step-by-step explanations, provide a problem-solving model, and illustrate required calculations. **Study Checks** associated with each Sample Problem allow you to check your problem-solving strategies.	Provide the intermediate steps to guide you successfully through each type of problem	173
UPDATED! Understanding the Concepts	**Understanding the Concepts** are questions with visual representations placed at the end of each chapter.	Build an understanding of newly learned chemical concepts	205
UPDATED! Additional Questions	**Additional Questions** at the end of chapter provide further study and application of the topics from the entire chapter.	Promote critical thinking	445
Challenge Questions	**Challenge Questions** at the end of each chapter provide complex questions.	Promote critical thinking, group work, and cooperative learning environments	208
UPDATED! Combining Ideas	**Combining Ideas** are sets of integrated problems that are placed after every 2-4 chapters.	Test your understanding of the concepts from previous chapters by integrating topics	211

Extend learning beyond the classroom

"One must learn by doing the thing; though you think you know it, you have no certainty until you try."
—Sophocles

MasteringCHEMISTRY®

www.masteringchemistry.com

MasteringChemistry is the most effective and widely used online tutorial, homework and assessment system for chemistry. It helps instructors maximize class time with customizable, easy-to-assign, and automatically graded assessments that motivate students to learn outside of class and arrive prepared for lecture. These assessments can easily be customized and personalized by instructors to suit their individual teaching style. To learn more visit: www.masteringchemistry.com

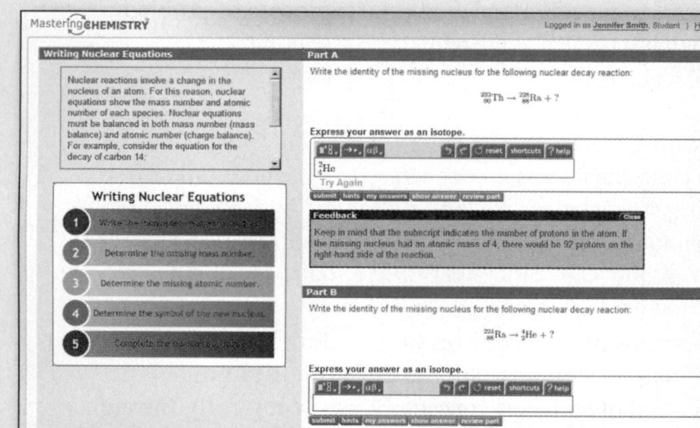

Student Tutorials

MasteringChemistry is the only system to provide instantaneous feedback specific to the most common wrong answers. Students can submit an answer and receive immediate, error-specific feedback. Simpler sub-problems—"hints"—are provided upon request.

Gradebook

Every assignment is automatically graded. Shades of red highlight vulnerable students and challenging assignments.

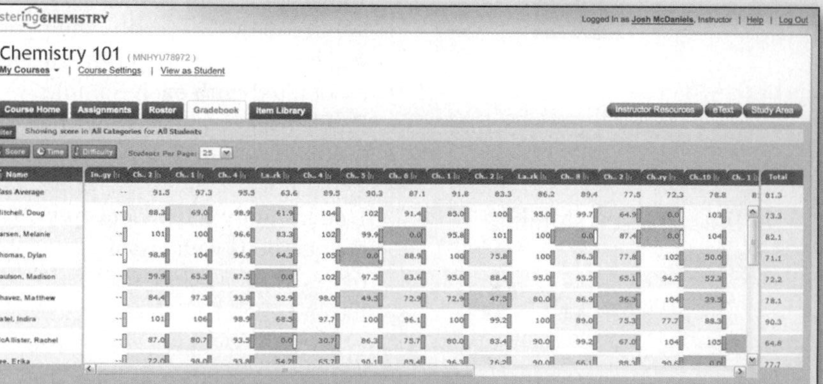

Gradebook Diagnostics

This screen provides weekly diagnostics. With a single click, charts summarize the most difficult problems, vulnerable students, grade distribution, and even score improvement over the course.

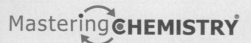

Tools for student self-study

The MasteringChemistry Study Area features in-depth resources for each of the health professions featured in the book.

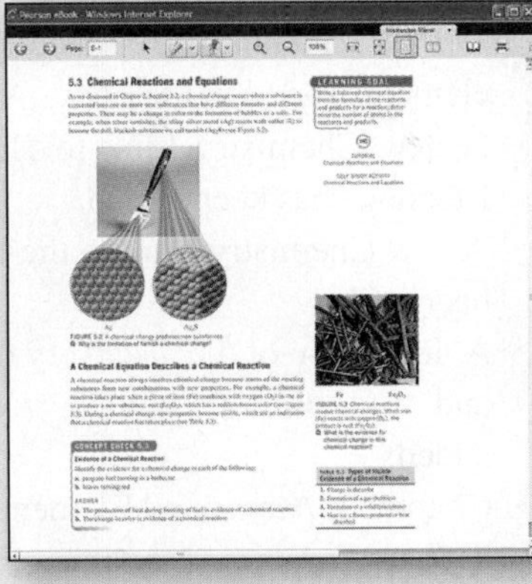

Pearson eText

The enhanced eText available in MasteringChemistry is fully searchable with note delivering capabilities. Content from the last two chapters of the text appears online only.

Reading Quizzes

Chapter-specific quizzes and activities focus on important, hard-to-grasp chemistry concepts. Icons throughout the textbook direct students to specific lessons related to the topics discussed in a given section.

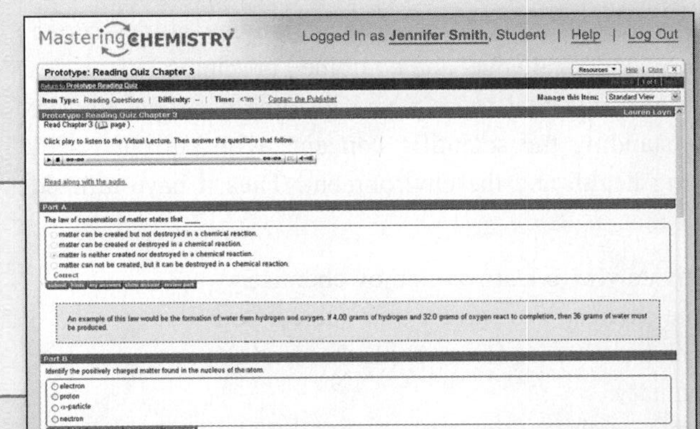

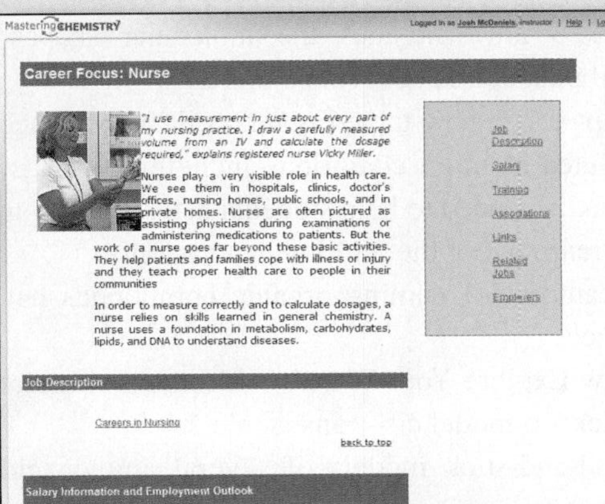

Career Focus

The career focus section includes additional information on careers that relates to the Chapter Openers and Career Focus boxes throughout the text.

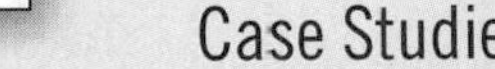

Case Studies

Case studies called out in the book correspond to activities within MasteringChemistry.

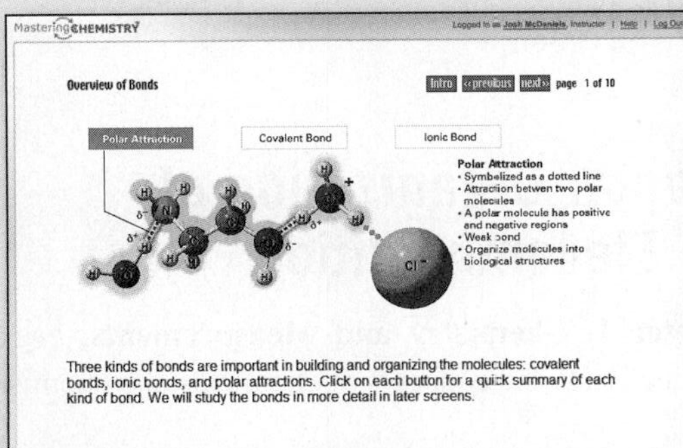

Self-Study Activities

These activities are called out within the book and appear in MasteringChemistry.

PREFACE

Welcome to the eleventh edition of *Chemistry: An Introduction to General, Organic, and Biological Chemistry.* This chemistry text was written and designed to help you prepare for a career in a health-related profession, such as nursing, dietetics, respiratory therapy, and environmental and agricultural science. This text assumes no prior knowledge of chemistry. My main objective in writing this text is to make the study of chemistry an engaging and positive experience for you by relating the structure and behavior of matter to its role in health and the environment. This new edition introduces more problem-solving strategies including new concept checks, more problem-solving guides, along with conceptual, challenge, and combined problems.

It is also my goal to help you become a critical thinker by understanding the scientific concepts with current issues concerning health and the environment. Thus, I have utilized materials that

- motivate you to learn and enjoy chemistry
- relate chemistry to careers that interest you
- develop problem-solving skills that lead to your success in chemistry
- promote your learning and success in chemistry

I hope that this textbook helps you discover exciting new ideas and gives you a rewarding experience as you develop an understanding and appreciation of the role of chemistry in your life.

New for the Eleventh Edition

New features have been added throughout this eleventh edition, including the following:

- The Prologue and Chapter 1 have been combined to form the new chapter 1, Chemistry and Measurements.
- New Concept Checks with Answers now appear in every chapter to help students focus on their understanding of newly introduced chemical terms and ideas.
- Captions have been added to photos to clarify content.
- Chapter Reviews now include art samples related to the content of each section.
- The interest boxes have been renamed Chemistry Link to Health and Chemistry Link to Environment. There are also two new types of boxes: Chemistry Link to Industry and Chemistry Link to History.
- A new Chemistry Link to Health feature, "Breathing Mixtures for Scuba," has been added.
- A new Chemistry Link to Industry, "Many Forms of Carbon," describes four forms of carbon: diamond, graphite, buckminsterfullerene, and nanotubes.
- A new Chemistry Link to Health, "Elements Essential to Health," includes a periodic table that highlights the elements found in the body.
- A new Chemistry Link to Health, "Hand Sanitizers and Ethanol," has been added.
- A new Chemistry Link to the Environment, "Plastics," has been added.
- A discussion of the discovery of electrons by J. J. Thomson and his "plum-pudding" model of the atom has been added.
- Chapter 3, Atoms and Elements, now includes electromagnetic radiation and atomic spectra as an introduction to energy levels and electron arrangements of electrons in atoms.
- New to Chapter 3 is a discussion of the metallic character of elements.
- There are now 30% more Guides to Problem Solving. The new guides include Calculating Density, Guide to Calculating the Atoms or Molecules of a Substance, Drawing Electron-Dot Formulas, Using Half-Lives, Naming Alcohols, Naming Ketones, Naming Carboxylic Acids, Naming Esters, Drawing Haworth Structures.
- Chapter 4 now includes the molecular shapes linear, trigonal planar, and bent shape of 120°.
- In Chapter 7, a new table organizes data for concentrated and diluted solution conditions including what is known and what is needed to help students determine an unknown before rearranging the dilution expression.
- The examples of naming organic compounds have been decreased.
- In a new Explore Your World, students use gumdrops and toothpicks to model cis–trans isomers.
- The ball-and-stick models of several amino acids have been added.
- Twenty-five percent of the Questions and Problems have been revised and/or are new.
- References to the tutorials available in *MasteringChemistry* have been placed in the *Study Guide* to point students to the relevant tutorials for each section and chapter in *MasteringChemistry*.

Chapter Organization of the Eleventh Edition

Chapter 1, Chemistry and Measurements, begins with an introduction to the concepts of chemicals and chemistry and asks students to develop a study plan for learning chemistry followed by types of measurements and the metric system in the sciences.

- New Concept Checks on metric prefixes, problem solving, and density are now included.
- Content was rewritten in the section on scientific notation, measured numbers, and problem solving to increase emphasis on metric and SI units.
- A new Sample Problem requires the use of percent and parts per million conversion factors.
- New photos, including laboratory technicians, the standard kilogram for the United States, chicken-pox virus, a terabyte hard disk drive, an ophthalmologist, and fertilizer application, were added to improve visual introduction to chemistry.
- More nutrients were added to daily values in Table 1.9.
- The Chemistry Link to Health "Bone Density" has been updated and is now included in Chapter 1.
- The content on specific gravity and the number of density sample problems were condensed.
- Content on the specific gravity of urine was removed.
- The apothecary unit of grains was deleted from clinical calculations.
- More Guides to Problem Solving (GPS) use color blocks as visual guides through the solution pathway.

Chapter 2, Matter and Energy, now includes the classification of matter and describes energy, temperature and its measurement, nutritional energy values, and states of matter.

- New Concept Checks on potential and kinetic energy, temperature, specific heat, states of matter, and physical and chemical properties clarify conceptual content.
- Section 2.1, Classification of Matter, and Section 2.2, States and Properties of Matter, now included in this chapter, contain more macro-to-micro diagrams and photos of the atomic arrangement in matter.
- A new discussion and photos of physical methods of separating mixtures, filtration, and chromatography have been added.
- A new Chemistry Link to Health, "Breathing Mixtures for Scuba," has been added.
- More heat problems now emphasize joules as the unit of energy.
- New problems relate the energy of fuel to the energy required to light a light bulb.
- Section 2.4, Temperature, was rewritten and several temperature equations deleted.
- Problems converting between temperature values in Fahrenheit and Kelvin were deleted.
- The Chemistry Link to the Environment "Carbon Dioxide and Global Warming" has been rewritten and the graph of atmospheric carbon dioxide levels has been extended to 2000 C.E.
- New art illustrates the atomic organization of solids, liquids, and gases and their changes of state.
- New to this edition is a discussion on combining energy calculations for heat changes and changes of state.
- The interchapter problem set, Combining Ideas from Chapters 1 and 2, completes the chapter.

Chapter 3, Atoms and Elements, now begins with an introduction to elements and atoms. The name and symbol of element 112 have been added to the periodic table as Copernicium, Cn. Using the masses of the naturally occurring isotopes and their abundances, atomic mass is calculated. A new discussion of the types of compounds formed by the elements has been added.

- New Concept Checks on symbols of elements, groups and periods on the periodic table, subatomic particles, average atomic mass, energy levels in atoms, and metallic character clarify conceptual content.
- A new Chemistry Link to Industry, "Many Forms of Carbon," describes four forms of carbon: diamond, graphite, buckminsterfullerene, and nanotubes.
- A new Chemistry Link to Health, "Elements Essential to Health," includes a periodic table that highlights the elements found in the body.
- The macro-to-micro art has been updated to improve visual representations of the atom and its structure.
- A discussion of the discovery of electrons by J. J. Thomson and his "plum-pudding" model of the atom has been added.
- Section 3.6, Electron Energy Levels, now includes electromagnetic radiation and atomic spectra as an introduction to energy levels and electron arrangements of electrons in atoms.
- The electron arrangement of electrons has been extended to the first 20 elements with a reference to electron arrangement in elements beyond atomic number 20.
- New to Section 3.7, Trends in Periodic Properties, is a discussion of the metallic character of elements.
- Updates, additions, and deletions to Questions and Problems were made.
- A new table, Summary of Trends in Periodic Properties of Representative Elements, has been added.
- New problems related to concepts of metallic character have been added.
- Table 3.8 now includes a column Most Prevalent Isotope.
- New problems with art have been added to Understanding the Concepts.
- Old Section 3.1, Classification of Matter, was moved to the beginning of Chapter 2.

Chapter 4, Compounds and Their Bonds, describes how atoms form ionic and covalent bonds and compounds. Students learn to write formulas and names of ionic compounds—including those with polyatomic ions—and covalent compounds.

Section 4.6, Electronegativity and Bond Polarity, discusses the polarity of bonds and Section 4.7 discusses the shapes and polarity of molecules. Section 4.8, Attractive Forces in Compounds, compares the attractive forces between particles and their impact on states of matter and changes of state.

- New Concept Checks including ions, drawing electron-dot formulas, naming covalent compounds, electronegativity, polarity, and shapes of molecules highlight conceptual understanding.
- New Understanding the Concepts problems have been added.
- The number of transition ions in the periodic table in Figure 4.2 has been increased.
- A Guide to Drawing Electron-Dot Formulas has been added.
- The names and formulas of the polyatomic ions perchlorate and hypochlorite have been added to Table 4.7.
- A new flowchart summarizes the naming of ionic and covalent compounds.
- New illustrations on the formation of a hydrogen molecule and elements that exist as diatomic, covalent molecules have been added.
- The material on writing electron-dot formulas has been expanded.
- Section 4.7, Shapes and Polarity of Molecules, now includes linear, trigonal planar, and bent shape of 120°.
- The material on sizes of atoms and their ions has been deleted.

Chapter 5, Chemical Quantities and Reactions, introduces moles and molar masses of compounds, which are used in calculations to determine the mass or number of particles in a given quantity. In Section 5.4, Types of Reactions, we look at collisions and interactions between atoms and molecules. Students learn to balance chemical equations and to recognize the types of chemical reactions. Section 5.6, Mole Relationships in Chemical Equations, and Section 5.7, Mass Calculations for Reactions, prepare students for the quantitative relationships in reactions. The chapter concludes with Section 5.8, Energy in Chemical Reactions, in which activation energy and energy changes in exothermic and endothermic reactions are discussed. Chapter 5 is followed by the interchapter problem set, Combining Ideas from Chapters 3 to 5.

- Matched statements are in all steps in Sample Problems to reflect Guides to Problem Solving (GPS) statements.
- New Concept Checks include moles, balancing chemical equations, types of reactions, quantitative relationships in reactions, and energy in reactions.
- A new GPS was written for calculating the atoms or molecules of a substance.
- New photos include sepia print, formation of patina on Cu, and pears.
- Combustion is now included in types of reactions.
- New problems of molar mass, grams, and moles have been added.
- Section 5.3, Chemical Reactions and Equations, now emphasizes chemical change.
- Sample Problems for conversion of grams to moles, moles to grams, moles to moles, or grams to grams were rewritten.
- In Section 5.8, units of joules and kilojoules have replaced units of calories and kilocalories.
- Material on chemical change has been combined with chemical equations to form a new Section 5.3, Chemical Reactions and Equations.

Chapter 6, Gases, discusses the properties of gases and calculates changes in gases using the gas laws. Problem-solving strategies enhance the discussion and calculations with gas laws.

- New Concept Checks include properties of gases, gas laws, and applications of gas laws to provide a transition between text information and problem solving.
- Tables for organization of data now include factors that remain constant.
- Many photos were updated and added.
- The properties of a gas that change in a gas law problem are clearly differentiated from properties that remain constant.
- More gas law problems now relate to real-world use of gases and gas mixtures.
- The units for answers to calculations of volume and pressure have been specified.
- New material on vapor pressure and boiling point has been added to Section 6.5, Temperature and Pressure (Gay-Lussac's Law).
- Material on gases in reactions at STP was deleted.

Chapter 7, Solutions, describes solutions, solubility, insoluble salts, concentrations, and colligative properties. New problem-solving strategies clarify the use of concentrations to determine volume or mass of solute. The volumes and molarities of solutions are used in calculations of reactants and products in chemical reactions as well as dilutions and titrations.

- New Concept Checks include solutes and solvents, types of solutions, electrolytes, solubility, dilution, and colligative properties.
- A new table describes possible combinations of solutes and solvents.
- Section 7.3, Solubility, now includes a discussion of solubility rules to identify a salt that is soluble or insoluble in water.
- New Table 7.10 gives a summary of the units used in percent concentration and molarity.
- A sample problem using volume/volume percent concentration has been added.
- Molarity has been combined with the section on solution concentration.

- The material on % concentration and molarity was combined to show the similar format of calculating any concentration.
- The table of concentration conversion factors for percent concentration is now combined with molarity.
- Section 7.5, Dilution of Solutions, is now a separate section that includes dilution of solutions using percent concentration and molarity.
- A new table was designed to organize data for concentrated and diluted solution conditions including what is known and what is needed to help students determine an unknown before rearranging the dilution expression.
- Section 7.6, Properties of Solutions, now includes the impact of particle concentration on boiling points and freezing points as well as on osmotic pressure.
- New information about insects and fish that produce biological antifreezes to survive subfreezing environments has been added.
- Questions and problems about freezing point and boiling point changes were added.
- Questions and problems about soluble and insoluble salts were added.
- Section 7.6, Solutions in Chemical Reactions, has been deleted.

Chapter 8, Acids and Bases, discusses acids and bases and their strengths, conjugate acid–base pairs, ionization of water, pH, and buffers. Section 8.1, Acid and Bases, now includes Brønsted–Lowry acids and bases. Acid–base titration uses neutralization reactions to calculate quantities of an acid in a Sample Problem. Chapter 8 is followed by the interchapter problem set, Combining Ideas from Chapters 6 to 8.

- New Concept Checks including names of acids and bases, strong and weak acids and bases, pH of solutions, and buffer solutions help clarify concepts. The list of acids and bases to emphasize most common acids has been updated.
- Table 8.1 of acids and bases has been updated to emphasize common acids.
- The use of the hydronium ion as the hydrated form of H^+ in acid solutions has been clarified.
- The order of weak acids in Table 8.3 has been revised.
- The number of problems on naming acids and bases has been reduced.
- The number of Sample Problems about balancing neutralization equations has been reduced.
- The redundant steps preceding Sample Problems have been deleted.

Chapter 9, Nuclear Radiation, looks at the type of radioactive particles that are emitted from the nuclei of radioactive atoms. Nuclear equations are written and balanced for both naturally occurring radioactivity and artificially produced radioactivity. A new Chemistry Link to Health, "Brachytherapy," was added along with an update of the discussion on radon in the home. The half-lives of radioisotopes are discussed, and the amount of time for a sample to decay is calculated. Radioisotopes important in the field of nuclear medicine are described.

- New Concept Checks include radiation particles, nuclear equations, half-lives, medical applications of radioisotopes, and fission and fusion.
- Table 9.2 now includes a column describing the change in the nucleus for each type of radiation emitted.
- Table 9.7 and Table 9.8 now include a column on type of radiation emitted by each isotope.
- The number of radioisotopes listed in Table 9.7 and Table 9.8 has been increased.
- New photos of a smoke detector and an excavation of bones for radiocarbon dating have been added.
- Questions and problems about positron decay were added.
- Material was added on the use of film-badges in radiology laboratories.

Chapter 10, Introduction to Organic Chemistry: Alkanes, discusses the structure, nomenclature, and reactions of alkanes. Guides to Problem Solving (GPS) clarify the rules for nomenclature. The chapter concludes with the introduction of functional groups for each family of organic compounds, which forms a basis for understanding the biomolecules of living systems.

- New Concept Checks include properties of organic compounds, naming alkanes, and functional groups.
- Skeletal formulas for alkanes and cycloalkanes to use for aromatic, carbohydrate, fatty acid, nucleic acid structures are introduced.
- Questions and problems now include the drawing of skeletal formulas.
- Several redundant Sample Problems on nomenclature were deleted.
- Material on uses of alkanes was placed before solubility and density of alkanes.
- The number of problems on naming and drawing condensed structural formulas was reduced.
- A new table lists the number of covalent bonds for elements in organic compounds.

Chapter 11, Unsaturated Hydrocarbons, discusses alkenes and alkynes, cis–trans isomers, addition reactions, polymers of alkenes used in everyday items, and aromatic compounds.

- New Concept Checks include unsaturated compounds and identifying cis–trans isomers.
- New visual art was added to illustrate cis–trans isomers.
- In a new Explore Your World, students use gumdrops and toothpicks to model cis–trans isomers.
- Addition reactions now include halogenation of alkenes.
- Redundant examples of alkene and alkyne nomenclature were deleted.
- Nomenclature of aromatic compounds was simplified.

Chapter 12, Organic Compounds with Oxygen and Sulfur, discusses alcohols, ethers, thiols, aldehydes, ketones, and chiral molecules. The discussions of Fischer projections and chiral molecules have been rewritten to better prepare students for the discussions on carbohydrates in Chapter 13 and amino acids in Chapter 16. The role of chiral drugs and their behavior in living systems is discussed. Chapter 12 is followed by the interchapter problem set, Combining Ideas from Chapters 9 to 12.

- New Concept Checks include isomers, classifying alcohols, solubility of alcohols, dehydration of alcohols, oxidation of alcohols, and chiral and achiral objects.
- Skeletal structures are included in structural formulas of alcohols, phenols, ethers, aldehydes, and ketones.
- New Guides to Naming Alcohols, Naming Aldehydes, and Naming Ketones have been added.
- The classification of alcohols has been moved to Section 12.2, Properties of Alcohols and Ethers.
- The discussion of boiling points for alcohols and ethers and aldehydes and ketones has been deleted.
- A new Chemistry Link to Health, "Hand Sanitizers and Ethanol," has been added.
- New tables for the solubility of alcohols and ethers and aldehydes and ketones have been added.
- New material on the oxidation of alcohols to aldehydes and carboxylic acids has been added.
- New material on the reduction of aldehydes and ketones has been added.
- A new Explore your World uses gumdrops and toothpicks to model chiral objects.

Chapter 13, Carbohydrates, which now follows Chapter 12, Organic Compounds with Oxygen and Sulfur, applies the organic chemistry of alcohols, aldehydes, and ketones to biomolecules.

- New Concept Checks include monosaccharides, D and L isomers, α and β isomers, and glycosidic bonds in disaccharides.
- Hydrogen atoms were added to all art of Haworth structures.
- Mutarotation of alpha and beta isomers of monosaccharides in aqueous solutions was added.
- A Guide to Drawing Haworth Structures was added.
- The —OH groups in open-chain and Haworth structures of D-glucose have been color coded.
- The C=O in carbonyl groups and the free hydroxyl group were color coded to highlight differences in alpha and beta isomers of monosaccharides.
- The steps for drawing Haworth structures were rewritten.
- The Chemistry Link to Health "Blood Types and Carbohydrates" was updated.
- Redundant structures of glucose, galactose, and fructose were removed.

Chapter 14, Carboxylic Acids, Esters, Amines, and Amides, completes the discussion of organic chemistry. The families and chemical reactions discussed are most applicable to biochemical systems. Sections 14.1, Carboxylic Acids, and 14.2, Properties of Carboxylic Acids, discuss the carboxyl group and carboxylic acids as weak acids. Sections 14.4, Amines, and 14.5, Amides, emphasize the nitrogen atom in their functional groups and their names. A Chemistry Link to Health feature on alkaloids discusses the naturally occurring amines in plants.

- New Concept Checks include carboxylic acids, solubility of carboxylic acids, acid hydrolysis, classifying amines, and heterocyclic amines.
- Skeletal formulas are now used in drawings of carboxylic acids, esters, amines, and amides.
- Examples of naming have been reduced.
- New Guides to Naming Carboxylic Acids and Naming Esters were added.
- The discussions of boiling points of carboxylic acids, amines, and amides have been deleted.
- Condensed structural formulas of carboxylic acids have been added to Table 14.2.
- A new Chemistry Link to the Environment, "Plastics," has been added.
- New material and questions on heterocyclic amines has been added.
- A new Chemistry Link to Health, "Synthesizing Drugs," has been added.

Chapter 15, Lipids, involves the functional groups of alcohols, carboxylic acids, and esters in fatty acids, triacylglycerols, glycerophospholipids, and steroids. Chemistry Links to Health include topics of interest to students such as olestra, trans fatty acids, and lipoproteins. A Chemistry Link to the Environment discusses biodiesel as an alternative nonpetroleum-based fuel. The role of lipids in cell membranes is discussed as well as the lipids that function as steroid hormones. Chapter 15 is followed by the interchapter problem set, Combining Ideas from Chapters 13 to 15.

- New Concept Checks include classes of lipids, formation and reactions of triacylglycerols, glycerophospholipids, steroid hormones, and cell membranes.
- The discussion of omega-3 fatty acids in fish oils has been rewritten.
- New Table 15.3 shows the similarities of organic and lipid reactions: esterification, hydrogenation, hydrolysis, and saponification.
- The condensed structural formulas of glycerophospholipids have been color coded to clarify glycerol, fatty acids, phosphate, and amino alcohol components,
- A discussion of bile salts has been added.

Chapter 16, Amino Acids, Proteins, and Enzymes, discusses amino acids, formation of proteins, structural levels of proteins, enzymes, and enzyme action. New items include an updated discussion of zwitterions and isoelectric points. Amino acids are written as their ionized forms in physiological solutions. Section 16.4, Levels of Protein Structure, describes the importance of protein structure at the primary, secondary, tertiary, and quaternary levels. Enzymes are discussed as biological catalysts, along with the impact of substrate concentration, cofactors, inhibitors, and denaturation on enzyme action.

- New Concept Checks include R groups of amino acids, ionized forms of amino acids, peptides, structures of proteins, denaturation of proteins, and enzyme action.
- New Section 16.1, Proteins and Amino Acids, combines old Sections 16.1 and 16.2.
- New descriptions of amino acids were written.
- The stereoisomers of amino acids were moved after the discussion on zwitterions.
- The ball-and-stick models of several amino acids have been added.
- The Chemistry Link to Health "Essential Amino Acids" was moved to Section 16.1.
- Section 16.2, Amino Acids as Acids and Bases, has been rewritten to clarify ionized forms of amino acids at pH values above and below pI.
- The names of di- and tripeptides were color coded to highlight the N terminal and C terminal.
- A discussion of the structures of the nonapeptides oxytocin and vasopressin has been added.
- The ribbon structures of myoglobin and hemoglobin have been added to the discussion of tertiary and quaternary structures of proteins.
- The illustration of pleated sheet has been added to secondary structures in Figure 16.11.
- A discussion of enzyme concentration was added to enzyme activity.
- A discussion of thermophiles and high-temperature resistance has been added to factors that affect enzyme activity.
- Irreversible inhibition is now discussed.

Chapter 17, Nucleic Acids and Protein Synthesis, describes the nucleic acids and their importance as biomolecules that store and direct information for the synthesis of cellular components. The role of complementary base pairing is highlighted in both DNA replication and the formation of mRNA during protein synthesis. The role of RNA is discussed in the relationship of the genetic code to the order of amino acids in a protein. Mutations describe ways in which the nucleotide sequences are altered in genetic diseases. We also look at how DNA or RNA in viruses utilizes host cells to produce more viruses.

- New Concept Checks include components of nucleic acids, DNA replication, types of RNA, genetic code, and viruses.
- Art for the structures of purines and pyrimidines was updated to show the N atom.
- The discussion of names of nucleosides and nucleotides was rewritten to clarify the components of each.
- A new table compares the components of DNA with those of RNA.
- Color coding was added to highlight the components of DNA, RNA, and amino acids.
- The term *chain elongation* was added to the discussion of translation.
- A new table shows complementary sequences in DNA, mRNA, tRNA, and peptides.
- Color coding has been added to highlight changes in mRNA codons and corresponding amino acids in mutations.
- The structure of Lexiva, protease inhibitor for AIDS, was added.
- A photo of an Epstein–Barr virus was added.
- The term *DNA nontemplate strand* was replaced with DNA coding strand.
- The subsections on initiation, chain elongation, and chain termination were rewritten for clarity.

Chapter 18, Metabolic Pathways and Energy Production, describes the metabolic pathways of biomolecules from the digestion of foodstuffs to the synthesis of ATP. Students look at the stages of metabolism and the digestion of carbohydrates along with the coenzymes required in metabolic pathways. The breakdown of glucose to pyruvate is described using glycolysis, which is followed by the decarboxylation of pyruvate to acetyl CoA. We look at the entry of acetyl CoA into the citric acid cycle and the production of reduced coenzymes. We describe electron transport, oxidative phosphorylation, and the synthesis of ATP. The oxidation of lipids and the degradation of amino acids are also discussed. Chapter 18 is followed by an interchapter problem set, Combining Ideas from Chapters 16 to 18.

- New Concept Checks include metabolism and cell structure, ATP, digestion, coenzymes, glycolysis, citric acid cycle, chemiosmotic model, and degradation of amino acids.
- Art for ATP and ADP has been updated with new symbols for phosphate.
- Added structures of ADP and AMP.
- The diagram for stages of metabolism in Figure 18.1 has been simplified.
- Figure 18.3 was updated by adding O to hexagons and pentagons in the representations of monosaccharides in polysaccharides.
- The Chemistry Link to Health "Lactose Intolerance" was moved to the discussion of the digestion of carbohydrates.
- The descriptions of NAD^+, FAD, and coenzyme A have been updated.
- A new Table 18.3 lists the enzymes and coenzymes utilized in metabolic reactions.
- The dark background in the glycolysis diagram was removed and symbols added to highlight ATP and ADP involvement.

- The names of enzymes in the discussion of steps 1–5 and 6–10 in glycolysis were deleted.
- A new section, Electron Transport and Oxidative Phosphorylation, combines old Sections 18.6 and 18.7.
- The names of the reaction steps are now named by type of reaction: oxidation, decarboxylation, hydration, dehydrogenation, isomerization, cleavage, and hydrolysis.
- The discussion on electron transport and complexes I, II, III, IV has been simplified.
- A new table summarizes the calculations of ATP from glycolysis, oxidative decarboxylation, and citric acid cycle.
- A new figure for beta oxidation that includes cycles of cleavage to acetyl CoA has been added.
- The structures and details of electron carriers FMN, FeS, cytochrome *c*, and Q were deleted.

Instructional Package

Chemistry: An Introduction to General, Organic, and Biological Chemistry, eleventh edition, is the nucleus of an integrated teaching and learning package of support material for both students and professors.

For Students

***Study Guide for Chemistry: An Introduction to General, Organic, and Biological Chemistry*, eleventh edition, by Karen Timberlake.** This manual is keyed to the learning goals in the text and designed to promote active learning through a variety of exercises with answers as well as practice tests. (ISBN: 0-321-71942-5)

***Selected Solutions Manual for Chemistry: An Introduction to General, Organic, and Biological Chemistry*, eleventh edition, by Karen Timberlake.** This manual contains the complete solutions to the odd-numbered problems. (ISBN: 0-321-76521-4)

MasteringChemistry® *(www.masteringchemistry.com)* The most advanced, most widely used online chemistry tutorial and homework program is available for the eleventh edition of *Chemistry: An Introduction to General, Organic, and Biological Chemistry*. MasteringChemistry® utilizes the Socratic method to coach students through problem-solving techniques, offering hints, and simpler questions on request to help students *learn*, not just practice. A powerful gradebook with diagnostics that gives instructors unprecedented insight into their students' learning is also available.

MC **Pearson eText** Pearson eText offers students the power to create notes, highlight text in different colors, create bookmarks, zoom, and view single or multiple pages. Access to the Pearson eText for *Chemistry: An Introduction to General, Organic, and Biological Chemistry*, eleventh edition, is available for purchase either as a stand-alone item or within MasteringChemistry®.

Media Icons **Media icons** in the margins throughout the text direct you to tutorials and case studies in the Study Area located within MasteringChemistry® for *Chemistry: An Introduction to General, Organic, and Biological Chemistry*, eleventh edition.

***Laboratory Manual* by Karen Timberlake** This best-selling lab manual coordinates 42 experiments with topics in *Chemistry: General, Organic, and Biological Chemistry*, eleventh edition, uses new terms during the lab, and explores chemical concepts. Laboratory investigations develop skills of manipulating equipment, reporting data, solving problems, making calculations, and drawing conclusions. (ISBN: 0-321-69529-1)

***Essential Laboratory Manual* by Karen Timberlake** This lab manual contains 25 experiments for the chemical topics in *Chemistry: General, Organic, and Biological Chemistry.* (ISBN: 0-13-605547-8)

For Teachers

MasteringChemistry® *(www.masteringchemistry.com)* Access to the Pearson eText for MasteringChemistry® for *Chemistry: An Introduction to General, Organic, and Biological Chemistry*, eleventh edition, is available for purchase either as a stand-alone item or within MasteringChemistry®. MasteringChemistry® is the first adaptive-learning online homework and tutorial system. Instructors can create online assignments for their students by choosing from a wide range of items, including end-of-chapter problems and research-enhanced tutorials. Assignments are automatically graded with up-to-date diagnostic information, helping instructors pinpoint where students struggle either individually or as a class as a whole.

Most of the teacher supplements and resources for this book are available electronically for download on the Instructor Resource Center (IRC). Please go to *www.PearsonSchool.com/Access_Request* and select "access to online instructor resources." You will be required to complete a one-time registration subject to verification before being emailed access information for download materials.

Instructor Solutions Manual by Mark Quirie The solutions manual highlights chapter topics, and includes answers and solutions for all questions and problems in the text. (ISBN: 0-321-71943-3)

Instructor Resource DVD: *An Introduction to General, Organic, and Biological Chemistry*, eleventh edition This DVD includes all the art, photos, and tables from the book in high-resolution format for use in classroom projection or when creating study materials and tests. In addition, the instructors can access over 2000 PowerPoint lecture outlines, written by Karen Timberlake, or create their own easily with the art provided in PowerPoint format. Also available on the DVD are downloadable files of the *Instructor Solutions Manual* and *Test Bank*, and a set of "clicker questions" designed for use with classroom-response systems. (ISBN: 0-321-72719-3)

Printed Test Bank The **Test Bank**, by William Timberlake, includes more than 1700 multiple-choice, matching, and true/false questions. (ISBN: 0-321-72718-5)

Online Instructor Manual for *Laboratory Manual* This manual contains answers to report pages for the *Laboratory Manual* and the *Essential Laboratory Manual*.

Acknowledgments

The preparation of a new text is a continuous effort of many people. As in my work on other textbooks, I am thankful for the support, encouragement, and dedication of many people who put in hours of tireless effort to produce a high-quality book that provides an outstanding learning package. The editorial team at Pearson has done an exceptional job. We want to thank Adam Jaworski, editor in chief, who supported our vision of this eleventh edition as well as the addition of new Concept Checks, more Guides to Problem Solving, new Chemistry Links to Health, History, Industry, and the Environment, and new problems in Understanding the Concepts and Combining Ideas. I am in awe and appreciate all the wonderful work of Jessica Neumann, associate editor, who was like an angel encouraging us at each step, while skillfully coordinating reviews, art, Web site materials, and all the things it takes to make a book come together. I appreciate the work of Beth Sweeten, project manager, and Rebecca Dunn of Preparé, who brilliantly coordinated all phases of the manuscript to the final pages of a beautiful book. Thanks to Mark Quirie, reviewer who precisely analyzed and edited the initial and final manuscripts and pages to make sure the words and problems were correct to help students learn chemistry.

I am especially proud of the art program in this text, which lends beauty and understanding to chemistry. I would like to thank Derek Bacchus and Tamara Newnam, art director and book designer, whose creative ideas provided the outstanding design for the cover and pages of the book. Eric Schrader, photo researcher, was invaluable in researching and selecting vivid photos for the text so that students can see the beauty of chemistry. Thanks also to Bio-Rad Laboratories for their courtesy and use of KnowItAll ChemWindows, drawing software that helped us produce chemical structures for the manuscript. The macro-to-micro illustrations designed by Production Solutions and Precision Graphics give students visual impressions of the atomic and molecular organization of everyday things and are a fantastic learning tool. I want to thank Donna Young for the hours of proofreading all the pages. I also appreciate all the hard work in the field put in by the marketing team and Erin Gardner, marketing manager.

I am extremely grateful to an incredible group of peers for their careful assessment of all the new ideas for the text; for their suggested additions, corrections, changes, and deletions; and for providing an incredible amount of feedback about improvements for the book. In addition, I appreciate the time scientists took to let us take photos and discuss their work with them. I admire and appreciate every one of you.

If you would like to share your experience with chemistry, or have questions and comments about this text, I would appreciate hearing from you.

Karen Timberlake
Email: khemist@aol.com

Reviewers

Eleventh Edition Reviewers

Gary Burce, *Ohlone College*
Pamela Goodman, *Moraine Valley Community College*
Emily Halvorson, *Pima Community College*
Mark Quirie, *Algonquin College*
Richard Triplett, *Des Moines Area Community College*

Tenth Edition Reviewers

Kristopher M. Baker, *SUNY, Rockland CC*
Daniel Bernier, *Riverside City College*
David Canoy, *Chemeketa Community College*
Ron Choppi, *Chaffey College*
Susan E. Cordova, *Central New Mexico Community College*
Alan D. Earhart, *Southeast Community College*
Coretta Fernandes, *Lansing Community College*
Eric Goll, *Brookdale Community College*
John G. Griggs, *Troy University*
Byron Howell, *Tyler Junior College*
Shelli Hull, *Tarrant County Community College, South*
Timothy Kreider, *University of Medicine & Dentistry of New Jersey*
Nancy Leigh, *Spoon River College*
Chunmei Li, *Stephen F. Austin State University*
Chris Massone *Molloy College*
Susanne McFadden, *Tarrant County Community College, NW*
Felix Ngassa, *Grand Valley State University*
Dr. MaryKay Orgill, *University of Nevada, Las Vegas*
Gordon A. Pomeroy, *York Technical College*
Lesley Putman, *Northern Michigan University*
Mark Quirie, *Algonquin College*
Kathleen Thrush Shaginaw, *Particular Solutions, Inc.*
Mark Thomson, *Ferris State University*
Rod Tracey, *College of the Desert*
Eileen Reilly-Wiedow, *Fairfield University*

1 Chemistry and Measurements

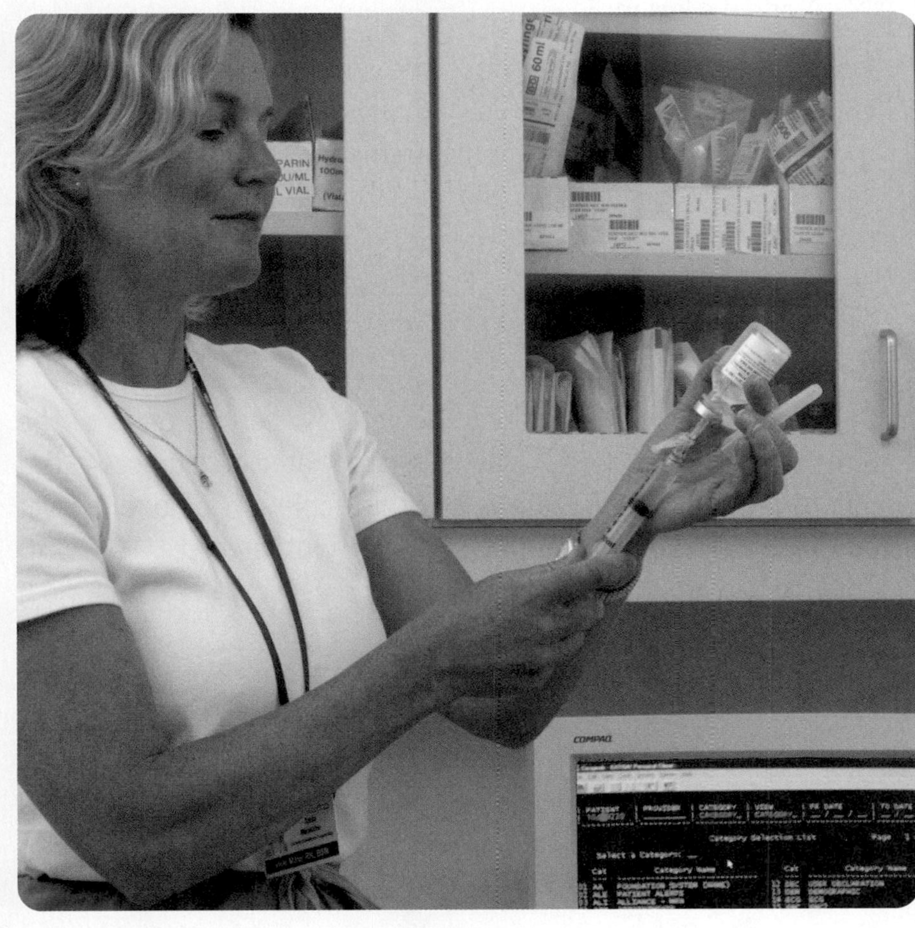

LOOKING AHEAD

"I use measurement in just about every part of my nursing practice," says registered nurse Vicki Miller. "When I receive a doctor's order for a medication, I have to verify that order. Then I draw a carefully measured volume from an IV or a vial to create that particular dose. Some dosage orders are specific to the size of the patient. I measure the patient's weight and calculate the dosage required for the weight of that patient."

Nurses use measurement each time they take a patient's temperature, height, weight, or blood pressure. Measurement is used to obtain the correct amounts for injections and medications and to determine the volumes of fluid intake and output. For each measurement, the amounts and units are recorded in the patient's records.

hat are some questions in chemistry you have been curious about? Perhaps you are interested in how smog is formed, what causes ozone depletion, or how aspirin relieves a headache. Just like you, chemists are curious about the world we live in.

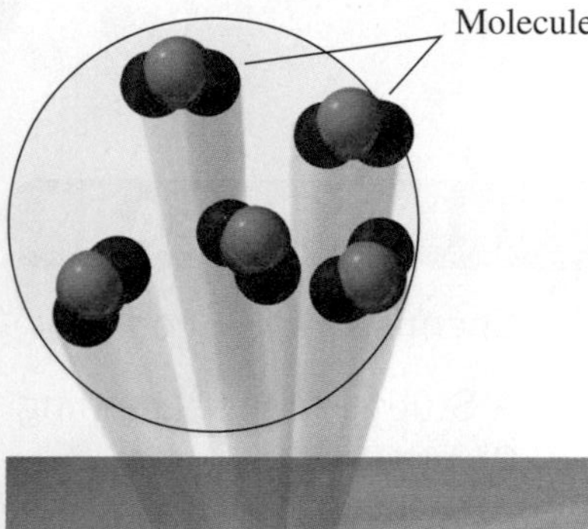

The chemical reaction of NO with oxygen in the air forms NO_2, which produces the reddish brown color of smog.

- How does car exhaust produce the smog that hangs over our cities? One component of car exhaust is nitrogen oxide (NO) formed in car engines, where high temperatures convert nitrogen gas (N_2) and oxygen gas (O_2) to NO gas. In chemistry, these reactions are written in the form of equations such as $N_2(g) + O_2(g) \longrightarrow 2NO(g)$. The reaction of NO with oxygen in the air produces NO_2, which gives the reddish brown color to smog.
- Why has the ozone layer been depleted in certain parts of the atmosphere? In the 1970s, scientists associated substances called chlorofluorocarbons (CFCs) with the depletion of ozone over Antarctica. As CFCs are broken down by ultraviolet (UV) light, chlorine (Cl) is released that causes the breakdown of ozone (O_3) molecules and destroys the ozone layer:

$$Cl + O_3 \longrightarrow ClO + O_2$$

- Why does aspirin relieve a headache? When a part of the body is injured, substances called prostaglandins are produced, which cause inflammation and pain. Aspirin acts to block the production of prostaglandins, thereby reducing inflammation, pain, and fever.

Chemists assess the impact of chemistry on our lives and our environment by making measurements. Levels of toxic materials in the air, soil, and water are measured. Measurements help us understand about radon in our homes, global warming, trans fatty acids, and DNA analysis. Understanding chemistry and measurement helps us make proper choices about our world.

Think about your day; you probably made some measurements. Perhaps you checked your weight by stepping on a scale. If you cooked some rice for dinner, you added 2 cups of water to 1 cup of rice. If you did not feel well, you may have taken your temperature. If you stopped at the gas station, you watched the gas pump measure the number of gallons of gasoline you put in the car.

Measurement is an essential part of health careers such as nursing, dental hygiene, respiratory therapy, nutrition, and veterinary technology. The temperature, height, and weight of a patient are measured and recorded. Samples of blood and urine are collected and sent to a laboratory where glucose, pH, urea, and protein are measured by the clinical technicians. By learning about measurement, you develop skills for solving problems and learn how to work with numbers.

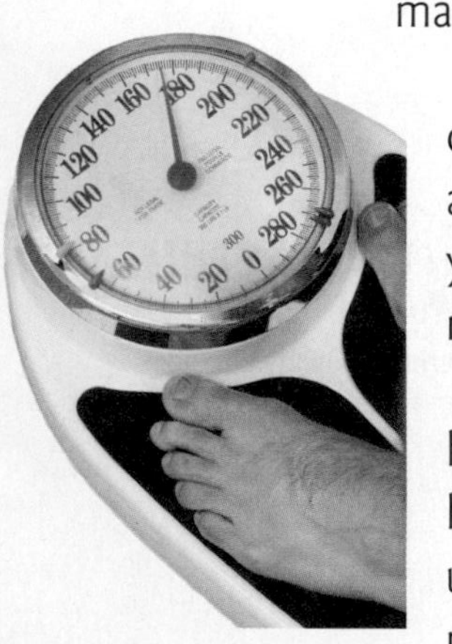

Your weight on a bathroom scale is a measurement.

1.1 Chemistry and Chemicals

LEARNING GOAL

Define the term *chemistry* and identify substances as chemicals.

Chemistry is the study of the composition, structure, properties, and reactions of matter. *Matter* is another word for all the substances that make up our world. Perhaps you imagine that chemistry is done only in a laboratory by a chemist wearing a lab coat and goggles. Actually, chemistry happens all around you every day and has a big impact on everything you use and do. You are doing chemistry when you cook food, add chlorine to a swimming pool, or drop an antacid tablet into water. Plants grow because chemical reactions convert carbon dioxide, water, and energy to carbohydrates. Chemical reactions take place when you digest food and break it down into substances that you need for energy and health.

Antacid tablets undergo a chemical reaction when dropped into water.

Branches of Chemistry

The field of chemistry is divided into branches such as organic, inorganic, and general chemistry. Organic chemistry is the study of substances that contain the element carbon. Inorganic chemistry is the study of all other substances except those that contain carbon. General chemistry is the study of the composition, properties, and reactions of matter.

Today, chemistry is often combined with other sciences such as geology, biology, and physics to form cross disciplines such as geochemistry, biochemistry, and physical chemistry. Geochemistry is the study of the chemical composition of ores, soils, and minerals of the surface of Earth and other planets. Biochemistry is the study of the chemical reactions that take place in biological systems. Physical chemistry is the study of the physical nature of chemical systems including energy changes.

Biochemists analyze laboratory samples.

Chemistry Link to History

EARLY CHEMISTS: THE ALCHEMISTS

For many centuries, chemists have studied changes in various substances. From the time of the ancient Greeks to about the sixteenth century, alchemists described a substance in terms of four components of nature: earth, air, fire, and water. By the eighth century, alchemists searched for an unknown substance called a philosopher's stone that they thought would turn metals into gold as well as prolong youth and postpone death. Although these efforts failed, the alchemists did provide information on the processes and chemical reactions involved in the extraction of metals from ores. The alchemists also designed some of the first laboratory equipment and developed early laboratory procedures.

The alchemist Paracelsus (1493–1541) thought that alchemy should be about preparing new medicines, not about producing gold. Using observation and experimentation, he proposed that a healthy body was regulated by a series of chemical processes that could be unbalanced by certain chemical compounds and rebalanced by using minerals and medicines. For example, he determined that inhaled dust, not underground spirits, caused lung disease in miners. He also thought that goiter was a problem caused by contaminated water, and he treated syphilis with compounds of mercury. His opinion of medicines was that the right dose makes the difference between a poison and a cure. Today, this idea is part of the risk analysis of medicines. Paracelsus changed alchemy in ways that helped to establish modern medicine and chemistry.

Alchemists in the Middle Ages developed laboratory procedures.

Swiss alchemist and physician Paracelsus (1493–1541) believed that chemicals could be used as medicines.

TABLE 1.1 Chemicals Commonly Used in Toothpaste

Chemical	Function
Calcium carbonate	An abrasive used to remove plaque
Sorbitol	Prevents loss of water and hardening of toothpaste
Glycerin	Makes toothpaste foam in the mouth
Sodium lauryl sulfate	A detergent used to loosen plaque
Titanium dioxide	Makes toothpaste white and opaque
Triclosan	Inhibits bacteria that cause plaque and gum disease
Sodium fluorophosphate	Prevents formation of cavities by strengthening tooth enamel with fluoride
Methyl salicylate	Gives toothpaste a pleasant wintergreen flavor

Toothpaste is a combination of many chemicals.

Chemicals

All the things you see around you are composed of one or more chemicals. A **chemical** is a substance that always has the same composition and properties wherever it is found. Chemical processes take place in chemistry laboratories, manufacturing plants, and pharmaceutical labs as well as every day in nature and in our bodies. Often the terms *chemical* and *substance* are used interchangeably to describe a specific type of material.

Every day you use products containing substances that were prepared by chemists. Soaps and shampoos contain chemicals that remove oils on your skin and scalp. When you brush your teeth, the chemicals in toothpaste clean your teeth, prevent plaque formation, and stop tooth decay. Some of the substances used to make toothpaste are listed in Table 1.1.

In cosmetics and lotions, chemicals are used to moisturize, prevent deterioration of the product, fight bacteria, and thicken the product. Your clothes may be made of natural materials such as cotton or synthetic substances such as nylon or polyester. Perhaps you wear a ring or watch made of gold, silver, or platinum. Your breakfast cereal is probably fortified with iron, calcium, and phosphorus, while the milk you drink is enriched with vitamins A and D. Antioxidants are chemicals added to your cereal to prevent it from spoiling. Some of the chemicals you may encounter when you cook in the kitchen are shown in Figure 1.1.

FIGURE 1.1 Many of the items found in a kitchen are obtained using chemical reactions.
Q What are some other chemicals found in a kitchen?

SAMPLE PROBLEM 1.1

Chemicals

Why is the copper in copper wire an example of a chemical, while sunlight is not?

SOLUTION

Copper is a chemical that has the same composition and properties wherever it is found. Sunlight is energy given off by the Sun. Thus, sunlight does not contain matter, which means it is not a chemical.

STUDY CHECK 1.1

Which of the following are chemicals?

a. iron **b.** tin
c. a low temperature **d.** water

The answers to all the Study Checks can be found at the end of this chapter.

QUESTIONS AND PROBLEMS

Chemistry and Chemicals

In every chapter, each magenta, odd-numbered exercise in *Questions and Problems* is paired with the next even-numbered exercise. The answers for the magenta, odd-numbered Questions and Problems are given at the end of this chapter. The complete solutions to the odd-numbered Questions and Problems are in the *Study Guide*.

1.1 Obtain a bottle of multivitamins, and observe the list of ingredients. List four. Which ones are chemicals?

1.2 Obtain a box of breakfast cereal, and observe the list of ingredients. List four. Which ones are chemicals?

1.3 A "chemical-free" shampoo includes the ingredients water, cocomide, glycerin, and citric acid. Is the shampoo truly "chemical-free"?

1.4 A "chemical-free" sunscreen includes the ingredients titanium dioxide, vitamin E, and vitamin C. Is the sunscreen truly "chemical-free"?

1.2 A Study Plan for Learning Chemistry

LEARNING GOAL

Develop a study plan for learning chemistry.

Here you are taking chemistry, perhaps for the first time. Whatever your reasons are for choosing to study chemistry, you can look forward to learning many new and exciting ideas.

Features in This Text Help You Study Chemistry

This text has been designed with a variety of study aids to complement your individual learning style. On the inside of the front cover is a periodic table of the elements. On the inside of the back cover are tables that summarize useful information needed throughout your study of chemistry. Each chapter begins with Looking Ahead, which outlines the topics in the chapter. A Learning Goal at the beginning of each section previews the concepts you are to learn. At the end of the text, there is a comprehensive Glossary and Index, which list and define key terms used in the text.

Before you begin reading, obtain an overview of a chapter by reviewing the topics in Looking Ahead. As you prepare to read a section of the chapter, look at the section title and turn it into a question. For example, for Section 1.1 "Chemistry and Chemicals," you could ask "What is chemistry?" or "What are chemicals?" When you are ready to read through that section, review the Learning Goal, which tells you what to expect in that section. As you read, try to answer your question. Throughout the chapter, you will find Concept Checks that will help you understand key ideas. When you come to a Sample Problem, take the time to work it through and try the associated Study Check. Many Sample Problems are accompanied by a visual Guide to Problem Solving (GPS). At the end of each section, you will find a set of Questions and Problems that allows you to apply problem solving immediately to the new concepts.

Throughout each chapter, boxes entitled Chemistry Link to Health, Chemistry Link to the Environment, Chemistry Link to Industry, and Chemistry Link to History help you connect the chemical concepts you are learning to real-life situations. Many of the figures and diagrams use macro-to-micro illustrations to depict the atomic level of organization of ordinary objects. These visual models illustrate the concepts described in the text and allow you to "see" the world in a microscopic way.

Studying in a group can be beneficial to learning.

At the end of each chapter, you will find several study aids that complete the chapter. *Concept Maps* show the connections between important concepts and *Chapter Reviews* provide a summary. The *Key Terms* are in boldface type in the text and listed with their definitions. *Understanding the Concepts*, a set of questions that uses art and structures, helps you visualize concepts. *Additional Questions and Problems* and *Challenge Problems* provide additional problems to test your understanding of the topics in the chapter. All the problems are paired, and the answers to all the *Study Checks* and to the odd-numbered *Questions and Problems* are provided. If your answers match, you most likely understand the topic; if not, you need to study the section again.

After some chapters, problem sets called *Combining Ideas* test your ability to solve problems containing material from more than one chapter.

Using Active Learning to Learn Chemistry

A student who is an active learner continually interacts with the chemical ideas while reading the text and attending lectures. Let's see how this is done.

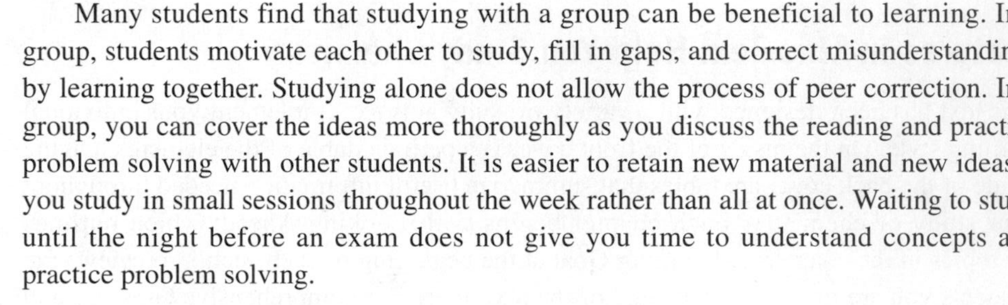

Students discuss a chemistry problem with their professor during office hours.

As you read and practice problem solving, you remain actively involved in studying, which enhances the learning process. In this way, you learn small bits of information at a time and establish the necessary foundation for understanding the next section. You should also note any questions you have about the reading to discuss with your professor and laboratory instructor. Table 1.2 summarizes these steps for active learning. The time you spend in lectures is also useful as a learning time. By keeping track of the class schedule and reading the assigned material before lectures, you become aware of the new terms and concepts you need to learn. Some questions that occur during your reading may be answered during the lectures. If not, you can ask for further clarification from your professor.

Many students find that studying with a group can be beneficial to learning. In a group, students motivate each other to study, fill in gaps, and correct misunderstandings by learning together. Studying alone does not allow the process of peer correction. In a group, you can cover the ideas more thoroughly as you discuss the reading and practice problem solving with other students. It is easier to retain new material and new ideas if you study in small sessions throughout the week rather than all at once. Waiting to study until the night before an exam does not give you time to understand concepts and practice problem solving.

Thinking about Your Study Plan

As you embark on your journey into the world of chemistry, think about your approach to studying and learning chemistry. You might consider some of the ideas in the following list. Check those ideas that will help you learn chemistry successfully. Commit to them now. *Your* success depends on *you*.

TABLE 1.2 Steps in Active Learning

1. Read each *Learning Goal* for an overview of the material.
2. Form a question from the title of the section you are going to read.
3. Read the section looking for answers to your question.
4. Self-test by working *Concept Checks*, *Sample Problems*, and *Study Checks*.
5. Complete the *Questions and Problems* that follow that section, and check the answers for the magenta odd-numbered problems.
6. Work the exercises in the *Study Guide* and go to *www.masteringchemistry.com* for self-study materials and instructor-assigned homework (optional).
7. Proceed to the next section, and repeat the previous steps.

My study of chemistry will include the following:

_______ reading the chapter before a lecture
_______ going to lectures
_______ reviewing the *Learning Goals*
_______ keeping a problem notebook
_______ reading the text as an active learner
_______ self-testing by working *Questions and Problems* following each section and checking solutions in the text
_______ being an active learner during lectures
_______ organizing a study group
_______ seeing the professor during office hours
_______ completing exercises in the *Study Guide*
_______ working through the tutorials at *www.masteringchemistry.com*
_______ attending review sessions
_______ organizing my own review sessions
_______ studying in small doses as often as I can

CONCEPT CHECK 1.1

A Study Plan for Learning Chemistry

Which of the following activities would you include in your study plan for learning chemistry successfully?

a. skipping lectures
b. forming a study group
c. keeping a problem notebook
d. waiting to study the night before the exam
e. becoming an active learner

ANSWER

Your success in chemistry can be improved by

b. forming a study group
c. keeping a problem notebook
e. becoming an active learner

QUESTIONS AND PROBLEMS

A Study Plan for Learning Chemistry

1.5 What are four things you can do to help yourself to succeed in chemistry?

1.6 What are four things that would make it difficult for you to succeed in chemistry?

1.7 A student in your class asks you for advice on learning chemistry. Which of the following might you suggest?
a. Form a study group.
b. Skip lectures.
c. Visit the professor during office hours.
d. Wait until the night before an exam to study.
e. Become an active learner.
f. Work the exercises in the *Study Guide*.

1.8 A student in your class asks you for advice on learning chemistry. Which of the following might you suggest?
a. Do the assigned problems.
b. Don't read the book; it's never on the test.
c. Attend review sessions.
d. Read the assignment before a lecture.
e. Keep a problem notebook.
f. Do the tutorials at *www.masteringchemistry.com*.

LEARNING GOAL

Write the names and abbreviations for the metric or SI units used in measurements of length, volume, mass, temperature, and time.

1.3 Units of Measurement

Scientists and health professionals throughout the world use the **metric system** of measurement. It is also the common measuring system in all but a few countries in the world. In 1960, scientists adopted a modification of the metric system called the **International System of Units (SI)**, or Système International, which is now the official system of measurement throughout the world except for the United States. In chemistry, we use metric units and SI units for length, volume, mass, temperature, and time, as listed in Table 1.3.

TABLE 1.3 Units of Measurement

Measurement	Metric	SI
Length	meter (m)	meter (m)
Volume	liter (L)	cubic meter (m^3)
Mass	gram (g)	kilogram (kg)
Temperature	degree Celsius (°C)	kelvin (K)
Time	second (s)	second (s)

Suppose today you walk 2.1 km to campus carrying a backpack that has a mass of 12 kg. Perhaps it is 8:30 A.M. and the temperature is 22 °C. You have a mass of 58.2 kg and a height of 165 cm. You may be more familiar with these measurements stated in the U.S. system of measurement; then you would walk 1.3 mi carrying a backpack that weighs 26 lb. The temperature at 8:30 A.M. would be 72 °F. You have a weight of 128 lb and a height of 65 in.

There are many measurements in everyday life.

Length

The metric and SI unit of length is the **meter (m)**. A meter is 39.4 inches (in.), which makes it slightly longer than a yard (yd). The **centimeter (cm)**, a smaller unit of length, is commonly used in chemistry and is about equal to the width of your little finger. For comparison, there are 2.54 cm in 1 in. (see Figure 1.2).

Some relationships between units for length are

$$1 \text{ m} = 100 \text{ cm}$$
$$1 \text{ m} = 39.4 \text{ in.}$$
$$1 \text{ m} = 1.09 \text{ yd}$$
$$2.54 \text{ cm} = 1 \text{ in.}$$

Volume

Volume is the amount of space a substance occupies. A **liter (L)** is slightly larger than the quart (qt), (1 L = 1.06 qt). In a laboratory or a hospital, chemists work with metric units

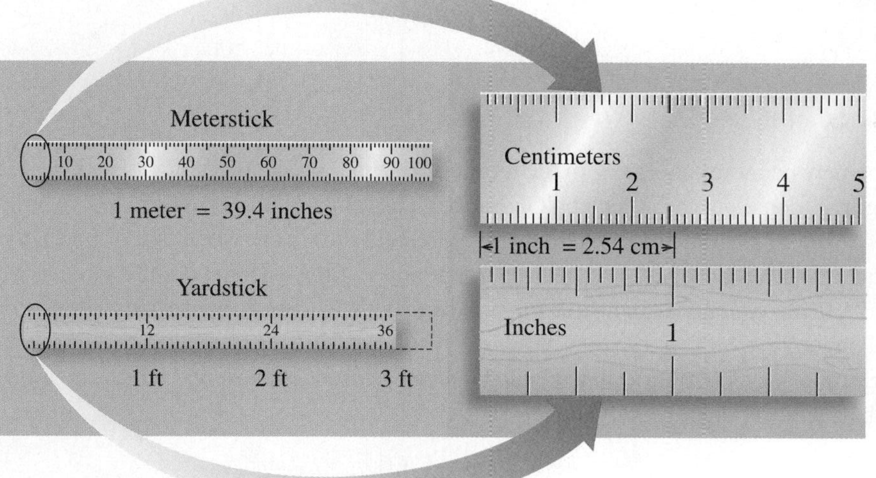

FIGURE 1.2 Length in the metric (SI) system is based on the meter, which is slightly longer than a yard.
Q How many centimeters are in a length of 1 inch?

of volume that are smaller and more convenient, such as the **milliliter (mL)**. There are 1000 mL in 1 L. A comparison of metric and U.S. units for volume appears in Figure 1.3.

Some relationships between units for volume are

$$1\ \text{L} = 1000\ \text{mL}$$
$$1\ \text{L} = 1.06\ \text{qt}$$
$$946\ \text{mL} = 1\ \text{qt}$$

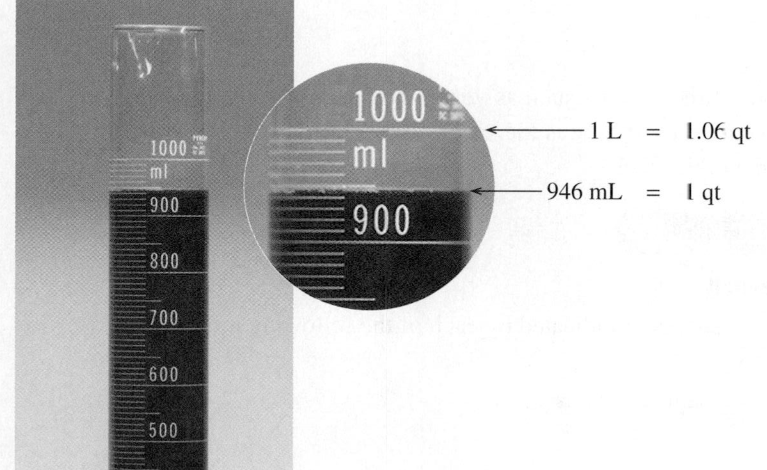

FIGURE 1.3 Volume is the space occupied by a substance. In the metric system, volume is based on the liter, which is slightly larger than a quart.
Q How many milliliters are in 1 quart?

The standard kilogram for the United States is stored at the National Institute of Standards and Technology (NIST).

Mass

The **mass** of an object is a measure of the quantity of material it contains. The SI unit of mass, the **kilogram (kg)**, is used for larger masses such as body mass. In the metric system, the unit for mass is the **gram (g)**, which is used for smaller masses. There are 1000 g in one kilogram. It takes 2.20 lb to make 1 kg, and 454 g is equal to 1 lb.

FIGURE 1.4 On an electronic balance, a nickel has a mass of 5.01 g in the digital readout.
Q What is the mass of 10 nickels?

FIGURE 1.5 A thermometer is used to determine temperature.
Q What kinds of temperature readings have you made today?

A stopwatch is used to measure the time of a race.

Some relationships between units for mass are

1 kg = 1000 g

1 kg = 2.20 lb

454 g = 1 lb

You may be more familiar with the term *weight* than with mass. Weight is a measure of the gravitational pull on an object. On Earth, an astronaut with a mass of 75.0 kg has a weight of 165 lb. On the moon where the gravitational pull is one-sixth that of Earth, the astronaut has a weight of 27.5 lb. However, the mass of the astronaut is the same as on Earth, 75.0 kg. Scientists measure mass rather than weight because mass does not depend on gravity.

In a chemistry laboratory, an analytical balance is used to measure the mass in grams of a substance (see Figure 1.4).

Temperature

Temperature tells us how hot something is, how cold it is outside, or helps us determine if we have a fever (see Figure 1.5). In the metric system, temperature is measured using Celsius temperature. On the **Celsius (°C) temperature scale**, water freezes at 0 °C and boils at 100 °C, while on the Fahrenheit (°F) scale, water freezes at 32 °F and boils at 212 °F. In the SI system, temperature is measured using the **Kelvin (K) temperature scale** on which the lowest possible temperature is 0 K. A unit on the Kelvin scale is called a kelvin (K) and is not written with a degree sign. We will discuss the relationships between these three temperature scales in Chapter 2.

Time

We typically measure time in units such as years, days, hours, minutes, or seconds. Of these, the SI and metric unit of time is the **second (s)**. The standard now used to determine a second is an atomic clock.

CONCEPT CHECK 1.2

Units of Measurement

State the type of measurement indicated by each of the following metric units:

a. gram **b.** liter
c. centimeter **d.** degree Celsius

ANSWER

a. A gram is a unit of mass.
b. A liter is a unit of volume.
c. A centimeter is a unit of length.
d. A degree Celsius is a unit of temperature.

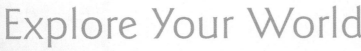

Explore Your World

UNITS LISTED ON LABELS

Read the labels on a variety of products such as sugar, salt, soft drinks, vitamins, and dental floss.

QUESTIONS

1. What metric or SI units of measurement are listed on the labels?
2. What type of measurement (mass, volume, etc.) do they indicate?
3. Write the metric or SI amounts in terms of a number plus a unit.

QUESTIONS AND PROBLEMS

Units of Measurement

1.9 Compare the units you would use and the units that a student in Mexico would use to measure each of the following:
a. your body mass
b. your height
c. amount of gasoline to fill a gas tank
d. temperature at noon

1.10 Why are each of the following statements confusing, and how would you make them clear using metric (SI) units?
a. I rode my bicycle for a distance of 15 today.
b. My dog weighs 15.
c. It is hot today. It is 30.
d. I lost 1.5 last week.

1.11 State the name of the unit and the type of measurement indicated for each of the following quantities:
a. 4.8 m **b.** 325 g **c.** 1.5 mL
d. 480 s **e.** 28 °C

1.12 State the name of the unit and the type of measurement indicated for each of the following quantities:
a. 0.8 L **b.** 3.6 cm **c.** 4 kg
d. 35 lb **e.** 373 K

1.4 Scientific Notation

LEARNING GOAL

Write a number in scientific notation.

TUTORIAL
Scientific Notation

Scientific Notation

In chemistry, we use numbers that are very large or very small. We might measure something as tiny as the width of a human hair, which is about 0.000 008 m. Or perhaps we want to count the number of hairs in the average human scalp, which is about 100 000 hairs (see Figure 1.6). In this text, we add spaces between sets of three digits when it helps make the places easier to count. However, it is more convenient to write large and small numbers in *scientific notation.*

Item	Standard Number	Scientific Notation
Width of a human hair	0.000 008 m	8×10^{-6} m
Hairs on a human scalp	100 000 hairs	1×10^{5} hairs

FIGURE 1.6 Humans have an average of 1×10^{5} hairs on their scalps. Each hair is about 8×10^{-6} m wide.
Q Why are large and small numbers written in scientific notation?

Writing a Number in Scientific Notation

A number written in **scientific notation** has three parts: a coefficient, a power of 10, and a measurement unit. For example, the number 2400 m is written in scientific notation as 2.4×10^{3} m. The coefficient is 2.4, the 3 is the power of 10, and m is the measurement unit of meters. The coefficient is determined by moving the decimal point to the left to give a coefficient that is at least 1 but less than 10. Because we moved the decimal point three places, the power of 10 is 3, which is written as 10^3. For any number greater than 1, the power of 10 is positive.

$$\underbrace{2400.}_{\leftarrow 3\text{ places}}\text{ m} = 2.4 \times 1000 = \underset{\text{Coefficient}}{2.4} \times \underset{\text{Power of 10}}{10^3}\text{ m}$$

When a number less than 1 is written in scientific notation, the power of 10 is a negative number. For example, the number 0.000 86 g is written in scientific notation by moving the decimal point four places to give a coefficient of 8.6. Because the decimal point was moved four places to the right, the power of 10 is a negative 4, written as 10^{-4}.

$$\underbrace{0.0008}_{4\text{ places}\rightarrow}6\text{ g} = \frac{8.6}{10\,000} = \frac{8.6}{10 \times 10 \times 10 \times 10} = \underset{\text{Coefficient}}{8.6} \times \underset{\text{Power of 10}}{10^{-4}}\text{ g}$$

Table 1.4 gives some examples of numbers written as positive and negative powers of 10. The powers of 10 are really a way of keeping track of the decimal point in the decimal number. Table 1.5 gives several examples of writing measurements in scientific notation.

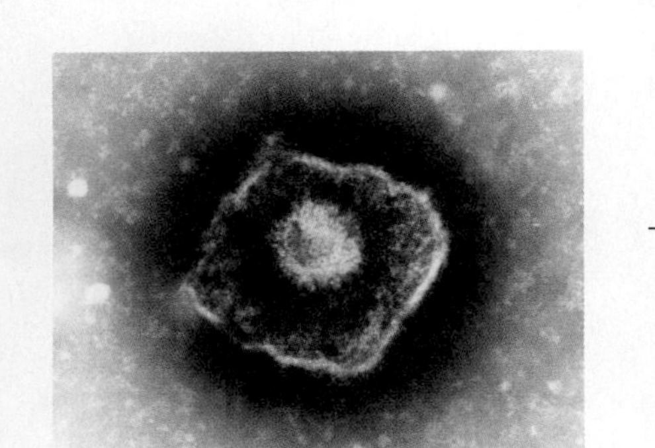

A chickenpox virus has a diameter of 3×10^{-7} m.

TABLE 1.4 Some Powers of 10

Number	Multiples of 10	Scientific Notation	
10 000	$10 \times 10 \times 10 \times 10$	1×10^4	Some positive powers of 10
1 000	$10 \times 10 \times 10$	1×10^3	
100	10×10	1×10^2	
10	10	1×10^1	
1	0	1×10^0	
0.1	$\frac{1}{10}$	1×10^{-1}	Some negative powers of 10
0.01	$\frac{1}{10} \times \frac{1}{10} = \frac{1}{100}$	1×10^{-2}	
0.001	$\frac{1}{10} \times \frac{1}{10} \times \frac{1}{10} = \frac{1}{1\,000}$	1×10^{-3}	
0.0001	$\frac{1}{10} \times \frac{1}{10} \times \frac{1}{10} \times \frac{1}{10} = \frac{1}{10\,000}$	1×10^{-4}	

TABLE 1.5 Some Measurements Written in Scientific Notation

Measured Quantity	Standard Number	Scientific Notation
Volume of gasoline used in United States each year	550 000 000 000 L	5.5×10^{11} L
Diameter of Earth	12 800 000 m	1.28×10^7 m
Time for light to travel from the Sun to Earth	500 s	5×10^2 s
Mass of a typical human	68 kg	6.8×10^1 kg
Mass of a hummingbird	0.002 kg	2×10^{-3} kg
Diameter of a chickenpox (varicella zoster) virus	0.000 000 3 m	3×10^{-7} m
Mass of bacterium (mycoplasma)	0.000 000 000 000 000 000 1 kg	1×10^{-19} kg

Scientific Notation and Calculators

You can enter a number in scientific notation on many calculators using the EE or EXP key. After you enter the coefficient, press the EXP (or EE) key and enter only the power of 10, because the EXP function key already includes the $\times$ 10 value. To enter a negative power of ten, press the plus/minus (+/−) key or the minus (−) key, depending on your calculator.

Number to Enter	Procedure	Calculator Display
4×10^6	4 EXP (EE) 6	4 06 or 4^{06} or 4 E06
2.5×10^{-4}	2.5 EXP (EE) +/− 4	2.5−04 or 2.5^{-04} or 2.5 E−04

When a calculator display appears in scientific notation, it is shown as a number between 1 and 10 followed by a space and the power of 10. To express this display in scientific notation, write the coefficient value, write × 10, and use the power of 10 as an exponent.

Calculator Display	Expressed in Scientific Notation
7.52 04 or 7.52^{04} or 7.52 E04	7.52×10^4
5.8−02 or 5.8^{-02} or 5.8 E−02	5.8×10^{-2}

SAMPLE PROBLEM 1.2

Scientific Notation

Write each of the following in scientific notation:

a. 75 000 m **b.** 0.0098 g **c.** 143 mL

SOLUTION

a. To write a coefficient of 7.5, which is more than 1 but less than 10, move the decimal point four places to the left to give 7.5×10^4 m.

b. To write a coefficient of 9.8, which is more than 1 but less than 10, move the decimal point three places to the right to give 9.8×10^{-3} g.

c. To write a coefficient of 1.43, which is more than 1 but less than 10, move the decimal point two places to the left to give 1.43×10^2 mL.

STUDY CHECK 1.2

Write each of the following in scientific notation:

a. 425 000 m **b.** 0.000 000 86 g

QUESTIONS AND PROBLEMS

Scientific Notation

1.13 Write each of the following in scientific notation:
a. 55 000 m **b.** 480 g **c.** 0.000 005 cm
d. 0.000 14 s **e.** 0.0072 L **f.** 670 000 kg

1.14 Write each of the following in scientific notation:
a. 180 000 000 g **b.** 0.000 06 m **c.** 750 °C
d. 0.15 mL **e.** 0.024 s **f.** 1500 cm

1.15 Which number in each of the following pairs is larger?
a. 7.2×10^3 cm or 8.2×10^2 cm
b. 4.5×10^{-4} kg or 3.2×10^{-2} kg
c. 1×10^4 L or 1×10^{-4} L
d. 0.000 52 m or 6.8×10^{-2} m

1.16 Which number in each of the following pairs is smaller?
a. 4.9×10^{-3} s or 5.5×10^{-9} s
b. 1250 kg or 3.4×10^2 kg
c. 0.000 000 4 m or 5.0×10^2 m
d. 2.50×10^2 g or 4×10^5 g

LEARNING GOAL

Identify a number as measured or exact; determine the number of significant figures in a measured number.

1.5 Measured Numbers and Significant Figures

When you make a measurement, you use some type of measuring device. For example, you may use a meterstick to measure your height, a scale to check your weight, or a thermometer to take your temperature. **Measured numbers** are the numbers you obtain when you measure a quantity such as your height, weight, or temperature.

Measured Numbers

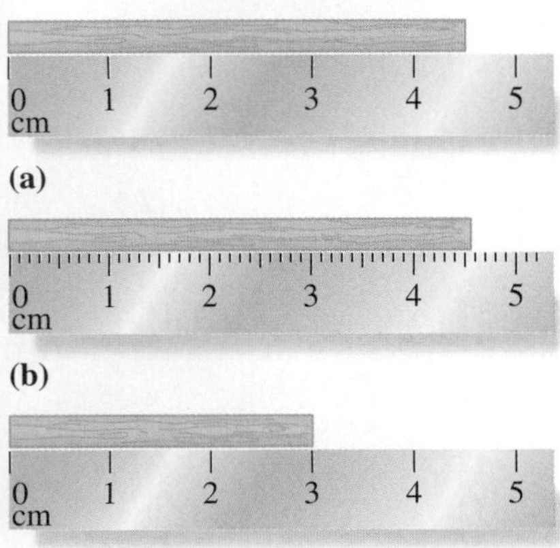

FIGURE 1.7 The lengths of the rectangular objects are measured as **(a)** 4.5 cm and **(b)** 4.55 cm.
Q What is the length of the object in (c)?

Suppose you are going to measure the lengths of the objects in Figure 1.7. You would select a metric ruler that may have lines marked in 1 cm divisions or perhaps in divisions of 0.1 cm. To report the length of the object, you observe the numerical values of the marked lines at the end of object. Then, you can *estimate* by visually dividing the space between the marked lines. This estimated value is the final digit in a measured number.

For example, in Figure 1.7a, the end of the object is between the marks of 4 cm and 5 cm, which means that the length is more than 4 cm but less than 5 cm. This is written as 4 cm plus an estimated digit. If you estimate that the end of the object is halfway between 4 cm and 5 cm, you would report its length as 4.5 cm. The last digit in a measured number may differ because people do not always estimate in the same way. Thus, someone else might report the length of the same object as 4.4 cm.

The metric ruler shown in Figure 1.7b is marked with lines at every 0.1 cm. With this ruler, you can estimate to the hundredths place (0.01 cm). Now you can determine that the end of the object is between 4.5 cm and 4.6 cm. Perhaps you report the length of the object as 4.55 cm, while another student reports its length as 4.56 cm. Both results are acceptable.

In Figure 1.7c, the end of the object appears to line up with the 3-cm mark. Because the divisions are marked in units of 1 cm, the length of the object is between 3 cm and 4 cm. Because the end of the object is on the 3-cm mark, the estimated digit is 0, which means the measurement is reported as 3.0 cm.

SELF STUDY ACTIVITY
Significant Figures

TUTORIAL
Counting Significant Figures

Significant Figures

In a measured number, the **significant figures (SFs)** are all the digits including the estimated digit. Nonzero numbers are always counted as significant figures. However, a zero may or may not be a significant figure depending on its position in a number. Table 1.6 gives the rules and examples of counting significant figures.

TABLE 1.6 Significant Figures in Measured Numbers

Rule	Measured Number	Number of Significant Figures
1. A number is a *significant figure* if it is		
a. not a zero	4.5 g	2
	122.35 m	5
b. a zero between nonzero digits	205 m	3
	5.082 kg	4
c. a zero at the end of a decimal number	50. L	2
	25.0 °C	3
	16.00 g	4
d. in the coefficient of a number written in scientific notation	4.8×10^5 m	2
	5.70×10^{-3} g	3
2. A zero is *not significant* if it is		
a. at the beginning of a decimal number	0.0004 s	1
	0.075 cm	2
b. used as a placeholder in a large number without a decimal point	850 000 m	2
	1 250 000 g	3

When one or more zeros in a large number are significant digits, they are shown more clearly by writing the number using scientific notation. For example, if the first zero in the measurement 500 m is significant, it is written as 5.0×10^2 m. In this text, we will place a decimal point at the end of a number if the zeros are significant. For example, a measurement written as 500. g indicates that all the zeros are significant. Thus 500. g has three significant figures. It could also be written as 5.00×10^2 g. We will assume that zeros at the end of large numbers without a decimal point are not significant. Therefore, we would write a value of 500 g as 5×10^2 g, which has only one significant figure.

CONCEPT CHECK 1.3

Significant Zeros

Identify the significant and nonsignificant zeros in each of the following measured numbers:

a. 0.000 250 m **b.** 70.040 g **c.** 1 020 000 L

ANSWER

a. The zeros preceding the first nonzero digit of 2 are not significant. The zero in the last decimal place following the 5 is significant.

b. Zeros between nonzero digits or at the end of decimal numbers are significant. All zeros in 70.040 g are significant.

c. Zeros between nonzero digits are significant. Zeros at the end of a large number with no decimal point are placeholders but not significant. The zero between 1 and 2 is significant, but the four zeros following the 2 are not significant.

Exact Numbers

Exact numbers are numbers obtained by counting items or using a definition that compares two units in the same measuring system. Suppose a friend asks you how many classes you are taking. You would answer by counting the number of classes in your schedule. It would not use any measuring tool. Suppose you are asked to state the number of seconds in one minute. Without using any measuring device, you would give the definition: There are 60 seconds in 1 minute. Exact numbers are not measured, do not have a limited number of significant figures, and do not affect the number of significant figures in a calculated answer. For more examples of exact numbers, see Table 1.7.

The number of baseballs is counted, which means 2 is an exact number.

TABLE 1.7 Examples of Some Exact Numbers

	Defined Equalities	
Counted Numbers	**U.S. System**	**Metric System**
8 doughnuts	1 ft = 12 in.	1 L = 1000 mL
2 baseballs	1 qt = 4 cups	1 m = 100 cm
5 capsules	1 lb = 16 oz	1 kg = 1000 g

CONCEPT CHECK 1.4

Measured Numbers and Significant Figures

Identify each of the following numbers as measured or exact, and give the number of significant figures in each of the measured numbers:

a. 42.2 g **b.** three eggs **c.** 5.0×10^{-3} cm
d. 450 000 km **e.** The value of 12 in. for 1 ft = 12 in.

ANSWER

a. The value 42.2 g is a measured number, which has three SFs because all nonzero digits are significant.

b. The value three eggs is obtained by counting, which makes it an exact number.

c. The value 5.0×10^{-3} cm is a measured number, which has two SFs because all numbers in the coefficient of a number written in scientific notation are significant.

d. The value 450 000 km is a measured number, which has two SFs because the zeros in a large number without a decimal point are not significant.

e. The value 12 in. is exact because the relationship 1 ft = 12 in. is a definition in the U.S. system of measurement.

QUESTIONS AND PROBLEMS

Measured Numbers and Significant Figures

1.17 Identify the numbers in each of the following statements as measured or exact:
a. A patient weighs 155 lb.
b. A patient is given 2 tablets of medication.
c. In the metric system, 1 kg is equal to 1000 g.
d. The distance from Denver, Colorado, to Houston, Texas, is 1720 km.

1.18 Identify the numbers in each of the following statements as measured or exact:
a. There are 31 students in the laboratory.
b. The oldest known flower lived 1.20×10^{8} years ago.
c. The largest gem ever found, an aquamarine, has a mass of 104 kg.
d. A laboratory test shows a blood cholesterol level of 184 mg/dL.

1.19 Identify the measured number(s), if any, in each of the following pairs of numbers:
a. 3 hamburgers and 6 oz of hamburger
b. 1 table and 4 chairs
c. 0.75 lb of grapes and 350 g of butter
d. 60 s = 1 min

1.20 Identify the exact number(s), if any, in each of the following pairs of numbers:
a. 5 pizzas and 50.0 g of cheese
b. 6 nickels and 16 g of nickel
c. 3 onions and 3 lb of onions
d. 5 miles and 5 cars

1.21 Indicate if the zeros are significant or not in each of the following measurements:
a. 0.0038 m **b.** 5.04 cm **c.** 800. L
d. 3.0×10^{-3} kg **e.** 85 000 g

1.22 Indicate if the zeros are significant or not in each of the following measurements:
a. 20.05 °C **b.** 5.00 m **c.** 0.000 02 g
d. 120 000 years **e.** 8.05×10^{2} L

1.23 How many significant figures are in each of the following measured quantities?
a. 11.005 g **b.** 0.000 32 m **c.** 36 000 000 km
d. 1.80×10^{4} kg **e.** 0.8250 L **f.** 30.0 °C

1.24 How many significant figures are in each of the following measured quantities?
a. 20.60 mL **b.** 1036.48 kg **c.** 4.00 m
d. 20.8 °C **e.** 60 800 000 g **f.** 5.0×10^{-3} L

1.25 In which of the following pairs do both numbers contain the same number of significant figures?
a. 11.0 m and 11.00 m
b. 405 K and 504.0 K
c. 0.000 12 s and 12 000 s
d. 250.0 L and 2.5×10^{-2} L

1.26 In which of the following pairs do both numbers contain the same number of significant figures?
a. 0.005 75 g and 5.75×10^{-3} g
b. 0.0250 m and 0.205 m
c. 150 000 s and 1.50×10^{4} s
d. 3.8×10^{-2} L and 7.5×10^{5} L

1.6 Significant Figures in Calculations

LEARNING GOAL

Adjust calculated answers to give the correct number of significant figures.

TUTORIAL
Significant Figures in Calculations

In the sciences, we measure many things: the length of a bacterium, the volume of a gas sample, the temperature of a reaction mixture, or the mass of iron in a sample. The numbers obtained from these types of measurements are often used in calculations. The number of significant figures in the measured numbers determines the number of significant figures in the reported or final answer.

Using a calculator will help you do calculations faster. However, calculators cannot think for you. It is up to you to enter the numbers correctly, press the correct function keys, and adjust the numbers obtained in the calculator display to give a final answer with the correct number of significant figures.

Rounding Off

Suppose you decide to buy carpeting for a room that measures 5.52 m by 3.58 m. Each of these measurements has three significant figures because the measuring tape limits your estimated place to 0.01 m. To determine how much carpeting you need, you would calculate the area of the room by multiplying 5.52 times 3.58 on your calculator. The calculator shows the numbers 19.7616 in its display. However, this is not the correct final answer because there are too many numbers, which is the result of the multiplication process. Because each of the original measurements has only three significant figures, the numbers 19.7616 in the calculator display must be *rounded off* to obtain an answer that also has three significant figures, which is 19.8. Therefore, you can order carpeting that will cover an area of 19.8 m^2.

A technician uses a calculator in the laboratory.

Each time you use a calculator, it is important to look at the measurements and determine the number of significant figures that can be used for the answer. You can use the following rules to round off the numbers shown in a calculator display.

TUTORIAL
Using Significant Figures

Rules for Rounding Off

1. If the first digit to be dropped is *4 or less*, then it and all following digits are simply dropped from the number.
2. If the first digit to be dropped is *5 or greater*, then the last retained digit of the number is increased by 1.

	Three Significant Figures	Two Significant Figures
Example 1: 8.4234 rounds off to	8.42	8.4
Example 2: 14.780 rounds off to	14.8	15
Example 3: 3256 rounds off to	3260* (3.26×10^3)	3300* (3.3×10^3)

*The value of a large number is retained by using placeholder zeros to replace dropped digits.

CONCEPT CHECK 1.5

Rounding Off

Select the correct value when 2.8456 m is rounded off to each of the following:

a. three significant figures:	2.84 m	2.85 m	2.8 m	2.90 m
b. two significant figures:	2.80 m	2.85 m	2.8 m	2.90 m

ANSWER

a. To round off 2.8456 m to three significant figures, drop the final digits 56 and increase the last retained digit by 1 to give 2.85 m.

b. To round off 2.8456 m to two significant figures, drop the final digits 456 to give 2.8 m.

SAMPLE PROBLEM 1.3

Rounding Off

Round off each of the following numbers to three significant figures:

a. 35.7823 m **b.** 0.002 625 L
c. 3.8268×10^3 g **d.** 1.2836 kg

SOLUTION

a. 35.8 m **b.** 0.002 63 L
c. 3.83×10^3 g **d.** 1.28 kg

STUDY CHECK 1.3

Round off each of the numbers in Sample Problem 1.3 to two significant figures.

A calculator is helpful in working problems and doing calculations faster.

Multiplication and Division

In multiplication or division, the final answer is written so it has the same number of significant figures as the measurement with the *fewest significant figures* (SFs).

Example 1

Multiply the following measured numbers: 24.65×0.67

24.65	×	0.67	=	16.5155	⟶	17
Four SFs		Two SFs		Calculator display		Final answer, rounded off to two SFs

The answer in the calculator display has more digits than the data allow. The measurement 0.67 has the fewer number of significant figures, two. Therefore, the calculator answer is rounded off to two significant figures.

Example 2

Solve the following:

$$\frac{2.85 \times 67.4}{4.39}$$

To do this problem on a calculator, we might press the keys in the following order:

2.85	×	67.4	÷	4.39	=	43.756264	⟶	43.8
Three SFs		Three SFs		Three SFs		Calculator display		Final answer, rounded off to three SFs

All the measurements in this problem have three significant figures. Therefore, the calculator result is rounded off to give an answer, 43.8, that has three significant figures.

Adding Significant Zeros

Sometimes, a calculator display gives a small whole number. To report an answer with the correct number of significant figures, you may need to write significant zeros after the calculator numbers. For example, suppose the calculator display is 4, but the measurements each have three significant figures. Then the display of 4 is written as 4.00 by adding two significant zeros.

$\frac{8.00}{2.00}$ =	4	⟶	4.00
Three SFs	Calculator display		Final answer, two zeros added to give three SFs

CONCEPT CHECK 1.6

Determination of Significant Figures for Final Answer

Perform the following calculation of measured numbers. Give the final answer with the correct number of significant figures.

$$\frac{8.560}{(2.84)(0.078)}$$

ANSWER

The calculation is performed as follows:

8.560	÷	2.84	÷	0.078	=	38.64210906	⟶	39
Four SFs		Three SFs		Two SFs		Calculator display		Final answer, rounded off to two SFs

The measurement with only two SFs limits the final answer to two SFs. To obtain the final answer, the calculator display must be rounded off by dropping all the digits after the 38. Because the first dropped digit 6 is greater than 5, the last retained digit is increased by one to give the final answer of 39.

SAMPLE PROBLEM 1.4

Significant Figures in Multiplication and Division

Perform the following calculations of measured numbers. Give the answers with the correct number of significant figures.

a. 56.8×0.37 **b.** $\frac{(2.075)(0.585)}{(8.42)(0.0045)}$ **c.** $\frac{25.0}{5.00}$

SOLUTION

a. 21 **b.** 32 **c.** 5.00 (add two significant zeros)

STUDY CHECK 1.4

Perform the following calculations of measured numbers and give the answers with the correct number of significant figures:

a. $45.26 \times 0.010\,88$ **b.** $2.6 \div 324$ **c.** $\frac{4.0 \times 8.00}{16}$

Addition and Subtraction

In addition or subtraction, the final answer is written so it has the same number of decimal places as the measurement with the *fewest decimal places.*

Example 3

Add:

2.045	Three decimal places
+ 34.1	One decimal place
36.145	Calculator display
36.1	Answer, rounded off to one decimal place

When numbers are added or subtracted to give an answer ending in zero, the zero does not appear after the decimal point in the calculator display. For example, 14.5 g − 2.5 g = 12.0 g. However, if you do the subtraction on your calculator, the display shows 12. To write the correct answer, a significant zero is written after the decimal point.

Example 4

Subtract:

14.5 g	One decimal place
− 2.5 g	One decimal place
12.	Calculator display
12.0 g	Answer, zero written after the decimal point

SAMPLE PROBLEM 1.5

Decimal Places in Addition and Subtraction

Perform the following calculations and give the answers with the correct number of decimal places:

a. 27.8 cm + 0.235 cm **b.** 153.247 g − 14.82 g

SOLUTION

a. 28.0 cm (rounded off to one decimal place)
b. 138.43 g (rounded off to two decimal places)

STUDY CHECK 1.5

Perform the following calculations and give the answers with the correct number of decimal places:

a. 82.45 mg + 1.245 mg + 0.000 56 mg **b.** 4.259 L − 3.8 L

QUESTIONS AND PROBLEMS

Significant Figures in Calculations

1.27 Why do we usually need to round off calculations that use measured numbers?

1.28 Why do we sometimes add a zero to a number in a calculator display?

1.29 Round off each of the following measurements to three significant figures:
a. 1.854 kg
b. 88.2038 L
c. 0.004 738 265 cm
d. 8807 m
e. 1.832×10^5 s

1.30 Round off each of the measurements in Problem 1.29 to two significant figures.

1.31 Round off or add zeros to each of the following to give an answer with three significant figures:
a. 5080 L **b.** 37 400 g
c. 104 720 m **d.** 0.000 250 82 s

1.32 Round off or add zeros to each of the following to give an answer with two significant figures:
a. 5 100 000 L **b.** 26 711 s
c. 0.003 378 m **d.** 56.982 g

1.33 For each of the following problems, give an answer with the correct number of significant figures:
a. 45.7×0.034 **b.** $0.002\,78 \times 5$
c. $\frac{34.56}{1.25}$ **d.** $\frac{(0.2465)(25)}{1.78}$

1.34 For each of the following problems, give an answer with the correct number of significant figures:
a. 400×185 **b.** $\frac{2.40}{(4)(125)}$
c. $0.825 \times 3.6 \times 5.1$ **d.** $\frac{3.5 \times 0.261}{8.24 \times 20.0}$

1.35 For each of the following, give an answer with the correct number of decimal places:
a. 45.48 cm + 8.057 cm
b. 23.45 g + 104.1 g + 0.025 g
c. 145.675 mL − 24.2 mL
d. 1.08 L − 0.585 L

1.36 For each of the following, give an answer with the correct number of decimal places:
a. 5.08 g + 25.1 g
b. 85.66 cm + 104.10 cm + 0.025 cm
c. 24.568 mL − 14.25 mL
d. 0.2654 L − 0.2585 L

1.7 Prefixes and Equalities

The special feature of the SI as well as the metric system is that a **prefix** can be placed in front of any unit to increase or decrease its size by some factor of ten. For example, the prefixes *milli* and *micro* are used to make the smaller units, milligram (mg) and microgram (μg). Table 1.8 lists some of the metric prefixes, their symbols, and their numerical values.

The prefix *centi* is like cents in a dollar. One cent would be a "centidollar" or 0.01 of a dollar. That also means that one dollar is the same as 100 cents. The prefix *deci* is like dimes in a dollar. One dime would be a "decidollar" or 0.1 of a dollar. That also means that one dollar is the same as 10 dimes.

The U.S. Food and Drug Administration has determined the daily values (DV) of nutrients for adults and children age four or older. Examples of these recommended daily values, some of which use prefixes, are listed in Table 1.9.

The relationship of a prefix to a unit can be expressed by replacing the prefix with its numerical value. For example, when the prefix *kilo* in kilometer is replaced with its value of 1000, we find that a kilometer is equal to 1000 meters. Other examples follow:

1 **kilo**meter (1 km) = **1000** meters (**10**^{3} m)

1 **kilo**liter (1 kL) = **1000** liters (**10**^{3} L)

1 **kilo**gram (1 kg) = **1000** grams (**10**^{3} g)

LEARNING GOAL

Use the numerical values of prefixes to write a metric equality.

SELF STUDY ACTIVITY
The Metric System

TUTORIAL
SI Prefixes and Units

TABLE 1.8 Metric and SI Prefixes

Prefix	Symbol	Numerical Value	Scientific Notation	Equality
Prefixes That Increase the Size of the Unit				
tera	T	1 000 000 000 000	10^{12}	1 Ts = 1×10^{12} s 1 s = 1×10^{-12} Ts
giga	G	1 000 000 000	10^{9}	1 Gm = 1×10^{9} m 1 m = 1×10^{-9} Gm
mega	M	1 000 000	10^{6}	1 Mg = 1×10^{6} g 1 g = 1×10^{-6} Mg
kilo	k	1 000	10^{3}	1 km = 1×10^{3} m 1 m = 1×10^{-3} km
Prefixes That Decrease the Size of the Unit				
deci	d	0.1	10^{-1}	1 dL = 1×10^{-1} L 1 L = 10 dL
centi	c	0.01	10^{-2}	1 cm = 1×10^{-2} m 1 m = 100 cm
milli	m	0.001	10^{-3}	1 ms = 1×10^{-3} s 1 s = 1×10^{3} ms
micro	μ	0.000 001	10^{-6}	1 μg = 1×10^{-6} g 1 g = 1×10^{6} μg
nano	n	0.000 000 001	10^{-9}	1 nm = 1×10^{-9} m 1 m = 1×10^{9} nm
pico	p	0.000 000 000 001	10^{-12}	1 ps = 1×10^{-12} s 1 s = 1×10^{12} ps

TABLE 1.9 Daily Values for Selected Nutrients

Nutrient	Amount Recommended
Protein	44 g
Vitamin C	60 mg
Vitamin B_{12}	6 μg
Calcium	1000 mg
Copper	2 mg
Iodine	150 μg
Iron	18 mg
Magnesium	400 mg
Niacin	20 mg
Potassium	3500 mg
Selenium	55 μg
Sodium	2400 mg
Zinc	15 mg

CONCEPT CHECK 1.7

Prefixes

Fill in the blanks with the correct symbol.

a. 1000 g = 1 ___ g **b.** 0.01 m = 1 ___ m
c. 1×10^6 L = 1 ___ L

ANSWER

a. The prefix for 1000 is *kilo*; 1000 g = 1 kg
b. The prefix for 0.01 is *centi*; 0.01 m = 1 cm
c. The prefix for 1×10^6 is *mega*; 1×10^6 L = 1 ML

A terabyte hard disk drive stores 10^{12} bytes of information.

SAMPLE PROBLEM 1.6

Prefixes

The storage capacity for a hard disk drive (HDD) is specified using prefixes: megabyte (MB), gigabyte (GB), or terabyte (TB). Indicate the storage capacity in bytes for each of the following hard disk drives. Suggest a reason for describing a HDD storage capacity in gigabytes or terabytes.

a. 5 MB **b.** 2 GB

SOLUTION

a. The prefix *mega* (M) in MB is equal to 1 000 000 or 1×10^6. Thus, 5 MB is equal to 5 000 000 (5×10^6) bytes.
b. The prefix *giga* (G) in GB is equal to 1 000 000 000 or 1×10^9. Thus, 2 GB is equal to 2 000 000 000 (2×10^9) bytes.

Expressing HDD capacity in gigabytes or terabytes gives a more reasonable number to work with than a number with many zeros or a large power of 10.

STUDY CHECK 1.6

Hard drives now have a storage capacity of 1.5 TB. How many bytes are stored?

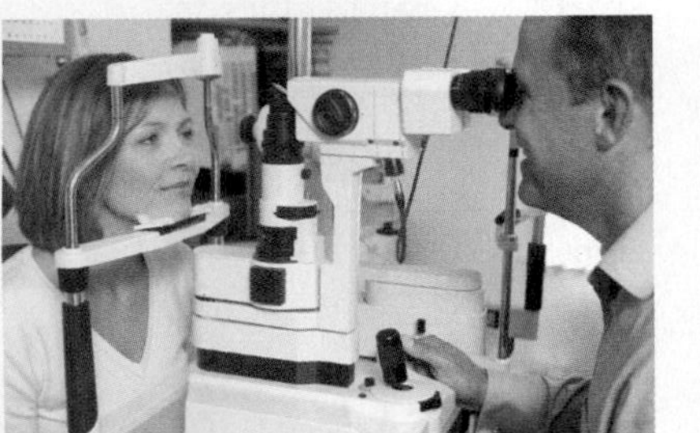

Using a retinal camera, an ophthalmologist photographs the retina of an eye.

First quantity			Second quantity	
1	m	=	100	cm
↑	↑		↑	↑
Number + unit			Number + unit	

This example of an equality shows the relationship between meters and centimeters.

Measuring Length

An ophthalmologist may measure the diameter of the retina of an eye in centimeters (cm), whereas a surgeon may need to know the length of a nerve in millimeters (mm). When the prefix *centi* is used with the unit meter, it becomes *centimeter*, a length that is one-hundredth of a meter (0.01 m). When the prefix *milli* is used with the unit meter, it becomes *millimeter*, and a length that is one-thousandth of a meter (0.001 m). There are 100 cm and 1000 mm in a meter.

If we compare the lengths of a millimeter and a centimeter, we find that 1 mm is 0.1 cm; there are 10 mm in 1 cm. These comparisons are examples of **equalities**, which show the relationship between two units that measure the same quantity. For example, in the equality 1 m = 100 cm, each quantity describes the same length but in a different unit. In every equality, each quantity has both a number and a unit.

Examples of Some Length Equalities

1 m = 100 cm = 1×10^2 cm
1 m = 1000 mm = 1×10^3 mm
1 cm = 10 mm = 1×10^1 mm

Some metric units for length are compared in Figure 1.8.

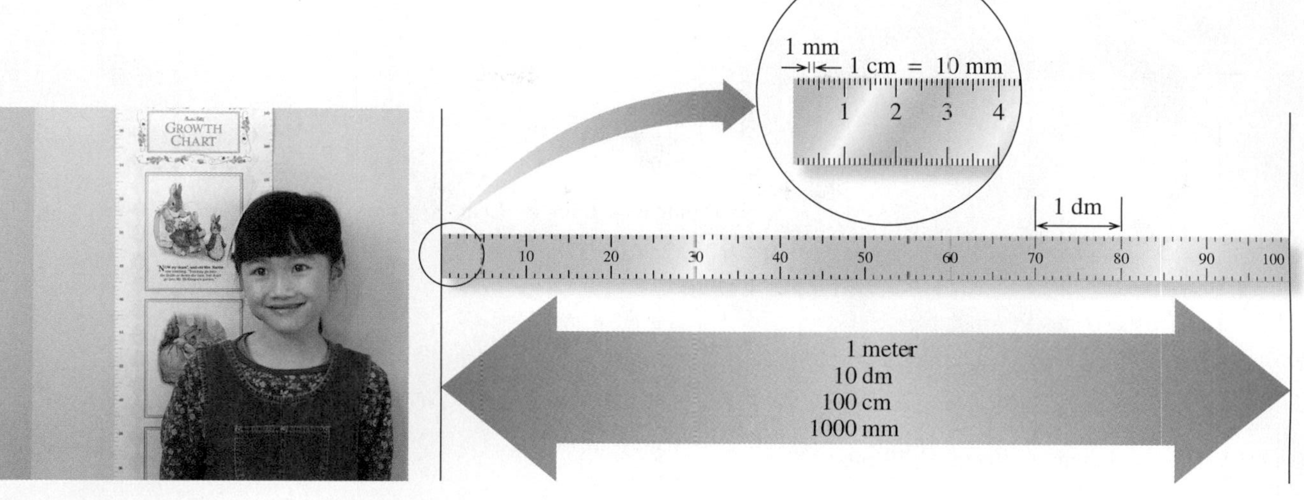

FIGURE 1.8 The metric length of 1 meter is the same length as 10 dm, 100 cm, or 1000 mm.
Q How many millimeters (mm) are in 1 centimeter (cm)?

Measuring Volume

Volumes of 1 L or smaller are common in the health sciences. When a liter is divided into 10 equal portions, each portion is a deciliter (dL). There are 10 dL in 1 L. Laboratory results for blood work are often reported in mass per deciliter. Table 1.10 lists typical laboratory test values for some substances in the blood.

When a liter is divided into a thousand parts, each of the smaller volumes is called a milliliter (mL). In a 1-L container of physiological saline, there are 1000 mL of solution (see Figure 1.9).

Examples of Some Volume Equalities

1 L = 10 dL = 1×10^1 dL

1 L = 1000 mL = 1×10^3 mL

1 dL = 100 mL = 1×10^2 mL

The **cubic centimeter** (abbreviated as $\mathbf{cm^3}$ or **cc**) is the volume of a cube whose dimensions are 1 cm on each side. A cubic centimeter has the same volume as a milliliter, and the units are often used interchangeably.

$1\ cm^3$ = 1 cc = 1 mL

When you see *1 cm*, you are reading about length; when you see *1 cc* or $1\ cm^3$ or *1 mL*, you are reading about volume. A comparison of units of volume is illustrated in Figure 1.10.

TABLE 1.10 Some Typical Laboratory Test Values

Substance in Blood	Typical Range
Albumin	3.5–5.0 g/dL
Ammonia	20–150 μg/dL
Calcium	8.5–10.5 mg/dL
Cholesterol	105–250 mg/dL
Iron (male)	80–160 μg/dL
Protein (total)	6.0–8.0 g/dL

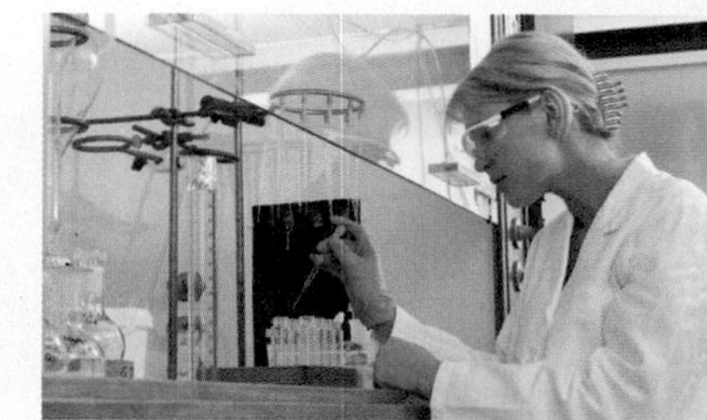

A laboratory technician transfers small volumes using a micropipette.

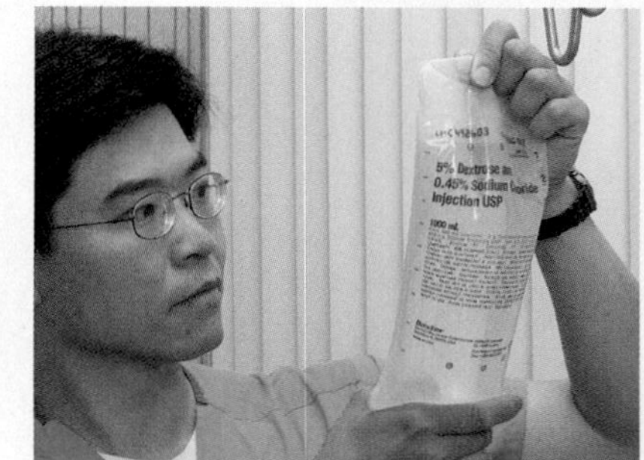

FIGURE 1.9 A plastic intravenous fluid container contains 1000 mL.
Q How many liters of solution are in the intravenous fluid container?

Measuring Mass

When you go to the doctor for a physical examination, your mass is recorded in kilograms, whereas the results of your laboratory tests are reported in grams, milligrams (mg), or micrograms (μg). A kilogram is equal to 1000 g. One gram represents the same mass as 1000 mg, and one mg equals 1000 μg.

Examples of Some Mass Equalities

1 kg = 1000 g = 1×10^3 g

1 g = 1000 mg = 1×10^3 mg

1 mg = 1000 μg = 1×10^3 μg

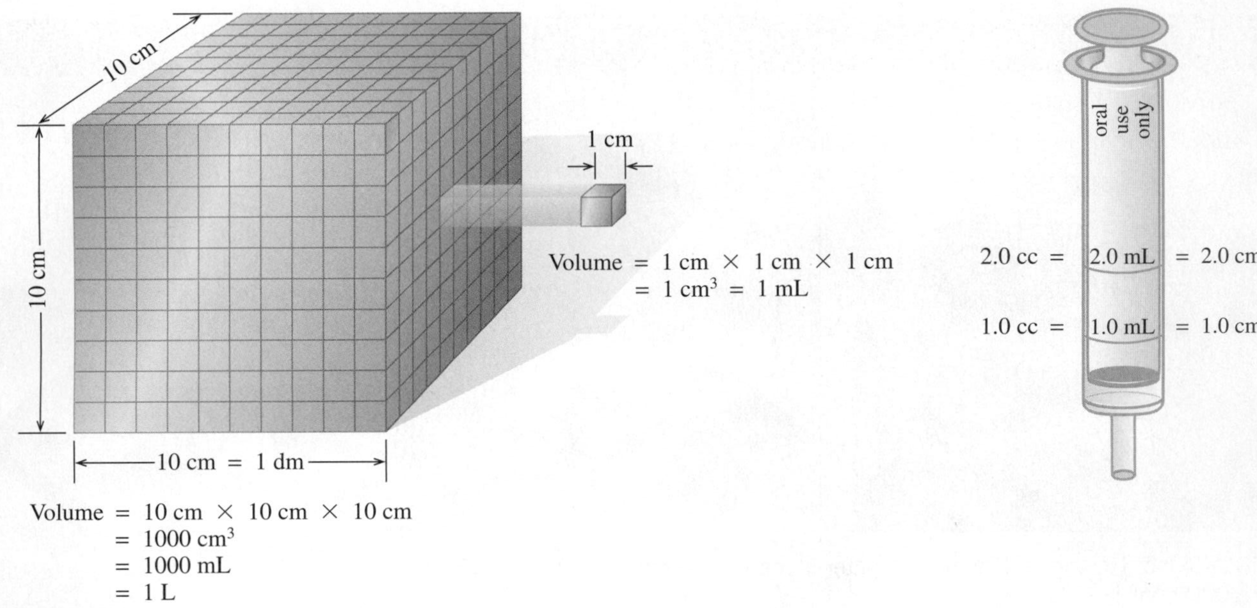

FIGURE 1.10 A cube measuring 10 cm on each side has a volume of 1000 cm^3, or 1 L; a cube measuring 1 cm on each side has a volume of 1 cm^3 (cc) or 1 mL.
Q What is the relationship between a milliliter (mL) and a cubic centimeter (cm^3)?

CONCEPT CHECK 1.8

Metric Relationships

Identify the larger unit in each of the following pairs:

a. centimeter or kilometer
b. mg or μg

ANSWER

a. A kilometer (1000 m) is larger than a centimeter (0.01 m).
b. A mg (0.001 g) is larger than a μg (0.000 001 g).

SAMPLE PROBLEM 1.7

Writing Metric Equalities

Complete the following list of metric equalities:

a. 1 L = ________ dL
b. 1 km = ________ m
c. 1 cm^3 = ________ mL

SOLUTION

a. 10 dL **b.** 1×10^3 m **c.** 1 mL

STUDY CHECK 1.7

Complete each of the following equalities:

a. 1 kg = ________ g
b. 1 mL = ________ L

QUESTIONS AND PROBLEMS

Prefixes and Equalities

1.37 The speedometer is marked in both km/h and mi/h, or mph. What is the meaning of each abbreviation?

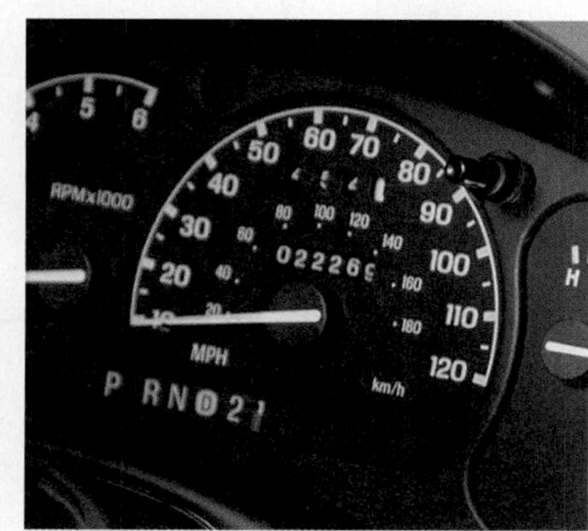

1.38 In a French car, the odometer reads 2250. What units would this be? What units would it be if this were an odometer in a car made for the United States?

1.39 How does the prefix *kilo* affect the gram unit in *kilogram*?

1.40 How does the prefix *centi* affect the meter unit in *centimeter*?

1.41 Write the abbreviation for each of the following units:
a. milligram **b.** deciliter
c. kilometer **d.** picogram
e. microliter **f.** nanosecond

1.42 Write the complete name for each of the following units:
a. cm **b.** ks
c. dL **d.** Gm
e. μg **f.** pg

1.43 Write the numerical values for each of the following prefixes:
a. centi **b.** tera
c. milli **d.** deci
e. mega **f.** pico

1.44 Write the complete name (prefix + unit) for each of the following numerical values:
a. 0.10 g **b.** 0.000 001 g
c. 1000 g **d.** 0.01 g
e. 0.001 g **f.** 1×10^9 g

1.45 Complete each of the following metric relationships:
a. 1 m = ________ cm **b.** 1 m = ________ nm
c. 1 mm = ________ m **d.** 1 L = ________ mL

1.46 Complete each of the following metric relationships:
a. 1 Mg = ________ g **b.** 1 mL = ________ μL
c. 1 g = ________ kg **d.** 1 g = ________ mg

1.47 For each of the following pairs, which is the larger unit?
a. milligram or kilogram **b.** milliliter or microliter
c. m or km **d.** kL or dL
e. nanometer or picometer

1.48 For each of the following pairs, which is the smaller unit?
a. mg or g **b.** centimeter or millimeter
c. mm or μm **d.** mL or dL
e. mg or Mg

1.8 Writing Conversion Factors

LEARNING GOAL

Write a conversion factor for two units that describe the same quantity.

Many problems in chemistry and the health sciences require a change of units. You make changes in units every day. For example, suppose you worked 2.0 hours (h) on your homework, and someone asked you how many minutes that was. You would answer 120 minutes (min). You must have multiplied 2.0 h × 60 min/h because you knew the equality (1 h = 60 min) that related the two units. When you expressed 2.0 h as 120 min, you did not change the amount of time you spent studying. You changed only the unit of measurement used to express the time. You can write any equality in the form of a fraction called a **conversion factor**, in which one of the quantities is the numerator and the other is the denominator. Be sure to include the units when you write the conversion factors. Two conversion factors are always possible from any equality.

TUTORIAL
Metric Conversions

Two Conversion Factors for the Equality 1 h = 60 min

$$\frac{\text{Numerator}}{\text{Denominator}} \begin{array}{c}\longrightarrow \\ \longrightarrow\end{array} \quad \frac{60\text{ min}}{1\text{ h}} \quad \text{and} \quad \frac{1\text{ h}}{60\text{ min}}$$

These factors are read as "60 minutes per 1 hour" and "1 hour per 60 minutes." The term *per* means "divided by." Some common relationships are given in Table 1.11. It is important that the equality you select to form a conversion factor is a true relationship.

When an equality shows the relationship for two units from the same system (metric or U.S.), it is considered a definition and exact. Thus, the numbers in that definition are not used to determine significant figures. When an equality shows the relationship of units from two different systems, the number is measured and counts toward the significant figures in a calculation. For example, in the equality 1 lb = 454 g, the measured number

TABLE 1.11 Some Common Equalities

Quantity	U.S.	Metric (SI)	Metric–U.S.
Length	1 ft = 12 in. 1 yd = 3 ft 1 mi = 5280 ft	1 km = 1000 m 1 m = 1000 mm 1 cm = 10 mm	2.54 cm = 1 in. (exact) 1 m = 39.4 in. 1 km = 0.621 mi
Volume	1 qt = 4 cups 1 qt = 2 pt 1 gal = 4 qt	1 L = 1000 mL 1 dL = 100 mL 1 mL = 1 cm^3	946 mL = 1 qt 1 L = 1.06 qt
Mass	1 lb = 16 oz	1 kg = 1000 g 1 g = 1000 mg	1 kg = 2.20 lb 454 g = 1 lb
Time		1 h = 60 min 1 min = 60 s	

454 has three significant figures. The number one in 1 lb is considered as exact. An exception is the relationship of 1 in. = 2.54 cm: The value 2.54 has been defined as exact.

Metric Conversion Factors

We can write metric conversion factors for any of the metric relationships. For example, from the equality for meters and centimeters, we can write the following factors:

Metric Equality	Conversion Factors
1 m = 100 cm	$\frac{100\text{ cm}}{1\text{ m}}$ and $\frac{1\text{ m}}{100\text{ cm}}$

Both are proper conversion factors for the relationship; one is just the inverse of the other. The usefulness of conversion factors is enhanced by the fact that we can turn a conversion factor over and use its inverse.

CONCEPT CHECK 1.9

Conversion Factors

Identify one or more correct conversion factors for the equality of gigagrams and grams.

a. $\frac{1\text{ Gg}}{1 \times 10^9\text{ g}}$ **b.** $\frac{1 \times 10^{-9}\text{ g}}{1\text{ Gg}}$ **c.** $\frac{1 \times 10^9\text{ Gg}}{1\text{ g}}$ **d.** $\frac{1 \times 10^9\text{ g}}{1\text{ Gg}}$

ANSWER

In Table 1.8, the equality for gigagrams and grams is 1 Gg = 1×10^9 g. Answers **a** and **d** are correctly written conversion factors.

FIGURE 1.11 In the United States, the contents of many packaged foods are listed in both U.S. and metric units.

Q What are some advantages of using the metric system?

Metric–U.S. System Conversion Factors

Suppose you need to convert from pounds, a unit in the U.S. system, to kilograms in the metric (or SI) system. A relationship you could use is

$$1\text{ kg} = 2.20\text{ lb}$$

The corresponding conversion factors would be

$$\frac{2.20\text{ lb}}{1\text{ kg}} \quad \text{and} \quad \frac{1\text{ kg}}{2.20\text{ lb}}$$

Figure 1.11 illustrates the contents of some packaged foods in both U.S. and metric units.

CONCEPT CHECK 1.10

Writing Conversion Factors from Equalities

Write conversion factors for the relationship between the following pairs of units:

a. millimeters and meters
b. quarts and milliliters

ANSWER

Equality	Conversion Factors		
a. 1 m = 1000 mm	$\frac{1\text{ m}}{1000\text{ mm}}$	and	$\frac{1000\text{ mm}}{1\text{ m}}$
b. 1 qt = 946 mL	$\frac{1\text{ qt}}{946\text{ mL}}$	and	$\frac{946\text{ mL}}{1\text{ qt}}$

Conversion Factors Stated within a Problem

Many times, an equality is specified within a problem that applies only to that problem. It might be the price of 1 kilogram of oranges or the speed of a car in kilometers per hour. However, we can write conversion factors for relationships stated within a problem.

1. The motorcycle was traveling at a speed of 85 km/h.
   Equality 1 h = 85 km
   Conversion factors $\frac{85\text{ km}}{1\text{ h}}$ and $\frac{1\text{ h}}{85\text{ km}}$
2. One vitamin tablet contains 500 mg of vitamin C.
   Equality 1 tablet = 500 mg of vitamin C
   Conversion factors $\frac{500\text{ mg vitamin C}}{1\text{ tablet}}$ and $\frac{1\text{ tablet}}{500\text{ mg vitamin C}}$

Conversion Factors from Percent, ppm, and ppb

When a *percent* (%) is given in a problem, it means parts per 100 parts. To write a percent as a conversion factor, we choose a unit and express the numerical relationship of the parts of this unit to 100 parts of the whole. For example, a person might have 18% body fat by mass. The percent quantity can be written as 18 mass units of body fat in every 100 mass units of body mass. Different mass units such as grams (g), kilograms (kg), or pounds (lb) can be used, but both units in the factor must be the same.

Percent quantity 18% body fat by mass
Equality 18 kg of body fat = 100 kg of body mass
Conversion factors $\frac{100\text{ kg body mass}}{18\text{ kg body fat}}$ and $\frac{18\text{ kg body fat}}{100\text{ kg body mass}}$

When scientists want to indicate ratios with particularly small percentage values, they use numerical relationships called *parts per million* (ppm) or *parts per billion* (ppb). The ratio of parts per million is the same as the milligrams of a substance per kilogram (mg/kg). The ratio of parts per billion equals the micrograms per kilogram (μg/kg). For example, the maximum amount of lead allowed by the FDA in glazed pottery bowls is 5 ppm, which is 5 mg/kg.

ppm quantity 5 ppm of lead in glaze
Equality 5 mg of lead = 1 kg of glaze
Conversion factors $\frac{5\text{ mg lead}}{1\text{ kg glaze}}$ and $\frac{1\text{ kg glaze}}{5\text{ mg lead}}$

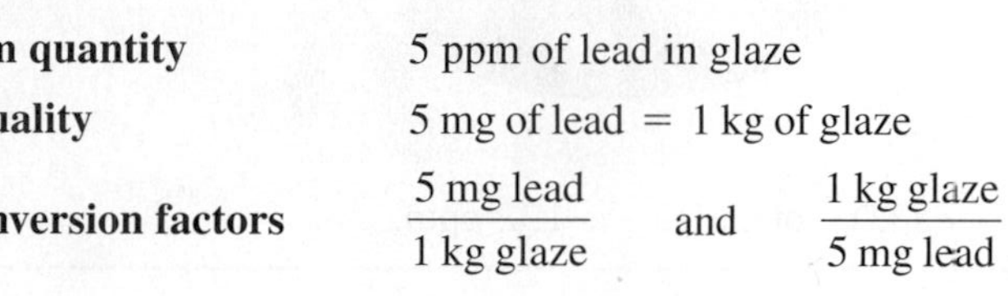

Explore Your World

SI AND METRIC EQUALITIES ON PRODUCT LABELS

Read the labels on some food products on your kitchen shelves and in your refrigerator, or refer to the labels in Figure 1.11. List the amount of product given in different units. Write a relationship for two of the amounts for the same product and container. Look for measurements of grams and pounds or quarts and milliliters.

QUESTIONS

1. Use the stated measurement to derive a metric–U.S. conversion factor.
2. How do your results compare with the conversion factors we have described in this text?

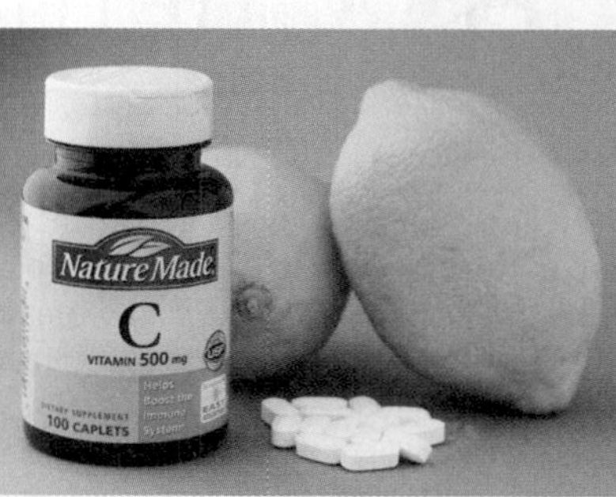

Vitamin C, an antioxidant needed by the body, is found in fruits such as lemons.

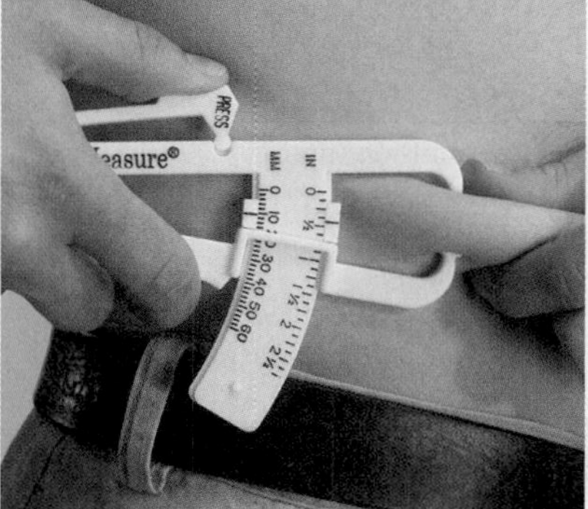

The thickness of the skin fold at the abdomen is used to determine percent body fat.

The maximum amount of mercury allowed in tuna is 0.1 ppm.

SAMPLE PROBLEM 1.8

Conversion Factors Stated in a Problem

Write two conversion factors for each of the following statements:

a. There are 325 mg of aspirin in 1 tablet.
b. The EPA has set the maximum level for mercury in tuna at 0.1 ppm.

SOLUTION

a. $\frac{325\text{ mg aspirin}}{1\text{ tablet}}$ and $\frac{1\text{ tablet}}{325\text{ mg aspirin}}$

b. $\frac{0.1\text{ mg mercury}}{1\text{ kg tuna}}$ and $\frac{1\text{ kg tuna}}{0.1\text{ mg mercury}}$

STUDY CHECK 1.8

What conversion factors can be written for the following statements?

a. A cyclist in the Tour de France bicycle race rides at the average speed of 62.2 km/h.
b. The permissible level of arsenic in water is 10 ppb.

Chemistry Link to Health

TOXICOLOGY AND RISK-BENEFIT ASSESSMENT

Each day we make choices about what we do or what we eat, often without thinking about the risks associated with these choices. We are aware of the risks of cancer from smoking or the risks of lead poisoning, and we know there is a greater risk of having an accident if we cross a street where there is no light or crosswalk.

A basic concept of toxicology is the statement of Paracelsus that the dose is the difference between a poison and a cure. To evaluate the level of danger from various substances, natural or synthetic, a risk assessment is made by exposing laboratory animals to the substances and monitoring the health effects. Often, doses very much greater than humans might ordinarily encounter are given to the test animals.

Many hazardous chemicals or substances have been identified by these tests. One measure of toxicity is the LD_{50}, or lethal dose, which is the concentration of the substance that causes death in 50% of the test animals. A dosage is typically measured in milligrams per kilogram (mg/kg) of body mass or micrograms per kilogram (μg/kg) of body mass.

Dosage	Units
parts per million (ppm)	milligrams per kilogram (mg/kg)
parts per billion (ppb)	micrograms per kilogram (μg/kg)

Other evaluations need to be made, but it is easy to compare LD_{50} values. Parathion, a pesticide, with an LD_{50} of 3 mg/kg would be highly toxic. That means that 3 mg of parathion per kg of body mass would be fatal to half the test animals. Table salt (sodium chloride) with an LD_{50} of 3000 mg/kg would have a much lower toxicity. You would need to ingest a huge amount of salt before any toxic effect would be observed. Although the risk to animals can be evaluated in the laboratory, it is more difficult to determine the impact in the environment since there is also a difference between continued exposure and a single, large dose of the substance.

Table 1.12 lists some LD_{50} values and compares substances in order of increasing toxicity.

TABLE 1.12 Some LD_{50} Values for Substances Tested in Rats

Substance	LD_{50} (mg/kg)
Table sugar	29 700
Baking soda	4220
Table salt	3000
Ethanol	2080
Aspirin	1100
Caffeine	192
DDT	113
Sodium cyanide	6
Parathion	3

The LD_{50} of caffeine is 192 ppm.

QUESTIONS AND PROBLEMS

Writing Conversion Factors

1.49 Why can two conversion factors be written for the equality 1 m = 100 cm?

1.50 How can you check that you have written the correct conversion factors for an equality?

1.51 Write the equality and conversion factors for each of the following:
a. centimeters and meters
b. milligrams and grams
c. liters and milliliters
d. deciliters and milliliters
e. days in 1 week

1.52 Write the equality and conversion factors for each of the following:
a. centimeters and inches **b.** pounds and kilograms
c. pounds and grams **d.** quarts and liters
e. dimes in 1 dollar

1.53 Write the conversion factors for each of the following statements:
a. A bee flies at an average speed of 3.5 m per second.
b. The daily value for potassium is 3500 mg.
c. An automobile traveled 46.0 km on 1.0 gal of gasoline.
d. The label on a bottle reads 50 mg of Atenolol per tablet.
e. The pesticide level in plums was 29 ppb.
f. A low-dose aspirin contains 81 mg of aspirin per tablet.

1.54 Write the conversion factors for each of the following statements:
a. The label on a bottle reads 10 mg of furosemide per mL.
b. The daily value for iodine is 150 μg.
c. The nitrate level in well water was 32 ppm.
d. Gold jewelry contains 58% by mass gold.
e. The price of a gallon of gas is \$3.19.
f. One capsule of fish oil contains 360 mg of omega-3 fatty acids.

1.9 Problem Solving

LEARNING GOAL
Use conversion factors to change from one unit to another.

TUTORIAL
Unit Conversions

TUTORIAL
Metric Conversions

The process of problem solving in chemistry often requires one or more conversion factors to change a given unit to the needed unit. For the problem, the unit of the given quantity and the unit of the needed quantity are identified. From there, the problem is set up with one or more conversion factors used to convert the given unit to the needed unit as seen in the following Sample Problem.

Given unit × one or more conversion factors = needed unit

SAMPLE PROBLEM 1.9

Problem Solving Using Metric Factors

In radiological imaging such as PET or CT scans, dosages of pharmaceuticals are based on body mass. If a person weighs 164 lb, what is the body mass in kilograms?

SOLUTION

Step 1 State the given and needed quantities.

Given 164 lb **Need** kilograms

Step 2 Write a plan to convert the given unit to the needed unit. We see that the given unit is in the U.S. system of measurement and the needed unit in the metric system. Therefore, the conversion factor must relate the U.S. unit lb to the metric unit kg.

pounds → U.S.–Metric factor → kilograms

Step 3 State the equalities and conversion factors needed to cancel units.

1 kg = 2.20 lb

$$\frac{2.20\text{ lb}}{1\text{ kg}} \quad \text{and} \quad \frac{1\text{ kg}}{2.20\text{ lb}}$$

Guide to Problem Solving Using Conversion Factors

1. State the given and needed quantities.
2. Write a plan to convert the given unit to the needed unit.
3. State the equalities and conversion factors needed to cancel units.
4. Set up problem to cancel units and calculate answer.

Step 4 **Set up problem to cancel units and calculate answer.** Write the given, 164 lb, and set up the conversion factor with the unit lb in the denominator (bottom number) to cancel out the given unit (lb) in the numerator.

Unit for answer goes here

$$164\ \cancel{\text{lb}} \times \frac{1\ \text{kg}}{2.20\ \cancel{\text{lb}}} = 74.5\ \text{kg}$$

Given — Conversion factor — Answer

Look at how the units cancel. The given unit lb cancels out and the needed unit kg is in the numerator. The unit you want in the final answer is the one that remains after all the other units have canceled out. This is a helpful way to check that you set up a problem properly.

$$\cancel{\text{lb}} \times \frac{\text{kg}}{\cancel{\text{lb}}} = \text{kg}$$ Unit needed for answer

The calculation gives the numerical answer, which is adjusted to give a final answer with the proper number of significant figures (SFs).

$$164 \times \frac{1}{2.20} = 164 \div 2.20 = 74.545454 55 = 74.5$$

Three SFs — Three SFs — Calculator display — Three SFs (rounded off)

The value of 74.5 combined with the unit, kg, gives the final answer of 74.5 kg. With few exceptions, answers to numerical problems contain a number and a unit.

STUDY CHECK 1.9

If 1890 mL of orange juice is prepared from orange juice concentrate, how many liters of orange juice is that?

Using Two or More Conversion Factors

In many problems, two or more conversion factors are needed to complete the change of units. In setting up these problems, one factor can follow the other. Each factor is arranged to cancel the preceding unit until the needed unit is obtained. Once the problem is set up to cancel units properly, the calculations can be done without writing intermediate results. The process is worth practicing until you understand unit cancellation, the steps on the calculator, and rounding off to give a final answer. In this text, when two or more conversion factors are required, the final answer will be based on obtaining a final calculator display and rounding to the correct number of significant figures.

CONCEPT CHECK 1.11

Canceling Units

Cancel the units in the following setup and give the unit of the final answer:

$$3.5\ \text{L} \times \frac{1 \times 10^3\ \text{mL}}{1\ \text{L}} \times \frac{0.48\ \text{g}}{1\ \text{mL}} \times \frac{1 \times 10^3\ \text{mg}}{1\ \text{g}} =$$

ANSWER

All matching units in the denominator and numerator cancel to give mg as the needed unit for the answer.

$$3.5\ \cancel{\text{L}} \times \frac{1 \times 10^3\ \cancel{\text{mL}}}{1\ \cancel{\text{L}}} \times \frac{0.48\ \cancel{\text{g}}}{1\ \cancel{\text{mL}}} \times \frac{1 \times 10^3\ \text{mg}}{1\ \cancel{\text{g}}} = \text{needed unit of mg}$$

SAMPLE PROBLEM 1.10

Problem Solving Using Two Factors

Synthroid is used as a replacement or supplemental therapy for diminished thyroid function. A doctor's order prescribed a dosage of 0.200 mg. If tablets in stock contain 50 μg of Synthroid, how many tablets are required to provide the prescribed medication?

SOLUTION

Step 1 **State the given and needed quantities.**

Given 0.200 mg of Synthroid **Need** number of tablets

Step 2 **Write a plan to convert the given unit to the needed unit.**

milligrams → [Metric factor] → micrograms → [Clinical factor] → number of tablets

Step 3 **State the equalities and conversion factors needed to cancel units.** In the problem, the information for the dosage is given as 50 μg per tablet. Using this as an equality along with the metric equality for milligrams and micrograms provides the following conversion factors:

$$1 \text{ mg} = 1000\ \mu\text{g} \qquad \frac{1 \text{ mg}}{1000\ \mu\text{g}} \text{ and } \frac{1000\ \mu\text{g}}{1 \text{ mg}}$$

$$1 \text{ tablet} = 50\ \mu\text{g} \qquad \frac{1 \text{ tablet}}{50\ \mu\text{g}} \text{ and } \frac{50\ \mu\text{g}}{1 \text{ tablet}}$$

Step 4 **Set up problem to cancel units and calculate answer.** The problem can be set up using the metric factor to cancel milligrams, and then the clinical factor to obtain tablets as the final unit.

$$0.200\ \cancel{\text{mg}} \times \frac{1000\ \cancel{\mu\text{g}}}{1\ \cancel{\text{mg}}} \times \frac{1 \text{ tablet}}{50\ \cancel{\mu\text{g}}} = 4 \text{ tablets}$$

STUDY CHECK 1.10

One medium bran muffin contains 4.2 g of fiber. How many ounces (oz) of fiber are obtained by eating three medium bran muffins if 1 lb = 16 oz?
(*Hint:* number of muffins ⟶ g of fiber ⟶ lb ⟶ oz)

VETERINARY TECHNICIAN (VT)

"I am checking this dog's ears for foxtails and her eyes for signs of conjunctivitis," says Joyce Rhodes, veterinary assistant at the Sonoma Animal Hospital. "We always check a dog's teeth for tartar, because dental care is very important to the well-being of the animal. When I do need to give a medication to an animal, I use my chemistry to prepare the proper dose that the pet should take. Dosages may be in milligrams, grams, or milliliters."

As a member of the veterinary health care team, a veterinary technician (VT) assists a veterinarian in the care and handling of animals. A VT takes medical histories, collects specimens, performs laboratory procedures, prepares an animal for surgery, assists in surgical procedures, takes X-rays, talks with animal owners, and cleans teeth.

SAMPLE PROBLEM 1.11

Using a Percent as a Conversion Factor

A person who exercises regularly has 16% body fat. If this person weighs 155 lb, what is the mass, in kilograms, of body fat?

SOLUTION

Step 1 **State the given and needed quantities.**

Given 155 lb of body weight; 16% body fat

Need kilograms of body fat

Step 2 **Write a plan to convert the given unit to the needed unit.**

lb of body weight → [U.S.–Metric factor] → kg of body mass → [Percent factor] → kg of body fat

Step 3 **State the equalities and conversion factors needed to cancel units.** One equality is the U.S.–metric factor for lb and kg. The second is the percent factor derived from the percentage information given in the problem.

TUTORIAL
Using Percentage as a Conversion Factor

1 kg of body fat = 2.20 lb of body fat

$$\frac{2.20 \text{ lb body fat}}{1 \text{ kg body fat}} \text{ and } \frac{1 \text{ kg body fat}}{2.20 \text{ lb body fat}}$$

16 kg of body fat = 100 kg of body mass

$$\frac{16 \text{ kg body fat}}{100 \text{ kg body mass}} \text{ and } \frac{100 \text{ kg body mass}}{16 \text{ kg body fat}}$$

Step 4 Set up problem to cancel units and calculate answer. We can set up the problem using conversion factors to cancel each unit, starting with lb of body weight, until we obtain the final unit, kg of body fat, in the numerator. After we count the significant figures in the measured quantities, we write the needed answer with the proper number of significant figures.

$$155 \cancel{\text{lb body weight}} \times \frac{1 \cancel{\text{kg body mass}}}{2.20 \cancel{\text{lb body weight}}} \times \frac{16 \text{ kg body fat}}{100 \cancel{\text{kg body mass}}} = 11 \text{ kg of body fat}$$

Three SFs — Three SFs — Two SFs — Two SFs

STUDY CHECK 1.11

Uncooked lean ground beef can contain up to 22% fat by mass. How many grams of fat would be contained in 0.25 lb of the ground beef?

QUESTIONS AND PROBLEMS

Problem Solving

1.55 When you convert one unit to another, how do you know which unit of the conversion factor to place in the denominator?

1.56 When you convert one unit to another, how do you know which unit of the conversion factor to place in the numerator?

1.57 Use metric conversion factors to solve each of the following problems:

a. The height of a student is 175 cm. How tall is the student in meters?

b. A cooler has a volume of 5500 mL. What is the capacity of the cooler in liters?

c. A Bee Hummingbird has a mass of 0.0018 kg. What is the mass of the hummingbird in grams?

1.58 Use metric conversion factors to solve each of the following problems:

a. The daily value of phosphorus is 800 mg. How many grams of phosphorus are recommended?

b. A glass of orange juice contains 0.85 dL of juice. How many milliliters of orange juice is that?

c. A package of chocolate instant pudding contains 2840 mg of sodium. How many grams of sodium is that?

1.59 Solve each of the following problems using one or more conversion factors:

a. A container holds 0.500 qt of liquid. How many milliliters of lemonade will it hold?

b. What is the mass, in kilograms, of a person who weighs 145 lb?

c. An athlete has 15% by mass body fat. What is the weight of fat, in pounds, of a 74-kg athlete?

d. A plant fertilizer contains 15% by mass nitrogen (N). In a container of soluble plant food, there are 10.0 oz of fertilizer. How many grams of nitrogen are in the container?

e. In a candy factory, the nutty chocolate bars contain 22.0% by mass pecans. If 5.0 kg of pecans were used for candy last Tuesday, how many pounds of nutty chocolate bars were made?

Agricultural fertilizers applied to a field provide nitrogen for plant growth.

1.60 Solve each of the following problems using one or more conversion factors:

a. You need 4.0 ounces of a steroid ointment. If there are 16 oz in 1 lb, how many grams of ointment does the pharmacist need to prepare?

b. During surgery, a patient receives 5.0 pints of plasma. How many milliliters of plasma were given? (1 quart = 2 pints)

c. Wine is 12% (by volume) alcohol. How many milliliters of alcohol are in a 0.750 L bottle of wine?

d. Blueberry high-fiber muffins contain 51% dietary fiber. If a package with a net weight of 12 oz contains six muffins, how many grams of fiber are in each muffin?

e. A jar of crunchy peanut butter contains 1.43 kg of peanut butter. If you use 8.0% of the peanut butter for a sandwich, how many ounces of peanut butter did you take out of the container?

1.61 Using conversion factors, solve each of the following clinical problems:

a. You have used 250 L of distilled water for a dialysis patient. How many gallons of water is that?

b. A patient needs 0.024 g of a sulfa drug. There are 8-mg tablets in stock. How many tablets should be given?

c. The daily dose of ampicillin for the treatment of an ear infection is 115 mg/kg of body weight. What is the daily dose for a 34-lb child?

1.62 Using conversion factors, solve each of the following clinical problems:

a. The physician has ordered 1.0 g of tetracycline to be given every 6 hours to a patient. If your stock on hand is 500-mg tablets, how many will you need for 1 day's treatment?

b. An intramuscular medication is given at 5.00 mg/kg of body weight. If you give 425 mg of medication to a patient, what is the patient's weight in pounds?

c. A physician has ordered 0.50 mg of atropine, intramuscularly. If atropine were available as 0.10 mg/mL of solution, how many milliliters would you need to give?

1.10 Density

LEARNING GOAL

Calculate the density of a substance; use the density to calculate the mass or volume of a substance.

The mass and volume of any object can be measured. If we compare the mass of the object to its volume, we obtain a relationship called **density**.

$$\text{Density} = \frac{\text{mass of substance}}{\text{volume of substance}}$$

Every substance has a unique density, which distinguishes it from other substances. For example, lead has a density of 11.3 g/mL whereas cork has a density of 0.26 g/mL. From these densities, we can predict if these substances will sink or float in water. If a substance, such as cork, is less dense than water, it will float. However, a lead object sinks because its density is greater than that of water (see Figure 1.12).

Metals such as gold and lead tend to have higher densities whereas gases have low densities. In the metric system, the densities of solids and liquids are usually expressed as grams per cubic centimeter (g/cm^3) or grams per milliliter (g/mL). The densities of gases are usually stated as grams per liter (g/L). Table 1.13 gives the densities of some common substances.

FIGURE 1.12 Objects that sink in water are more dense than water; objects float if they are less dense.
Q Why does an ice cube float and a piece of aluminum sink?

TABLE 1.13 Densities of Some Common Substances

Solids (at 25 °C)	Density (g/mL)	Liquids (at 25 °C)	Density (g/mL)	Gases (at 0 °C)	Density (g/L)
Cork	0.26	Gasoline	0.74	Hydrogen	0.090
Wood (maple)	0.75	Ethanol	0.79	Helium	0.179
Ice (at 0 °C)	0.92	Olive oil	0.92	Methane	0.714
Sugar	1.59	Water (at 4 °C)	1.00	Neon	0.90
Bone	1.80	Urine	1.003–1.030	Nitrogen	1.25
Salt (NaCl)	2.16	Plasma (blood)	1.03	Air (dry)	1.29
Aluminum	2.70	Milk	1.04	Oxygen	1.43
Diamond	3.52	Mercury	13.6	Carbon dioxide	1.96
Copper	8.92				
Silver	10.5				
Lead	11.3				
Gold	19.3				

SINK OR FLOAT?

1. Fill a large container or bucket with water. Place a can of diet soft drink and a can of a regular soft drink in the water. What happens? Using information on the label, how might you account for your observations?
2. Design an experiment to determine the substance that is the more dense in each of the following:
   a. water and vegetable oil
   b. water and ice
   c. rubbing alcohol and ice
   d. vegetable oil, water, and ice

CONCEPT CHECK 1.12

Density

(a) (b)

a. In drawing (a), the gray cube has a density of 4.5 g/cm^3. Is the density of the green cube the same, less than, or greater than that of the gray cube?

b. In drawing (b), the gray cube has a density of 4.5 g/cm^3. Is the density of the green cube the same, less than, or greater than that of the gray cube?

ANSWER

a. The green cube has the same volume as the gray cube but has a greater mass. Thus, the green cube has a density that is greater than that of the gray cube.

b. The green cube has the same mass as the gray cube, but the green cube has a greater volume. Thus, the green cube has a density that is less than that of the gray cube.

SAMPLE PROBLEM 1.12

Calculating Density

High-density lipoprotein (HDL) contains large amounts of proteins and small amounts of cholesterol. If a 0.258-g sample of HDL has a volume of 0.215 cm^3, what is the density of the HDL sample?

SOLUTION

Step 1 State the given and needed quantities.

Given mass of HDL sample = 0.258 g; volume = 0.215 cm^3

Need density (g/cm^3)

Step 2 Write the density expression.

$$\text{Density} = \frac{\text{mass of substance}}{\text{volume of substance}}$$

Step 3 Express mass in grams and volume in milliliters (mL) or cm^3.

Mass of HDL sample = 0.258 g

Volume of HDL sample = 0.215 cm^3

Step 4 Substitute mass and volume into the density expression and calculate the density.

$$\text{Density} = \frac{0.258\ \text{g}}{0.215\ \text{cm}^3} = \frac{1.20\ \text{g}}{1\ \text{cm}^3} = 1.20\ \text{g/cm}^3$$

(0.258 g: Three SFs; 0.215 cm^3: Three SFs; 1.20 g/cm^3: Three SFs)

STUDY CHECK 1.12

Low-density lipoprotein (LDL) contains small amounts of proteins and large amounts of cholesterol. If a 0.380-g sample of LDL has a volume of 0.362 cm^3, what is the density of the LDL sample?

Guide to Calculating Density

1. State the given and needed quantities.
2. Write the density expression.
3. Express mass in grams and volume in milliliters (mL) or cm^3.
4. Substitute mass and volume into the density expression and calculate the density.

Density Using Volume Displacement

The volume of a solid can be determined by volume displacement. When a solid is completely submerged in water, it displaces a volume that is equal to the volume of the solid. In Figure 1.13, the water level rises from 35.5 mL to 45.0 mL after the zinc object is added. This means that 9.5 mL of water is displaced and that the volume of the object is 9.5 mL. The density of the zinc is calculated as follows:

$$\text{Density} = \frac{\overset{\text{Four SFs}}{68.60\text{ g zinc}}}{\underset{\text{Two SFs}}{9.5\text{ mL}}} = \underset{\text{Two SFs}}{7.2\text{ g/mL}}$$

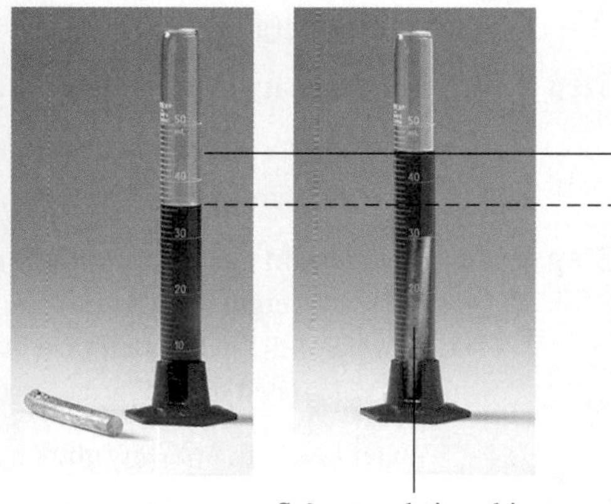

FIGURE 1.13 The density of a solid can be determined by volume displacement because a submerged object displaces a volume of water equal to its own volume.

Q How is the volume of the zinc object determined?

Chemistry Link to Health

BONE DENSITY

The density of our bones determines their health and strength. Our bones are constantly gaining and losing minerals such as calcium, magnesium, and phosphate. In childhood, bones form at a faster rate than they break down. As we age, the breakdown of bone occurs more rapidly than new bone forms. As the loss of bone minerals increases, bones begin to thin, causing a decrease in mass and density. Thinner bones lack strength, which increases the risk of fracture. Hormonal changes, disease, and certain medications can also contribute to the thinning of bone. Eventually, a condition of severe thinning of bone known as *osteoporosis* may occur. *Scanning electron micrographs* (SEMs) show (**a**) normal bone and (**b**) bone in osteoporosis due to loss of bone minerals.

Bone density is often determined by passing low-dose X-rays through the narrow part at the top of the femur (hip) and the spine (**c**). These locations are where fractures are more likely to occur, especially as we age. Bones with high density will block more of the X-rays compared to bones that are less dense. The results of a bone density test are compared to a healthy young adult as well as to other people of the same age.

Recommendations to improve bone strength include supplements of calcium and vitamin D and medications such as Fosamax, Evista, or Actonel. Weight-bearing exercise such as walking and lifting weights can also improve muscle strength, which in turn, increases bone strength.

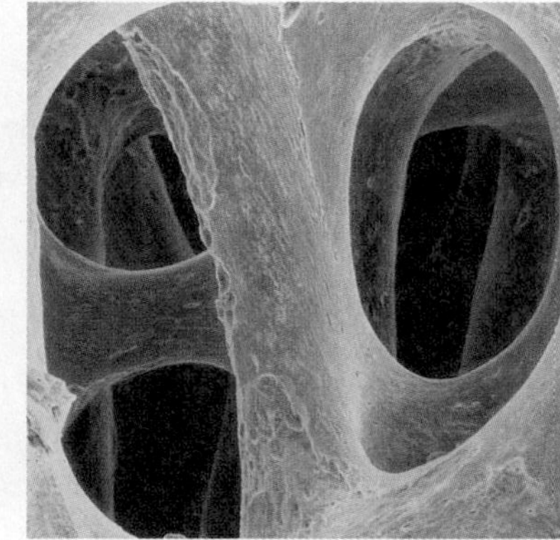

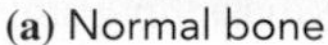
(**a**) Normal bone

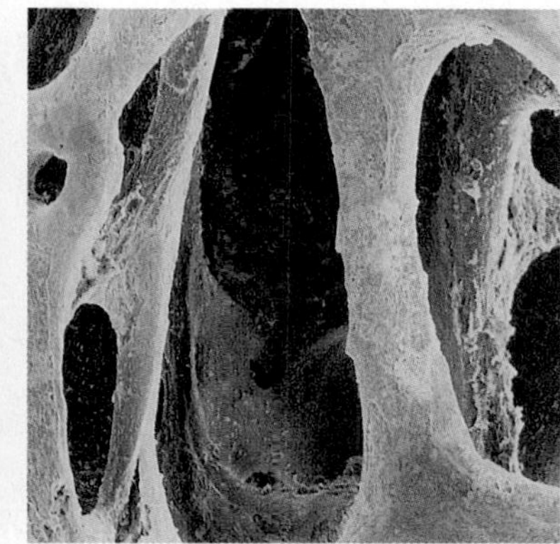
(**b**) Bone with osteoporosis

(**c**) Viewing a low-dose X-ray of the spine

Lead weights in a belt counteract the buoyancy of a scuba diver.

SAMPLE PROBLEM 1.13

Using Volume Displacement to Calculate Density

A lead weight used in the belt of a scuba diver has a mass of 226 g. When the lead weight is placed in a graduated cylinder containing 200.0 mL of water, the water level rises to 220.0 mL. What is the density of the lead weight (g/mL)?

SOLUTION

Step 1 State the given and needed quantities.

Given mass = 226 g; water level before object submerged = 200.0 mL; water level after object submerged = 220.0 mL

Need density (g/mL)

Step 2 Write the density expression.

$$\text{Density} = \frac{\text{mass of substance}}{\text{volume of substance}}$$

Step 3 Express mass in grams and volume in milliliters (mL) or cm³.

Mass of lead weight = 226 g

The volume of the lead weight is equal to the volume of water displaced, which is calculated as follows:

Water level after object submerged	=	220.0 mL
Water level before object submerged	=	−200.0 mL
Water displaced (volume of lead weight)	=	20.0 mL

Step 4 Substitute mass and volume into the density expression and calculate the density. The density is calculated by dividing the mass (g) by the volume (mL). Be sure to use the volume of water the object displaced and *not* the original volume of water.

$$\text{Density} = \frac{\overset{\text{Three SFs}}{226\text{ g}}}{\underset{\text{Three SFs}}{20.0\text{ mL}}} = \frac{11.3\text{ g}}{1\text{ mL}} = \underset{\text{Three SFs}}{11.3\text{ g/mL}}$$

STUDY CHECK 1.13

A total of 0.50 lb of glass marbles is added to 425 mL of water. The water level rises to a volume of 528 mL. What is the density (g/mL) of the glass marbles?

Guide to Using Density

1 State the given and needed quantities.

2 Write a plan to calculate the needed quantity.

3 Write equalities and their conversion factors including density.

4 Set up problem to calculate the needed quantity.

Problem Solving Using Density

Density can be used as a conversion factor. For example, if the volume and the density of a sample are known, the mass in grams of the sample can be calculated as shown in the following Sample Problem.

SAMPLE PROBLEM 1.14

Problem Solving with Density

John took 2.0 teaspoons (tsp) of cough syrup for a persistent cough. If the syrup had a density of 1.20 g/mL and there is 5.0 mL in 1 tsp, what was the mass, in grams, of the cough syrup?

SOLUTION

Step 1 State the given and needed quantities.

Given 2.0 tsp **Need** grams of syrup

Step 2 Write a plan to calculate the needed quantity.

Step 3 Write equalities and their conversion factors including density.

$$1\ \text{tsp} = 5.0\ \text{mL} \qquad \frac{5.0\ \text{mL}}{1\ \text{tsp}} \text{ and } \frac{1\ \text{tsp}}{5.0\ \text{mL}}$$

$$1\ \text{mL} = 1.20\ \text{g of syrup} \qquad \frac{1\ \text{mL}}{1.20\ \text{g syrup}} \text{ and } \frac{1.20\ \text{g syrup}}{1\ \text{mL}}$$

Step 4 Set up problem to calculate the needed quantity.

$$2.0\ \cancel{\text{tsp}} \times \frac{5.0\ \cancel{\text{mL}}}{1\ \cancel{\text{tsp}}} \times \frac{1.20\ \text{g syrup}}{1\ \cancel{\text{mL}}} = 12\ \text{g of syrup}$$

STUDY CHECK 1.14

How many milliliters of mercury are in a thermometer that contains 20.4 g of mercury? (See Table 1.13 for the density of mercury.)

Specific Gravity

Specific gravity (sp gr) is a relationship between the density of a substance and the density of water. Specific gravity is calculated by dividing the density of a sample by the density of water, which is 1.00 g/mL at 4 °C. A substance with a specific gravity of 1.00 has the same numerical value as the density of water (1.00 g/mL).

$$\text{Specific gravity} = \frac{\text{density of sample}}{\text{density of water}}$$

Specific gravity is one of the few unitless values you will encounter in chemistry. An instrument called a hydrometer is often used to measure the specific gravity of fluids such as battery fluid or a sample of urine. In Figure 1.14, a hydrometer is used to measure the specific gravity of beer.

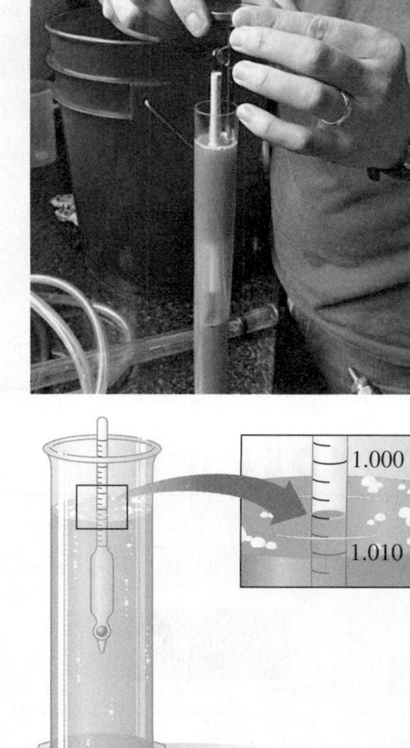

FIGURE 1.14 When the specific gravity of beer measures 1.010 or less with a hydrometer, the fermentation process is complete.

Q If the hydrometer reading is 1.006, what is the density of the liquid?

QUESTIONS AND PROBLEMS

Density

1.63 In an old trunk, you find a piece of metal that you think may be aluminum, silver, or lead. In lab, you find it has a mass of 217 g and a volume of 19.2 cm^3. Using Table 1.13, what is the metal you found?

1.64 Suppose you have two 100-mL graduated cylinders. In each cylinder there is 40.0 mL of water. You also have two cubes: One is lead, and the other is aluminum. Each cube measures 2.0 cm on each side. After you carefully lower each cube into the water of its own cylinder, what will the new water level be in each of the cylinders?

1.65 What is the density (g/mL) of each of the following samples?

a. A 20.0-mL sample of a salt solution that has a mass of 24.0 g.

b. A solid object with a mass of 1.65 lb and a volume of 170 mL.

c. A gem has a mass of 45.0 g. When the gem is placed in a graduated cylinder containing 20.0 mL of water, the water level rises to 34.5 mL.

d. A lightweight head on the driver of a golf club is made of titanium. If the volume of a sample of titanium is 114 cm^3 and the mass is 514.1 g, what is the density of titanium?

1.66 What is the density (g/mL) of each of the following samples?
a. A medication, if 3.00 mL has a mass of 3.85 g.
b. The fluid in a car battery, if it has a volume of 125 mL and a mass of 155 g.
c. A 5.00-mL urine sample from a patient suffering from symptoms resembling those of diabetes mellitus. The mass of the urine sample is 5.025 g.
d. A syrup is added to an empty container with a mass of 115.25 g. When 0.100 pint of syrup is added, the total mass of the container and syrup is 182.48 g. (1 qt = 2 pt)

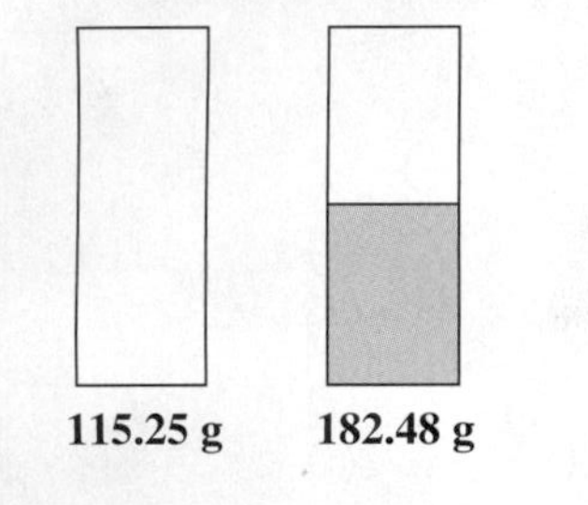

1.67 Use the density value to solve the following problems:
a. What is the mass, in grams, of 150 mL of a liquid with a density of 1.4 g/mL?
b. What is the mass of a glucose solution that fills a 0.500-L intravenous bottle if the density of the glucose solution is 1.15 g/mL?
c. A sculptor has prepared a mold for casting a bronze figure. The figure has a volume of 225 mL. If bronze has a density of 7.8 g/mL, how many ounces of bronze are needed in the preparation of the bronze figure?

1.68 Use the density value to solve the following problems:
a. A graduated cylinder contains 18.0 mL of water. What is the new water level after 35.6 g of silver metal with a density of 10.5 g/mL is submerged in the water?
b. A thermometer containing 8.3 g of mercury has broken. If mercury has a density of 13.6 g/mL, what volume spilled?
c. A fish tank holds 35 gal of water. Using the density of 1.00 g/mL for water, determine the number of pounds of water in the fish tank.

CONCEPT MAP

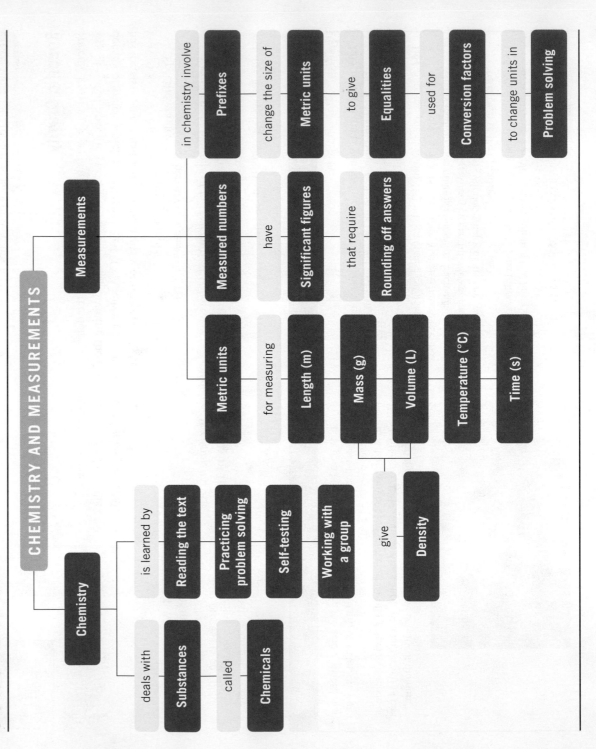

CHEMISTRY AND MEASUREMENTS

- **Chemistry**
  - deals with **Substances** called **Chemicals**
  - is learned by **Reading the text**, **Practicing problem solving**, **Self-testing**, **Working with a group**
- **Measurements**
  - **Metric units** for measuring **Length (m)**, **Mass (g)**, **Volume (L)**, **Temperature (°C)**, **Time (s)**
    - **Mass (g)** and **Volume (L)** give **Density**
  - **Measured numbers** have **Significant figures** that require **Rounding off answers**
  - in chemistry involve **Prefixes** change the size of **Metric units** to give **Equalities** used for **Conversion factors** to change units in **Problem solving**

CHAPTER REVIEW

1.1 Chemistry and Chemicals

Learning Goal: Define the term *chemistry* and identify substances as chemicals.

Chemistry is the study of the composition, structure, properties, and reactions of matter. A chemical is any substance that always has the same composition and properties wherever it is found.

1.2 Study Plan for Learning Chemistry

Learning Goal: Develop a study plan for learning chemistry.

A study plan for learning chemistry utilizes the features in this text and develops an active learning approach to study chemistry. By using the *Learning Goals* in the chapter and working the *Concept Checks*, *Sample Problems*, *Study Checks*, and the *Questions and Problems* at the end of each section, the student can successfully learn the concepts of chemistry.

1.3 Units of Measurement

Learning Goal: Write the names and abbreviations for metric or SI units used in measurements of length, volume, mass, temperature, and time.

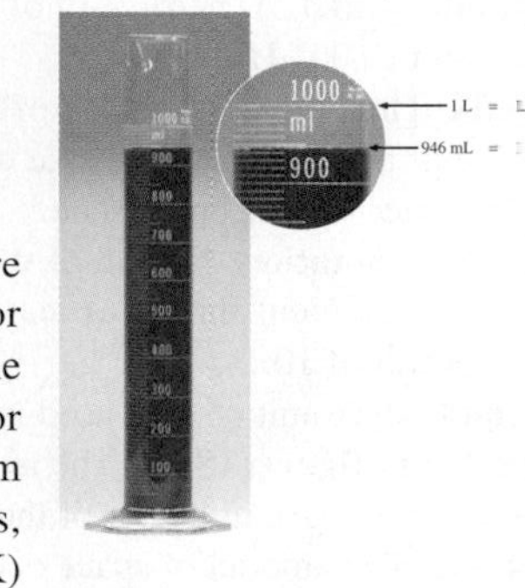

In science, physical quantities are described in units of the metric or International System (SI). Some important units are meter (m) for length, liter (L) for volume, gram (g) and kilogram (kg) for mass, degree Celsius (°C) and kelvin (K) for temperature, and second (s) for time.

1.4 Scientific Notation

Learning Goal: Write a number in scientific notation.

Large and small numbers can be written using scientific notation, in which the decimal point is moved to give a coefficient of at least 1 but less than 10 and the number of decimal places moved shown as a power of 10. A large number will have a positive power of 10, while a small number will have a negative power of 10.

1.5 Measured Numbers and Significant Figures

Learning Goal: Identify a number as measured or exact; determine the number of significant figures in a measured number.

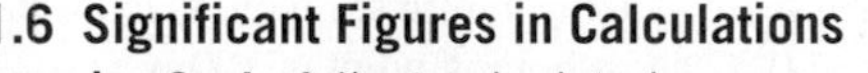

(a) (b) (c)

A measured number is any number obtained by using a measuring device. An exact number is obtained by counting items or from a definition; no measuring device is used. Significant figures are the numbers reported in a measurement including the last estimated digit. Zeros in front of a decimal number or at the end of a large number are not significant.

1.6 Significant Figures in Calculations

Learning Goal: Adjust calculated answers to give the correct number of significant figures.

In multiplication or division, the final answer is written so it has the same number of significant figures as the measurement with the fewest significant figures. In addition or subtraction, the final answer is written so it has the same number of decimal places as the measurement with the fewest decimal places.

1.7 Prefixes and Equalities

Learning Goal: Use the numerical values of prefixes to write a metric equality.

Prefixes placed in front of a unit change the size of the unit by factors of 10. Prefixes such as *centi*, *milli*, and *micro* provide smaller units; prefixes such as *kilo* provide larger units. An equality relates two units that measure the same quantity of length, volume, mass, or time. Examples of equalities are 1 m = 100 cm, 1 qt = 946 mL; 1 kg = 1000 g; 1 min = 60 s.

Examples of Some Mass Equalities

$1\ \text{kg} = 1000\ \text{g} = 1 \times 10^3\ \text{g}$

$1\ \text{g} = 1000\ \text{mg} = 1 \times 10^3\ \text{mg}$

$1\ \text{mg} = 1000\ \mu\text{g} = 1 \times 10^3\ \mu\text{g}$

1.8 Writing Conversion Factors

Learning Goal: Write a conversion factor for two units that describe the same quantity.

Conversion factors are used to express a relationship in the form of a fraction. Two factors can be written for any relationship in the metric or U.S. system.

1.9 Problem Solving

Learning Goal: Use conversion factors to change from one unit to another.

Conversion factors are useful when changing a quantity expressed in one unit to a quantity expressed in another unit. In the process, a given unit is multiplied by one or more conversion factors that cancel units until the needed answer is obtained.

$$\underset{\text{Given}}{164\ \cancel{\text{lb}}} \times \underset{\text{Conversion factor}}{\frac{1\ \text{kg}}{2.20\ \cancel{\text{lb}}}} = \underset{\text{Answer}}{74.5\ \text{kg}}$$

Unit for answer goes here

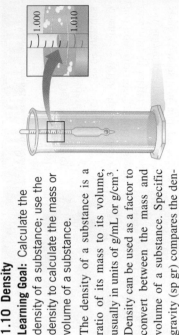

1.10 Density

Learning Goal: Calculate the density of a substance; use the density to calculate the mass or volume of a substance.

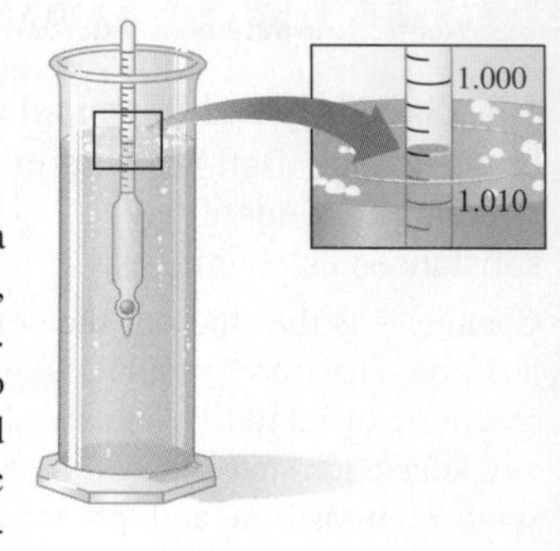

The density of a substance is a ratio of its mass to its volume, usually in units of g/mL or g/cm^3. Density can be used as a factor to convert between the mass and volume of a substance. Specific gravity (sp gr) compares the density of a substance to the density of water, 1.00 g/mL.

Key Terms

Celsius (°C) temperature scale A temperature scale on which water has a freezing point of 0 °C and a boiling point of 100 °C.

centimeter (cm) A unit of length in the metric system; there are 2.54 cm in 1 in.

chemical A substance that has the same composition and properties wherever it is found.

chemistry The science that studies the composition, structure, properties, and reactions of matter.

conversion factor A ratio in which the numerator and denominator are quantities from an equality or given relationship. For example, the conversion factors for the relationship 1 kg = 2.20 lb are written as

$$\frac{2.20\ \text{lb}}{1\ \text{kg}} \quad \text{and} \quad \frac{1\ \text{kg}}{2.20\ \text{lb}}$$

cubic centimeter (cm^3, cc) The volume of a cube that has 1-cm sides; 1 cm^3 is equal to 1 mL.

density The relationship of the mass of an object to its volume expressed as grams per cubic centimeter (g/cm^3), grams per milliliter (g/mL), or grams per liter (g/L).

equality A relationship between two units that measure the same quantity.

exact number A number obtained by counting or by definition.

gram (g) The metric unit used in measurements of mass.

International System of Units (SI) An international system of units that modifies the metric system.

Kelvin (K) temperature scale A temperature scale on which the lowest possible temperature is 0 K.

kilogram (kg) A metric mass of 1000 g, equal to 2.20 lb. The kilogram is the SI standard unit of mass.

liter (L) The metric unit for volume that is slightly larger than a quart.

mass A measure of the quantity of material in an object.

measured number A number obtained when a quantity is determined by using a measuring device.

meter (m) The metric unit for length that is slightly longer than a yard. The meter is the SI standard unit of length.

metric system A system of measurement used by scientists and in most countries of the world.

milliliter (mL) A metric unit of volume equal to one-thousandth of a liter (0.001 L).

prefix The part of the name of a metric unit that precedes the base unit and specifies the size of the measurement. All prefixes are related on a decimal scale.

scientific notation A form of writing large and small numbers using a coefficient that is at least 1 but less than 10, followed by a power of 10.

second (s) A unit of time used in both the SI and metric systems.

significant figures (SFs) The numbers recorded in a measurement.

temperature An indicator of the hotness or coldness of an object.

volume The amount of space occupied by a substance.

Understanding the Concepts

1.69 Which of the following will help you develop a successful study plan?
a. Skip lecture and just read the text.
b. Work the *Sample Problems* as you go through a chapter.
c. Go to your professor's office hours.
d. Read through the chapter, but work the problems later.

1.70 Which of the following will help you develop a successful study plan?
a. Study all night before the exam.
b. Form a study group and discuss the problems together.
c. Work problems in a notebook for easy reference.
d. Copy the answers to homework from a friend.

1.71 In which of the following pairs do both numbers contain the same number of significant figures?
a. 2.0500 m and 0.0205 m
b. 600.0 K and 60 K
c. 0.000 75 s and 75 000 s
d. 6.240 L and 6.240×10^{-2} L

1.72 In which of the following pairs do both numbers contain the same number of significant figures?
a. 3.44×10^{-3} g and 0.0344 g
b. 0.0098 s and 9.8×10^{4} s
c. 6.8×10^{3} mm and 68 000 m
d. 258.000 ng and 2.58×10^{-2} g

1.73 Indicate if each of the following is answered with an exact number or a measured number:

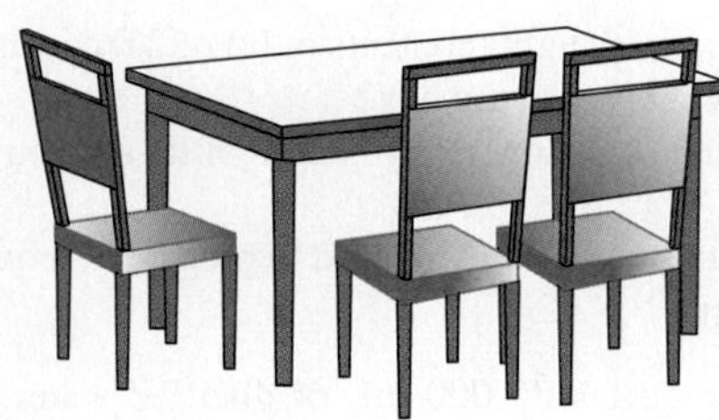

a. number of legs
b. height of table
c. number of chairs at the table
d. area of tabletop

1.74 Measure the length of each of the objects in diagrams **(a)**, **(b)**, and **(c)** using the metric ruler in the figure. Indicate the number of significant figures for each and the estimated digit for each.

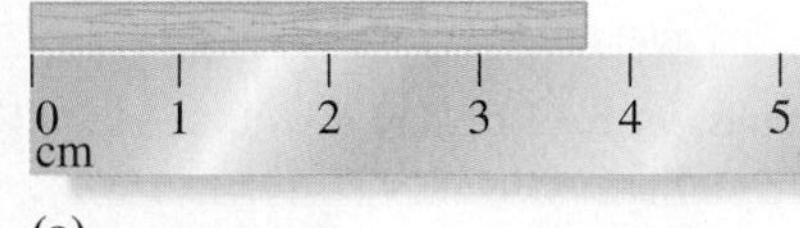

1.75 Measure the length and width of the rectangle, including the estimated digit, using a metric ruler.

a. What are the length and width of this rectangle measured in centimeters?
b. What are the length and width of this rectangle measured in millimeters?
c. How many significant figures are in the length measurement?
d. How many significant figures are in the width measurement?
e. What is the area of the rectangle in cm^2?
f. How many significant figures are in the calculated answer for area?

1.76 Each of the following diagrams represents a container of water and a cube. Some cubes float while others sink. Match diagrams 1, 2, 3, or 4 with one of the following descriptions and explain your choices:

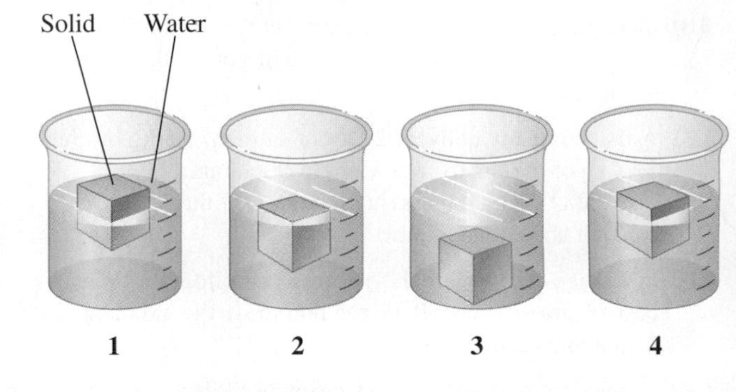

a. The cube has a greater density than water.
b. The cube has a density that is 0.60–0.80 g/mL.
c. The cube has a density that is one-half the density of water.
d. The cube has the same density as water.

1.77 What is the density of the solid object that is weighed and submerged in water?

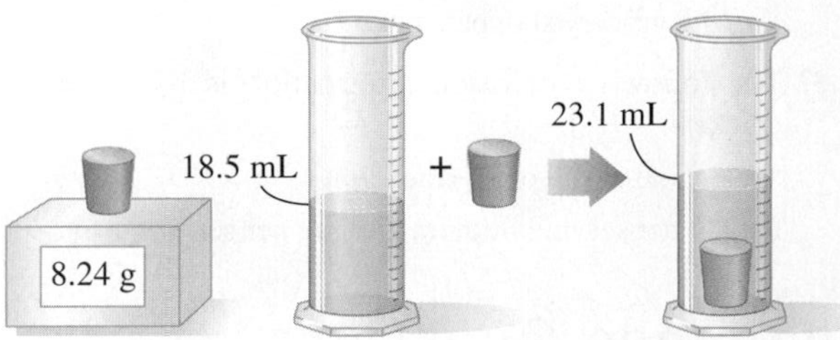

1.78 Consider the following solids. The solids A, B, and C represent aluminum, gold, and silver. If each has a mass of 10.0 g, what is the identity of each solid?

Density of aluminum = 2.70 g/mL
Density of gold = 19.3 g/mL
Density of silver = 10.5 g/mL

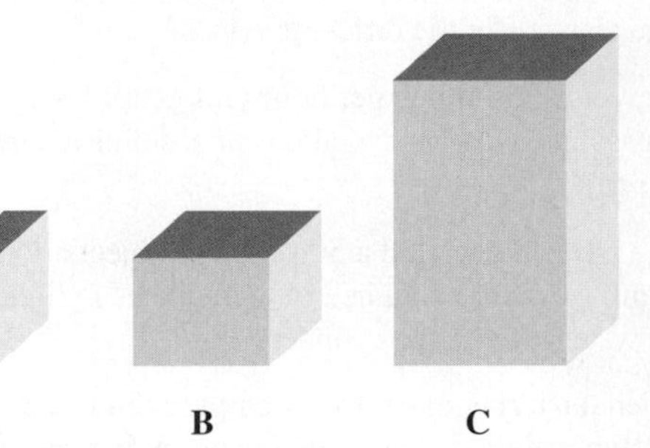

Additional Questions and Problems

For instructor-assigned homework, go to **www.masteringchemistry.com**.

1.79 Round off or add zeros to the following calculated answers to give a final answer with three significant figures:
a. 0.000 012 58 L **b.** 3.528×10^2 kg
c. 125 111 m **d.** 58.703 g
e. 3×10^{-3} s **f.** 0.010 826 g

1.80 What is the total mass, in grams, of a dessert containing 137.25 g of vanilla ice cream, 84 g of fudge sauce, and 43.7 g of nuts?

1.81 During a workout at the gym, you set the treadmill at a pace of 55.0 m/min. How many minutes will you walk if you cover a distance of 7500 ft?

1.82 A fish company delivers 22 kg of salmon, 5.5 kg of crab, and 3.48 kg of oysters to your seafood restaurant.
a. What is the total mass, in kilograms, of the seafood?
b. What is the total number of pounds?

1.83 In France, grapes are 1.75 Euros per kilogram. What is the cost of grapes, in dollars per pound, if the exchange rate is 1.36 dollars/Euro?

1.84 In Mexico, avocados are 48 pesos per kilogram. What is the cost, in cents, of an avocado that weighs 0.45 lb if the exchange rate is 13 pesos to the dollar?

1.85 Bill's recipe for onion soup calls for 4.0 lb of thinly sliced onions. If an onion has an average mass of 115 g, how many onions does Bill need?

1.86 The price of 1 lb of potatoes is $1.75. If all the potatoes sold today at the store bring in $1420, how many kilograms of potatoes did grocery shoppers buy?

1.87 The following nutrition information is listed on a box of crackers:

Serving size 0.50 oz (6 crackers)

Fat 4 g per serving Sodium 140 mg per serving

a. If the box has a net weight (contents only) of 8.0 oz, about how many crackers are in the box?
b. If you ate 10 crackers, how many ounces of fat are you consuming?
c. How many grams of sodium are used to prepare 50 boxes of crackers in part **a**?

1.88 A dialysis unit requires 75 000 mL of distilled water. How many gallons of water are needed? (1 gal = 4 qt)

1.89 To prevent bacterial infection, a doctor orders 4 tablets of amoxicillin per day for 10 days. If each tablet contains 250 mg of amoxicillin, how many ounces of the medication are given in 10 days?

1.90 Celeste's diet restricts her intake of protein to 24 g per day. If she eats 1.2 oz of protein, has she exceeded her protein limit for the day?

1.91 What is a cholesterol level of 1.85 g/L in the standard units of mg/dL?

1.92 An object has a mass of 3.15 oz and a volume of 0.1173 L. What is the density (g/mL) of the object?

1.93 The density of lead is 11.3 g/mL. The water level in a graduated cylinder initially at 215 mL rises to 285 mL after a piece of lead is submerged. What is the mass, in grams, of the lead?

1.94 A graduated cylinder contains 155 mL of water. A 15.0-g piece of iron (density = 7.86 g/cm^3) and a 20.0-g piece of lead (density = 11.3 g/cm^3) are added. What is the new water level, in milliliters, in the cylinder?

1.95 How many cubic centimeters (cm^3) of olive oil have the same mass as 1.50 L of gasoline (see Table 1.13)?

1.96 Ethyl alcohol has a density of 0.79 g/mL. What is the volume, in quarts, of 1.50 kg of alcohol?

Challenge Questions

1.97 A balance measures mass to 0.001 g. If you determine the mass of an object that weighs about 30 g, would you record the mass as 30 g, 32.5 g, 31.25 g, 31.075 g, or 3000 g? Explain your choice by writing two to three complete sentences that describe your thinking.

1.98 When three students use the same meterstick to measure the length of a paper clip, they obtain results of 5.8 cm, 5.75 cm, and 5.76 cm. If the meterstick has millimeter markings, what are some reasons for the different values?

1.99 A car travels at 55 miles per hour and gets 11 kilometers per liter of gasoline. How many gallons of gasoline are needed for a 3.0-hour trip?

1.100 A 50.0-g silver object and a 50.0-g gold object are both added to 75.5 mL of water contained in a graduated cylinder. What is the new water level in the cylinder?

1.101 In the manufacturing of computer chips, cylinders of silicon are cut into thin wafers that are 3.00 inches in diameter and have a mass of 1.50 g of silicon. How thick (mm) is each wafer if silicon has a density of 2.33 g/cm^3? (The volume of a cylinder is $V = \pi r^2 h$.)

1.102 A sunscreen preparation contains 2.50% by mass benzyl salicylate. If a tube contains 4.0 ounces of sunscreen, how many kilograms of benzyl salicylate are needed to manufacture 325 tubes of sunscreen?

1.103 For a 180-lb person, calculate the quantities of each of the following that must be ingested to provide the LD_{50} for caffeine given in Table 1.12:
a. 100. mg caffeine per 6.0 fl oz
b. 50. mg of caffeine
c. 100. mg of caffeine

1.104 The label on a 1-pint bottle of mineral water lists the following components. If the density is the same as pure water and you drink 3 bottles of water in one day, how many milligrams of each component will you obtain?
a. calcium 28 ppm **b.** fluoride 0.08 ppm
c. magnesium 12 ppm **d.** potassium 3.2 ppm
e. sodium 15 ppm

1.105 a. Some athletes have as little as 3.0% body fat. If such a person has a body mass of 65 kg, how many pounds of body fat does that person have?

b. In liposuction, a doctor removes fat deposits from a person's body. If body fat has a density of 0.94 g/mL and 3.0 L of fat is removed, how many pounds of fat were removed from the patient?

1.106 An 18-karat gold necklace is 75% gold by mass, 16% silver, and 9.0% copper.

a. What is the mass, in grams, of the necklace if it contains 0.24 oz of silver?

b. How many grams of copper are in the necklace?

c. If 18-karat gold has a density of 15.5 g/cm^3, what is the volume in cubic centimeters?

1.107 A graduated cylinder contains three liquids A, B, and C, which have different densities and do not mix: mercury (D = 13.6 g/mL), vegetable oil (D = 0.92 g/mL), and water (D = 1.00 g/mL). Identify the liquids A, B, and C in the cylinder.

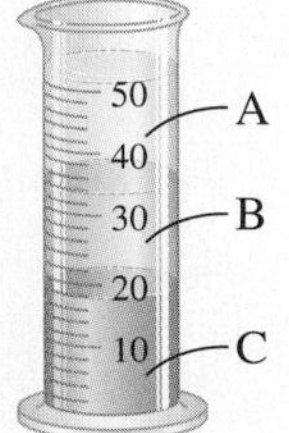

1.108 A mouthwash is 21.6% ethyl alcohol by mass. If each bottle contains 0.358 pt of mouthwash with a density of 0.876 g/mL, how many kilograms of ethyl alcohol are in 180 bottles of the mouthwash?

A mouthwash may contain ethyl alcohol.

Answers

Answers to Study Checks

1.1 a. iron, **b.** tin, and **d.** water are chemicals.

1.2 a. 4.25×10^5 m **b.** 8.6×10^{-7} g

1.3 a. 36 m **b.** 0.0026 L
c. 3.8×10^3 g **d.** 1.3 kg

1.4 a. 0.4924 **b.** 0.0080 or 8.0×10^{-3}
c. 2.0

1.5 a. 83.70 mg **b.** 0.5 L

1.6 1 500 000 000 000 (1.5×10^{12}) bytes

1.7 a. 1000 **b.** 0.001

1.8 a. $\frac{62.2\text{ km}}{1\text{ h}}$ and $\frac{1\text{ h}}{62.2\text{ km}}$ **b.** $\frac{10\ \mu g}{1\text{ kg}}$ and $\frac{1\text{ kg}}{10\ \mu g}$

1.9 1.89 L

1.10 0.44 oz

1.11 25 g of fat

1.12 1.05 g/cm^3

1.13 2.2 g/mL

1.14 1.50 mL

Answers to Selected Questions and Problems

1.1 Many chemicals are listed on a vitamin bottle such as vitamin A, vitamin B_3, vitamin B_{12}, folic acid, and so on.

1.3 No. All of the ingredients listed are chemicals.

1.5 Among the things you might do to help yourself succeed in chemistry: attend lecture regularly, review the *Learning Goals*, keep a problem notebook, read the text actively, read the chapter before lecture, join a study group, use your instructor's office hours, and others.

1.7 a, c, e, and **f**

1.9 In the United States, **a.** body mass is measured in pounds, **b.** height in feet and inches, **c.** amount of gasoline in gallons, and **d.** temperature in degrees Fahrenheit (°F). In Mexico, **a.** body mass is measured in kilograms, **b.** height in meters, **c.** amount of gasoline in liters, and **d.** temperature in degrees Celsius (°C).

1.11 a. meter, length **b.** gram, mass
c. milliliter, volume **d.** second, time
e. degree Celsius, temperature

1.13 a. 5.5×10^4 m **b.** 4.8×10^2 g
c. 5×10^{-6} cm **d.** 1.4×10^{-4} s
e. 7.2×10^{-3} L **f.** 6.7×10^5 kg

1.15 a. 7.2×10^3 cm **b.** 3.2×10^{-2} kg
c. 1×10^4 L **d.** 6.8×10^{-2} m

1.17 a. measured **b.** exact
c. exact **d.** measured

1.19 a. 6 oz of hamburger **b.** none
c. 0.75 lb, 350 g **d.** none (definitions are exact)

1.21 a. not significant **b.** significant
c. significant **d.** significant
e. not significant

1.23 **a.** 5 SFs **b.** 2 SFs **c.** 2 SFs
d. 3 SFs **e.** 4 SFs **f.** 3 SFs

1.25 **c**

1.27 A calculator often gives more digits than the number of significant figures allowed in the answer.

1.29 **a.** 1.85 kg **b.** 88.2 L
c. 0.004 74 cm **d.** 8810 m
e. 1.83×10^5 s

1.31 **a.** 5.08×10^3 L **b.** 3.74×10^4 g
c. 1.05×10^5 m **d.** 2.51×10^{-4} s

1.33 **a.** 1.6 **b.** 0.01
c. 27.6 **d.** 3.5

1.35 **a.** 53.54 cm **b.** 127.6 g
c. 121.5 mL **d.** 0.50 L

1.37 km/h is kilometers per hour; mi/h is miles per hour

1.39 The prefix *kilo* means to multiply by 1000. One kg is the same mass as 1000 g.

1.41 **a.** mg **b.** dL
c. km **d.** pg
e. μL **f.** ns

1.43 **a.** 0.01 **b.** 1×10^{12}
c. 1×10^{-3} **d.** 0.1
e. 1×10^6 **f.** 1×10^{-12}

1.45 **a.** 100 cm **b.** 1×10^9 nm
c. 0.001 m **d.** 1000 mL

1.47 **a.** kilogram **b.** milliliter
c. km **d.** kL
e. nanometer

1.49 A conversion factor can be inverted to give a second conversion factor.

1.51 **a.** 100 cm = 1 m, $\frac{100\text{ cm}}{1\text{ m}}$ and $\frac{1\text{ m}}{100\text{ cm}}$

b. 1000 mg = 1 g, $\frac{1000\text{ mg}}{1\text{ g}}$ and $\frac{1\text{ g}}{1000\text{ mg}}$

c. 1 L = 1000 mL, $\frac{1000\text{ mL}}{1\text{ L}}$ and $\frac{1\text{ L}}{1000\text{ mL}}$

d. 1 dL = 100 mL, $\frac{100\text{ mL}}{1\text{ dL}}$ and $\frac{1\text{ dL}}{100\text{ mL}}$

e. 1 week = 7 days, $\frac{1\text{ week}}{7\text{ days}}$ and $\frac{7\text{ days}}{1\text{ week}}$

1.53 **a.** $\frac{3.5\text{ m}}{1\text{ s}}$ and $\frac{1\text{ s}}{3.5\text{ m}}$

b. $\frac{3500\text{ mg potassium}}{1\text{ day}}$ and $\frac{1\text{ day}}{3500\text{ mg potassium}}$

c. $\frac{46.0\text{ km}}{1.0\text{ gal}}$ and $\frac{1.0\text{ gal}}{46.0\text{ km}}$

d. $\frac{50\text{ mg Atenolol}}{1\text{ tablet}}$ and $\frac{1\text{ tablet}}{50\text{ mg Atenolol}}$

e. $\frac{29\ \mu\text{g}}{1\text{ kg}}$ and $\frac{1\text{ kg}}{29\ \mu\text{g}}$

f. $\frac{81\text{ mg aspirin}}{1\text{ tablet}}$ and $\frac{1\text{ tablet}}{81\text{ mg aspirin}}$

1.55 The unit in the denominator must cancel with the preceding unit in the numerator.

1.57 **a.** 1.75 m **b.** 5.5 L **c.** 1.8 g

1.59 **a.** 473 mL **b.** 65.9 kg **c.** 24 lb
d. 43 g **e.** 50. lb

1.61 **a.** 66 gal **b.** 3 tablets **c.** 1800 mg

1.63 lead

1.65 **a.** 1.20 g/mL **b.** 4.4 g/mL
c. 3.10 g/mL **d.** 4.51 g/mL

1.67 **a.** 210 g **b.** 575 g **c.** 62 oz

1.69 **b** and **c**

1.71 **c** and **d**

1.73 **a.** exact **b.** measured
c. exact **d.** measured

1.75 **a.** length = 6.96 cm, width = 4.75 cm
b. length = 69.6 mm, width = 47.5 mm
c. 3 significant figures
d. 3 significant figures
e. 33.1 cm^2
f. 3 significant figures

1.77 1.8 g/mL

1.79 **a.** 0.000 012 6 L **b.** 3.53×10^2 kg
c. 125 000 m **d.** 58.7 g
e. 3.00×10^{-3} s **f.** 0.0108 g

1.81 42 min

1.83 $1.08 per lb

1.85 16 onions

1.87 **a.** 96 crackers **b.** 0.2 oz of fat
c. 110 g of sodium

1.89 0.35 oz

1.91 185 mg/dL

1.93 790 g

1.95 1200 cm^3

1.97 You would record the mass as 31.075 g. Since the balance will weigh to the nearest 0.001 g, the mass value would be reported to 0.001 g.

1.99 6.4 gal

1.101 0.141 mm

1.103 **a.** 79 cups **b.** 310 cans
c. 160 tablets

1.105 **a.** 4.3 lb of body fat **b.** 6.2 lb

1.107 A is vegetable oil, B is water, and C is mercury.

Matter and Energy

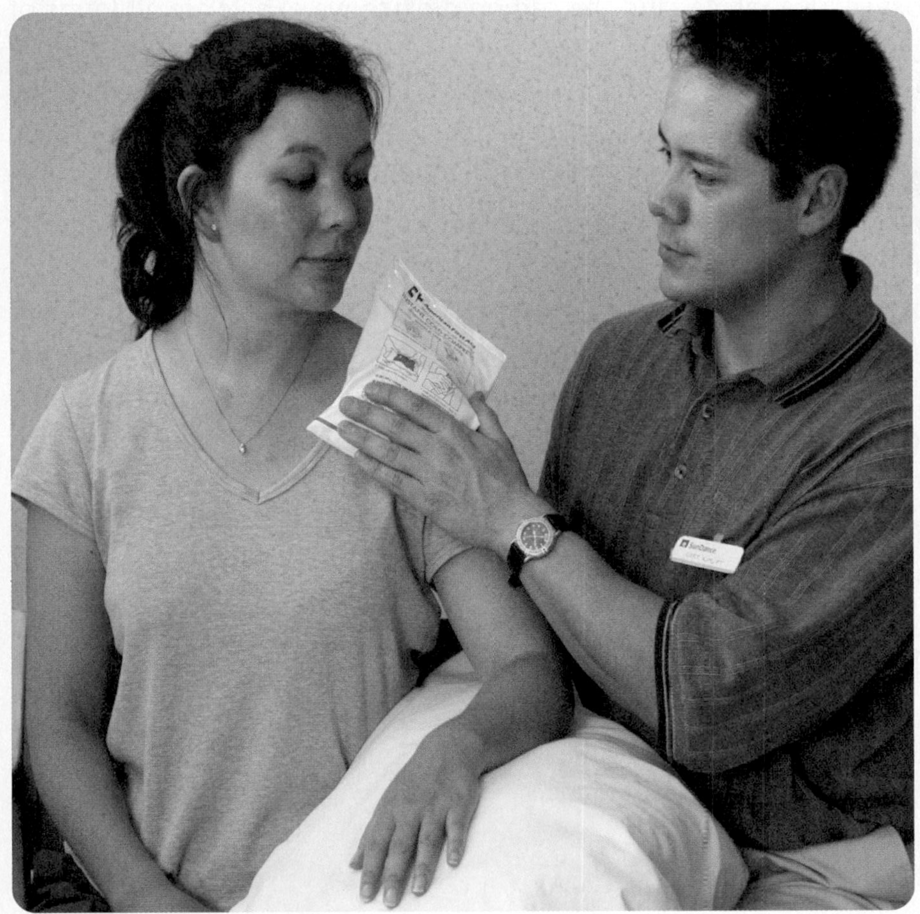

LOOKING AHEAD

"If you've had first aid for a sports injury," says Cort Kim, physical therapist at the Sunrise Sports Medicine Clinic, "you've likely been treated with a cold pack or hot pack. We use them for several kinds of injury. Here, I'm showing how I can use a cold pack to reduce swelling in my patient's shoulder."

A hot or cold pack is just a packaged chemical reaction. When you hit or open the pack to activate it, your action mixes chemicals together and thus initiates the reaction. In a cold pack, the reaction is one that absorbs heat energy, chills the pack, and draws heat from the injury. Hot packs use reactions that release energy, thus warming the pack. In both cases, the reaction proceeds at a moderate pace, so that the pack stays active for a long time and doesn't get too cold or hot.

Visit **www.masteringchemistry.com** for self-study materials and instructor-assigned homework.

Every day, we see a variety of materials with many different shapes and forms. To a scientist, all of this material is *matter*. Matter is everywhere around us: the orange juice we had for breakfast, the water we put in the coffee maker, the plastic bag we put our sandwich in, our toothbrush and toothpaste, the oxygen we inhale, and the carbon dioxide we exhale are all forms of matter.

Almost everything we do involves energy. We use energy when we walk, play tennis, study, and breathe. We use energy when we warm water, cook food, turn on lights, use computers, use a washing machine, or drive our cars. Of course, that energy has to come from somewhere. In our bodies, the food we eat provides us with energy. Energy from fossil fuels or the Sun is used to heat a home or water for a pool.

When we look around us, we see that matter takes the physical form of a solid, a liquid, or a gas. Water is a familiar example that we observe in all three states. In the solid state, water can be an ice cube or a snowflake. It is a liquid when it comes out of a faucet or fills a pool. Water forms a gas, or vapor, when it evaporates from wet clothes or boils in a pan. Energy is required to melt ice cubes and to boil water, whereas energy is removed to freeze water in an ice cube tray or to condense water vapor to liquid.

LEARNING GOAL

Classify examples of matter as pure substances or mixtures.

TUTORIAL
Classification of Matter

2.1 Classification of Matter

Matter is anything that has mass and occupies space. Matter makes up all things we use such as water, wood, plates, plastic bags, clothes, and shoes. The different types of matter are classified by their composition.

Pure Substances

A **pure substance** is matter that has a fixed or definite composition. There are two kinds of pure substances: elements and compounds. An **element**, the simplest type of a pure substance, is composed of only one type of material such as silver, iron, or aluminum. Every element is composed of *atoms*, which are extremely tiny particles that make up each type of matter. Silver is composed of silver atoms, iron of iron atoms, and aluminum of aluminum atoms. A complete list of the elements is found on the inside front cover of this text.

A molecule of water consists of two atoms of hydrogen (white) for one atom of oxygen (red) and has a formula of H_2O.

An aluminum can consists of many atoms of aluminum.

A **compound** is also a pure substance, but it consists of atoms of two or more elements always chemically combined in the same proportion. In compounds, the atoms are held together by attractions called *bonds*, which form small groups of atoms called molecules. For example, a molecule of the compound water has two hydrogen atoms for every one oxygen atom and is represented by the formula H_2O. The compound hydrogen peroxide is also a combination of hydrogen and oxygen, but it has two hydrogen atoms for every two oxygen atoms and is represented by the formula H_2O_2. Water (H_2O) and hydrogen peroxide (H_2O_2) are different compounds, which means they have different properties.

An important difference between compounds and elements is that compounds can be broken down by chemical processes into simpler substances, whereas elements cannot be broken down further. For example, ordinary table salt consists of the compound NaCl, which can be separated by chemical processes into sodium metal and chlorine gas, as seen in Figure 2.1. However, compounds such as NaCl cannot be separated into simpler substances by using physical methods such as boiling or sifting.

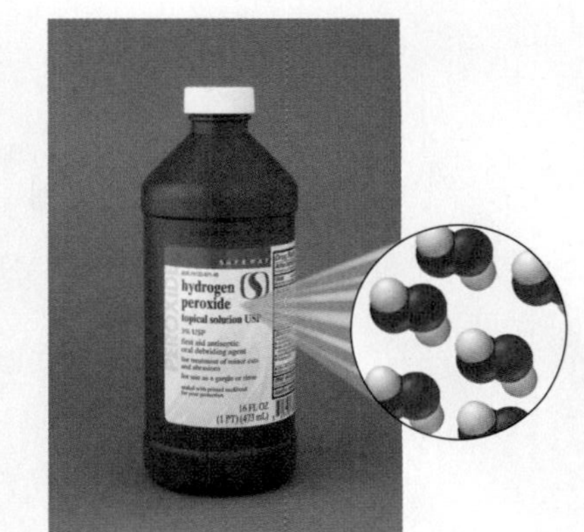

A molecule of hydrogen peroxide consists of two atoms of hydrogen (white) for two atoms of oxygen (red) and has a formula of H_2O_2.

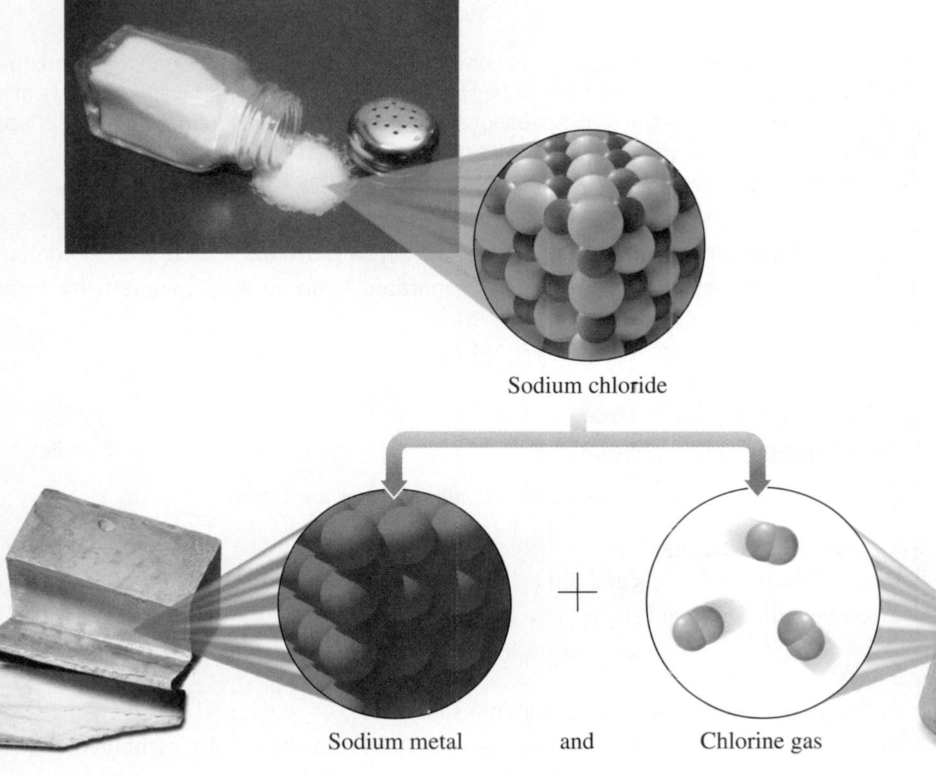

FIGURE 2.1 The decomposition of salt, NaCl, produces the elements sodium and chlorine.

Q How do elements and compounds differ?

Mixtures

Much of the matter in our everyday lives consists of mixtures. In a **mixture**, two or more substances are physically mixed, but not chemically combined. The air we breathe is a mixture of mostly oxygen and nitrogen gases. The steel in buildings and railroad tracks is a mixture of iron, nickel, carbon, and chromium. The brass in doorknobs and fixtures is a mixture of zinc and copper. Tea, coffee, and ocean water are mixtures too. In any mixture, the proportions of the components can vary. For example, two sugar–water mixtures may look the same, but the one with the higher ratio of sugar to water would taste sweeter. Different types of brass have different properties such as color or strength depending on the ratio of copper to zinc (see Figure 2.2).

Physical processes can be used to separate mixtures because there are no chemical interactions between the components. For example, different coins such as nickels,

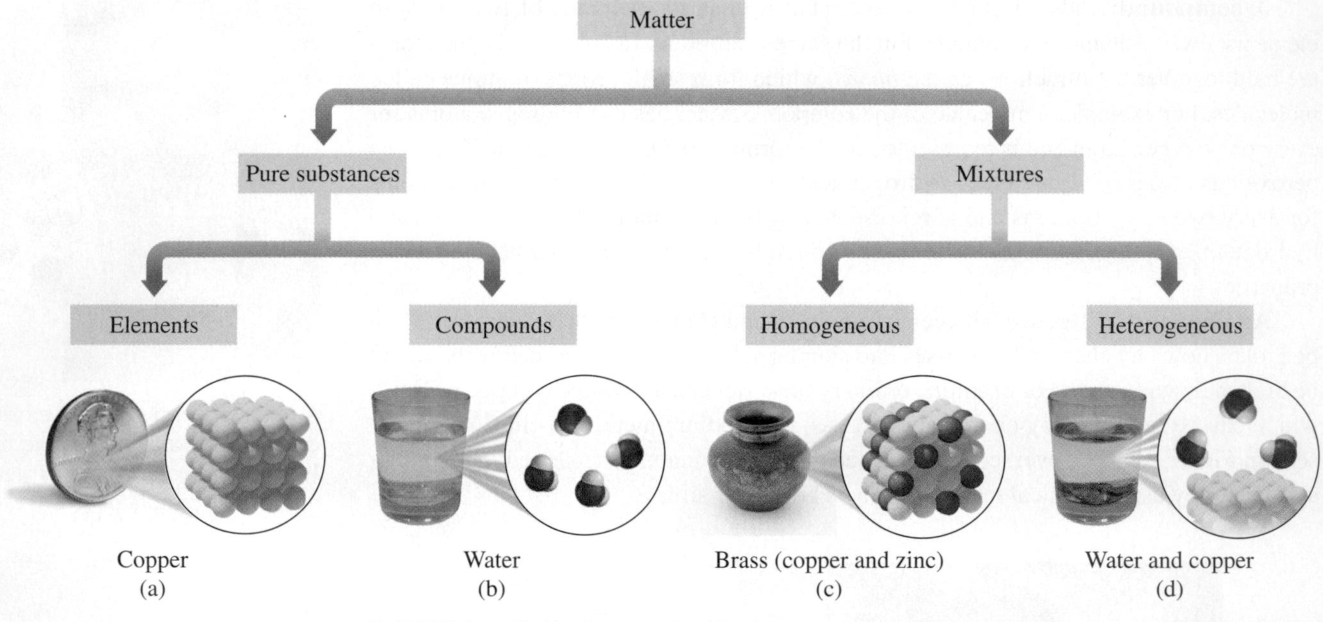

FIGURE 2.2 Matter is organized by its components: elements, compounds, and mixtures. (a) The element copper consists of copper atoms. (b) The compound water consists of H_2O molecules. (c) Brass is a homogeneous mixture of copper and zinc atoms. (d) Copper metal in water is a heterogeneous mixture of Cu atoms and H_2O molecules.
Q Why are copper and water pure substances, but brass is a mixture?

dimes, and quarters can be separated by size; iron particles mixed with sand can be picked up with a magnet; and water is separated from cooked spaghetti by using a strainer (see Figure 2.3).

Physical method of separation

FIGURE 2.3 A mixture of spaghetti and water is separated using a strainer, a physical method of separation.
Q Why can physical methods be used to separate mixtures but not compounds?

CONCEPT CHECK 2.1

Pure Substances and Mixtures

Classify each of the following as a pure substance or a mixture:

a. sugar in a sugar bowl
b. a collection of nickels and dimes
c. coffee with milk and sugar

ANSWER

a. Sugar is a compound, which is a pure substance.
b. The nickels and dimes are physically mixed, but not chemically combined, which makes the collection a mixture.
c. The coffee, milk, and sugar are physically mixed, but not chemically combined, which makes it a mixture.

Types of Mixtures

Mixtures are classified further as homogeneous or heterogeneous. In a *homogeneous mixture*, also called a *solution*, the composition is uniform throughout the sample. Familiar examples of homogeneous mixtures are air, which contains oxygen and nitrogen gases, and sea water, a solution of salt and water.

In a *heterogeneous mixture*, the components do not have a uniform composition throughout the sample. For example, a mixture of oil and water is heterogeneous because the oil floats on the surface of the water. Other examples of heterogeneous mixtures include the raisins in a cookie and the bubbles in a soda.

In the chemistry laboratory, mixtures are separated by various methods. Solids are separated from liquids by *filtration*, which involves pouring a mixture through a filter paper set in a funnel. In *chromatography*, different components of a liquid mixture separate as they move at different rates up the surface of a piece of chromatography paper.

A mixture of a liquid and a solid is separated by filtration.

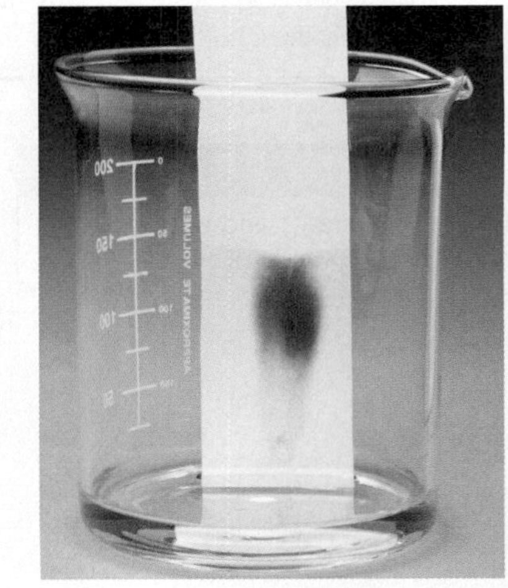

Different substances are separated as they travel at different rates up the surface of chromatography paper.

SAMPLE PROBLEM 2.1

Classifying Mixtures

Classify each of the following as a pure substance (element or compound) or a mixture (homogeneous or heterogeneous):

a. copper in copper wire
b. a chocolate chip cookie
c. Nitrox, a breathing mixture of oxygen and nitrogen for scuba diving

SOLUTION

a. Copper is an element, which is a pure substance.
b. A chocolate chip cookie does not have a uniform composition, which makes it a heterogeneous mixture.
c. The gases oxygen and nitrogen have a uniform composition in Nitrox, which makes it a homogeneous mixture.

STUDY CHECK 2.1

A salad dressing is prepared with oil, vinegar, and chunks of blue cheese. Is this a homogeneous or heterogeneous mixture?

Oil and water form a heterogeneous mixture.

Chemistry Link to Health

BREATHING MIXTURES FOR SCUBA

The air we breathe is composed mostly of the gases oxygen (21%) and nitrogen (79%). The homogeneous breathing mixtures used by scuba divers differ from the air we breathe depending on the depth of the dive. Nitrox is a mixture of oxygen and nitrogen, but with more oxygen gas (up to 32%) and less nitrogen gas (68%) than air. A breathing mixture with less nitrogen gas decreases the risk of nitrogen narcosis associated with breathing regular air while diving. Heliox is a breathing mixture of oxygen and helium gases typically used for diving to more than 200 feet. With deep dives, there is more chance of nitrogen narcosis, but by replacing nitrogen with helium, it does not occur. However, at dive depths over 300 ft, helium is associated with severe shaking and body temperature drop.

A breathing mixture used for dives over 400 ft is Trimix, which contains oxygen, helium, and some nitrogen. The addition of nitrogen lessens the problem of shaking that comes with breathing high levels of helium. Both Heliox and Trimix are used by only professional, military, or other highly trained divers.

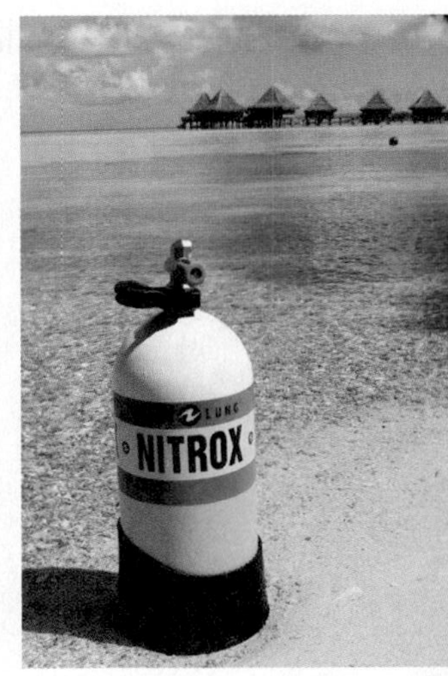

A Nitrox mixture is used to fill scuba tanks.

QUESTIONS AND PROBLEMS

Classification of Matter

2.1 Classify each of the following as a pure substance or a mixture:
a. baking soda ($NaHCO_3$) **b.** a blueberry muffin
c. ice (H_2O) **d.** zinc (Zn)
e. Trimix (oxygen, nitrogen, and helium) in a scuba tank

2.2 Classify each of the following as a pure substance or a mixture:
a. a soft drink **b.** propane (C_3H_8)
c. a cheese sandwich **d.** an iron (Fe) nail
e. salt substitute (KCl)

2.3 Classify each of the following pure substances as an element or a compound:
a. a silicon (Si) chip **b.** hydrogen peroxide (H_2O_2)
c. oxygen (O_2) **d.** rust (Fe_2O_3)
e. methane (CH_4) in natural gas

2.4 Classify each of the following pure substances as an element or a compound:
a. helium gas (He) **b.** mercury (Hg) in a thermometer
c. sugar ($C_{12}H_{22}O_{11}$) **d.** sulfur (S)
e. lye (NaOH)

2.5 Classify each of the following mixtures as homogeneous or heterogeneous:
a. vegetable soup **b.** sea water
c. tea **d.** tea with ice and lemon slices
e. fruit salad

2.6 Classify each of the following mixtures as homogeneous or heterogeneous:
a. nonfat milk **b.** chocolate-chip ice cream
c. gasoline **d.** peanut-butter-and-jelly sandwich
e. cranberry juice

LEARNING GOAL

Identify the states and physical and chemical properties of matter.

2.2 States and Properties of Matter

On Earth, matter exists in one of three *physical forms* called the **states of matter**: *solids*, *liquids*, and *gases*. A **solid**, such as a pebble or a baseball, has a definite shape and volume. You can probably recognize several solids within your reach right now such as books, pencils, or a computer mouse. In a *solid*, strong attractive forces hold the particles such as atoms or molecules close together. The particles are arranged in such a rigid pattern they can only vibrate slowly in fixed positions. For many solids, this rigid structure produces a crystal such as that seen in amethyst.

A **liquid** has a definite volume, but not a definite shape. In a *liquid*, the particles move in random directions but are sufficiently attracted to each other to maintain a definite volume, although not a rigid structure. Thus, when water, oil, or vinegar is poured from one container to another, the liquid maintains its own volume but takes the shape of the new container.

A **gas** does not have a definite shape or volume. In a *gas*, the particles are far apart, have little attraction to each other, and move at high speeds, taking the shape and volume of their container. When you inflate a bicycle tire, the air, which is a gas, fills the entire volume of the tire. The propane gas in a tank fills the entire volume of the tank. Table 2.1 compares the three states of matter.

Amethyst, a solid, is a purple form of quartz (SiO_2).

Water as a liquid takes the shape of its container.

A gas takes the shape and volume of its container.

TABLE 2.1 A Comparison of Solids, Liquids, and Gases

Characteristic	Solid	Liquid	Gas
Shape	Has a definite shape	Takes the shape of the container	Takes the shape of the container
Volume	Has a definite volume	Has a definite volume	Fills the volume of the container
Arrangement of particles	Fixed, very close	Random, close	Random, far apart
Interaction between particles	Very strong	Strong	Essentially none
Movement of particles	Very slow	Moderate	Very fast
Examples	Ice, salt, iron	Water, oil, vinegar	Water vapor, helium, air

Physical Properties and Physical Changes

One way to describe matter is to observe its properties. For example, if you were asked to describe yourself, you might list your characteristics such as the color of your eyes and skin or the length, color, and texture of your hair.

Physical properties are those characteristics that can be observed or measured without affecting the identity of a substance. In chemistry, typical physical properties include the shape, color, melting point, boiling point, and physical state of a substance. For example, a penny has the physical properties of a round shape, an orange-red color, solid state, and a shiny luster. Table 2.2 gives more examples of physical properties of copper found in pennies, electrical wiring, and copper pans.

Water is a substance that is commonly found in all three states: solid, liquid, and gas. When matter undergoes a **physical change**, its state or its appearance will change, but its composition remains the same. The solid state of water, snow or ice, has a different appearance than its liquid or gaseous state, but all three states are water (see Figure 2.4).

The physical appearance of a substance can change in other ways too. Suppose that you dissolve some salt in water. The appearance of the salt changes, but you could re-form the salt crystals by heating the mixture and evaporating the water. Thus in a physical change of state, no new substances are produced. Table 2.3 gives more examples of physical changes.

TUTORIAL
Properties and Changes of Matter

Copper, used in cookware, is a good conductor of heat.

TABLE 2.2 Some Physical Properties of Copper

State at 25 °C	Solid
Color	Orange-red
Odor	Odorless
Melting point	1083 °C
Boiling point	2567 °C
Luster	Shiny
Conduction of electricity	Excellent
Conduction of heat	Excellent

FIGURE 2.4 Water exists as a solid, as a liquid, or as a gas.
Q In what state of matter does water have a definite volume, but not a definite shape?

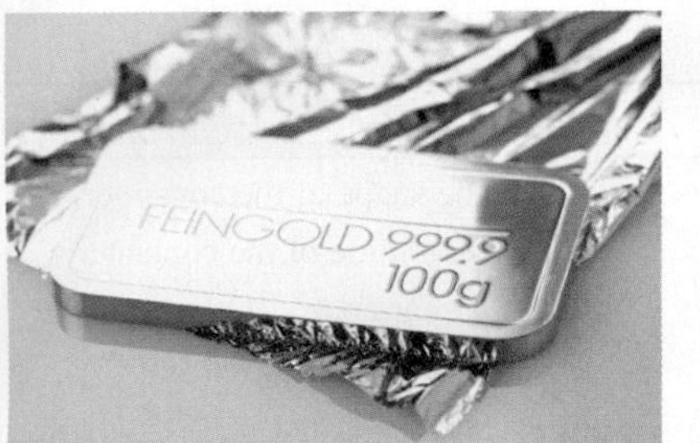

In a physical change, a gold ingot is hammered to form gold leaf.

TABLE 2.3 Examples of Some Physical Changes

Type of Physical Change	Example
Change of state	Water boiling
	Freezing of liquid water to solid water (ice)
Change of appearance	Dissolving sugar in water
Change of shape	Hammering a gold ingot into shiny gold leaf
	Drawing copper into thin copper wire
Change of size	Cutting paper into tiny pieces for confetti
	Grinding pepper into smaller particles

CONCEPT CHECK 2.2

States of Matter

Identify the state(s) of matter described by the substance in each of the following:

a. Its volume does not change in a different container.
b. Its shape depends on the container.
c. It has a definite shape and volume.

ANSWER

a. Both a solid and a liquid have their own volume that does not depend on the volume of their container.
b. Both a liquid and a gas take the shape of their containers.
c. A solid has a rigid arrangement of particles that gives it a definite shape and volume.

Chemical Properties and Chemical Changes

Chemical properties are those that describe the ability of a substance to change into a new substance. When a **chemical change** takes place, the original substance is converted into one or more new substances, which have different physical and chemical properties. For example, methane (CH_4) in natural gas can burn because it has the chemical property of being flammable. When methane burns in oxygen (O_2), it is converted to water (H_2O) and carbon dioxide (CO_2), which have different physical and chemical properties than the methane and oxygen. Rusting or corrosion is a chemical property of iron. In the rain, an iron nail undergoes a chemical change when it reacts with oxygen in the air to form rust (Fe_2O_3), a new substance. Table 2.4 gives some examples of chemical changes.

Flan has a topping of caramelized sugar.

TABLE 2.4 Examples of Some Chemical Changes

Type of Chemical Change	Change in Chemical Properties
Silver tarnishes	Shiny, silver metal reacts in air to give a black, grainy coating.
Methane burns	Methane burns with a bright flame, producing water vapor and carbon dioxide.
Sugar caramelizes	At high temperatures, white, granular sugar changes to a smooth, caramel-colored substance.
Iron rusts	Iron, which is gray and shiny, combines with oxygen to form orange-red rust.

CONCEPT CHECK 2.3

Physical and Chemical Properties

Classify each of the following as a physical or chemical property:

a. Gasoline is a liquid at room temperature.
b. Gasoline burns in air.
c. Gasoline has a pungent odor.

ANSWER

a. A liquid is a state of matter, which makes it a physical property.
b. When gasoline burns, it changes to different substances with new properties, which is a chemical property.
c. The odor of gasoline is a physical property.

SAMPLE PROBLEM 2.2

Physical and Chemical Changes

Classify each of the following as a physical or chemical change:

a. An ice cube melts to form liquid water.
b. An enzyme breaks down the lactose in milk.
c. Garlic is chopped into small pieces.

SOLUTION

a. A physical change occurs when the ice cube changes state from solid to liquid.
b. A chemical change occurs when an enzyme breaks down lactose into simpler substances.
c. A physical change occurs when the size of an object changes.

STUDY CHECK 2.2

Which of the following are chemical changes?

a. Water freezes on a pond.
b. Gas bubbles form when baking powder is placed in vinegar.
c. A log is chopped for firewood.
d. A log burns in a fireplace.

QUESTIONS AND PROBLEMS

States and Properties of Matter

2.7 Indicate whether each of the following describes a gas, a liquid, or a solid:
a. This substance has no definite volume or shape.
b. The particles in a substance do not interact with each other.
c. The particles in a substance are held in a rigid structure.

2.8 Indicate whether each of the following describes a gas, a liquid, or a solid:
a. This substance has a definite volume but takes the shape of its container.
b. The particles in a substance are very far apart.
c. This substance occupies the entire volume of the container.

2.9 Describe each of the following as a physical or chemical property:
a. Chromium is a steel-gray solid.
b. Hydrogen reacts readily with oxygen.
c. Nitrogen freezes at −210 °C.
d. Milk will sour when left in a warm room.
e. Butane gas in an igniter burns in oxygen.

2.10 Describe each of the following as a physical or chemical property:
a. Neon is a colorless gas at room temperature.
b. Apple slices turn brown when they are exposed to air.
c. Phosphorus will ignite when exposed to air.
d. At room temperature, mercury is a liquid.
e. Propane gas is compressed to a liquid for placement in a small cylinder.

2.11 What type of change, physical or chemical, takes place in each of the following?
a. Water vapor condenses to form rain.
b. Cesium metal reacts explosively with water.
c. Gold melts at 1064 °C.
d. A puzzle is cut into 1000 pieces.
e. Sugar dissolves in water.

2.12 What type of change, physical or chemical, takes place in each of the following?
a. Gold is hammered into thin sheets.
b. A silver pin tarnishes in the air.
c. A tree is cut into boards at a saw mill.
d. Food is digested.
e. A chocolate bar melts.

2.13 Describe each property of the element fluorine as physical or chemical.
a. is highly reactive
b. is a gas at room temperature
c. has a pale, yellow color
d. will explode in the presence of hydrogen
e. has a melting point of −220 °C

2.14 Describe each property of the element zirconium as physical or chemical.
a. melts at 1852 °C
b. is resistant to corrosion
c. has a grayish-white color
d. ignites spontaneously in air when finely divided
e. is a shiny metal

LEARNING GOAL

Identify energy as potential or kinetic; convert between units of energy.

2.3 Energy

When you are running, walking, dancing, or thinking, you are using energy to do *work*, which is any activity that requires energy. In fact, **energy** is defined as the ability to do work. Suppose you are climbing a steep hill and you become too tired to go on. At that moment, you do not have the energy to do any more work. Now suppose you sit down and have lunch. In a while, you will have obtained some energy from the food, and you will be able to do more work and complete the climb (see Figure 2.5).

FIGURE 2.5 Work is done as the rock climber moves up the cliff. At the top, the climber has more potential energy than when she started the climb.
Q What happens to the potential energy of the climber as she descends?

Potential and Kinetic Energy

Energy can be classified as potential energy or kinetic energy. **Kinetic energy** is the energy of motion. Any object that is moving has kinetic energy. **Potential energy** is determined by the position of an object or by the chemical composition of a substance. A boulder resting on top of a mountain has potential energy because of its location. If the boulder rolls down the mountain, the potential energy becomes kinetic energy. Water stored in a reservoir has potential energy. When the water goes over the dam and falls to the stream below, its potential energy becomes kinetic energy. Foods and fossil fuels have potential energy in their molecules. When you digest food or burn gasoline in your car, potential energy is converted to kinetic energy to do work.

When water flows from the top of a dam, potential energy is converted to kinetic energy.

CONCEPT CHECK 2.4

Potential and Kinetic Energy

Identify each of the following as an example of potential or kinetic energy:

a. gasoline
b. skating
c. a candy bar

ANSWER

a. Gasoline is burned to provide energy and heat; it contains potential energy (stored) in its molecules.
b. A skater uses energy to move; skating is kinetic energy (energy of motion).
c. A candy bar has potential energy. When digested, it provides energy for the body to do work.

Heat and Units of Energy

Heat, also known as *thermal energy*, is associated with the motion of particles. A frozen pizza feels cold because heat flows from your hand into the pizza. The faster the particles move, the greater the heat or thermal energy of the substance. In the frozen pizza, the particles are moving very slowly. As heat is added and the pizza becomes warmer, the motions of the particles in the pizza increase. Eventually, the particles have enough energy to make the pizza hot and ready to eat.

TABLE 2.5 A Comparison of Energy for Various Resources

Energy in joules

10^{27}	Energy radiated by Sun per second (10^{26})
10^{24}	World reserves of fossil fuel (10^{23})
10^{21}	Energy consumption for one year in US (10^{20})
10^{18}	Solar energy reaching the Earth per second (10^{17})
10^{15}	
10^{12}	Energy use per person in one year in US (10^{11})
10^{9}	Energy from one gallon of gasoline (10^{8})
10^{6}	Energy from one serving of pasta, a doughnut, or needed to bicycle one hour (10^{6})
10^{3}	Energy used to sleep one hour (10^{5})
10^{0}	

Units of Energy

The SI unit of energy and work is the **joule (J)** (pronounced "jewel"). The joule is a small amount of energy, so scientists often use the kilojoule (kJ), 1000 joules. To heat water for one cup of tea, you need about 75 000 J or 75 kJ of heat. Table 2.5 shows a comparison of energy in joules for several energy sources.

You may be more familiar with the unit **calorie (cal)**, from the Latin *calor*, meaning "heat." The calorie was originally defined as the amount of energy (heat) needed to raise the temperature of 1 g of water by 1 °C. Now one calorie is defined as *exactly* 4.184 J. This equality can be written as two conversion factors:

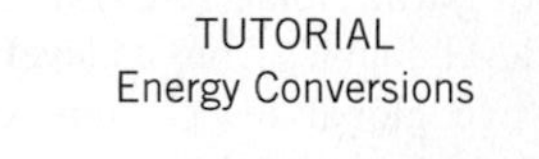

$$1 \text{ cal} = 4.184 \text{ J (exact)}$$

$$\frac{4.184 \text{ J}}{1 \text{ cal}} \quad \text{and} \quad \frac{1 \text{ cal}}{4.184 \text{ J}}$$

One *kilocalorie* (kcal) is equal to 1000 calories, and one *kilojoule* (kJ) is equal to 1000 joules.

$$1 \text{ kcal} = 1000 \text{ cal}$$

$$1 \text{ kJ} = 1000 \text{ J}$$

SAMPLE PROBLEM 2.3

Energy Units

When 1.0 g of diesel burns in an diesel car engine, 48 000 J are released. What is this quantity of energy in calories?

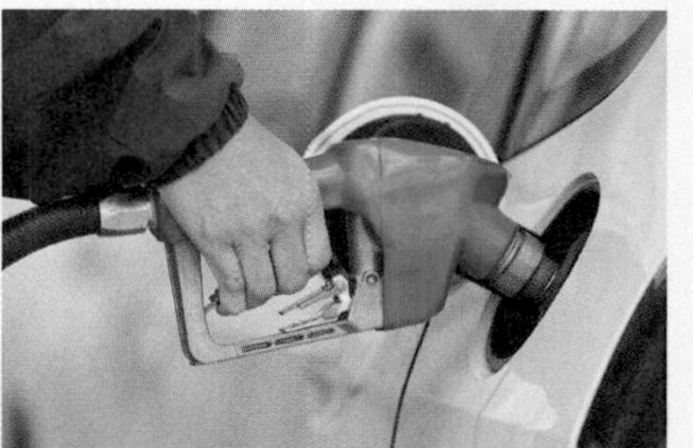

Diesel fuel reacts in a car engine to produce energy.

SOLUTION

Step 1 **Given** 48 000 J **Need** calories (cal)

Step 2 **Plan**

joules → Energy factor → calories

Step 3 **Equalities/Conversion Factors**

$$1 \text{ cal} = 4.184 \text{ J}$$

$$\frac{1 \text{ cal}}{4.184 \text{ J}} \quad \text{and} \quad \frac{4.184 \text{ J}}{1 \text{ cal}}$$

Step 4 Set Up Problem

$$\underset{\text{Two SFs}}{48\,000\ \cancel{J}} \times \underset{\text{Exact}}{\frac{1\ \text{cal}}{4.184\ \cancel{J}}} = \underset{\text{Two SFs}}{11\,000\ \text{cal}\ (1.1 \times 10^4\ \text{cal})}$$

STUDY CHECK 2.3

The burning of 1.0 g of coal produces 8.4 kcal. How many joules are produced?

Chemistry Link to the Environment

CARBON DIOXIDE AND GLOBAL WARMING

Earth's climate is a product of interactions between sunlight, the atmosphere, and the oceans. The Sun provides us with energy in the form of solar radiation. Some of this radiation is reflected back into space. The rest is absorbed by the clouds, atmospheric gases including carbon dioxide, and Earth's surface. For millions of years, concentrations of carbon dioxide (CO_2) have fluctuated. However in the last 100 years, the amount of carbon dioxide (CO_2) gas in our atmosphere has increased significantly. From the years 1000 to 1800, the atmospheric carbon dioxide averaged 280 ppm. But since the beginning of the Industrial Revolution in 1800 up until 2005, the level of atmospheric carbon dioxide has risen from about 280 ppm to about 390 ppm, a 40% increase.

As the atmospheric CO_2 level increases, more solar radiation is trapped by the atmospheric gases, which raises the temperature at the surface of Earth. Some scientists have estimated that if the carbon dioxide level doubles from its level before the Industrial Revolution, the average global temperature could increase by 2 to 4.4 °C. Although this seems to be a small temperature change, it could have dramatic impact worldwide. Even now, glaciers and snow cover in much of the world have diminished. Ice sheets in Antarctica and Greenland are melting faster and breaking apart. Although no one knows for sure how rapidly the ice in the polar regions is melting, this accelerating change will contribute to a rise in sea level. In the twentieth century, the sea level increased by about 20 cm. Some scientists predict the sea level will rise 1 m in this century. Such an increase will have a major impact on coastal areas.

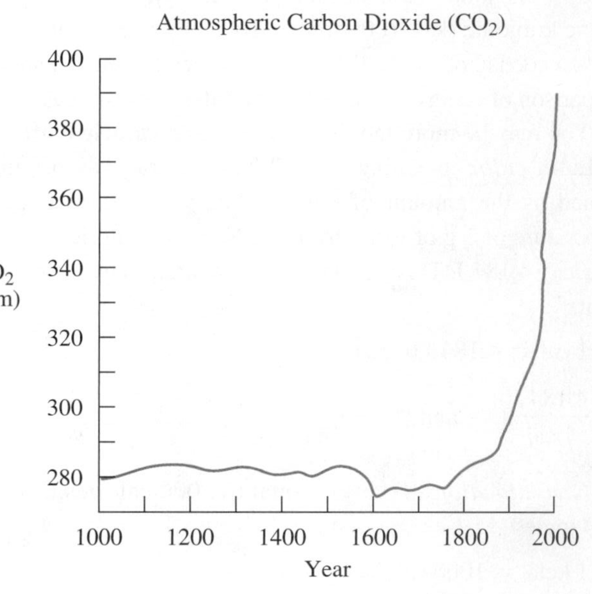

Until recently, the carbon dioxide level was maintained as algae in the oceans and the trees in the forests utilized the carbon dioxide. However, the ability of these and other forms of plant life to absorb carbon dioxide is not keeping up with the increase in carbon dioxide. Most scientists agree that the primary source of the increase of carbon dioxide is the burning of fossil fuels such as gasoline, coal, and natural gas. The cutting and burning of trees in the rain forests (deforestation) also reduces the amount of carbon dioxide removed from the atmosphere.

Worldwide efforts are being made to reduce the carbon dioxide produced by burning fossil fuels that heat our homes, run our cars, and provide energy for industries. Scientists are exploring ways to provide alternative energy sources and to reduce the effects of deforestation. Meanwhile, we can reduce energy use in our homes by using appliances that are more energy efficient such as replacing incandescent light bulbs with fluorescent lights. Such an effort worldwide will reduce the possible impact of global warming and at the same time save our fuel resources.

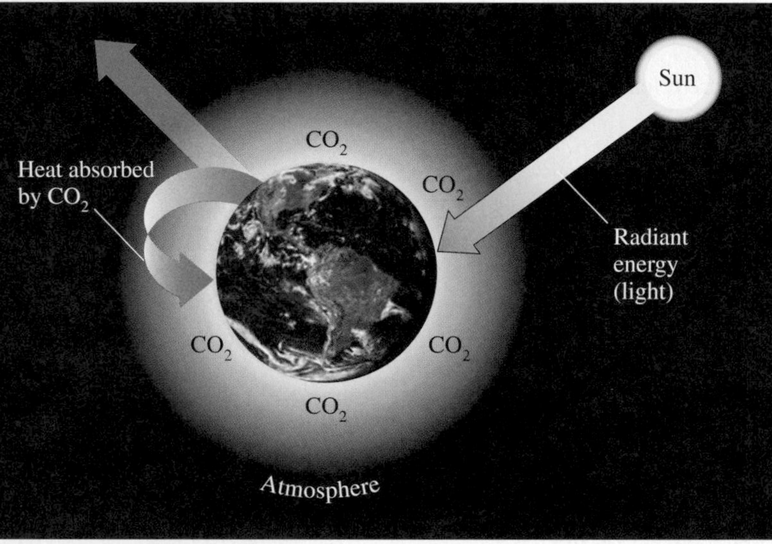

Heat from the Sun is trapped by the CO_2 layer in the atmosphere.

QUESTIONS AND PROBLEMS

Energy

2.15 Discuss the changes in the potential and kinetic energy of a roller-coaster ride as the roller-coaster car climbs to the top and goes down the other side.

2.16 Discuss the changes in the potential and kinetic energy of a ski jumper taking the elevator to the top of the jump and going down the ramp.

2.17 Indicate whether each of the following statements describes potential or kinetic energy:
a. water at the top of a waterfall
b. kicking a ball
c. the energy in a lump of coal
d. a skier at the top of a hill

2.18 Indicate whether each of the following statements describes potential or kinetic energy:
a. the energy in your food
b. a tightly wound spring
c. an earthquake
d. a car speeding down the freeway

2.19 The energy needed to keep a 75-watt lightbulb burning for 1.0 h is 270 kJ. Calculate the energy required to keep the lightbulb burning for 3.0 h in each of the following energy units:
a. joules **b.** kilocalories

2.20 A person uses 750 kcal on a long walk. Calculate the energy used for the walk in each of the following energy units:
a. joules **b.** kilojoules

2.4 Temperature

LEARNING GOAL

Given a temperature, calculate a corresponding temperature on another scale.

Temperatures in science, and in most of the world, are measured and reported in *Celsius* (°C) units. On the Celsius scale, the reference points are the freezing point of water, 0 °C, and the boiling point, 100 °C. In the United States, everyday temperatures are commonly reported in *Fahrenheit* (°F) units. On the Fahrenheit scale, water freezes at 32 °F and boils at 212 °F. A typical room temperature of 22 °C would be the same as 72 °F. A normal body temperature of 37.0 °C is 98.6 °F.

MC TUTORIAL Temperature Conversions

On the Celsius and Fahrenheit temperature scales, the temperature difference between freezing and boiling is divided into smaller units called *degrees*. The Celsius scale has 100 degrees between the freezing and boiling points of water whereas the Fahrenheit scale has 180 degrees between the freezing and boiling points. That makes a degree Celsius almost twice the size of a Fahrenheit degree: 1 °C = 1.8 °F (see Figure 2.6).

$$180 \text{ Fahrenheit degrees} = 100 \text{ degrees Celsius}$$

$$\frac{180 \text{ Fahrenheit degrees}}{100 \text{ degrees Celsius}} = \frac{1.8\ °\text{F}}{1\ °\text{C}}$$

We can write a temperature equation that relates a Fahrenheit temperature and its corresponding Celsius temperature.

$$T_F = \underbrace{1.8(T_C)}_{\text{Changes °C to °F}} + \underbrace{32}_{\text{Adjusts freezing point}}$$

In this equation, the Celsius temperature is multiplied by 1.8 to change °C to °F; then 32 is added to adjust the freezing point from 0 °C to 32 °F. Both values, 1.8 and 32, are exact numbers.

To convert from Fahrenheit to Celsius, the temperature equation is rearranged for T_C.

$$T_C = \frac{T_F - 32}{1.8}$$

A digital ear thermometer is used to measure body temperature.

Scientists have learned that the coldest temperature possible is −273 °C (more precisely, −273.15 °C). On the *Kelvin* scale, this temperature, called *absolute zero*, has

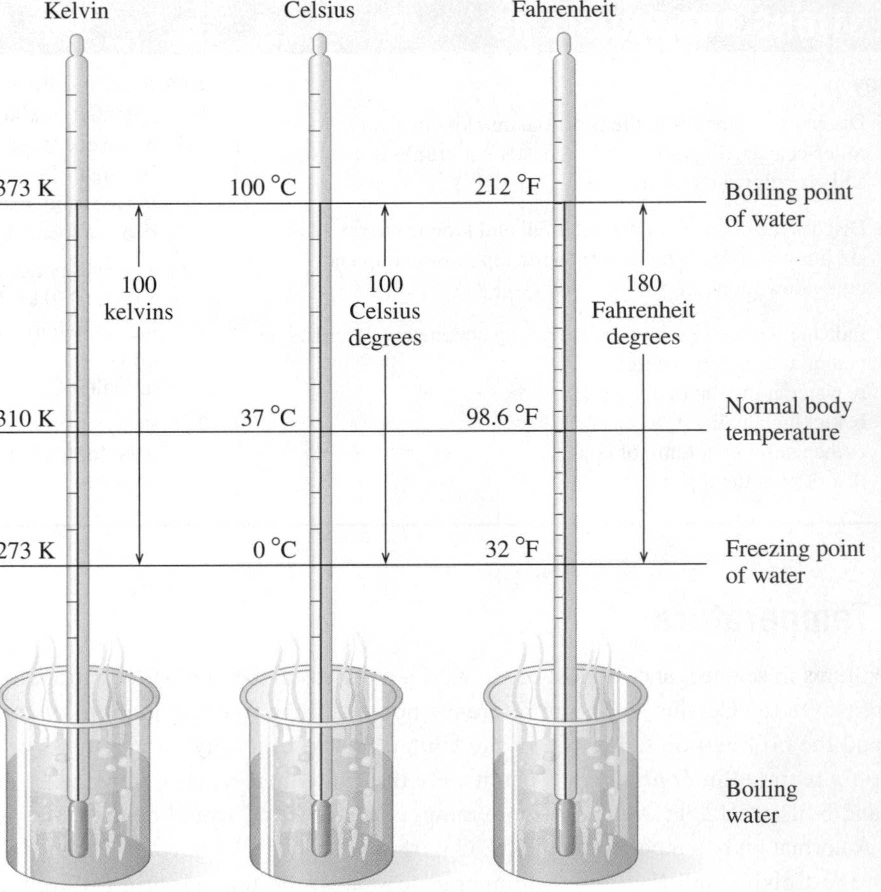

FIGURE 2.6 A comparison of the Fahrenheit, Celsius, and Kelvin temperature scales between the freezing and boiling points of water.
Q What is the difference in the values for freezing on the Celsius and Fahrenheit temperature scales?

the value of 0 K. Units on the Kelvin scale are called kelvins (K); no degree symbol is used. Because there are no lower temperatures, the Kelvin scale has no negative temperature values. Between the freezing point of water, 273 K, and the boiling point, 373 K, there are 100 kelvins, which makes a kelvin equal in size to a Celsius degree.

$$1\ \text{K} = 1\ °\text{C}$$

We can write an equation that relates a Celsius temperature to its corresponding Kelvin temperature by adding 273. Table 2.6 gives a comparison of some temperatures on the three scales.

$$T_K = T_C + 273$$

TABLE 2.6 A Comparison of Temperatures

Example	Fahrenheit (°F)	Celsius (°C)	Kelvin (K)
Sun	9937	5503	5776
A hot oven	450	232	505
A desert	120	49	322
A high fever	104	40	313
Room temperature	70	21	294
Water freezes	32	0	273
A northern winter	−66	−54	219
Helium boils	−452	−269	4
Absolute zero	−459	−273	0

SAMPLE PROBLEM 2.4

Converting Celsius to Fahrenheit

A room is heated to 22 °C. If that temperature is lowered by 1 °C, it can save as much as 5% in energy costs. What temperature, in Fahrenheit degrees, should be set to lower the temperature by 1 °C?

SOLUTION

Step 1 **Given** 22 °C − 1 °C = 21 °C **Need** T_F

Step 2 **Plan**

T_C → Temperature equation → T_F

Step 3 **Equality/Conversion Factor**

$$T_F = 1.8(T_C) + 32$$

Step 4 **Set Up Problem** Substitute the Celsius temperature into the equation and solve.

$$T_F = 1.8(21) + 32$$

Two SFs Exact

$$T_F = 38 + 32 \quad \text{1.8 is exact; 32 is exact}$$

$$= 70.\ °F \quad \text{Answer to the ones place}$$

In the equation, *the values of 1.8 and 32 are exact numbers*, which do not affect the number of SFs.

STUDY CHECK 2.4

In the process of making ice cream, rock salt is added to crushed ice to chill the ice cream mixture. If the temperature drops to −11 °C, what is it in Fahrenheit degrees?

SAMPLE PROBLEM 2.5

Converting Fahrenheit to Celsius

In a type of cancer treatment called *thermotherapy*, temperatures as high as 113 °F are used to destroy cancer cells. What is that temperature in degrees Celsius?

SOLUTION

Step 1 **Given** 113 °F **Need** T_C

Step 2 **Plan**

T_F → Temperature equation → T_C

Step 3 **Equality/Conversion Factor**

$$T_C = \frac{T_F - 32}{1.8}$$

Step 4 **Set Up Problem** Substitute the Fahrenheit temperature into the equation and solve.

$$T_C = \frac{T_F - 32}{1.8}$$

Career Focus

SURGICAL TECHNOLOGIST

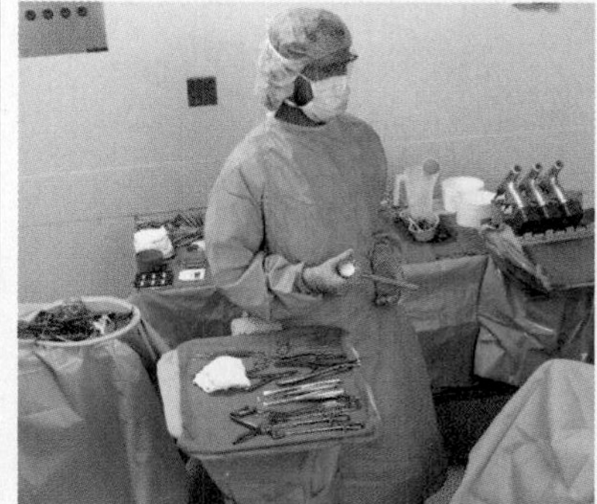

"As a surgical technologist, I assist the doctors during surgeries," says Christopher Ayars, surgical technologist, Kaiser Hospital. "I am there to help during general or orthopedic surgery by passing instruments, holding retractors, and maintaining the sterile field. Our equipment for surgery is sterilized by steam that is heated to 270 °F, which is the same as 130 °C."

Surgical technologists assist with surgical procedures by preparing and maintaining surgical equipment, instruments, and supplies; providing patient care in an operating room setting; preparing and maintaining a sterile field; and ensuring that there are no breaks in aseptic technique. Instruments, which have been sterilized, are wrapped and sent to surgery where they are checked again before they are opened.

$$T_C = \frac{(113 - 32)}{1.8} \quad \text{32 is exact; 1.8 is exact}$$

$$= \frac{81}{1.8} = 45\ °C \quad \text{Answer to the ones place}$$

STUDY CHECK 2.5

A child has a temperature of 103.6 °F. What is this temperature on a Celsius thermometer?

SAMPLE PROBLEM 2.6

Converting from Celsius to Kelvin Temperature

A dermatologist may use liquid cryogenic nitrogen at −196 °C to remove skin lesions and some skin cancers. What is the temperature of the liquid nitrogen in kelvins?

SOLUTION

Step 1 Given −196 °C **Need** T_K

Step 2 Plan

T_C → Temperature equation → T_K

Step 3 Equality/Conversion Factor

$$T_K = T_C + 273$$

Step 4 Set Up Problem Substitute the Celsius temperature into the equation and solve.

$$T_K = T_C + 273$$

$$T_K = -196 + 273$$

$$= 77\ K \quad \text{Answer to the ones place}$$

STUDY CHECK 2.6

On the planet Mercury, the average night temperature is 13 K, and the average day temperature is 683 K. What are these temperatures in degrees Celsius?

QUESTIONS AND PROBLEMS

Temperature

2.21 Your friend who is visiting from Canada just took her temperature. When she reads 99.8 °F, she becomes concerned that she is quite ill. How would you explain this temperature to your friend?

2.22 You have a friend who is using a recipe for flan from a Mexican cookbook. You notice that he set your oven temperature at 175 °F. What would you advise him to do?

2.23 Solve the following temperature conversions:
a. 37.0 °C = ________ °F **b.** 65.3 °F = ________ °C
c. −27 °C = ________ K **d.** 62 °C = ________ K
e. 114 °F = ________ °C

2.24 Solve the following temperature conversions:
a. 25 °C = ________ °F **b.** 155 °C = ________ °F
c. −25 °F = ________ °C **d.** 224 K = ________ °C
e. 145 °C = ________ K

2.25 a. A patient with hyperthermia has a temperature of 106 °F. What does this read on a Celsius thermometer?
b. Because high fevers can cause convulsions in children, the doctor wants to be called if the child's temperature goes over 40.0 °C. Should the doctor be called if a child has a temperature of 103 °F?

2.26 a. Hot compresses for a patient are prepared with water heated to 145 °F. What is the temperature of the hot water in degrees Celsius?
b. During extreme hypothermia, a boy's temperature dropped to 20.6 °C. What was his temperature on the Fahrenheit scale?

Chemistry Link to Health

VARIATION IN BODY TEMPERATURE

Normal body temperature is considered to be 37.0 °C, although it varies throughout the day and from person to person. Oral temperatures of 36.1 °C are common in the morning and climb to a high of 37.2 °C between 6 P.M. and 10 P.M. Temperatures above 37.2 °C for a person at rest are usually an indication of illness. Individuals who are involved in prolonged exercise may also experience elevated temperatures. Body temperatures of marathon runners can range from 39 °C to 41 °C as heat production during exercise exceeds the body's ability to lose heat.

Changes of more than 3.5 °C from the normal body temperature begin to interfere with bodily functions. At body temperatures above 41 °C, a condition called *hyperthermia* or heat stroke may occur in which sweat production stops, the pulse rate is elevated, and respiration becomes weak and rapid. A person with hyperthermia may become lethargic and lapse into a coma. In children, convulsions can occur, which may lead to permanent brain damage. Damage to internal organs is a major concern, and treatment, which must be immediate, may include immersing the person in an ice-water bath.

At the low temperature extreme of *hypothermia*, body temperature can drop as low as 28.5 °C. The person may appear cold and pale and have an irregular heartbeat. Unconsciousness can occur if the body temperature drops below 26.7 °C. Respiration becomes slow and shallow, and oxygenation of the tissues decreases. Treatment involves providing oxygen and increasing blood volume with glucose and saline fluids. Injecting warm fluids (37.0 °C) into the peritoneal cavity may restore the internal temperature.

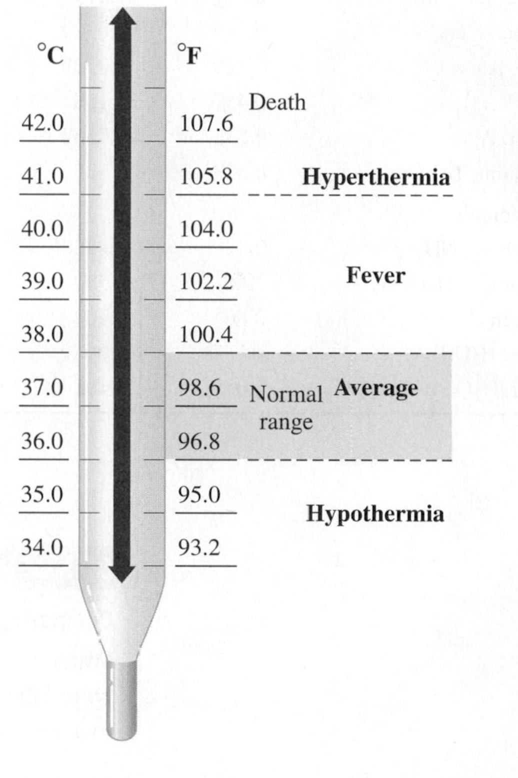

2.5 Specific Heat

LEARNING GOAL

Use specific heat to calculate heat loss or gain, temperature change, or mass of a sample.

TUTORIAL
Heat Capacity

Every substance can absorb heat. When you bake a potato, you place it in a hot oven. If you are cooking pasta, you add the pasta to boiling water. You already know that adding heat to water increases its temperature until it boils. Every substance has its own characteristic ability to absorb heat. Some substances must absorb more heat than others to reach a certain temperature. These energy requirements for different substances are described in terms of a physical property called *specific heat*. **Specific heat** (*SH*) is the amount of heat needed to raise the temperature of exactly 1 g of a substance by exactly 1 °C. This temperature change is written as ΔT (*delta T*), where the delta symbol means "a change in."

$$\text{Specific heat } (SH) = \frac{\text{heat}}{\text{grams} \times \Delta T} = \frac{\text{J (or cal)}}{1\text{ g} \times 1\text{ °C}}$$

Now we can write the specific heat for water using our definition of the calorie and joule.

$$\text{Specific heat } (SH) \text{ of } H_2O(l) = \frac{4.184\text{ J}}{\text{g} \times \text{°C}} = \frac{1.00\text{ cal}}{\text{g} \times \text{°C}}$$

If we look at Table 2.7, we see that 1 g of water requires 4.184 J to increase its temperature by 1 °C. Water has a large specific heat that is about five times the specific heat of aluminum. Aluminum has a specific heat that is about twice that of copper.

TABLE 2.7 Specific Heats of Some Substances

Substance	J/g °C	cal/g °C
Elements		
Aluminum, Al(*s*)	0.897	0.214
Copper, Cu(*s*)	0.385	0.0920
Gold, Au(*s*)	0.129	0.0308
Iron, Fe(*s*)	0.452	0.108
Silver, Ag(*s*)	0.235	0.0562
Titanium, Ti(*s*)	0.523	0.125
Compounds		
Ammonia, $NH_3(g)$	2.04	0.488
Ethanol, $C_2H_5OH(l)$	2.46	0.588
Sodium chloride, NaCl(*s*)	0.864	0.207
Water, $H_2O(l)$	4.184	1.00
Water, $H_2O(s)$	2.03	0.485

Therefore, 4.184 J (or 1.00 cal) will increase the temperature of 1 g of water by 1 °C. However, adding the same amount of heat (4.184 J or 1.00 cal) will raise the temperature of 1 g of aluminum by about 5 °C and 1 g of copper by 10 °C. The low specific heats of aluminum and copper mean they transfer heat efficiently, which makes them useful in cookware.

The high specific heat of water has a major impact on the temperatures in a coastal city compared to an inland city. A large mass of water near a coastal city can absorb or release five times the energy absorbed or released by the same mass of rock near an inland city. This means that in the summer a body of water absorbs large quantities of heat, which cools a coastal city, and then in the winter that same body of water releases large quantities of heat, which provide warmer temperatures. A similar effect happens with our bodies, which contain 70% by mass water. Water in the body absorbs or releases large quantities of heat in order to maintain an almost constant body temperature.

CONCEPT CHECK 2.5

Comparing Specific Heats

Water has a specific heat that is about six times larger than that of sandstone. How would the temperature change during the day and night if you live in a house next to a large lake compared to a house built in the desert on sandstone?

ANSWER

In the day, the water in the lake will absorb six times more energy than sandstone, which will keep the temperature in a house on a large lake more comfortable and cooler than a house in the desert. In the night, the water in the lake will release energy that warms the surrounding air so that the temperature will not drop as much as in the desert.

TUTORIAL
Specific Heat Calculations

Calculations Using Specific Heat

When we know the specific heat of a substance, we can calculate the heat lost or gained by measuring the mass of the substance and the initial and final temperature. We can substitute these measurements into the specific heat expression that is rearranged to solve for heat, which we call the *heat equation.*

$$\text{Heat} = \text{mass} \times \text{temperature change} \times \text{specific heat}$$

$$\text{Heat} = \text{mass} \times \Delta T \times SH$$

$$\text{cal} = \text{grams} \times {}^\circ\text{C} \times \frac{\text{cal}}{\text{g}\,{}^\circ\text{C}}$$

$$\text{J} = \text{grams} \times {}^\circ\text{C} \times \frac{\text{J}}{\text{g}\,{}^\circ\text{C}}$$

SAMPLE PROBLEM 2.7

Calculating Heat with Temperature Increase

How many joules are absorbed by 45.2 g of aluminum if its temperature rises from 12.5 °C to 76.8 °C (see Table 2.7)?

SOLUTION

Step 1 List given and needed data.

Given mass = 45.2 g

SH for aluminum = 0.897 J/g °C

Initial temperature = 12.5 °C

Final temperature = 76.8 °C

Need heat in joules (J)

Step 2 Calculate the temperature change (ΔT). The temperature change, ΔT, is the difference between the two temperatures.

$$\Delta T = T_{final} - T_{initial} = 76.8\ °C - 12.5\ °C = 64.3\ °C$$

Step 3 Write the heat equation and rearrange for unknown.

$$\text{Heat} = \text{mass} \times \Delta T \times SH$$

Step 4 Substitute the given values and solve, making sure units cancel.

$$\text{Heat} = 45.2\ \cancel{g} \times 64.3\ \cancel{°C} \times \frac{0.897\ J}{\cancel{g}\ \cancel{°C}} = 2610\ J\ (2.61 \times 10^3\ J)$$

STUDY CHECK 2.7

Some cooking pans have a layer of copper on the bottom. How many kilojoules are needed to raise the temperature of 125 g of copper from 22 °C to 325 °C (see Table 2.7)?

Guide to Calculations Using Specific Heat

1. List given and needed data.
2. Calculate the temperature change (ΔT).
3. Write the heat equation and rearrange for unknown.
4. Substitute the given values and solve, making sure units cancel.

The copper on a pan conducts heat rapidly to the food in the pan.

QUESTIONS AND PROBLEMS

Specific Heat

2.27 If the same amount of heat is supplied to samples of 10.0 g each of aluminum, iron, and copper all at 15 °C, which sample would reach the highest temperature (see Table 2.7)?

2.28 Substances **A** and **B** are the same mass and at the same initial temperature. When they are heated, the final temperature of **A** is 55 °C higher than the temperature of **B**. What does this tell you about the specific heats of **A** and **B**?

2.29 What is the amount of energy required in each of the following?
a. calories to heat 8.5 g of water from 15 °C to 36 °C
b. joules to heat 75 g of water from 22 °C to 66 °C
c. kilocalories to heat 150 g of water in a kettle from 15 °C to 77 °C
d. kilojoules to heat 175 g of copper from 28 °C to 188 °C

2.30 What is the amount of energy involved in each of the following?
a. calories given off when 85 g of water cools from 45 °C to 25 °C
b. joules given off when 25 g of water cools from 86 °C to 61 °C
c. kilocalories added when 5.0 kg of water warms from 22 °C to 28 °C
d. kilojoules to heat 224 g of gold from 18 °C to 185 °C

2.31 Calculate the energy, in joules and calories, for each of the following:
a. required to heat 25.0 g of water from 12.5 °C to 25.7 °C
b. required to heat 38.0 g of copper from 122 °C to 246 °C
c. lost when 15.0 g of ethanol, C_2H_5OH, cools from 60.5 °C to −42.0 °C
d. lost when 125 g of iron cools from 118 °C to 55 °C

2.32 Calculate the energy, in joules and calories, for each of the following:
a. required to heat 5.25 g of water from 5.5 °C to 64.8 °C
b. lost when 75.0 g of water cools from 86.4 °C to 2.1 °C
c. required to heat 10.0 g of silver from 112 °C to 275 °C
d. lost when 18.0 g of gold cools from 224 °C to 118 °C

LEARNING GOAL

Use the energy values to calculate the kilocalories (Cal) or kilojoules (kJ) in a food.

TUTORIAL
Nutritional Energy

CASE STUDY
Calories from Hidden Sugar

2.6 Energy and Nutrition

The food we eat provides energy to do work in the body, which includes the growth and repair of cells. Carbohydrates are the primary fuel for the body, but if the carbohydrate reserves are exhausted, fats, and then proteins are used for energy.

For many years in the field of nutrition, the energy from food was measured as Calories or kilocalories. The nutritional unit *Calorie*, *Cal* (with an uppercase C), is the same as 1000 cal, or 1 kcal. The international unit, kilojoule (kJ), is becoming more prevalent. For example, a baked potato has an energy value of 100 Calories, which is 100 kcal or 440 kJ. A typical diet that provides 2100 Cal (kcal) is the same as an 8800 kJ diet.

Energy Values in Nutrition

1 Cal = 1 kcal = 1000 cal

1 Cal = 4.184 kJ = 4184 J

The number of Calories in a food is determined by using an apparatus called a calorimeter, shown in Figure 2.7. A sample of food is placed in a steel container filled with oxygen with a measured amount of water which fills the surrounding chamber. The food sample is ignited, releasing heat that increases the temperature of the water. From the mass of the food and water as well as the temperature increase, the energy value of the food is calculated. We will assume that the energy absorbed by the calorimeter is negligible.

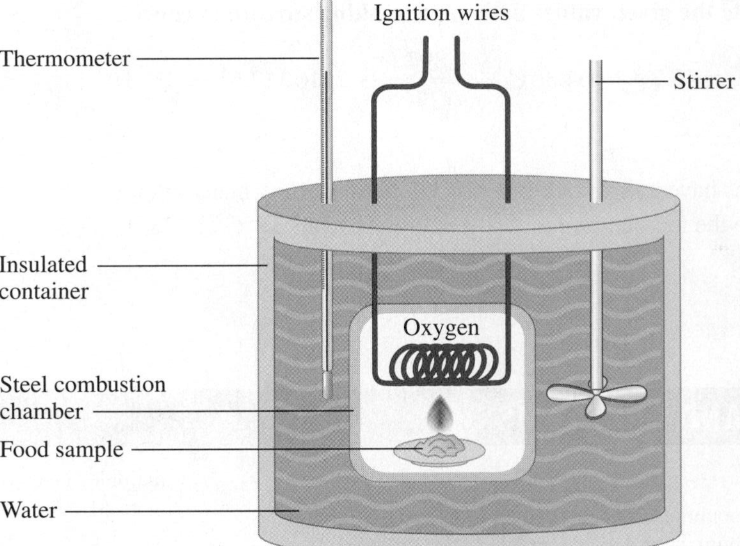

FIGURE 2.7 Heat released from burning a food sample in a calorimeter is used to determine the energy value of the food.
Q What happens to the temperature of water in a calorimeter during the combustion of a food sample?

CONCEPT CHECK 2.6

Energy Values of Food

When 55 g of pasta is burned in a calorimeter, 220 Cal of heat is released. What is the energy value of pasta in kcal/g?

ANSWER

Using the equality 1 Cal = 1 kcal, we can calculate the energy value of pasta.

$$\frac{220\ \cancel{\text{Cal}}}{55\ \text{g}} \times \frac{1\ \text{kcal}}{1\ \cancel{\text{Cal}}} = \frac{4.0\ \text{kcal}}{\text{g}}$$

Energy Values for Foods

The **energy (caloric) values** of food are the kilocalories or kilojoules obtained from burning 1 g of carbohydrate, fat, or protein. These values are listed in Table 2.8.

Using the energy values in Table 2.8, we can calculate the total energy of a food if the mass of each food type is known.

TABLE 2.8 Typical Energy (Caloric) Values for the Three Food Types

Food Type	kJ/g	kcal/g
Carbohydrate	17	4
Fat	38	9
Protein	17	4

$$\text{Kilojoules} = \cancel{g} \times \frac{\text{kJ}}{\cancel{g}} \qquad \text{Kilocalories} = \cancel{g} \times \frac{\text{kcal}}{\cancel{g}}$$

On packaged food, the energy content is listed in the Nutrition Facts label, usually in terms of the number of Calories for one serving. The general composition and energy content of some foods are given in Table 2.9.

TABLE 2.9 General Composition and Energy Content of Some Foods

Food	Carbohydrate (g)	Fat (g)	Protein (g)	Energy*
Banana, 1 medium	26	0	1	460 kJ (110 kcal)
Beef, ground, 3 oz	0	14	22	910 kJ (220 kcal)
Carrots, raw, 1 cup	11	0	1	200 kJ (50 kcal)
Chicken, no skin, 3 oz	0	3	20	460 kJ (110 kcal)
Egg, 1 large	0	6	6	330 kJ (80 kcal)
Milk, 4% fat, 1 cup	12	9	9	700 kJ (170 kcal)
Milk, nonfat, 1 cup	12	0	9	360 kJ (90 kcal)
Potato, baked	23	0	3	440 kJ (100 kcal)
Salmon, 3 oz	0	5	16	460 kJ (110 kcal)
Steak, 3 oz	0	27	19	1350 kJ (320 kcal)

*Energy values are rounded off to the tens place.

SAMPLE PROBLEM 2.8

Caloric Content for a Food

At a fast-food restaurant, a hamburger contains 37 g of carbohydrate, 19 g of fat, and 24 g of protein. What is the total energy content in kilocalories? Round off the kilocalories for each type of food to the tens place.

SOLUTION

Using the energy values for carbohydrate, fat, and protein (see Table 2.8), we can calculate the kilocalories for each type of food and the total kcal:

Food Type	Mass	Energy Value	Energy
Carbohydrate	$37\,\cancel{g} \times$	$\frac{4\text{ kcal}}{1\,\cancel{g}} =$	150 kcal*
Fat	$19\,\cancel{g} \times$	$\frac{9\text{ kcal}}{1\,\cancel{g}} =$	170 kcal*
Protein	$24\,\cancel{g} \times$	$\frac{4\text{ kcal}}{1\,\cancel{g}} =$	100 kcal*
	Total energy content	=	420 kcal

*Energy results are rounded off to the tens place.

STUDY CHECK 2.8

If you buy the same hamburger as in Sample Problem 2.8 at a fast-food restaurant in Canada, what is the energy content stated in kilojoules? Round off the kilojoules for each food type to the tens place.

Snack Crackers

Nutrition Facts
Serving Size 14 crackers (31g)
Servings Per Container About 7

Amount Per Serving
Calories 120 Calories from Fat 35
Kilojoules 500 kJ from Fat 150

% Daily Value*

Total Fat 4g 6%
Saturated Fat 0.5g 3%
Trans Fat 0g
Polyunsaturated Fat 0.5%
Monounsaturated Fat 1.5g
Cholesterol 0mg 0%
Sodium 310mg 13%
Total Carbohydrate 19g 6%
Dietary Fiber Less than 1g 4%
Sugars 2g
Proteins 2g

Vitamin A 0% • Vitamin C 0%
Calcium 4% • Iron 6%

*Percent Daily Values are based on a 2,000 calorie diet. Your daily values may be higher or lower depending on your calorie needs.

		Calories: 2,000	2,500
Total Fat	Less than	65g	80g
Sat Fat	Less than	20g	25g
Cholesterol	Less than	300mg	300mg
Sodium	Less than	2,400mg	2,400mg
Total Carbohydrate		300g	375g
Dietary Fiber		25g	30g

Calories per gram:
Fat 9 • Carbohydrate 4 • Protein 4

The nutrition facts include the total Calories, Calories from fat, and total grams of carbohydrate.

Explore Your World

COUNTING CALORIES

Obtain a food item that has a nutrition label. From the information on the label, determine the number of grams of carbohydrate, fat, and protein in one serving. Using the caloric values, calculate the total Calories for one serving. (For most products, the kilocalories for each food type are rounded off to the tens place.)

QUESTION

How does your total for the Calories in one serving compare to the Calories stated on the label for a single serving?

Chemistry Link to Health

LOSING AND GAINING WEIGHT

The number of kilocalories needed in the daily diet of an adult depends on gender and physical activity. Some general levels of energy needs are given in Table 2.10.

The amount of food a person eats is regulated by the hunger center in the hypothalamus, which is located in the brain. Food intake is normally proportional to the nutrient stores in the body. If these nutrient stores are low, you feel hungry; if they are high, you do not feel like eating.

A person gains weight when food intake exceeds energy output and loses weight when food intake is less than energy output. Many diet products contain cellulose, which has no nutritive value but provides bulk and makes you feel full. Some diet drugs depress the hunger center and must be used with caution, because they excite the nervous system and can elevate blood pressure. Because muscular exercise is an important way to expend energy, an increase in daily exercise aids weight loss. Table 2.11 lists some activities and the amount of energy they require.

One hour of swimming uses 2100 kJ of energy.

TABLE 2.10 Typical Energy Requirements for Adults

Gender	Age	Moderately active kcal (kJ)	Highly active kcal (kJ)
Female	19–30	2100 (8800)	2400 (10 000)
	31–50	2000 (8400)	2200 (9200)
Male	19–30	2700 (11 300)	3000 (12 600)
	31–50	2500 (10 500)	2900 (12 100)

TABLE 2.11 Energy Expended by a 70.0-kg (154-lb) Adult

Activity	Energy (kcal/h)	Energy (kJ/h)
Sleeping	60	250
Sitting	100	420
Walking	200	840
Swimming	500	2100
Running	750	3100

QUESTIONS AND PROBLEMS

Energy and Nutrition

2.33 Using the following data, calculate the kilocalories for each food burned in a calorimeter:

a. one stalk of celery that heats 505 g of water from 25.2 °C to 35.7 °C

b. a waffle that heats 4980 g of water from 20.6 °C to 62.4 °C

2.34 Using the following data, calculate the kilocalories for each food burned in a calorimeter:

a. 1 cup of popcorn that heats 1250 g of water from 25.5 °C to 50.8 °C

b. a sample of butter that heats 357 g of water from 22.7 °C to 38.8 °C

2.35 Using the energy values for foods (see Table 2.8), determine each of the following (round off the answers in kilocalories or kilojoules to the tens place):

a. the kilojoules for 1 cup of orange juice that has 26 g of carbohydrate, 2 g of protein, and no fat

b. the grams of carbohydrate in one apple if the apple has no fat and no protein and provides 72 Cal of energy

c. the kilocalories in 1 tablespoon of vegetable oil, which contains 14 g of fat and no carbohydrate or protein

d. the total kilocalories for a diet that consists of 68 g of carbohydrate, 150 g of protein, and 9.0 g of fat

2.36 Using the energy values for foods (see Table 2.8), determine each of the following (round off the answers in kilocalories or kilojoules to the tens place):

a. the kilojoules in 2 tablespoons of crunchy peanut butter that contains 6 g of carbohydrate, 16 g of fat, and 7 g of protein

b. the grams of protein in a cup of soup that has 110 kcal with 7 g of fat and 9 g of carbohydrate

c. the grams of sugar (carbohydrate) in a can of cola that has 140 Cal and no fat and no protein

d. the grams of fat in one avocado that has 405 kcal, 13 g of carbohydrate, and 5 g of protein

2.37 One cup of clam chowder contains 9 g of protein, 12 g of fat, and 16 g of carbohydrate. How much energy, in kilocalories and kilojoules, is in the clam chowder? (Round off the kilocalories or kilojoules to the tens place.)

2.38 A high-protein diet contains 70. g of carbohydrate, 150 g of protein, and 5.0 g of fat. How many kilocalories does this diet provide? How many kilojoules does this diet provide? (Round off the kilocalories and kilojoules to the tens place.)

2.7 Changes of State

LEARNING GOAL

Describe the changes of state between solids, liquids, and gases; calculate the energy involved.

In Section 2.2, we described the properties and states of matter: gases, liquids, and solids. We can now discuss how matter undergoes a **change of state** when it is converted from one state to another (see Figure 2.8).

When heat is added to a solid, the particles move faster. At a temperature called the **melting point (mp)**, the particles of a solid gain sufficient energy to overcome the attractive forces that hold them together. The particles in the solid separate and move about in random patterns. The substance is **melting**, changing from a solid to a liquid.

If the temperature is lowered, the reverse process takes place. Kinetic energy is lost, the particles slow down, and attractive forces pull the particles close together. The substance is **freezing**. A liquid changes to a solid at the **freezing point (fp)**, which is the same temperature as its melting point. Every substance has its own freezing (melting) point: water freezes (melts) at 0 °C; gold freezes (melts) at 1064 °C; nitrogen freezes (melts) at −210 °C.

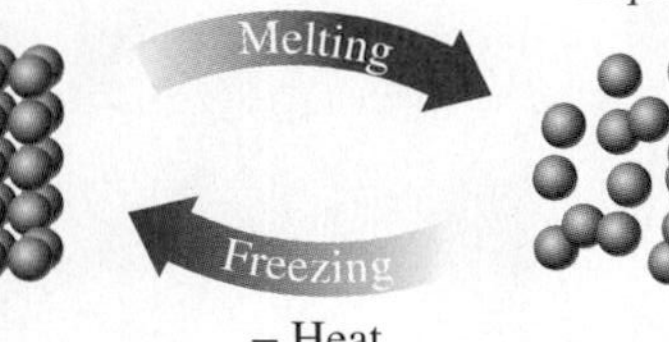

Melting and freezing are reversible processes.

During a change of state, the temperature of a substance remains constant. Suppose we have a glass containing ice and water. The ice melts when heat is added at 0 °C, forming more liquid. The liquid freezes when heat is removed at 0 °C. The processes of melting and freezing are reversible at 0 °C.

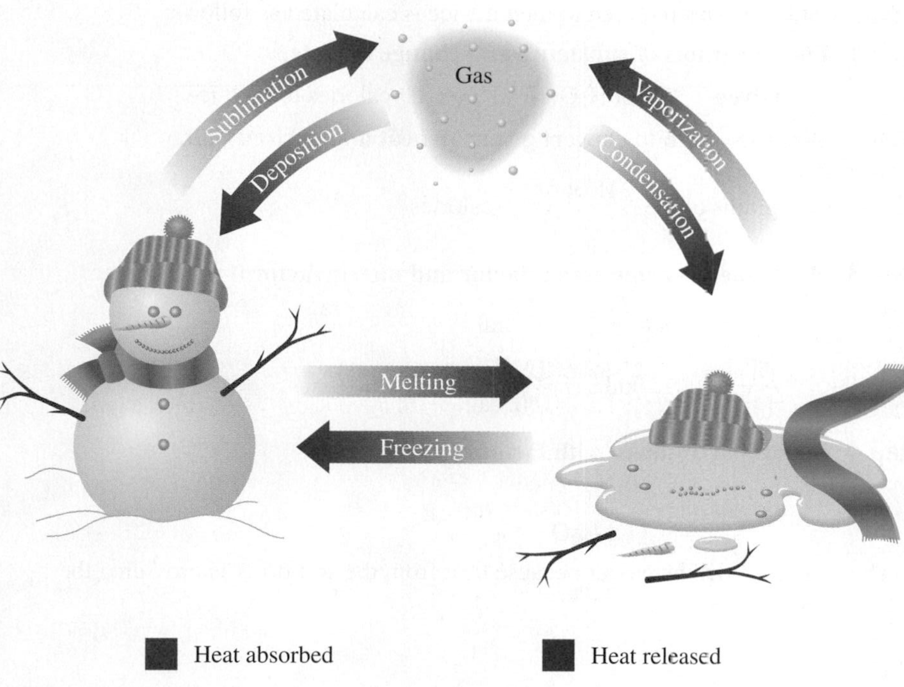

FIGURE 2.8 A summary of the changes of state.

Q Is heat added or released when liquid water freezes?

Heat of Fusion

During melting, energy called the **heat of fusion** is needed to separate the particles of a solid. For example, 80. cal (334 J) of heat are needed to melt exactly 1 g of ice at its melting point (0 °C).

Heat of Fusion for Water

$$\frac{80.\ \text{cal}}{1\ \text{g water}} \qquad \frac{334\ \text{J}}{1\ \text{g water}}$$

The heat of fusion (80. cal/g or 334 J/g) is also the heat that must be removed to freeze exactly 1 g of water at its freezing point (0 °C). Water is sometimes sprayed in fruit orchards during subfreezing weather. If the air temperature drops to 0 °C, the water begins to freeze. Heat is released as the water freezes, which warms the air and protects the fruit.

To determine the heat needed to melt a sample of ice, multiply the mass of the ice by its heat of fusion. There is no temperature change in the calculation because temperature remains constant as long as the ice is melting.

Calculating Heat to Melt (or Freeze) Water

Heat = mass × heat of fusion

$$\text{cal} = \cancel{g} \times \frac{80.\ \text{cal}}{\cancel{g}} \qquad \text{J} = \cancel{g} \times \frac{334\ \text{J}}{\cancel{g}}$$

Guide to Calculations Using Heat of Fusion

1 List the grams of substance and change of state.

2 Write the plan to convert grams to heat and desired unit.

3 Write the heat conversion factor and metric factor if needed.

4 Set up the problem with factors.

SAMPLE PROBLEM 2.9

Heat of Fusion

Ice cubes at 0 °C with a mass of 26 g are added to your soft drink.

a. How much heat (cal) must be added to melt all the ice at 0 °C?
b. What happens to the temperature of your soft drink? Why?

SOLUTION

a. The heat in calories required to melt the ice is calculated as follows:

Step 1 List the grams of substance and change of state.

Given 26 g of $H_2O(s)$ **Need** calories to melt ice

Step 2 Write the plan to convert grams to heat and desired unit.

grams of ice → Heat of fusion → calories

Step 3 Write the heat conversion factor and metric factor if needed.

$1\ \text{g of}\ H_2O(s \rightarrow l) = 80.\ \text{cal}$

$$\frac{80.\ \text{cal}}{1\ \text{g}\ H_2O} \quad \text{and} \quad \frac{1\ \text{g}\ H_2O}{80.\ \text{cal}}$$

Step 4 Set up the problem with factors.

$$26\ \cancel{\text{g}\ H_2O} \times \frac{80.\ \text{cal}}{1\ \cancel{\text{g}\ H_2O}} = 2100\ \text{cal}$$

b. The soft drink will be colder because heat from the soft drink is providing the energy to melt the ice.

STUDY CHECK 2.9

In a freezer, 150 g of water at 0 °C is placed in an ice cube tray. How much energy, in kilocalories, must be removed to form ice cubes at 0 °C?

Liquid + Heat Vaporization Gas
Condensation − Heat

Vaporization and condensation are reversible processes.

Boiling and Condensation

Water in a mud puddle disappears, unwrapped food dries out, and clothes hung on a line dry. **Evaporation** is taking place as water molecules with sufficient energy escape from the liquid surface and enter the gas phase (see Figure 2.9a). The loss of the "hot" water molecules removes heat, which cools the remaining liquid water. As heat is added, more and more water molecules evaporate. At the **boiling point (bp)**, the molecules of the liquid have the energy needed to change into a gas. The **boiling** of a liquid occurs as gas bubbles form throughout the liquid, then rise to the surface and escape (see Figure 2.9b).

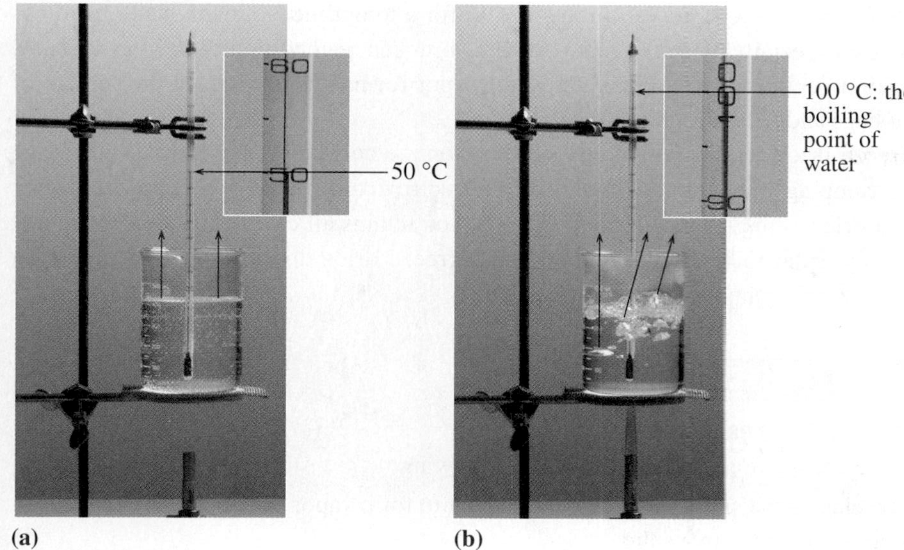

FIGURE 2.9 (a) Evaporation occurs at the surface of a liquid. (b) Boiling occurs as bubbles of gas form throughout the liquid.

Q Why does water evaporate faster at 80 °C than at 20 °C?

When heat is removed, a reverse process takes place. In **condensation**, water vapor is converted back to liquid as the water molecules lose kinetic energy and slow down. Condensation occurs at the same temperature as boiling but differs because heat is removed. You may have noticed that condensation occurs when you take a hot shower and the water vapor forms water droplets on a mirror. Because a substance loses heat as it condenses, its surroundings become warmer. That is why, when a rainstorm is approaching, you may notice a warming of the air as gaseous water molecules condense to rain.

Sublimation

In a process called **sublimation**, the particles on the surface of a solid change directly to a gas without going through the liquid state. In the process called **deposition**, gas changes directly to a solid. For example, dry ice, which is solid carbon dioxide, sublimes at −78 °C. It is called "dry" because it does not form a liquid as it warms. In extremely cold areas, snow does not melt but sublimes directly to water vapor.

Heat of Sublimation for Water

$$\frac{620.\ \text{cal}}{1\ \text{g water}} \qquad \frac{2590\ \text{J}}{1\ \text{g water}}$$

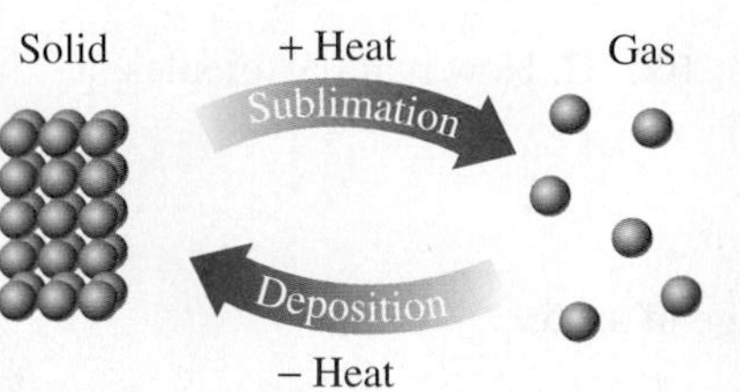

Sublimation and deposition are reversible processes.

Dry ice sublimes at −78 °C.

Career Focus

HISTOLOGIST

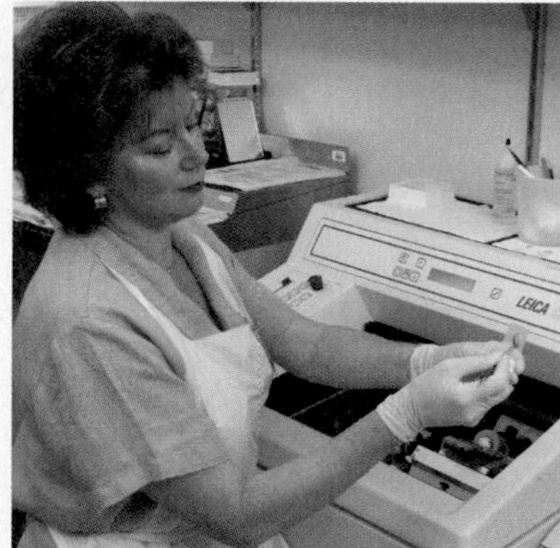

"While a patient is in surgery for skin cancer, some tissue around the cancer is sent to us," says Mary Ann Pipe, histology technician. "Using the Mohs surgical technique, we place the tissue block on a glass slide, chill it to −30 °C in a machine called a cryostat, and freeze it for longer in another machine called a heat extractor. From this frozen block of tissue, we cut extremely thin slices—one-thousandth of an inch—from different depths. We prepare three separate slides from skin at three different depths up to the surface of the skin. We stain the cells pink and blue by placing the slides in hemotoxin, and then in eosin. The slices are a tissue map that the doctor can easily read to determine if the margins around the skin cancer are clear or whether more tissue must be removed."

Water vapor will change to solid on contact with a cold surface.

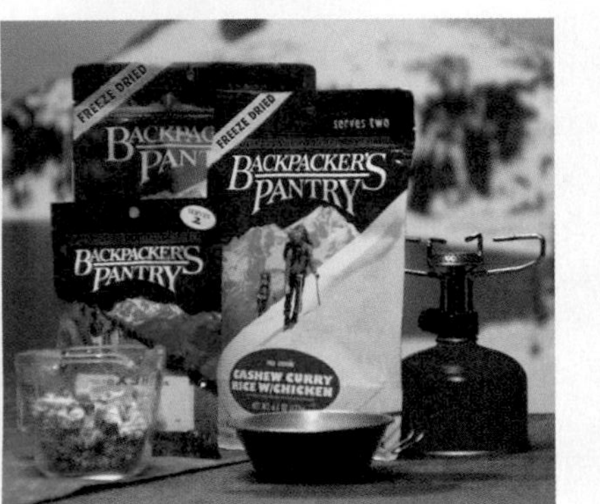

Freeze-dried foods have a long shelf life because they contain no water.

When frozen foods are left in the freezer for a long time, so much water sublimes that foods, especially meats, become dry and shrunken, a condition called *freezer burn*. Deposition occurs in a freezer when water vapor forms ice crystals on the surface of freezer bags and frozen food.

Freeze-dried foods prepared by sublimation are convenient for long-term storage and for camping and hiking. A food that has been frozen is placed in a vacuum chamber where it dries as the ice sublimes. The dried food retains all of its nutritional value and needs only water to be edible. A food that is freeze-dried does not need refrigeration because bacteria cannot grow without moisture.

CONCEPT CHECK 2.7

Identifying Changes of State

Give the change of state described in the following:

a. particles on the surface of a liquid escaping to form vapor
b. a liquid changing to a solid
c. gas bubbles forming throughout a liquid

ANSWER

a. evaporation **b.** freezing **c.** boiling

TUTORIAL
Heat of Vaporization and Heat of Fusion

Heat of Vaporization

The energy that must be added to convert exactly 1 g of liquid to gas at its boiling point is called the **heat of vaporization**. For water, 540 cal (2260 J) is needed to convert 1 g of water to vapor at 100 °C. This same amount of heat is released when 1 g of water vapor (gas) changes to liquid at 100 °C. Therefore, 2260 J or 540 cal/g is also the *heat of condensation* of water.

Heat of Vaporization for Water

$$\frac{540 \text{ cal}}{1 \text{ g water}} \qquad \frac{2260 \text{ J}}{1 \text{ g water}}$$

To calculate the amount of heat added to (or removed from) a sample of water, the mass of the sample is multiplied by the heat of vaporization. As before, no temperature change occurs during this change of state.

Calculating Heat to Vaporize (or Condense) Water

Heat = mass × heat of vaporization

$$\text{cal} = \cancel{\text{g}} \times \frac{540 \text{ cal}}{\cancel{\text{g}}} \qquad \text{J} = \cancel{\text{g}} \times \frac{2260 \text{ J}}{\cancel{\text{g}}}$$

Guide to Calculations Using Heat of Vaporization

1. List the grams of substance and change of state.
2. Write the plan to convert grams to heat and desired unit.
3. Write the heat conversion factor and metric factor if needed.
4. Set up the problem with factors.

SAMPLE PROBLEM 2.10

Using Heat of Vaporization

In a sauna, 122 g of water is converted to steam at 100 °C. How many kilojoules of heat are needed?

SOLUTION

Step 1 List the grams of substance and change of state.

Given 122 g of $H_2O(l)$ to $H_2O(g)$

Need kilojoules of heat to change state

Step 2 Write the plan to convert grams to heat and desired unit.

grams of water → Heat of vaporization → calories → Metric factor → kilocalories

Step 3 Write the heat conversion factor and metric factor if needed.

$$1 \text{ g of } H_2O\,(l \rightarrow g) = 2260 \text{ J} \qquad 1 \text{ kJ} = 1000 \text{ J}$$

$$\frac{2260 \text{ J}}{1 \text{ g } H_2O} \text{ and } \frac{1 \text{ g } H_2O}{2260 \text{ J}} \qquad \frac{1000 \text{ J}}{1 \text{ kJ}} \text{ and } \frac{1 \text{ kJ}}{1000 \text{ J}}$$

Step 4 Set up the problem with factors.

$$122 \cancel{\text{g } H_2O} \times \frac{2260 \cancel{\text{J}}}{1 \cancel{\text{g } H_2O}} \times \frac{1 \text{ kJ}}{1000 \cancel{\text{J}}} = 276 \text{ kJ}$$

STUDY CHECK 2.10

When steam from a pan of boiling water reaches a cool window, it condenses. How much heat, in kilojoules, is released when 25.0 g of steam condenses at 100 °C?

Chemistry Link to Health

STEAM BURNS

Hot water at 100 °C will cause burns and damage to the skin. However, getting steam on the skin is even more dangerous. If 25 g of hot water at 100 °C falls on a person's skin, the temperature of the water will drop to body temperature, 37 °C. The heat released during cooling can cause severe burns. The amount of heat can be calculated from the temperature change, 100 °C − 37 °C = 63 °C.

$$25 \cancel{\text{g}} \times 63 \cancel{°\text{C}} \times \frac{4.184 \text{ J}}{\cancel{\text{g}}\,\cancel{°\text{C}}} = 6600 \text{ J}$$

For comparison, we can calculate the amount of heat released when 25 g of steam at 100 °C hits the skin. First, the steam condenses to water (liquid) at 100 °C:

$$25 \cancel{\text{g}} \times \frac{2260 \text{ J}}{1 \cancel{\text{g}}} = 57\,000 \text{ J}$$

Then, the liquid water at 100 °C cools to body temperature 37 °C. The total amount of heat released from the condensation and cooling of the steam is calculated as follows:

Condensation (100 °C)	= 57 000 J
Cooling (100 °C to 37 °C)	= 6 600 J
Heat released	= 64 000 J (rounded off)

The amount of heat released from steam is almost ten times greater than the heat from the same amount of hot water.

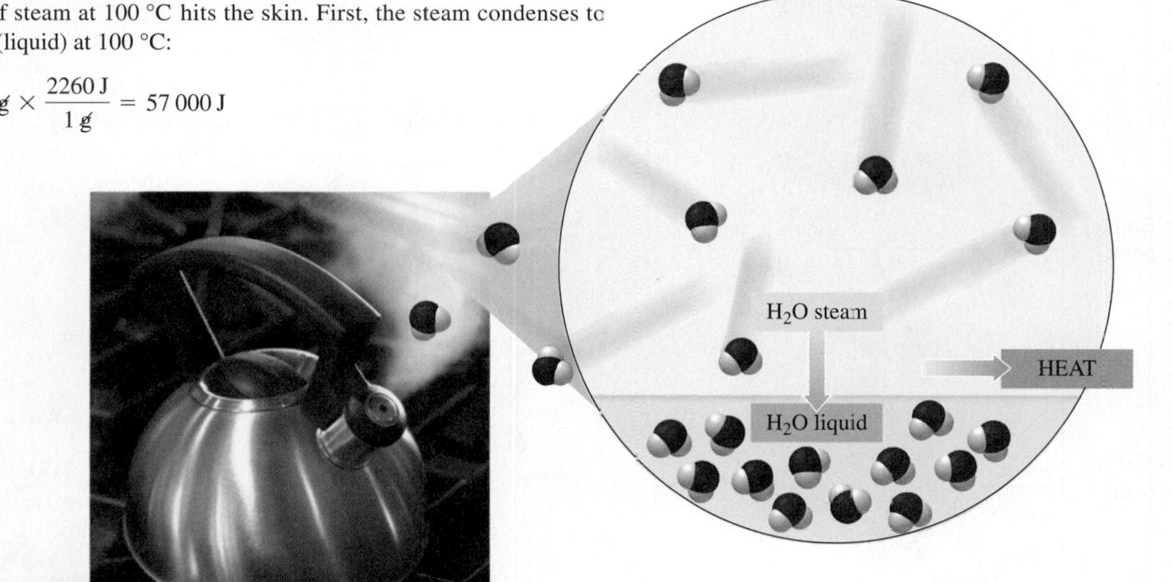

When a gram of steam condenses, 2260 J are released.

Heating and Cooling Curves

All the changes of state during the heating of a solid can be illustrated visually. On a **heating curve**, the temperature is shown on the vertical axis and the addition of heat is shown on the horizontal axis (see Figure 2.10a).

Steps on a Heating Curve

The first diagonal line indicates a warming of a solid as heat is added. When the melting temperature is reached, a horizontal line, or plateau, indicates that the solid is melting. As melting takes place, the solid is changing to liquid without any change in temperature (see Figure 2.10a).

Once all the particles are in the liquid state, adding more heat will increase the temperature of the liquid. This increase is shown as a diagonal line from the melting point temperature to the boiling point temperature. Once the liquid reaches its boiling point, a horizontal line indicates that the temperature is constant as liquid changes to gas. Because the heat of vaporization is greater than the heat of fusion, the horizontal line at the boiling point is longer than the line at the melting point. Once all the liquid becomes gas, adding more heat increases the temperature of the gas.

Steps on a Cooling Curve

A **cooling curve** is a diagram of the cooling process in which the temperature decreases as heat is removed (see Figure 2.10b). The cooling of the gas is shown as a diagonal line to the boiling (condensation) point. At the boiling (condensation) point, the horizontal line indicates a change of state as gas condenses to form a liquid. When all the gas has changed into liquid, the temperature decreases with further cooling of the liquid, which is shown as a diagonal line from the condensation point temperature to the freezing point temperature. At the freezing point, another horizontal line indicates that liquid is changing to solid at the freezing point temperature. Once all the substance is frozen, the removal of more heat lowers the temperature below its freezing point, which is shown as a diagonal line below the freezing point.

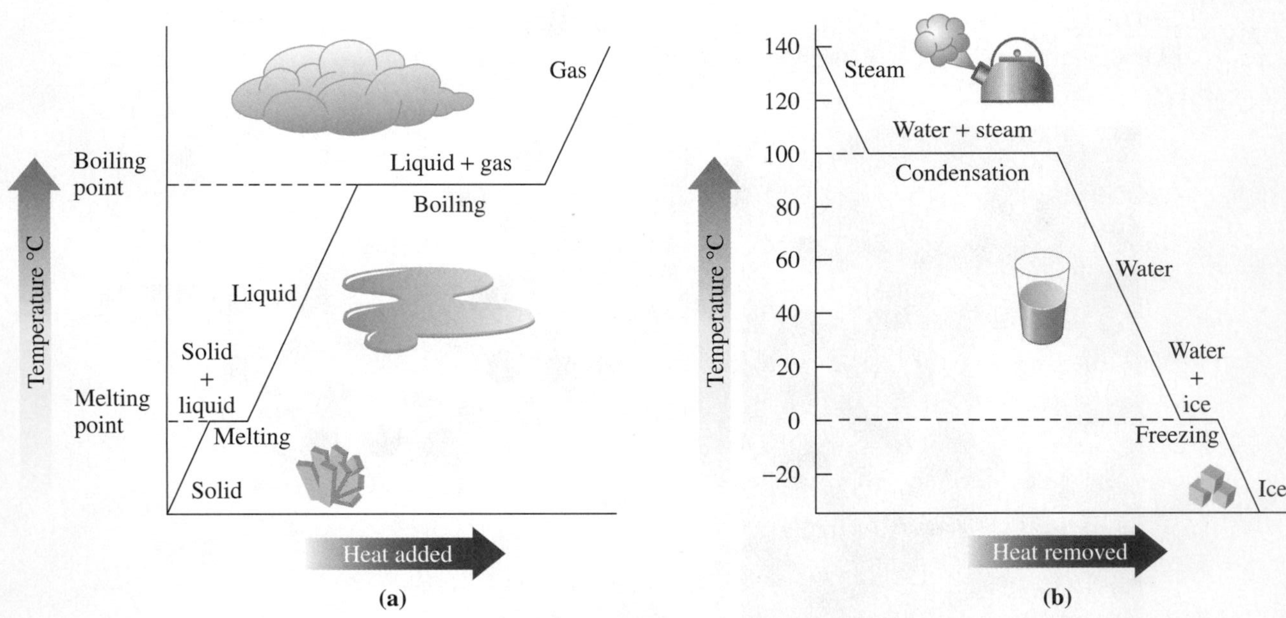

FIGURE 2.10 (a) A heating curve diagrams changes in state as temperature increases. (b) A cooling curve for water diagrams changes in state as temperature decreases.
Q What does the plateau at 100 °C represent on the cooling curve for water?

CONCEPT CHECK 2.8

Using a Cooling Curve

Using the cooling curve for water in Figure 2.10b, identify the state or change of state for water as solid, liquid, gas, condensation, or freezing.

a. at 120 °C **b.** at 100 °C
c. at 40 °C **d.** at 0 °C

ANSWER

a. A temperature of 120 °C occurs on the diagonal line above the boiling (condensation) point indicating that water is a gas.
b. A temperature of 100 °C, shown as a horizontal line, indicates that the water vapor is changing to liquid water, or condensing.
c. A temperature of 40 °C occurs on the diagonal line below the boiling point but above the freezing point, which indicates that the water is in the liquid state.
d. A temperature of 0 °C, shown as a horizontal line, indicates that the liquid water is changing to solid (ice) or freezing.

Combining Energy Calculations

Up to now, we have calculated one step in a heating or cooling curve. However, many problems require a combination of steps that include a temperature change as well as a change of state. The heat is calculated for each step separately and then added together to find the total energy, as seen in Sample Problem 2.11.

TUTORIAL
Heat, Energy, and Changes of State

SAMPLE PROBLEM 2.11

Combining Heat Calculations

Calculate the total heat, in joules, needed to convert 15.0 g of liquid ethanol at 25.0 °C to gas at its boiling point of 78.0 °C. Ethanol has a specific heat of 2.46 J/g °C, and a heat of vaporization of 841 J/g.

SOLUTION

Step 1 List the grams of substance and change of state.

Given 15.0 g of ethanol at 25.0 °C; boiling point of ethanol 78.0 °C
Specific heat 2.46 J/g °C; heat of vaporization 841 J/g

Need heat (J) to warm ethanol and change to gas at the boiling point

Step 2 Write the plan to convert grams to heat and desired unit. When several changes occur, draw a diagram of heating and changes of state.

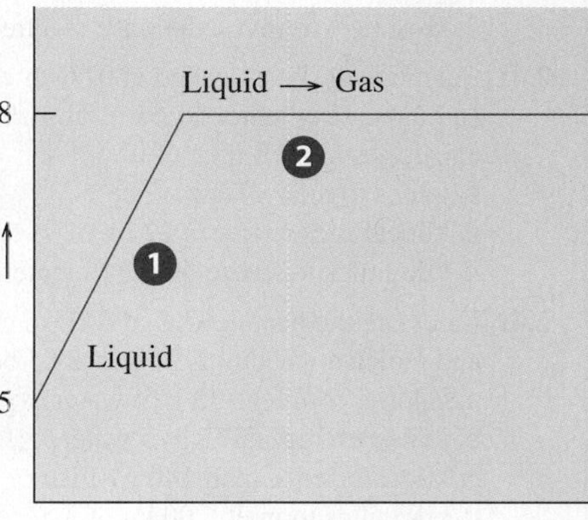

Total heat = joules needed to warm ethanol from 25.0 °C to 78.0 °C (boiling point)
+ joules to change liquid to gas at 78.0 °C (boiling point)

Step 3 Write the heat conversion factors needed.

$$SH_{\text{Ethanol}} = \frac{2.46\ \text{J}}{\text{g}\ ^\circ\text{C}}$$

$$\frac{2.46\ \text{J}}{\text{g}\ ^\circ\text{C}} \quad \text{and} \quad \frac{\text{g}\ ^\circ\text{C}}{2.46\ \text{J}}$$

$$1\ \text{g of ethanol}\ (l \rightarrow g) = 841\ \text{J}$$

$$\frac{841\ \text{J}}{1\ \text{g ethanol}} \quad \text{and} \quad \frac{1\ \text{g ethanol}}{841\ \text{J}}$$

Step 4 Set up problem with factors.

$$\Delta T = 78.0\ °C - 25.0\ °C = 53.0\ °C$$

Heat needed to warm ethanol (liquid) at 25 °C to ethanol (liquid) at 78.0 °C (boiling point):

$$15.0\ \cancel{g} \times 53.0\ \cancel{°C} \times \frac{2.46\ J}{\cancel{g}\ \cancel{°C}} = 1960\ J$$

Heat needed to change ethanol (liquid) to ethanol (gas) at 78.0 °C (boiling point):

$$15.0\ \cancel{g\ ethanol} \times \frac{841\ J}{1\ \cancel{g\ ethanol}} = 12\ 600\ J$$

Calculate the total heat:

Heating ethanol (25.0 °C to boiling point 78.0 °C)	1 960 J
Changing liquid to gas at boiling point (78.0 °C)	12 600 J
Total heat needed	14 600 J (rounded off)

STUDY CHECK 2.11

How many kilojoules are released when 75.0 g of steam at 100 °C condenses, cools to 0 °C, and freezes at 0 °C? (*Hint:* The solution will require three energy calculations.)

QUESTIONS AND PROBLEMS

Changes of State

2.39 Identify each of the following changes of state as melting, freezing, sublimation, or deposition:
- **a.** The solid structure of a substance breaks down as liquid forms.
- **b.** Coffee is freeze-dried.
- **c.** Water on the street turns to ice during a cold wintry night.
- **d.** Ice crystals form on a package of frozen corn.

2.40 Identify each of the following changes of state as melting, freezing, sublimation, or deposition:
- **a.** Dry ice in an ice-cream cart disappears.
- **b.** Snow on the ground turns to liquid water.
- **c.** Heat is removed from 125 g of liquid water at 0 °C.
- **d.** Frost (ice) forms on the walls of a freezer unit of a refrigerator.

2.41 Calculate the heat needed at 0 °C in each of the following and indicate whether heat was absorbed or released:
- **a.** calories to melt 65 g of ice
- **b.** joules to melt 17.0 g of ice
- **c.** kilocalories to freeze 225 g of water
- **d.** kilojoules to freeze 50.0 g of water

2.42 Calculate the heat needed at 0 °C in each of the following and indicate whether heat was absorbed or released:
- **a.** calories to freeze 35 g of water
- **b.** joules to freeze 275 g of water
- **c.** kilocalories to melt 140 g of ice
- **d.** kilojoules to melt 5.00 kg of ice

2.43 Identify each of the following changes of state as evaporation, boiling, or condensation:
- **a.** The water vapor in the clouds changes to rain.
- **b.** Wet clothes dry on a clothesline.
- **c.** Lava flows into the ocean and steam forms.
- **d.** After a hot shower, your bathroom mirror is covered with water.

2.44 Identify each of the following changes of state as evaporation, boiling, or condensation:
- **a.** At 100 °C, the water in a pan changes to steam.
- **b.** On a cool morning, the windows in your car fog up.
- **c.** A shallow pond dries up in the summer.
- **d.** Your teakettle whistles when the water is ready for tea.

2.45 Calculate the heat change at 100 °C in each of the following problems. Indicate whether heat was absorbed or released.
- **a.** calories to vaporize 10.0 g of water
- **b.** joules to vaporize 5.00 g of water
- **c.** kilocalories to condense 8.0 kg of steam
- **d.** kilojoules to condense 175 g of steam

2.46 Calculate the heat change at 100 °C in each of the following problems. Indicate whether heat was absorbed or released.
- **a.** calories to condense 10.0 g of steam
- **b.** joules to condense 7.60 g of steam
- **c.** kilocalories to vaporize 44 g of water
- **d.** kilojoules to vaporize 5.00 kg of water

2.47 Draw a heating curve for a sample of ice that is heated from −20 °C to 150 °C. Indicate the segment of the graph that corresponds to each of the following:
a. solid **b.** melting **c.** liquid
d. boiling **e.** gas

2.48 Draw a cooling curve for a sample of steam that cools from 110 °C to −10 °C. Indicate the segment of the graph that corresponds to each of the following:
- **a.** solid
- **b.** freezing
- **c.** liquid
- **d.** condensation (boiling)
- **e.** gas

2.49 Using the values for the heat of fusion, specific heat of water, and/or heat of vaporization, calculate the amount of heat energy in each of the following:

a. joules needed to melt 50.0 g of ice at 0 °C and to warm the liquid to 65.0 °C

b. kilocalories released when 15.0 g of steam condenses at 100 °C and the liquid cools to 0 °C

c. kilojoules needed to melt 24.0 g of ice at 0 °C, warm the liquid to 100 °C and change it to steam at 100 °C

2.50 Using the values for the heat of fusion, specific heat of water, and/or heat of vaporization, calculate the amount of heat energy in each of the following:

a. joules to condense 125 g of steam at 100 °C and to cool the liquid to 15.0 °C

b. kilocalories needed to melt a 525-g ice sculpture at 0 °C and to warm the liquid to 15.0 °C

c. kilojoules released when 85.0 g of steam condenses at 100 °C and to cool the liquid and freeze it at 0 °C

CONCEPT MAP

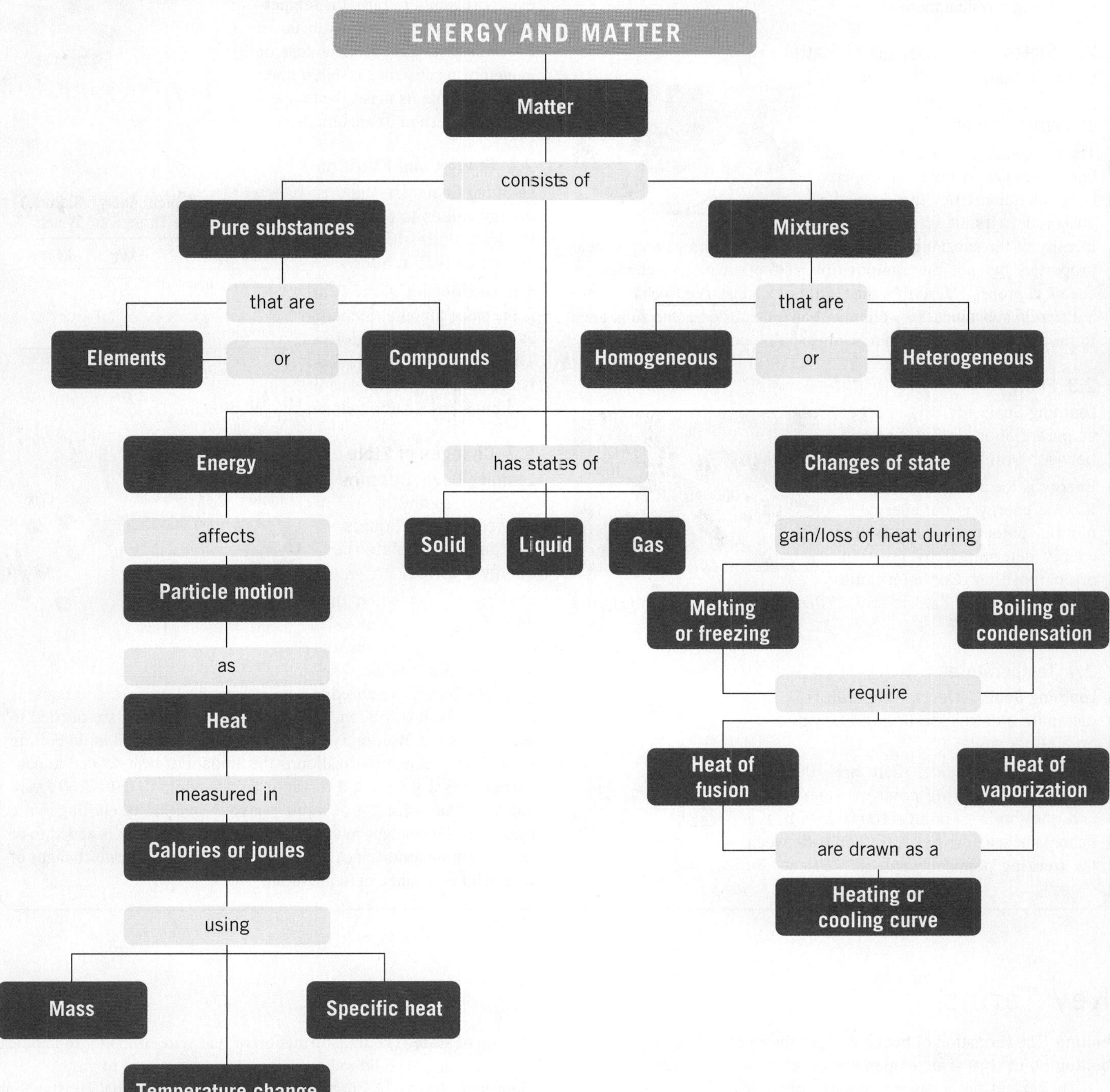

CHAPTER REVIEW

2.1 Classification of Matter

Learning Goal: Classify examples of matter as pure substances or mixtures.

Matter is classified as pure substances or mixtures. Pure substances, which are elements or compounds, have fixed compositions. Mixtures have variable compositions, which are classified further as homogeneous or heterogeneous. The substances in mixtures can be separated using physical methods.

2.2 States and Properties of Matter

Learning Goal: Identify the states and physical and chemical properties of matter.

The three states of matter are solid, liquid, and gas. A physical property is a characteristic that can be observed without changing the identity of the substance. A physical change occurs when physical properties but not the composition of the substance, change. A chemical property describes the ability of a substance to change into a different substance. In a chemical change, at least one substance forms a new substance with new physical properties.

2.3 Energy

Learning Goal: Identify energy as potential or kinetic; convert between units of energy.

Energy is the ability to do work. Kinetic energy is the energy of motion; potential energy is the energy determined by position or composition. Common units of energy are the calorie (cal), kilocalorie (kcal), joule (J), and kilojoule (kJ).

2.4 Temperature

Learning Goal: Given a temperature, calculate a corresponding temperature on another scale.

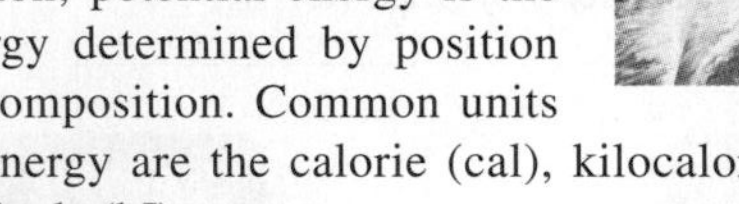

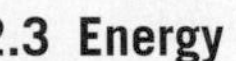

On the Celsius scale, there are 100 units between the freezing point of water (0 °C) and the boiling point (100 °C). On the Fahrenheit scale, there are 180 units between the freezing point of water (32 °F) and the boiling point (212 °F). A Fahrenheit temperature is related to its Celsius temperature by the equation $T_F = 1.8\,T_C + 32$. The SI temperature of Kelvin is related to the Celsius temperature by the equation $T_K = T_C + 273$.

2.5 Specific Heat

Learning Goal: Use specific heat to calculate heat loss or gain, temperature change, or mass of a sample.

Specific heat is the amount of energy required to raise the temperature of exactly 1 g of a substance by exactly 1 °C. The heat lost or gained by a substance is determined by multiplying its mass, the temperature change, and its specific heat.

2.6 Energy and Nutrition

Learning Goal: Use the energy values to calculate the kilocalories (Cal) or kilojoules (kJ) in a food.

The nutritional Calorie (Cal) is the same amount of energy as 1 kcal or 1000 calories. The energy content of a food is the sum of kilocalories or kilojoules from carbohydrate, fat, and protein.

TABLE 2.8 Typical Energy (Caloric) Values for the Three Food Types

Food Type	kJ/g	kcal/g
Carbohydrate	17	4
Fat	38	9
Protein	17	4

2.7 Changes of State

Learning Goal: Describe the changes of state between solids, liquids, and gases; calculate the energy involved.

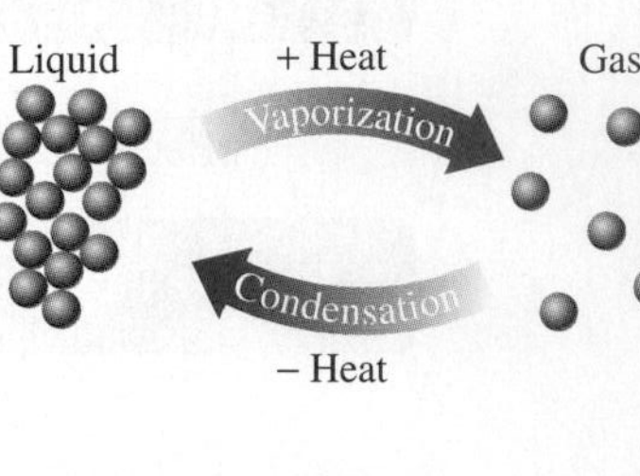

Melting occurs when the particles in a solid absorb enough energy to break apart and form a liquid. The amount of energy required to convert exactly 1 g of solid to liquid is called the heat of fusion. For water, 80. cal (334 J) are needed to melt 1 g of ice. Boiling is the vaporization of a liquid at its boiling point. The heat of vaporization is the amount of heat needed to convert exactly 1 g of liquid to vapor. For water, 540 cal (2260 J) are needed to vaporize 1 g of liquid water. A heating or cooling curve illustrates the changes in temperature and state as heat is added to or removed from a substance. Plateaus on the graph indicate changes of state with no change in temperature.

Key Terms

boiling The formation of bubbles of gas throughout a liquid.

boiling point (bp) The temperature at which a liquid changes to gas (boils) and gas changes to liquid (condenses).

calorie (cal) The amount of heat energy that raises the temperature of exactly 1 g of water exactly 1 °C.

change of state The transformation of one state of matter to another; for example, solid to liquid, liquid to solid, liquid to gas.

chemical change A change during which the original substance is converted into a new substance that has a different composition and new chemical and physical properties.

chemical properties The properties that indicate the ability of a substance to change into a new substance.

compound A pure substance consisting of two or more elements, with a definite composition, that can be broken down into simpler substances only by chemical methods.

condensation The change of state of a gas to a liquid.

cooling curve A diagram that illustrates temperature changes and changes of state for a substance as heat is removed.

deposition The change of a gas directly into a solid; the reverse of sublimation.

element A pure substance containing only one type of matter, which cannot be broken down by chemical methods.

energy The ability to do work.

energy (caloric) value The kilocalories (or kilojoules) obtained per gram of the food types: carbohydrate, fat, and protein.

evaporation The formation of a gas (vapor) by the escape of high-energy molecules from the surface of a liquid.

freezing The change of state from liquid to solid.

freezing point (fp) The temperature at which a liquid changes to a solid (freezes), a solid changes to a liquid (melts).

gas A state of matter that does not have a definite shape or volume.

heat The energy associated with the motion of particles in a substance.

heat of fusion The energy required to melt exactly 1 g of a substance at its melting point. For water, 80. cal (334 J) are needed to melt 1 g of ice; 80. cal (334 J) are released when 1 g of water freezes.

heat of vaporization The energy required to vaporize exactly 1 g of a substance at its boiling point. For water, 540 calories (2260 J) are needed to vaporize 1 g of liquid; 1 g of steam gives off 540 cal (2260 J) when it condenses.

heating curve A diagram that shows the temperature changes and changes of state of a substance as it is heated.

joule (J) The SI unit of heat energy; 4.184 J = 1 cal.

kinetic energy The energy of moving particles.

liquid A state of matter that takes the shape of its container but has a definite volume.

matter The material that makes up a substance and has mass and occupies space.

melting The change of state from a solid to a liquid.

melting point (mp) The temperature at which a solid becomes a liquid (melts). It is the same temperature as the freezing point.

mixture The physical combination of two or more substances that does not change the identities of the mixed substances.

physical change A change in which the physical properties of a substance change but its identity stays the same.

physical properties The properties that can be observed or measured without affecting the identity of a substance.

potential energy A type of energy related to position or composition of a substance.

pure substance A type of matter composed of elements and compounds that has a definite composition.

solid A state of matter that has its own shape and volume.

specific heat A quantity of heat that changes the temperature of exactly 1 g of a substance by exactly 1 °C.

states of matter Three forms of matter: solid, liquid, and gas.

sublimation The change of state in which a solid is transformed directly to a gas without forming a liquid first.

Understanding the Concepts

2.51 Identify each of the following as an element, a compound, or a mixture:

a.

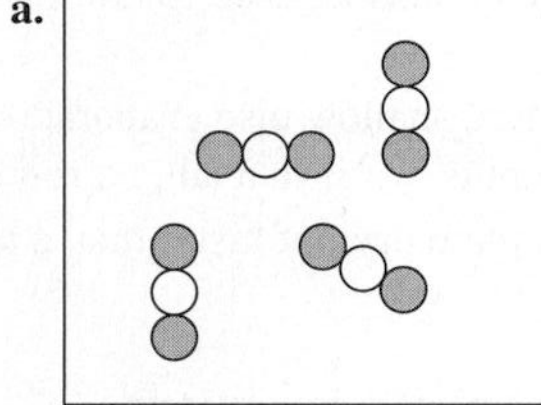

b.

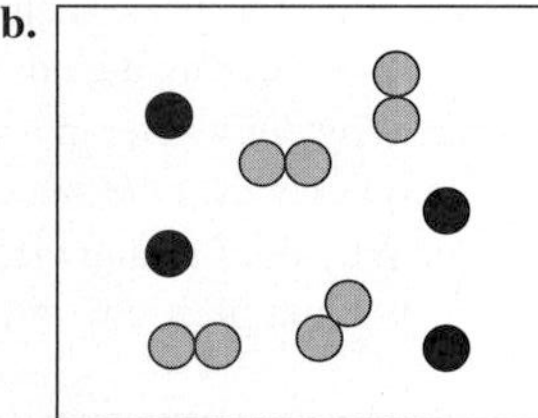

c.

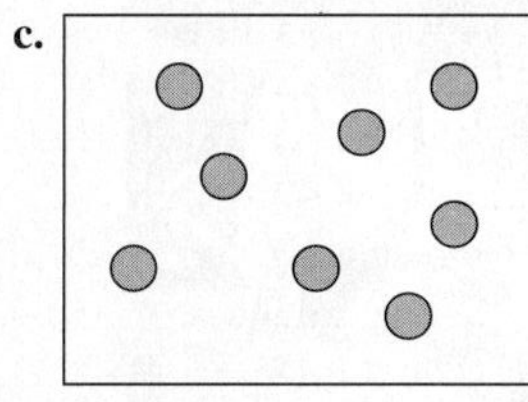

2.52 Which diagram illustrates a homogeneous mixture? Explain your choice. Which of the following diagrams illustrates heterogeneous mixtures? Explain your choice.

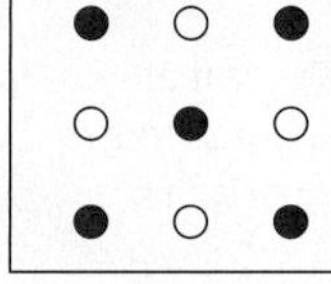

a.

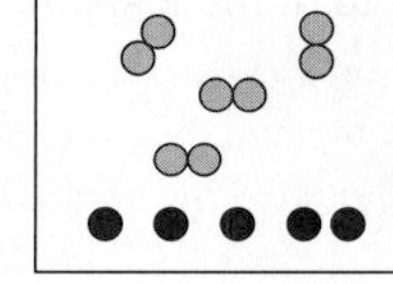

b.

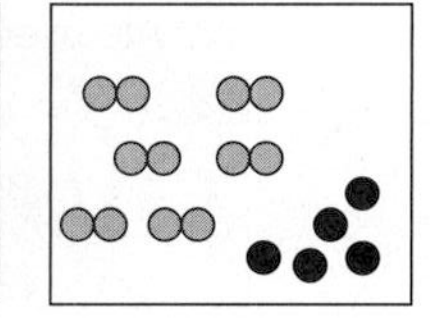

c.

2.53 Classify each of the following as a homogeneous or heterogeneous mixture:

a. lemon-flavored water **b.** stuffed mushrooms
c. tortilla soup

2.54 Classify each of the following as a homogeneous or heterogeneous mixture:

a. ketchup **b.** hard-boiled egg
c. eye drops

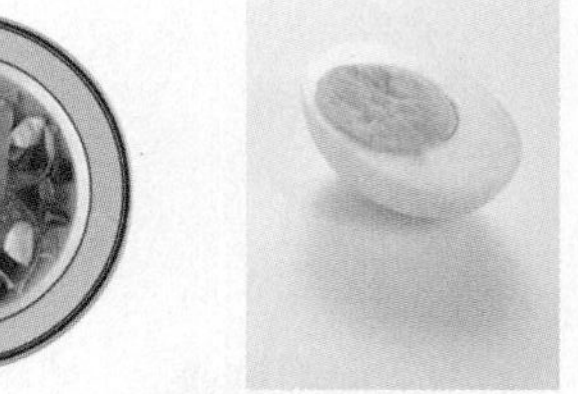

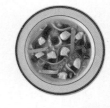

2.55 State the temperature, including the estimated digit, on each of the Celsius thermometers.

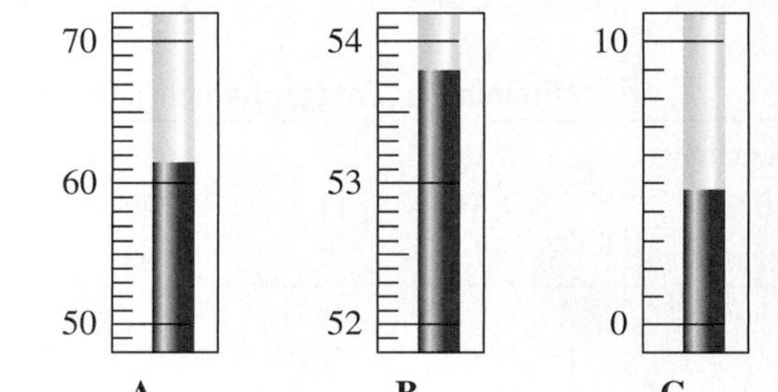

2.56 Select the warmer temperature in each pair.
a. 10 °C or 10 °F
b. 30 °C or 15 °F
c. −10 °C or 32 °F
d. 200 °C or 200 K

2.57 Compost can be made at home from grass clippings, some kitchen scraps, and dry leaves. As microbes break down organic matter, heat is generated and the compost can reach a temperature of 155 °F, which kills most pathogens. What is this initial temperature in Celsius degrees? In kelvins?

Compost produced from decayed plant material is used to enrich the soil.

2.58 After a week, biochemical reactions in compost slow, and the temperature drops to 45 °C. The dark brown organic-rich mixture is ready for use in the garden. What is this temperature in Fahrenheit degrees? In kelvins?

2.59 Calculate the energy to heat three cubes (gold, aluminum, and silver) each with a volume of $10.0\ cm^3$ from 15 °C to 25 °C. Refer to Tables 1.13 and 2.7. What do you notice about the energy needed for each?

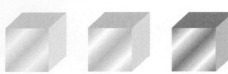

2.60 If you used the 8400 kilojoules you expend in energy in one day to heat 50 000 g of water at 20 °C, what would be the rise in temperature? What would be the new temperature of the water?

2.61 A 70.0-kg person has just eaten a quarter-pound cheeseburger, french fries, and a chocolate shake. Using Table 2.8, calculate the total kilocalories in this meal (round off the kilocalories to the tens place).

Item	Protein (g)	Fat (g)	Carbohydrate (g)
Cheeseburger	31	29	34
French fries	3	11	26
Chocolate shake	11	9	60

2.62 For the person in Problem 2.61, use Table 2.11 to determine each of the following:
a. The number of hours of sleeping needed to "burn off" the kilocalories in this meal.
b. The number of hours of running needed to "burn off" the kilocalories in this meal.

2.63 Use your knowledge of changes of state to explain the following:
a. How does perspiration during heavy exercise cool the body?
b. Why do towels dry more quickly on a hot summer day than on a cold winter day?
c. Why do wet clothes stay wet in a plastic bag?

Perspiration forms on the skin during heavy exercise.

2.64 Use your knowledge of changes of state to explain the following:
a. Why is a spray, such as ethyl chloride, used to numb a sports injury during a game?
b. Why does water in a wide, flat, shallow dish evaporate more quickly than the same amount of water in a tall, narrow glass?
c. Why does a sandwich on a plate dry out faster than a sandwich in plastic wrap?

A spray is used to numb a sports injury.

2.65 The following graph is a heating curve for chloroform, a solvent for fats, oils, and waxes:

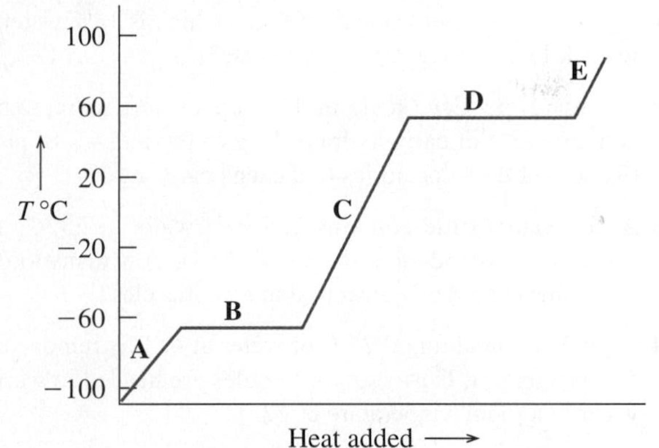

a. What is the approximate melting point of chloroform?
b. What is the approximate boiling point of chloroform?
c. On the heating curve, identify the segments **A**, **B**, **C**, **D**, and **E** as solid, liquid, gas, melting, or boiling.
d. At the following temperatures, is chloroform a solid, liquid, or gas?

−80 °C; −40 °C; 25 °C; 80 °C

2.66 Associate the contents of the beakers (**1–5**) with segments (**A–E**) on the following heating curve for water:

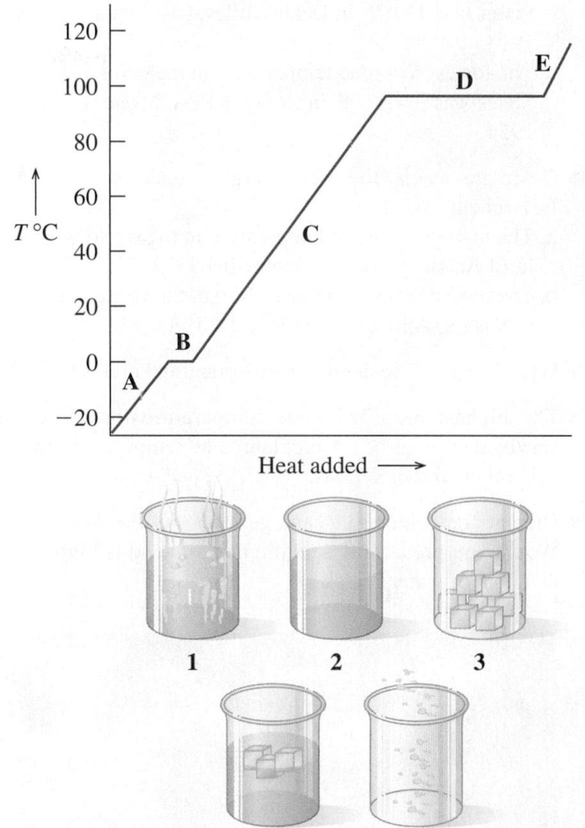

Additional Questions and Problems

For instructor-assigned homework, go to **www.masteringchemistry.com**.

2.67 Classify each of the following as an element, a compound, or a mixture:
a. carbon in pencils
b. carbon dioxide (CO_2) we exhale
c. orange juice
d. neon gas in lights
e. salad dressing of oil and vinegar

2.68 Classify each of the following as a homogeneous or heterogeneous mixture:
a. hot fudge sundae
b. herbal tea
c. vegetable oil
d. water and sand
e. mustard

2.69 Identify each of the following as a solid, a liquid, or a gas:
a. vitamin tablets in a bottle **b.** helium in a balloon
c. milk in a glass **d.** the air you breathe
e. charcoal briquettes on a barbecue

2.70 Identify each of the following as a solid, a liquid, or a gas:
a. popcorn in a bag **b.** water in a garden hose
c. a computer mouse **d.** air in a tire
e. hot tea

2.71 Identify each of the following as a physical or chemical property:
a. Gold is shiny.
b. Gold melts at 1064 °C.
c. Gold is a good conductor of electricity.
d. When gold reacts with yellow sulfur, a black sulfide compound forms.

2.72 Identify each of the following as a physical or chemical property of a candle:
a. The candle is 20 cm high with a diameter of 3 cm.
b. The candle burns.
c. The wax of the candle softens on a hot day.
d. The candle is blue.

2.73 Identify each of the following as a physical or chemical change:
a. A plant grows a new leaf.
b. Chocolate is melted for a dessert.
c. Wood is chopped for the fireplace.
d. Wood burns in a fireplace.

2.74 Identify each of the following as a physical or chemical change:
a. A medication tablet is broken in two.
b. Carrots are grated for use in a salad.
c. Malt undergoes fermentation to make beer.
d. A copper pipe reacts with air and turns green.

2.75 Calculate each of the following temperatures in degrees Celsius:
a. The highest recorded temperature in the continental United States was 134 °F in Death Valley, California, on July 10, 1913.
b. The lowest recorded temperature in the continental United States was −69.7 °F in Rodgers Pass, Montana, January 20, 1954.

2.76 Calculate each the following temperatures in degrees Fahrenheit:
a. The highest recorded temperature in the world was 58.0 °C in El Azizia, Libya, on September 13, 1922.
b. The lowest recorded temperature in the world was −89.2 °C in Vostok, Antarctica, on July 21, 1983.

2.77 What is −15 °F in degrees Celsius and in kelvins?

2.78 The highest recorded body temperature that a person has survived is 46.5 °C. Calculate that temperature in degrees Fahrenheit and in kelvins.

2.79 On a hot day, the beach sand gets hot, but the water stays cool. Would you predict the specific heat of sand is higher or lower than that of water? Explain.

The water, sand, and air gain energy from the Sun.

2.80 Why do drops of liquid water form on the outside of a glass of iced tea?

2.81 If you want to lose 1 pound of "fat," which is 15% water, how many kilocalories do you need to lose?

2.82 Calculate the Cal (kcal) in 1/2 cup of soft ice cream that contains 18 g of carbohydrate, 11 g of fat, and 4 g of protein. (Round off the kilocalories to the tens place.)

2.83 A hot-water bottle contains 725 g of water at 65 °C. If the water cools to body temperature (37 °C), how many kilocalories of heat could be transferred to sore muscles?

2.84 A pitcher containing 0.75 L of water at 4 °C is removed from the refrigerator. How many kilojoules are needed to warm the water to a room temperature of 22 °C?

2.85 The melting point of chloroform is −64 °C and its boiling point is 61 °C. Sketch a heating curve for chloroform from −100 °C to 100 °C.
a. What is the state of chloroform at −75 °C?
b. What happens on the curve at −64 °C?
c. What is the state of chloroform at −18 °C?
d. What is the state of chloroform at 80 °C?
e. At what temperature will both solid and liquid be present?

2.86 The melting point of benzene is 5.5 °C and its boiling point is 80.1 °C. Sketch a heating curve for benzene from 0 °C to 100 °C.
a. What is the state of benzene at 15 °C?
b. What happens on the curve at 5.5 °C?
c. What is the state of benzene at 63 °C?
d. What is the state of benzene at 98 °C?
e. At what temperature will both liquid and gas be present?

Challenge Questions

2.87 One liquid has a temperature of 140. °F and another liquid has a temperature of 60.0 °C. Are the liquids at the same temperature or at different temperatures?

2.88 A 0.50-g sample of vegetable oil is placed in a calorimeter. When the sample is burned, 18.9 kJ are given off. What is the caloric value, in kcal/g, of the oil?

2.89 How many kilocalories of heat are released when 75 g of steam at 100 °C is converted to ice at 0 °C? (*Hint*: The calculations include several steps.)

2.90 A 45-g piece of ice at 0.0 °C is added to a sample of water at 8.0 °C. All of the ice melts and the temperature of the water decreases to 0.0 °C. How many grams of water were in the sample?

2.91 In a large building, oil is used in a steam boiler heating system. The combustion of 1.0 lb of oil provides 2.4×10^7 J.
a. How many kilograms of oil are needed to heat 150 kg of water from 22 °C to 100 °C?
b. How many kilograms of oil are needed to change 150 kg of water to steam at 100 °C?

2.92 When 1.0 g of gasoline burns, it releases 11 kcal of heat. The density of gasoline is 0.74 g/mL.
a. How many megajoules are released when 1.0 gal of gasoline burns?
b. If a television requires 150 kJ/h to run, how many hours can the television run on the energy provided by 1.0 gal of gasoline?

2.93 An ice bag containing 275 g of ice at 0 °C was used to treat sore muscles. When the bag was removed, the ice had melted and the liquid water had a temperature of 24.0 °C. How many kilojoules of heat were absorbed?

2.94 A 115-g sample of steam at 100 °C is emitted from a volcano. It condenses, cools, and falls as snow at 0 °C. How many kilojoules of heat were released?

Answers

Answers to Study Checks

2.1 heterogeneous

2.2 **b** and **d** are chemical changes.

2.3 35 000 J

2.4 12 °F

2.5 39.8 °C

2.6 night −260. °C; day 410. °C

2.7 14.6 kJ

2.8 1760 kJ

2.9 12 kcal

2.10 56.5 kJ released

2.11 226 kJ

Answers to Selected Questions and Problems

2.1 **a.** pure substance **b.** mixture **c.** pure substance **d.** pure substance **e.** mixture

2.3 **a.** element **b.** compound **c.** element **d.** compound **e.** compound

2.5 **a.** heterogeneous **b.** homogeneous **c.** homogeneous **d.** heterogeneous **e.** heterogeneous

2.7 **a.** gas **b.** gas **c.** solid

2.9 **a.** physical **b.** chemical **c.** physical **d.** chemical **e.** chemical

2.11 **a.** physical **b.** chemical **c.** physical **d.** physical **e.** physical

2.13 **a.** chemical **b.** physical **c.** physical **d.** chemical **e.** physical

2.15 When the roller-coaster car is at the top of the ramp, it has its maximum potential energy. As it descends, potential energy changes to kinetic energy. At the bottom, all the energy is kinetic.

2.17 **a.** potential **b.** kinetic **c.** potential **d.** potential

2.19 **a.** 8.1×10^5 J **b.** 190 kcal

2.21 In the United States, we still use the Fahrenheit temperature scale. In °F, normal body temperature is 98.6. On the Celsius scale, her temperature would be 37.7 °C, a mild fever.

2.23 **a.** 98.6 °F **b.** 18.5 °C **c.** 246 K **d.** 335 K **e.** 46 °C

2.25 **a.** 41 °C **b.** No. The temperature is equivalent to 39 °C.

2.27 Copper has the lowest specific heat of the samples and will reach the highest temperature.

2.29 **a.** 180 cal **b.** 14 000 J **c.** 9.3 kcal **d.** 10.8 kJ

2.31 **a.** 1380 J, 330. cal **b.** 1810 J, 434 cal **c.** 3780 J, 904 cal **d.** 3600 J, 850 cal

2.33 **a.** 5.30 kcal **b.** 208 kcal

2.35 **a.** 470 kJ **b.** 18 g **c.** 130 kcal **d.** 950 kcal

2.37 210 kcal, 880 kJ

2.39 **a.** melting **b.** sublimation **c.** freezing **d.** deposition

2.41 **a.** 5200 cal absorbed
b. 5680 J absorbed
c. 18 kcal released
d. 16.7 kJ released

2.43 **a.** condensation **b.** evaporation **c.** boiling **d.** condensation

2.45 **a.** 5400 cal absorbed
b. 11 300 J absorbed
c. 4300 kcal released
d. 396 kJ released

2.47

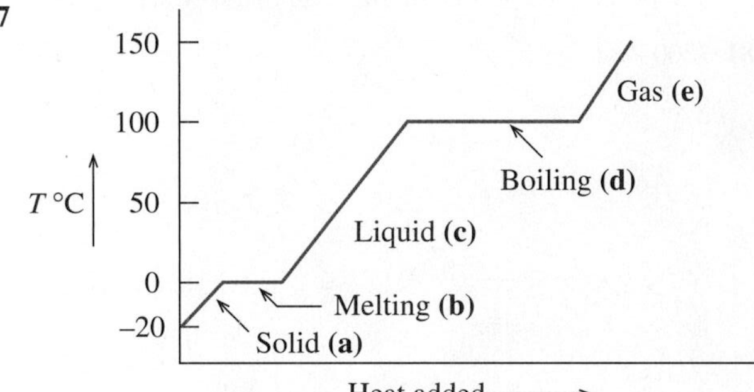

2.49 **a.** 30 300 J **b.** 9.6 kcal **c.** 72.2 kJ

2.51 **a.** compound **b.** mixture **c.** element

2.53 **a.** homogeneous **b.** heterogeneous **c.** heterogeneous

2.55 **a.** 61.4 °C **b.** 53.80 °C **c.** 4.8 °C

2.57 68.3 °C, 341 K

2.59 gold, 250 J or 59 cal; aluminum, 240 J or 58 cal; silver, 250 J or 59 cal. The heat needed for all the metal samples is almost the same.

2.61 1100 kcal (rounded off)

2.63 **a.** The heat from the skin is used to evaporate the water (perspiration). Therefore, the skin is cooled.
b. On a hot day, there are more molecules with sufficient energy to become water vapor.
c. In a closed plastic bag, some water molecules evaporate, but they cannot escape and will condense back to liquid; the clothes will not dry.

2.65 **a.** about −60 °C
b. about 60 °C
c. The diagonal line A represents the solid state as temperature increases. The horizontal line B represents the change from

solid to liquid or melting of the substance. The diagonal line C represents the liquid state as temperature increases. The horizontal line D represents the change from liquid to gas or boiling of the liquid. The diagonal line E represents the gas state as temperature increases.

d. At −80 °C, solid; at −40 °C, liquid; at 25 °C, liquid; at 80 °C, gas

2.67 **a.** element **b.** compound **c.** mixture
d. element **e.** mixture

2.69 **a.** solid **b.** gas **c.** liquid
d. gas **e.** solid

2.71 **a.** physical **b.** physical
c. physical **d.** chemical

2.73 **a.** chemical **b.** physical
c. physical **d.** chemical

2.75 **a.** 57 °C (or 56.7 °C using 3 SFs)
b. −56.5 °C

2.77 −26 °C; 247 K

2.79 Sand must have a lower specific heat than water. When both substances absorb the same amount of heat, the final temperature of the sand will be higher than that of water.

2.81 3500 kcal

2.83 20. kcal

2.85

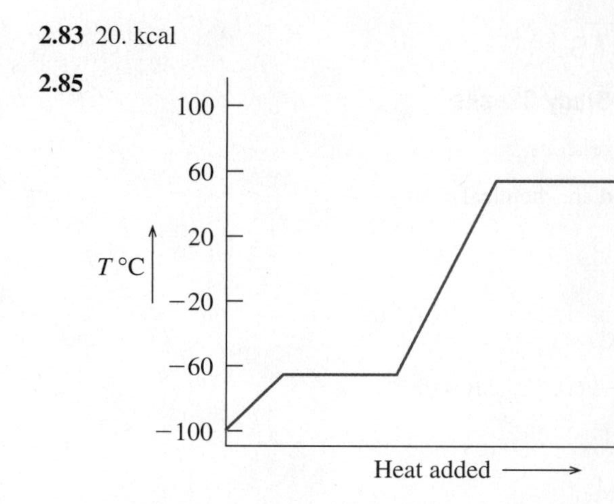

a. solid **b.** solid chloroform melts
c. liquid **d.** gas
e. −64 °C

2.87 The two liquids are at the same temperature.

2.89 55 kcal

2.91 **a.** 0.93 kg **b.** 6.4 kg

2.93 119.5 kJ

Combining Ideas from Chapters 1 and 2

CI.1 Gold, one of the most sought-after metals in the world, has a density of 19.3 g/cm^3, a melting point of 1064 °C, a specific heat of 0.129 J/g °C, and a heat of fusion of 63.6 J/g. A gold nugget found in Alaska in 1998 weighs 20.17 lb.

Gold nuggets, also called native gold, can be found in streams and mines.

a. How many significant figures are in the measurement of the weight of the nugget?
b. Which is the mass of the nugget in kilograms?
c. If the nugget were pure gold, what would its volume be in cm^3?
d. What is the melting point of gold in Fahrenheit degrees and kelvins?
e. How many kilocalories are required to raise the temperature of the nugget from 500. °C to 1064 °C and melt all the gold to liquid at 1064 °C?
f. If the price of gold is $35.10 per gram, what is the nugget worth, in dollars?

CI.2 The mileage for a motorcycle with a fuel-tank capacity of 22 L is 35 mi/gal. The density of gasoline is 0.74 g/mL.

a. How long a trip, in kilometers, can be made on one full tank of gasoline?
b. If the price of gasoline is $2.67 per gallon, what would be the cost of fuel for the trip?
c. If the average speed during the trip is 44 mi/h, how many hours will it take to reach the destination?
d. If the density of gasoline is 0.74 g/mL, what is the mass, in grams, of the fuel in the tank?
e. When 1.00 g of gasoline burns, 47 kJ of energy is released. How many kilojoules are produced when the fuel in one full tank is burned?

CI.3 Answer the following questions for the water samples A and B shown in the diagrams:

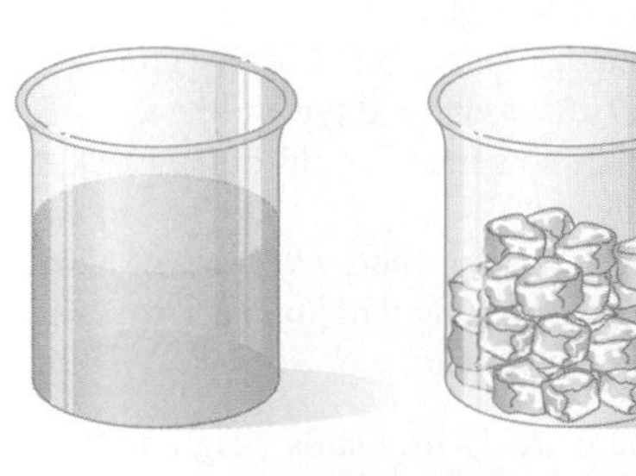

a. When each sample is transferred to another container, what happens to the shape and volume of each?
b. Match the diagrams (1, 2, or 3) that represent the water particles with sample **A** or **B**. Give a reason for your choice.

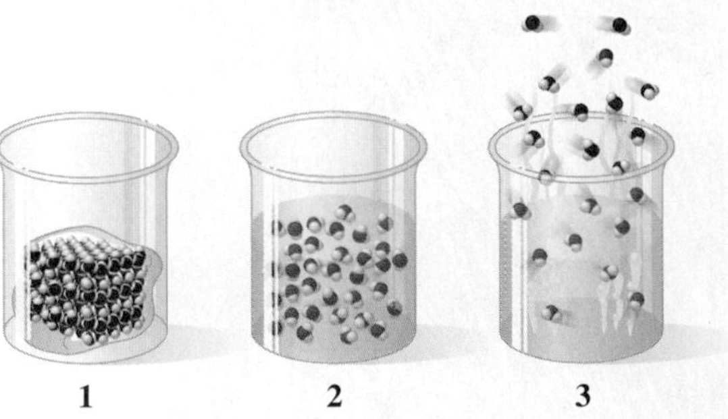

c. The state of matter indicated in diagram 1 is a _________; in diagram 2 it is a _________; and in diagram 3 it is a _________.
d. When the water in diagram 1 changes to the water in diagram 2, the process is called _________, which occurs at a temperature called the _________ point. This is an example of a _________ change.
e. When the water in diagram 2 changes to the water in diagram 3, the process is called _________, which occurs at a temperature called the _________ point. This is an example of a _________ change.
f. If the water in diagram 2 has a mass of 19 g and a temperature of 45 °C, how much heat, in kilojoules, is removed to cool the liquid and form solid at 0 °C?

CI.4 The label of a black cherry almond energy bar with a mass of 68 g lists the "nutrition facts" as 5 g of fat, 39 g of carbohydrate, and 10 g of protein.

a. Using the energy values of carbohydrates, fats, and proteins (see Table 2.8), what are the total kilocalories (Calories) listed for a black cherry almond bar? (Round off answers for each food type to the tens place.)
b. What are the kilojoules for the black cherry almond bar? (Round off answers for each food type to the tens place.)
c. If you obtain 160 kJ, how many grams of the black cherry almond bar did you eat?
d. If you are walking and using energy at a rate of 840 kJ/h, how many minutes of walking will you need to walk to expend the energy of two bars?

CI.5 In a box of nails, there are 75 iron nails weighing 0.25 lb. The density of iron is 7.86 g/cm^3. The specific heat of iron is 0.452 J/g °C. The melting point of iron is 1535 °C. The heat of fusion for iron is 272 J/g.

a. What is the volume, in cm^3, of the iron nails in the box?
b. If 30 nails are added to a graduated cylinder containing 17.6 mL of water, what is the new level of water in the cylinder?
c. How many joules must be added to the nails in the box to raise the temperature from 16 °C to 125 °C?
d. How many joules are required to heat one nail from 25 °C to its melting point and change it to liquid iron?

CI.6 A hot tub is filled with 450 gal of water.

a. What is the volume of water, in liters, in the tub?
b. What is the mass, in kilograms, of water in the tub?
c. How many kilocalories are needed to heat the water from 62 °F to 105 °F?
d. If the hot-tub heater provides 5900 kJ/min, how long, in hours, will it take to heat the water in the hot tub from 62 °F to 105 °F?

Answers

CI.1 **a.** 4 significant figures
b. 9.17 kg
c. 474 cm^3
d. 1947 °F; 1337 K
e. 298 kcal
f. $322 000

CI.3 **a.** The shape of A changes to the shape of the new container, while the shape of B remains the same. The volumes of both A and B remain the same.
b. A is liquid water represented by diagram 2. In liquid water, the water particles are in a random arrangement, but close together. B is solid water represented by diagram 1. In solid water, the water particles are fixed in a definite arrangement.
c. solid; liquid; gas
d. melting; melting point; physical
e. boiling; boiling point; physical
f. 9.9 kJ

CI.5 **a.** 14 cm^3
b. 23.4 mL
c. 5600 J or 5.6×10^3 J
d. 1400 J

Atoms and Elements

LOOKING AHEAD

"Many of my patients have diabetes, ulcers, hypertension, and cardiovascular problems," says Sylvia Lau, registered dietitian. "If a patient has diabetes, I discuss foods that raise blood sugar such as fruit, milk, and starches. I talk about how dietary fat contributes to weight gain and complications from diabetes. For stroke patients, I suggest diets low in fat and cholesterol because high blood pressure increases the risk of another stroke."

If a lab test shows low levels of iron, zinc, iodine, magnesium, or calcium, a dietitian discusses foods that provide those essential elements. For instance, she may recommend more beef for an iron deficiency, whole grain for zinc, leafy green vegetables for magnesium, dairy products for calcium, and iodized table salt and seafood for iodine.

Visit **www.masteringchemistry.com** for self-study materials and instructor-assigned homework.

All matter is composed of *elements*, of which there are 117 different kinds. Of these, 88 elements occur naturally and make up all the substances in our world. Many elements are already familiar to you. You may have a ring or necklace made of gold, silver, or perhaps platinum. If you play tennis or golf, then you may have noticed that your racket or clubs may be made from the elements titanium or carbon. In our bodies, calcium and phosphorus form the structure of bones and teeth, iron and copper are needed in the formation of red blood cells, and iodine is required for the proper functioning of the thyroid.

The correct amounts of certain elements are crucial to the proper growth and function of the body. Low levels of iron can lead to anemia, while lack of iodine can cause hypothyroidism and goiter. Some elements known as microminerals, such as chromium, cobalt, and selenium, are needed in our bodies in very small amounts. Laboratory tests are used to confirm that these elements are within normal ranges in our bodies.

LEARNING GOAL

Given the name of an element, write its correct symbol; from the symbol, write the correct name.

TUTORIAL
Elements and Symbols in the Periodic Table

3.1 Elements and Symbols

Elements are pure substances from which all other things are built. As we discussed in Chapter 2, elements cannot be broken down into simpler substances. Over the centuries, elements have been named for planets, mythological figures, minerals, colors, geographic locations, and famous people. Some sources of names of elements are listed in Table 3.1. A complete list of all the elements and their symbols appears on the inside front cover of this text.

TABLE 3.1 Some Elements and Their Names

Element	Source of Name
Uranium	The planet Uranus
Titanium	Titans (mythology)
Chlorine	*Chloros*, "greenish-yellow" (Greek)
Iodine	*Ioeides*, "violet" (Greek)
Magnesium	Magnesia, a mineral
Californium	California
Curium	Marie and Pierre Curie
Copernicium	Nicolaus Copernicus

Chemical Symbols

Chemical symbols are one- or two-letter abbreviations for the names of the elements. Only the first letter of an element's symbol is capitalized. If the symbol has a second letter, it is lowercase so that we know when a different element is indicated. If two letters are capitalized, they represent the symbols of two different elements. For example, the element cobalt has the symbol Co. However, the two capital letters CO specify two elements, carbon (C) and oxygen (O).

One-Letter Symbols	Two-Letter Symbols
C carbon	Co cobalt
S sulfur	Si silicon
N nitrogen	Ne neon
I iodine	Ni nickel

Although most of the symbols use letters from the current names, some are derived from their ancient names. For example, Na, the symbol for sodium, comes from the Latin word *natrium*. The symbol for iron, Fe, is derived from the Latin name *ferrum*. Table 3.2 lists the names and symbols of some common elements. Learning their names and symbols will greatly help your learning of chemistry.

TABLE 3.2 Names and Symbols of Some Common Elements

Name*	Symbol	Name*	Symbol	Name*	Symbol
Aluminum	Al	Gold (*aurum*)	Au	Oxygen	O
Argon	Ar	Helium	He	Phosphorus	P
Arsenic	As	Hydrogen	H	Platinum	Pt
Barium	Ba	Iodine	I	Potassium (*kalium*)	K
Boron	B	Iron (*ferrum*)	Fe	Radium	Ra
Bromine	Br	Lead (*plumbum*)	Pb	Silicon	Si
Cadmium	Cd	Lithium	Li	Silver (*argentum*)	Ag
Calcium	Ca	Magnesium	Mg	Sodium (*natrium*)	Na
Carbon	C	Manganese	Mn	Strontium	Sr
Chlorine	Cl	Mercury (*hydrargyrum*)	Hg	Sulfur	S
Chromium	Cr	Neon	Ne	Tin (*stannum*)	Sn
Cobalt	Co	Nickel	Ni	Titanium	Ti
Copper (*cuprum*)	Cu	Nitrogen	N	Uranium	U
Fluorine	F			Zinc	Zn

*Names given in parentheses are ancient Latin or Greek words from which the symbols are derived.

Aluminum

Carbon

Gold

Silver

Sulfur

CONCEPT CHECK 3.1

Symbols of the Elements

The symbol for carbon is C, and the symbol for sulfur is S. However, the symbol for cesium is Cs, not CS. Why?

ANSWER

When the symbol for an element has two letters, the first letter is capitalized, but the second letter is lowercase. If both letters are capitalized such as in CS, two elements—carbon and sulfur—are indicated.

SAMPLE PROBLEM 3.1

Writing Chemical Symbols

What are the chemical symbols for the following elements?

a. nickel **b.** nitrogen **c.** neon

SOLUTION

a. Ni **b.** N **c.** Ne

STUDY CHECK 3.1

What are the chemical symbols for the elements silicon, sulfur, and silver?

Chemistry Link to Health

TOXICITY OF MERCURY

Mercury is a silvery, shiny element that is a liquid at room temperature. Mercury can enter the body through inhaled mercury vapor, contact with the skin, or ingestion of foods or water contaminated with mercury. In the body, mercury destroys proteins and disrupts cell function. Long-term exposure to mercury can damage the brain and kidneys, cause mental retardation, and decrease physical development. Blood, urine, and hair samples are used to test for mercury.

In both freshwater and seawater, bacteria convert mercury into toxic methylmercury, which attacks the central nervous system (CNS). Because fish absorb methylmercury, we are exposed to mercury by consuming mercury-contaminated fish. As levels of mercury ingested from fish became a concern, the Food and Drug Administration (FDA) set a maximum level of one part mercury per million parts seafood (1 ppm), which is the same as 1 mg of mercury in every kilogram of seafood. Fish higher in the food chain, such as swordfish and shark, can have such high levels of mercury that the Environmental Protection Agency (EPA) recommends they be consumed no more than once a week.

One of the worst incidents of mercury poisoning occurred in Minamata and Niigata, Japan, in 1950. At that time, the ocean was polluted with high levels of mercury from industrial wastes. Because fish were a major food in the diet, more than 2000 people were affected with mercury poisoning and died or developed neural damage. In the United States between 1988 and 1997, the use of mercury decreased by 75% when the use of mercury was banned in paint and pesticides, and regulated in batteries and other products. Mercury batteries come with warnings on the label and should be disposed of properly.

This mercury fountain, housed in glass, was designed by Alexander Calder for the 1937 World's Fair in Paris.

SAMPLE PROBLEM 3.2

Names and Symbols of Chemical Elements

Give the name of the element that corresponds to each of the following chemical symbols:

a. Zn **b.** K **c.** H **d.** Fe

SOLUTION

a. zinc **b.** potassium **c.** hydrogen **d.** iron

STUDY CHECK 3.2

What are the names of the elements with the chemical symbols Mg, Al, and F?

QUESTIONS AND PROBLEMS

Elements and Symbols

3.1 Write the symbols for the following elements:
a. copper **b.** platinum **c.** calcium **d.** manganese
e. iron **f.** barium **g.** lead **h.** strontium

3.2 Write the symbols for the following elements:
a. oxygen **b.** lithium **c.** uranium **d.** titanium
e. hydrogen **f.** chromium **g.** tin **h.** gold

3.3 Write the name of the element for each symbol.
a. C **b.** Cl **c.** I **d.** Hg
e. Ag **f.** Ar **g.** B **h.** Ni

3.4 Write the name of the element for each symbol.
a. He **b.** P **c.** Na **d.** As
e. Ca **f.** Br **g.** Cd **h.** Si

3.5 What elements are in the following substances?
a. table salt, $NaCl$ **b.** plaster casts, $CaSO_4$
c. Demerol, $C_{15}H_{22}ClNO_2$ **d.** antacid, $CaCO_3$

3.6 What elements are in the following substances?
a. water, H_2O **b.** baking soda, $NaHCO_3$
c. lye, $NaOH$ **d.** sugar, $C_{12}H_{22}O_{11}$

Chemistry Link to Industry

MANY FORMS OF CARBON

Carbon, which has the symbol C and atomic number 6, is located in Group 4A (14) in Period 2 on the periodic table. However, its atoms can be arranged in different ways to give several different substances. Two forms of carbon—diamond and graphite—have been known since prehistoric times. In diamond, carbon atoms are arranged in a rigid structure. A diamond is transparent and harder than any other substance, whereas graphite is black and soft. In graphite, carbon atoms are arranged in sheets that slide over each other. Graphite is used as pencil lead, as lubricants, and as carbon fibers for the manufacture of light weight golf clubs and tennis rackets.

Two other forms of carbon have been discovered more recently. In the form called buckminsterfullerene or buckyball, 60 carbon atoms are arranged as rings of 5 and 6 atoms to give a spherical, cage-like structure. When a fullerene structure is stretched out, it produces a cylinder with a diameter of only a few nanometers called a nanotube. Practical uses for buckyballs and nanotubes are not yet developed, but they are expected to find use in light weight structural materials, heat conductors, computer parts, and medicine.

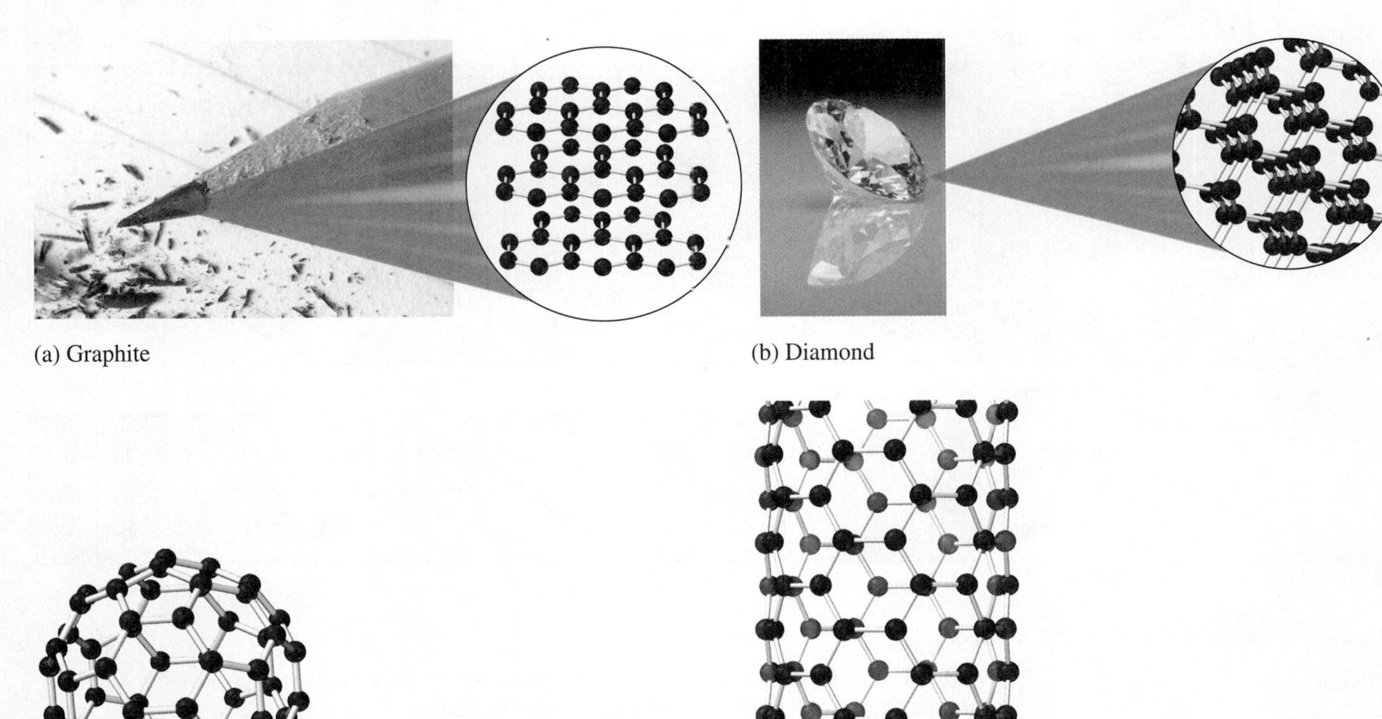

(a) Graphite

(b) Diamond

(c) Buckminsterfullerene

(d) Nanotubes

Carbon atoms can form different types of structures.

3.2 The Periodic Table

LEARNING GOAL

Use the periodic table to identify the group and the period of an element; identify the element as a metal, nonmetal, or metalloid.

As more and more elements were discovered, it became necessary to organize them into some type of classification system. By the late 1800s, scientists recognized that certain elements looked alike and behaved much the same way. In 1872, a Russian chemist, Dmitri Mendeleev, arranged the 60 elements known at that time into groups with similar properties and placed them in order of increasing atomic masses. Today, this arrangement of 117 elements is known as the **periodic table** (see Figure 3.1).

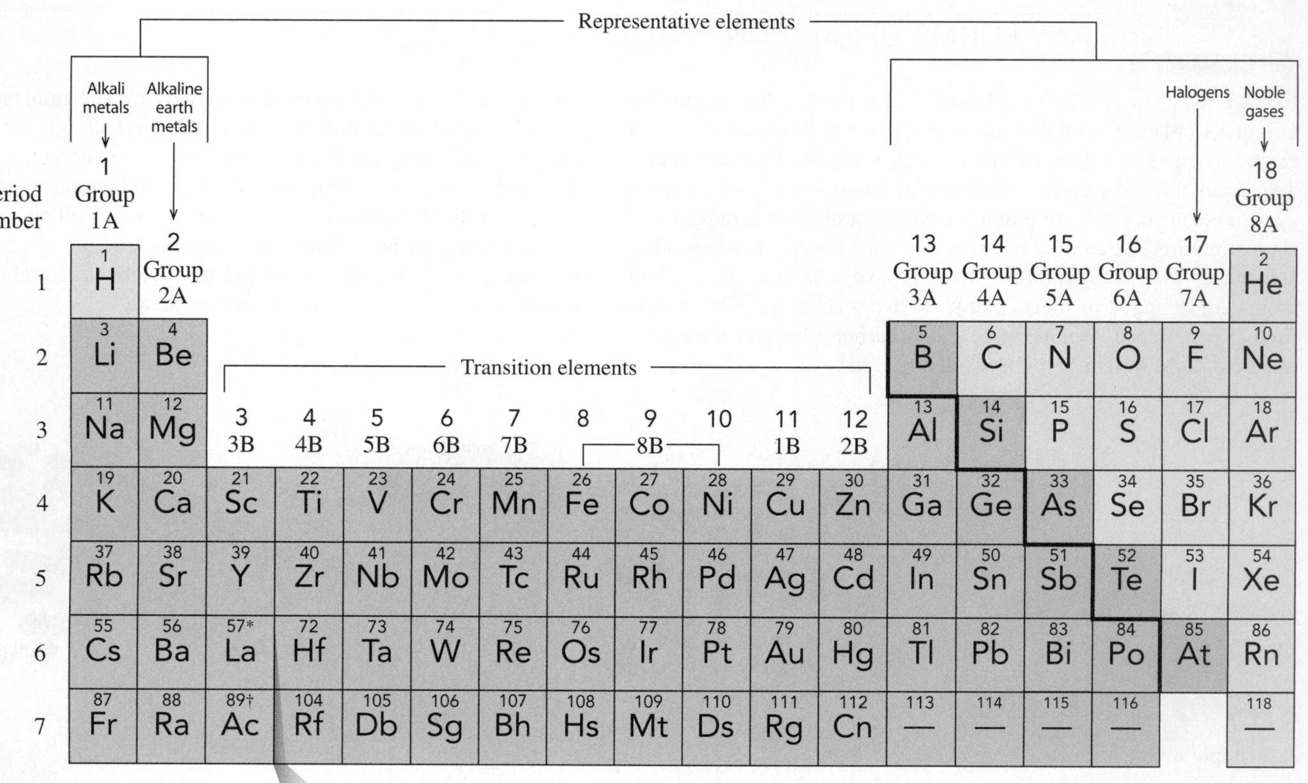

FIGURE 3.1 On the periodic table, groups are elements arranged as vertical columns, and periods are the elements in each horizontal row.
Q What is the symbol of the alkali metal in Period 3?

TUTORIAL
Elements and Symbols in the Periodic Table

Periods and Groups

Each horizontal row in the periodic table is called a **period** (see Figure 3.2). Each period is counted from the top of the table as Period 1 to Period 7. The first period contains 2 elements: hydrogen (H) and helium (He). The second period contains 8 elements: lithium (Li), beryllium (Be), boron (B), carbon (C), nitrogen (N), oxygen (O), fluorine (F), and neon (Ne). The third period also contains 8 elements beginning with sodium (Na) and ending with argon (Ar). The fourth period, which begins with potassium (K), and the fifth period, which begins with rubidium (Rb), have 18 elements each. The sixth period, which begins with cesium (Cs), has 32 elements. The seventh period, as of today, contains the 31 remaining elements, although it could go up to 32.

Each vertical column on the periodic table contains a **group** (or family) of elements that have similar properties. At the top of each column is a number that is assigned to each group. The elements in the first two columns on the left and the last six columns on the right of the periodic table are called the **representative elements**. For many years, they have been given group numbers 1A–8A. In the center of the periodic table is a block

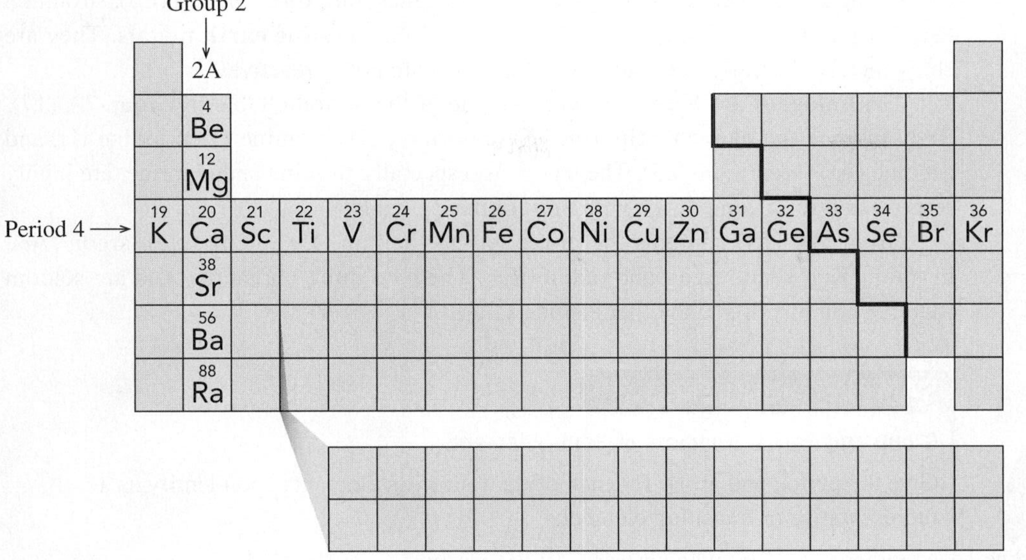

FIGURE 3.2 On the periodic table, each vertical column represents a group of elements and each horizontal row of elements represents a period.
Q Are the elements Si, P, and S part of a group or a period?

of elements known as the **transition elements**, which are designated with the letter "B." A newer numbering system assigns group numbers of 1–18 going across the periodic table. Because both systems of group numbers are currently in use, they are both indicated on the periodic table in this text and are included in our discussions of elements and group numbers. The lanthanides and actinides that are part of Periods 6 and 7 are placed at the bottom of the periodic table to allow it fit on a page.

Classification of Groups

Several groups in the periodic table have special names (see Figure 3.3). Group 1A (1) elements—lithium (Li), sodium (Na), potassium (K), rubidium (Rb), cesium (Cs), and francium (Fr)—are a family of elements known as the **alkali metals** (see Figure 3.4). The elements within this group are soft, shiny metals that are good conductors of heat and electricity and have relatively low melting points. Alkali metals react vigorously with water and form white products when they combine with oxygen.

Although hydrogen (H) is at the top of Group 1A (1), hydrogen is not an alkali metal and has very different properties than the rest of the elements in this group. Thus hydrogen is not included in the classification of alkali metals.

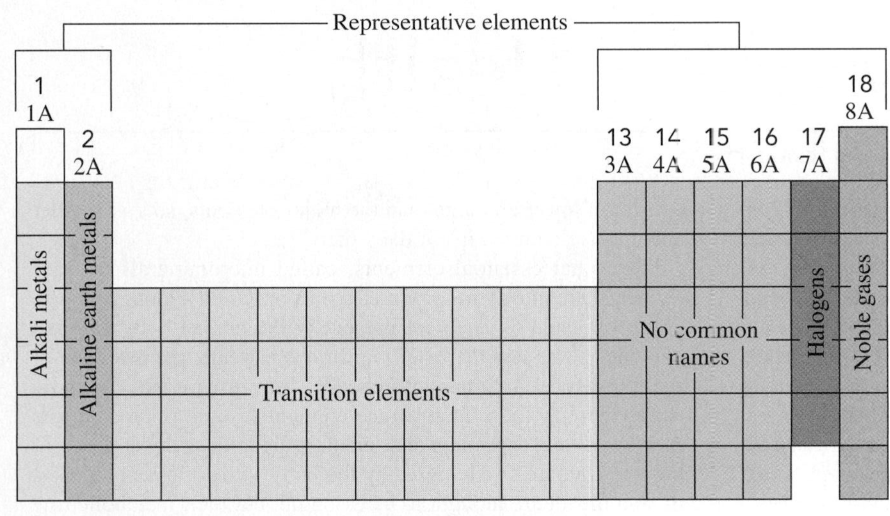

FIGURE 3.3 Certain groups on the periodic table have common names.
Q What is the common name for the group of elements that includes helium and argon?

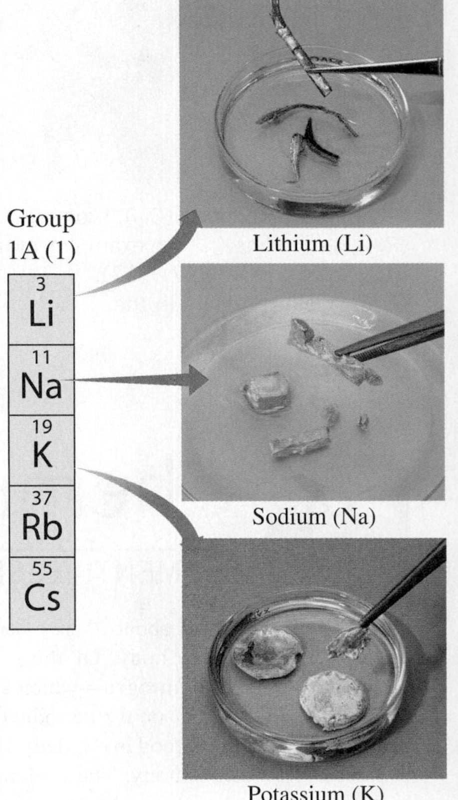

FIGURE 3.4 Lithium (Li), sodium (Na), and potassium (K) are some alkali metals from Group 1A (1).
Q What physical properties do these alkali metals have in common?

Group 2A (2) elements—beryllium (Be), magnesium (Mg), calcium (Ca), strontium (Sr), barium (Ba), and radium (Ra)—are called the **alkaline earth metals**. They are shiny metals like those in Group 1A (1), but they are not as reactive.

The **halogens** are found on the right side of the periodic table in Group 7A (17). They include the elements fluorine (F), chlorine (Cl), bromine (Br), iodine (I), and astatine (At) (see Figure 3.5). The halogens, especially fluorine and chlorine, are highly reactive and form compounds with most of the elements.

Group 8A (18) contains the **noble gases**—helium (He), neon (Ne), argon (Ar), krypton (Kr), xenon (Xe), and radon (Rn). They are quite unreactive and are seldom found in combination with other elements.

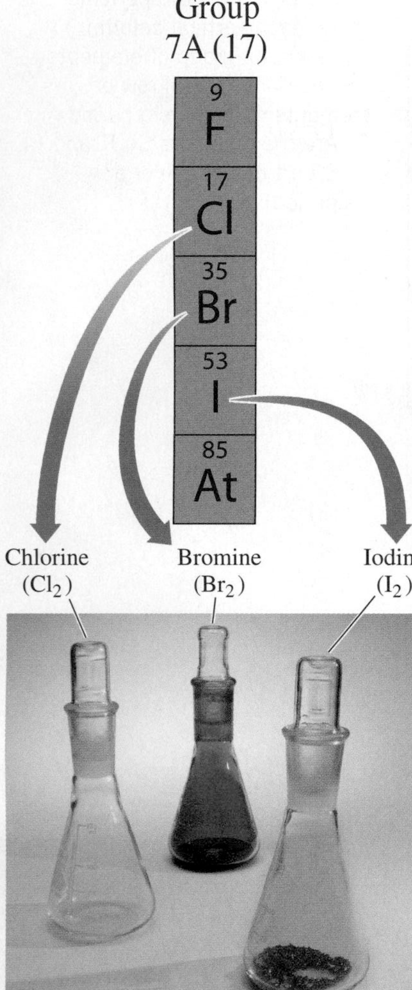

FIGURE 3.5 Chlorine (Cl_2), bromine (Br_2), and iodine (I_2) are examples of halogens from Group 7A (17).

Q What elements are in the halogen group?

SAMPLE PROBLEM 3.3

Group and Period Numbers of Some Elements

Give the period and group for each of the following elements and identify as a representative or transition element:

a. iodine **b.** manganese **c.** barium **d.** gold

SOLUTION

a. Iodine (I), Period 5, Group 7A (17), is a representative element.
b. Manganese (Mn), Period 4, Group 7B (7), is a transition element.
c. Barium (Ba), Period 6, Group 2A (2), is a representative element.
d. Gold (Au), Period 6, Group 1B (11), is a transition element.

STUDY CHECK 3.3

Strontium is an element that gives a brilliant red color to fireworks.

a. In what group is strontium found?
b. In what chemical family is strontium found?
c. In what period is strontium found?
d. What are the name and symbol of the element in Period 3 that is in the same group as strontium?
e. What alkali metal, halogen, and noble gas are in the same period as strontium?

Strontium provides the red color in fireworks.

Chemistry Link to Health

ELEMENTS ESSENTIAL TO HEALTH

Of all the elements, only about 20 are essential for the well-being and survival of the human body. Of those, four elements—oxygen, carbon, hydrogen, and nitrogen—which are representative elements in Period 1 and Period 2 on the periodic table, make up 96% of our body mass. Most of the food in our daily diet provides these elements to maintain a healthy body. These elements are found in carbohydrates, fats, and proteins. Most of the hydrogen and oxygen is found in water, which makes up 55–60% of our body mass.

The macrominerals—Ca, P, K, Cl, S, Na, and Mg—are located in Period 3 and Period 4 of the periodic table. They are involved in the formation of bones and teeth, maintenance of heart and blood vessels, muscle contraction, nerve impulses, acid–base balance of body fluids, and regulation of cellular metabolism. The macrominerals are present in lower amounts than the major elements, so that smaller amounts are required in our daily diets.

The other essential elements, called microminerals or trace elements, are mostly transition elements in Period 4 along with Mo and I in Period 5. They are present in the human body in small amounts, some less than 100 mg. In recent years, the detection of such small amounts has improved so that researchers can more easily identify the roles of trace elements. Some trace elements such as arsenic, chromium and selenium are toxic at higher levels in the body but are still required by the body. Other elements such as tin and nickel are thought to be essential, but their metabolic role has not yet been determined. Some examples and the amounts present in a 60-kg person are listed in Table 3.3.

TABLE 3.3 Typical Amounts of Essential Elements in a 60-kg Adult

Element	Quantity	Function
Major Elements		
Oxygen (O)	39 kg	Building block of biomolecules and water (H_2O)
Carbon (C)	11 kg	Building block of organic and biomolecules
Hydrogen (H)	6 kg	Component of biomolecules, water (H_2O), and pH of body fluids, stomach acid (HCl)
Nitrogen (N)	1.5 kg	Component of proteins and nucleic acids
Macrominerals		
Calcium (Ca)	1000 g	Bone and teeth, muscle contraction, nerve impulses
Phosphorus (P)	600 g	Bone and teeth, nucleic acids, ATP
Potassium (K)	120 g	Most abundant positive ion (K^+) in cells, muscle contraction, nerve impulses
Chlorine (Cl)	100 g	Most abundant negative ion (Cl^-) in fluids outside cells, stomach acid (HCl)
Sulfur (S)	86 g	Proteins, liver, vitamin B_1, insulin
Sodium (Na)	60 g	Most abundant positive ion (Na^+) in fluids outside cells, water balance, muscle contraction, nerve impulses
Magnesium (Mg)	36 g	Bone, required for metabolic reactions
Microminerals (trace elements)		
Iron (Fe)	3600 mg	Component of oxygen carrier hemoglobin
Silicon (Si)	3000 mg	Growth and maintenance of bone and teeth, tendons and ligaments, hair and skin
Zinc (Zn)	2000 mg	Metabolic reactions in cells, DNA synthesis, growth of bone, teeth, connective tissue, immune system
Copper (Cu)	240 mg	Blood vessels, blood pressure, immune system
Manganese (Mn)	60 mg	Bone growth, blood clotting, necessary for metabolic reactions
Iodine (I)	20 mg	Proper thyroid function
Molybdenum (Mo)	12 mg	Needed to process Fe and N from diets
Arsenic (As)	3 mg	Growth and reproduction
Chromium (Cr)	3 mg	Maintenance of blood sugar levels, synthesis of biomolecules
Cobalt (Co)	3 mg	Vitamin B_{12}, red blood cells
Selenium (Se)	2 mg	Immune system, health of heart and pancreas
Vanadium (V)	2 mg	Formation of bone and teeth, energy from food

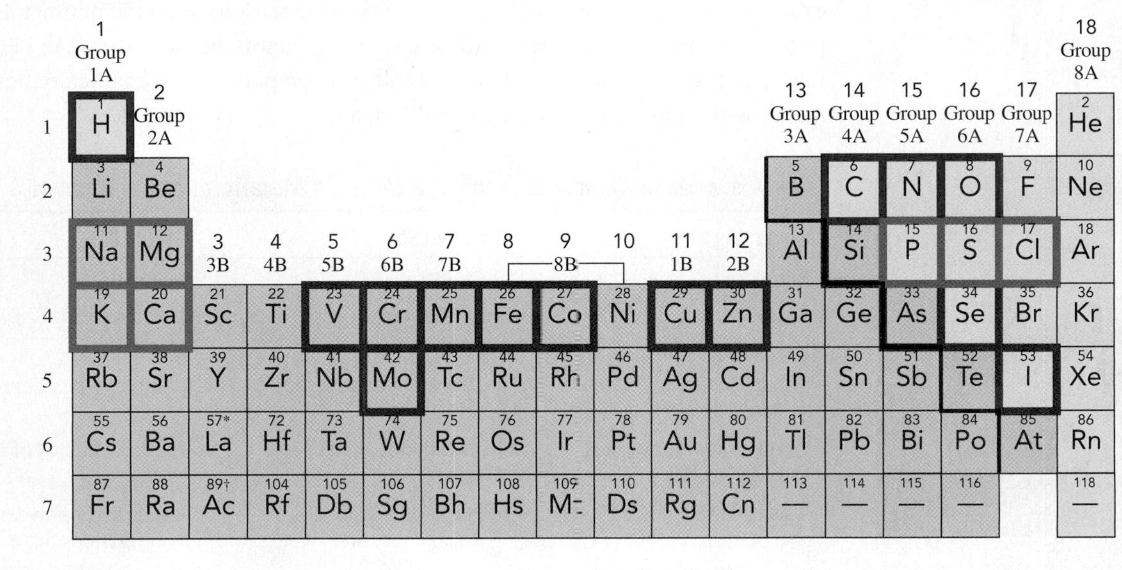

Metals, Nonmetals, and Metalloids

TUTORIAL
Metals, Nonmetals, and Metalloids

Another feature of the periodic table is the heavy zigzag line that separates the elements into the *metals* and the *nonmetals*. The metals are those elements on the left of the line *except for hydrogen*, and the nonmetals are the elements on the right (see Figure 3.6).

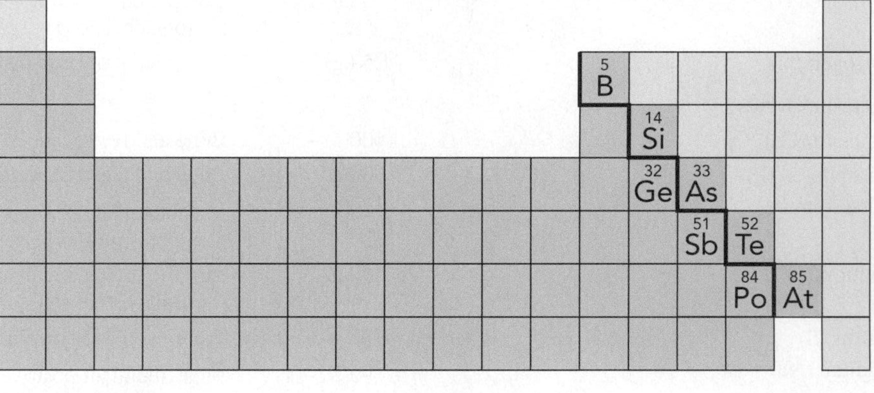

FIGURE 3.6 Along the heavy zigzag line on the periodic table that separates the metals and nonmetals are metalloids, which exhibit characteristics of both metals and nonmetals.
Q On which side of the heavy zigzag line are the nonmetals located?

In general, most **metals** are shiny solids, such as copper (Cu), gold (Au), and silver (Ag). Metals can be shaped into wires (ductile) or hammered into a flat sheet (malleable). Metals are good conductors of heat and electricity. They usually melt at higher temperatures than nonmetals. All the metals are solids at room temperature, except for mercury (Hg), which is a liquid.

Nonmetals are not especially shiny, ductile, or malleable, and they are often poor conductors of heat and electricity. They typically have low melting points and low densities. Some examples of nonmetals are hydrogen (H), carbon (C), nitrogen (N), oxygen (O), chlorine (Cl), and sulfur (S).

Except for aluminum, the elements located along the heavy line are **metalloids**: B, Si, Ge, As, Sb, Te, Po, and At. Metalloids are elements that exhibit some properties that are typical of the metals and other properties that are characteristic of the nonmetals. For example, they are better conductors of heat and electricity than the nonmetals, but not as good as the metals. The metalloids are semiconductors because they can be modified to function as conductors or insulators. Table 3.4 compares some characteristics of silver, a metal, with those of antimony, a metalloid, and sulfur, a nonmetal.

A silver cup is shiny, antimony is a blue-gray solid, and sulfur is a dull, yellow color.

TABLE 3.4 Some Characteristics of a Metal, a Metalloid, and a Nonmetal

Silver (Ag)	Antimony (Sb)	Sulfur (S)
Metal	Metalloid	Nonmetal
Shiny	Blue-gray, shiny	Dull, yellow
Extremely ductile	Brittle	Brittle
Can be hammered into sheets (malleable)	Shatters when hammered	Shatters when hammered
Good conductor of heat and electricity	Poor conductor of heat and electricity	Poor conductor of heat and electricity, good insulator
Used in coins, jewelry, tableware	Used to harden lead, color glass and plastics	Used in gunpowder, rubber, fungicides
Density 10.5 g/mL	Density 6.7 g/mL	Density 2.1 g/mL
Melting point 962 °C	Melting point 630 °C	Melting point 113 °C

CONCEPT CHECK 3.2

Groups and Periods on the Periodic Table

Consider the elements aluminum, germanium, and phosphorus.

a. In what group and period are they found?
b. Identify each as a metal, nonmetal, or metalloid.

ANSWER

a. Aluminum is in Group 3A (13), Period 3. Germanium is in Group 4A (14), Period 4. Phosphorus is in Group 5A (15), Period 3.
b. Aluminum is a metal, germanium is a metalloid, and phosphorus is a nonmetal.

SAMPLE PROBLEM 3.4

Metals, Nonmetals, or Metalloids

Use the periodic table to classify each of the following elements by its group and period, group name (if any), and as a metal, nonmetal, or metalloid:

a. Na **b.** I **c.** B

SOLUTION

a. Na (sodium), Group 1A (1), Period 3, is an alkali metal.
b. I (iodine), Group 7A (17), Period 5, halogen, is a nonmetal.
c. B (boron), Group 3A (13), Period 2, is a metalloid.

STUDY CHECK 3.4

Give the name and symbol of the following elements:

a. Group 5A (15), Period 4
b. A noble gas in Period 3
c. A metalloid in Period 3

Career Focus

MATERIALS SCIENTIST

"The unique qualities of semiconducting metals make it possible for us to create sophisticated electronic circuits," says Tysen Streib, Global Product Manager, Applied Materials. "Elements from Groups 3A (13), 4A (14), and 5A (15) of the periodic table make good semiconductors because they readily form covalently bonded crystals. When small amounts of impurities are added, free-flowing electrons or holes can travel through the crystal with very little interference. Without these covalent bonds and loosely bound electrons, we wouldn't have any of the microchips that we use in computers, cell phones, and thousands of other devices."

Materials scientists study the chemical properties of materials to find new uses for them in products such as cars, bridges, and clothing. They also develop materials that can be used as superconductors or in integrated-circuit chips and fuel cells. Chemistry is important in materials science because it provides information about structure and composition.

QUESTIONS AND PROBLEMS

The Periodic Table

3.7 Identify the group or period number described by each of the following:
a. contains the elements C, N, and O
b. begins with helium
c. the alkali metals
d. ends with neon

3.8 Identify the group or period number described by each of the following:
a. contains Na, K, and Rb
b. the row that begins with Li
c. the noble gases
d. contains F, Cl, Br, and I

3.9 Classify each of the following as an alkali metal, alkaline earth metal, transition element, halogen, or noble gas:
a. Ca **b.** Fe **c.** Xe
d. K **e.** Cl

3.10 Classify each of the following as an alkali metal, alkaline earth metal, transition element, halogen, or noble gas:
a. Ne **b.** Mg **c.** Cu
d. Br **e.** Ba

3.11 Give the symbol of the element described by each of the following:
a. Group 4A (14), Period 2
b. a noble gas in Period 1
c. an alkali metal in Period 3
d. Group 2A (2), Period 4
e. Group 3A (13), Period 3

3.12 Give the symbol of the element described by each of the following:
a. an alkaline earth metal in Period 2
b. Group 5A (15), Period 3
c. a noble gas in Period 4
d. a halogen in Period 5
e. Group 4A (14), Period 4

3.13 Is each of the following elements a metal, nonmetal, or metalloid?
a. calcium
b. sulfur
c. a shiny element
d. an element that is a gas at room temperature
e. located in Group 8A (18)
f. bromine
g. tellurium
h. silver

3.14 Is each of the following elements a metal, nonmetal, or metalloid?
a. located in Group 2A (2)
b. a good conductor of electricity
c. chlorine
d. arsenic
e. an element that is not shiny
f. oxygen
g. nitrogen
h. tin

LEARNING GOAL

Describe the electrical charge and location in an atom for a proton, a neutron, and an electron.

MC

SELF STUDY ACTIVITY
Atoms and Isotopes

TUTORIAL
The Anatomy of Atoms

3.3 The Atom

All the elements listed on the periodic table are made up of atoms. In Chapter 2, we described an **atom** as the smallest particle of an element that retains the characteristics of that element. Imagine that you are tearing a piece of aluminum foil into smaller and smaller pieces. Now imagine that you have a piece so small that you cannot tear it apart any further. Then you would have a single atom of aluminum.

The concept of the atom is relatively recent. Although the Greek philosophers in 500 B.C.E. reasoned that everything must contain minute particles they called *atomos*, the idea of atoms did not become a scientific theory until 1808. Then John Dalton (1766–1844) developed an atomic theory that proposed that atoms were responsible for the combinations of elements found in compounds.

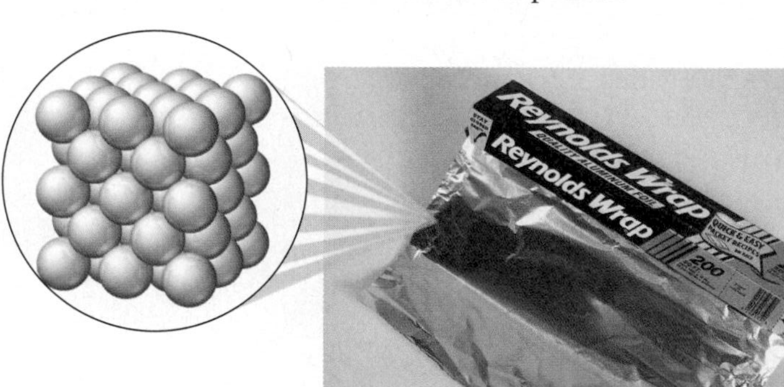

Aluminum foil consists of atoms of aluminum.

Dalton's Atomic Theory

1. All matter is made up of tiny particles called atoms.
2. All atoms of a given element are similar to one another and different from atoms of other elements.
3. Atoms of two or more different elements combine to form compounds. A particular compound is always made up of the same kinds of atoms and always has the same number of each kind of atom.
4. A chemical reaction involves the rearrangement, separation, or combination of atoms. Atoms are never created or destroyed during a chemical reaction.

Dalton's atomic theory formed the basis of current atomic theory, although we have modified some of Dalton's statements. We now know that atoms of the same element are not completely identical to each other and consist of even smaller particles. However, an atom is still the smallest particle that retains the properties of an element.

Although atoms are the building blocks of everything we see around us, we cannot see an atom or even a billion atoms with the naked eye. However, when billions and billions of atoms are packed together, the characteristics of each atom are added to those of the next until we can see the characteristics we associate with the element. For example, a small piece of the shiny element nickel consists of many, many nickel atoms. A special kind of microscope called a *scanning tunneling microscope* (STM) produces images of individual atoms (see Figure 3.7).

FIGURE 3.7 Images of nickel atoms are produced when nickel is magnified millions of times by a scanning tunneling microscope (STM).
Q Why is a microscope with extremely high magnification needed to see these atoms?

Electrical Charges in an Atom

By the end of the 1800s, experiments with electricity showed that atoms were not solid spheres but were composed of even smaller bits of matter called *subatomic particles*, three of which are the *proton*, *neutron*, and *electron*. Two of these subatomic particles were discovered because they have electrical charges.

An electrical charge can be positive or negative. Experiments show that like charges repel, or push away from each other. When you brush your hair on a dry day, electrical charges that are alike build up on the brush and in your hair. As a result, your hair flies away from the brush. However, opposite or unlike charges attract. The clinginess of the clothing taken from the clothes dryer is due to the attraction of opposite, unlike charges, as shown in Figure 3.8.

Positive charges repel

Negative charges repel

Unlike charges attract

FIGURE 3.8 Like charges repel and unlike charges attract.
Q Why are the electrons attracted to the protons in the nucleus of an atom?

Structure of the Atom

In 1897, J. J. Thomson, an English physicist, applied electricity to a glass tube, which produced streams of small particles called *cathode rays*. Because these rays were attracted to a positively charged electrode, Thomson realized that the particles in the rays must be negatively charged. In further experiments, these particles called **electrons** were found to be much smaller than the atom and to have extremely small masses. Because atoms are neutral, scientists soon discovered that atoms contained positively charged particles called **protons** that were much heavier than the electrons.

Thomson proposed the "plum-pudding" model for the atom in which the electrons and protons were randomly distributed through the atom. In 1911, Ernest Rutherford worked with Thomson to test this model. In Rutherford's experiment, positively charged particles were aimed at a thin sheet of gold foil (see Figure 3.9). If the Thomson model were correct, the particles would travel in straight paths through the gold foil. Rutherford was greatly surprised to find that some of the particles were deflected as they passed through the gold foil, and a few particles were deflected so much that they went back in the opposite direction. According to Rutherford, it was as though he had shot a cannonball at a piece of tissue paper, and it bounced back at him.

From the gold-foil experiments, Rutherford realized that the protons must be contained in a small, positively charged region at the center of the atom, which he called the **nucleus**. He proposed that the electrons in the atom occupy the space surrounding the nucleus through which most of the particles traveled undisturbed. Only the particles

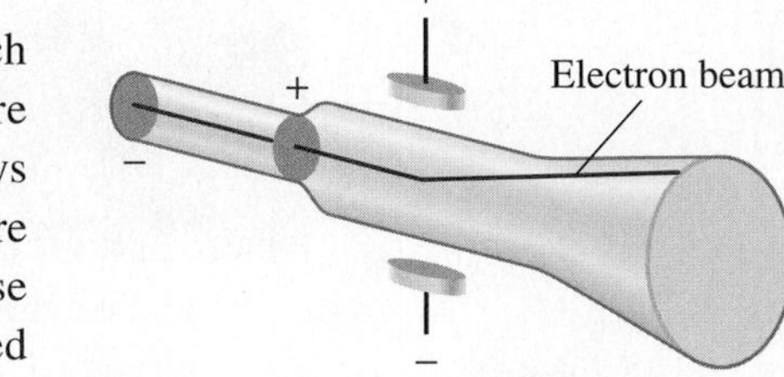

Cathode ray tube

Negatively charged cathode rays (electrons) are attracted to the positive electrode.

Negative electron "plums"

Positive "pudding"

Thomson proposed the "plum-pudding" model of the atom.

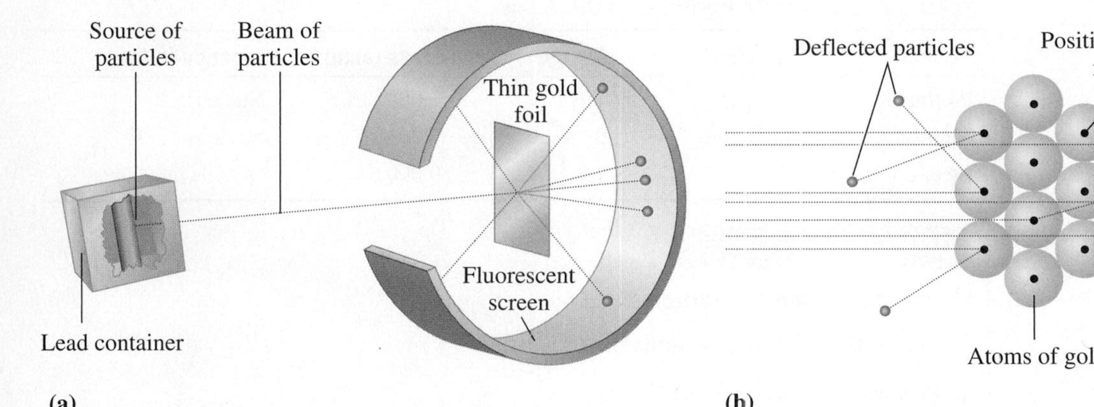

(a)

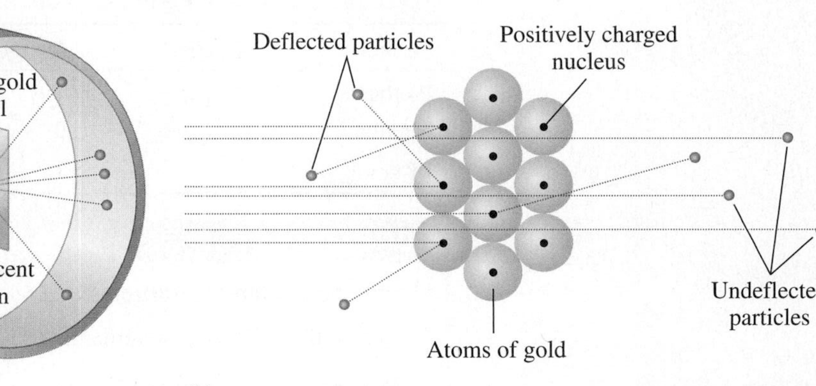

(b)

FIGURE 3.9 **(a)** Positive particles are aimed at a piece of gold foil. **(b)** Particles that come close to the atomic nuclei are deflected from their straight path.
Q Why are some particles deflected while most pass through the gold foil undeflected?

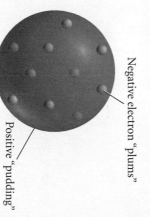

that came near this dense, positive center were deflected. If an atom were the size of a football stadium, the nucleus would be about the size of a golf ball placed in the center of the field.

Scientists knew that the nucleus was heavier than the mass of the protons, so they looked for another subatomic particle. Eventually, they discovered that the nucleus also contained a particle that is neutral, which they called a **neutron**. Thus, the masses of the protons and neutrons in the nucleus determine the mass of an atom (see Figure 3.10).

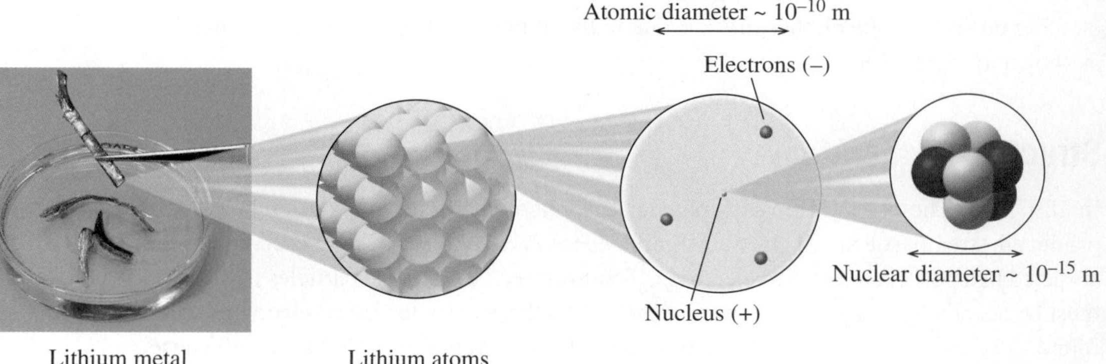

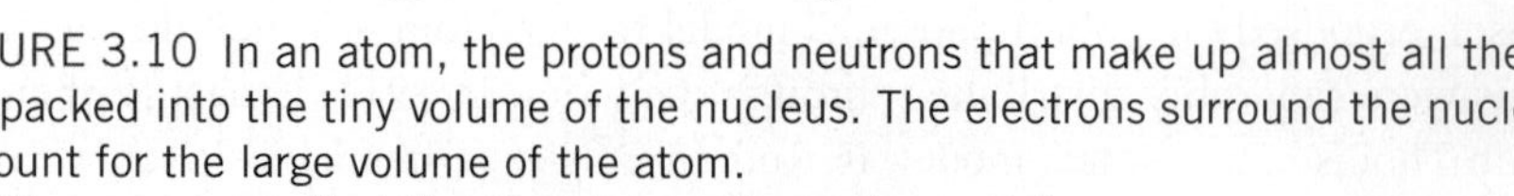

FIGURE 3.10 In an atom, the protons and neutrons that make up almost all the mass are packed into the tiny volume of the nucleus. The electrons surround the nucleus and account for the large volume of the atom.
Q Why can we say that the atom is mostly empty space?

TUTORIAL
Atomic Structure and Properties of Subatomic Particles

Mass of the Atom

All the subatomic particles are extremely small compared with the things you see around you. One proton has a mass of 1.7×10^{-24} g, and the neutron is about the same. However, the electron has a mass 9.1×10^{-28} g, which is about 1/2000th of the mass of either a proton or neutron. Because the masses of subatomic particles are so small, chemists use a very small unit of mass called an **atomic mass unit (amu)**. An amu is defined as one-twelfth of the mass of a carbon atom which has a nucleus containing six protons and six neutrons. In biology, the atomic mass unit is called a *Dalton* (Da) in honor of John Dalton. On the amu scale, the proton and neutron each have a mass of about 1 amu. Table 3.5 summarizes some information about the subatomic particles in an atom.

TABLE 3.5 Subatomic Particles in the Atom

Particle	Symbol	Charge	Mass (amu)	Location in Atom
Proton	p or p^+	1+	1.007	Nucleus
Neutron	n or n^0	0	1.008	Nucleus
Electron	e^-	1−	0.000 55	Outside nucleus

CONCEPT CHECK 3.3

Identifying Subatomic Particles

Is each of the following statements true or false?

a. Protons are heavier than electrons.
b. Protons are attracted to neutrons.
c. Electrons are so small that they have no electrical charge.
d. The nucleus contains all the protons and neutrons of an atom.

ANSWER

a. True
b. False; protons are attracted to electrons.
c. False; electrons have a 1− charge.
d. True

SAMPLE PROBLEM 3.5

Identifying Subatomic Particles

Identify the subatomic particle that has the following characteristics:

a. no charge
b. a mass of 0.000 55 amu
c. a mass about the same as a neutron

SOLUTION

a. neutron **b.** electron **c.** proton

STUDY CHECK 3.5

Is the following statement true or false?

The nucleus occupies a large volume in an atom.

Explore Your World

REPULSION AND ATTRACTION

Tear a small piece of paper into bits. Brush your hair several times, and place the brush just above the bits of paper. Use your knowledge of electrical charges to give an explanation for your observations. Try the same experiment with a comb.

QUESTIONS

1. What happens when objects with like charges are placed close together?
2. What happens when objects with unlike charges are placed close together?

QUESTIONS AND PROBLEMS

The Atom

3.15 Identify each of the following as describing either a proton, neutron, or electron:
a. has the smallest mass
b. has a 1+ charge
c. is found outside the nucleus
d. is electrically neutral

3.16 Identify each of the following as describing either a proton, neutron, or electron:
a. has a mass about the same as a proton
b. is found in the nucleus
c. is attracted to the protons
d. has a 1− charge

3.17 What did Rutherford determine about the structure of the atom from his gold-foil experiment?

3.18 Why does the nucleus in every atom have a positive charge?

3.19 Identify each of the following statements as true or false:
a. A proton and an electron have opposite charges.
b. The nucleus contains most of the mass of an atom.
c. Electrons repel each other.
d. A proton is attracted to a neutron.

3.20 Identify each of the following statements as true or false:
a. A proton is attracted to an electron.
b. A neutron has twice the mass of a proton.
c. Neutrons repel each other.
d. Electrons and neutrons have opposite charges.

3.21 On a dry day, your hair flies away when you brush it. How would you explain this?

3.22 Sometimes clothes removed from the dryer cling together. What kinds of charges are on the clothes?

3.4 Atomic Number and Mass Number

LEARNING GOAL

Given the atomic number and the mass number of an atom, state the number of protons, neutrons, and electrons.

All the atoms of the same element always have the same number of protons. This feature distinguishes atoms of one element from atoms of all the other elements.

Atomic Number

An **atomic number**, which is equal to the number of protons in the nucleus of an atom, is used to identify and define each element.

Atomic number = number of protons in an atom

TUTORIAL
Element Names, Symbols, and Atomic Numbers

On the inside front cover of this text is a periodic table, which gives all the elements in order of increasing atomic number. The atomic number is the whole number that appears above the symbol of each element. For example, a hydrogen atom, with atomic number 1, has 1 proton; a lithium atom, with atomic number 3, has 3 protons; an atom of carbon, with atomic number 6, has 6 protons; and gold, with atomic number 79, has 79 protons; and so forth.

An atom is electrically neutral. That means that the number of protons in an atom is equal to the number of electrons. This electrical balance gives an atom an overall charge of zero. Thus, in every atom, the atomic number also gives the number of electrons.

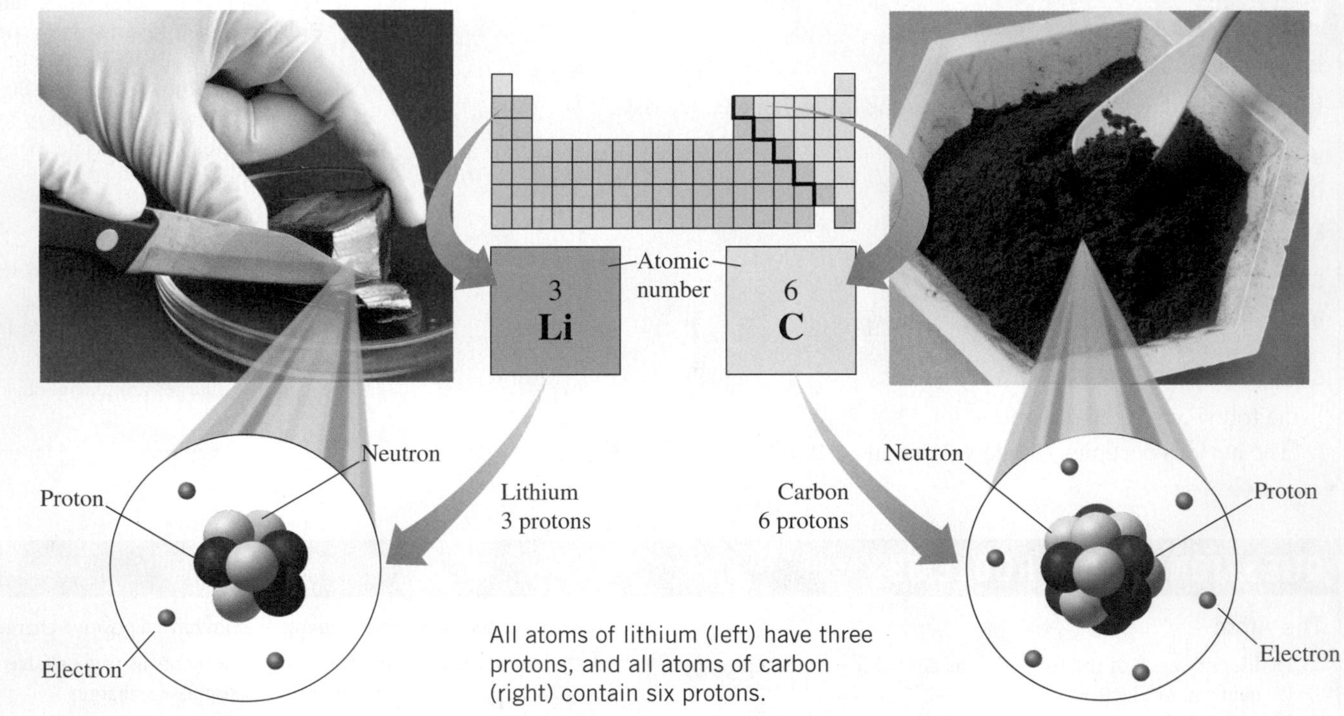

All atoms of lithium (left) have three protons, and all atoms of carbon (right) contain six protons.

SAMPLE PROBLEM 3.6

Using Atomic Number to Find the Number of Protons and Electrons

Using the periodic table in Figure 3.1, state the atomic number, number of protons, and number of electrons for an atom of each of the following elements:

a. nitrogen **b.** magnesium **c.** bromine

SOLUTION

a. atomic number 7; 7 protons and 7 electrons
b. atomic number 12; 12 protons and 12 electrons
c. atomic number 35; 35 protons and 35 electrons

STUDY CHECK 3.6

Consider an atom that has 79 electrons.

a. How many protons are in its nucleus?
b. What is its atomic number?
c. What is its name, and what is its symbol?

Mass Number

TUTORIAL
Atomic Number and Mass Number

We now know that the protons and neutrons determine the mass of the nucleus. For any atom, the **mass number** is the sum of the number of protons and neutrons in the nucleus of a single atom. Thus, the mass number is a counting number, which is always a whole number. Because mass number represents the particles in the nucleus of a single atom, it does not appear on the periodic table.

Mass number = number of protons + number of neutrons

For example, an atom of oxygen that contains 8 protons and 8 neutrons has a mass number of 16. An atom of iron that contains 26 protons and 30 neutrons has a mass number of 56. Table 3.6 illustrates the relationship between atomic number, mass number, and the number of protons, neutrons, and electrons in examples of atoms for different elements.

TABLE 3.6 Composition of Some Atoms of Different Elements

Element	Symbol	Atomic Number	Mass Number	Number of Protons	Number of Neutrons	Number of Electrons
Hydrogen	H	1	1	1	0	1
Nitrogen	N	7	14	7	7	7
Chlorine	Cl	17	37	17	20	17
Iron	Fe	26	57	26	31	26
Gold	Au	79	197	79	118	79

CONCEPT CHECK 3.4

Counting Subatomic Particles in Atoms

An atom of silver has a mass number of 109.

a. How many protons are in the nucleus?
b. How many neutrons are in the nucleus?
c. How many electrons are in the atom?

ANSWER

a. Silver (Ag) with atomic number 47 has 47 protons.
b. The number of neutrons is calculated by subtracting the number of protons from the mass number. 109 − 47 = 62 neutrons for Ag with a mass number of 109.
c. An atom is neutral, which means that the number of electrons is equal to the number of protons. An atom of silver with 47 protons has 47 electrons.

SAMPLE PROBLEM 3.7

Calculating Numbers of Protons, Neutrons, and Electrons

For an atom of zinc that has a mass number of 68, determine the following:

a. the number of protons
b. the number of neutrons
c. the number of electrons

SOLUTION

a. Zinc (Zn), with an atomic number of 30, has 30 protons.
b. The number of neutrons in this atom is found by subtracting the atomic number from the mass number.

Mass number − atomic number = number of neutrons
68 − 30 = 38

c. Because the zinc atom is neutral, the number of electrons is equal to the number of protons. A zinc atom has 30 electrons.

STUDY CHECK 3.7

How many neutrons are in the nucleus of a bromine atom that has a mass number of 80?

Career Focus

OPTICIAN

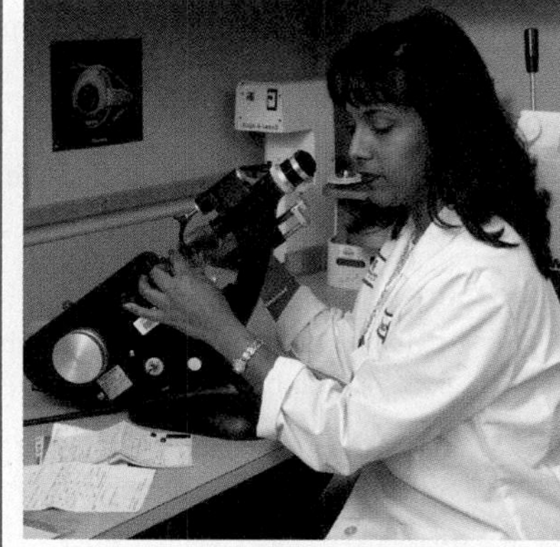

"When a patient brings in a prescription, I help select the proper lenses, put them into a frame, and fit them properly on the patient's face," says Suranda Lara, optician, Kaiser Hospital. "If a prescription requires a thinner and lighter-weight lens, we formulate that lens. So we have to understand the different materials used to make lenses. Sometimes patients come in with their own glasses that they want to convert to sunglasses. We remove the lenses and put them into a tint bath, which turns them into sunglasses." Opticians fit and adjust eyewear for patients who have had their eyesight tested by an ophthalmologist or optometrist. Optics and mathematics are used to select materials for frames and lenses that are compatible with patients' facial measurements and lifestyles.

QUESTIONS AND PROBLEMS

Atomic Number and Mass Number

3.23 Would you use the atomic number, mass number, or both to determine each of the following?
a. number of protons in an atom
b. number of neutrons in an atom
c. number of particles in the nucleus
d. number of electrons in a neutral atom

3.24 What do you know about the subatomic particles from each of the following?
a. atomic number
b. mass number
c. mass number − atomic number
d. mass number + atomic number

3.25 Write the names and symbols of the elements with the following atomic numbers:
a. 3 **b.** 9 **c.** 20 **d.** 30
e. 10 **f.** 14 **g.** 53 **h.** 8

3.26 Write the names and symbols of the elements with the following atomic numbers:
a. 1 **b.** 11 **c.** 19 **d.** 82
e. 35 **f.** 47 **g.** 15 **h.** 2

3.27 How many protons and electrons are there in a neutral atom of each of the following elements?
a. argon
b. zinc
c. iodine
d. potassium

3.28 How many protons and electrons are there in a neutral atom of each of the following elements?
a. carbon **b.** fluorine
c. calcium **d.** sulfur

3.29 Complete the following table for a neutral atom of each element:

Name of the Element	Symbol	Atomic Number	Mass Number	Number of Protons	Number of Neutrons	Number of Electrons
	Al		27			
		12			12	
Potassium					20	
				16	15	
			56			26

3.30 Complete the following table for a neutral atom of each element:

Name of the Element	Symbol	Atomic Number	Mass Number	Number of Protons	Number of Neutrons	Number of Electrons
	N		15			
Calcium			42			
				38	50	
		14			16	
		56	138			

LEARNING GOAL

Give the number of protons, electrons, and neutrons in an isotope of an element; calculate the atomic mass of an element using the abundance and mass of its naturally occurring isotopes.

SELF STUDY ACTIVITY
Atoms and Isotopes

TUTORIAL
Isotopes

3.5 Isotopes and Atomic Mass

We have seen that all atoms of the same element have the same number of protons and electrons. However, the atoms of any one element are not completely identical because they can have different numbers of neutrons.

Isotopes are atoms of the same element that have the same number of protons, but different numbers of neutrons. For example, in a large sample of magnesium atoms, there are three different types of atoms or isotopes. We already know that all the isotopes of magnesium have 12 protons. However, one isotope has 12 neutrons, another has 13 neutrons, and yet another isotope has 14 neutrons. Because the mass number of a single atom is the sum of the protons and neutrons, the three isotopes of magnesium, which have the same atomic number, will have different mass numbers.

Atomic Symbols for the Isotopes of Magnesium

To distinguish between the different isotopes of the same element, we can write an **atomic symbol** that indicates the mass number of each isotope in the upper left corner and the atomic number of the element in the lower left corner.

Mass number → $^{24}_{12}Mg$ ← Symbol of element
Atomic number

Atomic symbol for an isotope of magnesium.

An isotope may be referred to by its name or symbol followed by the mass number, such as magnesium-24 or Mg-24. Magnesium has three naturally occurring isotopes, as shown in Table 3.7. In a large sample of naturally occurring magnesium atoms, each type of isotope can be present as a low percentage or a high percentage. For example, the Mg-24 isotope makes up almost 80% of the total sample, whereas Mg-25 and Mg-26 each make up only about 10% of the total number of magnesium atoms.

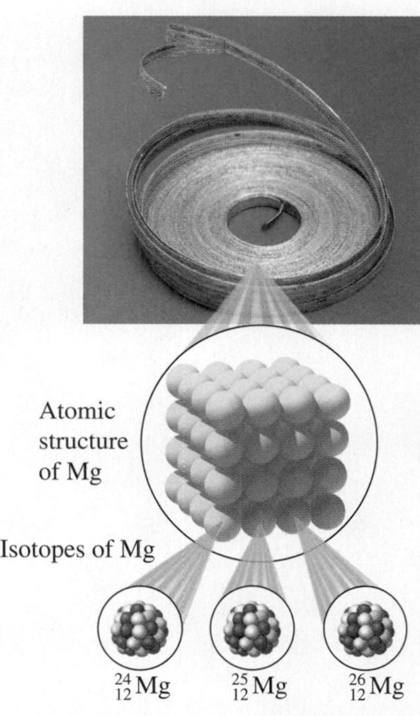

The nuclei of three naturally occurring magnesium isotopes have different numbers of neutrons.

TABLE 3.7 Isotopes of Magnesium

Atomic Symbol	$^{24}_{12}Mg$	$^{25}_{12}Mg$	$^{26}_{12}Mg$
Number of protons	12	12	12
Number of electrons	12	12	12
Mass number	**24**	**25**	**26**
Number of neutrons	**12**	**13**	**14**
Mass of isotope (amu)	23.99	24.99	25.98
% abundance	78.70%	10.13%	11.17%

SAMPLE PROBLEM 3.8

Identifying Protons and Neutrons in Isotopes

State the number of protons and neutrons for each of the following isotopes of neon (Ne):

a. $^{20}_{10}Ne$ **b.** $^{21}_{10}Ne$ **c.** $^{22}_{10}Ne$

SOLUTION

The atomic number of Ne is 10, which means that the nucleus of each isotope has 10 protons. The number of neutrons in each isotope is found by subtracting the atomic number (10) from each of their mass numbers.

a. 10 protons; 10 neutrons (20 – 10)
b. 10 protons; 11 neutrons (21 – 10)
c. 10 protons; 12 neutrons (22 – 10)

STUDY CHECK 3.8

Write an atomic symbol for each of the following:

a. a nitrogen atom with 8 neutrons
b. an atom with 20 protons and 22 neutrons
c. an atom with mass number 27 and 14 neutrons

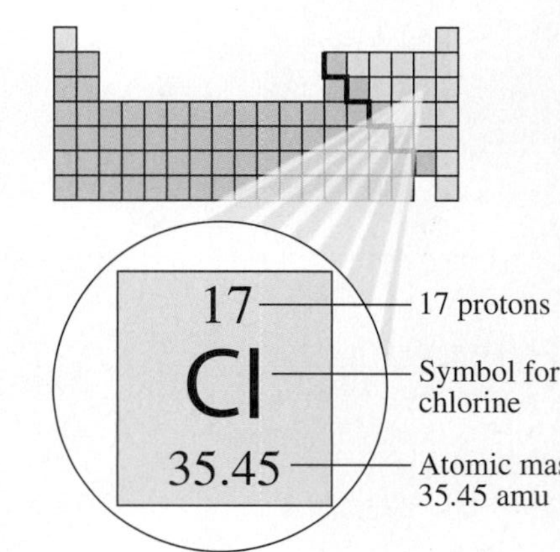

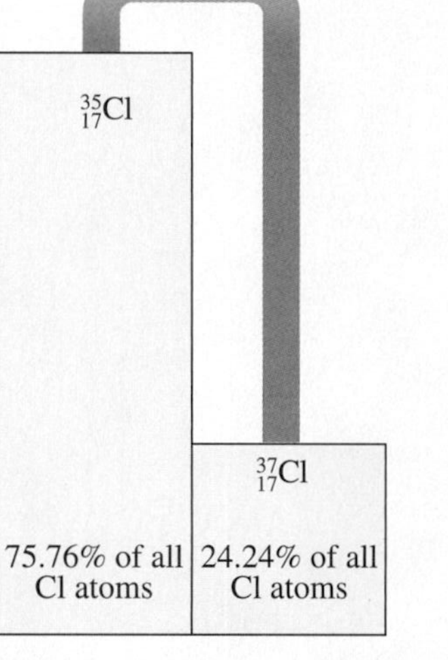

Chlorine, with two naturally occurring isotopes, has an atomic mass of 35.45 amu.

Atomic Mass

In laboratory work, a chemist generally uses samples with many atoms that contain all the different atoms or isotopes of an element. Because each isotope has a different mass, chemists have calculated an **atomic mass** for an "average atom," which is a *weighted average* of the masses of all the naturally occurring isotopes of that element. On the periodic table, the atomic mass is the number including decimal places that is given below the symbol of each element. Most elements consist of two or more isotopes, which is one reason that the atomic masses on the periodic table are seldom whole numbers.

TUTORIAL
Atomic Mass Calculations

Calculating Atomic Mass

To calculate the atomic mass of an element, the percentage abundance of each isotope and its mass must be determined experimentally. For example, a large sample of naturally occurring chlorine atoms consists of 75.76% of ${}^{35}_{17}Cl$ atoms and 24.24% of ${}^{37}_{17}Cl$ atoms. The atomic mass is calculated using the percentage of each isotope and its mass: The ${}^{35}_{17}Cl$ isotope has a mass of 34.97 amu and the ${}^{37}_{17}Cl$ isotope has a mass of 36.97 amu.

$$\text{Atomic mass of Cl} = \underbrace{\text{mass of } {}^{35}_{17}\text{Cl} \times \frac{{}^{35}_{17}\text{Cl}\%}{100\%}}_{\text{amu from } {}^{35}_{17}\text{Cl}} + \underbrace{\text{mass of } {}^{37}_{17}\text{Cl} \times \frac{{}^{37}_{17}\text{Cl}\%}{100\%}}_{\text{amu from } {}^{37}_{17}\text{Cl}}$$

Isotope	Mass (amu)	×	Abundance (%)	=	Contribution to Average Cl Atom
${}^{35}_{17}Cl$	34.97	×	$\frac{75.76}{100}$	=	26.49 amu
${}^{37}_{17}Cl$	36.97	×	$\frac{24.24}{100}$	=	8.962 amu
			Atomic mass of Cl	=	35.45 amu

The atomic mass of 35.45 amu is the weighted average mass of a sample of Cl atoms, although no individual Cl atom actually has this mass. An atomic mass of 35.45 for chlorine also indicates that there is a higher percentage of ${}^{35}_{17}Cl$ atoms because the atomic mass of 35.45 is closer to the mass number of Cl-35. In fact, there are about three atoms of ${}^{35}_{17}Cl$ for every one atom of ${}^{37}_{17}Cl$ in a sample of chlorine atoms.

Table 3.8 lists the naturally occurring isotopes of some selected elements and their atomic masses.

TABLE 3.8 The Atomic Mass of Some Elements

Element	Isotopes	Atomic Mass (weighted average)	Most Prevalent Isotope
Lithium	${}^{6}_{3}Li$, ${}^{7}_{3}Li$	6.941 amu	${}^{7}_{3}Li$
Carbon	${}^{12}_{6}C$, ${}^{13}_{6}C$, ${}^{14}_{6}C$	12.01 amu	${}^{12}_{6}C$
Oxygen	${}^{16}_{8}O$, ${}^{17}_{8}O$, ${}^{18}_{8}O$	16.00 amu	${}^{16}_{8}O$
Fluorine	${}^{19}_{9}F$	19.00 amu	${}^{19}_{9}F$
Sulfur	${}^{32}_{16}S$, ${}^{33}_{16}S$, ${}^{34}_{16}S$, ${}^{36}_{16}S$	32.07 amu	${}^{32}_{16}S$
Copper	${}^{63}_{29}Cu$, ${}^{65}_{29}Cu$	63.55 amu	${}^{63}_{29}Cu$

CONCEPT CHECK 3.5

Average Atomic Mass

Neon consists of three naturally occurring isotopes: ${}^{20}_{10}Ne$, ${}^{21}_{10}Ne$, and ${}^{22}_{10}Ne$. Using the atomic mass on the periodic table, which isotope of neon is likely to be the most prevalent?

ANSWER

Using the periodic table, we find that the atomic mass for all the naturally occurring isotopes of neon is 20.18 amu. Since this number is very close to 20, the isotope Ne-20 that has a mass number of 20 is the most prevalent isotope in a naturally occurring sample of neon atoms.

SAMPLE PROBLEM 3.9

Calculating Atomic Mass

Using Table 3.7, calculate the atomic mass for magnesium using the weighted average mass method.

SOLUTION

Isotope	Mass (amu)		Abundance (%)		Contribution to the Atomic Mass
$^{24}_{12}\text{Mg}$	23.99	×	$\frac{78.70}{100}$	=	18.88 amu
$^{25}_{12}\text{Mg}$	24.99	×	$\frac{10.13}{100}$	=	2.531 amu
$^{26}_{12}\text{Mg}$	25.98	×	$\frac{11.17}{100}$	=	2.902 amu
	Atomic mass of Mg			=	24.31 amu (weighted average mass)

STUDY CHECK 3.9

There are two naturally occurring isotopes of boron. The isotcpe $^{10}_{5}\text{B}$ has a mass of 10.01 amu with an abundance of 19.80%, and the isotope $^{11}_{5}\text{B}$ has a mass of 11.01 amu with an abundance of 80.20%. What is the atomic mass of boron?

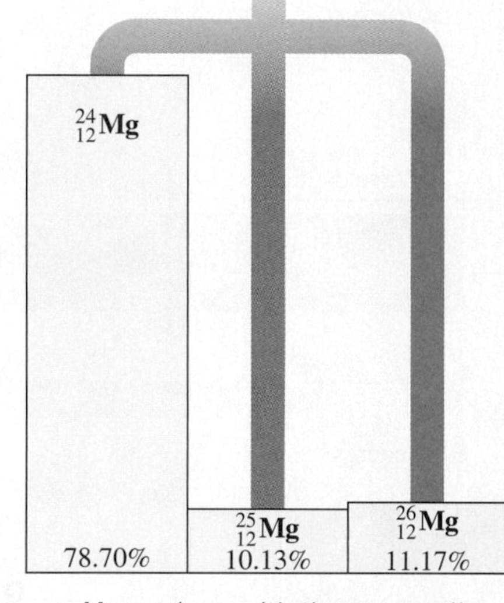

Magnesium, with three naturally occurring isotopes, has an atomic mass of 24.31 amu.

QUESTIONS AND PROBLEMS

Isotopes and Atomic Mass

3.31 What are the number of protons, neutrons, and electrons in the following isotopes?
a. $^{27}_{13}\text{Al}$ **b.** $^{52}_{24}\text{Cr}$ **c.** $^{34}_{16}\text{S}$ **d.** $^{81}_{35}\text{Br}$

3.32 What are the number of protons, neutrons, and electrons in the following isotopes?
a. $^{2}_{1}\text{H}$ **b.** $^{14}_{7}\text{N}$ **c.** $^{26}_{14}\text{Si}$ **d.** $^{70}_{30}\text{Zn}$

3.33 Write the atomic symbol for the isotope with each of the following characteristics:
a. 15 protons and 16 neutrons
b. 35 protons and 45 neutrons
c. 50 electrons and 72 neutrons
d. a chlorine atom with 18 neutrons
e. a mercury atom with 122 neutrons

3.34 Write the atomic symbol for the isotope with each of the following characteristics:
a. an oxygen atom with 10 neutrons
b. 4 protons and 5 neutrons
c. 25 electrons and 28 neutrons
d. a mass number of 24 and 13 neutrons
e. a nickel atom with 32 neutrons

3.35 There are three naturally occurring isotopes of argon, with mass numbers 36, 38, and 40.
a. Write the atomic symbol for each of these atoms.
b. How are these isotopes alike?
c. How are they different?
d. Why is the atomic mass of argon listed on the periodic table not a whole number?
e. Which isotope is the most prevalent in a sample of argon?

3.36 There are four isotopes of strontium with mass numbers 84, 86, 87, and 88.
a. Write the atomic symbol for each of these atoms.
b. How are these isotopes alike?
c. How are they different?
d. Why is the atomic mass of strontium listed on the periodic table not a whole number?
e. Which isotope is the most prevalent in a sample of strontium?

3.37 Two isotopes of gallium are naturally occurring, with $^{69}_{31}\text{Ga}$ at 60.11% (68.93 amu) and $^{71}_{31}\text{Ga}$ at 39.89% (70.92 amu). What is the atomic mass of gallium?

3.38 Two isotopes of copper are naturally occurring, with $^{63}_{29}\text{Cu}$ at 69.09% (62.93 amu) and $^{65}_{29}\text{Cu}$ at 30.91% (64.93 amu). What is the atomic mass of copper?

LEARNING GOAL

Given the name or symbol of one of the first 20 elements in the periodic table, write the electron arrangement.

TUTORIAL
Electromagnetic Radiation

TUTORIAL
Energy Levels

3.6 Electron Energy Levels

When we listen to a radio, use a microwave oven, turn on a light, see the colors of a rainbow, or have an X-ray, we are using various forms of *electromagnetic radiation.* Light and other electromagnetic radiation consist of energy particles that move as waves of energy. In a wave, just like the waves in an ocean, the distance between the peaks is called the *wavelength.* All forms of electromagnetic radiation travel in space at the speed of light, 3.0×10^8 m/s but differ in energy and wavelength. High-energy radiation has short wavelengths compared to low-energy radiation, which has longer wavelengths. The *electromagnetic spectrum* shows the arrangement of different types of electromagnetic radiation in order of increasing energy (see Figure 3.11).

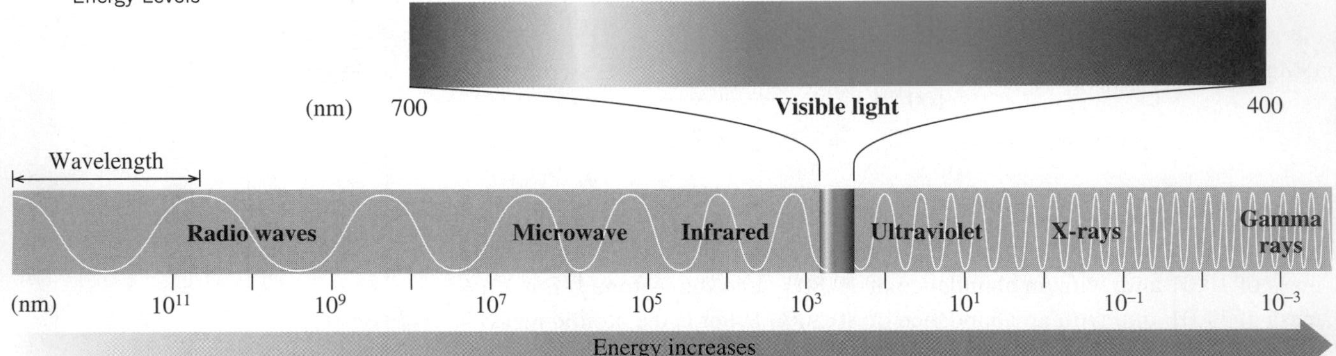

FIGURE 3.11 The electromagnetic spectrum shows the arrangement of wavelengths of electromagnetic radiation. The visible portion consists of wavelengths from 700 nm to 400 nm.

Q How does the wavelength of red light compare to that of blue light?

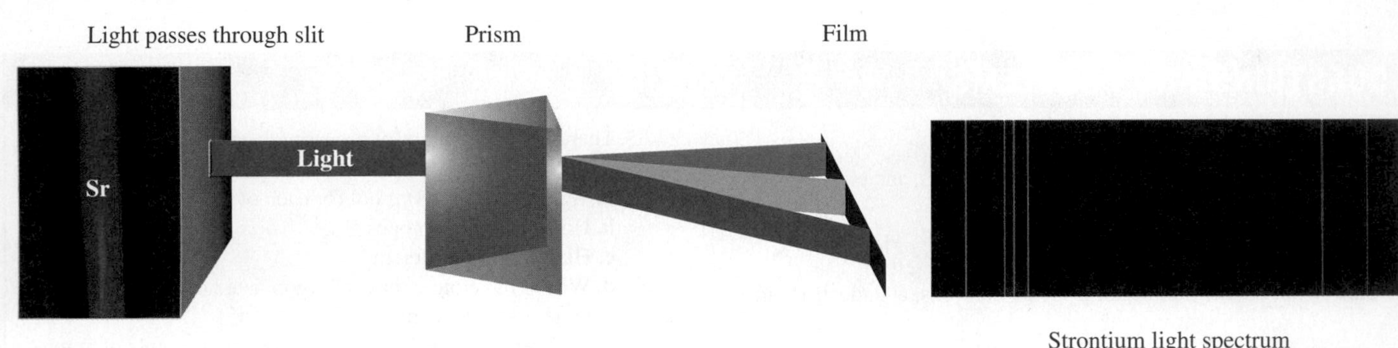

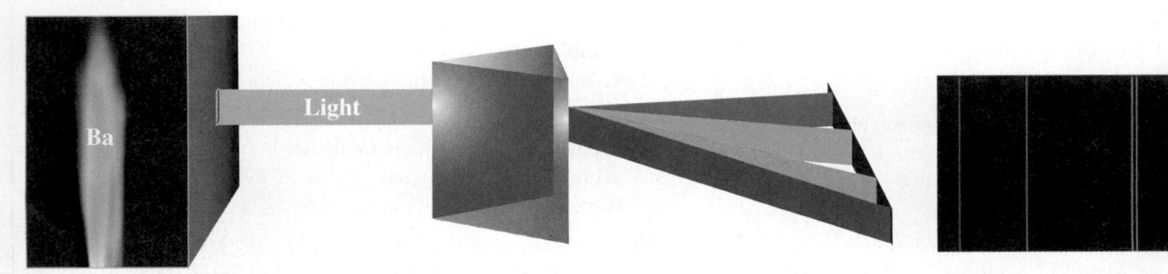

In an atomic spectrum, light from a heated element separates into distinct lines.

When the light from the Sun passes through a prism, the light separates into a continuous color spectrum, which consists of the colors we see in a rainbow. In contrast, when light from an element that is heated passes through a prism, it separates into distinct lines of color called an *atomic spectrum*. Each element has its own unique atomic spectrum.

A rainbow forms when light passes through water droplets.

Electron Energy Levels

Scientists have now determined that the lines in the atomic spectra of elements are caused by changes in the energies of the electrons. In an atom, each electron has a specific energy known as its **energy level**. All the electrons with the same energy are grouped in the same energy level. The energy levels are assigned numbers (n) beginning with $n = 1$, $n = 2$, up to $n = 7$. Electrons in the lower energy levels are closer to the nucleus, whereas electrons in the higher energy levels are farther away.

As an analogy, we can think of the energy levels of an atom as similar to the shelves in a bookcase. The first shelf is the lowest energy level; the second shelf is the second energy level, and so on. If we are arranging books on the shelves, it would take less energy to fill the bottom shelf first, and then the second shelf, and so on. However, we could never get any book to stay in the space between any of the shelves. Similarly, the energy of an electron must be at specific energy levels, and not between.

Unlike bookcases, however, there is a large difference between the energy of the first and second levels, but then the higher energy levels are closer together. Another difference is that the lower electron energy levels hold fewer electrons than the higher energy levels.

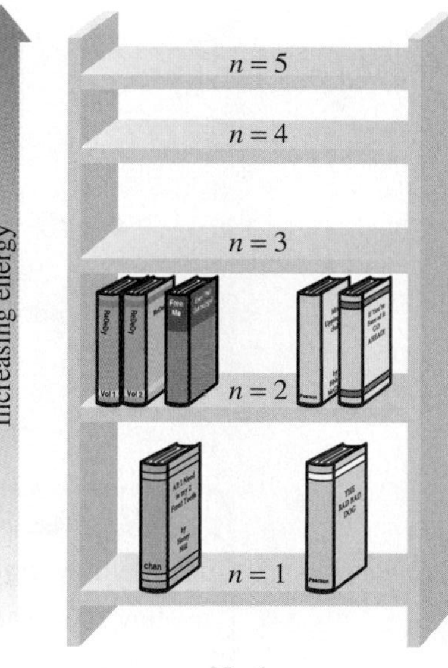

An electron can have the energy of only one of the energy levels in an atom.

Chemistry Link to the Environment

ENERGY-SAVING FLUORESCENT BULBS

A compact fluorescent light bulb (CFL) is replacing the standard light bulb we use in our homes and workplaces. Compared to a standard light bulb, the CFL has a longer life and uses less electricity. Within about 20 days of use, the CFL saves enough money in electricity costs to pay for its higher initial cost.

A standard incandescent light bulb has a thin tungsten filament inside a sealed glass bulb. When the light is switched on, electricity flows through this filament, and electrical energy is converted to heat energy. When the filament reaches a temperature of about 2300 °C, we see white light.

A fluorescent bulb produces light in a different way. When the switch is turned on, electrons move between two electrodes and collide with mercury atoms in a mixture of mercury and argon gas inside the light. When the electrons in the mercury atoms absorb energy from the collisions, they are raised to higher energy levels. As electrons fall to lower energy levels, energy in the ultraviolet range is emitted. This ultraviolet light strikes the phosphor coating inside the tube, and fluorescence occurs as visible light is emitted.

The production of light in a fluorescent bulb is more efficient than in an incandescent light bulb. A 75-watt incandescent bulb can be replaced by a 20-watt CFL that gives the same amount of light, providing a 70% reduction in electricity costs. A typical incandescent light bulb lasts for one to two months, whereas a CFL lasts from one to two years. One drawback of the CFL is that each contains about 4 mg of mercury. As long as the bulb stays intact, no mercury is released. However, used CFL bulbs should not be disposed of in household trash but rather should be taken to a recycling center.

A compact fluorescent light bulb (CFL) uses up to 70% less energy.

Changes in Electron Energy Level

An electron can change from one energy level to a higher level only if it absorbs the energy equal to the difference between two levels. When an electron changes to a lower energy level, it emits energy equal to the difference between the two levels (see Figure 3.12). If the energy emitted is in the visible range, we see one of the colors of visible light. The yellow color of sodium streetlights and the red color of neon lights are examples of electrons emitting energy in the visible color range.

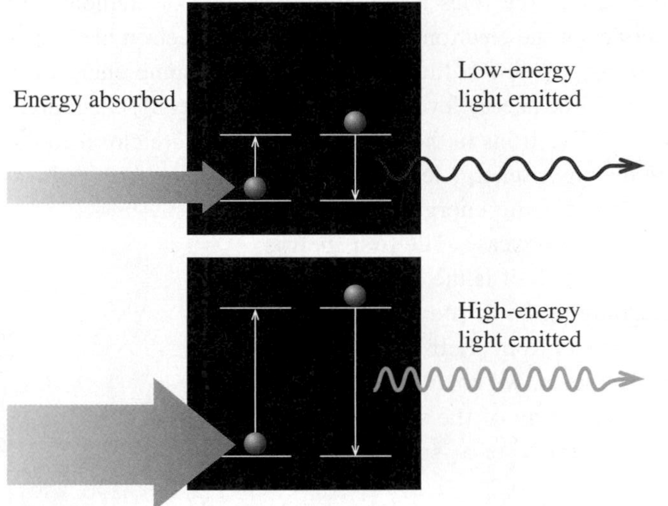

FIGURE 3.12 Electrons absorb a specific amount of energy to move to a higher energy level. When electrons lose energy, a specific quantity of energy is emitted.
Q What causes electrons to move to higher energy levels?

CONCEPT CHECK 3.6

Change in Energy Levels

a. How does an electron move to a higher energy level?
b. When an electron drops to a lower energy level, how is energy lost?

ANSWER

a. An electron moves to a higher energy level when it absorbs an amount of energy equal to the difference in energy levels.
b. Energy equal to the difference in energy levels is emitted when an electron drops to a lower energy level.

Electron Arrangements for the First 20 Elements

SELF STUDY ACTIVITY
Bohr's Shell Model of the Atom

The *electron arrangement* of an atom gives the number of electrons in each energy level. We can write the electron arrangements for the first 20 elements by placing electrons in energy levels beginning with the lowest. There is a limit to the number of electrons allowed in each energy level. Only a few electrons can occupy the lower energy levels, while more electrons can be accommodated in higher energy levels. We can now look at the numbers of electrons in the first four energy levels for the first 20 elements as shown in Table 3.9.

Period 1

H	1
He	2

The single electron of hydrogen goes into energy level 1, and the two electrons of helium fill energy level 1. Thus, energy level 1 can hold just two electrons. The electron arrangements for H and He are shown in the margin.

TABLE 3.9 Electron Arrangements for the First 20 Elements

Element	Symbol	Atomic Number	Number of Electrons in Energy Level			
			1	2	3	4
Hydrogen	H	1	1			
Helium	He	2	2			
Lithium	Li	3	2	1		
Beryllium	Be	4	2	2		
Boron	B	5	2	3		
Carbon	C	6	2	4		
Nitrogen	N	7	2	5		
Oxygen	O	8	2	6		
Fluorine	F	9	2	7		
Neon	Ne	10	2	8		
Sodium	Na	11	2	8	1	
Magnesium	Mg	12	2	8	2	
Aluminum	Al	13	2	8	3	
Silicon	Si	14	2	8	4	
Phosphorus	P	15	2	8	5	
Sulfur	S	16	2	8	6	
Chlorine	Cl	17	2	8	7	
Argon	Ar	18	2	8	8	
Potassium	K	19	2	8	8	1
Calcium	Ca	20	2	8	8	2

Period 2

For the elements of the second period (lithium, Li, to neon, Ne), we fill the first energy level with two electrons, and place the remaining electrons in the second energy level. For example, lithium has three electrons. Two of those electrons fill energy level 1. The remaining electron goes into the second energy level. We can write this electron arrangement as 2,1. Going across Period 2, more electrons are added to the second energy level. For example, an atom of carbon, with a total of six electrons, fills energy level 1, which leaves the four remaining electrons in the second energy level. The electron arrangement for carbon can be written 2,4. For neon, the last element in Period 2, both the first and second energy levels are filled to give an electron arrangement of 2,8.

Lithium	2,1
Carbon	2,4
Neon	2,8

Period 3

For sodium, the first element in Period 3, 10 electrons fill the first and second energy levels, which means that the remaining electron must go into the third energy level. The electron arrangement for sodium can be written 2,8,1. The elements that follow sodium in the third period continue to add electrons, one at a time, until the third level is complete. For example, a sulfur atom with 16 electrons fills the first and second energy levels, which leaves 6 electrons in the third level. The electron arrangement for sulfur can be written 2,8,6. At the end of Period 3, argon has 8 electrons in the third level. Once energy level 3 has 8 electrons, it stops filling. However, 10 more electrons will be added later.

Sodium	2,8,1
Sulfur	2,8,6
Argon	2,8,8

Period 4

For potassium, the first element in Period 4, electrons fill the first and second energy levels, and 8 electrons are in the third energy level. The remaining electron of potassium enters energy level 4, which gives the electron arrangement of 2,8,8,1. For calcium, the process is similar, except that calcium has 2 electrons in the fourth energy level. The electron arrangement for calcium can be written 2,8,8,2.

Potassium	2,8,8,1
Calcium	2,8,8,2

Elements Beyond 20

Following calcium, the next 10 electrons (scandium to zinc) are added to the third energy level, which already has 8 electrons, until it is complete with 18 electrons. The higher energy levels 5, 6, and 7 can theoretically accommodate up to 50, 72, and 98 electrons, but they are not completely filled. Beyond the first 20 elements, the electron arrangements become more complicated and will not be covered in this text.

Energy Level (n)	1	2	3	4	5	6	7
Number of Electrons	2	8	18	32	32	18	8

SAMPLE PROBLEM 3.10

Writing Electron Arrangements

Write the electron arrangement for each of the following:

a. oxygen **b.** chlorine

SOLUTION

a. Oxygen with atomic number 8, has eight electrons, which are arranged with two electrons in energy level 1 and six electrons in energy level 2.

2,6

b. Chlorine with atomic number 17, has 17 electrons, which are arranged with 2 electrons in energy level 1, 8 electrons in energy level 2, and 7 electrons in energy level 3.

2,8,7

STUDY CHECK 3.10

What element has an electron arrangement of 2,8,5?

QUESTIONS AND PROBLEMS

Electron Energy Levels

3.39 Electrons can move to higher energy levels when they ________ (absorb/emit) energy.

3.40 Electrons drop to lower energy levels when they ________ (absorb/emit) energy.

3.41 Identify the form of electromagnetic radiation in each pair that has the greater energy:
a. green light or yellow light
b. microwaves or blue light

3.42 Identify the form of electromagnetic radiation in each pair that has the greater energy:
a. radio waves or violet light
b. infrared light or ultraviolet light

3.43 Write the electron arrangement for each of the following elements:
(*Example*: sodium 2,8,1)
a. carbon **b.** argon **c.** potassium
d. silicon **e.** helium **f.** nitrogen

3.44 Write the electron arrangement for each of the following atoms:
(*Example*: sodium 2,8,1)
a. phosphorus **b.** neon **c.** sulfur
d. magnesium **e.** aluminum **f.** fluorine

3.45 Identify the elements that have the following electron arrangements:

	Energy Level			
	1	**2**	**3**	**4**
a.	2	1		
b.	2	8	2	
c.	1			
d.	2	8	7	
e.	2	6		

3.46 Identify the elements that have the following electron arrangements:

	Energy Level			
	1	**2**	**3**	**4**
a.	2	5		
b.	2	8	6	
c.	2	4		
d.	2	8	8	1
e.	2	8	3	

Chemistry Link to Health

BIOLOGICAL REACTIONS TO UV LIGHT

Our everyday life depends on sunlight, but exposure to sunlight can have damaging effects on living cells, and too much exposure can even cause their death. The light energy, especially ultraviolet (UV), excites electrons and may lead to unwanted chemical reactions. The list of damaging effects of sunlight includes sunburn; wrinkling; premature aging of the skin; changes in the DNA of the cells, which can lead to skin cancers; inflammation of the eyes; and perhaps cataracts. Some drugs, like the acne medications Accutane and Retin-A, as well as antibiotics, diuretics, sulfonamides, and estrogen, make the skin extremely sensitive to light.

However, medicine does take advantage of the beneficial effect of sunlight. Phototherapy can be used to treat certain skin conditions including psoriasis, eczema, and dermatitis. In the treatment of psoriasis, for example, oral drugs are given to make the skin more photosensitive; exposure to UV follows. Low-energy light is used to break down bilirubin in neonatal jaundice. Sunlight is also a factor in stimulating the immune system.

In cutaneous T-cell lymphoma, an abnormal increase in T cells causes painful ulceration of the skin. The skin is treated by photophoresis, in which the patient receives a photosensitive chemical, and then blood is removed from the body and exposed to ultraviolet light. The blood is returned to the patient, and the treated T cells stimulate the immune system to respond to the cancer cells.

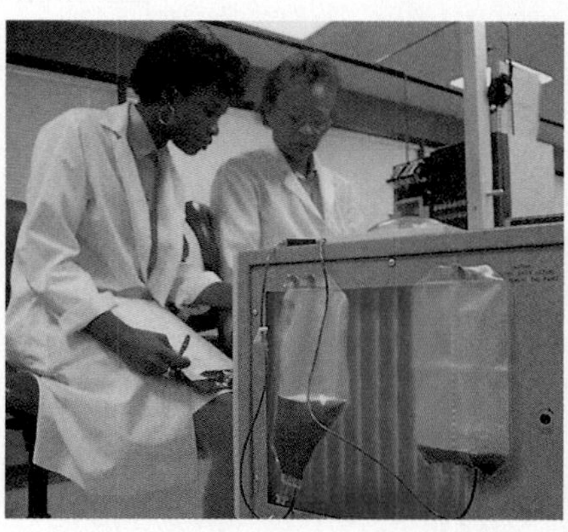

In a disorder called seasonal affective disorder or SAD, people experience mood swings and depression during the winter. Some research suggests that SAD is the result of a decrease of serotonin, or an increase in melatonin, when there are fewer hours of sunlight. One treatment for SAD is therapy using bright light provided by a lamp called a light box. A daily exposure to intense light for 30 to 60 minutes seems to reduce symptoms of SAD.

3.7 Trends in Periodic Properties

LEARNING GOAL

Use the electron arrangement of elements to explain the trends in periodic properties.

TUTORIAL
Periodic Trends

The electron arrangements of atoms are an important factor in the physical and chemical properties of the elements. Now we will look at the *valence electrons* in atoms, *atomic size*, and *ionization energy*. Known as *periodic properties*, each increases or decreases across a period, and then the trend is repeated again in each successive period.

Group Number and Valence Electrons

The chemical properties of representative elements in Groups 1A (1) to 8A (18) are mostly due to the **valence electrons**, which are the electrons in the outermost energy level. The **group number** gives the number of valence electrons for each group of representative elements. For example, all the elements in Group 1A (1) have one valence electron. All the elements in Group 2A (2) have two valence electrons. The halogens in Group 7A (17) all have seven valence electrons. Table 3.10 shows how the number of valence electrons for common representative elements is consistent with the group number.

CONCEPT CHECK 3.7

Using Group Numbers

How can we determine that strontium has two valence electrons without writing its electron arrangement?

ANSWER

Strontium is in Group 2A (2). Because the group number is the same as the number of electrons in the outermost energy level, strontium would have two valence electrons.

TABLE 3.10 Comparison of Electron Arrangements, by Group, for Some Representative Elements

			Number of Electrons in Energy Level			
Group Number	**Element**	**Symbol**	**1**	**2**	**3**	**4**
1A (1)	Lithium	Li	2	**1**		
	Sodium	Na	2	8	**1**	
	Potassium	K	2	8	8	**1**
2A (2)	Beryllium	Be	2	**2**		
	Magnesium	Mg	2	8	**2**	
	Calcium	Ca	2	8	8	**2**
3A (13)	Boron	B	2	**3**		
	Aluminum	Al	2	8	**3**	
	Gallium	Ga	2	8	18	**3**
4A (14)	Carbon	C	2	**4**		
	Silicon	Si	2	8	**4**	
	Germanium	Ge	2	8	18	**4**
5A (15)	Nitrogen	N	2	**5**		
	Phosphorus	P	2	8	**5**	
	Arsenic	As	2	8	18	**5**
6A (16)	Oxygen	O	2	**6**		
	Sulfur	S	2	8	**6**	
	Selenium	Se	2	8	18	**6**
7A (17)	Fluorine	F	2	**7**		
	Chlorine	Cl	2	8	**7**	
	Bromine	Br	2	8	18	**7**
8A (18)	Helium	He	**2**			
	Neon	Ne	2	**8**		
	Argon	Ar	2	8	**8**	
	Krypton	Kr	2	8	18	**8**

SAMPLE PROBLEM 3.11

Using Group Numbers

Using the periodic table, write the group number and the number of valence electrons for each of the following elements:

a. cesium **b.** iodine

SOLUTION

a. Cesium (Cs) is in Group 1A (1); cesium has one valence electron.
b. Iodine (I) is in Group 7A (17); iodine has seven valence electrons.

STUDY CHECK 3.11

What is the group number of elements with atoms that have five valence electrons?

Electron-Dot Symbols

TUTORIAL
Electron-Dot Symbols for Elements

An **electron-dot symbol**, also known as a Lewis structure, represents the valence electrons as dots that are placed on the sides, top, or bottom of the symbol for the element. One to four valence electrons are arranged as single dots. When an atom has five to eight valence electrons, one or more electrons are paired. Any of the following would be an acceptable electron-dot symbol for magnesium, which has two valence electrons:

Possible Electron-Dot Symbols for the Two Valence Electrons in Magnesium

Ṁg· Ṁg ·Mg ·Mg· Mg· ·Mg

Electron-dot symbols for selected elements are given in Table 3.11.

Increasing number of valence electrons

Same number of valence electrons

TABLE 3.11 Electron-Dot Symbols for Selected Elements in Periods 1–4

	Group Number							
	1A (1)	2A (2)	3A (13)	4A (14)	5A (15)	6A (16)	7A (17)	8A (18)
Number of Valence Electrons	1	2	3	4	5	6	7	8*
Electron-Dot Symbol	H·							He:
	Li·	Be·	·Ḃ·	·Ċ·	·N̈·	·Ö:	·F̤̈:	:N̤̈e:
	Na·	Ṁg·	·Ȧl·	·Ṡi·	·P̈·	·S̈:	·C̤̈l:	:Ä̤r:
	K·	Ċa·	·Ġa·	·Ġe·	·Äs·	·S̈e:	·B̤̈r:	:K̤̈r:

*Helium (He) is stable with 2 valence electrons.

SAMPLE PROBLEM 3.12

Writing Electron-Dot Symbols

Write the electron-dot symbol for each of the following:

a. bromine **b.** aluminum

SOLUTION

a. Because the group number for bromine is 7A (17), bromine has seven valence electrons, which are drawn as seven dots, three pairs and one single dot, around the symbol Br.

·B̤̈r:

b. Aluminum, in Group 3A (13), has three valence electrons, which are shown as three single dots around the symbol Al.

·Ȧl·

STUDY CHECK 3.12

What is the electron-dot symbol for phosphorus?

Atomic Size

The size of an atom is determined by its *atomic radius*, which is the distance of the outermost electrons from the nucleus. For each group of representative elements, the atomic size *increases* from the top to the bottom because the outermost electrons in each energy level are farther from the nucleus. For example, in Group 1A (1), Li has a valence electron in energy level 2; Na has a valence electron in energy level 3; and K has a valence electron in energy level 4. This means that a K atom is larger than a Na atom, and a Na atom is larger than a Li atom (see Figure 3.13).

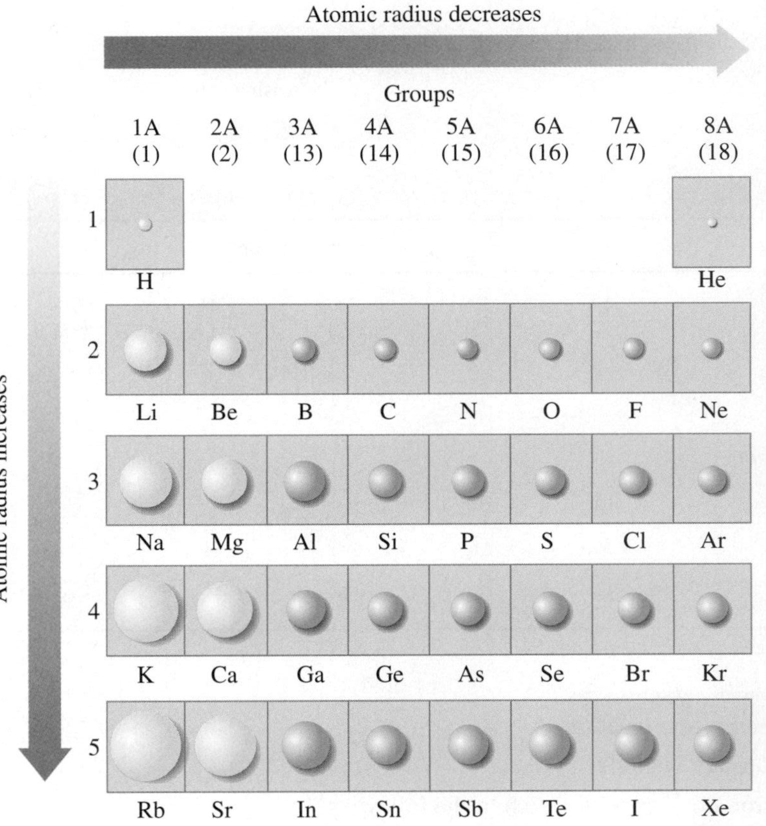

FIGURE 3.13 For representative elements, the atomic size increases going down a group but decreases going from left to right across a period.

Q Why does the atomic size increase going down a group of representative elements?

For the elements in a period, an increase in the number of protons in the nucleus increases the attraction for the outermost electrons. As a result, the outer electrons are pulled closer to the nucleus, which means that the size of representative elements decreases going from left to right across a period.

TUTORIAL
Ionization Energy

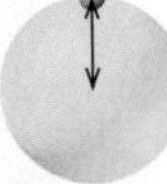

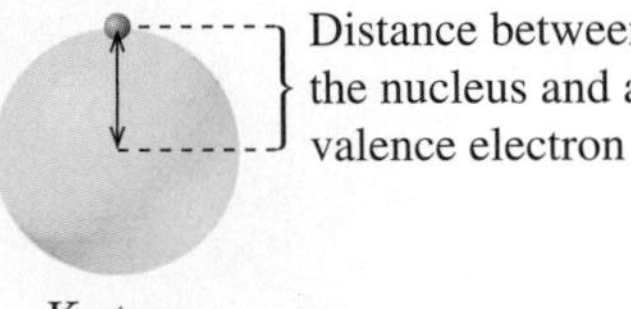

As the distance from a valence electron to the nucleus increases in Group 1A (1), the ionization energy decreases.

Ionization Energy

In an atom, negatively charged electrons are attracted to the positive charge of the protons in the nucleus. Thus, a quantity of energy known as the **ionization energy** is required to remove one of the outermost electrons. When an electron is removed from a neutral atom, a positive particle called a cation with a 1+ charge is formed.

$$Na(g) + \text{energy (ionization)} \longrightarrow Na^+(g) + e^-$$

The ionization energy *decreases* going down a group. Less energy is needed to remove an electron because nuclear attraction decreases when electrons are farther from the nucleus. Going across a period from left to right, the ionization energy *increases*. As the positive charge of the nucleus increases, more energy is needed to remove an electron.

In Period 1, the valence electrons are close to the nucleus and strongly held. H and He have high ionization energies because a large amount of energy is required to remove

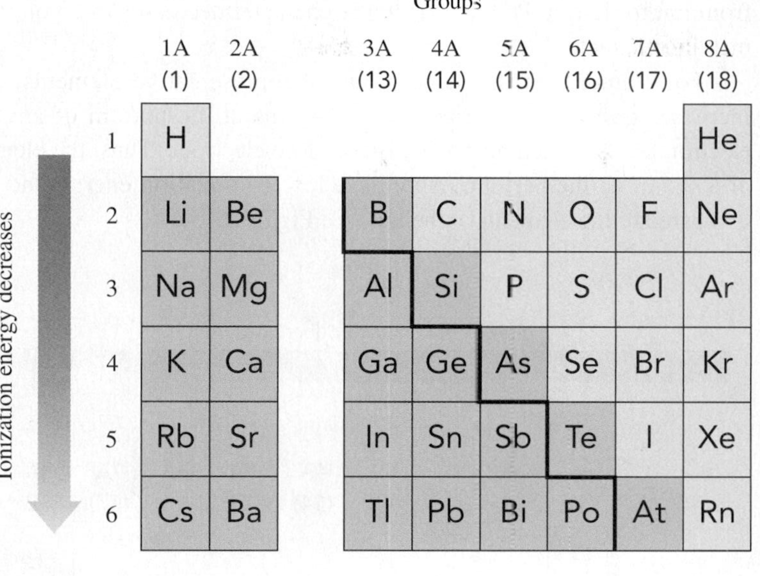

FIGURE 3.14 Ionization energies for the representative elements tend to decrease going down a group and increase going across a period.

Q Why is the ionization energy for F greater than for Cl?

an electron. The ionization energy for He is the highest of any element because He has a full, stable, energy level which is disrupted by removing an electron. The high ionization energies of the noble gases indicate that their electron arrangements are especially stable. In general, the ionization energy is low for the metals and high for the nonmetals (see Figure 3.14).

SAMPLE PROBLEM 3.13

Ionization Energy

Indicate the element in each group that has the higher ionization energy and explain your choice.

a. K or Na **b.** Mg or Cl **c.** F, N, or C

SOLUTION

a. Na. In Na, the valence electron is closer to the nucleus.

b. Cl. Attraction for the valence electrons increases going from left to right across a period.

c. F. Attraction for the valence electrons increases going from left to right across a period.

STUDY CHECK 3.13

Arrange Sn, Sr, and I in order of increasing ionization energy.

Metallic Character

TUTORIAL
Patterns in the Periodic Table

In Section 3.2, we identified elements as metals, nonmetals, and metalloids. An element that has **metallic character** is an element that loses valence electrons easily. Metallic character is more prevalent in the elements (metals) on the left side of the period table and decreases going from the left side to the right side of the periodic table. The elements (nonmetals) on the right side of the periodic table do not easily lose electrons, which means they are the least metallic. Most of the metalloids between the metals and nonmetals tend to lose electrons, but not as easily as the metals. Thus, in Period 3, sodium, which loses electrons most easily, would be the most metallic. Going across from left to right in Period 3, metallic character decreases to argon, which has the least metallic character.

For elements in the same group of representative elements, metallic character increases going from top to bottom. Atoms at the bottom of any group have more electron levels, which makes it easier to lose electrons. Thus, the elements at the bottom of a group on the periodic table have lower ionization energy and are more metallic compared to the elements at the top (see Figure 3.15).

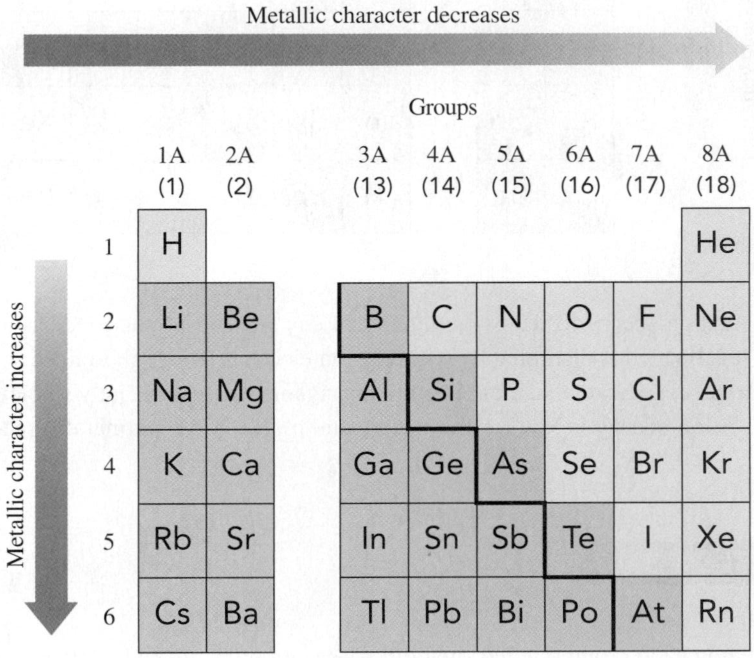

Metallic character decreases

Groups

Metallic character increases

	1A (1)	2A (2)	3A (13)	4A (14)	5A (15)	6A (16)	7A (17)	8A (18)
1	H							He
2	Li	Be	B	C	N	O	F	Ne
3	Na	Mg	Al	Si	P	S	Cl	Ar
4	K	Ca	Ga	Ge	As	Se	Br	Kr
5	Rb	Sr	In	Sn	Sb	Te	I	Xe
6	Cs	Ba	Tl	Pb	Bi	Po	At	Rn

FIGURE 3.15 Metallic character of the representative elements increases going down a group and decreases going from left to right across a period.

Q Why is the metallic character greater for Rb than for Li?

A summary of the trends in periodic properties we have discussed is given in Table 3.12.

TABLE 3.12 Summary of Trends in Periodic Properties of Representative Elements

Periodic Property	Top to Bottom of a Group	Left to Right across a Period
Valence Electrons	Remains the same	Increases
Atomic Radius	Increases due to the increase in number of energy levels	Decreases due to the increase of protons in the nucleus that pull electrons closer
Ionization Energy	Decreases because outer electrons are easier to remove when they are farther away from the nucleus	Increases as the attraction of the protons for outer electrons requires more energy to remove an electron
Metallic Character	Increases because outer electrons are easier to remove when they are farther away from the nucleus	Decreases as the attraction of the protons makes it more difficult to remove an electron

CONCEPT CHECK 3.8

Metallic Character

Identify the element that has more metallic character in each of the following:

a. Mg or Al **b.** Na or K

ANSWER

a. Mg is more metallic than Al because metallic character decreases from left to right across a period.

b. K is more metallic than Na because metallic character increases going down a group.

QUESTIONS AND PROBLEMS

Trends in Periodic Properties

3.47 What is the group number and number of valence electrons for each of the following elements?
a. magnesium **b.** chlorine **c.** oxygen
d. nitrogen **e.** barium **f.** bromine

3.48 What is the group number and number of valence electrons for each of the following elements?
a. lithium **b.** silicon **c.** neon
d. argon **e.** tin **f.** boron

3.49 Write the group number and electron-dot symbol for each of the following elements:
a. sulfur **b.** nitrogen **c.** calcium
d. sodium **e.** gallium

3.50 Write the group number and electron-dot symbol for each of the following elements:
a. carbon **b.** oxygen **c.** argon
d. lithium **e.** chlorine

3.51 Place the elements in each set in order of decreasing atomic radius.
a. Al, Si, Mg **b.** Cl, I, Br **c.** Sr, Sb, I **d.** P, Si, Na

3.52 Place the elements in each set in order of decreasing atomic radius.
a. Cl, S, P **b.** Ge, Si, C **c.** Ba, Ca, Sr **d.** S, O, Se

3.53 Select the larger atom in each pair.
a. Na or Cl **b.** Na or Rb **c.** Na or Mg **d.** Rb or I

3.54 Select the larger atom in each pair.
a. S or Cl **b.** S or O **c.** S or Se **d.** S or Mg

3.55 Arrange each set of elements in order of increasing ionization energy.
a. F, Cl, Br **b.** Na, Cl, Al **c.** Na, K, Cs **d.** As, Ca, Br

3.56 Arrange each set of elements in order of increasing ionization energy.
a. O, N, C **b.** S, P, Cl **c.** As, P, N **d.** Al, Si, P

3.57 Select the element in each pair with the higher ionization energy.
a. Br or I **b.** Mg or Sr **c.** Si or P **d.** I or Xe

3.58 Select the element in each pair with the higher ionization energy.
a. O or Ne **b.** K or Br **c.** Ca or Ba **d.** N or O

3.59 Fill in the following blanks using larger or smaller, more metallic or less metallic. Na has a ________ atomic size and is ________ than P.

3.60 Fill in the following blanks using larger or smaller, lower or higher. Mg has a ________ atomic size and a ________ ionization energy than Cs.

3.61 Place the following in order of decreasing metallic character:
Br, Ge, Ca, Ga

3.62 Place the following in order of increasing metallic character:
Mg, P, Al, Ar

3.63 Fill in each of the following blanks using higher or lower, more or less:
Sr has a ________ ionization energy and ________ metallic character than Sb.

3.64 Fill in each of the following blanks using higher or lower, more or less:
N has a ________ ionization energy and ________ metallic character than As.

3.65 Complete each of the statements **a** through **d** using 1, 2, or 3:
1. decreases **2.** increases **3.** remains the same
Going down Group 6A (16),
a. the ionization energy ________
b. the atomic size ________
c. the metallic character ________
d. the number of valence electrons ________

3.66 Complete each of the statements **a** through **d** using 1, 2, or 3:
1. decreases **2.** increases **3.** remains the same
Going from left to right across Period 4,
a. the ionization energy ________
b. the atomic size ________
c. the metallic character ________
d. the number of valence electrons ________

3.67 Which statements completed with **a** through **e** will be true?

In Period 2, the nonmetals compared to the metals have larger (greater)

a. atomic size
b. ionization energies
c. number of protons
d. metallic character
e. number of valence electrons

3.68 Which statements completed with **a** through **e** will be true?

In Group 4A (14), an atom of C compared to an atom of Sn has a larger (greater)

a. atomic size
b. ionization energy
c. number of protons
d. metallic character
e. number of valence electrons

CONCEPT MAP

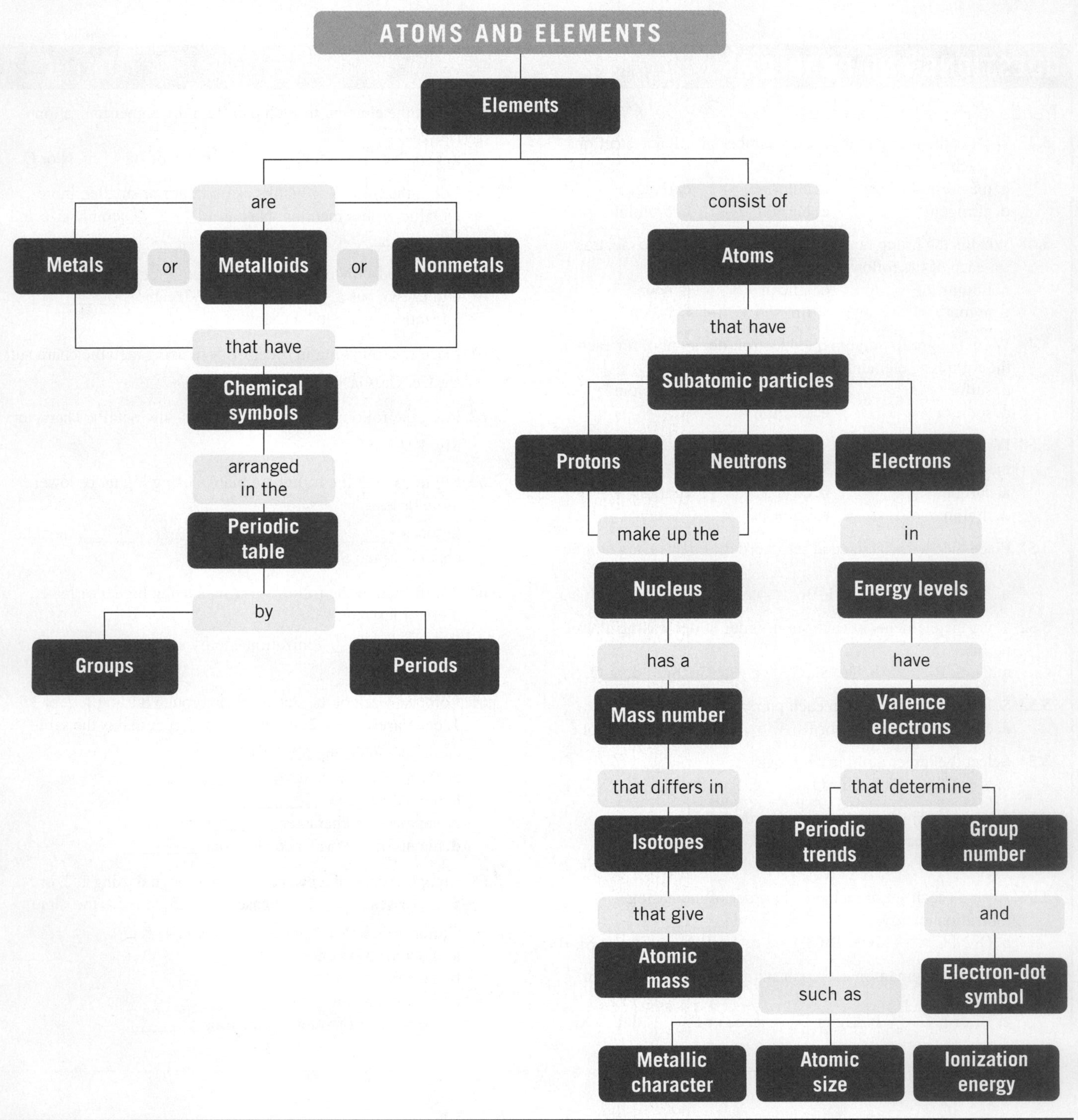

CHAPTER REVIEW

3.1 Elements and Symbols

Learning Goal: Given the name of an element, write its correct symbol; from the symbol, write the correct name.

Elements are the primary substances of matter. Chemical symbols are one- or two-letter abbreviations of the names of the elements.

3.2 The Periodic Table

Learning Goal: Use the periodic table to identify the group and the period of an element; identify the element as a metal, nonmetal, or metalloid.

5 B
14 Si
32 Ge 33 As
51 Sb 52 Te
84 Po 85 At

The periodic table is an arrangement of the elements by increasing atomic number. A vertical column or group on the periodic table contains elements with similar properties. A horizontal row is called a period. Elements in Group 1A (1) are called the alkali metals; Group 2A (2), alkaline earth metals; Group 7A (17), the halogens; and Group 8A (18), the noble gases. On the periodic table, metals are located on the left of the heavy zigzag line, and nonmetals are to the right of the heavy zigzag line. Except for aluminum, elements located on the heavy line are called metalloids.

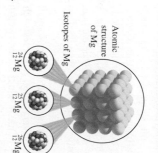

3.3 The Atom

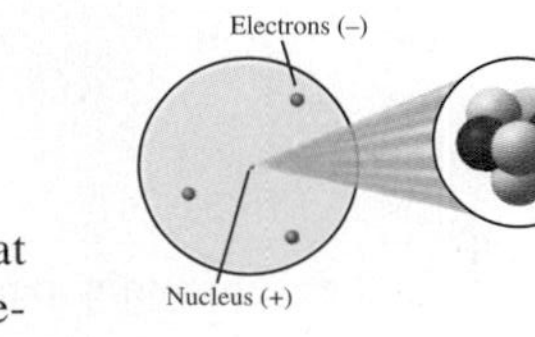

Learning Goal: Describe the electrical charge and location in an atom for a proton, a neutron, and an electron.

An atom is the smallest particle that retains the characteristics of an element. Within atoms, there are three subatomic particles. Protons have a positive charge (+), electrons carry a negative charge (−), and neutrons are electrically neutral. The protons and neutrons are found in the tiny, dense nucleus whereas electrons are located in a large space surrounding the nucleus.

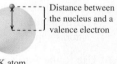

3.4 Atomic Number and Mass Number

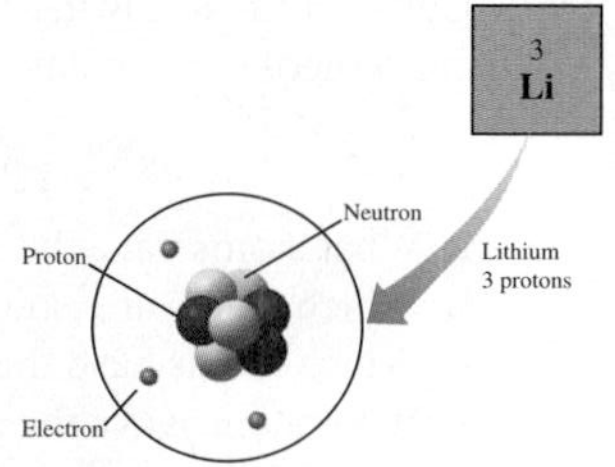

Learning Goal: Given the atomic number and the mass number of an atom, state the number of protons, neutrons, and electrons.

The atomic number gives the number of protons in all the atoms of the same element. In a neutral atom, the number of protons and the number of electrons are equal. The mass number is the total number of protons and neutrons in an atom.

3.5 Isotopes and Atomic Mass

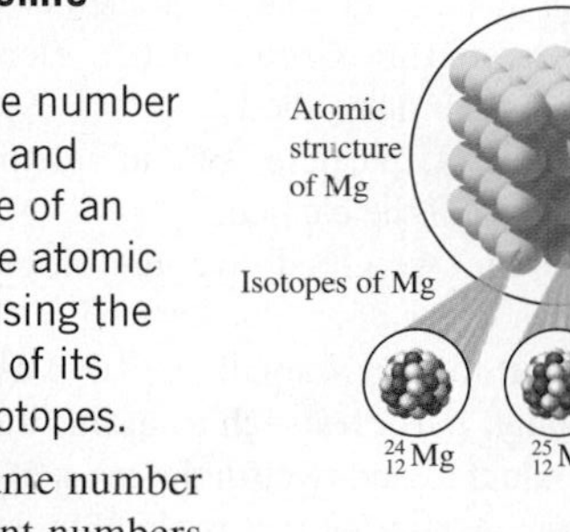

Learning Goal: Give the number of protons, electrons, and neutrons in an isotope of an element; calculate the atomic mass of an element using the abundance and mass of its naturally occurring isotopes.

Atoms that have the same number of protons, but different numbers of neutrons, are called isotopes. The atomic mass of an element is the weighted average mass of all the isotopes in a naturally occurring sample of that element.

3.6 Electron Energy Levels

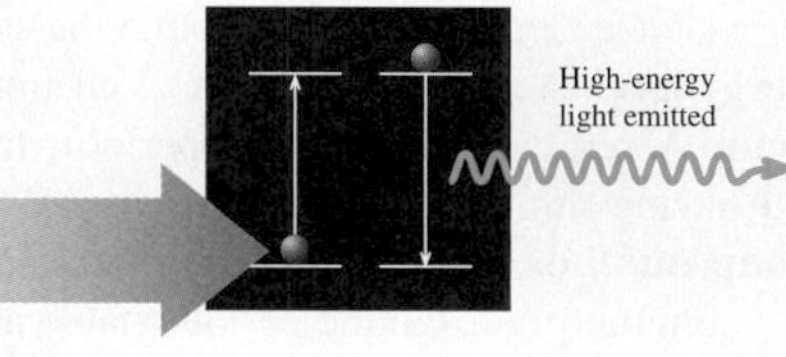

Learning Goal: Given the name or symbol of one of the first 20 elements in the periodic table, write the electron arrangement.

Every electron has a specific amount of energy. In an atom, the electrons of similar energy are grouped in specific energy levels. The first level nearest the nucleus can hold 2 electrons, the second level can hold 8 electrons, and the third level will take up to 18 electrons. The electron arrangement is written by placing the number of electrons in that atom in order from the lowest energy levels and filling to higher levels.

3.7 Trends in Periodic Properties

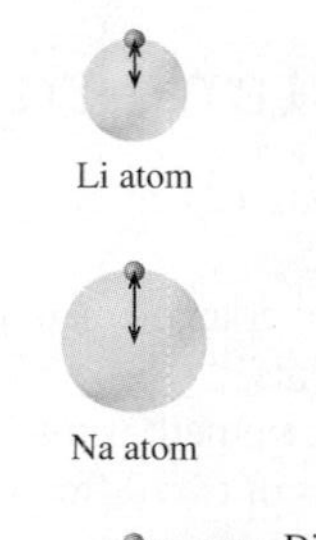

Learning Goal: Use the electron arrangement of elements to explain the trends in periodic properties.

The similarity of behavior for the elements in a group is related to having the same number of valence electrons, which are the electrons in the outermost energy level. The group number for an element gives the number of valence electrons, which means that each group of elements has the same arrangement of valence electrons differing only in the energy level. The size of an atom increases going down a group and decreases going across a period. The energy required to remove a valence electron is the ionization energy, which decreases going down a group and generally increases going from left to right across a period. The metallic character increases going down a group and decreases going across a period.

Key Terms

alkali metals Elements of Group 1A (1) except hydrogen; these are soft, shiny metals with one outer shell electron.

alkaline earth metals Group 2A (2) elements, which have 2 electrons in their outer shells.

atom The smallest particle of an element that retains the characteristics of the element.

atomic mass The weighted average mass of all the naturally occurring isotopes of an element.

atomic mass unit (amu) A small mass unit used to describe the mass of very small particles such as atoms and subatomic particles; 1 amu is equal to one-twelfth the mass of a $^{12}_{6}C$ atom.

atomic number A number that is equal to the number of protons in an atom.

atomic symbol An abbreviation used to indicate the mass number and atomic number of an isotope.

chemical symbol An abbreviation that represents the name of an element.

electron A negatively charged subatomic particle having a very small mass that is usually ignored in calculations; its symbol is e^-.

electron-dot symbol The representation of an atom that shows valence electrons as dots around the symbol of the element.

energy level A group of electrons with similar energy.

group A vertical column in the periodic table that contains elements having similar physical and chemical properties.

group number A number that appears at the top of each vertical column (group) in the periodic table and indicates the number of electrons in the outermost energy level.

halogens Group 7A (17) elements fluorine, chlorine, bromine, iodine, and astatine.

ionization energy The energy needed to remove the least tightly bound electron from the outermost energy level of an atom.

isotope An atom that differs only in mass number from another atom of the same element. Isotopes have the same atomic number (number of protons), but different numbers of neutrons.

mass number The total number of neutrons and protons in the nucleus of an atom.

metal An element that is shiny, malleable, ductile, and a good conductor of heat and electricity. The metals are located to the left of the zigzag line in the periodic table.

metallic character A measure of how easily an element loses a valence electron.

metalloid Elements with properties of both metals and nonmetals, located along the zigzag line on the periodic table.

neutron A neutral subatomic particle having a mass of about 1 amu and found in the nucleus of an atom; its symbol is n or n^0.

noble gas An element in Group 8A (18) of the periodic table, generally unreactive and seldom found in combination with other elements.

nonmetal An element with little or no luster that is a poor conductor of heat and electricity. The nonmetals are located to the right of the zigzag line in the periodic table.

nucleus The compact, very dense center of an atom, containing the protons and neutrons of the atom.

period A horizontal row of elements in the periodic table.

periodic table An arrangement of elements by increasing atomic number such that elements having similar chemical behavior are grouped in vertical columns.

proton A positively charged subatomic particle having a mass of about 1 amu and found in the nucleus of an atom; its symbol is p or p^+.

representative elements Elements found in Groups 1A (1) through 8A (18) excluding B groups (3–12) of the periodic table.

transition elements Elements located between Groups 2A (2) and 3A (13) on the periodic table.

valence electrons Electrons in the outermost energy level of an atom.

Understanding the Concepts

3.69 According to Dalton's atomic theory, which of the following are true?

a. Atoms of an element are identical to atoms of other elements.
b. Every element is made of atoms.
c. Atoms of two different elements combine to form compounds.
d. In a chemical reaction, some atoms disappear and new atoms appear.

3.70 Use Rutherford's gold-foil experiment to answer each of the following:

a. What did Rutherford expect to happen when he aimed particles at the gold foil?
b. How did the results differ from what he expected?
c. How did he use the results to propose a model of the atom?

3.71 Use the subatomic particles (**1–3**) to define each of the following:

1. protons **2.** neutrons **3.** electrons

a. atomic mass
b. atomic number
c. positive charge
d. negative charge
e. mass number – atomic number

3.72 Use the subatomic particles (**1–3**) to define each of the following:

1. protons **2.** neutrons **3.** electrons

a. mass number
b. surround the nucleus
c. in the nucleus
d. charge of 0
e. equal to number of electrons

3.73 Consider the following atoms in which X represents the chemical symbol of the element:

$$^{16}_{8}X \quad ^{16}_{9}X \quad ^{18}_{10}X \quad ^{17}_{8}X \quad ^{18}_{8}X$$

a. What atoms have the same number of protons?
b. Which atoms are isotopes? Of what element?
c. Which atoms have the same mass number?
d. What atoms have the same number of neutrons?

3.74 Cadmium, atomic number 48, consists of eight naturally occurring isotopes. Do you expect any of the isotopes to have the atomic mass listed on the periodic table for cadmium? Explain.

3.75 Indicate if the atoms in each pair have the same number of protons? Neutrons? Electrons?

a. $^{37}_{17}Cl$, $^{38}_{18}Ar$ **b.** $^{36}_{14}Si$, $^{35}_{14}Si$ **c.** $^{40}_{18}Ar$, $^{39}_{17}Cl$

3.76 Complete the following table for the three naturally occurring isotopes of silicon, the major component in computer chips:

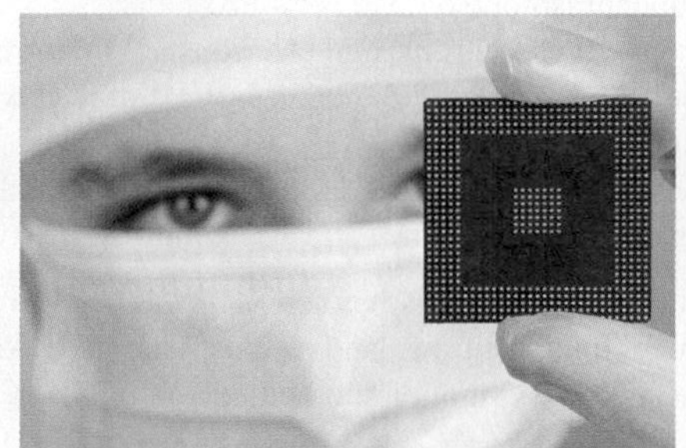

Computer chips consist primarily of the element silicon.

	Isotope		
	$^{28}_{14}Si$	$^{29}_{14}Si$	$^{30}_{14}Si$
Number of protons			
Number of neutrons			
Number of electrons			
Atomic number			
Mass number			

3.77 For each representation of a nucleus **A** through **E**, write the atomic symbol, and identify which are isotopes.

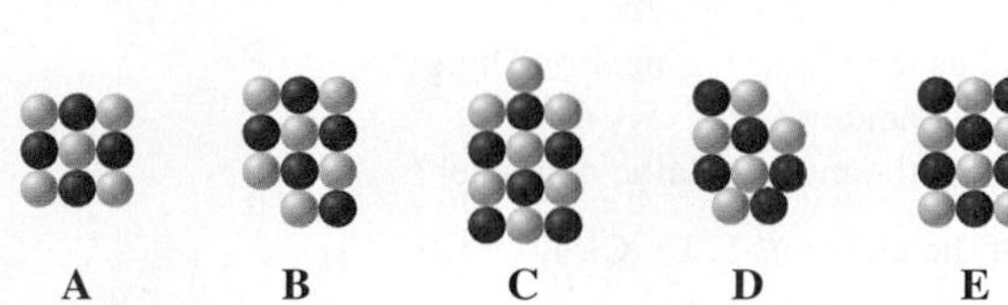

3.78 Identify the element represented by each nucleus **A** through **E** in Problem 3.77 as a metal, nonmetal, or metalloid.

3.79 Match the spheres **A** through **D** with atoms of Li, Na, K, and Rb.

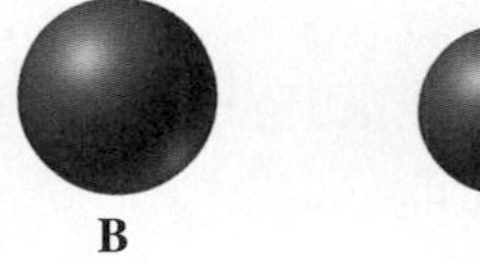

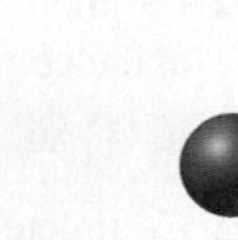

3.80 Match the spheres **A** through **D** with atoms of K, Ge, Ca, and Kr.

3.81 Of the elements Na, Mg, Si, S, Cl, and Ar, identify one that fits each of the following:
a. largest atomic size
b. a halogen
c. electron arrangement 2,8,4
d. highest ionization energy
e. in Group 6A (16)
f. most metallic character
g. two valence electrons

3.82 Of the elements Sn, Xe, Te, Sr, I, and Rb, identify one that fits each of the following:
a. smallest atomic size
b. an alkaline earth metal
c. a metalloid
d. lowest ionization energy
e. in Group 4A (14)
f. least metallic character
g. seven valence electrons

Additional Questions and Problems

For instructor-assigned homework, go to **www.masteringchemistry.com**.

3.83 Give the group number and period number for each of the following elements:
a. bromine **b.** argon **c.** potassium **d.** radium

3.84 Give the group number and period number for each of the following elements:
a. radon **b.** arsenic **c.** carbon **d.** neon

3.85 Indicate if each of the following statements is true or false:
a. The proton is a negatively charged particle.
b. The neutron is 2000 times as heavy as a proton.
c. The atomic mass unit is based on a carbon atom with 6 protons and 6 neutrons.
d. The nucleus is the largest part of the atom.
e. The electrons are located outside the nucleus.

3.86 Indicate if each of the following statements is true or false:
a. The neutron is electrically neutral.
b. Most of the mass of an atom is due to the protons and neutrons.
c. The charge of an electron is equal, but opposite, to the charge of a neutron.
d. The proton and the electron have about the same mass.
e. The mass number is the number of protons.

3.87 Complete the following statements:
a. The atomic number gives the number of ________ in the nucleus.
b. In an atom, the number of electrons is equal to the number of ________.
c. Sodium and potassium are examples of elements called ________.

3.88 Complete the following statements:
a. The number of protons and neutrons in an atom is also the ________ number.
b. The elements in Group 7A (17) are called the ________.
c. Elements that are shiny and conduct heat are called ________.

3.89 Write the names and symbols of the elements with the following atomic numbers:
a. 28 **b.** 56 **c.** 88 **d.** 33
e. 50 **f.** 55 **g.** 79 **h.** 80

3.90 Write the names and symbols of the elements with the following atomic numbers:
a. 10 **b.** 22 **c.** 48 **d.** 26
e. 54 **f.** 78 **g.** 83 **h.** 92

3.91 For the following atoms, give the number of protons, neutrons, and electrons:

a. $^{107}_{47}Ag$ **b.** $^{98}_{43}Tc$ **c.** $^{208}_{82}Pb$ **d.** $^{222}_{86}Rn$ **e.** $^{136}_{54}Xe$

3.92 For the following atoms, give the number of protons, neutrons, and electrons:

a. $^{198}_{79}Au$ **b.** $^{127}_{53}I$ **c.** $^{75}_{35}Br$ **d.** $^{133}_{55}Cs$ **e.** $^{195}_{78}Pt$

3.93 Complete the following table:

Name	Atomic Symbol	Number of Protons	Number of Neutrons	Number of Electrons
	$^{34}_{16}S$			
		28	34	
Magnesium			14	
	$^{220}_{86}Rn$			

3.94 Complete the following table:

Name	Atomic Symbol	Number of Protons	Number of Neutrons	Number of Electrons
Potassium			22	
	$^{51}_{23}V$			
		48	64	
Barium			82	

3.95 Write the atomic symbol for each of the following:
a. an atom with 4 protons and 5 neutrons
b. an atom with 12 protons and 14 neutrons
c. a calcium atom with a mass number of 46
d. an atom with 30 electrons and 40 neutrons
e. a copper atom with 34 neutrons

3.96 Write the atomic symbol for each of the following:
a. an aluminum atom with 14 neutrons
b. an atom with atomic number 26 and 32 neutrons

c. a strontium atom with 50 neutrons
d. an atom with a mass number of 72 and atomic number 33

3.97 The most abundant isotope of lead is $^{208}_{82}Pb$.
a. How many protons, neutrons, and electrons are in $^{208}_{82}Pb$?
b. What is the atomic symbol of another isotope of lead with 132 neutrons?
c. What is the atomic symbol and name of an atom with the same mass number as in part **b** and 131 neutrons?

3.98 The most abundant isotope of silver is $^{107}_{47}Ag$.
a. How many protons, neutrons, and electrons are in $^{107}_{47}Ag$?
b. What is the atomic symbol of another isotope of silver with 62 neutrons?
c. What is the atomic symbol and name of an atom with the same mass number as in part **b** and 61 neutrons?

3.99 Write the group number and the electron arrangement for each of the following:
a. oxygen **b.** sodium
c. neon **d.** boron

3.100 Write the group number and the electron arrangement for each of the following:
a. magnesium **b.** chlorine
c. beryllium **d.** argon

3.101 Why is the ionization energy of Ca higher than K, but lower than Mg?

3.102 Why is the ionization energy of Cl lower than F, but higher than S?

3.103 Of the elements Li, Be, N, and F, which
a. is an alkaline earth metal?
b. has the largest atomic radius?
c. has the highest ionization energy?
d. is found in Group 5A (15)?
e. has the most metallic character?

3.104 Of the elements F, Br, Cl, I, which
a. is the largest atom?
b. is the smallest atom?
c. has the lowest ionization energy?
d. requires the most energy to remove an electron?
e. is found in Period 4?

Challenge Questions

3.105 For each of the following, write the symbol and name for X and the number of protons and neutrons. Which are isotopes of each other?

a. $^{37}_{17}X$ **b.** $^{56}_{27}X$ **c.** $^{116}_{50}X$ **d.** $^{124}_{50}X$ **e.** $^{116}_{48}X$

3.106 There are four naturally occurring isotopes of iron: $^{54}_{26}Fe$, $^{56}_{26}Fe$, $^{57}_{26}Fe$, and $^{58}_{26}Fe$.
a. How many protons, neutrons, and electrons are in $^{58}_{26}Fe$?
b. What is the most prevalent isotope in an iron sample?
c. How many neutrons are in $^{57}_{26}Fe$?
d. Why don't any of the isotopes of iron have the atomic mass of 55.85 amu listed on the periodic table?

3.107 Give the symbol of the element that has the
a. smallest atomic size in Group 6A (16)
b. smallest atomic size in Period 3
c. highest ionization energy in Group 5A (15)
d. lowest ionization energy in Period 3
e. most metallic character in Group 2A (2)

3.108 Give the symbol of the element that has the
a. largest atomic size in Group 1A (1)
b. largest atomic size in Period 4
c. highest ionization energy in Group 2A (2)
d. lowest ionization energy in Group 7A (17)
e. least metallic character in Group 4A (14)

3.109 A lead atom has a mass of 3.4×10^{-22} g. How many lead atoms are in a cube of lead that has a volume of 2.00 cm^3 if the density of lead is 11.3 g/cm^3?

3.110 If the diameter of a sodium atom is 3.14×10^{-8} cm, how many sodium atoms would fit along a line exactly 1 inch long?

3.111 Silicon has three naturally occurring isotopes: Si-28 (27.977 amu) with an abundance of 92.23%, Si-29 (28.976 amu) with a 4.68% abundance, and Si-30 (29.974 amu) with a 3.09% abundance. What is the atomic mass of silicon?

3.112 Antimony (Sb), which has an atomic weight of 121.75 amu, has two naturally occurring isotopes: Sb-121 and Sb-123. If a sample of antimony is 42.70% Sb-123, which has a mass of 122.90 amu, what is the mass of Sb-121?

Answers

Answers to Study Checks

3.1 Si, S, and Ag

3.2 magnesium, aluminum, and fluorine

3.3 **a.** Strontium is in Group 2A (2).
b. Strontium is an alkaline earth metal.
c. Strontium is in Period 5.
d. Magnesium, Mg
e. Alkali metal, Rb; halogen, I; noble gas, Xe

3.4 **a.** arsenic, As **b.** argon, Ar **c.** silicon, Si

3.5 False; the nucleus occupies a very small volume in an atom.

3.6 **a.** 79 **b.** 79 **c.** gold, Au

3.7 45 neutrons

3.8 **a.** ${}^{15}_{7}N$ **b.** ${}^{42}_{20}Ca$ **c.** ${}^{27}_{13}Al$

3.9 10.81 amu

3.10 phosphorus

3.11 Group 5A (15)

3.12 $\cdot\ddot{\underset{\cdot}{P}}\cdot$

3.13 Ionization energy increases going left to right across a period: Sr is lowest, Sn is higher, and I is the highest of this set.

Answers to Selected Questions and Problems

3.1 **a.** Cu **b.** Pt **c.** Ca **d.** Mn
e. Fe **f.** Ba **g.** Pb **h.** Sr

3.3 **a.** carbon **b.** chlorine **c.** iodine **d.** mercury
e. silver **f.** argon **g.** boron **h.** nickel

3.5 **a.** sodium, chlorine
b. calcium, sulfur, oxygen
c. carbon, hydrogen, chlorine, nitrogen, oxygen
d. calcium, carbon, oxygen

3.7 **a.** Period 2 **b.** Group 8A (18)
c. Group 1A (1) **d.** Period 2

3.9 **a.** alkaline earth metal **b.** transition element
c. noble gas **d.** alkali metal
e. halogen

3.11 **a.** C **b.** He **c.** Na **d.** Ca **e.** Al

3.13 **a.** metal **b.** nonmetal **c.** metal **d.** nonmetal
e. nonmetal **f.** nonmetal **g.** metalloid **h.** metal

3.15 **a.** electron **b.** proton **c.** electron **d.** neutron

3.17 Rutherford determined that an atom contains a small, compact nucleus that is positively charged.

3.19 **a.** True **b.** True
c. True **d.** False; a proton is attracted to an electron

3.21 In the process of brushing hair, strands of hair become charged with like charges that repel each other.

3.23 **a.** atomic number **b.** both
c. mass number **d.** atomic number

3.25 **a.** lithium, Li **b.** fluorine, F
c. calcium, Ca **d.** zinc, Zn
e. neon, Ne **f.** silicon, Si
g. iodine, I **h.** oxygen, O

3.27 **a.** 18 protons and 18 electrons
b. 30 protons and 30 electrons
c. 53 protons and 53 electrons
d. 19 protons and 19 electrons

3.29 See Table 3.13.

3.31 **a.** 13 protons, 14 neutrons, 13 electrons
b. 24 protons, 28 neutrons, 24 electrons
c. 16 protons, 18 neutrons, 16 electrons
d. 35 protons, 46 neutrons, 35 electrons

3.33 **a.** ${}^{31}_{15}P$ **b.** ${}^{80}_{35}Br$ **c.** ${}^{122}_{50}Sn$
d. ${}^{35}_{17}Cl$ **e.** ${}^{202}_{80}Hg$

3.35 **a.** ${}^{36}_{18}Ar$ ${}^{38}_{18}Ar$ ${}^{40}_{18}Ar$
b. They all have the same number of protons and electrons.
c. They have different numbers of neutrons, which gives them different mass numbers.
d. The atomic mass of Ar listed on the periodic table is the average atomic mass of all the isotopes.
e. Because argon has an atomic mass of 39.95, the isotope of ${}^{40}_{18}Ar$ would be the most prevalent.

3.37 69.72 amu

3.39 absorb

3.41 **a.** green light **b.** blue light

3.43 **a.** 2,4 **b.** 2,8,8 **c.** 2,8,8,1
d. 2,8,4 **e.** 2 **f.** 2,5

3.45 **a.** Li **b.** Mg **c.** H
d. Cl **e.** O

TABLE 3.13

Name of the Element	Symbol	Atomic Number	Mass Number	Number of Protons	Number of Neutrons	Number of Electrons
Aluminum	Al	13	27	13	14	13
Magnesium	Mg	12	24	12	12	12
Potassium	K	19	39	19	20	19
Sulfur	S	16	31	16	15	16
Iron	Fe	26	56	26	30	26

3.47 **a.** Group 2A (2), 2 e^-
b. Group 7A (17), 7 e^-
c. Group 6A (16), 6 e^-
d. Group 5A (15), 5 e^-
e. Group 2A (2), 2 e^-
f. Group 7A (17), 7 e^-

3.49 **a.** Group 6A (16)
$\cdot\ddot{\underset{\cdot\cdot}{S}}\cdot$
b. Group 5A (15)
$\cdot\ddot{\underset{\cdot}{N}}\cdot$
c. Group 2A (2)
·Ca·
d. Group 1A (1)
Na·
e. Group 3A (13)
$\cdot\dot{Ga}\cdot$

3.51 **a.** Mg, Al, Si **b.** I, Br, Cl **c.** Sr, Sb, I **d.** Na, Si, P

3.53 **a.** Na **b.** Rb **c.** Na **d.** Rb

3.55 **a.** Br, Cl, F **b.** Na, Al, Cl **c.** Cs, K, Na **d.** Ca, As, Br

3.57 **a.** Br **b.** Mg **c.** P **d.** Xe

3.59 larger, more metallic

3.61 Ca, Ga, Ge, Br

3.63 lower, more

3.65 **a.** decreases **b.** increases
c. increases **d.** remains the same

3.67 **b**, **c**, and **e** are true.

3.69 **b** and **c** are true.

3.71 **a.** 1 + 2 **b.** 1 **c.** 1 **d.** 3 **e.** 2

3.73 **a.** ${}^{16}_{8}X$, ${}^{17}_{8}X$, ${}^{18}_{8}X$ all have 8 protons.
b. ${}^{16}_{8}X$, ${}^{17}_{8}X$, ${}^{18}_{8}X$ are all isotopes of O.
c. ${}^{16}_{8}X$ and ${}^{16}_{9}X$ have mass numbers of 16, and ${}^{18}_{10}X$ and ${}^{18}_{8}X$ have mass numbers of 18.
d. ${}^{16}_{8}X$ and ${}^{18}_{10}X$ have 8 neutrons each.

3.75 **a.** Both have 20 neutrons.
b. Both have 14 protons and 14 electrons.
c. Both have 22 neutrons.

3.77 **a.** ${}^{9}_{4}Be$ **b.** ${}^{11}_{5}B$ **c.** ${}^{13}_{6}C$ **d.** ${}^{10}_{5}B$ **e.** ${}^{12}_{6}C$
Representations **B** and **D** are isotopes of boron; **C** and **E** are isotopes of carbon.

3.79 Li is **D**, Na is **A**, K is **C**, and Rb is **B**.

3.81 **a.** Na **b.** Cl **c.** Si **d.** Ar
e. S **f.** Na **g.** Mg

3.83 **a.** Group 7A (17), Period 4
b. Group 8A (18), Period 3
c. Group 1A (1), Period 4
d. Group 2A (2), Period 7

3.85 **a.** False **b.** False **c.** True
d. False **e.** True

3.87 **a.** protons **b.** protons **c.** alkali metals

3.89 **a.** nickel, Ni **b.** barium, Ba
c. radium, Ra **d.** arsenic, As
e. tin, Sn **f.** cesium, Cs
g. gold, Au **h.** mercury, Hg

3.91 **a.** 47 protons, 60 neutrons, 47 electrons
b. 43 protons, 55 neutrons, 43 electrons
c. 82 protons, 126 neutrons, 82 electrons
d. 86 protons, 136 neutrons, 86 electrons
e. 54 protons, 82 neutrons, 54 electrons

3.93

Name	Atomic Symbol	Number of Protons	Number of Neutrons	Number of Electrons
Sulfur	${}^{34}_{16}S$	16	18	16
Nickel	${}^{62}_{28}Ni$	28	34	28
Magnesium	${}^{26}_{12}Mg$	12	14	12
Radon	${}^{220}_{86}Rn$	86	134	86

3.95 **a.** ${}^{9}_{4}Be$ **b.** ${}^{26}_{12}Mg$ **c.** ${}^{46}_{20}Ca$
d. ${}^{70}_{30}Zn$ **e.** ${}^{63}_{29}Cu$

3.97 **a.** 82 protons, 126 neutrons, 82 electrons
b. ${}^{214}_{82}Pb$
c. ${}^{214}_{83}Bi$, bismuth

3.99 **a.** O; Group 6A (16); 2,6
b. Na; Group 1A (1); 2,8,1
c. Ne; Group 8A (18); 2,8
d. B; Group 3A (13); 2,3

3.101 Calcium has a greater number of protons than K. The increase in positive charge increases the attraction for electrons, which means that more energy is required to remove the valence electrons. The valence electrons are farther from the nucleus in Ca than in Mg and less energy is needed to remove them.

3.103 **a.** Be **b.** Li **c.** F
d. N **e.** Li

3.105 **a.** Cl, chlorine; 17 protons and 20 neutrons
b. Co, cobalt; 27 protons and 29 neutrons
c. Sn, tin; 50 protons and 66 neutrons
d. Sn, tin; 50 protons and 74 neutrons
e. Cd, cadmium; 48 protons and 68 neutrons
The symbols **c** and **d** are isotopes of Sn, tin.

3.107 **a.** O **b.** Ar **c.** N
d. Na **e.** Ra

3.109 6.6×10^{22} atoms of Pb

3.111 28.09 amu

Compounds and Their Bonds

4

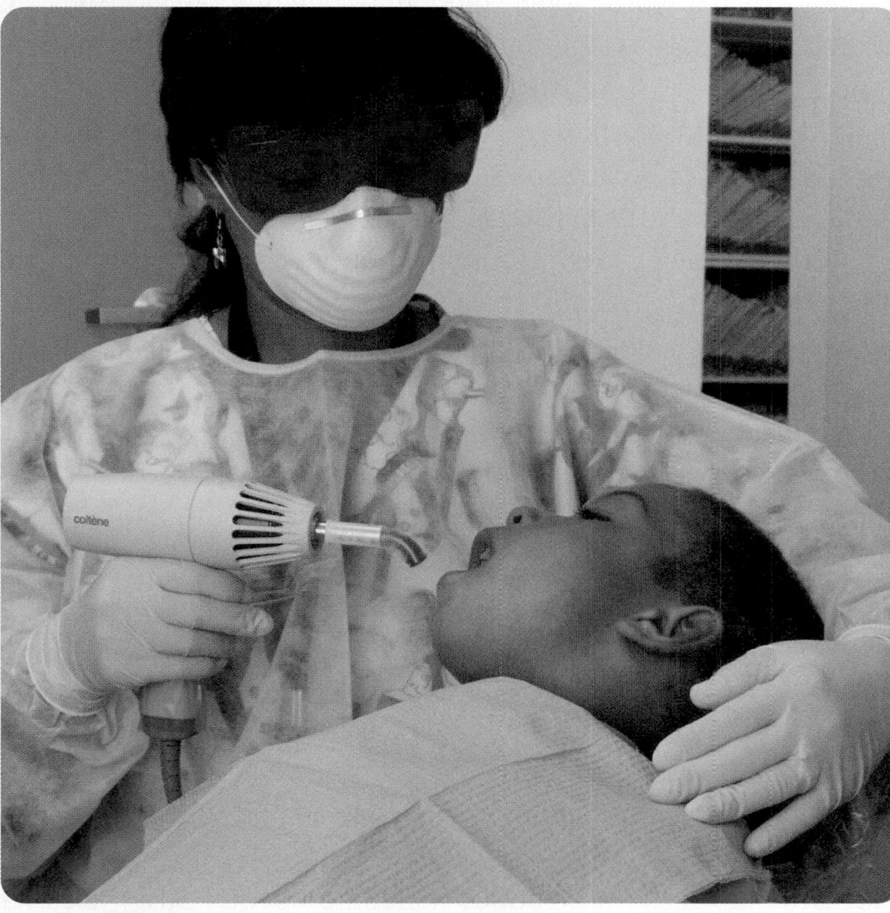

LOOKING AHEAD

"One way to prevent cavities in children is to apply a thin, plastic coating called a sealant to their teeth," says Dr. Pam Alston, a dentist in private practice. "We look for teeth with deep grooves and pits that trap food. We clean the teeth and apply an etching agent, which helps the sealant bind to the teeth. Then we apply the liquid sealant, which fills in the grooves and pits, and use ultraviolet light to solidify the coating."

The use of fluoride compounds, such as SnF_2 in toothpaste and NaF in water and mouth rinses have greatly reduced tooth decay. The fluoride ion replaces the hydroxide ion to form $Ca_{10}(PO_4)_6F_2$, which strengthens the enamel and makes it less susceptible to decay. Other compounds used in dentistry are the anesthetic known as laughing gas, N_2O, and Novocain, $C_{13}H_{20}N_2O_2$.

Visit **www.masteringchemistry.com** for self-study materials and instructor-assigned homework.

In nature, atoms of almost all the elements on the periodic table are found in combination with other atoms. Only the atoms of the noble gases—He, Ne, Ar, Kr, Xe, and Rn—do not combine in nature with other atoms. As discussed in Chapter 2, a compound is a pure substance, composed of two or more elements, with a definite composition. In a typical ionic compound, one or more electrons are transferred from the atoms of metals to atoms of nonmetals. The attraction that results is called an ionic bond.

We use many ionic compounds every day. When we cook or bake, we use ionic compounds such as salt, NaCl, and baking soda, $NaHCO_3$. Epsom salts, $MgSO_4$, may be used to soak sore feet. Milk of magnesia, $Mg(OH)_2$, or calcium carbonate, $CaCO_3$, may be taken to settle an upset stomach. In a mineral supplement, iron may be present as iron(II) sulfate, $FeSO_4$, iodine as potassium iodide, KI, and manganese as manganese(II) sulfate, $MnSO_4$. Some sunscreens contain zinc oxide, ZnO, and the tin(II) fluoride, SnF_2, in toothpaste provides fluoride to help prevent tooth decay.

Small amounts of metals cause the different colors of gemstones.

The structures of ionic crystals result in the beautiful facets seen in gems. Sapphires and rubies are made of aluminum oxide, Al_2O_3. Impurities of chromium make rubies red, and iron and titanium make sapphires blue.

In compounds of nonmetals, covalent bonding occurs by atoms sharing one or more valence electrons. There are many more covalent compounds than there are ionic ones and many simple covalent compounds are present in our everyday lives. For example, water (H_2O), oxygen (O_2), and carbon dioxide (CO_2) are all covalent compounds.

Covalent compounds consist of molecules, which are discrete groups of atoms. A molecule of water (H_2O) consists of two atoms of hydrogen and one atom of oxygen. When you have iced tea, perhaps you add molecules of sugar ($C_{12}H_{22}O_{11}$), which is a covalent compound. Other covalent compounds include propane (C_3H_8), alcohol (C_2H_6O), the antibiotic amoxicillin ($C_{16}H_{19}N_3O_5S$), and the antidepressant Prozac ($C_{17}H_{18}F_3NO$).

LEARNING GOAL

Using the octet rule, write the symbols of the simple ions for the representative elements.

TUTORIAL
Octet Rule and Ions

4.1 Octet Rule and Ions

Compounds are the result of the formation of chemical bonds between two or more different elements. Ionic bonds occur when atoms of one element lose valence electrons and the atoms of another element gain valence electrons. Ionic compounds typically occur between metals and nonmetals. For example, sodium atoms lose electrons and chlorine atoms gain electrons to form the ionic compound NaCl. Covalent bonds form when atoms of nonmetals share valence electrons. In the covalent compound NCl_3, atoms of nitrogen and chlorine share electrons.

Most of the elements, except the noble gases, are found in nature combined as compounds. The noble gases are so stable that they form compounds only under extreme conditions. One explanation for the stability of noble gases is that they have eight valence electrons. A few elements achieve the stability of helium with two valence electrons. This tendency for atoms to attain a noble gas electron arrangement is known as the **octet rule** and provides a key to our understanding of the ways in which atoms bond and form compounds.

M• ⌒ •Nm: :Nm• ⊂⊃ •Nm:

Loss and gain of electrons   Sharing electrons

M^+ :Nm:$^-$:Nm:Nm:

Ionic bond   Covalent bond

M is a metal
Nm is a nonmetal

Positive Ions

In ionic bonding, ions, which have electrical charges, form when atoms lose or gain electrons to form **octets**. As we have seen in Chapter 3, Section 3.7, the ionization energy of a metal in Group 1A (1), 2A (2), or 3A (13) is low. Thus, a metal atom readily loses valence electrons. In doing so, a metal atom obtains the same electron arrangement as its nearest noble gas (usually eight valence electrons). By losing electrons, a metal atom forms an **ion**, which has a positive charge. In the following diagram, a sodium atom loses its single valence electron, which leaves 10 electrons that have the same electron arrangement as neon. Because there are 11 protons in its nucleus, the atom is no longer neutral. Then the ion has an **ionic charge**, which is the difference between the number of protons (positive charge) and the number of electrons (negative charge). For the sodium ion, the ionic charge is 1+, which is written in the upper right-hand corner as Na^+.

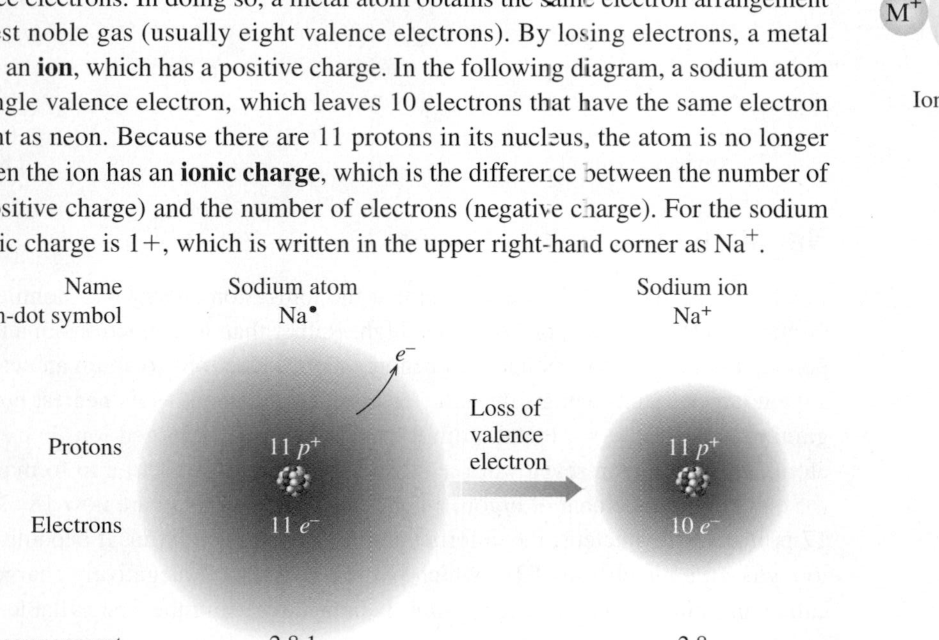

Metals in ionic compounds lose their valence electrons to form positively charged ions called **cations** (pronounced *cat'-i-ons*). Magnesium, a metal in Group 2A (2), attains a noble gas electron arrangement like neon by losing two valence electrons to

Chemistry Link to Industry

SOME USES FOR NOBLE GASES

Noble gases may be used when it is necessary to have a substance that is unreactive. Scuba divers normally use a pressurized mixture of nitrogen and oxygen gases for breathing under water. However, when the air mixture is used at depths where pressure is high, the nitrogen gas is absorbed into the blood, where it can cause mental disorientation. To avoid this problem, a breathing mixture of oxygen and helium may be substituted (see *Breathing Mixtures for Scuba* in Chapter 2). The diver still obtains the necessary oxygen, but the unreactive helium that dissolves in the blood does not cause mental disorientation. However, its lower density does change the vibrations of the vocal cords, and the diver will sound like Donald Duck.

Helium is also used to fill blimps and balloons. When dirigibles were first designed, they were filled with hydrogen, a very light gas. However, when they came in contact with any type of spark or heating source, they exploded violently because of the extreme reactivity of hydrogen gas with oxygen present in the air. Today blimps are filled with unreactive helium gas, which presents no danger of explosion.

Lighting tubes are generally filled with a noble gas such as neon or argon. While the electrically heated filaments that produce the light get very hot, the surrounding noble gases do not react with the hot filament. If heated in air, the elements that constitute the filament will quickly burn out when oxygen is present.

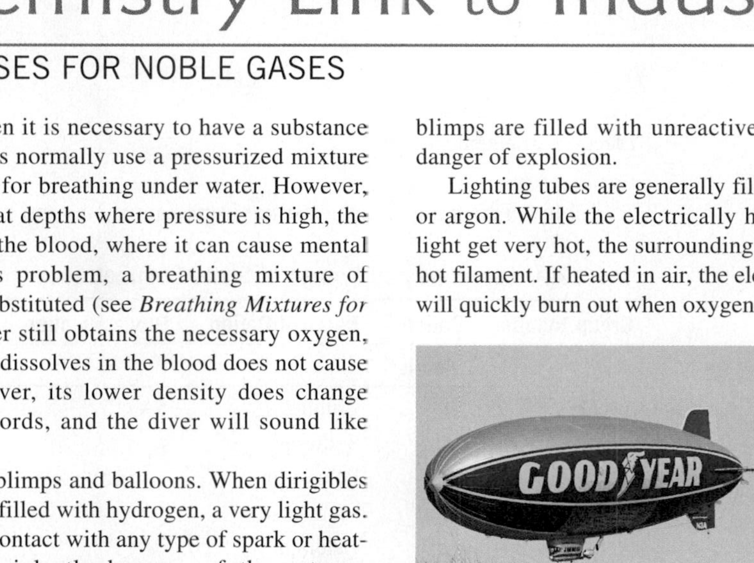

The helium in a blimp is much less dense than air, which allows the blimp to fly at about 150 m above ground.

form a positive ion with a 2+ ionic charge. A metal ion is named by its element name. Thus, Mg^{2+} is named the *magnesium ion*.

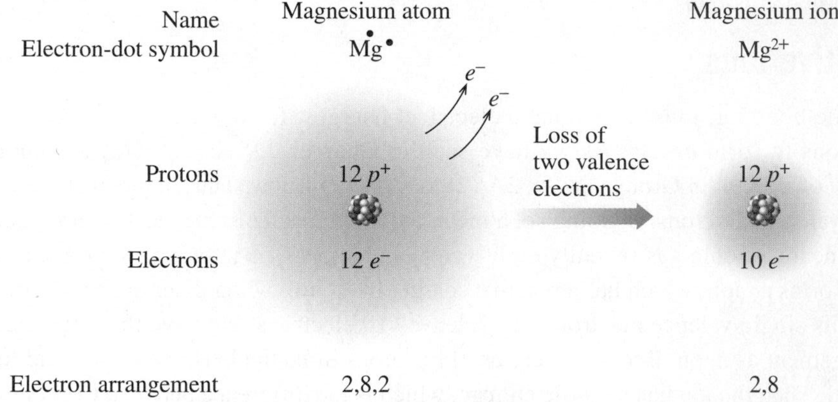

Negative Ions

In Chapter 3, Section 3.7, we learned that the ionization energy of a nonmetal atom in Group 5A (15), 6A (16), or 7A (17) is high. Rather than lose electrons in an ionic compound, a nonmetal atom gains one or more valence electrons to attain an octet. In doing so, a nonmetal atom obtains the same electron arrangement as its nearest noble gas. By gaining electrons, a nonmetal atom forms a negatively charged ion. For example, an atom of chlorine with seven valence electrons obtains one electron to form an octet and the electron arrangement of argon, a noble gas. Because there are now 18 electrons and 17 protons in its nucleus, the chlorine atom is no longer neutral. It becomes a chloride ion with an ionic charge of 1−, which is written as Cl^-. A negatively charged ion, also called an **anion** (pronounced *an′-i-on*), is named by using the first syllable of its name followed by *ide*.

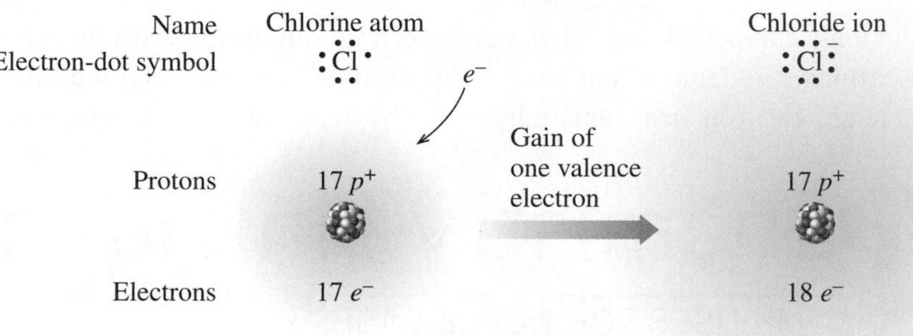

Table 4.1 lists the names of some important metal and nonmetal ions.

TABLE 4.1 Formulas and Names of Some Common Ions

Group Number	Cation	Name of Cation	Group Number	Anion	Name of Anion
	Metals			**Nonmetals**	
1A (1)	Li^+	Lithium	5A (15)	N^{3-}	Nitride
	Na^+	Sodium		P^{3-}	Phosphide
	K^+	Potassium	6A (16)	O^{2-}	Oxide
2A (2)	Mg^{2+}	Magnesium		S^{2-}	Sulfide
	Ca^{2+}	Calcium	7A (17)	F^-	Fluoride
	Ba^{2+}	Barium		Cl^-	Chloride
3A (13)	Al^{3+}	Aluminum		Br^-	Bromide
				I^-	Iodide

CONCEPT CHECK 4.1

Ions

a. Write the symbol and name of the ion that has 7 protons and 10 electrons.

b. Write the symbol and name of the ion that has 20 protons and 18 electrons.

ANSWER

a. The element with 7 protons is nitrogen. In an ion of nitrogen with 10 electrons, the ionic charge would be 3−, (7+) + (10−) = 3−. The ion, written as N^{3-}, is the *nitride* ion.

b. The element with 20 protons is calcium. In an ion of calcium with 18 electrons, the ionic charge would be 2+, (20+) + (18−) = 2+. The ion, written as Ca^{2+}, is the *calcium* ion.

TUTORIAL
Ions

Ionic Charges from Group Numbers

As we learned in Chapter 3, Section 3.7, we can obtain the number of valence electrons of the representative elements from the group numbers on the periodic table. Now we can use group numbers to determine the ionic charges for most of the ions. The atoms of the elements in Group 1A (1) lose one electron to form ions with a 1+ charge. The atoms of the elements in Group 2A (2) lose two electrons to form ions with a 2+ charge. The atoms of the elements in Group 3A (13) lose three electrons to form ions with a 3+ charge.

In ionic compounds, the atoms of the elements in Group 7A (17) gain one electron to form ions with a 1− charge. The atoms of the elements in Group 6A (16) gain two electrons to form ions with a 2− charge. The atoms of the elements in Group 5A (15) gain three electrons to form ions with a 3− charge.

The nonmetals of Group 4A (14) do not typically form ions. However, the metals Sn and Pb in Group 4A (14) lose electrons to form positive ions. Table 4.2 lists the ionic charges for some common monatomic ions of representative elements.

TABLE 4.2 Examples of Monatomic Ions and Their Nearest Noble Gases

		Metals Lose Valence Electrons			Nonmetals Gain Valence Electrons				
Noble Gases		1A (1)	2A (2)	3A (13)	5A (15)	6A (16)	7A (17)		Noble Gases
He	⇦	Li^+							
Ne	⇦	Na^+	Mg^{2+}	Al^{3+}	N^{3-}	O^{2-}	F^-	⇨	Ne
Ar	⇦	K^+	Ca^{2+}		P^{3-}	S^{2-}	Cl^-	⇨	Ar
Kr	⇦	Rb^+	Sr^{2+}				Br^-	⇨	Kr
Xe	⇦	Cs^+	Ba^{2+}				I^-	⇨	Xe

SAMPLE PROBLEM 4.1

Writing Ions

Consider the elements aluminum and oxygen.

a. Identify each as a metal or a nonmetal.
b. State the number of valence electrons for each.
c. State the number of electrons that must be lost or gained for each to acquire an octet.
d. Write the symbol, including its ionic charge, and name of each resulting ion.

SOLUTION

	Aluminum	Oxygen
a.	metal	nonmetal
b.	three valence electrons	six valence electrons
c.	loses $3\ e^-$	gains $2\ e^-$
d.	Al^{3+}, aluminum ion	O^{2-}, oxide ion

STUDY CHECK 4.1

What are the symbols for the ions formed by potassium and sulfur?

Chemistry Link to Health

SOME IMPORTANT IONS IN THE BODY

Several ions in body fluids have important physiological and metabolic functions. Some are listed in Table 4.3.

Foods such as bananas, milk, cheese, and potatoes provide the body with ions that are important in regulating body functions.

TABLE 4.3 Ions in the Body

Ion	Occurrence	Function	Source	Result of Too Little	Result of Too Much
Na^+	Principal cation outside the cell	Regulation and control of body fluids	Salt, cheese, pickles	Hyponatremia, anxiety, diarrhea, circulatory failure, decrease in body fluid	Hypernatremia, little urine, thirst, edema
K^+	Principal cation inside the cell	Regulation of body fluids and cellular functions	Bananas, orange juice, milk, prunes, potatoes	Hypokalemia (hypopotassemia), lethargy, muscle weakness, failure of neurological impulses	Hyperkalemia (hyperpotassemia), irritability, nausea, little urine, cardiac arrest
Ca^{2+}	Cation outside the cell; 90% of calcium in the body in bone as $Ca_3(PO_4)_2$ or $CaCO_3$	Major cation of bone; needed for muscle contraction	Milk, yogurt, cheese, greens, spinach	Hypocalcemia, tingling fingertips, muscle cramps, osteoporosis	Hypercalcemia, relaxed muscles, kidney stones, deep bone pain
Mg^{2+}	Cation outside the cell; 70% of magnesium in the body in bone structure	Essential for certain enzymes, muscles, nerve control	Widely distributed (part of chlorophyll of all green plants), nuts, whole grains	Disorientation, hypertension, tremors, slow pulse	Drowsiness
Cl^-	Principal anion outside the cell	Gastric juice, regulation of body fluids	Salt	Same as for Na^+	Same as for Na^+

QUESTIONS AND PROBLEMS

Octet Rule and Ions

4.1 State the number of electrons that must be lost by atoms of each of the following to acquire a noble gas electron arrangement:
a. Li **b.** Ca **c.** Ga **d.** Cs **e.** Ba

4.2 State the number of electrons that must be gained by atoms of each of the following to acquire a noble gas electron arrangement:
a. Cl **b.** Se **c.** N **d.** I **e.** S

4.3 Write the symbols of the ions with the following number of protons and electrons:
a. 3 protons, 2 electrons **b.** 9 protons, 10 electrons
c. 12 protons, 10 electrons **d.** 26 protons, 23 electrons

4.4 Write the symbols of the ions with the following number of protons and electrons:
a. 8 protons, 10 electrons **b.** 19 protons, 18 electrons
c. 35 protons, 36 electrons **d.** 50 protons, 46 electrons

4.5 State the number of electrons lost or gained when the following elements form ions:
a. Sr **b.** P **c.** Group 7A (17)
d. Na **e.** Br

4.6 State the number of electrons lost or gained when the following elements form ions:
a. O **b.** Group 2A (2) **c.** F
d. K **e.** Rb

4.7 Write the symbol for the ion of each of the following:
a. chlorine **b.** cesium
c. sulfur **d.** radium

4.8 Write the symbol for the ion of each of the following:
a. fluorine **b.** calcium
c. sodium **d.** iodine

4.2 Ionic Compounds

LEARNING GOAL

Using charge balance, write the correct formula for an ionic compound.

TUTORIAL
Ionic Compounds

Ionic compounds consist of positive and negative ions. The ions are held together by strong attractions between the oppositely charged ions, called *ionic bonds*.

Properties of Ionic Compounds

The physical and chemical properties of an ionic compound such as NaCl are very different from those of the original elements. For example, the original elements of NaCl were sodium, which is a soft, shiny metal, and chlorine, which is a yellow-green poisonous gas. However, when they react and form positive and negative ions, they produce NaCl, which is ordinary table salt, a hard, white, crystalline substance that is important in our diet.

In a crystal of NaCl, every Na^+ ion is surrounded by six Cl^- ions, and every Cl^- ion is surrounded by six Na^+ ions (see Figure 4.1). Thus, there are many strong attractions between the positive and negative ions, which account for the high melting points of ionic compounds. For example, the melting point of NaCl is 801 °C. At room temperature, ionic compounds are solids.

Formulas of Ionic Compounds

The **formula** of an ionic compound indicates the number and kinds of ions that make up the ionic compound. The sum of the ionic charges in the formula is always zero, which means that the total amount of positive charge is equal to the total amount of negative charge. For example, the formula NaCl indicates that the compound consists of one sodium ion, Na^+, for every chloride ion, Cl^-. Although the ions are positively or negatively charged, their ionic charges are not shown in the formula of the compound.

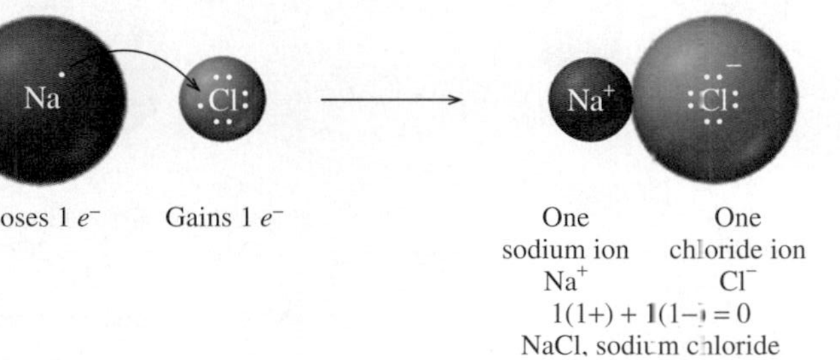

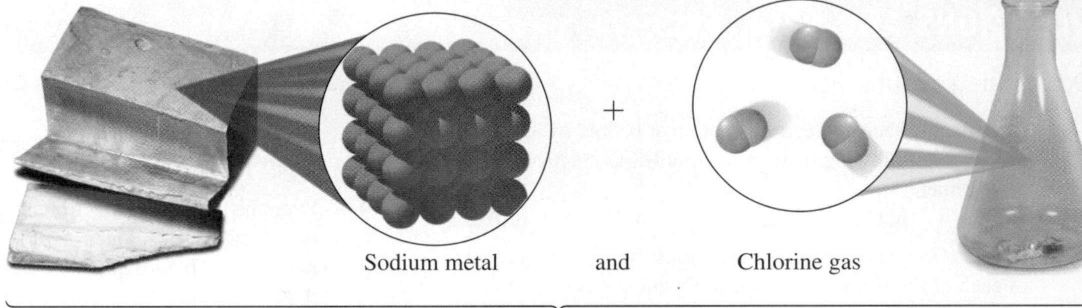

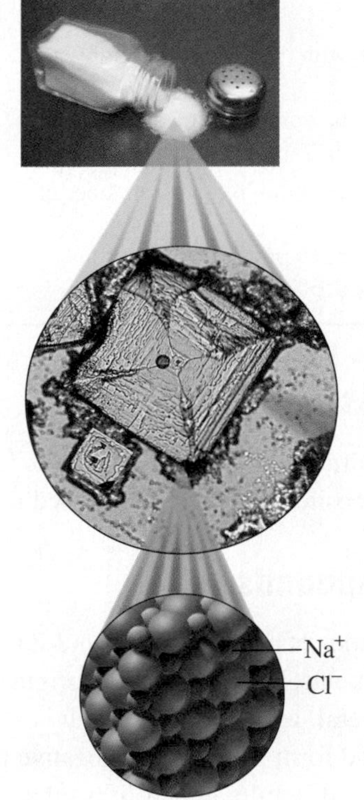

FIGURE 4.1 The elements sodium and chlorine react to form the ionic compound sodium chloride, the compound that makes up table salt. The magnification of NaCl crystals shows the arrangements of Na^+ and Cl^- ions in a NaCl crystal.

Q What is the type of bonding between Na^+ and Cl^- ions in NaCl?

PHYSICAL THERAPIST

"Physical therapists need to understand how the body, muscles, and joints function in order to recognize when something is not working and which area needs strengthening," says Vincent Leddy, physical therapist. "Chemistry is important to understanding the body's physiology and how chemical changes in the body affect movement. I went into physical therapy because I enjoy teaching movement to children. I am guiding Maggie into her chair but allowing her to move as much as she can by herself. I help her by giving gentle pressure and reassurance that she'll be safe in the transition. I work on getting Maggie to use her body, and the occupational therapist works on Maggie's fine motor skills and how she uses her hand on the switch or picks up objects. By using both physical and occupational therapy, we enhance the child's performance."

Subscripts in Formulas

Consider a compound of magnesium and chlorine. To achieve an octet, a Mg atom loses its two valence electrons to form Mg^{2+}. Two Cl atoms each gain one electron to form two Cl^- ions. The two Cl^- ions are needed to balance the positive charge of Mg^{2+}. This gives the formula $MgCl_2$, magnesium chloride, in which a subscript of 2 shows that two Cl^- are needed for charge balance.

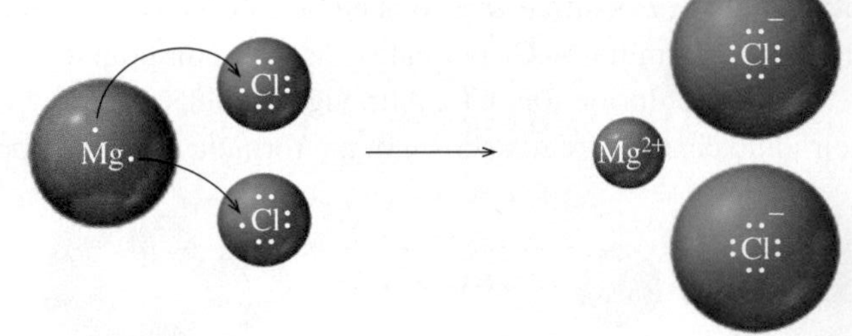

Writing Ionic Formulas from Ionic Charges

TUTORIAL
Writing Ionic Formulas

The subscripts in the formula of an ionic compound represent the number of positive and negative ions that give an overall charge of zero. Thus, we can now write a formula directly from the ionic charges of the positive and negative ions. Suppose we wish to write the formula of the ionic compound containing Na^+ and S^{2-} ions. To balance the ionic charge of the S^{2-} ion, we will need to place two Na^+ ions in the formula. This gives the formula Na_2S, which has an overall charge of zero. In the formula of an ionic compound, the cation is written first followed by the anion. Appropriate subscripts are used to show the number of each of the ions. This formula, which is the lowest ratio of the ions in an ionic compound, is called a *formula unit*.

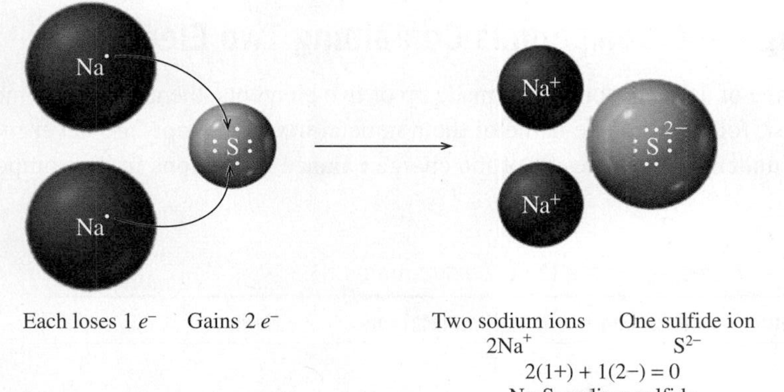

SAMPLE PROBLEM 4.2

Writing Formulas from Ionic Charges

Determine the ionic charges, and write the formula for the ionic compound formed when lithium and nitrogen react.

SOLUTION

Lithium in Group 1A (1) forms Li^+; nitrogen in Group 5A (15) forms N^{3-}. The charge of 3− for N^{3-} is balanced by three Li^+ ions. Writing the positive ion first gives the formula Li_3N.

STUDY CHECK 4.2

Determine the ionic charges, and write the formula of the compound that would form when calcium and oxygen react.

QUESTIONS AND PROBLEMS

Ionic Compounds

4.9 Which of the following pairs of elements are likely to form an ionic compound?
a. lithium and chlorine **b.** oxygen and bromine
c. potassium and oxygen **d.** sodium and neon
e. cesium and magnesium **f.** nitrogen and fluorine

4.10 Which of the following pairs of elements are likely to form an ionic compound?
a. helium and oxygen **b.** magnesium and chlorine
c. chlorine and bromine **d.** potassium and sulfur
e. sodium and potassium **f.** nitrogen and iodine

4.11 Write the correct ionic formula for the compound formed between each of the following pairs of ions:
a. Na^+ and O^{2-} **b.** Al^{3+} and Br^-
c. Ba^{2+} and N^{3-} **d.** Mg^{2+} and F^-
e. Al^{3+} and S^{2-}

4.12 Write the correct ionic formula for compounds formed between each of the following pairs of ions:
a. Al^{3+} and Cl^- **b.** Ca^{2+} and S^{2-}
c. Li^+ and S^{2-} **d.** Rb^+ and P^{3-}
e. Cs^+ and I^-

4.13 Determine the ions, and write the correct formula for the ionic compound formed by each of the following:
a. potassium and sulfur **b.** sodium and nitrogen
c. aluminum and iodine **d.** gallium and oxygen

4.14 Determine the ions, and write the correct formula for the ionic compound formed by each of the following:
a. calcium and chlorine **b.** rubidium and sulfur
c. sodium and phosphorus **d.** magnesium and oxygen

LEARNING GOAL

Given the formula of an ionic compound, write the correct name; given the name of an ionic compound, write the correct formula.

4.3 Naming and Writing Ionic Formulas

In naming an ionic compound, the name of a metal ion is the same as its elemental name. The name of a nonmetal ion is obtained by using the first syllable of its elemental name followed by *ide*. In the name of any compound, a space separates the name of the metal ion from the name of the nonmetal ion.

Iodized salt contains KI to prevent iodine deficiency.

Naming Ionic Compounds Containing Two Elements

In the name of an ionic compound made up of two elements, the name of the metal ion is given first, followed by the name of the nonmetal ion. Subscripts are never mentioned; they are understood as a result of the charge balance of the ions in the compound (see Table 4.4).

TABLE 4.4 Names of Some Ionic Compounds

Compound	Metal Ion	Nonmetal Ion	Name
KI	K^+	I^-	
	Potassium	Iodide	Potassium iodide
$MgBr_2$	Mg^{2+}	Br^-	
	Magnesium	Bromide	Magnesium bromide
Al_2O_3	Al^{3+}	O^{2-}	
	Aluminum	Oxide	Aluminum oxide

Guide to Naming Ionic Compounds with Metals That Form a Single Ion

1 Identify the cation and anion.

2 Name the cation by its element name.

3 Name the anion by using the first syllable of its element name followed by *ide*.

4 Write the name of the cation first and the name of the anion second.

SAMPLE PROBLEM 4.3

Naming Ionic Compounds

Write the name of the ionic compound Mg_3N_2.

SOLUTION

Step 1 **Identify the cation and anion.** The cation, Mg^{2+}, is from Group 2A (2), and the anion, N^{3-}, is from Group 5A (15).

Step 2 **Name the cation by its element name.** The cation Mg^{2+} is magnesium.

Step 3 **Name the anion by using the first syllable of its element name followed by *ide*.** The anion N^{3-} is nitride.

Step 4 **Write the name of the cation first and the name of the anion second.**
Mg_3N_2 magnesium nitride

STUDY CHECK 4.3

Name the compound Cs_2S.

Metals with Variable Charge

We have seen that the charge of an ion of a representative element can be obtained from its group number. However, it is not as easy to determine the charge of a transition element because they typically form two or more positive ions. The transition elements lose electrons, but they are lost from the highest energy level and sometimes from a lower energy level as well. This is also true for metals of representative elements in Groups 4A (14) and 5A (15), such as Pb, Sn, and Bi.

In some ionic compounds, iron is in the Fe^{2+} form, but in other compounds, it has the Fe^{3+} form. Copper also forms two different ions, Cu^{+} and Cu^{2+}. When a metal can form two or more types of ions, it has *variable charge*. Then we cannot predict the ionic charge from the group number.

When different ions are possible, a naming system is used to identify the particular cation. To do this, a Roman numeral that is equal to the ionic charge is placed in parentheses immediately after the elemental name. For example, Fe^{2+} is iron(II), and Fe^{3+} is iron(III). Table 4.5 lists the ions of some elements that produce more than one ion.

Figure 4.2 shows some ions and their location on the periodic table. The transition elements form more than one positive ion except for zinc (Zn^{2+}), cadmium (Cd^{2+}), and silver (Ag^{+}), which form only one ion. Thus, only the elemental names of zinc, cadmium, and silver are sufficient when naming their cations in ionic compounds. Metals in Groups 4A (14) and 5A (15) also form more than one type of positive ion. For example, lead and tin in Group 4A (14) form cations with charges of 2+ and 4+.

TABLE 4.5 Some Elements That Form More Than One Positive Ion

Element	Possible Ions	Name of Ion
Chromium	Cr^{2+}	Chromium(II)
	Cr^{3+}	Chromium(III)
Cobalt	Co^{2+}	Cobalt(II)
	Co^{3+}	Cobalt(III)
Copper	Cu^{+}	Copper(I)
	Cu^{2+}	Copper(II)
Gold	Au^{+}	Gold(I)
	Au^{3+}	Gold(III)
Iron	Fe^{2+}	Iron(II)
	Fe^{3+}	Iron(III)
Lead	Pb^{2+}	Lead(II)
	Pb^{4+}	Lead(IV)
Manganese	Mn^{2+}	Manganese(II)
	Mn^{3+}	Manganese(III)
Mercury	Hg_2^{2+}	Mercury(I)*
	Hg^{2+}	Mercury(II)
Nickel	Ni^{2+}	Nickel(II)
	Ni^{3+}	Nickel(III)
Tin	Sn^{2+}	Tin(II)
	Sn^{4+}	Tin(IV)

*Mercury(I) ions form an ion pair with a 2+ charge.

Determination of Variable Charge

When you name an ionic compound, you need to determine if the metal is a representative element or a transition element. If it is a transition element, except for zinc, cadmium, or silver, you will need to use its ionic charge as a Roman numeral as part of its name. The calculation of ionic charge depends on the negative charge of the anions in the formula. For example, we use charge balance to determine the charge of a copper cation in the ionic compound $CuCl_2$. Because there are two chloride ions, each with a 1− charge, the total negative charge is 2−. To balance this 2− charge, the copper ion must have a charge of 2+, or Cu^{2+}:

TUTORIAL
Writing Ionic Formulas

$CuCl_2$

Cu charge	+ 2 Cl^- charge	= 0
?	+ 2(1−)	= 0
2+	+ 2−	= 0

1 Group 1A	2 Group 2A	3 3B	4 4B	5 5B	6 6B	7 7B	8 8B	9 8B	10 8B	11 1B	12 2B	13 Group 3A	14 Group 4A	15 Group 5A	16 Group 6A	17 Group 7A	18 Group 8A
H^+																	
Li^+														N^{3-}	O^{2-}	F^-	
Na^+	Mg^{2+}											Al^{3+}		P^{3-}	S^{2-}	Cl^-	
K^+	Ca^{2+}				Cr^{2+} Cr^{3+}	Mn^{2+} Mn^{3+}	Fe^{2+} Fe^{3+}	Co^{2+} Co^{3+}	Ni^{2+} Ni^{3+}	Cu^{+} Cu^{2+}	Zn^{2+}					Br^-	
Rb^+	Sr^{2+}									Ag^+	Cd^{2+}		Sn^{2+} Sn^{4+}			I^-	
Cs^+	Ba^{2+}									Au^+ Au^{3+}	Hg_2^{2+} Hg^{2+}		Pb^{2+} Pb^{4+}				

Metals Metalloids Nonmetals

FIGURE 4.2 Metals form positive ions, and nonmetals form negat ve ions.
Q What are the typical ions of calcium, copper, and oxygen in ion c compounds?

TABLE 4.6 Some Ionic Compounds of Elements That Form Two Kinds of Positive Ions

Compound	Systematic Name
$FeCl_2$	Iron(II) chloride
Fe_2O_3	Iron(III) oxide
Cu_3P	Copper(I) phosphide
$CrBr_2$	Chromium(II) bromide
$SnCl_2$	Tin(II) chloride
PbS_2	Lead(IV) sulfide
BiF_3	Bismuth(III) fluoride

Guide to Naming Ionic Compounds with Variable Charge Metals

1. Determine the charge of the cation from the anion.
2. Name the cation by its element name and use a Roman numeral in parentheses for the charge.
3. Name the anion by using the first syllable of its element name followed by *ide*.
4. Write the name of the cation first and the name of the anion second.

To indicate the 2+ charge for the copper ion Cu^{2+}, we place the Roman numeral (II) immediately after copper when naming this compound: copper(II) chloride.

Table 4.6 lists the names of some ionic compounds in which the transition elements and elements from Groups 4A (14) and 5A (15) have more than one positive ion.

SAMPLE PROBLEM 4.4

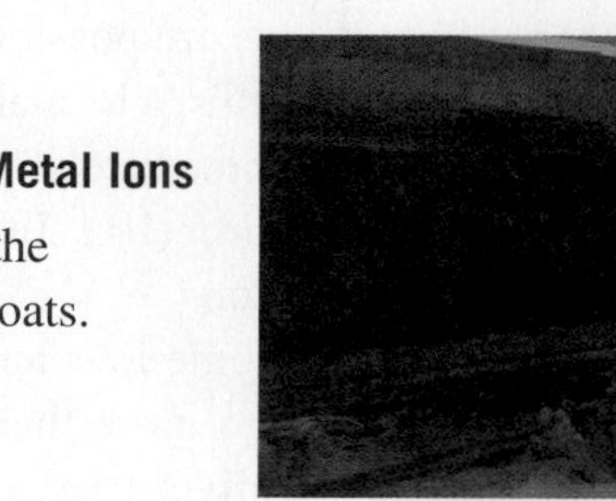

Naming Ionic Compounds with Variable Charge Metal Ions

Antifouling paint contains Cu_2O, which prevents the growth of barnacles and algae on the bottoms of boats. What is the name of Cu_2O?

SOLUTION

Step 1 Determine the charge of the cation from the anion. The nonmetal O in Group 6A (16) forms the O^{2-} ion. Because two Cu ions are in the formula to balance the charge of O^{2-}, the ionic charge of each Cu ion must be 1+.

	Metal		Nonmetal
Element	Copper		Oxygen
Group	Transition		6A (16)
Ion	Cu?		O^{2-}
Charge balance	2(1+)	+	(2−) = 0
Ion	Cu^+		O^{2-}

Step 2 Name the cation by its element name and use a Roman numeral in parentheses for the charge. copper(I)

Step 3 Name the anion by using the first syllable of its element name followed by *ide*. oxide

Step 4 Write the name of the cation first and the name of the anion second. copper(I) oxide

STUDY CHECK 4.4

Write the name of the compound whose formula is MnS.

Writing Formulas from the Name of an Ionic Compound

The formula of an ionic compound is written from the first part of the name for the ionic compound, which is the metal ion, and the second part of the name, which is the nonmetal. Subscripts are added, as needed, to balance the charge. The steps for writing a formula from the name of an ionic compound are shown in Sample Problem 4.5.

Guide to Writing Formulas from the Name of an Ionic Compound

1. Identify the cation and anion.
2. Balance the charges.
3. Write the formula, cation first, using subscripts from the charge balance.

SAMPLE PROBLEM 4.5

Writing Formulas of Ionic Compounds

Write the formula for iron(III) chloride.

SOLUTION

Step 1 Identify the cation and anion. The Roman numeral (III) indicates that the charge of the iron ion is 3+, Fe^{3+}.

	Cation	Anion
Ion	Iron(III)	Chloride
Group	Transition	7A (17)
Symbol	Fe^{3+}	Cl^-

Step 2 Balance the charges.

Fe^{3+} Cl^-
Cl^-
Cl^-

$1(3+) + \mathbf{3}(1-) = 0$

Becomes a subscript in the formula

Step 3 Write the formula, cation first, using subscripts from the charge balance.

$FeCl_3$

STUDY CHECK 4.5

Write the correct formula for chromium(III) oxide.

The pigment chrome oxide green contains chromium(III) oxide.

QUESTIONS AND PROBLEMS

Naming and Writing Ionic Formulas

4.15 Write the name for each of the following:
a. Al_2O_3 **b.** $CaCl_2$ **c.** Na_2O
d. Mg_3P_2 **e.** KI **f.** BaF_2

4.16 Write the name for each of the following:
a. $MgCl_2$ **b.** K_3P **c.** Li_2S
d. CsF **e.** MgO **f.** $SrBr_2$

4.17 Why is a Roman numeral placed after the name of most transition element ions?

4.18 The compound $CaCl_2$ is named calcium chloride; the compound $CuCl_2$ is named copper(II) chloride. Explain why a Roman numeral is used in one name but not the other.

4.19 Write the name for each of the following (include the Roman numeral when necessary):
a. Fe^{2+} **b.** Cu^{2+} **c.** Zn^{2+}
d. Pb^{4+} **e.** Cr^{3+} **f.** Mn^{2+}

4.20 Write the name for each of the following (include the Roman numeral when necessary):
a. Ag^+ **b.** Cu^+ **c.** Fe^{3+}
d. Sn^{2+} **e.** Au^{3+} **f.** Ni^{2+}

4.21 Write the name for each of the following:
a. $SnCl_2$ **b.** FeO **c.** Cu_2S
d. CuS **e.** $CdBr_2$ **f.** $ZnCl_2$

4.22 Write the name for each of the following:
a. Ag_3P **b.** PbS **c.** SnO_2
d. $MnCl_3$ **e.** FeS **f.** $CoCl_2$

4.23 Give the symbol of the cation in each of the following:
a. $AuCl_3$ **b.** Fe_2O_3 **c.** PbI_4 **d.** $SnCl_2$

4.24 Give the symbol of the cation in each of the following:
a. $FeCl_2$ **b.** CrO **c.** Ni_2S_3 **d.** AlP

4.25 Write the formula for each of the following:
a. magnesium chloride **b.** sodium sulfide
c. copper(I) oxide **d.** zinc phosphide
e. gold(III) nitride **f.** chromium(II) chloride

4.26 Write the formula for each of the following:
a. nickel(III) oxide **b.** barium fluoride
c. tin(IV) chloride **d.** silver sulfide
e. copper(II) iodide **f.** lithium nitride

4.4 Polyatomic Ions

LEARNING GOAL

Write the name and formula of a compound containing a polyatomic ion.

A **polyatomic ion** is a group of atoms that has an overall ionic charge. Most polyatomic ions consist of a nonmetal such as phosphorus, sulfur, carbon, or nitrogen bonded to oxygen atoms.

Almost all the polyatomic ions are anions with charges of 1−, 2−, or 3−, which indicate that the group of atoms has gained 1, 2, or 3 electrons to complete octets. Only one common polyatomic ion, NH_4^+, has a positive charge. Some models of common polyatomic ions are shown in Figure 4.3.

TUTORIAL
Polyatomic Ions

Names of Polyatomic Ions

The names of the most common polyatomic ions end in *ate*. When a related ion has one less oxygen atom, the *ite* ending is used for its name. Recognizing these endings will help you identify polyatomic ions in the name of a compound. The hydroxide ion (OH^-) and cyanide ion (CN^-) are exceptions to this naming pattern.

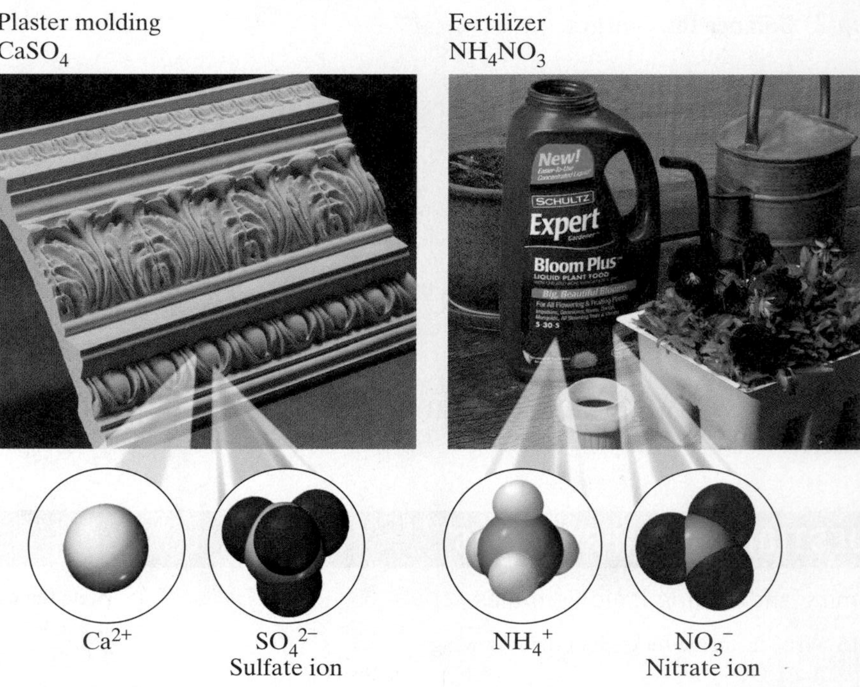

FIGURE 4.3 Many products contain polyatomic ions, which are groups of atoms that carry an ionic charge.
Q What is the charge of a sulfate ion?

By recognizing the formulas, charges, and names of the ions shown in bold type in Table 4.7, you can derive the related ions. Note that both the *ate* ion and *ite* ion of a particular nonmetal have the same ionic charge. For example, the sulfate ion is SO_4^{2-}, and the sulfite ion, which has one less oxygen atom, is SO_3^{2-}. Phosphate and phosphite ions each have a 3− charge; nitrate and nitrite have a 1− charge; and perchlorate, chlorate, chlorite, and hypochlorite all have a 1− charge. The halogens form four different polyatomic ions with oxygen. The one with one more oxygen than chlorate uses the prefix *per*; the one with one oxygen less than chlorite uses the prefix *hypo*.

TABLE 4.7 Names and Formulas of Some Common Polyatomic Ions

Nonmetal	Formula of Ion*	Name of Ion
Hydrogen	OH^-	Hydroxide
Nitrogen	NH_4^+	Ammonium
	NO_3^-	**Nitrate**
	NO_2^-	Nitrite
Chlorine	ClO_4^-	Perchlorate
	ClO_3^-	**Chlorate**
	ClO_2^-	Chlorite
	ClO^-	Hypochlorite
Carbon	CO_3^{2-}	**Carbonate**
	HCO_3^-	Hydrogen carbonate (or bicarbonate)
	CN^-	Cyanide
	$C_2H_3O_2^-$	Acetate
Sulfur	SO_4^{2-}	**Sulfate**
	HSO_4^-	Hydrogen sulfate (or bisulfate)
	SO_3^{2-}	Sulfite
	HSO_3^-	Hydrogen sulfite (or bisulfite)
Phosphorus	PO_4^{3-}	**Phosphate**
	HPO_4^{2-}	Hydrogen phosphate
	$H_2PO_4^-$	Dihydrogen phosphate
	PO_3^{3-}	Phosphite

*Formulas in bold type indicate the most common polyatomic ion for that element.

The formula of hydrogen carbonate, or *bicarbonate*, is written with a hydrogen in front of the polyatomic formula for carbonate and decreasing the charge to 1− to give HCO_3^-.

$$CO_3^{2-} + H^+ = HCO_3^-$$

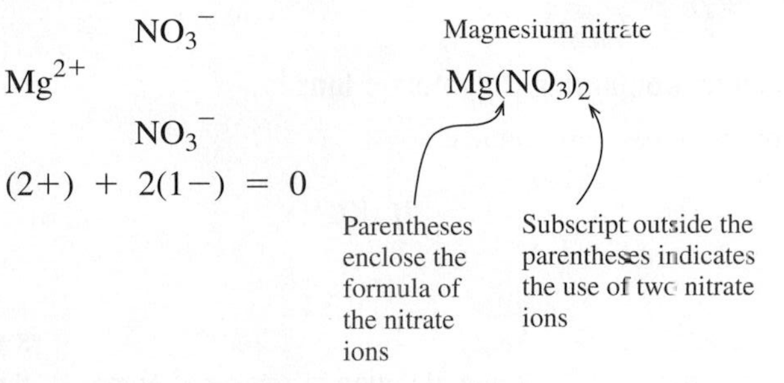
Sodium chlorite is used in the processing and bleaching of pulp from wood fibers and recycled cardboard.

Compounds Containing Polyatomic Ions

No polyatomic ion exists by itself. Like any ion, a polyatomic ion must be associated with ions of opposite charge. The bonding between polyatomic ions and other ions is one of electrical attraction. For example, the compound sodium sulfate consists of sodium ions (Na^+) and sulfate ions (SO_4^{2-}) held together by ionic bonds.

To write correct formulas for compounds containing polyatomic ions, we follow the same rules of charge balance that we used for writing the formulas of simple ionic compounds. The total negative and positive charges must equal zero. For example, consider the formula for a compound containing calcium ions and carbonate ions. The ions are written as

Ca^{2+} (Calcium ion) CO_3^{2-} (Carbonate ion)

Ionic charge: $(2+) + (2-) = 0$

Because one ion of each balances the charge, the formula is written as

$CaCO_3$
Calcium carbonate

When more than one polyatomic ion is needed for charge balance, parentheses are used to enclose the formula of the ion. A subscript is written outside the closing parenthesis to indicate the number of polyatomic ions. Consider the formula for magnesium nitrate. The ions in this compound are the magnesium ion and the nitrate ion, a polyatomic ion.

Mg^{2+} (Magnesium ion) NO_3^- (Nitrate ion)

To balance the positive charge of 2+, two nitrate ions are needed. The formula, including the parentheses around the nitrate ion, is as follows:

Mg^{2+} NO_3^- NO_3^-

$(2+) + 2(1-) = 0$

Magnesium nitrate
$Mg(NO_3)_2$

Parentheses enclose the formula of the nitrate ions

Subscript outside the parentheses indicates the use of two nitrate ions

CONCEPT CHECK 4.2

Polyatomic Ions in Bones and Teeth

Bones and teeth contain a solid mineral substance called hydroxyapatite, $Ca_{10}(PO_4)_6(OH)_2$. What polyatomic ions are contained in the mineral substance of bone and teeth?

ANSWER

The polyatomic ions are phosphate (PO_4^{3-}) and hydroxide (OH^-).

The mineral substance in teeth contains phosphate and hydroxide ions.

Naming Compounds Containing Polyatomic Ions

When naming ionic compounds containing polyatomic ions, we first write the positive ion, usually a metal, and then we write the name of the polyatomic ion. It is important that you learn to recognize the polyatomic ion in the formula and name it correctly. As with other ionic compounds, no prefixes are used.

Na_2SO_4	$FePO_4$	$Al_2(CO_3)_3$
$Na_2\boxed{SO_4}$	$Fe\boxed{PO_4}$	$Al_2(\boxed{CO_3})_3$
Sodium sulfate	Iron(III) phosphate	Aluminum carbonate

Table 4.8 lists the formulas and names of some ionic compounds that include polyatomic ions and also gives their uses in medicine and industry.

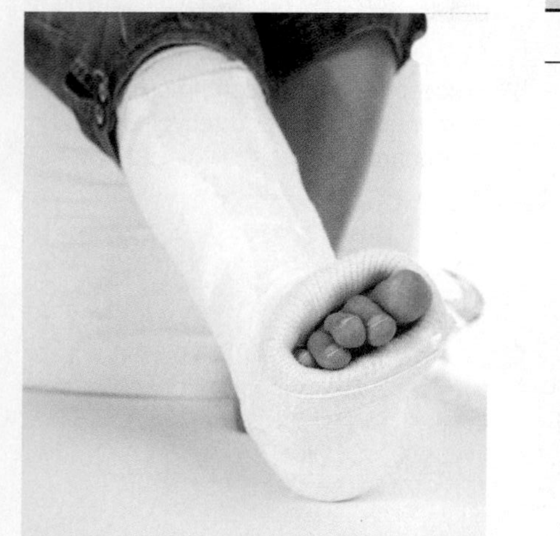

A plaster cast made of $CaSO_4$ immobilizes a broken leg.

TABLE 4.8 Some Compounds That Contain Polyatomic Ions

Formula	Name	Use
$BaSO_4$	Barium sulfate	Contrast medium for X-rays
$CaCO_3$	Calcium carbonate	Antacid, calcium supplement
$Ca_3(PO_4)_2$	Calcium phosphate	Calcium dietary supplement
$CaSO_3$	Calcium sulfite	Preservative in cider and fruit juices
$CaSO_4$	Calcium sulfate	Plaster casts
$AgNO_3$	Silver nitrate	Topical anti-infective
$NaHCO_3$	Sodium bicarbonate *or* Sodium hydrogen carbonate	Antacid
$Zn_3(PO_4)_2$	Zinc phosphate	Dental cement
$FePO_4$	Iron(III) phosphate	Food additive
K_2CO_3	Potassium carbonate	Alkalizer, diuretic
$Al_2(SO_4)_3$	Aluminum sulfate	Antiperspirant, anti-infective
$AlPO_4$	Aluminum phosphate	Antacid
$MgSO_4$	Magnesium sulfate	Cathartic, Epsom salts

Guide to Naming Ionic Compounds with Polyatomic Ions

1. Identify the cation and polyatomic ion (anion).
2. Name the cation using a Roman numeral, if needed.
3. Name the polyatomic ion usually ending with *ite* or *ate*.
4. Write the name of the compound, cation first and the polyatomic ion second.

SAMPLE PROBLEM 4.6

Naming Compounds Containing Polyatomic Ions

Name the following ionic compounds:

a. $Cu(NO_2)_2$ **b.** $KClO_3$

SOLUTION

Formula	Step 1 Cation/Anion	Step 2 Name of Cation	Step 3 Name of Anion	Step 4 Name of Compound
a. $Cu(NO_2)_2$	Cu^{2+} NO_2^-	Copper(II) ion	Nitrite ion	Copper(II) nitrite
b. $KClO_3$	K^+ ClO_3^-	Potassium ion	Chlorate ion	Potassium chlorate

STUDY CHECK 4.6

What is the name of $Co_3(PO_4)_2$?

QUESTIONS AND PROBLEMS

Polyatomic Ions

4.27 Write the formulas, including the charge, for the following polyatomic ions:
a. hydrogen carbonate (bicarbonate)
b. ammonium
c. phosphate
d. hydrogen sulfate

4.28 Write the formulas, including the charge, for the following polyatomic ions:
a. nitrite **b.** sulfite
c. hydroxide **d.** phosphite

4.29 Name each of the following polyatomic ions:
a. SO_4^{2-} **b.** CO_3^{2-} **c.** PO_4^{3-} **d.** NO_3^-

4.30 Name each of the following polyatomic ions:
a. OH^- **b.** HSO_3^- **c.** CN^- **d.** NO_2^-

4.31 Complete the following table with the formula of the compound that forms between each pair of ions:

	NO_2^-	CO_3^{2-}	HSO_4^-	PO_4^{3-}
Li^+				
Cu^{2+}				
Ba^{2+}				

4.32 Complete the following table with the formula of the compound that forms between each pair of ions:

	NO_3^-	HCO_3^-	SO_3^{2-}	HPO_4^{2-}
NH_4^+				
Al^{3+}				
Pb^{4+}				

4.33 Write the formula for the polyatomic ion in each of the following and name each compound:
a. Na_2CO_3 **b.** NH_4Cl **c.** K_3PO_4
d. $Cr(NO_2)_2$ **e.** $FeSO_3$

4.34 Write the formula for the polyatomic ion in each of the following and name each compound:
a. KOH **b.** $NaNO_3$ **c.** Au_2CO_3
d. $NaHCO_3$ **e.** $BaSO_4$

4.35 Write the correct formula for the following compounds:
a. barium hydroxide **b.** sodium sulfate
c. iron(II) nitrate **d.** zinc phosphate
e. iron(III) carbonate

4.36 Write the correct formula for the following compounds:
a. aluminum chlorate **b.** ammonium oxide
c. magnesium bicarbonate **d.** sodium nitrite
e. copper(I) sulfate

4.5 Covalent Compounds and Their Names

LEARNING GOAL

Given the formula of a covalent compound, write its correct name; given the name of a covalent compound, write its formula.

A **covalent compound** forms when atoms of two nonmetals share electrons. Because of the high ionization energies of the nonmetals, electrons are not transferred between atoms but are shared to achieve stability. When atoms share electrons, the bond is a *covalent bond*. When two or more atoms share electrons, they form **molecules**.

SELF STUDY ACTIVITY
Covalent Bonds

Formation of a Hydrogen Molecule

The simplest covalent molecule is hydrogen gas, H_2. When two H atoms are far apart, there is no attraction between them. As the H atoms move closer, the positive charge of each nucleus attracts the electron of the other atom. This attraction, which is greater than the repulsion between the valence electrons, pulls the H atoms closer until they share a pair of valence electrons (see Figure 4.4). The result is called a *covalent bond*, in which the shared electrons give the noble gas arrangement of He to *each* of the H atoms. When the H atoms form H_2, they are more stable than two individual H atoms.

Formation of Octets in Covalent Molecules

TUTORIAL
Covalent Molecules and the Octet Rules

The valence electrons in covalent molecules are shown using an electron-dot formula, or Lewis structure. The shared electrons, or *bonding pairs*, are shown as two dots or a single line between atoms. The nonbonding pairs of electrons, or *lone pairs*, are placed on the outside. For example, a fluorine molecule, F_2, consists of two fluorine atoms,

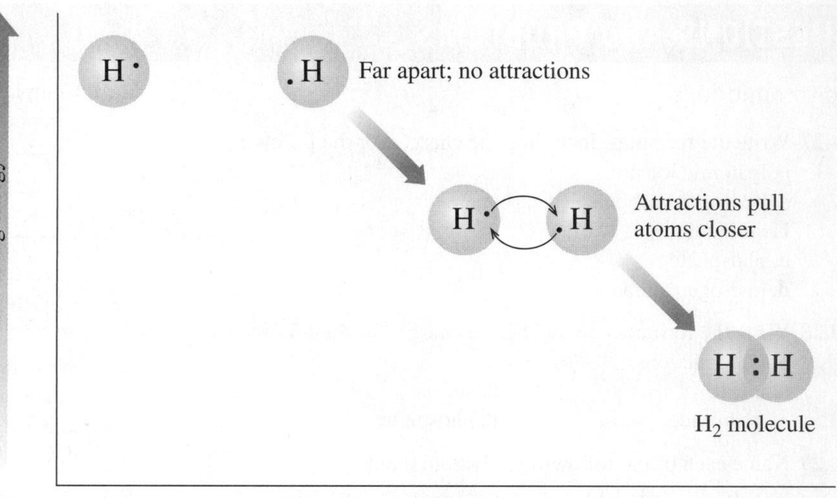

FIGURE 4.4 A covalent bond forms as H atoms move close together to share electrons.
Q What determines the attraction between two H atoms?

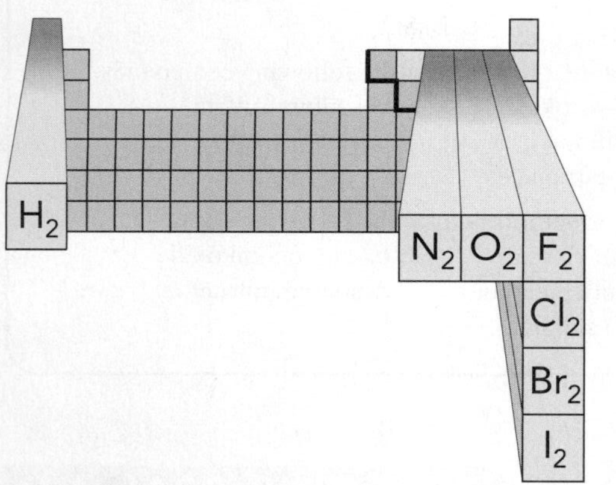

Group 7A (17), each with seven valence electrons. Both F atoms achieve octets by sharing their unpaired valence electrons. In the F_2 molecule, each F atom has the noble gas arrangement of neon.

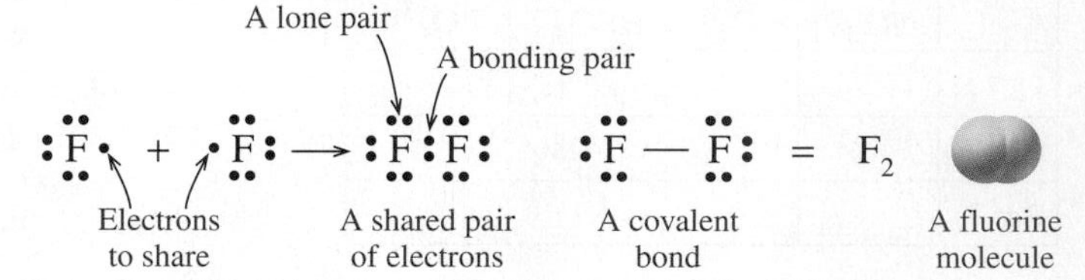

Hydrogen (H_2) and fluorine (F_2) are examples of nonmetal elements whose natural state is diatomic; that is, they contain two like atoms. The elements that exist as diatomic molecules are listed in Table 4.9.

TABLE 4.9 Elements That Exist as Diatomic, Covalent Molecules

Diatomic Molecule	Name
H_2	Hydrogen
N_2	Nitrogen
O_2	Oxygen
F_2	Fluorine
Cl_2	Chlorine
Br_2	Bromine
I_2	Iodine

Sharing Electrons between Atoms of Different Elements

The number of electrons that a nonmetal atom shares and the number of covalent bonds it forms are usually equal to the number of electrons it needs to acquire a noble gas arrangement. Table 4.10 gives the most typical bonding patterns for some nonmetals. For example, the element carbon combines with hydrogen to form a covalent compound, CH_4, methane, which is a component of natural gas.

Drawing Electron-Dot Formulas

To draw the electron-dot formula of CH_4, we first draw the electron-dot symbols of carbon and hydrogen (see Chapter 3, Section 3.7).

$$\cdot\dot{\underset{\cdot}{C}}\cdot \qquad \cdot H$$

TABLE 4.10 Typical Bonding Patterns of Some Nonmetals in Covalent Compounds

1A (1)	3A (13)	4A (14)	5A (15)	6A (16)	7A (17)
*H 1 bond					
	*B 3 bonds	C 4 bonds	N 3 bonds	O 2 bonds	F 1 bond
		Si 4 bonds	P 3 bonds	S 2 bonds	Cl, Br, I 1 bond

*H and B do not form octets. H atoms share one electron pair; B atoms share three electron pairs for a set of six electrons.

TABLE 4.11 Electron-Dot Formulas for Some Covalent Compounds

CH_4	NH_3	H_2O
Formulas Using Electron Dots		
H H:C:H H	H:N:H H	:O:H H
Formulas Using Bonds and Electron Dots		
H | H—C—H | H	H—N—H | H	:O—H | H
Molecular Models		
Methane molecule	Ammonia molecule	Water molecule

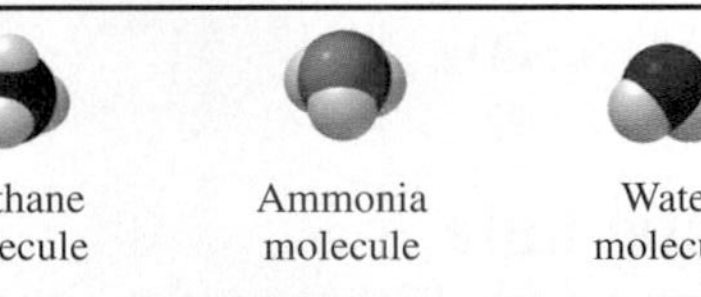

Then we can determine the number of valence electrons needed for carbon and hydrogen. When a carbon atom shares its four electrons with four hydrogen atoms, carbon obtains an octet and each hydrogen atom is complete with two shared electrons. The electron-dot formula is drawn with the carbon atom as the center atom with the hydrogen atoms on each of the sides. The bonding pairs of electrons, which are single covalent bonds, may also be shown as single lines between the carbon atom and each of the hydrogen atoms.

```
   H           H
H:C:H        H—C—H
   H           H
```

(MC) TUTORIAL Writing Electron-Dot Formulas

TUTORIAL Covalent Lewis-Dot Structures

Table 4.11 gives the formulas and three-dimensional models of some covalent molecules.

CONCEPT CHECK 4.3

Drawing Electron-Dot Formulas

Use the electron-dot symbols of S and F to draw the electron-dot formula for SF_2, sulfur difluoride, in which sulfur is the central atom.

ANSWER

To draw the electron-dot formula of SF_2, we first draw the electron-dot symbols of sulfur with six valence electrons and fluorine with seven valence electrons.

:S·   ·F:

Then we can determine the number of valence electrons needed for sulfur and fluorine. A sulfur atom will share each of its two unpaired electrons with the unpaired electrons of two fluorine atoms. In this way, both the S atom and the two F atoms obtain octets. The electron-dot formula for SF_2 is drawn with the sulfur atom with two fluorine atoms attached with paired electrons.

```
:S:F:      :S—F:
:F:         |
           :F:
```

SAMPLE PROBLEM 4.7

Drawing Electron-Dot Formulas for Covalent Compounds

Draw an electron-dot formula for H_2S.

SOLUTION

Sulfur, which has six valence electrons, shares two electrons to form an octet. Two atoms of hydrogen each share one valence electron to be stable. The shared pairs of electrons representing two covalent bonds can be written as single lines.

```
    H
    |
H—S:
    ··
```

Electron-dot formula for H_2S

STUDY CHECK 4.7

Draw the electron-dot formula for PH_3.

Exceptions to the Octet Rule

While the octet rule is useful, there are exceptions. We have already seen that a hydrogen (H_2) molecule requires just two electrons or a single bond to achieve the stability of the nearest noble gas, helium. Although the nonmetals typically form octets, atoms such as P, S, Cl, Br, and I can form compounds with 10 or 12 valence electrons. For example, in PCl_3, the P atom has an octet, but in PCl_5, the P atom has 10 valence electrons or 5 bonds. In H_2S, the S atom has an octet, but in SF_6, there are 12 valence electrons or 6 bonds to the sulfur atom. In this text, we will encounter formulas with expanded octets, but we do not represent them with electron-dot formulas.

Double and Triple Covalent Bonds

Up to now, we have looked at covalent bonding in molecules having only single bonds. In many covalent compounds, atoms share two or three pairs of electrons to complete their octets. A **double bond** occurs when two pairs of electrons are shared; in a **triple bond**, three pairs of electrons are shared. Atoms of carbon, oxygen, nitrogen, and sulfur are most likely to form multiple bonds.

Double and triple bonds form when the number of valence electrons is not enough to complete the octets of all the atoms in the molecule. Then one or more lone pairs from the atoms attached to the central atom are shared with the central atom.

For example, there are double bonds in CO_2 because two pairs of electrons are shared between the carbon atom and each oxygen atom to give octets. Atoms of hydrogen and the halogens do not form double or triple bonds. The process of drawing an electron-dot formula is shown in Sample Problem 4.8.

Guide to Drawing Electron-Dot Formulas

1. Determine the arrangement of atoms.
2. Determine the total number of valence electrons.
3. Attach each bonded atom to the central atom with a pair of electrons.
4. Place the remaining electrons using single or multiple bonds to complete octets (two for H, six for B).

SAMPLE PROBLEM 4.8

Drawing Electron-Dot Formulas with Multiple Bonds

Draw the electron-dot formula for carbon dioxide, CO_2, in which the central atom is C.

SOLUTION

Step 1 **Determine the arrangement of atoms.** O C O

Step 2 **Determine the total number of valence electrons.** Using the group number to determine the number of valence electrons, one carbon atom has four valence electrons, and each oxygen atom has six valence electrons, which gives a total of 16 valence electrons.

Valence electrons

C 2(O) CO_2

$4e^- + 2(6e^-) = 16e^-$

Step 3 **Attach each bonded atom to the central atom with a pair of electrons.** A pair of bonding electrons (single bond) is placed between each O atom and the central C atom.

O:C:O or O—C—O

Step 4 **Place the remaining electrons using single or multiple bonds to complete octets.** Because we used four valence electrons to attach the C atom to two O atoms, there are 12 valence electrons remaining:

16 valence e^- − 4 bonding e^- = 12 e^- remaining

The 12 remaining electrons are placed as six lone pairs on the outside O atoms. However, this does not complete the octet of the C atom.

:Ö:C:Ö: or :Ö—C—Ö:

To obtain an octet, the C atom must share lone pairs from each of the O atoms. When two bonding pairs occur between atoms, it is known as a double bond.

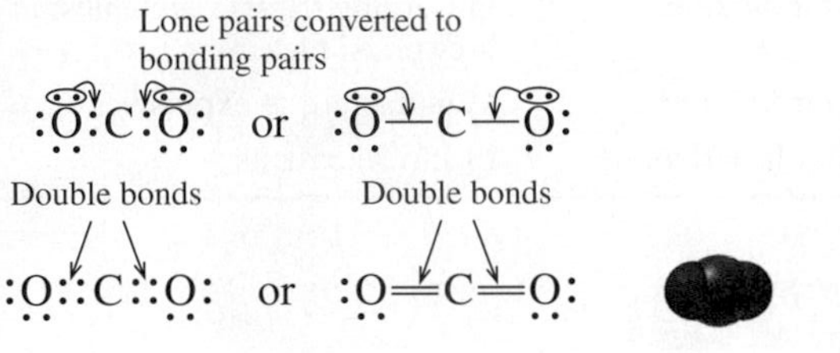

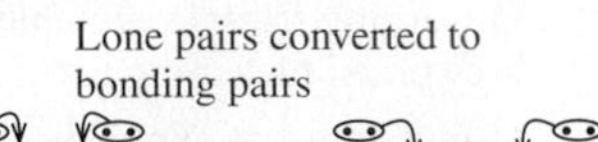

Molecule of carbon dioxide

STUDY CHECK 4.8

Draw the electron-dot formula for SO_2. The atoms are arranged as O S O.

CONCEPT CHECK 4.4

Forming Triple Bonds in Covalent Molecules

The covalent molecule N_2 contains a triple bond. Describe how the atoms of N achieve octets by forming a triple bond.

ANSWER

Because nitrogen is in Group 5A (15), each N atom has five valence electrons. An octet for each N atom cannot be achieved by sharing only one or two pairs of electrons. However, each N atom achieves an octet by sharing three bonding pairs of electrons to form a triple bond.

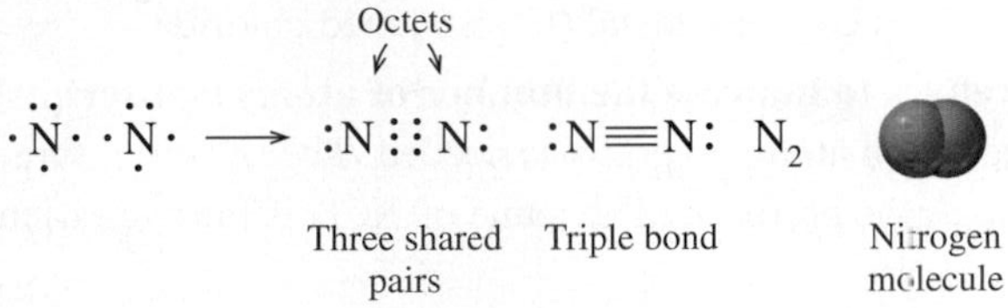

Octets

·N̈· ·N̈· ⟶ :N⋮⋮N: :N≡N: N_2

Three shared pairs Triple bond

Nitrogen molecule

TUTORIAL
Naming Covalent Compounds

TUTORIAL
Naming Molecular Compounds

Names and Formulas of Covalent Compounds

When naming a covalent compound, the first nonmetal in the formula is named by its element name; the second nonmetal is named using the first syllable of its name, followed by *ide*. When a subscript indicates two or more atoms of an element, a prefix is used in front of its name. Table 4.12 lists prefixes used in naming covalent compounds. The names of covalent compounds need prefixes because several different compounds can be formed from the same two nonmetals. For example, carbon and oxygen can form two different compounds, carbon monoxide, CO, and carbon dioxide, CO_2.

When the vowels *o* and *o* or *a* and *o* appear together, the first vowel is omitted as in carbon monoxide. In the name of a covalent compound, the prefix *mono* is usually omitted, as in NO, nitrogen oxide. Traditionally, however, CO is named carbon monoxide. Table 4.13 lists the formulas, names, and commercial uses of some covalent compounds.

TABLE 4.12 Prefixes Used in Naming Covalent Compounds

1 mono	6 hexa
2 di	7 hepta
3 tri	8 octa
4 tetra	9 nona
5 penta	10 deca

TABLE 4.13 Some Common Covalent Compounds

Formula	Name	Commercial Uses
CS_2	Carbon disulfide	Manufacture of rayon
CO_2	Carbon dioxide	Fire extinguishers, dry ice, propellant in aerosols, carbonation of beverages
NO	Nitrogen oxide	Stabilizer
N_2O	Dinitrogen oxide	Inhalation anesthetic, "laughing gas"
SO_2	Sulfur dioxide	Preserving fruits, vegetables; disinfectant in breweries; bleaching textiles
SO_3	Sulfur trioxide	Manufacture of explosives
SF_6	Sulfur hexafluoride	Electrical circuits

CONCEPT CHECK 4.5

Naming Covalent Compounds

Why does the name of the covalent compound BrCl, bromine chloride, not include a prefix, but the name of OCl_2, oxygen dichloride, does?

ANSWER

When a formula has one atom of any element, the prefix *mono* is not used in the name. Thus, the name of BrCl is bromine chloride. However, two or more atoms of an element are indicated by using a prefix. Thus, the name of OCl_2 is oxygen dichloride.

Guide to Naming Covalent Compounds

1 Name the first nonmetal by its element name.

2 Name the second element by using the first syllable of its name followed by *ide*.

3 Add prefixes to indicate the number of atoms (subscripts).

SAMPLE PROBLEM 4.9

Naming Covalent Compounds

Name the covalent compound NCl_3.

SOLUTION

Step 1 **Name the first nonmetal by its element name.** In NCl_3, the first nonmetal (N) is nitrogen.

Step 2 **Name the second element by using the first syllable of its name followed by *ide*.** The second nonmetal (Cl) is named chloride.

Step 3 **Add prefixes to indicate the number of atoms (subscripts).** Because there is one nitrogen atom, no prefix is needed. The subscript three for the Cl atoms is shown as the prefix *tri*. The name of NCl_3 is nitrogen trichloride.

STUDY CHECK 4.9

Write the name of each of the following compounds:

a. $SiBr_4$ **b.** Br_2O

Writing Formulas from the Names of Covalent Compounds

In the name of a covalent compound, the names of two nonmetals are given along with prefixes for the number of atoms of each. To obtain a formula, we write the symbol for each element and a subscript if a prefix indicates two or more atoms.

SAMPLE PROBLEM 4.10

Writing Formulas for Covalent Compounds

Write the formula for diboron trioxide.

SOLUTION

Step 1 **Write the symbols in the order of the elements in the name.** In this covalent compound of two nonmetals, the first nonmetal is boron (B), and the second nonmetal is oxygen (O).

B O

Step 2 **Write any prefixes as subscripts.** The prefix *di* in *di*boron indicates that there are two atoms of boron, shown as a subscript 2 in the formula. The prefix *tri* in *tri*oxide indicates that there are three atoms of oxygen, shown as a subscript 3 in the formula.

B_2O_3

STUDY CHECK 4.10

What is the formula of iodine pentafluoride?

Guide to Writing Formulas for Covalent Compounds

1 Write the symbols in the order of the elements in the name.

2 Write any prefixes as subscripts.

Summary of Naming Ionic and Covalent Compounds

We have now examined strategies for naming ionic and covalent compounds. In general, compounds having two elements are named by stating the first element name, followed by the name of the second element with an *ide* ending. If the first element is a metal, the compound is usually ionic; if the first element is a nonmetal, the compound is usually covalent. For ionic compounds, it is necessary to determine whether the metal can form more than one type of positive ion; if so, a Roman numeral following the name of the metal indicates the particular ionic charge. One exception is the ammonium ion, NH_4^+, which is also written first as a positively charged polyatomic ion. Ionic compounds having three or more elements include some type of polyatomic ion. They are named by ionic rules but have an *ate* or *ite* ending when the polyatomic ion has a negative charge.

In naming covalent compounds having two elements, prefixes are necessary to indicate two or more atoms of each nonmetal as shown in that particular formula (see Figure 4.5).

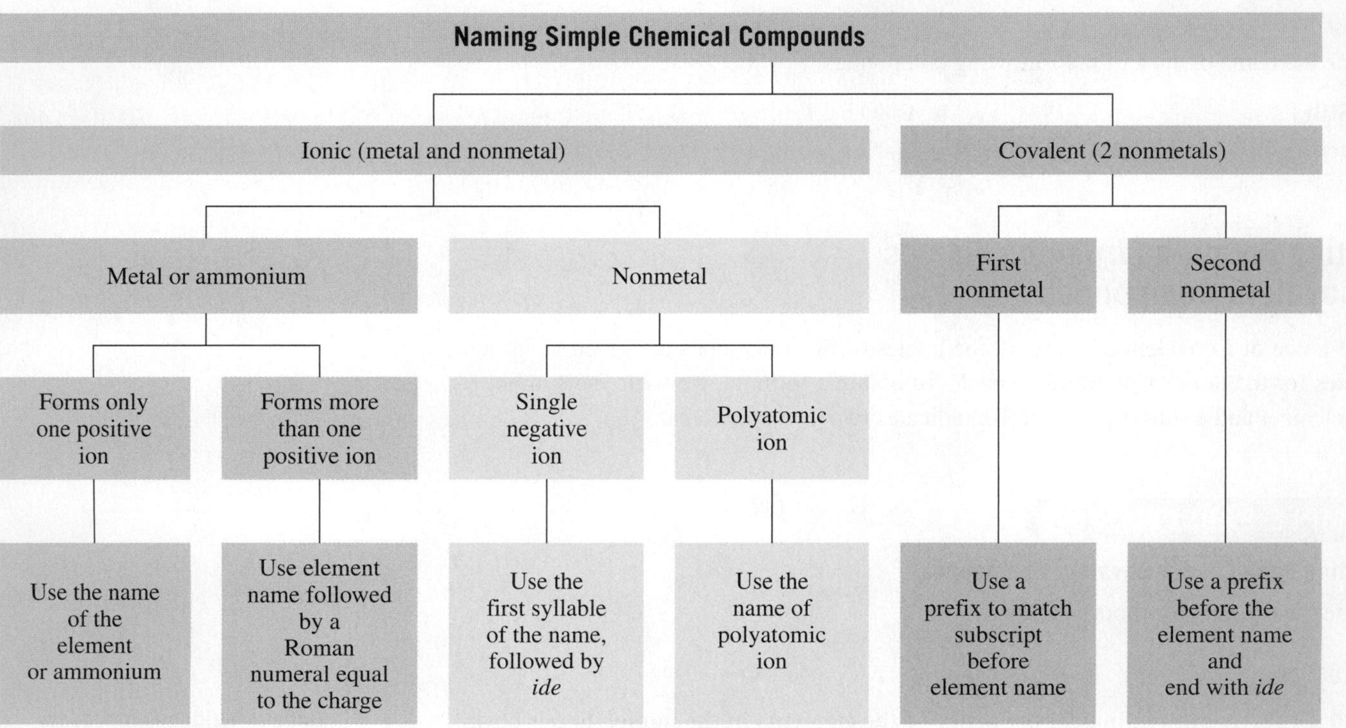

FIGURE 4.5 A flowchart for naming ionic and covalent compounds.

Q Why are the names of some metal ions followed by a Roman numeral in the name of a compound?

SAMPLE PROBLEM 4.11

Naming Ionic and Covalent Compounds

Identify each of the following compounds as ionic or covalent and give its name:

a. Na_3P **b.** $NiSO_4$ **c.** SO_3

SOLUTION

a. Na_3P, consisting of a metal and nonmetal, is an ionic compound. As a representative element in Group 1A (1), Na forms the sodium ion, Na^+. Phosphorus, as a representative element in Group 5A (15), forms a phosphide ion, P^{3-}. Writing the name of the cation followed by the name of the anion gives the name sodium phosphide.

b. $NiSO_4$, consisting of a cation of a transition element and a polyatomic ion SO_4^{2-}, is an ionic compound. As a transition element, Ni forms more than one type of ion. In this formula, the 2− charge of SO_4^{2-} is balanced by one nickel ion, Ni^{2+}. In the name, a Roman numeral written after the metal name, nickel(II), specifies the 2+ charge. The anion SO_4^{2-} is a polyatomic ion named sulfate. The compound is named nickel(II) sulfate.

c. SO_3 consists of two nonmetals, which indicates that it is a covalent compound. The first element S is *sulfur* (no prefix is needed). The second element O, *oxide*, has subscript 3, which requires a prefix *tri* in the name. The compound is named sulfur trioxide.

STUDY CHECK 4.11

What is the name of $Fe(NO_3)_3$?

QUESTIONS AND PROBLEMS

Covalent Compounds and Their Names

4.37 What elements on the periodic table are most likely to form covalent compounds?

4.38 How does the bond that forms between Na and Cl differ from a bond that forms between N and Cl?

4.39 Draw the electron-dot formula for each of the following covalent molecules:

a. Br_2 **b.** H_2 **c.** HF **d.** OF_2

4.40 Draw the electron-dot formula for each of the following covalent molecules:

a. NI_3 **b.** CCl_4 **c.** Cl_2 **d.** SiF_4

4.41 Name each of the following:
a. PBr_3 **b.** CBr_4 **c.** SiO_2 **d.** HF **e.** NI_3

4.42 Name each of the following:
a. CS_2 **b.** P_2O_5 **c.** Cl_2O **d.** PCl_3 **e.** CO

4.43 Name each of the following:
a. N_2O_3 **b.** Si_2Br_6 **c.** P_4S_3 **d.** PCl_5 **e.** N_2S_3

4.44 Name each of the following:
a. SiF_4 **b.** IBr_3 **c.** CO_2 **d.** N_2F_2 **e.** SeF_6

4.45 Write the formula of each of the following:
a. carbon tetrachloride **b.** carbon monoxide
c. phosphorus trichloride **d.** dinitrogen tetroxide

4.46 Write the formula of each of the following:
a. sulfur dioxide **b.** silicon tetrachloride
c. iodine trifluoride **d.** dinitrogen oxide

4.47 Write the formula of each of the following:
a. oxygen difluoride **b.** boron trichloride
c. dinitrogen trioxide **d.** sulfur hexafluoride

4.48 Write the formula of each of the following:
a. sulfur dibromide **b.** carbon disulfide
c. tetraphosphorus hexoxide **d.** dinitrogen pentoxide

4.49 Name each of the following
a. $Al_2(SO_4)_3$ antiperspirant **b.** $CaCO_3$ antacid
c. N_2O "laughing gas," inhaled anesthetic
d. Na_3PO_4 cathartic **e.** $(NH_4)_2SO_4$ fertilizer
f. Fe_2O_3 pigment

4.50 Name each of the following
a. N_2 Earth's atmosphere **b.** $Mg_3(PO_4)_2$ antacid
c. $FeSO_4$ iron supplement in vitamins
d. N_2O_4 rocket fuel **e.** Cu_2O fungicide
f. NI_3 contact explosive

4.6 Electronegativity and Bond Polarity

LEARNING GOAL

Use electronegativity to determine the polarity of a bond.

We can learn more about the chemistry of compounds by looking at how the bonding electrons are shared between atoms. Although we have discussed covalent bonds as one or more bonding pairs of electrons, we do not know if those bonding electrons are shared equally or unequally.

To do this, we use **electronegativity**, which is the ability of an atom to attract the shared electrons in a bond. Nonmetals have higher electronegativities than do metals, because nonmetals have a greater attraction for electrons than metals. On the electronegativity scale, fluorine was assigned a value of 4.0, and the electronegativities for all other elements were determined relative to the attraction of fluorine for shared electrons. The nonmetals fluorine (4.0) and oxygen (3.5), which are located in the upper right corner of the periodic table, have the highest electronegativities. The metals cesium and francium, which have the lowest electronegativity (0.7), are located in the lower left corner of the periodic table. The electronegativities for the representative elements are shown in Figure 4.6. Note that there are no electronegativity values for the noble gases because they do not typically form bonds. The electronegativity values for transition elements are also low, but we have not included them in our discussion.

SELF STUDY ACTIVITY
Electronegativity

Electronegativity increases →

Electronegativity decreases ↓

H 2.1

1 Group 1A	2 Group 2A	13 Group 3A	14 Group 4A	15 Group 5A	16 Group 6A	17 Group 7A	18 Group 8A
Li 1.0	Be 1.5	B 2.0	C 2.5	N 3.0	O 3.5	F 4.0	
Na 0.9	Mg 1.2	Al 1.5	Si 1.8	P 2.1	S 2.5	Cl 3.0	
K 0.8	Ca 1.0	Ga 1.6	Ge 1.8	As 2.0	Se 2.4	Br 2.8	
Rb 0.8	Sr 1.0	In 1.7	Sn 1.8	Sb 1.9	Te 2.1	I 2.5	
Cs 0.7	Ba 0.9	Tl 1.8	Pb 1.9	Bi 1.9	Po 2.0	At 2.1	

FIGURE 4.6 The electronegativities of representative elements in Group 1A (1) to Group 7A (17), which indicate the ability of atoms to attract shared electrons, increase across a period and decrease going down a group.

Q What element on the periodic table has the strongest attraction for shared electrons?

CONCEPT CHECK 4.6

Electronegativity

Without referring to the electronegativity scale, predict the order of increasing electronegativity for the elements Cl, F, P, Mg.

ANSWER

We know that electronegativities increase from the lower left corner to the upper right corner of the periodic table. Because Mg is on the left side of Period 3, Mg has the lowest electronegativity of these elements. The element P, which is on the right of Mg, but on the left of Cl, would have the next lowest electronegativity. The Cl on the right end of Period 3 would have a higher electronegativity than P. The highest electronegativity is F, which is above Cl in the upper right corner of the periodic table. The order of increasing electronegativity is: Mg, P, Cl, F.

SELF STUDY ACTIVITY
Bonds and Bond Polarities

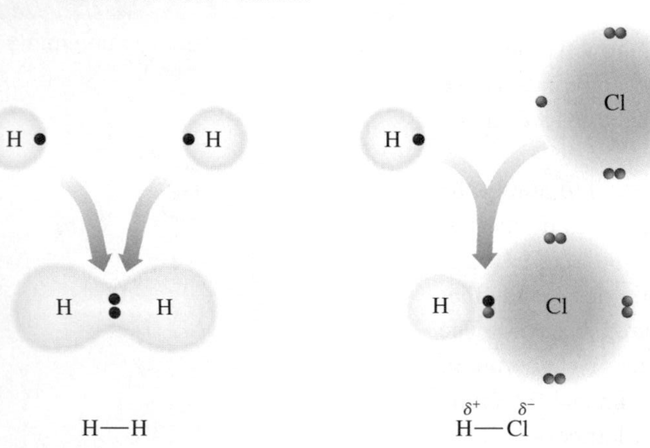

FIGURE 4.7 In the nonpolar covalent bond of H_2, electrons are shared equally. In the polar covalent bond of HCl, electrons are shared unequally.

Q H_2 has a nonpolar covalent bond, but HCl has a polar covalent bond. Explain.

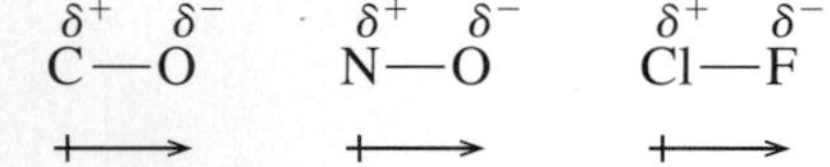

Examples of Dipoles in Polar Covalent Bonds

$\overset{\delta^+}{C}—\overset{\delta^-}{O}$ $\overset{\delta^+}{N}—\overset{\delta^-}{O}$ $\overset{\delta^+}{Cl}—\overset{\delta^-}{F}$

Polarity of Bonds

The difference in the electronegativity of two atoms can be used to predict the type of bond, ionic or covalent, that forms. For the H—H bond, the electronegativity difference is zero $(2.1 - 2.1 = 0)$, which means the bonding electrons are shared equally. A bond between atoms with identical or very similar electronegativity values is a **nonpolar covalent bond**. However, when bonds are between atoms with different electronegativity values, the electrons are shared unequally; the bond is a **polar covalent bond**. For the H—Cl bond, there is an electronegativity difference of $3.0\,(\text{Cl}) - 2.1(\text{H}) = 0.9$, which means that the H—Cl bond is polar covalent (see Figure 4.7).

Dipoles and Bond Polarity

The **polarity** of a bond depends on the electronegativity difference. In a polar covalent bond, the shared electrons are attracted to the more electronegative atom, which makes it partially negative due to the negatively charged electrons around that atom. At the other end of the bond, the atom with the lower electronegativity becomes partially positive due to the lack of the electrons at that atom. A bond becomes more *polar* as the electronegativity difference increases. A polar covalent bond that has a separation of charges is called a **dipole**. The positive and negative ends of the dipole are indicated by the lowercase Greek letter delta with a positive or negative sign, δ^+ and δ^-. Sometimes we use an arrow that points from the positive charge to the negative charge $\leftarrow\!\!\!+\!\!\!\rightarrow$ to indicate the dipole.

Variations in Bonding

The variations in bonding are continuous; there is no definite point at which one type of bond stops and the next starts. When the electronegativity difference is 0.0 to 0.4, the electrons are considered to share equally in a *nonpolar covalent bond*. For example, the H—H bond $(2.1 - 2.1 = 0)$ and the C—H bond $(2.5 - 2.1 = 0.4)$ are classified as nonpolar covalent bonds. As the electronegativity difference increases, the shared electrons are attracted more strongly to the more electronegative atom, which increases the

TABLE 4.14 Electronegativity Difference and Types of Bonds

Electronegativity difference	0	0.4	1.8	3.3
Bond type	Covalent nonpolar	Covalent polar	Ionic	
Electron bonding	Electrons shared equally	Electrons shared unequally	Electron transfer	
		δ^+ δ^-	+ −	

SELF STUDY ACTIVITY
Bonds and Bond Polarities

polarity of the bond. When the electronegativity difference is from 0.5 to 1.8, the bond is a *polar covalent bond*. For example, the O—H bond ($3.5 - 2.1 = 1.4$) is classified as a *polar covalent bond* (see Table 4.14).

When the electronegativity difference is greater than 1.8, electrons are transferred from one atom to another, which results in an ionic bond. For example, the electronegativity difference for the ionic compound NaCl is $3.0 - 0.9 = 2.1$. Thus for large differences in electronegativity, we would predict an ionic bond (see Table 4.15).

TABLE 4.15 Predicting Bond Type from Electronegativity Differences

Molecule		Electronegativity Difference	Bond Type	Reason
H_2	H—H	$2.1 - 2.1 = 0$	Nonpolar covalent	Less than 0.4
Cl_2	Cl—Cl	$3.0 - 3.0 = 0$	Nonpolar covalent	Less than 0.4
HBr	$\overset{\delta^+}{H}—\overset{\delta^-}{Br}$	$2.8 - 2.1 = 0.7$	Polar covalent	Greater than 0.4, but less than 1.8
HCl	$\overset{\delta^+}{H}—\overset{\delta^-}{Cl}$	$3.0 - 2.1 = 0.9$	Polar covalent	Greater than 0.4, but less than 1.8
NaCl	Na^+Cl^-	$3.0 - 0.9 = 2.1$	Ionic	Greater than 1.8
MgO	$Mg^{2+}O^{2-}$	$3.5 - 1.2 = 2.3$	Ionic	Greater than 1.8

CONCEPT CHECK 4.7

Using Electronegativity to Determine Polarity of Bonds

Complete the following table for each of the bonds indicated:

Bond	Electronegativity Difference	Type of Bond	Reason
Si—P			
Si—S			
Cs—Cl			

ANSWER

Bond	Electronegativity Difference	Type of Bond	Reason
Si—P	$2.1 - 1.8 = 0.3$	Nonpolar covalent	Less than 0.4
Si—S	$2.5 - 1.8 = 0.7$	Polar covalent	Greater than 0.4, but less than 1.8
Cs—Cl	$3.0 - 0.7 = 2.3$	Ionic	Greater than 1.8

SAMPLE PROBLEM 4.12

Bond Polarity

Using electronegativity values, classify each bond as nonpolar covalent, polar covalent, or ionic.

N—N, O—H, Cl—As, O—K

SOLUTION

For each bond, we calculate the difference in electronegativity.

Bond	Electronegativity Difference	Type of Bond
N—N	3.0 − 3.0 = 0.0	Nonpolar covalent
O—H	3.5 − 2.1 = 1.4	Polar covalent
Cl—As	3.0 − 2.0 = 1.0	Polar covalent
O—K	3.5 − 0.8 = 2.7	Ionic

STUDY CHECK 4.12

Using electronegativity values, classify each of the following bonds as nonpolar covalent, polar covalent, or ionic:

a. P—Cl **b.** Br—Br **c.** Na—O

QUESTIONS AND PROBLEMS

Electronegativity and Bond Polarity

4.51 Describe the trend in electronegativity going from left to right across a period.

4.52 Describe the trend in electronegativity going down a group.

4.53 Using the periodic table, arrange the atoms in each set in order of increasing electronegativity.
a. Li, Na, K **b.** Na, Cl, P **c.** Se, Ca, O

4.54 Using the periodic table, arrange the atoms in each set in order of increasing electronegativity.
a. Cl, F, Br **b.** B, O, N **c.** Mg, F, S

4.55 Predict whether each of the following bonds is ionic, polar covalent, or nonpolar covalent:
a. Si—Br **b.** Li—F **c.** Br—F
d. I—I **e.** N—P **f.** C—O

4.56 Predict whether each of the following bonds is ionic, polar covalent, or nonpolar covalent:
a. Si—O **b.** K—Cl **c.** S—F
d. P—Br **e.** Li—O **f.** N—S

4.57 For each of the following bonds, indicate the positive end with δ^+ and the negative end with δ^-. Draw an arrow to show the dipole for each.
a. N—F **b.** Si—Br **c.** C—O
d. P—Br **e.** N—P

4.58 For each of the following bonds, indicate the positive end with δ^+ and the negative end with δ^-. Draw an arrow to show the dipole for each.
a. P—Cl **b.** Se—F **c.** Br—F
d. N—H **e.** B—Cl

LEARNING GOAL

Predict the three-dimensional structure of a molecule, and classify it as polar or nonpolar.

4.7 Shapes and Polarity of Molecules

Using the information in Section 4.5 about electron-dot formulas, we can predict the three-dimensional shapes of many molecules. The shape is important in our understanding of how molecules interact with enzymes or certain antibiotics or produce our sense of taste and smell.

The three-dimensional shape is determined by drawing an electron-dot formula and identifying the number of electron groups around the central atom. In the **valence shell electron-pair repulsion (VSEPR) theory**, the electron groups are arranged as far apart as possible around the central atom to minimize the repulsion of the electron groups. The specific shape of a molecule is determined by the number of atoms attached to the central atom.

TUTORIAL
Molecular Shape

TUTORIAL
Shapes of Molecules

SELF STUDY ACTIVITY
The Shape of Molecules

Central Atoms with Two Electron Groups

In $BeCl_2$, two chlorine atoms are bonded to a central beryllium atom. Because an atom of Be has a strong attraction for valence electrons, it forms a covalent rather than ionic compound. With only two electron groups (two electron pairs) around the central atom, the electron-dot formula of $BeCl_2$ is an exception to the octet rule. The best arrangement of two electron groups for minimal repulsion is to place them on opposite sides of the Be atom. This gives a *linear* shape and a bond angle of 180° to the $BeCl_2$ molecule.

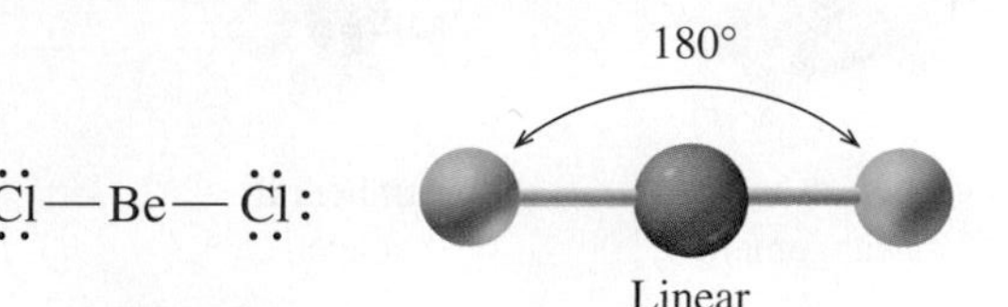

Linear

Another example of a linear molecule is CO_2. In predicting shapes, we count a *double* or *triple* bond (two or three electron pairs) as one electron group. In the electron-dot formula of CO_2, two electron groups (two double bonds) are on opposite sides of the C atom as far apart as possible, which gives a bond angle of 180°. The shape of the CO_2 molecule is *linear*.

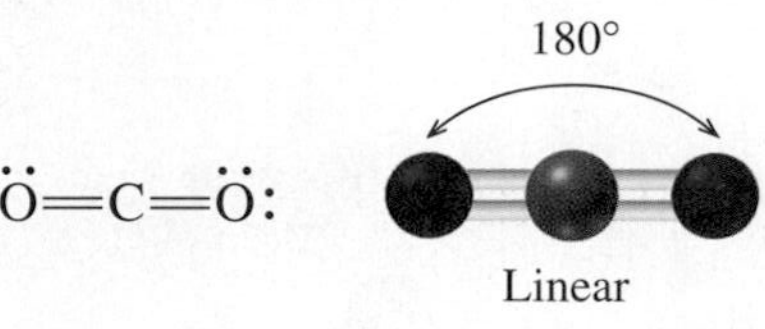

Linear

Central Atom with Three Electron Groups

In the electron-dot formula of BF_3, the central atom B has three electron groups attached to three fluorine atoms (another exception to the octet rule). With the three electron groups as far apart as possible, the bond angles are 120° and the shape is *trigonal planar*. In the BF_3 molecule, all the atoms are in the same plane and have bond angles of 120°.

:F:
:F—B—F:

Electron-dot formula

:F:
B
:F: :F:

Electron arrangement

120°

Trigonal planar

In the electron-dot formula for SO_2, there are also three electron groups around the sulfur atom: a single-bonded O atom, a double-bonded O atom, and a lone pair of electrons. As in BF_3, three electron groups have minimal repulsion by forming a trigonal planar arrangement. However, in SO_2 one of the electron groups is a lone pair. Therefore its shape is determined only by the two oxygen atoms bonded to the central S atom to give the SO_2 molecule a *bent* shape.

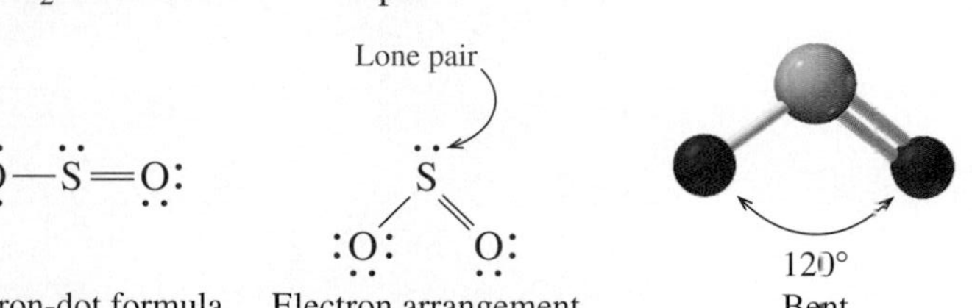

Electron-dot formula Electron arrangement Bent

Central Atom with Four Electron Groups

In a molecule of CH_4, the central carbon atom is bonded to four electron groups attached to four hydrogen atoms. From the electron-dot formula, you may think that CH_4 is planar with 90° angles. However, the best arrangement for minimal repulsion is the one with the electron groups at the corners of a tetrahedron, which gives bond angles of 109°. When there are four atoms attached to the four electron groups, the shape of the molecule is *tetrahedral*.

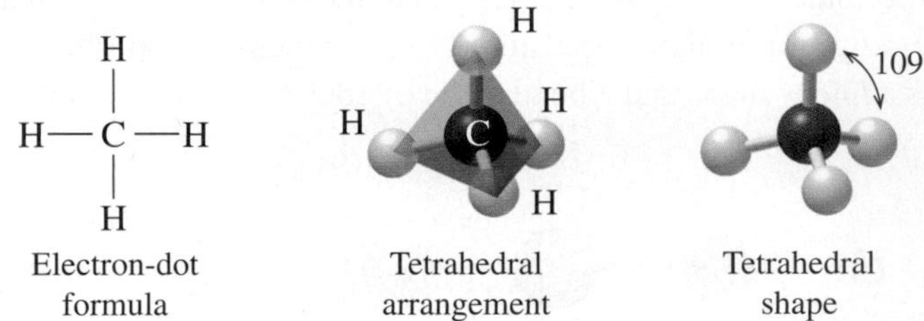

Now we can look at molecules that have four electron groups, but only two or three attached atoms. For example, ammonia, NH_3, has four electron groups, which have minimal repulsion by forming a tetrahedron. However, in NH_3 one of the electron groups is a lone pair. Therefore its shape is determined by only the three hydrogen atoms bonded to the central N atom to give the NH_3 molecule a *trigonal pyramidal* shape.

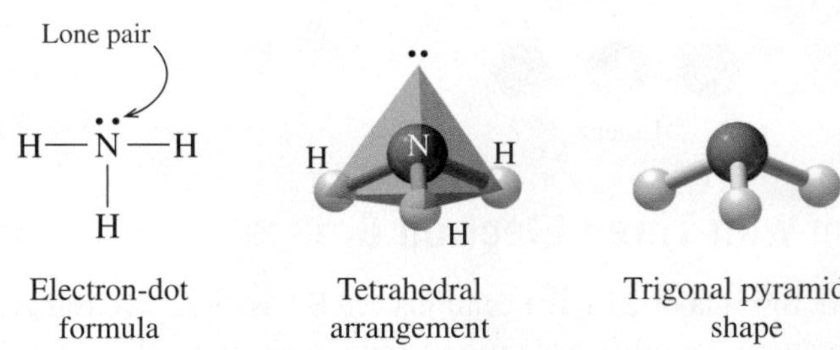

In the electron-dot formula of water, H_2O, there are four electron groups, which have minimal repulsion by forming a tetrahedron. However, in H_2O, two of the electron groups are lone pairs. Therefore its shape is determined by only the two hydrogen atoms bonded to the central O atom to give the H_2O molecule a *bent* shape. Table 4.16 gives the electron geometry and shape of molecules with two, three, or four bonded atoms.

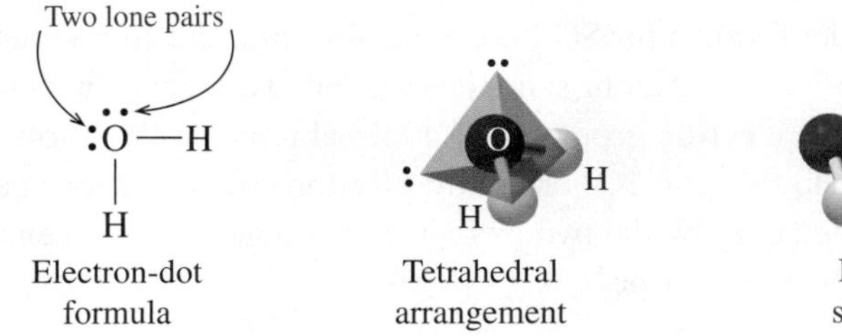

TABLE 4.16 Molecular Shapes for a Central Atom with Two, Three, and Four Bonded Atoms

Electron Groups	Electron-Group Geometry	Bonded Atoms	Lone Pairs	Bond Angle	Molecular Shape	Example	Three-Dimensional Model
2	Linear	2	0	180°	Linear	$BeCl_2$	
3	Trigonal planar	3	0	120°	Trigonal planar	BF_3	
		2	1	120°	Bent	SO_2	
4	Tetrahedral	4	0	109°	Tetrahedral	CH_4	
		3	1	109°	Trigonal pyramidal	NH_3	
		2	2	109°	Bent	H_2O	

CONCEPT CHECK 4.8

Shapes of Molecules

If the four electron groups in a PH_3 molecule form a tetrahedron, why does a PH_3 molecule have a trigonal pyramidal shape?

ANSWER

In the electron-dot formula, the four electron groups, one of which is a lone pair, achieve minimal repulsion when they are in the shape of a tetrahedron. Because the shape of the PH_3 molecule is determined by only the three atoms attached to the central atom, the shape of PH_3 is *trigonal pyramidal*.

$$\begin{array}{c} \cdot\cdot \\ \mathrm{H-P-H} \\ | \\ \mathrm{H} \end{array}$$

Guide to Predicting Molecular Shape (VSEPR Theory)

1. Draw the electron-dot formula.
2. Arrange the electron groups around the central atom to minimize repulsion.
3. Use the atoms bonded to the central atom to determine the molecular shape.

SAMPLE PROBLEM 4.13

Shapes of Molecules

Predict the shape of a molecule of $SiCl_4$.

SOLUTION

Step 1 **Draw the electron-dot formula.**

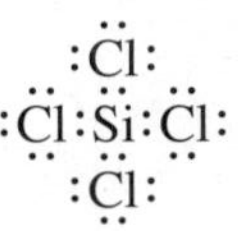

Step 2 **Arrange the electron groups around the central atom to minimize repulsion.** To minimize repulsion, four electron groups have a tetrahedral arrangement.

Step 3 **Use the atoms bonded to the central atom to determine the molecular shape.** With four bonded atoms and no lone pairs, the $SiCl_4$ molecule has a tetrahedral shape.

STUDY CHECK 4.13

Predict the shape of SCl_2.

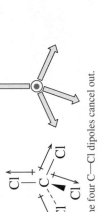

TUTORIAL
Distinguishing Polar and Nonpolar Molecules

Polarity of Molecules

We learned in Section 4.6 that covalent bonds can be polar or nonpolar. Molecules with covalent bonds can also be polar or nonpolar, which depends on their bond polarity and their shape.

Nonpolar Molecules

Molecules such as H_2, Cl_2, or PH_3 are nonpolar because they contain nonpolar covalent bonds. Molecules with polar bonds can be nonpolar if the polar bonds (dipoles) cancel each other in a symmetrical arrangement. For example, CO_2, a linear molecule, contains two polar covalent bonds, whose dipoles point in opposite directions. As a result, the dipoles cancel out, which makes a CO_2 molecule nonpolar. Thus a **nonpolar molecule** contains nonpolar bonds or has polar bonds with dipoles that cancel.

When the polar bonds or dipoles in a molecule cancel each other, the molecule is nonpolar. For example, CO_2 and CCl_4 contain polar bonds. However, the symmetrical arrangement of the polar bonds cancels the dipoles, which makes CO_2, CCl_4, and BF_3 molecules nonpolar.

O=C=O

Dipoles cancel

CO_2 is a nonpolar molecule.

Examples of Nonpolar Molecules with Polar Bonds

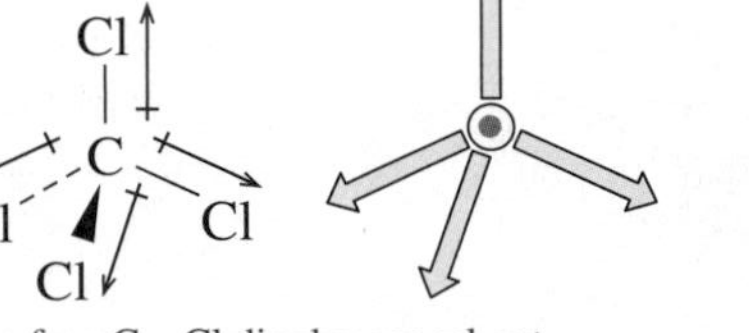

The four C—Cl dipoles cancel out.

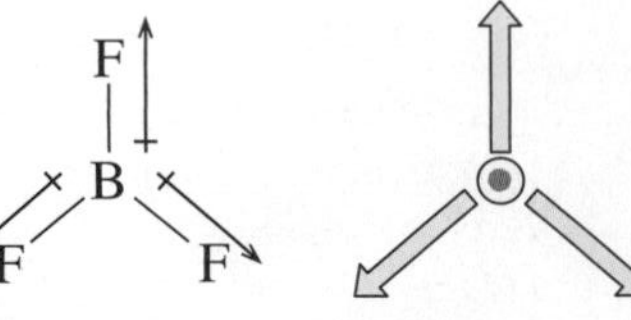

BF_3 is a nonpolar molecule.

Polar Molecules

In a **polar molecule**, one end of the molecule is more negatively charged than another end. Polarity in a molecule occurs when the polar bonds do not cancel each other. This cancellation depends on the type of atoms, the electron pairs around the central atom, and the shape of the molecule. For example, the HCl molecule is polar because electrons are shared unequally in a polar covalent bond.

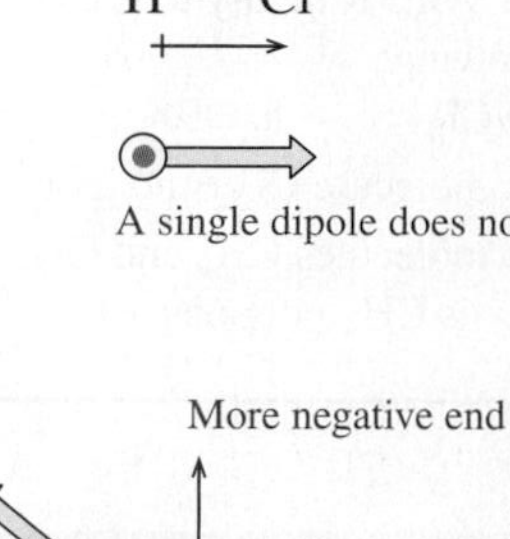

A single dipole does not cancel.

In polar molecules with three or more atoms, the shape of the molecule determines whether the dipoles cancel or not. Often there are lone pairs around the central atom. In H_2O, the dipoles do not cancel, which makes the molecule positive at one end and negative at the other end. Thus, water is a polar molecule.

More negative end of molecule

More positive end of molecule

H_2O is a polar molecule because its dipoles do not cancel.

In the molecule NH_3, there are three dipoles, but they do not cancel.

More negative end of molecule

More positive end of molecule

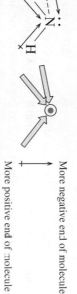

NH_3 is a polar molecule because its dipoles do not cancel.

In the molecule CH_3F, the CF bond is polar but the CH bonds are nonpolar, which makes CH_3F a polar molecule.

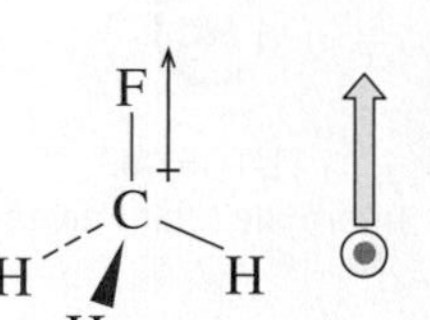

CH_3F is a polar molecule.

SAMPLE PROBLEM 4.14

Polarity of Molecules

Determine whether a molecule of OF_2 is polar or nonpolar.

OF_2

SOLUTION

The electron-dot formula for OF_2 shows four electron groups with two bonded atoms and two lone pairs.

:Ö:F̤:
:F̤:

The molecule would have a bent shape. The two polar O—F bonds do not cancel, which makes OF_2 a polar molecule.

STUDY CHECK 4.14

Would PCl_3 be a polar or nonpolar molecule?

QUESTIONS AND PROBLEMS

Shapes and Polarity of Molecules

4.59 Predict the shape of a molecule with each of the following:
a. two bonded atoms and two lone pairs
b. three bonded atoms and one lone pair

4.60 Predict the shape of a molecule with each of the following:
a. three bonded atoms and no lone pairs
b. two bonded atoms and one lone pair

4.61 In the molecule PCl_3, the four electron pairs around the phosphorus atom are arranged in a tetrahedral geometry. However, the shape of the molecule is called *trigonal pyramidal*. Why does the shape of the molecule have a different name from the name of the electron pair geometry?

4.62 In the molecule H_2S, the four electron pairs around the sulfur atom are arranged in a tetrahedral geometry. However, the shape of the molecule is called *bent*. Why does the shape of the molecule have a different name from the name of the electron pair geometry?

4.63 Compare the electron-dot formulas of PH_3 and NH_3. Why do these molecules have the same shape?

4.64 Compare the electron-dot formulas CH_4 and H_2O. Why do these molecules have approximately the same bond angles but different shapes?

4.65 Use VSEPR theory to predict the shape of each of the following:
a. $SeBr_2$ **b.** CCl_4 **c.** $GaCl_3$ **d.** SeO_2

4.66 Use VSEPR theory to predict the shape of each of the following:
a. NCl_3 **b.** OBr_2 **c.** SiF_2Cl_2 **d.** $BeBr_2$

4.67 The molecule Cl_2 is nonpolar, but HCl is polar. Explain.

4.68 The molecules CH_4 and CH_3Cl both contain four bonds. Why is CH_4 nonpolar whereas CH_3Cl is polar?

4.69 Identify each of the following molecules as polar or nonpolar:
a. HBr **b.** NF_3 **c.** CHF_3

4.70 Identify each of the following molecules as polar or nonpolar:
a. SeF_2 **b.** PBr_3 **c.** $SiCl_4$

LEARNING GOAL

Describe the attractive forces between ions, polar molecules, and nonpolar molecules.

TUTORIAL
Intermolecular Forces

4.8 Attractive Forces in Compounds

In gases, the interactions between particles are minimal, which allows gas molecules to move far apart from each other. In solids and liquids, there are sufficient interactions between the particles to hold them close together, although some solids have low melting points while others have very high melting points. Such differences in properties are explained by looking at the various kinds of attractive forces between particles.

Ionic compounds have high melting points. For example, solid NaCl melts at 801 °C. Large amounts of energy are needed to overcome the strong attractive forces between positive and negative ions. In solids containing molecules with covalent bonds, there are attractive forces too, but they are weaker than those of an ionic compound.

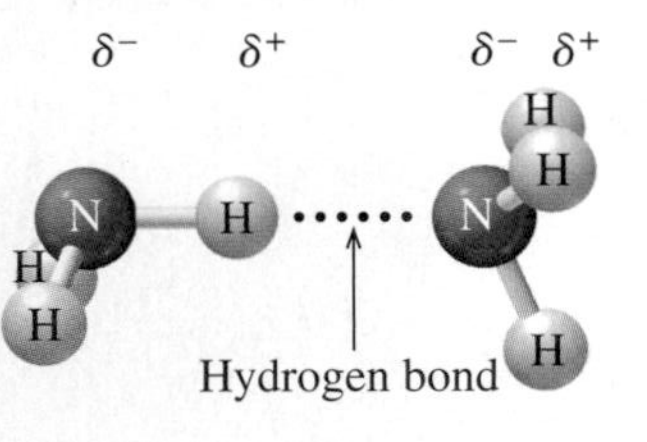

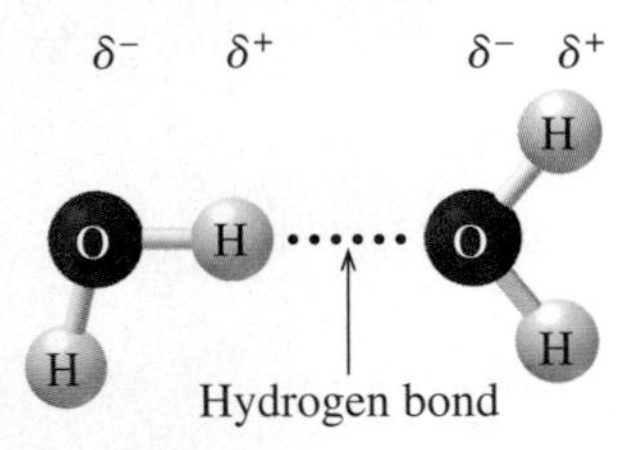

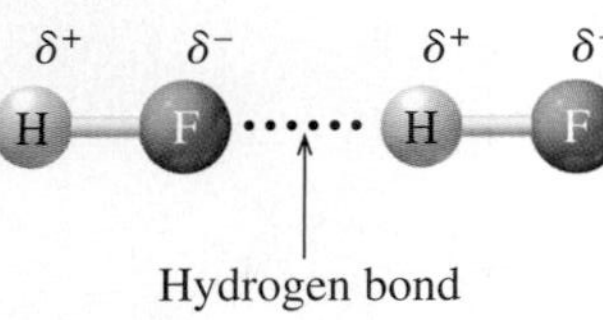

Dipole–Dipole Attractions and Hydrogen Bonds

For polar molecules, attractive forces called **dipole–dipole attractions** occur between the positive end of one molecule and the negative end of another. For a polar molecule with a dipole such as HCl, the partially positive H atom of one HCl molecule attracts the partially negative Cl atom in another molecule.

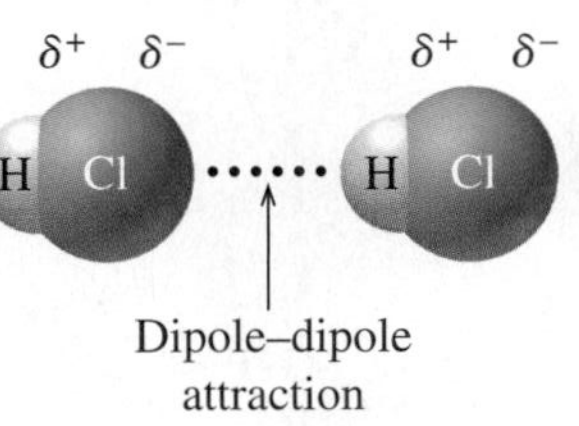

When a hydrogen atom is attached to highly electronegative atoms of fluorine, oxygen, or nitrogen, there are strong dipole–dipole attractions between the polar molecules. This type of attraction, called a **hydrogen bond**, occurs between the partially positive hydrogen atom of one molecule and a lone pair of electrons on a nitrogen, oxygen, or fluorine atom in another molecule. Hydrogen bonds are the strongest type of attractive forces between polar molecules. They are a major factor in the formation and structure of biological molecules such as proteins and DNA.

Dispersion Forces

Nonpolar compounds do form solids, but at low temperatures. Very weak attractions called **dispersion forces** occur between nonpolar molecules. Usually, the electrons in a nonpolar molecule are distributed symmetrically. However, electrons may accumulate more in one part of the molecule than another, which forms a temporary dipole. Although dispersion forces are very weak, they make it possible for nonpolar molecules to form liquids and solids.

The melting points of substances are related to the strength of the attractive forces. Compounds with weak attractive forces such as dispersion forces have low melting points because only a small amount of energy is needed to separate the molecules and form a liquid. Compounds with hydrogen bonds and dipole–dipole attractions require more energy to break the attractive forces between the molecules. The highest melting points are seen with ionic compounds that have the very strong attractions between ions. Table 4.17 compares the melting points of some substances with various kinds of attractive forces. The various types of attractions between particles in solids and liquids are summarized in Table 4.18.

TABLE 4.17 Melting Points of Selected Substances

Substance	Melting Point (°C)
Ionic bonds	
MgF_2	1248
NaCl	801
Hydrogen bonds	
H_2O	0
NH_3	−78
Dipole–dipole attractions	
HBr	−89
HCl	−115
Dispersion forces	
Cl_2	−101
F_2	−220

SAMPLE PROBLEM 4.15

Attractive Forces between Particles

Indicate the major type of molecular interaction expected of each of the following:

1. dipole–dipole attractions **2.** hydrogen bonding **3.** dispersion forces

a. HF **b.** Br_2 **c.** PCl_3

SOLUTION

a. (2) HF is a polar molecule that interacts with other HF molecules by hydrogen bonding.

b. (3) Br_2 is nonpolar; only dispersion forces provide attractive forces.

c. (1) The polarity of the PCl_3 molecules provides dipole–dipole attractions.

STUDY CHECK 4.15

Why is the boiling point of H_2S lower than that of H_2O?

TABLE 4.18 Comparison of Bonding and Attractive Forces

Type of Force	Particle Arrangement	Example	Strength
Between atoms or ions		Na^+Cl^-	**Strong**
Ionic bond	+ − + / − + −		
Covalent bond (X = nonmetal)	X:X	Cl—Cl	
Between molecules			
Hydrogen bond (X = F, O, or N)	$\delta^+\ \delta^-$ H X ··· $\delta^+\ \delta^-$ H X	$\overset{\delta^+}{H}—\overset{\delta^-}{F}\cdots\overset{\delta^+}{H}—\overset{\delta^-}{F}$	
Dipole–dipole attractions (X and Y = nonmetals)	$\delta^+\ \delta^-$ Y X ··· $\delta^+\ \delta^-$ Y X	$\overset{\delta^+}{Br}—\overset{\delta^-}{Cl}\cdots\overset{\delta^+}{Br}—\overset{\delta^-}{Cl}$	
Dispersion forces (Temporary shift of electrons in nonpolar bonds)	$\delta^+\ \delta^-$ X:X ··· $\delta^+\ \delta^-$ X:X (temporary dipoles)	$\overset{\delta^+}{F}—\overset{\delta^-}{F}\cdots\overset{\delta^+}{F}—\overset{\delta^-}{F}$	**Weak**

QUESTIONS AND PROBLEMS

Attractive Forces in Compounds

4.71 Identify the major type of attractive force between particles of each of the following substances:
a. BrF **b.** KCl **c.** Cl_2 **d.** CH_4

4.72 Identify the major type of attractive force between particles of each of the following substances:
a. OF_2 **b.** MgF_2 **c.** NH_3 **d.** HCl

4.73 Identify the strongest attractive force between particles of each of the following:
a. H_2O **b.** Ar **c.** HBr **d.** NF_3 **e.** CO

4.74 Identify the strongest attractive force between particles of each of the following:
a. O_2 **b.** HI **c.** NaF **d.** $CH_3—OH$ **e.** Ne

CONCEPT MAP

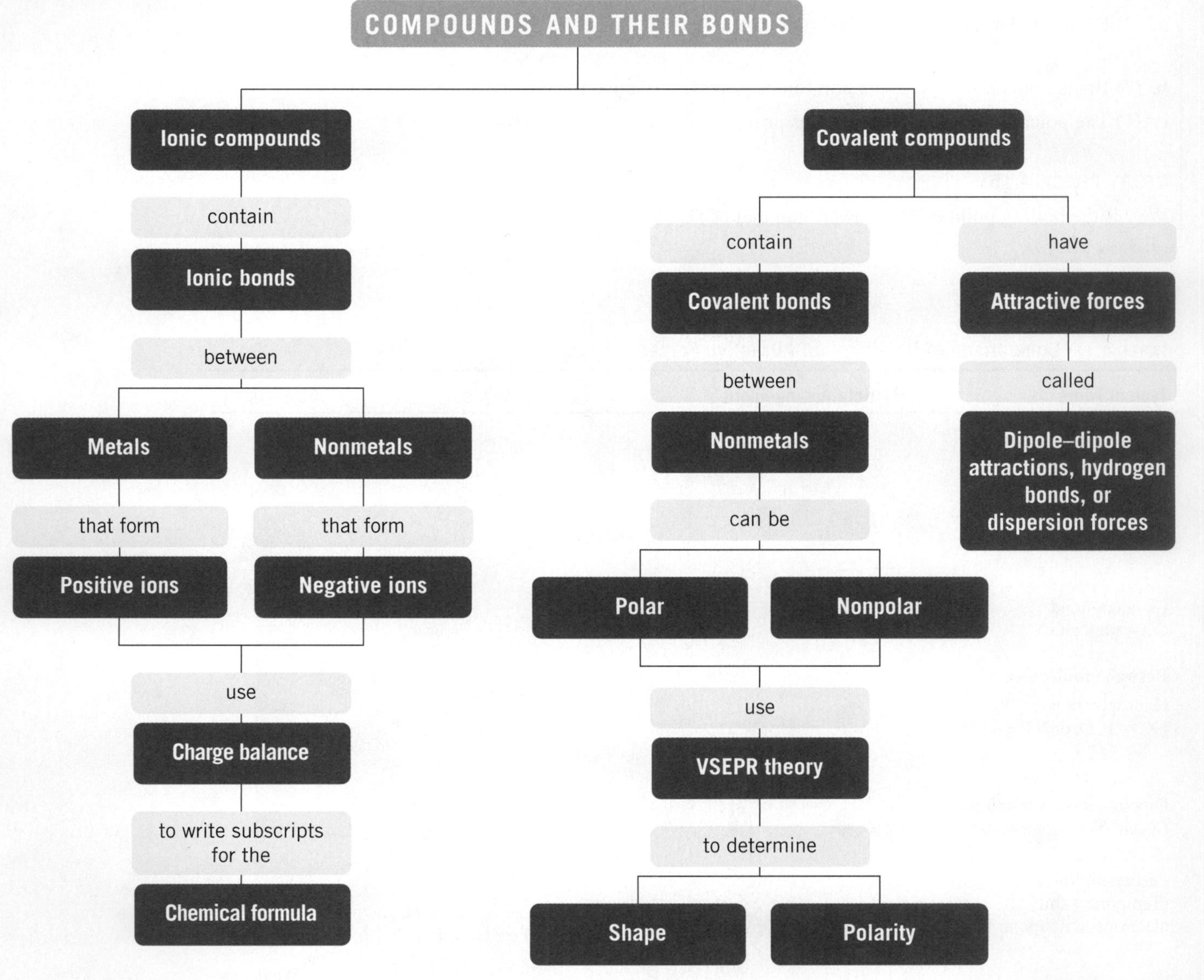

CHAPTER REVIEW

4.1 Octet Rule and Ions

Learning Goal: Using the octet rule, write the symbols of the simple ions for the representative elements.

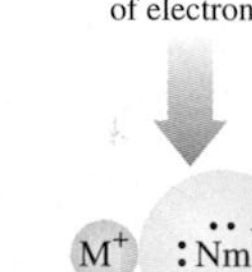

Loss and gain of electrons

Ionic bond

The stability of the noble gases is associated with an octet of 8 electrons in their valence shells; helium needs 2 electrons for stability. Atoms of elements in Groups 1A–7A (1, 2, 13–17) achieve stability by losing, gaining, or sharing their valence electrons in the formation of compounds. Metals of the representative elements form octets by losing valence electrons to form positively charged ions (cations): Group 1A (1), 1+, Group 2A (2), 2+, and Group 3A (13), 3+. When reacting with metals, nonmetals gain electrons to form octets and form negatively charged ions (anions): Group 5A (15), 3−, Group 6A (16), 2−, Group 7A (17), 1−.

4.2 Ionic Compounds

Learning Goal: Using charge balance, write the correct formula for an ionic compound.

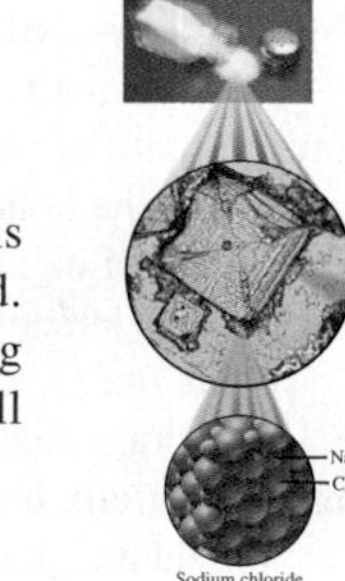

Sodium chloride

The total positive and negative ionic charge is balanced in the formula of an ionic compound. Charge balance in a formula is achieved by using subscripts after each symbol so that the overall charge is zero.

4.3 Naming and Writing Ionic Formulas

Learning Goal: Given the formula of an ionic compound, write the correct name; given the name of an ionic compound, write the correct formula.

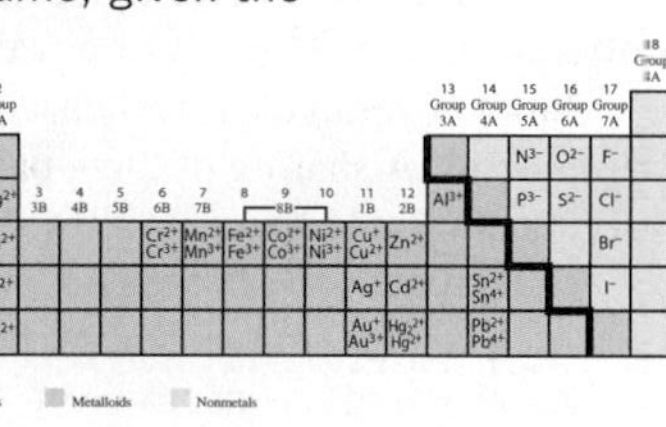

In naming ionic compounds, the name of the positive ion is given first, followed by the name of the negative ion. Ionic compounds containing two elements end with *ide*. Except for Ag, Cd, and Zn, transition elements form cations with two or more ionic charges. Then the charge of the cation is determined from the total negative charge in the formula and included as a Roman numeral following the name.

4.4 Polyatomic Ions

Learning Goal: Write the name and formula of a compound containing a polyatomic ion.

Fertilizer NH_4NO_3

NH_4^+

NO_3^- Nitrate ion

A polyatomic ion is a group of nonmetal atoms that carries an electrical charge; for example, the carbonate ion has the formula CO_3^{2-}. Most polyatomic ions have names that end with *ate* or *ite*.

4.5 Covalent Compounds and Their Names

Learning Goal: Given the formula of a covalent compound, write its correct name; given the name of a covalent compound, write its formula.

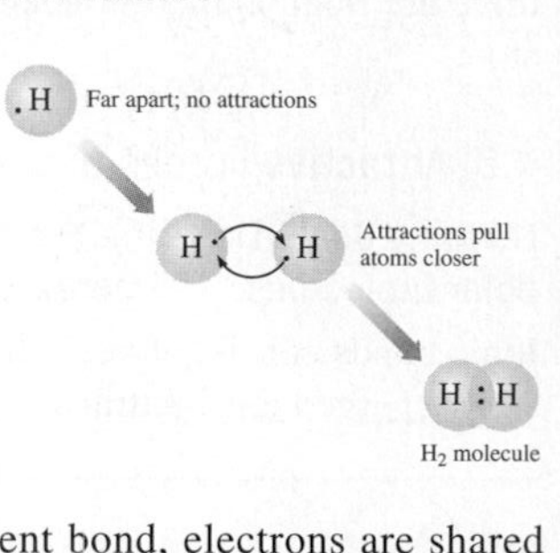

H_2 molecule

In a covalent bond, electrons are shared by atoms of two nonmetals such that each of the atoms achieves an octet (or two for hydrogen). In a nonpolar covalent bond, electrons are shared equally by atoms. In a polar covalent bond, electrons are unequally shared due to their attraction to the more electronegative atom. In some covalent compounds, double or triple bonds are needed to provide an octet. Their names use prefixes to indicate the subscripts of each type of atom in the formula. The ending of the second nonmetal is changed to *ide*.

4.6 Electronegativity and Bond Polarity

Learning Goal: Use electronegativity to determine the polarity of a bond.

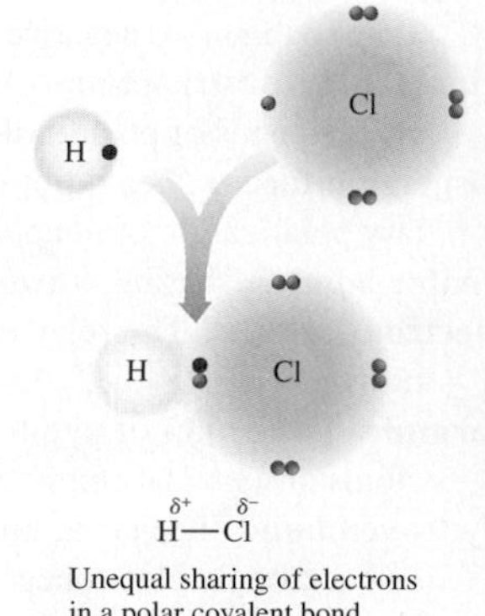

Unequal sharing of electrons in a polar covalent bond

Electronegativity is the ability of an atom to attract shared pairs of electrons. The electronegativity values of the metals are low, while nonmetals have high electronegativities. If atoms share the bonding pair of electrons equally, it is called a nonpolar covalent bond. If the bonding electrons are unequally shared, it is called a polar covalent bond. In polar covalent bonds, the atom with the lower electronegativity is partially positive (δ^+) and the atom with the higher electronegativity is partially negative (δ^-). Atoms that form ionic bonds have large differences in electronegativity.

4.7 Shapes and Polarity of Molecules

Learning Goal: Predict the three-dimensional structure of a molecule, and classify it as polar or nonpolar.

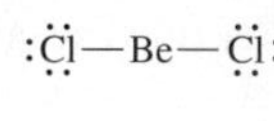

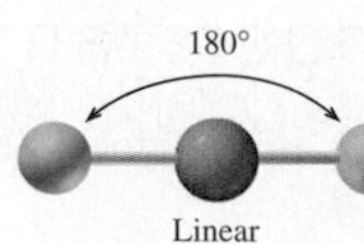

Linear

The VSEPR theory indicates that the repulsion of electrons around a central atom pushes the electron groups as far apart as possible. The shape of a molecule is predicted from the arrangement of the bonded atoms and lone pairs around the central atom. The shape of two atoms around a central atom with no lone pairs is linear. The shape of three atoms around a central atom with no lone pairs is trigonal planar. The shape of two atoms around a central atom with one lone pair is bent, 120°. The shape of four atoms around a central atom with no lone pairs is tetrahedral. The shape of three atoms around a central atom with one lone pair is trigonal pyramidal. The shape of two bonded atoms around a central atom with two lone pairs is bent, 109°.

Molecules are nonpolar if they contain nonpolar covalent bonds or have an arrangement of polar covalent bonds with dipoles that cancel out. In polar molecules, the dipoles do not cancel because there are nonidentical bonded atoms or lone pairs on the central atom.

4.8 Attractive Forces in Compounds

Learning Goal: Describe the attractive forces between ions, polar molecules, and nonpolar molecules.

Ionic bonds consist of very strong attractive forces between oppositely charged ions. Attractive forces in polar covalent compounds are weaker than ionic bonds and include dipole–dipole attractions and hydrogen bonds. Nonpolar covalent compounds form solids using temporary dipoles called dispersion forces.

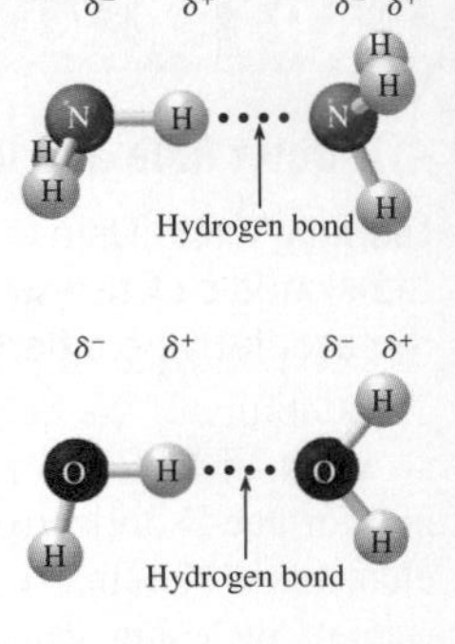

Key Terms

anion A negatively charged ion such as Cl^-, O^{2-}, or SO_4^{2-}.

cation A positively charged ion such as Na^+, Mg^{2+}, Al^{3+}, or NH_4^+.

covalent compound A combination of atoms in which noble gas arrangements are attained by sharing electrons.

dipole The separation of positive and negative charges in a polar bond indicated by an arrow that is drawn from the more positive atom to the more negative atom.

dipole–dipole attractions Attractive forces between oppositely charged ends of polar molecules.

dispersion forces Weak dipole bonding that results from a momentary polarization of nonpolar molecules.

double bond A sharing of two pairs of electrons by two atoms.

electronegativity The relative ability of an element to attract electrons in a bond.

formula The group of symbols and subscripts that represents the atoms or ions in a compound.

hydrogen bond The attraction between a partially positive H atom and a strongly electronegative atom of F, O, or N.

ion An atom or group of atoms having an electrical charge because of a loss or gain of electrons.

ionic charge The difference between the number of protons (positive) and the number of electrons (negative) written in the upper right corner of the symbol for the element or polyatomic ion.

ionic compound A compound of positive and negative ions held together by ionic bonds.

molecule The smallest unit of two or more atoms held together by covalent bonds.

nonpolar covalent bond A covalent bond in which the electrons are shared equally between atoms.

nonpolar molecule A molecule that has only nonpolar bonds or in which the bond dipoles cancel.

octet A set of eight valence electrons.

octet rule Elements in Groups 1A–7A (1, 2, 13–17) react with other elements by forming ionic or covalent bonds to produce a noble gas arrangement, usually eight electrons in the outer shell.

polar covalent bond A covalent bond in which the electrons are shared unequally between atoms.

polar molecule A molecule containing bond dipoles that do not cancel.

polarity A measure of the unequal sharing of electrons indicated by the difference in electronegativities.

polyatomic ion A group of covalently bonded nonmetal atoms that has an overall electrical charge.

triple bond A sharing of three pairs of electrons by two atoms.

valence-shell electron-pair repulsion (VSEPR) theory A theory that predicts the shape of a molecule by placing the electron pairs on a central atom as far apart as possible to minimize the mutual repulsion of the electrons.

Understanding the Concepts

4.75 **a.** How does the octet rule explain the formation of a sodium ion?
b. What noble gas has the same electron arrangement as the sodium ion?
c. Why are Group 1A (1) and Group 2A (2) elements found in many compounds, but not Group 8A (18) elements?

4.76 **a.** How does the octet rule explain the formation of a chloride ion?
b. What noble gas has the same electron arrangement as the chloride ion?
c. Why are Group 7A (17) elements found in many compounds, but not Group 8A (18) elements?

4.77 Identify each of the following atoms or ions:

A	B	C	D
$18\,e^-$	$8\,e^-$	$28\,e^-$	$23\,e^-$
$15\,p^+$, $16\,n$	$8\,p^+$, $8\,n$	$30\,p^+$, $35\,n$	$26\,p^+$, $28\,n$

4.78 Identify each of the following atoms or ions:

A	B	C	D	E
$2\,e^-$	$0\,e^-$	$3\,e^-$	$10\,e^-$	$1\,e^-$
$3\,p^+$, $4\,n$	$1\,p^+$	$3\,p^+$, $4\,n$	$7\,p^+$, $8\,n$	$1\,p^+$, $2\,n$

4.79 Consider the following electron-dot symbols for elements X and Y:

X· $\cdot\ddot{\underset{\cdot\cdot}{Y}}\cdot$

a. What are the group numbers of X and Y?
b. Will a compound of X and Y be ionic or covalent?

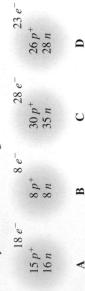

c. What ions would be formed by X and Y?
d. What would be the formula of a compound of X and Y?
e. What would be the formula of a compound of X and sulfur?
f. What would be the formula of a compound of Y and chlorine?
g. Is the compound in part **f** ionic or covalent?

4.80 Consider the following electron-dot symbols for elements X and Y:

$\dot{X}\cdot \qquad \cdot\ddot{Y}\cdot$

a. What are the group numbers of X and Y?
b. Will a compound of X and Y be ionic or covalent?
c. What ions would be formed by X and Y?
d. What would be the formula of a compound of X and Y?
e. What would be the formula of a compound of X and sulfur?
f. What would be the formula of a compound of Y and chlorine?
g. Is the compound in part **f** ionic or covalent?

4.81 Match each of the electron-dot formulas (**a–c**) with the correct diagram (**1–3**) of its shape, and name the shape; indicate if each molecule is polar or nonpolar. (Assume X and Y are nonmetals and all bonds are polar covalent.)

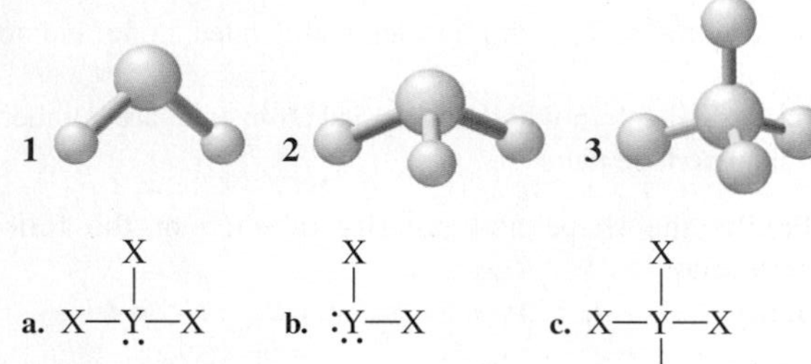

a. X—Ÿ—X (X above Y, lone pair below Y) **b.** :Ÿ—X (X above Y, lone pair below Y) **c.** X—Y—X (X above and below Y)

4.82 Match each of the formulas (**a-c**) with the correct diagram (**1-3**) of its shape, and name the shape; indicate if each molecule is polar or nonpolar.

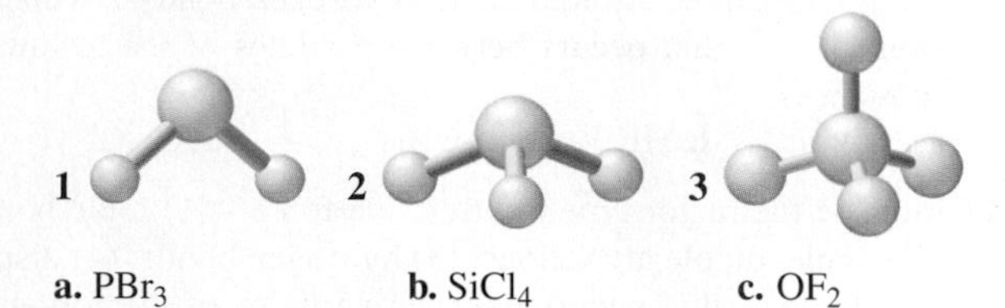

a. PBr_3 **b.** $SiCl_4$ **c.** OF_2

4.83 Using each of the following electron arrangements, give the formulas of the cation and anion that form, the formula of the compound they form, and its name.

Electron Arrangements		Cation	Anion	Formula of Compound	Name of Compound
2,8,2	2,5				
2,8,8,1	2,6				
2,8,3	2,8,7				

4.84 Using each of the following electron arrangements, give the formulas of the cation and anion that form, the formula of the compound they form, and its name.

Electron Arrangements		Cation	Anion	Formula of Compound	Name of Compound
2,8,1	2,7				
2,8,8,2	2,8,6				
2,8,3	2,8,5				

4.85 State the number of valence electrons, bonding pairs, and lone pairs in each of the following electron-dot formulas:
a. H:H **b.** H:B̈r: **c.** :B̈r:B̈r:

4.86 State the number of valence electrons, bonding pairs, and lone pairs in each of the following electron-dot formulas:
a. H:Ö: (with H above O) **b.** H:N̈:H (with H below N) **c.** :B̈r:Ö:B̈r:

Additional Questions and Problems

For instructor-assigned homework, go to **www.masteringchemistry.com**.

4.87 Write the electron arrangement of the following:
a. N^{3-} **b.** Mg^{2+} **c.** P^{3-} **d.** Al^{3+} **e.** Li^{+}

4.88 Write the electron arrangement of the following:
a. K^{+} **b.** Na^{+} **c.** S^{2-} **d.** Cl^{-} **e.** Ca^{2+}

4.89 Consider an ion with the symbol X^{2+} formed from a representative element.
a. What is the group number of the element?
b. What is the electron-dot symbol of the element?
c. If X is in Period 3, what is the element?
d. What is the formula of the compound formed from X and the nitride ion?

4.90 Consider an ion with the symbol Y^{3-} formed from a representative element.
a. What is the group number of the element?
b. What is the electron-dot symbol of the element?
c. If Y is in Period 3, what is the element?
d. What is the formula of the compound formed from the barium ion and Y?

4.91 One of the ions of tin is tin(IV).
a. What is the symbol for this ion?
b. How many protons and electrons are in the ion?
c. What is the formula of tin(IV) oxide?
d. What is the formula of tin(IV) phosphate?

4.92 One of the ions of gold is gold(III).
a. What is the symbol for this ion?
b. How many protons and electrons are in the ion?
c. What is the formula of gold(III) sulfate?
d. What is the formula of gold(III) nitrate?

4.93 Write the formula of the following ionic compounds:
a. tin(II) sulfide **b.** lead(IV) oxide
c. silver chloride **d.** calcium nitride
e. copper(I) phosphide **f.** chromium(II) bromide

4.94 Write the formula of the following ionic compounds:
a. nickel(III) oxide **b.** iron(III) sulfide
c. lead(II) sulfate **d.** chromium(III) iodide
e. lithium nitride **f.** gold(I) oxide

4.95 Name each of the following:
a. NCl_3 **b.** N_2S_3 **c.** N_2O
d. F_2 **e.** PCl_5 **f.** P_2O_5

4.96 Name each of the following:
a. CBr_4 **b.** SF_6 **c.** Br_2
d. N_2O_4 **e.** SO_2 **f.** CS_2

4.97 Give the formula for each of the following:
a. carbon monoxide **b.** diphosphorus pentoxide
c. dihydrogen sulfide **d.** sulfur dichloride

4.98 Give the formula for each of the following:
a. silicon dioxide **b.** carbon tetrabromide
c. diphosphorus tetraiodide **d.** dinitrogen oxide

4.99 Classify each of the following as ionic or covalent, and give its name:
a. $FeCl_3$ **b.** Na_2SO_4 **c.** NO_2
d. N_2 **e.** PF_5 **f.** CF_4

4.100 Classify each of the following as ionic or covalent, and give its name:
a. $Al_2(CO_3)_3$ **b.** ClF_5 **c.** H_2
d. Mg_3N_2 **e.** ClO_2 **f.** $CrPO_4$

4.101 Write the formulas for the following:
a. tin(II) carbonate **b.** lithium phosphide
c. silicon tetrachloride **d.** manganese(III) oxide
e. bromine **f.** calcium bromide

4.102 Write the formulas for the following:
a. sodium carbonate **b.** nitrogen dioxide
c. aluminum nitrate **d.** copper(I) nitride
e. potassium phosphate **f.** cobalt(III) sulfate

4.103 Select the more polar bond in each of the following pairs:
a. C—N or C—O **b.** N—F or N—Br
c. Br—Cl or S—Cl **d.** Br—Cl or Br—I
e. N—F or N—O

4.104 Select the more polar bond in each of the following pairs:
a. C—C or C—O **b.** P—Cl or P—Br
c. Si—S or Si—Cl **d.** F—Cl or F—Br
e. P—O or P—S

4.105 Calculate the electronegativity difference and classify each of the following bonds as nonpolar covalent, polar covalent, or ionic:
a. Si—Cl **b.** C—C **c.** Na—Cl
d. C—H **e.** F—F

4.106 Calculate the electronegativity difference and classify each of the following bonds as nonpolar covalent, polar covalent, or ionic:
a. C—N **b.** Cl—Cl **c.** K—Br
d. H—H **e.** N—F

4.107 Classify the following molecules as polar or nonpolar:
a. NH_3 **b.** CH_3Cl **c.** SiF_4

4.108 Classify the following molecules as polar or nonpolar:
a. GeH_4 **b.** SeO_2 **c.** SCl_2

4.109 Predict the shape and polarity of each of the following molecules:
a. A central atom with three identical bonded atoms and one lone pair.
b. A central atom with two bonded atoms and two lone pairs.

4.110 Predict the shape and polarity of each of the following molecules:
a. A central atom with four identical bonded atoms and no lone pairs.
b. A central atom with four bonded atoms that are not identical and no lone pairs.

4.111 Predict the shape and polarity of each of the following molecules:
a. SI_2 **b.** PBr_3

4.112 Predict the shape and polarity of each of the following molecules:
a. H_2O **b.** CF_4

4.113 Indicate the major type of attractive force—(1) ionic bonds, (2) dipole–dipole attractions, (3) hydrogen bonds, (4) dispersion forces—that occurs between particles of the following substances:
a. NH_3 **b.** HI **c.** Br_2 **d.** Cs_2O

4.114 Indicate the major type of attractive force—(1) ionic bonds, (2) dipole–dipole attractions, (3) hydrogen bonds, (4) dispersion forces—that occurs between particles of the following substances:
a. $CHCl_3$ **b.** H_2O **c.** LiCl **d.** Cl_2

Challenge Questions

4.115 Complete the following table for atoms or ions:

Atom or Ion	Number of Protons	Number of Electrons	Electrons Lost/Gained
K^+			
	$12\ p^+$	$10\ e^-$	
	$8\ p^+$		$2\ e^-$ gained
		$10\ e^-$	$3\ e^-$ lost

4.116 Complete the following table for atoms or ions:

Atom or Ion	Number of Protons	Number of Electrons	Electrons Lost/Gained
	$30\ p^+$		$2\ e^-$ lost
	$36\ p^+$	$36\ e^-$	
	$16\ p^+$		$2\ e^-$ gained
		$46\ e^-$	$4\ e^-$ lost

4.117 Identify the group number in the periodic table of X, a representative element, in each of the following ionic compounds:
a. XCl_3 **b.** Al_2X_3 **c.** XCO_3

4.118 Identify the group number in the periodic table of X, a representative element, in each of the following ionic compounds:
a. X_2O_3 **b.** X_2SO_3 **c.** Na_3X

4.119 Classify each of the following as ionic or covalent, and name each:
a. Li_2O **b.** $Cr(NO_3)_2$ **c.** $Mg(HCO_3)_2$
d. NF_3 **e.** $CaCl_2$ **f.** K_3PO_4
g. $Au_2(SO_3)_3$ **h.** I_2

4.120 Classify each of the following as ionic or covalent, and name each:
a. $FeCl_2$ **b.** Cl_2O_7 **c.** Br_2
d. $Ca_3(PO_4)_2$ **e.** PCl_3 **f.** $Al(NO_3)_3$
g. $PbCl_4$ **h.** $MgCO_3$

Answers

Answers to Study Checks

4.1 K^+ and S^{2-}

4.2 Ca^{2+} and O^{2-}, CaO

4.3 cesium sulfide

4.4 manganese(II) sulfide

4.5 Cr_2O_3

4.6 cobalt(II) phosphate

4.7 H:P̤̈:H H—P̈—H (with H below P in each)

4.8 :Ö—S̈=Ö: or :Ö=S̈—Ö:

4.9 a. silicon tetrabromide **b.** dibromine oxide

4.10 IF_5

4.11 iron(III) nitrate

4.12 a. polar covalent (0.9) **b.** nonpolar covalent (0)
c. ionic (2.6)

4.13 The central atom sulfur has four electron pairs: two bonded atoms and two lone pairs. The geometry of SCl_2 would be bent, 109°.

4.14 polar

4.15 The attractive forces between H_2S molecules are dipole–dipole attractions, whereas H_2O molecules have hydrogen bonds, which are stronger and require more energy to break.

Answers to Selected Questions and Problems

4.1 a. 1 **b.** 2 **c.** 3 **d.** 1 **e.** 2

4.3 a. Li^+ **b.** F^- **c.** Mg^{2+} **d.** Fe^{3+}

4.5 a. 2 e^- lost **b.** 3 e^- gained **c.** 1 e^- gained
d. 1 e^- lost **e.** 1 e^- gained

4.7 a. Cl^- **b.** Cs^+ **c.** S^{2-} **d.** Ra^{2+}

4.9 a, c

4.11 a. Na_2O **b.** $AlBr_3$ **c.** Ba_3N_2
d. MgF_2 **e.** Al_2S_3

4.13 a. K^+ and S^{2-}, K_2S **b.** Na^+ and N^{3-}, Na_3N
c. Al^{3+} and I^-, AlI_3 **d.** Ga^{3+} and O^{2-}, Ga_2O_3

4.15 a. aluminum oxide **b.** calcium chloride
c. sodium oxide **d.** magnesium phosphide
e. potassium iodide **f.** barium fluoride

4.17 Most of the transition elements form more than one positive ion. The specific ion is indicated in a name by writing a Roman numeral that is the same as the ionic charge. For example, iron forms Fe^{2+} and Fe^{3+} ions, which are named iron(II) and iron(III).

4.19 a. iron(II) **b.** copper(II) **c.** zinc
d. lead(IV) **e.** chromium(III) **f.** manganese(II)

4.21 a. tin(II) chloride **b.** iron(II) oxide
c. copper(I) sulfide **d.** copper(II) sulfide
e. cadmium bromide **f.** zinc chloride

4.23 a. Au^{3+} **b.** Fe^{3+} **c.** Pb^{4+} **d.** Sn^{2+}

4.25 a. $MgCl_2$ **b.** Na_2S **c.** Cu_2O **d.** Zn_3P_2
e. AuN **f.** $CrCl_2$

4.27 a. HCO_3^- **b.** NH_4^+ **c.** PO_4^{3-} **d.** HSO_4^-

4.29 a. sulfate **b.** carbonate **c.** phosphate **d.** nitrate

4.31

	NO_2^-	CO_3^{2-}	HSO_4^-	PO_4^{3-}
Li^+	$LiNO_2$	Li_2CO_3	$LiHSO_4$	Li_3PO_4
Cu^{2+}	$Cu(NO_2)_2$	$CuCO_3$	$Cu(HSO_4)_2$	$Cu_3(PO_4)_2$
Ba^{2+}	$Ba(NO_2)_2$	$BaCO_3$	$Ba(HSO_4)_2$	$Ba_3(PO_4)_2$

4.33 a. CO_3^{2-}, sodium carbonate
b. NH_4^+, ammonium chloride
c. PO_4^{3-}, potassium phosphate
d. NO_2^-, chromium(II) nitrite
e. SO_3^{2-}, iron(II) sulfite

4.35 a. $Ba(OH)_2$ **b.** Na_2SO_4 **c.** $Fe(NO_3)_2$
d. $Zn_3(PO_4)_2$ **e.** $Fe_2(CO_3)_3$

4.37 Nonmetals share electrons to form covalent compounds.

4.39 a. :B̤̈r:B̤̈r: or :B̤̈r—B̤̈r: **b.** H:H or H—H
c. H:F̤̈: or H—F̤̈: **d.** :F̤̈:Ö:(:F̤̈:) or :F̤̈—Ö:(—F̤̈:)

4.41 a. phosphorus tribromide **b.** carbon tetrabromide
c. silicon dioxide **d.** hydrogen fluoride
e. nitrogen triiodide

4.43 a. dinitrogen trioxide **b.** disilicon hexabromide
c. tetraphosphorus trisulfide **d.** phosphorus pentachloride
e. dinitrogen trisulfide

4.45 a. CCl_4 **b.** CO **c.** PCl_3 **d.** N_2O_4

4.47 a. OF_2 **b.** BCl_3 **c.** N_2O_3 **d.** SF_6

4.49 a. aluminum sulfate **b.** calcium carbonate
c. dinitrogen oxide **d.** sodium phosphate
e. ammonium sulfate **f.** iron(III) oxide

4.51 The electronegativity increases going from left to right across a period.

4.53 **a.** K, Na, Li **b.** Na, P, Cl **c.** Ca, Se, O

4.55 **a.** polar covalent **b.** ionic
c. polar covalent **d.** nonpolar covalent
e. polar covalent **f.** polar covalent

4.57 **a.** $\overset{\delta^+}{N}-\overset{\delta^-}{F}$ (↦) **b.** $\overset{\delta^+}{Si}-\overset{\delta^-}{Br}$ (↦) **c.** $\overset{\delta^+}{C}-\overset{\delta^-}{O}$ (↦)
d. $\overset{\delta^+}{P}-\overset{\delta^-}{Br}$ (↦) **e.** $\overset{\delta^-}{N}-\overset{\delta^+}{P}$ (↤)

4.59 **a.** bent, 109° **b.** trigonal pyramidal

4.61 The four electron groups in PCl_3 have a tetrahedral arrangement, but three bonded atoms with a lone pair around a central atom give a trigonal pyramidal shape.

4.63 In both PH_3 and NH_3, there are three bonded atoms and one lone pair on the central atoms. The shapes of both are trigonal pyramidal.

4.65 **a.** bent, 109° **b.** tetrahedral
c. trigonal planar **d.** bent, 120°

4.67 Cl_2 is a nonpolar molecule because there is a nonpolar covalent bond between Cl atoms, which have identical electronegativity values. In HCl, the bond is a polar bond, which makes HCl a polar molecule.

4.69 **a.** polar **b.** polar **c.** polar

4.71 **a.** dipole–dipole attractions **b.** ionic bonds
c. dispersion forces **d.** dispersion forces

4.73 **a.** hydrogen bonds **b.** dispersion forces
c. dipole–dipole attractions **d.** dipole–dipole attractions
e. dipole–dipole attractions

4.75 **a.** By losing one valence electron from the third energy level, sodium achieves an octet in the second energy level.
b. The sodium ion Na^+ has the same electron arrangement as Ne (2,8).
c. Group 1A (1) and 2A (2) elements acquire octets by losing electrons to form compounds. Group 8A (18) elements are stable with octets (or two electrons for helium).

4.77 **a.** P^{3-} ion **b.** O atom
c. Zn^{2+} ion **d.** Fe^{3+} ion

4.79 **a.** X = Group 1A (1), Y = Group 6A (16)
b. ionic **c.** X^+ and Y^{2-} **d.** X_2Y
e. X_2S **f.** YCl_2 **g.** covalent

4.81 **a.** 2. trigonal pyramidal, polar
b. 1. bent, 109°, polar **c.** 3. tetrahedral, nonpolar

4.83

Electron Arrangements		Cation	Anion	Formula of Compound	Name of Compound
2,8,2	2,5	Mg^{2+}	N^{3-}	Mg_3N_2	magnesium nitride
2,8,8,1	2,6	K^+	O^{2-}	K_2O	potassium oxide
2,8,3	2,8,7	Al^{3+}	Cl^-	$AlCl_3$	aluminum chloride

4.85 **a.** two valence electrons, one bonding pair, no lone pairs
b. eight valence electrons, one bonding pair, three lone pairs
c. 14 valence electrons, one bonding pair, six lone pairs

4.87 **a.** 2,8 **b.** 2,8 **c.** 2,8,8 **d.** 2,8 **e.** 2

4.89 **a.** 2A (2) **b.** $\dot{X}\cdot$ **c.** Mg **d.** X_3N_2

4.91 **a.** Sn^{4+} **b.** 50 protons and 46 electrons
c. SnO_2 **d.** $Sn_3(PO_4)_4$

4.93 **a.** SnS **b.** PbO_2 **c.** AgCl
d. Ca_3N_2 **e.** Cu_3P **f.** $CrBr_2$

4.95 **a.** nitrogen trichloride **b.** dinitrogen trisulfide
c. dinitrogen oxide **d.** fluorine
e. phosphorus pentachloride **f.** diphosphorus pentoxide

4.97 **a.** CO **b.** P_2O_5 **c.** H_2S **d.** SCl_2

4.99 **a.** ionic, iron(III) chloride **b.** ionic, sodium sulfate
c. covalent, nitrogen dioxide **d.** covalent, nitrogen
e. covalent, phosphorus pentafluoride
f. covalent, carbon tetrafluoride

4.101 **a.** $SnCO_3$ **b.** Li_3P **c.** $SiCl_4$
d. Mn_2O_3 **e.** Br_2 **f.** $CaBr_2$

4.103 **a.** C—O **b.** N—F **c.** S—Cl
d. Br—I **e.** N—F

4.105 **a.** polar covalent (Cl 3.0 − Si 1.8 = 1.2)
b. nonpolar covalent (C 2.5 − C 2.5 = 0)
c. ionic (Cl 3.0 − Na 0.9 = 2.1)
d. nonpolar covalent (C 2.5 − H 2.1 = 0.4)
e. nonpolar covalent (F 4.0 − F 4.0 = 0)

4.107 **a.** polar **b.** polar **c.** nonpolar

4.109 **a.** trigonal pyramidal, polar **b.** bent 109°, polar

4.111 **a.** bent 109°, nonpolar **b.** trigonal pyramidal, polar

4.113 **a.** (3) hydrogen bonds **b.** (2) dipole–dipole attractions
c. (4) dispersion forces **d.** (1) ionic bonds

4.115

Atom or Ion	Number of Protons	Number of Electrons	Electrons Lost/Gained
K^+	19 p^+	18 e^-	1 e^- lost
Mg^{2+}	12 p^+	10 e^-	2 e^- lost
O^{2-}	8 p^+	10 e^-	2 e^- gained
Al^{3+}	13 p^+	10 e^-	3 e^- lost

4.117 **a.** Group 3A (13) **b.** Group 6A (16) **c.** Group 2A (2)

4.119 **a.** ionic, lithium oxide **b.** ionic, chromium(II) nitrate
c. ionic, magnesium bicarbonate or magnesium hydrogen carbonate
d. covalent, nitrogen trifluoride
e. ionic, calcium chloride **f.** ionic, potassium phosphate
g. ionic, gold(III) sulfite **h.** covalent, iodine

Chemical Quantities and Reactions

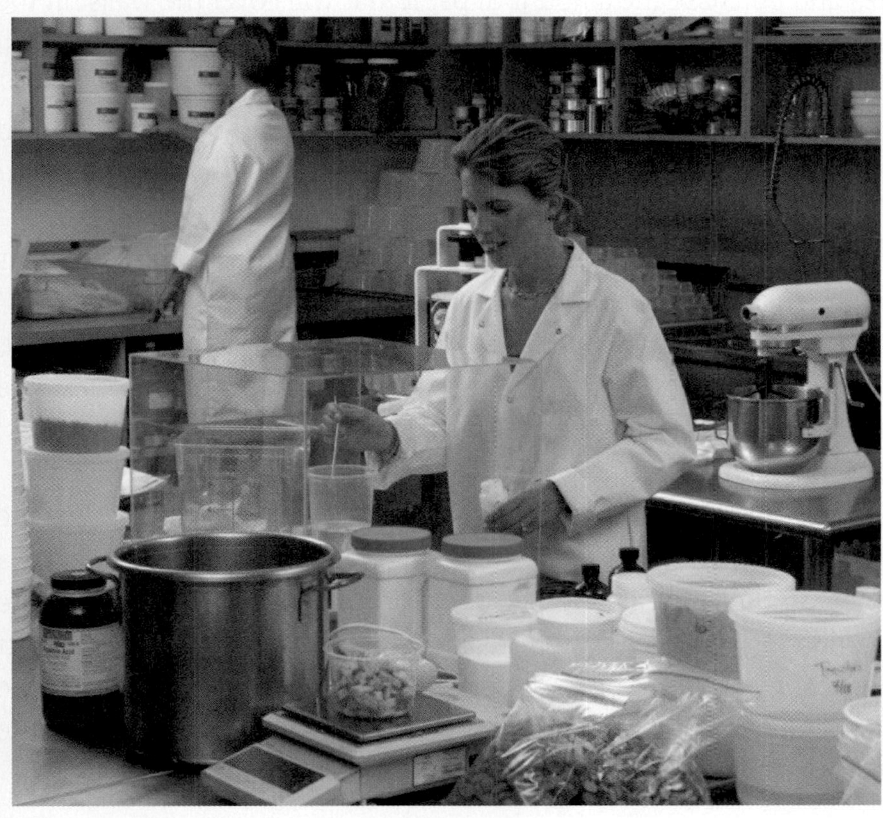

LOOKING AHEAD

"In our food science laboratory I develop a variety of food products, from cake doughnuts to energy beverages," says Anne Cristofano, senior food technologist at Mattson & Company. "When I started the doughnut project, I researched the ingredients and then weighed them out in the lab. I added water to make a batter and cooked the doughnuts in a fryer. The batter and the oil temperature make a big difference. If I don't get the right taste or texture, I adjust the ingredients, such as sugar and flour, or adjust the temperature."

A food technologist studies the physical and chemical properties of food and develops scientific ways to process and preserve it for extended shelf life. The food products are tested for texture, color, and flavor. The results of these tests help improve the quality and safety of food.

In chemistry, we calculate and measure the amounts of substances to use in the lab. Actually, measuring the amount of a substance is something you do every day. When you cook, you measure out the proper amounts of ingredients so you don't have too much of one or too little of another. At the gas station, you measure out a certain amount of fuel in your gas tank. If you paint the walls of a room, you measure the area and purchase the amount of paint that will cover the walls. In the lab, the chemical formula of a substance tells us the number and kinds of atoms it has, which we use to determine the amount of a substance you need.

Chemical reactions occur everywhere. The fuel in our cars burns with oxygen to provide energy to make the car move or run the air conditioner. When we cook our food or bleach our hair, chemical reactions take place. In our bodies, chemical reactions convert food into molecules to build muscles and move them. In the leaves of trees and plants, carbon dioxide and water are converted into carbohydrates.

Some chemical reactions are simple, whereas others are quite complex. However, they can all be written with the chemical equations that chemists use to describe chemical reactions. In every chemical reaction, the atoms in the reacting substances, called reactants, are rearranged to give new substances called products.

In this chapter, we will see how equations are written and how we can determine the amount of reactant or product involved. We do the same thing at home when we use a recipe to make cookies. At the automotive repair shop, a mechanic does essentially the same thing when adjusting the fuel system of an engine to allow for the correct amounts of fuel and oxygen. In the body, a certain amount of O_2 must reach the tissues for efficient metabolic reactions. If the oxygenation of the blood is low, the therapist will oxygenate the patient and recheck the blood levels.

LEARNING GOAL

Use Avogadro's number to determine the number of particles in a given number of moles.

TUTORIAL
Using Avogadro's Number

5.1 The Mole

At the grocery store, you buy eggs by the dozen. In an office-supply store, pencils are ordered by the gross and paper by the ream. In a restaurant, soda is ordered by the case. The terms dozen, gross, ream, and case are used to count the number of items present. For example, when you buy a dozen eggs, you know you will get 12 eggs in the carton.

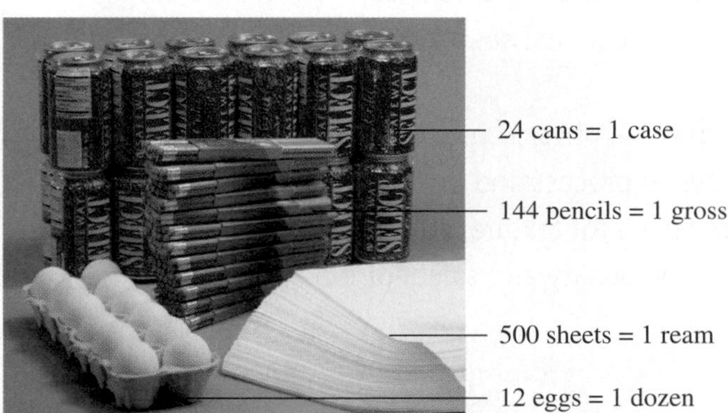

Collections of items include dozen, gross, and mole.

Avogadro's Number

In chemistry, particles such as atoms, molecules, and ions are counted by the **mole**, a unit that contains 6.02×10^{23} items. This very large number, called **Avogadro's number** after Amedeo Avogadro, an Italian physicist, looks like this when written with 3 significant figures:

One mole of sulfur contains 6.02×10^{23} sulfur atoms.

Avogadro's number

$$602\,000\,000\,000\,000\,000\,000\,000 = 6.02 \times 10^{23}$$

One mole of an element contains Avogadro's number of atoms. For example, 1 mole of carbon contains 6.02×10^{23} carbon atoms; 1 mole of aluminum contains 6.02×10^{23} aluminum atoms; 1 mole of sulfur contains 6.02×10^{23} sulfur atoms.

$$1 \text{ mole of an element} = 6.02 \times 10^{23} \text{ atoms of that element}$$

Avogadro's number tells us that one mole of a compound contains 6.02×10^{23} of the particular type of particles that make up that compound. One mole of a covalent compound contains Avogadro's number of molecules. For example, 1 mole of CO_2 contains 6.02×10^{23} molecules of CO_2. One mole of an ionic compound contains Avogadro's number of **formula units**, which are the groups of ions represented by the formula of an ionic compound. One mole of NaCl contains 6.02×10^{23} formula units of NaCl (Na^+, Cl^-). Table 5.1 gives examples of the number of particles in some 1-mole quantities.

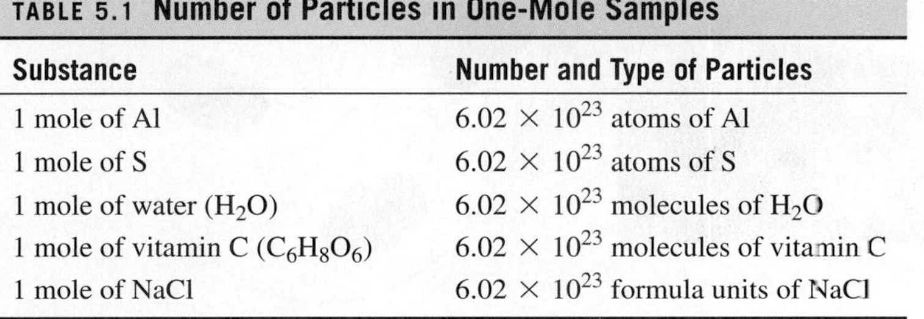

TABLE 5.1 Number of Particles in One-Mole Samples

Substance	Number and Type of Particles
1 mole of Al	6.02×10^{23} atoms of Al
1 mole of S	6.02×10^{23} atoms of S
1 mole of water (H_2O)	6.02×10^{23} molecules of H_2O
1 mole of vitamin C ($C_6H_8O_6$)	6.02×10^{23} molecules of vitamin C
1 mole of NaCl	6.02×10^{23} formula units of NaCl

We can use Avogadro's number as a conversion factor to convert between the moles of a substance and the number of particles it contains.

$$\frac{6.02 \times 10^{23} \text{ particles}}{1 \text{ mole}} \quad \text{and} \quad \frac{1 \text{ mole}}{6.02 \times 10^{23} \text{ particles}}$$

For example, we use Avogadro's number to convert 4.00 moles of sulfur to atoms of sulfur.

$$4.00 \cancel{\text{moles S atoms}} \times \underbrace{\frac{6.02 \times 10^{23} \text{ S atoms}}{1 \cancel{\text{mole S atoms}}}}_{\text{Avogadro's number as a conversion factor}} = 2.41 \times 10^{24} \text{ S atoms}$$

We can also use Avogadro's number to convert 3.01×10^{24} molecules of CO_2 to moles of CO_2.

$$3.01 \times 10^{24} \cancel{CO_2 \text{ molecules}} \times \underbrace{\frac{1 \text{ mole } CO_2 \text{ molecules}}{6.02 \times 10^{23} \cancel{CO_2 \text{ molecules}}}}_{\text{Avogadro's number as a conversion factor}} = 5.00 \text{ moles of } CO_2 \text{ molecules}$$

In calculations that convert between moles and particles, the number of moles will be a small number compared to the very large number of atoms or molecules.

CONCEPT CHECK 5.1

Moles and Particles

Explain why 0.50 mole of aluminum is a small number, but the number of atoms in 0.50 mole is a large number: 3.0×10^{23} atoms of aluminum.

ANSWER

The term mole is a unit that represents 6.02×10^{23} particles. Because atoms are submicroscopic particles, a large number of atoms are present in 1 mole of aluminum.

SAMPLE PROBLEM 5.1

Calculating the Number of Molecules

How many molecules are present in 1.75 moles of carbon dioxide, CO_2?

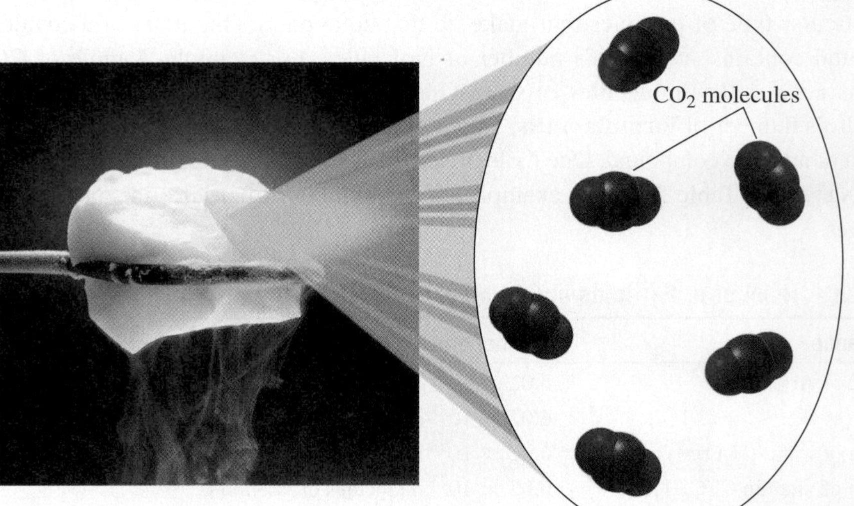

The solid form of carbon dioxide is known as "dry ice."

SOLUTION

Step 1 State the given and needed quantities.

Given 1.75 moles of CO_2 **Need** molecules of CO_2

Step 2 Write a plan to convert moles to atoms or molecules.

moles of CO_2 → Avogadro's number → molecules of CO_2

Step 3 Use Avogadro's number to write conversion factors.

$$1 \text{ mole of } CO_2 = 6.02 \times 10^{23} \text{ molecules of } CO_2$$

$$\frac{6.02 \times 10^{23} \text{ molecules } CO_2}{1 \text{ mole } CO_2} \quad \text{and} \quad \frac{1 \text{ mole } CO_2}{6.02 \times 10^{23} \text{ molecules } CO_2}$$

Step 4 Set up the problem to calculate the number of particles.

$$1.75 \cancel{\text{moles } CO_2} \times \frac{6.02 \times 10^{23} \text{ molecules } CO_2}{1 \cancel{\text{mole } CO_2}} = 1.05 \times 10^{24} \text{ molecules of } CO_2$$

STUDY CHECK 5.1

How many moles of water, H_2O, contain 2.60×10^{23} molecules of water?

Guide to Calculating the Atoms or Molecules of a Substance

1 State the given and needed quantities.

2 Write a plan to convert moles to atoms or molecules.

3 Use Avogadro's number to write conversion factors.

4 Set up the problem to calculate the number of particles.

Moles of Elements in a Formula

TUTORIAL
Moles and the Chemical Formula

We have seen that the subscripts in a chemical formula of a compound indicate the number of atoms of each type of element. For example, in a molecule of aspirin, chemical formula $C_9H_8O_4$, there are 9 carbon atoms, 8 hydrogen atoms, and 4 oxygen atoms. The subscripts also state the number of moles of each element in 1 mole of aspirin: 9 moles of C atoms, 8 moles of H atoms, and 4 moles of O atoms.

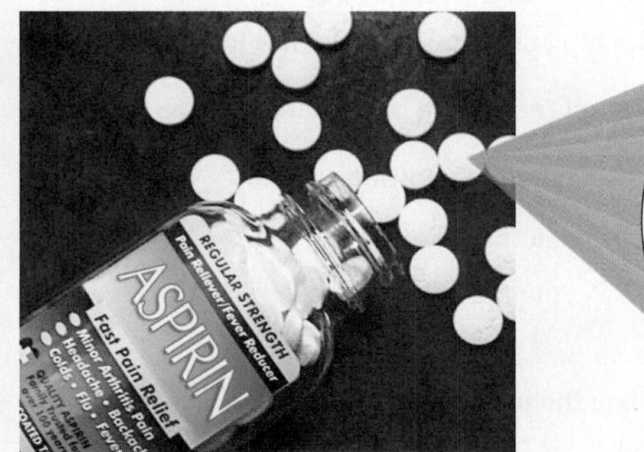

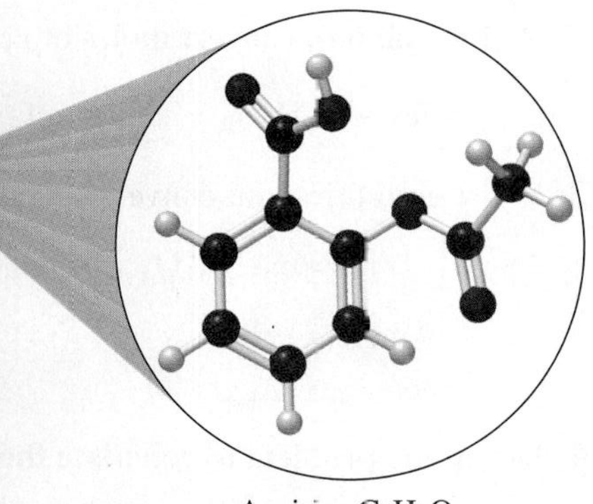

Aspirin $C_9H_8O_4$

Number of atoms in 1 molecule

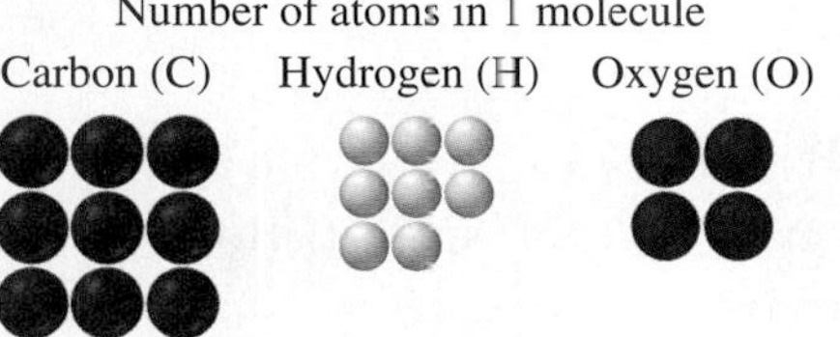

The formula subscript specifies the $C_9H_8O_4$

	Carbon	**Hydrogen**	**Oxygen**
Atoms in 1 molecule	9 atoms of C	8 atoms of H	4 atoms of O
Moles of each element in 1 mole	9 moles of C	8 moles of H	4 moles of O

Using the subscripts from the formula, $C_9H_8O_4$, we can write the conversion factors for each of the elements in 1 mole of aspirin.

$$\frac{9\text{ moles C}}{1\text{ mole }C_9H_8O_4} \quad \frac{8\text{ moles H}}{1\text{ mole }C_9H_8O_4} \quad \frac{4\text{ moles O}}{1\text{ mole }C_9H_8O_4}$$

$$\frac{1\text{ mole }C_9H_8O_4}{9\text{ moles C}} \quad \frac{1\text{ mole }C_9H_8O_4}{8\text{ moles H}} \quad \frac{1\text{ mole }C_9H_8O_4}{4\text{ moles O}}$$

CONCEPT CHECK 5.2

Using Subscripts of a Formula

Indicate the moles of each type of atom in 1 mole of propyl acetate, $C_5H_{10}O_2$, which gives the odor and taste to pears.

ANSWER

The subscripts in the formula indicate that there are 5 moles of C atoms, 10 moles of H atoms, and 2 moles of O atoms in 1 mole of propyl acetate.

The compound propyl acetate provides the odor and taste of pears.

SAMPLE PROBLEM 5.2

Calculating the Moles of an Element in a Compound

For 1.50 moles of aspirin, $C_9H_8O_4$, how many moles of carbon are present?

SOLUTION

Step 1 State the given and needed quantities.

Given 1.50 moles of $C_9H_8O_4$ **Need** moles of C

Step 2 Write a plan to convert moles of compound to moles of an element.

moles of $C_9H_8O_4$ → Subscript → moles of C atoms

Step 3 Write equalities and conversion factors using subscripts.

$$1 \text{ mole of } C_9H_8O_4 = 9 \text{ moles of C}$$

$$\frac{9 \text{ moles C}}{1 \text{ mole } C_9H_8O_4} \quad \text{and} \quad \frac{1 \text{ mole } C_9H_8O_4}{9 \text{ moles C}}$$

Step 4 Set up the problem to calculate the moles of an element.

$$1.50 \cancel{\text{moles } C_9H_8O_4} \times \frac{9 \text{ moles C}}{1 \cancel{\text{mole } C_9H_8O_4}} = 13.5 \text{ moles of C}$$

STUDY CHECK 5.2

How many moles of aspirin, $C_9H_8O_4$, contain 0.480 mole of O atoms?

Guide to Calculating the Moles of an Element in a Compound

1. State the given and needed quantities.
2. Write a plan to convert moles of compound to moles of an element.
3. Write equalities and conversion factors using subscripts.
4. Set up the problem to calculate the moles of an element.

QUESTIONS AND PROBLEMS

The Mole

5.1 What is a mole?

5.2 What is Avogadro's number?

5.3 Calculate each of the following:
a. number of C atoms in 0.500 mole of C
b. number of SO_2 molecules in 1.28 moles of SO_2
c. moles of Fe in 5.22×10^{22} atoms of Fe
d. moles of C_2H_6O in 8.50×10^{24} molecules of C_2H_6O

5.4 Calculate each of the following:
a. number of Li atoms in 4.5 moles of Li
b. number of CO_2 molecules in 0.0180 mole of CO_2
c. moles of Cu in 7.8×10^{21} atoms of Cu
d. moles of C_2H_6 in 3.75×10^{23} molecules of C_2H_6

5.5 Calculate each of the following quantities in 2.00 moles of H_3PO_4:
a. moles of H **b.** moles of O
c. atoms of P **d.** atoms of O

5.6 Calculate each of the following quantities in 0.185 mole of $C_6H_{14}O$:
a. moles of C **b.** moles of O
c. atoms of H **d.** atoms of C

5.7 Quinine, $C_{20}H_{24}N_2O_2$, is a component of tonic water and bitter lemon.
a. How many moles of hydrogen are in 1.0 mole of quinine?
b. How many moles of carbon are in 5.0 moles of quinine?
c. How many moles of nitrogen are in 0.020 mole of quinine?

5.8 Aluminum sulfate, $Al_2(SO_4)_3$, is used in some antiperspirants.
a. How many moles of sulfur are present in 3.0 moles of $Al_2(SO_4)_3$?
b. How many moles of aluminum ions are present in 0.40 mole of $Al_2(SO_4)_3$?
c. How many moles of sulfate ions (SO_4^{2-}) are present in 1.5 moles of $Al_2(SO_4)_3$?

LEARNING GOAL

Determine the molar mass of a substance, and use molar mass to convert between grams and moles.

5.2 Molar Mass

A single atom or molecule is much too small to weigh, even on the most sensitive balance. In fact, it takes a huge number of atoms or molecules to make enough of a substance for you to see. An amount of water that contains Avogadro's number of water molecules is only a few sips. In the laboratory, we can use a balance to weigh out Avogadro's number of particles or 1 mole of a substance.

For any element, the quantity called **molar mass** is the quantity in grams that equals the atomic mass of that element. We are counting 6.02×10^{23} atoms of an element when we weigh out the number of grams equal to its molar mass. For example, if we need 1 mole of carbon (C) atoms, we would first find the atomic mass of 12.01 on the periodic table. Then to obtain 1 mole of carbon atoms, we would weigh out 12.01 g of carbon. Thus, the molar mass of carbon is found by looking at the atomic mass on the periodic table.

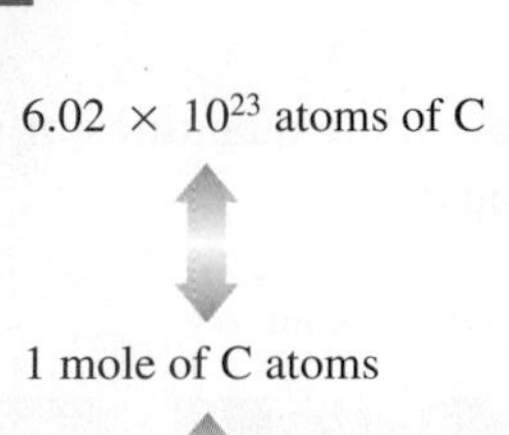

6.02×10^{23} atoms of C

1 mole of C atoms

12.01 g of C atoms

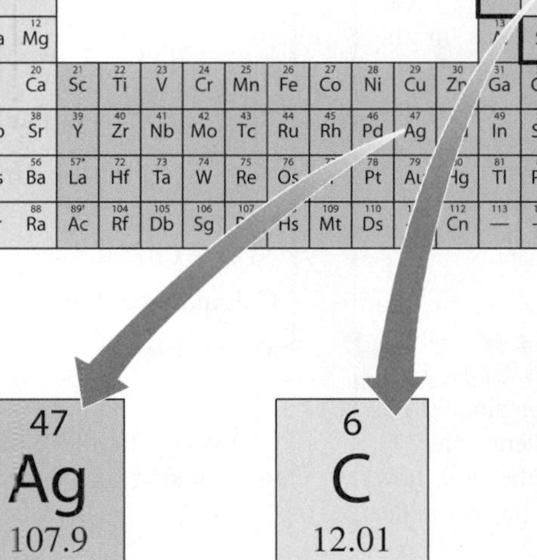

47 Ag 107.9	6 C 12.01	16 S 32.07
1 mole of silver atoms has a mass of 107.9 g.	1 mole of carbon atoms has a mass of 12.01 g.	1 mole of sulfur atoms has a mass of 32.07 g.

Molar Mass of a Compound

To determine the molar mass of a compound, multiply the molar mass of each element by its subscript in the formula, and add the results as shown in Sample Problem 5.3. ***In this text, we round molar mass to the tenths (0.1 g) place or use at least three significant figures for calculations.***

Lithium carbonate produces a red color in fireworks.

SAMPLE PROBLEM 5.3

Calculating the Molar Mass of a Compound

Find the molar mass of Li_2CO_3 used to produce red color in fireworks.

SOLUTION

Step 1 Obtain the molar mass of each element.

$$\frac{6.94\text{ g Li}}{1\text{ mole Li}} \qquad \frac{12.0\text{ g C}}{1\text{ mole C}} \qquad \frac{16.0\text{ g O}}{1\text{ mole O}}$$

Step 2 Multiply each molar mass by the number of moles (subscript) in the formula.

Grams from 2 moles of Li

$$2\ \cancel{\text{moles Li}} \times \frac{6.94\text{ g Li}}{1\ \cancel{\text{mole Li}}} = 13.8\text{ g of Li}$$

Guide to Calculating Molar Mass

1. Obtain the molar mass of each element.
2. Multiply each molar mass by the number of moles (subscript) in the formula.
3. Calculate the molar mass by adding the masses of the elements.

CALCULATING MOLES IN THE KITCHEN

The labels on food products list the components in grams and milligrams. It is possible to convert those values to moles, which is also interesting. Read the labels of some products in the kitchen and convert the amounts given in grams or milligrams to moles using molar mass. Here are some examples.

QUESTIONS

1. How many moles of NaCl are in a box of salt?
2. How many moles of sugar are contained in a 5-lb bag of sugar if sugar has the formula $C_{12}H_{22}O_{11}$?
3. A serving of cereal contains 90 mg of potassium. If there are 11 servings of cereal in the box, how many moles of K^+ are present in the cereal in the box?

Grams from 1 mole of C

$$1\ \cancel{\text{mole C}} \times \frac{12.0\text{ g C}}{1\ \cancel{\text{mole C}}} = 12.0\text{ g of C}$$

Grams from 3 moles of O

$$3\ \cancel{\text{moles O}} \times \frac{16.0\text{ g O}}{1\ \cancel{\text{mole O}}} = 48.0\text{ g of O}$$

Step 3 Calculate the molar mass by adding the masses of the elements.

2 moles of Li	=	13.8 g of Li
1 mole of C	=	12.0 g of C
3 moles of O	=	+48.0 g of O
Molar mass of Li_2CO_3	=	73.8 g

STUDY CHECK 5.3

Calculate the molar mass of salicylic acid, $C_7H_6O_3$.

Figure 5.1 shows some 1-mole quantities of substances. Table 5.2 lists the molar mass for several 1-mole samples.

One–Mole Quantities

S

Fe

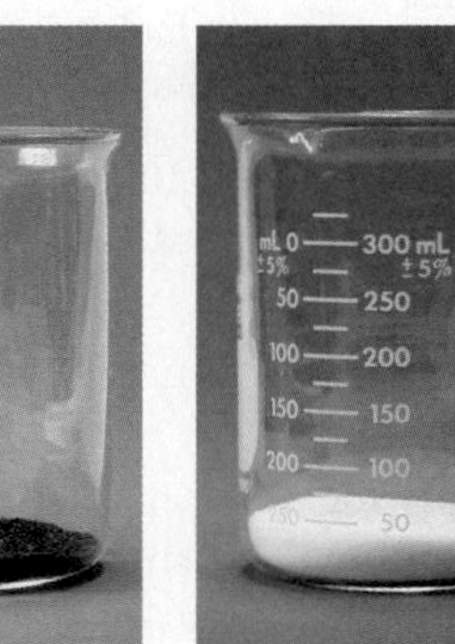
NaCl

$K_2Cr_2O_7$

$C_{12}H_{22}O_{11}$

FIGURE 5.1 One-mole samples: sulfur, S (32.1 g); iron, Fe (55.9 g); salt, NaCl (58.5 g); potassium dichromate, $K_2Cr_2O_7$ (294.2 g); and sucrose, $C_{12}H_{22}O_{11}$ (342.2 g).
Q How is the molar mass for $K_2Cr_2O_7$ obtained?

TABLE 5.2 The Molar Mass of Selected Elements and Compounds

Substance	Molar Mass
1 mole of C	12.0 g
1 mole of Na	23.0 g
1 mole of Fe	55.9 g
1 mole of NaF	42.0 g
1 mole of $C_6H_{12}O_6$ (glucose)	180.1 g
1 mole of $C_8H_{10}N_4O_2$ (caffeine)	194.1 g

Calculations Using Molar Mass

The molar mass of an element or a compound is one of the most useful conversion factors in chemistry. Molar mass is used to change from moles of a substance to grams, or from grams to moles. For example, 1 mole of magnesium has a mass of 24.3 g. To express molar mass as an equality, we can write

$$1\text{ mole of Mg} = 24.3\text{ g of Mg}$$

From this equality, two conversion factors can be written.

$$\frac{24.3\text{ g Mg}}{1\text{ mole Mg}} \quad \text{and} \quad \frac{1\text{ mole Mg}}{24.3\text{ g Mg}}$$

Conversion factors for the molar mass of a compound are written in the same way. For example, the equality for the molar mass of the compound H_2O is written

$$1\text{ mole of }H_2O = 18.0\text{ g of }H_2O$$

From this equality, the conversion factors from the molar mass of H_2O are written as

$$\frac{18.0\ \text{g}\ H_2O}{1\ \text{mole}\ H_2O} \quad \text{and} \quad \frac{1\ \text{mole}\ H_2O}{18.0\ \text{g}\ H_2O}$$

MC TUTORIAL
Converting Between Grams and Moles

We can now change from moles to grams, or grams to moles, using the conversion factors derived from the molar mass as shown in Sample Problem 5.4. (Remember, you must determine the molar mass of the substance first.)

Table salt is sodium chloride, NaCl.

Guide to Calculating the Moles (or Grams) of a Substance from Grams (or Moles)

1. State the given and needed quantities.
2. Write a plan to convert moles to grams (or grams to moles).
3. Determine the molar mass and write conversion factors.
4. Set up the problem to convert moles to grams (or grams to moles).

SAMPLE PROBLEM 5.4

Converting the Mass of a Compound to Moles

A box of salt contains 737 g of NaCl. How many moles of NaCl are present in the box?

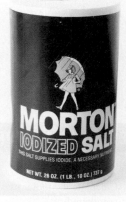

SOLUTION

Step 1 State the given and needed quantities.

Given 737 g of NaCl **Need** moles of NaCl

Step 2 Write a plan to convert grams to moles.

grams of NaCl → Molar mass → moles of NaCl

Step 3 Determine the molar mass and write conversion factors.

$$1\ \text{mole of NaCl} = 58.5\ \text{g of NaCl}$$

$$\frac{58.5\ \text{g NaCl}}{1\ \text{mole NaCl}} \quad \text{and} \quad \frac{1\ \text{mole NaCl}}{58.5\ \text{g NaCl}}$$

Step 4 Set up the problem to convert given grams to moles.

$$737\ \cancel{\text{g NaCl}} \times \frac{1\ \text{mole NaCl}}{58.5\ \cancel{\text{g NaCl}}} = 12.6\ \text{moles of NaCl}$$

STUDY CHECK 5.4

Silver metal is used to make jewelry.

Silver metal is used in the manufacture of tableware, mirrors, jewelry, and dental alloys. If the design for a piece of jewelry requires 0.750 mole of silver, how many grams of silver are needed?

We can summarize the calculations to show the connections between the moles of a compound, its mass in grams, number of molecules (or formula units if ionic), and the moles and atoms of each element in that compound in the following flowchart:

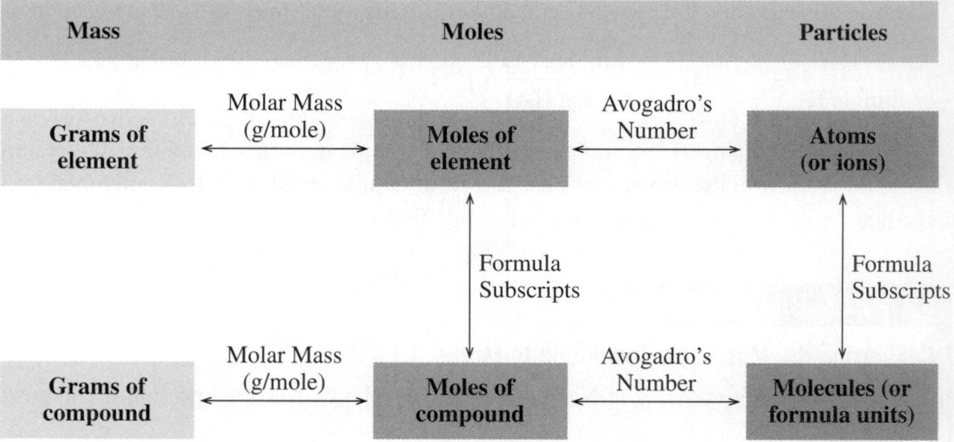

QUESTIONS AND PROBLEMS

Molar Mass

5.9 Calculate the molar mass for each of the following compounds:
a. $KC_4H_5O_6$ (cream of tartar)
b. Fe_2O_3 (rust)
c. $C_{19}H_{20}FNO_3$ (Paxil, an antidepressant)
d. $Al_2(SO_4)_3$ (antiperspirant)
e. $Mg(OH)_2$ (antacid)
f. $C_{16}H_{19}N_3O_5S$ (amoxicillin, an antibiotic)

5.10 Calculate the molar mass for each of the following compounds:
a. $FeSO_4$ (iron supplement)
b. Al_2O_3 (absorbent and abrasive)
c. $C_7H_5NO_3S$ (saccharin)
d. C_3H_8O (rubbing alcohol)
e. $(NH_4)_2CO_3$ (baking powder)
f. $Zn(C_2H_3O_2)_2$ (zinc dietary supplement)

5.11 Calculate the molar mass of each of the following compounds:
a. Cl_2 **b.** $C_3H_6O_3$ **c.** $Mg_3(PO_4)_2$
d. AlF_3 **e.** $C_2H_4Cl_2$ **f.** SnF_2

5.12 Calculate the molar mass of each of the following compounds:
a. O_2 **b.** KH_2PO_4 **c.** $Fe(ClO_4)_3$
d. $C_4H_8O_4$ **e.** $Ga_2(CO_3)_3$ **f.** $KBrO_4$

5.13 Calculate the mass, in grams, for each of the following:
a. 2.00 moles of Na **b.** 2.80 moles of Ca
c. 0.125 mole of Sn **d.** 1.76 moles of Cu

5.14 Calculate the mass, in grams, for each of the following:
a. 1.50 moles of K **b.** 2.5 moles of C
c. 0.25 mole of P **d.** 12.5 moles of He

5.15 Calculate the number of grams in each of the following:
a. 0.500 mole of NaCl **b.** 1.75 moles of Na_2O
c. 0.225 mole of H_2O **d.** 4.42 moles of CO_2

5.16 Calculate the number of grams in each of the following:
a. 2.0 moles of $MgCl_2$ **b.** 3.5 moles of C_6H_6
c. 5.00 moles of C_2H_6O **d.** 0.488 mole of $C_3H_6O_3$

5.17 **a.** The compound $MgSO_4$ is called Epsom salts. How many grams will you need to prepare a bath containing 5.00 moles of Epsom salts?
b. In a bottle of soda, there is 0.25 mole of CO_2. How many grams of CO_2 are in the bottle?

5.18 **a.** Cyclopropane, C_3H_6, is an anesthetic given by inhalation. How many grams are in 0.25 mole of cyclopropane?
b. The sedative Demerol hydrochloride has the formula $C_{15}H_{22}ClNO_2$. How many grams are in 0.025 mole of Demerol hydrochloride?

5.19 How many moles are contained in each of the following?
a. 50.0 g of Ag **b.** 0.200 g of C
c. 15.0 g of NH_3 **d.** 75.0 g of SO_2

5.20 How many moles are contained in each of the following?
a. 25.0 g of Ca **b.** 5.00 g of S
c. 40.0 g of H_2O **d.** 100.0 g of O_2

5.21 Calculate the number of moles in 25.0 g of each of the following:
a. Ne **b.** O_2 **c.** $Al(OH)_3$ **d.** Ga_2S_3

5.22 Calculate the number of moles in 4.00 g of each of the following:
a. He **b.** SnO_2 **c.** $Cr(OH)_3$ **d.** Ca_3N_2

5.23 How many moles of S are in each of the following quantities?
a. 25 g of S **b.** 125 g of SO_2 **c.** 2.0 moles of Al_2S_3

5.24 How many moles of C are in each of the following quantities?
a. 75 g of C **b.** 0.25 mole of C_2H_6 **c.** 88 g of CO_2

5.25 Propane gas, C_3H_8, a hydrocarbon, is used as a fuel for many barbecues.
a. How many grams of the compound are in 1.50 moles of propane?
b. How many moles of the compound are in 34.0 g of propane?
c. How many grams of carbon are in 34.0 g of propane?

5.26 Allyl sulfide, $C_6H_{10}S$, is the substance that gives garlic, onions, and leeks their characteristic odor.
a. How many moles of sulfur are in 23.2 g of $C_6H_{10}S$?
b. How many moles of hydrogen are in 0.75 mole of $C_6H_{10}S$?
c. How many grams of carbon are in 44.0 g of $C_6H_{10}S$?

The characteristic odor of garlic is a sulfur-containing compound.

5.3 Chemical Reactions and Equations

As we discussed in Chapter 2, Section 2.2, a *chemical change* occurs when a substance is converted into one or more new substances that have different formulas and different properties. There may be a change in color or the formation of bubbles or a solid. For example, when silver tarnishes, the shiny silver metal (Ag) reacts with sulfur (S) to become the dull, blackish substance we call tarnish (Ag_2S) (see Figure 5.2).

LEARNING GOAL

Write a balanced chemical equation from the formulas of the reactants and products for a reaction; determine the number of atoms in the reactants and products.

TUTORIAL
Chemical Reactions and Equations

SELF STUDY ACTIVITY
Chemical Reactions and Equations

TUTORIAL
Signs of a Chemical Reaction

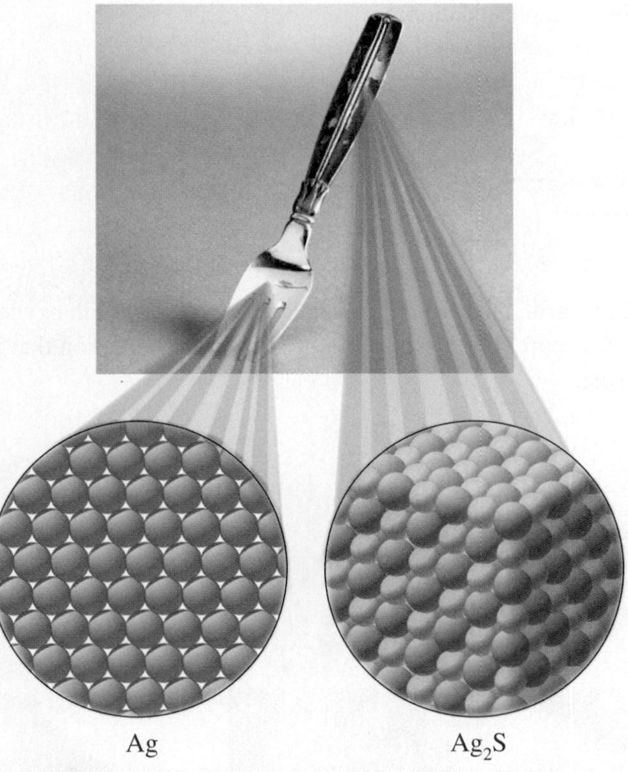

FIGURE 5.2 A chemical change produces new substances.
Q Why is the formation of tarnish a chemical change?

A Chemical Equation Describes a Chemical Reaction

A *chemical reaction* always involves chemical change because atoms of the reacting substances form new combinations with new properties. For example, a chemical reaction takes place when a piece of iron (Fe) combines with oxygen (O_2) in the air to produce a new substance, rust (Fe_2O_3), which has a reddish-brown color (see Figure 5.3). During a chemical change, new properties become visible, which are an indication that a chemical reaction has taken place (see Table 5.3).

FIGURE 5.3 Chemical reactions involve chemical changes. When iron (Fe) reacts with oxygen (O_2), the product is rust (Fe_2O_3).
Q What is the evidence for chemical change in this chemical reaction?

CONCEPT CHECK 5.3

Evidence of a Chemical Reaction

Identify the evidence for a chemical change in each of the following:

a. propane fuel burning in a barbecue
b. leaves turning red

ANSWER

a. The production of heat during burning of fuel is evidence of a chemical reaction.
b. The change in color is evidence of a chemical reaction.

TABLE 5.3 Types of Visible Evidence of a Chemical Reaction

1. Change in the color
2. Formation of a gas (bubbles)
3. Formation of a solid (precipitate)
4. Heat (or a flame) produced or heat absorbed

When you build a model airplane, prepare a new recipe, or mix a medication, you follow a set of directions. These directions tell you what materials to use and the products you will obtain. In chemistry, a *chemical equation* tells us the materials we need and the products that will form in a chemical reaction.

Writing a Chemical Equation

Suppose you work in a bicycle shop, assembling wheels and frames into bicycles. You could represent this process by a simple equation:

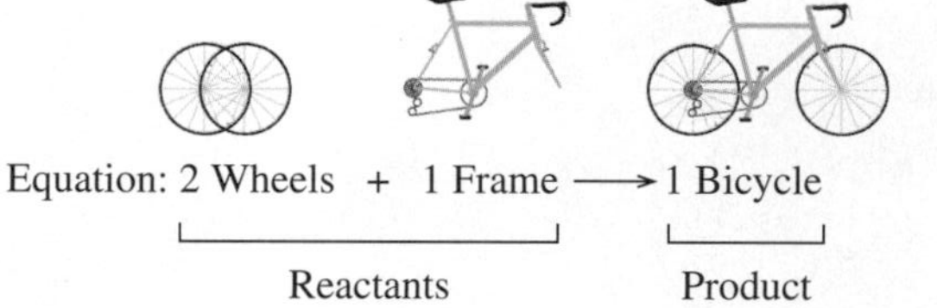

Equation: 2 Wheels + 1 Frame ⟶ 1 Bicycle

Reactants — Product

When you burn charcoal in a grill, the carbon in the charcoal combines with oxygen to form carbon dioxide. We can represent this reaction by a chemical equation that is much like the one for the bicycle:

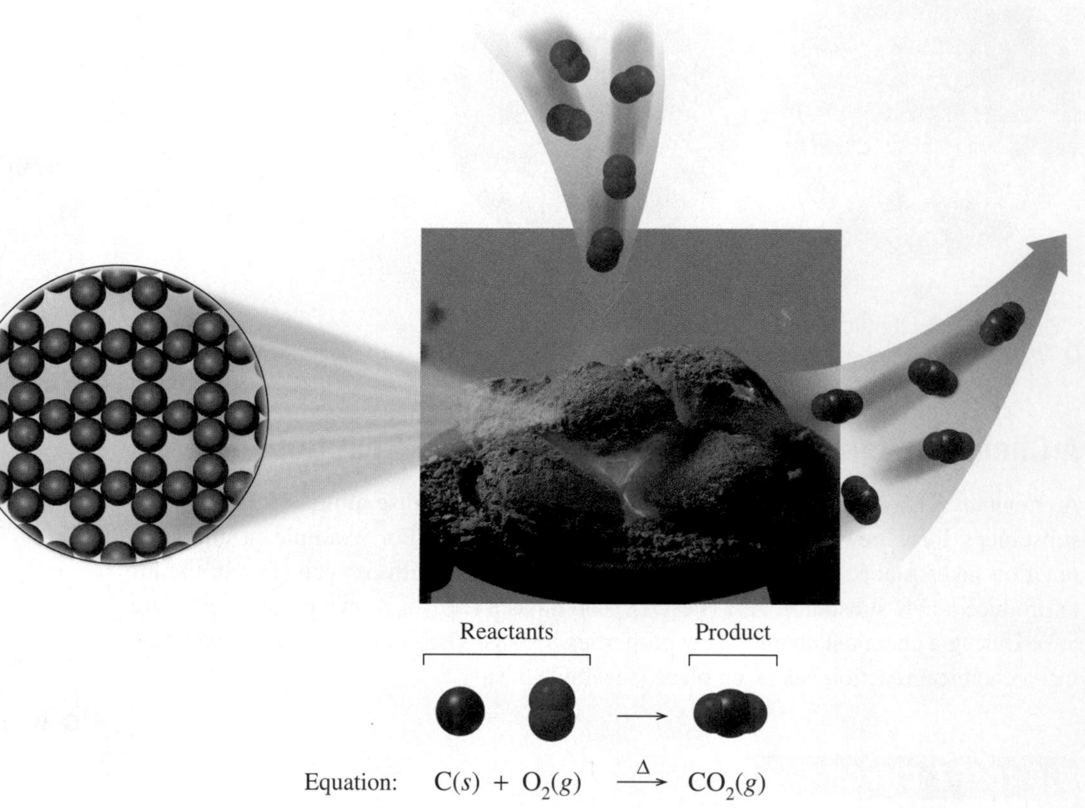

Reactants — Product

Equation: $C(s) + O_2(g) \xrightarrow{\Delta} CO_2(g)$

TUTORIAL
Chemical Reactions and Equations

In a **chemical equation**, the formulas of the **reactants** are written on the left of the arrow and the formulas of the **products** on the right. When there are two or more formulas on the same side, they are separated by plus (+) signs. The delta sign (Δ) indicates that heat was used to start the reaction.

Generally, each formula in an equation is followed by an abbreviation, in parentheses, that gives the physical state of the substance: solid (*s*), liquid (*l*), or gas (*g*). If a substance is dissolved in water, it is an aqueous (*aq*) solution. Table 5.4 summarizes some of the symbols used in equations.

TABLE 5.4 Some Symbols Used in Writing Equations

Symbol	Meaning
+	Separates two or more formulas
$\longrightarrow$	Reacts to form products
Δ	Reactants are heated
(*s*)	Solid
(*l*)	Liquid
(*g*)	Gas
(*aq*)	Aqueous

Identifying a Balanced Chemical Equation

When a reaction takes place, the bonds between the atoms of the reactants are broken and new bonds are formed to give the products. All atoms are conserved, which means that atoms cannot be gained, lost, or changed into other types of atoms during a chemical reaction. Every chemical reaction must be written as a **balanced equation**, which shows the same number of atoms for each element in the reactants as well as in the products. For example, the chemical equation we wrote previously for burning carbon is balanced because there is one carbon atom and two oxygen atoms in both the reactants and the products:

$$C(s) + O_2(g) \longrightarrow CO_2(g)$$

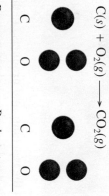

C O C O

Reactant atoms = Product atoms

Now consider the reaction in which hydrogen reacts with oxygen to form water. The formulas of the reactants and products are written as follows:

$$H_2(g) + O_2(g) \longrightarrow H_2O(g)$$

When we add up the atoms of each element on each side, we find that the equation is *not balanced*. There are two oxygen atoms to the left of the arrow, but only one to the right. To balance this equation, we place whole numbers called **coefficients** in front of the formulas. If we write a coefficient of 2 in front of the H_2O formula, it represents two molecules of water. Because the coefficient multiplies all the atoms in H_2O, there are now four hydrogen atoms and two oxygen atoms in the products. To obtain four atoms of hydrogen in the reactants, we must write a coefficient of 2 in front of the formula H_2. However, we *do not change subscripts*, which would alter the chemical identity of a reactant or product. Now the number of hydrogen atoms and the number of oxygen atoms are the same in the reactants as in the products. The equation is *balanced*.

$$2H_2(g) + O_2(g) \longrightarrow 2H_2O(g)$$

H O H O

Reactant atoms = Product atoms

Balancing a Chemical Equation

The chemical reaction that occurs in the flame of a gas burner you use in the laboratory or a gas cooktop is the reaction of methane gas, CH_4, and oxygen to produce carbon dioxide and water. We now show the process of balancing a chemical equation in Sample Problem 5.5.

TUTORIAL
Balancing Chemical Equations

Guide to Balancing a Chemical Equation

1. Write an equation using the correct formulas of the reactants and products.
2. Count the atoms of each element in the reactants and products.
3. Use coefficients to balance each element.
4. Check the final equation to confirm it is balanced.

SAMPLE PROBLEM 5.5

Balancing a Chemical Equation

The chemical reaction of methane, CH_4, and oxygen gas, O_2, produces carbon dioxide (CO_2) and water (H_2O). Write a balanced chemical equation for this reaction.

SOLUTION

Step 1 Write an equation using the correct formulas of the reactants and products.

$$CH_4(g) + O_2(g) \longrightarrow CO_2(g) + H_2O(g)$$

Step 2 Count the atoms of each element in the reactants and products. In the initial unbalanced equation, a coefficient of 1 is understood and not usually written. When we compare the atoms on the reactant side with the atoms on the product side, we see that there are more H atoms in the reactants and more O atoms in the products.

$$CH_4(g) + O_2(g) \longrightarrow CO_2(g) + H_2O(g)$$

Reactants	Products	
1 C atom	1 C atom	Balanced
4 H atoms	2 H atoms	Not balanced
2 O atoms	3 O atoms	Not balanced

Step 3 Use coefficients to balance each element. We will start by balancing the H atoms in CH_4 because it has the most atoms. By placing a coefficient of 2 in front of the formula for water, a total of 4 H atoms in the products is obtained.

$$CH_4(g) + O_2(g) \longrightarrow CO_2(g) + 2H_2O(g)$$

Reactants	Products	
1 C atom	1 C atom	Balanced
4 H atoms	4 H atoms	Balanced
2 O atoms	4 O atoms	Not balanced

We can balance the O atoms on the reactant side by placing a coefficient of 2 in front of the formula O_2. There are now 4 O atoms and 4 H atoms in both the reactants and products.

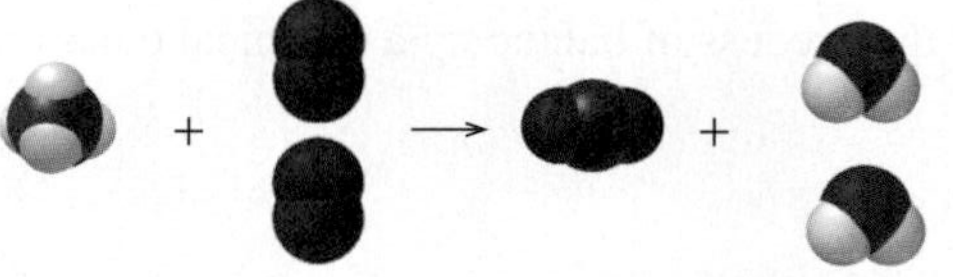

$$CH_4(g) + 2O_2(g) \longrightarrow CO_2(g) + 2H_2O(g) \quad \text{Balanced}$$

Step 4 **Check the final equation to confirm it is balanced.** In the final equation, the numbers of atoms of C, H, and O are the same in both the reactants and the products. The equation is balanced.

$$CH_4(g) + 2O_2(g) \longrightarrow CO_2(g) + 2H_2O(g)$$

Reactants	Products	
1 C atom	1 C atom	Balanced
4 H atoms	4 H atoms	Balanced
4 O atoms	4 O atoms	Balanced

In a balanced equation, the coefficients must be the lowest whole numbers possible. Suppose you had obtained the following:

$$2CH_4(g) + 4O_2(g) \longrightarrow 2CO_2(g) + 4H_2O(g) \quad \text{Incorrect}$$

Although there are equal numbers of atoms on both sides of the equation, this is not written correctly. The correctly written equation is obtained by dividing all the coefficients by 2.

STUDY CHECK 5.5

Balance the following equation:

$$Al(s) + Cl_2(g) \longrightarrow AlCl_3(s)$$

CONCEPT CHECK 5.4

Atoms in a Balanced Chemical Equation

Indicate the number of each atom in the reactants and products in the following equation:

$$Fe_2S_3(s) + 6HCl(aq) \longrightarrow 2FeCl_3(aq) + 3H_2S(g)$$

	Reactants	Products
Atoms of Fe		
Atoms of S		
Atoms of H		
Atoms of Cl		

ANSWER

The total number of atoms of each element is obtained by multiplying its subscript by the coefficient.

	Reactants	Products
Atoms of Fe	2	2
Atoms of S	3	3
Atoms of H	6	6
Atoms of Cl	6	6

SAMPLE PROBLEM 5.6

Balancing Chemical Equations

Balance the following equation:

$$Na_3PO_4(aq) + MgCl_2(aq) \longrightarrow Mg_3(PO_4)_2(s) + NaCl(aq)$$

SOLUTION

Step 1 **Write an equation using the correct formulas of the reactants and products.**

Step 2 **Count the atoms of each element in the reactants and products.** When we compare the number of ions on the reactant and product sides, we find they are not balanced. In this equation, we balance the phosphate ion as a group because it appears on both sides of the equation.

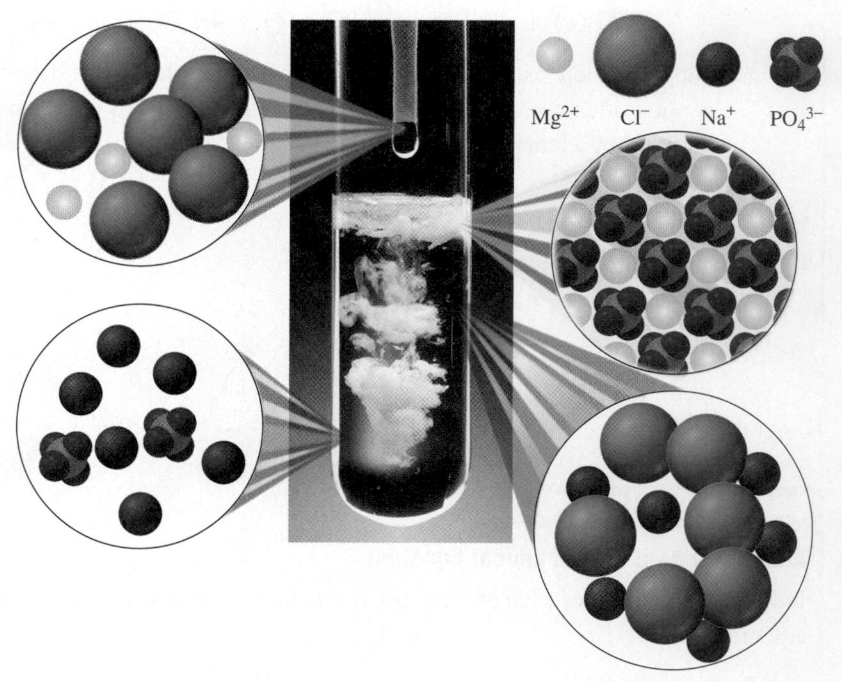

Reactants	Products	
$Na_3PO_4(aq) + MgCl_2(aq) \longrightarrow$	$Mg_3(PO_4)_2(s) + NaCl(aq)$	
3 Na^+	1 Na^+	Not balanced
1 PO_4^{3-}	2 PO_4^{3-}	Not balanced
1 Mg^{2+}	3 Mg^{2+}	Not balanced
2 Cl^-	1 Cl^-	Not balanced

Step 3 **Use coefficients to balance each element.** We begin with the formula that has the highest subscript values, which in this equation is $Mg_3(PO_4)_2$. The subscript 3 in $Mg_3(PO_4)_2$ is used as a coefficient for $MgCl_2$ to balance magnesium. The subscript 2 in $Mg_3(PO_4)_2$ is used as a coefficient for Na_3PO_4 to balance the phosphate ion.

$$2Na_3PO_4(aq) + 3MgCl_2(aq) \longrightarrow Mg_3(PO_4)_2(s) + NaCl(aq)$$

Reactants	Products	
6 Na^+	1 Na^+	Not balanced
2 PO_4^{3-}	2 PO_4^{3-}	Balanced
3 Mg^{2+}	3 Mg^{2+}	Balanced
6 Cl^-	1 Cl^-	Not balanced

Looking again at each of the ions in the reactants and products, we see that the sodium and chloride ions are not yet equal. A coefficient of 6 for the NaCl balances the equation.

$$2Na_3PO_4(aq) + 3MgCl_2(aq) \longrightarrow Mg_3(PO_4)_2(s) + 6NaCl(aq)$$

Step 4 **Check the final equation to confirm it is balanced.** A check of the total number of ions confirms the equation is balanced. A coefficient of 1 is understood and not usually written.

$$2Na_3PO_4(aq) + 3MgCl_2(aq) \longrightarrow Mg_3(PO_4)_2(s) + 6NaCl(aq) \quad \text{Balanced}$$

Reactants	Products	
6 Na^+	6 Na^+	Balanced
2 $PO_4{}^{3-}$	2 $PO_4{}^{3-}$	Balanced
3 Mg^{2+}	3 Mg^{2+}	Balanced
6 Cl^-	6 Cl^-	Balanced

STUDY CHECK 5.6

Balance the following equation:

$$Sb_2S_3(s) + HCl(aq) \longrightarrow SbCl_3(s) + H_2S(g)$$

QUESTIONS AND PROBLEMS

Chemical Reactions and Equations

5.27 Determine whether each of the following equations is balanced or not balanced:
a. $S(s) + O_2(g) \longrightarrow SO_3(g)$
b. $2Al(s) + 3Cl_2(g) \longrightarrow 2AlCl_3(s)$
c. $H_2(g) + O_2(g) \longrightarrow H_2O(g)$
d. $C_3H_8(g) + 5O_2(g) \longrightarrow 3CO_2(g) + 4H_2O(g)$

5.28 Determine whether each of the following equations is balanced or not balanced:
a. $PCl_3(s) + Cl_2(g) \longrightarrow PCl_5(s)$
b. $CO(g) + 2H_2(g) \longrightarrow CH_3OH(g)$
c. $2KClO_3(s) \longrightarrow 2KCl(s) + O_2(g)$
d. $Mg(s) + N_2(g) \longrightarrow Mg_3N_2(s)$

5.29 Balance each of the following equations:
a. $N_2(g) + O_2(g) \longrightarrow NO(g)$
b. $HgO(s) \longrightarrow Hg(l) + O_2(g)$
c. $Fe(s) + O_2(g) \longrightarrow Fe_2O_3(s)$
d. $Na(s) + Cl_2(g) \longrightarrow NaCl(s)$

5.30 Balance each of the following equations:
a. $Ca(s) + Br_2(l) \longrightarrow CaBr_2(s)$
b. $P_4(s) + O_2(g) \longrightarrow P_4O_{10}(s)$
c. $Sb_2S_3(s) + HCl(aq) \longrightarrow SbCl_3(s) + H_2S(g)$
d. $Fe_2O_3(s) + C(s) \longrightarrow Fe(s) + CO(g)$

5.31 Balance each of the following equations:
a. $Mg(s) + AgNO_3(aq) \longrightarrow Mg(NO_3)_2(aq) + Ag(s)$
b. $Al(s) + CuSO_4(aq) \longrightarrow Cu(s) + Al_2(SO_4)_3(aq)$
c. $Pb(NO_3)_2(aq) + NaCl(aq) \longrightarrow PbCl_2(s) + NaNO_3(aq)$
d. $Al(s) + HCl(aq) \longrightarrow AlCl_3(aq) + H_2(g)$

5.32 Balance each of the following equations:
a. $Zn(s) + H_2SO_4(aq) \longrightarrow ZnSO_4(aq) + H_2(g)$
b. $Al(s) + H_2SO_4(aq) \longrightarrow Al_2(SO_4)_3(aq) + H_2(g)$
c. $K_2SO_4(aq) + BaCl_2(aq) \longrightarrow BaSO_4(s) + KCl(aq)$
d. $CaCO_3(s) \longrightarrow CaO(s) + CO_2(g)$

5.4 Types of Reactions

LEARNING GOAL

Identify a reaction as a combination, decomposition, single replacement, double replacement, or combustion.

A great number of reactions occur in nature, in biological systems, and in the laboratory. However, there are some general patterns that help us classify most reactions into five general types.

Combination Reactions

In a **combination reaction**, two or more elements or compounds bond to form one product. For example, sulfur and oxygen combine to form the product sulfur dioxide.

Combination

Two or more reactants combine to yield a single product

A + B ⟶ A B

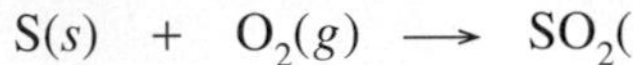

$$S(s) + O_2(g) \longrightarrow SO_2(g)$$

In Figure 5.4, the elements magnesium and oxygen combine to form a single product, magnesium oxide.

$$2Mg(s) + O_2(g) \longrightarrow 2MgO(s)$$

In other examples of combination reactions, elements or compounds combine to form a single product.

$$N_2(g) + 3H_2(g) \longrightarrow 2NH_3(g) \quad \text{Ammonia}$$

$$Cu(s) + S(s) \longrightarrow CuS(s)$$

$$MgO(s) + CO_2(g) \longrightarrow MgCO_3(s)$$

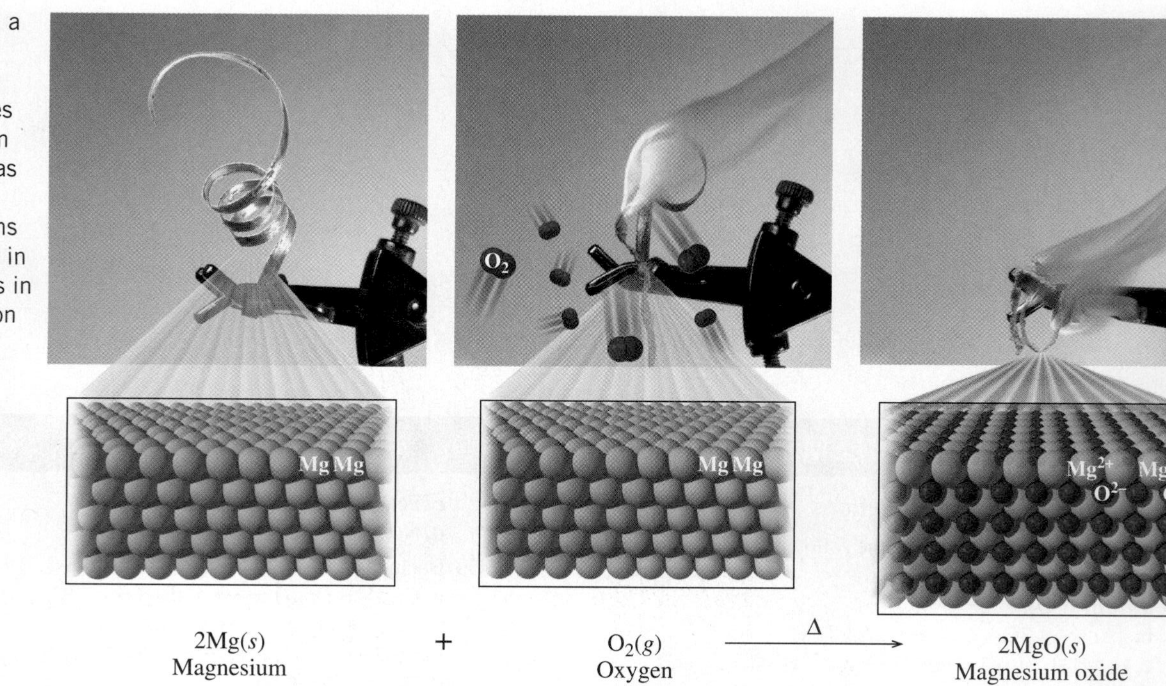

FIGURE 5.4 In a combination reaction, two or more substances combine to form one substance as product.
Q What happens to the atoms in the reactants in a combination reaction?

Decomposition Reactions

Decomposition

One reactant splits into two or more products

A B ⟶ A + B

In a **decomposition reaction**, a reactant splits into two or more simpler products. For example, when mercury(II) oxide is heated, the compound breaks apart into mercury atoms and oxygen (see Figure 5.5).

$$2HgO(s) \xrightarrow{\Delta} 2Hg(l) + O_2(g)$$

In another example of a decomposition reaction, calcium carbonate breaks apart into simpler compounds of calcium oxide and carbon dioxide.

$$CaCO_3(s) \xrightarrow{\Delta} CaO(s) + CO_2(g)$$

Replacement Reactions

In a replacement reaction, elements in a compound are replaced by other elements. In a **single replacement reaction**, a reacting element switches place with an element in the other reacting compound.

Single replacement

One element replaces another element

A + B C ⟶ A C + B

In the single replacement reaction shown in Figure 5.6, zinc replaces hydrogen in hydrochloric acid, HCl(*aq*).

$$Zn(s) + 2HCl(aq) \longrightarrow ZnCl_2(aq) + H_2(g)$$

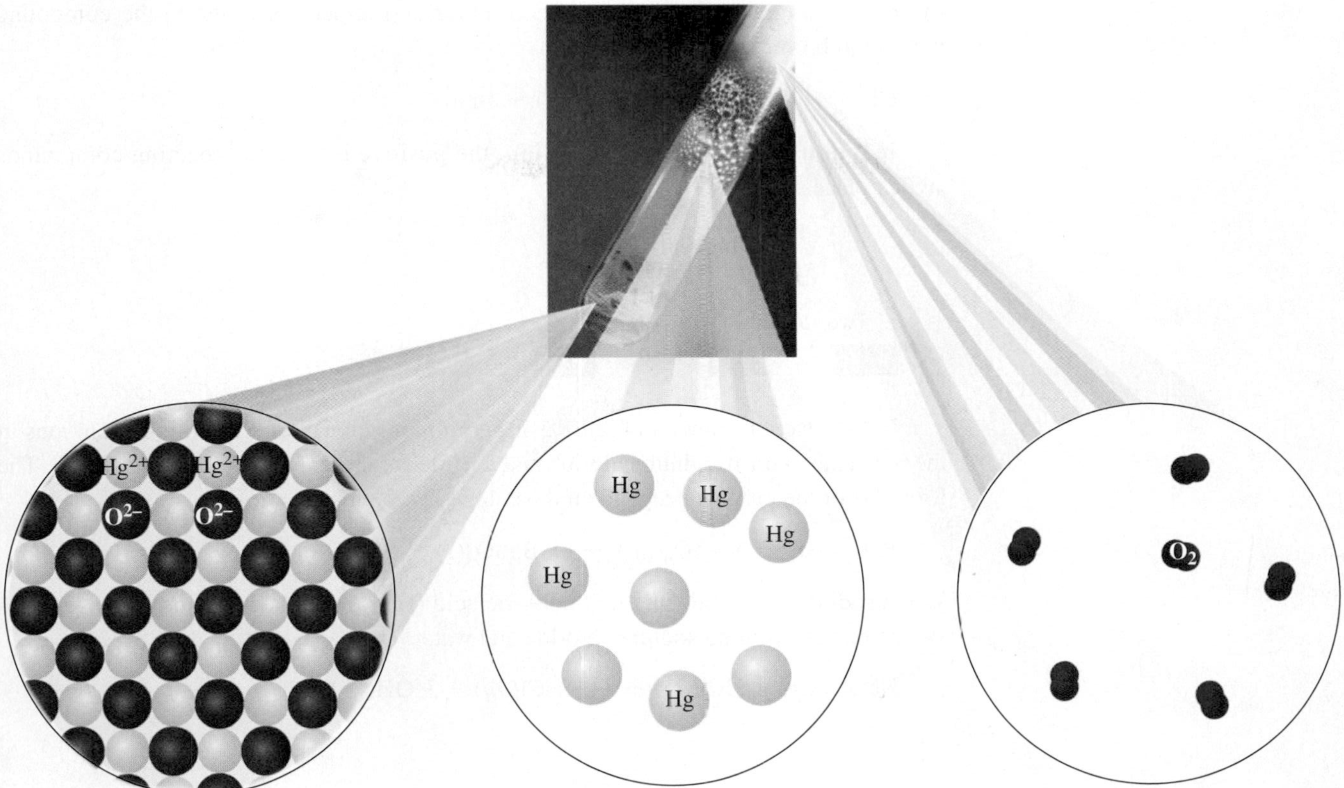

$$2HgO(s) \xrightarrow{\Delta} 2Hg(l) + O_2(g)$$

Mercury(II) oxide → Mercury + Oxygen

FIGURE 5.5 In a decomposition reaction, one reactant breaks down into two or more products.
Q How do the differences in the reactant and products classify this as a decomposition reaction?

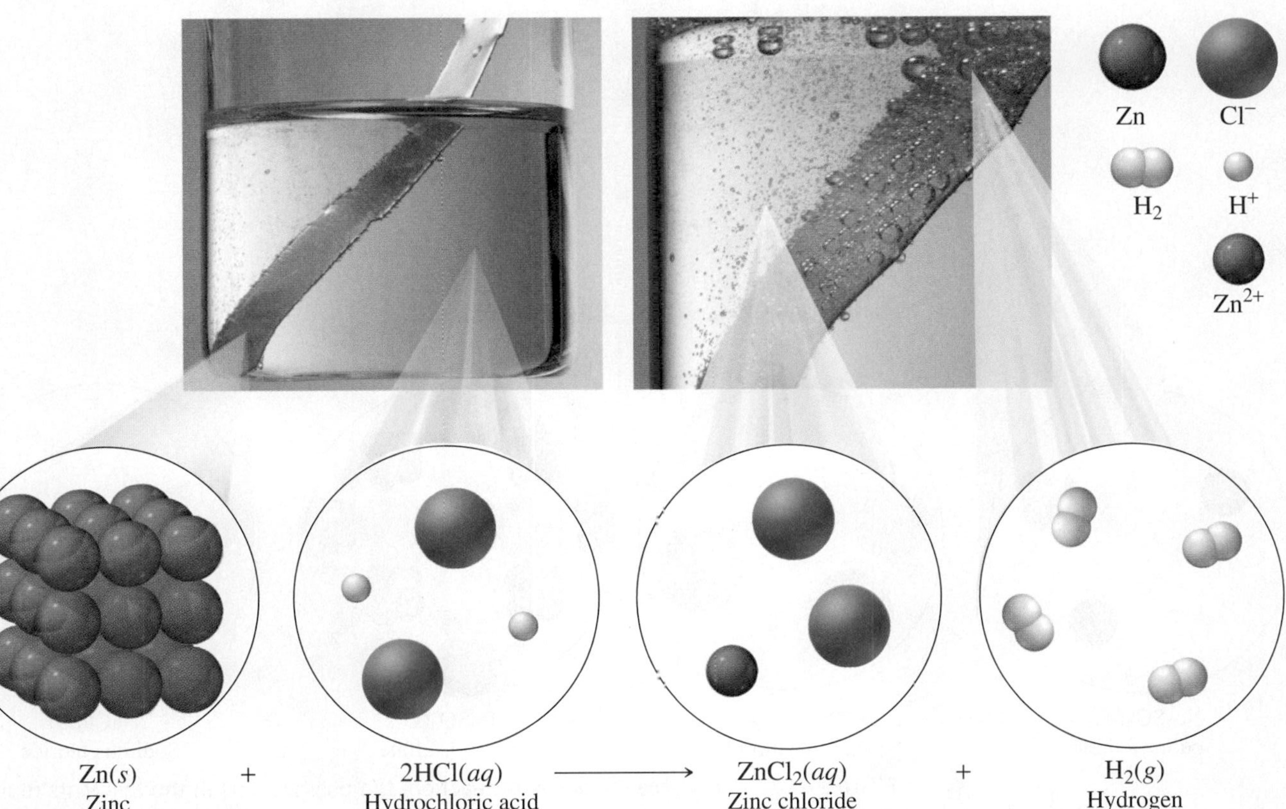

$$Zn(s) + 2HCl(aq) \longrightarrow ZnCl_2(aq) + H_2(g)$$

Zinc + Hydrochloric acid → Zinc chloride + Hydrogen

FIGURE 5.6 In a single replacement reaction, an atom or ion replaces an atom or ion in a compound.
Q What changes in the formulas of the reactants identify this equation as a single replacement?

In another single replacement reaction, chlorine replaces bromine in the compound potassium bromide.

$$Cl_2(g) + 2KBr(s) \longrightarrow 2KCl(s) + Br_2(l)$$

In a **double replacement reaction**, the positive ions in the reacting compounds switch places.

Double replacement

Two elements replace each other

A B + C D ⟶ A D + C B

In the reaction shown in Figure 5.7, barium ions change places with sodium ions in the reactants to form sodium chloride and a white solid precipitate of barium sulfate. The formulas of the products depend on the charges of the ions.

$$BaCl_2(aq) + Na_2SO_4(aq) \longrightarrow BaSO_4(s) + 2NaCl(aq)$$

When sodium hydroxide and hydrochloric acid (HCl) react, sodium and hydrogen ions switch places, forming sodium chloride and water.

$$NaOH(aq) + HCl(aq) \longrightarrow NaCl(aq) + HOH(l)$$

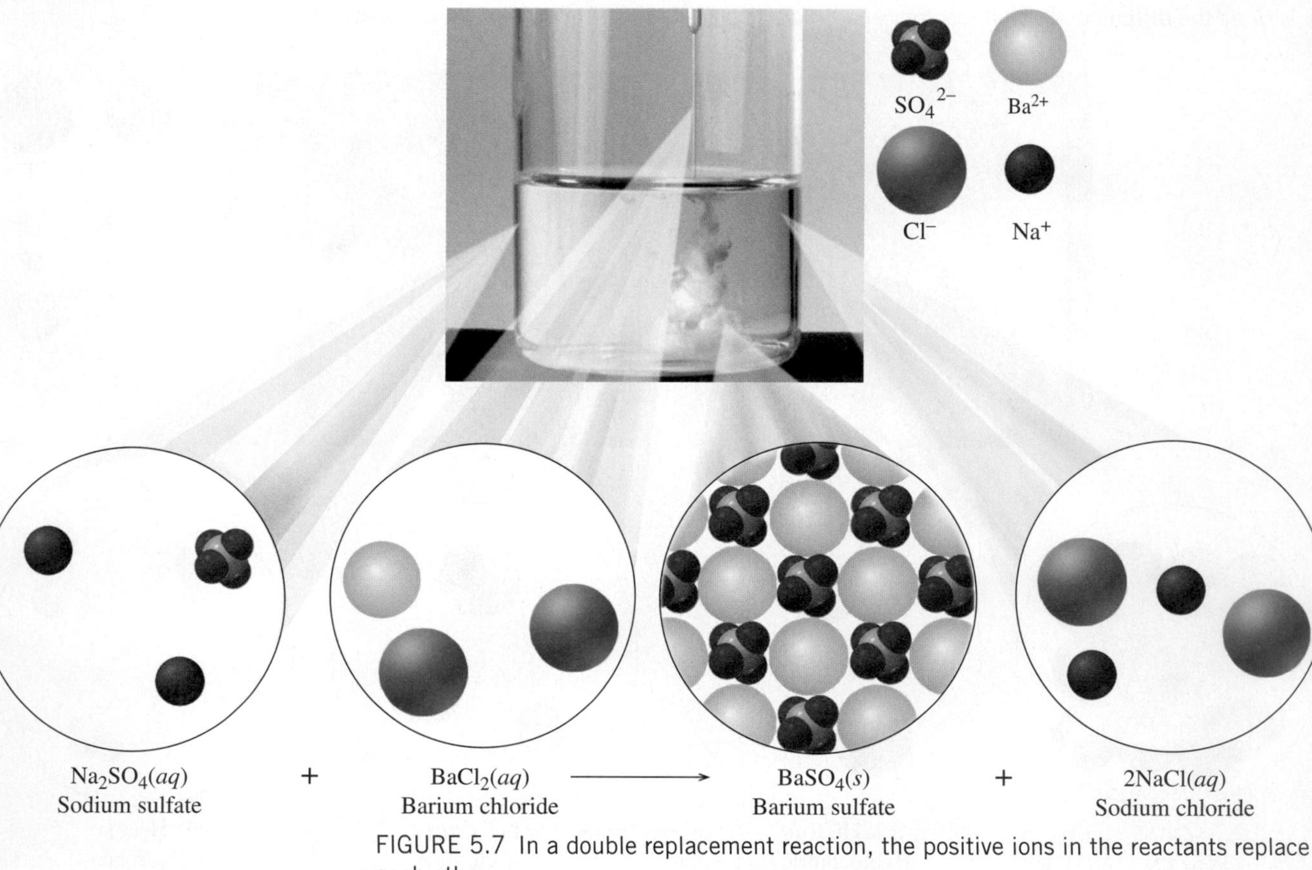

FIGURE 5.7 In a double replacement reaction, the positive ions in the reactants replace each other.

Q How do the changes in the formulas of the reactants identify this equation as a double replacement reaction?

Combustion Reactions

The burning of a candle and the burning of fuel in the engine of a car are examples of combustion reactions. In a **combustion reaction**, a carbon-containing compound, usually a fuel, burns in oxygen from the air to produce carbon dioxide (CO_2), water (H_2O), and energy in the form of heat or a flame. For example, methane gas (CH_4) undergoes combustion when used to cook our food on a gas cooktop and to heat our homes. In the equation for the combustion of methane, each element in the fuel (CH_4) forms a compound with oxygen.

$$\underset{\text{Methane}}{CH_4(g)} + 2O_2(g) \xrightarrow{\Delta} CO_2(g) + 2H_2O(g) + \text{energy}$$

The balanced equation for the combustion of propane (C_3H_8) is

$$C_3H_8(g) + 5O_2(g) \xrightarrow{\Delta} 3CO_2(g) + 4H_2O(g) + \text{energy}$$

Propane is the fuel used in portable heaters and gas barbecues. Gasoline, a mixture of liquid hydrocarbons, is the fuel that powers our cars, lawn mowers, and snow blowers.

Table 5.5 summarizes the reaction types and gives examples.

In a combustion reaction, a candle burns using the oxygen in the air.

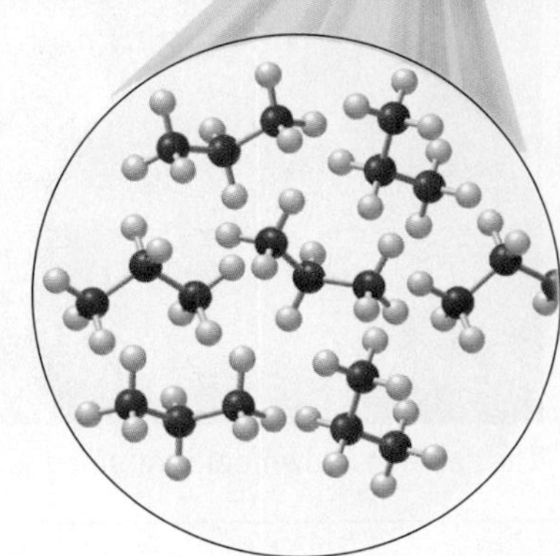

The propane from the torch undergoes combustion, which provides energy to solder metals.

CONCEPT CHECK 5.5

Identifying Reactions and Predicting Products

Classify the following reactions as combination, decomposition, single replacement, double replacement, or combustion:

a. $2Fe_2O_3(s) + 3C(s) \longrightarrow 3CO_2(g) + 4Fe(s)$

b. $2KClO_3(s) \xrightarrow{\Delta} 2KCl(s) + 3O_2(g)$

c. $C_2H_4(g) + 3O_2(g) \xrightarrow{\Delta} 2CO_2(g) + 2H_2O(g) + \text{energy}$

ANSWER

a. In this single replacement reaction, a C atom replaces Fe in Fe_2O_3 to form the compound CO_2 and Fe atoms.

b. When one reactant breaks down to produce two products, the reaction is decomposition.

c. The reaction of a carbon compound with oxygen to produce carbon dioxide, water, and energy makes this a combustion reaction.

TABLE 5.5 Summary of Reaction Types

Reaction Type	Example
Combination	
$A + B \longrightarrow AB$	$Ca(s) + Cl_2(g) \longrightarrow CaCl_2(s)$
Decomposition	
$AB \longrightarrow A + B$	$Fe_2S_3(s) \longrightarrow 2Fe(s) + 3S(s)$
Single Replacement	
$A + BC \longrightarrow AC + B$	$Cu(s) + 2AgNO_3(aq) \longrightarrow Cu(NO_3)_2(aq) + 2Ag(s)$
Double Replacement	
$AB + CD \longrightarrow AD + CB$	$BaCl_2(aq) + K_2SO_4(aq) \longrightarrow BaSO_4(s) + 2KCl(aq)$
Combustion	
$C_XH_Y + ZO_2(g) \xrightarrow{\Delta} XCO_2(g) + Y/2\ H_2O(g) + \text{energy}$	$CH_4(g) + 2O_2(g) \xrightarrow{\Delta} CO_2(g) + 2H_2O(g) + \text{energy}$

Chemistry Link to Health

SMOG AND HEALTH CONCERNS

There are two types of smog. One, photochemical smog, requires sunlight to initiate reactions that produce pollutants such as nitrogen oxides and ozone. The other type of smog, industrial or London smog, occurs in areas where coal containing sulfur is burned and the unwanted product sulfur dioxide is emitted.

Photochemical smog is most prevalent in cities where people are dependent on cars for transportation. On a typical day in Los Angeles, for example, nitrogen oxide (NO) emissions from car exhausts increase as traffic increases on the roads. When N_2 and O_2 react at high temperatures in car and truck engines, the product is nitrogen oxide.

$$N_2(g) + O_2(g) \xrightarrow{\text{Heat}} 2NO(g)$$

Then, NO reacts with oxygen in the air to produce NO_2, a reddish-brown gas that is irritating to the eyes and damaging to the respiratory tract.

$$2NO(g) + O_2(g) \longrightarrow 2NO_2(g)$$

The reddish-brown color of smog is due to nitrogen dioxide.

When NO_2 molecules are exposed to sunlight, they are converted into NO molecules and oxygen atoms.

$$NO_2(g) \xrightarrow{\text{Sunlight}} NO(g) + \underset{\text{Oxygen atoms}}{O(g)}$$

Oxygen atoms are so reactive that they combine with oxygen molecules in the atmosphere, forming ozone.

$$O(g) + O_2(g) \longrightarrow \underset{\text{Ozone}}{O_3(g)}$$

In the upper atmosphere (the stratosphere), ozone is beneficial because it protects us from harmful ultraviolet radiation that comes from the Sun. However, in the lower atmosphere, ozone irritates the eyes and respiratory tract, where it causes coughing, decreased lung function, and fatigue. It also causes deterioration of fabrics, cracks rubber, and damages trees and crops.

Industrial smog is produced when sulfur is converted to sulfur dioxide during the burning of coal or other sulfur-containing fuels.

$$S(s) + O_2(g) \longrightarrow SO_2(g)$$

The SO_2 is damaging to plants and is corrosive to metals such as steel. The SO_2 is also damaging to humans and can cause lung impairment and respiratory difficulties. In the air, SO_2 reacts with more oxygen to form SO_3, which can combine with water to form sulfuric acid. When rain falls, it absorbs the sulfuric acid, which makes acid rain.

$$2SO_2(g) + O_2(g) \longrightarrow 2SO_3(g)$$

$$SO_3(g) + H_2O(l) \longrightarrow \underset{\text{Sulfuric acid}}{H_2SO_4(aq)}$$

The presence of sulfuric acid in rivers and lakes causes an increase in the acidity of the water, reducing the ability of animals and plants to survive.

QUESTIONS AND PROBLEMS

Types of Reactions

5.33 **a.** Why is the following reaction called a decomposition reaction?

$$2Al_2O_3(s) \xrightarrow{\Delta} 4Al(s) + 3O_2(g)$$

b. Why is the following reaction called a single replacement reaction?

$$Br_2(g) + BaI_2(s) \longrightarrow BaBr_2(s) + I_2(g)$$

5.34 **a.** Why is the following reaction called a combination reaction?

$$H_2(g) + Br_2(g) \longrightarrow 2HBr(g)$$

b. Why is the following reaction called a double replacement reaction?

$$AgNO_3(aq) + NaCl(aq) \longrightarrow AgCl(s) + NaNO_3(aq)$$

5.35 Classify each of the following reactions as a combination, decomposition, single replacement, double replacement, or combustion:

a. $4Fe(s) + 3O_2(g) \longrightarrow 2Fe_2O_3(s)$

b. $Mg(s) + 2AgNO_3(aq) \longrightarrow Mg(NO_3)_2(aq) + 2Ag(s)$

c. $CuCO_3(s) \xrightarrow{\Delta} CuO(s) + CO_2(g)$

d. $NaOH(aq) + HCl(aq) \longrightarrow NaCl(aq) + H_2O(l)$

e. $ZnCO_3(s) \xrightarrow{\Delta} CO_2(g) + ZnO(s)$

f. $Al_2(SO_4)_3(aq) + 6KOH(aq) \longrightarrow 2Al(OH)_3(s) + 3K_2SO_4(aq)$

g. $Pb(s) + O_2(g) \longrightarrow PbO_2(s)$

h. $C_4H_8(g) + 6O_2(g) \xrightarrow{\Delta} 4CO_2(g) + 4H_2O(g)$

5.36 Classify each of the following reactions as a combination, decomposition, single replacement, double replacement, or combustion:

a. $CuO(s) + 2HCl(aq) \longrightarrow CuCl_2(aq) + H_2O(l)$
b. $2Al(s) + 3Br_2(g) \longrightarrow 2AlBr_3(s)$
c. $Pb(NO_3)_2(aq) + 2NaCl(aq) \longrightarrow PbCl_2(s) + 2NaNO_3(aq)$
d. $2Mg(s) + O_2(g) \xrightarrow{\Delta} 2MgO(s)$
e. $2C_2H_2(g) + 5O_2(g) \xrightarrow{\Delta} 4CO_2(g) + 2H_2O(g)$
f. $Fe_2O_3(s) + 3C(s) \longrightarrow 2Fe(s) + 3CO(g)$
g. $C_6H_{12}O_6(aq) \longrightarrow 2C_2H_6O(aq) + 2CO_2(g)$
h. $BaCl_2(aq) + K_2CO_3(aq) \longrightarrow BaCO_3(s) + 2KCl(aq)$

5.37 Predict the products that would result from each of the following reactions and balance:

a. combination: $Mg(s) + Cl_2(g) \longrightarrow$
b. decomposition: $HBr(g) \xrightarrow{\Delta}$
c. single replacement: $Mg(s) + Zn(NO_3)_2(aq) \longrightarrow$
d. double replacement: $K_2S(aq) + Pb(NO_3)_2(aq) \longrightarrow$
e. combustion: $C_2H_6(g) + O_2(g) \xrightarrow{\Delta}$

5.38 Predict the products that would result from each of the following reactions and balance:

a. combination: $Ca(s) + O_2(g) \longrightarrow$
b. combustion: $C_6H_6(g) + O_2(g) \xrightarrow{\Delta}$
c. decomposition: $PbO_2(s) \xrightarrow{\Delta}$
d. single replacement: $KI(s) + Cl_2(g) \longrightarrow$
e. double replacement: $CuCl_2(aq) + Na_2S(aq) \longrightarrow$

5.5 Oxidation–Reduction Reactions

LEARNING GOAL

Define the terms oxidation and reduction; identify the reactants oxidized and reduced.

Perhaps you have never heard of an oxidation and reduction reaction. However, this type of reaction has many important applications in your everyday life. When you see a rusty nail, tarnish on a silver spoon, or corrosion on metal, you are observing oxidation.

$$4Fe(s) + 3O_2(g) \longrightarrow \underset{\text{Rust}}{2Fe_2O_3(s)} \quad \text{Fe is oxidized}$$

Rust forms when the oxygen in the air reacts with iron.

When we turn the lights on in our automobiles, an oxidation–reduction reaction within the car battery provides the electricity. On a cold, wintry day, we might build a fire. As the wood burns, oxygen combines with carbon and hydrogen to produce carbon dioxide, water, and heat. In the previous section, we called this a combustion reaction, but it is also an oxidation–reduction reaction. When we eat foods with starches in them, the starches break down to give glucose, which is oxidized in our cells to give us energy along with carbon dioxide and water. Every breath we take provides oxygen to carry out oxidation in our cells.

$$\underset{\text{Glucose}}{C_6H_{12}O_6(aq)} + 6O_2(g) \longrightarrow 6CO_2(g) + 6H_2O(l) + \text{energy}$$

Oxidation–Reduction Reactions

TUTORIAL
Identifying Oxidation–Reduction Reactions

In an **oxidation–reduction reaction** (*redox*), electrons are transferred from one substance to another. If one substance loses electrons, another substance must gain electrons. **Oxidation** is defined as the *loss* of electrons; **reduction** is the *gain* of electrons.

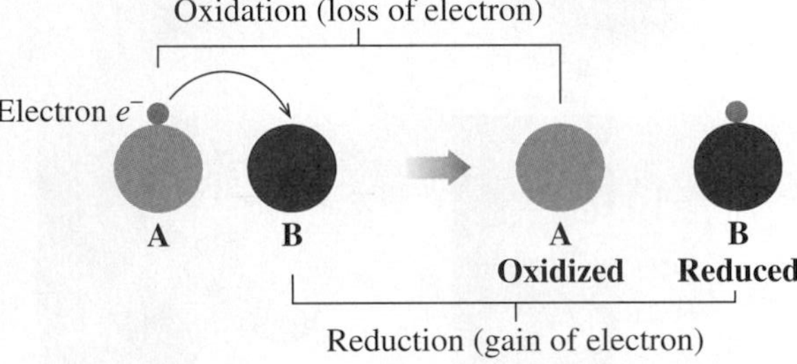

One way to remember these definitions is to use the following mnemonic:

OIL RIG

Oxidation **I**s **L**oss of electrons

Reduction **I**s **G**ain of electrons

The green patina on copper is due to oxidation.

Reduced		Oxidized
Na	Oxidation: lose e^- →	$Na^+ + e^-$
Ca		$Ca^{2+} + 2\,e^-$
$2Br^-$	← Reduction: gain e^-	$Br_2 + 2\,e^-$
Fe^{2+}		$Fe^{3+} + e^-$

Oxidation is a loss of electrons; reduction is a gain of electrons.

Oxidation–Reduction

In general, atoms of metals lose electrons to form positive ions, whereas nonmetals gain electrons to form negative ions. Now we can say that metals are oxidized and nonmetals are reduced.

The green color that appears on copper surfaces from weathering, known as patina, is a mixture of $CuCO_3$ and CuO. We can now look at the oxidation and reduction reactions that take place when copper metal reacts with oxygen in the air to produce copper(II) oxide.

$$2Cu(s) + O_2(g) \longrightarrow 2CuO(s)$$

Because the copper ion (Cu^{2+}) has a 2+ charge in CuO, each copper atom has lost two electrons. Therefore, copper metal (Cu) has been oxidized in the reaction.

$$Cu(s) \longrightarrow Cu^{2+}(s) + 2\,e^- \quad \text{Oxidation: loss of electrons}$$

Because the oxide ion (O^{2-}) has a 2− charge in CuO, each oxygen atom has gained two electrons. Therefore, oxygen has been reduced.

$$O_2(g) + 4\,e^- \longrightarrow 2O^{2-}(s) \quad \text{Reduction: gain of electrons}$$

The equation for the formation of CuO involves an oxidation and a reduction that occur simultaneously. In every oxidation and reduction, the number of electrons lost must be equal to the number of electrons gained. Therefore, we multiply the oxidation reaction of Cu by 2. Canceling the $4\,e^-$ on each side, we obtain the overall oxidation–reduction equation for the formation of CuO.

$$2Cu(s) \longrightarrow 2Cu^{2+}(s) + 4\,e^- \quad \text{Oxidation}$$
$$O_2(g) + 4\,e^- \longrightarrow 2O^{2-}(s) \quad \text{Reduction}$$
$$2Cu(s) + O_2(g) \longrightarrow 2CuO(s) \quad \text{Oxidation–reduction equation}$$

As we see in the next reaction between zinc and copper(II) sulfate, there is always an oxidation with every reduction (see Figure 5.8).

We write the equation to show the atoms and ions that react:

$$\mathbf{Zn}(s) + \mathbf{Cu^{2+}}(aq) + SO_4^{2-}(aq) \longrightarrow \mathbf{Zn^{2+}}(aq) + SO_4^{2-}(aq) + \mathbf{Cu}(s)$$

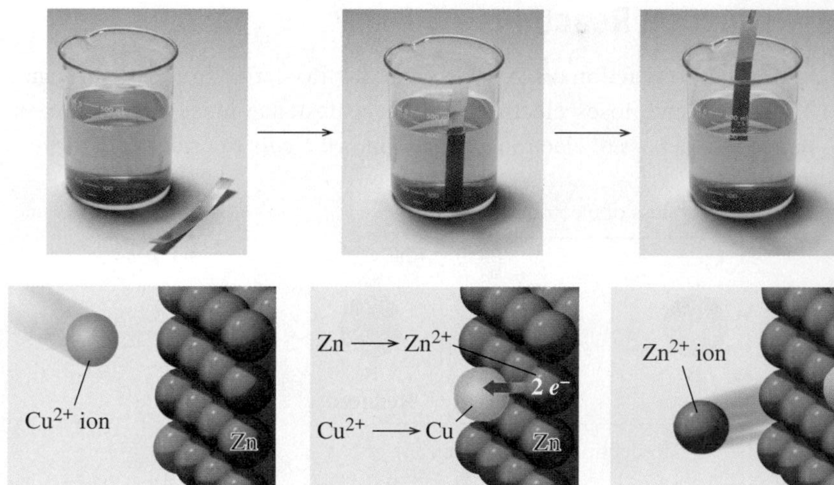

FIGURE 5.8 In this single replacement reaction, Zn(*s*) is oxidized to $Zn^{2+}(aq)$ when it provides two electrons to reduce $Cu^{2+}(aq)$ to Cu(*s*):

$$Zn(s) + Cu^{2+}(aq) \longrightarrow Cu(s) + Zn^{2+}(aq)$$

Q In the oxidation, does Zn(*s*) lose or gain electrons?

In this reaction, Zn atoms lose two electrons to form Zn^{2+}. At the same time, Cu^{2+} gains two electrons. The SO_4^{2-} ions are *spectator ions* and do not change.

$$Zn(s) \longrightarrow Zn^{2+}(aq) + 2\,e^- \quad \text{Oxidation of Zn}$$

$$Cu^{2+}(aq) + 2\,e^- \longrightarrow Cu(s) \quad \text{Reduction of } Cu^{2+}$$

In this single replacement reaction, zinc was oxidized and copper(II) ion was reduced.

A vintage photograph has a sepia tone that is caused by the reaction of light with silver in the film.

CONCEPT CHECK 5.6

Identifying Oxidation–Reduction Reactions

In photographic film, the following decomposition reaction occurs in the presence of light. What is oxidized, and what is reduced?

$$2AgBr(s) \xrightarrow{\text{Light}} 2Ag(s) + Br_2(g)$$

ANSWER

To identify the substances oxidized and reduced, we need to determine the ions and charges in the reactants and products. In AgBr, there is a silver ion (Ag^+) with a 1+ charge and a bromide ion (Br^-) with a charge of 1−. By writing AgBr as ions, we obtain the following equation:

$$2Ag^+(s) + 2Br^-(s) \longrightarrow 2Ag(s) + Br_2(g)$$

Now we can compare Ag^+ in the reactant with the Ag atom in the product. We see that each Ag^+ gained an electron; Ag^+ is reduced.

$$2Ag^+(s) + 2\,e^- \longrightarrow 2Ag(s) \quad \text{Reduction}$$

When we compare Br^- in the reactant with the Br atom in the product Br_2, we see that each Br^- lost an electron; Br^- is oxidized.

$$2Br^-(s) \longrightarrow Br_2(g) + 2\,e^- \quad \text{Oxidation}$$

Oxidation and Reduction in Biological Systems

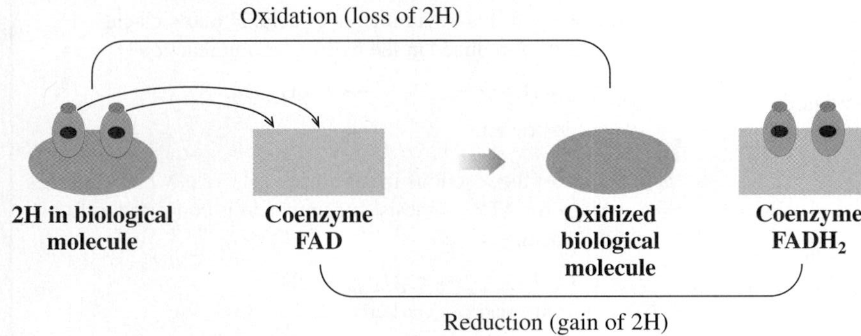

Oxidation may also involve the addition of oxygen or the loss of hydrogen, and reduction may involve the loss of oxygen or the gain of hydrogen. In the cells of the body, oxidation of organic (carbon) compounds involves the transfer of hydrogen atoms (H), which are composed of electrons and protons. For example, the oxidation of a typical biochemical molecule can involve the transfer of two hydrogen atoms (or $2H^+$ and $2\,e^-$) to a proton acceptor such as the coenzyme FAD (flavin adenine dinucleotide). The

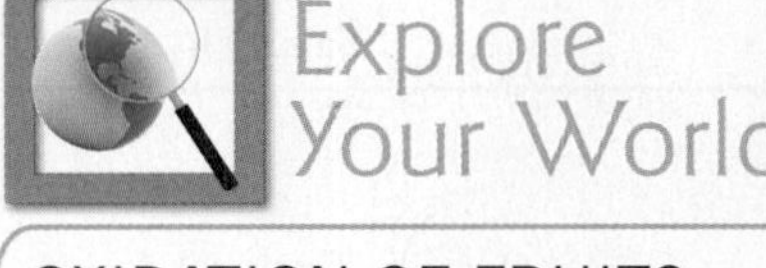

OXIDATION OF FRUITS AND VEGETABLES

Freshly cut surfaces of fruits and vegetables discolor when exposed to oxygen in the air. Cut three slices of a fruit or vegetable such as apple, potato, avocado, or banana. Leave one piece on the kitchen counter (uncovered). Wrap one piece in plastic wrap and leave it on the kitchen counter. Dip one piece in lemon juice and leave it uncovered.

QUESTIONS

1. What changes take place in each sample after 1–2 h?
2. Why would wrapping fruits and vegetables slow the rate of discoloration?
3. If lemon juice contains vitamin C (an antioxidant), why would dipping a fruit or vegetable in lemon juice affect the oxidation reaction on the surface of the fruit or vegetable?
4. Other kinds of antioxidants are vitamin E, citric acid, and BHT. Look for these antioxidants on the labels of cereals, potato chips, and other foods in your kitchen. Why are antioxidants added to food products that will be stored on our kitchen shelves?

coenzyme is reduced to $FADH_2$. In many biochemical oxidation–reduction reactions, the transfer of hydrogen atoms is necessary for the production of energy in the cells. For example, methyl alcohol (CH_3OH), a poisonous substance, is metabolized in the body by the following reactions:

$$\underset{\text{Methyl alcohol}}{CH_3OH} \longrightarrow \underset{\text{Formaldehyde}}{H_2CO} + 2H \quad \text{Oxidation: loss of H atoms}$$

The formaldehyde can be oxidized further, this time by the addition of oxygen, to produce formic acid.

$$\underset{\text{Formaldehyde}}{2H_2CO} + O_2 \longrightarrow \underset{\text{Formic acid}}{2H_2CO_2} \quad \text{Oxidation: addition of O atoms}$$

Finally, formic acid is oxidized to carbon dioxide and water.

$$\underset{\text{Formic acid}}{2H_2CO_2} + O_2 \longrightarrow 2CO_2 + 2H_2O \quad \text{Oxidation: addition of O atoms}$$

The intermediate products of the oxidation of methyl alcohol are quite toxic, causing blindness and possibly death as they interfere with key reactions in the cells of the body.

In summary, we find that the particular definition of oxidation and reduction we use depends on the process that occurs in the reaction. All these definitions are summarized in Table 5.6. Oxidation always involves a loss of electrons, but it may also be seen as an addition of oxygen or the loss of hydrogen atoms. A reduction always involves a gain of electrons and may also be seen as the loss of oxygen or the gain of hydrogen.

TABLE 5.6 Characteristics of Oxidation and Reduction

Oxidation	
Always Involves	**May Involve**
Loss of electrons	Addition of oxygen
	Loss of hydrogen
Reduction	
Always Involves	**May Involve**
Gain of electrons	Loss of oxygen
	Gain of hydrogen

QUESTIONS AND PROBLEMS

Oxidation–Reduction Reactions

5.39 Indicate each of the following as an oxidation or a reduction:
a. $Na^+(aq) + e^- \longrightarrow Na(s)$
b. $Ni(s) \longrightarrow Ni^{2+}(aq) + 2\,e^-$
c. $Cr^{3+}(aq) + 3\,e^- \longrightarrow Cr(s)$
d. $2H^+(aq) + 2\,e^- \longrightarrow H_2(g)$

5.40 Indicate each of the following as an oxidation or a reduction:
a. $O_2(g) + 4\,e^- \longrightarrow 2O^{2-}(aq)$
b. $Al(s) \longrightarrow Al^{3+}(aq) + 3\,e^-$
c. $Fe^{3+}(aq) + e^- \longrightarrow Fe^{2+}(aq)$
d. $2Br^-(aq) \longrightarrow Br_2(l) + 2\,e^-$

5.41 In each of the following reactions, identify the reactant that is oxidized and the reactant that is reduced:
a. $Zn(s) + Cl_2(g) \longrightarrow ZnCl_2(s)$
b. $Cl_2(g) + 2NaBr(aq) \longrightarrow 2NaCl(aq) + Br_2(l)$
c. $2PbO(s) \longrightarrow 2Pb(s) + O_2(g)$
d. $2Fe^{3+}(aq) + Sn^{2+}(aq) \longrightarrow 2Fe^{2+}(aq) + Sn^{4+}(aq)$

5.42 In each of the following reactions, identify the reactant that is oxidized and the reactant that is reduced:
a. $2Li(s) + F_2(g) \longrightarrow 2LiF(s)$
b. $Cl_2(g) + 2KI(aq) \longrightarrow 2KCl(aq) + I_2(s)$
c. $2Al(s) + 3Sn^{2+}(aq) \longrightarrow 2Al^{3+}(aq) + 3Sn(s)$
d. $Fe(s) + CuSO_4(aq) \longrightarrow FeSO_4(aq) + Cu(s)$

5.43 In the mitochondria of human cells, energy for the production of ATP is provided by the oxidation and reduction reactions of the iron ions in the cytochromes in electron transport. Identify each of the following reactions as an oxidation or reduction:
a. $Fe^{3+} + e^- \longrightarrow Fe^{2+}$ **b.** $Fe^{2+} \longrightarrow Fe^{3+} + e^-$

5.44 Chlorine (Cl_2) is a strong germicide used to disinfect drinking water and to kill microbes in swimming pools. If the product is Cl^-, was the elemental chlorine oxidized or reduced?

5.45 When linoleic acid, an unsaturated fatty acid, reacts with hydrogen, it forms a saturated fatty acid. Is linoleic acid oxidized or reduced in the hydrogenation reaction?

$$\underset{\text{Linoleic acid}}{C_{18}H_{32}O_2} + 2H_2 \longrightarrow C_{18}H_{36}O_2$$

5.46 In one of the reactions in the citric acid cycle, which provides energy for ATP synthesis, succinic acid is converted to fumaric acid.

$$\underset{\text{Succinic acid}}{C_4H_6O_4} \longrightarrow \underset{\text{Fumaric acid}}{C_4H_4O_4} + 2H$$

The reaction is accompanied by a coenzyme, flavin adenine dinucleotide (FAD).

$$FAD + 2H \longrightarrow FADH_2$$

a. Is succinic acid oxidized or reduced?
b. Is FAD oxidized or reduced?
c. Why would the two reactions occur together?

Chemistry Link to the Environment

FUEL CELLS: CLEAN ENERGY FOR THE FUTURE

Fuel cells are of interest to scientists because they provide an alternative source of electrical energy that is more efficient, does not use up oil reserves, and generates products that do not pollute the atmosphere. Fuel cells are considered to be a clean way to produce energy.

In a fuel cell, the reactants continuously enter the cell, which generates an electrical current. One type of hydrogen–oxygen fuel cell has been used in automobile prototypes. In this cell, hydrogen gas enters the fuel cell and comes in contact with a platinum catalyst embedded in a plastic membrane. The catalyst assists in the oxidation of hydrogen atoms to hydrogen ions and electrons.

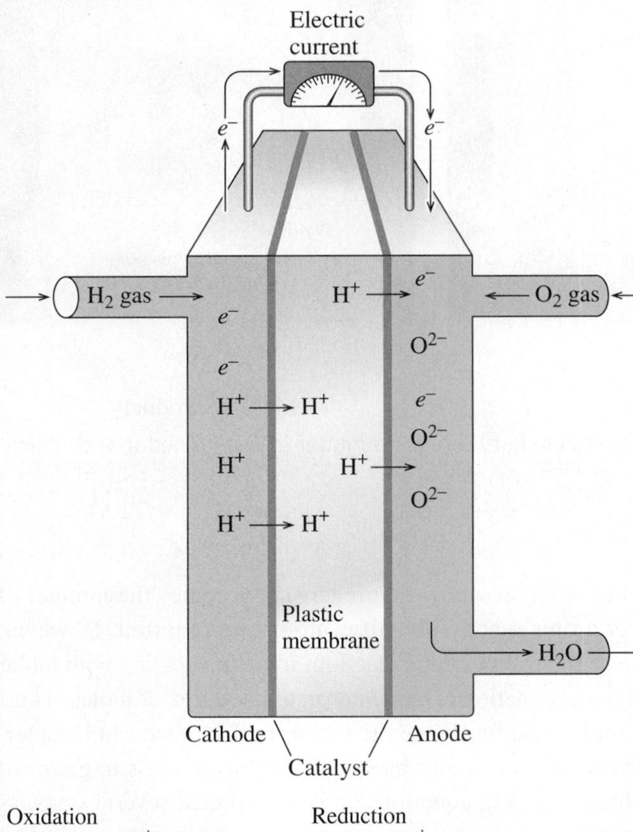

A fuel cell uses a continuous supply of hydrogen and oxygen to generate electricity.

The electrons produce an electric current as they travel through the wire. The protons move through the plastic membrane to react with oxygen molecules. The oxygen molecules are reduced to oxide ions that combine with the hydrogen ions to form water. The overall hydrogen–oxygen fuel cell reaction can be written as

$$2H_2(g) + O_2(g) \longrightarrow 2H_2O(l)$$

Fuel cells have already been used for power on the space shuttle. A major drawback to the practical use of fuel cells is the economic impact of converting cars to fuel cell operation. The storage and cost of producing hydrogen are also problems. Some manufacturers are experimenting with systems that convert gasoline or methanol to hydrogen for use in fuel cells.

In homes, fuel cells may one day replace the batteries currently used to provide electrical power for cell phones, CD and DVD players, and laptop computers. Fuel cell design is still in the prototype phase, although there is much interest in development of these cells. We already know they can work, but modifications must still be made before they become reasonably priced and part of our everyday lives.

Fuel cells are used to supply power on the space shuttle orbiter.

5.6 Mole Relationships in Chemical Equations

LEARNING GOAL

Given a quantity in moles of reactant or product, use a mole–mole factor from the balanced equation to calculate the moles of another substance in the reaction.

In Section 5.2, we saw that equations are balanced in terms of the numbers of each type of atom in the reactants and products. However, when we do experiments in the laboratory or prepare medications in the pharmacy, the samples we use contain billions of atoms or molecules, which makes it impossible to count them. What we can measure is their mass, using a balance. Because mass is related to the number of particles through the molar mass, measuring the mass is equivalent to counting the number of particles or moles.

SELF STUDY ACTIVITY
Stoichiometry

TUTORIAL
Law of Conservation of Mass

Conservation of Mass

In any chemical reaction, the total amount of matter in the reactants is equal to the total amount of matter in the products. If all the reactants were weighed, they would have a total mass equal to the total mass of the products. This is known as the *law of conservation of mass*, which says that there is no change in the total mass of the substances reacting in a chemical reaction. Thus, no material is lost or gained when original substances are changed to new substances.

For example, tarnish forms when silver reacts with sulfur to form silver sulfide.

$$2Ag(s) + S(s) \longrightarrow Ag_2S(s)$$

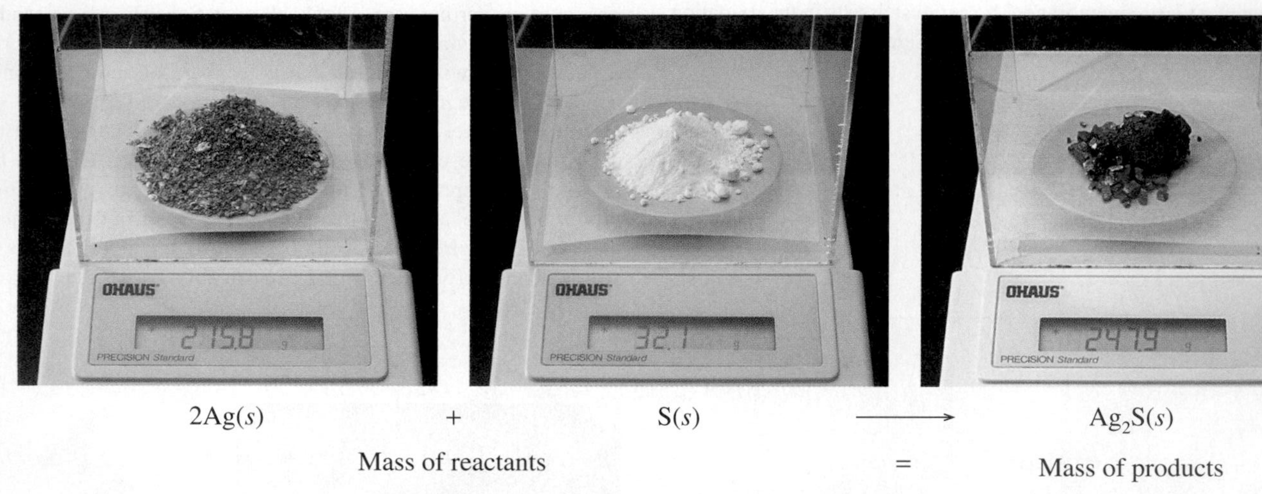

$2Ag(s)$ + $S(s)$ ⟶ $Ag_2S(s)$

Mass of reactants = Mass of products

The law of conservation of mass states that there is no matter lost or gained in a chemical reaction.

In this reaction, the number of silver atoms that reacts is two times the number of sulfur atoms. When 200 silver atoms react, 100 sulfur atoms are required. However, many more atoms would actually be present in this reaction. If we are dealing with molar amounts, then the coefficients in the equation can be interpreted in terms of moles. Thus, 2 moles of silver react with 1 mole of sulfur. Since the molar mass of each can be determined, the quantities of silver and sulfur can also be stated in terms of mass in grams of each. Therefore, an equation for a chemical equation can be interpreted several ways, as seen in Table 5.7.

TABLE 5.7 Information Available from a Balanced Equation

	Reactants			Products
Equation	**$2Ag(s)$**	**$+ S(s)$**	⟶	**$Ag_2S(s)$**
Atoms	2 Ag atoms	+ 1 S atom	⟶	1 Ag_2S formula unit
	200 Ag atoms	+ 100 S atoms	⟶	100 Ag_2S formula units
Avogadro's number of atoms	$2(6.02 \times 10^{23})$ Ag atoms	$+ 1(6.02 \times 10^{23})$ S atoms	⟶	$1(6.02 \times 10^{23})$ Ag_2S formula units
Moles	2 moles of Ag	+ 1 mole of S	⟶	1 mole of Ag_2S
Mass (g)	2(107.9 g) of Ag	+ 1(32.1 g) of S	⟶	1(247.9 g) of Ag_2S
Total mass (g)	247.9 g		⟶	247.9 g

Iron (Fe) + Sulfur (S) ⟶ Iron(III) sulfide (Fe_2S_3)

$2Fe(s) + 3S(s) \longrightarrow Fe_2S_3(s)$

In the chemical reaction of Fe and S, the mass of the reactants is the same as the mass of the product, Fe_2S_3.

Mole–Mole Factors from an Equation

When iron reacts with sulfur, the product is iron(III) sulfide.

$$2Fe(s) + 3S(s) \longrightarrow Fe_2S_3(s)$$

Because the equation is balanced, we know the proportions of iron and sulfur in the reaction. For this reaction, we see that 2 moles of iron reacts with 3 moles of sulfur to form 1 mole of iron(III) sulfide. Actually, any amount of iron or sulfur may be used, but the *ratio* of iron reacting with sulfur will be the same. From the coefficients, we can write **mole–mole factors** between reactants and between reactants and products. The coefficients used in the mole–mole factors are exact numbers; they do not limit the number of significant figures.

Fe and S: $\dfrac{2 \text{ moles Fe}}{3 \text{ moles S}}$ and $\dfrac{3 \text{ moles S}}{2 \text{ moles Fe}}$

Fe and Fe_2S_3: $\dfrac{2 \text{ moles Fe}}{1 \text{ mole } Fe_2S_3}$ and $\dfrac{1 \text{ mole } Fe_2S_3}{2 \text{ moles Fe}}$

S and Fe_2S_3: $\dfrac{3 \text{ moles S}}{1 \text{ mole } Fe_2S_3}$ and $\dfrac{1 \text{ mole } Fe_2S_3}{3 \text{ moles S}}$

CONCEPT CHECK 5.7

Writing Mole–Mole Factors

Consider the following balanced equation:

$$4Na(s) + O_2(g) \longrightarrow 2Na_2O(s)$$

Write the mole–mole factors for

a. Na and O_2 **b.** Na and Na_2O

ANSWER

a. The mole–mole factors for Na and O_2 use the coefficient of Na to write 4 moles of Na, and the coefficient of 1 (understood) to write 1 mole of O_2.

4 moles of Na = 1 mole of O_2

$\dfrac{4 \text{ moles Na}}{1 \text{ mole } O_2}$ and $\dfrac{1 \text{ mole } O_2}{4 \text{ moles Na}}$

b. The mole–mole factors for Na and Na_2O use the coefficient of Na to write 4 moles of Na and the coefficient of Na_2O to write 2 moles of Na_2O.

4 moles of Na = 2 moles of Na_2O

$\dfrac{4 \text{ moles Na}}{2 \text{ moles } Na_2O}$ and $\dfrac{2 \text{ moles } Na_2O}{4 \text{ moles Na}}$

TUTORIAL
Moles of Reactants and Products

TUTORIAL
Conversions Involving Moles

Using Mole–Mole Factors in Calculations

Whenever you prepare a recipe, adjust an engine for the proper mixture of fuel and air, or prepare medicines in a pharmaceutical laboratory, you need to know the proper amounts of reactants to use and how much of the product will form. Earlier we wrote all the possible conversion factors that can be obtained from the balanced equation $2Fe(s) + 3S(s) \longrightarrow Fe_2S_3(s)$. Now we can use mole–mole factors in chemical calculations.

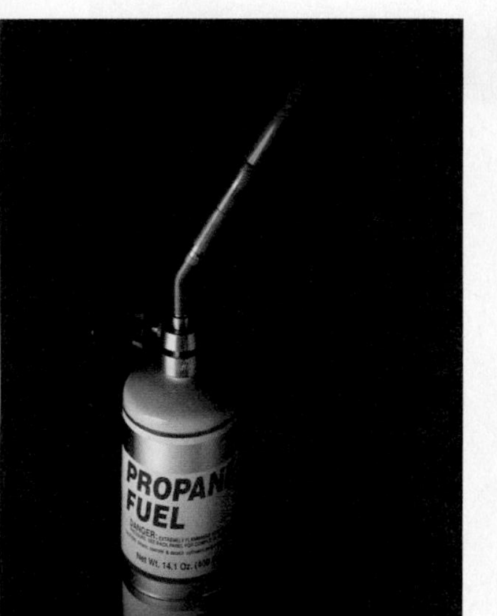

Propane fuel reacts with O_2 in the air to produce CO_2, H_2O, and energy.

Guide to Using Mole–Mole Factors

1. State the given and needed quantities.
2. Write a plan to convert the given to the needed moles.
3. Use coefficients to write relationships and mole–mole factors.
4. Set up the problem using the mole–mole factor that cancels given moles.

SAMPLE PROBLEM 5.7

Calculating Moles of a Product

Propane gas (C_3H_8), a fuel for camp stoves and specially equipped automobiles, reacts with oxygen to produce carbon dioxide, water, and energy. How many moles of CO_2 can be produced when 2.25 moles of C_3H_8 reacts?

$$\underset{\text{Propane}}{C_3H_8(g)} + 5O_2(g) \xrightarrow{\Delta} 3CO_2(g) + 4H_2O(g) + \text{energy}$$

SOLUTION

Step 1 State the given and needed quantities.

Given moles of C_3H_8 **Need** moles of CO_2

Step 2 Write a plan to convert the given to the needed moles.

moles of C_3H_8 → Mole–mole factor → moles of CO_2

Step 3 Use coefficients to write relationships and mole–mole factors.

$$1 \text{ mole of } C_3H_8 = 3 \text{ moles of } CO_2$$

$$\frac{1 \text{ mole } C_3H_8}{3 \text{ moles } CO_2} \quad \text{and} \quad \frac{3 \text{ moles } CO_2}{1 \text{ mole } C_3H_8}$$

Step 4 Set up the problem using the mole–mole factor that cancels given moles.

$$2.25 \cancel{\text{moles } C_3H_8} \times \frac{3 \text{ moles } CO_2}{1 \cancel{\text{mole } C_3H_8}} = 6.75 \text{ moles of } CO_2$$

The answer is given with three SFs because the given quantity, 2.25 moles of C_3H_8, has three SFs. The values in the mole–mole factor are exact.

STUDY CHECK 5.7

Using the equation in Sample Problem 5.7, calculate the number of moles of oxygen that must react to produce 0.756 mole of water.

QUESTIONS AND PROBLEMS

Mole Relationships in Chemical Equations

5.47 Write all the mole–mole factors for each of the following equations:

a. $2SO_2(g) + O_2(g) \longrightarrow 2SO_3(g)$

b. $4P(s) + 5O_2(g) \longrightarrow 2P_2O_5(s)$

5.48 Write all the mole–mole factors for each of the following equations:

a. $2Al(s) + 3Cl_2(g) \longrightarrow 2AlCl_3(s)$

b. $4HCl(g) + O_2(g) \longrightarrow 2Cl_2(g) + 2H_2O(g)$

5.49 The reaction of hydrogen with oxygen produces water.

$$2H_2(g) + O_2(g) \longrightarrow 2H_2O(g)$$

a. How many moles of O_2 are required to react with 2.0 moles of H_2?

b. How many moles of H_2 are needed to react with 5.0 moles of O_2?

c. How many moles of H_2O form when 2.5 moles of O_2 reacts?

5.50 Ammonia is produced by the reaction of hydrogen and nitrogen.

$$N_2(g) + 3H_2(g) \longrightarrow \underset{\text{Ammonia}}{2NH_3(g)}$$

a. How many moles of H_2 are needed to react with 1.0 mole of N_2?
b. How many moles of N_2 reacted if 0.60 mole of NH_3 is produced?
c. How many moles of NH_3 are produced when 1.4 moles of H_2 reacts?

5.51 Carbon disulfide and carbon monoxide are produced when carbon is heated with sulfur dioxide.

$$5C(s) + 2SO_2(g) \xrightarrow{\Delta} CS_2(l) + 4CO(g)$$

a. How many moles of C are needed to react with 0.500 mole of SO_2?
b. How many moles of CO are produced when 1.2 moles of C reacts?
c. How many moles of SO_2 are needed to produce 0.50 mole of CS_2?
d. How many moles of CS_2 are produced when 2.5 moles of C reacts?

5.52 In the acetylene torch, acetylene gas (C_2H_2) burns in oxygen to produce carbon dioxide and water.

$$2C_2H_2(g) + 5O_2(g) \xrightarrow{\Delta} 4CO_2(g) + 2H_2O(g)$$

a. How many moles of O_2 are needed to react with 2.00 moles of C_2H_2?
b. How many moles of CO_2 are produced when 3.5 moles of C_2H_2 reacts?
c. How many moles of C_2H_2 are required to produce 0.50 mole of H_2O?
d. How many moles of CO_2 are produced from 0.100 mole of O_2?

5.7 Mass Calculations for Reactions

LEARNING GOAL

Given the mass in grams of a substance in a reaction, calculate the mass in grams of another substance in the reaction.

When you perform a chemistry experiment in the laboratory, you measure a specific mass of the reactant. From the mass in grams, you can determine the number of moles of reactant. By using mole–mole factors, you can predict the moles of product that can be produced. Then the molar mass of the product is used to convert the moles back into mass in grams as seen in Sample Problem 5.8.

TUTORIAL
Masses of Reactants and Products

SELF STUDY ACTIVITY
Stoichiometry

SAMPLE PROBLEM 5.8

Mass of Product from Mass of Reactant

When acetylene (C_2H_2) burns in oxygen, high temperatures are produced that are used for welding metals.

$$2C_2H_2(g) + 5O_2(g) \xrightarrow{\Delta} 4CO_2(g) + 2H_2O(g) + \text{energy}$$

How many grams of CO_2 are produced when 54.6 g of C_2H_2 is burned?

SOLUTION

Step 1 Use molar mass to convert grams of given to moles.

$$1 \text{ mole of } C_2H_2 = 26.0 \text{ g of } C_2H_2$$

$$\frac{26.0 \text{ g } C_2H_2}{1 \text{ mole } C_2H_2} \quad \text{and} \quad \frac{1 \text{ mole } C_2H_2}{26.0 \text{ g } C_2H_2}$$

$$54.6 \text{ g } \cancel{C_2H_2} \times \frac{1 \text{ mole } C_2H_2}{26.0 \text{ g } \cancel{C_2H_2}} = 2.10 \text{ moles of } C_2H_2$$

Step 2 Write a mole–mole factor from the coefficients in the equation.

$$2 \text{ moles of } C_2H_2 = 4 \text{ moles of } CO_2$$

$$\frac{2 \text{ moles } C_2H_2}{4 \text{ moles } CO_2} \quad \text{and} \quad \frac{4 \text{ moles } CO_2}{2 \text{ moles } C_2H_2}$$

Guide to Calculating the Masses of Reactants and Products in a Chemical Reaction

1. Use molar mass to convert grams of given to moles (if necessary).
2. Write a mole–mole factor from the coefficients in the equation.
3. Convert moles of given to moles of needed substance using the mole–mole factor.
4. Convert moles of needed substance to grams using molar mass.

A mixture of acetylene and oxygen undergoes combustion during the welding of metals.

Step 3 Convert moles of given to moles of needed substance using the mole–mole factor.

$$2.10\ \cancel{\text{moles } C_2H_2} \times \frac{4\ \text{moles } CO_2}{2\ \cancel{\text{moles } C_2H_2}} = 4.20\ \text{moles of } CO_2$$

Step 4 Convert moles of needed substance to grams using molar mass.

$$1\ \text{mole of } CO_2 = 44.0\ \text{g of } CO_2$$

$$\frac{44.0\ \text{g } CO_2}{1\ \text{mole } CO_2} \quad \text{and} \quad \frac{1\ \text{mole } CO_2}{44.0\ \text{g } CO_2}$$

$$4.20\ \cancel{\text{moles } CO_2} \times \frac{44.0\ \text{g } CO_2}{1\ \cancel{\text{mole } CO_2}} = 185\ \text{g of } CO_2$$

The solution can also be obtained by placing the conversion factors in sequence using the following plan:

grams of C_2H_2 → (Molar mass) → moles of C_2H_2 → (Mole–mole factor) → moles of CO_2 → (Molar mass) → grams of CO_2

$$54.6\ \cancel{\text{g } C_2H_2} \times \frac{1\ \cancel{\text{mole } C_2H_2}}{26.0\ \cancel{\text{g } C_2H_2}} \times \frac{4\ \cancel{\text{moles } CO_2}}{2\ \cancel{\text{moles } C_2H_2}} \times \frac{44.0\ \text{g } CO_2}{1\ \cancel{\text{mole } CO_2}} = 185\ \text{g of } CO_2$$

STUDY CHECK 5.8

Using the equation in Sample Problem 5.8, calculate the grams of CO_2 that can be produced when 25.0 g of O_2 reacts.

QUESTIONS AND PROBLEMS

Mass Calculations for Reactions

5.53 Sodium reacts with oxygen to produce sodium oxide.

$$4Na(s) + O_2(g) \longrightarrow 2Na_2O(s)$$

a. How many grams of Na_2O are produced when 57.5 g of Na reacts?

b. If you have 18.0 g of Na, how many grams of O_2 are needed for the reaction?

c. How many grams of O_2 are needed in a reaction that produces 75.0 g of Na_2O?

5.54 Nitrogen gas reacts with hydrogen gas to produce ammonia by the following equation:

$$N_2(g) + 3H_2(g) \longrightarrow 2NH_3(g)$$

a. If you have 3.64 g of H_2, how many grams of NH_3 can be produced?

b. How many grams of H_2 are needed to react with 2.80 g of N_2?

c. How many grams of NH_3 can be produced from 12.0 g of H_2?

5.55 Ammonia and oxygen react to form nitrogen and water.

$$4NH_3(g) + 3O_2(g) \longrightarrow 2N_2(g) + 6H_2O(g)$$

a. How many grams of O_2 are needed to react with 13.6 g of NH_3?

b. How many grams of N_2 can be produced when 6.50 g of O_2 reacts?

c. How many grams of water are formed from the reaction of 34.0 g of NH_3?

5.56 Iron(III) oxide reacts with carbon to give iron and carbon monoxide.

$$Fe_2O_3(s) + 3C(s) \longrightarrow 2Fe(s) + 3CO(g)$$

a. How many grams of C are required to react with 16.5 g of Fe_2O_3?

b. How many grams of CO are produced when 36.0 g of C reacts?

c. How many grams of Fe can be produced when 6.00 g of Fe_2O_3 reacts?

5.57 Nitrogen dioxide and water react to produce nitric acid, HNO_3, and nitrogen oxide.

$$3NO_2(g) + H_2O(l) \longrightarrow 2HNO_3(aq) + NO(g)$$

a. How many grams of H_2O are needed to react with 28.0 g of NO_2?

b. How many grams of NO are obtained from 15.8 g of NO_2?

c. How many grams of HNO_3 are produced from 8.25 g of NO_2?

5.58 Calcium cyanamide reacts with water to form calcium carbonate and ammonia.

$$CaCN_2(s) + 3H_2O(l) \longrightarrow CaCO_3(s) + 2NH_3(g)$$

a. How many grams of water are needed to react with 75.0 g of $CaCN_2$?

b. How many grams of NH_3 are produced from 5.24 g of $CaCN_2$?

c. How many grams of $CaCO_3$ form if 155 g of water reacts?

5.59 When lead(II) sulfide ore burns in oxygen, the products are solid lead(II) oxide and sulfur dioxide gas.
a. Write the balanced equation for the reaction.
b. How many grams of oxygen are required to react with 0.125 mole of lead(II) sulfide?
c. How many grams of sulfur dioxide can be produced when 65.0 g of lead(II) sulfide reacts?
d. How many grams of lead(II) sulfide are used to produce 128 g of lead(II) oxide?

5.60 When the gases dihydrogen sulfide and oxygen react, they form the gases sulfur dioxide and water vapor.
a. Write the balanced equation for the reaction.
b. How many grams of oxygen are needed to react with 2.50 g of dihydrogen sulfide?
c. How many grams of sulfur dioxide can be produced when 38.5 g of oxygen reacts?
d. How many grams of oxygen are needed to produce 55.8 g of water vapor?

5.8 Energy in Chemical Reactions

LEARNING GOAL

Describe exothermic and endothermic reactions and factors that affect the rate of a reaction.

TUTORIAL
Heat of Reaction

For a chemical reaction to take place, the molecules of the reactants must collide with each other and have the proper orientation and energy. Even when a collision has the proper orientation, there still must be sufficient energy to break the bonds of the reactants. The **activation energy** is the amount of energy required to break the bonds between atoms of the reactants. If the energy of a collision is less than the activation energy, the molecules bounce apart without reacting. Many collisions occur, but only a few actually lead to the formation of product.

The concept of activation energy is analogous to climbing over a hill. To reach a destination on the other side, we must expend energy to climb to the top of the hill. Once we are at the top, we can easily run down the other side. The energy needed to get us from our starting point to the top of the hill would be the activation energy.

Three Conditions Required for a Reaction to Occur

1. **Collision** The reactants must collide.
2. **Orientation** The reactants must align properly to break and form bonds.
3. **Energy** The collision must provide the energy of activation.

The activation energy is the energy needed to convert reacting molecules into products.

Exothermic Reactions

In every chemical reaction, heat is absorbed or released as reactants are converted to products. The *heat of reaction* is the difference between the energy of the reactants and the energy of the products. In an **exothermic reaction**, the energy of the reactants is greater than that of the products. Thus, heat is released in exothermic reactions. For example, in the thermite reaction, the reaction of aluminum and iron(III) oxide produces

The high temperature of the thermite reaction has been used to cut or weld railroad tracks.

so much heat that temperatures of 2500 °C can be reached. The thermite reaction has been used to cut or weld railroad tracks.

Exothermic, Heat Released (Given Off)

$$2Al(s) + Fe_2O_3(s) \longrightarrow 2Fe(l) + Al_2O_3(s) + 850\ kJ \quad \text{Heat is a product}$$

Endothermic Reactions

In **endothermic reactions**, the energy of the reactants is lower than that of the products. Thus, heat is absorbed in endothermic reactions. For example, when hydrogen and iodine react to form hydrogen iodide, heat must be absorbed. For an endothermic reaction, the heat of reaction is written on the same side as the reactants.

Endothermic, Heat Absorbed (Taken In)

$$H_2(g) + I_2(g) + 55\ kJ \longrightarrow 2HI(g) \quad \text{Heat is a reactant}$$

Reaction	Energy Change	Heat in the Equation
Exothermic	Heat released	Product side
Endothermic	Heat absorbed	Reactant side

CONCEPT CHECK 5.8

Exothermic and Endothermic Reactions

In the reaction of one mole of solid carbon with oxygen gas, the energy of the carbon dioxide product is 393 kJ lower than the energy of the reactants.

a. Is the reaction exothermic or endothermic?

b. Write the balanced equation for the reaction, including the heat of reaction.

ANSWER

a. When the products have a lower energy than the reactants, the reaction is exothermic.

b. $C(s) + O_2(g) \longrightarrow CO_2(g) + 393\ kJ$

Chemistry Link to Health

HOT PACKS AND COLD PACKS

In a hospital, at a first-aid station, or at an athletic event, a *cold pack* may be used to reduce swelling from an injury, remove heat from inflammation, or decrease capillary size to lessen the effect of hemorrhaging. Inside the plastic container of a cold pack, there is a compartment containing solid ammonium nitrate (NH_4NO_3) that is separated from a compartment containing water. The pack is activated when it is hit or squeezed hard enough to break the walls between the compartments and cause the ammonium nitrate to mix with the water (shown as H_2O over the reaction arrow). In an endothermic process, one mole of NH_4NO_3 that dissolves in water absorbs 26 kJ. The temperature drops to about 4–5 °C to give a cold pack ready to use.

Endothermic Reaction in a Cold Pack

$$26\ kJ + NH_4NO_3(s) \xrightarrow{H_2O} NH_4NO_3(aq)$$

Hot packs are used to relax muscles, lessen aches and cramps, and increase circulation by expanding capillary size. Constructed in the same way as cold packs, a hot pack contains a salt such as $CaCl_2$. When one mole of $CaCl_2$ dissolves in water, 81 kJ are released. The temperature rises to as much as 66 °C to give a hot pack ready to use.

Exothermic Reaction in a Hot Pack

$$CaCl_2(s) \xrightarrow{H_2O} CaCl_2(aq) + 81\ kJ$$

Cold packs use an endothermic reaction.

Rate of Reaction

TUTORIAL
Factors That Affect Rate

The *rate* (or speed) *of reaction* is measured by the amount of reactant used up, or the amount of product formed, in a certain period of time. Reactions with low activation energies go faster than reactions with high activation energies. Some reactions go very fast, while others are very slow. For any reaction, the rate is affected by changes in temperature, changes in the concentration of the reactants, and the addition of catalysts.

Temperature

At higher temperatures, the increase in kinetic energy of the reactants makes them move faster and collide more often, and it provides more collisions with the required energy of activation. Reactions almost always go faster at higher temperatures. For every 10 °C increase in temperature, most reaction rates approximately double. If we want food to cook faster, we raise the temperature. When body temperature rises, there is an increase in the pulse rate, rate of breathing, and metabolic rate. On the other hand, we slow down a reaction by lowering the temperature. For example, we refrigerate perishable foods to make them last longer. In some cardiac surgeries, body temperature is lowered to 28 °C so the heart can be stopped and less oxygen is required by the brain. This is also the reason why some people have survived submersion in icy lakes for long periods of time.

Concentrations of Reactants

The rate of a reaction also increases when reactants are added. Then there are more collisions between the reactants and the reaction goes faster. For example, a patient having difficulty breathing may be given a breathing mixture with a higher oxygen content than the atmosphere. The increase in the number of oxygen molecules in the lungs increases the rate at which oxygen combines with hemoglobin. The increased rate of oxygenation of the blood means that the patient can breathe more easily.

$$\underset{\text{Hemoglobin}}{Hb(aq)} + \underset{\text{Oxygen}}{O_2(g)} \longrightarrow \underset{\text{Oxyhemoglobin}}{HbO_2(aq)}$$

Catalysts

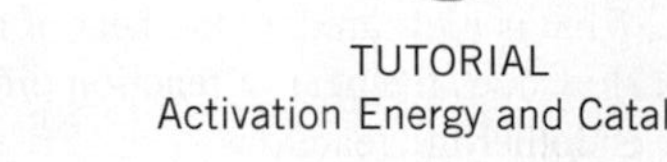
TUTORIAL
Activation Energy and Catalysis

Another way to speed up a reaction is to lower the energy of activation. This can be done by adding a **catalyst**. Earlier, we discussed the energy required to climb a hill. If instead we find a tunnel through the hill, we do not need as much energy to get to the other side. A catalyst acts by providing an alternate pathway with a lower energy requirement. As a result, more collisions form product successfully. Catalysts have found many uses in industry. In the production of margarine, the reaction of hydrogen with vegetable oils is normally very slow. However, when finely divided platinum is present as a catalyst, the reaction occurs rapidly. In the body, biocatalysts called enzymes make most metabolic reactions go at the rates necessary for proper cellular activity.

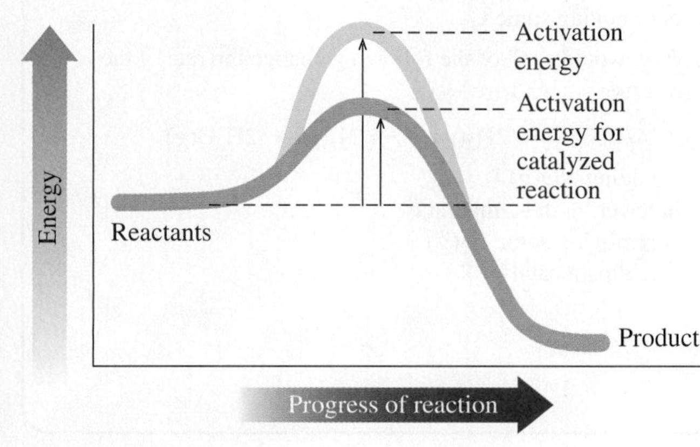

A catalyst lowers the activation energy of a chemical reaction.

TABLE 5.8 Factors That Increase Reaction Rate

Factor	Reason
Increase reactant concentration	More collisions
Increase temperature	More collisions, more collisions with energy of activation
Add a catalyst	Lowers energy of activation

A summary of the factors affecting reaction rate is given in Table 5.8.

SAMPLE PROBLEM 5.9

Factors That Affect the Rate of Reaction

Indicate whether the following changes will increase, decrease, or have no effect upon the rate of reaction:

a. increase in temperature
b. decrease in the number of reactants
c. addition of a catalyst

SOLUTION

a. increase
b. decrease
c. increase

STUDY CHECK 5.9

How does lowering the temperature affect the rate of reaction?

QUESTIONS AND PROBLEMS

Energy in Chemical Reactions

5.61 **a.** Why do chemical reactions require energy of activation?
b. What is the function of a catalyst?
c. In an exothermic reaction, is the energy of the products higher or lower than that of the reactants?
d. Draw an energy diagram for an exothermic reaction.

5.62 **a.** What is measured by the heat of reaction?
b. How does the heat of reaction differ in exothermic and endothermic reactions?
c. In an endothermic reaction, is the energy of the products higher or lower than that of the reactants?
d. Draw an energy diagram for an endothermic reaction.

5.63 Classify each of the following as an exothermic or endothermic reaction:
a. A reaction releases 550 kJ.
b. The energy level of the reactants is lower than that of the products.
c. The metabolism of glucose in the body provides energy.

5.64 Classify each of the following as an exothermic or endothermic reaction:
a. The energy level of the reactants is higher than that of the products.
b. In the body, the synthesis of proteins requires energy.
c. A reaction absorbs 125 kJ.

5.65 Classify each of the following as an exothermic or endothermic reaction:

a. $CH_4(g) + 2O_2(g) \xrightarrow{\Delta} CO_2(g) + 2H_2O(g) + 890\text{ kJ}$
b. $Ca(OH)_2(s) + 65.3\text{ kJ} \longrightarrow CaO(s) + H_2O(l)$
c. $2Al(s) + Fe_2O_3(s) \longrightarrow Al_2O_3(s) + 2Fe(s) + 850\text{ kJ}$

5.66 Classify each of the following as an exothermic or endothermic reaction:

a. $C_3H_8(g) + 5O_2(g) \xrightarrow{\Delta} 3CO_2(g) + 4H_2O(g) + 2220\text{ kJ}$
b. $2Na(s) + Cl_2(g) \longrightarrow 2NaCl(s) + 819\text{ kJ}$
c. $PCl_5(g) + 67\text{ kJ} \longrightarrow PCl_3(g) + Cl_2(g)$

5.67 **a.** What is meant by the rate of a reaction?
b. Why does bread grow mold more quickly at room temperature than in the refrigerator?

5.68 **a.** How does a catalyst affect the activation energy?
b. Why is pure oxygen used in respiratory distress?

5.69 How would each of the following change the rate of the reaction shown here?

$$2SO_2(g) + O_2(g) \longrightarrow 2SO_3(g)$$

a. adding $SO_2(g)$
b. increasing the temperature
c. adding a catalyst
d. removing some O_2

5.70 How would each of the following change the rate of the reaction shown here?

$$2NO(g) + 2H_2(g) \longrightarrow N_2(g) + 2H_2O(g)$$

a. adding $NO(g)$
b. lowering the temperature
c. removing some $H_2(g)$
d. adding a catalyst

CONCEPT MAP

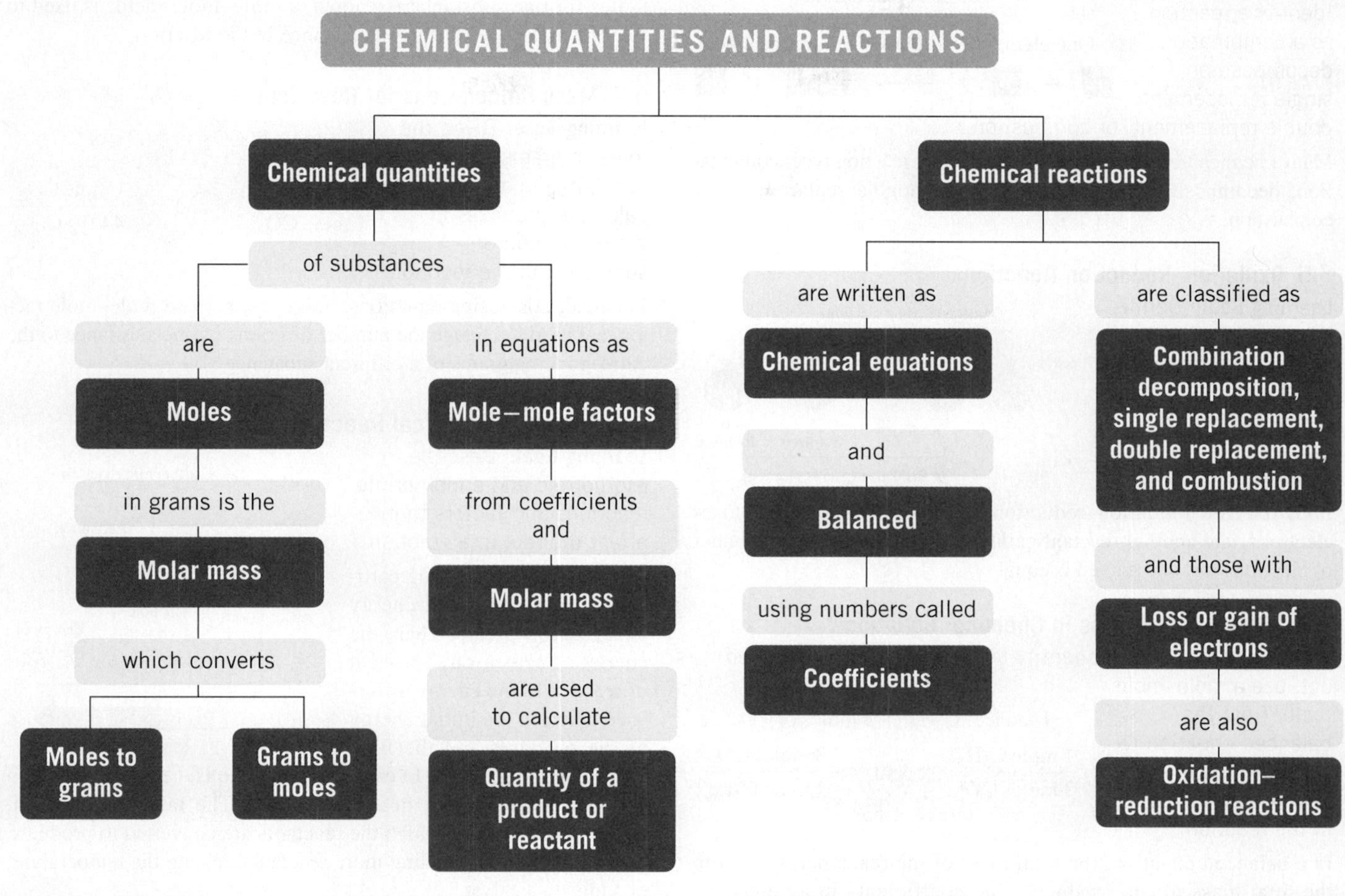

CHAPTER REVIEW

5.1 The Mole

Learning Goal: Use Avogadro's number to determine the number of particles in a given number of moles.

One mole of an element contains 6.02×10^{23} atoms; a mole of a compound contains 6.02×10^{23} molecules or formula units.

TABLE 5.1 Number of Particles in One-Mole Samples

Substance	Number and Type of Particles
1 mole of Al	6.02×10^{23} atoms of Al
1 mole of S	6.02×10^{23} atoms of S
1 mole of water (H_2O)	6.02×10^{23} molecules of H_2O
1 mole of vitamin C ($C_6H_8O_6$)	6.02×10^{23} molecules of vitamin C
1 mole of NaCl	6.02×10^{23} formula units of NaCl

5.2 Molar Mass

Learning Goal: Determine the molar mass of a substance, and use molar mass to convert between grams and moles.

The molar mass (g/mole) of any substance is the mass in grams equal numerically to its atomic mass, or the sum of the atomic masses, which have been multiplied by their subscripts in a formula. The molar mass is used as a conversion factor to change a quantity in grams to moles or to change a given number of moles to grams.

TABLE 5.2 The Molar Mass of Selected Elements and Compounds

Substance	Molar Mass
1 mole of C	12.0 g
1 mole of Na	23.0 g
1 mole of Fe	55.9 g
1 mole of NaF	42.0 g
1 mole of $C_6H_{12}O_6$ (glucose)	180.1 g
1 mole of $C_8H_{10}N_4O_2$ (caffeine)	194.1 g

5.3 Chemical Reactions and Equations

Learning Goal: Write a balanced chemical equation from the formulas of the reactants and products for a reaction; determine the number of atoms in the reactants and products.

A chemical change occurs when the atoms of the initial substances rearrange to form new substances. A chemical equation shows the formulas of the substances that react on the left side of a reaction arrow and the products that form on the right side of the reaction arrow. A chemical equation is balanced by writing coefficients, small whole numbers, in front of formulas to equalize the atoms of each of the elements in the reactants and the products.

$$C(s) + O_2(g) \longrightarrow CO_2(g)$$

C O C O

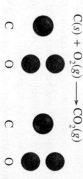

Reactant atoms = Product atoms

5.4 Types of Reactions

Learning Goal: Identify a reaction as a combination, decomposition, single replacement, double replacement, or combustion.

Single replacement

One element replaces another element

A + BC ⟶ AC + B

Many chemical reactions can be organized by reaction type: combination, decomposition, single replacement, double replacement, or combustion.

5.5 Oxidation–Reduction Reactions

Learning Goal: Define the terms oxidation and reduction; identify the reactants oxidized and reduced.

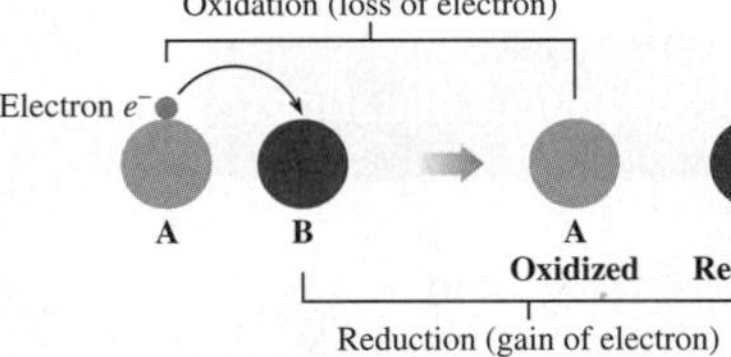

When electrons are transferred in a reaction, it is an oxidation–reduction reaction. One reactant loses electrons, and another reactant gains electrons. Overall, the number of electrons lost and gained is equal.

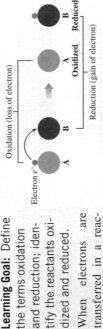

5.6 Mole Relationships in Chemical Equations

Learning Goal: Given a quantity in moles of reactant or product, use a mole–mole factor from the balanced equation to calculate the moles of another substance in the reaction.

$$1 \text{ mole of } C_3H_8 = 3 \text{ moles of } CO_2$$

$$\frac{1 \text{ mole } C_3H_8}{3 \text{ moles } CO_2} \quad \text{and} \quad \frac{3 \text{ moles } CO_2}{1 \text{ mole } C_3H_8}$$

In a balanced equation, the total mass of the reactants is equal to the total mass of the products. The coefficients in an equation describing the relationship between the moles of any two components are used to write mole–mole factors. When the number of moles for one substance is known, a mole–mole factor is used to find the moles of a different substance in the reaction.

5.7 Mass Calculations for Reactions

Learning Goal: Given the mass in grams of a substance in a reaction, calculate the mass in grams of another substance in the reaction.

$$1 \text{ mole of } CO_2 = 44.0 \text{ g of } CO_2$$

$$\frac{44.0 \text{ g } CO_2}{1 \text{ mole } CO_2} \quad \text{and} \quad \frac{1 \text{ mole } CO_2}{44.0 \text{ g } CO_2}$$

In calculations using equations, molar masses and mole–mole factors are used to change the number of grams of one substance to the corresponding grams of a different substance.

5.8 Energy in Chemical Reactions

Learning Goal: Describe exothermic and endothermic reactions and factors that affect the rate of a reaction.

Exothermic reaction
Energy
Energy of reactants
Activation energy
Heat of reaction (released)
Energy of products
Progress of reaction

In a reaction, the reacting particles must collide with energy equal to or greater than the energy of activation. The heat of reaction is the energy difference between the initial energy of the reactants and the final energy of the products. In exothermic reactions, heat is released. In endothermic reactions, heat is absorbed. The rate of a reaction, which is the speed at which the reactants are converted to products, can be increased by adding more reactants, raising the temperature, or adding a catalyst.

Key Terms

activation energy The energy needed upon collision to break apart the bonds of the reacting molecules.

Avogadro's number The number of items in a mole, equal to 6.02×10^{23}.

balanced equation The final form of a chemical equation that shows the same number of atoms of each element in the reactants and products.

catalyst A substance that increases the rate of reaction by lowering the activation energy.

chemical equation A shorthand way to represent a chemical reaction using chemical formulas to indicate the reactants and products and coefficients to show reacting ratios.

coefficients Whole numbers placed in front of the formulas to balance the number of atoms or moles of atoms of each element on both sides of an equation.

combination reaction A chemical reaction in which reactants combine to form a single product.

combustion reaction A chemical reaction in which a fuel containing carbon and hydrogen reacts with oxygen to produce CO_2, H_2O, and energy.

decomposition reaction A reaction in which a single reactant splits into two or more simpler substances.

double replacement reaction A reaction in which parts of two different reactants exchange places.

endothermic reaction A reaction that requires heat; the energy of the products is higher than the energy of the reactants.

exothermic reaction A reaction that releases heat; the energy of the products is lower than the energy of the reactants.

formula unit The group of ions represented by the formula of an ionic compound.

molar mass The mass in grams of 1 mole of an element equal numerically to its atomic mass. The molar mass of a compound is equal to the sum of the masses of the elements multiplied by their subscripts in the formula.

mole A group of atoms, molecules, or formula units that contains 6.02×10^{23} of these items.

mole–mole factor A conversion factor that relates the number of moles of two compounds derived from the coefficients in an equation.

oxidation The loss of electrons by a substance. Biological oxidation may involve the addition of oxygen or the loss of hydrogen.

oxidation–reduction reaction A reaction in which the oxidation of one reactant is always accompanied by the reduction of another reactant.

products The substances formed as a result of a chemical reaction.

reactants The initial substances that undergo change in a chemical reaction.

reduction The gain of electrons by a substance. Biological reduction may involve the loss of oxygen or the gain of hydrogen.

single replacement reaction A reaction in which an element replaces a different element in a compound.

Understanding the Concepts

5.71 Using the following models of the molecules (black = C, white = H, yellow = S, green = Cl), determine each of the following:

1. 

2. 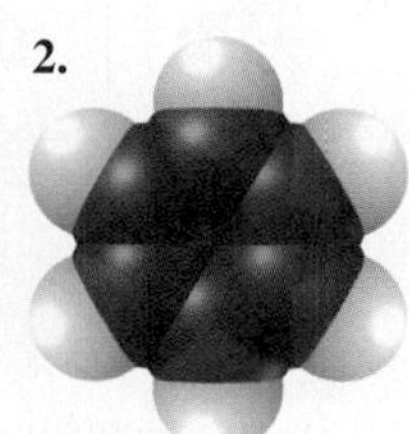

a. molecular formula
b. molar mass
c. number of moles in 10.0 g

5.72 Using the following models of the molecules (black = C, white = H, yellow = S, red = O), determine each of the following:

1. 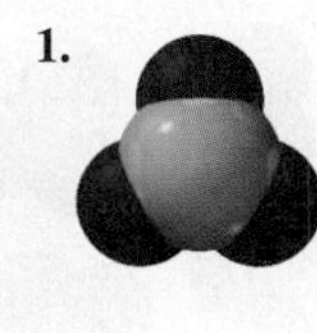

2. 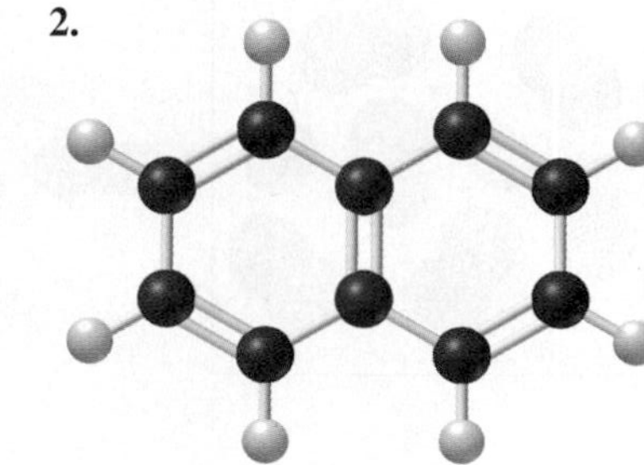

a. molecular formula
b. molar mass
c. number of moles in 10.0 g

5.73 A dandruff shampoo contains dipyrithione, $C_{10}H_8N_2O_2S_2$, which acts as an antibacterial and antifungal agent.
a. What is the molar mass of dipyrithione?
b. How many moles of dipyrithione are in 25.0 g?
c. How many moles of carbon are in 25.0 g of dipyrithione?
d. How many moles of dipyrithione contain 8.2×10^{24} atoms of nitrogen?

Dandruff shampoo contains dipyrithione.

5.74 Ibuprofen, an anti-inflammatory, has the formula $C_{13}H_{18}O_2$.

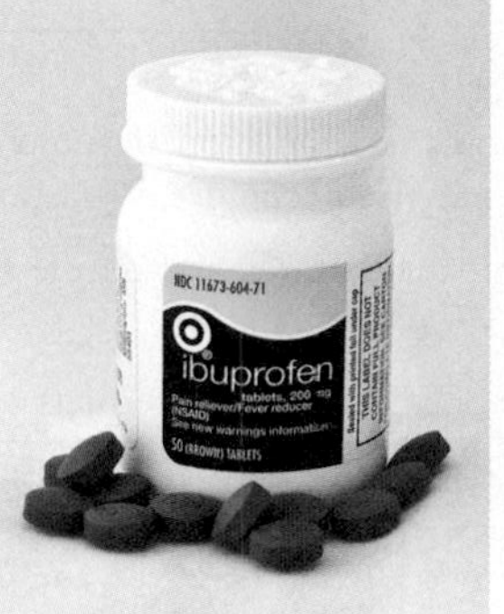

Ibuprofen is an anti-inflammatory.

a. What is the molar mass of ibuprofen?
b. How many grams of ibuprofen are in 0.525 mole?
c. How many moles of carbon are in 12.0 g of ibuprofen?
d. How many moles of ibuprofen contain 1.22×10^{23} atoms of carbon?

5.75 Balance each of the following by adding coefficients, and identify the type of reaction for each:

a. 

b.   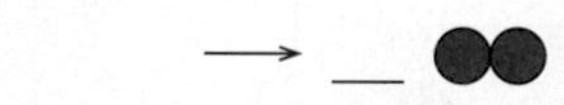

5.76 Balance each of the following by adding coefficients, and identify the type of reaction for each:

a. 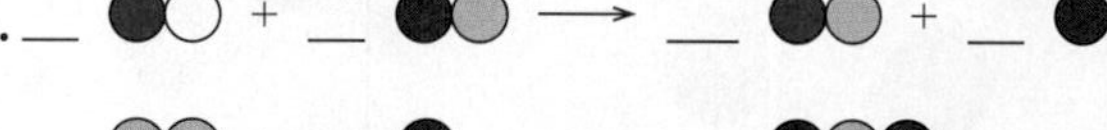

b. 

5.77 If red spheres represent oxygen atoms and blue spheres represent nitrogen atoms,

Reactants **Products**

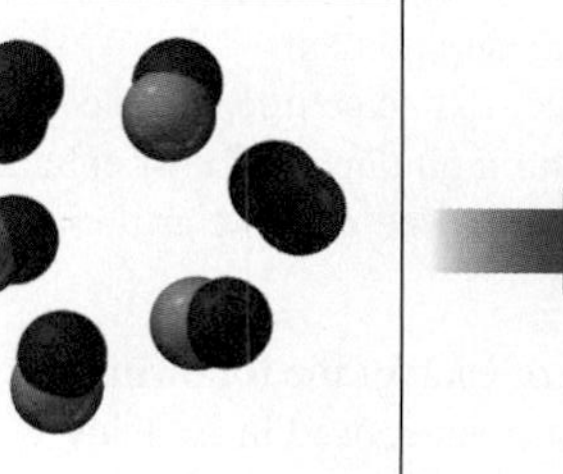

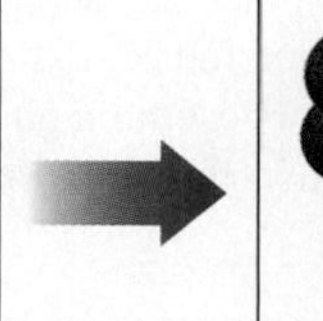

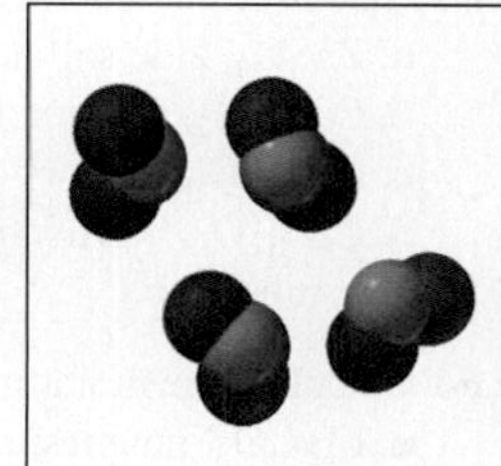

a. write the balanced equation for the reaction.
b. indicate the type of reaction as combination, decomposition, single replacement, double replacement, or combustion.

5.78 If purple spheres represent iodine atoms and white spheres represent hydrogen atoms,

Reactants **Products**

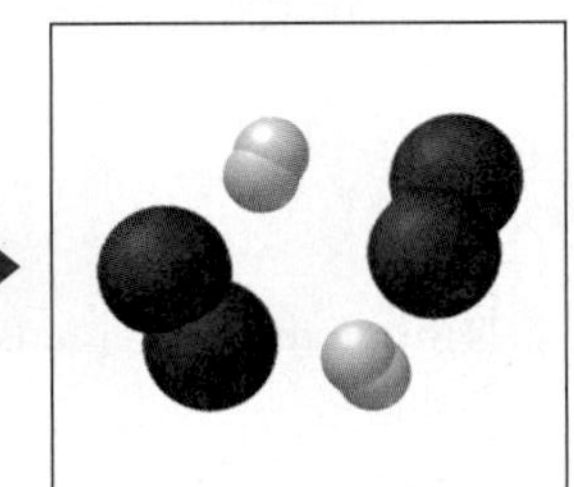

a. write the balanced equation for the reaction.
b. indicate the type of reaction as combination, decomposition, single replacement, double replacement, or combustion.

5.79 If blue spheres represent nitrogen atoms and purple spheres represent iodine atoms,

Reactants **Products**

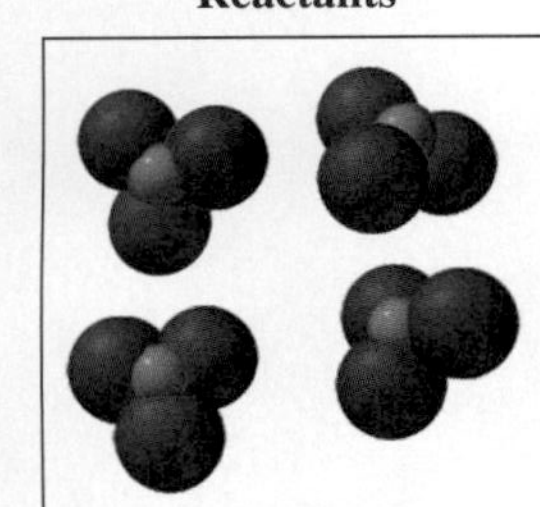

a. write the balanced equation for the reaction.
b. indicate the type of reaction as combination, decomposition, single replacement, double replacement, or combustion.

5.80 If green spheres represent chlorine atoms, yellow-green spheres represent fluorine, and white spheres represent hydrogen atoms,

Reactants **Products**

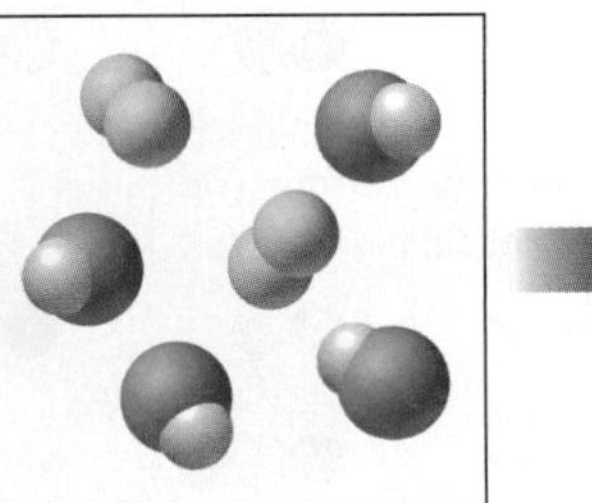

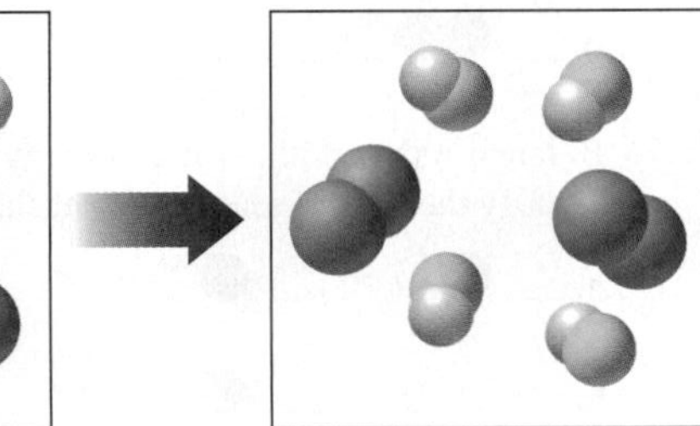

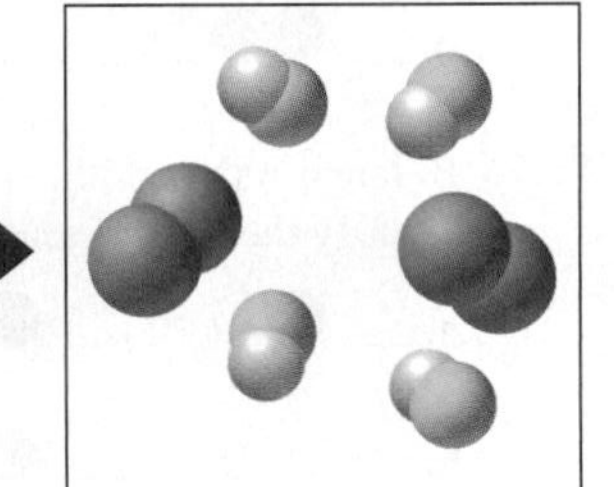

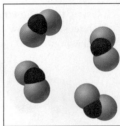

a. write the balanced equation for the reaction.
b. indicate the type of reaction as combination, decomposition, single replacement, double replacement, or combustion.

5.81 If green spheres represent chlorine atoms and red spheres represent oxygen atoms,

Reactants **Products**

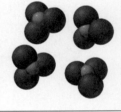

a. write the balanced equation for the reaction.
b. indicate the type of reaction as combination, decomposition, single replacement, double replacement, or combustion.

5.82 If blue spheres represent nitrogen atoms and purple spheres represent iodine atoms,

Reactants **Products**

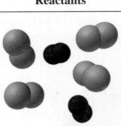

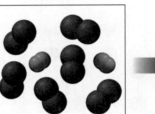

a. write the balanced equation for the reaction.
b. indicate the type of reaction as combination, decomposition, single replacement, double replacement, or combustion.

Additional Questions and Problems

For instructor-assigned homework, go to **www.masteringchemistry.com**.

5.83 Calculate the molar mass of each of the following:
a. $ZnSO_4$, zinc sulfate, zinc supplement
b. $Ca(IO_3)_2$, calcium iodate, iodine source in table salt
c. $C_5H_8NNaO_4$, monosodium glutamate, flavor enhancer
d. $C_6H_{12}O_2$, isoamyl formate, used to make artificial fruit syrups

5.84 Calculate the molar mass of each of the following:
a. $MgCO_3$, magnesium carbonate, used in antacids
b. $Au(OH)_3$, gold(III) hydroxide, used in gold plating
c. $C_{18}H_{34}O_2$, oleic acid, from olive oil
d. $C_{21}H_{26}O_5$, prednisone, anti-inflammatory

5.85 How many grams are in 0.150 mole of each of the following?
a. K **b.** Cl_2 **c.** Na_2CO_3

5.86 How many grams are in 2.25 moles of each of the following?
a. N_2 **b.** NaBr **c.** C_6H_{14}

5.87 How many moles are in 25.0 g of each of the following compounds?
a. CO_2 **b.** Al_2O_3 **c.** $MgCl_2$

5.88 How many moles are in 4.00 g of each of the following compounds?
a. NH_3 **b.** $Ca(NO_3)_2$ **c.** SO_3

5.89 Identify the type of reaction for each of the following as combination, decomposition, single replacement, double replacement, or combustion:
a. Atoms of a metal and a nonmetal form an ionic compound.
b. A compound of hydrogen and carbon reacts with oxygen.

c. Heating calcium carbonate produces calcium oxide and carbon dioxide.
d. Zinc replaces copper in $Cu(NO_3)_2$.

5.90 Identify the type of reaction for each of the following as combination, decomposition, single replacement, double replacement, or combustion:
a. A compound breaks apart into its elements.
b. Copper and bromine form copper(II) bromide.
c. Iron(II) sulfite breaks down to iron(II) oxide and sulfur dioxide.
d. Silver ion from $AgNO_3(aq)$ forms a solid with bromide ion from $KBr(aq)$.

5.91 Balance each of the following equations, and identify the type of reaction:
a. $NH_3(g) + HCl(g) \longrightarrow NH_4Cl(s)$
b. $C_5H_{12}(g) + O_2(g) \xrightarrow{\Delta} CO_2(g) + H_2O(g)$
c. $Sb(s) + Cl_2(g) \longrightarrow SbCl_3(s)$
d. $NI_3(s) \longrightarrow N_2(g) + I_2(g)$
e. $KBr(aq) + Cl_2(aq) \longrightarrow KCl(aq) + Br_2(l)$
f. $Fe(s) + H_2SO_4(aq) \longrightarrow Fe_2(SO_4)_3(aq) + H_2(g)$
g. $Al_2(SO_4)_3(aq) + NaOH(aq) \longrightarrow Na_2SO_4(aq) + Al(OH)_3(s)$

5.92 Balance each of the following equations, and identify the type of reaction:
a. $Li_3N(s) \longrightarrow Li(s) + N_2(g)$
b. $Mg(s) + N_2(g) \longrightarrow Mg_3N_2(s)$
c. $Al(s) + HCl(aq) \longrightarrow AlCl_3(aq) + H_2(g)$
d. $C_3H_4(g) + O_2(g) \xrightarrow{\Delta} CO_2(g) + H_2O(g)$
e. $Cr_2O_3(s) + H_2(g) \longrightarrow Cr(s) + H_2O(g)$
f. $Al(s) + Cl_2(g) \longrightarrow AlCl_3(s)$
g. $MgCl_2(aq) + AgNO_3(aq) \longrightarrow Mg(NO_3)_2(aq) + AgCl(s)$

5.93 Predict the products and write a balanced equation for each of the following:
a. Single replacement:
$Zn(s) + HCl(aq) \longrightarrow$ ________ + ________
b. Decomposition:
$BaCO_3(s) \xrightarrow{\Delta}$ ________ + ________
c. Double replacement:
$NaOH(aq) + HCl(aq) \longrightarrow$ ________ + ________
d. Combination:
$Al(s) + F_2(g) \longrightarrow$ ________

5.94 Predict the products and write a balanced equation for each of the following:
a. Decomposition:
$NaCl(s) \xrightarrow{\text{Electricity}}$ ________ + ________
b. Combination:
$Ca(s) + Br_2(g) \longrightarrow$ ________
c. Combustion:
$C_2H_4(g) + O_2(g) \xrightarrow{\Delta}$ ________ + ________
d. Double replacement:
$NiCl_2(aq) + NaOH(aq) \longrightarrow$ ________ + ________

5.95 Identify each of the following as an oxidation or a reduction reaction:
a. $Zn^{2+} + 2\,e^- \longrightarrow Zn$ **b.** $Al \longrightarrow Al^{3+} + 3\,e^-$
c. $Pb \longrightarrow Pb^{2+} + 2\,e^-$ **d.** $Cl_2 + 2\,e^- \longrightarrow 2Cl^-$

5.96 Write a balanced chemical equation for each of the following oxidation–reduction reactions:
a. Sulfur reacts with molecular chlorine to form sulfur dichloride.
b. Molecular chlorine and sodium bromide react to form molecular bromine and sodium chloride.
c. Aluminum metal and iron(III) oxide react to produce aluminum oxide and elemental iron.
d. Copper(II) oxide reacts with carbon to form elemental copper and carbon dioxide.

5.97 When ammonia (NH_3) reacts with fluorine, the products are dinitrogen tetrafluoride (N_2F_4) and hydrogen fluoride (HF).
a. Write the balanced chemical equation.
b. How many moles of each reactant are needed to produce 4.00 moles of HF?
c. How many grams of F_2 are required to react with 25.5 g of NH_3?
d. How many grams of N_2F_4 can be produced when 3.40 g of NH_3 reacts?

5.98 When nitrogen dioxide (NO_2) from car exhaust combines with water in the air, it forms nitric acid (HNO_3), which causes acid rain, and nitrogen oxide.
a. Write the balanced chemical equation.
b. How many molecules of NO_2 are needed to react with 0.250 mole of H_2O?
c. How many grams of HNO_3 are produced when 60.0 g of NO_2 completely reacts?

5.99 Pentane gas, C_5H_{12}, undergoes combustion with oxygen to produce carbon dioxide and water.
a. Write the balanced chemical equation.
b. How many grams of pentane are needed to produce 72 g of water?
c. How many grams of CO_2 are produced from 32.0 g of oxygen?

5.100 Propane gas, C_3H_8, reacts with oxygen to produce water and carbon dioxide. Propane has a density of 2.02 g/L at room temperature.
a. Write the balanced chemical equation.
b. How many grams of water form when 5.00 L of propane gas (C_3H_8) reacts?
c. How many grams of CO_2 are produced from 18.5 g of oxygen gas?
d. How many grams of H_2O can be produced from the reaction of 8.50×10^{22} molecules of propane gas, C_3H_8?

5.101 The equation for the formation of silicon tetrachloride from silicon and chlorine is

$$Si(s) + 2Cl_2(g) \longrightarrow SiCl_4(g) + 157\text{ kcal}$$

a. Is the formation of $SiCl_4$ an endothermic or exothermic reaction?
b. Is the energy of the product higher or lower than the energy of the reactants?

5.102 The equation for the formation of nitrogen oxide is

$$N_2(g) + O_2(g) + 90.2\text{ kJ} \longrightarrow 2NO(g)$$

a. Is the formation of NO an endothermic or exothermic reaction?
b. Is the energy of the product higher or lower than the energy of the reactants?

Challenge Questions

5.103 At a winery, glucose ($C_6H_{12}O_6$) in grapes undergoes fermentation to produce ethanol (C_2H_6O) and carbon dioxide.

$$\underset{\text{Glucose}}{C_6H_{12}O_6} \longrightarrow \underset{\text{Ethanol}}{2C_2H_6O} + 2CO_2$$

Glucose in grapes ferments to produce ethanol.

a. How many grams of glucose are required to form 124 g of ethanol?
b. How many grams of ethanol would be formed from the reaction of 0.240 kg of glucose?

5.104 Gasohol is a fuel containing ethanol (C_2H_6O) that burns in oxygen (O_2) to give carbon dioxide and water.
a. Write the balanced chemical equation.
b. How many moles of O_2 are needed to completely react with 4.0 moles of C_2H_6O?
c. If a car produces 88 g of CO_2, how many grams of O_2 are used up in the reaction?
d. If you burn 125 g of C_2H_6O, how many grams of CO_2 and H_2O can be produced?

5.105 Consider the following *unbalanced* equation:

$$Al(s) + O_2(g) \longrightarrow Al_2O_3(s)$$

a. Write the balanced chemical equation.
b. Identify the type of reaction.
c. How many moles of oxygen are needed to react with 4.50 moles of aluminum?
d. How many grams of aluminum oxide are produced when 50.2 g of aluminum reacts?
e. When aluminum is reacted in a closed container with 8.00 g of oxygen, how many grams of aluminum oxide can form?

5.106 A toothpaste contains 0.240% by mass sodium fluoride used to prevent dental caries and 0.30% by mass triclosan $C_{12}H_7Cl_3O_2$, an antigingivitis agent. One tube contains 119 g of toothpaste.

Components in toothpaste include triclosan and NaF.

a. How many moles of NaF are in the tube of toothpaste?
b. How many fluoride ions, F^-, are in the tube of toothpaste?
c. How many grams of sodium ion, Na^+, are in 1.50 g of toothpaste?
d. How many molecules of triclosan are in the tube of toothpaste?

5.107 During heavy exercise and workouts, lactic acid, $C_3H_6O_3$, accumulates in the muscles where it can cause pain and soreness.

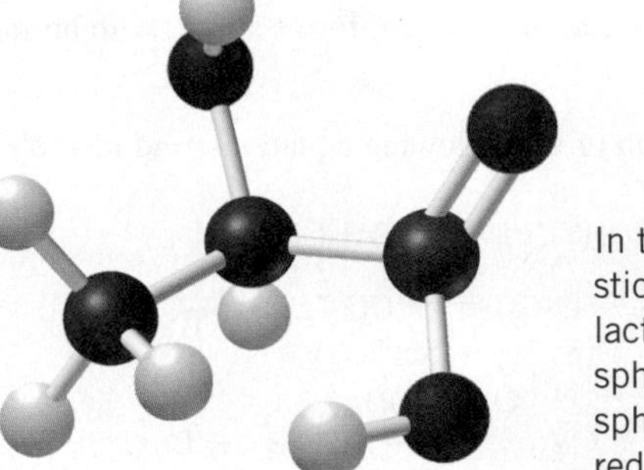

In the ball-and-stick model of lactic acid, black spheres = C, white spheres = H, and red spheres = O.

a. How many molecules are in 0.500 mole of lactic acid?
b. How many atoms of carbon are in 1.50 moles of lactic acid?
c. How many moles of lactic acid contain 4.5×10^{24} atoms of O?
d. What is the molar mass of lactic acid?

5.108 Ammonium sulfate, $(NH_4)_2SO_4$, is used in fertilizers to provide nitrogen for the soil.
a. How many formula units are in 0.200 mole of ammonium sulfate?
b. How many H atoms are in 0.100 mole of ammonium sulfate?
c. How many moles of ammonium sulfate contain 7.4×10^{25} atoms of N?
d. What is the molar mass of ammonium sulfate?

5.109 The gaseous hydrocarbon acetylene, C_2H_2, used in welders' torches, releases 1300 kJ when 1 mole of C_2H_2 undergoes combustion.
a. Write a balanced equation for the reaction, including the heat of reaction.
b. Is the reaction endothermic or exothermic?
c. How many moles of water are produced when 64.0 g of oxygen reacts?
d. How many moles of oxygen are needed to react with 2.25×10^{24} molecules of acetylene?

5.110 Methanol (CH_3OH), which is used as a cooking fuel, burns with oxygen gas (O_2) to produce the gases carbon dioxide and water. The reaction produces 363 kJ of heat per mole of methanol.
a. Write a balanced equation for the reaction, including the heat of reaction.
b. Is the reaction endothermic or exothermic?
c. How many moles of oxygen must react with 25.0 g of methanol?
d. How many grams of carbon dioxide are produced when 78.0 g of methanol reacts?

Answers

Answers to Study Checks

5.1 0.432 mole of H_2O

5.2 0.120 mole of aspirin

5.3 138.1 g of salicylic acid

5.4 80.9 g of Ag

5.5 $2Al(s) + 3Cl_2(g) \longrightarrow 2AlCl_3(s)$

5.6 $Sb_2S_3(s) + 6HCl(aq) \longrightarrow 2SbCl_3(s) + 3H_2S(g)$

5.7 0.945 mole of O_2

5.8 27.5 g of CO_2

5.9 The rate of reaction will decrease because the number of collisions between reacting particles will be fewer, and a smaller number of the collisions that do occur will have sufficient activation energy.

Answers to Selected Questions and Problems

5.1 One mole contains 6.02×10^{23} atoms of an element, molecules of a covalent substance, or formula units of an ionic substance.

5.3 **a.** 3.01×10^{23} atoms of C
b. 7.71×10^{23} molecules of SO_2
c. 0.0867 mole of Fe
d. 14.1 moles of C_2H_6O

5.5 **a.** 6.00 moles of H
b. 8.00 moles of O
c. 1.20×10^{24} atoms of P
d. 4.82×10^{24} atoms of O

5.7 **a.** 24 moles of H
b. 1.0×10^2 moles of C
c. 0.040 mole of N

5.9 **a.** 188.2 g **b.** 159.8 g **c.** 329.2 g
d. 342.3 g **e.** 58.3 g **f.** 365.3 g

5.11 **a.** 71.0 g **b.** 90.1 g **c.** 262.9 g
d. 84.0 g **e.** 99.0 g **f.** 156.7 g

5.13 **a.** 46.0 g **b.** 112 g
c. 14.8 g **d.** 112 g

5.15 **a.** 29.3 g **b.** 109 g
c. 4.05 g **d.** 194 g

5.17 **a.** 602 g **b.** 11 g

5.19 **a.** 0.463 mole of Ag **b.** 0.0167 mole of C
c. 0.882 mole of NH_3 **d.** 1.17 moles of SO_2

5.21 **a.** 1.24 moles of Ne **b.** 0.781 mole of O_2
c. 0.321 mole of $Al(OH)_3$ **d.** 0.106 mole of Ga_2S_3

5.23 **a.** 0.78 mole of S **b.** 1.95 moles of S
c. 6.0 moles of S

5.25 **a.** 66.2 g of propane **b.** 0.771 mole of propane
c. 27.8 g of C

5.27 **a.** not balanced **b.** balanced
c. not balanced **d.** balanced

5.29 **a.** $N_2(g) + O_2(g) \longrightarrow 2NO(g)$
b. $2HgO(s) \longrightarrow 2Hg(l) + O_2(g)$
c. $4Fe(s) + 3O_2(g) \longrightarrow 2Fe_2O_3(s)$
d. $2Na(s) + Cl_2(g) \longrightarrow 2NaCl(s)$

5.31 **a.** $Mg(s) + 2AgNO_3(aq) \longrightarrow Mg(NO_3)_2(aq) + 2Ag(s)$
b. $2Al(s) + 3CuSO_4(aq) \longrightarrow 3Cu(s) + Al_2(SO_4)_3(aq)$
c. $Pb(NO_3)_2(aq) + 2NaCl(aq) \longrightarrow PbCl_2(s) + 2NaNO_3(aq)$
d. $2Al(s) + 6HCl(aq) \longrightarrow 2AlCl_3(aq) + 3H_2(g)$

5.33 **a.** A single reactant splits into two simpler substances (elements).
b. One element in the reacting compound is replaced by the other reactant.

5.35 **a.** combination **b.** single replacement
c. decomposition **d.** double replacement
e. decomposition **f.** double replacement
g. combination **h.** combustion

5.37 **a.** $Mg(s) + Cl_2(g) \longrightarrow MgCl_2(s)$
b. $2HBr(g) \longrightarrow H_2(g) + Br_2(g)$
c. $Mg(s) + Zn(NO_3)_2(aq) \longrightarrow Zn(s) + Mg(NO_3)_2(aq)$
d. $K_2S(aq) + Pb(NO_3)_2(aq) \longrightarrow 2KNO_3(aq) + PbS(s)$
e. $2C_2H_6(g) + 7O_2(g) \xrightarrow{\Delta} 4CO_2(g) + 6H_2O(g)$

5.39 **a.** reduction **b.** oxidation
c. reduction **d.** reduction

5.41 **a.** Zn is oxidized, Cl_2 is reduced.
b. Br^- in NaBr is oxidized, Cl_2 is reduced.
c. The O^{2-} in PbO is oxidized, the Pb^{2+} in PbO is reduced.
d. Sn^{2+} is oxidized, Fe^{3+} is reduced.

5.43 **a.** reduction **b.** oxidation

5.45 Linoleic acid gains hydrogen atoms and is reduced.

5.47 **a.** $\frac{2 \text{ moles } SO_2}{1 \text{ mole } O_2}$ and $\frac{1 \text{ mole } O_2}{2 \text{ moles } SO_2}$

$\frac{2 \text{ moles } SO_2}{2 \text{ moles } SO_3}$ and $\frac{2 \text{ moles } SO_3}{2 \text{ moles } SO_2}$

$\frac{2 \text{ moles } SO_3}{1 \text{ mole } O_2}$ and $\frac{1 \text{ mole } O_2}{2 \text{ moles } SO_3}$

b. $\frac{4 \text{ moles P}}{5 \text{ moles } O_2}$ and $\frac{5 \text{ moles } O_2}{4 \text{ moles P}}$

$\frac{4 \text{ moles P}}{2 \text{ moles } P_2O_5}$ and $\frac{2 \text{ moles } P_2O_5}{4 \text{ moles P}}$

$\frac{5 \text{ moles } O_2}{2 \text{ moles } P_2O_5}$ and $\frac{2 \text{ moles } P_2O_5}{5 \text{ moles } O_2}$

5.49 **a.** 1.0 mole of O_2 **b.** 10. moles of H_2
c. 5.0 moles of H_2O

5.51 **a.** 1.25 moles of C **b.** 0.96 mole of CO
c. 1.0 mole of SO_2 **d.** 0.50 mole of CS_2

5.53 **a.** 77.5 g of Na_2O **b.** 6.26 g of O_2
c. 19.4 g of O_2

5.55 **a.** 19.2 g of O_2 **b.** 3.79 g of N_2
c. 54.0 g of H_2O

5.57 **a.** 3.65 g of H_2O **b.** 3.43 g of NO
c. 7.53 g of HNO_3

5.59 **a.** $2PbS(s) + 3O_2(g) \longrightarrow 2PbO(s) + 2SO_2(g)$
b. 6.00 g of O_2
c. 17.4 g of SO_2
d. 137 g of PbS

5.61 **a.** The energy of activation is the energy required to break the bonds of the reacting molecules.
b. A catalyst provides a pathway that lowers the activation energy and speeds up a reaction.
c. In exothermic reactions, the energy of the products is lower than the energy of the reactants.
d.

Energy
Reactants
Products
Reaction progress →

5.63 **a.** exothermic **b.** endothermic
c. exothermic

5.65 **a.** exothermic **b.** endothermic
c. exothermic

5.67 **a.** The rate of a reaction tells how fast the products are formed or how fast the reactants are consumed.
b. Reactions go faster at higher temperatures.

5.69 **a.** increase **b.** increase
c. increase **d.** decrease

5.71 **1. a.** S_2Cl_2 **b.** 135.2 g/mole
c. 0.0740 mole
2. a. C_6H_6 **b.** 78.1 g/mole
c. 0.128 mole

5.73 **a.** 252.3 g/mole **b.** 0.0991 mole of dipyrithione
c. 0.991 mole of C **d.** 6.8 moles of dipyrithione

5.75 **a.** 1,1,2 combination **b.** 2,2,1 decomposition

5.77 **a.** $2NO(g) + O_2(g) \longrightarrow 2NO_2(g)$
b. combination

5.79 **a.** $2NI_3(g) \longrightarrow N_2(g) + 3I_2(g)$
b. decomposition

5.81 **a.** $2Cl_2(g) + O_2(g) \longrightarrow 2OCl_2(g)$
b. combination

5.83 **a.** 161.5 g/mole **b.** 389.9 g/mole
c. 169.1 g/mole **d.** 116.1 g/mole

5.85 **a.** 5.87 g **b.** 10.7 g **c.** 15.9 g

5.87 **a.** 0.568 mole **b.** 0.245 mole **c.** 0.262 mole

5.89 **a.** combination **b.** combustion
c. decomposition **d.** single replacement

5.91 **a.** $NH_3(g) + HCl(g) \longrightarrow NH_4Cl(s)$ combination
b. $C_5H_{12}(g) + 8O_2(g) \xrightarrow{\Delta} 5CO_2(g) + 6H_2O(g)$ combustion
c. $2Sb(s) + 3Cl_2(g) \longrightarrow 2SbCl_3(s)$ combination
d. $2NI_3(s) \longrightarrow N_2(g) + 3I_2(g)$ decomposition
e. $2KBr(aq) + Cl_2(aq) \longrightarrow 2KCl(aq) + Br_2(l)$ single replacement
f. $2Fe(s) + 3H_2SO_4(aq) \longrightarrow Fe_2(SO_4)_3(aq) + 3H_2(g)$ single replacement
g. $Al_2(SO_4)_3(aq) + 6NaOH(aq) \longrightarrow 3Na_2SO_4(aq) + 2Al(OH)_3(s)$ double replacement

5.93 **a.** $Zn(s) + 2HCl(aq) \longrightarrow ZnCl_2(aq) + H_2(g)$
b. $BaCO_3(s) \xrightarrow{\Delta} BaO(s) + CO_2(g)$
c. $NaOH(aq) + HCl(aq) \longrightarrow NaCl(aq) + H_2O(l)$
d. $2Al(s) + 3F_2(g) \longrightarrow 2AlF_3(s)$

5.95 **a.** reduction **b.** oxidation
c. oxidation **d.** reduction

5.97 **a.** $2NH_3(g) + 5F_2(g) \longrightarrow N_2F_4(g) + 6HF(g)$
b. 1.33 moles of NH_3 and 3.33 moles of F_2
c. 143 g of F_2
d. 10.4 g of N_2F_4

5.99 **a.** $C_5H_{12}(g) + 8O_2(g) \xrightarrow{\Delta} 5CO_2(g) + 6H_2O(g) +$ energy
b. 48 g of pentane
c. 27.5 g of CO_2

5.101 **a.** exothermic **b.** lower

5.103 **a.** 242 g of glucose **b.** 123 g of ethanol

5.105 **a.** $4Al(s) + 3O_2(g) \longrightarrow 2Al_2O_3(s)$
b. combination
c. 3.38 moles of oxygen
d. 94.8 g of Al_2O_3
e. 17.0 g of Al_2O_3

5.107 **a.** 3.01×10^{23} molecules **b.** 2.71×10^{24} atoms of C
c. 2.5 moles of lactic acid **d.** 90.1 g

5.109 **a.** $2C_2H_2(g) + 5O_2(g) \xrightarrow{\Delta} 4CO_2(g) + 2H_2O(g) + 2600$ kJ
b. exothermic
c. 0.800 mole of H_2O
d. 9.34 moles of O_2

Combining Ideas from Chapters 3 to 5

CI.7 For parts **a** to **f**, consider the loss of electrons by atoms of the element X, and a gain of electrons by atoms of the element Y, if X is in Group 2A (2), Period 3, and Y is in Group 7A (17), Period 3.

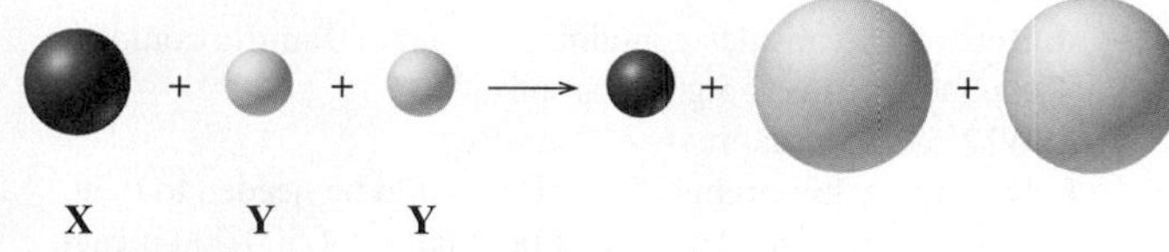

a. Which reactant has the higher electronegativity?
b. What are the ionic charges of X and Y in the product?
c. Write the electron arrangements of the atoms X and Y.
d. Write the electron arrangements of the ions of X and Y.
e. Give the names of the noble gases with the same electron arrangements as each of these ions.
f. Write the formula and name of the ionic compound indicated by the ions of X and Y.

CI.8 A sterling silver bracelet, which is 92.5% silver by mass, has a volume of 25.6 cm^3 and a density of 10.2 g/cm^3.

Sterling silver is 92.5% silver by mass.

a. What is the mass, in kilograms, of the bracelet?
b. How many atoms of silver are in the bracelet?
c. Determine the number of protons and neutrons in each of the two stable isotopes of silver:

$^{107}_{47}Ag$ $^{109}_{47}Ag$

CI.9 In the following diagram, blue spheres represent the element A and yellow spheres represent the element B:

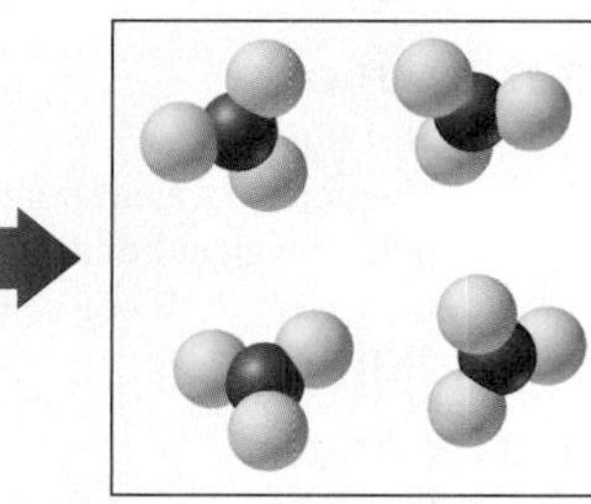

a. Write the formulas of the reactants and the products.
b. Write the balanced equation for the reaction.
c. Identify the type of reaction as combination, decomposition, single replacement, double replacement, or combustion.

CI.10 The active ingredient in Tums is calcium carbonate. One Tums tablet contains 500. mg of calcium carbonate.

The active ingredient in Tums neutralizes excess stomach acid.

a. What is the formula for calcium carbonate?
b. What is the molar mass of calcium carbonate?
c. How many moles of calcium carbonate are in 1 roll of Tums that contains 12 tablets?
d. If a person takes two Tums tablets, how many grams of calcium are obtained?
e. If the daily recommended quantity of Ca^{2+} to maintain bone strength in older women is 1500 mg, how many Tums tablets are needed each day to supply the needed calcium?

CI.11 Acetone (propanone), a clear liquid solvent with an acrid odor, is used to remove nail polish, paints, and resins. It has a low boiling point and is highly flammable. Acetone has a density of 0.786 g/mL.

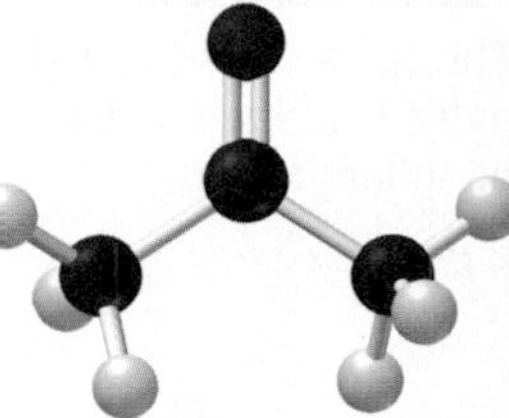

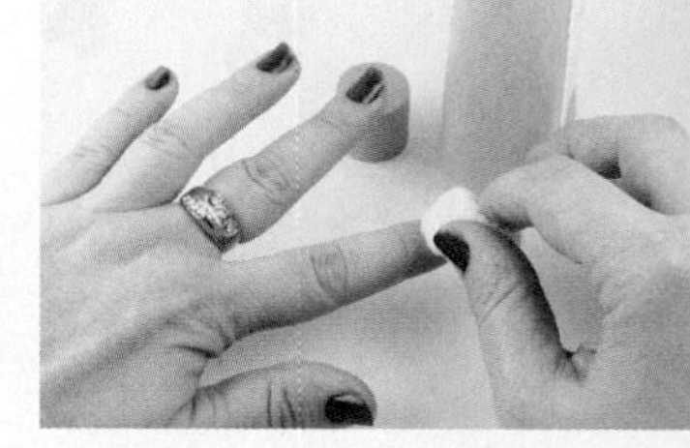

Acetone consists of carbon atoms (black), hydrogen atoms (white), and an oxygen atom (red).

a. What is the molecular formula of acetone?
b. What is the molar mass of acetone?
c. Identify the bonds C—C, C—H, and C—O in a molecule of acetone as polar covalent or nonpolar covalent.
d. Write the balanced equation for the combustion of acetone.
e. How many grams of oxygen gas are needed to react with 15.0 mL of acetone?

CI.12 The compound butyric acid gives rancid butter its characteristic odor.

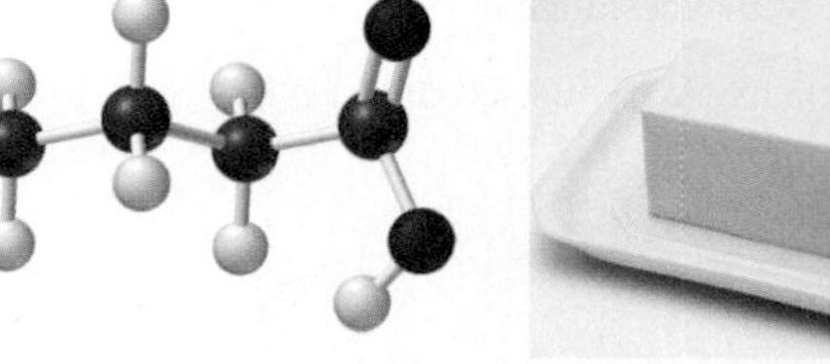

Butyric acid produces the characteristic odor of rancid butter.

a. If black spheres are carbon atoms, white spheres are hydrogen atoms, and red spheres are oxygen atoms, what is the formula of butyric acid?
b. What is the molar mass of butyric acid?
c. How many grams of butyric acid contain 3.28×10^{23} atoms of oxygen?
d. How many grams of carbon are in 5.28 g of butyric acid?
e. Butyric acid has a density of 0.959 g/mL at 20 °C. How many moles of butyric acid are contained in 1.56 mL of butyric acid?

CI.13 Tamiflu (Oseltamivir), $C_{16}H_{28}N_2O_4$, is an antiviral drug used to treat influenza. The preparation of Tamiflu begins with the extraction of shikimic acid from the seedpods of star anise. From 2.6 g of star anise, 0.13 g of shikimic acid can be obtained and used to produce one capsule containing 75 mg of Tamiflu. The usual adult dosage for treatment of influenza is two capsules of Tamiflu daily for 5 days.

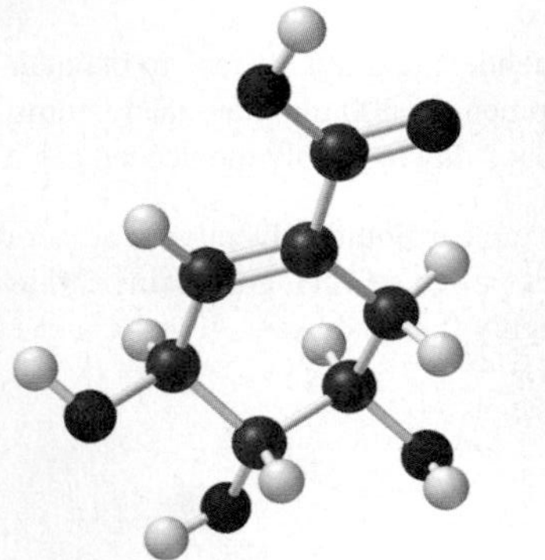

Shikimic acid is the basis for the antiviral drug in Tamiflu.

The spice called star anise is a plant source of shikimic acid.

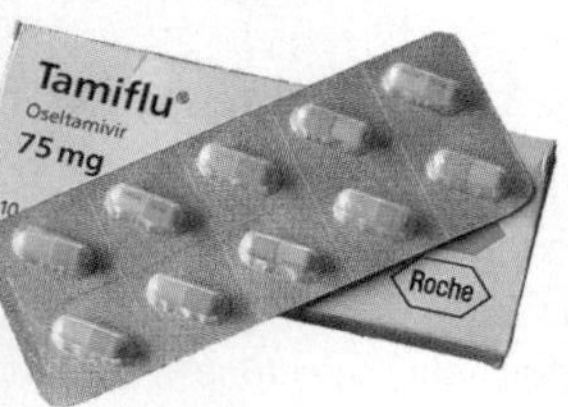

Each capsule contains 75 mg of Tamiflu.

a. What is the formula of shikimic acid? (Black spheres = C, white spheres = H, and red spheres = O)
b. What is the molar mass of shikimic acid?
c. How many moles of shikimic acid are contained in 130 g of shikimic acid?
d. How many capsules containing 75 mg of Tamiflu could be produced from 155 g of star anise?
e. What is the molar mass of Tamiflu?
f. How many kilograms of Tamiflu would be needed to treat all the people in a city with a population of 500 000 if each person consumes two Tamiflu capsules a day for 5 days?

CI.14 One of the components of gasoline is octane, C_8H_{18}, which has a density of 0.803 g/cm^3. The combustion of 1 mole of octane provides 5510 kJ. A hybrid car has a fuel tank with a capacity of 11.9 gal and a gas mileage of 45 mi/gal.

Octane is one of the components of gasoline.

a. Write the balanced chemical equation for the combustion of octane.
b. Is the combustion of octane endothermic or exothermic?
c. How many moles of octane are in one tank of fuel, assuming it is all octane?
d. If the total mileage of this hybrid car for one year is 24 500 mi, how many kilograms of carbon dioxide would be produced from the combustion of the fuel, assuming it is all octane?

Answers

CI.7 **a.** Y has the higher electronegativity.
b. X^{2+}, Y^-
c. X = 2,8,2 Y = 2,8,7
d. X^{2+} = 2,8 Y^- = 2,8,8
e. X^{2+} has the same electron arrangement as Ne. Y^- has the same electron arrangement as Ar.
f. $MgCl_2$, magnesium chloride

CI.9 **a.** Reactants: A and B_2; Products: AB_3
b. $2A + 3B_2 \longrightarrow 2AB_3$
c. combination

CI.11 **a.** C_3H_6O
b. 58.1 g/mole
c. Nonpolar covalent bonds: C—C, C—H; polar covalent bond: C—O
d. $C_3H_6O(l) + 4O_2(g) \xrightarrow{\Delta} 3CO_2(g) + 3H_2O(g)$ + energy
e. 26.0 g of O_2

CI.13 **a.** $C_7H_{10}O_5$ **b.** 174.1 g/mole
c. 0.75 mole **d.** 59 capsules
e. 312.3 g/mole **f.** 400 kg

6 Gases

LOOKING AHEAD

"When oxygen levels in the blood are low, the cells in the body don't get enough oxygen," says Sunanda Tripathi, registered nurse, Santa Clara Valley Medical Center. "We use a nasal cannula to give supplemental oxygen to a patient; a patient breathes in a gaseous mixture that is about 28% oxygen compared to 21% in ambient air."

When a patient has a breathing disorder, the flow and volume of oxygen into and out of the lungs are measured. A ventilator may be used if a patient has difficulty breathing. When pressure is increased, the lungs expand. When the pressure of the incoming gas is reduced, the lung volume contracts to expel carbon dioxide. These relationships—known as gas laws—are an important part of ventilation and breathing.

Visit **www.masteringchemistry.com** for self-study materials and instructor-assigned homework.

We all live at the bottom of a sea of gases called the atmosphere. The most important of these gases is oxygen, which constitutes about 21% of the atmosphere. Without oxygen, life on this planet would be impossible—oxygen is vital to all life processes of plants and animals. Ozone (O_3), formed in the upper atmosphere by the interaction of oxygen with ultraviolet light, absorbs some of the harmful radiation before it can strike Earth's surface. The other gases in the atmosphere include nitrogen (78% of the atmosphere), argon, carbon dioxide (CO_2), and water vapor. Carbon dioxide gas, a product of combustion and metabolism, is used by plants in photosynthesis, a process that produces the oxygen that is essential for humans and animals.

The atmosphere has become a dumping ground for other gases, such as methane, chlorofluorocarbons (CFCs), and nitrogen oxides as well as volatile organic compounds (VOCs), which are gases from paints, paint thinners, and cleaning supplies. The chemical reactions of these gases with sunlight and oxygen in the air are contributing to air pollution, ozone depletion, global warming, and acid rain. Such chemical changes can seriously affect our health and our lifestyle. An understanding of gases and some of the laws that govern gas behavior can help us understand the nature of matter and allow us to make decisions concerning important environmental and health issues.

LEARNING GOAL

Use the kinetic molecular theory of gases to describe the properties of gases.

TUTORIAL
Properties of Gases

SELF STUDY ACTIVITY
Properties of Gases

TUTORIAL
Kinetic Molecular Theory of Gases

6.1 Properties of Gases

We are surrounded by gases, but not often aware of their presence. Of the elements on the periodic table, only a handful exist as gases at room temperature: H_2, N_2, O_2, F_2, Cl_2, and the noble gases. Another group of gases includes the oxides of the nonmetals on the upper right corner of the periodic table, such as CO, CO_2, NO, NO_2, SO_2, and SO_3. Generally molecules that are gases at room temperature have fewer than five atoms from the first or second period.

The behavior of gases is quite different from that of liquids and solids. As we learned in Chapter 2, gas particles are far apart, whereas particles of both liquids and solids are held close together. A gas has no definite shape or volume and will completely fill any container. As we learned in Chapter 4, the attractive forces between gas particles are minimal. Thus, there are great distances between gas particles, which make a gas less dense than a solid or liquid, and easy to compress. A model for the behavior of a gas, called the **kinetic molecular theory of gases**, helps us understand gas behavior.

Kinetic Molecular Theory of Gases

1. **A gas consists of small particles (atoms or molecules) that move randomly with high velocities.** Gas molecules moving in random directions at high speeds cause a gas to fill the entire volume of a container.
2. **The attractive forces between the particles of a gas are usually very small.** Gas particles are far apart and fill a container of any size and shape.
3. **The actual volume occupied by gas molecules is extremely small compared with the volume that the gas occupies.** The volume of the gas is considered equal

to the volume of the container. Most of the volume of a gas is empty space, which allows gases to be easily compressed.

4. **Gas particles are in constant motion, moving rapidly in straight paths.** When gas particles collide, they rebound and travel in new directions. Every time they hit the walls of the container, they exert pressure. An increase in the number or force of collisions against the walls of the container causes an increase in the pressure of the gas.
5. **The average kinetic energy of gas molecules is proportional to the Kelvin temperature.** Gas particles move faster as the temperature increases. At higher temperatures, gas particles hit the walls of the container more often, with more force, producing higher pressures.

The kinetic molecular theory helps explain some of the characteristics of gases. For example, we can quickly smell perfume when a bottle is opened on the other side of a room, because its particles move rapidly in all directions. At room temperature, the molecules of air are moving at about 1000 miles per hour. They move faster at higher temperatures and more slowly at lower temperatures. Sometimes tires or gas-filled containers explode when temperatures are too high. From the kinetic molecular theory, we know that gas particles move faster when heated, hit the walls of a container with more force, and cause a buildup of pressure inside a container.

CONCEPT CHECK 6.1

Kinetic Molecular Theory of Gases

Use the kinetic molecular theory of gases to explain each of the following:

a. You can smell the odor of cooking onions from far away.
b. The volume of a balloon filled with helium gas increases when left in the sun.

ANSWER

a. Molecules of gas, which carry the aroma of cooking food, move at high speeds in random directions and great distances to reach you in a different location.
b. Raising the temperature of a gas causes the gas particles to move faster, hitting the walls of the balloon more often and with more force, which increases its volume.

When we talk about a gas, we describe it in terms of four properties: pressure, volume, temperature, and the amount of gas.

Pressure (*P*)

Gas particles are extremely small and move rapidly. When they hit the walls of a container, they exert *pressure* (see Figure 6.1). If we heat the container, the molecules move faster and smash into the walls more often and with increased force, thus increasing the pressure. The gas particles in the air, mostly oxygen and nitrogen, exert a pressure on us called **atmospheric pressure** (see Figure 6.2). As you go to higher altitudes, the atmospheric pressure is less because there are fewer particles in the air. The most common units used for gas pressure measurement are the atmosphere (atm) and millimeters of mercury (mmHg). On the TV weather report, you may hear or see the atmospheric pressure given in inches of mercury, or kilopascals in countries other than the United States. In a hospital, the unit torr may be used.

FIGURE 6.1 Gas particles move in straight lines within a container. The gas particles exert pressure when they collide with the walls of the container.
Q Why does heating the container increase the pressure of the gas within it?

Volume (*V*)

The volume of gas equals the size of the container in which the gas is placed. When you inflate a tire or a basketball, you are adding more gas particles. The increase in the

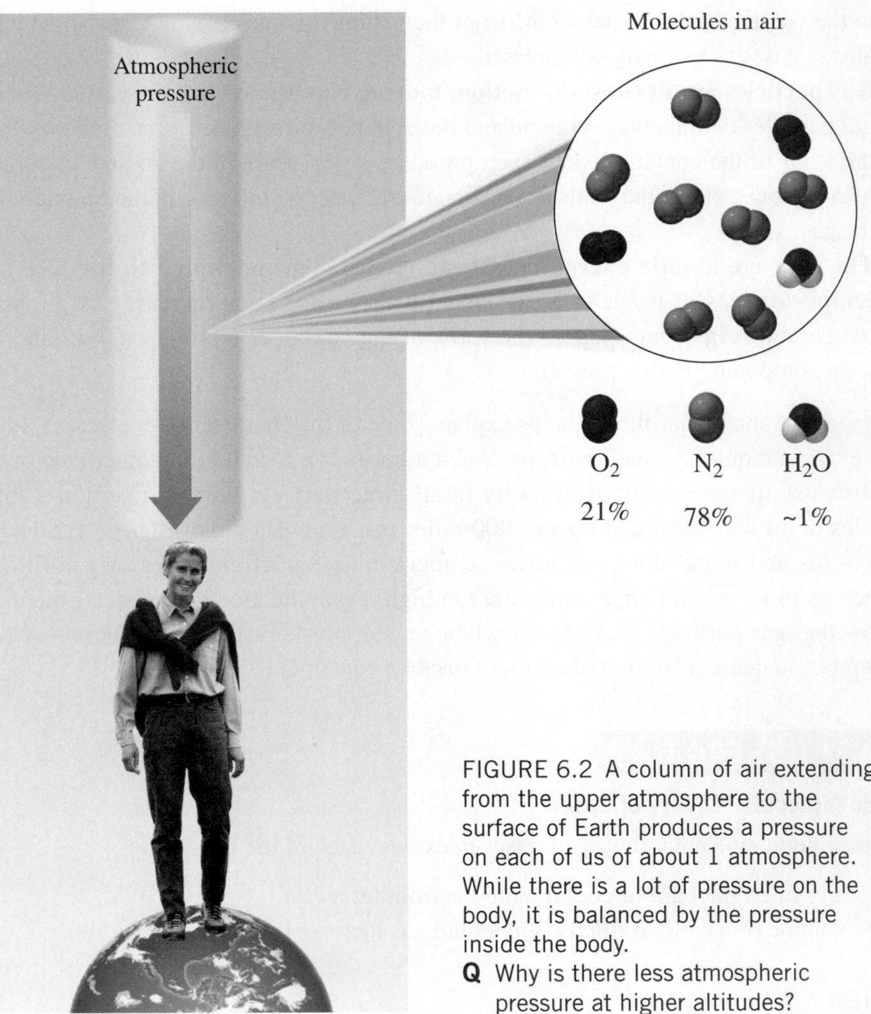

FIGURE 6.2 A column of air extending from the upper atmosphere to the surface of Earth produces a pressure on each of us of about 1 atmosphere. While there is a lot of pressure on the body, it is balanced by the pressure inside the body.

Q Why is there less atmospheric pressure at higher altitudes?

number of particles hitting the walls of the tire or basketball increases its volume. Sometimes, on a cold morning, a tire looks flat. The volume of the tire has decreased because a lower temperature decreases the speed of the molecules, which in turn reduces the force of their impacts on the walls of the tire. The most common units for volume measurement are liters (L) and milliliters (mL).

Temperature (*T*)

The temperature of a gas is related to the kinetic energy of its particles. For example, if we have a gas at 200 K in a rigid container and heat it to a temperature of 400 K, the gas particles will have twice the kinetic energy that they did at 200 K. This also means that the gas at 400 K exerts twice the pressure of the gas at 200 K. Although you measure gas temperature using a Celsius thermometer, all comparisons of gas behavior and all calculations related to temperature must use the Kelvin temperature scale. No one has quite created the conditions for absolute zero (0 K), but we predict that the particles will have zero kinetic energy and the gas will exert zero pressure at absolute zero.

Amount of Gas (*n*)

When you add air to a bicycle tire, you increase the amount of gas, which results in a higher pressure in the tire. Usually we measure the amount of gas by its mass in grams. In gas law calculations, we need to change the grams of gas to moles.

A summary of the four properties of a gas are given in Table 6.1.

TABLE 6.1 Properties That Describe a Gas

Property	Description	Units of Measurement
Pressure (P)	The force exerted by a gas against the walls of the container	atmosphere (atm); millimeters of mercury (mmHg); torr; pascal (Pa)
Volume (V)	The space occupied by a gas	liter (L); milliliter (mL)
Temperature (T)	The factor that determines kinetic energy and rate of motion of gas particles	degree Celsius (°C); Kelvin (K) *is required in calculations*
Amount (n)	The quantity of gas present in a container	grams (g); moles (n) *is required in calculations*

SAMPLE PROBLEM 6.1

Properties of Gases

Identify the property of a gas that is described by each of the following:

a. increases the kinetic energy of gas particles
b. the force of the gas particles hitting the walls of the container
c. the space that is occupied by a gas

SOLUTION

a. temperature **b.** pressure **c.** volume

STUDY CHECK 6.1

As more helium gas is added to a balloon, the number of grams of helium increases. What property of a gas is described?

Explore Your World

FORMING A GAS

Obtain baking soda and a jar or a plastic bottle. You will also need an elastic glove that fits over the mouth of the jar or a balloon that fits snugly over the top of the plastic bottle. Place a cup of vinegar in the jar or bottle. Sprinkle some baking soda into the fingertips of the glove or into the balloon. Carefully fit the glove or balloon over the top of the jar or bottle. Slowly lift the fingers of the glove or the balloon so that the baking soda falls into the vinegar. Watch what happens. Squeeze the glove or balloon.

QUESTIONS

1. Describe the properties of gas that you observe as the reaction takes place between vinegar and baking soda.
2. How do you know that a gas was formed?

QUESTIONS AND PROBLEMS

Properties of Gases

6.1 Use the kinetic molecular theory of gases to explain each of the following:
a. Gases move faster at higher temperatures.
b. Gases can be compressed much more than liquids or solids.

6.2 Use the kinetic molecular theory of gases to explain each of the following:
a. A container of nonstick cooking spray explodes when thrown into a fire.
b. The air in a hot-air balloon is heated to make the balloon rise.

6.3 Identify the property of a gas that is measured in each of the following:
a. 350 K
b. 125 mL
c. 2.00 g of O_2
d. 755 mmHg

6.4 Identify the property of a gas that is measured in each of the following:
a. 425 K
b. 1.0 atm
c. 10.0 L
d. 0.50 mole of He

6.2 Gas Pressure

LEARNING GOAL

Describe the units of measurement used for pressure, and change from one unit to another.

When billions and billions of gas particles hit against the walls of a container, they exert **pressure**, which is defined as a force acting on a certain area.

$$\text{Pressure } (P) = \frac{\text{force}}{\text{area}}$$

The atmospheric pressure can be measured using a barometer (see Figure 6.3). At a pressure of exactly 1 atmosphere (atm), the mercury column would be exactly 760 mm high. One **atmosphere (atm)** is defined as *exactly* 760 mmHg (millimeters of mercury).

TUTORIAL
Converting Between Units of Pressure

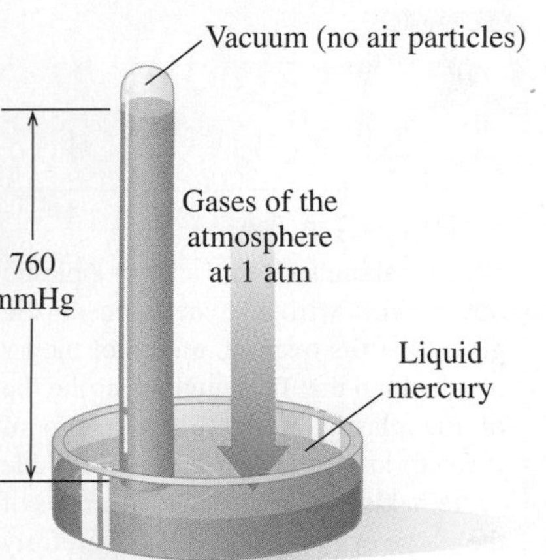

FIGURE 6.3 **A barometer:** The pressure exerted by the gases in the atmosphere is equal to the downward pressure of a mercury column in a closed glass tube. The height of the mercury column measured in mmHg is called atmospheric pressure.

Q Why does the height of the mercury column change from day to day?

One atmosphere is also 760 torr, a pressure unit named to honor Evangelista Torricelli, the inventor of the barometer. Because they are equal, units of torr and mmHg are used interchangeably.

$$1\ \text{atm} = 760\ \text{mmHg} = 760\ \text{torr (exact)}$$

$$1\ \text{mmHg} = 1\ \text{torr (exact)}$$

In SI units, pressure is measured in pascals (Pa); 1 atm is equal to 101 325 Pa. Because a pascal is a very small unit, it is likely that pressures would be reported in kilopascals.

$$1\ \text{atm} = 1.01325 \times 10^5\ \text{Pa} = 101.325\ \text{kPa}$$

The U.S. equivalent of 1 atm is 14.7 lb/in.^{2} (psi). When you use a pressure gauge to check the air pressure in the tires of a car, it may read 30–35 psi. This measurement is actually 30–35 psi above the pressure that the atmosphere exerts on the outside of the tire.

$$1\ \text{atm} = 14.7\ \text{lb/in.}^2$$

Table 6.2 summarizes the various units used in the measurement of pressure.

TABLE 6.2 Units for Measuring Pressure

Unit	Abbreviation	Unit Equivalent to 1 atm
Atmosphere	atm	1 atm
Millimeters of Hg	mmHg	760 mmHg (exact)
Torr	torr	760 torr (exact)
Inches of Hg	in. Hg	29.9 in. Hg
Pounds per square inch	lb/in.^{2} (psi)	14.7 lb/in.^{2}
Pascal	Pa	101 325 Pa
Kilopascal	kPa	101.325 kPa

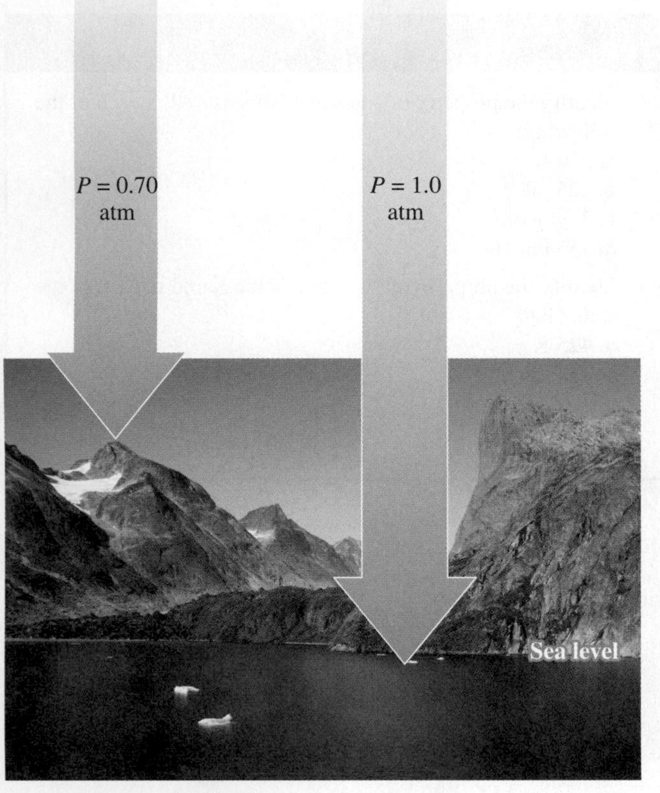

The atmospheric pressure decreases as the altitude increases.

If you have a barometer in your home, it probably gives pressure in inches of mercury. Atmospheric pressure changes with variations in weather and altitude. On a hot, sunny day, a column of air has more molecules, which increase the pressure on the mercury surface. The mercury column rises, indicating a higher atmospheric pressure. On a rainy day, the atmosphere exerts less pressure, which causes the mercury column to fall. In the weather report, this type of weather is called a low-pressure system. Above sea level, the density of the gases in the air decreases, which causes lower atmospheric pressures; the atmospheric pressure is greater than 760 mmHg at the Dead Sea because it is below sea level (see Table 6.3).

TABLE 6.3 Altitude and Atmospheric Pressure

Location	Altitude (km)	Atmospheric Pressure (mmHg)
Dead Sea	−0.40	800
Sea level	0	760
Los Angeles	0.09	750
Las Vegas	0.70	700
Denver	1.60	630
Mount Whitney	4.50	440
Mount Everest	8.90	250

Divers must be concerned about increasing pressures on their ears and lungs when they dive below the surface of the ocean. Because water is more dense than air, the pressure on a diver increases rapidly as the diver descends. At a depth of 33 ft below the surface of the ocean, an additional 1 atm of pressure is exerted by the water on a diver, for a total of 2 atm. At 100 ft below the surface, there is a total pressure of 4 atm on a diver. The regulator that a diver uses continuously adjusts the pressure of the breathing mixture to match the increase in pressure.

CASE STUDY
Scuba Diving and Blood Gases

CONCEPT CHECK 6.2

Units of Pressure

A sample of neon gas has a pressure of 0.50 atm. Calculate the pressure, in mmHg, of the neon.

ANSWER

The equality 1 atm = 760 mmHg can be written as two conversion factors:

$$\frac{760\ \text{mmHg}}{1\ \text{atm}} \quad \text{or} \quad \frac{1\ \text{atm}}{760\ \text{mmHg}}$$

Using the conversion factor that cancels atm and gives mmHg, we can set up the problem as

$$0.50\ \cancel{\text{atm}} \times \frac{760\ \text{mmHg}}{1\ \cancel{\text{atm}}} = 380\ \text{mmHg}$$

Chemistry Link to Health

MEASURING BLOOD PRESSURE

The measurement of your blood pressure is one of the important measurements a doctor or nurse makes during a physical examination. Acting like a pump, the heart contracts to create the pressure that pushes blood through the circulatory system. During contraction, the blood pressure is called *systolic* and is at its highest. When the heart muscles relax, the blood pressure is called *diastolic* and falls. The normal range for systolic pressure is 100–120 mmHg, and for diastolic pressure, 60–80 mmHg, usually expressed as a ratio such as 100/80. These values are somewhat higher in older people. When blood pressures are elevated, such as 140/90, there is a greater risk of stroke, heart attack, or kidney damage. Low blood pressure prevents the brain from receiving adequate oxygen, causing dizziness and fainting.

The blood pressures are measured by a sphygmomanometer, an instrument consisting of a stethoscope and an inflatable cuff connected to a tube of mercury called a manometer. After the cuff is wrapped around the upper arm, it is pumped up with air until it cuts off the flow of blood through the arm. With the stethoscope over the artery, the air is slowly released from the cuff. When the pressure equals the systolic pressure, blood starts to flow again, and the noise it makes is heard through the stethoscope. As air continues to be released, the cuff deflates until no sound is heard in the artery. That second pressure reading is noted as the diastolic pressure, the pressure when the heart is not contracting.

The use of digital blood pressure monitors is becoming more common. However, they have not been validated for use in all situations and can sometimes give inaccurate readings.

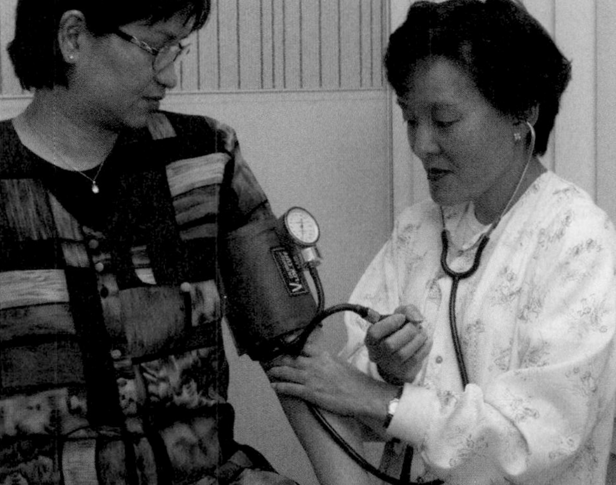

The measurement of blood pressure is part of a routine physical exam.

QUESTIONS AND PROBLEMS

Gas Pressure

6.5 What units are used to measure the pressure of a gas?

6.6 Which of the following statement(s) describes the pressure of a gas?
a. the force of the gas particles on the walls of the container
b. the number of gas particles in a container
c. the volume of the container
d. 3.00 atm
e. 750 torr

6.7 An oxygen tank contains oxygen (O_2) at a pressure of 2.00 atm. What is the pressure in the tank in terms of the following units?
a. torr
b. mmHg

6.8 On a climb up Mt. Whitney, the atmospheric pressure drops to 467 mmHg. What is the pressure in terms of the following units?
a. atm
b. torr

LEARNING GOAL

Use the pressure–volume relationship (Boyle's law) to determine the new pressure or volume of a certain amount of gas at a constant temperature.

TUTORIAL
Pressure and Volume

6.3 Pressure and Volume (Boyle's Law)

Imagine that you can see air particles hitting the walls inside a bicycle tire pump. What happens to the pressure inside the pump as we push down on the handle? As the volume decreases, there is a decrease in the surface area of the container. The air particles are crowded together, more collisions occur, and the pressure increases within the container.

When a change in one property (in this case, volume) causes a change in another property (in this case, pressure), the properties are related. If the changes occur in opposite directions, the properties have an *inverse relationship*. The inverse relationship between the pressure and volume of a gas is known as **Boyle's law**. The law states that the volume (V) of a sample of gas changes inversely with the pressure (P) of the gas as long as there is no change in the temperature (T) or amount of gas (n) (see Figure 6.4).

If the volume or pressure of a gas sample changes without any change in the temperature or in the amount of the gas, the new pressure and volume will give the same PV product as the initial pressure and volume. Then we can set the initial and final PV products equal to each other.

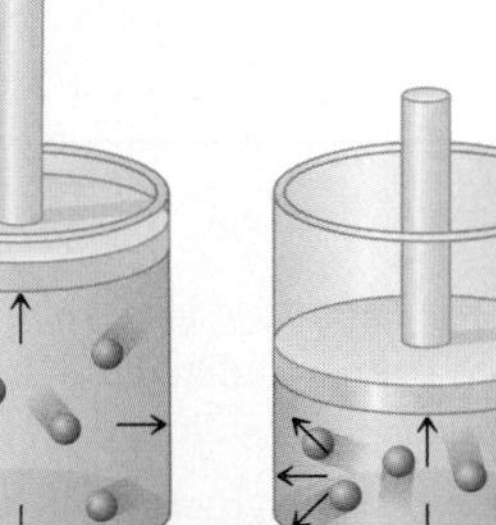

FIGURE 6.4 **Boyle's law:** As volume decreases, gas molecules become more crowded, which causes the pressure to increase. Pressure and volume are inversely related.
Q If the volume of a gas increases, what will happen to its pressure?

Boyle's Law

$P_1V_1 = P_2V_2$ No change in number of moles and temperature

CONCEPT CHECK 6.3

Boyle's Law

State and explain the reason for the change (*increases*, *decreases*) in the pressure of a gas that occurs in each of the following when n and T do not change:

	Pressure (*P*)	Volume (*V*)	Amount (*n*)	Temperature (*T*)
a.		decreases	constant	constant
b.		increases	constant	constant

ANSWER

a. When the volume of a gas decreases at constant n and T, the gas particles are closer together, which increases the number of collisions with the container walls. Therefore, pressure increases when volume decreases with no change in n and T.

b. When the volume of a gas increases at constant n and T, the gas particles move farther apart, which decreases the number of collisions with the container walls. Therefore, pressure decreases when the volume increases with no change in n and T.

	Pressure (*P*)	Volume (*V*)	Amount (*n*)	Temperature (*T*)
a.	increases	decreases	constant	constant
b.	decreases	increases	constant	constant

SAMPLE PROBLEM 6.2

Calculating Pressure When Volume Changes

A sample of hydrogen gas (H_2) has a volume of 5.0 L and a pressure of 1.0 atm. What is the new pressure, in atmospheres, if its volume decreases to 2.0 L with no change in temperature or amount of gas?

SOLUTION

Step 1 Organize the data in a table of initial and final conditions. In this problem, we want to know the final pressure (P_2) for the change in volume. In calculations with gas laws, it is helpful to organize the data in a table. The only properties listed in the table are those that change, which are the volume and pressure. Because we are given the initial and final volume of the gas, we know that the volume decreases. We predict that the pressure will increase. The properties that remain constant, in this case, are temperature (T) and the amount of gas (n).

Conditions 1	Conditions 2	Know	Predict
$V_1 = 5.0\ \text{L}$	$V_2 = 2.0\ \text{L}$	V decreases	
$P_1 = 1.0\ \text{atm}$	$P_2 = ?\ \text{atm}$		P increases

Factors that remain constant: T and n

Step 2 Rearrange the gas law equation to solve for the unknown quantity. For a PV relationship, we use Boyle's law and solve for P_2 by dividing both sides by V_2.

$$P_1V_1 = P_2V_2$$

$$\frac{P_1V_1}{V_2} = \frac{P_2\cancel{V_2}}{\cancel{V_2}}$$

$$P_2 = P_1 \times \frac{V_1}{V_2}$$

Step 3 Substitute values into the gas law equation and calculate. When we substitute in the values, we see that the ratio of the volumes is greater than 1, which increases the pressure as we predicted in Step 1. Note that the units of volume (L) cancel to give the final pressure in atmospheres.

$$P_2 = 1.0\ \text{atm} \times \frac{5.0\ \cancel{\text{L}}}{2.0\ \cancel{\text{L}}} = 2.5\ \text{atm}$$

Volume factor increases pressure

STUDY CHECK 6.2

A sample of helium gas has a volume of 150 mL at 750 torr. If the volume expands to 450 mL at constant temperature, what is the new pressure, in torr, when temperature and the amount of gas do not change?

Guide to Using the Gas Laws

1 Organize the data in a table of initial and final conditions.

2 Rearrange the gas law equation to solve for the unknown quantity.

3 Substitute values into the gas law equation and calculate.

A gauge indicates the pressure of a gas in a tank.

SAMPLE PROBLEM 6.3

Calculating Volume When Pressure Changes

The gauge on a 12-L tank of compressed oxygen reads 3800 mmHg. How many liters would this same gas occupy at a pressure of 0.75 atm at constant temperature and amount of gas?

SOLUTION

Step 1 **Organize the data in a table of initial and final conditions.** To match the units for the initial and final pressures, we can either convert atm to mmHg, or mmHg to atm.

$$0.75\ \cancel{\text{atm}} \times \frac{760\ \text{mmHg}}{1\ \cancel{\text{atm}}} = 570\ \text{mmHg}$$

or

$$3800\ \cancel{\text{mmHg}} \times \frac{1\ \text{atm}}{760\ \cancel{\text{mmHg}}} = 5.0\ \text{atm}$$

We can place the information using units of mmHg for pressure and volume in liters in a table. (We could have both pressures in atm as well.) We know that pressure decreases, and can predict that the volume increases.

Conditions 1	Conditions 2	Know	Predict
$P_1 = 3800$ mmHg (5.0 atm)	$P_2 = 570$ mmHg (0.75 atm)	P decreases	
$V_1 = 12$ L	$V_2 = ?$ L		V increases

Factors that remain constant: T and n

Step 2 **Rearrange the gas law equation to solve for the unknown quantity.** For a PV relationship, we use Boyle's law and solve for V_2 by dividing both sides by P_2. According to Boyle's law, a decrease in the pressure will cause an increase in the volume.

$$P_1V_1 = P_2V_2$$

$$\frac{P_1V_1}{P_2} = \frac{\cancel{P_2}V_2}{\cancel{P_2}}$$

$$V_2 = V_1 \times \frac{P_1}{P_2}$$

Step 3 **Substitute values into the gas law equation and calculate.** When we substitute in the values with pressures in units of mmHg or atm, the ratio of pressures (pressure factor) is greater than 1, which increases the volume as predicted in Step 1.

$$V_2 = 12\ \text{L} \times \frac{3800\ \cancel{\text{mmHg}}}{570\ \cancel{\text{mmHg}}} = 80.\ \text{L}$$

Pressure factor increases volume

or

$$V_2 = 12\ \text{L} \times \frac{5.0\ \cancel{\text{atm}}}{0.75\ \cancel{\text{atm}}} = 80.\ \text{L}$$

Pressure factor increases volume

STUDY CHECK 6.3

In an underground natural gas reserve, a bubble of methane gas, CH_4, has a volume of 45.0 mL at 1.60 atm. What volume, in milliliters, will the gas bubble occupy when it reaches the surface where the atmospheric pressure is 745 mmHg, if there is no change in the temperature or amount of gas?

Chemistry Link to Health

PRESSURE–VOLUME RELATIONSHIP IN BREATHING

The importance of Boyle's law becomes more apparent when you consider the mechanics of breathing. Our lungs are elastic, balloon-like structures contained within an airtight chamber called the thoracic cavity. The diaphragm, a muscle, forms the flexible floor of the cavity.

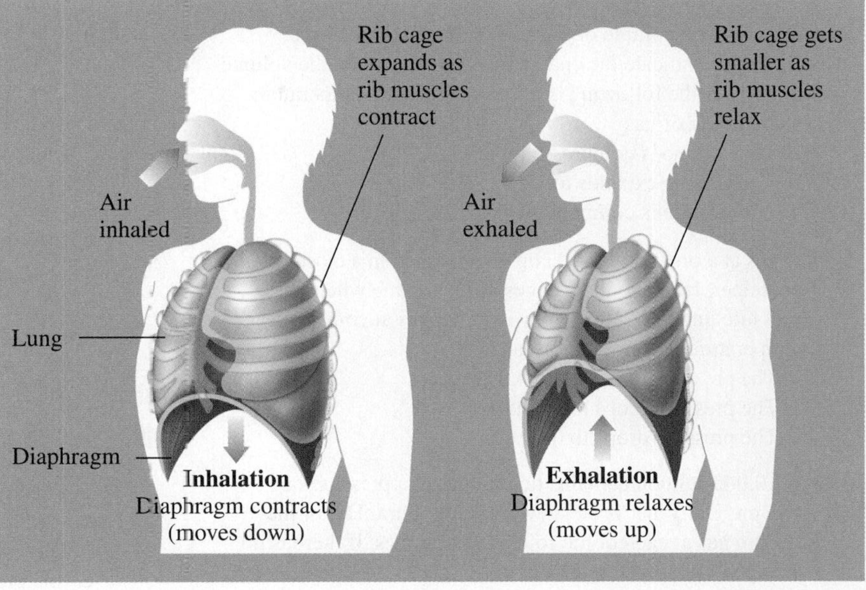

Inspiration

The process of taking a breath of air begins when the diaphragm contracts and the rib cage expands, causing an increase in the volume of the thoracic cavity. The elasticity of the lungs allows them to expand when the thoracic cavity expands. According to Boyle's law, the pressure inside the lungs decreases when their volume increases, causing the pressure inside the lungs to fall below the pressure of the atmosphere. This difference in pressures produces a *pressure gradient* between the lungs and the atmosphere. In a pressure gradient, molecules flow from an area of higher pressure to an area of lower pressure. During the inhalation phase of breathing, air flows into the lungs (*inspiration*), until the pressure within the lungs becomes equal to the pressure of the atmosphere.

Expiration

Expiration, or the exhalation phase of breathing, occurs when the diaphragm relaxes and moves back up into the thoracic cavity to its resting position. The volume of the thoracic cavity decreases, which squeezes the lungs and decreases their volume. Now the pressure in the lungs is higher than the pressure of the atmosphere, so air flows out of the lungs. Thus, breathing is a process in which pressure gradients are continuously created between the lungs and the environment because of the changes in the volume and pressure.

QUESTIONS AND PROBLEMS

Pressure and Volume (Boyle's Law)

6.9 Why do scuba divers need to exhale air when they ascend to the surface of the water?

6.10 Why does a sealed bag of chips expand when you take it to a higher altitude?

6.11 The air in a cylinder with a piston has a volume of 220 mL and a pressure of 650 mmHg.

a. To obtain a higher pressure inside the cylinder at constant temperature and amount of gas, would the cylinder change as shown in **A** or **B**? Explain your choice.

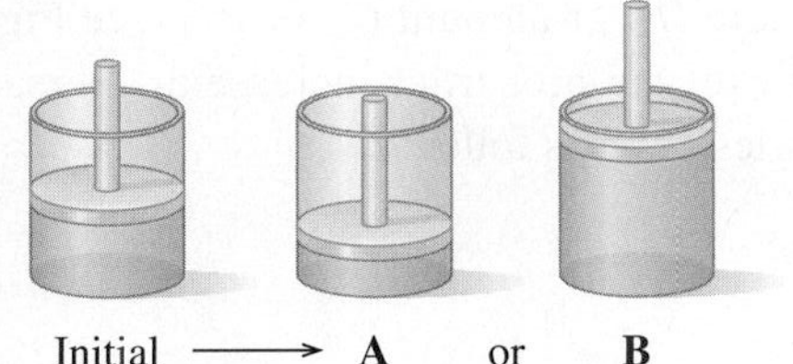

b. If the pressure inside the cylinder increases to 1.2 atm, what is the final volume, in milliliters, of the cylinder? Complete the following data table:

Property	Conditions 1	Conditions 2	Know	Predict
Pressure (P)				
Volume (V)				

6.12 A balloon is filled with helium gas. When the following changes are made at constant temperature, which of the following diagrams (**A**, **B**, or **C**) shows the new volume of the balloon?

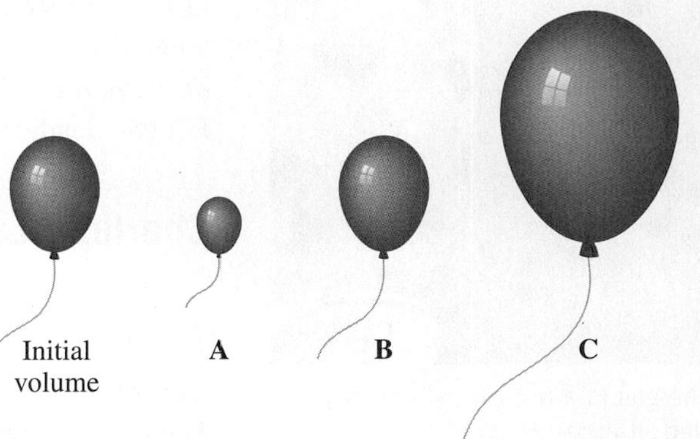

a. The balloon floats to a higher altitude where the outside pressure is lower.
b. The balloon is taken inside the house, but the atmospheric pressure remains the same.
c. The balloon is put in a hyperbaric chamber in which the pressure is increased.

6.13 A gas with a volume of 4.0 L is contained in a closed container. Indicate the changes in pressure when the volume undergoes the following changes at constant temperature and amount of gas:
a. The volume is compressed to 2.0 L.
b. The volume expands to 12 L.
c. The volume is compressed to 0.40 L.

6.14 A gas at a pressure of 2.0 atm is contained in a closed container. Indicate the changes in its volume when the pressure undergoes the following changes at constant temperature and amount of gas:
a. The pressure increases to 6.0 atm.
b. The pressure drops to 1.0 atm.
c. The pressure drops to 0.40 atm.

6.15 A 10.0-L balloon contains helium gas at a pressure of 655 mmHg. What is the new pressure, in mmHg, of the helium gas at each of the following volumes, if there is no change in temperature or amount of gas?
a. 20.0 L
b. 2.50 L
c. 1500. mL

6.16 The air in a 5.00-L tank has a pressure of 1.20 atm. What is the new pressure, in atm, when the air is placed in tanks that have the following volumes, if there is no change in temperature and amount of gas?
a. 1.00 L **b.** 2500. mL **c.** 750. mL

6.17 A sample of nitrogen (N_2) has a volume of 50.0 L at a pressure of 760. mmHg. What is the volume, in liters, of the gas at each of the following pressures, if there is no change in temperature and amount of gas?
a. 1500 mmHg **b.** 4.00 atm **c.** 0.500 atm

6.18 A sample of methane (CH_4) has a volume of 25 mL at a pressure of 0.80 atm. What is the volume, in milliliters, of the gas at each of the following pressures, if there is no change in temperature and amount of gas?
a. 0.40 atm **b.** 2.00 atm **c.** 2500 mmHg

6.19 Cyclopropane, C_3H_6, is a general anesthetic. A 5.0-L sample has a pressure of 5.0 atm. What is the volume, in liters, of the anesthetic given to a patient at a pressure of 1.0 atm with no change in temperature and amount of gas?

6.20 The volume of air in a person's lungs is 615 mL at a pressure of 760. mmHg. Inhalation occurs as the pressure in the lungs drops to 752 mmHg with no change in temperature and amount of gas. To what volume, in milliliters, did the lungs expand?

6.21 Use the words *inspiration* or *expiration* to describe the part of the breathing cycle that occurs as a result of each of the following:
a. The diaphragm contracts (flattens out).
b. The volume of the lungs decreases.
c. The pressure within the lungs is less than that of the atmosphere.

6.22 Use the words *inspiration* or *expiration* to describe the part of the breathing cycle that occurs as a result of each of the following:
a. The diaphragm relaxes, moving up into the thoracic cavity.
b. The volume of the lungs expands.
c. The pressure within the lungs is greater than that of the atmosphere.

LEARNING GOAL

Use the temperature–volume relationship (Charles's law) to determine the new temperature or volume of a certain amount of gas at a constant pressure.

As the gas in a hot-air balloon is heated, it expands.

6.4 Temperature and Volume (Charles's Law)

Suppose that you are going to take a ride in a hot-air balloon. The captain turns on a propane burner to heat the air inside the balloon. As the temperature rises, the air particles move faster and spread out, causing the volume of the balloon to increase. As the air is heated, it becomes less dense than the air outside, causing the balloon and its passengers to lift off. In 1787, Jacques Charles, a balloonist as well as a physicist, proposed that the volume of a gas is related to the temperature. This became **Charles's law**, which states that the volume (V) of a gas is directly related to the temperature (T) when there is no change in the pressure (P) or amount (n) of gas (see Figure 6.5). A *direct relationship* is one in which the related properties increase or decrease together. For two conditions, we can write Charles's law as follows:

Charles's Law

$$\frac{V_1}{T_1} = \frac{V_2}{T_2}$$ No change in number of moles and pressure

All temperatures used in gas law calculations must be converted to their corresponding Kelvin (K) temperatures.

To determine the effect of changing temperature on the volume of a gas, the pressure and the amount of gas are kept constant. If we increase the temperature of a gas sample, we know from the kinetic molecular theory that the motion (kinetic energy) of the gas particles will also increase. To keep the pressure constant, the volume of the container must increase (see Figure 6.5). If the temperature of the gas decreases, the volume of the container must decrease to maintain the same pressure when the amount of gas is constant.

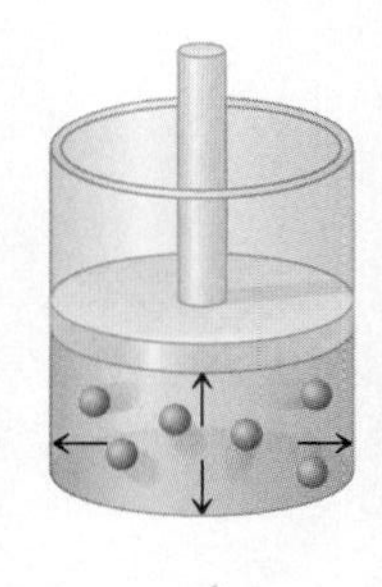
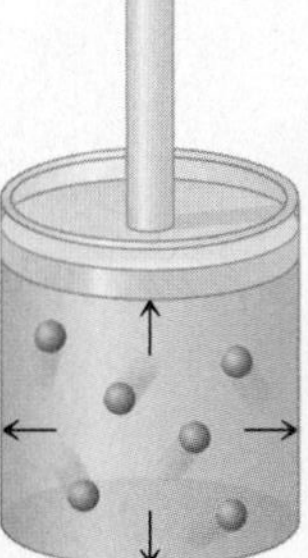

FIGURE 6.5 **Charles's law:** The Kelvin temperature of a gas is directly related to the volume of the gas when there is no change in the pressure and amount. When the temperature increases, making the molecules move faster, the volume must increase to maintain constant pressure.

Q If the temperature of a gas decreases at constant pressure and amount of gas, how will the volume change?

CONCEPT CHECK 6.4

Charles's Law

State and explain the reason for the change (*increases*, *decreases*) in the volume of a gas that occurs in each of the following when n and P do not change:

	Temperature (T)	Volume (V)	Pressure (P)	Amount (n)
a.	increases		constant	constant
b.	decreases		constant	constant

ANSWER

a. When the temperature of a gas increases at constant n and P, the gas particles move faster. To keep the pressure constant, the volume of the container must increase when temperature increases with no change in n and P.

b. When the temperature of a gas decreases at constant n and P, the gas particles move more slowly. To keep the pressure constant, the volume of the container must decrease when the temperature decreases with no change in n and P.

	Temperature (T)	Volume (V)	Pressure (P)	Amount (n)
a.	increases	increases	constant	constant
b.	decreases	decreases	constant	constant

SAMPLE PROBLEM 6.4

TUTORIAL
Temperature and Volume

Calculating Volume When Temperature Changes

A sample of argon gas has a volume of 5.40 L and a temperature of 15 °C. Find the new volume, in liters, of the gas if the temperature increases to 42 °C at constant pressure and amount of gas.

SOLUTION

Step 1 **Organize the data in a table of initial and final conditions.** The only properties listed in the table are those that change, the temperature and volume. When the temperatures are given in degrees Celsius, they must be changed to kelvins. Because we know the initial and final temperatures of the gas, we know that the temperature increases. Thus, we can predict that the volume increases.

$$T_1 = 15\ °\text{C} + 273 = 288\ \text{K}$$
$$T_2 = 42\ °\text{C} + 273 = 315\ \text{K}$$

Conditions 1	Conditions 2	Know	Predict
$T_1 = 288$ K	$T_2 = 315$ K	T increases	
$V_1 = 5.40$ L	$V_2 = ?$ L		V increases

Factors that remain constant: P and n

Step 2 Rearrange the gas law equation to solve for the unknown quantity. In this problem, we want to know the final volume (V_2) when the temperature increases. Using Charles's law, we solve for V_2 by multiplying both sides by T_2.

$$\frac{V_1}{T_1} = \frac{V_2}{T_2}$$

$$\frac{V_1}{T_1} \times T_2 = \frac{V_2}{\cancel{T_2}} \times \cancel{T_2}$$

$$V_2 = V_1 \times \frac{T_2}{T_1}$$

Step 3 Substitute the values into the gas law equation and calculate. From the table, we see that the temperature has increased. Because temperature is directly related to volume, the volume must increase. When we substitute in the values, we see that the ratio of the temperatures (temperature factor) is greater than 1, which increases the volume, as predicted in Step 1.

$$V_2 = 5.40\text{ L} \times \frac{315\text{ }\cancel{K}}{288\text{ }\cancel{K}} = 5.91\text{ L}$$

Temperature factor increases volume

STUDY CHECK 6.4

A mountain climber with a body temperature of 37 °C inhales 486 mL of air at a temperature of −8 °C. What volume, in milliliters, will the air occupy in the lungs, if the pressure and amount of gas do not change?

QUESTIONS AND PROBLEMS

Temperature and Volume (Charles's Law)

6.23 Select the diagram that shows the new volume of a balloon when the following changes are made at constant pressure and amount of gas:

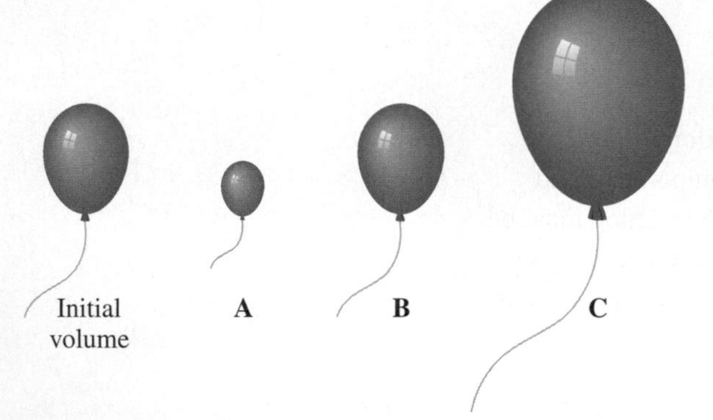

a. The temperature changes from 100 K to 300 K.
b. The balloon is placed in a freezer.
c. The balloon is first warmed and then returned to its starting temperature.

6.24 Indicate whether the final volume in each of the following is the same, larger, or smaller than the initial volume, if pressure and amount of gas do not change:

a. A volume of 505 mL of air on a cold winter day at −15 °C is breathed into the lungs, where body temperature is 37 °C.
b. The heater used to heat the air in a hot-air balloon is turned off.
c. A balloon filled with helium at the amusement park is left in a car on a hot day.

6.25 A sample of neon initially has a volume of 2.50 L at 15 °C. What is the new temperature, in degrees Celsius, when the volume of the sample is changed at constant pressure and amount of gas to each of the following?
a. 5.00 L **b.** 1250 mL **c.** 7.50 L **d.** 3550 mL

6.26 A gas has a volume of 4.00 L at 0 °C. What final temperature, in degrees Celsius, is needed to change the volume of the gas to each of the following, if n and P do not change?
a. 1.50 L **b.** 1200 mL **c.** 250 L **d.** 50.0 mL

6.27 A balloon contains 2500 mL of helium gas at 75 °C. What is the new volume, in milliliters, of the gas when the temperature changes to each of the following, if n and P do not change?
a. 55 °C **b.** 680. K **c.** −25 °C **d.** 240. K

6.28 An air bubble has a volume of 0.500 L at 18 °C. What is the volume, in liters, at each of the following temperatures, if n and P do not change?
a. 0 °C **b.** 425 K **c.** −12 °C **d.** 575 K

Chemistry Link to the Environment

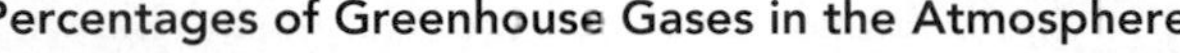

GREENHOUSE GASES

Percentages of Greenhouse Gases in the Atmosphere

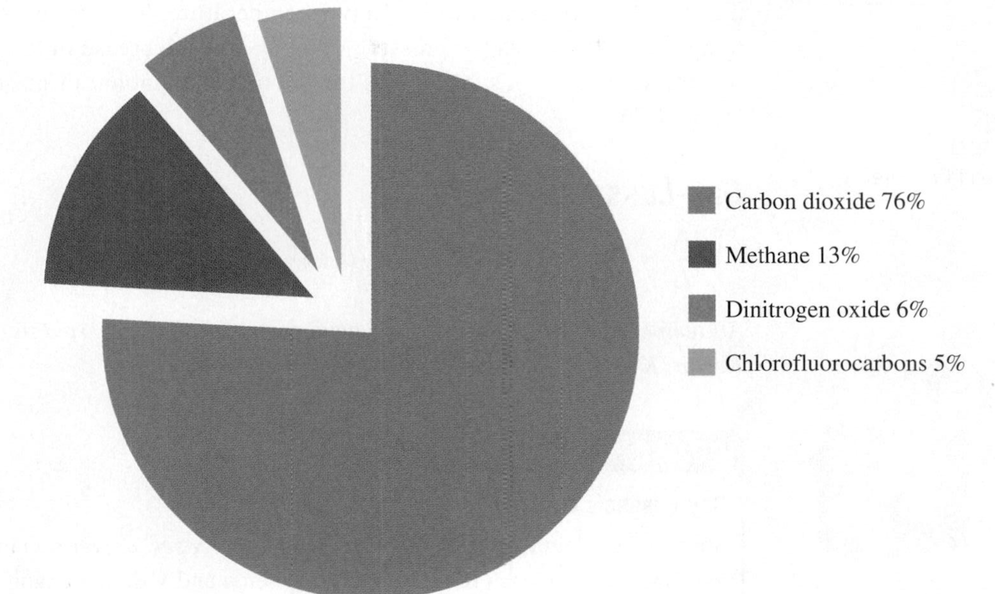

The term *greenhouse gases* was first used in early 1800s for the gases in the atmosphere that trap heat. Among the *greenhouse gases* are carbon dioxide (CO_2), methane (CH_4), dinitrogen oxide (N_2O), and chlorofluorocarbons (CFCs). The molecules of greenhouse gases consist of more than two atoms that vibrate when heat is absorbed. By contrast, oxygen and nitrogen are not greenhouse gases because the two atoms in their molecules are so tightly bonded, they do not absorb heat.

Greenhouse gases are beneficial in keeping the average surface temperature for Earth at 15 °C. Without greenhouse gases, it is estimated that the average surface temperature of Earth would be −18 °C. Most scientists say that the concentration of greenhouse gases in the atmosphere, and therefore the surface temperature of Earth, is increasing as a result of human activities. As we discussed in Chapter 2, the increase in atmospheric carbon dioxide is mostly a result of the burning of fossil fuels and wood.

Methane (CH_4) is a colorless, odorless gas that is released by livestock, rice farming, the decomposition of organic plant material in landfills, and the mining, drilling, and transport of coal and oil. The contribution from livestock comes from the breakdown of organic material in the digestive tracts of cows, sheep, and camels. The level of methane in the atmosphere has increased about 150% since industrialization. In one year, as much as 5×10^{11} kg of methane is added to the atmosphere. Livestock produces about 20% of the greenhouse gases. In one day, one cow emits about 200 g of methane. For a global population of 1.5 billion livestock, a total of 3×10^{8} kg of methane is produced every day. In the past few years, methane levels have stabilized due to improvements in the recovery of methane. Methane remains in the atmosphere for about ten years, but its molecular structure causes it to trap 20 times more heat than does carbon dioxide.

Dinitrogen oxide (N_2O), commonly called nitrous oxide, is a colorless greenhouse gas that has a sweet odor. Most people recognize it as an anesthetic used in dentistry called "laughing gas." Although some dinitrogen oxide is released naturally from soil bacteria, the major sources are agricultural and industrial processes. Atmospheric dinitrogen oxide has increased by about 15% since industrialization from the extensive use of fertilizers, sewage treatment plants, and car exhaust. Each year, 1×10^{10} kg of dinitrogen oxide is added to the atmosphere. Dinitrogen oxide released today will remain in the atmosphere for about 150–180 years, where it has a greenhouse effect that is 300 times greater than that of carbon dioxide.

Chlorofluorinated gases (CFCs) are synthetic compounds containing chlorine, fluorine, and carbon. Chlorofluorocarbons were used as propellants in aerosol cans and in refrigerants in refrigerators and air conditioners. During the 1970s, scientists determined that CFCs in the atmosphere were destroying the protective ozone layer. Since then, many countries banned the production and use of CFCs, and their levels in the atmosphere have declined slightly. Hydrofluorocarbons (HFCs), in which hydrogen atoms replace chlorine atoms, are now used as refrigerants. Although HFCs do not destroy the ozone layer, they are greenhouse gases because they trap heat in the atmosphere.

Based on current trends and climate models, scientists estimate that levels of atmospheric carbon dioxide will increase by about 2% each year up to 2025. As long as more heat is trapped by the greenhouse gases than is reflected back into space, average surface temperatures on Earth will continue to rise. Efforts are taking place around the world to slow or decrease the emissions of greenhouse gases into the atmosphere. It is anticipated that temperature will stabilize only when the amount of energy that reaches the surface of Earth is equal to the heat that is reflected back into space.

In 2007, former U.S. Vice-president Al Gore and the United Nations Panel on Climate Change were awarded the Nobel Peace Prize for increasing global awareness of the relationship between human activities and global warming.

LEARNING GOAL

Use the temperature–pressure relationship (Gay-Lussac's law) to determine the new temperature or pressure of a certain amount of gas at a constant volume.

TUTORIAL
Temperature and Pressure

FIGURE 6.6 **Gay-Lussac's law:** When the Kelvin temperature of a gas doubles at constant volume and amount of gas, the pressure also doubles.

Q How does a decrease in the temperature of a gas affect its pressure at constant volume and amount of gas?

6.5 Temperature and Pressure (Gay-Lussac's Law)

If we could observe the molecules of a gas as the temperature rises, we would notice that they move faster and hit the sides of the container more often and with greater force. If we maintain a constant volume and amount of gas, the pressure will increase. In the temperature–pressure relationship known as **Gay-Lussac's law**, the pressure of a gas is directly related to its Kelvin temperature. This means that an increase in temperature increases the pressure of a gas and a decrease in temperature decreases the pressure of the gas, as long as the volume and amount of gas do not change (see Figure 6.6).

Gay-Lussac's Law

$$\frac{P_1}{T_1} = \frac{P_2}{T_2}$$ No change in number of moles and volume

All temperatures used in gas law calculations must be converted to their corresponding Kelvin (K) temperatures.

CONCEPT CHECK 6.5

Gay-Lussac's Law

State and explain the reason for the change (*increases*, *decreases*) in the pressure of a gas that occurs in each of the following when n and V do not change:

	Temperature (T)	Pressure (P)	Volume (V)	Amount (n)
a.	increases		constant	constant
b.	decreases		constant	constant

ANSWER

a. When the temperature of a gas increases with no change in n and V, the particles of gas move faster. When the volume does not change, the gas particles collide more often with the container walls, and with more force, increasing the pressure.

b. When the temperature of a gas decreases, the particles of gas move slower. When the volume does not change, the gas particles do not collide as often with the container walls, and with less force, decreasing the pressure.

	Temperature (T)	Pressure (P)	Volume (V)	Amount (n)
a.	increases	increases	constant	constant
b.	decreases	decreases	constant	constant

SAMPLE PROBLEM 6.5

Calculating Pressure When Temperature Changes

Aerosol containers can be dangerous if they are heated, because they can explode. Suppose a container of hair spray with a pressure of 4.0 atm at a room temperature of 25 °C is thrown into a fire. If the temperature of the gas inside the aerosol can reaches 402 °C, what will be its pressure in atmospheres? The aerosol container may explode if the pressure inside exceeds 8.0 atm. Would you expect it to explode?

SOLUTION

Step 1 **Organize the data in a table of initial and final conditions.** The properties listed in the table, the pressure and temperature, are those that change. The temperatures given in degrees Celsius must be changed to kelvins. Because we know the initial and final temperatures of the gas, we know that the temperature increases. Thus, we can predict that the pressure increases.

$$T_1 = 25\ °C + 273 = 298\ K$$

$$T_2 = 402\ °C + 273 = 675\ K$$

Conditions 1	Conditions 2	Know	Predict
$P_1 = 4.0$ atm	$P_2 = ?$ atm		P increases
$T_1 = 298$ K	$T_2 = 675$ K	T increases	

Factors that remain constant: V and n

Step 2 **Rearrange the gas law equation to solve for the unknown quantity.** Using Gay-Lussac's law, we can solve for P_2 by multiplying both sides by T_2.

$$\frac{P_1}{T_1} = \frac{P_2}{T_2}$$

$$\frac{P_1}{T_1} \times T_2 = \frac{P_2}{\cancel{T_2}} \times \cancel{T_2}$$

$$P_2 = P_1 \times \frac{T_2}{T_1}$$

Step 3 **Substitute values into the gas law equation and calculate.** When we substitute in the known temperatures, we see that their ratio is greater than 1, which increases pressure as predicted in Step 1.

$$P_2 = 4.0\ \text{atm} \times \frac{675\ \cancel{K}}{298\ \cancel{K}} = 9.1\ \text{atm}$$

Temperature factor increases pressure

Because the calculated pressure of 9.1 atm of the gas exceeds the limit of 8.0 atm for the can, we would expect it to explode.

STUDY CHECK 6.5

In a storage area where the temperature has reached 55 °C, the pressure of oxygen gas in a 15.0-L steel cylinder is 965 torr. To what temperature, in degrees Celsius, would the gas have to be cooled to reduce the pressure to 850. torr?

NURSE ANESTHETIST

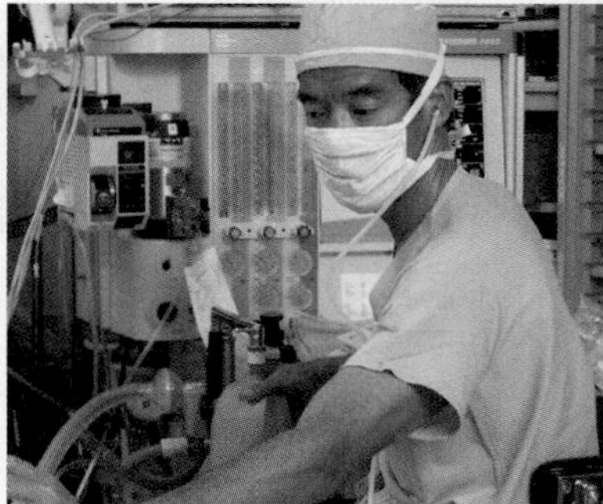

"During surgery, I work with the surgeon to provide a safe level of anesthetics that renders the patient free from pain," says Mark Noguchi, nurse anesthetist (CRNA), Kaiser Hospital. "We do spinal and epidural blocks as well as general anesthetics, which means the patient is totally asleep. We use a variety of pharmaceutical agents including halothane ($C_2HBrClF_3$) and bupivacaine ($C_{18}H_{28}N_2O$), as well as muscle relaxants such as midazolam ($C_{18}H_{13}ClFN_3$) to achieve the results we want for the surgical situation. We also assess the patient's overall hemodynamic status. If blood is lost, we replace components such as plasma, platelets, and coagulation factors. We also monitor the heart rate and run EKGs to determine cardiac function."

Vapor Pressure and Boiling Point

In Chapter 2, we learned that liquid molecules with sufficient kinetic energy can break away from the surface of the liquid as they become gas particles or vapor. In an open container, all the liquid will eventually evaporate. In a closed container, the vapor accumulates and creates pressure called *vapor pressure*. Each liquid exerts its own vapor pressure at a given temperature. As temperature increases, more vapor forms, and vapor pressure increases. Table 6.4 lists the vapor pressure of water at various temperatures.

A liquid reaches its boiling point when its vapor pressure becomes equal to the external pressure. As boiling occurs, bubbles of the gas form within the liquid and quickly rise to the surface. For example, at an atmospheric pressure of 760 mmHg, water will boil at 100 °C, the temperature at which its vapor pressure reaches 760 mmHg.

At higher altitudes, atmospheric pressures are lower and the boiling point of water is lower than 100 °C. For example, a typical atmospheric pressure in Denver is 634 mmHg, which means that a vapor pressure of 634 mmHg is required for water to boil in Denver. From Table 6.4, we see that water has a vapor pressure of 634 mmHg at 95 °C. Therefore, water boils at 95 °C in Denver. A boiling temperature below 100 °C means

TABLE 6.4 Vapor Pressure of Water

Temperature (°C)	Vapor Pressure (mmHg)
0	5
10	9
20	18
30	32
37*	47
40	55
50	93
60	149
70	234
80	355
90	528
95	634
100	760

*Body temperature.

TUTORIAL
Vapor Pressure and Boiling Point

An autoclave used to sterilize equipment attains a temperature higher than 100 °C.

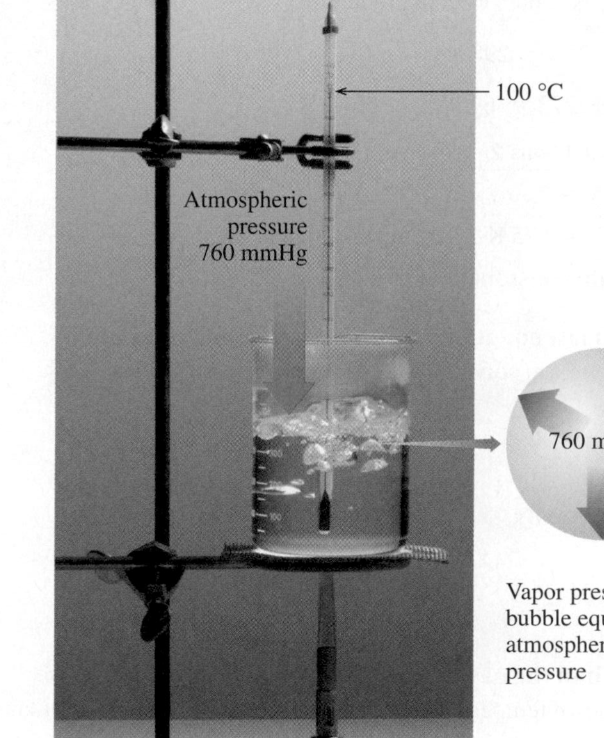

Water boils when its vapor pressure is equal to the pressure of the atmosphere.

TABLE 6.5 Pressure and the Boiling Point of Water

Pressure (mmHg)	Boiling Point (°C)
270	70
467	87
634	95
752	99
760	100
800	100.4
1075	110
1520 (2 atm)	120
3800 (5 atm)	160
7600 (10 atm)	200

that it takes longer to cook food. People who live at high altitudes often use pressure cookers to obtain higher temperatures when preparing food. In a pressure cooker, water is heated in a closed container so that pressures above 1 atm are obtained, which raises the boiling temperature of water above 100 °C.

Laboratories and hospitals use devices called autoclaves to sterilize laboratory and surgical equipment. An autoclave, like a pressure cooker, is a closed container that increases the total pressure. Then water boils at higher temperatures, which is more effective at killing bacteria. Table 6.5 shows how the boiling point of water increases as pressure increases.

QUESTIONS AND PROBLEMS

Temperature and Pressure (Gay-Lussac's Law)

6.29 Calculate the new temperature, in degrees Celsius, for each of the following with n and V constant:

a. A sample of xenon at 25 °C and 740 mmHg is cooled to give a pressure of 620 mmHg.

b. A tank of argon gas with a pressure of 0.950 atm at −18 °C is heated to give a pressure of 1250 torr.

6.30 Calculate the new temperature, in degrees Celsius, for each of the following with n and V constant:

a. A tank of helium gas with a pressure of 250 torr at 0 °C is heated to give a pressure of 1500 torr.

b. A sample of air at 40. °C and 740 mmHg is cooled to give a pressure of 680 mmHg.

6.31 Solve for the new pressure, in torr, for each of the following, if n and V are constant:

a. A gas with an initial pressure of 1200 torr at 155 °C is cooled to 0 °C.

b. A gas in an aerosol can with an initial pressure of 1.40 atm at 12 °C is heated to 35 °C.

6.32 Solve for the new pressure, in atm, for each of the following, if n and V are constant:

a. A gas with an initial pressure of 1.20 atm at 75 °C is cooled to −32 °C.

b. A sample of N_2 with an initial pressure of 780. mmHg at −75 °C is heated to 28 °C.

6.33 Explain each of the following observations:
a. Water boils at 87 °C on the top of Mt. Whitney.
b. Food cooks more quickly in a pressure cooker than in an open pan.

6.34 Explain each of the following observations:
a. Boiling water at sea level is hotter than boiling water in the mountains.
b. Water used to sterilize surgical equipment is heated to 120 °C at 2.0 atm in an autoclave.

6.6 The Combined Gas Law

LEARNING GOAL

Use the combined gas law to find the new pressure, volume, or temperature of a gas when changes in two of these properties are given.

MC TUTORIAL The Combined Gas Law

All of the pressure–volume–temperature relationships for gases that we have studied may be combined into a single relationship called the **combined gas law**. This expression is useful for studying the effect of changes in two of these variables on the third as long as the amount of gas (number of moles) remains constant.

Combined Gas Law

$$\frac{P_1V_1}{T_1} = \frac{P_2V_2}{T_2} \quad \text{No change in moles of gas}$$

By using the combined gas law, we can derive any of the gas laws by omitting those properties that do not change as seen in Table 6.6.

TABLE 6.6 Summary of Gas Laws

Combined Gas Law	Properties Held Constant	Relationship	Name of Gas Law
$\frac{P_1V_1}{\cancel{T_1}} = \frac{P_2V_2}{\cancel{T_2}}$	T, n	$P_1V_1 = P_2V_2$	Boyle's law
$\frac{\cancel{P_1}V_1}{T_1} = \frac{\cancel{P_2}V_2}{T_2}$	P, n	$\frac{V_1}{T_1} = \frac{V_2}{T_2}$	Charles's law
$\frac{P_1\cancel{V_1}}{T_1} = \frac{P_2\cancel{V_2}}{T_2}$	V, n	$\frac{P_1}{T_1} = \frac{P_2}{T_2}$	Gay-Lussac's law

CONCEPT CHECK 6.6

The Combined Gas Law

State and explain the reason for the change (*increases*, *decreases*, *no change*) in the pressure or volume of a gas that occurs for the following when n does not change:

	Pressure (P)	Volume (V)	Temperature (K)	Amount (n)
a.		twice as large	half the Kelvin temperature	constant
b.	twice as large		twice as large	constant

ANSWER

a. Pressure decreases by one-half when the volume (at constant n) doubles. If the temperature in Kelvin is halved, the pressure is also halved. The changes in both V and T decrease the pressure to one-fourth its initial value.

b. No change. When the Kelvin temperature of a gas (at constant n) is doubled, the volume is doubled. But when the pressure is twice as much, the volume must decrease to one-half. The changes offset each other, and no change occurs in the volume.

	Pressure (P)	Volume (V)	Temperature (K)	Amount (n)
a.	one-fourth as large	twice as large	half the Kelvin temperature	constant
b.	twice as large	no change	twice as large	constant

Under water, the pressure on a diver is greater than the atmospheric pressure.

SAMPLE PROBLEM 6.6

Using the Combined Gas Law

A 25.0-mL bubble is released from a diver's air tank at a pressure of 4.00 atm and a temperature of 11 °C. What is the volume, in milliliters, of the bubble when it reaches the ocean surface, where the pressure is 1.00 atm and the temperature is 18 °C (Assume the amount of gas in the bubble remains the same)?

SOLUTION

Step 1 **Organize the data in a table of initial and final conditions.** The properties listed in the table, the pressure, volume, and temperature, are those that change. We do not list the amount of gas because it remains constant. The temperatures in degrees Celsius must be changed to kelvins.

$$T_1 = 11\ °C + 273 = 284\ K$$
$$T_2 = 18\ °C + 273 = 291\ K$$

Conditions 1	Conditions 2
$P_1 = 4.00$ atm	$P_2 = 1.00$ atm
$V_1 = 25.0$ mL	$V_2 = ?$ mL
$T_1 = 284$ K	$T_2 = 291$ K

Factor that remains constant: n

Step 2 **Rearrange the gas law equation to solve for the unknown quantity.** For changes in two conditions, we rearrange the combined gas law to solve for V_2.

$$\frac{P_1V_1}{T_1} = \frac{P_2V_2}{T_2}$$

$$\frac{P_1V_1}{T_1} \times \frac{T_2}{P_2} = \frac{\cancel{P_2}V_2 \times \cancel{T_2}}{\cancel{T_2} \times \cancel{P_2}}$$

$$V_2 = V_1 \times \frac{P_1}{P_2} \times \frac{T_2}{T_1}$$

Step 3 **Substitute values into the gas law equation and calculate.** From the data table, we determine that both the pressure decrease and the temperature increase will increase the volume.

$$V_2 = 25.0\ \text{mL} \times \frac{4.00\ \cancel{\text{atm}}}{1.00\ \cancel{\text{atm}}} \times \frac{291\ \cancel{K}}{284\ \cancel{K}} = 102\ \text{mL}$$

Pressure factor increases volume; Temperature factor increases volume

However, when the unknown value is decreased by one change but increased by the second change, it is not possible to predict the overall change.

STUDY CHECK 6.6

A weather balloon is filled with 15.0 L of helium at a temperature of 25 °C at a pressure of 685 mmHg. What is the pressure, in mmHg, of the helium in the balloon in the upper atmosphere when the temperature is −35 °C and the volume becomes 34.0 L, if the amount of He is constant?

QUESTIONS AND PROBLEMS

The Combined Gas Law

6.35 A sample of helium gas has a volume of 6.50 L at a pressure of 845 mmHg and a temperature of 25 °C. What is the pressure of the gas, in atmospheres, when the volume and temperature of the gas sample are changed to the following, if the amount of gas is constant?
a. 1850 mL and 325 K **b.** 2.25 L and 12 °C
c. 12.8 L and 47 °C

6.36 A sample of argon gas has a volume of 735 mL at a pressure of 1.20 atm and a temperature of 112 °C. What is the volume of the gas, in milliliters, when the pressure and temperature of the gas sample are changed to the following, if the amount of gas remains the same?
a. 658 mmHg and 281 K **b.** 0.55 atm and 75 °C
c. 15.4 atm and −15 °C

6.37 A 100.0-mL bubble of hot gases at 225 °C and 1.80 atm is emitted from an active volcano. What is the new volume, in milliliters, of the bubble outside the volcano where the temperature is −25 °C and the pressure is 0.80 atm, if the amount of gas remains the same?

6.38 A scuba diver 60 ft below the ocean surface inhales 50.0 mL of compressed air from a scuba tank at a pressure of 3.00 atm and a temperature of 8 °C. What is the pressure of the air, in atm, in the lungs when the gas expands to 150.0 mL at a body temperature of 37 °C, and the amount of gas remains constant?

6.7 Volume and Moles (Avogadro's Law)

LEARNING GOAL

Use Avogadro's law to describe the relationship between the amount of a gas and its volume, and use this relationship in calculations.

TUTORIAL
Volume and Moles

In our study of the gas laws, we have looked at changes in properties for a specified amount (n) of gas. Now we will consider how the properties of a gas change when there is a change in the number of moles or grams.

When you blow up a balloon, its volume increases because you add more air molecules. If the balloon has a hole in it, air leaks out, causing its volume to decrease. In 1811, Amedeo Avogadro formulated **Avogadro's law**, which states that the volume of a gas is directly related to the number of moles of a gas when temperature and pressure do not change. For example, if the number of moles of a gas is doubled, then the volume will also double as long as we do not change the pressure or the temperature (see Figure 6.7). At constant pressure and temperature, we can write Avogadro's law:

Avogadro's Law

$$\frac{V_1}{n_1} = \frac{V_2}{n_2} \quad \text{No change in pressure and temperature}$$

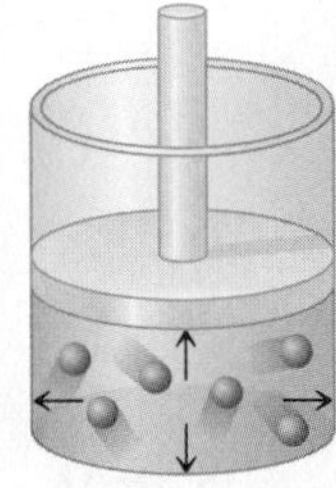

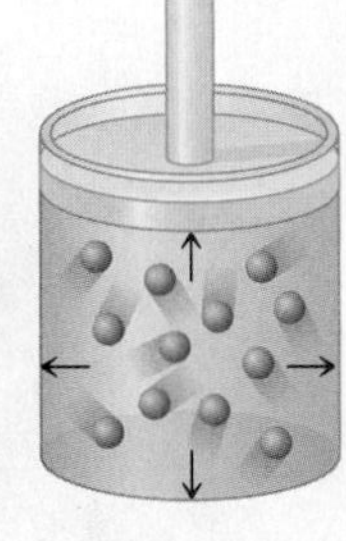

FIGURE 6.7 **Avogadro's law:** The volume of a gas is directly related to the number of moles of the gas. If the number of moles is doubled, the volume must double at constant temperature and pressure.
Q If a balloon has a leak, what happens to its volume?

SAMPLE PROBLEM 6.7

Calculating Volume for a Change in Moles

A weather balloon with a volume of 44 L is filled with 2.0 moles of helium. To what volume, in liters, will the balloon expand if 3.0 moles of helium are added, to give a total of 5.0 moles of helium, if pressure and temperature do not change?

SOLUTION

Step 1 **Organize the data in a table of initial and final conditions.** A data table can be set up by listing the initial and final conditions for those properties that change. Because there is an increase in the number of moles of gas, we can predict that the volume increases.

Conditions 1	Conditions 2	Know	Predict
$V_1 = 44$ L	$V_2 = ?$ L		V increases
$n_1 = 2.0$ moles	$n_2 = 5.0$ moles	n increases	

Factors that remain constant: P and T

Step 2 Rearrange the gas law equation to solve for the unknown quantity. Using Avogadro's law, we can solve for V_2.

$$\frac{V_1}{n_1} = \frac{V_2}{n_2}$$

$$n_2 \times \frac{V_1}{n_1} = \frac{V_2}{\cancel{n_2}} \times \cancel{n_2}$$

$$V_2 = V_1 \times \frac{n_2}{n_1}$$

Step 3 Substitute values into the gas law equation and calculate. When we substitute in the moles of gas, we see that their ratio is greater than 1, which increases volume as predicted in Step 1.

$$V_2 = 44\ \text{L} \times \frac{5.0\ \cancel{\text{moles}}}{2.0\ \cancel{\text{moles}}} = 110\ \text{L}$$

Mole factor increases volume

STUDY CHECK 6.7

A sample containing 8.00 g of oxygen gas has a volume of 5.00 L. What is the volume, in liters, after 4.00 g of oxygen gas is added to the 8.00 g of oxygen in the balloon, if the temperature and pressure do not change?

STP and Molar Volume

Using Avogadro's law, we can say that any two gases will have equal volumes if they contain the same number of moles of gas at the same temperature and pressure. To help us make comparisons between different gases, arbitrary conditions called *standard temperature* (273 K) and *standard pressure* (1 atm) together abbreviated **STP**, were selected by scientists:

STP Conditions

Standard temperature is *exactly* 0 °C (273 K).

Standard pressure is *exactly* 1 atm (760 mmHg).

The molar volume of a gas is about the same as the volume of three basketballs.

At STP, one mole of any gas occupies a volume of 22.4 L, which is about the same as the volume of three basketballs. This volume, 22.4 L, of any gas at STP is called its **molar volume** (see Figure 6.8).

When a gas is at STP conditions (0 °C and 1 atm), its molar volume can be used to write conversion factors between the number of moles of gas and its volume, in liters.

Molar Volume Conversion Factors

1 mole of gas at STP = 22.4 L of gas

$$\frac{1\ \text{mole gas (STP)}}{22.4\ \text{L gas}} \quad \text{and} \quad \frac{22.4\ \text{L gas}}{1\ \text{mole gas (STP)}}$$

The molar volume conversion factors are used to convert between the number of moles of gas and its volume.

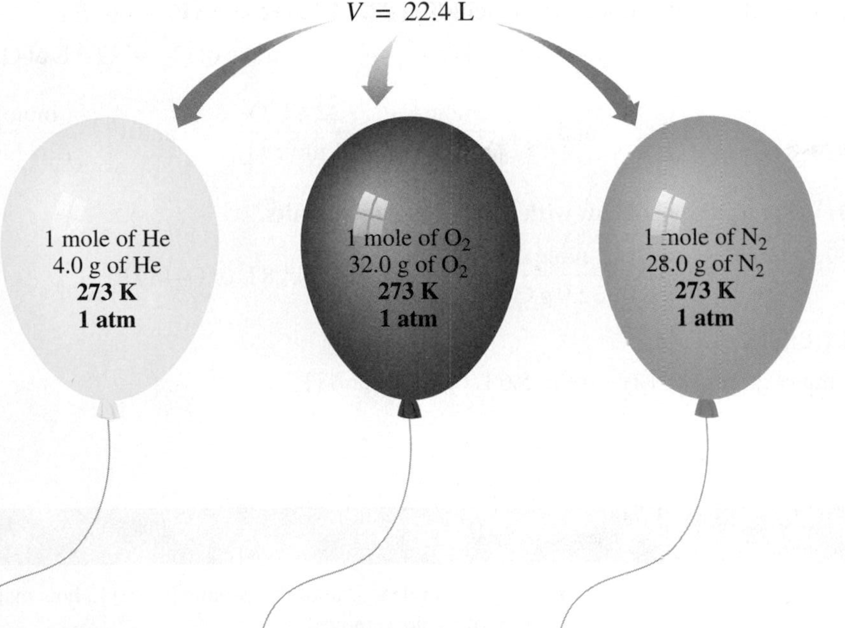

FIGURE 6.8 Avogadro's law indicates that 1 mole of any gas at STP has a volume of 22.4 L.

Q What volume of gas, in liters, is occupied by 16.0 g of methane gas, CH_4, at STP?

CONCEPT CHECK 6.7

Molar Volume

Write the equality and conversion factors for molar volume of helium at STP.

ANSWER

The equality for the molar volume of helium at STP is

1 mole of He = 22.4 L of He

Because an equality has two conversion factors, we write the factors of molar volume for helium as

$$\frac{22.4 \text{ L He}}{1 \text{ mole He (STP)}} \quad \text{and} \quad \frac{1 \text{ mole He (STP)}}{22.4 \text{ L He}}$$

Guide to Using Molar Volume

1. State the given and needed quantities.
2. Write a plan.
3. Write conversion factors including 22.4 L/mole at STP.
4. Set up the problem with factors to cancel units.

SAMPLE PROBLEM 6.8

Using Molar Volume to Find Volume at STP

What is the volume, in liters, of 64.0 g of O_2 gas at STP?

SOLUTION

Step 1 State the given and needed quantities.

Given 64.0 g of $O_2(g)$ at STP **Need** volume in liters (L)

Step 2 Write a plan. The grams of O_2 are converted to moles using molar mass. Then a molar volume conversion factor is used to convert the number of moles to volume (L).

grams of O_2 → Molar mass → moles of O_2 → Molar volume → liters of O_2

Step 3 Write conversion factors including 22.4 L/mole at STP.

$1 \text{ mole of } O_2 = 32.0 \text{ g of } O_2$ $\quad$ $1 \text{ mole of } O_2 = 22.4 \text{ L of } O_2$

$$\frac{32.0 \text{ g } O_2}{1 \text{ mole } O_2} \text{ and } \frac{1 \text{ mole } O_2}{32.0 \text{ g } O_2} \qquad \frac{22.4 \text{ L } O_2}{1 \text{ mole } O_2} \text{ and } \frac{1 \text{ mole } O_2}{22.4 \text{ L } O_2}$$

Step 4 Set up the problem with factors to cancel units.

$$64.0 \cancel{\text{g } O_2} \times \frac{1 \cancel{\text{mole } O_2}}{32.0 \cancel{\text{g } O_2}} \times \frac{22.4 \text{ L } O_2}{1 \cancel{\text{mole } O_2}} = 44.8 \text{ L of } O_2 \text{ (STP)}$$

STUDY CHECK 6.8

How many grams of $N_2(g)$ are in 5.6 L of $N_2(g)$ at STP?

QUESTIONS AND PROBLEMS

Volume and Moles (Avogadro's Law)

6.39 What happens to the volume of a bicycle tire or a basketball when you use an air pump to add air?

6.40 Sometimes when you blow up a balloon and release it, it flies around the room. What is happening to the air that was in the balloon and its volume?

6.41 A sample containing 1.50 moles of neon gas has a volume of 8.00 L. What is the new volume of the gas, in liters, when each of the following changes occurs in the quantity of the gas at constant pressure and temperature?
a. A leak allows one half of the neon atoms to escape.
b. A sample of 25.0 g of neon is added to the 1.50 moles of neon gas in the container.
c. A sample of 3.50 moles of O_2 is added to the 1.50 moles of neon gas in the container.

6.42 A sample containing 4.80 g of O_2 gas has a volume of 15.0 L at constant pressure and temperature.
a. What is the new volume, in liters, after 0.500 mole of O_2 gas is added to the initial sample of 4.80 g of O_2?
b. If oxygen is released until the volume is 10.0 L, how many moles of O_2 are removed?
c. What is the volume, in liters, after 4.00 g of He is added to the O_2 gas in the container?

6.43 Use the molar volume to solve each of the following at STP:
a. the number of moles of O_2 in 44.8 L of O_2 gas
b. the number of moles of CO_2 in 4.00 L of CO_2 gas

6.44 Use molar volume to solve each of the following at STP:
a. the volume, in liters, occupied by 2.50 moles of N_2
b. the volume, in milliliters, occupied by 0.420 mole of He

6.45 Use the molar volume to solve each of the following at STP:
a. the volume, in liters, of 6.40 g of O_2
b. the volume, in milliliters, occupied by 50.0 g of neon

6.46 Use molar volume to solve each of the following at STP:
a. the number of grams of neon contained in 11.2 L of Ne gas
b. the number of grams of H_2 in 1620 mL of H_2 gas

LEARNING GOAL

Use partial pressures to calculate the total pressure of a mixture of gases.

TUTORIAL
Mixtures of Gases

6.8 Partial Pressures (Dalton's Law)

Many gas samples are a mixture of gases. For example, the air you breathe is a mixture of mostly oxygen and nitrogen gases. In gas mixtures, scientists observed that all gas particles behave in the same way. Therefore, the total pressure of the gases in a mixture is a result of the collisions of the gas particles regardless of what type of gas they are.

In a gas mixture, each gas exerts its **partial pressure**, which is the pressure it would exert if it were the only gas in the container. **Dalton's law** states that the total pressure of a gas mixture is the sum of the partial pressures of the gases in the mixture.

Dalton's Law

$$P_{total} = P_1 + P_2 + P_3 + \cdots$$

Total pressure of a gas mixture = Sum of the partial pressures of the gases in the mixture

Suppose we have two separate tanks, one filled with helium at 2.0 atm and the other filled with argon at 4.0 atm. When the gases are combined in a single tank at the same volume and temperature, the number of gas molecules, not the type of gas, determines the pressure in a container. Then the pressure of the gases in the gas mixture would be 6.0 atm, which is the sum of their individual or partial pressures.

$P_{He} = 2.0$ atm + $P_{Ar} = 4.0$ atm

$$P_{total} = P_{He} + P_{Ar} = 2.0 \text{ atm} + 4.0 \text{ atm} = 6.0 \text{ atm}$$

The total pressure of two gases is the sum of their partial pressures.

CONCEPT CHECK 6.8

Pressure of a Gas Mixture

A scuba tank is filled with Trimix, a breathing gas mixture for deep scuba diving. The tank contains oxygen with a partial pressure of 20. atm, nitrogen with a partial pressure of 40. atm, and helium with a partial pressure of 140. atm. What is the total pressure of the breathing mixture, in atmospheres?

ANSWER

Using Dalton's law of partial pressures, we add together the partial pressures of oxygen, nitrogen, and helium present in the mixture.

$$P_{total} = P_{oxygen} + P_{nitrogen} + P_{helium}$$

$$P_{total} = 20.\text{ atm} + 40.\text{ atm} + 140.\text{ atm}$$

$$= 200.\text{ atm}$$

Therefore, when oxygen, nitrogen, and helium are placed in the same container, the sum of their partial pressures is the total pressure of the mixture, which is 200. atm.

Air Is a Gas Mixture

The air you breathe is a mixture of gases. What we call the atmospheric pressure is actually the sum of the partial pressures of the gases in the air. Table 6.7 lists partial pressures for the gases in air on a typical day.

TABLE 6.7 Typical Composition of Air

Gas	Partial Pressure (mmHg)	Percentage (%)
Nitrogen, N_2	594	78.2
Oxygen, O_2	160.	21.0
Carbon dioxide, CO_2; Argon, Ar; Water, H_2O	6	0.8
Total air	760.	100

Guide to Solving for Partial Pressure

1. Write the equation for the sum of the partial pressures.
2. Solve for the unknown pressure.
3. Substitute known pressures and calculate the unknown partial pressure.

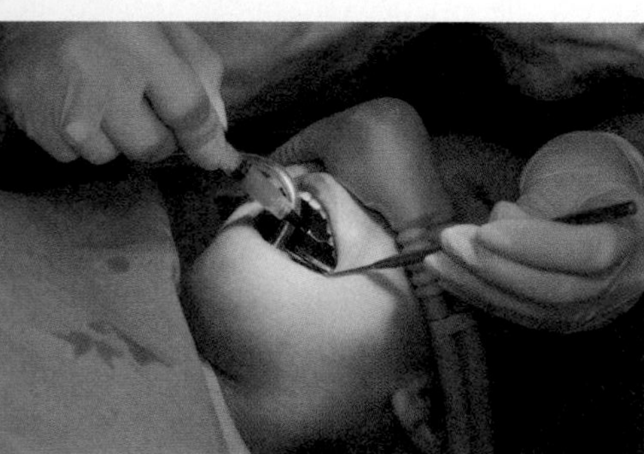

Dinitrogen oxide is used as an anesthetic in dentistry.

SAMPLE PROBLEM 6.9

Partial Pressure of a Gas in a Mixture

A Heliox breathing mixture of oxygen and helium is prepared for a scuba diver who is going to descend 200 ft below the ocean surface. At that depth, the diver breathes a gas mixture that has a total pressure of 7.00 atm. If the partial pressure of the oxygen in the tank at that depth is 1140 mmHg, what is the partial pressure, in atm, of the helium in the breathing mixture?

SOLUTION

Step 1 Write the equation for the sum of the partial pressures. From Dalton's law of partial pressures, we know that the total pressure is equal to the sum of the partial pressures.

$$P_{\text{total}} = P_{O_2} + P_{\text{He}}$$

Step 2 Solve for the unknown pressure.

$$P_{\text{total}} = P_{O_2} + P_{\text{He}}$$

$$P_{\text{He}} = P_{\text{total}} - P_{O_2}$$

Convert units to match.

$$P_{O_2} = 1140\ \cancel{\text{mmHg}} \times \frac{1\ \text{atm}}{760\ \cancel{\text{mmHg}}} = 1.50\ \text{atm}$$

Step 3 Substitute known pressures and calculate the unknown partial pressure.

$$P_{\text{He}} = P_{\text{total}} - P_{O_2}$$

$$P_{\text{He}} = 7.00\ \text{atm} - 1.50\ \text{atm} = 5.50\ \text{atm}$$

STUDY CHECK 6.9

A common anesthetic in dentistry known as Nitronox consists of a mixture of dinitrogen oxide, N_2O, and oxygen gas, O_2, which is administered through an inhaler over the nose. If the anesthetic mixture has a total pressure of 740 mmHg, and the partial pressure of the oxygen is 370 mmHg, what is the partial pressure, in torr, of the dinitrogen oxide?

Chemistry Link to Health

BLOOD GASES

Our cells continuously use oxygen and produce carbon dioxide. Both gases move in and out of the lungs through the membranes of the alveoli, the tiny air sacs at the ends of the airways in the lungs. An exchange of gases occurs in which oxygen from the air diffuses into the lungs and into the blood, while carbon dioxide produced in the cells is carried to the lungs to be exhaled. In Table 6.8, partial pressures are given for the gases in air that we inhale (inspired air), air that we exhale (expired air), and air in the alveoli.

At sea level, oxygen normally has a partial pressure of 100 mmHg in the alveoli of the lungs. Because the partial pressure of oxygen in venous blood is 40 mmHg, oxygen diffuses from the alveoli into the bloodstream. The oxygen combines with hemoglobin, which carries it to the tissues of the body where the partial pressure of oxygen can be very low, less than 30 mmHg. Oxygen diffuses from the blood where the partial pressure of O_2 is high into the tissues where O_2 pressure is low.

TABLE 6.8 Partial Pressures of Gases During Breathing

Gas	Partial Pressure (mmHg)		
	Inspired Air	Expired Air	Alveolar Air
Nitrogen, N_2	594	569	573
Oxygen, O_2	160.	116	100.
Carbon dioxide, CO_2	0.3	28	40.
Water vapor, H_2O	5.7	47	47
Total	760.	760.	760.

As oxygen is used in the cells of the body during metabolic processes, carbon dioxide is produced, so the partial pressure of CO_2 may be as high as 50 mmHg or more. Carbon dioxide diffuses from the tissues into the bloodstream and is carried to the lungs. There it diffuses out of the blood, where CO_2 has a partial pressure of 46 mmHg into the alveoli, where the CO_2 is at 40 mmHg and is exhaled. Table 6.9 gives the partial pressures of blood gases in the tissues and in oxygenated and deoxygenated blood.

TABLE 6.9 Partial Pressures of Oxygen and Carbon Dioxide in Blood and Tissues

	Partial Pressure (mmHg)		
Gas	**Oxygenated Blood**	**Deoxygenated Blood**	**Tissues**
O_2	100	40	30 or less
CO_2	40	46	50 or greater

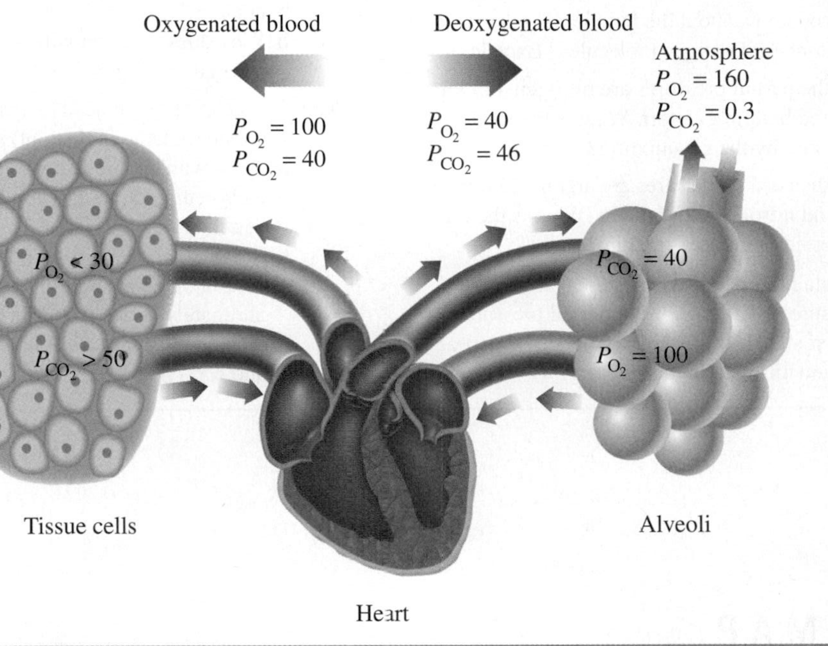

Chemistry Link to Health

HYPERBARIC CHAMBERS

A burn patient may undergo treatment for burns and infections in a hyperbaric chamber, a device in which pressures that are two to three times greater than atmospheric pressure can be obtained. A greater oxygen pressure increases the level of dissolved oxygen in the blood and tissues, where it fights bacterial infections. High levels of oxygen are toxic to many strains of bacteria. The hyperbaric chamber may also be used during surgery, to help counteract carbon monoxide (CO) poisoning, and to treat some cancers.

The blood is normally capable of dissolving up to 95% of the oxygen. Thus, if the partial pressure of the oxygen is 2280 mmHg (3 atm), about 2170 mmHg of oxygen can dissolve in the blood where it saturates the tissues. In the case of carbon monoxide poisoning, this oxygen can replace the carbon monoxide that has attached to the hemoglobin.

A patient undergoing treatment in a hyperbaric chamber must also undergo decompression (reduction of pressure) at a rate that slowly reduces the concentration of dissolved oxygen in the blood. If decompression is too rapid, the oxygen dissolved in the blood may form gas bubbles in the circulatory system.

If a scuba diver does not decompress slowly, a condition called the "bends" may occur. While below the surface of the ocean, a diver uses breathing mixtures with higher pressures. If there is nitrogen in the mixture, higher quantities of nitrogen gas dissolve in the blood. If the diver ascends to the surface too quickly, the dissolved nitrogen becomes less soluble. Then the nitrogen forms gas bubbles that can produce life-threatening blood clots, or form in the joints and tissues of the body, which can be quite painful. A diver suffering from the bends is placed immediately into a decompression chamber where pressure is first increased and then slowly decreased. The dissolved nitrogen can then diffuse through the lungs until atmospheric pressure is reached. See also Chapter 2, Chemistry Link to Health "Breathing Mixtures for Scuba."

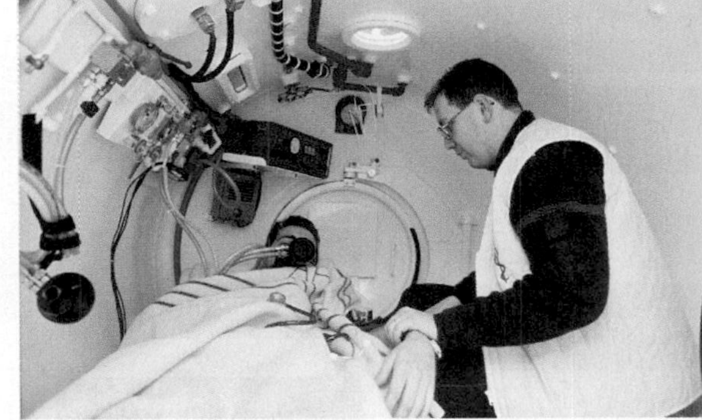

A hyperbaric chamber is used in the treatment of certain diseases.

QUESTIONS AND PROBLEMS

Partial Pressures (Dalton's Law)

6.47 A typical air sample in the lungs contains oxygen at 100 mmHg, nitrogen at 573 mmHg, carbon dioxide at 40 mmHg, and water vapor at 47 mmHg. Why are these pressures called partial pressures?

6.48 Suppose a mixture contains helium and oxygen gases. If the partial pressure of helium is the same as the partial pressure of oxygen, what do you know about the number of helium atoms compared to the number of oxygen molecules? Explain.

6.49 In a gas mixture, the partial pressures are nitrogen 425 torr, oxygen 115 torr, and helium 225 torr. What is the total pressure (torr) exerted by the gas mixture?

6.50 In a gas mixture, the partial pressures are argon 415 mmHg, neon 75 mmHg, and nitrogen 125 mmHg. What is the total pressure (atm) exerted by the gas mixture?

6.51 A gas mixture containing oxygen, nitrogen, and helium exerts a total pressure of 925 torr. If the partial pressures are oxygen 425 torr and helium 75 torr, what is the partial pressure (torr) of the nitrogen in the mixture?

6.52 A gas mixture containing oxygen, nitrogen, and neon exerts a total pressure of 1.20 atm. If helium added to the mixture increases the pressure to 1.50 atm, what is the partial pressure (atm) of the helium?

6.53 In certain lung ailments such as emphysema, there is a decrease in the ability of oxygen to diffuse into the blood.
a. How would the partial pressure of oxygen in the blood change?
b. Why does a person with severe emphysema sometimes use a portable oxygen tank?

6.54 An accident to the head can affect the ability of a person to ventilate (breathe in and out).
a. What would happen to the partial pressures of oxygen and carbon dioxide in the blood if a person cannot properly ventilate?
b. When a person who cannot breathe properly is placed on a ventilator, an air mixture is delivered at pressures that are alternately above the air pressure in the person's lung, and then below. How will this move oxygen gas into the lungs, and carbon dioxide out?

CONCEPT MAP

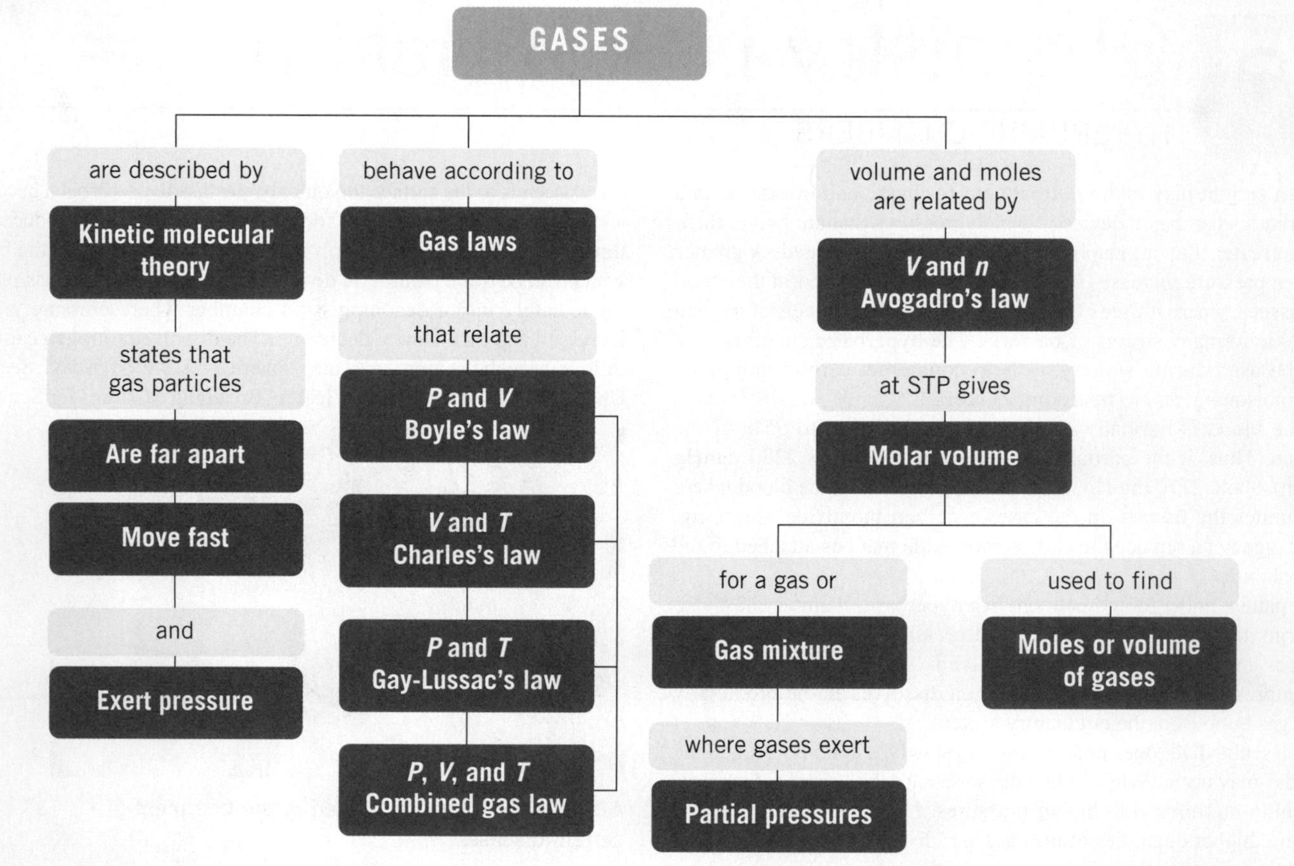

CHAPTER REVIEW

6.1 Properties of Gases

Learning Goal: Use the kinetic molecular theory of gases to describe the properties of gases.

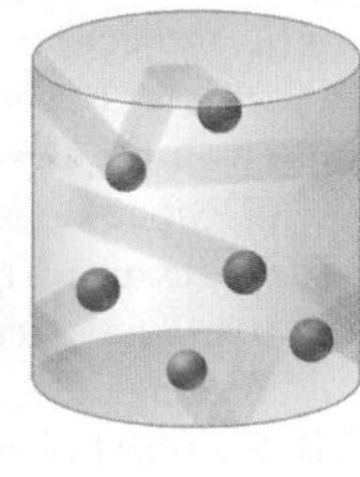

In a gas, particles are so far apart and moving so fast that their attractions are negligible. A gas is described by the physical properties of pressure (P), volume (V), temperature (T) in kelvins (K), and the amount in moles (n).

6.2 Gas Pressure

Learning Goal: Describe the units of measurement used for pressure and change from one unit to another.

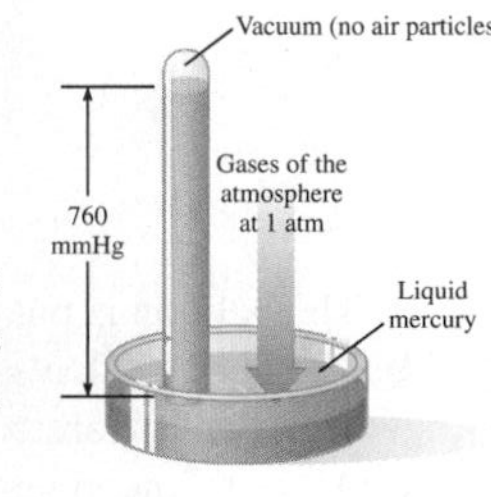

A gas exerts pressure, the force of the gas particles striking the walls of a container. Gas pressure is measured in units such as torr, mmHg, atm, and Pa.

6.3 Pressure and Volume (Boyle's Law)

Learning Goal: Use the pressure–volume relationship (Boyle's law) to determine the new pressure or volume of a certain amount of gas at a constant temperature.

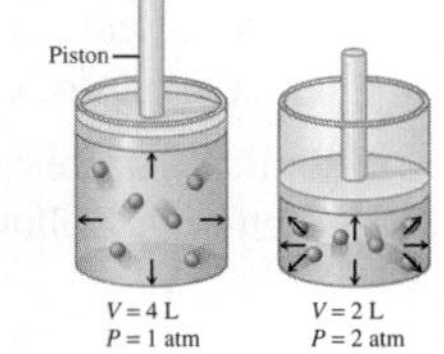

The volume (V) of a gas changes inversely with the pressure (P) of the gas if there is no change in the temperature and the amount of gas: $P_1V_1 = P_2V_2$. This means that the pressure increases if volume decreases; pressure decreases if volume increases.

6.4 Temperature and Volume (Charles's Law)

Learning Goal: Use the temperature–volume relationship (Charles's law) to determine the new temperature or volume of a certain amount of gas at a constant pressure.

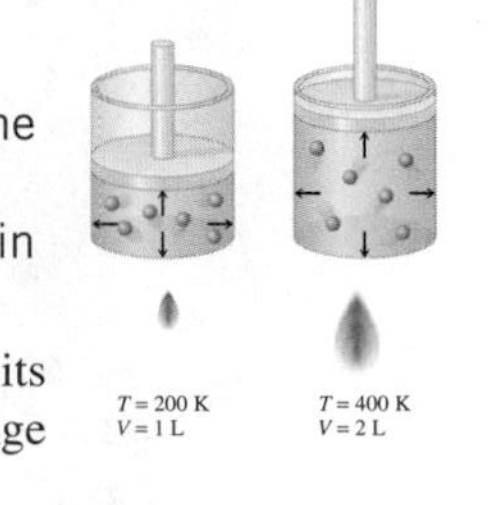

The volume (V) of a gas is directly related to its Kelvin temperature (T) when there is no change in the pressure and amount of the gas.

$$\frac{V_1}{T_1} = \frac{V_2}{T_2}$$

Therefore, if temperature increases, the volume of the gas increases; if temperature decreases, volume decreases.

6.5 Temperature and Pressure (Gay-Lussac's Law)

Learning Goal: Use the temperature–pressure relationship (Gay-Lussac's law) to determine the new temperature or pressure of a certain amount of gas at a constant volume.

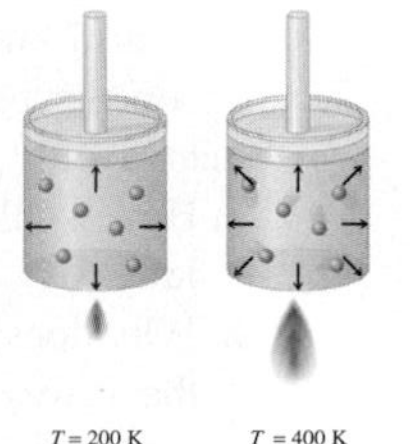

The pressure (P) of a gas is directly related to its Kelvin temperature (T).

$$\frac{P_1}{T_1} = \frac{P_2}{T_2}$$

This means that an increase in temperature increases the pressure of a gas, or a decrease in temperature decreases the pressure as long as the volume and the amount of gas remain constant. Vapor pressure is the pressure of the gas that forms when a liquid evaporates. At the boiling point of a liquid, the vapor pressure equals the external pressure.

6.6 The Combined Gas Law

Learning Goal: Use the combined gas law to find the new pressure, volume, or temperature of a gas when changes in two of these properties are given.

Combined Gas Law

$$\frac{P_1V_1}{T_1} = \frac{P_2V_2}{T_2}$$

The combined gas law is the relationship of pressure (P), volume (V), and temperature (T) for a constant amount (n) of gas.

$$\frac{P_1V_1}{T_1} = \frac{P_2V_2}{T_2}$$

This expression is used to determine the effect of changes in two of the variables on the third.

6.7 Volume and Moles (Avogadro's Law)

Learning Goal: Use Avogadro's law to describe the relationship between the amount of a gas and its volume, and use this relationship in calculations.

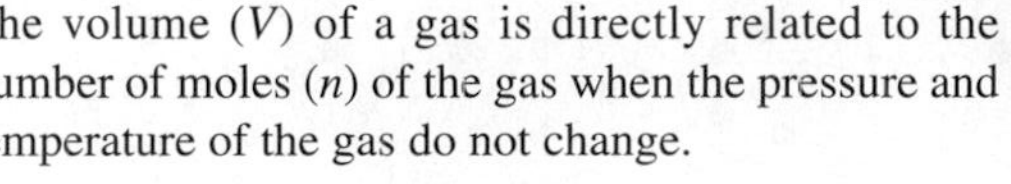

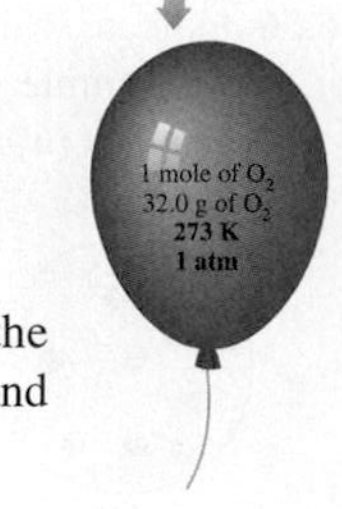

The volume (V) of a gas is directly related to the number of moles (n) of the gas when the pressure and temperature of the gas do not change.

$$\frac{V_1}{n_1} = \frac{V_2}{n_2}$$

If the moles of gas are increased, the volume must increase; if the moles of gas are decreased, the volume must decrease. At standard temperature (273 K) and standard pressure (1 atm), abbreviated STP, one mole of any gas has a volume of 22.4 L.

6.8 Partial Pressures (Dalton's Law)

Learning Goal: Use partial pressures to calculate the total pressure of a mixture of gases.

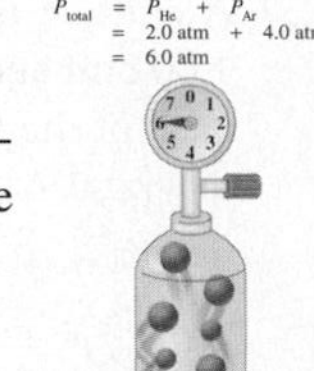

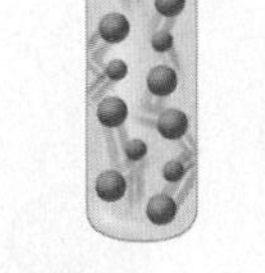

In a mixture of two or more gases, the total pressure is the sum of the partial pressures of the individual gases.

$$P_{\text{total}} = P_1 + P_2 + P_3 + \ldots$$

The partial pressure of a gas in a mixture is the pressure it would exert if it were the only gas in the container.

Key Terms

atmosphere (atm) The pressure exerted by a column of mercury 760 mm high.

atmospheric pressure The pressure exerted by the atmosphere.

Avogadro's law A gas law stating that the volume of gas is directly related to the number of moles of gas when pressure and temperature do not change.

Boyle's law A gas law stating that the pressure of a gas is inversely related to the volume when temperature and moles of the gas do not change.

Charles's law A gas law stating that the volume of a gas changes directly with a change in Kelvin temperature when pressure and moles of the gas do not change.

combined gas law A relationship that combines several gas laws relating pressure, volume, and temperature when the amount of gas does not change.

$$\frac{P_1V_1}{T_1} = \frac{P_2V_2}{T_2}$$

Dalton's law A gas law stating that the total pressure exerted by a mixture of gases in a container is the sum of the partial pressures that each gas would exert alone.

Gay-Lussac's law A gas law stating that the pressure of a gas changes directly with a change in temperature when the number of moles of a gas and its volume do not change.

kinetic molecular theory of gases A model used to explain the behavior of gases.

molar volume A volume of 22.4 L occupied by 1 mole of a gas at STP conditions of 0 °C (273 K) and 1 atm.

partial pressure The pressure exerted by a single gas in a gas mixture.

pressure The force exerted by gas particles that hit the walls of a container.

STP Standard conditions of exactly 0 °C (273 K) temperature and 1 atm pressure used for the comparison of gases.

Understanding the Concepts

6.55 At 100 °C, which of the following diagrams (**1**, **2**, or **3**) represents a gas sample that exerts the

a. lowest pressure? **b.** highest pressure?

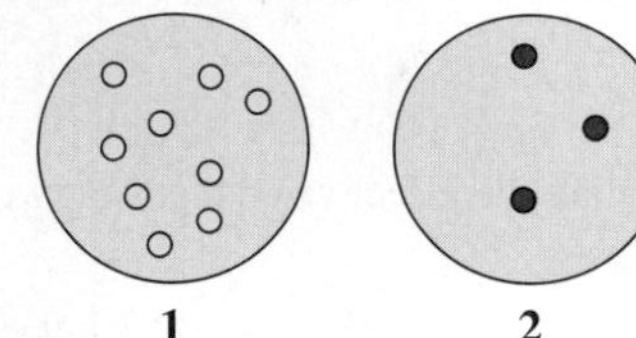

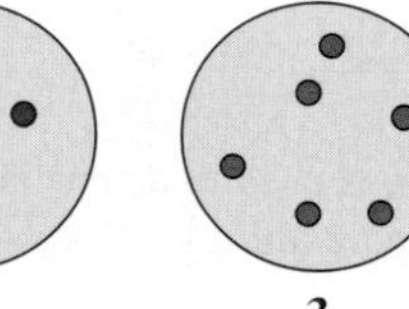

6.56 Indicate which diagram (**1**, **2**, or **3**) represents the volume of the gas sample in a flexible container when each of the following changes (**a–e**) takes place:

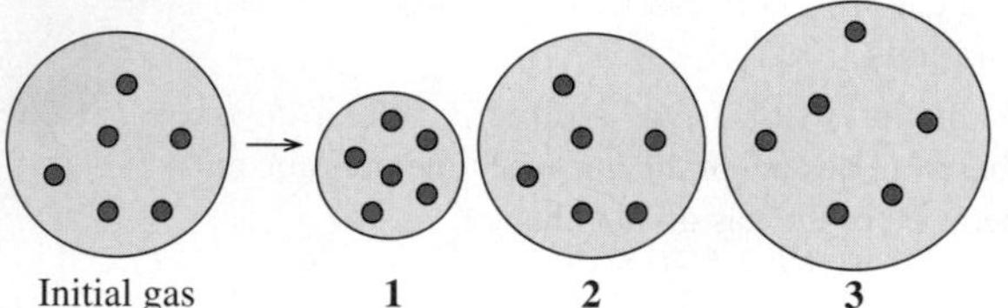

a. Temperature increases at constant pressure.
b. Temperature decreases at constant pressure.
c. Atmospheric pressure increases at constant temperature.
d. Atmospheric pressure decreases at constant temperature.
e. Doubling the atmospheric pressure and doubling the Kelvin temperature.

6.57 A balloon is filled with helium gas with a partial pressure of 1.00 atm and neon gas with a partial pressure of 0.50 atm. For each of the following changes of the initial balloon, select the diagram (**A**, **B**, or **C**) that shows the final volume of the balloon:

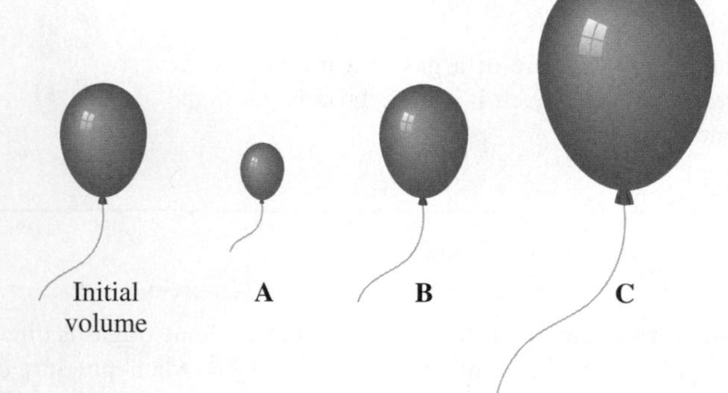

a. The balloon is put in a cold storage unit (P and n constant).
b. The balloon floats to a higher altitude where the pressure is less (n and T are constant).
c. All of the neon gas is removed (T constant).
d. The Kelvin temperature doubles and one half of the gas atoms leak out (P is constant).
e. 2.0 moles of O_2 gas is added at constant T and P.

6.58 Indicate if pressure increases, decreases, or stays the same in each of the following:

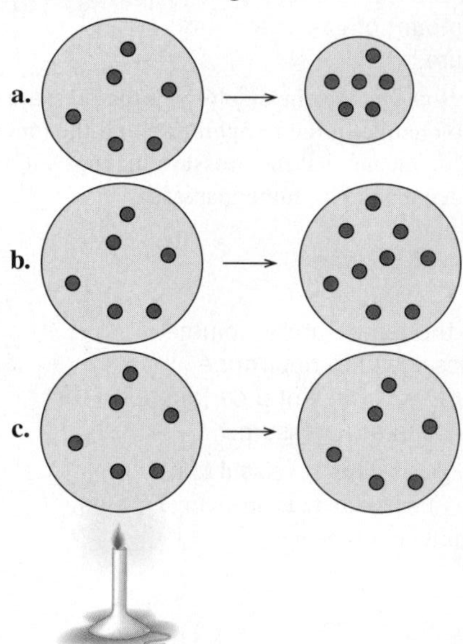

6.59 At a restaurant, a customer chokes on a piece of food. You put your arms around the person's waist and use your fists to push up on the person's abdomen, an action called the Heimlich maneuver.
a. How would this action change the volume of the chest and lungs?
b. Why does it cause the person to expel the food item from the airway?

6.60 An airplane is pressurized with air to 650 mmHg.
a. If air is 21% oxygen, what is the partial pressure of oxygen on the plane?
b. If the partial pressure of oxygen drops below 100 mmHg, passengers become drowsy. If this happens, oxygen masks are released. What is the total cabin pressure at which oxygen masks are dropped?

Additional Questions and Problems

For instructor-assigned homework, go to **www.masteringchemistry.com**.

6.61 In 1783, Jacques Charles launched his first balloon filled with hydrogen gas because it was lighter than air. The balloon had a volume of 31 000 L when it reached an altitude of 1000 m where the pressure was 658 mmHg and the temperature was −8 °C. How many kilograms of hydrogen were used to fill the balloon at STP?

Jacques Charles used hydrogen to launch his balloon in 1783.

6.62 A weather balloon has a volume of 750 L when filled with helium at 8 °C at a pressure of 380 torr. What is the new volume of the balloon, where the pressure is 0.20 atm and the temperature is −45 °C?

6.63 A sample of hydrogen (H_2) gas at 127 °C has a pressure of 2.00 atm. At what temperature (°C) will the pressure of the H_2 decrease to 0.25 atm, if V and n are constant?

6.64 A mixture of nitrogen (N_2) and helium has a volume of 250 mL at 30 °C and a total pressure of 745 mmHg.
a. If the partial pressure of helium is 32 mmHg, what is the partial pressure of the nitrogen?
b. What is the volume of the nitrogen at STP?

6.65 A weather balloon is partially filled with helium to allow for expansion at high altitudes. At STP, a weather balloon is filled with enough helium to give a volume of 25.0 L. At an altitude of 30.0 km and −35 °C, it has expanded to 2460 L. The increase in volume causes it to burst and a small parachute returns the instruments to Earth.
a. How many grams of helium are added to the balloon?
b. What is the pressure, in mmHg, of the helium inside the balloon when it bursts?

6.66 What is the total pressure, in mmHg, of a gas mixture containing argon gas at 0.25 atm, helium gas at 350 mmHg, and nitrogen gas at 360 torr?

6.67 A gas mixture contains oxygen and argon at partial pressures of 0.60 atm and 425 mmHg. If nitrogen gas added to the sample increases the total pressure to 1250 torr, what is the partial pressure, in torr, of the nitrogen added?

6.68 A gas mixture contains helium and oxygen at partial pressures of 255 torr and 0.450 atm. What is the total pressure, in mmHg, of the mixture after it is placed in a container one-half the volume of the original container?

Challenge Questions

6.69 Your spaceship has docked at a space station above Mars. The temperature inside the space station is a carefully controlled 24 °C at a pressure of 745 mmHg. A balloon with a volume of 425 mL drifts into the airlock where the temperature is −95 °C and the pressure is 0.115 atm. What is the new volume of the balloon (n remains constant)? Assume that the balloon is very elastic.

6.70 You are doing research on planet X. The temperature inside the space station is a carefully controlled 24 °C and the pressure is 755 mmHg. Suppose that a balloon, which has a volume of 850. mL inside the space station, is placed into the airlock, and floats out to planet X. If planet X has an atmospheric pressure of 0.150 atm and the volume of the balloon changes to 3.22 L, what is the temperature (°C) on planet X (n remains constant)?

6.71 Two flasks of equal volume and at the same temperature contain different gases. One flask contains 1.00 g of Ne, and the other flask contains 1.00 g of He. Which of the following statements are correct? Explain your answers.
a. Both flasks contain the same number of atoms.
b. The pressures in the flasks are the same.
c. The flask that contains helium has a higher pressure than the flask that contains neon.
d. The densities of the gases are the same.

6.72 In the fermentation of glucose (wine making), 780 mL of CO_2 gas was produced at 37 °C and 1.00 atm. What is the volume (L) of the gas when measured at 22 °C and 675 mmHg?

6.73 A gas sample has a volume of 4250 mL at 15 °C and 745 mmHg. What is the new temperature (°C) after the sample is transferred to a new container with a volume of 2.50 L and a pressure of 1.20 atm?

6.74 A liquid is placed in a 25.0-L flask. At 140 °C, the liquid evaporates completely to give a pressure of 0.900 atm. If the flask can withstand pressures up to 1.30 atm, calculate the maximum temperature to which the gas can be heated without breaking the flask.

6.75 As seen in Chapter 1, one teragram (Tg) is equal to 10^{12} g. In 2000, CO_2 emissions from the generation of electricity for use in homes in the United States was 780 Tg. In 2020, it is estimated that CO_2 emissions from the generation of electricity for use in homes will be 990 Tg.
a. Calculate the number of kilograms of CO_2 emitted in the years 2000 and 2020.
b. Calculate the number of moles of CO_2 emitted in the years 2000 and 2020.
c. What is the increase, in megagrams, for the CO_2 emissions between the years 2000 and 2020?

6.76 As seen in Chapter 1, one teragram (Tg) is equal to 10^{12} g. In 2000, CO_2 emissions from fuels used for transportation in the United States was 1990 Tg. In 2020, it is estimated that CO_2 emissions from the fuels used for transportation in the United States will be 2760 Tg.

a. Calculate the number of kilograms of CO_2 emitted in the years 2000 and 2020.
b. Calculate the number of moles of CO_2 emitted in the years 2000 and 2020.
c. What is the increase, in megagrams, for the CO_2 emissions between the years 2000 and 2020?

Answers

Answers to Study Checks

6.1 The mass, in grams, gives the amount of gas.

6.2 250 torr

6.3 73.4 mL

6.4 569 mL

6.5 16 °C

6.6 241 mmHg

6.7 7.50 L

6.8 7.0 g of N_2

6.9 370 torr

Answers to Selected Questions and Problems

6.1 **a.** At a higher temperature, gas particles have greater kinetic energy, which makes them move faster.
b. Because there are great distances between the particles of a gas, they can be pushed closer together and still remain a gas.

6.3 **a.** temperature **b.** volume **c.** amount **d.** pressure

6.5 atmospheres (atm), mmHg, torr, lb/in.^{2}, kPa

6.7 **a.** 1520 torr **b.** 1520 mmHg

6.9 As a diver ascends to the surface, external pressure decreases. If the air in the lungs were not exhaled, its volume would expand and severely damage the lungs. The pressure in the lungs must adjust to changes in the external pressure.

6.11 **a.** The pressure is greater in cylinder A. According to Boyle's law, a decrease in volume pushes the gas particles closer together, which will cause an increase in the pressure.

Property	Conditions 1	Conditions 2	Know	Predict
Pressure (P)	650 mmHg	1.2 atm (910 mmHg)	P increases	
Volume (V)	220 mL	160 mL		V decreases

6.13 **a.** The pressure doubles.
b. The pressure falls to one-third the initial pressure.
c. The pressure increases to ten times the original pressure.

6.15 **a.** 328 mmHg **b.** 2620 mmHg **c.** 4370 mmHg

6.17 **a.** 25 L **b.** 12.5 L **c.** 100. L

6.19 25 L of cyclopropane

6.21 **a.** inspiration **b.** expiration **c.** inspiration

6.23 **a.** C **b.** A **c.** B

6.25 **a.** 303 °C **b.** −129 °C **c.** 591°C **d.** 136 °C

6.27 **a.** 2400 mL **b.** 4900 mL **c.** 1800 mL **d.** 1700 mL

6.29 **a.** −23 °C **b.** 168 °C

6.31 **a.** 770 torr **b.** 1150 torr

6.33 **a.** On top of a mountain, water boils below 100 °C because the atmospheric pressure is less than 1 atm.
b. Because the pressure inside a pressure cooker is greater than 1 atm, water boils above 100 °C. At a higher temperature, food cooks faster.

6.35 **a.** 4.26 atm **b.** 3.07 atm **c.** 0.606 atm

6.37 110 mL

6.39 The volume increases because the number of gas particles is increased.

6.41 **a.** 4.00 L **b.** 14.6 L **c.** 26.7 L

6.43 **a.** 2.00 moles of O_2 **b.** 0.179 mole of CO_2

6.45 **a.** 4.48 L of O_2 **b.** 55 400 mL of Ne

6.47 In a gas mixture, the pressure that each gas exerts as part of the total pressure is called the partial pressure of that gas. Because the air sample is a mixture of gases, the total pressure is the sum of the partial pressures of each gas in the sample.

6.49 765 torr

6.51 425 torr

6.53 **a.** The partial pressure of oxygen will be lower than normal.
b. Breathing a higher concentration of oxygen will help to increase the supply of oxygen in the lungs and blood and raise the partial pressure of oxygen in the blood.

6.55 **a.** 2 **b.** 1

6.57 **a.** A **b.** C **c.** A
d. B **e.** C

6.59 **a.** The volume of the chest and lungs is decreased.
b. The decrease in volume increases the pressure, which can dislodge the food in the trachea.

6.61 2.5 kg of H_2

6.63 −223 °C

6.65 **a.** 4.46 g of helium **b.** 6.73 mmHg

6.67 370 torr

6.69 2170 mL

6.71 **a.** False. The flask containing helium gas contains more atoms because 1 gram of helium contains more moles of helium and therefore more atoms than 1 gram of neon.
b. False. There are different numbers of moles of gas in the flasks, which means that the pressures are different.
c. True. There are more moles of helium, which makes the pressure of helium greater than that of neon.
d. True. Density is mass divided by volume. If gases have the same mass and volume, they have the same density.

6.73 −66 °C

6.75 **a.** 7.8×10^{11} kg of CO_2 (2000); 9.9×10^{11} kg of CO_2 (2020)
b. 1.8×10^{13} moles of CO_2 (2000); 2.3×10^{13} moles of CO_2 (2020)
c. 2.1×10^{8} Mg of CO_2 increase

7 Solutions

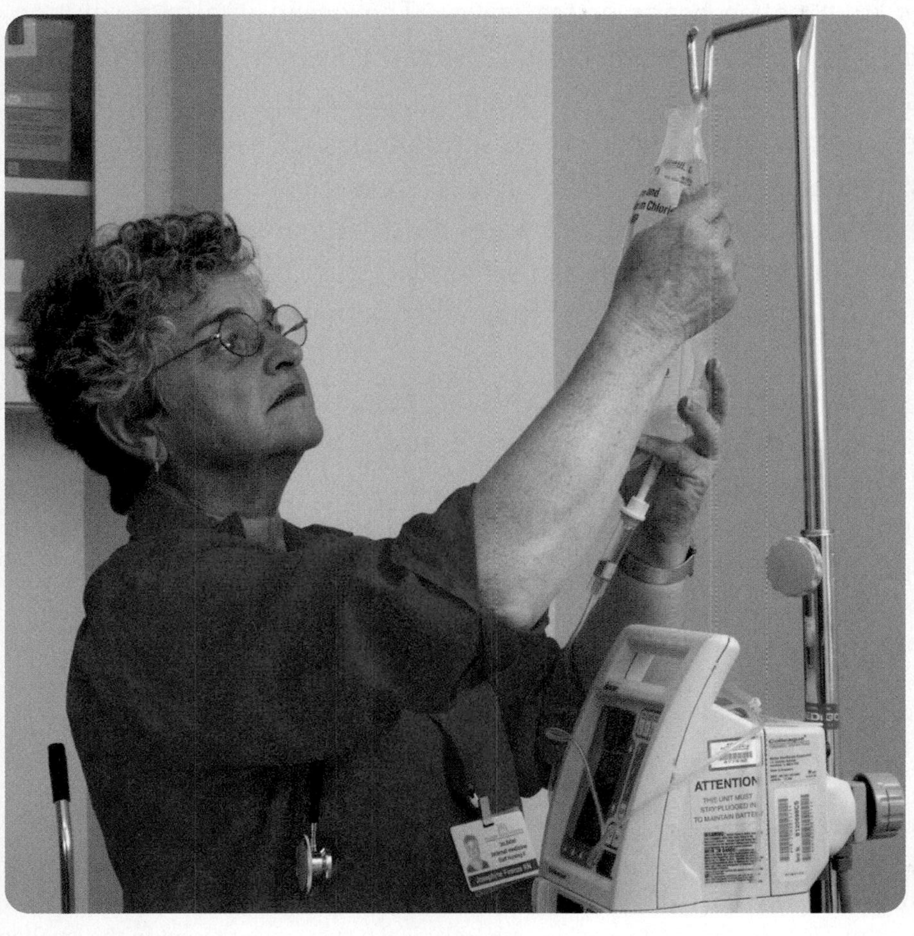

LOOKING AHEAD

"There is a lot of chemistry going on in the body, including drug interactions," says Josephine Firenze, registered nurse, Kaiser Hospital.

Normally, the body maintains a homeostasis of fluids and electrolytes. Conditions that alter the composition of body fluids can lead to convulsions, coma, or death. To halt the disease process and to establish homeostasis, a patient may be given intravenous fluid therapy. Solutions that are compatible with body fluids such as a 5% glucose or a 0.9% saline are used. An infusion pump delivers the desired number of milliliters per hour to the patient. During IV therapy, a patient is checked for fluid overload as indicated by edema, which is swelling, or a greater fluid input than output.

Visit **www.masteringchemistry.com** for self-study materials and instructor-assigned homework.

Solutions are everywhere around us. Most consist of one substance dissolved in another. The air we breathe is a solution that is primarily oxygen and nitrogen gases. Carbon dioxide gas dissolved in water makes carbonated drinks. When we make solutions of coffee or tea, we use hot water to dissolve substances from coffee beans or tea leaves. The ocean is also a solution, consisting of many salts such as sodium chloride dissolved in water. In your medicine cabinet, the antiseptic tincture of iodine is a solution of iodine dissolved in alcohol.

Our body fluids contain water and dissolved substances such as glucose and urea and electrolytes such as K^+, Na^+, Cl^-, Mg^{2+}, HCO_3^-, and HPO_4^{2-}. Proper amounts of each of these dissolved substances and water must be maintained in the body fluids. Small changes in electrolyte levels can seriously disrupt cellular processes and endanger our health. Therefore, the measurement of their concentrations is a valuable diagnostic tool.

In the processes of osmosis and dialysis, water, essential nutrients, and waste products enter and leave the cells of the body. In osmosis, water flows in and out of the cells of the body. In dialysis, small particles in solution as well as water diffuse through semipermeable membranes. The kidneys utilize osmosis and dialysis to regulate the amount of water and electrolytes that are excreted.

LEARNING GOAL

Identify the solute and solvent in a solution; describe the formation of a solution.

A solution consists of at least one solute dispersed in a solvent.

7.1 Solutions

A **solution** is a homogeneous mixture in which one substance called the **solute** is uniformly dispersed in another substance called the **solvent**. Because the solute and the solvent do not react with each other, they can be mixed in varying proportions. A little salt dissolved in water tastes slightly salty. When more salt dissolves, the water tastes very salty. Usually, the solute (in this case, salt) is the substance present in the smaller amount, whereas the solvent (in this case, water) is present in the larger amount. In a solution, the particles of the solute are evenly dispersed among the molecules within the solvent (see Figure 7.1).

CONCEPT CHECK 7.1

Identifying a Solute and a Solvent

Identify the solute and the solvent in each of the following solutions:

a. 15 g of sugar dissolved in 100 mL of water
b. 75 mL of water mixed with 25 mL of isopropyl alcohol
c. a tincture of iodine prepared with 0.10 g of I_2 and 10.0 mL of ethanol

ANSWER

a. Sugar, the smaller quantity, is the solute; water is the solvent.
b. Isopropyl alcohol, which has the smaller volume, is the solute; water is the solvent.
c. Iodine, the smaller quantity, is the solute; ethanol is the solvent.

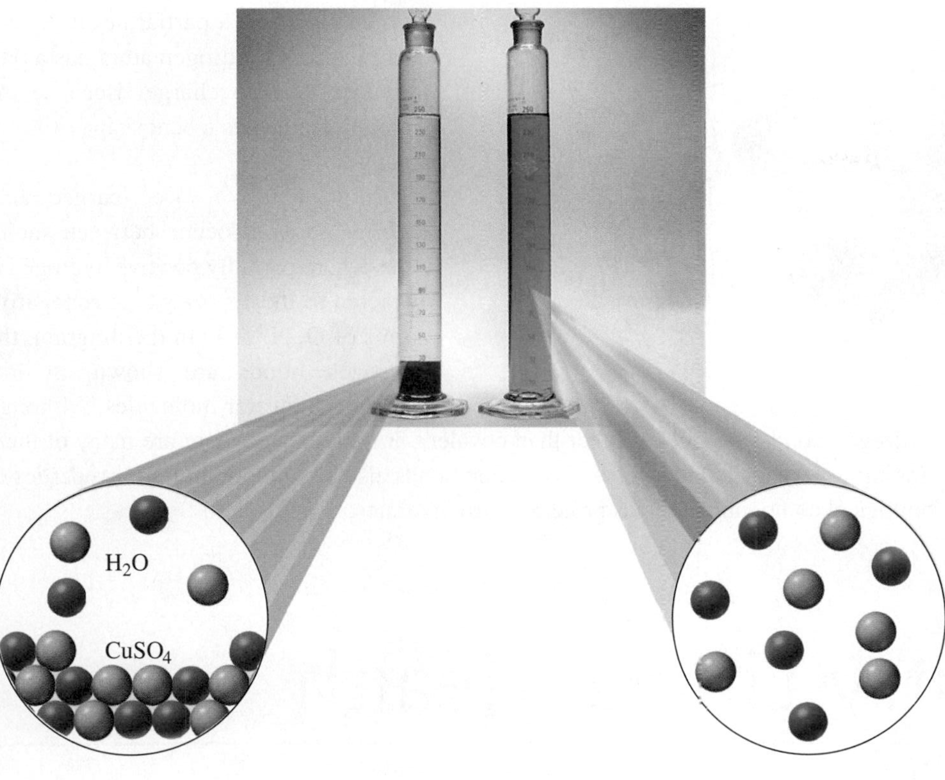

FIGURE 7.1 A solution of copper(II) sulfate ($CuSO_4$) forms as particles of solute dissolve and become evenly dispersed among the solvent (water) molecules.

Q What does the uniform blue color in the graduated cylinder on the right indicate?

Types of Solutes and Solvents

Solutes and solvents may be solids, liquids, or gases. The solution that forms has the same physical state as the solvent. When sugar crystals are dissolved in water, the resulting sugar solution is liquid. Sugar is the solute, and water is the solvent. Soda water and soft drinks are prepared by dissolving carbon dioxide gas in water. The carbon dioxide gas is the solute, and water is the solvent. Table 7.1 lists some solutes and solvents and their solutions.

TABLE 7.1 Some Examples of Solutions

Type	Example	Primary Solute	Solvent
Gas Solutions			
Gas in a gas	Air	Oxygen (gas)	Nitrogen (gas)
Liquid Solutions			
Gas in a liquid	Soda water	Carbon dioxide (gas)	Water (liquid)
	Household ammonia	Ammonia (gas)	Water (liquid)
Liquid in a liquid	Vinegar	Acetic acid (liquid)	Water (liquid)
Solid in a liquid	Seawater	Sodium chloride (solid)	Water (liquid)
	Tincture of iodine	Iodine (solid)	Ethanol (liquid)
Solid Solutions			
Liquid in a solid	Dental amalgam	Mercury (liquid)	Silver (solid)
Solid in a solid	Brass	Zinc (solid)	Copper (solid)
	Steel	Carbon (solid)	Iron (solid)

Water as a Solvent

Water is one of the most common solvents in nature. In the H_2O molecule, an oxygen atom shares electrons with two hydrogen atoms. Because oxygen is much more electronegative than hydrogen, the O—H bonds are polar. In each polar bond, the

SELF STUDY ACTIVITY
Hydrogen Bonding

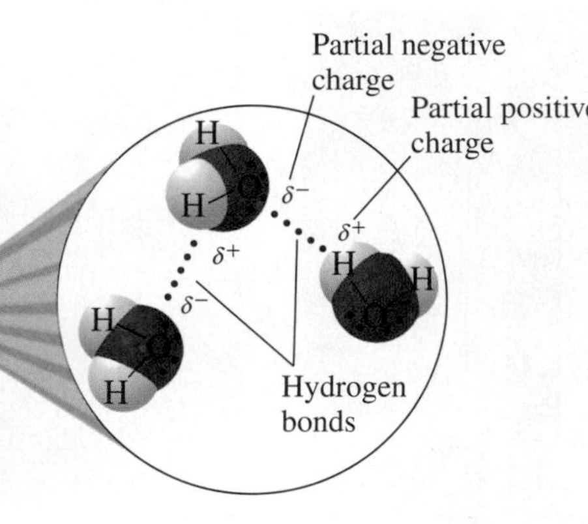

In water, hydrogen bonds form between an oxygen atom in one water molecule and the hydrogen atom in another.

oxygen atom has a partial negative (δ^-) charge and the hydrogen atom has a partial positive (δ^+) charge. Because the water molecule has a bent shape, water is a *polar solvent.*

In Chapter 4, we learned that *hydrogen bonds* occur between molecules where partially positive hydrogen is attracted to the strongly electronegative atoms of O, N, or F. In the diagram, the hydrogen bonds are shown as dots between the water molecules. Although hydrogen bonds are much weaker than covalent or ionic bonds, there are many of them linking water molecules together. Hydrogen bonds also are important in the properties of biological compounds such as proteins, carbohydrates, and DNA.

Chemistry Link to Health

WATER IN THE BODY

The average adult contains about 60% water by weight, and the average infant about 75%. About 60% of the body's water is contained within the cells as intracellular fluids; the other 40% makes up extracellular fluids, which include the interstitial fluid in tissue and the plasma in the blood. These external fluids carry nutrients and waste materials between the cells and the circulatory system.

24 Hours

Water gain	
Liquid	1000 mL
Food	1200 mL
Metabolism	300 mL
Total	2500 mL

Water loss	
Urine	1500 mL
Perspiration	300 mL
Breath	600 mL
Feces	100 mL
Total	2500 mL

Every day you lose between 1500 and 3000 mL of water from the kidneys as urine, from the skin as perspiration, from the lungs as you exhale, and from the gastrointestinal tract. Serious dehydration can occur in an adult if there is a 10% net loss in total body fluid, and a 20% loss of fluid can be fatal. An infant suffers severe dehydration with a 5–10% loss in body fluid.

Water loss is continually replaced by the liquids and foods in the diet and from metabolic processes that produce water in the cells of the body. Table 7.2 lists the % by mass of water contained in some foods.

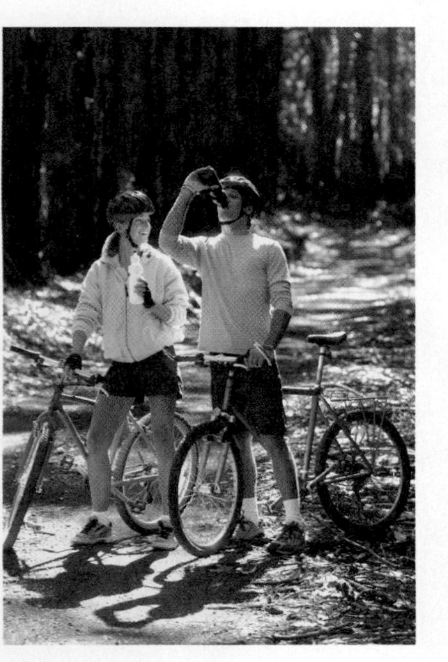

The water lost from the body is replaced by the intake of fluids.

TABLE 7.2 Percentage of Water in Some Foods

Food	Water (% by mass)	Food	Water (% by mass)
Vegetables		**Meats/Fish**	
Carrot	88	Chicken, cooked	71
Celery	94	Hamburger, broiled	60
Cucumber	96	Salmon	71
Tomato	94		
Fruits		**Milk Products**	
Apple	85	Cottage cheese	78
Cantaloupe	91	Milk, whole	87
Orange	86	Yogurt	88
Strawberry	90		
Watermelon	93		

Formation of Solutions

The interactions between solute and solvent will determine whether or not a solution will form. A solution forms when there is sufficient attraction between the particles of the solute and the solvent, which provides energy to separate the particles. Such attractions only occur when the solute and the solvent have similar polarities.

Solutions with Ionic and Polar Solutes

In ionic solutes such as sodium chloride, NaCl, there are strong solute–solute attractions between positively charged Na^+ ions and negatively charged Cl^- ions. When NaCl crystals are placed in water, the process of dissolution begins as the partially negative oxygen atoms in water molecules attract positive Na^+ ions and the partially positive hydrogen atoms in other water molecules attract negative Cl^- ions (see Figure 7.2). As soon as the Na^+ and Cl^- ions form a solution, they undergo **hydration** as water molecules surround each ion. Hydration of the ions diminishes their attraction to other ions and keeps them in solution. The strong solute–solvent attractions between Na^+ and Cl^- ions and the polar water molecules provide energy needed to form the solution.

In the equation for the formation of the NaCl solution, the solid and aqueous NaCl are shown with the formula H_2O over the arrow, which indicates that water is needed for the dissociation process but is not a reactant.

$$\mathrm{NaCl}(s) \xrightarrow{\mathrm{H_2O}} \mathrm{Na^+}(aq) + \mathrm{Cl^-}(aq)$$

In another example, we find that a polar covalent compound such as methanol, CH_3—OH, is soluble in water because methanol has a polar —OH group that forms hydrogen bonds with water (see Figure 7.3).

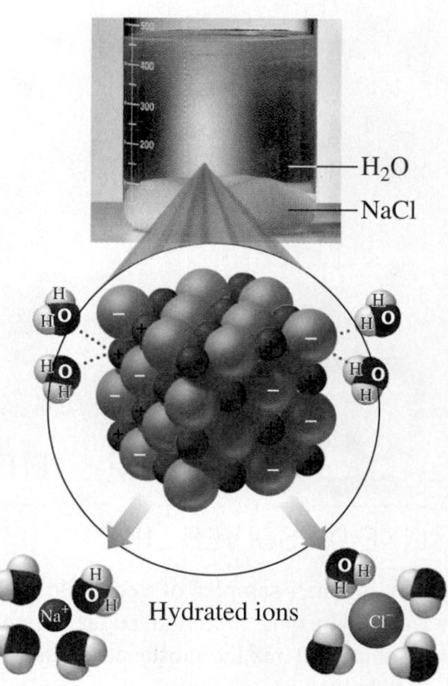

FIGURE 7.2 Ions on the surface of a crystal of NaCl dissolve in water as they are attracted to the polar water molecules that pull the ions into solution and surround them.

Q What helps keep the Na^+ and Cl^- ions in solution?

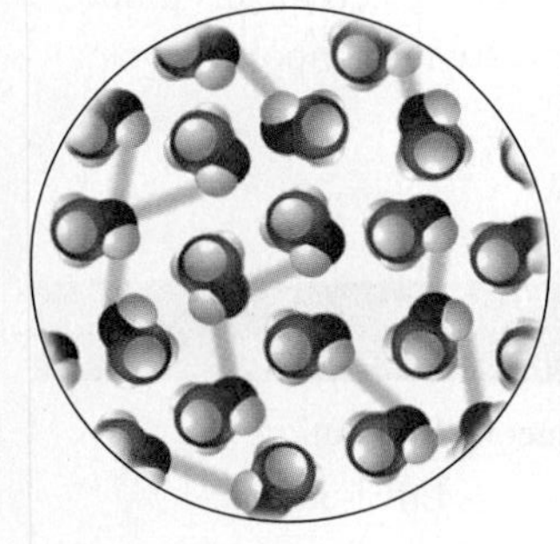

Methanol (CH_3—OH) solute

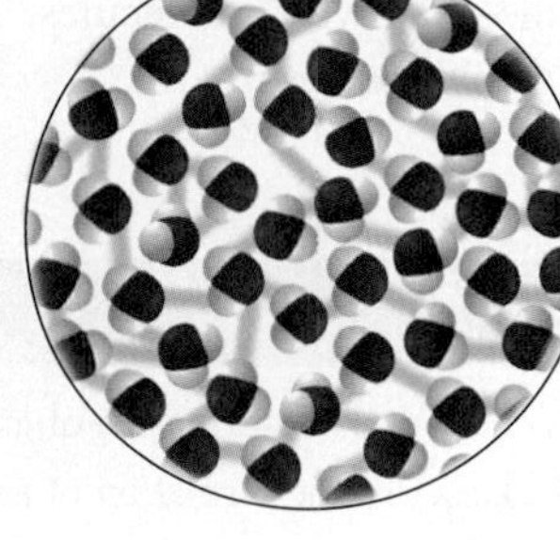

Water solvent

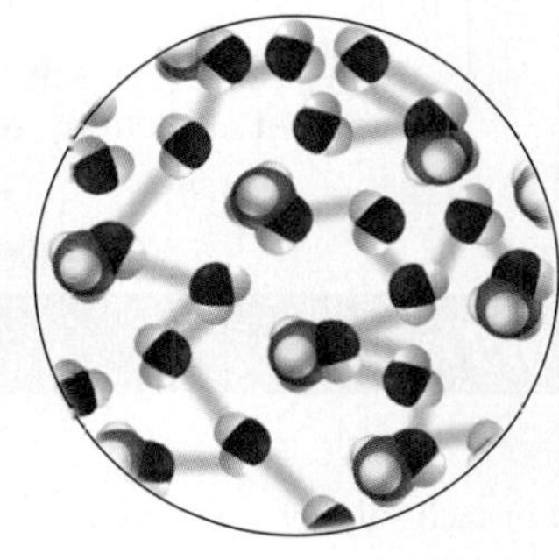

Methanol–water solution with hydrogen bonding

FIGURE 7.3 Molecules of polar covalent compound methanol, CH_3—OH, form hydrogen bonds with polar molecules to form a methanol–water solution.

Q Why are there attractions between the solute methanol and the solvent water?

Solutions with Nonpolar Solutes

Compounds containing nonpolar molecules such as iodine (I_2), oil, or grease do not dissolve in water because there are essentially no attractions between the nonpolar solute and the polar solvent. Nonpolar solutes require nonpolar solvents for a solution to form. The expression "*like dissolves like*" is a way of saying that the polarities of a solute and a solvent must be similar in order to form a solution (see Table 7.3).

TABLE 7.3 Possible Combinations of Solutes and Solvents

Solutions Will Form		Solutions Will Not Form	
Solute	**Solvent**	**Solute**	**Solvent**
Polar/ionic	Polar	Polar/ionic	Nonpolar
Nonpolar	Nonpolar	Nonpolar	Polar

Figure 7.4 illustrates the formation of some solutions.

FIGURE 7.4 Like dissolves like. **(a)** The test tubes contain an upper layer of the polar solvent water and a lower layer of the nonpolar solvent CH_2Cl_2. **(b)** The nonpolar solute I_2 (purple) is soluble in the CH_2Cl_2 layer. **(c)** The ionic solute $Ni(NO_3)_2$ (green) is soluble in the water layer.

Q In which solvent would polar molecules of sugar be soluble?

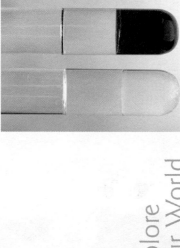

Explore Your World

LIKE DISSOLVES LIKE

Obtain small samples of vegetable oil, water, vinegar, salt, and sugar. Then combine and mix the substances as listed in **a** through **e**.

a. oil and water
b. water and vinegar
c. salt and water
d. sugar and water
e. salt and oil

QUESTIONS

1. Which of the mixtures formed a solution?
2. Which of the mixtures did not form a solution?
3. Why do some mixtures form solutions, but others do not?

CONCEPT CHECK 7.2

Polar and Nonpolar Solutes

Indicate whether each of the following substances will dissolve in water. Explain.

a. KCl
b. octane, C_8H_{18}, a compound in gasoline

ANSWER

a. Yes. KCl is an ionic compound. The solute–solvent attractions between the ions K^+ and Cl^- and polar water molecules provide the energy to break solute–solute and solvent–solvent bonds. Thus, a KCl solution will form.

b. No. Octane is a nonpolar compound of carbon and hydrogen, which means it does not form a solution with the polar water molecules. There are no nonpolar–polar solvent attractions, and no solution forms.

QUESTIONS AND PROBLEMS

Solutions

7.1 Identify the solute and the solvent in each solution composed of the following:
a. 10.0 g of NaCl and 100.0 g of H_2O
b. 50.0 mL of ethanol, C_2H_5OH, and 10.0 mL of H_2O
c. 0.20 L of O_2 and 0.80 L of N_2

7.2 Identify the solute and the solvent in each solution composed of the following:
a. 50.0 g of silver and 4.0 g of mercury
b. 100.0 mL of water and 5.0 g of sugar
c. 1.0 g of Br_2 and 50.0 mL of methylene chloride

7.3 Describe the formation of an aqueous KI solution.

7.4 Describe the formation of an aqueous LiBr solution.

7.5 Water is a polar solvent and carbon tetrachloride (CCl_4) is a nonpolar solvent. In which solvent is each of the following more likely to be soluble?
a. $NaNO_3$, ionic
b. I_2, nonpolar
c. sucrose (table sugar), polar
d. gasoline, nonpolar

7.6 Water is a polar solvent and hexane is a nonpolar solvent. In which solvent is each of the following more likely to be soluble?
a. vegetable oil, nonpolar
b. benzene, nonpolar
c. LiCl, ionic
d. Na_2SO_4, ionic

LEARNING GOAL

Identify solutes as electrolytes or nonelectrolytes.

7.2 Electrolytes and Nonelectrolytes

Solutes can be classified by their ability to conduct an electrical current. When solutes called **electrolytes** dissolve in water, they separate into ions forming solutions that are able to conduct electricity. When **nonelectrolytes** dissolve in water, they dissolve as molecules, not as ions. Thus, solutions of nonelectrolytes do not conduct electricity.

To test solutions for the presence of ions, we can use an apparatus that consists of a battery and a pair of electrodes connected by wires to a light bulb. The light bulb glows when electricity can flow, which can only happen when electrolytes provide ions that move between the electrodes to complete the circuit.

Electrolytes

Electrolytes can be further classified as *strong electrolytes* or *weak electrolytes*. All electrolytes undergo *dissociation*, in which some or all of the solute that dissolves produces ions. For a **strong electrolyte**, such as sodium chloride (NaCl), there is 100 percent dissociation of the solute into ions. When the electrodes from the light bulb apparatus are placed in a NaCl solution, the light bulb is very bright.

In an equation for dissociation of a compound in water, the charges must balance. For example, magnesium nitrate dissociates to give one magnesium ion for every two nitrate ions. However, only the ionic bonds between Mg^{2+} and NO_3^- are broken, not the covalent bonds within the polyatomic ion. The dissociation for $Mg(NO_3)_2$ is written as follows:

$$Mg(NO_3)_2(s) \xrightarrow{H_2O} Mg^{2+}(aq) + 2NO_3^-(aq)$$

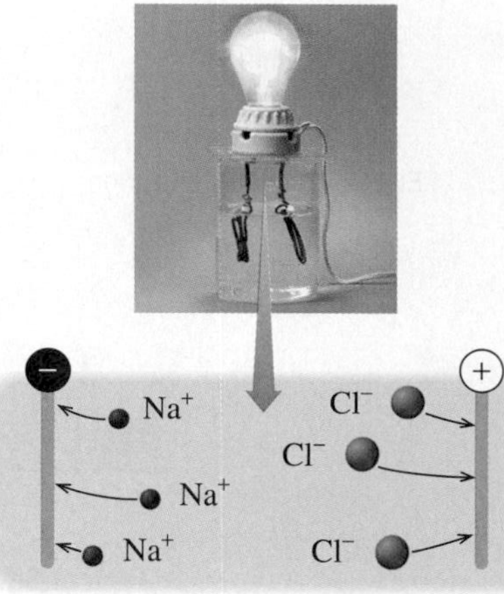

(a) A strong electrolyte completely dissociates into ions in an aqueous solution.

Weak Electrolytes

A **weak electrolyte** is a compound that dissolves in water mostly as molecules. Only a few of the dissolved molecules separate, producing a small number of ions in solution. Thus solutions of weak electrolytes do not conduct electrical current as well as solutions of strong electrolytes. For example, an aqueous solution of the weak electrolyte HF consists of mostly HF molecules and only a few H^+ and F^- ions. When the electrodes are placed in a solution of a weak electrolyte, the glow of the light bulb is very dim. As more H^+ and F^- ions form, some recombine to give HF molecules. These forward and reverse reactions of molecules to ions and back again are indicated by two arrows that point in opposite directions between reactant and products:

$$HF(aq) \underset{\text{Recombination}}{\overset{\text{Dissociation}}{\rightleftharpoons}} H^+(aq) + F^-(aq)$$

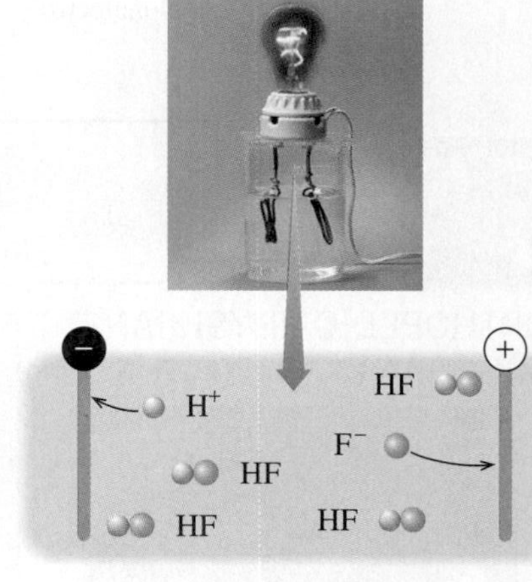

(b) A weak electrolyte forms mostly molecules and a few ions in an aqueous solution.

Nonelectrolytes

A nonelectrolyte such as sucrose (sugar) dissolves in water only as molecules, which do not separate into ions. Thus, solutions of nonelectrolytes do not conduct electricity. When the electrodes of the light bulb apparatus are placed in a solution of a nonelectrolyte, the light bulb does not glow because the solution does not conduct electricity.

$$\underset{\text{Sucrose}}{C_{12}H_{22}O_{11}(s)} \xrightarrow{H_2O} \underset{\text{Solution of sucrose molecules}}{C_{12}H_{22}O_{11}(aq)}$$

Table 7.4 summarizes the classification of solutes in aqueous solutions.

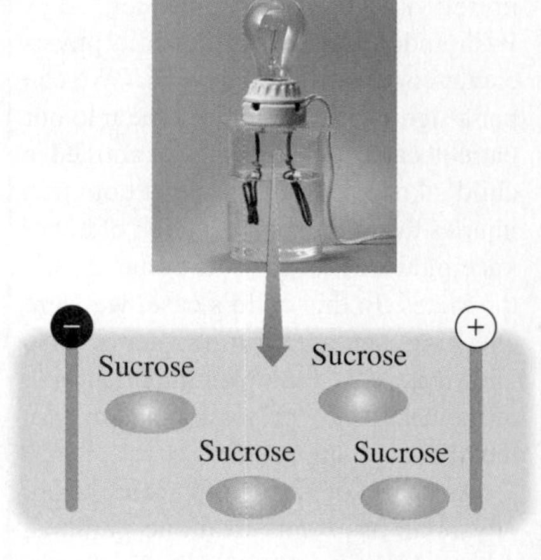

(c) A nonelectrolyte dissolves as molecules in an aqueous solution.

SAMPLE PROBLEM 7.1

Solutions of Electrolytes and Nonelectrolytes

Indicate whether solutions of each of the following contain only ions, only molecules, or mostly molecules and a few ions:

a. Na_2SO_4, a strong electrolyte
b. CH_3OH, a nonelectrolyte

TUTORIAL
Electrolytes and Ionization

SOLUTION

a. An aqueous solution of Na_2SO_4 contains only the ions Na^+ and SO_4^{2-}.

b. A nonelectrolyte such as CH_3OH produces only molecules when it dissolves in water.

STUDY CHECK 7.1

Boric acid, H_3BO_3, is a weak electrolyte. Would you expect a boric acid solution to contain only ions, only molecules, or mostly molecules and a few ions?

TABLE 7.4 Classification of Solutes in Aqueous Solutions

Type of Solute	Dissociates	Type(s) of Particles in Solution	Conducts Electricity?	Examples
Strong electrolyte	Completely	Ions only	Yes	Ionic compounds such as $NaCl$, KBr, $MgCl_2$, $NaNO_3$; bases such as $NaOH$, KOH; acids such as HCl, HBr, HI, HNO_3, $HClO_4$, H_2SO_4
Weak electrolyte	Partially	Mostly molecules and a few ions	Weakly	HF, H_2O, NH_3, $HC_2H_3O_2$ (acetic acid)
Nonelectrolyte	None	Molecules only	No	Carbon compounds such as CH_3OH (methanol), C_2H_5OH (ethanol), $C_{12}H_{22}O_{11}$ (sucrose), CH_4N_2O (urea)

Career Focus

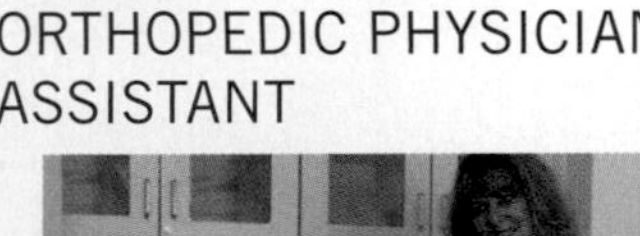

ORTHOPEDIC PHYSICIAN ASSISTANT

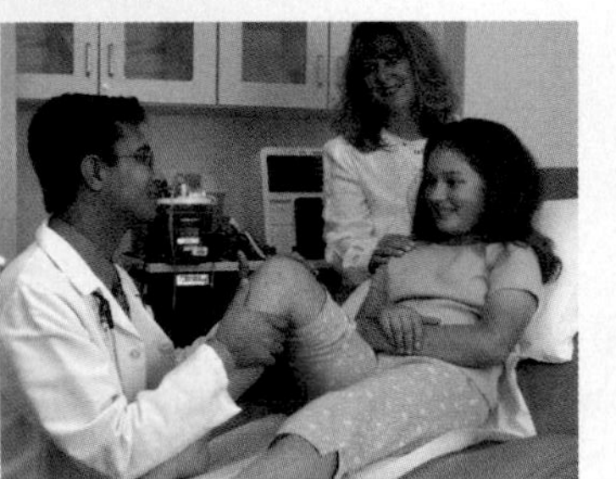

"At a time when we have a shortage of health care professionals, I think of myself as a physician extender," says Pushpinder Beasley, orthopedic physician assistant, Kaiser Hospital. "We can put a significant amount of time into our patient care. Just today, I examined a child's knee. One of the most common injuries to children is disruption of either knee ligaments or the soft tissue around the knees. In this child's case, we were checking her anterior ligaments, also known as ACL. I think an important role of the health care professional is to earn the trust of young people."

As part of a health care team, physician assistants examine patients, order laboratory tests, make diagnoses, report patient progress, order therapeutic procedures and, in most states, prescribe medications.

Equivalents

Body fluids contain a mixture of several electrolytes, such as Na^+, Cl^-, K^+, and Ca^{2+}. We measure each individual ion in terms of an **equivalent (Eq)**, which is the amount of that ion equal to 1 mole of positive or negative electrical charge. For example, 1 mole of Na^+ ions and 1 mole of Cl^- ions are each 1 equivalent because they each contain 1 mole of charge. For an ion with a charge of 2+ or 2−, there are 2 equivalents for each mole. Some examples of ions and equivalents are shown in Table 7.5.

TABLE 7.5 Equivalents of Electrolytes

Ion	Electrical Charge	Number of Equivalents in 1 Mole
Na^+	1+	1 Eq
Ca^{2+}	2+	2 Eq
Fe^{3+}	3+	3 Eq
Cl^-	1−	1 Eq
SO_4^{2-}	2−	2 Eq

In any solution, the charge of the positive ions is always balanced by the charge of the negative ions. The concentrations of electrolytes in intravenous fluids are expressed in milliequivalents per liter (mEq/L); 1 Eq = 1000 mEq. For example, a solution containing 25 mEq/L of Na^+ and 4 mEq/L of K^+ has a total positive charge of 29 mEq/L. If Cl^- is the only anion, its concentration must be 29 mEq/L.

SAMPLE PROBLEM 7.2

Electrolyte Concentration

The laboratory tests for a patient indicate a blood calcium level of 8.8 mEq/L.

a. How many moles of calcium ion are in 0.50 L of blood?

b. If chloride ion is the only other ion present, what is its concentration in mEq/L?

SOLUTION

a. Using the volume and the electrolyte concentration in mEq/L, we can find the number of equivalents in 0.50 L of blood.

$$0.50\,\cancel{L} \times \frac{8.8\,\cancel{mEq}}{1\,\cancel{L}} \times \frac{1\text{ Eq}}{1000\,\cancel{mEq}} = 0.0044\text{ Eq of }Ca^{2+}$$

We can then convert equivalents to moles (for Ca^{2+} there are 2 Eq per mole).

$$0.0044\,\cancel{Eq\,Ca^{2+}} \times \frac{1\text{ mole }Ca^{2+}}{2\,\cancel{Eq\,Ca^{2+}}} = 0.0022\text{ mole of }Ca^{2+}$$

b. If the concentration of Ca^{2+} is 8.8 mEq/L, then the concentration of Cl^- must be 8.8 mEq/L to balance the charge.

STUDY CHECK 7.2

A Ringer's solution for intravenous fluid replacement contains 155 mEq of Cl^- per liter of solution. If a patient receives 1250 mL of Ringer's solution, how many moles of chloride ion were given?

QUESTIONS AND PROBLEMS

Electrolytes and Nonelectrolytes

7.7 KF is a strong electrolyte, and HF is a weak electrolyte. How does their dissociation in water differ?

7.8 NaOH is a strong electrolyte, and CH_3OH is a nonelectrolyte. How does their dissociation in water differ?

7.9 Write a balanced equation for the dissociation of each of the following strong electrolytes in water:
a. KCl **b.** $CaCl_2$
c. K_3PO_4 **d.** $Fe(NO_3)_3$

7.10 Write a balanced equation for the dissociation of each of the following strong electrolytes in water:
a. LiBr **b.** $NaNO_3$
c. $CuCl_2$ **d.** K_2CO_3

7.11 Indicate whether aqueous solutions of each of the following will contain only ions, only molecules, or mostly molecules and a few ions:
a. acetic acid ($HC_2H_3O_2$), a weak electrolyte
b. NaBr, a strong electrolyte
c. fructose ($C_6H_{12}O_6$), a nonelectrolyte

7.12 Indicate whether aqueous solutions of each of the following will contain only ions, only molecules, or mostly molecules and a few ions:
a. NH_4Cl, a strong electrolyte
b. ethanol, C_2H_5OH, a nonelectrolyte
c. HCN, hydrocyanic acid, a weak electrolyte

7.13 Classify each solute represented in the following equations as a strong, weak, or nonelectrolyte:

a. $K_2SO_4(s) \xrightarrow{H_2O} 2K^+(aq) + SO_4^{2-}(aq)$

b. $NH_4OH(aq) \overset{H_2O}{\rightleftharpoons} NH_4^+(aq) + OH^-(aq)$

c. $C_6H_{12}O_6(s) \xrightarrow{H_2O} C_6H_{12}O_6(aq)$

7.14 Classify each solute represented in the following equations as a strong, weak, or nonelectrolyte:

a. $CH_3OH(l) \xrightarrow{H_2O} CH_3OH(aq)$

b. $MgCl_2(s) \xrightarrow{H_2O} Mg^{2+}(aq) + 2Cl^-(aq)$

c. $HClO(aq) \overset{H_2O}{\rightleftharpoons} H^+(aq) + ClO^-(aq)$

7.15 Indicate the number of equivalents in each of the following:
a. 1 mole of K^+
b. 2 moles of OH^-
c. 1 mole of Ca^{2+}
d. 3 moles of CO_3^{2-}

7.16 Indicate the number of equivalents in each of the following:
a. 1 mole of Mg^{2+}
b. 0.5 mole of H^+
c. 4 moles of Cl^-
d. 2 moles of Fe^{3+}

7.17 A physiological saline solution contains 154 mEq/L each of Na^+ and Cl^-. How many moles each of Na^+ and Cl^- are in 1.00 L of the saline solution?

7.18 A solution to replace potassium loss contains 40. mEq/L each of K^+ and Cl^-. How many moles each of K^+ and Cl^- are in 1.5 L of the solution?

7.19 A solution contains 40. mEq/L of Cl^- and 15 mEq/L of HPO_4^{2-}. If Na^+ is the only cation in the solution, what is the Na^+ concentration, in milliequivalents per liter?

7.20 A sample of Ringer's solution contains the following concentrations (mEq/L) of cations: Na^+ 147, K^+ 4, and Ca^{2+} 4. If Cl^- is the only anion in the solution, what is the Cl^- concentration, in milliequivalents per liter?

Chemistry Link to Health

ELECTROLYTES IN BODY FLUIDS

Table 7.6 gives the concentrations of some typical electrolytes in blood plasma. There is a charge balance because the total number of positive charges is equal to the total number of negative charges. The use of a specific intravenous solution depends on the nutritional, electrolyte, and fluid needs of the individual patient. Examples of various types of solutions are given in Table 7.7.

TABLE 7.6 Some Typical Concentrations Electrolytes in Blood Plasma

Electrolyte	Concentration (mEq/L)
Cations	
Na^+	138
K^+	5
Mg^{2+}	3
Ca^{2+}	4
Total	150
Anions	
Cl^-	110
HCO_3^-	30
HPO_4^{2-}	4
Proteins	6
Total	150

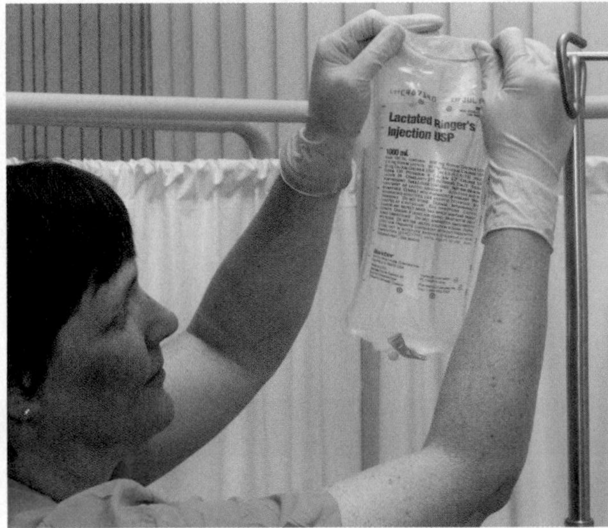

An intravenous solution is used to replace electrolytes in the body.

TABLE 7.7 Electrolyte Concentrations in Intravenous Replacement Solutions

Solution	Electrolytes (mEq/L)	Use
Sodium chloride (0.9%)	Na^+ 154, Cl^- 154	Replacement of fluid loss
Potassium chloride with 5.0% dextrose	K^+ 40, Cl^- 40	Treatment of malnutrition (low potassium levels)
Ringer's solution	Na^+ 147, K^+ 4, Ca^{2+} 4, Cl^- 155	Replacement of fluids and electrolytes lost through dehydration
Maintenance solution with 5.0% dextrose	Na^+ 40, K^+ 35, Cl^- 40, lactate$^-$ 20, HPO_4^{2-} 15	Maintenance of fluid and electrolyte levels
Replacement solution (extracellular)	Na^+ 140, K^+ 10, Ca^{2+} 5, Mg^{2+} 3, Cl^- 103, acetate$^-$ 47, citrate^{3-} 8	Replacement of electrolytes in extracellular fluids

LEARNING GOAL

Define solubility; distinguish between an unsaturated and a saturated solution; identify a salt as soluble or insoluble.

7.3 Solubility

The term **solubility** is used to describe the amount of a solute that can dissolve in a given amount of solvent. Many factors such as the type of solute, the type of solvent, and the temperature, affect the solubility of a solute. Solubility, usually expressed in grams of solute in 100 grams of solvent, is the maximum amount of solute that can be dissolved at a certain temperature. If a solute readily dissolves when added to the solvent, the solution does not contain the maximum amount of solute. We call this solution an **unsaturated solution**.

A solution that contains all the solute that can dissolve is a **saturated solution**. When a solution is saturated, the rate at which the solute dissolves becomes equal to the rate at which solid forms, a process known as recrystallization. Then there is no further change in the amount of dissolved solute in solution.

CASE STUDY
Kidney Stones and Saturated Solutions

$$\text{Solute} + \text{solvent} \underset{\text{Solute recrystallizes}}{\overset{\text{Solute dissolves}}{\rightleftharpoons}} \text{saturated solution}$$

We can prepare a saturated solution by adding an amount of solute greater than that needed to reach solubility. Stirring the solution will dissolve the maximum amount of solute and leave the excess on the bottom of the container. Once we have a saturated solution, the addition of more solute will only increase the amount of undissolved solute.

More solute can dissolve in an unsaturated solution, but not in a saturated solution.

SAMPLE PROBLEM 7.3

Saturated Solutions

At 20 °C, the solubility of KCl is 34 g/100 g of water. In the laboratory, a student mixes 75 g of KCl with 200. g of water at a temperature of 20 °C.

a. How much of the KCl will dissolve?
b. Is the solution saturated or unsaturated?
c. What is the mass, in grams, of any solid KCl on the bottom of the container?

SOLUTION

a. KCl has a solubility of 34 g of KCl in 100 g of water. Using the solubility as a conversion factor, we can calculate the maximum amount of KCl that can dissolve in 200. g of water as follows:

$$200.\ \cancel{\text{g } H_2O} \times \frac{34\text{ g KCl}}{100\ \cancel{\text{g } H_2O}} = 68\text{ g of KCl}$$

b. Because 75 g of KCl exceeds the amount (68 g) that can dissolve in 200. g of water, the KCl solution is saturated.

c. If we add 75 g of KCl to 200. g of water and only 68 g of KCl can dissolve, there is 7 g (75 g − 68 g) of solid (undissolved) KCl on the bottom of the container.

STUDY CHECK 7.3

At 40 °C, the solubility of KNO_3 is 65 g of KNO_3/100 g of H_2O. How many grams of KNO_3 will dissolve in 120 g of H_2O at 40 °C?

Chemistry Link to Health

GOUT AND KIDNEY STONES: A PROBLEM OF SATURATION IN BODY FLUIDS

The conditions of gout and kidney stones involve compounds in the body that exceed their solubility levels and form solid products. Gout affects adults, primarily men, over the age of 40. Attacks of gout may occur when the concentration of uric acid in blood plasma exceeds its solubility, which is 7 mg/100 mL of plasma at 37 °C. Insoluble deposits of needle-like crystals of uric acid can form in the cartilage, tendons, and soft tissues where they cause painful gout attacks. They may also form in the tissues of the kidneys, where they can cause renal damage. High levels of uric acid in the body can be caused by an increase in uric acid production, failure of the kidneys to remove uric acid, or by a diet with an overabundance of foods containing purines, which are metabolized to uric acid in the body. Foods in the diet that contribute to high levels of uric acid include certain meats, sardines, mushrooms, asparagus, and beans. Drinking alcoholic beverages may also significantly increase uric acid levels and bring about gout attacks.

Treatment for gout involves diet changes and drugs. Depending on the levels of uric acid, a medication, such as probenecid, can be used to help the kidneys eliminate uric acid, or allopurinol, which blocks the production of uric acid by the body.

Kidney stones are solid materials that form in the urinary tract. Most kidney stones are composed of calcium phosphate and calcium oxalate, although they can be solid uric acid. The excessive ingestion of minerals and insufficient water intake can cause the concentration of mineral salts to exceed their solubility and lead to the formation of kidney stones. When a kidney stone passes through the urinary tract, it causes considerable pain and discomfort, necessitating the use of painkillers and surgery. Sometimes ultrasound is used to break up kidney stones. Persons prone to kidney stones are advised to drink six to eight glasses of water every day to prevent saturation levels of minerals in the urine.

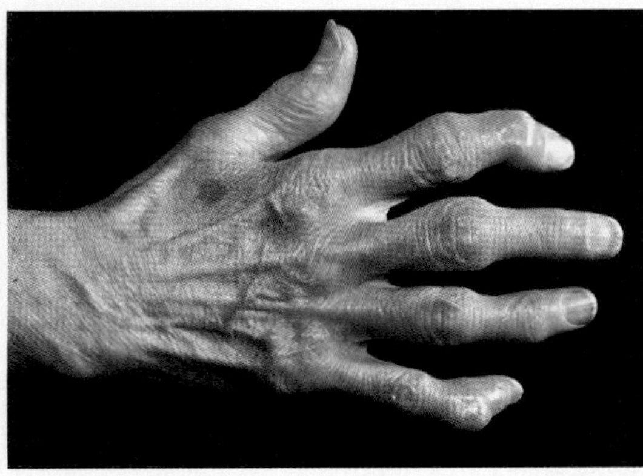

Gout occurs when uric acid exceeds its solubility.

Kidney stones form when calcium phosphate exceeds its solubility.

TUTORIAL
Solubility

SELF STUDY ACTIVITY
Solubility

TUTORIAL
Solubility of Gases and Solids in Water

Effect of Temperature on Solubility

The solubility of most solids is greater as temperature increases, which means that solutions usually contain more dissolved solute at higher temperature. A few substances show little change in solubility at higher temperatures, and a few are less soluble (see Figure 7.5). For example, when you add sugar to iced tea, some undissolved sugar may form on the bottom of the glass. But if you add sugar to hot tea, many teaspoons of sugar are needed before solid sugar appears. Hot tea dissolves more sugar than does cold tea because the solubility of sugar is much greater at a higher temperature.

When a saturated solution is carefully cooled, it becomes a *supersaturated solution* because it contains more solute than the solubility allows. Such a solution is unstable, and if the solution is agitated or if a solute crystal is added, the excess solute will recrystallize to give a saturated solution again.

The solubility of a gas in water decreases as the temperature increases. At higher temperatures, more gas molecules have the energy to escape from the solution. Perhaps you have observed the bubbles escaping from a cold carbonated soft drink as it warms. At high temperatures, bottles containing carbonated solutions may burst as more gas molecules leave the solution and increase the gas pressure inside the bottle. Biologists have found that increased temperatures in rivers and lakes cause the amount of dissolved oxygen to decrease until the warm water can no longer support a biological community. Electricity-generating plants are required to have their own ponds to use with their cooling towers to lessen the threat of thermal pollution in surrounding waterways.

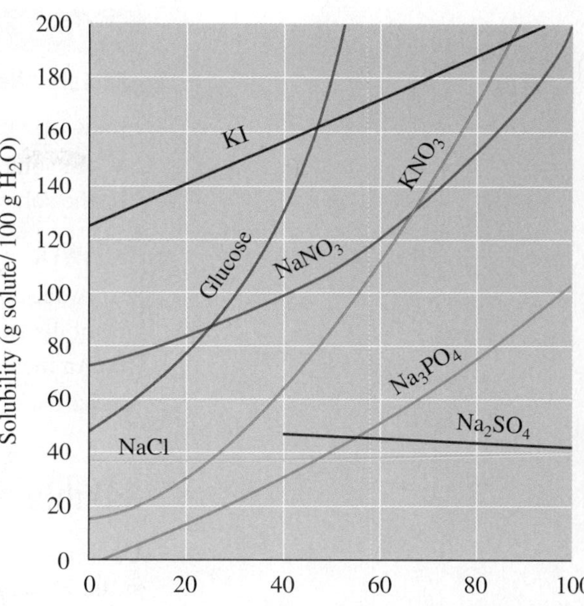

FIGURE 7.5 In water, most common solids are more soluble as the temperature increases.

Q Compare the solubility of $NaNO_3$ at 20 °C and 60 °C.

Henry's Law

Henry's law states that the solubility of gas in a liquid is directly related to the pressure of that gas above the liquid. At higher pressures, there are more gas molecules available to enter and dissolve in the liquid. A can of soda is carbonated by using CO_2 gas at high pressure to increase the solubility of the CO_2 in the beverage. When you open the can at atmospheric pressure, the pressure on the CO_2 drops, which decreases the solubility of CO_2. As a result, bubbles of CO_2 rapidly escape from the solution. The burst of bubbles is even more noticeable when you open a warm can of soda.

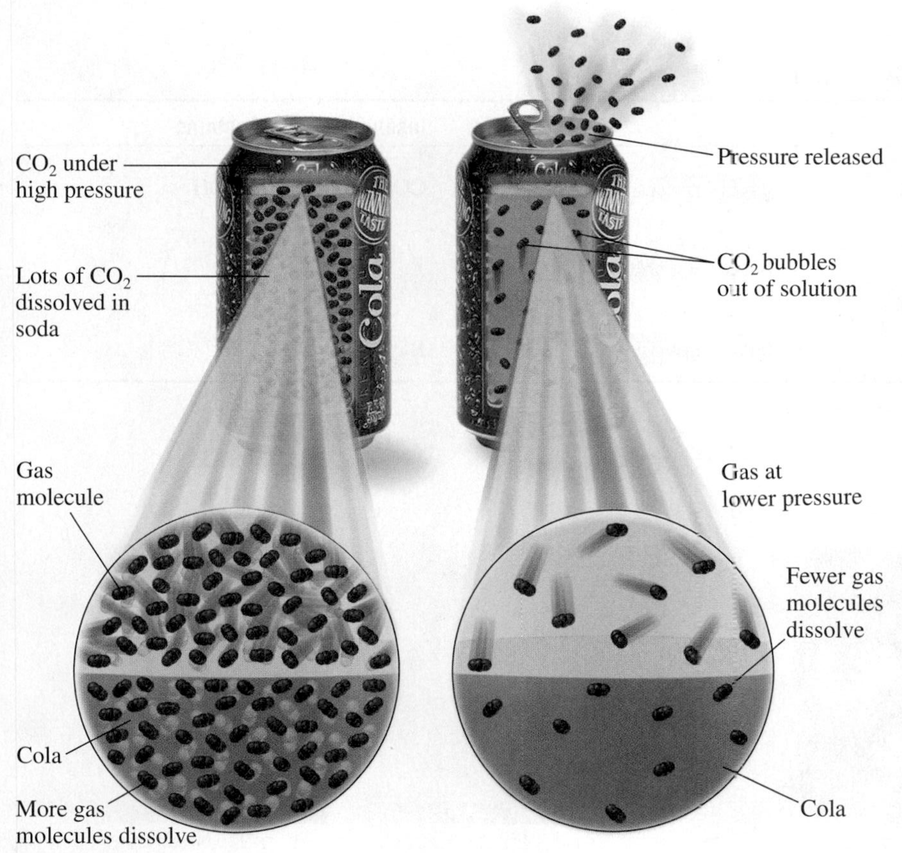

When the pressure of a gas above a solution decreases, the solubility of that gas in the solution also decreases.

Explore Your World

PREPARING SOLUTIONS

Obtain a drinking glass and water and place $\frac{1}{4}$ or $\frac{1}{2}$ cup of cold water in a glass. Begin adding 1 tablespoon of sugar at a time and stir thoroughly. Take a sip of the liquid in the glass as you proceed. As the sugar solution becomes more concentrated, you may need to stir for a few minutes until all the sugar dissolves. Each time, observe the solution after several minutes to determine when it is saturated.

Repeat the above activity with $\frac{1}{4}$ or $\frac{1}{2}$ cup of warm water. Count the number of tablespoons of sugar you need to form a saturated solution.

QUESTIONS

1. What do you notice about how sweet each sugar solution tastes?
2. How did you know when you obtained a saturated solution?
3. How much sugar dissolved in the warm water compared to the cold water?

CONCEPT CHECK 7.3

Factors Affecting Solubility

Indicate whether there is an increase or decrease in each of the following:

a. the solubility of sugar in water at 45 °C compared to its solubility in water at 25 °C
b. the solubility of O_2 in a lake as the water warms

ANSWER

a. A decrease in the temperature from 45 °C to 25 °C decreases the solubility of the sugar.
b. An increase in the temperature decreases the solubility of O_2 gas.

Soluble and Insoluble Salts

Up to now, we have considered ionic compounds that dissolve in water; they are **soluble salts**. However, some ionic compounds do not separate into ions in water. They are **insoluble salts** that remain as solids even in contact with water.

Salts that are soluble in water typically contain at least one of the following ions: Li^+, Na^+, K^+, NH_4^+, NO_3^- or $C_2H_3O_2^-$ (acetate). Most salts containing Cl^- are soluble, but AgCl, $PbCl_2$, and Hg_2Cl_2 are not; they are insoluble chloride salts. Similarly, most salts containing SO_4^{2-} are soluble, but a few are insoluble, as shown in Table 7.8. Most other salts are insoluble and do not dissolve in water (see Figure 7.6). In an insoluble salt, attractions between its positive and negative ions are too strong for the polar water molecules to break. We can use the solubility rules to predict whether a salt (a solid ionic compound) would be expected to dissolve in water. Table 7.9 illustrates the use of these rules.

TABLE 7.8 Solubility Rules for Ionic Solids in Water

Soluble If Salt Contains		Insoluble If Salt Contains
NH_4^+, Li^+, Na^+, K^+, NO_3^-, $C_2H_3O_2^-$ (acetate)	← but are soluble with	CO_3^{2-}, S^{2-}, PO_4^{3-}, OH^-
Cl^-, Br^-, I^-	but are not soluble with →	Ag^+, Pb^{2+}, Hg_2^{2+}
SO_4^{2-}	but are not soluble with →	Ba^{2+}, Pb^{2+}, Ca^{2+}, Sr^{2+}

CdS

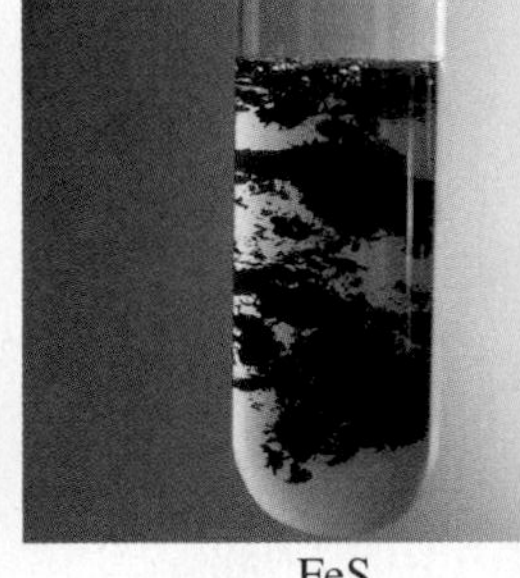

FeS

PbI_2

$Ni(OH)_2$

FIGURE 7.6 Mixing certain aqueous solutions produces insoluble salts.
Q What ions make each of these salts insoluble in water?

TABLE 7.9 Using Solubility Rules

Ionic Compound	Solubility in Water	Reasoning
K_2S	Soluble	Contains K^+
$Ca(NO_3)_2$	Soluble	Contains NO_3^-
$PbCl_2$	Insoluble	Is an insoluble chloride
NaOH	Soluble	Contains Na^+
$AlPO_4$	Insoluble	Contains no soluble ions

In medicine, the insoluble salt $BaSO_4$ is used as an opaque substance to enhance X-rays of the gastrointestinal tract. $BaSO_4$ is so insoluble that it does not dissolve in gastric fluids (see Figure 7.7). Other barium salts cannot be used because they would dissolve in water, releasing Ba^{2+} which is poisonous.

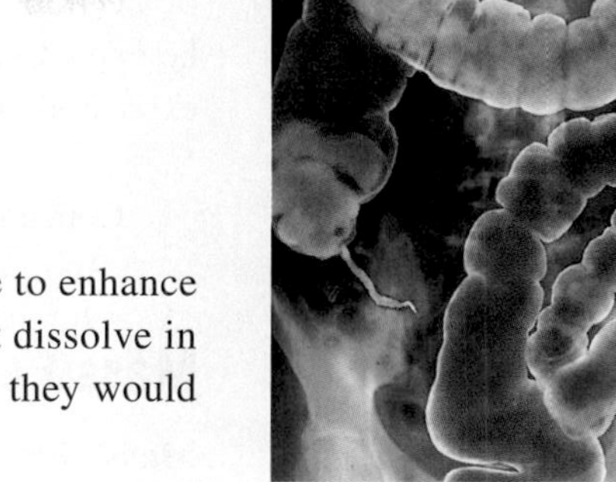

FIGURE 7.7 A barium sulfate enhanced X-ray of the abdomen shows the large intestine.
Q Is $BaSO_4$ a soluble or an insoluble substance?

CONCEPT CHECK 7.4

Soluble and Insoluble Salts

Predict whether each of the following salts is soluble in water and explain why:

a. Na_3PO_4 **b.** $CaCO_3$

ANSWER

a. The salt Na_3PO_4 is soluble in water because any compound that contains Na^+ is soluble.

b. The salt $CaCO_3$ is not soluble. The compound does not contain a soluble positive ion, which means that a calcium salt containing CO_3^{2-} is not soluble.

QUESTIONS AND PROBLEMS

Solubility

7.21 State whether each of the following refers to a saturated or unsaturated solution:
a. A crystal added to a solution does not change in size.
b. A sugar cube completely dissolves when added to a cup of coffee.

7.22 State whether each of the following refers to a saturated or unsaturated solution:
a. A spoonful of salt added to boiling water dissolves.
b. A layer of sugar forms on the bottom of a glass of tea as ice is added.

Use the following table for Problems 7.23–7.26:

	Solubility (g/100 g H_2O)	
Substance	**20 °C**	**50 °C**
KCl	34	43
$NaNO_3$	88	110
$C_{12}H_{22}O_{11}$ (sugar)	204	260

7.23 Use the previous table to determine whether each of the following solutions will be saturated or unsaturated at 20 °C:
a. Adding 25 g of KCl to 100. g of H_2O
b. Adding 11 g of $NaNO_3$ to 25 g of H_2O
c. Adding 400. g of sugar to 125 g of H_2O

7.24 Using the above table, determine whether each of the following solutions will be saturated or unsaturated at 50 °C:
a. Adding 25 g of KCl to 50. g of H_2O
b. Adding 150. g of $NaNO_3$ to 75 g of H_2O
c. Adding 80. g of sugar to 25 g of H_2O

7.25 A solution containing 80. g of KCl in 200. g of H_2O at 50 °C is cooled to 20 °C.
a. How many grams of KCl remain in solution at 20 °C?
b. How many grams of solid KCl crystallized after cooling?

7.26 A solution containing 80. g of $NaNO_3$ in 75 g of H_2O at 50 °C is cooled to 20 °C.
a. How many grams of $NaNO_3$ remain in solution at 20 °C?
b. How many grams of solid $NaNO_3$ crystallized after cooling?

7.27 Explain the following observations:
a. More sugar dissolves in hot tea than in iced tea.
b. Champagne in a warm room goes flat.
c. A warm can of soda has more spray when opened than a cold one.

7.28 Explain the following observations:
a. An open can of soda loses its "fizz" faster at room temperature than in the refrigerator.
b. Chlorine gas in tap water escapes as the sample warms to room temperature.
c. Less sugar dissolves in iced coffee than in hot coffee.

7.29 Predict whether each of the following ionic compounds is soluble in water:
a. LiCl **b.** AgCl **c.** $BaCO_3$
d. K_2O **e.** $Fe(NO_3)_3$

7.30 Predict whether each of the following ionic compounds is soluble in water:
a. PbS **b.** KI **c.** Na_2S
d. Ag_2O **e.** $CaSO_4$

LEARNING GOAL

Calculate the concentration of a solute in a solution; use concentration to calculate the amount of solute or solution.

7.4 Concentration of a Solution

The amount of solute dissolved in a certain amount of solution is called the **concentration** of the solution. We will look at ways to express a concentration as a ratio of a certain amount of solute in a given amount of solution. The amount of a solute may be expressed in units of grams, milliliters, or moles. The amount of a solution may be expressed in units of grams, milliliters, or liters.

$$\text{Concentration of a solution} = \frac{\text{amount of solute}}{\text{amount of solution}}$$

TUTORIAL
Calculating Mass Percent Concentration

Mass Percent (m/m) Concentration

Mass percent (m/m) describes the mass of the solute in grams for exactly 100 g of solution. In the calculation of mass percent (m/m), the units of mass of the solute and solution must be the same. If the mass of the solute is given as grams, then the mass of the solution must also be grams. The mass of the solution is the sum of the mass of the solute and the mass of the solvent.

$$\text{Mass percent (m/m)} = \frac{\text{mass of solute (g)}}{\text{mass of solute (g)} + \text{mass of solvent (g)}} \times 100\%$$

$$= \frac{\text{mass of solute (g)}}{\text{mass of solution (g)}} \times 100\%$$

Suppose we prepared a solution by mixing 8.00 g of KCl (solute) with 42.00 g of water (solvent). Together, the mass of the solute and mass of solvent give the mass of the solution (8.00 g + 42.00 g = 50.00 g). Mass percent is calculated by substituting in the mass of the solute and the mass of the solution into the mass percent expression.

$$\frac{8.00 \text{ g KCl}}{50.00 \text{ g solution}} \times 100\% = 16.0\% \text{ (m/m)}$$

$$\overbrace{\underset{\text{Solute}}{8.00 \text{ g KCl}} \underset{+}{+} \underset{\text{Solvent}}{42.00 \text{ g } H_2O}}$$

Add 8.00 g of KCl

Add water until the solution weighs 50.00 g

When water is added to 8.00 g of KCl to form 50.00 g of a KCl solution, the mass percent concentration is 16.0% (m/m).

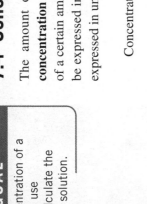

CONCEPT CHECK 7.5

Mass Percent (m/m) Concentration

A NaBr solution is prepared by dissolving 4.0 g of NaBr in 50.0 g of H_2O.

a. What is the mass of the solution?

b. Is the final concentration of the NaBr solution equal to 7.4% (m/m), 8.0% (m/m) or 80.% (m/m)?

ANSWER

a. The mass of the NaBr solution is the sum of 4.0 g of NaBr and the mass of the H_2O solvent, which is 54.0 g (4.0 g of NaBr + 50.0 g of H_2O).

b. The mass percent of the NaBr is equal to 7.4% (m/m).

$$\frac{4.0 \text{ g NaBr}}{54.0 \text{ g solution}} \times 100\% = 7.4\% \text{ (m/m) NaBr solution}$$

SAMPLE PROBLEM 7.4

Calculating Mass Percent (m/m) Concentration

What is the mass percent of a solution prepared by dissolving 30.0 g of NaOH in 120.0 g of H_2O?

SOLUTION

Step 1 Determine the quantities of solute and solution.

$$\begin{aligned}\text{Mass of solute} &= 30.0\text{ g of NaOH}\\ \text{Mass of solvent} &= 120.0\text{ g of } H_2O\\ \text{Mass of solution} &= 150.0\text{ g of NaOH solution}\end{aligned}$$

Step 2 Write the concentration expression.

$$\text{Mass percent (m/m)} = \frac{\text{grams of solute}}{\text{grams of solution}} \times 100\%$$

Step 3 Substitute solute and solution quantities into the expression.

$$\text{Mass percent (m/m)} = \frac{30.0\text{ g NaOH}}{150.0\text{ g solution}} \times 100\%$$
$$= 20.0\%\text{ (m/m) NaOH solution}$$

STUDY CHECK 7.4

What is the mass percent (m/m) of NaCl solution made by dissolving 2.0 g of NaCl in 56.0 g of H_2O?

Guide to Calculating Solution Concentration

1 Determine the quantities of solute and solution.

2 Write the concentration expression.

3 Substitute solute and solution quantities into the expression.

Volume Percent (v/v) Concentration

Because the volumes of liquids or gases are easily measured, the concentrations of their solutions are often expressed as **volume percent (v/v)**. The units of volume used in the ratio must be the same, for example, both in milliliters or both in liters.

$$\text{Volume percent (v/v)} = \frac{\text{volume of solute}}{\text{volume of solution}} \times 100\%$$

We interpret a volume percent as the volume of solute in exactly 100 mL of solution. In the wine industry, a label that reads 12% (v/v) means 12 mL of ethanol, C_2H_5OH, in 100 mL of wine.

SAMPLE PROBLEM 7.5

Calculating Volume Percent (v/v) Concentration

A student prepared a solution by adding water to 18 mL of ethanol (C_2H_5OH) to give a final solution volume of 150 mL. What is the volume percent (v/v) of the ethanol solution?

SOLUTION

Step 1 Determine the quantities of solute and solution.

$$\begin{aligned}\text{Volume of solute} &= 18\text{ mL of } C_2H_5OH\\ \text{Volume of solution} &= 150\text{ mL of } C_2H_5OH\text{ solution}\end{aligned}$$

Step 2 Write the concentration expression.

$$\text{Volume percent (v/v)} = \frac{\text{volume of solute}}{\text{volume of solution}} \times 100\%$$

Step 3 Substitute solute and solution quantities into the expression.

$$\text{Volume percent (v/v)} = \frac{18\text{ mL } C_2H_5OH}{150\text{ mL solution}} \times 100\%$$
$$= 12\%\text{ (v/v) } C_2H_5OH\text{ solution}$$

STUDY CHECK 7.5

What is the volume percent (v/v) of Br_2 in a solution prepared by dissolving 12 mL of bromine (Br_2) in the solvent carbon tetrachloride to make 250 mL of solution?

Mass/Volume Percent (m/v) Concentration

Mass/volume percent (m/v) describes the mass of the solute in grams for exactly 100 mL of solution. In the calculation of mass/volume percent, the unit of mass of the solute is grams and the unit of volume is milliliters.

$$\text{Mass/volume percent (m/v)} = \frac{\text{grams of solute}}{\text{milliliters of solution}} \times 100\%$$

The mass/volume percent is widely used in hospitals and pharmacies for the preparation of intravenous solutions and medicines. For example, a 5% (m/v) glucose solution contains 5 g of glucose in 100 mL of solution. The volume of solution represents the combination of the glucose and H_2O.

5.0 g of KI 2.0% (m/v) KI solution

SAMPLE PROBLEM 7.6

Calculating Mass/Volume Percent (m/v) Concentration

A student prepared a solution by dissolving 5.0 g of KI in enough water to give a final volume of 250 mL. What is the mass/volume percent of the KI solution?

SOLUTION

Step 1 Determine the quantities of solute and solution.

Mass of solute = 5.0 g of KI

Volume of solution = 250 mL of KI solution

Step 2 Write the concentration expression.

$$\text{Mass/volume percent (m/v)} = \frac{\text{mass of solute}}{\text{volume of solution}} \times 100\%$$

Step 3 Substitute solute and solution quantities into the expression.

$$\text{Mass/volume percent (m/v)} = \frac{5.0 \text{ g KI}}{250 \text{ mL solution}} \times 100\%$$

$$= 2.0\% \text{ (m/v) KI solution}$$

STUDY CHECK 7.6

What is the mass/volume percent (m/v) of NaOH in a solution prepared by dissolving 12 g of NaOH in enough water to make 220 mL of solution?

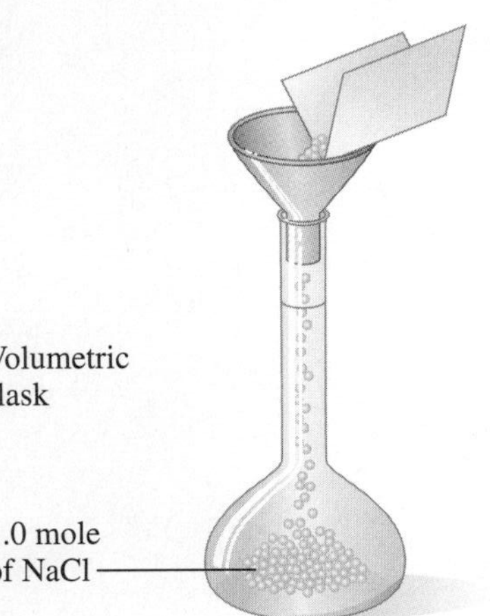

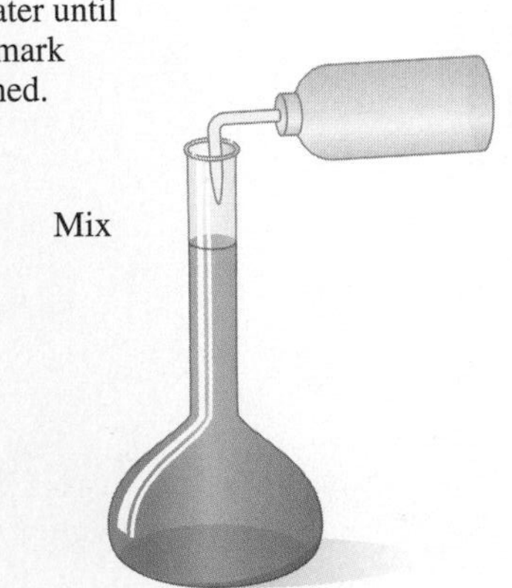

A 1.0 molar (M) NaCl solution

Molarity (M) Concentration

When chemists work with solutions, they often use **molarity (M)**, a concentration that states the number of moles of solute in exactly one liter of solution.

$$\text{Molarity (M)} = \frac{\text{moles of solute}}{\text{liters of solution}}$$

The molarity of a solution can be calculated knowing the moles of solute and the volume of solution in liters. For example, if 1.0 mole of NaCl were dissolved in enough water to prepare 1.0 L of solution, the resulting NaCl solution has a molarity of 1.0 M. The abbreviation M indicates the units of moles per liter (moles/L).

$$\text{M} = \frac{\text{moles of solute}}{\text{liters of solution}} = \frac{1.0 \text{ mole NaCl}}{1 \text{ L solution}} = 1.0 \text{ M NaCl solution}$$

SAMPLE PROBLEM 7.7

Calculating Molarity

What is the molarity (M) of 60.0 g of NaOH in 0.250 L of solution?

SOLUTION

Step 1 **Determine the quantities of solute and solution.** For molarity, we need the quantity in moles, and the volume of the solution in liters. To calculate the moles of NaOH, we need to write the equality and conversion factors for the molar mass of NaOH. Then the moles in 60.0 g of NaOH can be determined.

$$1 \text{ mole of NaOH} = 40.0 \text{ g of NaOH}$$

$$\frac{1 \text{ mole NaOH}}{40.0 \text{ g NaOH}} \quad \text{and} \quad \frac{40.0 \text{ g NaOH}}{1 \text{ mole NaOH}}$$

$$\text{Moles of NaOH} = 60.0 \cancel{\text{g NaOH}} \times \frac{1 \text{ mole NaOH}}{40.0 \cancel{\text{g NaOH}}} = 1.50 \text{ moles of NaOH}$$

$$\text{Volume of solution} = 0.250 \text{ L of NaOH solution}$$

Step 2 **Write the concentration expression.**

$$\text{Molarity (M)} = \frac{\text{moles of solute}}{\text{liters of solution}}$$

Step 3 **Substitute solute and solution quantities into the expression.**

$$\text{M} = \frac{1.50 \text{ moles NaOH}}{0.250 \text{ L solution}} = \frac{6.00 \text{ moles NaOH}}{1 \text{ L solution}} = 6.00 \text{ M NaOH solution}$$

STUDY CHECK 7.7

What is the molarity of a solution that contains 75.0 g of KNO_3 dissolved in 0.350 L of solution?

Table 7.10 summarizes the types of units used in the various types of concentration expressions for solution.

TABLE 7.10 Summary of Types of Concentration Expressions and Their Units

Concentration Units	Mass Percent (m/m)	Volume Percent (v/v)	Mass/Volume Percent (m/v)	Molarity (M)
Solute	g	mL	g	mole
Solution	g	mL	mL	L

Using Concentrations as Conversion Factors

In the preparation of solutions, we often need to calculate the amount of solute or volume of solution. Then the concentration is useful as a conversion factor. Some examples of percent concentrations and molarity, their meanings, and conversion factors are given in Table 7.11. Some examples of using percent concentration or molarity as conversion factors are given in Sample Problems 7.8 and 7.9.

TUTORIAL
Percent Concentration as a Conversion Factor

TABLE 7.11 Conversion Factors from Concentrations

Percent Concentration	Meaning	Conversion Factors		
10% (m/m) KCl solution	10 g of KCl in 100 g of KCl solution	$\frac{10\text{ g KCl}}{100\text{ g solution}}$	and	$\frac{100\text{ g solution}}{10\text{ g KCl}}$
12% (v/v) ethanol solution	12 mL of ethanol in 100 mL of ethanol solution	$\frac{12\text{ mL ethanol}}{100\text{ mL solution}}$	and	$\frac{100\text{ mL solution}}{12\text{ mL ethanol}}$
5% (m/v) glucose solution	5 g of glucose in 100 mL of glucose solution	$\frac{5\text{ g glucose}}{100\text{ mL solution}}$	and	$\frac{100\text{ mL solution}}{5\text{ g glucose}}$
Molarity				
6.0 M HCl solution	6.0 moles of HCl in 1 liter of HCl solution	$\frac{6.0\text{ moles HCl}}{1\text{ L solution}}$	and	$\frac{1\text{ L solution}}{6.0\text{ moles HCl}}$

Guide to Using Concentration to Calculate Mass or Volume

1. State the given and needed quantities.
2. Write a plan to calculate mass or volume.
3. Write equalities and conversion factors.
4. Set up the problem to calculate mass or volume.

SAMPLE PROBLEM 7.8

Using Mass/Volume Percent to Find Mass of Solute

A topical antibiotic is 1.0% (m/v) Clindamycin. How many grams of Clindamycin are in 60. mL of the 1.0% (m/v) solution?

SOLUTION

Step 1 State the given and needed quantities.

Given 1.0% (m/v) Clindamycin solution **Need** grams of Clindamycin

Step 2 Write a plan to calculate mass or volume.

milliliters of solution → % (m/v) factor → grams of Clindamycin

Step 3 Write equalities and conversion factors. The percent (m/v) indicates the grams of a solute in every 100 mL of a solution. The 1.0% (m/v) can be written as two conversion factors.

100 mL of solution = 1.0 g of Clindamycin

$$\frac{1.0\text{ g Clindamycin}}{100\text{ mL solution}} \quad \text{and} \quad \frac{100\text{ mL solution}}{1.0\text{ g Clindamycin}}$$

Step 4 Set up the problem to calculate mass or volume. The volume of the solution is converted to mass of solute using the conversion factor that cancels mL.

$$60.\ \cancel{\text{mL solution}} \times \frac{1.0\text{ g Clindamycin}}{100\ \cancel{\text{mL solution}}} = 0.60\text{ g of Clindamycin}$$

STUDY CHECK 7.8

Calculate the grams of KCl in 225 g of an 8.00% (m/m) KCl solution.

SAMPLE PROBLEM 7.9

Using Molarity to Calculate Volume of Solution

How many liters of a 2.00 M NaCl solution are needed to provide 67.3 g of NaCl?

SOLUTION

Step 1 State the given and needed quantities.

Given 67.3 g of NaCl from a 2.00 M NaCl solution

Need liters of NaCl solution

Step 2 Write a plan to calculate mass or volume.

grams of NaCl → Molar mass → moles of NaCl → Molarity → liters of NaCl solution

Step 3 Write equalities and conversion factors.

$$1 \text{ mole of NaCl} = 58.5 \text{ g of NaCl}$$

$$\frac{1 \text{ mole NaCl}}{58.5 \text{ g NaCl}} \quad \text{and} \quad \frac{58.5 \text{ g NaCl}}{1 \text{ mole NaCl}}$$

The molarity of any solution can be written as two conversion factors.

$$1 \text{ L of NaCl solution} = 2.00 \text{ moles of NaCl}$$

$$\frac{1 \text{ L NaCl solution}}{2.00 \text{ moles NaCl}} \quad \text{and} \quad \frac{2.00 \text{ moles NaCl}}{1 \text{ L NaCl solution}}$$

Step 4 Set up the problem to calculate mass or volume.

$$\text{Liters of NaCl solution} = 67.3 \cancel{\text{g NaCl}} \times \frac{1 \cancel{\text{mole NaCl}}}{58.5 \cancel{\text{g NaCl}}} \times \frac{1 \text{ L NaCl solution}}{2.00 \cancel{\text{moles NaCl}}}$$

$$= 0.575 \text{ L of NaCl solution}$$

STUDY CHECK 7.9

How many milliliters of a 6.0 M HCl solution will provide 4.5 moles of HCl?

QUESTIONS AND PROBLEMS

Concentration of a Solution

7.31 What is the difference between a 5% (m/m) glucose solution and a 5% (m/v) glucose solution?

7.32 What is the difference between a 10% (v/v) methyl alcohol (CH_3OH) solution and a 10% (m/m) methyl alcohol solution?

7.33 Calculate the mass percent (m/m) for the solute in each of the following solutions:
a. 25 g of KCl and 125 g of H_2O
b. 12 g of sugar in 225 g of tea solution with sugar
c. 8.0 g of $CaCl_2$ in 80.0 g of $CaCl_2$ solution

7.34 Calculate the mass percent (m/m) for the solute in each of the following solutions:
a. 75 g of NaOH in 325 g of NaOH solution
b. 2.0 g of KOH and 20.0 g of H_2O
c. 48.5 g of Na_2CO_3 in 250.0 g of Na_2CO_3 solution

7.35 Calculate the mass/volume percent (m/v) for the solute in each of the following solutions:
a. 75 g of Na_2SO_4 in 250 mL of Na_2SO_4 solution
b. 39 g of sucrose in 355 mL of a carbonated drink

7.36 Calculate the mass/volume percent (m/v) for the solute in each of the following solutions:
a. 2.50 g of LiCl in 40.0 mL of LiCl solution
b. 7.5 g of casein in 120 mL of low-fat milk

7.37 Calculate the grams or milliliters of solute needed to prepare the following solutions:
a. 50.0 mL of a 5.0% (m/v) KCl solution
b. 1250 mL of a 4.0% (m/v) NH_4Cl solution
c. 250. mL of a 10.0% (v/v) acetic acid solution

7.38 Calculate the grams or milliliters of solute needed to prepare the following solutions:
a. 150 mL of a 40.0% (m/v) $LiNO_3$ solution
b. 450 mL of a 2.0% (m/v) KOH solution
c. 225 mL of a 15% (v/v) isopropyl alcohol solution

7.39 A mouthwash contains 22.5% (v/v) alcohol. If the bottle of mouthwash contains 355 mL, what is the volume, in milliliters, of alcohol?

7.40 A bottle of champagne is 11% (v/v) alcohol. If there are 750 mL of champagne in the bottle, how many milliliters of alcohol are present?

7.41 A patient receives 100. mL of 20.% (m/v) mannitol solution every hour.
a. How many grams of mannitol are given in 1 h?
b. How many grams of mannitol does the patient receive in 12 h?

7.42 A patient receives 250 mL of a 4.0% (m/v) amino acid solution twice a day.
a. How many grams of amino acids are in 250 mL of solution?
b. How many grams of amino acids does the patient receive in 1 day?

7.43 A patient needs 100. g of glucose in the next 12 h. How many liters of a 5% (m/v) glucose solution must be given?

7.44 A patient received 2.0 g of NaCl in 8 h. How many milliliters of a 0.90% (m/v) NaCl (saline) solution were delivered?

7.45 Calculate the amount of solution (g or mL) that contains each of the following amounts of solute:
a. 5.0 g of $LiNO_3$ from a 25% (m/m) $LiNO_3$ solution
b. 40.0 g of KOH from a 10.0% (m/v) KOH solution
c. 2.0 mL of formic acid from a 10.0% (v/v) formic acid solution

7.46 Calculate the amount of solution (g or mL) that contains each of the following amounts of solute:
a. 7.50 g of NaCl from a 2.0% (m/m) NaCl solution
b. 4.0 g of NaF from a 25% (m/v) NaF solution
c. 20.0 g of KBr from an 8.0% (m/m) KBr solution

7.47 Calculate the molarity of each of the following solutions:
a. 2.00 moles of glucose in 4.00 L of solution
b. 4.00 g of KOH in 2.00 L of solution
c. 5.85 g of NaCl in 400. mL of solution

7.48 Calculate the molarity of each of the following solutions:
a. 0.500 mole of glucose in 0.200 L of solution
b. 36.5 g of HCl in 1.00 L of solution
c. 30.0 g of NaOH in 350. mL of solution

7.49 Calculate the grams of solute needed to prepare each of the following solutions:
a. 2.00 L of a 1.50 M NaOH solution
b. 4.00 L of a 0.200 M KCl solution
c. 25.0 mL of a 6.00 M HCl solution

7.50 Calculate the grams of solute needed to prepare each of the following solutions:
a. 2.00 L of a 6.00 M NaOH solution
b. 5.00 L of a 0.100 M $CaCl_2$ solution
c. 175 mL of a 3.00 M $NaNO_3$ solution

7.51 What volume is needed to obtain each of the following amounts of solute?
a. liters of a 2.00 M KBr solution to obtain 3.00 moles of KBr
b. liters of a 1.50 M NaCl solution to obtain 15.0 moles of NaCl
c. milliliters of a 0.800 M $Ca(NO_3)_2$ solution to obtain 0.0500 mole of $Ca(NO_3)_2$

7.52 What volume is needed to obtain each of the following amounts of solute?
a. liters of 4.00 M KCl solution to obtain 0.100 mole of KCl
b. liters of a 6.00 M HCl solution to obtain 5.00 moles of HCl
c. milliliters of a 2.50 M K_2SO_4 solution to obtain 1.20 moles of K_2SO_4

LEARNING GOAL

Describe the dilution of a solution; calculate the new concentration or volume of a diluted solution.

7.5 Dilution of Solutions

In chemistry and biology, we often prepare diluted solutions from more concentrated solutions. In a process called **dilution**, a solvent, usually water, is added to a solution, which increases the volume. As a result, the concentration of the solution decreases. In an everyday example, you are making a dilution when you add three cans of water to a can of concentrated orange juice.

1 can of orange juice concentrate + 3 cans of water = 4 cans of orange juice

MC

TUTORIAL
Dilution

TUTORIAL
Solution Dilution

Although the addition of solvent increases the volume, the amount of solute in the diluted solution is equal to the amount of solute present before dilution (see Figure 7.8).

Amount of solute in the concentrated solution = amount of solute in the diluted solution

Grams or moles of solute (Concentrated solution) = grams or moles of solute (Diluted solution)

We can write this equality in terms of the concentration, C, and the volume, V. The concentration, C, may be percent concentration or molarity.

$$\underset{\text{Concentrated solution}}{C_1V_1} = \underset{\text{Diluted solution}}{C_2V_2}$$

If we are given any 3 of the 4 variables, we can rearrange the dilution expression to solve for the unknown quantity as seen in Sample Problem 7.10.

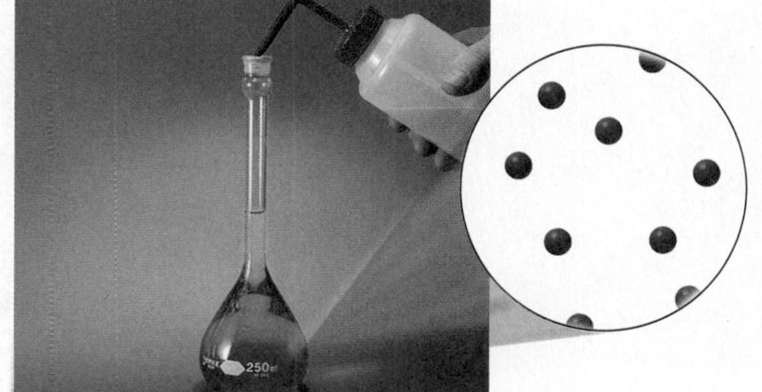

FIGURE 7.8 When water is added to a concentrated solution, there is no change in the number of particles. However, the solute particles spread out as the volume of the diluted solution increases.

Q What is the concentration of the diluted solution after an equal volume of water is added to a sample of a 6 M HCl solution?

CONCEPT CHECK 7.6

Volume of a Diluted Solution

A 50.0-mL sample of a 20.0% (m/v) $SrCl_2$ solution is diluted with water to give a 5.00% (m/v) $SrCl_2$ solution. Use this data to complete the table with the given concentrations and volumes of the solutions. Indicate what we know (increases or decreases), and predict the change in the unknown (increases or decreases).

Concentrated Solution	Diluted Solution	Know	Predict
C_1 =	C_2 =		
V_1 =	V_2 = ? mL		

ANSWER

Concentrated Solution	Diluted Solution	Know	Predict
C_1 = 20.0% (m/v)	C_2 = 5.00% (m/v)	C decreases	
V_1 = 50.0 mL	V_2 = ? mL		*V* increases

SAMPLE PROBLEM 7.10

Volume of a Diluted Solution

What volume (mL) of a 2.5% (m/v) KOH solution can be prepared by diluting 50.0 mL of a 12% (m/v) KOH solution?

SOLUTION

Step 1 Prepare a table of the concentrations and volumes of the solutions. Organize the data in a table making sure that the units of concentration and volume are the same.

Concentrated Solution	Diluted Solution	Know	Predict
C_1 = 12% (m/v) KOH	C_1 = 2.5% (m/v) KOH	C decreases	
V_1 = 50.0 mL	V_2 = ? mL		*V* increases

Guide to Calculating Dilution Quantities

1. Prepare a table of the concentrations and volumes of the solutions.
2. Rearrange the dilution expression to solve for the unknown quantity.
3. Substitute the known quantities into the dilution expression and solve.

Step 2 Rearrange the dilution expression to solve for the unknown quantity.

$$C_1V_1 = C_2V_2$$

$$\frac{C_1V_1}{C_2} = \frac{\cancel{C_2}V_2}{\cancel{C_2}}$$

$$V_2 = V_1 \times \frac{C_1}{C_2}$$

Step 3 Substitute the known quantities into the dilution expression and solve.

$$V_2 = 50.0\text{ mL} \times \frac{12.0\cancel{\%}}{2.50\cancel{\%}} = 240\text{ mL of diluted KOH solution}$$

When the initial volume (V_1) is multiplied by a ratio of the percent concentrations that is greater than 1, the volume of the diluted solution increases as predicted in Step 1.

STUDY CHECK 7.10

What is the percent concentration (m/v) of the diluted solution when 25.0 mL of a 15% (m/v) HCl solution is diluted to 125 mL?

SAMPLE PROBLEM 7.11

Molarity of a Diluted Solution

What is the molarity of a solution prepared when 75.0 mL of a 4.00 M KCl solution is diluted to a volume of 500. mL?

SOLUTION

Step 1 Prepare a table of the concentrations and volumes of the solutions.

Concentrated Solution	Diluted Solution	Know	Predict
M_1 = 4.00 M KCl	M_2 = ? M KCl		M decreases
V_1 = 75.0 mL	V_2 = 500. mL	V increases	

Step 2 Rearrange the dilution expression to solve for the unknown quantity.

$$M_1V_1 = M_2V_2$$

$$\frac{M_1V_1}{V_2} = \frac{M_2\cancel{V_2}}{\cancel{V_2}}$$

$$M_2 = M_1 \times \frac{V_1}{V_2}$$

Step 3 Substitute the known quantities into the dilution expression and solve.

$$M_2 = 4.00\text{ M} \times \frac{75.0\ \cancel{\text{mL}}}{500.\ \cancel{\text{mL}}} = 0.600\text{ M (diluted KCl solution)}$$

When the initial molarity (M_1) is multiplied by a ratio of the volumes that is less than 1, the molarity of the diluted solution decreases as predicted in Step 1.

STUDY CHECK 7.11

You need to prepare 600. mL of a 2.00 M NaOH solution from a 10.0 M NaOH solution. What volume, in milliliters, of the 10.0 M NaOH solution do you need?

QUESTIONS AND PROBLEMS

Dilution of Solutions

7.53 To make tomato soup, you add one can of water to the condensed soup. Why is this a dilution?

7.54 A can of frozen lemonade calls for the addition of three cans of water to make a pitcher of the beverage. Why is this a dilution?

7.55 Calculate the concentration of each of the following diluted solutions:

a. 2.0 L of a 6.0 M HCl solution is added to water so that the final volume is 6.0 L.

b. Water is added to 0.50 L of a 12 M NaOH solution to make 3.0 L of a diluted NaOH solution.

c. A 10.0-mL sample of a 25% (m/v) KOH solution is diluted with water so that the final volume is 100.0 mL.

d. A 50.0-mL sample of a 15% (m/v) H_2SO_4 solution is added to water to give a final volume of 250 mL.

7.56 Calculate the concentration of each of the following diluted solutions:

a. 1.0 L of a 4.0 M HNO_3 solution is added to water so that the final volume is 8.0 L.

b. Water is added to 0.25 L of a 6.0 M NaF solution to make 2.0 L of a diluted NaF solution.

c. A 50.0-mL sample of an 8.0% (m/v) KBr solution is diluted with water so that the final volume is 200.0 mL.

d. A 5.0-mL sample of a 50.0% (m/v) acetic acid ($HC_2H_3O_2$) solution is added to water to give a final volume of 25 mL.

7.57 What is the volume, in milliliters, of each of the following diluted solutions?

a. A 1.5 M HCl solution prepared from 20.0 mL of a 6.0 M HCl solution

b. A 2.0% (m/v) LiCl solution prepared from 50.0 mL of a 10.0% (m/v) LiCl solution

c. A 0.500 M H_3PO_4 solution prepared from 50.0 mL of a 6.00 M H_3PO_4 solution

d. A 5.0% (m/v) glucose solution prepared from 75 mL of a 12% (m/v) glucose solution

7.58 What is the volume, in milliliters, of each of the following diluted solutions?

a. A 1.0% (m/v) H_2SO_4 solution prepared from 10.0 mL of a 20.0% H_2SO_4 solution

b. A 0.10 M HCl solution prepared from 25 mL of a 6.0 M HCl solution

c. A 1.0 M NaOH solution prepared from 50.0 mL of a 12 M NaOH solution

d. A 1.0% (m/v) $CaCl_2$ solution prepared from 18 mL of a 4.0% (m/v) $CaCl_2$ solution

7.59 Determine the volume, in milliliters, required to prepare each of the following diluted solutions:

a. 255 mL of a 0.200 M HNO_3 solution from a 4.00 M HNO_3 solution

b. 715 mL of a 0.100 M $MgCl_2$ solution using a 6.00 M $MgCl_2$ solution

c. 0.100 L of a 0.150 M KCl solution using an 8.00 M KCl solution

7.60 Determine the volume, in milliliters, required to prepare each of the following diluted solutions:

a. 20.0 mL of a 0.250 M KNO_3 solution from a 6.00 M KNO_3 solution

b. 25.0 mL of 2.50 M H_2SO_4 solution using a 12.0 M H_2SO_4 solution

c. 0.500 L of a 1.50 M NH_4Cl solution using a 10.0 M NH_4Cl solution

7.6 Properties of Solutions

LEARNING GOAL

Identify a mixture as a solution, a colloid, or a suspension. Describe how the number of particles in a solution affects the freezing point, boiling point, and osmotic pressure of a solution.

The size and number of solute particles in different types of mixtures play an important role in determining the properties of that mixture.

Solutions

In the solutions discussed up to now, the solute was dissolved as small particles that are uniformly dispersed throughout the solvent to give a homogeneous solution. When you observe a solution, such as salt water, you cannot visually distinguish the solute from the solvent. The solution appears transparent. The particles are so small that they go through filters and through *semipermeable membranes* such as the cell membranes in the body.

Colloids

In **colloids**, the solute particles are large molecules, such as proteins, or groups of molecules or ions. Colloids, similar to solutions, are homogeneous mixtures that do not separate or settle out. Colloidal particles are small enough to pass through filters, but too large to pass through semipermeable membranes. Table 7.12 lists several examples of colloids.

TABLE 7.12 Examples of Colloids

Colloid	Substance Dispersed	Dispersing Medium
Fog, clouds, hair sprays	Liquid	Gas
Dust, smoke	Solid	Gas
Shaving cream, whipped cream, soapsuds	Gas	Liquid
Styrofoam, marshmallows	Gas	Solid
Mayonnaise, milk	Liquid	Liquid
Gelatin, agar	Liquid	Solid
Blood plasma, paints (latex)	Solid	Liquid

Suspensions

Suspensions are heterogeneous, nonuniform mixtures that are very different from solutions or colloids. The particles of a suspension are so large that they can often be seen with the naked eye. They are trapped by filters and semipermeable membranes.

The weight of the suspended solute particles causes them to settle out soon after mixing. If you stir muddy water, it mixes but then quickly separates as the suspended particles settle to the bottom and leave clear liquid at the top. You can find suspensions among the medications in a hospital or in your medicine cabinet. These include Kaopectate, calamine lotion, antacid mixtures, and liquid penicillin. It is important to read the label that states "shake well before using" so that the particles form a suspension.

Water-treatment plants make use of the properties of suspensions to purify water. When flocculants such as aluminum sulfate or ferric sulfate are added to untreated water, they react with impurities to form large suspended particles called floc. In the water-treatment plant, a system of filters traps the suspended particles but clean water passes through.

Table 7.13 compares the different types of mixtures and Figure 7.9 illustrates some properties of solutions, colloids, and suspensions.

TABLE 7.13 Comparison of Solutions, Colloids, and Suspensions

Type of Mixture	Type of Particle	Settling	Separation
Solution	Small particles such as atoms, ions, or small molecules	Particles do not settle	Particles cannot be separated by filters or semipermeable membranes
Colloid	Larger molecules or groups of molecules or ions	Particles do not settle	Particles can be separated by semipermeable membranes but not by filters
Suspension	Very large particles that may be visible	Particles settle rapidly	Particles can be separated by filters

Chemistry Link to Health

COLLOIDS AND SOLUTIONS IN THE BODY

In the body, colloids are retained by semipermeable membranes. For example, the intestinal lining allows solution particles to pass into the blood and lymph circulatory systems. However, the colloids from foods are too large to pass through the membrane, and they remain in the intestinal tract. Digestion breaks down large colloidal particles, such as starch and protein, into smaller particles, such as glucose and amino acids that can pass through the intestinal membrane and enter the circulatory system. Certain foods, such as bran, a fiber, cannot be broken down by human digestive processes, and they move through the intestine intact.

Because large proteins, such as enzymes, are colloids, they remain inside cells. However, many of the substances that must be obtained by cells, such as oxygen, amino acids, electrolytes, glucose, and minerals, can pass through cellular membranes. Waste products, such as urea and carbon dioxide, pass out of the cell to be excreted.

CONCEPT CHECK 7.7

Classifying Types of Mixtures

Classify each of the following as a solution, a colloid, or a suspension:

a. a mixture that has particles that settle upon standing
b. a mixture whose solute particles pass through both filters and membranes
c. an enzyme, which is a large protein molecule, that cannot pass through cellular membranes, but does pass through a filter

ANSWER

a. A suspension has very large particles that settle upon standing.
b. A solution contains particles small enough to pass through both filters and membranes.
c. A colloid contains particles that are small enough to pass through a filter, but too large to pass through a membrane.

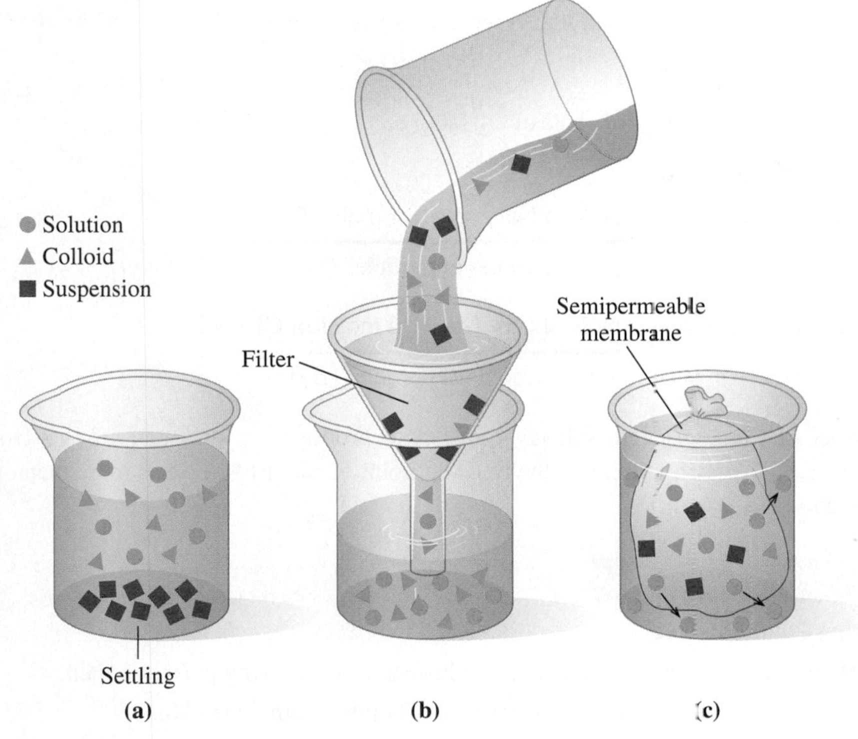

FIGURE 7.9 Properties of different types of mixtures: **(a)** suspensions settle out; **(b)** suspensions are separated by a filter; **(c)** solution particles go through a semipermeable membrane, but colloids and suspensions do not.

Q A filter can be used to separate suspension particles from a solution, but a semipermeable membrane is needed to separate colloids from a solution. Explain.

Freezing Point Lowering and Boiling Point Elevation

When a solute dissolves in water, the particles in the solution affect the physical properties such as freezing point and boiling point. Any aqueous solution will have a lower freezing point and a higher boiling point than pure water. These types of changes in physical properties are examples of *colligative properties*, which depend only on the number of solute particles in the solution.

Probably one familiar example is the process of spreading salt on icy sidewalks and roads when temperatures drop below freezing. The particles from the salt combine with water to lower the freezing point, which causes the ice to melt. Another example is the addition of antifreeze, such as ethylene glycol, $C_2H_6O_2$, to the water in a car radiator. If the ethylene glycol and water mixture is about 50% water by mass, it does not freeze until the temperature drops to about -34 °F and does not boil unless the temperature goes above 255 °F. The solution in the radiator prevents the water from forming ice in cold weather and boiling over on a hot desert highway.

Ethylene glycol is added to a radiator to form an aqueous solution that has a lower freezing point and a higher boiling point than water.

The Alaskan Upis beetle produces a biological antifreeze to survive subfreezing temperatures.

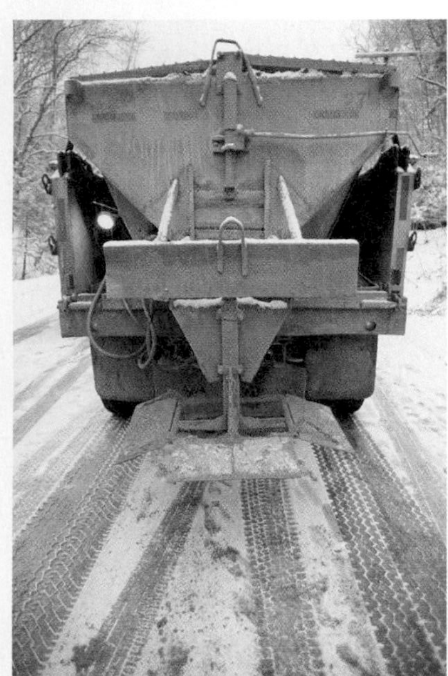

A truck spreads calcium chloride on the road to melt ice and snow.

Insects and fish in climates with subfreezing temperatures control ice formation by producing biological antifreezes made of glycerol, proteins, and sugars such as glucose, within their bodies. Some insects can survive temperatures below −60 °C. These forms of biological antifreezes may one day be applied to the long-term preservation of human organs.

The lowering of the freezing point of water occurs because the solute particles disrupt the formation of the solid ice structure. Thus, a lower temperature is required to freeze the water in the solution. The greater the solute concentration, the lower the freezing point will be. When there is 1 mole of particles in 1.0 kg (1.0 L) of water, the freezing point drops from 0 °C to −1.86 °C. If there are 2 moles of particles in 1.0 kg (1.0 L) of water, the freezing point drops twice as much.

A similar change occurs with the boiling point of water. Then the solute particles disrupt the formation of water vapor. One mole of particles in 1.0 kg (1.0 L) of water raises its boiling point from 100. °C to 100.52 °C.

As we discussed in Section 7.2, a solute that is a nonelectrolyte dissolves as molecules, whereas a solute that is a strong electrolyte dissolves entirely as ions. The solute in antifreeze, which is ethylene glycol, dissolves as molecules.

Nonelectrolyte:

$$1 \text{ mole of } C_2H_6O_2(l) = 1 \text{ mole of } C_2H_6O_2(aq)$$

Strong electrolytes:

$$1 \text{ mole of } NaCl(s) = \underbrace{1 \text{ mole of } Na^+(aq) + 1 \text{ mole of } Cl^-(aq)}_{2 \text{ moles of particles } (aq)}$$

$$1 \text{ mole of } CaCl_2(s) = \underbrace{1 \text{ mole of } Ca^{2+}(aq) + 2 \text{ moles of } Cl^-(aq)}_{3 \text{ moles of particles } (aq)}$$

The effect of $CaCl_2$ used to salt icy roads is to produce 3 moles of particles from 1 mole of $CaCl_2$, which will lower the freezing point of water three times more than one mole of ethylene glycol.

CONCEPT CHECK 7.8

Freezing Point Changes

In each pair, identify the solution that will have a lower freezing point. Explain.

a. 1.0 mole of NaOH (strong electrolyte) or 1.0 mole of ethylene glycol (nonelectrolyte) each in 1.0 L of water

b. 0.20 mole of KNO_3 (strong electrolyte) or 0.20 mole of $Ca(NO_3)_2$ (strong electrolyte) each in 1.0 L of water

ANSWER

a. When 1.0 mole of NaOH dissolves in water, it will produce 2.0 moles of particles because each NaOH dissociates to give two particles, Na^+ and OH^-. However, 1.0 mole of ethylene glycol dissolves as molecules to produce only 1.0 mole of particles. Thus, 1.0 mole of NaOH in 1.0 L of water will have the lower freezing point.

b. When 0.20 mole of KNO_3 dissolves in water, it will produce 0.40 mole of particles because each KNO_3 dissociates to give two particles, K^+ and NO_3^-. When 0.20 mole of $Ca(NO_3)_2$ dissolves in water, it will produce 0.60 mole of particles because each $Ca(NO_3)_2$ dissociates to give three particles, Ca^{2+} and $2NO_3^-$. Thus, 0.20 mole of $Ca(NO_3)_2$ in 1.0 L of water will have the lower freezing point.

Osmotic Pressure

The movement of water into and out of the cells of plants as well as our own bodies is an important biological process that also depends on the solute concentration. In a process called **osmosis**, water molecules move through a semipermeable membrane from the solution with the lower concentration of solute into a solution with the higher solute concentration. In an osmosis apparatus, water is placed on one side of a semipermeable membrane and a sucrose (sugar) solution on the other side. The semipermeable membrane allows water molecules to flow back and forth but blocks the sucrose molecules because they are too large to pass through the membrane. Because the sucrose solution has a higher solute concentration, more water molecules flow into the sucrose solution than out of the sucrose solution. The volume level of the sucrose solution rises as the volume level on the water side falls. The increase of water dilutes the sucrose solution to equalize (or attempt to equalize) the concentrations on both sides of the membrane.

Eventually the height of the sucrose solution creates sufficient pressure to equalize the flow of water between the two compartments. This pressure, called **osmotic pressure**, prevents the flow of additional water into the more concentrated solution. Then there is no further change in the volumes of the two solutions. The osmotic pressure depends on the concentration of solute particles in the solution. The greater the number of particles dissolved, the higher its osmotic pressure. In this example, the sucrose solution has a higher osmotic pressure than pure water, which has an osmotic pressure of zero.

In a process called *reverse osmosis*, a pressure greater than the osmotic pressure is applied to a solution. The flow of water is reversed so that water flows out of the solution with the higher solute concentration. The process of reverse osmosis is used in desalination plants to obtain pure water from sea (salt) water.

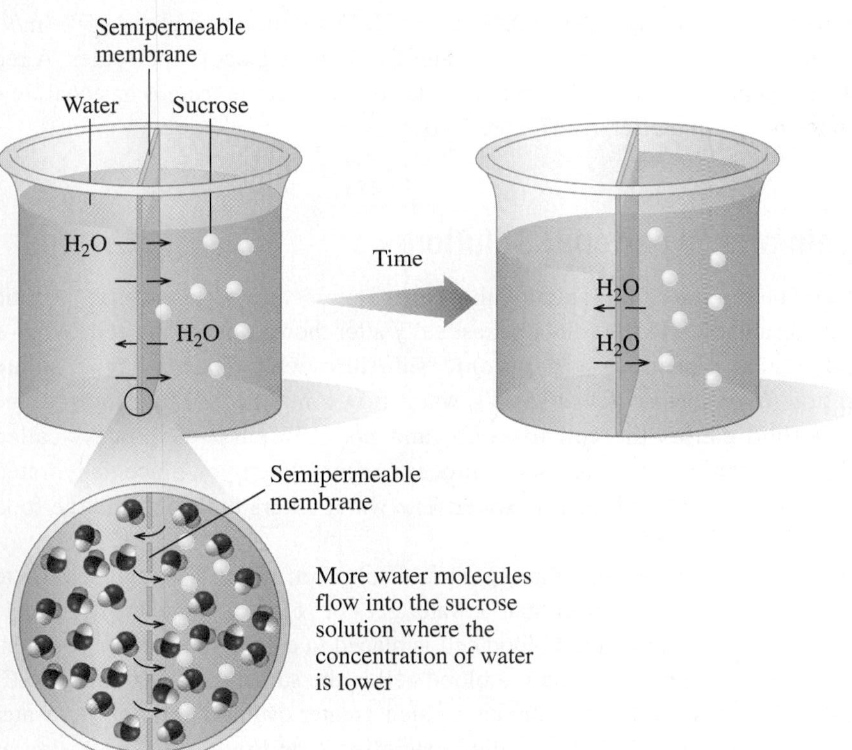

Water flows into the solution with a higher solute concentration until the flow of water becomes equal in both directions.

TUTORIAL
Osmosis

EVERYDAY OSMOSIS

1. Place a few pieces of dried fruit such as raisins, prunes, or banana chips in water. Observe them after 1 hour or more. Look at them again the next day.
2. Place some grapes in a concentrated salt-water solution. Observe them after 1 hour or more. Look at them again the next day.
3. Place one potato slice in water and another slice in a concentrated salt-water solution. After 1–2 hours, observe the shapes and size of the slices. Look at them again the next day.

QUESTIONS

1. How did the shape of the dried fruit change after being in water? Explain.
2. How did the appearance of the grapes change after being in a concentrated salt solution? Explain.
3. How does the appearance of the potato slice that was placed in water compare to the appearance of the potato slice placed in salt water? Explain.
4. At the grocery store, why are sprinklers used to spray water on fresh produce such as lettuce, carrots, and cucumbers?

CONCEPT CHECK 7.9

Osmotic Pressure

A 2% (m/m) sucrose solution and an 8% (m/m) sucrose solution are separated by a semipermeable membrane.

a. Which sucrose solution exerts the greater osmotic pressure?
b. In what direction does water flow initially?
c. When the flow of water is equal in both directions, which solution has the higher level?

ANSWER

a. The 8% (m/m) sucrose solution has the higher solute concentration, more solute particles, and the greater osmotic pressure.
b. Initially, water will flow out of the 2% (m/m) solution into the more concentrated 8% (m/m) solution.
c. The level of the 8% (m/m) solution will be higher.

Isotonic Solutions

Because the cell membranes in biological systems are semipermeable, osmosis is an ongoing process. The solutes in body solutions such as blood, tissue fluids, lymph, and plasma all exert osmotic pressure. Most intravenous solutions used in a hospital are **isotonic solutions**, which exert the same osmotic pressure as body fluids such as blood. The percent concentration typically used in IV solutions is mass/volume percent (m/v), which is a type of percent concentration we have already discussed. The most typical isotonic solutions are 0.9% (m/v) NaCl solution, or 0.9 g of NaCl/100 mL of solution, and 5% (m/v) glucose, or 5 g of glucose/100 mL of solution. Although they do not contain the same kinds of particles, a 0.9% (m/v) NaCl solution as well as a 5% (m/v) glucose solution is a 0.3 M solution of Na^+ and Cl^- ions or glucose molecules. A red blood cell placed in an isotonic solution retains its volume because there is an equal flow of water into and out of the cell (see Figure 7.10a).

Hypotonic and Hypertonic Solutions

If a red blood cell is placed in a solution that is not isotonic, the differences in osmotic pressure inside and outside the cell can drastically alter the volume of the cell. When a red blood cell is placed in a **hypotonic solution**, which has a lower solute concentration (*hypo* means "lower than"), water flows into the cell by osmosis. The increase in fluid causes the cell to swell, and possibly burst—a process called **hemolysis** (see Figure 7.10b). A similar process occurs when you place dehydrated food, such as raisins or dried fruit, in water. The water enters the cells, and the food becomes plump and smooth.

If a red blood cell is placed in a **hypertonic solution**, which has a higher solute concentration (*hyper* means "greater than"), water goes out of the cell into the hypertonic solution by osmosis. Suppose a red blood cell is placed in a 10% (m/v) NaCl solution. Because the osmotic pressure in the red blood cell is the same as a 0.9% (m/v) NaCl solution, the 10% (m/v) NaCl solution has a much greater osmotic pressure. As water leaves the cell, it shrinks, a process called **crenation** (see Figure 7.10c). A similar process occurs when making pickles when a hypertonic salt solution causes the cucumbers to shrivel as they lose water.

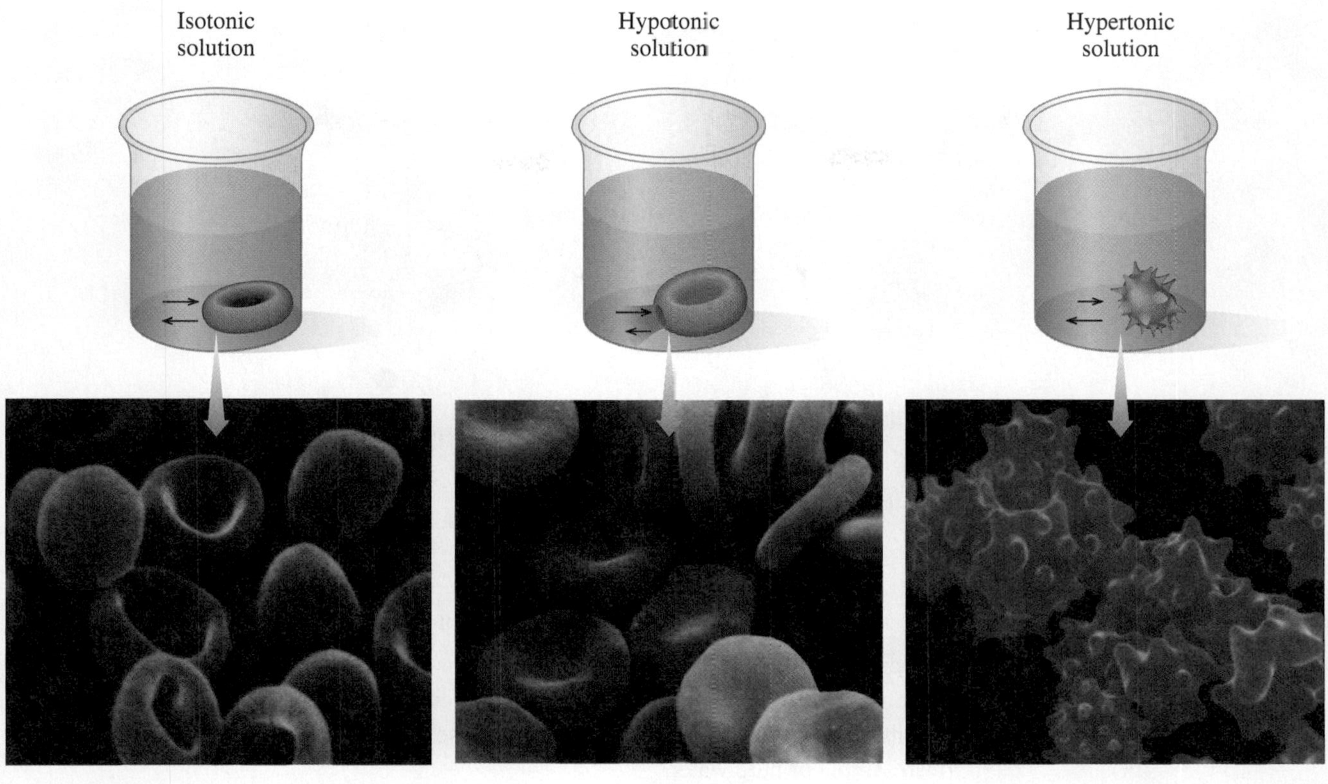

FIGURE 7.10 **(a)** In an isotonic solution, a red blood cell retains its normal volume. **(b)** Hemolysis: In a hypotonic solution, water flows into a red blood cell, causing it to swell and burst. **(c)** Crenation: In a hypertonic solution, water leaves the red blood cell, causing it to shrink.

Q What happens to a red blood cell placed in a 4% (m/v) NaCl solution?

SAMPLE PROBLEM 7.12

Isotonic, Hypotonic, and Hypertonic Solutions

Describe each of the following solutions as isotonic, hypotonic, or hypertonic. Indicate whether a red blood cell placed in each solution will undergo hemolysis, crenation, or no change.

a. a 5% (m/v) glucose solution
b. a 0.2% (m/v) NaCl solution

SOLUTION

a. A 5% (m/v) glucose solution is isotonic. A red blood cell will not undergo any change.
b. A 0.2% (m/v) NaCl solution is hypotonic. A red blood cell will undergo hemolysis.

STUDY CHECK 7.12

What will happen to a red blood cell placed in a 10% (m/v) glucose solution?

Dialysis

TUTORIAL
Dialysis

Dialysis is a process that is similar to osmosis. In dialysis, a semipermeable membrane, called a dialyzing membrane, permits small solute molecules and ions as well as solvent water molecules to pass through, but it retains large particles, such as colloids. Dialysis is a way to separate solution particles from colloids.

Suppose we fill a cellophane bag with a solution containing NaCl, glucose, starch, and protein and place it in pure water. Cellophane is a dialyzing membrane, and the sodium ions, chloride ions, and glucose molecules will pass through it into the surrounding water. However, starch and protein remain inside because they are colloids. Water

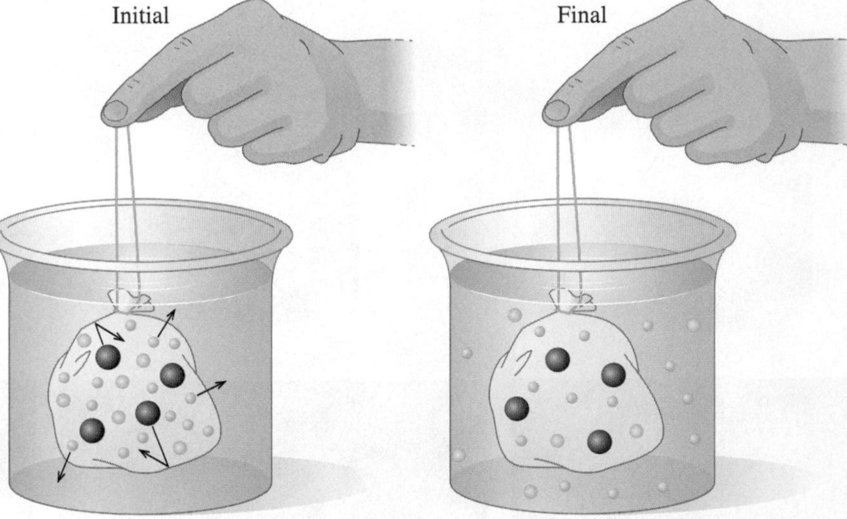

Solution particles pass through a dialyzing membrane but colloidal particles are retained.

molecules will flow by osmosis into the cellophane bag. Eventually the concentrations of sodium ions, chloride ions, and glucose molecules inside and outside the dialysis bag become equal. To remove more NaCl or glucose, the cellophane bag must be placed in a fresh sample of pure water.

Chemistry Link to Health

DIALYSIS BY THE KIDNEYS AND THE ARTIFICIAL KIDNEY

The fluids of the body undergo dialysis by the membranes of the kidneys, which remove waste materials, excess salts, and water. In an adult, each kidney contains about 2 million nephrons. At the top of each nephron, there is a network of arterial capillaries called the glomerulus.

As blood flows into the glomerulus, small particles, such as amino acids, glucose, urea, water, and certain ions, will move through the capillary membranes into the nephron. As this solution moves through the nephron, substances still of value to the body (such as amino acids, glucose, certain ions, and 99% of the water) are reabsorbed. The major waste product, urea, is excreted in the urine.

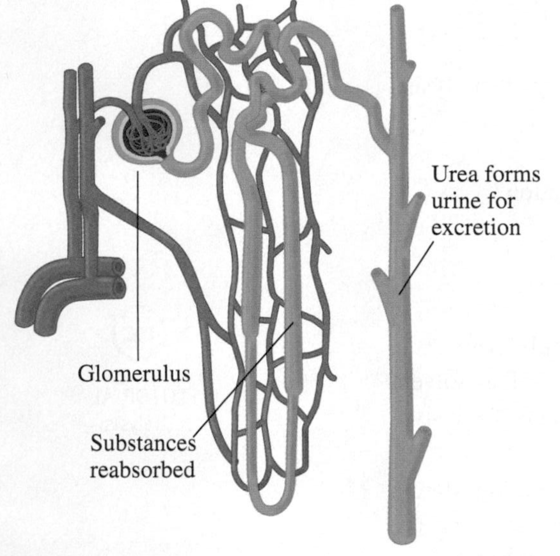

In the nephron, substances such as glucose, ions, water are reabsorbed from the blood; the waste product, urea, is excreted as urine.

Hemodialysis

If the kidneys fail to dialyze waste products, increased levels of urea can become life-threatening in a relatively short time. A person with kidney failure must use an artificial kidney, which cleanses the blood by **hemodialysis**.

A typical artificial kidney machine contains a large tank filled with about 100 L of water containing selected electrolytes. In the center of this dialyzing bath (dialysate), there is a dialyzing coil or membrane made of cellulose tubing. As the patient's blood flows through the dialyzing coil, the highly concentrated waste products dialyze out of the blood. No blood is lost because the membrane is not permeable to large particles such as red blood cells.

Dialysis patients do not produce much urine. As a result, they retain large amounts of water between dialysis treatments, which produces a strain on the heart. The intake of fluids for a dialysis patient may be restricted to as little as a few teaspoons of water a day. In the dialysis procedure, the pressure of the blood is increased as it circulates through the dialyzing coil so water can be squeezed out of the blood. For some dialysis patients, 2–10 L of water may be removed during one treatment. Dialysis patients have from two to three treatments a week, each treatment requiring about 5–7 hours. Some of the newer treatments require less time. For many patients, dialysis is done at home with a home dialysis unit.

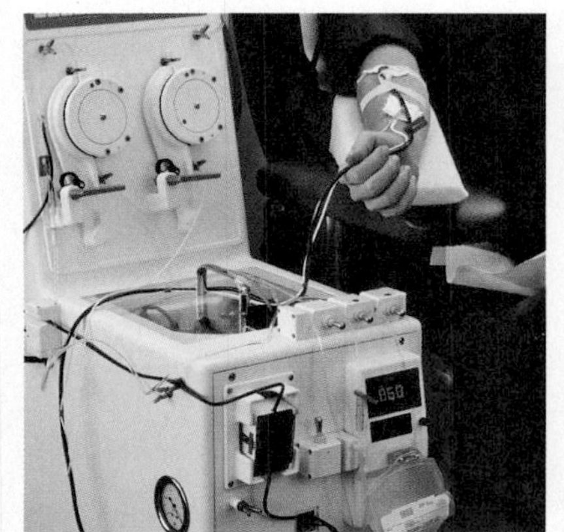

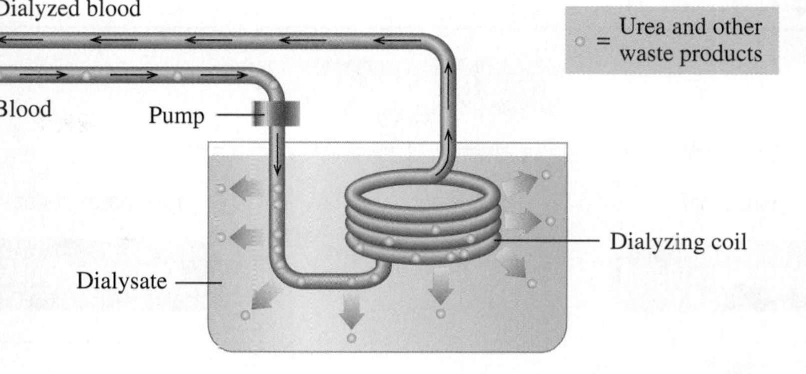

During dialysis, waste products and excess water are removed from the blood.

QUESTIONS AND PROBLEMS

Properties of Solutions

7.61 Identify the following as characteristic of a solution, a colloid, or a suspension:
a. a mixture that cannot be separated by a semipermeable membrane
b. a mixture that settles out upon standing

7.62 Identify the following as characteristic of a solution, a colloid, or a suspension:
a. Particles of this mixture remain inside a semipermeable membrane but pass through filters.
b. The particles of solute in this solution are very large and visible.

7.63 In each pair, identify the solution that will have a lower freezing point. Explain.
a. 1.0 mole of glycerol (nonelectrolyte) or 2.0 moles of ethylene glycol (nonelectrolyte) each in 1.0 L of water.
b. 0.50 mole of KCl (strong electrolyte) or 0.50 mole of $MgCl_2$ (strong electrolyte) each in 2.0 L of water.

7.64 In each pair, identify the solution that will have a higher boiling point. Explain.
a. 1.50 moles of LiOH (strong electrolyte) or 3.00 moles of LiOH (strong electrolyte) each in 0.50 L of water.
b. 0.40 mole of $Al(NO_3)_3$ (strong electrolyte) or 0.40 mole of CsCl (strong electrolyte) each in 0.50 L of water.

7.65 A 10% (m/v) starch solution is separated from a 1% (m/v) starch solution by a semipermeable membrane. (Starch is a colloid.)
a. Which compartment has the higher osmotic pressure?
b. In which direction will water flow initially?
c. In which compartment will the volume level rise?

7.66 Two solutions, a 0.1% (m/v) albumin solution and a 2% (m/v) albumin solution, are separated by a semipermeable membrane. (Albumin is a colloid.)
a. Which compartment has the higher osmotic pressure?
b. In which direction will water flow initially?
c. In which compartment will the volume level rise?

7.67 Indicate the compartment (A or B) that will increase in volume for each of the following pairs of solutions separated by a semipermeable membrane:

A	B
a. 5% (m/v) sucrose	10% (m/v) sucrose
b. 8% (m/v) albumin	4% (m/v) albumin
c. 0.1% (m/v) starch	10% (m/v) starch

7.68 Indicate the compartment (A or B) that will increase in volume for each of the following pairs of solutions separated by semipermeable membrane:

A	B
a. 20% (m/v) starch	10% (m/v) starch
b. 10% (m/v) albumin	2% (m/v) albumin
c. 0.5% (m/v) sucrose	5% (m/v) sucrose

7.69 Are the following solutions isotonic, hypotonic, or hypertonic compared with a red blood cell?
a. distilled H_2O **b.** 1% (m/v) glucose
c. 0.9% (m/v) NaCl **d.** 15% (m/v) glucose

7.70 Will a red blood cell undergo crenation, hemolysis, or no change in each of the following solutions?
a. 1% (m/v) glucose **b.** 2% (m/v) NaCl
c. 5% (m/v) glucose **d.** 0.1% (m/v) NaCl

7.71 Each of the following mixtures is placed in a dialyzing bag and immersed in distilled water. Which substances will be found outside the bag in the distilled water?
a. NaCl solution
b. starch solution (colloid) and alanine, an amino acid, solution
c. NaCl solution and starch solution (colloid)
d. urea solution

7.72 Each of the following mixtures is placed in a dialyzing bag and immersed in distilled water. Which substances will be found outside the bag in the distilled water?
a. KCl solution and glucose solution
b. albumin solution (colloid)
c. an albumin solution (colloid), KCl solution, and glucose solution
d. urea solution and NaCl solution

CONCEPT MAP

SOLUTIONS

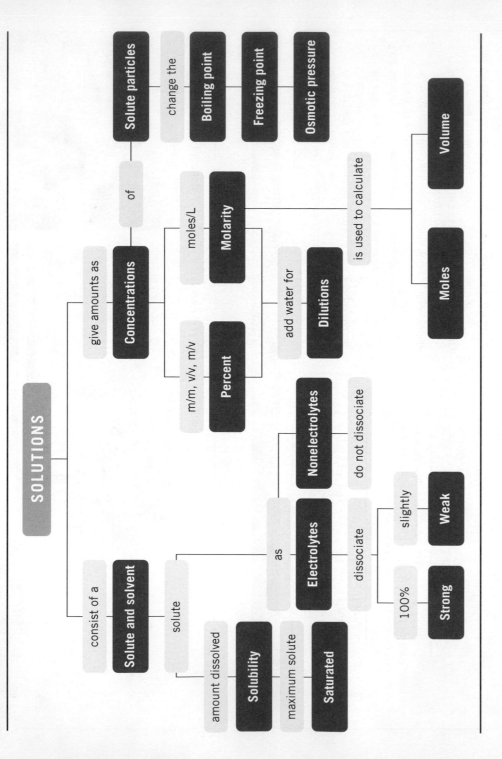

- consist of a **Solute and solvent**
  - solute
    - amount dissolved **Solubility**
      - maximum solute **Saturated**
    - as **Electrolytes**
      - dissociate
        - 100% **Strong**
        - slightly **Weak**
    - as **Nonelectrolytes**
      - do not dissociate
- give amounts as **Concentrations** of **Solute particles**
  - m/m, v/v, m/v **Percent**
  - moles/L **Molarity**
  - add water for **Dilutions**
  - Molarity is used to calculate **Moles** and **Volume**
  - Solute particles change the **Boiling point**, **Freezing point**, **Osmotic pressure**

CHAPTER REVIEW

7.1 Solutions

Learning Goal: Identify the solute and solvent in a solution; describe the formation of a solution.

Solute: The substance present in lesser amount

Salt

Water

Solvent: The substance present in greater amount

A solution forms when a solute dissolves in a solvent. In a solution, the particles of solute are evenly dispersed in the solvent. The solute and solvent may be solid, liquid, or gas. The polar O—H bond leads to hydrogen bonding between water molecules. An ionic solute dissolves in water, a polar solvent, because the polar water molecules attract and pull the ions into solution, where they become hydrated. The expression "like dissolves like" means that a polar or an ionic solute dissolves in a polar solvent while a nonpolar solute dissolves in a nonpolar solvent.

7.2 Electrolytes and Nonelectrolytes

Learning Goal: Identify solutes as electrolytes or nonelectrolytes.

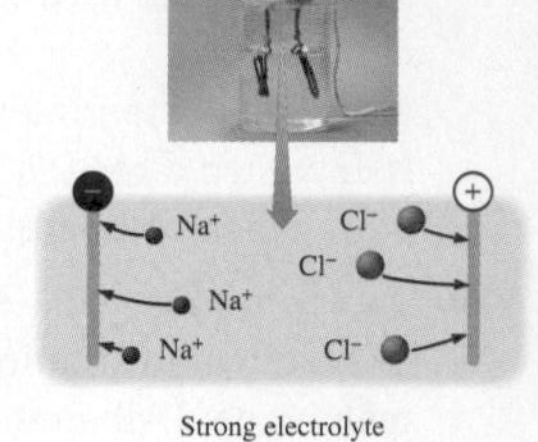

Strong electrolyte

Substances that release ions in water are called electrolytes because their solutions will conduct an electrical current. Strong electrolytes are completely ionized, whereas weak electrolytes are only partially ionized. Nonelectrolytes are substances that dissolve in water to produce molecules and cannot conduct electrical currents. An equivalent (Eq) is the amount of an electrolyte that carries one mole of positive or negative charge. One mole of Na^+ is 1 Eq. One mole of Ca^{2+} has 2 Eq. In fluid replacement solutions, the concentrations of electrolytes are expressed as mEq/L of solution.

7.3 Solubility

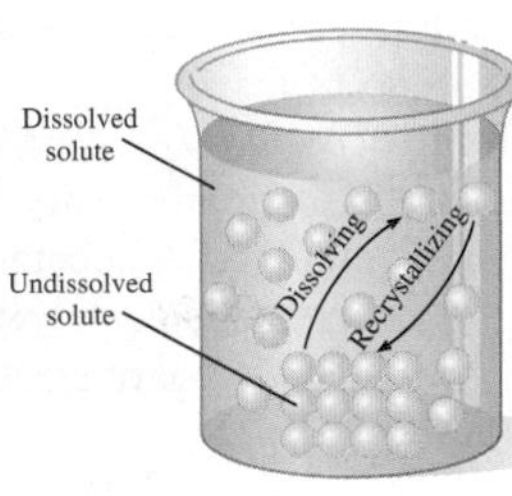

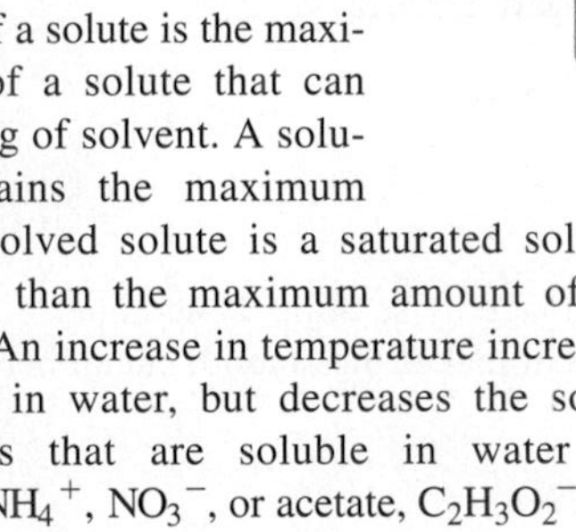

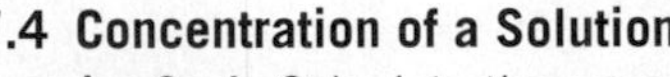

Learning Goal: Define solubility; distinguish between an unsaturated and a saturated solution; identify a salt as soluble or insoluble.

The solubility of a solute is the maximum amount of a solute that can dissolve in 100 g of solvent. A solution that contains the maximum amount of dissolved solute is a saturated solution. A solution containing less than the maximum amount of dissolved solute is unsaturated. An increase in temperature increases the solubility of most solids in water, but decreases the solubility of gases in water. Salts that are soluble in water usually contain Li^+, Na^+, K^+, NH_4^+, NO_3^-, or acetate, $C_2H_3O_2^-$.

7.4 Concentration of a Solution

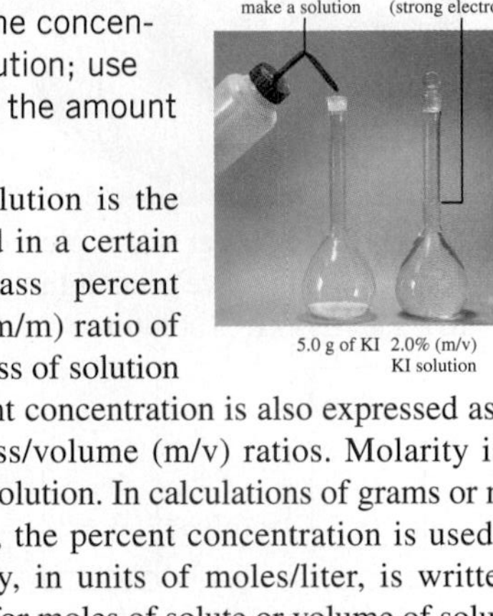

Learning Goal: Calculate the concentration of a solute in a solution; use concentration to calculate the amount of solute or solution.

The concentration of a solution is the amount of solute dissolved in a certain amount of solution. Mass percent expresses the mass/mass (m/m) ratio of the mass of solute to the mass of solution multiplied by 100%. Percent concentration is also expressed as volume/volume (v/v) and mass/volume (m/v) ratios. Molarity is the moles of solute per liter of solution. In calculations of grams or milliliters of solute or solution, the percent concentration is used as a conversion factor. Molarity, in units of moles/liter, is written as conversion factors to solve for moles of solute or volume of solution.

7.5 Dilution of Solutions

Learning Goal: Describe the dilution of a solution; calculate the new concentration or volume of a diluted solution.

In dilution, a solvent such as water is added to a solution, which increases its volume and decreases its concentration.

7.6 Properties of Solutions

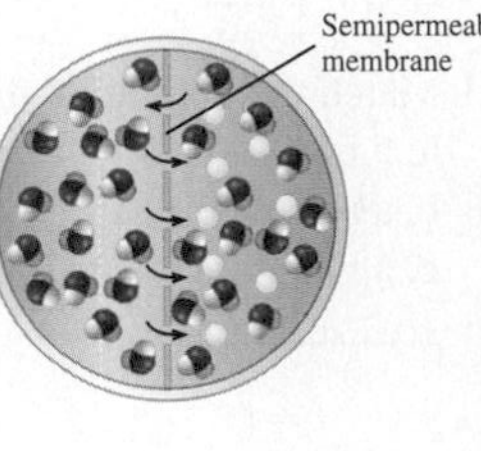

Learning Goal: Identify a mixture as a solution, a colloid, or a suspension. Describe how the number of particles in a solution affects the freezing point, boiling point, and osmotic pressure of a solution.

Colloids contain particles that do not settle out; they pass through filters, but not through semipermeable membranes. Suspensions have very large particles that settle out of solution. The particles in a solution lower the freezing point, raise the boiling point, and increase the osmotic pressure.

In osmosis, solvent (water) passes through a semipermeable membrane from a solution of a lower solute concentration to a solution of a higher concentration. Isotonic solutions have osmotic pressures equal to that of body fluids. A red blood cell maintains its volume in an isotonic solution but swells and may burst (hemolyzes) in a hypotonic solution, and shrinks (crenates) in a hypertonic solution. In dialysis, water and small solute particles pass through a dialyzing membrane, while larger particles are retained.

Key Terms

colloid A mixture having particles that are moderately large. Colloids pass through filters but cannot pass through semipermeable membranes.

concentration A measure of the amount of solute that is dissolved in a specified amount of solution.

crenation The shriveling of a cell because water leaves the cell when the cell is placed in a hypertonic solution.

dialysis A process in which water and small solute particles pass through a semipermeable membrane.

dilution A process by which water (solvent) is added to a solution to increase the volume and decrease (dilute) the concentration of the solute.

electrolyte A substance that produces ions when dissolved in water; its solution conducts electricity.

equivalent (Eq) The amount of a positive or negative ion that supplies 1 mole of electrical charge.

hemodialysis A mechanical cleansing of the blood by an artificial kidney using the principle of dialysis.

hemolysis A swelling and bursting of red blood cells in a hypotonic solution because of an increase in fluid volume.

Henry's law The solubility of a gas in a liquid is directly related to the pressure of that gas above the liquid.

hydration The process of surrounding dissolved ions by water molecules.

hypertonic solution A solution that has a higher particle concentration and higher osmotic pressure than the cells of the body.

hypotonic solution A solution that has a lower particle concentration and lower osmotic pressure than the cells of the body.

insoluble salt An ionic compound that does not dissolve in water.

isotonic solution A solution that has the same particle concentration and osmotic pressure as that of the cells of the body.

mass percent (m/m) The grams of solute in exactly 100 g of solution.

mass/volume percent (m/v) The grams of solute in exactly 100 mL of solution.

molarity (M) The number of moles of solute in exactly 1 L of solution.

nonelectrolyte A substance that dissolves in water as molecules; its solution does not conduct an electrical current.

osmosis The flow of a solvent, usually water, through a semipermeable membrane into a solution of higher solute concentration.

osmotic pressure The pressure that prevents the flow of water into the more concentrated solution.

saturated solution A solution containing the maximum amount of solute that can dissolve at a given temperature. Any additional solute will remain undissolved in the container.

soluble salt An ionic compound that dissolves in water.

solubility The maximum amount of solute that can dissolve in exactly 100 g of solvent, usually water, at a given temperature.

solute The component in a solution that is present in the smaller quantity.

solution A homogeneous mixture in which the solute is made up of small particles (ions or molecules) that can pass through filters and semipermeable membranes.

solvent The substance in which the solute dissolves; usually the component present in greatest amount.

strong electrolyte A polar or ionic compound that ionizes completely when it dissolves in water. Its solution is a good conductor of electricity.

suspension A mixture in which the solute particles are large enough and heavy enough to settle out and be retained by both filters and semipermeable membranes.

unsaturated solution A solution that contains less solute than can be dissolved.

volume percent (v/v) A percent concentration that relates the volume of the solute in exactly 100 ml of solution.

weak electrolyte A substance that produces only a few ions along with many molecules when it dissolves in water. Its solution is a weak conductor of electricity.

Understanding the Concepts

7.73 Match the diagram with each of the following:
- **a.** a polar solute and a polar solvent
- **b.** a nonpolar solute and a polar solvent
- **c.** a nonpolar solute and a nonpolar solvent

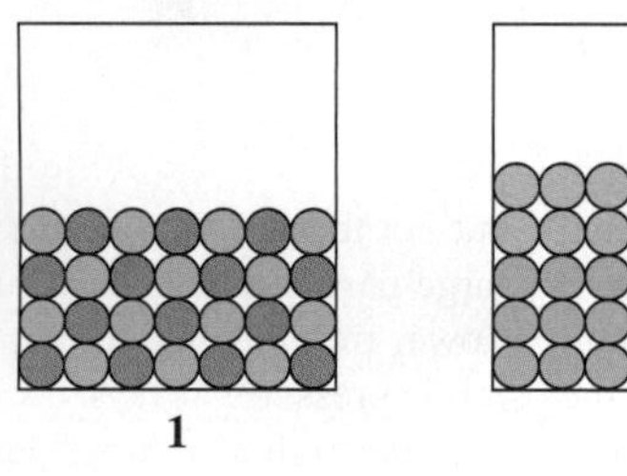
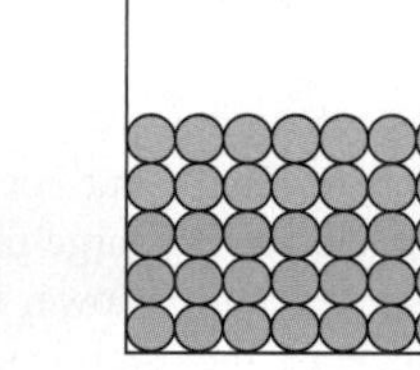

1 2

7.74 Would heating or cooling cause each of the following changes?
- **a.** 2 to 3
- **b.** 2 to 1

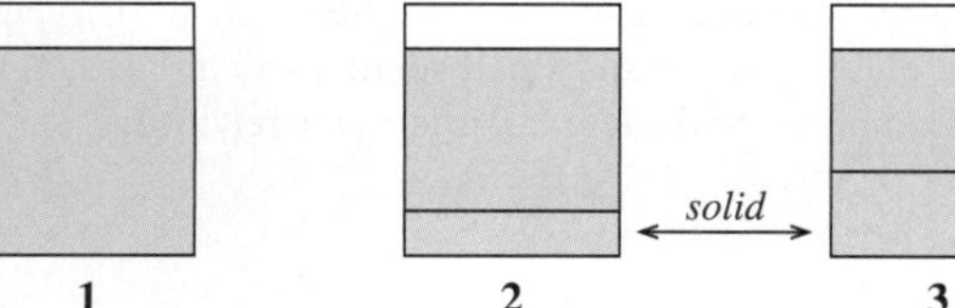

1 2 3

7.75 Select the diagram that represents the solution formed by a solute ⚫⚫ that is a
- **a.** nonelectrolyte
- **b.** weak electrolyte
- **c.** strong electrolyte

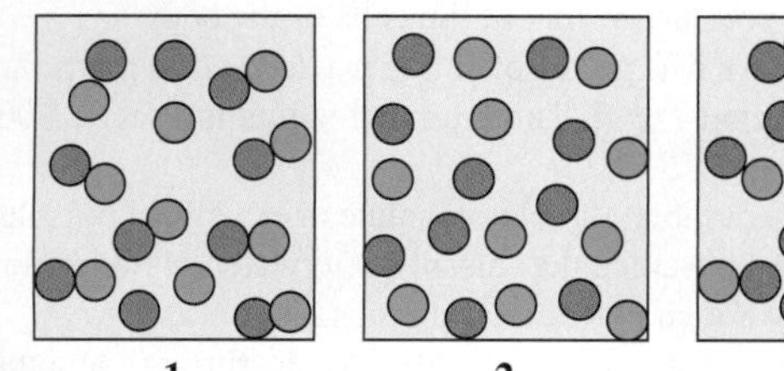

1 2 3

7.76 Why do lettuce leaves in a salad wilt after a vinaigrette dressing containing salt is added?

7.77 A pickle is made by soaking a cucumber in brine, a saltwater solution. What makes the smooth cucumber become wrinkled like a prune?

7.78 Select the container that represents the dilution of a 4% (m/v) KCl solution to give each of the following:
a. a 2% (m/v) KCl solution **b.** a 1% (m/v) KCl solution

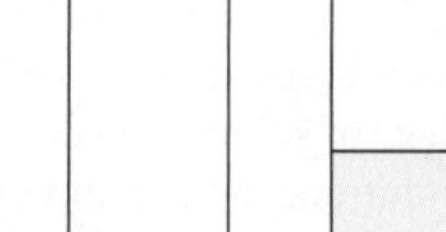
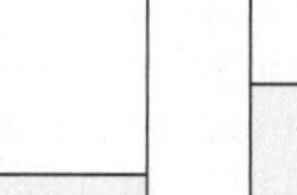
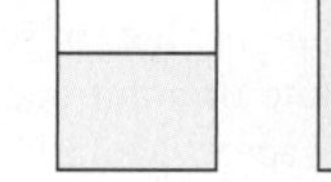

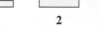

4% (m/v) KCl 1 2 3

7.79 A semipermeable membrane separates compartments A and B. If the levels of the following solutions in A and B are equal initially, select the diagram that illustrates the final levels:

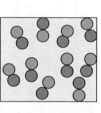

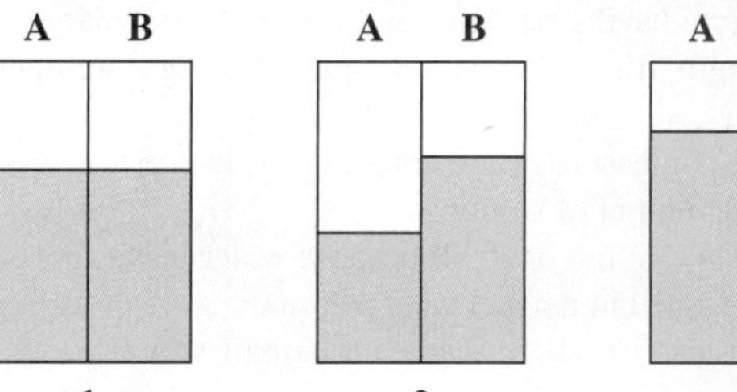

1 2 3

	Solution in A	Solution in B
a.	2% (m/v) starch	8% (m/v) starch
b.	1% (m/v) starch	1% (m/v) starch
c.	5% (m/v) sucrose	1% (m/v) sucrose
d.	0.1% (m/v) sucrose	1% (m/v) sucrose

7.80 Select the diagram that represents the shape of a red blood cell when placed in each of the following solutions:

1 2 3

Normal red blood cell

- **a.** 0.9% (m/v) NaCl
- **b.** 10% (m/v) glucose
- **c.** 0.01% (m/v) NaCl
- **d.** 5% (m/v) glucose
- **e.** 1% (m/v) glucose

Additional Questions and Problems

For instructor-assigned homework, go to **www.masteringchemistry.com**.

7.81 Why does iodine dissolve in hexane, but not in water?

7.82 How does temperature and pressure affect the solubility of solids and gases in water?

7.83 Potassium nitrate has a solubility of 32 g of KNO_3 in 100 g of H_2O at 20 °C. State if each of the following forms an unsaturated or saturated solution at 20 °C:
a. adding 2 g of KNO_3 to 200. g of H_2O
b. adding 19 g of KNO_3 to 50. g of H_2O
c. adding 68 g of KNO_3 to 150. g of H_2O

7.84 Potassium chloride has a solubility of 43 g of KCl in 100 g of H_2O at 50 °C. State if each of the following forms an unsaturated or saturated solution at 50 °C:
a. adding 25 g of KCl to 100. g of H_2O
b. adding 25 g of KCl to 50. g of H_2O
c. adding 86 g of KCl to 150. g of H_2O

7.85 Indicate whether each of the following ionic compounds is soluble in water:
a. KCl
b. $MgSO_4$
c. CuS
d. $AgNO_3$
e. $Ca(OH)_2$

7.86 Indicate whether each of the following ionic compounds is soluble in water:
a. $CuCO_3$
b. FeO
c. $Mg_3(PO_4)_2$
d. $(NH_4)_2SO_4$
e. $NaHCO_3$

7.87 Calculate the mass percent (m/m) of a solution containing 15.5 g of Na_2SO_4 and 75.5 g of H_2O.

7.88 How many grams of K_2CO_3 are in 750 mL of a 3.5% (m/v) K_2CO_3 solution?

7.89 A patient receives all her nutrition from fluids given through the vena cava. Every 12 hours, 500 mL of a solution that is 5.0% (m/v) amino acids (protein) and 20% (m/v) glucose (carbohydrate) is given along with 500 mL of a 10% (m/v) lipid (fat).
a. In 1 day, how many grams each of amino acids, glucose, and lipid are given to the patient?
b. How many kilocalories does she obtain in 1 day?

7.90 An 80-proof brandy is a 40.% (v/v) ethanol solution. The "proof" is twice the percent concentration of alcohol in the beverage. How many milliliters of alcohol are present in 750 mL of brandy?

7.91 How many milliliters of a 12% (v/v) propyl alcohol solution would you need to obtain 4.5 mL of propyl alcohol?

7.92 How many liters of a 5.0% (m/v) glucose solution would you need to obtain 75 g of glucose?

7.93 If you were in the laboratory, how would you prepare 0.250 L of a 2.00 M KCl solution?

7.94 What is the molarity of a solution containing 15.6 g of KCl in 274 mL of KCl solution?

7.95 A solution is prepared with 70.0 g of HNO_3 and 130.0 g of H_2O. It has a density of 1.21 g/mL.
a. What is the mass percent (m/m) of the HNO_3 solution?
b. What is the total volume of the solution?
c. What is the mass/volume percent (m/v)?
d. What is its molarity (M)?

7.96 What is the molarity of a 15% (m/v) NaOH solution?

7.97 How many grams of solute are in each of the following solutions?
a. 2.5 L of a 3.0 M $Al(NO_3)_3$ solution
b. 75 mL of a 0.50 M $C_6H_{12}O_6$ solution
c. 235 mL of a 1.80 M LiCl solution

7.98 How many milliliters of each of the following solutions will provide 25.0 g of KOH?
a. 2.50 M KOH solution
b. 0.750 M KOH solution
c. 5.60 M KOH solution

7.99 Calculate the molarity of the solution when water is added to prepare each of the following:
a. 25.0 mL of a 0.200 M NaBr solution is diluted to 50.0 mL
b. 15.0 mL of a 1.20 M K_2SO_4 solution is diluted to 40.0 mL
c. 75.0 mL of a 6.00 M NaOH solution is diluted to 255 mL

7.100 Calculate the molarity of the solution when water is added to prepare each of the following:
a. 25.0 mL of a 18.0 M HCl solution is diluted to 500. mL
b. 50.0 mL of a 1.50 M NaCl solution is diluted to 125 mL
c. 4.50 mL of a 8.50 M KOH solution is diluted to 75.0 mL

7.101 What is the final volume, in mL, when 25.0 mL of a 5.00 M HCl solution is diluted to each of the following concentrations?
a. 2.50 M HCl solution
b. 1.00 M HCl solution
c. 0.500 M HCl solution

7.102 What is the final volume, in mL, when 5.00 mL of a 12.0 M NaOH solution is diluted to each of the following concentrations?
a. 0.600 M NaOH solution
b. 1.00 M NaOH solution
c. 2.50 M NaOH solution

7.103 Why would solutions with high salt content be used to prepare dried flowers?

7.104 A patient on dialysis has a high level of urea, a high level of sodium, and a low level of potassium in the blood. Why is the dialyzing solution prepared with a high level of potassium but no sodium or urea?

7.105 Why can't you drink seawater even if you are stranded on a desert island?

7.106 Why would a dialysis unit (artificial kidney) use isotonic concentrations of NaCl, KCl, $NaHCO_3$, and glucose in the dialysate?

Challenge Questions

7.107 In a laboratory experiment, a 10.0-mL sample of NaCl solution is poured into an evaporating dish with a mass of 24.10 g. The combined mass of the evaporating dish and NaCl solution is 36.15 g. After heating, the evaporating dish and dry NaCl have a combined mass of 25.50 g.
a. What is the mass percent (m/m) of the NaCl solution?
b. What is the molarity (M) of the NaCl solution?
c. If water is added to 10.0 mL of the initial NaCl solution to give a final volume of 60.0 mL, what is the molarity of the diluted NaCl solution?

7.108 A solution contains 4.56 g of KCl in 175 mL of solution. If the density of the KCl solution is 1.12 g/mL, what are the mass percent (m/m) and molarity (M) for the potassium chloride solution?

7.109 Potassium fluoride has a solubility of 92 g of KF in 100 g of H_2O at 18 °C. State if each of the following mixtures forms an unsaturated or saturated solution at 18 °C:
a. 35 g of KF and 25 g of H_2O
b. 42 g of KF and 50. g of H_2O
c. 145 g of KF and 150. g of H_2O

7.110 A solution is prepared by dissolving 22.0 g of NaOH in 118.0 g of water. The NaOH solution has a density of 1.15 g/mL.
a. What is the mass percent (m/m) of the NaOH solution?
b. What is the total volume (mL) of the solution?
c. What is the molarity (M) of the solution?

7.111 How many milliliters of a 1.75 M LiCl solution contain 15.2 g of LiCl?

7.112 How many grams of NaBr are contained in 75.0 mL of a 1.50 M NaBr solution?

7.113 Identify the solution that has the lower freezing point and explain.
a. a 2.0 M KF solution (strong electrolyte) or a 1.0 M $CaCl_2$ solution (strong electrolyte)
b. a 0.50 M glucose solution (nonelectrolyte) or a 0.25 M $CaCl_2$ solution (strong electrolyte)

7.114 Identify the solution that has the higher boiling point and explain.
a. a 2.0 M glucose solution (nonelectrolyte) or a 1.0 M glycerol solution (nonelectrolyte)
b. a 0.50 M $MgCl_2$ solution (strong electrolyte) or a 0.50 M $Fe(NO_3)_3$ solution (strong electrolyte)

Answers

Answers to Study Checks

7.1 A solution of a weak electrolyte would contain mostly molecules and a few ions.

7.2 0.194 mole of Cl^-

7.3 78 g of KNO_3

7.4 3.4% (m/m) NaCl solution

7.5 4.8% (v/v) Br_2 in CCl_4

7.6 5.5% (m/v) NaOH solution

7.7 2.12 M KNO_3 solution

7.8 18.0 g of KCl

7.9 750 mL of HCl solution

7.10 3.0% (m/v) HCl solution

7.11 120. mL

7.12 The red blood cell will shrink (crenate).

Answers to Selected Questions and Problems

7.1 **a.** NaCl, solute; water, solvent
b. water, solute; ethanol, solvent
c. oxygen, solute; nitrogen, solvent

7.3 The polar water molecules pull the K^+ and I^- ions away from the solid and into solution, where they are hydrated.

7.5 **a.** water
b. CCl_4
c. water
d. CCl_4

7.7 In a solution of KF, only the ions of K^+ and F^- are present in the solvent. In an HF solution, there are a few ions of H^+ and F^- present but mostly dissolved HF molecules.

7.9 **a.** $KCl(s) \xrightarrow{H_2O} K^+(aq) + Cl^-(aq)$

b. $CaCl_2(s) \xrightarrow{H_2O} Ca^{2+}(aq) + 2Cl^-(aq)$

c. $K_3PO_4(s) \xrightarrow{H_2O} 3K^+(aq) + PO_4^{3-}(aq)$

d. $Fe(NO_3)_3(s) \xrightarrow{H_2O} Fe^{3+}(aq) + 3NO_3^-(aq)$

7.11 **a.** mostly molecules and a few ions
b. ions only
c. molecules only

7.13 **a.** strong electrolyte
b. weak electrolyte
c. nonelectrolyte

7.15 **a.** 1 Eq
b. 2 Eq
c. 2 Eq
d. 6 Eq

7.17 0.154 mole of Na^+, 0.154 mole of Cl^-

7.19 55 mEq/L

7.21 **a.** saturated
b. unsaturated

7.23 **a.** unsaturated
b. unsaturated
c. saturated

7.25 **a.** 68 g of KCl
b. 12 g of KCl

7.27 **a.** The solubility of solid solutes typically increases as temperature increases.
b. The solubility of a gas is less at a higher temperature.
c. Gas solubility is less at a higher temperature and the CO_2 pressure in the can is increased.

7.29 **a.** soluble
b. insoluble
c. insoluble
d. soluble
e. soluble

7.31 5% (m/m) is 5 g of glucose in 100 g of solution, whereas 5% (m/v) is 5 g of glucose in 100 mL of solution.

7.33 **a.** 17% (m/m) KCl solution
b. 5.3% (m/m) sugar solution
c. 10.% (m/m) $CaCl_2$ solution

7.35 **a.** 30.% (m/v) Na_2SO_4 solution
b. 11% (m/v) sucrose solution

7.37 **a.** 2.5 g of KCl
b. 50. g of NH_4Cl
c. 25.0 mL of acetic acid solution

7.39 79.9 mL of alcohol

7.41 **a.** 20. g of mannitol
b. 240 g of mannitol

7.43 2 L of glucose solution

7.45 **a.** 20. g of $LiNO_3$ solution
b. 400. mL of KOH solution
c. 20. mL of formic acid solution

7.47 **a.** 0.500 M glucose solution
b. 0.0357 M KOH solution
c. 0.250 M NaCl solution

7.49 **a.** 120. g of NaOH
b. 59.7 g of KCl
c. 5.48 g of HCl

7.51 **a.** 1.50 L of solution
b. 10.0 L of solution
c. 62.5 mL of solution

7.53 Adding water (solvent) to the soup increases the volume and dilutes the tomato soup concentration.

7.55 **a.** 2.0 M HCl solution
b. 2.0 M NaOH solution
c. 2.5% (m/v) KOH solution
d. 3.0% (m/v) H_2SO_4 solution

7.57 **a.** 80. mL of HCl solution
b. 250 mL of LiCl solution
c. 600. mL of H_3PO_4 solution
d. 180 mL of glucose solution

7.59 **a.** 12.8 mL of the HNO_3 solution
b. 11.9 mL of the $MgCl_2$ solution
c. 1.88 mL of the KCl solution

7.61 **a.** solution
b. suspension

7.63 **a.** 2.0 moles of ethylene glycol in 1.0 L of water will have a lower freezing point because it has more particles in solution.
b. 0.50 mole of $MgCl_2$ in 2.0 L of water has a lower freezing point because each formula unit of $MgCl_2$ dissociates in water to give three particles whereas each formula unit of KCl dissociates to give only two particles.

7.65 **a.** 10% (m/v) starch solution
b. from the 1% (m/v) starch solution into the 10% (m/v) starch solution
c. 10% (m/v) starch solution

7.67 **a.** B 10% (m/v) sucrose solution
b. A 8% (m/v) albumin solution
c. B 10% (m/v) starch solution

7.69 **a.** hypotonic
b. hypotonic
c. isotonic
d. hypertonic

7.71 **a.** NaCl
b. alanine
c. NaCl
d. urea

7.73 **a.** 1
b. 2
c. 1

7.75 **a.** 3
b. 1
c. 2

7.77 The skin of the cucumber acts like a semipermeable membrane, and the more dilute solution inside flows into the brine solution.

7.79 **a.** 2
b. 1
c. 3
d. 2

7.81 Because iodine is a nonpolar molecule, it will dissolve in hexane, a nonpolar solvent. Iodine does not dissolve in water because water is a polar solvent.

7.83 **a.** unsaturated solution
b. saturated solution
c. saturated solution

7.85 **a.** soluble
b. soluble
c. insoluble
d. soluble
e. insoluble

7.87 17.0% (m/m) Na_2SO_4 solution

7.89 **a.** 50 g of amino acids, 200 g of glucose, and 100 g of lipid
b. 1900 kcal

7.91 38 mL of solution

7.93 To make a 2.00 M KCl solution, weigh out 37.3 g of KCl (0.500 mole) and place in a volumetric flask. Add water to dissolve the KCl and give a final volume of 0.250 L.

7.95 **a.** 35.0% (m/m) HNO_3 solution
b. 165 mL
c. 42.4% (m/v) HNO_3 solution
d. 6.73 M HNO_3 solution

7.97 **a.** 1600 g of $Al(NO_3)_3$
b. 6.8 g of $C_6H_{12}O_6$
c. 17.9 g of LiCl

7.99 **a.** 0.100 M NaBr solution
b. 0.450 M K_2SO_4 solution
c. 1.76 M NaOH solution

7.101 **a.** 50.0 mL of HCl solution
b. 125 mL of HCl solution
c. 250. mL of HCl solution

7.103 The solution will dehydrate the flowers because water will flow out of the cells of the flowers into the more concentrated (hypertonic) salt solution.

7.105 Drinking seawater will cause water to flow out of the body cells and further dehydrate a person.

7.107 **a.** 11.6% (m/m) NaCl solution
b. 2.39 M NaCl solution
c. 0.398 M NaCl solution

7.109 **a.** saturated
b. unsaturated
c. saturated

7.111 205 mL of LiCl solution

7.113 **a.** 2.0 M KF solution
b. 0.25 M $CaCl_2$ solution

Acids and Bases

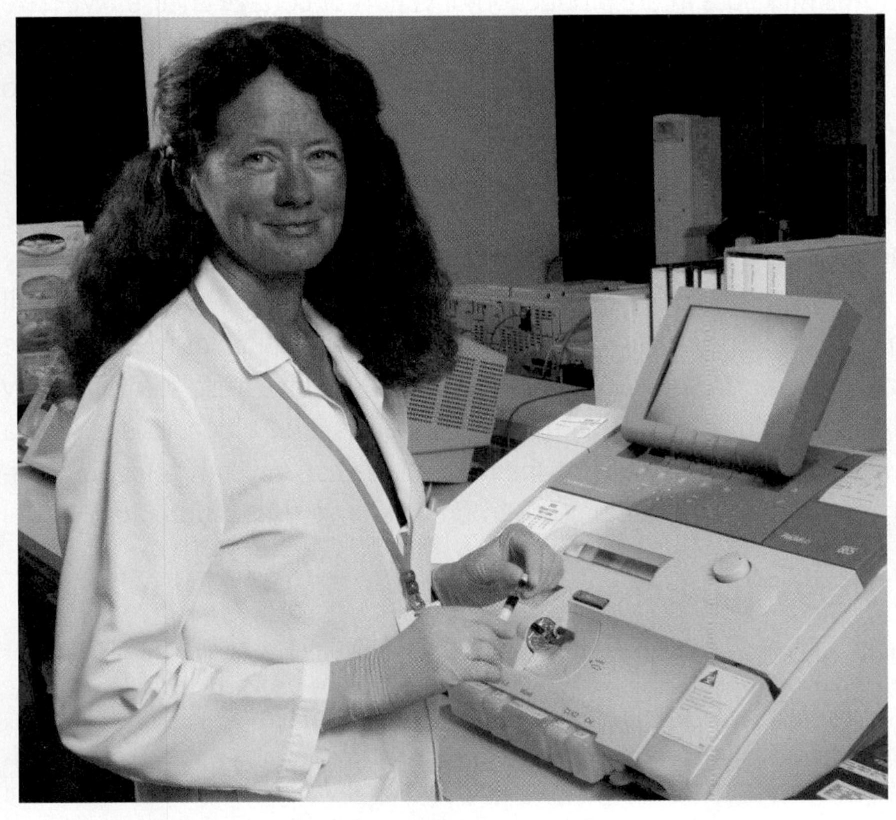

LOOKING AHEAD

"In a stat lab, we are sent blood samples of patients in emergency situations," says Audrey Trautwein, clinical laboratory technician, Stat Lab, Santa Clara Valley Medical Center. "We may need to assess the status of a trauma patient in ER or a patient who is in surgery. For example, an acidic blood pH diminishes cardiac function and affects the actions of certain drugs. In a stat situation, it is critical that we obtain our results fast. This is done using a blood gas analyzer. As I put a blood sample into the analyzer, a small probe draws out a measured volume, which is tested simultaneously for pH, P_{O_2}, and P_{CO_2} as well as electrolytes, glucose, and hemoglobin. In about one minute, we have our test results, which are sent to the doctor's computer."

Visit **www.masteringchemistry.com** for self-study materials and instructor-assigned homework.

Acids and bases are important substances in health, industry, and the environment. One of the most common characteristics of acids is their sour taste. Lemons and grapefruits are sour because they contain organic acids such as citric and ascorbic acid (vitamin C). Vinegar tastes sour because it contains acetic acid. We produce lactic acid in our muscles when we exercise. Acid from bacteria turns milk sour in the production of yogurt or cottage cheese. We have hydrochloric acid in our stomachs that helps us digest food. Sometimes we take antacids, which are bases such as sodium bicarbonate or milk of magnesia, to neutralize the effects of too much stomach acid.

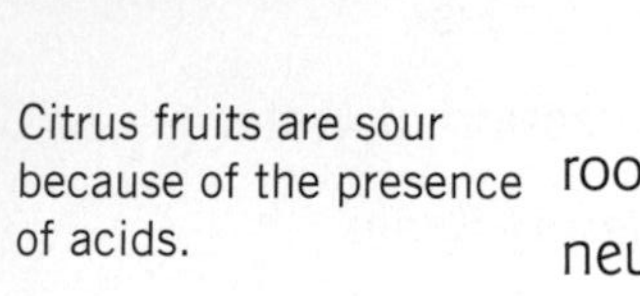

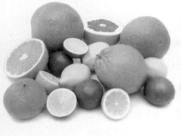

Citrus fruits are sour because of the presence of acids.

In the environment, the acidity, or pH, of rain, water, and soil can have significant effects. When rain becomes too acidic, it can dissolve marble statues and accelerate the corrosion of metals. In lakes and ponds, the acidity of water can affect the ability of plants and fish to survive. The acidity of soil around plants affects their growth. If the soil pH is too acidic or too basic, the roots of the plant cannot take up some nutrients. Most plants thrive in soil with a nearly neutral pH, although certain plants such as orchids, camellias, and blueberries require a more acidic soil. Major changes in the pH of the body fluids can severely affect biological activities within the cells. Buffers are present to prevent large fluctuations in pH.

LEARNING GOAL

Describe and name acids and bases; identify Brønsted–Lowry acids and bases.

TUTORIAL
Definitions of Acids and Bases

TUTORIAL
Naming Acids and Bases

TUTORIAL
Acid and Base Formulas

8.1 Acids and Bases

The term *acid* comes from the Latin word *acidus*, which means "sour." We are familiar with the sour tastes of vinegar and lemons and other common acids in foods.

In 1887, the Swedish chemist Svante Arrhenius was the first to describe **acids** as substances that produce hydrogen ions (H^+) when they dissolve in water. Because acids produce ions in water, they are also electrolytes (see Section 7.2). For example, hydrogen chloride ionizes completely in water to give hydrogen ions, H^+, and chloride ions, Cl^-. It is the hydrogen ions that give acids a sour taste, change blue litmus indicator to red, and corrode some metals.

$$\mathbf{HCl}(g) \xrightarrow[\text{Ionization}]{H_2O} \mathbf{H^+}(aq) + \mathbf{Cl^-}(aq)$$

Polar covalent compound — Ionization — Hydrogen ion

Naming Acids

When an acid dissolves in water to produce a hydrogen ion and a simple nonmetal anion, the prefix *hydro* is used before the name of the nonmetal, and its *ide* ending is changed to *ic acid*. For example, hydrogen chloride (HCl) dissolves in water to form HCl(*aq*), which is named hydrochloric acid. An exception is hydrogen cyanide (HCN), which dissolves in water to form hydrocyanic acid, HCN(*aq*).

When an acid contains oxygen, it dissolves in water to produce a hydrogen ion and an oxygen-containing polyatomic anion. The most common form of an oxygen-containing acid has a name that ends with *ic acid*. The name of its polyatomic anion ends in *ate*. An acid that contains one less oxygen atom than the common form is named as an *ous acid*. The name of its polyatomic anion ends with *ite* (see Table 8.1). By learning the names of the most common acids, we can derive the names of the corresponding *ous acids* and their polyatomic anions.

TABLE 8.1 Names of Common Acids and Their Anions

	Acid	Name of Acid	Anion	Name of Anion
Acids without oxygen				
	HCl	**Hydro**chlor**ic acid**	Cl^-	Chlor**ide**
	HBr	**Hydro**brom**ic acid**	Br^-	Brom**ide**
Acids with oxygen				
Nonmetal				
N	HNO_3	Nitr**ic acid**	NO_3^-	Nitr**ate**
	HNO_2	Nitr**ous acid**	NO_2^-	Nitr**ite**
S	H_2SO_4	Sulfur**ic acid**	SO_4^{2-}	Sulf**ate**
	H_2SO_3	Sulfur**ous acid**	SO_3^{2-}	Sulf**ite**
C	H_2CO_3	Carbon**ic acid**	CO_3^{2-}	Carbon**ate**
	$HC_2H_3O_2$	Acet**ic acid**	$C_2H_3O_2^-$	Acet**ate**
P	H_3PO_4	Phosphor**ic acid**	PO_4^{3-}	Phosph**ate**
	H_3PO_3	Phosphor**ous acid**	PO_3^{3-}	Phosph**ite**
Cl	$HClO_3$	Chlor**ic acid**	ClO_3^-	Chlor**ate**
	$HClO_2$	Chlor**ous acid**	ClO_2^-	Chlor**ite**

Bases

You may be familiar with some bases such as antacids, drain openers, and oven cleaners. According to the Arrhenius theory, **bases** are ionic compounds that dissociate into a metal ion and hydroxide ions (OH^-) when they dissolve in water. Thus, Arrhenius bases are also electrolytes. For example, sodium hydroxide is an Arrhenius base that dissociates in water to give sodium ions, Na^+, and hydroxide ions, OH^-.

Most Arrhenius bases are formed from Groups 1A (1) and 2A (2) metals, such as NaOH, KOH, LiOH, and $Ca(OH)_2$. The hydroxide ions (OH^-) give Arrhenius bases common characteristics such as a bitter taste and soapy, slippery feel. A base turns litmus indicator blue and phenolphthalein indicator pink.

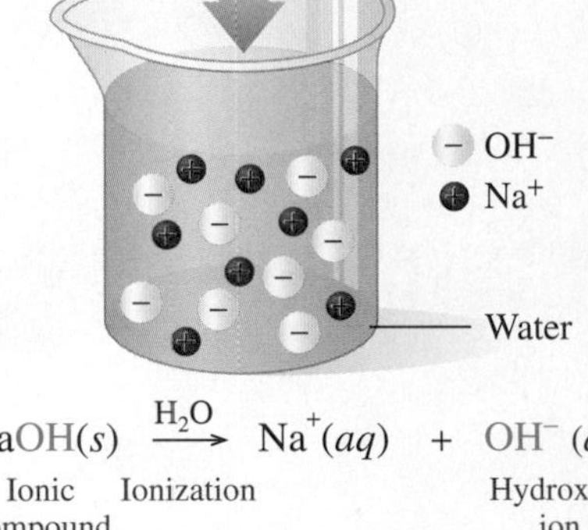

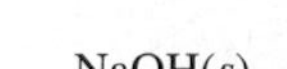

$$NaOH(s) \xrightarrow{H_2O} Na^+(aq) + OH^-(aq)$$

Ionic compound — Ionization — Hydroxide ion

An Arrhenius base produces a cation and an OH^- anion in an aqueous solution.

Naming Bases

Typical Arrhenius bases are named as hydroxides.

Bases	Name
LiOH	Lithium **hydroxide**
NaOH	Sodium **hydroxide**
KOH	Potassium **hydroxide**
$Ca(OH)_2$	Calcium **hydroxide**
$Al(OH)_3$	Aluminum **hydroxide**

CONCEPT CHECK 8.1

Names and Formulas of Acids and Bases

a. Name each of the following as an acid or a base:
1. H_3PO_4 **2.** NaOH

b. Write the formula of each of the following acid or base:
1. nitrous acid **2.** hydrobromic acid

ANSWER

a. 1. phosphoric acid **2.** sodium hydroxide

b. 1. HNO_2 **2.** HBr

SAMPLE PROBLEM 8.1

Dissociation of an Arrhenius Base

Write the equation for the dissociation of $Ca(OH)_2(s)$ in water.

SOLUTION

The base $Ca(OH)_2$ dissociates in water to give a solution of calcium ions (Ca^{2+}) and twice as many hydroxide ions (OH^-).

$$Ca(OH)_2(s) \xrightarrow{H_2O} Ca^{2+}(aq) + 2OH^-(aq)$$

STUDY CHECK 8.1

Write the equation for the dissociation of lithium hydroxide in water.

Brønsted–Lowry Acids and Bases

In 1923, J. N. Brønsted in Denmark and T. M. Lowry in Great Britain expanded the definition of acids and bases. A **Brønsted–Lowry acid** donates a proton (hydrogen ion, H^+) to another substance, and a **Brønsted–Lowry base** accepts a proton.

A Brønsted–Lowry acid is a proton (H^+) donor.

A Brønsted–Lowry base is a proton (H^+) acceptor.

A free, dissociated proton (H^+) does not actually exist in water. It undergoes hydration just like other cations because it has a strong attraction to polar water molecules. The hydrated H^+ is written as H_3O^+ and called a **hydronium ion**.

$$H{-}\ddot{O}{:}(H) + H^+ \longrightarrow \left[H{-}\ddot{O}{-}H(H)\right]^+$$

Water Proton Hydronium ion

We can write the formation of a hydrochloric acid solution as a transfer of a proton from hydrogen chloride to water. By accepting a proton in the reaction, water is acting as a base according to the Brønsted–Lowry concept.

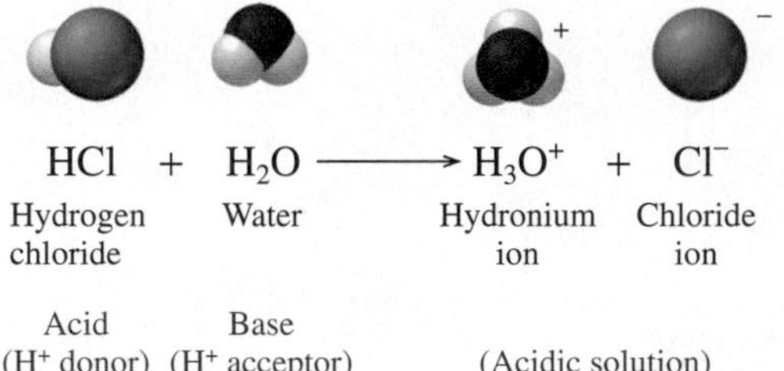

$$HCl + H_2O \longrightarrow H_3O^+ + Cl^-$$

Hydrogen chloride (Acid, H^+ donor) + Water (Base, H^+ acceptor) → Hydronium ion + Chloride ion (Acidic solution)

Ammonia, NH_3, acts as a base by accepting a proton when it reacts with water. Because the nitrogen of NH_3 has a stronger attraction for a proton than the oxygen of water, water acts as an acid and donates a proton.

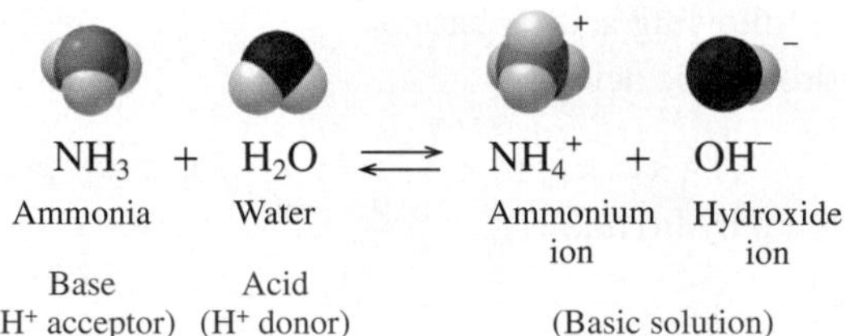

$$NH_3 + H_2O \rightleftarrows NH_4^+ + OH^-$$

Ammonia (Base, H^+ acceptor) + Water (Acid, H^+ donor) ⇌ Ammonium ion + Hydroxide ion (Basic solution)

Table 8.2 compares some characteristics of acids and bases.

TABLE 8.2 Some Characteristics of Acids and Bases

Characteristic	Acids	Bases
Arrhenius	Produce H^+	Produce OH^-
Brønsted–Lowry	Donate H^+	Accept H^+
Electrolytes	Yes	Yes
Taste	Sour	Bitter, chalky
Feel	May sting	Soapy, slippery
Litmus	Red	Blue
Phenolphthalein	Colorless	Pink
Neutralization	Neutralize bases	Neutralize acids

MC SELF STUDY ACTIVITY
Nature of Acids and Bases

TUTORIAL
Properties of Acids and Bases

SAMPLE PROBLEM 8.2

Acids and Bases

In each of the following equations, identify the reactant that is a Brønsted–Lowry acid and the reactant that is a Brønsted–Lowry base:

a. $HBr(aq) + H_2O(l) \longrightarrow H_3O^+(aq) + Br^-(aq)$
b. $H_2O(l) + CN^-(aq) \rightleftharpoons HCN(aq) + OH^-(aq)$

SOLUTION

a. HBr, acid; H_2O, base **b.** H_2O, acid; CN^-, base

STUDY CHECK 8.2

When HNO_3 reacts with water, water acts as a base (H^+ acceptor). Write the equation for the reaction.

Conjugate Acid–Base Pairs

According to the Brønsted–Lowry theory, a **conjugate acid–base pair** consists of molecules or ions related by the loss or gain of one H^+. Every acid–base reaction contains two conjugate acid–base pairs because protons are transferred in both the forward and the reverse reactions. When the acid HA donates H^+, the conjugate base A^- forms. When the base B accepts the H^+, it forms the conjugate acid BH^+. We can write this as a general equation for a Brønsted–Lowry acid–base reaction as follows:

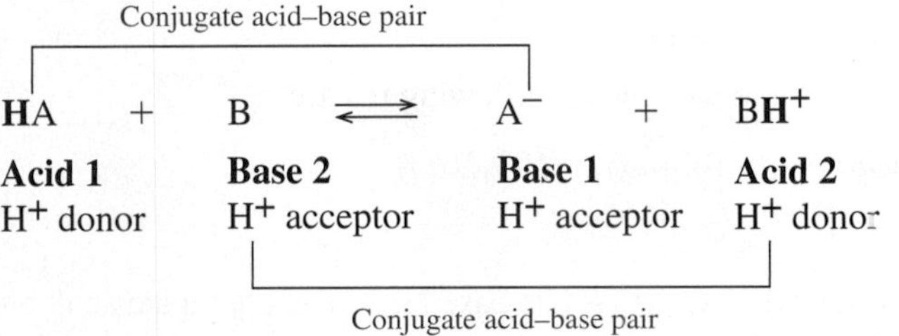

Now we can identify the conjugate acid–base pairs in a reaction between hydrofluoric acid, HF, and water. Because HF is a weak acid, the reaction is written with two arrows to show that the conjugate acid H_3O^+ in the products can transfer a proton to re-form the acid HF. We can now identify the conjugate acid–base pairs as HF/F^- and H_3O^+/H_2O.

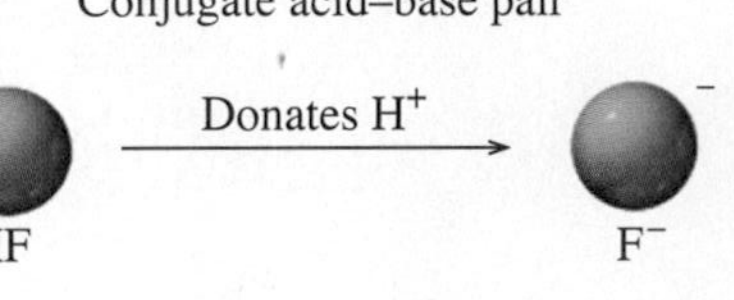

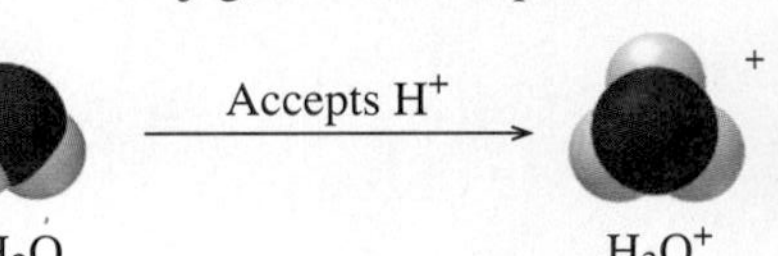

TUTORIAL
Identifying Conjugate Acid–Base Pairs

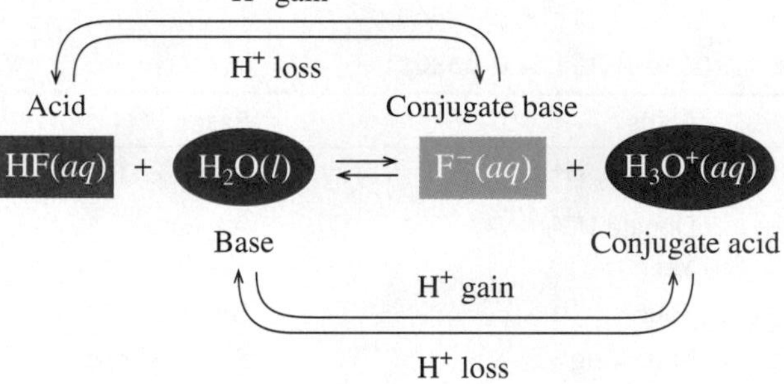

In the reaction between ammonia, NH_3, and water, NH_3 accepts one H^+ from H_2O to form the conjugate acid NH_4^+ and conjugate base OH^-. Both conjugate acid–base pairs, NH_4^+/NH_3 and H_2O/OH^-, are related by the loss and gain of one H^+.

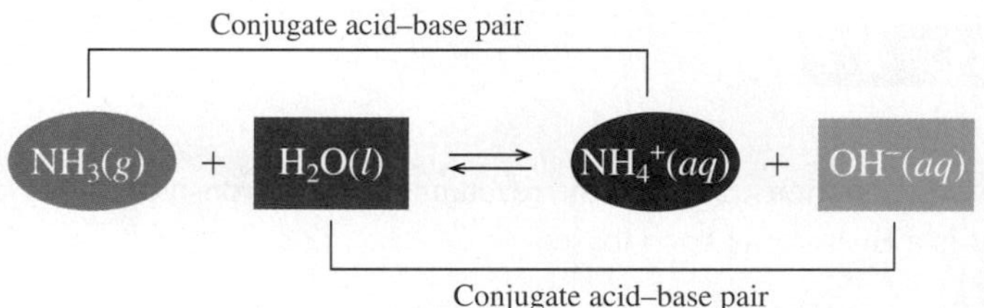

We have now seen that water can act as an acid or a base. Substances that can act as both acids and bases are *amphoteric*. For water, the most common amphoteric substance, the acidic or basic behavior depends on whether the other reactant is a stronger acid or base.

CONCEPT CHECK 8.2

Conjugate Acid–Base Pairs

Write the formula of the conjugate base of each of the following:

a. $HBrO_3$ **b.** HCO_3^-

ANSWER

a. When $HBrO_3$ loses H^+, it forms its conjugate base BrO_3^-.
b. When HCO_3^- loses H^+, it forms its conjugate base CO_3^{2-}.

SAMPLE PROBLEM 8.3

Identifying Conjugate Acid–Base Pairs

Identify the conjugate acid–base pairs in the following reaction:

$$HBr(aq) + NH_3(aq) \longrightarrow Br^-(aq) + NH_4^+(aq)$$

SOLUTION

In the reaction, the acid HBr donates H^+ to the base NH_3. The conjugate acid–base pairs are HBr/Br^- and NH_4^+/NH_3.

STUDY CHECK 8.3

In the following reaction, identify the conjugate acid–base pairs:

$$HCN(aq) + SO_4^{2-}(aq) \rightleftharpoons CN^-(aq) + HSO_4^-(aq)$$

QUESTIONS AND PROBLEMS

Acids and Bases

8.1 Indicate whether each of the following statements is characteristic of an acid or a base:
a. has a sour taste
b. neutralizes bases
c. produces H^+ ions in water
d. is named potassium hydroxide

8.2 Indicate whether each of the following statements is characteristic of an acid or a base:
a. neutralizes acids
b. produces OH^- ions in water
c. has a soapy feel
d. turns litmus red

8.3 Name each of the following acids or bases:
a. HCl **b.** $Ca(OH)_2$ **c.** H_2CO_3
d. HNO_3 **e.** H_2SO_3

8.4 Name each of the following acids or bases:
a. $Al(OH)_3$ **b.** HBr **c.** H_2SO_4
d. KOH **e.** HNO_2

8.5 Write the formula for each of the following acids and bases:
a. magnesium hydroxide **b.** hydrofluoric acid
c. phosphoric acid **d.** lithium hydroxide
e. ammonium hydroxide **f.** sulfuric acid

8.6 Write the formula for each of the following acids and bases:
a. barium hydroxide **b.** hydroiodic acid
c. nitric acid **d.** strontium hydroxide
e. sodium hydroxide **f.** chloric acid

8.7 Identify the acid (proton donor) and the base (proton acceptor) for the reactants in each of the following:
a. $HI(aq) + H_2O(l) \longrightarrow H_3O^+(aq) + I^-(aq)$
b. $F^-(aq) + H_2O(l) \rightleftharpoons HF(aq) + OH^-(aq)$

8.8 Identify the acid (proton donor) and the base (proton acceptor) for the reactants in each of the following:
a. $CO_3^{2-}(aq) + H_2O(l) \rightleftharpoons HCO_3^-(aq) + OH^-(aq)$
b. $H_2SO_4(aq) + H_2O(l) \longrightarrow H_3O^+(aq) + HSO_4^-(aq)$

8.9 Write the formula of the conjugate base for each of the following:
a. HF **b.** H_2O **c.** H_2CO_3 **d.** HSO_4^-

8.10 Write the formula of the conjugate base for each of the following:
a. HSO_3^- **b.** H_3O^+ **c.** HPO_4^{2-} **d.** HNO_2

8.11 Write the formula of the conjugate acid for each of the following:
a. CO_3^{2-} **b.** H_2O **c.** $H_2PO_4^-$ **d.** Br^-

8.12 Write the formula of the conjugate acid for each of the following:
a. SO_4^{2-} **b.** CN^- **c.** OH^- **d.** ClO_2^-

8.13 Identify the Brønsted–Lowry acid–base pairs in each of the following equations:
a. $H_2CO_3(aq) + H_2O(l) \rightleftharpoons H_3O^+(aq) + HCO_3^-(aq)$
b. $NH_4^+(aq) + H_2O(l) \rightleftharpoons H_3O^+(aq) + NH_3(aq)$
c. $HCN(aq) + NO_2^-(aq) \rightleftharpoons HNO_2(aq) + CN^-(aq)$

8.14 Identify the Brønsted–Lowry acid–base pairs in each of the following equations:
a. $H_3PO_4(aq) + H_2O(l) \rightleftharpoons H_3O^+(aq) + H_2PO_4^-(aq)$
b. $CO_3^{2-}(aq) + H_2O(l) \rightleftharpoons OH^-(aq) + HCO_3^-(aq)$
c. $H_3PO_4(aq) + NH_3(aq) \rightleftharpoons NH_4^+(aq) + H_2PO_4^-(aq)$

8.2 Strengths of Acids and Bases

LEARNING GOAL
Write equations for the dissociation of strong and weak acids.

The *strength* of an acid is determined by the moles of H_3O^+ that are produced for each mole of acid that dissolves. The *strength* of a base is determined by the moles of OH^- that are produced for each mole of base that dissolves. In the process called *dissociation*, an acid or base produces ions in water. However, they vary greatly in their ability to produce H_3O^+ or OH^-. Strong acids and strong bases dissociate completely in water. Weak acids and weak bases dissolve mostly as molecules with only a few dissociating into ions.

Acids produce hydrogen ions in aqueous solutions.

Strong and Weak Acids

Strong acids are examples of strong electrolytes because they donate protons so easily that their dissociation in water is virtually complete. For example, when HCl, a strong acid, dissociates in water, H^+ is transferred to H_2O; the resulting solution contains only the ions H_3O^+ and Cl^-. We consider the reaction of HCl in H_2O as going 100% to products. The equation for the dissociation of a strong acid, such as HCl, is written with a single arrow to the products.

$$HCl(g) + H_2O(l) \longrightarrow H_3O^+(aq) + Cl^-(aq)$$

There are only six common strong acids. All other acids are weak (see Table 8.3).

TABLE 8.3 Common Strong and Weak Acids

Increasing acid strength ↑

Strong Acids	
Hydroiodic acid	HI
Hydrobromic acid	HBr
Perchloric acid	$HClO_4$
Hydrochloric acid	HCl
Sulfuric acid	H_2SO_4
Nitric acid	HNO_3
Weak Acids	
Hydronium ion	H_3O^+
Hydrogen sulfate ion	HSO_4^-
Phosphoric acid	H_3PO_4
Hydrofluoric acid	HF
Nitrous acid	HNO_2
Acetic acid	$HC_2H_3O_2$
Carbonic acid	H_2CO_3
Hydrosulfuric acid	H_2S
Dihydrogen phosphate	$H_2PO_4^{2-}$
Ammonium ion	NH_4^+
Hydrocyanic acid	HCN
Bicarbonate ion	HCO_3^-
Hydrogen sulfide ion	HS^-
Water	H_2O

With the exception of the six strong acids, all other acids are weak. **Weak acids** are weak electrolytes because they produce only a few ions in water. A solution of a strong acid contains all ions, whereas a solution of a weak acid contains mostly molecules and few ions (see Figure 8.1). Many of the products you use at home

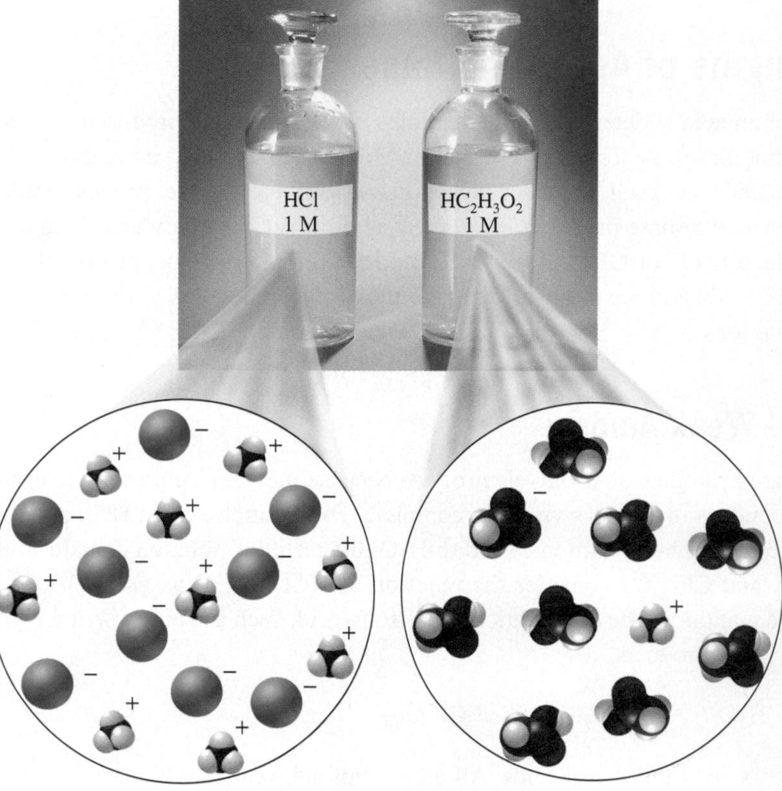

FIGURE 8.1 A strong acid such as HCl is completely dissociated (100%), whereas a weak acid such as $HC_2H_3O_2$ contains mostly dissolved molecules and only a few ions.

Q What is the difference between a strong acid and a weak acid?

contain weak acids. In carbonated soft drinks, CO_2 dissolves in water to form carbonic acid, H_2CO_3, a weak acid.

$$H_2CO_3(aq) + H_2O(l) \rightleftharpoons H_3O^+(aq) + HCO_3^-(aq)$$

A weak acid such as carbonic acid, H_2CO_3, is written with a double arrow. A longer reverse arrow may be used to indicate that a solution of the weak carbonic acid contains mostly undissociated acid molecules of the reactants and a few H_3O^+ and bicarbonate ions of the products.

Citric acid is a weak acid found in fruits and fruit juices such as lemons, oranges, and grapefruit. In the vinegar used in salad dressings, the weak acid, acetic acid, $HC_2H_3O_2$, is present as a 5% (m/v) acetic acid solution.

$$\underset{\text{Acetic acid}}{HC_2H_3O_2(l)} + H_2O(l) \rightleftharpoons H_3O^+(aq) + \underset{\text{Acetate ion}}{C_2H_3O_2^-(aq)}$$

In summary, if HA is a strong acid in water, its aqueous solution consists of the ions H_3O^+ and A^-. However, if HA is a weak acid, its aqueous solution consists of mostly undissociated HA and only a few H_3O^+ and A^- ions (see Figure 8.2).

Strong acid: $HA(aq) + H_2O(l) \longrightarrow H_3O^+(aq) + A^-(aq)$ (~100% dissociated)

Weak acid: $HA(aq) + H_2O(l) \rightleftharpoons H_3O^+(aq) + A^-(aq)$ (small % dissociated)

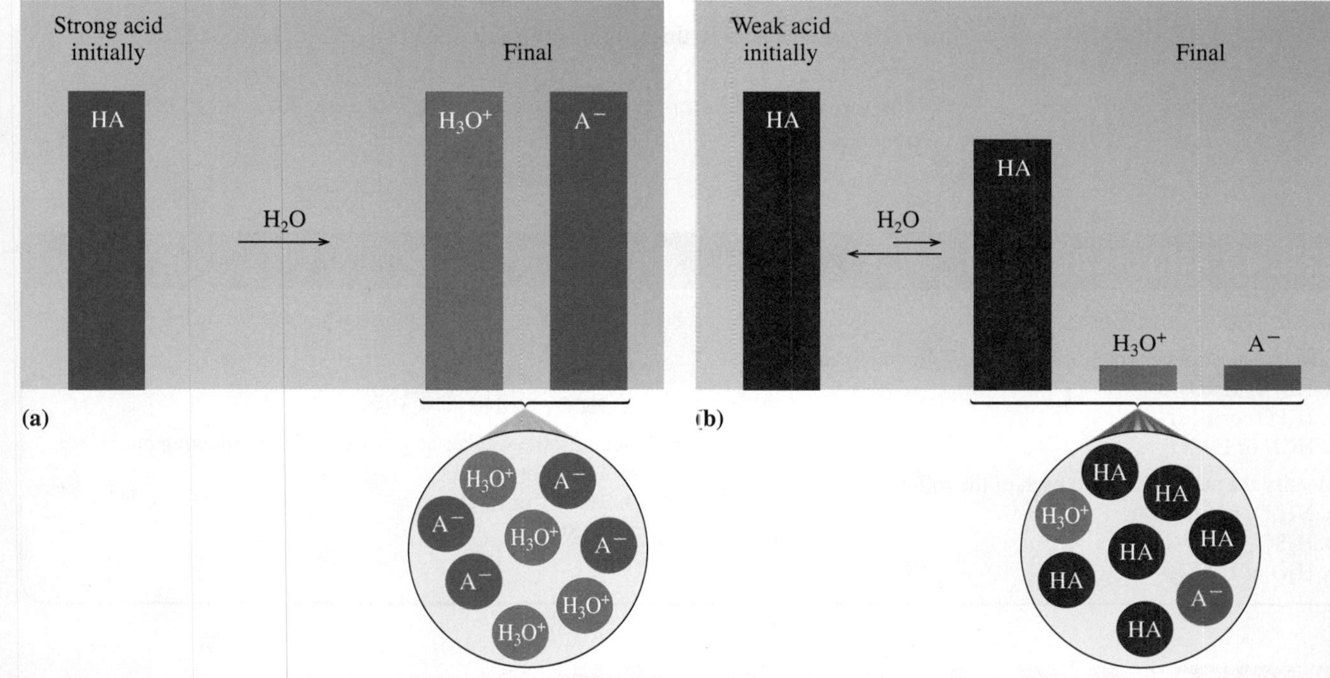

FIGURE 8.2 (a) A solution of a strong acid (HA) contains only H_3O^+ and A^- ions. (b) A solution of a weak acid (HA) contains mostly undissociated HA molecules and only a few H_3O^+ and A^- ions.

Q Why do the heights of the H_3O^+ and A^- ions in the bar diagrams differ for a strong acid and a weak acid?

Strong and Weak Bases

A **strong base** is a strong electrolyte that dissociates completely in water to give an aqueous solution of a metal ion and hydroxide ions. The Group 1A (1) hydroxides are very soluble in water, which can give high concentrations of OH^- ions. The Group 2A (2) bases are less soluble in water, but they are strong bases because they dissociate completely as ions. For example, when KOH forms a KOH solution, it contains only K^+ and OH^- ions.

$$KOH(s) \xrightarrow{H_2O} K^+(aq) + OH^-(aq)$$

Strong Bases

Group 1A (1)
- LiOH
- NaOH
- KOH

Group 2A (2)
- $Ca(OH)_2$
- $Sr(OH)_2$
- $Ba(OH)_2$

Bases in household products are used to remove grease and to open drains.

Sodium hydroxide, NaOH (also known as lye), is used in household products to remove grease in ovens and to clean drains. Because high concentrations of hydroxide ions cause severe damage to the skin and eyes, directions must be followed carefully when such products are used in the home, and use in the chemistry laboratory should be carefully supervised. If you spill an acid or a base on your skin or get some in your eyes, be sure to flood the area immediately with water for at least 10 minutes and seek medical attention.

A **weak base** is an electrolyte that produces only a few ions in solution. A typical weak base, ammonia, NH_3, is found in window cleaners. In an aqueous solution of NH_3, only a few molecules accept protons to form ammonium hydroxide, which is NH_4^+ and OH^- ions.

$$\underset{\text{Ammonia}}{NH_3(g)} + H_2O(l) \rightleftarrows \underset{\text{Ammonium hydroxide}}{NH_4^+(aq) + OH^-(aq)}$$

CONCEPT CHECK 8.3

Strengths of Acids and Bases

For each of the following questions, select from HCO_3^-, H_2SO_4, or HNO_2:

a. Which is the strongest acid?
b. Which acid is the weakest?

ANSWER

a. The strongest acid in this group is the acid listed closest to the top of Table 8.3, which is H_2SO_4.
b. The weakest acid in this group is the acid listed closest to the bottom of Table 8.3, which is HCO_3^-.

QUESTIONS AND PROBLEMS

Strengths of Acids and Bases

8.15 Identify the stronger acid in each of the following pairs:
a. HBr or HNO_2
b. H_3PO_4 or HSO_4^-
c. HCN or H_2CO_3

8.16 Identify the stronger acid in each of the following pairs:
a. NH_4^+ or H_3O^+
b. H_2SO_4 or HCN
c. H_2O or H_2CO_3

8.17 Identify the weaker acid in each of the following pairs:
a. HCl or HSO_4^-
b. HNO_2 or HF
c. HCO_3^- or NH_4^+

8.18 Identify the weaker acid in each of the following pairs:
a. HNO_3 or HCO_3^-
b. HSO_4^- or H_2O
c. H_2SO_4 or H_2CO_3

LEARNING GOAL

Use the ion product of water to calculate the $[H_3O^+]$ and $[OH^-]$ in an aqueous solution.

TUTORIAL
Ionization of Water

8.3 Ionization of Water

We have seen that in some acid–base reactions, water is amphoteric, which means that water can be both an acid and a base. One water molecule acts as an acid by donating H^+ to another water molecule, which acts as a base. Every time a H^+ is transferred between two water molecules, the products are one H_3O^+ and one OH^-.

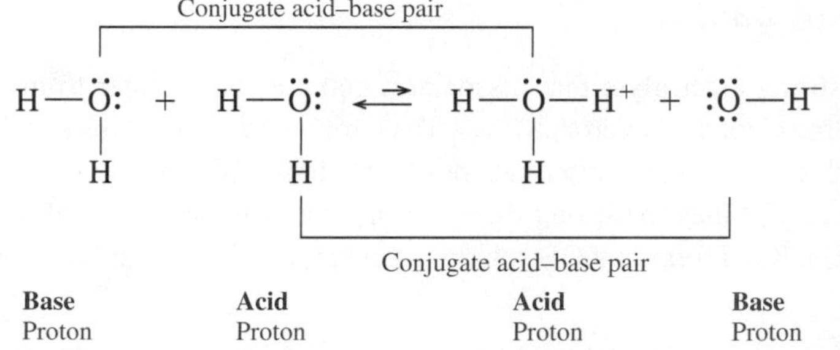

In the equation for the ionization of water, there is both a forward and reverse reaction:

$$\underset{\text{Base}}{H_2O(l)} + \underset{\text{Acid}}{H_2O(l)} \rightleftarrows \underset{\text{Conjugate acid}}{H_3O^+(aq)} + \underset{\text{Conjugate base}}{OH^-(aq)}$$

Experiments have determined that, in pure water, the concentrations of H_3O^+ and OH^- at 25 °C are each 1.0×10^{-7} M. Square brackets are used to indicate the concentrations in moles per liter (M).

$$\text{Pure water} \quad [H_3O^+] = [OH^-] = 1.0 \times 10^{-7}\,\text{M}$$

When these concentrations are multiplied, we obtain the expression and value called the **ion product constant of water, K_w**, which is 1.0×10^{-14}. The concentration units are omitted in the K_w value.

$$K_w = [H_3O^+][OH^-]$$
$$= (1.0 \times 10^{-7}\,\text{M})(1.0 \times 10^{-7}\,\text{M}) = 1.0 \times 10^{-14}$$

The K_w value (1.0×10^{-14}) applies to any aqueous solution at 25 °C because all aqueous solutions contain both H_3O^+ and OH^- (see Figure 8.3).

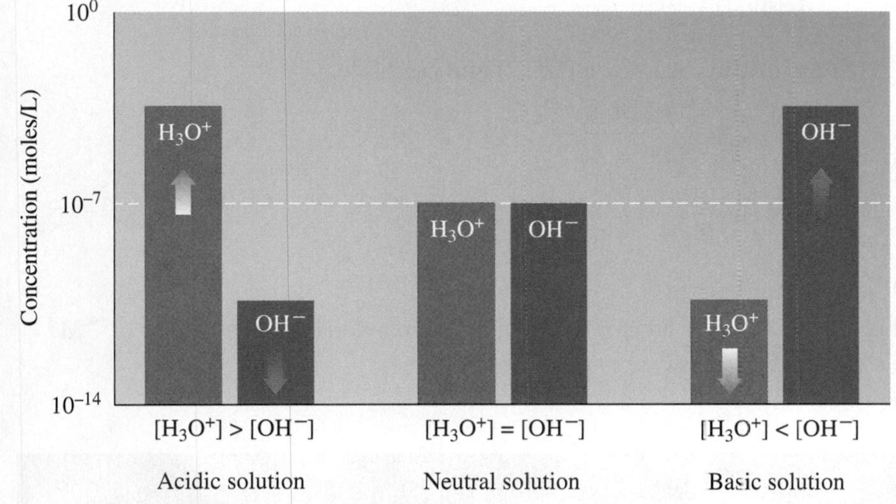

FIGURE 8.3 In a neutral solution, $[H_3O^+]$ and $[OH^-]$ are equal. In acidic solutions, the $[H_3O^+]$ is greater than the $[OH^-]$. In basic solutions, the $[OH^-]$ is greater than the $[H_3O^+]$.

Q Is a solution that has a $[H_3O^+]$ of 1.0×10^{-3} M acidic, basic, or neutral?

When the $[H_3O^+]$ and $[OH^-]$ in a solution are equal, the solution is **neutral**. However, most solutions are not neutral; they have different concentrations of $[H_3O^+]$ and $[OH^-]$. If acid is added to water, there is an increase in $[H_3O^+]$ and a decrease in $[OH^-]$, which makes an acidic solution. If base is added, $[OH^-]$ increases and $[H_3O^+]$ decreases, which makes a basic solution. Thus, for any aqueous solution, whether it is neutral, acidic, or basic, the product $[H_3O^+][OH^-]$ is equal to K_w (1.0×10^{-14}) (see Table 8.4).

TABLE 8.4 Examples of $[H_3O^+]$ and $[OH^-]$ in Neutral, Acidic, and Basic Solutions

K_w	$[H_3O^+]$	$[OH^-]$	Type of Solution
1.0×10^{-14}	1.0×10^{-7} M	1.0×10^{-7} M	Neutral
1.0×10^{-14}	1.0×10^{-2} M	1.0×10^{-12} M	Acidic
1.0×10^{-14}	2.5×10^{-5} M	4.0×10^{-10} M	Acidic
1.0×10^{-14}	1.0×10^{-8} M	1.0×10^{-6} M	Basic
1.0×10^{-14}	5.0×10^{-11} M	2.0×10^{-4} M	Basic

If the $[H_3O^+]$ is given, K_w can be rearranged to calculate $[OH^-]$. If the $[OH^-]$ is given, K_w can be rearranged to calculate $[H_3O^+]$ as seen in Sample Problem 8.4.

$$K_w = [H_3O^+][OH^-]$$

$$[OH^-] = \frac{K_w}{[H_3O^+]} \qquad [H_3O^+] = \frac{K_w}{[OH^-]}$$

Guide to Calculating $[H_3O^+]$ and $[OH^-]$ in Aqueous Solutions

1 Write the K_w for water.

2 Solve the K_w for the unknown $[H_3O^+]$ or $[OH^-]$.

3 Substitute in the known $[H_3O^+]$ or $[OH^-]$ and calculate.

SAMPLE PROBLEM 8.4

Calculating the $[H_3O^+]$ of a Solution

A vinegar solution has a $[OH^-] = 5.0 \times 10^{-12}$ M at 25 °C. What is the $[H_3O^+]$ of the vinegar solution? Is the solution acidic, basic, or neutral?

SOLUTION

Step 1 Write the K_w for water.

$$K_w = [H_3O^+][OH^-] = 1.0 \times 10^{-14}$$

Step 2 Solve the K_w for the unknown $[H_3O^+]$. Rearrange the ion product expression by dividing through by the $[OH^-]$.

$$\frac{K_w}{[OH^-]} = \frac{[H_3O^+]\cancel{[OH^-]}}{\cancel{[OH^-]}}$$

$$[H_3O^+] = \frac{1.0 \times 10^{-14}}{[OH^-]}$$

Step 3 Substitute in the known $[OH^-]$ and calculate.

$$[H_3O^+] = \frac{1.0 \times 10^{-14}}{[5.0 \times 10^{-12}]} = 2.0 \times 10^{-3}\ \text{M}$$

Because the $[H_3O^+]$ of 2.0×10^{-3} M is larger than the $[OH^-]$ of 5.0×10^{-12} M, the solution is acidic.

STUDY CHECK 8.4

What is the $[H_3O^+]$ of an ammonia cleaning solution with $[OH^-] = 4.0 \times 10^{-4}$ M? Is the solution acidic, basic, or neutral?

QUESTIONS AND PROBLEMS

Ionization of Water

8.19 Why are the concentrations of H_3O^+ and OH^- equal in pure water?

8.20 What is the meaning and value of K_w?

8.21 In an acidic solution, how does the concentration of H_3O^+ compare to the concentration of OH^-?

8.22 If a base is added to pure water, why does the concentration of H_3O^+ decrease?

8.23 Indicate whether each of the following is acidic, basic, or neutral:
a. $[H_3O^+] = 2.0 \times 10^{-5}$ M **b.** $[H_3O^+] = 1.4 \times 10^{-9}$ M
c. $[OH^-] = 8.0 \times 10^{-3}$ M **d.** $[OH^-] = 3.5 \times 10^{-10}$ M

8.24 Indicate whether each of the following is acidic, basic, or neutral:
a. $[H_3O^+] = 6.0 \times 10^{-12}$ M **b.** $[H_3O^+] = 1.4 \times 10^{-4}$ M
c. $[OH^-] = 5.0 \times 10^{-12}$ M **d.** $[OH^-] = 4.5 \times 10^{-2}$ M

8.25 Calculate the $[OH^-]$ of each aqueous solution with the following $[H_3O^+]$:
a. coffee, 1.0×10^{-5} M **b.** soap, 1.0×10^{-8} M
c. cleanser, 5.0×10^{-10} M **d.** lemon juice, 2.5×10^{-2} M

8.26 Calculate the $[OH^-]$ of each aqueous solution with the following $[H_3O^+]$:
a. NaOH solution, 1.0×10^{-12} M
b. aspirin, 6.0×10^{-4} M
c. milk of magnesia, 1.0×10^{-9} M
d. stomach acid, 5.2×10^{-2} M

8.27 Calculate the $[H_3O^+]$ of each aqueous solution with the following $[OH^-]$:
a. vinegar, 1.0×10^{-11} M **b.** urine, 2.0×10^{-9} M
c. ammonia, 5.6×10^{-3} M **d.** NaOH solution, 2.5×10^{-2} M

8.28 Calculate the $[H_3O^+]$ of each aqueous solution with the following $[OH^-]$:
a. baking soda, 1.0×10^{-6} M **b.** orange juice, 5.0×10^{-11} M
c. milk, 2.0×10^{-8} M **d.** bleach, 2.1×10^{-3} M

8.4 The pH Scale

LEARNING GOAL

Calculate the pH of a solution from $[H_3O^+]$; given the pH, calculate $[H_3O^+]$.

TUTORIAL
The pH Scale

CASE STUDY
Hyperventilation and Blood pH

Many kinds of careers such as respiratory therapy, food processing, medicine, agriculture, and wine making, require the measurement of the $[H_3O^+]$ and $[OH^-]$ of solutions. The proper level of acidity is necessary to evaluate the functioning of the lungs and kidneys, to control bacterial growth in foods, and to prevent the growth of pests in food crops.

On the pH scale, a number between 0 and 14 represents the $[H_3O^+]$ for most solutions. A neutral solution has a pH of 7.0. An acidic solution has a pH less than 7.0, and a basic solution has a pH greater than 7.0 (see Figure 8.4).

Neutral solution	pH = 7.0	$[H_3O^+] = 1 \times 10^{-7}$ M
Acidic solution	pH < 7.0	$[H_3O^+] > 1 \times 10^{-7}$ M
Basic solution	pH > 7.0	$[H_3O^+] < 1 \times 10^{-7}$ M

pH Value

0 1 2 3 4 5 6 7 8 9 10 11 12 13 14

Acidic

Neutral
$[H^+] = [OH^-]$

Basic

1 M HCl 0.0
Gastric juice 1.6
Lemon juice 2.2
Vinegar 2.8
Carbonated beverage 3.0
Orange 3.5
Apple juice 3.8
Tomato 4.2
Coffee 5.0
Bread 5.5
Potato 5.8
Urine 6.0
Milk 6.4
Water (pure) 7.0
Drinking water 7.2
Blood 7.4
Bile 8.0
Detergents 8.0–9.0
Seawater 8.5
Milk of magnesia 10.5
Ammonia 11.0
Bleach 12.0
1 M NaOH (lye) 14.0

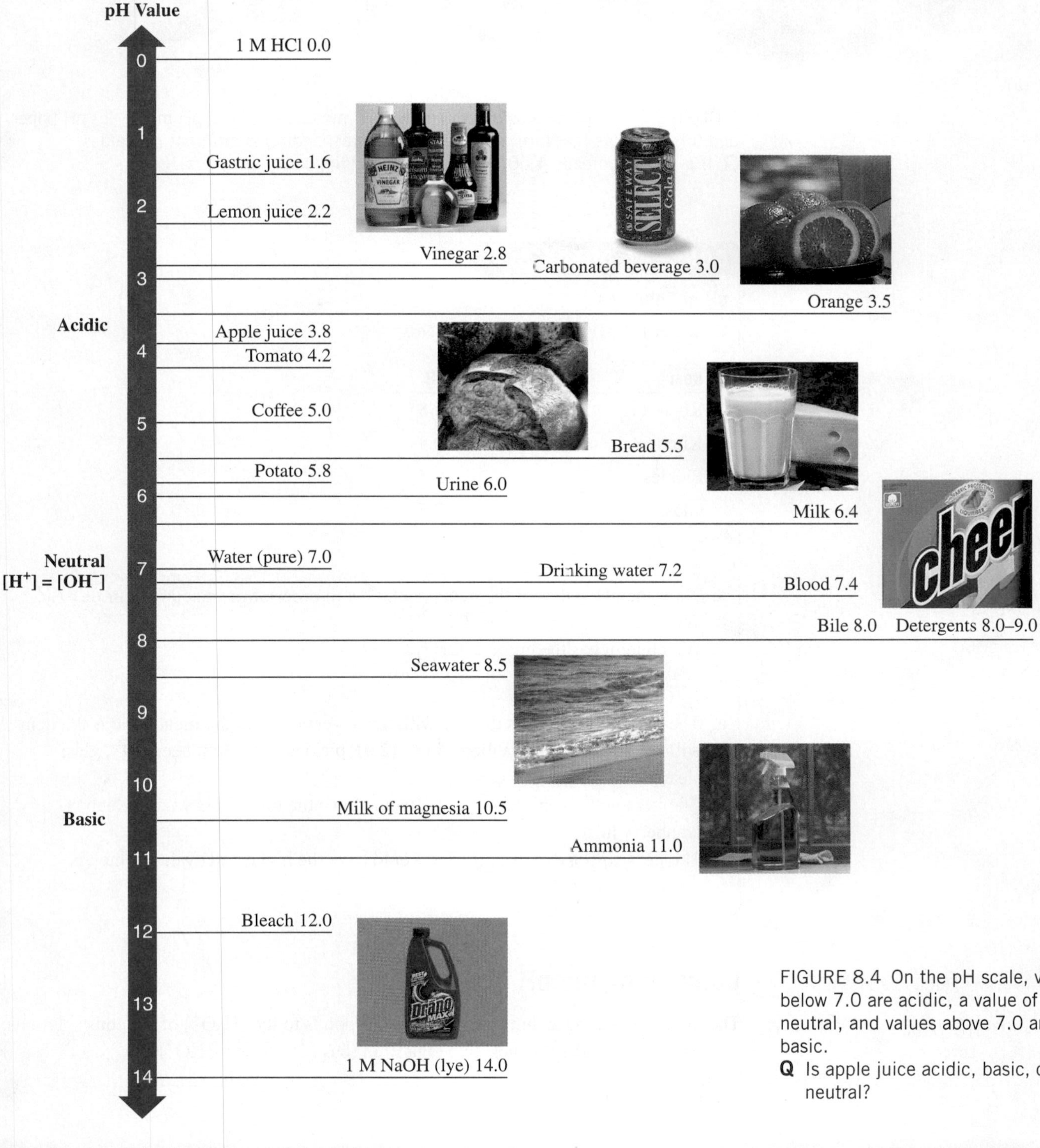

FIGURE 8.4 On the pH scale, values below 7.0 are acidic, a value of 7.0 is neutral, and values above 7.0 are basic.

Q Is apple juice acidic, basic, or neutral?

In the laboratory, a pH meter is commonly used to determine the pH of a solution. There are also indicators and pH papers that turn specific colors when placed in solutions of different pH values. The pH is found by comparing the colors to a chart (see Figure 8.5).

(a)

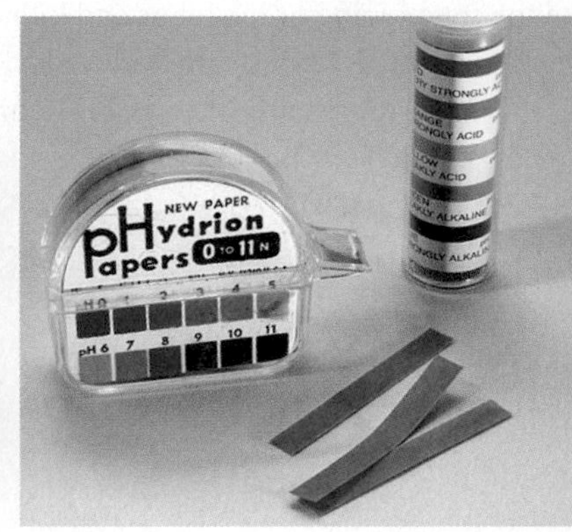

(b)

(c)

FIGURE 8.5 The pH of a solution can be determined using (a) a pH meter, (b) pH paper, and (c) indicators that turn different colors corresponding to different pH values.
Q If a pH meter reads 4.00, is the solution acidic, basic, or neutral?

CONCEPT CHECK 8.4

pH of Solutions

Consider the pH of the following items:

Item	pH
Root beer	5.8
Kitchen cleaner	10.9
Pickles	3.5
Glass cleaner	7.6
Cranberry juice	2.9

a. Place the pH values of the items on the list in order of most acidic to most basic.
b. Which item has the highest $[H_3O^+]$?
c. Which item has the highest $[OH^-]$?

ANSWER

a. The most acidic item is the one with the lowest pH, and the most basic is the item with the highest pH: cranberry juice (2.9), pickles (3.5), root beer (5.8), glass cleaner (7.6), kitchen cleaner (10.9).
b. The item with the highest $[H_3O^+]$ would have the lowest pH value, which is cranberry juice.
c. The item with the highest $[OH^-]$ would have the highest pH value, which is kitchen cleaner.

TUTORIAL
Logarithms

Calculating the pH of Solutions

The pH scale is a logarithmic scale that corresponds to the $[H_3O^+]$ of aqueous solutions. Mathematically, **pH** is the negative logarithm (base 10) of the $[H_3O^+]$.

$$pH = -\log[H_3O^+]$$

Essentially, the negative powers of 10 in the molar concentrations are converted to positive numbers. For example, a lemon juice solution with the $[H_3O^+] = 1.0 \times 10^{-2}$ M has a pH of 2.00. This can be calculated using the pH equation:

$$\begin{aligned} pH &= -\log[1.0 \times 10^{-2}] \\ pH &= -(-2.00) \\ &= 2.00 \end{aligned}$$

The number of *decimal places* in the pH value is the same as the number of significant figures in the coefficient of $[H_3O^+]$. The number on the left of the decimal point in the pH value is the power of 10.

$$[H_3O^+] = \mathbf{1.0} \times 10^{-2}\text{ M} \qquad pH = 2.\mathbf{00}$$

Two significant figures — Two decimal places

Because pH is a log scale, a change of one pH unit corresponds to a tenfold change in $[H_3O^+]$. It is important to note that the pH decreases as the $[H_3O^+]$ increases. For example, a solution with a pH of 2.00 has a $[H_3O^+]$ 10 times higher than a solution with a pH of 3.00 and 100 times higher than a solution with a pH of 4.00.

CONCEPT CHECK 8.5

Calculating pH

Indicate if the pH values given for each the following are correct or incorrect and why:

a. $[H_3O^+] = 1 \times 10^{-6}$ M pH = −6.0
b. $[H_3O^+] = 1.0 \times 10^{-10}$ M pH = 10
c. $[H_3O^+] = 1.0 \times 10^{-6}$ M pH = 6.00

ANSWER

a. Incorrect. The pH of this solution is 6.0, which has a positive value, not negative.
b. Incorrect. This solution has a $[H_3O^+]$ of 1.0×10^{-10} M, which has a pH of 10.00. Two zeros are needed after the decimal point to match the two significant figures in the coefficient of the $[H_3O^+]$.
c. The pH is correctly calculated from the $[H_3O^+]$ because there are two zeros after the decimal point to match the two significant figures in the coefficient of the molarity.

SAMPLE PROBLEM 8.5

Calculating pH Using the $[H_3O^+]$

Determine the pH for the following solutions:

a. $[H_3O^+] = 1.0 \times 10^{-5}$ M **b.** $[H_3O^+] = 5 \times 10^{-8}$ M

SOLUTION

a. Step 1 Enter the $[H_3O^+]$.

Calculator Display

$pH = -\log[1.0 \times 10^{-5}] = 1.0$ [EE or EXP] 5 [+/−] 1.0⁻⁰⁵ or 1.0E−05

Step 2 Press the *log* key and change the sign.

[log] [+/−] 5

Guide to Calculating pH of an Aqueous Solution

1 Enter the $[H_3O^+]$.

2 Press the *log* key and change the sign.

3 Adjust the number of SFs on the *right* of the decimal point to equal the SFs in the coefficient.

Step 3 Adjust the number of SFs on the *right* of the decimal point to equal the SFs in the coefficient.

1.0×10^{-5} M (Two SFs) ⟶ pH = 5.00 (Two SFs on the *right* of the decimal point)

b. Step 1 Enter the $[H_3O^+]$.

	Calculator Display
pH = $-\log[5 \times 10^{-8}]$ = 5 EE or EXP 8 +/–	5^{-08} or 5E–08

Step 2 Press the *log* key and change the sign.

log +/–	7.301029

Step 3 Adjust the number of SFs on the *right* of the decimal point to equal the SFs in the coefficient.

5×10^{-8} M (One SF) ⟶ pH = 7.3 (One SF on the *right* of the decimal point)

STUDY CHECK 8.5

What is the pH of bleach with $[H_3O^+] = 4.2 \times 10^{-12}$ M?

Chemistry Link to Health

STOMACH ACID, HCl

When a person sees, smells, thinks about, or tastes food, the gastric glands in the stomach begin to secrete a strongly acidic solution of HCl. In a single day, a person may secrete as much as 2000 mL of gastric juice.

The HCl in the gastric juice activates a digestive enzyme called *pepsin* that breaks down proteins in food entering the stomach. The secretion of HCl continues until the stomach has a pH of about 2, without ulcerating the stomach lining. Normally, large quantities of viscous mucus are secreted within the stomach to protect its lining from acid and enzyme damage.

SAMPLE PROBLEM 8.6

Calculating pH Using the $[OH^-]$

What is the pH of an ammonia solution with $[OH^-] = 3.7 \times 10^{-3}$ M?

SOLUTION

Step 1 Enter the $[H_3O^+]$. Because $[OH^-]$ is given for the ammonia solution, we have to calculate $[H_3O^+]$ using the K_w. When we divide through by the $[OH^-]$, we obtain the $[H_3O^+]$.

$$\frac{K_w}{[OH^-]} = \frac{[H_3O^+]\cancel{[OH^-]}}{\cancel{[OH^-]}}$$

$$[H_3O^+] = \frac{1.0 \times 10^{-14}}{[3.7 \times 10^{-3}]} = 2.7 \times 10^{-12}\text{ M}$$

	Calculator Display
pH = $-\log[2.7 \times 10^{-12}]$ = 2.7 EE or EXP 12 +/–	2.7^{-12} or 2.7E–12

Step 2 Press the *log* key and change the sign.

log +/–	11.56863

Step 3 Adjust the number of SFs on the *right* of the decimal point to equal the SFs in the coefficient.

2.7×10^{-12} M (Two SFs) ⟶ pH = 11.57 (Two SFs on the *right* of the decimal point)

STUDY CHECK 8.6

Calculate the pH of a sample of acid rain with $[OH^-] = 2 \times 10^{-10}$ M.

Calculating the $[H_3O^+]$ from pH

In another type of calculation, we are given the pH of a solution and asked to determine the $[H_3O^+]$. This is a reverse of the pH calculation. For whole number pH values, the negative pH value is the power of 10 in the H_3O^+ concentration.

$$[H_3O^+] = 1 \times 10^{-pH}$$

For pH values that are not whole numbers, the calculation requires the use of the *10^x* key, which is usually a *2nd function* key. On some calculators, this operation is done using the *inverse* key and the *log* key.

SAMPLE PROBLEM 8.7

Calculating $[H_3O^+]$ from pH

Determine the $[H_3O^+]$ for solutions having each of the following pH values:

a. pH = 3.0
b. pH = 8.2

SOLUTION

a. For pH values that are whole numbers, the $[H_3O^+]$ can be written 1×10^{-pH}.

pH →

$$[H_3O^+] = 1 \times 10^{-3}\,M$$

b. For pH values that are not whole numbers, the $[H_3O^+]$ is calculated as follows:

Step 1 Enter the pH value and press the *change sign* key.

	Calculator Display
8.2 [+/–]	−8.2

Step 2 Convert −pH to concentration. Press the *2nd function* key and then the *10^x* key. Or press the *inverse* key and then the *log* key.

	Calculator Display
[2nd] [10^x] or [inv] [log]	6.3095^{-09} or 6.3095 E−09

Step 3 Adjust the significant figures in the coefficient. Because the pH value of 8.2 has one digit on the right of the decimal point, the $[H_3O^+]$ is written with one significant figure.

$$[H_3O^+] = 6 \times 10^{-9}\,M$$

STUDY CHECK 8.7

What is the $[H_3O^+]$ of a beer that has a pH of 4.5?

A comparison of $[H_3O^+]$, $[OH^-]$, and their corresponding pH values is given in Table 8.5.

TABLE 8.5 A Comparison of $[H_3O^+]$, $[OH^-]$, and Corresponding pH Values

$[H_3O^+]$	pH	$[OH^-]$	
10^0	0	10^{-14}	
10^{-1}	1	10^{-13}	
10^{-2}	2	10^{-12}	
10^{-3}	3	10^{-11}	Acidic
10^{-4}	4	10^{-10}	
10^{-5}	5	10^{-9}	
10^{-6}	6	10^{-8}	
10^{-7}	7	10^{-7}	Neutral
10^{-8}	8	10^{-6}	
10^{-9}	9	10^{-5}	
10^{-10}	10	10^{-4}	Basic
10^{-11}	11	10^{-3}	
10^{-12}	12	10^{-2}	
10^{-13}	13	10^{-1}	
10^{-14}	14	10^0	

QUESTIONS AND PROBLEMS

The pH Scale

8.29 Why does a neutral solution have a pH of 7.0?

8.30 If you know the $[OH^-]$, how can you determine the pH of a solution?

8.31 State whether each of the following solutions is acidic, basic, or neutral:
a. blood, pH 7.38 **b.** vinegar, pH 2.8
c. drain cleaner, pH 11.2 **d.** coffee, pH 5.54
e. tomatoes, pH 4.2 **f.** chocolate cake, pH 7.6

8.32 State whether each of the following solutions is acidic, basic, or neutral:
a. soda, pH 3.22 **b.** shampoo, pH 5.7
c. laundry detergent, pH 9.4 **d.** rain, pH 5.83
e. honey, pH 3.9 **f.** cheese, pH 5.2

8.33 Calculate the pH of each solution given the following $[H_3O^+]$ or $[OH^-]$ values:
a. $[H_3O^+] = 1 \times 10^{-4}$ M **b.** $[H_3O^+] = 3 \times 10^{-9}$ M
c. $[OH^-] = 1 \times 10^{-5}$ M **d.** $[OH^-] = 2.5 \times 10^{-11}$ M
e. $[H_3O^+] = 6.7 \times 10^{-8}$ M **f.** $[OH^-] = 8.2 \times 10^{-4}$ M

8.34 Calculate the pH of each solution given the following $[H_3O^+]$ or $[OH^-]$ values:
a. $[H_3O^+] = 1 \times 10^{-8}$ M **b.** $[H_3O^+] = 5 \times 10^{-6}$ M
c. $[OH^-] = 4 \times 10^{-2}$ M **d.** $[OH^-] = 8 \times 10^{-3}$ M
e. $[H_3O^+] = 4.7 \times 10^{-2}$ M **f.** $[OH^-] = 3.9 \times 10^{-6}$ M

8.35 Complete the following table:

$[H_3O^+]$	$[OH^-]$	pH	Acidic, Basic, or Neutral?
	1×10^{-6} M		
		3.0	
2×10^{-5} M			
1×10^{-12} M			
		4.62	

8.36 Complete the following table:

$[H_3O^+]$	$[OH^-]$	pH	Acidic, Basic, or Neutral?
		10.0	
			Neutral
	1×10^{-5} M		
1×10^{-2} M			
		11.3	

Chemistry Link to the Environment

ACID RAIN

Natural rain is slightly acidic, with a pH of 5.6. In the atmosphere, carbon dioxide combines with water to form carbonic acid, a weak acid that dissociates to give hydronium ions and bicarbonate.

$$CO_2(g) + H_2O(l) \rightleftharpoons H_2CO_3(aq)$$

$$H_2CO_3(aq) + H_2O(l) \rightleftharpoons H_3O^+(aq) + HCO_3^-(aq)$$

However, in many parts of the world, rain has become considerably more acidic. *Acid rain* is a term given to precipitation such as rain, snow, hail, or fog when the water has a pH less than 5.6. In the United States, pH values of rain have decreased to about 4–4.5. In some parts of the world, pH values of rain have been reported as low as 2.6, which is about as acidic as lemon juice or vinegar. Because the calculation of pH involves powers of 10, a pH value of 2.6 would be 1000 times more acidic than natural rain.

Although natural sources such as volcanoes and forest fires release SO_2, the primary sources of acid rain today are from the burning of fossil fuels in automobiles and coal in industrial plants. When coal and oil are burned, the sulfur impurities combine with the oxygen in the air to produce SO_2 and SO_3. The reaction of SO_3 with water forms sulfuric acid, H_2SO_4, a strong acid.

$$S(s) + O_2(g) \longrightarrow SO_2(g)$$

$$2SO_2(g) + O_2(g) \longrightarrow 2SO_3(g)$$

$$SO_3(g) + H_2O(l) \longrightarrow H_2SO_4(aq)$$

In an effort to decrease the formation of acid rain, legislation has required a reduction in SO_2 emissions. Coal-burning plants have installed equipment called "scrubbers" that absorb SO_2 before it is emitted. In a smokestack, "scrubbing" removes 95% of the SO_2 as the flue gases containing SO_2 pass through limestone ($CaCO_3$) and water. The end product, $CaSO_4$, also called "gypsum," is used in agriculture and to prepare cement products.

Nitrogen oxide forms at high temperatures in the engines of automobiles. As nitrogen oxide is emitted into the air, it combines with more oxygen to form nitrogen dioxide, which is responsible for the brown color of smog. When nitrogen dioxide dissolves in water in the atmosphere, nitric acid, a strong acid, forms.

$$N_2(g) + O_2(g) \longrightarrow 2NO(g)$$

$$2NO(g) + O_2(g) \longrightarrow 2NO_2(g)$$

$$3NO_2(g) + H_2O(g) \longrightarrow 2HNO_3(aq) + NO(g)$$

Air currents in the atmosphere carry the sulfuric acid and nitric acid many thousands of kilometers before they precipitate in areas far away from the site of the contamination. The acids in acid rain have detrimental effects on marble and limestone structures, lakes, and forests. Throughout the world, monuments made of marble (a form of $CaCO_3$) are deteriorating as acid rain dissolves the marble.

$$CaCO_3(s) + H_2SO_4(aq) \longrightarrow CaSO_4(aq) + CO_2(g) + H_2O(l)$$

Acid rain is changing the pH of many lakes and streams in parts of the United States and Europe. When the pH of a lake falls below 4.5–5, most fish and plant life cannot survive. As the soil near a lake becomes more acidic, aluminum becomes more soluble. Increased levels of aluminum ions in lakes are toxic to fish and water animals.

Trees and forests are susceptible to acid rain too. Acid rain breaks down the protective waxy coating on leaves and interferes with photosynthesis. Tree growth is impaired as nutrients and minerals in the soil dissolve and wash away. In Eastern Europe, acid rain is causing an environmental disaster. Nearly 70% of the forests in the Czech Republic have been severely damaged, and some parts of the land are so acidic that crops will not grow.

A marble statue in Washington Square Park has been eroded by acid rain.

Acid rain has severely damaged forests in Eastern Europe.

Explore Your World

USING VEGETABLES AND FLOWERS AS pH INDICATORS

Many flowers and vegetables with strong color, especially reds and purples, contain compounds that change color with changes in pH. Some examples are red cabbage, cranberry juice, and other highly colored drinks.

Materials Needed

Red cabbage or cranberry juice or drink, water, and a saucepan

Several glasses or small glass containers and some tape and a pen or pencil to mark the containers

Several colorless household solutions such as vinegar, lemon juice, other fruit juices, window cleaners, soaps, shampoos, and detergents and common household products such as baking soda, antacids, aspirin, salt, and sugar

Procedure

1. Obtain a bottle of cranberry juice or cranberry drink, or use a red cabbage to prepare the red cabbage pH indicator, as follows: Tear up several red cabbage leaves and place them in a saucepan and cover with water. Boil for about 5 minutes. Cool and collect the purple solution. Alternatively, place red cabbage leaves in a blender and cover with water. Blend until thoroughly mixed, remove, and pour off the red liquid.
2. Place small amounts of each household solution into separate clear glass containers and mark what each one is. If the sample is a solid or a thick liquid, add a small amount of water. Add some cranberry juice or red cabbage indicator until you obtain a color.
3. Observe the colors of the various samples. The colors that indicate acidic solutions are the pink and orange colors (pH 1–4) and the pink to lavender colors (pH 5–6). A neutral solution has about the same purple color as the indicator. Bases will give blue to green color (pH 8–11) or a yellow color (pH 12–13).
4. Arrange your samples by color and pH. Classify each of the solutions as acidic (1–6), neutral (7), or basic (8–13).
5. Try to make an indicator using other colorful fruits or flowers.

QUESTIONS

1. Which products that tested acidic listed an acid on their labels?
2. Which products that tested basic listed a base on their labels?
3. Which products were neutral?
4. Which flowers or vegetables behaved as indicators?

LEARNING GOAL

Write balanced equations for reactions of acids and bases; calculate the molarity or volume of an acid from titration information.

SELF STUDY ACTIVITY
Nature of Acids and Bases

8.5 Reactions of Acids and Bases

Typical reactions of acids and bases include the reactions of acids with metals, bases, and carbonate or bicarbonate ions. For example, when you drop an antacid tablet in water, the bicarbonate ion and citric acid in the tablet react to produce carbon dioxide bubbles, a salt, and water. A *salt* is an ionic compound that does not have H^+ as the cation or OH^- as the anion.

Acids and Metals

Acids react with certain metals to produce a salt and hydrogen gas (H_2). Metals that react with acids include potassium, sodium, calcium, magnesium, aluminum, zinc, iron, and tin. In reactions that are single replacement reactions, the metal ion replaces the hydrogen in the acid.

$$\underset{\text{Metal}}{Mg(s)} + \underset{\text{Acid}}{2HCl(aq)} \longrightarrow \underset{\text{Salt}}{MgCl_2(aq)} + \underset{\text{Hydrogen}}{H_2(g)}$$

$$\underset{\text{Metal}}{Zn(s)} + \underset{\text{Acid}}{2HNO_3(aq)} \longrightarrow \underset{\text{Salt}}{Zn(NO_3)_2(aq)} + \underset{\text{Hydrogen}}{H_2(g)}$$

CONCEPT CHECK 8.6

Equations for Metals and Acids

Write a balanced equation for the reaction of $Al(s)$ with $HCl(aq)$.

ANSWER

Write the reactants and products. When a metal reacts with an acid, the products are a salt and hydrogen gas.

$$Al(s) + HCl(aq) \longrightarrow \text{salt} + H_2(g)$$

Determine the formula of the salt. When Al(*s*) reacts, it forms Al^{3+}, which is balanced by $3Cl^-$ from HCl.

$$Al(s) + HCl(aq) \longrightarrow AlCl_3(aq) + H_2(g)$$

Balance the equation.

$$2Al(s) + 6HCl(aq) \longrightarrow 2AlCl_3(aq) + 3H_2(g)$$

Acids, Carbonates, and Bicarbonates

When an acid is added to a carbonate or bicarbonate (hydrogen carbonate), the products are carbon dioxide gas, water, and an ionic compound (salt). The acid reacts with CO_3^{2-} or HCO_3^- to produce carbonic acid, H_2CO_3, which breaks down rapidly to CO_2 and H_2O.

$$\underset{\text{Acid}}{HBr(aq)} + \underset{\text{Bicarbonate}}{NaHCO_3(aq)} \longrightarrow \underset{\text{Carbon dioxide}}{CO_2(g)} + \underset{\text{Water}}{H_2O(l)} + \underset{\text{Salt}}{NaBr(aq)}$$

$$\underset{\text{Acid}}{2HCl(aq)} + \underset{\text{Carbonate}}{Na_2CO_3(aq)} \longrightarrow \underset{\text{Carbon dioxide}}{CO_2(g)} + \underset{\text{Water}}{H_2O(l)} + \underset{\text{Salt}}{2NaCl(aq)}$$

Acids and Hydroxides: Neutralization

Neutralization is a reaction between an acid and a base to produce a salt and water. The H^+ of an acid that can be strong or weak and the OH^- of a strong base combine to form water as one product. The salt is the cation from the base and the anion from the acid. We can write the following equation for the neutralization reaction between HCl and NaOH:

$$\underset{\text{Acid}}{HCl(aq)} + \underset{\text{Base}}{NaOH(aq)} \longrightarrow \underset{\text{Salt}}{NaCl(aq)} + \underset{\text{Water}}{H_2O(l)}$$

If we write the strong acid HCl and the strong base NaOH as ions, we see that H^+ reacts with OH^- to form water, leaving the ions Na^+ and Cl^- in solution.

$$\mathbf{H^+}(aq) + Cl^-(aq) + Na^+(aq) + \mathbf{OH^-}(aq) \longrightarrow Na^+(aq) + Cl^-(aq) + \mathbf{H_2O}(l)$$

When we omit the ions that do not change during the reaction (Na^+ and Cl^-), we see that the equation for neutralization is the formation of H_2O from H^+ and OH^-.

$$\mathbf{H^+}(aq) + \cancel{Cl^-(aq)} + \cancel{Na^+(aq)} + \mathbf{OH^-}(aq) \longrightarrow \cancel{Na^+(aq)} + \cancel{Cl^-(aq)} + \mathbf{H_2O}(l)$$

$$H^+(aq) + OH^-(aq) \longrightarrow H_2O(l)$$

In a neutralization reaction, one H^+ always combines with one OH^-. Therefore, a neutralization equation uses coefficients to balance H^+ in the acid with the OH^- in the base. Sample Problem 8.8 shows how to balance the equation for the neutralization of HCl and $Ba(OH)_2$.

Guide to Balancing an Equation for Neutralization

1. Write the reactants and products.
2. Balance the H^+ in the acid with the OH^- in the base.
3. Balance the H_2O with the H^+ and the OH^-.
4. Write the salt from the remaining ions.

SAMPLE PROBLEM 8.8

Balancing Equations of Acids

Write the balanced equation for the neutralization of HCl(*aq*) and $Ba(OH)_2(s)$.

SOLUTION

Step 1 **Write the reactants and products.**

$$HCl(aq) + Ba(OH)_2(s) \longrightarrow \text{salt} + H_2O(l)$$

Step 2 **Balance the H^+ in the acid with the OH^- in the base.** Placing a coefficient of 2 in front of HCl provides $2H^+$ for the $2OH^-$ in $Ba(OH)_2$.

$$2HCl(aq) + Ba(OH)_2(s) \longrightarrow \text{salt} + H_2O(l)$$

Step 3 **Balance the H_2O with the H^+ and the OH^-.** Use a coefficient of 2 in front of H_2O to balance $2H^+$ and $2OH^-$.

$$2HCl(aq) + Ba(OH)_2(s) \longrightarrow \text{salt} + 2H_2O(l)$$

Step 4 **Write the salt from the remaining ions.** Use the ions Ba^{2+} and $2Cl^-$ to write the formula of the salt, $BaCl_2$.

$$2\mathbf{HCl}(aq) + \mathbf{Ba}(OH)_2(s) \longrightarrow \mathbf{BaCl_2}(aq) + 2H_2O(l)$$

STUDY CHECK 8.8

Write the balanced equation for the reaction between $H_2SO_4(aq)$ and $NaHCO_3(aq)$.

TUTORIAL
Acid–Base Titrations

Acid–Base Titration

Suppose we need to find the molarity of a solution of HCl, which has an unknown concentration. We can do this by a laboratory procedure called **titration** in which we neutralize an acid sample with a known amount of base. In a titration, we place a measured volume of the acid in a flask and add a few drops of an *indicator* such as phenolphthalein. In an acidic solution, phenolphthalein is colorless. Then we fill a buret with a NaOH solution of known molarity and carefully add NaOH solution to the acid in the flask (see Figure 8.6).

In the titration, we neutralize the acid by adding a volume of base that contains a matching number of moles of OH^-. We know that neutralization has taken place when the phenolphthalein in the solution changes from colorless to pink. This is called the neutralization endpoint. From the volume added and molarity of the NaOH solution, we can calculate the number of moles of NaOH and then the concentration of the acid.

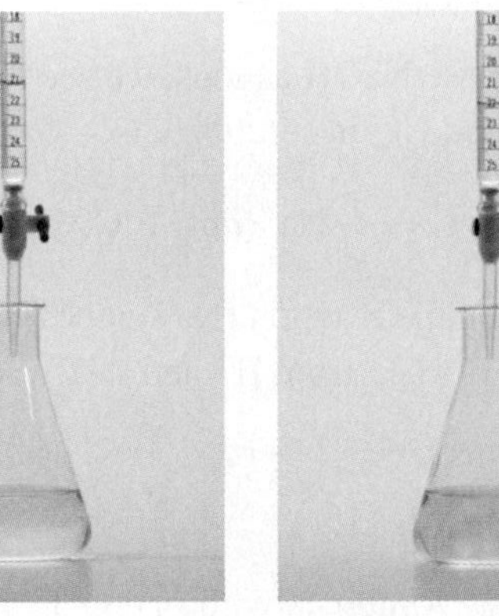

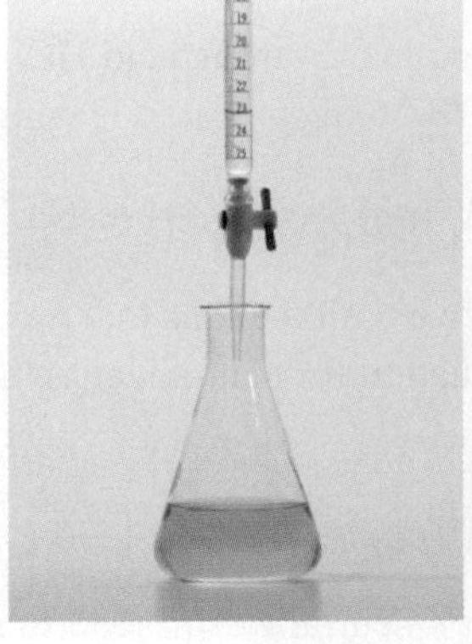

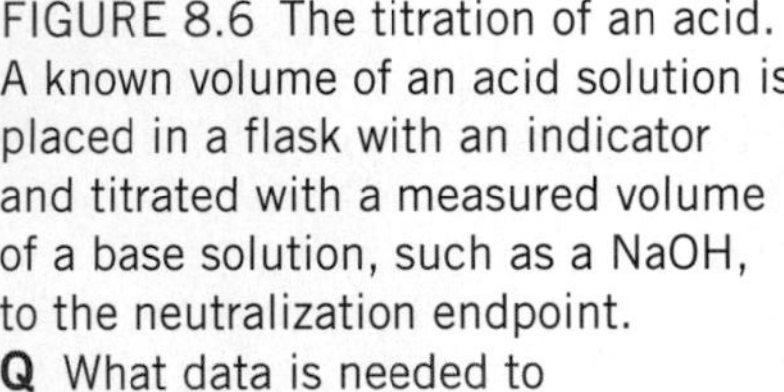

FIGURE 8.6 The titration of an acid. A known volume of an acid solution is placed in a flask with an indicator and titrated with a measured volume of a base solution, such as a NaOH, to the neutralization endpoint.

Q What data is needed to determine the molarity of the acid in the flask?

SAMPLE PROBLEM 8.9

Titration of an Acid

A 25.0-mL (0.025 L) sample of HCl solution is placed in a flask with a few drops of phenolphthalein (indicator). If 32.6 mL of a 0.185 M NaOH solution is needed to reach the endpoint, what is the molarity of the HCl solution?

$$NaOH(aq) + HCl(aq) \longrightarrow NaCl(aq) + H_2O(l)$$

SOLUTION

Step 1 **State the given and needed quantities.**

Given 25.0 mL (0.025 L) of a HCl solution; 32.6 mL of a 0.185 M NaOH solution

Need molarity of HCl solution

Step 2 Write a plan to calculate molarity.

mL of solution → (Metric factor) → L → (Molarity) → moles of NaOH → (Mole–mole factor) → moles of HCl → (Divide by liters) → molarity of HCl solution

Step 3 State equalities and conversion factors including concentration.

$$1\text{ L of solution} = 1000\text{ mL of solution}$$
$$\frac{1\text{ L solution}}{1000\text{ mL NaOH}} \text{ and } \frac{1000\text{ mL NaOH}}{1\text{ L solution}}$$

$$0.185\text{ mole of NaOH} = 1\text{ L of solution}$$
$$\frac{0.185\text{ mole NaOH}}{1\text{ L of solution}} \text{ and } \frac{1\text{ L of solution}}{0.185\text{ mole NaOH}}$$

$$1\text{ mole of HCl} = 1\text{ mole of NaOH}$$
$$\frac{1\text{ mole HCl}}{1\text{ mole NaOH}} \text{ and } \frac{1\text{ mole NaOH}}{1\text{ mole HCl}}$$

Step 4 Set up the problem to calculate the needed quantity.

$$32.6\ \cancel{\text{mL solution}} \times \frac{1\ \cancel{\text{L solution}}}{1000\ \cancel{\text{mL solution}}} \times \frac{0.185\ \cancel{\text{mole NaOH}}}{1\ \cancel{\text{L solution}}} \times \frac{1\text{ mole HCl}}{1\ \cancel{\text{mole NaOH}}} = 0.00603\text{ mole of HCl}$$

$$\text{Molarity of HCl} = \frac{0.00603\text{ mole HCl}}{0.0250\text{ L solution}} = 0.241\text{ M HCl solution}$$

Guide to Calculations for an Acid–Base Titration

1. State the given and needed quantities.
2. Write a plan to calculate molarity or volume.
3. State equalities and conversion factors including concentration.
4. Set up the problem to calculate the needed quantity.

STUDY CHECK 8.9

What is the molarity of a HCl solution, if 28.6 mL of a 0.175 M NaOH solution is needed to neutralize a 25.0-mL sample of the HCl solution?

Chemistry Link to Health

ANTACIDS

Antacids are substances used to neutralize excess stomach acid (HCl). Some antacids are mixtures of aluminum hydroxide and magnesium hydroxide. These hydroxides are not very soluble in water, so the levels of available OH^- are not damaging to the intestinal tract. However, aluminum hydroxide has the side effects of producing constipation and binding phosphate in the intestinal tract, which may cause weakness and loss of appetite. Magnesium hydroxide has a laxative effect. These side effects are less likely when a combination of the antacids is used.

$$Al(OH)_3(s) + 3HCl(aq) \longrightarrow AlCl_3(aq) + 3H_2O(l)$$
$$Mg(OH)_2(s) + 2HCl(aq) \longrightarrow MgCl_2(aq) + 2H_2O(l)$$

Some antacids use calcium carbonate to neutralize excess stomach acid. About 10% of the calcium is absorbed into the bloodstream, where it elevates the levels of serum calcium. Calcium carbonate is not recommended for patients who have peptic ulcers or a tendency to form kidney stones.

$$CaCO_3(s) + 2HCl(aq) \longrightarrow CaCl_2(aq) + CO_2(g) + H_2O(l)$$

Still other antacids contain sodium bicarbonate. This type of antacid has a tendency to increase blood pH and elevate sodium levels in the body fluids. It also is not recommended in the treatment of peptic ulcers.

$$NaHCO_3(s) + HCl(aq) \longrightarrow NaCl(aq) + CO_2(g) + H_2O(l)$$

The neutralizing substances in some antacid preparations are given in Table 8.6.

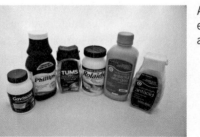

Antacids neutralize excess stomach acid.

TABLE 8.6 Basic Compounds in Some Antacids

Antacid	Base(s)
Amphojel	$Al(OH)_3$
Milk of magnesia	$Mg(OH)_2$
Mylanta, Maalox, Di-Gel, Gelusil, Riopan	$Mg(OH)_2$, $Al(OH)_3$
Bisodol, Rolaids	$CaCO_3$, $Mg(OH)_2$
Titralac, Tums, Pepto-Bismol	$CaCO_3$
Alka-Seltzer	$NaHCO_3$, $KHCO_3$

QUESTIONS AND PROBLEMS

Reactions of Acids and Bases

8.37 Complete and balance the equation for each of the following reactions:

a. $ZnCO_3(s) + HBr(aq) \longrightarrow$
b. $Zn(s) + HCl(aq) \longrightarrow$
c. $HCl(aq) + NaHCO_3(s) \longrightarrow$
d. $H_2SO_4(aq) + Mg(OH)_2(s) \longrightarrow$

8.38 Complete and balance the equation for each of the following reactions:

a. $KHCO_3(s) + HCl(aq) \longrightarrow$
b. $Ca(s) + H_2SO_4(aq) \longrightarrow$
c. $H_2SO_4(aq) + Al(OH)_3(s) \longrightarrow$
d. $Na_2CO_3(s) + H_2SO_4(aq) \longrightarrow$

8.39 Balance each of the following neutralization reactions:

a. $HCl(aq) + Mg(OH)_2(s) \longrightarrow MgCl_2(aq) + H_2O(l)$
b. $H_3PO_4(aq) + LiOH(aq) \longrightarrow Li_3PO_4(aq) + H_2O(l)$

8.40 Balance each of the following neutralization reactions:

a. $HNO_3(aq) + Ba(OH)_2(s) \longrightarrow Ba(NO_3)_2(aq) + H_2O(l)$
b. $H_2SO_4(aq) + Al(OH)_3(aq) \longrightarrow Al_2(SO_4)_3(aq) + H_2O(l)$

8.41 Write a balanced equation for the neutralization of each of the following:

a. $H_2SO_4(aq)$ and $NaOH(aq)$ **b.** $HCl(aq)$ and $Fe(OH)_3(s)$
c. $H_2CO_3(aq)$ and $Mg(OH)_2(s)$

8.42 Write a balanced equation for the neutralization of each of the following:

a. $H_3PO_4(aq)$ and $NaOH(aq)$ **b.** $HI(aq)$ and $LiOH(aq)$
c. $HNO_3(aq)$ and $Ca(OH)_2(s)$

8.43 What is the molarity of a HCl solution if 5.00 mL is neutralized by 28.6 mL of a 0.145 M NaOH solution?

$$HCl(aq) + NaOH(aq) \longrightarrow NaCl(aq) + H_2O(l)$$

8.44 If 29.7 mL of a 0.205 M KOH solution is required to neutralize 25.0 mL of a $HC_2H_3O_2$ solution, what is the molarity of the acetic acid solution?

$$HC_2H_3O_2(aq) + KOH(aq) \longrightarrow KC_2H_3O_2(aq) + H_2O(l)$$

8.45 If 38.2 mL of a 0.163 M KOH solution is required to neutralize 25.0 mL of a H_2SO_4 solution, what is the molarity of the H_2SO_4 solution?

$$H_2SO_4(aq) + 2KOH(aq) \longrightarrow K_2SO_4(aq) + 2H_2O(l)$$

8.46 A solution of 0.162 M NaOH is used to neutralize 25.0 mL of a H_2SO_4 solution. If 32.8 mL of the NaOH solution is required, what is the molarity of the H_2SO_4 solution?

$$H_2SO_4(aq) + 2NaOH(aq) \longrightarrow Na_2SO_4(aq) + 2H_2O(l)$$

LEARNING GOAL

Describe the role of buffers in maintaining the pH of a solution.

TUTORIAL
Buffer Solutions

SELF STUDY ACTIVITY
pH and Buffers

8.6 Buffers

The pH of water and most solutions changes drastically when a small amount of acid or base is added. However, if a solution is buffered, there is little change in pH. A **buffer** is a solution that maintains pH by neutralizing added acid or base. For example, blood contains buffers that maintain a consistent pH of about 7.4. If the pH of the blood goes slightly above or below 7.4, changes in oxygen uptake and metabolic processes can be drastic enough to cause death. Even though we obtain acids and bases from foods and cellular reactions, the buffers in the body absorb those compounds so effectively that the pH of the blood remains essentially unchanged (see Figure 8.7).

In a buffer, an acid must be present to react with any OH^- that is added, and a base must be available to react with any added H_3O^+. However, that acid and base must not be able to neutralize each other. Therefore, a combination of an acid–base conjugate pair is used to prepare a buffer. Most buffer solutions consist of nearly equal concentrations of a weak acid and a salt containing its conjugate base (see Figure 8.8). Buffers may also contain a weak base and a salt containing its conjugate acid.

For example, a typical buffer contains acetic acid ($HC_2H_3O_2$) and a salt such as sodium acetate ($NaC_2H_3O_2$). As a weak acid, acetic acid dissociates slightly in water to form H_3O^+ and a very small amount of $C_2H_3O_2^-$. The addition of its salt provides a much larger concentration of the acetate ion ($C_2H_3O_2^-$), which is necessary for its buffering capability.

$$\underset{\text{Large amount}}{HC_2H_3O_2(aq)} + H_2O(l) \rightleftarrows H_3O^+(aq) + \underset{\text{Large amount}}{C_2H_3O_2^-(aq)}$$

We can now describe how this buffer solution maintains the $[H_3O^+]$. When a small amount of acid is added, it combines with the acetate ion $C_2H_3O_2^-$ (conjugate base) to produce more $HC_2H_3O_2$. There will be a slight decrease in the $[C_2H_3O_2^-]$ and a slight increase in $[HC_2H_3O_2]$, but the $[H_3O^+]$ and thus the pH will not change very much.

$$HC_2H_3O_2(aq) + H_2O(l) \longleftarrow H_3O^+(aq) + C_2H_3O_2^-(aq)$$

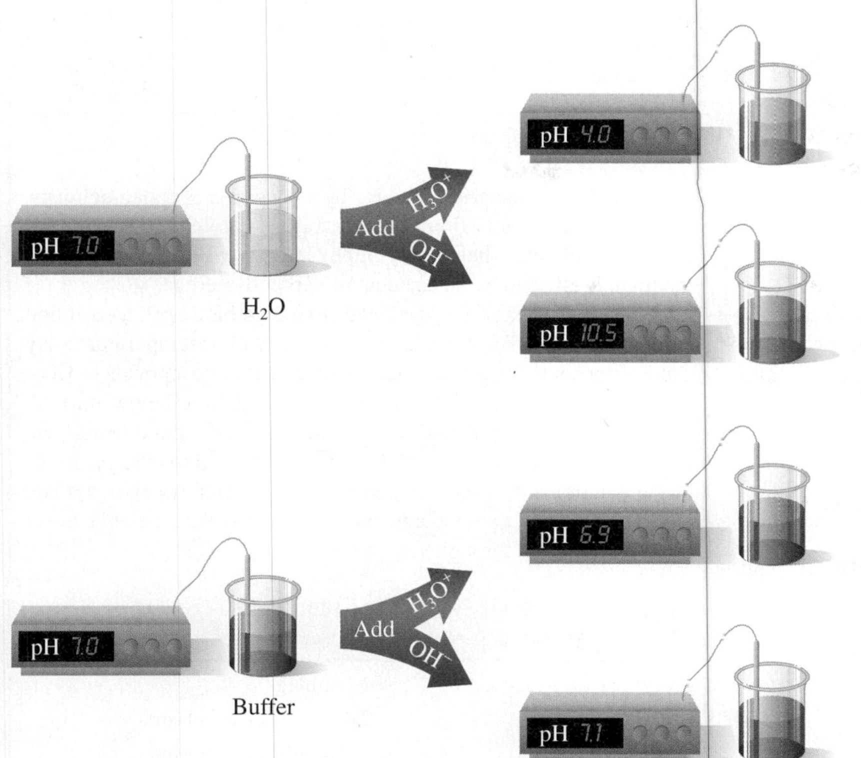

FIGURE 8.7 Adding an acid or a base to water changes the pH drastically, but a buffer resists pH change when small amounts of acid or base are added.

Q Why does the pH change several pH units when acid is added to water, but not when acid is added to a buffer?

If a small amount of base is added to this buffer solution, it is neutralized by the acetic acid, $HC_2H_3O_2$. The products are water and $C_2H_3O_2^-$. The $[HC_2H_3O_2]$ decreases slightly and the $[C_2H_3O_2^-]$ increases slightly, but again the $[H_3O^+]$ does not change very much. Thus, pH of the buffer solution is maintained (see Figure 8.8).

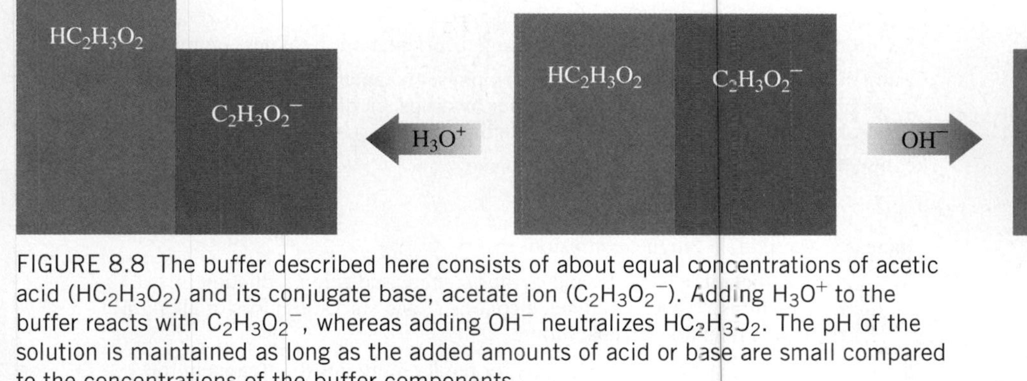

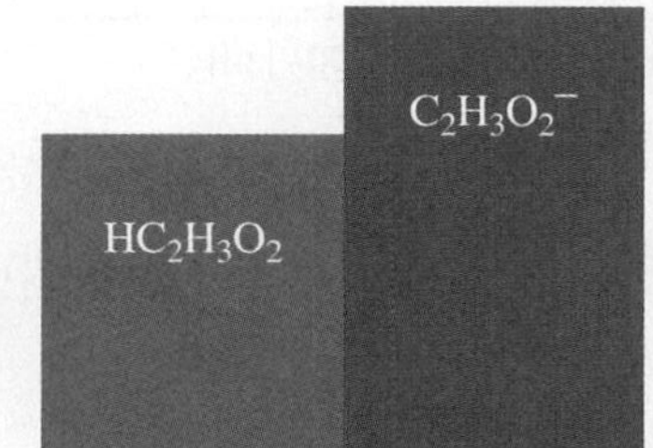

FIGURE 8.8 The buffer described here consists of about equal concentrations of acetic acid ($HC_2H_3O_2$) and its conjugate base, acetate ion ($C_2H_3O_2^-$). Adding H_3O^+ to the buffer reacts with $C_2H_3O_2^-$, whereas adding OH^- neutralizes $HC_2H_3O_2$. The pH of the solution is maintained as long as the added amounts of acid or base are small compared to the concentrations of the buffer components.

Q How does this acetic acid–acetate ion buffer maintain pH?

CONCEPT CHECK 8.6

Identifying Buffer Solutions

Indicate whether each of the following would make a buffer solution:

a. HCl, a strong acid, and NaCl

b. H_3PO_4, a weak acid

c. HF, a weak acid, and NaF

ANSWER

a. No. A strong acid will change the pH of a solution. A buffer requires a weak acid and its salt.

b. No. A weak acid is part of a buffer, but the salt of the weak acid is also needed.

c. Yes. This mixture would be a buffer since it contains a weak acid and its salt.

Chemistry Link to Health

BUFFERS IN THE BLOOD

The arterial blood has a normal pH of 7.35 – 7.45. If changes in H_3O^+ lower the pH below 6.8 or raise it above 8.0, cells cannot function properly and death may result. In our cells, CO_2 is continually produced as an end product of cellular metabolism. Some CO_2 is carried to the lungs for elimination, and the rest dissolves in body fluids such as plasma and saliva, forming carbonic acid. As a weak acid, carbonic acid dissociates to give bicarbonate, HCO_3^-, and H_3O^+. More of the anion HCO_3^- is supplied by the kidneys to give an important buffer system in the body fluid: the H_2CO_3/HCO_3^- buffer.

$$CO_2(g) + H_2O(l) \rightleftharpoons H_2CO_3(aq) \underset{}{\overset{H_2O}{\rightleftharpoons}} H_3O^+(aq) + HCO_3^-(aq)$$

Excess H_3O^+ entering the body fluids reacts with the HCO_3^- and excess OH^- reacts with the carbonic acid.

$$H_2CO_3(aq) + H_2O(l) \longleftarrow H_3O^+(aq) + HCO_3^-(aq)$$

$$H_2CO_3(aq) + OH^-(aq) \longrightarrow H_2O(l) + HCO_3^-(aq)$$

In the body, the concentration of carbonic acid is closely associated with the partial pressure of CO_2. Table 8.7 lists the normal values for arterial blood. If the CO_2 level increases, it produces more H_2CO_3 and more H_3O^+, lowering the pH. This condition is called **acidosis**. Difficulty with ventilation or gas diffusion can lead to respiratory acidosis, which can happen in emphysema or when the medulla of the brain is affected by an accident or depressive drugs.

A decrease in the CO_2 level leads to a high blood pH, a condition called **alkalosis**. Excitement, trauma, or a high temperature may cause a person to hyperventilate, which expels large amounts of CO_2. As the partial pressure of CO_2 in the blood falls below normal, H_2CO_3 forms CO_2 and H_2O, decreasing the $[H_3O^+]$ and raising the pH. Table 8.8 lists some of the conditions that lead to changes in the blood pH and some possible treatments. The kidneys also regulate H_3O^+ and HCO_3^- components, but more slowly than the adjustment made by the lungs through ventilation.

TABLE 8.7 Normal Values for Blood Buffer in Arterial Blood

P_{CO_2}	40 mmHg
H_2CO_3	2.4 mmoles/L of plasma
HCO_3^-	24 mmoles/L of plasma
pH	7.35 – 7.45

TABLE 8.8 Acidosis and Alkalosis: Symptoms, Causes, and Treatments

Respiratory Acidosis: $CO_2\uparrow$ pH$\downarrow$	
Symptoms:	Failure to ventilate, suppression of breathing, disorientation, weakness, coma
Causes:	Lung disease blocking gas diffusion (e.g. emphysema, pneumonia, bronchitis, asthma); depression of respiratory center by drugs, cardiopulmonary arrest, stroke, poliomyelitis, or nervous system disorders
Treatment:	Correction of disorder, infusion of bicarbonate
Metabolic Acidosis: $H^+\uparrow$ pH$\downarrow$	
Symptoms:	Increased ventilation, fatigue, confusion
Causes:	Renal disease, including hepatitis and cirrhosis; increased acid production in diabetes mellitus, hyperthyroidism, alcoholism, and starvation; loss of alkali in diarrhea; acid retention in renal failure
Treatment:	Sodium bicarbonate given orally, dialysis for renal failure, insulin treatment for diabetic ketosis
Respiratory Alkalosis: $CO_2\downarrow$ pH$\uparrow$	
Symptoms:	Increased rate and depth of breathing, numbness, light-headedness, tetany
Causes:	Hyperventilation because of anxiety, hysteria, fever, exercise; reaction to drugs such as salicylate, quinine, and antihistamines; conditions causing hypoxia (e.g. pneumonia, pulmonary edema, heart disease)
Treatment:	Elimination of anxiety-producing state, rebreathing into a paper bag
Metabolic Alkalosis: $H^+\downarrow$ pH$\uparrow$	
Symptoms:	Depressed breathing, apathy, confusion
Causes:	Vomiting, diseases of the adrenal glands, ingestion of excess alkali
Treatment:	Infusion of saline solution, treatment of underlying diseases

QUESTIONS AND PROBLEMS

Buffers

8.47 Which of the following represent a buffer system? Explain.
a. NaOH and NaCl
b. H_2CO_3 and $NaHCO_3$
c. HF and KF
d. KCl and NaCl

8.48 Which of the following represent a buffer system? Explain.
a. H_3PO_3
b. $NaNO_3$
c. $HC_2H_3O_2$ and $NaC_2H_3O_2$
d. HCl and NaOH

8.49 Consider the buffer system of hydrofluoric acid, HF, and its salt, NaF.

$$HF(aq) + H_2O(l) \rightleftharpoons H_3O^+(aq) + F^-(aq)$$

a. What is the purpose of a buffer system?
b. Why does a buffer require a salt that contains the conjugate base of the weak acid?
c. How does the buffer react when some H_3O^+ is added?
d. How does the buffer react when some OH^- is added?

8.50 Consider the buffer system of nitrous acid, HNO_2, and its salt, $NaNO_2$.

$$HNO_2(aq) + H_2O(l) \rightleftharpoons H_3O^+(aq) + NO_2^-(aq)$$

a. What is the purpose of a buffer system?
b. What is the purpose of $NaNO_2$ in the buffer?
c. How does the buffer react when some H_3O^+ is added?
d. How does the buffer react when some OH^- is added?

CONCEPT MAP

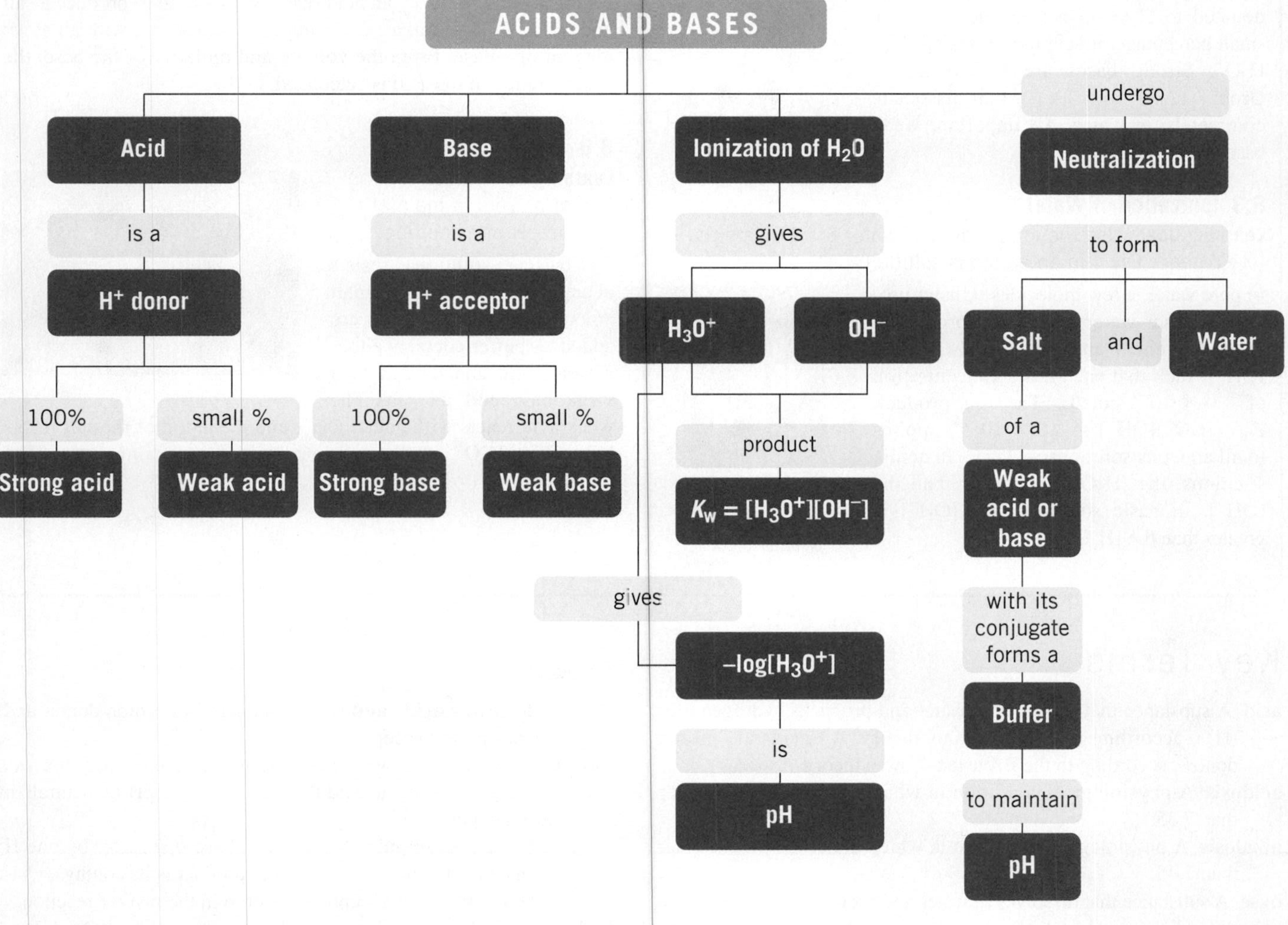

CHAPTER REVIEW

8.1 Acids and Bases

Learning Goal: Describe and name acids and bases; identify Brønsted–Lowry acids and bases.

An Arrhenius acid produces H^+ and an Arrhenius base produces OH^- in aqueous solutions. Acids taste sour, may sting, and neutralize bases. Bases taste bitter, feel slippery, and neutralize acids. Acids containing a simple anion are named using a *hydro* prefix and an *ic acid* ending. Acids with oxygen-containing polyatomic anions are named as *ic* or *ous acids.*

According to the Brønsted–Lowry theory, acids are proton (H^+) donors and bases are proton acceptors. Two conjugate acid–base pairs are present in an acid–base reaction. Each acid–base pair is related by the loss or gain of one H^+. For example, when the acid HF donates one H^+, the F^- that forms is its conjugate base because F^- can accept one H^+. The other acid–base pair would be H_3O^+/H_2O.

$$HF(aq) + H_2O(l) \rightleftharpoons H_3O^+(aq) + F^-(aq)$$

8.2 Strengths of Acids and Bases

Learning Goal: Write equations for the dissociation of strong and weak acids.

In strong acids, all the H^+ in the acid is donated to H_2O; in a weak acid, only a small percentage of acid molecules produce H_3O^+. Strong bases are hydroxides of Groups 1A (1) and 2A (2) that dissociate completely in water. An important weak base is ammonia, NH_3.

8.3 Ionization of Water

Learning Goal: Use the ion product of water to calculate the $[H_3O^+]$ and $[OH^-]$ in an aqueous solution.

In pure water, a few molecules transfer protons to other water molecules, producing small but equal amounts of $[H_3O^+]$ and $[OH^-]$, such that each has a concentration of 1.0×10^{-7} mole/L. The ion product, K_w, $[H_3O^+][OH^-] = 1.0 \times 10^{-14}$, applies to all aqueous solutions at 25 °C. In acidic solutions, the $[H_3O^+]$ is greater than the $[OH^-]$. In basic solutions, the $[OH^-]$ is greater than the $[H_3O^+]$.

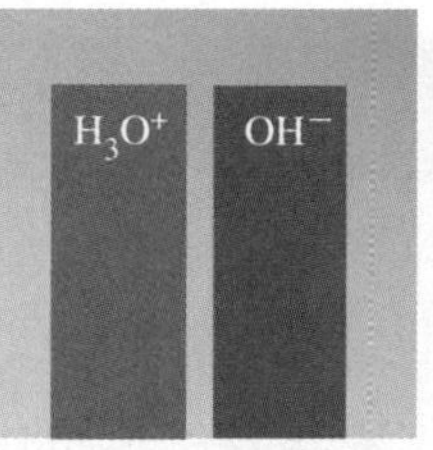

$[H_3O^+] = [OH^-]$

Neutral solution

8.4 The pH Scale

Learning Goal: Calculate the pH of a solution from $[H_3O^+]$; given the pH, calculate $[H_3O^+]$.

The pH scale is a range of numbers from 0 to 14 related to the $[H_3O^+]$ of the solution. A neutral solution has a pH of 7.0. In acidic solutions, the pH is below 7.0, and in basic solutions, the pH is above 7.0. Mathematically, pH is the negative logarithm of the hydronium ion concentration ($-\log[H_3O^+]$).

8.5 Reactions of Acids and Bases

Learning Goal: Write balanced equations for reactions of acids and bases; calculate the molarity or volume of an acid from titration information.

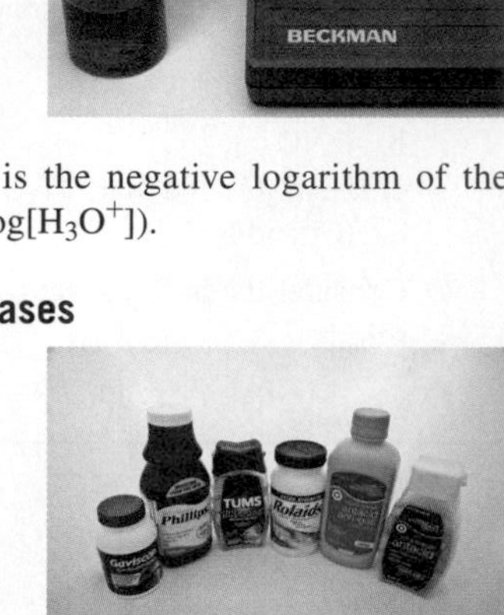

When an acid reacts with a metal, hydrogen gas and a salt are produced. The reaction of an acid with a carbonate or bicarbonate produces carbon dioxide, a salt, and water. In neutralization, an acid reacts with a base to produce a salt and water. In titration, an acid sample is neutralized with a known amount of a base. From the volume and molarity of the base, the concentration of the acid is calculated.

8.6 Buffers

Learning Goal: Describe the role of buffers in maintaining the pH of a solution.

A buffer solution resists changes in pH when small amounts of acid or base are added. A buffer contains either a weak acid and its salt or a weak base and its salt. The weak acid reacts with added OH^-, and the anion of the salt reacts with added H_3O^+. Buffers are important in maintaining the pH of the blood.

Key Terms

acid A substance that dissolves in water and produces hydrogen ions (H^+), according to the Arrhenius theory. All acids are proton donors, according to the Brønsted–Lowry theory.

acidosis A physiological condition in which the blood pH is lower than 7.35.

alkalosis A physiological condition in which the blood pH is higher than 7.45.

base A substance that dissolves in water and produces hydroxide ions according to the Arrhenius theory. All bases are proton acceptors, according to the Brønsted–Lowry theory.

Brønsted–Lowry acids and bases An acid is a proton donor, and a base is a proton acceptor.

buffer A solution of a weak acid and its conjugate base or a weak base and its conjugate acid that maintains the pH by neutralizing added acid or base.

conjugate acid–base pair An acid and base that differ by one H^+. When an acid donates a proton, the product is its conjugate base, which is capable of accepting a proton in the reverse reaction.

hydronium ion The ion formed by the attraction of a proton (H^+) to a H_2O molecule written as H_3O^+.

ion product constant of water, K_w The product of $[H_3O^+]$ and $[OH^-]$ in solution; $K_w = [H_3O^+][OH^-]$.

neutral The term that describes a solution with equal concentrations of H_3O^+ and OH^-.

neutralization A reaction between an acid and a base to form a salt and water.

pH A measure of the $[H_3O^+]$ in a solution; $pH = -\log[H_3O^+]$.

strong acid An acid that completely ionizes in water.

strong base A base that completely ionizes in water.

titration The addition of base to an acid sample to determine the concentration of the acid.

weak acid An acid that is a poor donor of H^+ and dissociates only slightly in water.

weak base A base that is a poor acceptor of H^+ and dissociates only slightly in water.

Understanding the Concepts

8.51 Identify each of the following as an acid, a base, or a salt, and give its name:

a. LiOH **b.** $Ca(NO_3)_2$ **c.** HBr
d. $Ba(OH)_2$ **e.** H_2CO_3

8.52 Identify each of the following as an acid, a base, or a salt, and give its name:

a. H_3PO_4 **b.** $MgBr_2$ **c.** NH_3
d. $HClO_2$ **e.** NaCl

8.53 Complete the following table:

Acid	Conjugate Base
H_2O	
	CN^-
HNO_2	
	$H_2PO_4^-$

8.54 Complete the following table:

Base	Conjugate Acid
	HS^-
	H_3O^+
NH_3	
HCO_3^-	

8.55 In each of the following diagrams of acid solutions, determine if the diagram represents a strong acid or a weak acid. The acid has the formula HX.

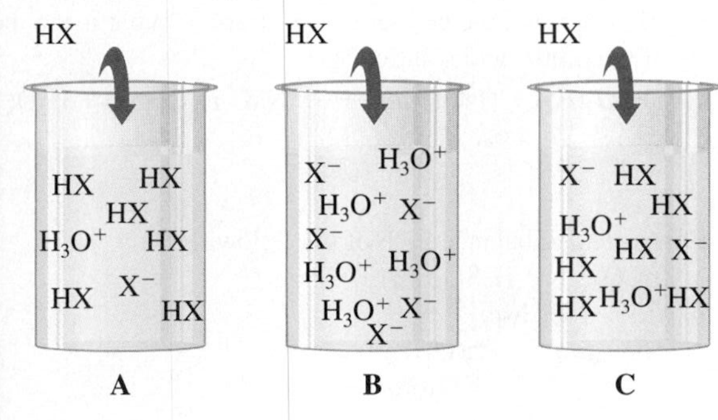

8.56 Adding a few drops of a strong acid to water will lower the pH appreciably. However, adding the same number of drops to a buffer does not appreciably alter the pH. Explain.

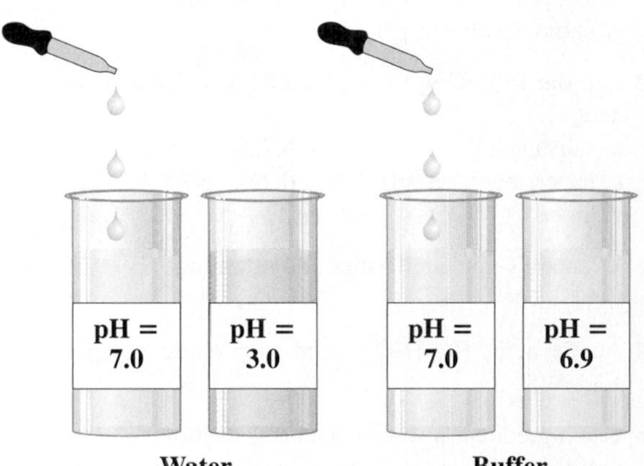

8.57 Sometimes, during stress or trauma, a person can start to hyperventilate. To avoid fainting, a person might breathe into a paper bag.

a. What changes occur in the blood pH during hyperventilation?

b. How does breathing into a paper bag help return blood pH to normal?

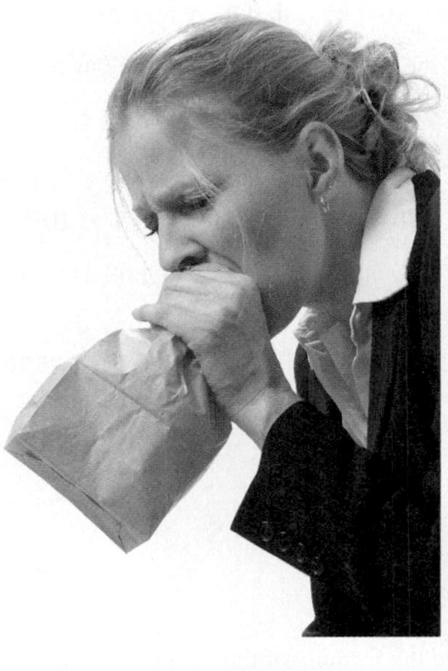

Breathing into a paper bag can help a person who is hyperventilating.

8.58 In the blood plasma, pH is maintained by the carbonic acid–bicarbonate buffer system.

a. How is pH maintained when acid is added to the buffer system?

b. How is pH maintained when base is added to the buffer system?

Additional Questions and Problems

For instructor-assigned homework, go to **www.masteringchemistry.com**.

8.59 Name each of the following as an acid or base:
a. H_2SO_4 **b.** RbOH
c. $Ca(OH)_2$ **d.** HI

8.60 Name each of the following as an acid or base:
a. $Sr(OH)_2$ **b.** H_2SO_3
c. $HC_2H_3O_2$ **d.** CsOH

8.61 Are each of the following examples acidic, basic, or neutral?
a. rain, pH 5.2 **b.** tears, pH 7.5
c. tea, pH 3.8 **d.** cola, pH 2.5
e. photo developer, pH 12.0

8.62 Are the following examples of body fluids acidic, basic, or neutral?
a. saliva, pH 6.8 **b.** urine, pH 5.9
c. pancreatic juice, pH 8.0 **d.** bile, pH 8.4
e. blood, pH 7.45

8.63 What are some similarities and differences between strong and weak acids?

8.64 Acetic acid, $HC_2H_3O_2$, used to prepare vinegar, is a weak acid. Why?

8.65 One ingredient in some antacids is $Mg(OH)_2$.
a. If the base is not very soluble in water, why is it considered a strong base?
b. What is the neutralization reaction of $Mg(OH)_2$ with stomach acid, HCl?

8.66 What are some ingredients found in antacids? What do they do?

8.67 Determine the pH for the following solutions:
a. $[H_3O^+] = 1.0 \times 10^{-8}$ M **b.** $[H_3O^+] = 5.0 \times 10^{-2}$ M
c. $[OH^-] = 3.5 \times 10^{-4}$ M **d.** $[OH^-] = 0.005$ M

8.68 Identify each of the solutions in Problem 8.67 as acidic, basic, or neutral.

8.69 Determine the pH for the following solutions:
a. $[OH^-] = 1.0 \times 10^{-7}$ M **b.** $[H_3O^+] = 4.2 \times 10^{-3}$ M
c. $[H_3O^+] = 0.0001$ M **d.** $[OH^-] = 8.5 \times 10^{-9}$ M

8.70 Identify each of the solutions in Problem 8.69 as acidic, basic, or neutral.

8.71 What are the $[H_3O^+]$ and $[OH^-]$ for a solution with the following pH values?
a. 3.0 **b.** 6.00 **c.** 8.0
d. 11.0 **e.** 9.20

8.72 What are the $[H_3O^+]$ and $[OH^-]$ for a solution with the following pH values?
a. 10.0 **b.** 5.0 **c.** 7.00
d. 6.5 **e.** 1.82

8.73 Solution A has a pH of 4.0, and solution B has a pH of 6.0.
a. Which solution is more acidic?
b. What is the $[H_3O^+]$ in each?
c. What is the $[OH^-]$ in each?

8.74 Solution X has a pH of 9.5, and solution Y has a pH of 7.5.
a. Which solution is more acidic?
b. What is the $[H_3O^+]$ in each?
c. What is the $[OH^-]$ in each?

8.75 A buffer is made by dissolving H_3PO_4 and NaH_2PO_4 in water.
a. Write an equation that shows how this buffer neutralizes added acid.
b. Write an equation that shows how this buffer neutralizes added base.

8.76 A buffer is made by dissolving $HC_2H_3O_2$ and $NaC_2H_3O_2$ in water.
a. Write an equation that shows how this buffer neutralizes added acid.
b. Write an equation that shows how this buffer neutralizes added base.

8.77 Calculate the volume (mL) of a 0.150 M NaOH solution needed to neutralize each of the following:
a. 25.0 mL of a 0.288 M HCl solution
b. 10.0 mL of a 0.560 M H_2SO_4 solution
c. 5.00 mL of a 0.618 M HBr solution

8.78 Calculate the volume (mL) of a 0.215 M KOH solution that will completely neutralize each of the following:
a. 2.50 mL of a 0.825 M H_2SO_4 solution
b. 18.5 mL of a 0.560 M HNO_3 solution
c. 5.00 mL of a 3.18 M HCl solution

8.79 A solution of 0.205 M NaOH solution is used to neutralize 20.0 mL of a H_2SO_4 solution. If 45.6 mL of the NaOH solution is required, what is the molarity of the H_2SO_4 solution?

$$H_2SO_4(aq) + 2NaOH(aq) \longrightarrow Na_2SO_4(aq) + 2H_2O(l)$$

8.80 A 10.0-mL sample of vinegar, which is an aqueous solution of acetic acid, $HC_2H_3O_2$, requires 16.5 mL of 0.500 M NaOH solution to reach the endpoint in a titration. What is the molarity of the acetic acid solution?

$$HC_2H_3O_2(aq) + NaOH(aq) \longrightarrow NaC_2H_3O_2(aq) + H_2O(l)$$

Challenge Questions

8.81 Consider the following:
1. H_2S **2.** H_3PO_4 **3.** HCO_3^-
a. For each, write the formula of the conjugate base.
b. Write the formula of the weakest acid.
c. Write the formula of the strongest acid.

8.82 Identify the conjugate acid–base pairs in each of the following equations:
a. $NH_3(aq) + HNO_3(aq) \rightleftharpoons NH_4^+(aq) + NO_3^-(aq)$
b. $H_2O(l) + HBr(aq) \rightleftharpoons H_3O^+(aq) + Br^-(aq)$
c. $HNO_2(aq) + HS^-(aq) \rightleftharpoons H_2S(g) + NO_2^-(aq)$
d. $Cl^-(aq) + H_2O(l) \rightleftharpoons OH^-(aq) + HCl(aq)$

8.83 Complete and balance each of the following:
a. $ZnCO_3(s) + H_2SO_4(aq) \longrightarrow$
b. $Al(s) + HCl(aq) \longrightarrow$
c. $H_3PO_4(aq) + Ca(OH)_2(s) \longrightarrow$
d. $KHCO_3(s) + HNO_3(aq) \longrightarrow$

8.84 Determine each of the following for a 0.050 M KOH solution:
a. $[H_3O^+]$
b. pH
c. products when allowed to react with H_3PO_4
d. milliliters required to neutralize 40.0 mL of a 0.035 M H_2SO_4 solution

8.85 What is the pH of a solution prepared by dissolving 2.5 g of HCl in water to make 425 mL of HCl solution?

8.86 Consider the reaction of KOH and HNO_2.
a. Write the balanced chemical equation.
b. Calculate the milliliters of 0.122 M KOH solution required to neutralize 36.0 mL of a 0.250 M HNO_2 solution.

8.87 A 0.204 M NaOH solution is used to titrate 50.0 mL of a H_3PO_4 solution.
a. Write the balanced chemical equation.
b. What is the molarity of the H_3PO_4 solution if 16.4 mL of the NaOH solution is required?

8.88 A solution of 0.312 M KOH is used to titrate 15.0 mL of a H_2SO_4 solution.
a. Write the balanced chemical equation.
b. What is the molarity of the H_2SO_4 solution if 28.2 mL of the KOH solution is required?

8.89 One of the most acidic lakes in the United States is Little Echo Pond in the Adirondacks in New York. Recently, this lake had a pH of 4.2, well below the recommended pH of 6.5.

A helicopter drops calcium carbonate on an acidic lake to increase its pH.

a. What are the $[H_3O^+]$ and $[OH^-]$ of Little Echo Pond?
b. What are the $[H_3O^+]$ and $[OH^-]$ of a lake that has a pH of 6.5?
c. One way to raise the pH and restore aquatic life of an acidic lake is to add limestone ($CaCO_3$). How many grams of $CaCO_3$ are needed to neutralize 1.0 kL of the acidic water from Little Echo Pond if the acid is written as HA?

$$2HA + CaCO_3(s) \longrightarrow CaA_2 + CO_2(g) + H_2O(l)$$

8.90 The daily output of stomach acid (gastric juices) is 1000 mL to 1400 mL. Prior to a meal, stomach acid (HCl) typically has a pH of 1.42.
a. What is the $[H_3O^+]$ of stomach acid?
b. The antacid Maalox contains 200. mg of $Al(OH)_3$ per tablet. Write the balanced equation for the neutralization, and calculate the milliliters of stomach acid neutralized by two tablets of Maalox.
c. The antacid milk of magnesia contains 400. mg of $Mg(OH)_2$ per teaspoon. Write the neutralization equation, and calculate the number of milliliters of stomach acid neutralized by 1 tablespoon of milk of magnesia (1 tablespoon = 3 teaspoons).

Answers

Answers to Study Checks

8.1 $LiOH(s) \longrightarrow Li^+(aq) + OH^-(aq)$

8.2 $HNO_3(aq) + H_2O(l) \longrightarrow H_3O^+(aq) + NO_3^-(aq)$

8.3 The conjugate acid–base pairs are HCN/CN^- and HSO_4^-/SO_4^{2-}.

8.4 $[H_3O^+] = 2.5 \times 10^{-11}$ M, basic

8.5 11.38

8.6 4.3

8.7 $[H_3O^+] = 3 \times 10^{-5}$ M

8.8 $H_2SO_4(aq) + 2NaHCO_3(s) \longrightarrow Na_2SO_4(aq) + 2CO_2(g) + 2H_2O(l)$

8.9 0.200 M HCl solution

Answers to Selected Questions and Problems

8.1 **a.** acid **b.** acid **c.** acid **d.** base

8.3 **a.** hydrochloric acid **b.** calcium hydroxide
c. carbonic acid **d.** nitric acid
e. sulfurous acid

8.5 **a.** $Mg(OH)_2$ **b.** HF
c. H_3PO_4 **d.** LiOH
e. NH_4OH **f.** H_2SO_4

8.7 **a.** HI is the acid (proton donor) and H_2O is the base (proton acceptor).
b. H_2O is the acid (proton donor) and F^- is the base (proton acceptor).

8.9 **a.** F^- **b.** OH^-
c. HCO_3^- **d.** SO_4^{2-}

8.11 **a.** HCO_3^- **b.** H_3O^+
c. H_3PO_4 **d.** HBr

8.13 **a.** The conjugate acid–base pairs are H_2CO_3/HCO_3^- and H_3O^+/H_2O.
b. The conjugate acid–base pairs are NH_4^+/NH_3 and H_3O^+/H_2O.
c. The conjugate acid–base pairs are HCN/CN^- and HNO_2/NO_2^-.

8.15 **a.** HBr **b.** HSO_4^- **c.** H_2CO_3

8.17 **a.** HSO_4^- **b.** HNO_2 **c.** HCO_3^-

8.19 In pure water, $[H_3O^+] = [OH^-]$ because one of each is produced every time a proton transfers from one water molecule to another.

8.21 In an acidic solution, the $[H_3O^+]$ is greater than the $[OH^-]$.

8.23 **a.** acidic **b.** basic **c.** basic **d.** acidic

8.25 **a.** 1.0×10^{-9} M **b.** 1.0×10^{-6} M **c.** 2.0×10^{-5} M **d.** 4.0×10^{-13} M

8.27 **a.** 1.0×10^{-3} M **b.** 5.0×10^{-6} M **c.** 1.8×10^{-12} M **d.** 4.0×10^{-13} M

8.29 In a neutral solution, the $[H_3O^+]$ is 1×10^{-7} M and the pH is 7.0, which is the negative value of the power of 10.

8.31 **a.** basic **b.** acidic **c.** basic **d.** acidic **e.** acidic **f.** basic

8.33 **a.** 4.0 **b.** 8.5 **c.** 9.0 **d.** 3.40 **e.** 7.17 **f.** 10.92

8.35

$[H_3O^+]$	$[OH^-]$	pH	Acidic, Basic, or Neutral?
1×10^{-8} M	1×10^{-6} M	8.0	Basic
1×10^{-3} M	1×10^{-11} M	3.0	Acidic
2×10^{-5} M	5×10^{-10} M	4.7	Acidic
1×10^{-12} M	1×10^{-2} M	12.0	Basic
2.4×10^{-5} M	4.2×10^{-10} M	4.62	Acidic

8.37 **a.** $ZnCO_3(s) + 2HBr(aq) \longrightarrow ZnBr_2(aq) + CO_2(g) + H_2O(l)$
b. $Zn(s) + 2HCl(aq) \longrightarrow ZnCl_2(aq) + H_2(g)$
c. $HCl(g) + NaHCO_3(s) \longrightarrow NaCl(aq) + CO_2(g) + H_2O(l)$
d. $H_2SO_4(aq) + Mg(OH)_2(s) \longrightarrow MgSO_4(aq) + 2H_2O(l)$

8.39 **a.** $2HCl(aq) + Mg(OH)_2(s) \longrightarrow MgCl_2(aq) + 2H_2O(l)$
b. $H_3PO_4(aq) + 3LiOH(aq) \longrightarrow Li_3PO_4(aq) + 3H_2O(l)$

8.41 **a.** $H_2SO_4(aq) + 2NaOH(aq) \longrightarrow Na_2SO_4(aq) + 2H_2O(l)$
b. $3HCl(aq) + Fe(OH)_3(aq) \longrightarrow FeCl_3(aq) + 3H_2O(l)$
c. $H_2CO_3(aq) + Mg(OH)_2(s) \longrightarrow MgCO_3(s) + 2H_2O(l)$

8.43 0.830 M HCl solution

8.45 0.124 M H_2SO_4 solution

8.47 **b** and **c** are buffer systems; **b** contains the weak acid H_2CO_3 and its salt $NaHCO_3$; **c** contains HF, a weak acid, and its salt KF.

8.49 **a.** A buffer system keeps the pH constant.
b. A buffer needs a salt to supply more of the conjugate base of the weak acid to neutralize any H_3O^+ added.
c. The added H_3O^+ reacts with F^- from NaF.
d. The HF of the buffer neutralizes added OH^-.

8.51 **a.** base, lithium hydroxide
b. salt, calcium nitrate
c. acid, hydrobromic acid
d. base, barium hydroxide
e. acid, carbonic acid

8.53

Acid	Conjugate Base
H_2O	OH^-
HCN	CN^-
HNO_2	NO_2^-
H_3PO_4	$H_2PO_4^-$

8.55 **a.** weak acid **b.** strong acid **c.** weak acid

8.57 **a.** Hyperventilation will lower the CO_2 level in the blood, which lowers the $[H_2CO_3]$, which decreases the $[H_3O^+]$ and increases the blood pH.
b. Breathing into a bag increases the CO_2 level, increases the $[H_2CO_3]$, and increases the $[H_3O^+]$, which lowers the blood pH.

8.59 **a.** sulfuric acid **b.** rubidium hydroxide **c.** calcium hydroxide **d.** hydroiodic acid

8.61 **a.** acidic **b.** basic **c.** acidic **d.** acidic **e.** basic

8.63 Both strong and weak acids produce H_3O^+ in water. Weak acids are only slightly dissociated, to give an aqueous solution of only a few ions, whereas a strong acid is completely dissociated, to give only ions in solution.

8.65 **a.** The $Mg(OH)_2$ that dissolves is completely ionized, which makes it a strong base.
b. $Mg(OH)_2(s) + 2HCl(aq) \longrightarrow MgCl_2(aq) + 2H_2O(l)$

8.67 **a.** 8.00 **b.** 1.30 **c.** 10.54 **d.** 11.7

8.69 **a.** 7.00 **b.** 2.38 **c.** 4.0 **d.** 5.92

8.71 **a.** $[H_3O^+] = 1 \times 10^{-3}$ M, $[OH^-] = 1 \times 10^{-11}$ M
b. $[H_3O^+] = 1.0 \times 10^{-6}$ M, $[OH^-] = 1.0 \times 10^{-8}$ M
c. $[H_3O^+] = 1 \times 10^{-8}$ M, $[OH^-] = 1 \times 10^{-6}$ M
d. $[H_3O^+] = 1 \times 10^{-11}$ M, $[OH^-] = 1 \times 10^{-3}$ M
e. $[H_3O^+] = 6.3 \times 10^{-10}$ M, $[OH^-] = 1.6 \times 10^{-5}$ M

8.73 **a.** Solution A with a pH of 4.0 is more acidic than solution B.
b. Solution A: $[H_3O^+] = 1 \times 10^{-4}$ M, Solution B: $[H_3O^+] = 1 \times 10^{-6}$ M
c. Solution A: $[OH^-] = 1 \times 10^{-10}$ M, Solution B: $[OH^-] = 1 \times 10^{-8}$ M

8.75 **a.** Acid added: $H_2PO_4^-(aq) + H_3O^+(aq) \longrightarrow H_3PO_4(aq) + H_2O(l)$
b. Base added: $H_3PO_4(aq) + OH^-(aq) \longrightarrow H_2PO_4^-(aq) + H_2O(l)$

8.77 **a.** 48.0 mL **b.** 74.7 mL **c.** 20.6 mL

8.79 0.234 M H_2SO_4 solution

8.81 **a.** **1.** HS^- **2.** $H_2PO_4^-$ **3.** CO_3^{2-}
b. HCO_3^-
c. H_3PO_4

8.83 **a.** $ZnCO_3(s) + H_2SO_4(aq) \longrightarrow ZnSO_4(aq) + CO_2(g) + H_2O(l)$
b. $2Al(s) + 6HCl(aq) \longrightarrow 2AlCl_3(aq) + 3H_2(g)$
c. $2H_3PO_4(aq) + 3Ca(OH)_2(s) \longrightarrow Ca_3(PO_4)_2(aq) + 6H_2O(l)$
d. $KHCO_3(s) + HNO_3(aq) \longrightarrow KNO_3(aq) + CO_2(g) + H_2O(l)$

8.85 0.80

8.87 **a.** $H_3PO_4(aq) + 3NaOH(aq) \longrightarrow Na_3PO_4(aq) + 3H_2O(l)$
b. 0.0224 M H_3PO_4 solution

8.89 **a.** $[H_3O^+] = 6 \times 10^{-5}$ M; $[OH^-] = 2 \times 10^{-10}$ M
b. $[H_3O^+] = 3 \times 10^{-7}$ M; $[OH^-] = 3 \times 10^{-8}$ M
c. 3 g of $CaCO_3$

Combining Ideas from Chapters 6 to 8

CI.15 Methane is a major component of purified natural gas used for heating and cooking. When 1 mole of methane gas burns with oxygen to produce carbon dioxide and water, 883 kJ is produced. Methane gas has a density of 0.715 g/L at STP. For transport, the natural gas is cooled to −163 °C to form liquefied natural gas (LNG) with a density of 0.45 g/mL. A tank on a ship can hold 7.0 million gallons of LNG.

An LNG carrier transports liquefied natural gas.

a. Draw the electron-dot formula of methane, which has the formula CH_4.
b. What is the mass, in kilograms, of LNG (assume that LNG is all methane) transported in one tank on a ship?
c. What is the volume, in liters, of LNG (methane) from one tank when the LNG is converted to gas at STP?
d. Write the balanced equation for the combustion of methane in a gas burner, including the heat of reaction.

Methane is the fuel burned in a gas cooktop.

e. How many kilograms of oxygen are needed to react with all of the methane provided by one tank of LNG?
f. How much heat, in kilojoules, is released from burning all of the methane in one tank of LNG?

CI.16 Automobile exhaust is a major cause of air pollution. The pollutants formed from gasoline include nitrogen oxide that is produced at high temperatures in an automobile engine from nitrogen and oxygen gases in the air. Once emitted into the air, nitrogen oxide reacts with oxygen to produce nitrogen dioxide, a reddish brown gas with a sharp, pungent odor that makes up smog. One component of gasoline is heptane, C_7H_{16}, which has a density of 0.684 g/mL. In one year, a typical automobile uses 550 gal of gasoline and produces 41 lb of nitrogen oxide.

Two gases found in automobile exhaust are carbon dioxide and nitrogen oxide.

a. Write balanced equations for the production of nitrogen oxide and nitrogen dioxide.
b. If all the nitrogen oxide emitted by one automobile is converted to nitrogen dioxide in the atmosphere, how many kilograms of nitrogen dioxide are produced in one year by a single automobile?
c. Write a balanced equation for the combustion of heptane.
d. How many moles of CO_2 are produced from the gasoline used by the typical automobile in one year, assuming the gasoline is all heptane?
e. How many liters of carbon dioxide at STP are produced in one year from the gasoline used by the typical automobile?

CI.17 When clothes have stains, bleach is often added to the wash to react with the soil and make the stains colorless. The active ingredient in bleach is sodium hypochlorite (NaClO). A bleach solution can be prepared by bubbling chlorine gas into a solution of sodium hydroxide to produce sodium hypochlorite, sodium chloride, and water. A typical bottle of bleach has a volume of 1.42 gal, with a density of 1.08 g/mL and contains 282 g of NaClO.

The active component of bleach is sodium hypochlorite.

a. Is sodium hypochlorite an ionic or a covalent compound?
b. What is the mass/volume percent (m/v) of sodium hypochlorite in the bleach solution?
c. Write the equation for the preparation of a bleach solution.
d. How many liters of chlorine gas at STP are required to produce 1.42 gal of bleach for one bottle of bleach?
e. If the pH of the bleach solution is 10.3, what are the $[H_3O^+]$ and $[OH^-]$?

CI.18 In wine making, glucose ($C_6H_{12}O_6$) from grapes undergoes fermentation in the absence of oxygen to produce ethanol and carbon dioxide. A bottle of vintage port wine has a volume of 750 mL and contains 135 mL of ethanol (C_2H_6O). Ethanol has a density of 0.789 g/mL. In 1.5 lb of grapes, there is 26 g of glucose.

When the sugar in grapes is fermented, ethanol is produced.

Port is a type of fortified wine that is produced in Portugal.

a. Calculate the volume percent (v/v) of ethanol in the port wine.
b. What is the molarity (M) of ethanol in the port wine?
c. Write the balanced equation for the fermentation reaction of glucose.
d. How many grams of glucose are required to produce one bottle of port wine?
e. How many bottles of port wine can be produced from 1.0 ton of grapes (1 ton = 2000 lb)?

CI.19 A metal completely reacts with 34.8 mL of a 0.520 M HCl solution.

When a metal reacts with a strong acid, hydrogen gas forms.

a. Write a balanced equation for the reaction of the metal M and HCl(aq) to form $MCl_3(aq)$ and H_2 gas.
b. What volume, in milliliters, of H_2 is produced at STP?
c. How many moles of metal M reacted?
d. If the metal has a mass of 0.420 g, use your results from part **c** to determine the molar mass of the metal M.
e. What are the name and symbol of metal M in part **d**?
f. Write the balanced equation using the symbol of the metal from part **e**.

CI.20 In a teaspoon (5.0 mL) of a common liquid antacid, there are 200. mg of $Mg(OH)_2$ and 200. mg of $Al(OH)_3$. A 0.080 M HCl solution, which is similar to stomach acid, is used to neutralize 5.0 mL of the liquid antacid.

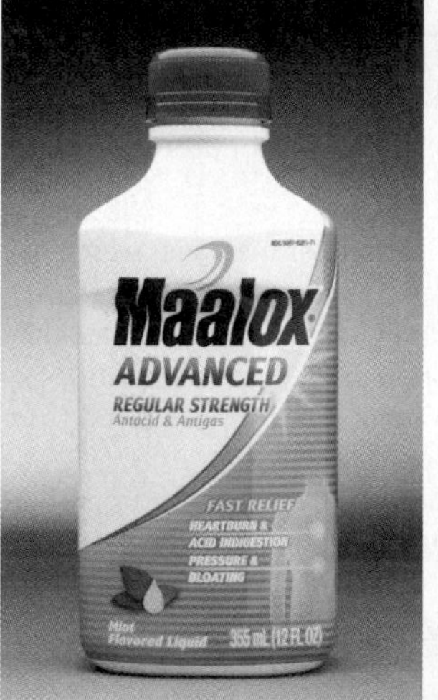

An antacid neutralizes stomach acid.

a. Write the balanced equation for the neutralization reaction of HCl and $Mg(OH)_2$.
b. Write the balanced equation for the neutralization reaction of HCl and $Al(OH)_3$.
c. What is the pH of the HCl solution?
d. How many milliliters of the HCl solution are needed to neutralize the $Mg(OH)_2$?
e. How many milliliters of the HCl solution are needed to neutralize the $Al(OH)_3$?

Answers

CI.15 a. 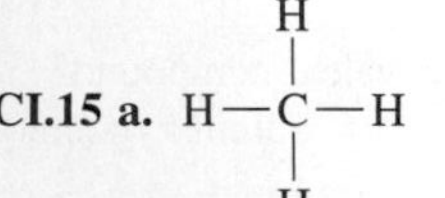

b. 1.2×10^7 kg of LNG (methane)
c. 1.7×10^{10} L of LNG (methane) at STP
d. $CH_4(g) + 2O_2(g) \xrightarrow{\Delta} CO_2(g) + 2H_2O(g) + 883$ kJ
e. 4.8×10^7 kg of O_2
f. 6.6×10^{11} kJ

CI.17 a. ionic **b.** 5.25% (m/v)
c. $2NaOH(aq) + Cl_2(g) \longrightarrow NaClO(aq) + NaCl(aq) + H_2O(l)$
d. 84.8 L of chlorine gas
e. $[H_3O^+] = 5 \times 10^{-11}$ M; $[OH^-] = 2 \times 10^{-4}$ M

CI.19 a. $2M(s) + 6HCl(aq) \longrightarrow 2MCl_3(aq) + 3H_2(g)$
b. 203 mL of H_2
c. 6.03×10^{-3} mole of M
d. 69.7 g/mole
e. Gallium; Ga
f. $2Ga(s) + 6HCl(aq) \longrightarrow 2GaCl_3(aq) + 3H_2(g)$

Nuclear Radiation

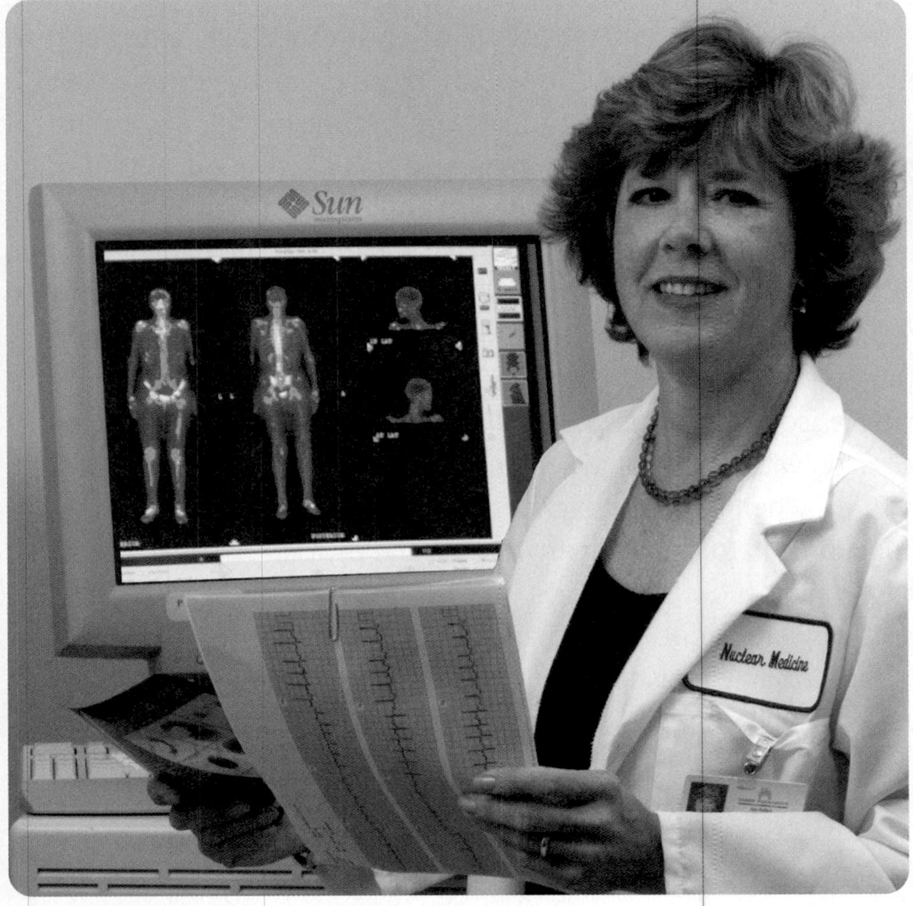

LOOKING AHEAD

"Everything we do in this department involves radioactive materials," says Julie Goudak, nuclear medicine technologist at Kaiser Hospital. "The radioisotopes are given in several ways. The patient may ingest an isotope, breathe it in, or receive it by an IV injection. We do many diagnostic tests, particularly of the heart function, to determine if a patient needs a cardiac CT scan."

A nuclear medicine technologist administers isotopes that emit radiation to determine the level of function of an organ such as the thyroid or heart, to detect the presence and size of a tumor, or to treat disease. A radioisotope locates in a specific organ, and its radiation is used by a computer to create an image of that organ. From this data, a physician can make a diagnosis and design a treatment program.

Visit **www.masteringchemistry.com** for self-study materials and instructor-assigned homework.

A female patient, age 50, complains of nervousness, irritability, increased perspiration, brittle hair, and muscle weakness. Her hands are shaky at times and her heart often beats rapidly. She has been experiencing weight loss. The doctor decides to test her thyroid. To get a detailed look, a thyroid scan is ordered. The patient is given a small amount of an iodine radioisotope, which will be taken up by the thyroid. The scan shows a higher than normal rate of uptake of the radioactive iodine, indicating an overactive thyroid gland, a condition called hyperthyroidism. Treatment for hyperthyroidism includes the use of drugs to lower the level of thyroid hormone, the use of radioactive iodine to destroy thyroid cells, or surgical removal of part of or the entire thyroid. In our case, the physician decides to use radioactive iodine. To begin treatment, the patient drinks a solution containing radioactive iodine. In the following few weeks, the cells that take up the radioactive iodine are destroyed by the radiation. Tests show that the patient's thyroid is smaller and the blood level of thyroid hormone is once again normal.

The first radioactive isotope was used, in 1937, at the University of California at Berkeley to treat a person with leukemia. Major strides in the use of radioactivity in medicine occurred in 1946, when a radioactive iodine isotope was successfully used to diagnose thyroid function and to treat hyperthyroidism and thyroid cancer. In the 1970s and 1980s, a variety of radioactive substances were used to produce images of organs such as liver, spleen, thyroid, kidney, and brain, and to detect heart disease. Today, procedures in nuclear medicine provide information about the function and structure of every organ in the body, allowing the physician to diagnose and treat diseases early.

LEARNING GOAL

Describe alpha, beta, positron, and gamma radiation.

9.1 Natural Radioactivity

Most naturally occurring isotopes of elements up to atomic number 19 have stable nuclei. An atom has a stable nucleus when the forces of attraction and repulsion between the protons and neutrons are balanced. Elements with atomic numbers 20 and higher usually have one or more isotopes that have unstable nuclei. An unstable atom has a nucleus with too many or too few protons compared to the number of neutrons, which means the forces between protons and neutrons are unbalanced. An unstable nucleus is *radioactive*, which means that it spontaneously emits small particles of energy, called **radiation**, to become more stable. Radiation may take the form of particles such as alpha (α) and beta (β) particles, positrons (β^+), or pure energy such as gamma (γ) rays. An isotope that emits radiation is called a *radioisotope*. For most types of radiation, there is a change in the number of protons in the nucleus, which means that an atom is converted into an atom of a different element. In 1803, this kind of nuclear change was not evident to Dalton when he made his predictions about atoms. Elements with atomic numbers of 93 and higher are produced artificially in nuclear laboratories and consist only of radioactive isotopes.

In Chapter 3, we wrote symbols for the different isotopes of an element. These symbols had their mass number written in the upper left corner and the atomic number written in the lower left corner. Recall that the mass number is equal to the number of protons and neutrons in the nucleus, and the atomic number is equal to the number of protons. For example, a radioactive isotope of iodine used in the diagnosis and treatment of thyroid conditions has a symbol with a mass number of 131 and an atomic number of 53, as shown.

Mass number (protons and neutrons)
Element symbol
Atomic number (protons)

$$^{131}_{53}\mathrm{I}$$

Radioactive isotopes are identified by writing the mass number after the element's name or symbol. Thus, in this example, the isotope is called iodine-131 or I-131. Table 9.1 compares some stable, nonradioactive isotopes with some radioactive isotopes.

TABLE 9.1 Stable and Radioactive Isotopes of Some Elements

Magnesium	Iodine	Uranium
Stable Isotopes		
$^{24}_{12}\mathrm{Mg}$	$^{127}_{53}\mathrm{I}$	None
Magnesium-24	Iodine-127	
Radioactive Isotopes		
$^{23}_{12}\mathrm{Mg}$	$^{125}_{53}\mathrm{I}$	$^{235}_{92}\mathrm{U}$
Magnesium-23	Iodine-125	Uranium-235
$^{27}_{12}\mathrm{Mg}$	$^{131}_{53}\mathrm{I}$	$^{238}_{92}\mathrm{U}$
Magnesium-27	Iodine-131	Uranium-238

Types of Radiation

By emitting radiation, an unstable nucleus forms a more stable, lower energy nucleus. One type of radiation consists of *alpha particles*. An **alpha particle** is identical to a helium (He) nucleus, which has 2 protons and 2 neutrons. An alpha particle has a mass number of 4, an atomic number of 2, and a charge of 2+. The symbol for an alpha particle is the Greek letter alpha (α) or the symbol of a helium nucleus except that the 2+ charge is omitted.

$^{4}_{2}\mathrm{He}$

Alpha (α) particle

$^{0}_{-1}e$

Beta (β) particle

Another type of radiation occurs when a radioisotope emits a *beta particle*. A **beta particle** is a high-energy electron, with a charge of 1− and, because its mass is so much less than the mass of a proton, it has a mass number of 0. It is represented by the Greek letter beta (β) or by the symbol for the electron (e) including the mass number and the charge. A beta particle forms when a neutron in an unstable nucleus changes to a proton and an electron.

$$^{1}_{0}n \longrightarrow {}^{1}_{1}\mathrm{H} + {}^{0}_{-1}e \text{ (or } \beta)$$

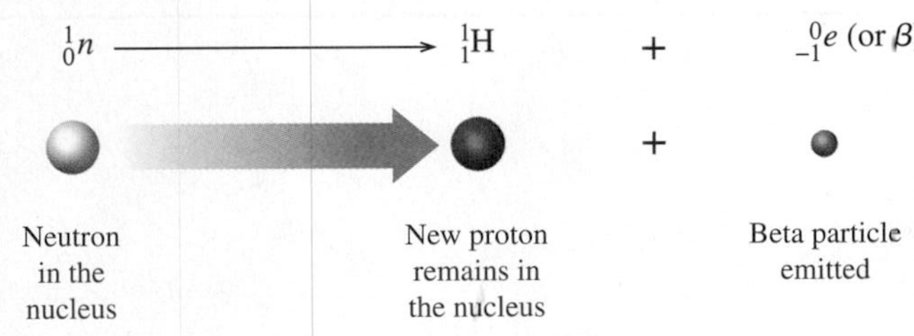

A **positron**, represented as β^{+}, has a positive (1+) charge with a mass number of 0, which makes it similar to a beta (β) particle. When the symbol β is used with no charge, it represents a beta particle rather than a positron. We write the symbols of a beta particle and a positron as follows:

	Electron	Positron
Mass number Charge	$^{0}_{-1}e$	$^{0}_{+1}e$

MC TUTORIAL
Types of Radiation

A positron is produced by an unstable nucleus when a proton is transformed into a neutron and a positron.

$$^{1}_{1}\mathrm{H} \longrightarrow {}^{1}_{0}n + {}^{0}_{+1}e \text{ (or } \beta^{+})$$

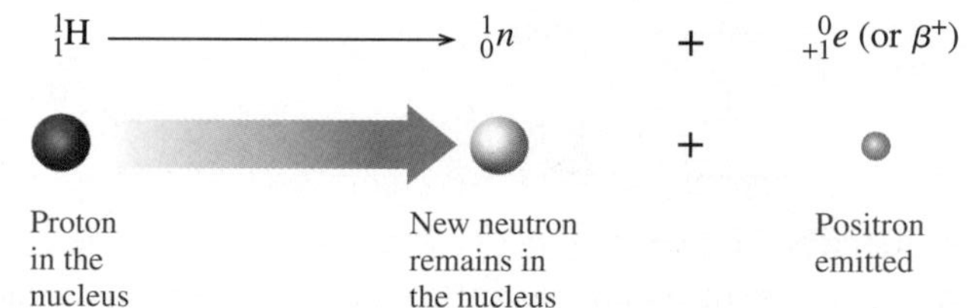

A positron is an example of *antimatter*, a term physicists use to describe a particle that is the exact opposite of another particle, in this case, an electron. When an electron and a positron collide, their minute masses are completely converted to energy in the form of *gamma rays*.

$$^{0}_{-1}e + {}^{0}_{+1}e \longrightarrow 2\,^{0}_{0}\gamma$$

Gamma rays are high-energy radiation, released when an unstable nucleus undergoes a rearrangement of its particles to give a more stable, lower energy nucleus. Gamma rays are often emitted along with other types of radiation. A gamma ray is written as the Greek letter gamma (γ). Because gamma rays are energy only, zeros are used to show that a gamma ray has no mass or charge.

$^{0}_{0}\gamma$

Gamma (γ) ray

Table 9.2 summarizes the types of radiation we will use in nuclear equations.

TABLE 9.2 Some Forms of Radiation

Type of Radiation	Symbol		Change in Nucleus	Mass Number	Charge
Alpha particle	α	$^{4}_{2}\mathrm{He}$	Two protons and two neutrons are emitted as an alpha particle.	4	2+
Beta particle	β	$^{0}_{-1}e$	A neutron changes to a proton and an electron is emitted.	0	1−
Positron	β^{+}	$^{0}_{+1}e$	A proton changes to a neutron and a positron is emitted.	0	1+
Gamma ray	γ	$^{0}_{0}\gamma$	Energy is lost to stabilize the nucleus.	0	0
Proton	p	$^{1}_{1}\mathrm{H}$	A proton is emitted.	1	1+
Neutron	n	$^{1}_{0}n$	A neutron is emitted.	1	0

CONCEPT CHECK 9.1

Radiation Particles

Give the name and write the symbol for each of the following types of radiation:

a. contains two protons and two neutrons

b. has a mass number of 0 and a 1− charge

ANSWER

a. An alpha (α) particle, $^{4}_{2}\mathrm{He}$, has two protons and two neutrons.

b. A beta (β) particle, $^{0}_{-1}e$, is an electron with a mass number of 0 and a 1− charge.

Chemistry Link to Health

BIOLOGICAL EFFECTS OF RADIATION

When radiation strikes molecules in its path, electrons may be knocked away, forming unstable ions. If this *ionizing radiation* passes through the human body, it may interact with water molecules, removing electrons and producing H_2O^+, which can cause undesirable chemical reactions.

The cells most sensitive to ionizing radiation are the ones undergoing rapid division—those of the bone marrow, skin, reproductive organs, and intestinal lining, as well as all cells of growing children. Damaged cells may lose their ability to produce necessary materials. For example, if radiation damages cells of the bone marrow, red blood cells may no longer be produced. If sperm cells, ova, or the cells of a fetus are damaged, birth defects may result. In contrast, cells of the nerves, muscles, liver, and adult bones are much less sensitive to radiation because they undergo little or no cellular division.

Cancer cells are another example of rapidly dividing cells. Because cancer cells are highly sensitive to radiation, large doses of radiation are used to destroy them. The normal tissue that surrounds cancer cells divides at a slower rate and suffers less damage from radiation. However, radiation, due to its high penetrating energy, may itself cause malignant tumors, leukemia, anemia, and genetic mutations.

Radiation Protection

Radiologists, doctors, and nurses working with radioactive isotopes must use proper radiation protection. Proper **shielding** is necessary to prevent exposure. Alpha particles, which have the greatest mass and highest charge of the radiation particles, travel only a few centimeters in the air before they collide with air molecules, acquire electrons, and become helium atoms. External sources of protection from alpha particles are provided by a piece of paper, lab coats, and the skin. However, if ingested or inhaled, alpha particles can cause serious internal damage because of the large mass and high charge.

Beta particles have a very small mass and move much faster and farther than alpha particles, traveling as much as several meters through air. They can pass through paper and penetrate as far as 4–5 mm into body tissue. External exposure to beta particles can burn the surface of the skin, but they are stopped before they reach the internal organs. Heavy clothing such as lab coats and gloves are needed to protect the skin from beta particles.

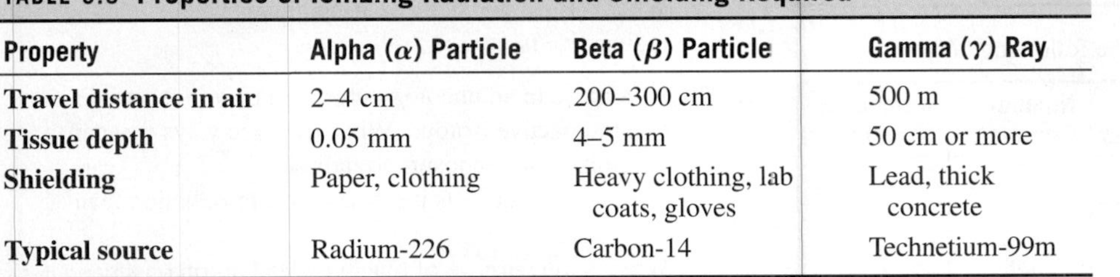

Different types of shielding are needed for different types of radiation.

Gamma rays travel great distances through the air and pass through many materials, including body tissues. Because gamma rays can penetrate so deeply, exposure to these rays is extremely hazardous. Only very dense shielding, such as lead or concrete, will stop them. Syringes used for injections of radioactive isotopes use shielding made of lead or heavy-weight materials such as tungsten and plastic composites.

When working with radioactive materials, medical personnel wear protective clothing and gloves and stand behind a shield (see Figure 9.1). Long tongs may be used to pick up vials of radioactive material, keeping them away from the hands and body.

Table 9.3 summarizes the shielding materials required for the various types of radiation.

TABLE 9.3 Properties of Ionizing Radiation and Shielding Required

Property	Alpha (α) Particle	Beta (β) Particle	Gamma (γ) Ray
Travel distance in air	2–4 cm	200–300 cm	500 m
Tissue depth	0.05 mm	4–5 mm	50 cm or more
Shielding	Paper, clothing	Heavy clothing, lab coats, gloves	Lead, thick concrete
Typical source	Radium-226	Carbon-14	Technetium-99m

TUTORIAL
Radiation and Its Biological Effects

If you work in an environment where radioactive materials are present, such as a nuclear medicine facility, try to keep the time you must spend in a radioactive area to a minimum. Remaining in a radioactive area twice as long exposes you to twice as much radiation.

Keep your distance! The greater the distance from the radioactive source, the lower the intensity of radiation you receive. Just by doubling your distance from the radiation source, the intensity of radiation drops to $\left(\frac{1}{2}\right)^2$ or one-fourth of its previous value.

SAMPLE PROBLEM 9.1

Radiation Protection

How does the type of shielding for alpha radiation differ from that used for gamma radiation?

SOLUTION

Alpha radiation is stopped by paper and clothing. However, lead or concrete is needed for protection from gamma radiation.

STUDY CHECK 9.1

Besides shielding, what other methods help reduce exposure to radiation?

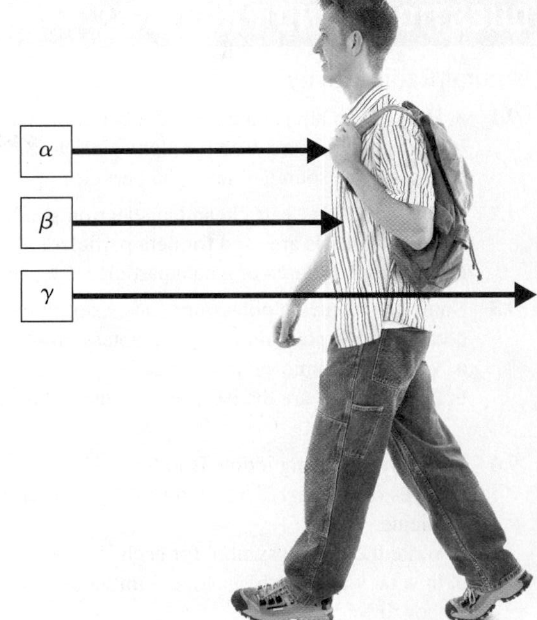

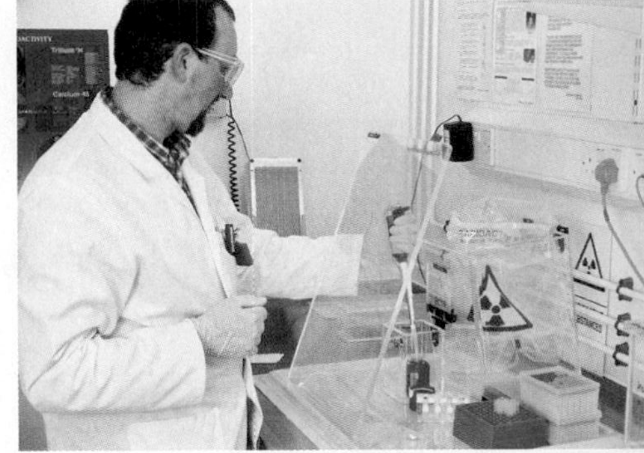

FIGURE 9.1 A person working with radioisotopes wears protective clothing and gloves and stands behind a shield.

Q What types of radiation would a lead shield block?

QUESTIONS AND PROBLEMS

Natural Radioactivity

9.1 **a.** How are an alpha particle and a helium nucleus similar?
b. What symbols are used for alpha particles?
c. What is the source of an alpha particle?

9.2 **a.** How are a beta particle and an electron similar?
b. What symbols are used for beta particles?
c. What is the source of a beta particle?

9.3 Naturally occurring potassium consists of three isotopes: potassium-39, potassium-40, and potassium-41.
a. Write the atomic symbol for each isotope.
b. In what ways are the isotopes similar and in what ways do they differ?

9.4 Naturally occurring iodine is iodine-127. The radioactive isotopes of iodine-125 and iodine-130 are used in nuclear medicine.
a. Write the atomic symbol for each isotope.
b. In what ways are the isotopes similar and in what ways do they differ?

9.5 Supply the missing information in the following table:

Medical Use	Atomic Symbol	Mass Number	Number of Protons	Number of Neutrons
Heart imaging	$^{201}_{81}Tl$			
Radiation therapy		60	27	
Abdominal scan			31	36
Hyperthyroidism	$^{131}_{53}I$			
Leukemia treatment		32		17

9.6 Supply the missing information in the following table:

Medical Use	Atomic Symbol	Mass Number	Number of Protons	Number of Neutrons
Cancer treatment	$^{131}_{55}Cs$			
Brain scan		99	43	
Blood flow		141	58	
Bone scan		85		47
Lung function	$^{133}_{54}Xe$			

9.7 Write a symbol for each of the following:
a. alpha particle
b. neutron
c. beta particle
d. nitrogen-15
e. iodine-125

9.8 Write a symbol for each of the following:
a. proton
b. gamma ray
c. barium-131
d. positron
e. palladium-103

9.9 Identify each of the following:
a. $^{0}_{-1}X$ **b.** $^{4}_{2}X$
c. $^{1}_{0}X$ **d.** $^{24}_{11}X$
e. $^{14}_{6}X$

9.10 Identify each of the following:
a. $^{1}_{1}X$ **b.** $^{32}_{15}X$
c. $^{0}_{0}X$ **d.** $^{59}_{26}X$
e. $^{0}_{+1}X$

9.11 **a.** Why does beta radiation penetrate farther into body tissue than alpha radiation?
b. How does radiation cause damage to cells of the body?
c. Why does the radiation technician leave the room when you receive an X-ray?
d. What is the purpose of wearing gloves when handling radioactive isotopes?

9.12 **a.** As a nurse in an oncology unit, you may give an injection of a radioactive isotope. What are three ways you can minimize your exposure to radiation?
b. Why are cancer cells more sensitive to radiation than nerve cells?
c. What is the purpose of placing a lead apron on a patient who is receiving dental X-rays?
d. Why are the walls in a radiology treatment room built of thick concrete blocks?

LEARNING GOAL

Write an equation showing mass numbers and atomic numbers for radioactive decay.

TUTORIAL
Writing Nuclear Equations

9.2 Nuclear Equations

In a process called **radioactive decay**, a nucleus spontaneously breaks down by emitting radiation. This process can be written using the atomic symbols of the original radioactive nucleus, the new nucleus, and the radiation emitted. An arrow between the atomic symbols indicates that this is a *nuclear equation.*

$$\text{Radioactive nucleus} \longrightarrow \text{new nucleus} + \text{radiation } (\alpha, \beta, \beta^+, \gamma)$$

In Chapter 4, we balanced chemical equations to give the same number of atoms in the reactants and products. In a nuclear equation, the mass numbers and the atomic numbers must balance so the number of protons and neutrons are equal on both sides. However, in a nuclear equation, there is often a change in the number of protons, which gives a different element.

Alpha Decay

An unstable nucleus undergoes alpha decay by emitting an alpha particle. Because an alpha particle consists of two protons and two neutrons, the mass number decreases by four, and the atomic number decreases by two. For example, uranium-238 emits an alpha particle to form a nucleus with a mass number of 234. Compared to uranium with 92 protons, the new nucleus has 90 protons, which is thorium.

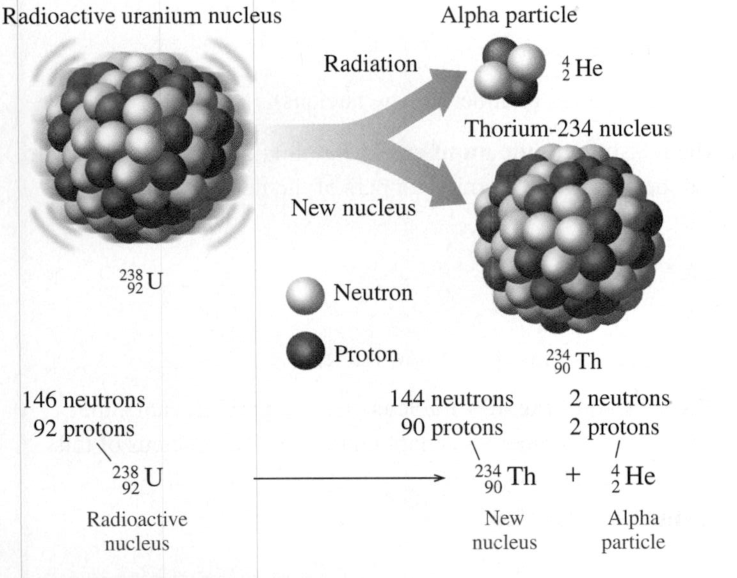

146 neutrons, 92 protons; 144 neutrons, 90 protons; 2 neutrons, 2 protons

$$\underset{\text{Radioactive nucleus}}{^{238}_{92}U} \longrightarrow \underset{\text{New nucleus}}{^{234}_{90}Th} + \underset{\text{Alpha particle}}{^{4}_{2}He}$$

CONCEPT CHECK 9.2

Alpha Decay

Francium-221 emits an alpha particle, which forms a new nucleus.

a. Does the new nucleus have a larger or smaller mass number? By how much?
b. Does the new nucleus have a larger or smaller atomic number? By how much?

ANSWER

a. The loss of an alpha particle will give the new nucleus with a smaller mass number. Because an alpha particle is a helium nucleus, $^{4}_{2}He$, the mass number of the new nucleus will decrease by four from 221 to 217.
b. The loss of an alpha particle will give the new nucleus with a smaller atomic number. Because an alpha particle is a helium nucleus, $^{4}_{2}He$, the atomic number of the new nucleus will decrease by two from 87 to 85.

We can look at writing a balanced nuclear equation for americium-241, which undergoes alpha decay as shown in Sample Problem 9.2.

SAMPLE PROBLEM 9.2

Writing an Equation for Alpha Decay

Smoke detectors used in homes and apartments contain americium-241, which undergoes alpha decay. When alpha particles collide with air molecules, charged particles are produced that generate an electrical current. If smoke particles enter the detector, they interfere with the formation of charged particles in the air, and the electric current is interrupted. This causes the alarm to sound, and warns the occupants of the danger of fire. Complete the following nuclear equation for the decay of americium-241:

$$^{241}_{95}Am \longrightarrow ? + ^{4}_{2}He$$

A smoke detector sounds an alarm when smoke enters its ionization chamber.

Guide to Completing a Nuclear Equation

1. Write the incomplete nuclear equation.
2. Determine the missing mass number.
3. Determine the missing atomic number.
4. Determine the symbol of the new nucleus.
5. Complete the nuclear equation.

SOLUTION

Step 1 Write the incomplete nuclear equation.

$$^{241}_{95}Am \longrightarrow ? + ^{4}_{2}He$$

Step 2 Determine the missing mass number. In the equation, the mass number of americium, 241, is equal to the sum of the mass numbers of the new nucleus and the alpha particle.

$$241 = ? + 4$$
$$241 - 4 = ?$$
$$241 - 4 = 237 \text{ (mass mumber of new nucleus)}$$

Step 3 Determine the missing atomic number. The atomic number of americium, 95, must equal the sum of the atomic numbers of the new nucleus and the alpha particle.

$$95 = ? + 2$$
$$95 - 2 = ?$$
$$95 - 2 = 93 \text{ (atomic number of new nucleus)}$$

Step 4 Determine the symbol of the new nucleus. On the periodic table, the element that has atomic number 93 is neptunium, Np. The nucleus of this isotope of Np is written as $^{237}_{93}Np$.

Step 5 Complete the nuclear equation.

$$^{241}_{95}Am \longrightarrow ^{237}_{93}Np + ^{4}_{2}He$$

STUDY CHECK 9.2

Write a balanced nuclear equation for the alpha decay of polonium Po-214.

Chemistry Link to the Environment

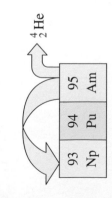

RADON IN OUR HOMES

The presence of radon has become a much publicized environmental and health issue because of the radiation danger it poses. Radioactive isotopes such as radium-226 are naturally present in many types of rocks and soils. Radium-226 emits an alpha particle and is converted into radon gas, which diffuses out of the rocks and soil.

$$^{226}_{88}Ra \longrightarrow ^{222}_{86}Rn + ^{4}_{2}He$$

Outdoors, radon gas poses little danger because it disperses in the air. However, if the radioactive source is under a house or building, the radon gas can enter the house through cracks in the foundation or other openings. Those who live or work there may inhale the radon. Inside the lungs, radon-222 emits alpha particles to form polonium-218, which is known to cause lung cancer.

$$^{222}_{86}Rn \longrightarrow ^{218}_{84}Po + ^{4}_{2}He$$

Some researchers have estimated that 10 percent of all lung cancer deaths in the United States result from radon gas exposure. The Environmental Protection Agency (EPA) recommends that the maximum level of radon not exceed 4 picocuries (pCi) per liter of air in a home. One picocurie (pCi) is equal to 10^{-12} curies (Ci); curies are described in Section 9.3. In California, 1 percent of all the houses surveyed exceeded the EPA's recommended maximum radon level.

A radon gas detector is used to determine radon levels in buildings.

Beta Decay

As we learned in Section 9.1, the formation of a beta particle is the result of the breakdown of a neutron into a proton and an electron (beta particle). Because the proton remains in the nucleus, the number of protons increases by one, while the number of neutrons decreases by one. Therefore, in a nuclear equation for beta decay, the mass number of the radioactive nucleus and the mass number of the new nucleus are the same. However, the atomic number of the new nucleus increases by one, indicating a change of one element into another. For example, the beta decay of a carbon-14 nucleus produces a nitrogen-14 nucleus.

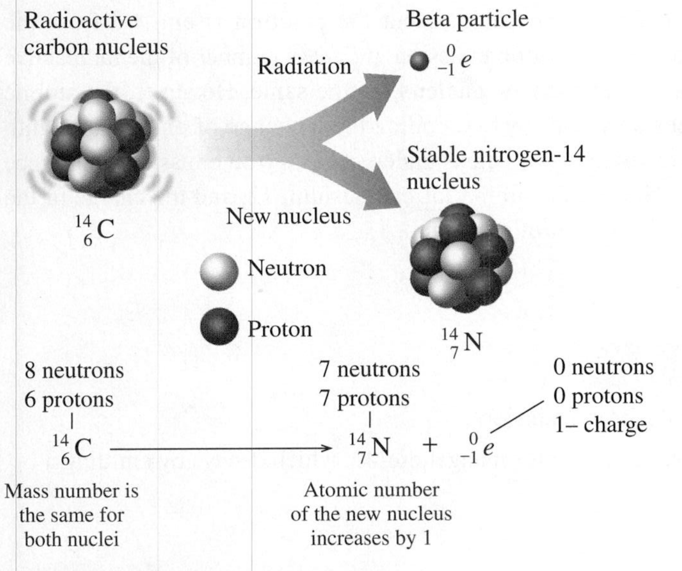

SAMPLE PROBLEM 9.3

Writing an Equation for Beta Decay

Write the balanced nuclear equation for the beta decay of cobalt-60.

SOLUTION

Step 1 Write the incomplete nuclear equation.

$$^{60}_{27}Co \longrightarrow ? + ^{0}_{-1}e$$

Step 2 Determine the missing mass number. In the equation for beta decay, the mass number of cobalt, 60, is equal to the sum of the mass numbers of the new nucleus and the beta particle.

$$60 = ? + 0$$
$$60 - 0 = ?$$
$$60 = ? \text{ (mass number of new nucleus)}$$

Step 3 Determine the missing atomic number. The atomic number of cobalt, 27, must equal the sum of the atomic numbers of the new nucleus and the beta particle.

$$27 = ? - 1$$
$$27 + 1 = ?$$
$$28 = ? \text{ (atomic number of new nucleus)}$$

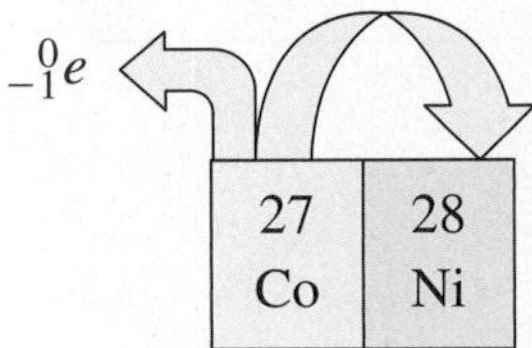

Step 4 Determine the symbol of the new nucleus. On the periodic table, the element that has atomic number 28 is nickel (Ni). The nucleus of this isotope of Ni is written as $^{60}_{28}Ni$.

BETA EMITTERS IN MEDICINE

The radioactive isotopes of several biologically important elements are beta emitters. When a radiologist wants to treat a malignancy within the body, a beta emitter may be used. The short range of penetration into the tissue by beta particles is advantageous for certain conditions. For example, some malignant tumors increase the fluid within the body tissues. A compound containing phosphorus-32, a beta emitter, is injected into the tumor. The beta particles travel only a few millimeters through the tissue, so only the malignancy and any tissue within that range are affected. The growth of the tumor is slowed or stopped, and the production of fluid decreases. Phosphorus-32 is also used to treat leukemia, polycythemia vera (excessive production of red blood cells), and lymphomas.

$$^{32}_{15}\text{P} \longrightarrow\ ^{32}_{16}\text{S} +\ ^{0}_{-1}e$$

Another beta emitter, iron-59, is used in blood tests to determine the level of iron in the blood and the rate of production of red blood cells by the bone marrow.

$$^{59}_{26}\text{Fe} \longrightarrow\ ^{59}_{27}\text{Co} +\ ^{0}_{-1}e$$

Step 5 **Complete the nuclear equation.**

$$^{60}_{27}\text{Co} \longrightarrow\ ^{60}_{28}\text{Ni} +\ ^{0}_{-1}e$$

STUDY CHECK 9.3

Write the balanced nuclear equation for the beta decay of chromium-51.

Positron Emission

In positron emission, a proton in an unstable nucleus is converted to a neutron and a positron. The neutron remains in the nucleus, but the positron is emitted from the nucleus. In a nuclear equation for positron emission, the mass number of the radioactive nucleus and the mass number of the new nucleus are the same. However, the atomic number of the new nucleus decreases by one, indicating a change of one element into another. For example, an aluminum-24 nucleus undergoes positron emission to produce a magnesium-24 nucleus. The atomic number of magnesium, 12, and the charge of the positron (1+) give the atomic number of aluminum, 13.

$$^{24}_{13}\text{Al} \longrightarrow\ ^{24}_{12}\text{Mg} +\ ^{0}_{+1}e$$

SAMPLE PROBLEM 9.4

Writing an Equation for Positron Emission

Write the balanced nuclear equation for manganese-49, which decays by emitting a positron.

SOLUTION

Step 1 **Write the incomplete nuclear equation.**

$$^{49}_{25}\text{Mn} \longrightarrow ? +\ ^{0}_{+1}e$$

Step 2 **Determine the missing mass number.** In the equation, the mass number of manganese, 49, is equal to the sum of the mass numbers of the new nucleus and the positron.

$$49 = ? + 0$$
$$49 - 0 = ?$$
$$49 = ?\ \text{(mass number of new nucleus)}$$

Step 3 **Determine the missing atomic number.** The atomic number of manganese, 25, must equal the sum of the atomic numbers of the new nucleus and the positron.

$$25 = ? + 1$$
$$25 - 1 = ?$$
$$24 = ?\ \text{(atomic number of new nucleus)}$$

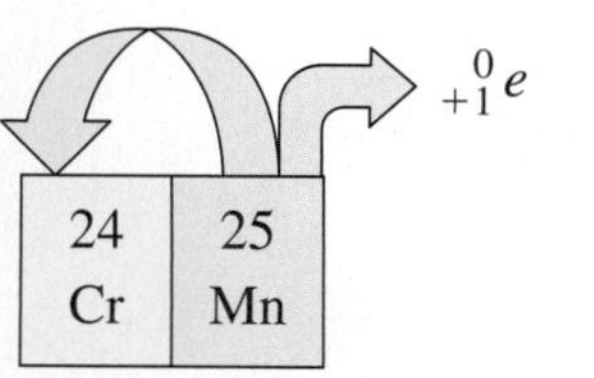

Step 4 **Determine the symbol of the new nucleus.** On the periodic table, the element that has atomic number 24 is chromium, Cr. The nucleus of this isotope of Cr is written as $^{49}_{24}\text{Cr}$.

Step 5 **Complete the nuclear equation.**

$$^{49}_{25}\text{Mn} \longrightarrow\ ^{49}_{24}\text{Cr} +\ ^{0}_{+1}e$$

STUDY CHECK 9.4

Write the balanced nuclear equation for xenon-118, which decays by emitting a positron.

Gamma Emission

Pure gamma emitters are rare, although some gamma radiation accompanies most alpha and beta radiation. In radiology, one of the most commonly used gamma emitters is technetium (Tc). Because the unstable isotope of technetium decays quickly, it is written as the *metastable* (symbol m) isotope: technetium-99m, Tc-99m, or $^{99m}_{43}\text{Tc}$. By emitting energy in the form of gamma rays, the unstable nucleus becomes more stable.

$$^{99m}_{43}\text{Tc} \longrightarrow {}^{99}_{43}\text{Tc} + {}^{0}_{0}\gamma$$

Figure 9.2 summarizes the changes in the nucleus for alpha, beta, positron, and gamma radiation.

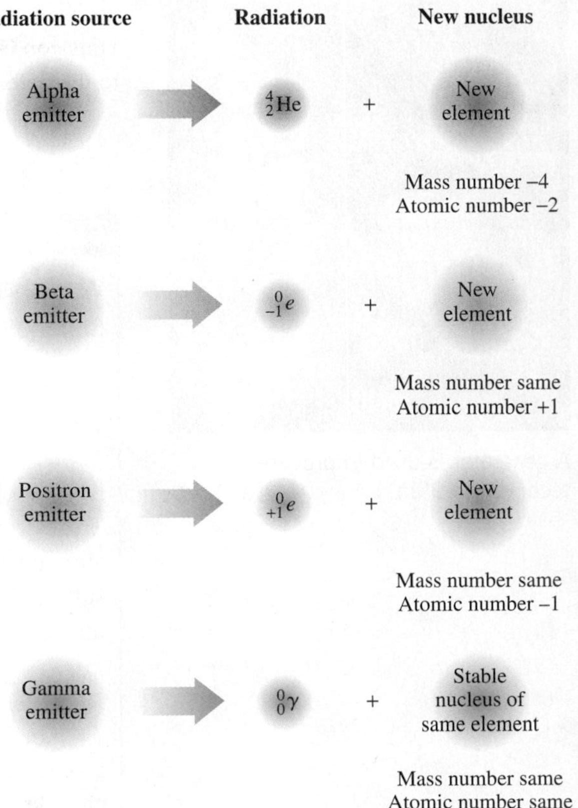

FIGURE 9.2 When the nuclei of alpha, beta, positron, and gamma emitters emit radiation, new, more stable nuclei are produced.

Q What changes occur in the number of protons and neutrons of an unstable nucleus that undergoes alpha decay?

Producing Radioactive Isotopes

Today, many radioisotopes are produced in small amounts by converting stable, nonradioactive isotopes into radioactive ones. In a process called *transmutation*, a stable nucleus is bombarded by high-speed particles such as alpha particles, protons, neutrons, and small nuclei. When one of these particles is absorbed, the nucleus becomes a radioactive isotope.

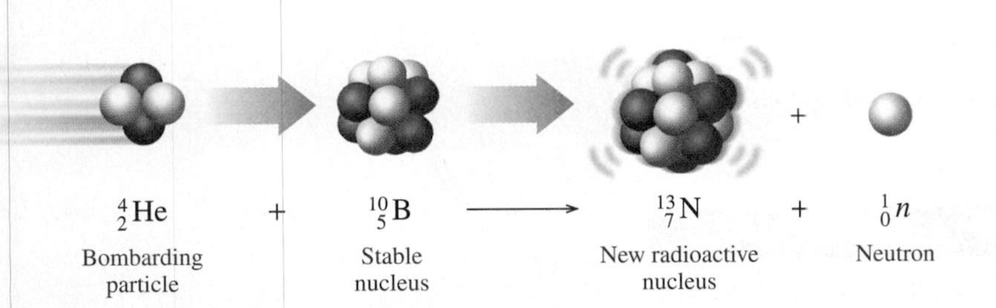

When nonradioactive B-10 is bombarded by an alpha particle, the products are radioactive N-13 and a neutron.

All elements that have an atomic number greater than 92 have been produced by bombardment; none of these elements occurs naturally. Most have been produced in small amounts and exist for such a short time that it is difficult to study their properties. An example is element 105, dubnium (Db), which is produced when californium-249 is bombarded with nitrogen-15.

$$^{15}_{7}\text{N} + {}^{249}_{98}\text{Cf} \longrightarrow {}^{260}_{105}\text{Db} + 4\,{}^{1}_{0}n$$

Technetium-99m is a radioisotope used in nuclear medicine for several diagnostic procedures, including the detection of brain tumors and examinations of the liver and spleen. The source of technetium-99m is molybdenum-99, which is produced in a nuclear reactor by neutron bombardment of molybdenum-98.

$$^{1}_{0}n + {}^{98}_{42}\text{Mo} \longrightarrow {}^{99}_{42}\text{Mo}$$

Many radiology laboratories have small generators containing molybdenum-99, which decays to give technetium-99m.

$$^{99}_{42}\text{Mo} \longrightarrow {}^{99m}_{43}\text{Tc} + {}^{0}_{-1}e$$

TUTORIAL
Alpha, Beta, and Gamma Emitters

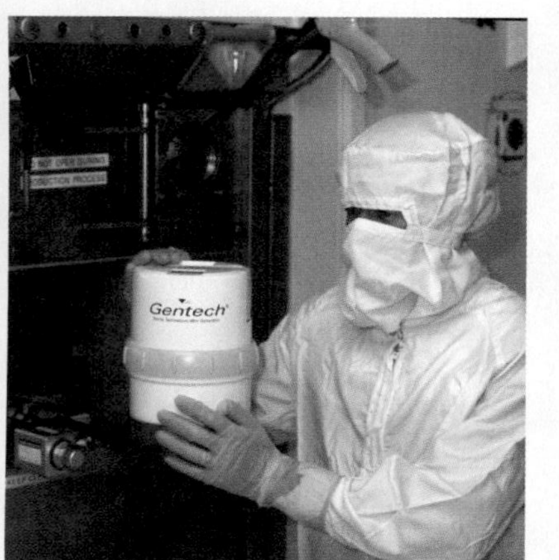

A generator is used to prepare technetium-99m.

The technetium-99m radioisotope decays by emitting gamma rays. Gamma emission is desirable for diagnostic work because the gamma rays pass through the body to the detection equipment.

$${}^{99m}_{43}\text{Tc} \longrightarrow {}^{99}_{43}\text{Tc} + {}^{0}_{0}\gamma$$

CONCEPT CHECK 9.3

Writing an Isotope Produced by Bombardment

Sulfur-32 is bombarded with a neutron to produce a new isotope and an alpha particle.

$${}^{1}_{0}n + {}^{32}_{16}\text{S} \longrightarrow ? + {}^{4}_{2}\text{He}$$

What is the new isotope?

ANSWER

To determine the name of the new isotope, we need to calculate its mass number and atomic number. On the left side, the sum of the mass numbers of one neutron, 1, and sulfur, 32, gives a total of 33. On the right side, we subtract the mass number of the alpha particle, 4, which gives a mass number of 29 to the new nucleus.

$${}^{1}_{0}n + {}^{32}_{16}\text{S} \longrightarrow {}^{29}_{?}? + {}^{4}_{2}\text{He}$$

On the left side, the sum of the atomic numbers of a neutron, 0, and sulfur, 16, gives a total of 16. On the right side, we subtract the atomic number of the alpha particle, 2, to give an atomic number of 14 to the new nucleus. The element with the atomic number of 14 is silicon.

$${}^{1}_{0}n + {}^{32}_{16}\text{S} \longrightarrow {}^{29}_{14}\text{Si} + {}^{4}_{2}\text{He}$$

Thus, the new isotope is silicon-29.

SAMPLE PROBLEM 9.5

Producing Radioactive Isotopes

Write the balanced nuclear equation for the bombardment of nickel-58 by a proton (${}^{1}_{1}\text{H}$), which produces a radioactive isotope and an alpha particle.

SOLUTION

Step 1 Write the incomplete nuclear equation.

$${}^{1}_{1}\text{H} + {}^{58}_{28}\text{Ni} \longrightarrow ? + {}^{4}_{2}\text{He}$$

Step 2 Determine the missing mass number. In the equation, the sum of the mass numbers of the proton, 1, and the nickel, 58, must be equal to the sum of the mass numbers of the new nucleus and the alpha particle.

$$1 + 58 = ? + 4$$
$$59 - 4 = ?$$
$$55 = ? \text{ (mass number of new nucleus)}$$

Step 3 Determine the missing atomic number. The sum of the atomic numbers of the proton, 1, and nickel, 28, must equal the sum of the atomic numbers of the new nucleus and the alpha particle, 2.

$$1 + 28 = ? + 2$$
$$29 - 2 = ?$$
$$27 = ? \text{ (atomic number of new nucleus)}$$

Step 4 **Determine the symbol of the new nucleus.** On the periodic table, the element that has atomic number 27 is cobalt, Co. The nucleus of this isotope of Co is written as $^{55}_{27}Co$.

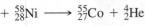

Step 5 **Complete the nuclear equation.**

$$^{1}_{1}H + ^{58}_{28}Ni \longrightarrow ^{55}_{27}Co + ^{4}_{2}He$$

$^{4}_{2}He$ $^{1}_{1}H$

27	28
Co	Ni

STUDY CHECK 9.5

The first radioactive isotope was produced in 1934 by the bombardment of aluminum-27 by an alpha particle to produce a radioactive isotope and one neutron. What is the balanced nuclear equation for this transmutation?

QUESTIONS AND PROBLEMS

Nuclear Equations

9.13 Write a balanced nuclear equation for the alpha decay of each of the following radioactive isotopes:
a. $^{208}_{84}Po$ **b.** $^{232}_{90}Th$
c. $^{251}_{102}No$ **d.** radon-220

9.14 Write a balanced nuclear equation for the alpha decay of each of the following radioactive isotopes:
a. curium-243 **b.** $^{252}_{99}Es$
c. $^{251}_{98}Cf$ **d.** $^{261}_{107}Bh$

9.15 Write a balanced nuclear equation for the beta decay of each of the following radioactive isotopes:
a. $^{25}_{11}Na$ **b.** $^{20}_{8}O$
c. strontium-92 **d.** iron-60

9.16 Write a balanced nuclear equation for the beta decay of each of the following radioactive isotopes:
a. $^{44}_{19}K$ **b.** iron-59
c. potassium-42 **d.** $^{141}_{56}Ba$

9.17 Write a balanced nuclear equation for the positron emission of each of the following radioactive isotopes:
a. silicon-26 **b.** cobalt-54
c. $^{77}_{37}Rb$ **d.** $^{93}_{45}Rh$

9.18 Write a balanced nuclear equation for the positron emission of each of the following radioactive isotopes:
a. boron-8 **b.** $^{15}_{8}O$
c. $^{40}_{19}K$ **d.** carbon-11

9.19 Complete each of the following nuclear equations, and describe the type of radiation:
a. $^{28}_{13}Al \longrightarrow ? + ^{0}_{-1}e$
b. $^{180m}_{73}Ta \longrightarrow ^{180}_{73}Ta + ?$
c. $^{66}_{29}Cu \longrightarrow ^{66}_{30}Zn + ?$
d. $? \longrightarrow ^{4}_{2}He + ^{234}_{90}Th$
e. $^{188}_{80}Hg \longrightarrow ? + ^{0}_{+1}e$

9.20 Complete each of the following nuclear equations, and describe the type of radiation:
a. $^{11}_{6}C \longrightarrow ^{11}_{5}B + ?$
b. $^{35}_{16}S \longrightarrow ? + ^{0}_{-1}e$
c. $? \longrightarrow ^{90}_{39}Y + ^{0}_{-1}e$
d. $^{210}_{83}Bi \longrightarrow ? + ^{4}_{2}He$
e. $? \longrightarrow ^{89}_{39}Y + ^{0}_{+1}e$

9.21 Complete each of the following bombardment reactions:
a. $^{1}_{0}n + ^{9}_{4}Be \longrightarrow ?$
b. $^{1}_{0}n + ^{131}_{52}Te \longrightarrow ? + ^{0}_{-1}e$
c. $^{1}_{0}n + ? \longrightarrow ^{24}_{11}Na + ^{4}_{2}He$
d. $^{4}_{2}He + ^{27}_{13}Al \longrightarrow ? + ^{1}_{0}n$

9.22 Complete each of the following bombardment reactions:
a. $? + ^{40}_{18}Ar \longrightarrow ^{43}_{19}K + ^{1}_{1}H$
b. $^{1}_{0}n + ^{238}_{92}U \longrightarrow ?$
c. $^{1}_{0}n + ? \longrightarrow ^{14}_{6}C + ^{1}_{1}H$
d. $? + ^{64}_{28}Ni \longrightarrow ^{272}_{111}Rg + ^{1}_{0}n$

9.3 Radiation Measurement

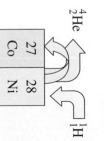

LEARNING GOAL

Describe the detection and measurement of radiation.

TUTORIAL
Measuring Radiation

One of the most common instruments for detecting beta and gamma radiation is the Geiger counter. It consists of a metal tube filled with a gas such as argon. When radiation enters a window on the end of the tube, argon atoms form ions, which produce an electrical current. Each burst of current is amplified to give a click and a reading on a meter.

$$Ar + \text{radiation} \longrightarrow Ar^{+} + e^{-}$$

Radiation is measured in several different ways. We can measure the activity of a radioactive sample or determine the impact of radiation on biological tissue.

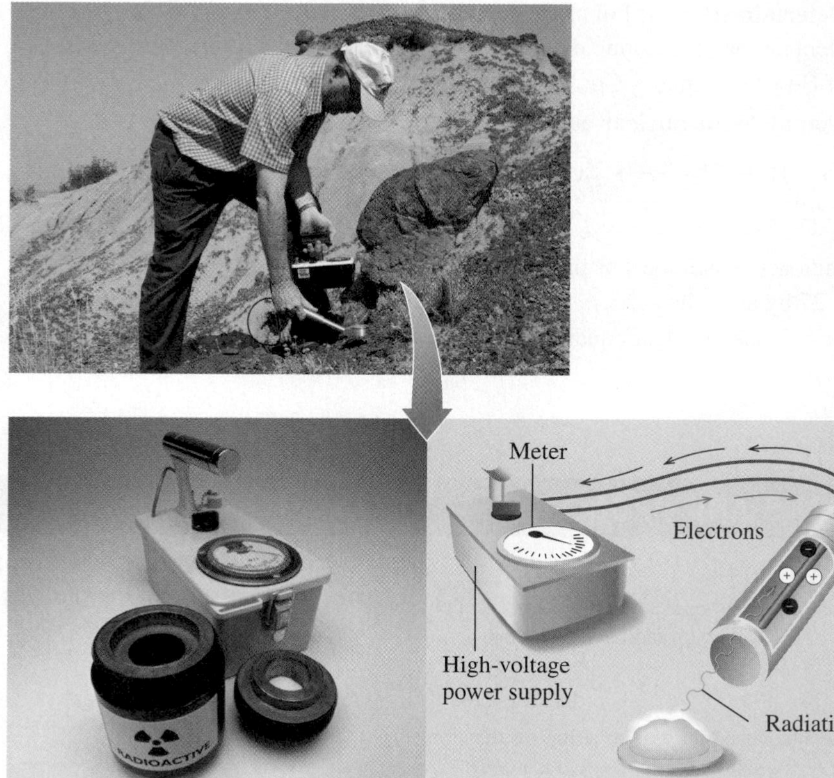

A radiation technician uses a Geiger counter to check radiation levels.

Measuring Radiation

SELF STUDY ACTIVITY
Nuclear Chemistry

When a radiology laboratory obtains a radioisotope, the *activity* of the sample is measured in terms of the number of nuclear disintegrations per second. The **curie (Ci)**, the original unit of activity, was defined as the number of disintegrations that occur in one second for 1 gram of radium, which is equal to 3.7×10^{10} disintegrations per second. The curie was named for the Polish scientist Marie Curie, who along with her husband, Pierre, discovered the radioactive elements radium and polonium. The SI unit of radiation activity is the **becquerel (Bq)**, which is one disintegration per second.

The **rad (radiation absorbed dose)** is a unit that measures the amount of radiation absorbed by a gram of material such as body tissue. The SI unit for absorbed dose is the **gray (Gy)**, which is defined as the joules of energy absorbed by 1 kilogram of body tissue. The gray is equal to 100 rad.

The **rem (radiation equivalent in humans)** measures the biological effects of different kinds of radiation. Although alpha particles do not penetrate the skin, if they should enter the body by some other route, they can cause extensive damage within a short distance. High-energy radiation such as beta particles and high-energy protons and neutrons that penetrate the skin and travel into tissue cause more damage. Gamma rays are damaging because they travel a long way through body tissue.

To determine the **equivalent dose** or rem dose, the absorbed dose (rad) is multiplied by a factor that adjusts for biological damage caused by a particular form of radiation. For beta and gamma radiation the factor is 1, so the biological damage in rems is the same as the absorbed radiation (rad). For high-energy protons and neutrons, the factor is about 10, and for alpha particles it is 20.

$$\text{Biological damage (rem)} = \text{absorbed dose (rad)} \times \text{factor}$$

TABLE 9.4 Units of Radiation Measurement

Measurement	Common Unit	SI Unit	Relationship
Activity	curie (Ci) 1 Ci = 3.7×10^{10} disintegrations/s	becquerel (Bq) 1 Bq = 1 disintegration/s	1 Ci = 3.7×10^{10} Bq
Absorbed dose	rad	gray (Gy) 1 Gy = 1 J/kg of tissue	1 Gy = 100 rad
Biological damage	rem = rad × factor	sievert (Sv)	1 Sv = 100 rem

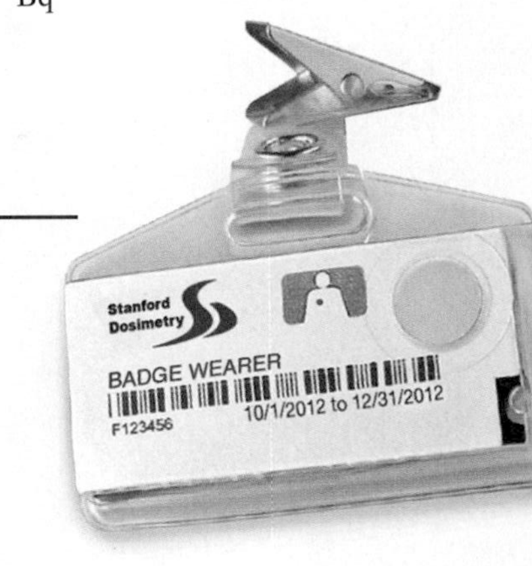

A film badge measures radiation exposure.

Often the measurement for an equivalent dose is in millirems (mrem). One rem is equal to 1000 mrem. The SI unit is the **sievert (Sv)**. One sievert is equal to 100 rem. Table 9.4 summarizes the units used to measure radiation.

People who work in radiology laboratories wear film badges to determine their exposure to radiation. A film badge consists of a piece of photographic film in a container that is attached to clothing. Periodically, the film badges are collected and developed to determine the level of exposure to radiation.

CASE STUDY
Food Irradiation

Chemistry Link to Health

RADIATION AND FOOD

Food-borne illnesses caused by pathogenic bacteria such as *Salmonella*, *Listeria*, and *Escherichia coli* (*E. coli*) have become a major health concern in the United States. The Centers for Disease Control and Prevention estimates that each year *E. coli* in contaminated foods infects 70 000 people in the United States, and that 60 people die. *E. coli* has been responsible for outbreaks of illness from contaminated ground beef, eggs, fruit juices, lettuce, spinach, and alfalfa sprouts.

The Food and Drug Administration (FDA) has approved the use of 0.3 kGy to 1 kGy of ionizing radiation produced by cobalt-60 or cesium-137 for the treatment of foods. The irradiation technology is much like that used to sterilize medical supplies. Cobalt pellets are placed in stainless steel tubes, which are arranged in racks. When food moves through the series of racks, the gamma rays pass through the food and kill the bacteria.

It is important for consumers to understand that when food is irradiated, it never comes in contact with the radioactive source. The gamma rays pass through the food to kill bacteria, but that does not make the food radioactive. The radiation kills bacteria because it stops their ability to divide and grow. We cook or heat food thoroughly for the same purpose. Radiation, as well as heat, has little effect on the food itself because its cells are no longer dividing or growing. Thus, irradiated food is not harmed although a small amount of vitamin B_1 and C may be lost.

Currently, tomatoes, blueberries, strawberries, and mushrooms are irradiated to allow them to be harvested when completely ripe and extend their shelf life (see Figure 9.3). The FDA has also approved the irradiation of pork, poultry, and beef in order to decrease potential infections and to extend shelf life. Currently, irradiated food is available in over 40 countries. In the United States, irradiated foods such as tropical fruit, spinach, and ground meat are found in some supermarkets. *Apollo 17* astronauts ate irradiated foods on the moon, and some U.S. hospitals and nursing homes now use irradiated poultry to reduce the possibility of infections among patients. The extended shelf life of irradiated food also makes it useful for campers and military personnel. Soon consumers concerned about food safety will have a choice of irradiated meats, fruits, and vegetables at the market.

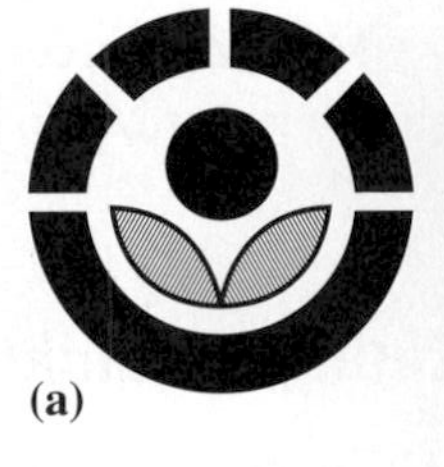
(a)

(b)

FIGURE 9.3 **(a)** The FDA requires this symbol to appear on irradiated retail foods. **(b)** After 2 weeks, the irradiated strawberries on the right show no spoilage. Mold is growing on the nonirradiated ones on the left.

Q Why are irradiated foods used on spaceships and in nursing homes?

SAMPLE PROBLEM 9.6

Radiation Measurement

One treatment for bone pain involves intravenous administration of the radioisotope phosphorus-32, which is incorporated into bone. A typical dose of 7 mCi can produce up to 450 rad in the bone. What is the difference between the units of mCi and rad?

SOLUTION

The millicuries (mCi) indicate the activity of the P-32 in terms of the number of nuclei that break down in 1 second. The radiation absorbed dose (rad) is a measure of amount of radiation absorbed by the bone.

STUDY CHECK 9.6

If phosphorus-32 is a beta emitter, how do the number of rems compare to the rads?

TABLE 9.5 Average Annual Radiation Received by a Person in the United States

Source	Dose (mrem)
Natural	
The ground	20
Air, water, food	30
Cosmic rays	40
Wood, concrete, brick	50
Medical	
Chest X-ray	20
Dental X-ray	20
Mammogram	40
Lumbar spine X-ray	70
Upper gastrointestinal tract X-ray	200
Other	
Nuclear power plants	0.1
Air travel	10
Television	20
Radon	200*

*Varies widely.

TABLE 9.6 Lethal Doses of Radiation for Some Life-Forms

Life-Form	LD_{50} (rem)
Insect	100 000
Bacterium	50 000
Rat	800
Human	500
Dog	300

Exposure to Radiation

Every day, we are exposed to low levels of radiation from naturally occurring radioactive isotopes in the buildings where we live and work, in our food and water, and in the air we breathe. For example, potassium-40 is a naturally occurring isotope that is present in any potassium-containing food. Other naturally occurring radioisotopes in air and food are carbon-14, radon-222, strontium-90, and iodine-131. The average person in the United States is exposed to about 360 mrem of radiation annually. Table 9.5 lists some common sources of radiation.

Another source of background radiation is cosmic radiation produced in space by the Sun. People who live at high altitudes or travel by airplane receive a greater amount of cosmic radiation because there are fewer molecules in the atmosphere to absorb the radiation. For example, a person living in Denver receives about twice the cosmic radiation as a person living in Los Angeles. A person living close to a nuclear power plant normally does not receive much additional radiation, perhaps 0.1 mrem in one year. However, in the accident at the Chernobyl nuclear power plant in 1986 in Ukraine, it is estimated that people in a nearby town received as much as 1 rem/h.

Medical sources of radiation including dental, hip, spine and chest X-rays, and mammograms add to our radiation exposure.

Radiation Sickness

The larger the dose of radiation received at one time, the greater the effect on the body. Exposure to radiation less than 25 rem is not usually detected. Whole-body exposure of 100 rem produces a temporary decrease in the number of white blood cells. If the exposure to radiation is greater than 100 rem, a person may experience symptoms of radiation sickness: nausea, vomiting, fatigue, and a reduction in white-cell count. A whole-body dosage greater than 300 rem can decrease the white-cell count to zero. The victim may have diarrhea, hair loss, and infection. Exposure to radiation of 500 rem is expected to cause death in 50% of the people receiving that dose. This amount of radiation to the whole body is called the *lethal dose for one-half the population*, or the LD_{50}. The LD_{50} varies for different life-forms, as Table 9.6 shows. Radiation dosages of 600 rem or greater would be fatal to all humans within a few weeks.

QUESTIONS AND PROBLEMS

Radiation Measurement

9.23 **a.** How does a Geiger counter detect radiation?
b. What are the SI and common units that describe the activity of a radioactive sample?
c. What are the SI and common units that describe the radiation dose absorbed by tissue?
d. What is meant by the term kilogray?

9.24 **a.** What is background radiation?
b. What are the SI and common units that describe the biological effect of radiation?
c. What is meant by the abbreviations mCi and mrem?
d. Why is a factor used to determine the equivalent dose?

9.25 The recommended dosage of iodine-131 is 4.20 μCi/kg of body weight. How many microcuries of iodine-131 are needed for a 70.0-kg patient with hyperthyroidism?

9.26 **a.** The dosage of technetium-99m for a lung scan is 20. μCi/kg of body mass. How many millicuries of technetium-99m should be given to a 50.0-kg patient? (1 mCi = 1000 μCi)
b. A person receives 50 mrad of gamma radiation. What is that amount in grays? What would be the equivalent dose in millirems?
c. Suppose a person absorbed 50 mrad of alpha radiation. What would be the equivalent dose in millirems? How does it compare with the millirems in part **b**?

9.4 Half-Life of a Radioisotope

LEARNING GOAL

Given the half-life of a radioisotope, calculate the amount of radioisotope remaining after one or more half-lives.

SELF STUDY ACTIVITY
Nuclear Chemistry

The **half-life** of a radioisotope is the amount of time it takes for one-half of a sample to decay. For example, $^{131}_{53}I$ has a half-life of 8.0 days. As $^{131}_{53}I$ decays, it produces a beta particle and the nonradioactive isotope $^{131}_{54}Xe$.

$$^{131}_{53}I \longrightarrow {}^{131}_{54}Xe + {}^{0}_{-1}e$$

Suppose we have a sample that initially contains 20. mg of $^{131}_{53}I$. In 8.0 days, one-half (10. mg) of all the I-131 nuclei in the sample will decay, which leaves 10. mg of I-131. After 16 days (two half-lives), 5.0 mg of the remaining I-131 decays, which leaves 5.0 mg of I-131. After 24 days (three half-lives), 2.5 mg of the remaining I-131 decays, which leaves 2.5 mg of I-131 still capable of producing radiation.

As the I-131 undergoes beta decay, there is also the buildup of the decay product Xe-131 with each half-life. That means that after the first half-life, the decay process produces 10. mg of $^{131}_{54}Xe$ and after a second half-life, there is a total of 15 mg of the product $^{131}_{54}Xe$. After the third half-life, there is a total of 17.5 mg of $^{131}_{54}Xe$.

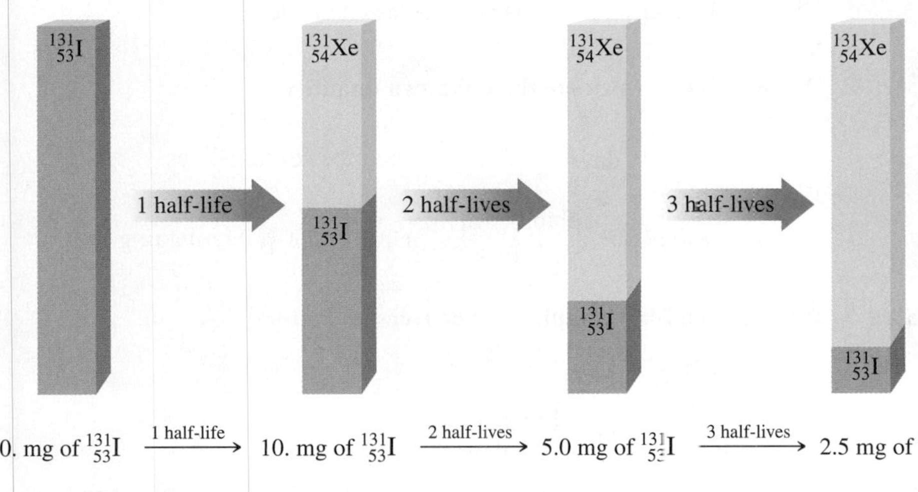

A **decay curve** is a diagram that shows the decay of a radioactive isotope for each half-life. Figure 9.4 shows the decay curve for the $^{131}_{53}I$ we have discussed.

Explore Your World

MODELING HALF-LIVES

Obtain a piece of paper and a licorice stick or celery stalk. Draw a vertical and a horizontal axis on the paper. Label the vertical axis as number of radioactive atoms and the horizontal axis as minutes. Place the licorice stick or celery against the vertical axis and mark its height for zero minutes. In the next minute, cut the licorice stick or celery in two. (You can eat the half if you are hungry.) Place the shortened licorice stick or celery at 1 minute on the horizontal axis and mark its height. Every minute, cut the licorice stick or celery in half again and mark the shorter height at the corresponding time. Keep reducing the length by half until you cannot divide the licorice or celery in half any more. Connect the marks you made for each minute. What does the curve look like? How does this curve represent the concept of a half-life for a radioisotope?

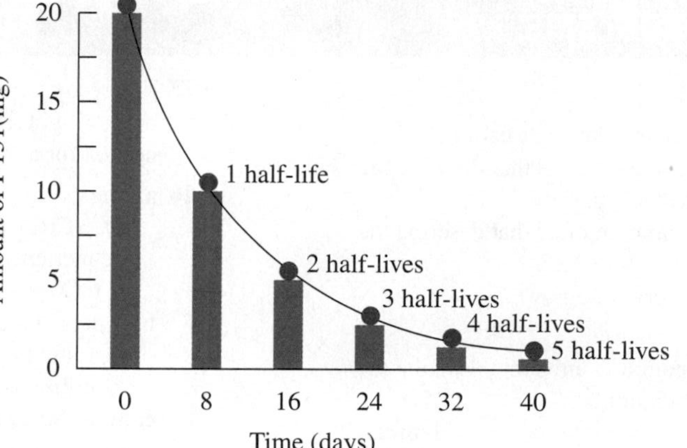

FIGURE 9.4 The decay curve for iodine-131 shows that one-half of the radioactive sample decays and one-half remains radioactive after each half-life of 8 days.

Q How many milligrams of the 20.-mg sample remain radioactive after 2 half-lives?

TUTORIAL
Radioactive Half-Lives

CONCEPT CHECK 9.4

Half-Lives

Iridium-192, used to treat breast cancer, has a half-life of 74 days. What is the activity of the Ir-192 after 74 days if the activity of the initial sample of Ir-192 is 8×10^4 Bq?

ANSWER

In 74 days, which is one half-life of iridium-192, one-half of all of the iridium-192 atoms will decay. Thus, after 74 days, the activity is half of the initial activity, which is 4×10^4 Bq.

Guide to Using Half-Lives

1. State the given and needed quantities.
2. Write a plan to calculate the unknown quantity.
3. Write the half-life equality and conversion factors.
4. Set up the problem to calculate amount of active radioisotope.

SAMPLE PROBLEM 9.7

Using Half-Lives of a Radioisotope

Phosphorus-32, a radioisotope used in the treatment of leukemia, has a half-life of 14.3 days. If a sample contains 8.0 mg of phosphorus-32, how many milligrams of phosphorus-32 remain after 42.9 days?

SOLUTION

Step 1 State the given and needed quantities.

Given 8.0 mg of ${}^{32}_{15}\text{P}$; 42.9 days; 14.3 days/half-life

Need mg of ${}^{32}_{15}\text{P}$ remaining

Step 2 Write a plan to calculate the unknown quantity.

days → Half-life → number of half-lives

milligrams of ${}^{32}_{15}\text{P}$ → Number of half-lives → milligrams of ${}^{32}_{15}\text{P}$ remaining

Step 3 Write the half-life equality and conversion factors.

$$1 \text{ half-life} = 14.3 \text{ days}$$

$$\frac{14.3 \text{ days}}{1 \text{ half-life}} \text{ and } \frac{1 \text{ half-life}}{14.3 \text{ days}}$$

Step 4 Set up the problem to calculate amount of active radioisotope. First, we determine the number of half-lives in the amount of time that has elapsed.

$$\text{Number of half-lives} = 42.9 \cancel{\text{days}} \times \frac{1 \text{ half-life}}{14.3 \cancel{\text{days}}} = 3 \text{ half-lives}$$

Now, we can calculate how much of the sample decays in 3 half-lives and how many milligrams of the phosphorus remain.

$$8.0 \text{ mg of } {}^{32}_{15}\text{P} \xrightarrow{\text{1 half-life}} 4.0 \text{ mg of } {}^{32}_{15}\text{P} \xrightarrow{\text{2 half-lives}} 2.0 \text{ mg of } {}^{32}_{15}\text{P} \xrightarrow{\text{3 half-lives}} 1.0 \text{ mg of } {}^{32}_{15}\text{P}$$

STUDY CHECK 9.7

Iron-59 has a half-life of 44 days. If the laboratory received a sample of 8.0 μg of iron-59, how many micrograms of iron-59 remain active after 176 days?

Naturally occurring isotopes of the elements usually have long half-lives, as shown in Table 9.7. They disintegrate slowly and produce radiation over a long period of time, even hundreds or millions of years. In contrast, many of the radioisotopes used in nuclear medicine have much shorter half-lives. They disintegrate rapidly and produce almost all their radiation in a short period of time. For example, technetium-99m emits half of its radiation in the first 6 h. This means that a small amount of the radioisotope given to a patient is essentially gone within 2 days. The decay products of technetium-99m are totally eliminated from the body.

TABLE 9.7 Half-Lives of Some Radioisotopes

Element	Radioisotope	Half-Life	Radiation
Naturally Occurring			
Carbon-14	${}^{14}_{6}\text{C}$	5730 y	Beta
Potassium-40	${}^{40}_{19}\text{K}$	1.3×10^9 y	Beta, gamma
Radium-226	${}^{226}_{88}\text{Ra}$	1600 y	Alpha
Strontium-90	${}^{90}_{38}\text{Sr}$	38.1 y	Alpha
Uranium-238	${}^{238}_{92}\text{U}$	4.5×10^9 y	Alpha
Medical			
Chromium-51	${}^{51}_{24}\text{Cr}$	28 d	Gamma
Iodine-131	${}^{131}_{53}\text{I}$	8 d	Beta, gamma
Iron-59	${}^{59}_{26}\text{Fe}$	44 d	Beta, gamma
Radon-222	${}^{222}_{86}\text{Rn}$	3.8 d	Alpha
Technetium-99m	${}^{99m}_{43}\text{Tc}$	6.0 h	Gamma

MC TUTORIAL
Radiocarbon Dating

CONCEPT CHECK 9.5

Dating Using Half-Lives

Carbon material in the bones of humans and animals assimilates carbon until death. Using carbon dating, the number of half-lives of carbon-14 from a bone sample determines the age of the bone. Suppose a bone sample is obtained from a prehistoric animal and used for carbon dating. We can calculate the age of the bone or the years elapsed since the animal died by using the half-life of carbon-14 (5730 y). If carbon dating shows that four half-lives have passed, how much time has elapsed since the animal died?

ANSWER

Because we know that four half-lives have passed, we can use the half-life of carbon-14 (5730 y) to calculate the age of the bone sample.

$$1 \text{ half-life} = 5730 \text{ y}$$

$$\frac{5730 \text{ y}}{1 \text{ half-life}} \quad \text{and} \quad \frac{1 \text{ half-life}}{5730 \text{ y}}$$

$$\text{Years elapsed} = 4 \cancel{\text{half-lives}} \times \frac{5730 \text{ y}}{1 \cancel{\text{half-life}}} = 22\,900 \text{ y}$$

We would estimate that the animal lived 22 900 years ago.

The age of a bone sample from a skeleton can be determined by carbon dating.

Chemistry Link to the Environment

DATING ANCIENT OBJECTS

A technique known as radiological dating is used by geologists, archaeologists, and historians as a way to determine the age of ancient objects. The age of an object derived from plants or animals (such as wood, fiber, natural pigments, bone, and cotton or woolen clothing) is determined by measuring the amount of carbon-14, a naturally occurring radioactive form of carbon. In 1960, Willard Libby received the Nobel Prize in Chemistry for the work he did developing carbon-14 dating techniques during the 1940s. Carbon-14 is produced in the upper atmosphere by the bombardment of ${}^{14}_{7}N$ by high-energy neutrons from cosmic rays.

$$\underset{\text{Neutron from cosmic rays}}{{}^{1}_{0}n} + \underset{\text{Nitrogen in atmosphere}}{{}^{14}_{7}N} \longrightarrow \underset{\text{Radioactive carbon-14}}{{}^{14}_{6}C} + \underset{\text{Proton}}{{}^{1}_{1}H}$$

The carbon-14 reacts with oxygen to form radioactive carbon dioxide, ${}^{14}_{6}CO_2$. Living plants continuously absorb carbon dioxide, which incorporates carbon-14 into the plant material. The uptake of carbon-14 stops when the plant dies.

As the carbon-14 decays, the amount of radioactive carbon-14 in the plant material steadily decreases.

$${}^{14}_{6}C \longrightarrow {}^{14}_{7}N + {}^{0}_{-1}e$$

In a process called **carbon dating**, scientists use the half-life of carbon-14 (5730 y) to calculate the length of time since the plant died. For example, a wooden beam found in an ancient Indian dwelling might have one-half of the carbon-14 found in living plants today. Because one half-life of carbon-14 is 5730 y, the dwelling was constructed about 5730 y ago. Carbon-14 dating was used to determine that the Dead Sea Scrolls are about 2000 y old.

A radiological dating method used for determining the age of much older items is based on uranium-238, which decays through a series of reactions to lead-206. The uranium-238 has an incredibly long half-life, about 4×10^9 (4 billion) y. Measurements of the amounts of uranium-238 and lead-206 enable geologists to determine the age of rock samples. The older rocks will have a higher percentage of lead-206 because more of the uranium-238 has decayed. The age of rocks brought back from the moon by the *Apollo* missions, for example, was determined using uranium-238. They were found to be about 4×10^9 y old, approximately the same age calculated for Earth.

The age of the Dead Sea scrolls was determined using carbon-14 dating.

QUESTIONS AND PROBLEMS

Half-Life of a Radioisotope

9.27 What is meant by the term half-life?

9.28 Why are radioisotopes with short half-lives used for diagnosis in nuclear medicine?

9.29 Technetium-99m is an ideal radioisotope for scanning organs because it has a half-life of 6.0 h and is a pure gamma emitter. Suppose that 80.0 mg were prepared in the technetium generator this morning. How many milligrams of technetium-99m would remain active after the following intervals?
a. one half-life **b.** two half-lives
c. 18 h **d.** 24 h

9.30 A sample of sodium-24 with an activity of 12 mCi is used to study the rate of blood flow in the circulatory system. If sodium-24 has a half-life of 15 h, what is the activity of the sodium after 2.5 days?

9.31 Strontium-85, used for bone scans, has a half-life of 65 days. How long will it take for the radiation level of strontium-85 to drop to one-fourth of its original level? To one-eighth?

9.32 Fluorine-18, which has a half-life of 110 min, is used in PET scans. If 100. mg of fluorine-18 is shipped at 8:00 A.M., how many milligrams of the radioisotope are still active if the sample arrives at the radiology laboratory at 1:30 P.M.?

LEARNING GOAL

Describe the use of radioisotopes in medicine.

9.5 Medical Applications Using Radioactivity

To determine the condition of an organ in the body, a radiologist may use a radioisotope that concentrates in that organ. The cells in the body do not differentiate between a nonradioactive atom and a radioactive one. However, radioactive atoms can be detected because they emit radiation. Some radioisotopes used in nuclear medicine are listed in Table 9.8.

TABLE 9.8 Medical Applications of Some Common Radioisotopes

Isotope	Half-Life	Radiation	Medical Application
Au-198	2.7 d	Beta	Liver imaging; treatment of abdominal carcinoma
Ce-141	32.5 d	Gamma	Gastrointestinal tract diagnosis; measuring blood flow to the heart
Cs-131	9.7 d	Gamma	Prostate brachytherapy
F-18	110 min	Positron	Positron emission tomography (PET)
Ga-67	78 h	Gamma	Abdominal imaging; tumor detection
Ga-68	68 min	Gamma	Detection of pancreatic cancer
I-125	60 d	Gamma	Treatment of thyroid, brain, and prostate cancer
I-131	8 d	Beta	Treatment of Graves' disease, goiter, hyperthyroidism, thyroid and prostate cancer
Ir-192	74 d	Gamma	Treatment of breast and prostate cancer
P-32	14.3 d	Beta	Treatment of leukemia, excess red blood cells, pancreatic cancer
Pd-103	17 d	Gamma	Prostate brachytherapy
Sr-85	65 d	Gamma	Detection of bone lesions; brain scans
Tc-99m	6 h	Gamma	Imaging of skeleton and heart muscle, brain, liver, heart, lungs, bone, spleen, kidney, and thyroid; most widely used radioisotope in nuclear medicine

Scans with Radioisotopes

After a person receives a radioisotope, the radiologist determines the level and location of radioactivity emitted by the radioisotope. An apparatus called a scanner is used to produce an image of the organ. The scanner moves slowly across the patient's body above the region where the organ containing the radioisotope is located. The gamma rays emitted from the radioisotope in the organ can be used to expose a photographic plate, producing a **scan** of the organ. On a scan, an area of decreased or increased radiation can indicate such conditions as a disease of the organ, a tumor, a blood clot, or edema.

A common method of determining thyroid function is the use of *radioactive iodine uptake* (RAIU). Taken orally, the radioisotope iodine-131 mixes with the iodine already present in the thyroid. Twenty-four hours later, the amount of iodine taken up by the thyroid is determined. A detection tube held up to the area of the thyroid gland detects the radiation coming from the iodine-131 that has located there (see Figure 9.5).

(a)

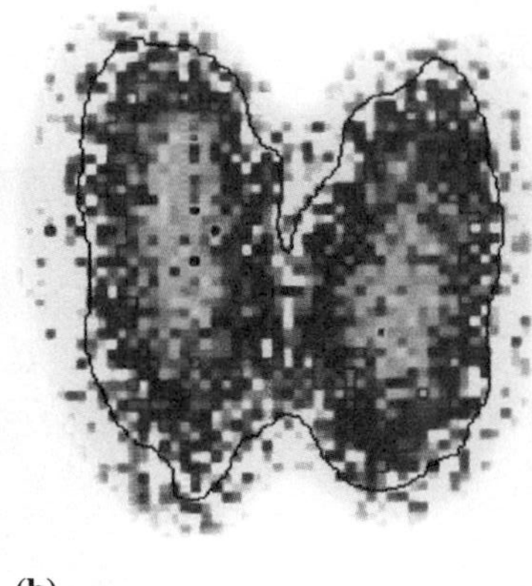

(b)

FIGURE 9.5 **(a)** A scanner is used to detect radiation from a radioisotope that has accumulated in an organ. **(b)** A scan of the thyroid shows the accumulation of radioactive iodine-131 in the thyroid.

Q What type of radiation would move through body tissues to create a scan?

A person with a hyperactive thyroid will have a higher than normal level of radioactive iodine, whereas a patient with a hypoactive thyroid will record low values. If a patient has hyperthyroidism, treatment is begun to lower the activity of the thyroid. One treatment involves giving the patient a therapeutic dosage of radioactive iodine, which has a higher radiation count than the diagnostic dose. The radioactive iodine goes to the thyroid where its radiation destroys some of the thyroid cells. The thyroid produces less thyroid hormone, bringing the hyperthyroid condition under control.

Positron Emission Tomography (PET)

Positron emitters with short half-lives, such as carbon-11, nitrogen-13, oxygen-15, and fluorine-18, are used in an imaging method called *positron emission tomography* (PET). A positron-emitting isotope such as fluorine-18 combined with substances in the body such as glucose is used to study brain function, metabolism, and blood flow.

$${}^{18}_{9}\text{F} \longrightarrow {}^{18}_{8}\text{O} + {}^{0}_{+1}e$$

As positrons are emitted, they combine with electrons to produce gamma rays that are detected by computerized equipment to create a three-dimensional image of the organ (see Figure 9.6).

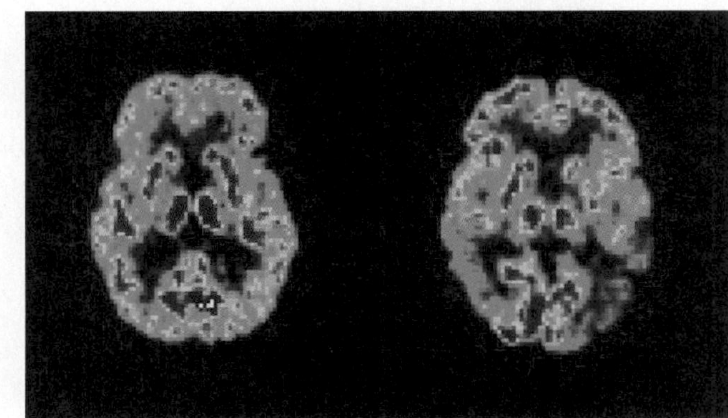

FIGURE 9.6 These PET scans of the brain show a normal brain on the left and a brain affected by Alzheimer's disease on the right.
Q When positrons collide with electrons, what type of radiation is produced that gives an image of an organ?

Chemistry Link to Health

OTHER IMAGING METHODS

Computed Tomography (CT)
Another imaging method used to detect changes within the body is computed tomography (CT). A computer monitors the degree of absorption of 30 000 X-ray beams directed at the brain at successive layers. The differences in absorption based upon the densities of the tissues and fluids in the brain provide a series of images of the brain. This technique is successful in the identification of brain hemorrhages, tumors, and atrophy (see Figure 9.7).

Magnetic Resonance Imaging (MRI)
Magnetic resonance imaging (MRI) is a powerful imaging technique that does not involve X-ray radiation. It is the least invasive imaging method available. MRI is based on the absorption of energy when the protons in hydrogen atoms are excited by a strong magnetic field. Hydrogen atoms make up 63% of all the atoms in the body. In the nucleus, the spin of the single proton acts like a tiny magnet. With no external field, the spins of the protons have random orientations. However, when placed within a large magnet, the spins of the

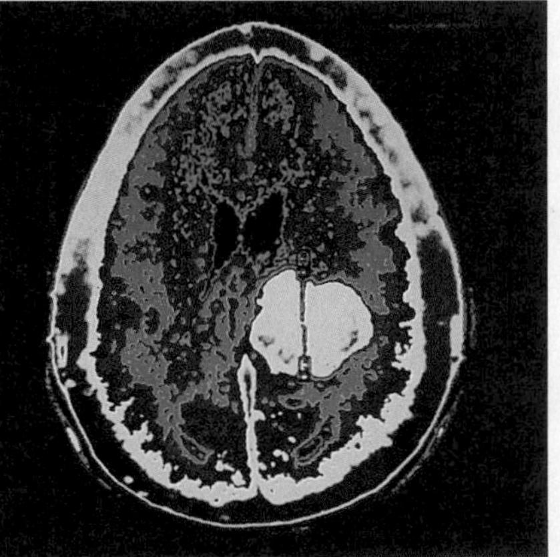

FIGURE 9.7 A CT scan shows a tumor (yellow) in the brain.
Q What is the type of radiation used to give a CT scan?

protons align with the magnetic field. A magnet aligned with the field has a lower energy than one that is aligned against the field. As the MRI scan proceeds, radiofrequency pulses of energy are applied. When a nucleus absorbs certain energy, its proton "flips" and becomes aligned against the field. Because hydrogen atoms in the body are in different chemical environments, energies of different frequencies are absorbed. The energies absorbed are calculated and converted to color images of the body. MRI is particularly useful to image soft tissues because soft tissues contain large amounts of water (see Figure 9.8).

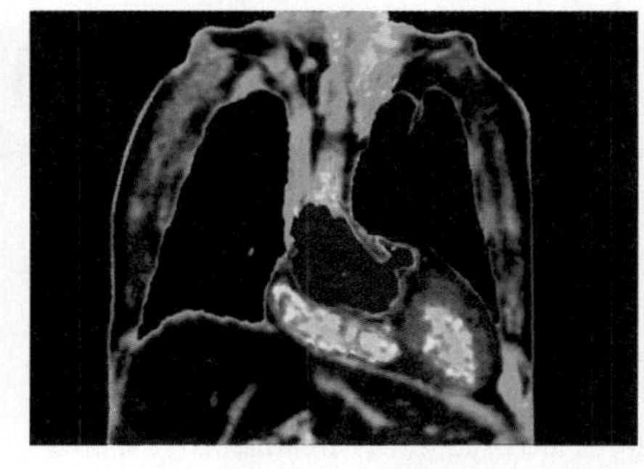

FIGURE 9.8 An MRI scan of the heart and lungs, with the left ventricle shown in red.

Q What is the source of energy in MRI?

CONCEPT CHECK 9.6

Medical Application of Radioactivity

In the determination of thyroid function, a patient receives an oral dose of sodium iodide that contains 10 μCi of iodine-131, which is a beta emitter. Write the balanced nuclear equation for the beta decay of iodine-131.

ANSWER

We can write the incomplete nuclear equation starting with iodine-131, which has the atomic number 53.

$$^{131}_{53}\text{I} \longrightarrow ? + {}^{0}_{-1}e$$

In beta decay, the mass number (131) does not change, but the atomic number of the new nucleus increases by 1. The new atomic number is 54, which is xenon (Xe).

$$^{131}_{53}\text{I} \longrightarrow {}^{131}_{54}\text{Xe} + {}^{0}_{-1}e$$

SAMPLE PROBLEM 9.8

Medical Application of Radioisotopes

In the treatment of abdominal carcinoma, a person is treated with seeds of gold-198, which is a beta emitter. Write the balanced nuclear equation for the beta decay of gold-198.

SOLUTION

We can write the incomplete nuclear equation starting with gold-198, which has the atomic number 79.

$$^{198}_{79}\text{Au} \longrightarrow ? + {}^{0}_{-1}e$$

In beta decay, the mass number, 198, does not change, but the atomic number of the new nucleus increases by one. The new atomic number is 80, which is mercury, Hg.

$$^{198}_{79}\text{Au} \longrightarrow {}^{198}_{80}\text{Hg} + {}^{0}_{-1}e$$

STUDY CHECK 9.8

In an experimental treatment, a person is given boron-10, which is taken up by malignant tumors. When bombarded with neutrons, boron-10 decays by emitting alpha particles that destroy the surrounding tumor cells. Write the balanced nuclear equation for this experimental procedure.

Chemistry Link to Health

BRACHYTHERAPY

The process called *brachytherapy*, or seed implantation, is an internal form of radiation therapy. The prefix *brachy* is from the Greek word for "short distance." With internal radiation, a high dose of radiation is delivered to a cancerous area, while normal tissue sustains minimal damage. Because higher doses are used, fewer treatments of shorter duration are needed. Conventional external treatment delivers a lower dose per treatment, but requires six to eight weeks of treatments.

Permanent Brachytherapy

One of the most common forms of cancer in males is prostate cancer. In addition to surgery and chemotherapy, one treatment option is to place 40 or more titanium capsules or "seeds" in the malignant area. Each seed, which is the size of a small grain of rice, contains radioactive iodine-125, palladium-103, or cesium-131, which decay by gamma emission. The radiation from the seeds destroys the cancer by interfering with the reproduction of cancer cells. Because the radiation targets the cancer cells, there is minimal damage to normal tissues. Ninety percent (90%) of the radioisotopes decay within a few months because they have short half-lives.

Isotope	I-125	Pd-103	Cs-131
Half-life	60 days	17 days	10 days
Time to deliver 90% of radiation	7 months	2 months	1 month

Almost no radiation passes out of the patient's body. The amount of radiation received by a family member is no greater than that received on a long plane flight. The titanium capsules are left in the body permanently, but the products of decay are not radioactive and cause no further damage.

Temporary Brachytherapy

In another type of treatment for prostate cancer, long needles containing iridium-192 are placed in the tumor. However, the needles are removed after 5 to 10 minutes depending on the activity of the iridium isotope. Compared to permanent brachytherapy, temporary brachytherapy can deliver a higher dose of radiation over a shorter time. The procedure may be repeated in a few days.

Brachytherapy is also used following breast cancer lumpectomy. An iridium-192 isotope is inserted into the catheter implanted in the space left by the removal of the tumor. The isotope is removed after 5 to 10 minutes depending on the activity of the iridium source. Radiation is delivered primarily to the tissue surrounding the cavity that contained the tumor and where the cancer is most likely to reoccur. The procedure is repeated twice a day for five days to give an absorbed dose of 34 Gy (3400 rads). The catheter is removed and no radioactive material remains in the body.

In conventional external beam therapy for breast cancer, a patient receives 2 Gy/treatment once a day for 35 days or about 7 weeks, which gives a total absorbed dose of about 70 Gy or 7000 rads. The external beam therapy irradiates the entire breast including the tumor cavity.

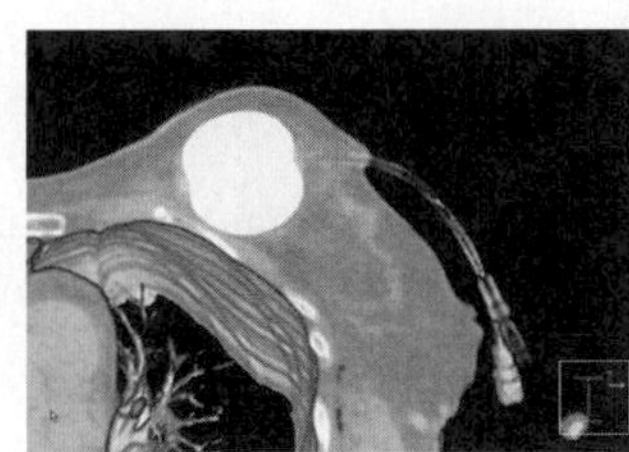

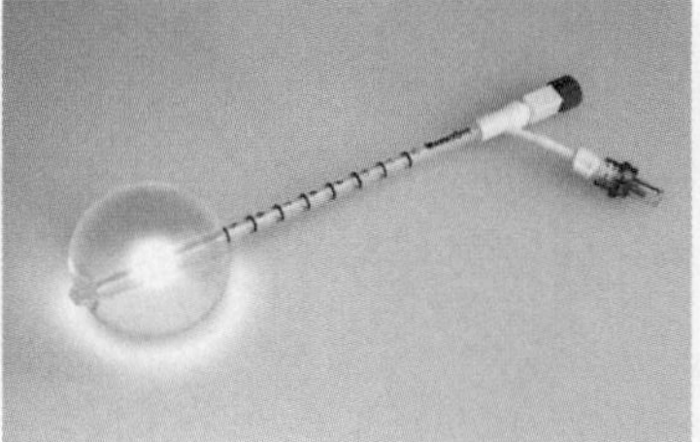

A catheter is placed temporarily in the breast for radiation from Ir-192.

Chemistry Link to Health

RADIATION DOSES IN DIAGNOSTIC AND THERAPEUTIC PROCEDURES

We can compare the levels of radiation exposure commonly used during diagnostic and therapeutic procedures in nuclear medicine. In diagnostic procedures, the radiologist minimizes radiation damage by using the minimum amount of radioisotope needed to evaluate the condition of an organ or tissue. The doses used in radiation therapy are much greater than those used for diagnostic procedures. For example, a therapeutic dose would be used to destroy the cells in a malignant tumor. Although there will be some damage to surrounding tissue, the healthy cells are more resistant to radiation and can repair themselves (see Table 9.9).

TABLE 9.9 Radiation Doses Used for Diagnostic and Therapeutic Procedures

Organ/Condition	Dose (rem)
Diagnostic	
Liver	0.3
Lung	2.0
Thyroid	50.0
Therapeutic	
Lymphoma	4500
Skin cancer	5000–6000
Lung cancer	6000
Brain tumor	6000–7000

QUESTIONS AND PROBLEMS

Medical Applications Using Radioactivity

9.33 Bone and bony structures contain calcium and phosphorus.
a. Why would the radioisotopes calcium-47 and phosphorus-32 be used in the diagnosis and treatment of bone diseases?
b. The radioisotope strontium-89, a beta emitter, is used to treat bone cancer. Write the balanced nuclear equation, and explain why a strontium radioisotope would be used to treat bone cancer.

9.34 **a.** Technetium-99m emits only gamma radiation. Why would this type of radiation be used in diagnostic imaging rather than an isotope that also emits beta or alpha radiation?
b. A person with polycythemia vera (excess production of red blood cells) receives radioactive phosphorus-32. Why would this treatment reduce the production of red blood cells in the bone marrow of the patient?

9.35 In a diagnostic test for leukemia, a person receives 4.0 mL of a solution containing selenium-75. If the activity of the selenium-75 is 45 μCi/mL, what dose, in μCi, is given?

9.36 A vial contains radioactive iodine-131 with an activity of 2.0 mCi/mL. If the thyroid test requires 3.0 mCi in an "atomic cocktail," how many milliliters are used to prepare the iodine-131 solution?

9.6 Nuclear Fission and Fusion

LEARNING GOAL

Describe the processes of nuclear fission and fusion.

TUTORIAL
Fission and Fusion

TUTORIAL
Nuclear Fission and Fusion Reactions

In the 1930s, scientists bombarding uranium-235 with neutrons discovered that the U-235 nucleus splits into two medium-weight nuclei and produces a great amount of energy. This was the discovery of nuclear **fission**. The energy generated by splitting the atom was called atomic energy. When uranium-235 absorbs a neutron, it breaks apart to form two smaller nuclei, several neutrons, and a great amount of energy. A typical equation for nuclear fission is

$$^{1}_{0}n + ^{235}_{92}U \longrightarrow ^{236}_{92}U \longrightarrow ^{91}_{35}Kr + ^{142}_{56}Ba + 3\,^{1}_{0}n + \text{energy}$$

If we could weigh the initial reactants and the products with great accuracy, we would find that their total mass is slightly less than the mass of the starting materials. The missing mass has been converted into energy, consistent with the famous equation derived by Albert Einstein:

$$E = mc^2$$

E is the energy released, m is the mass lost, and c is the speed of light, 3×10^8 m/s. Even though the mass loss is very small, when it is multiplied by the speed of light squared, the result is a large value for the energy released. The fission of one gram of uranium-235 produces about as much energy as the burning of three tons of coal.

Chain Reaction

Fission begins when a neutron collides with the nucleus of a uranium atom. The resulting nucleus is unstable and splits into smaller nuclei. This fission process also releases several neutrons and large amounts of gamma radiation and energy. The neutrons emitted have high energies and bombard more uranium-235 nuclei. As fission continues, there is a rapid increase in the number of high-energy neutrons capable of splitting more

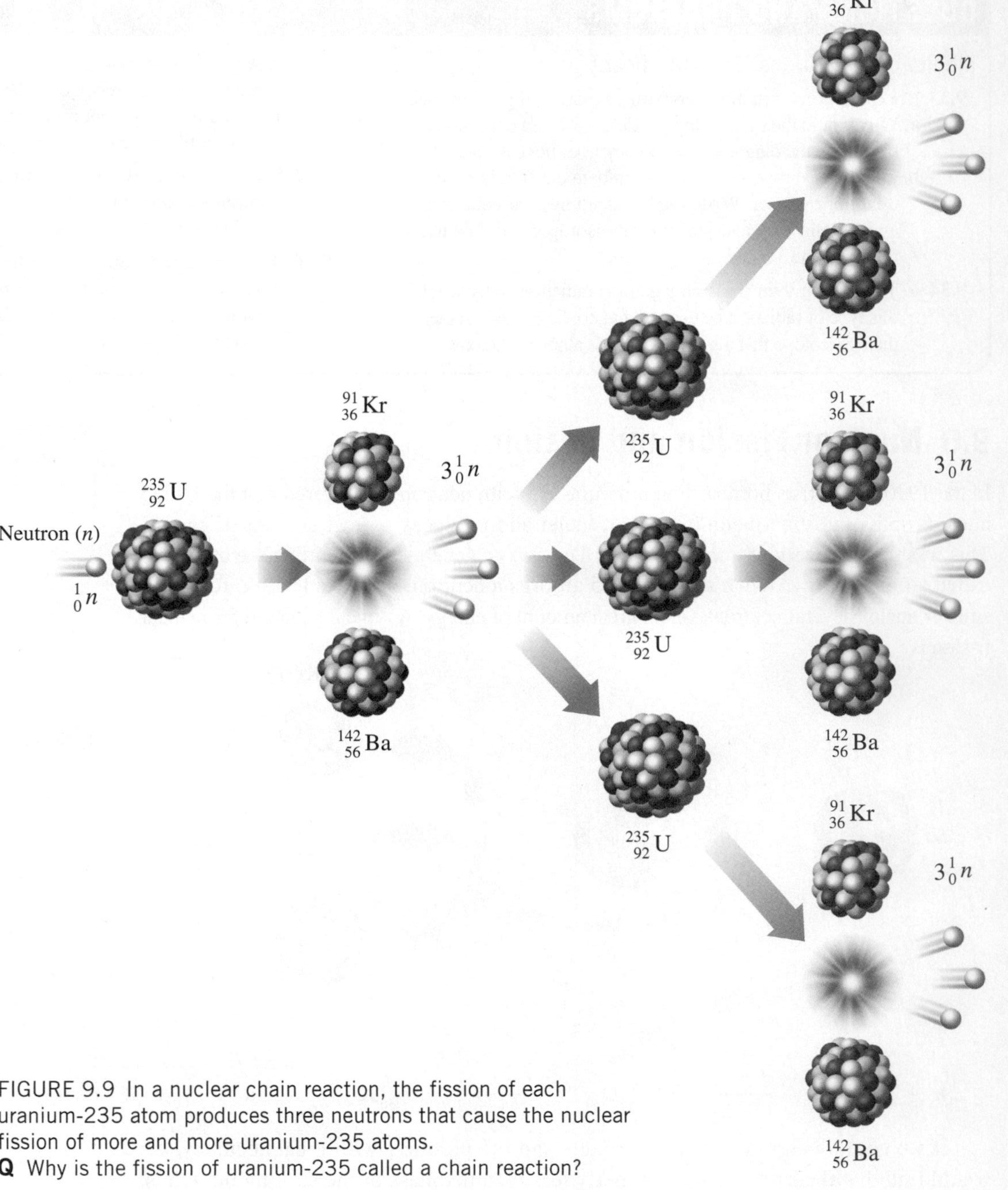

FIGURE 9.9 In a nuclear chain reaction, the fission of each uranium-235 atom produces three neutrons that cause the nuclear fission of more and more uranium-235 atoms.
Q Why is the fission of uranium-235 called a chain reaction?

uranium atoms, a process called a **chain reaction**. To sustain a nuclear chain reaction, sufficient quantities of uranium-235 must be brought together to provide a *critical mass* in which almost all the neutrons immediately collide with more uranium-235 nuclei. So much heat and energy are released that an atomic explosion can occur (see Figure 9.9).

Nuclear Fusion

In **fusion**, two small nuclei such as those in hydrogen combine to form a larger nucleus. Mass is lost, and a tremendous amount of energy is released, even more than the energy released from nuclear fission. However, a fusion reaction requires a temperature of 100 000 000 °C to overcome the repulsion of the hydrogen nuclei and cause

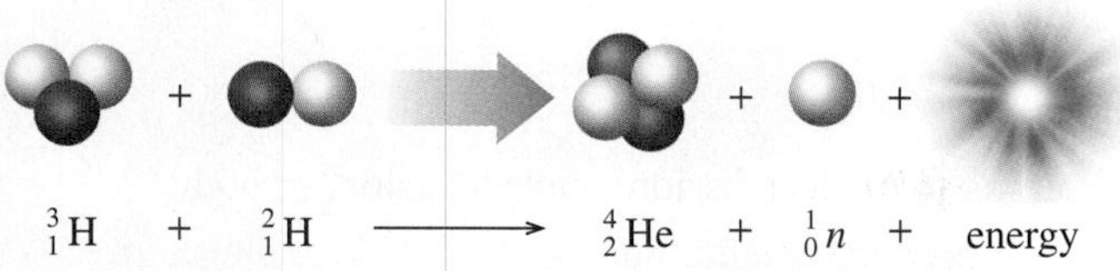

$$^{3}_{1}\text{H} + ^{2}_{1}\text{H} \longrightarrow ^{4}_{2}\text{He} + ^{1}_{0}n + \text{energy}$$

Hydrogen atoms combine in a fusion reactor at very high temperatures.

them to undergo fusion. Fusion reactions occur continuously in the Sun and other stars, providing us with heat and light. The huge amounts of energy produced by our Sun come from the fusion of 6×10^{11} kg of hydrogen every second. In a fusion reaction, isotopes of hydrogen combine to form helium and large amounts of energy.

Scientists expect less radioactive waste with shorter half-lives from fusion reactors. However, fusion is still in the experimental stage because the extremely high temperatures needed have been difficult to attain and even more difficult to maintain. Research groups around the world are attempting to develop the technology needed to make the harnessing of the fusion reaction a reality in our lifetime.

Chemistry Link to the Environment

NUCLEAR POWER PLANTS

In a nuclear power plant, the quantity of uranium-235 is held below a critical mass, so it cannot sustain a chain reaction. The fission reactions are slowed by placing control rods, which absorb some of the fast-moving neutrons, among the uranium samples. In this way, less fission occurs, and there is a slower, controlled production of energy. The heat from the controlled fission is used to produce steam. The steam drives a generator, which produces electricity. Approximately 10% of the electrical energy produced in the United States is generated in nuclear power plants.

Although nuclear power plants help meet some of our energy needs, there are some problems. One of the most serious problems is the production of radioactive by-products that have very long half-lives. It is essential that these waste products be stored safely for a very long time in a place where they do not contaminate the environment. Early in 1990, the EPA gave its approval for the storage of radioactive hazardous wastes in chambers 2150 ft underground. In 1998, the Waste Isolation Pilot Plant (WIPP) repository site in New Mexico was ready to receive plutonium waste from former U.S. bomb factories. Although authorities claim the caverns are safe, some people are concerned with the safe transport of the radioactive waste by trucks on the highways.

Nuclear power plants supply about 10% of the electricity in the United States.

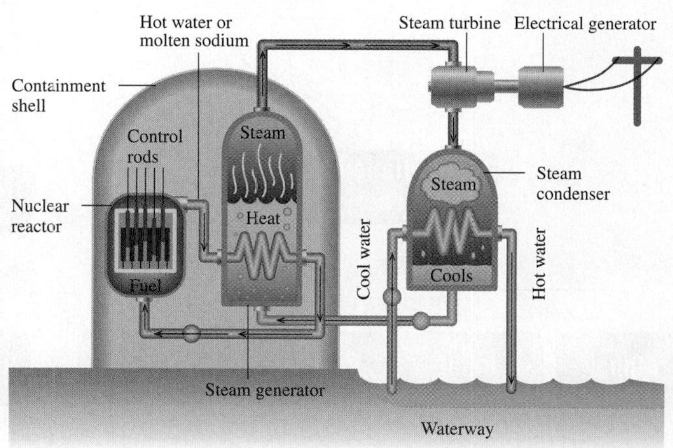

Heat from nuclear fission is used to generate electricity.

CONCEPT CHECK 9.7

Identifying Fission and Fusion

Classify the following as pertaining to nuclear fission, nuclear fusion, or both:

a. A large nucleus breaks apart to produce smaller nuclei.
b. Large amounts of energy are released.
c. Very high temperatures are needed for reaction.
d. $^{3}_{1}H + ^{2}_{1}H \longrightarrow ^{4}_{2}He + ^{1}_{0}n + \text{energy}$

ANSWER

a. When a large nucleus breaks apart to produce smaller nuclei, the process is fission.
b. Large amounts of energy are generated in both the fusion and fission processes.
c. An extremely high temperature is required for fusion.
d. When small nuclei of hydrogen atoms combine, the process is fusion.

QUESTIONS AND PROBLEMS

Nuclear Fission and Fusion

9.37 What is nuclear fission?

9.38 How does a chain reaction occur in nuclear fission?

9.39 Complete the following fission reaction:

$$^{1}_{0}n + ^{235}_{92}U \longrightarrow ^{131}_{50}Sn + ? + 2^{1}_{0}n + \text{energy}$$

9.40 In another fission reaction, uranium-235 bombarded with a neutron produces strontium-94, another small nucleus, and three neutrons. Write the complete equation for the fission reaction.

9.41 Indicate whether each of the following is characteristic of the fission or fusion process or both:

a. Neutrons bombard a nucleus.
b. This nuclear process occurs in the Sun.
c. A large nucleus splits into smaller nuclei.
d. Small nuclei combine to form larger nuclei.

9.42 Indicate whether each of the following is characteristic of the fission or fusion process or both:

a. Very high temperatures are required to initiate the reaction.
b. Less radioactive waste is produced.
c. Hydrogen nuclei are the reactants.
d. Large amounts of energy are released when the nuclear reaction occurs.

CONCEPT MAP

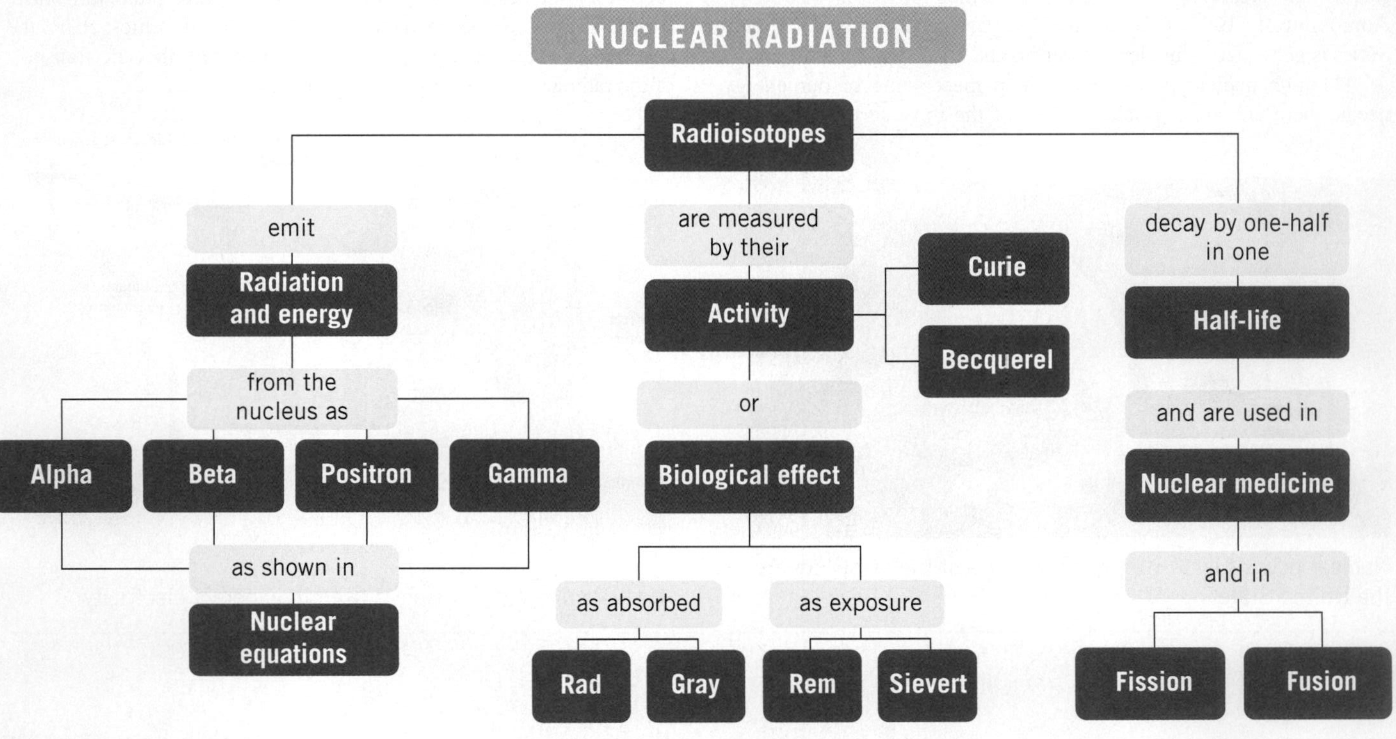

CHAPTER REVIEW

9.1 Natural Radioactivity

Learning Goal: Describe alpha, beta, positron, and gamma radiation.

Radioactive isotopes have unstable nuclei that break down (decay), spontaneously emitting alpha (α), beta (β), positron (β^+), and gamma (γ) radiation. Because radiation can damage the cells in the body, proper protection must be used: shielding, limiting the time of exposure, and distance.

9.2 Nuclear Equations

Learning Goal: Write an equation showing mass numbers and atomic numbers for radioactive decay.

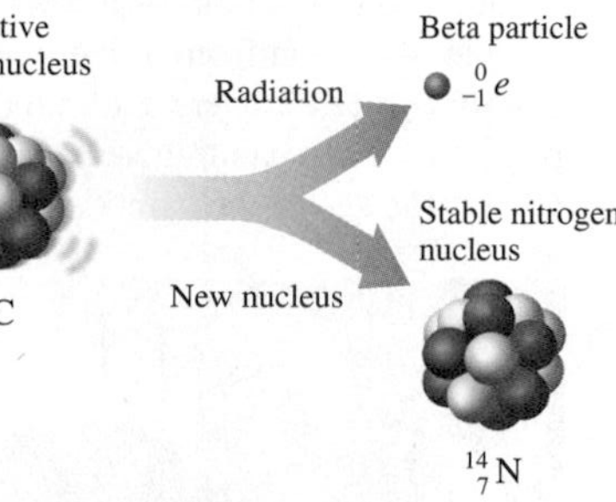

A balanced equation is used to represent the changes that take place in the nuclei of the reactants and products. The new isotopes and the type of radiation emitted can be determined from the symbols that show the mass numbers and atomic numbers of the isotopes in the nuclear equation. A radioisotope is produced artificially when a nonradioactive isotope is bombarded by a small particle. Many radioactive isotopes used in nuclear medicine are produced in this way.

9.3 Radiation Measurement

Learning Goal: Describe the detection and measurement of radiation.

In a Geiger counter, radiation produces charged particles in the gas in a metal tube, which produces an electrical current. The curie (Ci) and the becquerel (Bq) measure the activity, which is the number of nuclear transformations per second. The amount of radiation absorbed by a substance is measured in rads or the gray (Gy). The rem and the sievert (Sv) are units used to determine the biological damage from the different types of radiation.

9.4 Half-Life of a Radioisotope

Learning Goal: Given the half-life of a radioisotope, calculate the amount of radioisotope remaining after one or more half-lives.

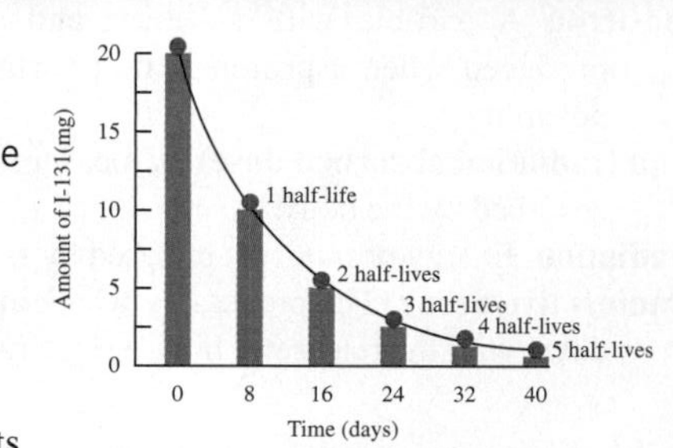

Every radioisotope has its own rate of emitting radiation. The time it takes for one-half of a radioactive sample to decay is called its half-life. For many medical radioisotopes, such as Tc-99m and I-131, half-lives are short. For other isotopes, usually naturally occurring ones such as C-14, Ra-226, and U-238, half-lives are extremely long.

9.5 Medical Applications Using Radioactivity

Learning Goal: Describe the use of radioisotopes in medicine.

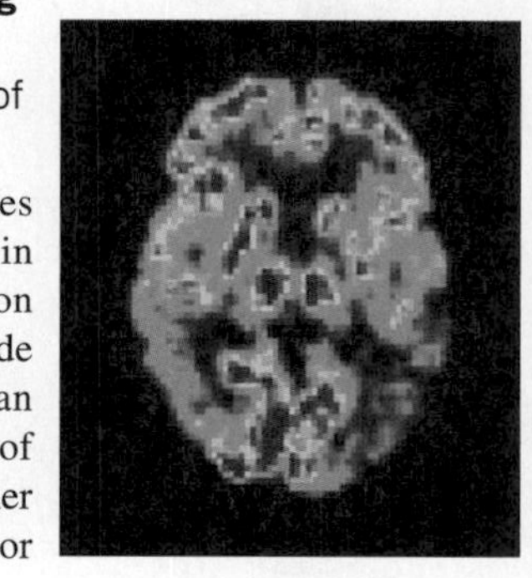

In nuclear medicine, radioisotopes are given that go to specific sites in the body. By detecting the radiation they emit, an evaluation can be made about the location and extent of an injury, disease, tumor, or the level of function of a particular organ. Higher levels of radiation are used to treat or destroy tumors.

9.6 Nuclear Fission and Fusion

Learning Goal: Describe the processes of nuclear fission and fusion.

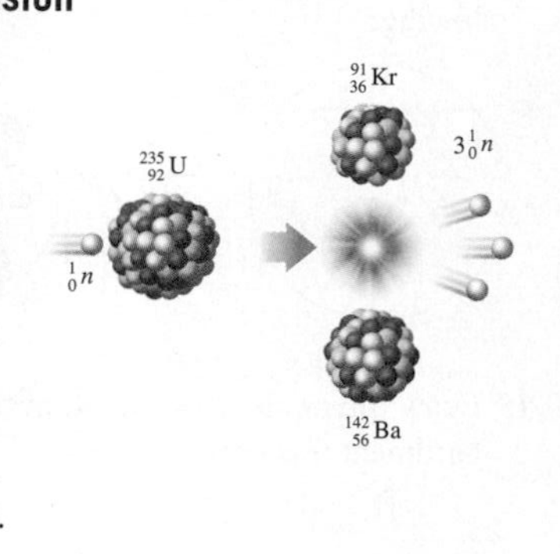

In fission, a large nucleus breaks apart into smaller pieces, releasing one or more types of radiation and a great amount of energy. In fusion, small nuclei combine to form a larger nucleus, while great amounts of energy are released.

Key Terms

alpha particle A nuclear particle identical to a helium nucleus (two protons and two neutrons) with symbol α or $^{4}_{2}He$.

becquerel (Bq) A unit of activity of a radioactive sample equal to one disintegration per second.

beta particle A particle identical to an electron with symbol β or $^{0}_{-1}e$ that forms in the nucleus when a neutron changes to a proton and an electron.

carbon dating A technique used to date ancient specimens that contain carbon. The age is determined by the amount of active carbon-14 that remains in the sample.

chain reaction A fission reaction that will continue once it has been initiated by a high-energy neutron bombarding a heavy nucleus such as U-235.

curie (Ci) A unit of the activity of a radioactive sample equal to 3.7×10^{10} disintegrations/s.

decay curve A diagram of the decay of a radioactive element.

equivalent dose The measure of biological damage from an absorbed dose that has been adjusted for the type of radiation.

fission A process in which large nuclei are split into smaller pieces, releasing large amounts of energy.

fusion A reaction in which large amounts of energy are released when small nuclei combine to form larger nuclei.

gamma ray High-energy radiation with symbol $^{0}_{0}\gamma$ emitted by an unstable nucleus.

gray (Gy) A unit of absorbed dose equal to 100 rad.

half-life The length of time it takes for one-half of a radioactive sample to decay.

positron A particle with no mass and a positive charge ($^{0}_{+1}e$) produced when a proton is transformed into a neutron and a positron.

rad (radiation absorbed dose) A measure of an amount of radiation absorbed by the body.

radiation Energy or particles released by radioactive atoms.

radioactive decay The process by which an unstable nucleus breaks down with the release of high-energy radiation.

rem (radiation equivalent in humans) A measure of the biological damage caused by the various kinds of radiation (rad × radiation biological factor).

scan The image of a site in the body created by the detection of radiation from radioactive isotopes that have accumulated in that site.

shielding Materials used to provide protection from radioactive sources.

sievert (Sv) A unit of biological damage (equivalent dose) equal to 100 rem.

Understanding the Concepts

9.43 Consider the following nucleus of a radioactive isotope:

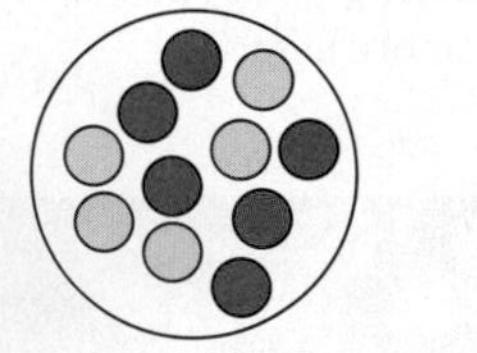

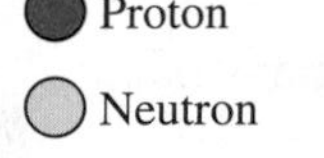

a. What is the nuclear symbol for this isotope?
b. Draw the resulting nucleus if this isotope emits a positron.

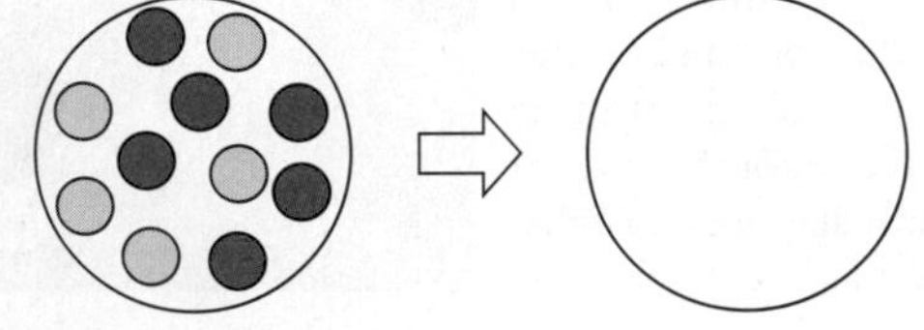

9.44 Draw the nucleus that emits a beta particle to complete the following:

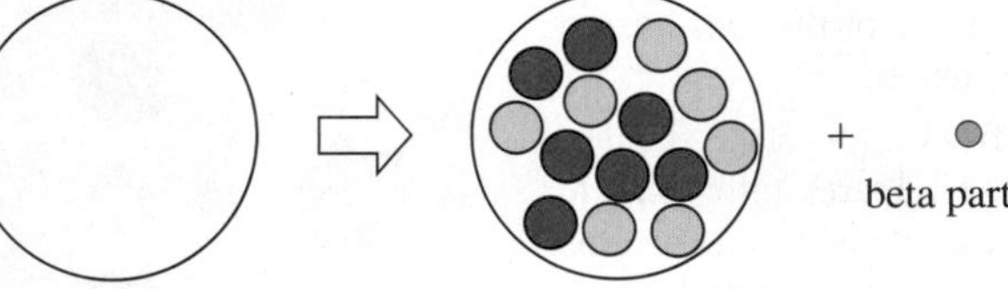

9.45 Draw the nucleus of the atom to complete the following bombardment reaction:

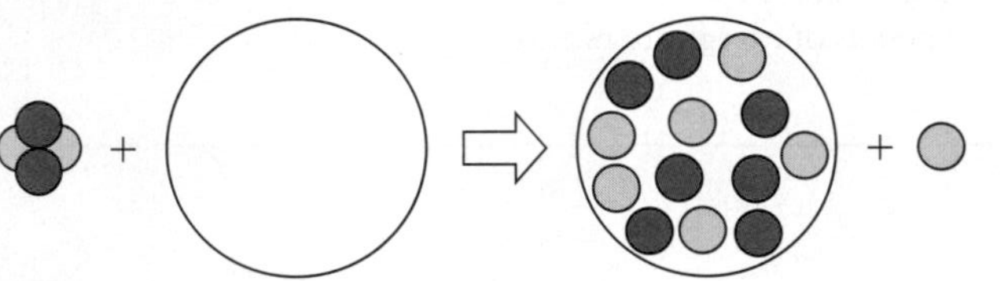

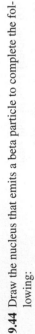

9.46 Complete the following bombardment reaction by drawing the nucleus of the atom produced:

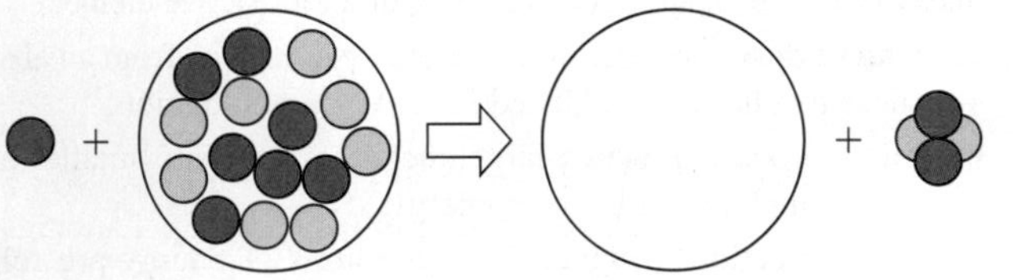

9.47 Carbon dating of small bits of charcoal used in cave paintings has determined that some of the paintings are from 10 000 to 30 000 y old. Carbon-14 has a half-life of 5730 y. In a 1-μg sample of carbon from a live tree, the activity of carbon-14 is 6.4 μCi. If researchers determine that 1 μg of charcoal from a prehistoric cave painting in France has an activity of 0.80 μCi, what is the age of the painting?

The technique of carbon dating is used to determine the age of ancient cave paintings.

9.48 Using the decay curve for iodine-131, determine the following:

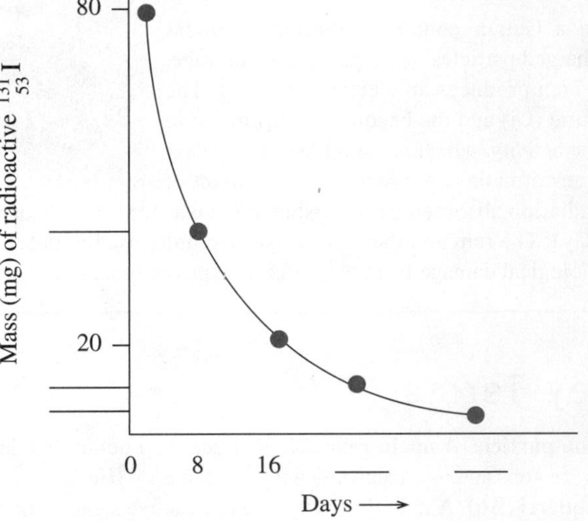

a. Complete the values for the mass of radioactive I-131 on the vertical axis.
b. Complete the number of days on the horizontal axis.
c. Calculate the half-life, in days, of iodine-131.

Additional Questions and Problems

For instructor-assigned homework, go to **www.masteringchemistry.com**.

9.49 State the number of protons and the number of neutrons in the nucleus of each of the following:
a. sodium-25 **b.** nickel-61
c. rubidium-84 **d.** silver-110

9.50 State the number of protons and the number of neutrons in the nucleus of each of the following:
a. boron-10 **b.** zinc-72
c. iron-59 **d.** gold-198

9.51 Describe alpha, beta, and gamma radiation in terms of the following:
a. radiation **b.** symbols

9.52 Describe alpha, beta, and gamma radiation in terms of the following:
a. depth of tissue penetration
b. shielding needed for protection

9.53 Identify each of the following as alpha decay, beta decay, positron emission, or gamma emission:
a. $^{27m}_{13}Al \longrightarrow ^{27}_{13}Al + ^{0}_{0}\gamma$
b. $^{8}_{5}B \longrightarrow ^{8}_{4}Be + ^{0}_{+1}e$
c. $^{220}_{86}Rn \longrightarrow ^{216}_{84}Po + ^{4}_{2}He$

9.54 Identify each of the following as alpha decay, beta decay, positron emission, or gamma emission:
a. $^{127}_{55}Cs \longrightarrow ^{127}_{54}Xe + ^{0}_{+1}e$
b. $^{90}_{38}Sr \longrightarrow ^{90}_{39}Y + ^{0}_{-1}e$
c. $^{218}_{85}At \longrightarrow ^{214}_{83}Bi + ^{4}_{2}He$

9.55 Write a balanced nuclear equation for each of the following:
a. Th-225 (α decay) **b.** Bi-210 (α decay)
c. cesium-137 (β decay) **d.** tin-126 (β decay)
e. F-18 (β^+ emission)

9.56 Write a balanced nuclear equation for each of the following:
a. potassium-40 (β decay) **b.** sulfur-35 (β decay)
c. platinum-190 (α decay) **d.** Ra-210 (α decay)
e. In-113m (γ emission)

9.57 Complete each of the following nuclear equations:
a. $^{4}_{2}He + ^{14}_{7}N \longrightarrow ? + ^{1}_{1}H$
b. $^{4}_{2}He + ^{27}_{13}Al \longrightarrow ^{30}_{14}Si + ?$
c. $^{1}_{0}n + ^{235}_{92}U \longrightarrow ^{90}_{38}Sr + 3^{1}_{0}n + ?$

9.58 Complete each of the following nuclear equations:
a. $? + ^{59}_{27}Co \longrightarrow ^{56}_{25}Mn + ^{4}_{2}He$
b. $? \longrightarrow ^{14}_{7}N + ^{0}_{-1}e$
c. $^{0}_{-1}e + ^{76}_{36}Kr \longrightarrow ?$

9.59 Write the balanced nuclear equation for the following:
a. When two oxygen-16 atoms collide, one of the products is an alpha particle.
b. When californium-249 is bombarded by oxygen-18, a new isotope and four neutrons are produced.
c. Radon-222 emits an alpha particle.

9.60 Write the balanced nuclear equation for the following:
a. Polonium-210 decays to give lead-206.
b. Bismuth-211 emits an alpha particle.
c. A radioisotope emits a positron to form titanium-48.

9.61 If the amount of radioactive phosphorus-32 in a sample decreases from 1.2 mg to 0.30 mg in 28 days, what is the half-life, in days, of phosphorus-32?

9.62 If the amount of radioactive iodine-123 in a sample decreases from 0.4 mg to 0.1 mg in 26.2 h, what is the half-life, in hours, of iodine-123?

9.63 Calcium-47, a beta emitter, has a half-life of 4.5 days.
a. Write the balanced nuclear equation for the beta decay of calcium-47.
b. How many milligrams of a 16-mg sample of calcium-47 would remain after 18 days?
c. How many days have passed if 4.8 mg of calcium-47 decayed to 1.2 mg of calcium-47?

9.64 Cesium-137, a beta emitter, has a half-life of 30 y.
a. Write the balanced nuclear equation for the beta decay of cesium-137.
b. How many milligrams of a 16-mg sample of cesium-137 would remain after 90 y?
c. How many years are required for 28 mg of cesium-137 to decay to 3.5 mg of cesium-137?

9.65 A nuclear technician was accidentally exposed to potassium-42 while doing some brain scans for possible tumors. The error was not discovered until 36 h later when the activity of the potassium-42 sample was 2.0 μCi. If potassium-42 has a half-life of 12 h, what was the activity of the sample at the time the technician was exposed?

9.66 A wooden object from the site of an ancient temple has a carbon-14 activity of 10 counts/min, compared with a reference piece of wood cut today that has an activity of 40 counts/min. If the half-life for carbon-14 is 5730 y, what is the age of the ancient wood object?

9.67 A 120-mg sample of technetium-99m is used for a diagnostic test. If technetium-99m has a half-life of 6.0 h, how much of the technetium-99m sample remains 24 h after the test?

9.68 The half-life of oxygen-15 is 124 s. If a sample of oxygen-15 has an activity of 4000 Bq, how many minutes will elapse before it reaches an activity of 500 Bq?

9.69 What is the purpose of irradiating meats, fruits, and vegetables?

9.70 The irradiation of foods was approved in the United States in the 1980s.
a. Why have we not seen many irradiated products in our markets?
b. Would you buy foods that have been irradiated? Why or why not?

9.71 What is the difference between fission and fusion?

9.72 **a.** What are the products in the fission of uranium-235 that make possible a nuclear chain reaction?
b. What is the purpose of placing control rods among uranium samples in a nuclear reactor?

9.73 Where does fusion occur naturally?

9.74 Why are scientists continuing to try to build a fusion reactor even though very high temperatures required have been difficult to reach and maintain?

Challenge Questions

9.75 Uranium-238 decays in a series of nuclear changes until stable lead-206 is produced. Complete the following nuclear equations that are part of the uranium-238 decay series:

a. $^{238}_{92}U \longrightarrow ^{234}_{90}Th + ?$

b. $^{234}_{90}Th \longrightarrow ? + ^{0}_{-1}e$

c. $? \longrightarrow ^{222}_{86}Rn + ^{4}_{2}He$

9.76 The iceman known as "Ötzi" was discovered in a high mountain pass on the Austrian–Italian border. Samples of his hair and bones had carbon-14 activity that was about 50% of that present in new hair or bone. Carbon-14 is a beta emitter.

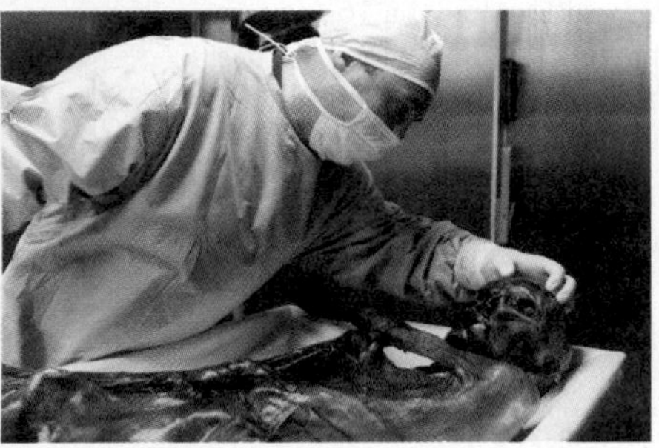

The mummified remains of "Ötzi" were discovered in 1991.

a. How long ago did "Ötzi" live if the half-life for C-14 is 5730 y?

b. Write a balanced nuclear equation for the decay of carbon-14.

9.77 Complete and balance each of the following nuclear equations:

a. $^{23m}_{12}Mg \longrightarrow ? + ^{0}_{0}\gamma$

b. $^{61}_{30}Zn \longrightarrow ^{61}_{29}Cu + ?$

c. C-12 + Cf-249 $\longrightarrow$? + 4 neutrons

9.78 Complete and balance each of the following nuclear equations:

a. Bi-209 + Cr-54 $\longrightarrow ? + ^{1}_{0}n$

b. Np-237 undergoes alpha decay.

c. $^{4}_{2}He + ^{241}_{95}Am \longrightarrow ? + 2^{1}_{0}n$

9.79 The half-life for the radioactive decay of calcium-47 is 4.5 days. If a sample has an activity of 4.0 μCi after 18 days, what was the initial activity of the sample?

9.80 A 16-μg sample of sodium-24 decays to 2.0 μg in 45 h. What is the half-life of sodium-24?

Answers

Answers to Study Checks

9.1 Limiting the time one spends near a radioactive source and staying as far away as possible will reduce exposure to radiation.

9.2 $^{214}_{84}Po \longrightarrow ^{210}_{82}Pb + ^{4}_{2}He$

9.3 $^{51}_{24}Cr \longrightarrow ^{51}_{25}Mn + ^{0}_{-1}e$

9.4 $^{118}_{54}Xe \longrightarrow ^{118}_{53}I + ^{0}_{+1}e$

9.5 $^{4}_{2}He + ^{27}_{13}Al \longrightarrow ^{30}_{15}P + ^{1}_{0}n$

9.6 For β, the factor is 1; rads and rems are equal.

9.7 0.50 μg

9.8 $^{1}_{0}n + ^{10}_{5}B \longrightarrow ^{7}_{3}Li + ^{4}_{2}He$

Answers to Selected Questions and Problems

9.1 a. Both an alpha particle and a helium nucleus have two protons and two neutrons.

b. α, $^{4}_{2}He$

c. An alpha particle is emitted from an unstable nucleus during radioactive decay.

9.3 a. $^{39}_{19}K$, $^{40}_{19}K$, $^{41}_{19}K$

b. They all have 19 protons and 19 electrons, but they differ in the number of neutrons.

9.5

Medical Use	Atomic Symbol	Mass Number	Number of Protons	Number of Neutrons
Heart imaging	$^{201}_{81}Tl$	201	81	120
Radiation therapy	$^{60}_{27}Co$	60	27	33
Abdominal scan	$^{67}_{31}Ga$	67	31	36
Hyperthyroidism	$^{131}_{53}I$	131	53	78
Leukemia treatment	$^{32}_{15}P$	32	15	17

9.7 a. α, $^{4}_{2}He$

b. $^{1}_{0}n$

c. β, $^{0}_{-1}e$

d. $^{15}_{7}N$

e. $^{125}_{53}I$

9.9 a. β, $^{0}_{-1}e$

b. α, $^{4}_{2}He$

c. $^{1}_{0}n$

d. $^{24}_{11}Na$

e. $^{14}_{6}C$

9.11 a. Because β particles have less mass and move faster than α particles, they can penetrate farther into body tissue.
b. Ionizing radiation produces ions that cause undesirable reactions in the cells.
c. Radiation technicians leave the room to increase the distance between themselves and the radiation. Also, a wall that contains lead shields them.
d. Wearing gloves shields the skin from α and β radiation.

9.13 a. ${}^{208}_{84}Po \longrightarrow {}^{204}_{82}Pb + {}^{4}_{2}He$
b. ${}^{232}_{90}Th \longrightarrow {}^{228}_{88}Ra + {}^{4}_{2}He$
c. ${}^{251}_{102}No \longrightarrow {}^{247}_{100}Fm + {}^{4}_{2}He$
d. ${}^{220}_{86}Rn \longrightarrow {}^{216}_{84}Po + {}^{4}_{2}He$

9.15 a. ${}^{25}_{11}Na \longrightarrow {}^{25}_{12}Mg + {}^{0}_{-1}e$
b. ${}^{20}_{8}O \longrightarrow {}^{20}_{9}F + {}^{0}_{-1}e$
c. ${}^{92}_{38}Sr \longrightarrow {}^{92}_{39}Y + {}^{0}_{-1}e$
d. ${}^{60}_{26}Fe \longrightarrow {}^{60}_{27}Co + {}^{0}_{-1}e$

9.17 a. ${}^{26}_{14}Si \longrightarrow {}^{26}_{13}Al + {}^{0}_{+1}e$
b. ${}^{54}_{27}Co \longrightarrow {}^{54}_{26}Fe + {}^{0}_{+1}e$
c. ${}^{77}_{37}Rb \longrightarrow {}^{77}_{36}Kr + {}^{0}_{+1}e$
d. ${}^{93}_{45}Rh \longrightarrow {}^{93}_{44}Ru + {}^{0}_{+1}e$

9.19 a. ${}^{28}_{14}Si$, beta decay
b. ${}^{0}_{0}\gamma$, gamma emission
c. ${}^{0}_{-1}e$, beta decay
d. ${}^{298}_{92}U$, alpha decay
e. ${}^{188}_{79}Au$, positron emission

9.21 a. ${}^{10}_{4}Be$
b. ${}^{132}_{53}I$
c. ${}^{27}_{13}Al$
d. ${}^{30}_{15}P$

9.23 a. When radiation enters the Geiger counter, it ionizes a gas in the detection tube, which produces an electrical current that is detected by the instrument.
b. becquerel (Bq), curie (Ci)
c. gray (Gy), rad
d. 1000 Gy

9.25 294 μCi

9.27 A half-life is the time it takes for one-half of a radioactive sample to decay.

9.29 a. 40.0 mg
b. 20.0 mg
c. 10.0 mg
d. 5.00 mg

9.31 130 days, 195 days

9.33 a. Since the elements Ca and P are part of the bone, the radioactive isotopes of Ca and P will also become part of the bony structures of the body where their radiation can be used to diagnose or treat bone diseases.
b. ${}^{89}_{38}Sr \longrightarrow {}^{89}_{39}Y + {}^{0}_{-1}e$

Strontium (Sr) acts much like calcium (Ca) because both are Group 2A (2) elements. The body will accumulate radioactive strontium in bones in the same way that it incorporates calcium. Once the strontium isotope is absorbed by the bone, the beta radiation will destroy cancer cells.

9.35 180 μCi

9.37 Nuclear fission is the splitting of a large atom into smaller fragments with the release of large amounts of energy.

9.39 ${}^{103}_{42}Mo$

9.41 a. fission
b. fusion
c. fission
d. fusion

9.43 a. ${}^{11}_{6}C$
b.

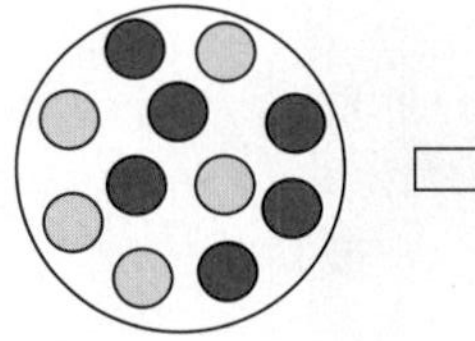
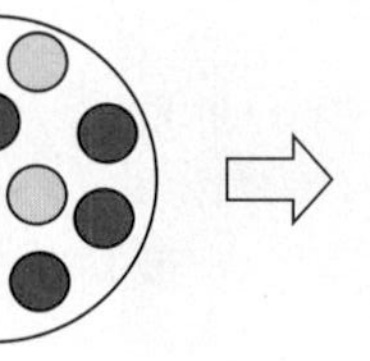
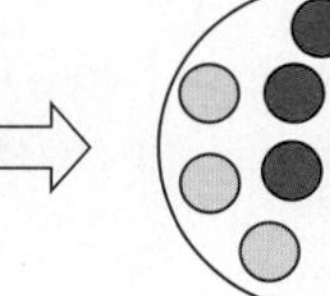
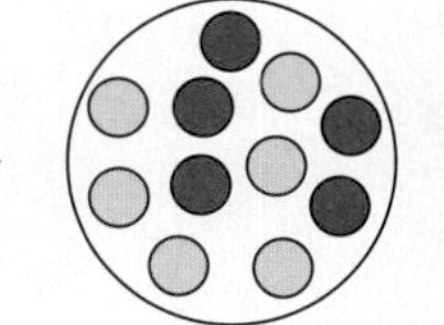

9.45

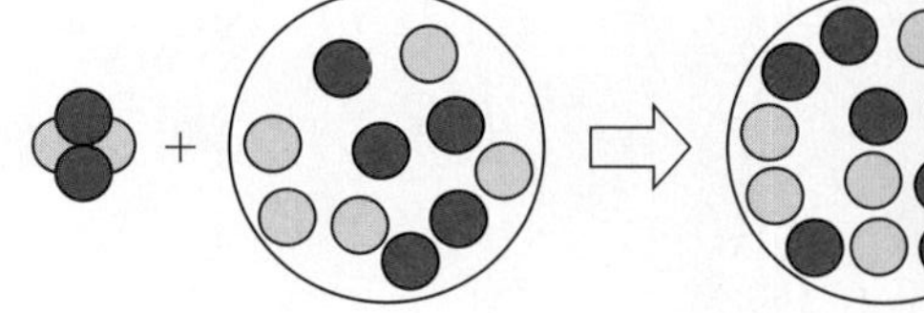

9.47 17 200 y

9.49 a. 11 protons and 14 neutrons
b. 28 protons and 33 neutrons
c. 37 protons and 47 neutrons
d. 47 protons and 63 neutrons

9.51 a. In alpha decay, a helium nucleus or alpha particle is emitted from the nucleus of a radioisotope. In beta decay, a neutron in an unstable nucleus is converted to a proton and an electron, which is emitted as a beta particle. In gamma emission, high-energy radiation is emitted from the nucleus of a radioisotope.
b. alpha radiation: α or ${}^{4}_{2}He$
beta radiation: β or ${}^{0}_{-1}e$
gamma radiation: γ or ${}^{0}_{0}\gamma$

9.53 a. gamma emission
b. positron emission
c. alpha decay

9.55 a. ${}^{225}_{90}Th \longrightarrow {}^{221}_{88}Ra + {}^{4}_{2}He$
b. ${}^{210}_{83}Bi \longrightarrow {}^{206}_{81}Tl + {}^{4}_{2}He$
c. ${}^{137}_{55}Cs \longrightarrow {}^{137}_{56}Ba + {}^{0}_{-1}e$
d. ${}^{126}_{50}Sn \longrightarrow {}^{126}_{51}Sb + {}^{0}_{-1}e$
e. ${}^{18}_{9}F \longrightarrow {}^{18}_{8}O + {}^{0}_{+1}e$

9.57 **a.** ${}^{17}_{8}O$
b. ${}^{1}_{1}H$
c. ${}^{143}_{54}Xe$

9.59 **a.** ${}^{16}_{8}O + {}^{16}_{8}O \longrightarrow {}^{4}_{2}He + {}^{28}_{14}Si$
b. ${}^{249}_{98}Cf + {}^{18}_{8}O \longrightarrow {}^{263}_{106}Sg + 4{}^{1}_{0}n$
c. ${}^{222}_{86}Rn \longrightarrow {}^{218}_{84}Po + {}^{4}_{2}He$

9.61 14 days

9.63 **a.** ${}^{47}_{20}Ca \longrightarrow {}^{47}_{21}Sc + {}^{0}_{-1}e$
b. 1.0 mg of Ca-47
c. 9.0 days

9.65 16 μCi

9.67 7.5 mg

9.69 The irradiation of meats, fruits, and vegetables kills bacteria such as *E. coli* that can cause food-borne illnesses. In addition, spoilage is deterred, and shelf life is extended.

9.71 In the fission process, an atom splits into smaller nuclei. In fusion, small nuclei combine (fuse) to form a larger nucleus.

9.73 Fusion occurs naturally in the Sun and other stars.

9.75 **a.** ${}^{238}_{92}U \longrightarrow {}^{234}_{90}Th + {}^{4}_{2}He$
b. ${}^{234}_{90}Th \longrightarrow {}^{234}_{91}Pa + {}^{0}_{-1}e$
c. ${}^{226}_{88}Ra \longrightarrow {}^{222}_{86}Rn + {}^{4}_{2}He$

9.77 **a.** ${}^{23}_{12}Mg$
b. ${}^{0}_{+1}e$
c. ${}^{257}_{104}Rf$

9.79 64 μCi

Introduction to Organic Chemistry: Alkanes

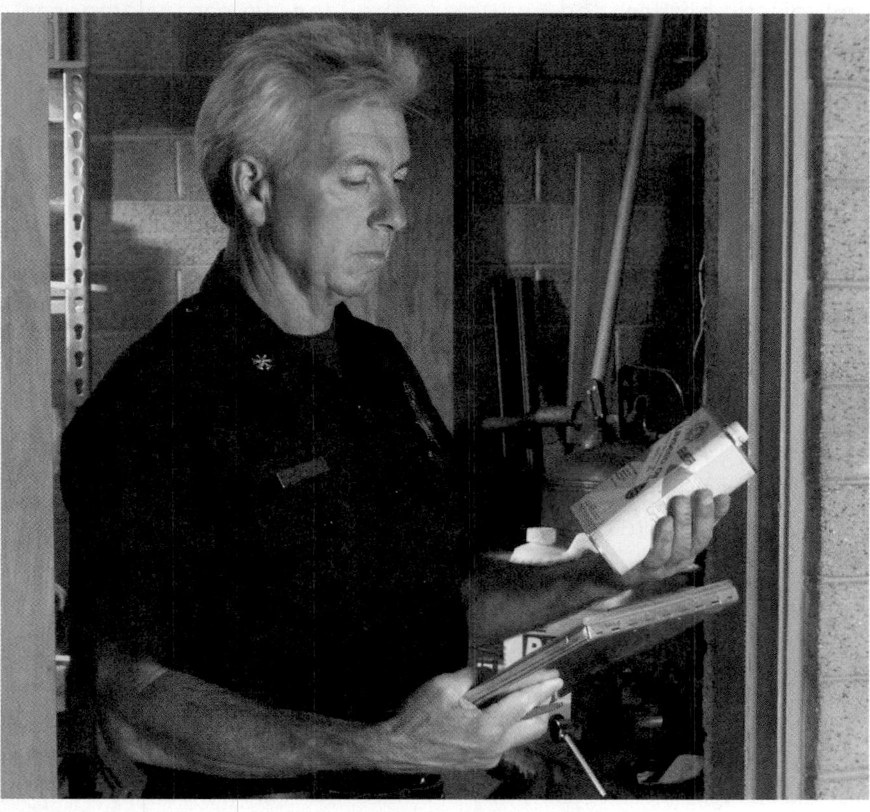

LOOKING AHEAD

"When we have a hazardous materials spill, the first thing we do is isolate it," says Don Dornell, assistant fire chief, Burlingame Fire Station. "Then our technicians and a county chemist identify the product from its flammability and solubility in water so we can use the proper materials to clean up the spill. We use different methods for alcohol, which mixes with water, than for gasoline, which floats. Because hydrocarbons are volatile, we use foam to cover them and trap the vapors. At oil refineries, we will use foams, but many times we squirt water on the tanks to cool the contents below their boiling points, too. By knowing the boiling point of the product and its density and vapor density, we know if it floats or sinks in water and where its vapors will go."

Visit **www.masteringchemistry.com** for self-study materials and instructor-assigned homework.

Organic chemistry is the chemistry of carbon compounds that contain, primarily, carbon and hydrogen. The element carbon has a special role in chemistry because it bonds with other carbon atoms to give a vast array of molecules. The variety of molecules is so great that we find organic compounds in many common products we use, such as gasoline, medicines, shampoos, plastic bottles, and perfumes. The food we eat is composed of different organic compounds that supply us with fuel for energy and the carbon atoms needed to build and repair the cells of our bodies.

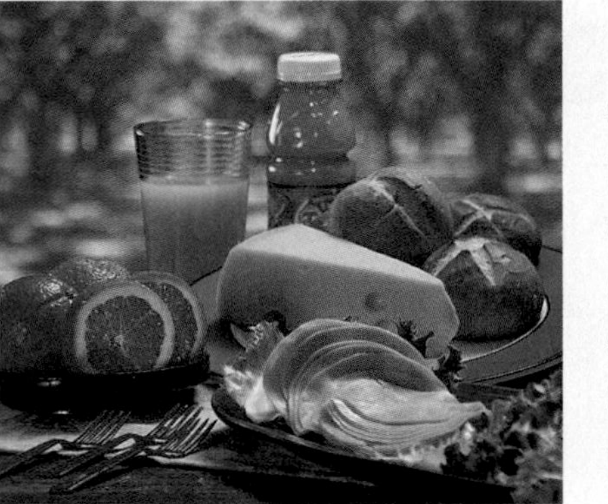

Foods in the diet provide energy and materials for the cells of the body.

Although many organic compounds occur in nature, chemists have synthesized even more. The cotton, wool, or silk in your clothes contain naturally occurring organic compounds, whereas materials such as polyester, nylon, and plastic have been synthesized through organic reactions. Sometimes it is convenient to synthesize a molecule in the lab even though that molecule is also found in nature. For example, vitamin C synthesized in a laboratory has the same structure and properties as the vitamin C in oranges and lemons. Learning about the structures and reactions of organic molecules will provide you with a foundation for understanding the more complex molecules of biochemistry.

LEARNING GOAL

Identify properties characteristic of organic or inorganic compounds.

SELF STUDY ACTIVITY
Introduction to Organic Molecules

10.1 Organic Compounds

At the beginning of the nineteenth century, scientists classified chemical compounds as inorganic and organic. An inorganic compound was a substance that was composed of minerals, and an organic compound was a substance that came from an organism, thus the use of the word "organic." Early scientists thought that some type of "vital force," which could be found only in living cells, was required to synthesize an organic compound. This idea was shown to be incorrect in 1828 when the German chemist Friedrich Wöhler synthesized urea, a product of protein metabolism, by heating an inorganic compound, ammonium cyanate.

$$NH_4CNO \xrightarrow{\text{Heat}} H_2N-\overset{\overset{\displaystyle O}{\|}}{C}-NH_2$$

Ammonium cyanate (inorganic) → Urea (organic)

Vegetable oil, an organic compound, is not soluble in water.

Organic chemistry is the study of carbon compounds. **Organic compounds** typically contain carbon (C) and hydrogen (H), and sometimes oxygen (O), sulfur (S), nitrogen (N), or a halogen (F, Cl, Br, and I). The formulas of organic compounds are written with carbon first, followed by hydrogen and then any other elements.

Many organic compounds are nonpolar molecules with weak attractions between molecules. As a result, they typically have low melting and boiling points, are not soluble in water, and are less dense than water. For example, vegetable oil, which is a mixture of organic compounds, does not dissolve in water but floats on top. A very typical reaction of organic compounds is that they burn vigorously in air.

TABLE 10.1 Some Properties of Organic and Inorganic Compounds

Property	Organic	Example: C_3H_8	Inorganic	Example: NaCl
Elements present	C and H, sometimes O, S, N, or Cl (F, Br, I)	C and H	Most metals and nonmetals	Na and Cl
Particles	Molecules	C_3H_8	Mostly ions	Na^+ and Cl^-
Bonding	Mostly covalent	Covalent	Many are ionic, some covalent	Ionic
Polarity of bonds	Nonpolar, unless a strongly electronegative atom is present	Nonpolar	Most are ionic or polar covalent, a few are nonpolar covalent	Ionic
Melting point	Usually low	−188 °C	Usually high	801 °C
Boiling point	Usually low	−42 °C	Usually high	1413 °C
Flammability	High	Burns in air	Low	Does not burn
Solubility in water	Not soluble, unless a polar group is present	No	Most are soluble, unless nonpolar	Yes

In contrast, inorganic compounds contain elements other than carbon and hydrogen and are ionic with high melting and boiling points. Inorganic compounds that are ionic or polar covalent are usually soluble in water and most do not burn in air. Table 10.1 contrasts some of the properties typically associated with organic and inorganic compounds such as propane, C_3H_8, and sodium chloride, NaCl (see Figure 10.1).

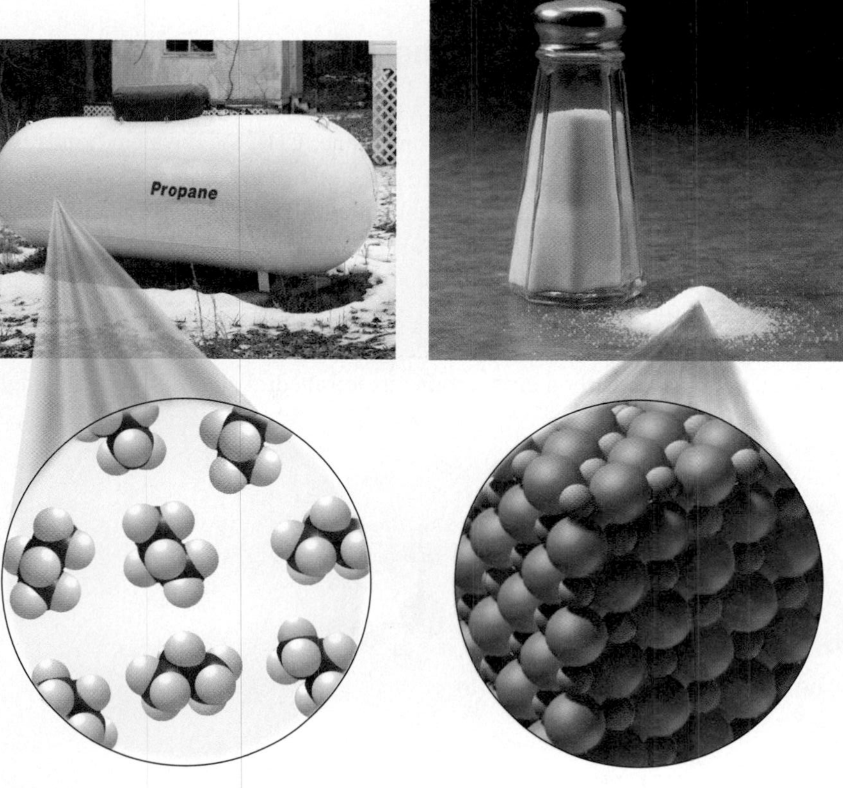

FIGURE 10.1 Propane, C_3H_8, is an organic compound, whereas sodium chloride, NaCl, is an inorganic compound.

Q Why is propane used as a fuel?

CONCEPT CHECK 10.1

Properties of Organic Compounds

Indicate whether the following properties are more typical of organic or inorganic compounds:

a. is not soluble in water

b. has a high melting point

c. burns in air

ANSWER

a. Many organic compounds are not soluble in water.
b. Inorganic compounds are more likely to have high melting points.
c. Organic compounds are more likely to be flammable.

Bonding in Organic Compounds

Hydrocarbons, as the name suggests, are organic compounds that consist of carbon and hydrogen. In the simplest hydrocarbon, methane (CH_4), the carbon atom forms an octet by sharing its four valence electrons with the valence electrons of four hydrogen atoms. In all organic molecules, every carbon atom has four bonds. A hydrocarbon is referred to as a *saturated hydrocarbon* when all the bonds in the molecules are single bonds. We can draw an **expanded structural formula** of a compound by showing all the bonds between atoms.

$$\cdot\dot{\underset{\cdot}{C}}\cdot + 4H\cdot \longrightarrow H:\overset{H}{\overset{\cdot\cdot}{\underset{\underset{H}{\cdot\cdot}}{C}}}:H = H-\overset{H}{\overset{|}{\underset{\underset{H}{|}}{C}}}-H$$

Methane

The Tetrahedral Structure of Carbon

The VSEPR theory (Chapter 4) predicts that a molecule with four atoms bonded to a central atom has a tetrahedral shape. Thus, for methane, CH_4, the covalent bonds from the carbon atom to the four hydrogen atoms are directed to the corners of a tetrahedron with bond angles of 109°. The three-dimensional structure of methane can be illustrated as a ball-and-stick model or a space-filling model. The expanded structural formula is a two-dimensional representation in which the bonds from carbon to each hydrogen atom are shown (see Figure 10.2).

In ethane, C_2H_6, each carbon atom is bonded to another carbon and three hydrogen atoms. As in methane, each carbon atom retains its tetrahedral shape, with bond angles close to 109° (see Figure 10.3).

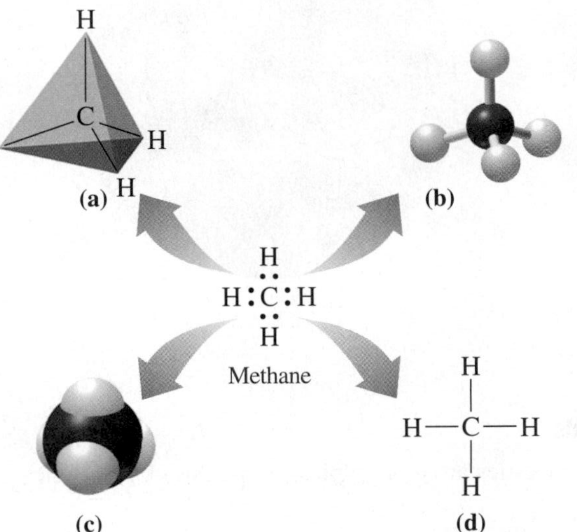

FIGURE 10.2 Representations of methane, CH_4: **(a)** tetrahedron, **(b)** ball-and-stick model, **(c)** space-filling model, **(d)** expanded structural formula.
Q Why does methane have a tetrahedral shape and not a flat shape?

FIGURE 10.3 Representations of ethane, C_2H_6: **(a)** tetrahedral shape of each carbon, **(b)** ball-and-stick model, **(c)** space-filling model, **(d)** expanded structural formula.
Q How is the tetrahedral shape maintained in a molecule with two carbon atoms?

QUESTIONS AND PROBLEMS

Organic Compounds

10.1 Identify the following as formulas of organic or inorganic compounds:
a. KCl **b.** C_4H_{10} **c.** C_2H_6O
d. H_2SO_4 **e.** $CaCl_2$ **f.** C_3H_7Cl

10.2 Identify the following as formulas of organic or inorganic compounds:
a. $C_6H_{12}O_6$ **b.** K_3PO_4 **c.** I_2
d. C_2H_6S **e.** $C_{10}H_{22}$ **f.** CH_4

10.3 Identify the following properties as more typical of organic or inorganic compounds:
a. is soluble in water
b. has a low boiling point
c. contains carbon and hydrogen
d. contains ionic bonds

10.4 Identify the following properties as more typical of organic or inorganic compounds:
a. contains Na and Br
b. is a gas at room temperature
c. contains covalent bonds
d. is an electrolyte

10.5 Match the following physical and chemical properties with ethane, C_2H_6, or sodium bromide, NaBr:
a. boils at −89 °C **b.** burns vigorously in air
c. is a solid at 250 °C **d.** dissolves in water

10.6 Match the following physical and chemical properties with cyclohexane, C_6H_{12}, or calcium nitrate, $Ca(NO_3)_2$:
a. melts at 500 °C
b. is insoluble in water
c. does not burn in air
d. is a liquid at room temperature

10.7 How are the hydrogen atoms arranged in space in methane, CH_4?

10.8 In a propane molecule with three carbon atoms, what is the shape around each carbon atom?

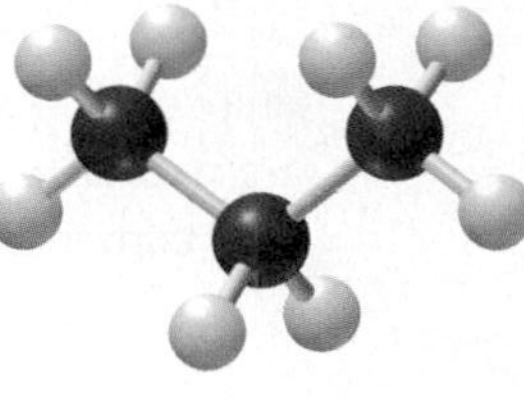

Propane

10.2 Alkanes

LEARNING GOAL

Write the IUPAC names and draw the condensed structural formulas and skeletal formulas for alkanes.

TUTORIAL
IUPAC Naming of Alkanes

More than 90% of the compounds in the world are organic compounds. The large number of carbon compounds is possible because the covalent bond between carbon atoms (C—C) is very strong, allowing carbon atoms to form long, stable chains. To help us study this large group of compounds, we organize them into classes that have similar structures and chemical properties.

The **alkanes** are a class of hydrocarbons in which the atoms are connected only by single bonds. One of the most common uses of alkanes is as fuels. Methane, used in gas

TABLE 10.2 IUPAC Names of the First Ten Alkanes

Number of Carbon Atoms	Prefix	Name	Molecular Formula	Condensed Structural Formula
1	Meth	Methane	CH_4	CH_4
2	Eth	Ethane	C_2H_6	CH_3-CH_3
3	Prop	Propane	C_3H_8	$CH_3-CH_2-CH_3$
4	But	Butane	C_4H_{10}	$CH_3-CH_2-CH_2-CH_3$
5	Pent	Pentane	C_5H_{12}	$CH_3-CH_2-CH_2-CH_2-CH_3$
6	Hex	Hexane	C_6H_{14}	$CH_3-CH_2-CH_2-CH_2-CH_2-CH_3$
7	Hept	Heptane	C_7H_{16}	$CH_3-CH_2-CH_2-CH_2-CH_2-CH_2-CH_3$
8	Oct	Octane	C_8H_{18}	$CH_3-CH_2-CH_2-CH_2-CH_2-CH_2-CH_2-CH_3$
9	Non	Nonane	C_9H_{20}	$CH_3-CH_2-CH_2-CH_2-CH_2-CH_2-CH_2-CH_2-CH_3$
10	Dec	Decane	$C_{10}H_{22}$	$CH_3-CH_2-CH_2-CH_2-CH_2-CH_2-CH_2-CH_2-CH_2-CH_3$

heaters and gas cooktops, is an alkane with one carbon atom. Ethane, propane, and butane contain two, three, and four carbon atoms, respectively, connected in a row or a *continuous* chain. As we can see, all the names for alkanes end in *ane*. Such names are part of the **IUPAC** (International Union of Pure and Applied Chemistry) **system** used by chemists to name organic compounds. Alkanes with five or more carbon atoms in a chain are named using Greek prefixes: *pent* (5), *hex* (6), *hept* (7), *oct* (8), *non* (9), and *dec* (10) (see Table 10.2).

CONCEPT CHECK 10.2

Naming Alkanes

Give the IUPAC name for each of the following:

a. $CH_3-CH_2-CH_3$ **b.** C_6H_{14}

ANSWER

a. A chain with three carbon atoms is propane.
b. An alkane with six carbon atoms is hexane.

Condensed Structural Formulas

In a **condensed structural formula**, each carbon atom and its attached hydrogen atoms are written as a group. A subscript indicates the number of hydrogen atoms bonded to each carbon atom.

$$\underset{\text{Expanded}}{H-\overset{\displaystyle H}{\underset{\displaystyle H}{\overset{|}{\underset{|}{C}}}}-} = \underset{\text{Condensed}}{CH_3-} \qquad \underset{\text{Expanded}}{-\overset{\displaystyle H}{\underset{\displaystyle H}{\overset{|}{\underset{|}{C}}}}-} = \underset{\text{Condensed}}{-CH_2-}$$

By contrast, the *molecular formula* gives the total number of each kind of atom, but does not indicate their arrangement in the molecule.

When an organic molecule consists of a chain of three or more carbon atoms, the carbon atoms do not lie in a straight line. The tetrahedral shape of carbon arranges the carbon bonds in a zigzag pattern. A simplified structure called the *line-bond* or *skeletal formula* represents only the carbon skeleton in which carbon atoms are represented as the ends of each line or as corners in a zigzag line. The hydrogen atoms are not shown, but each carbon is understood to have the proper number of atoms including hydrogen atoms to give four bonds. In the skeletal formula of hexane, each line in the zigzag drawing represents a single bond. The carbon atoms on the ends would have three hydrogen atoms. However the other carbon atoms in the middle of the carbon chain are each bonded to two carbons and therefore two hydrogen atoms. Figure 10.4 shows the molecular formula, expanded structural formula, condensed structural formula, and skeletal formula for hexane.

Because an alkane has only C—C single bonds, the groups attached to each C are not in fixed positions. They can rotate freely about the bond connecting the carbon atoms. This motion is analogous to the independent rotation of the wheels of a toy car. Thus, different arrangements, known as *conformations*, occur during the rotation about a single bond.

Alkane Name Hexane
Molecular Formula C_6H_{14}
Ball-and-Stick Model

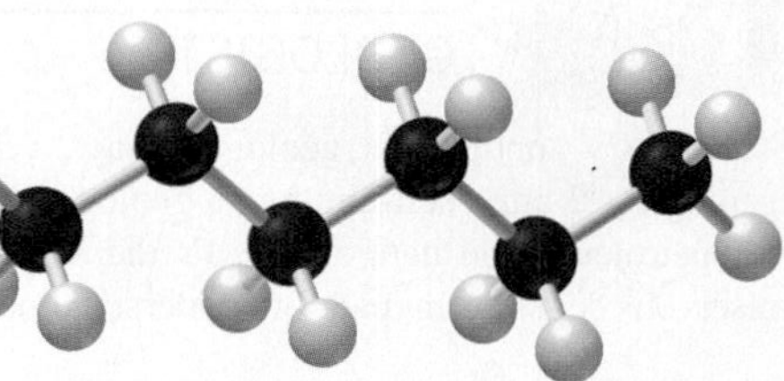

Expanded Structural Formula

Condensed Structural Formulas

$CH_3-CH_2-CH_2-CH_2-CH_2-CH_3$

Skeletal Formula

FIGURE 10.4 A hexane molecule can be represented in several ways: molecular formula, ball-and-stick model, expanded structural formula, condensed structural formula, and skeletal formula.

Q Why do the carbon atoms in hexane appear to be arranged in a zigzag chain?

Suppose we could look at butane, C_4H_{10}, as it rotates. Sometimes the $-CH_3$ groups line up in front of each other, and at other times they are opposite each other. As the $-CH_3$ groups turn around the single bond, the carbon chain in the condensed structural formulas may appear at different angles. For example, butane can be drawn using a variety of two-dimensional condensed structural formulas as shown in Table 10.3. All these condensed structural formulas represent the same compound with four carbon atoms.

TABLE 10.3 Some Structural Formulas for Butane, C_4H_{10}

Expanded Structural Formula

Condensed Structural Formulas

$CH_3-CH_2-CH_2-CH_3$

Skeletal Formulas

Career Focus

GEOLOGIST

"Chemistry underpins geology," says Vic Abadie, consulting geologist. "I am a self-employed geologist consulting in exploration for petroleum and natural gas. Predicting the occurrence of an oil reservoir depends in part on understanding chemical reactions of minerals. This is because over geologic time, such reactions create or destroy pore spaces that host crude oil or gas in a reservoir rock formation. Chemical analysis can match oil in known reservoir formations with distant formations that generated oil and from which oil migrated into the reservoirs in the geologic past. This can help identify target areas to explore for new reservoirs."

"I evaluate proposals to drill for undiscovered oil and gas. I recommend that my clients invest in proposed wells that my analysis suggests have strong geologic and economic merit. This is a commercial application of the scientific method: The proposal to drill is the hypothesis, and the drill bit tests it. A successful well validates the hypothesis and generates oil or gas production and revenue for my clients and me. I do this and other consulting for private and corporate clients. The risk is high, the work is exciting, and my time is flexible."

SAMPLE PROBLEM 10.1

Drawing Expanded and Condensed Structural Formulas for Alkanes

Draw the expanded structural formula, condensed structural formula, and skeletal formula for pentane.

SOLUTION

In the expanded structural formula, five carbon atoms are connected to each other and to hydrogen atoms using single bonds to give each carbon atom a total of four bonds. In the condensed structural formula, each carbon atom and its attached hydrogen atoms are written as CH_3— or —CH_2—. The skeletal formula shows the carbon skeleton as a zigzag line where the ends and corners represent C atoms.

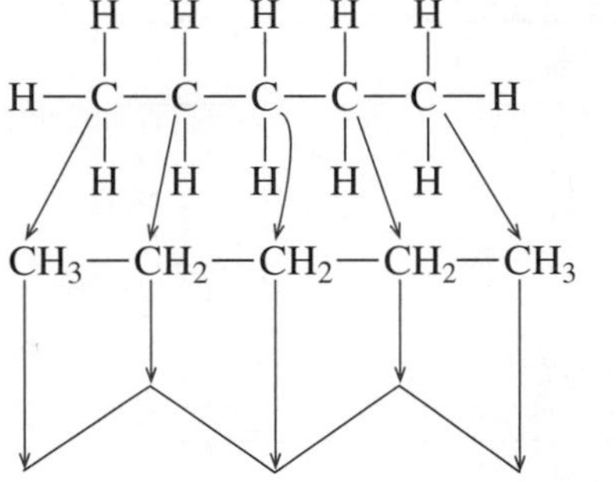

Expanded structural formula

Condensed structural formula

Skeletal formula

STUDY CHECK 10.1

Draw the condensed structural formula and give the name for the following skeletal formula:

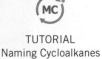

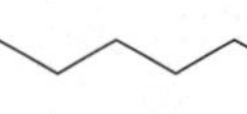

Cycloalkanes

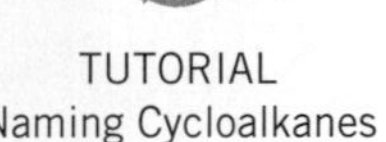

TUTORIAL
Naming Cycloalkanes

Hydrocarbons can also form cyclic structures called **cycloalkanes**, which have two fewer hydrogen atoms than the corresponding alkanes. The simplest cycloalkane, cyclopropane (C_3H_6) has a ring of three carbon atoms bonded to six hydrogen atoms.

TABLE 10.4 Formulas of Some Common Cycloalkanes

Ball-and-Stick Models			
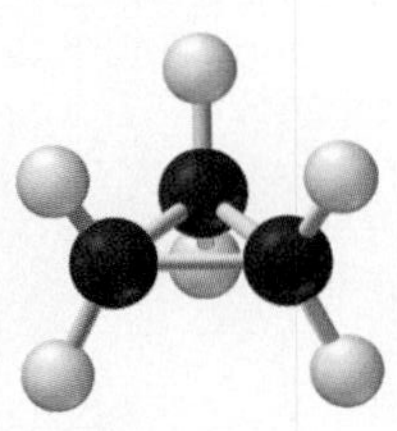	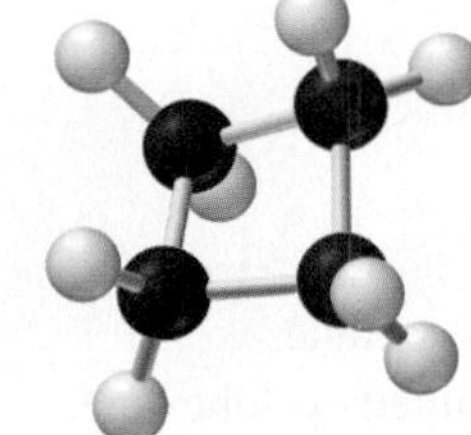	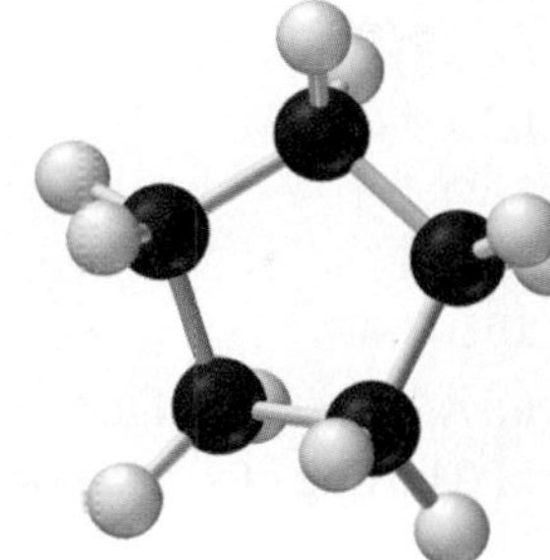	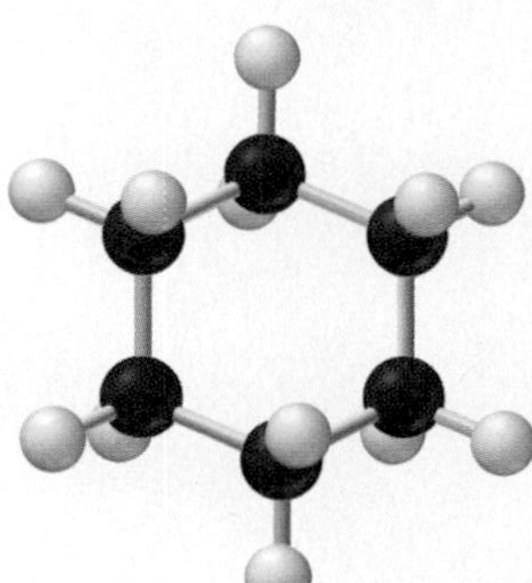
Condensed Structural Formulas			
CH_2 / $H_2C—CH_2$	$H_2C—CH_2$ / $H_2C—CH_2$	CH_2 / H_2C CH_2 / $H_2C—CH_2$	CH_2 / H_2C CH_2 / H_2C CH_2 / CH_2
Skeletal Formulas			

Names			
Cyclopropane	Cyclobutane	Cyclopentane	Cyclohexane

Most often cycloalkanes are drawn using their skeletal formulas, which appear as simple geometric figures. As seen for alkanes, each corner of the skeletal formula for a cycloalkane represents a carbon atom.

The ball-and-stick models, condensed structural formulas, and skeletal formulas for several cycloalkanes are shown in Table 10.4. A cycloalkane is named by adding the prefix *cyclo* to the name of the alkane with the same number of carbon atoms.

CONCEPT CHECK 10.3

Identifying Cycloalkanes

Name the cycloalkanes represented in the skeletal formula for cholesterol, an important steroid in the body.

CH_3 CH_3 CH_3 CH_3 CH_3 HO

Cholesterol

ANSWER

Cholesterol contains the cycloalkanes cyclopentane and cyclohexane.

SAMPLE PROBLEM 10.2

Naming Alkanes

Give the IUPAC name for each of the following:

a.

b. 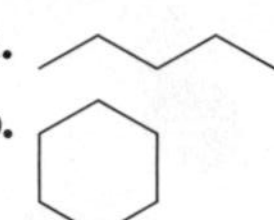

SOLUTION

a. A chain with eight carbon atoms is octane.
b. The ring of six carbon atoms is named cyclohexane.

STUDY CHECK 10.2

What is the IUPAC name of the following compound?

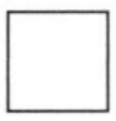

QUESTIONS AND PROBLEMS

Alkanes

10.9 Give the IUPAC name for each of the following alkanes:

a.
```
CH3
|
CH2—CH2—CH2
         |
         CH3
```

b. $CH_3—CH_3$

c.
```
          CH2—CH3
          |
          CH2
          |
CH3—CH2—CH2
```

10.10 Give the IUPAC name for each of the following alkanes:

a. CH_4
b. $CH_3—CH_2—CH_2—CH_3$

c.
```
CH3
|
CH2
|
CH3
```

10.11 Draw the condensed structural formula for alkanes or a skeletal formula for cycloalkanes for each of the following:

a. methane
b. ethane
c. pentane
d. cyclopropane

10.12 Draw the condensed structural formula for alkanes or a skeletal formula for cycloalkanes for each of the following:

a. propane **b.** hexane
c. heptane **d.** cyclopentane

LEARNING GOAL

Write the IUPAC names for alkanes with substituents.

TUTORIAL
Naming Alkanes with Substituents

10.3 Alkanes with Substituents

When an alkane has four or more carbon atoms, the atoms can be arranged so that a side group called a **branch** or **substituent** is attached to a carbon chain. For example, there are different ball-and-stick models for two compounds that have the molecular formula C_4H_{10}. One model is shown as a chain of four carbon atoms. In the other model, a carbon atom is attached as a branch or substituent to a carbon in a chain of three atoms (see Figure 10.5). An alkane with at least one branch is called a **branched alkane**. When two compounds have the same molecular formula but different arrangements of atoms, they are called **isomers**.

In another example, we can draw the condensed structural formulas of three different isomers with the molecular formula C_5H_{12} as follows:

Isomers of C_5H_{12}

Alkane	Branched Alkanes	
$CH_3—CH_2—CH_2—CH_2—CH_3$	$CH_3—CH(CH_3)—CH_2—CH_3$	$CH_3—C(CH_3)_2—CH_3$

SAMPLE PROBLEM 10.3

Isomers

Identify each pair of condensed structural formulas as isomers or the same molecule.

a. $CH_2(CH_3)—CH_2(CH_3)$ and $CH_2(CH_3)—CH_2—CH_3$

b. $CH_3—CH(CH_3)—CH_2—CH_2—CH_3$ and $CH_3—CH(CH_3)—CH(CH_3)—CH_3$

SOLUTION

a. Although the carbon chain has different angles due to rotation, these condensed structural formulas represent the same molecule. The two formulas represent the same molecule because each has four C atoms in a continuous chain.

b. The molecular formula of both these condensed structural formulas is C_6H_{14}. However, they represent isomers because the C atoms are bonded in a different order; they have different arrangements. In one, there is a $—CH_3$ group attached to a five-carbon chain, and in the other there are two $—CH_3$ groups attached to a four-carbon chain.

STUDY CHECK 10.3

Why is the following condensed structural formula an isomer of the molecules in Sample Problem 10.3, part **b**?

$CH_3—CH_2—CH(CH_3)—CH_2—CH_3$

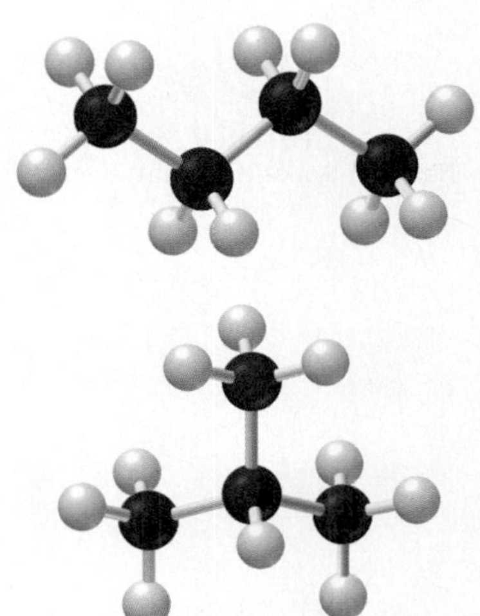

FIGURE 10.5 The isomers of C_4H_{10} have the same number and type of atoms but are bonded in a different order.

Q What makes these molecules isomers?

TUTORIAL
Drawing Organic Compounds with Functional Groups

SELF STUDY ACTIVITY
Organic Molecules and Isomers

Substituents in Alkanes

In the IUPAC names for alkanes, a carbon branch is named as an **alkyl group**, which is an alkane that is missing one hydrogen atom. The alkyl group is named by replacing the *ane* ending of the corresponding alkane name with *yl*. Alkyl groups cannot exist on their own: they must be attached to a carbon chain. When a halogen atom is attached to a carbon chain, it is named as a *halo* group: *fluoro*, *chloro*, *bromo*, or *iodo*. Some of the common groups attached to carbon chains are illustrated in Table 10.5.

TABLE 10.5 Names and Formulas of Some Common Substituents

Substituent	Name
$CH_3—$	Methyl
$CH_3—CH_2—$	Ethyl
$CH_3—CH_2—CH_2—$	Propyl
$CH_3—CH(—)—CH_3$	Isopropyl
$F—, Cl—, Br—, I—$	Fluoro, chloro, bromo, iodo

TUTORIAL
Naming Alkanes with Substituents

Naming Alkanes with Substituents

In the IUPAC system of naming, a carbon chain is numbered to give the location of one or more substituents. Let's take a look at how we use the IUPAC system to name the alkane shown in Sample Problem 10.4.

Guide to Naming Alkanes

1. Write the alkane name of the longest chain of carbon atoms.
2. Number the carbon atoms starting from the end nearer a substituent.
3. Give the location and name of each substituent (alphabetical order) as a prefix to the name of the main chain.

SAMPLE PROBLEM 10.4

Writing IUPAC Names for Alkanes with Substituents

Give the IUPAC name for the following alkane:

$$CH_3-\underset{\displaystyle |}{\overset{\displaystyle CH_3}{CH}}-CH_2-\underset{\displaystyle CH_3}{\overset{\displaystyle Br}{C}}-CH_2-CH_3$$

SOLUTION

Step 1 **Write the alkane name of the longest chain of carbon atoms.** In this alkane, the longest chain has six carbon atoms, which is hexane.

$$CH_3-\overset{\displaystyle CH_3}{CH}-CH_2-\underset{\displaystyle CH_3}{\overset{\displaystyle Br}{C}}-CH_2-CH_3 \qquad \text{hexane}$$

Step 2 **Number the carbon atoms starting from the end nearer a substituent.** The numbering of the carbon chain begins at carbon 1 at the end of the carbon chain closer to the methyl group (CH_3-) and continues in the same direction.

$$\underset{1}{CH_3}-\underset{2}{\overset{\displaystyle CH_3}{CH}}-\underset{3}{CH_2}-\underset{4}{\underset{\displaystyle CH_3}{\overset{\displaystyle Br}{C}}}-\underset{5}{CH_2}-\underset{6}{CH_3} \qquad \text{hexane}$$

Step 3 **Give the location and name of each substituent (alphabetical order) as a prefix to the name of the main chain.** The substituents, which are bromo and methyl groups, are listed in alphabetical order. A hyphen is placed between the number on the carbon chain and the substituent name. When there are two or more of the same substituent, a prefix (*di*, *tri*, *tetra*) is used in front of the name. Then commas are used to separate the numbers for the locations of the substituents.

$$\underset{1}{CH_3}-\underset{2}{\overset{\displaystyle CH_3}{CH}}-\underset{3}{CH_2}-\underset{4}{\underset{\displaystyle CH_3}{\overset{\displaystyle Br}{C}}}-\underset{5}{CH_2}-\underset{6}{CH_3} \qquad \text{4-bromo-2,4-dimethyl hexane}$$

STUDY CHECK 10.4

Give the IUPAC name for the following compound:

$$CH_3-CH_2-\overset{\displaystyle CH_3}{CH}-CH_2-\overset{\displaystyle CH_3}{CH}-CH_2-Cl$$

TUTORIAL
Naming Cycloalkanes

Naming Cycloalkanes with Substituents

When one substituent is attached to a carbon atom in a cycloalkane, the name of the substituent is placed in front of the cycloalkane name. No number is needed for a single alkyl group or halogen atom because the carbon atoms in the cycloalkane are equivalent.

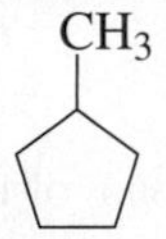

CH_3

methylcyclopentane

Haloalkanes

In a *haloalkane*, halogen atoms replace hydrogen atoms in an alkane. The halo substituents are numbered and arranged alphabetically, just as we did with the alkyl groups. Many times chemists use the common, traditional name for these compounds rather than the systematic IUPAC name. Simple haloalkanes are commonly named as alkyl halides; the carbon group is named as an alkyl group followed by the halide name.

	$CH_3—Cl$	$CH_3—CH_2—Br$	$CH_3—CH_2—CH_2—F$
IUPAC	Chloromethane	Bromoethane	1-Fluoropropane
Common	Methyl chloride	Ethyl bromide	Propyl fluoride

SAMPLE PROBLEM 10.5

Naming Haloalkanes

Freon-11 and Freon-12 are compounds, known as *chlorofluorocarbons (CFCs)*, previously used as refrigerants and aerosol propellants. What are their IUPAC names?

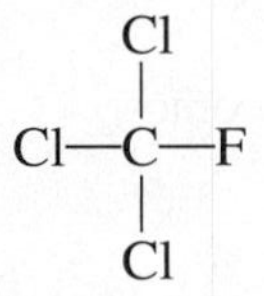

```
     Cl               Cl
     |                |
 Cl—C—F           F—C—F
     |                |
     Cl               Cl
  Freon-11         Freon-12
```

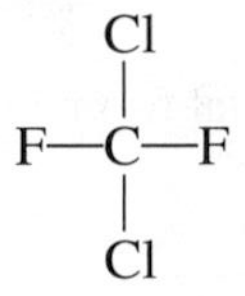

SOLUTION

Freon-11, trichlorofluoromethane; Freon-12, dichlorodifluoromethane

STUDY CHECK 10.5

A haloalkane used as a fumigant has two carbon atoms with one bromine atom attached to each carbon. What is its IUPAC name?

Drawing Condensed Structural Formulas for Alkanes

TUTORIAL
Drawing Haloalkanes and Branched Alkanes

The IUPAC name gives all the information needed to draw the condensed structural formula of an alkane. Suppose you are asked to draw the condensed structural formula of 2,3-dimethylbutane. The alkane name gives the number of carbon atoms in the longest chain. The names in the beginning indicate the substituents and where they are attached. We can break down the name in the following way:

2,3-Dimethylbutane

2,3-	Di	methyl	but	ane
Substituents on carbons 2 and 3	Two identical groups	$CH_3—$ alkyl groups	Four carbon atoms in the main chain	Single C—C bonds

Guide to Drawing Alkane Formulas

1. Draw the main chain of carbon atoms.
2. Number the chain and place the substituents on the carbons indicated by the numbers.
3. Add the correct number of hydrogen atoms to give four bonds to each C atom.

SAMPLE PROBLEM 10.6

Drawing Condensed Structures from IUPAC Names

Draw the condensed structural formula and skeletal formula for 2,3-dimethylbutane.

SOLUTION

We can use the following guide to draw the condensed structural formula:

Step 1 **Draw the main chain of carbon atoms.** For butane, we draw a chain of four carbon atoms.

$$C-C-C-C$$

Step 2 **Number the chain and place the substituents on the carbons indicated by the numbers.** The first part of the name indicates two methyl groups (CH_3-), one on carbon 2 and one on carbon 3.

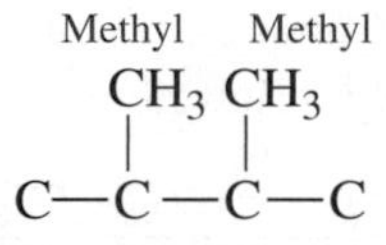

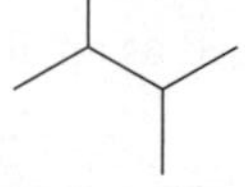

Methyl Methyl

$$\underset{1}{C}-\underset{2}{\overset{CH_3}{\overset{|}{C}}}-\underset{3}{\overset{CH_3}{\overset{|}{C}}}-\underset{4}{C}$$

1 2 3 4

Step 3 **Add the correct number of hydrogen atoms to give four bonds to each C atom.**

$$CH_3-\overset{CH_3}{\overset{|}{CH}}-\overset{CH_3}{\overset{|}{CH}}-CH_3$$

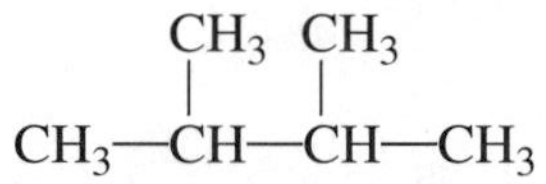

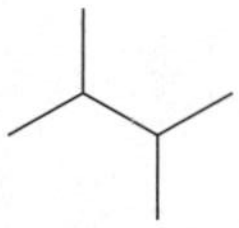

2,3-Dimethylbutane

STUDY CHECK 10.6

What is the condensed structural formula and skeletal formula for 2-bromo-4-methylpentane?

QUESTIONS AND PROBLEMS

Alkanes with Substituents

10.13 Indicate whether each of the following pairs of condensed structural formulas represent isomers or the same molecule:

a. 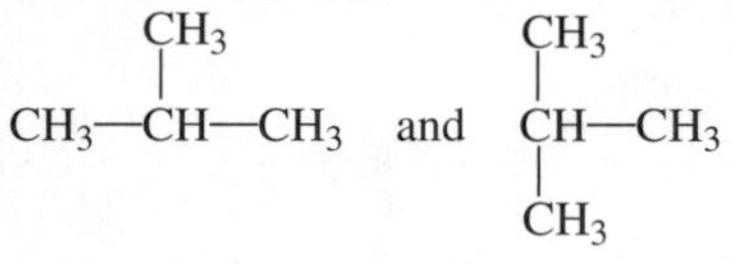 $CH_3-\overset{CH_3}{\overset{|}{CH}}-CH_3$ and $\overset{CH_3}{\overset{|}{\underset{\underset{CH_3}{|}}{CH}}}-CH_3$

b. 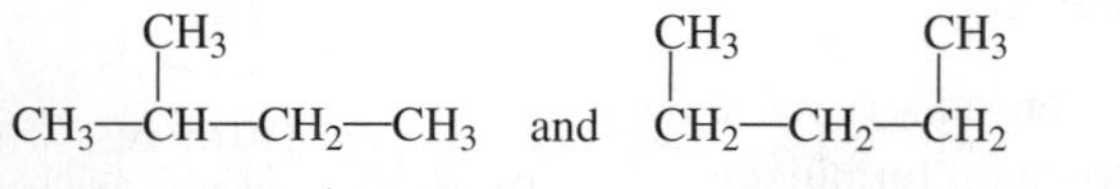 $CH_3-\overset{CH_3}{\overset{|}{CH}}-CH_2-CH_3$ and $\overset{CH_3}{\overset{|}{CH_2}}-CH_2-\overset{CH_3}{\overset{|}{CH_2}}$

c. 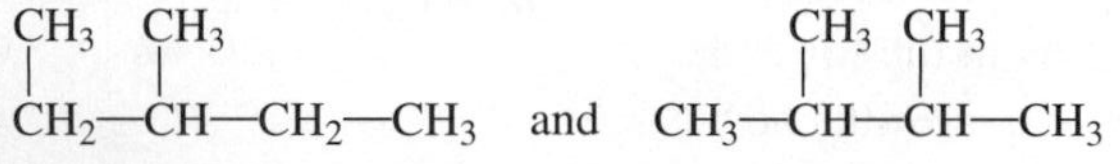 $\overset{CH_3}{\overset{|}{CH_2}}-\overset{CH_3}{\overset{|}{CH}}-CH_2-CH_3$ and $CH_3-\overset{CH_3}{\overset{|}{CH}}-\overset{CH_3}{\overset{|}{CH}}-CH_3$

10.14 Indicate whether each of the following pairs of condensed structural formulas represent isomers or the same molecule:

a. 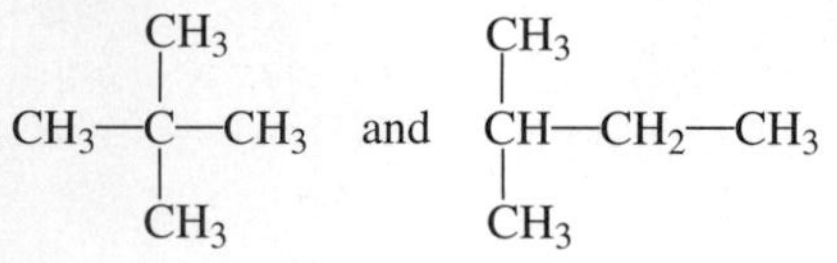 $CH_3-\overset{CH_3}{\overset{|}{\underset{\underset{CH_3}{|}}{C}}}-CH_3$ and $\overset{CH_3}{\overset{|}{\underset{\underset{CH_3}{|}}{CH}}}-CH_2-CH_3$

b. $CH_3-\overset{CH_3}{\overset{|}{CH}}-\overset{CH_3}{\overset{|}{CH}}-\overset{CH_3}{\overset{|}{CH_2}}$ and $CH_3-\overset{CH_3}{\overset{|}{CH}}-CH_2-\overset{CH_3}{\overset{|}{CH}}-CH_3$

c. $CH_3-\overset{CH_3}{\overset{|}{CH}}-CH_2-CH_3$ and $CH_3-CH_2-\overset{CH_3}{\overset{|}{CH}}-CH_3$

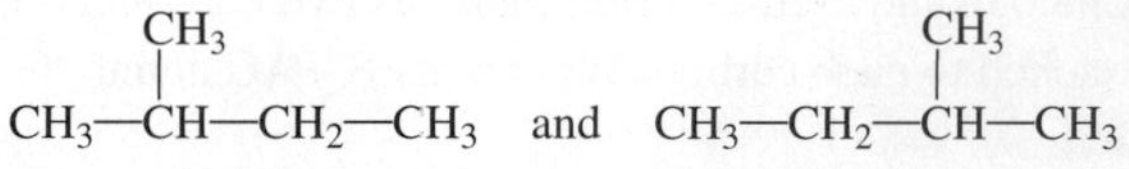

10.15 Give the IUPAC name for each of the following:

a. 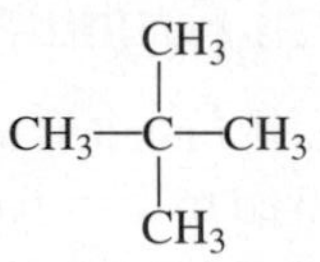 $CH_3-\overset{CH_3}{\overset{|}{\underset{\underset{CH_3}{|}}{C}}}-CH_3$

b. [skeletal formula]

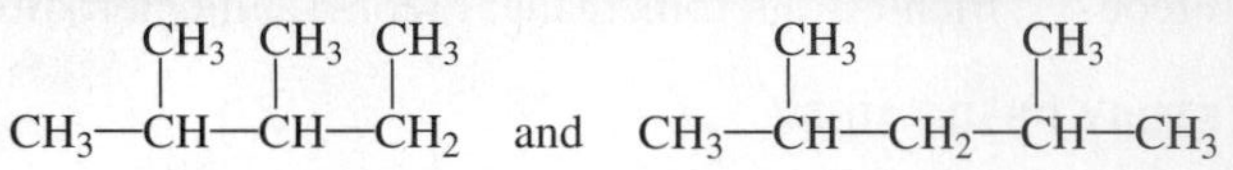

c. $CH_3-\overset{CH_3}{\overset{|}{\underset{\underset{CH_3}{|}}{C}}}-CH_2-\overset{CH_2-CH_3}{\overset{|}{CH}}-CH_2-CH_3$

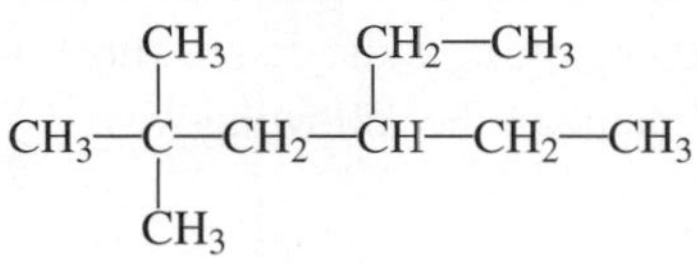

d. 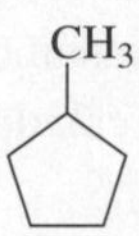 [cyclopentane ring with CH_3]

e. $CH_3-CH_2-CH_2-F$

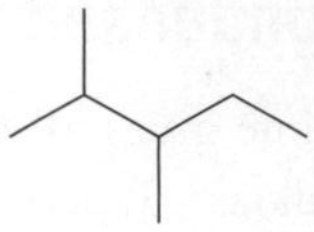

f. 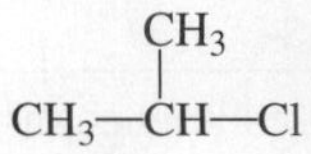 $CH_3-\overset{CH_3}{\overset{|}{CH}}-Cl$

10.16 Give the IUPAC name for each of the following:

a. $CH_3-CH(CH_3)-CH_2-CH_2-CH_3$

b.

c. $CH_3-CH_2-CH(CH_2-CH_3)-CH(CH_2-CH_3)-CH_2-CH_3$

d. cyclohexane with CH_3

e. $CH_3-CH_2-CH(Cl)-CH_3$

f. $CH_3-C(CH_3)_2-I$

10.17 Draw a condensed structural formula for each of the following alkanes:
a. 3,3-dimethylpentane
b. 2,3,5-trimethylhexane
c. 3-ethyl-2,5-dimethyloctane
d. 1-bromo-2-chloroethane
e. 2-bromo-3-chlorobutane

10.18 Draw a condensed structural formula for each of the following alkanes:
a. 3-ethylpentane
b. 3-ethyl-2-methylpentane
c. 4-ethyl-2,2-dimethyloctane
d. 2-bromopropane
e. 2,3-dichloro-2-methylbutane

10.19 Draw the skeletal formula for each of the following:
a. 3-methylheptane
b. ethylcyclopentane
c. bromocyclobutane
d. 2,3-dichlorohexane

10.20 Draw the skeletal formula for each of the following:
a. 1-bromo-2-methylpentane
b. methylcyclopropane
c. ethylcyclohexane
d. 4-chlorooctane

Chemistry Link to Health

COMMON USES OF HALOGENATED ALKANES

Some common uses of haloalkanes include solvents and anesthetics. For many years, carbon tetrachloride was widely used in dry cleaners, and in home spot removers to take oils and grease out of clothes. However, its use was discontinued when carbon tetrachloride was found to be toxic to the liver, where it can cause cancer. Today, dry cleaners use other halogenated compounds such as dichloromethane, 1,1,1-trichloroethane, and 1,1,2-trichloro-1,2,2-trifluoroethane.

CH_2Cl_2	Cl_3C-CH_3	$FCl_2C-CClF_2$
Dichloromethane	1,1,1-Trichloroethane	1,1,2-Trichloro-1,2,2-trifluoroethane

General anesthetics are compounds that are inhaled or injected to cause a loss of sensation so that surgery or other procedures can be done without causing pain to the patient. As nonpolar compounds, they are soluble in the nonpolar nerve membranes, where they decrease the ability of the nerve cells to conduct the sensation of pain. Trichloromethane, commonly called chloroform ($CHCl_3$), was once used as an anesthetic, but it is toxic and may be carcinogenic. One of the most widely used general anesthetics is halothane (2-bromo-2-chloro-1,1,1-trifluoroethane), also called Fluothane. It has a pleasant odor, is nonexplosive, has few side effects, undergoes few reactions within the body, and is eliminated quickly from the body.

$F_3C-CHClBr$ Halothane (Fluothane)

For minor surgeries, a local anesthetic such as chloroethane (ethyl chloride), CH_3-CH_2-Cl, is applied to an area of the skin. Chloroethane evaporates quickly, cooling the skin and causing a loss of sensation.

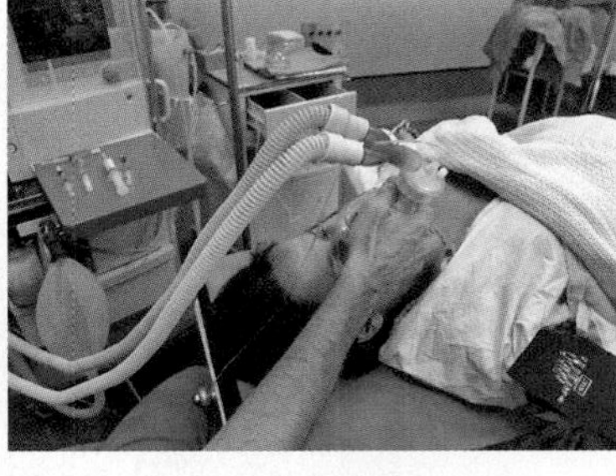

Anesthetics, such as halothane, decrease the sensation of pain.

A local anesthetic evaporates quickly to reduce the sensation of pain.

LEARNING GOAL

Identify the properties of alkanes and write a balanced equation for combustion.

CASE STUDY
Hazardous Materials

10.4 Properties of Alkanes

Many types of alkanes are the components of fuels that power our cars and oil that heats our homes. You may have used a mixture of hydrocarbons such as mineral oil as a laxative or petrolatum to soften your skin. The differences in uses of many of the alkanes result from their physical properties including solubility, density, and boiling point.

The solid alkanes that make up waxy coatings on fruits and vegetables help retain moisture, inhibit mold, and enhance appearance.

Some Uses of Alkanes

The first four alkanes with 1 to 4 carbon atoms—methane, ethane, propane, and butane—are gases at room temperature and are widely used as heating fuels.

Alkanes having 5 to 8 carbon atoms (pentane, hexane, heptane, and octane) are liquids at room temperature. They are highly volatile, which makes them useful in fuels such as gasoline.

Liquid alkanes with 9 to 17 carbon atoms have higher boiling points and are found in kerosene, diesel, and jet fuels. Motor oil is a mixture of high-molecular-weight liquid hydrocarbons and is used to lubricate the internal components of engines. Mineral oil is a mixture of liquid hydrocarbons and is used as a laxative and a lubricant. Larger alkanes, known as paraffins, are waxy solids used to coat fruits and vegetables to retain moisture, inhibit mold growth, and enhance appearance. Petrolatum, or Vaseline, is a semisolid mixture of hydrocarbons with more than 25 carbon atoms used in ointments and cosmetics and as a lubricant and a solvent.

Solubility and Density

Alkanes are nonpolar, which makes them insoluble in water. However, they are soluble in nonpolar solvents such as other alkanes. Alkanes have densities from 0.62 g/mL to about 0.79 g/mL, which is less dense than water (1.0 g/mL).

If there is an oil spill in the ocean, the alkanes in the oil, which do not mix with water, form a thin layer on the surface that spreads over a large area. In April 2010, an explosion on an oil-drilling rig in the Gulf of Mexico caused the largest oil spill in United States history. At its maximum, an estimated 10 million liters of oil was leaked every day. Other major oil spills occurred in Queensland, Australia (2009), the coast of Wales (1996), the Shetland Islands (1993), and Alaska, the *Exxon Valdez*, in 1989 (see Figure 10.6). If the

FIGURE 10.6 In oil spills, large quantities of oil spread over the water.

Q What physical properties cause oil to remain on the surface of water?

crude oil reaches land, there can be considerable damage to beaches, shellfish, fish, birds, and wildlife habitats. When animals such as birds are covered with oil, they must be cleaned quickly because ingestion of the hydrocarbons when they try to clean themselves is fatal.

Cleanup of oil spills includes mechanical, chemical, and microbiological methods. A boom may be placed around the leaking oil until it can be removed. Boats called skimmers then scoop up the oil and place it in tanks. In a chemical method, a substance that attracts oil is used to pick up oil, which is then scraped off into recovery tanks. Certain bacteria that ingest oil are also used to break oil down into less harmful products.

Combustion of Alkanes

TUTORIAL
Writing Balanced Equations for the Combustion of Alkanes

CASE STUDY
Poison in the Home: Carbon Monoxide

The carbon–carbon single bonds in alkanes are difficult to break, which makes them the least reactive family of organic compounds. However, alkanes burn readily in oxygen. As we discussed in Section 5.4, a carbon-containing compound, such as an alkane, undergoes *combustion* when it completely reacts with oxygen to produce carbon dioxide, water, and energy. For example, methane is the natural gas we use to cook our food on a gas cooktop and heat our homes.

$$CH_4(g) + 2O_2(g) \xrightarrow{\Delta} CO_2(g) + 2H_2O(g) + \text{energy}$$

CONCEPT CHECK 10.4

Combustion

Write a balanced equation for the complete combustion of butane.

ANSWER

The balanced equation for the complete combustion of butane can be written

$$2C_4H_{10}(g) + 13O_2(g) \xrightarrow{\Delta} 8CO_2(g) + 10H_2O(g) + \text{energy}$$

Chemistry Link to Health

TOXICITY OF CARBON MONOXIDE

When a propane heater, fireplace, or woodstove is used in a closed room, there must be adequate ventilation. If the supply of oxygen is limited, *incomplete combustion* from burning gas, oil, or wood produces carbon monoxide. The incomplete combustion of methane in natural gas is written

$$\underset{\text{Limited oxygen supply}}{2CH_4(g) + 3O_2(g)} \xrightarrow{\Delta} \underset{\text{Carbon monoxide}}{2CO(g)} + 4H_2O(g) + \text{energy}$$

Carbon monoxide (CO) is a colorless, odorless, poisonous gas. When inhaled, CO passes into the bloodstream, where it attaches to hemoglobin. When CO binds to the hemoglobin, it reduces the amount of oxygen (O_2) reaching the organs and cells. As a result, a healthy person can experience a reduction in exercise capability, visual perception, and manual dexterity.

When the amount of hemoglobin (Hb) bound to CO as COHb is about 10%, a person may experience shortness of breath, mild headache, and drowsiness. Heavy smokers can have as high as 9 percent COHb in their blood. When as much as 30 percent of the hemoglobin is COHb, a person may experience more severe symptoms including dizziness, mental confusion, severe headache, and nausea. If 50 percent or more of the hemoglobin is bound to CO, a person could become unconscious and die if not treated immediately with oxygen.

Chemistry Link to Industry

CRUDE OIL

Crude oil or petroleum contains a wide variety of hydrocarbons. At an oil refinery, the components in crude oil are separated by fractional distillation, a process that removes groups or fractions of hydrocarbons by continually heating the mixture to higher temperatures (see Table 10.6). Fractions containing alkanes with longer carbon chains require higher temperatures before they reach their boiling temperature and form gases. The gases are removed and passed through a distillation column where they cool and condense back to liquids. The major use of crude oil is to obtain gasoline, but a barrel of crude oil is only about 35% gasoline. To increase the production of gasoline, heating oils are broken down to give the lower-weight alkanes.

TABLE 10.6 Typical Alkane Mixtures Obtained by Distillation of Crude Oil

Distillation Temperatures (°C)	Number of Carbon Atoms	Product
Below 30	1–4	Natural gas
30–200	5–12	Gasoline
200–250	12–16	Kerosene, jet fuel
250–350	15–18	Diesel fuel, heating oil
350–450	18–25	Lubricating oil
Nonvolatile residue	Over 25	Asphalt, tar

A refinery converts crude oil into gasoline, heating oil, and other organic products.

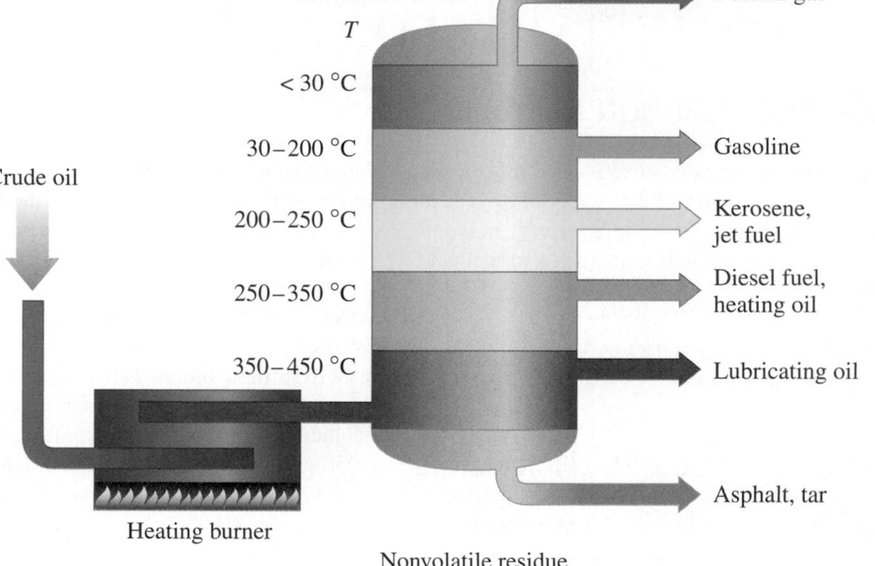

QUESTIONS AND PROBLEMS

Properties of Alkanes

10.21 Heptane, used as a solvent for rubber cement, has a density of 0.68 g/mL and boils at 98 °C.
a. Draw the condensed structural and skeletal formulas of heptane.
b. Is it a solid, liquid, or gas at room temperature?
c. Is it soluble in water?
d. Will it float on water or sink?
e. Write the balanced chemical equation for the complete combustion of heptane.

10.22 Nonane has a density of 0.79 g/mL and boils at 151 °C.
a. Draw the condensed structural and skeletal formulas of nonane.
b. Is it a solid, liquid, or gas at room temperature?
c. Is it soluble in water?
d. Will it float on water or sink?
e. Write the balanced chemical equation for the complete combustion of nonane.

10.23 Write a balanced chemical equation for the complete combustion of each of the following compounds:
a. ethane
b. cyclopropane, C_3H_6
c. octane

10.24 Write a balanced chemical equation for the complete combustion of each of the following compounds:
a. hexane
b. cyclopentane, C_5H_{10}
c. 2-methylbutane

10.5 Functional Groups

LEARNING GOAL
Classify organic molecules according to their functional groups.

TUTORIAL
Identifying Functional Groups

SELF STUDY ACTIVITY
Functional Groups

We have discussed the organic compounds that contain carbon and hydrogen. Now we can look at those organic compounds in which carbon atoms bond with the nonmetals oxygen, nitrogen, sulfur, and halogens. Table 10.7 lists the number of covalent bonds formed by elements found in organic compounds. In a typical organic compound, carbon forms four covalent bonds, hydrogen forms one covalent bond, nitrogen forms three covalent bonds, and oxygen and sulfur each form two covalent bonds.

TABLE 10.7 Covalent Bonds for Elements in Organic Compounds

Element	Group	Covalent Bonds	Structure of Atoms
H	1A (1)	1	H—
C	4A (14)	4	—C— (with bonds above and below)
N	5A (15)	3	—N̈— (with bond above)
O, S	6A (16)	2	—Ö— —S̈—
F, Cl, Br, I	7A (17)	1	—X: (X = F, Cl, Br, I)

There are millions of organic compounds and new ones are synthesized every day. We organize the organic compounds according to specific groups of atoms called **functional groups**. Compounds that have the same functional groups have similar properties. By identifying functional groups, we can classify organic compounds according to their structure and predict some of their common reactions.

Alkenes, Alkynes, and Aromatic Compounds

The hydrocarbon family also includes *alkenes*, *alkynes*, and *aromatic compounds*. An **alkene** contains at least one double bond, which forms when two adjacent carbon atoms share two pairs of valence electrons. To give carbon four bonds, each carbon

atom (black) in the double bond of ethene is attached to two hydrogen atoms (white). For an alkene, the ending of the name of the corresponding alkane is changed from *ane* to *ene*.

The simplest alkene, ethene (or ethylene), is $CH_2{=}CH_2$. Ethene is an important plant hormone involved in promoting the ripening of fruit. Commercially grown fruit, such as avocados, bananas, and tomatoes, are often picked before they are ripe. Before the fruit is brought to market, it is exposed to ethene to accelerate the ripening process. Ethene also accelerates the breakdown of cellulose in plants, which causes flowers to wilt and leaves to fall from trees.

In an **alkyne**, a triple bond forms when two carbon atoms share three pairs of valence electrons. The simplest alkyne is called ethyne, $HC{\equiv}CH$, commonly known as acetylene. In ethyne, each carbon atom in the triple bond is attached to one hydrogen atom. Alkenes and alkynes are known as *unsaturated hydrocarbons* because they contain double or triple bonds.

In 1825, Michael Faraday isolated a hydrocarbon called *benzene*, which had the formula, C_6H_6. Because compounds containing benzene often had fragrant odors, the family became known as **aromatic compounds**. A benzene molecule consists of a ring of six carbon atoms with one hydrogen atom attached to each carbon. Later, scientists proposed that the carbon atoms were arranged in a ring with alternating single and double bonds. Today, the benzene structure is typically represented as a hexagon with a circle in the center.

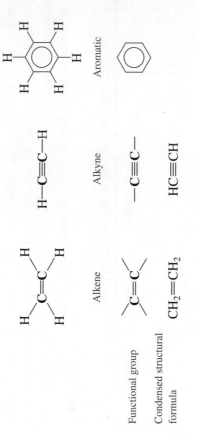

	Alkene	Alkyne	Aromatic
	$H_2C{=}CH_2$ (H, H / C=C / H, H)	$H{-}C{\equiv}C{-}H$	(benzene ring with six H)
Functional group	$>C{=}C<$	$-C{\equiv}C-$	(hexagon with circle)
Condensed structural formula	$CH_2{=}CH_2$	$HC{\equiv}CH$	

CONCEPT CHECK 10.5

Identifying Alkanes, Alkenes, and Alkynes

Classify each of the following condensed structural formulas as an alkane, alkene, or alkyne:

a. $CH_3{-}C{\equiv}C{-}CH_3$
b. $CH_3{-}CH_2{-}CH_3$
c. $CH_3{-}CH_2{-}CH_2{-}CH{=}CH_2$

ANSWER

a. A condensed structural formula with a carbon–carbon triple bond is an alkyne.
b. A condensed structural formula with only single bonds between carbon atoms is an alkane.
c. A condensed structural formula with a carbon–carbon double bond is an alkene.

Alcohols, Thiols, and Ethers

In an **alcohol**, a **hydroxyl** (—OH) **group** replaces a hydrogen atom in a hydrocarbon. The oxygen (O) atom is shown in red in the ball-and-stick models. A **thiol** contains the functional group —SH bonded to a carbon atom. Many thiols, such as those found in garlic and cheese, have strong, often unpleasant, odors. The odor of onions is due to the thiol called 1-propanethiol, $CH_3—CH_2—CH_2—SH$.

	Alcohol	Thiol	Ether
	$CH_3—CH_2—$**OH**	$CH_3—CH_2—$**SH**	$CH_3—$**O**$—CH_3$
Functional group	—OH	—SH	—O—

Alcohol

Thiol

An **ether** contains an oxygen atom (red) bonded to two carbon atoms. The oxygen atom also has two unshared pairs of electrons, but they are not shown in the condensed structural formulas.

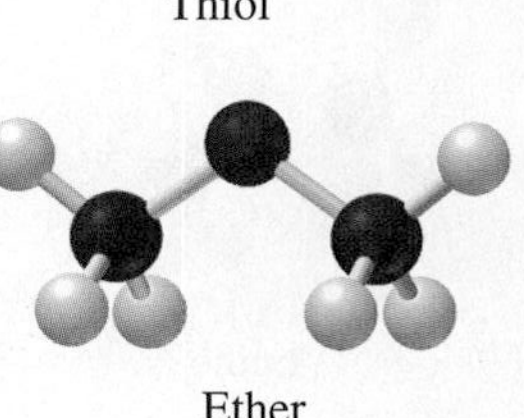

Ether

Aldehydes and Ketones

Aldehydes and ketones contain a **carbonyl group** (C=O), which consists of a carbon atom attached to an oxygen atom with a double bond. In an **aldehyde**, the carbon atom of the carbonyl group is bonded to a hydrogen atom. Only the simplest aldehyde, CH_2O, has a carbonyl group attached to two hydrogen atoms. In a **ketone**, the carbon atom of the carbonyl group is bonded to two other carbon atoms.

	Aldehyde	Ketone
	$CH_3—C(=O)—H$	$CH_3—C(=O)—CH_3$
Functional group	$—C(=O)—H$	$—C(=O)—$

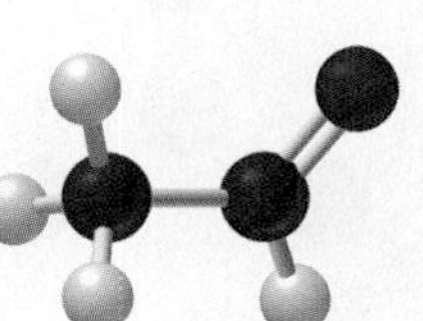

Aldehyde

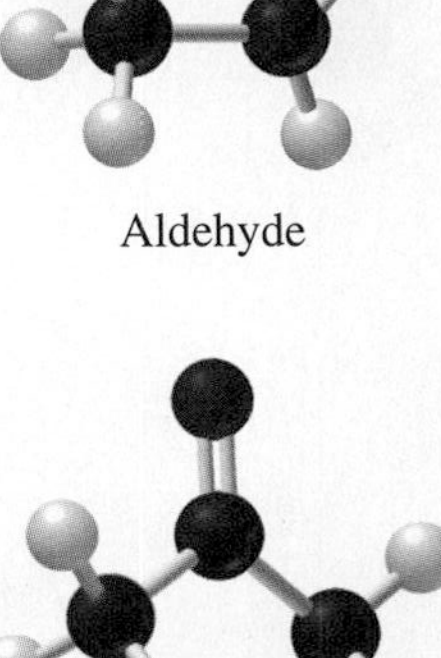

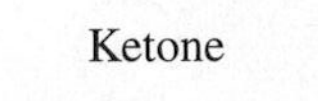

Ketone

Carboxylic Acids and Esters

In **carboxylic acids**, the functional group is the *carboxyl group*, which is a combination of the *carb*onyl and hydr*oxyl* groups. Carboxylic acids are common in nature. Formic acid, HCOOH, is injected under the skin from bee and red ant stings. A solution of acetic acid ($CH_3—COOH$) and water is the vinegar used in food preparation and salad dressings.

$CH_3—C(=O)—O—H$ or $CH_3—$**COOH**

Carboxylic acid

Functional group $—C(=O)—O—H$ or —**COOH**

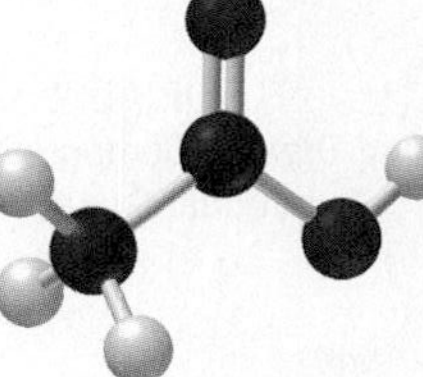

Carboxylic acid

An **ester** is similar to a carboxylic acid, except that the oxygen of the carboxyl group is attached to a carbon and not to hydrogen. Many of the fragrances of perfumes and flowers and the flavors of fruits are due to esters. Small esters are volatile, so we can smell them, and soluble in water, so we can taste them. Esters are responsible for the odor and flavor of oranges, bananas, pears, pineapples, and strawberries.

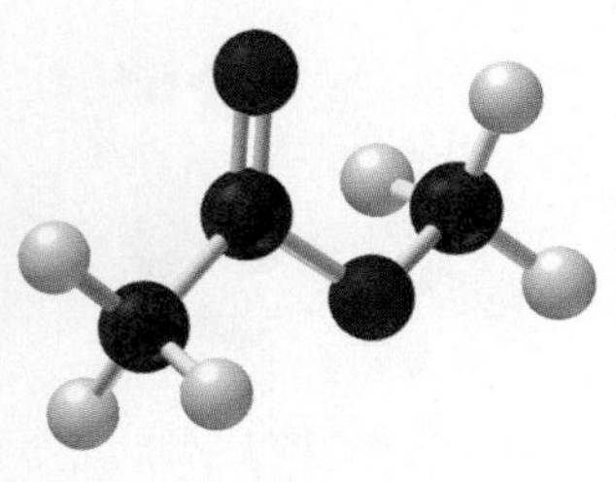
Ester

$$CH_3{-}\overset{\overset{\large O}{\|}}{C}{-}O{-}CH_3 \quad \text{or} \quad CH_3{-}\mathbf{COO}{-}CH_3$$

Ester

$$\text{Functional group} \quad {-}\overset{\overset{\large O}{\|}}{C}{-}O{-} \quad \text{or} \quad {-}\mathbf{COO}{-}$$

Amines and Amides

Amines are derivatives of ammonia, NH_3, in which carbon groups replace one, two, or three of the hydrogen atoms. From Table 10.7, we know that a nitrogen atom forms three bonds. One characteristic of fish is their odor, which is due to amines.

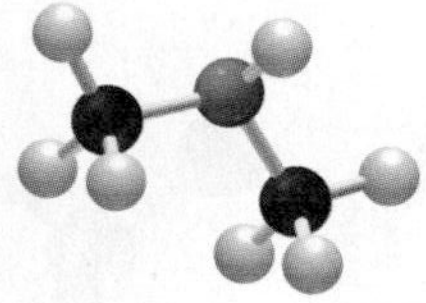

$$NH_3 \qquad CH_3{-}NH_2 \qquad CH_3{-}\underset{\underset{CH_3}{|}}{NH} \qquad CH_3{-}\underset{\underset{CH_3}{|}}{N}{-}CH_3$$

Ammonia — Example of amines

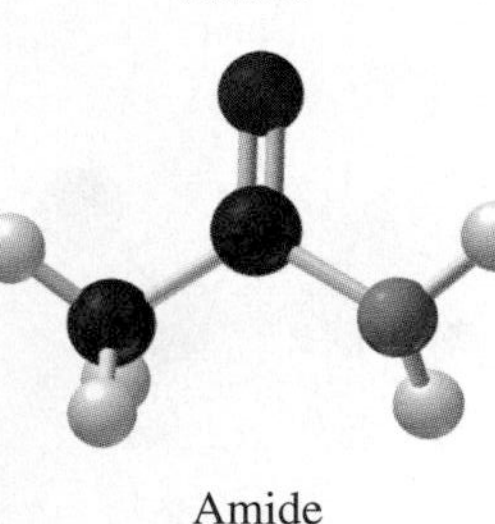
Amine

In an **amide**, the hydroxyl group of a carboxylic acid is replaced by a nitrogen group.

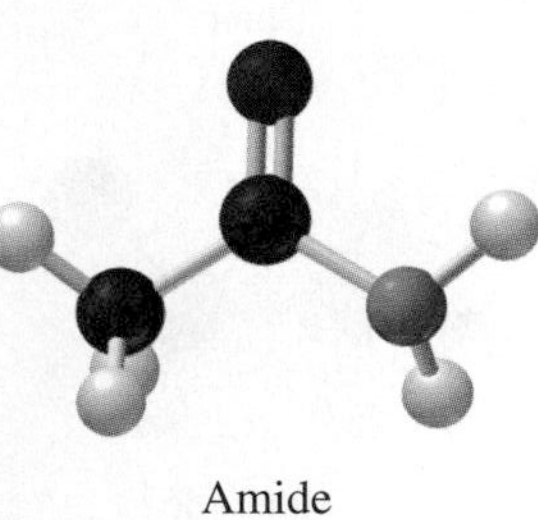
Amide

$$CH_3{-}\overset{\overset{\large O}{\|}}{C}{-}NH_2$$

Amide

The simplest natural amide is urea, an end product of protein metabolism in the body. The kidneys remove urea from the blood and provide for its excretion in urine. Urea is also used as a component of fertilizer to increase nitrogen in the soil.

$$H_2N{-}\overset{\overset{\large O}{\|}}{C}{-}NH_2$$

Urea

CASE STUDY
Death by Chocolate

TUTORIAL
Drawing Organic Compounds with Functional Groups

When we look at a molecule with functional groups, we need to isolate the functional group before we can classify the compound. In the following examples, we have highlighted the functional groups and classified the compound.

$$CH_3{-}CH_2{-}O{-}CH_3 \qquad CH_3{-}CH_2{-}CH_2{-}OH \qquad CH_3{-}NH{-}CH_3$$

Ether — Alcohol — Amine

$$CH_3{-}\overset{\overset{\large O}{\|}}{C}{-}CH_2{-}CH_3 \qquad CH_3{-}CH{=}CH{-}CH_3$$

Ketone — Alkene

SAMPLE PROBLEM 10.7

Classifying Organic Compounds

Classify the following organic compounds according to their functional groups:

a. $CH_3{-}CH_2{-}CH_2{-}CH_2{-}OH$
b. $CH_3{-}CH_2{-}NH{-}CH_3$

c. $CH_3-CH_2-\overset{\overset{\displaystyle O}{\|}}{C}-CH_2-CH_3$

d. $CH_3-CH_2-\overset{\overset{\displaystyle O}{\|}}{C}-OH$

SOLUTION

a. alcohol **b.** amine **c.** ketone **d.** carboxylic acid

STUDY CHECK 10.7

Classify the following organic compound according to its functional group:

$$CH_3-\overset{\overset{\displaystyle O}{\|}}{C}-O-CH_2-CH_3$$

A list of the common functional groups in organic compounds is shown in Table 10.8.

TABLE 10.8 Classification of Organic Compounds

Class	Functional Group	Example
Alkene	$>C=C<$	$H_2C=CH_2$
Alkyne	$-C\equiv C-$	$HC\equiv CH$
Aromatic	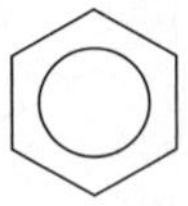 benzene ring	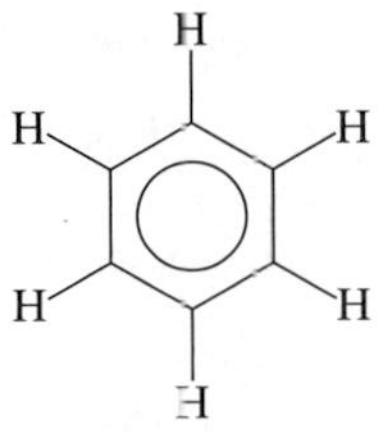 C_6H_6 (benzene ring with an H on each carbon)
Alcohol	$-OH$	CH_3-CH_2-OH
Ether	$-O-$	CH_3-O-CH_3
Thiol	$-SH$	CH_3-SH
Aldehyde	$-\overset{\overset{\displaystyle O}{\|}}{C}-H$	$CH_3-\overset{\overset{\displaystyle O}{\|}}{C}-H$
Ketone	$-\overset{\overset{\displaystyle O}{\|}}{C}-$	$CH_3-\overset{\overset{\displaystyle O}{\|}}{C}-CH_3$
Carboxylic acid	$-\overset{\overset{\displaystyle O}{\|}}{C}-OH$	$CH_3-\overset{\overset{\displaystyle O}{\|}}{C}-OH$
Ester	$-\overset{\overset{\displaystyle O}{\|}}{C}-O-$	$CH_3-\overset{\overset{\displaystyle O}{\|}}{C}-O-CH_3$
Amine	$-\overset{\vert}{N}-$	CH_3-NH_2
Amide	$-\overset{\overset{\displaystyle O}{\|}}{C}-\overset{\vert}{N}-$	$CH_3-\overset{\overset{\displaystyle O}{\|}}{C}-NH_2$

Chemistry Link to the Environment

FUNCTIONAL GROUPS IN FAMILIAR COMPOUNDS

The flavors and odors of foods and many household products can be attributed to the functional groups of organic compounds. As we discuss these familiar products, look for the functional groups we have described.

Ethyl alcohol is the alcohol found in alcoholic beverages. Isopropyl alcohol is another alcohol commonly used to disinfect skin before giving injections and to treat cuts.

$$CH_3-CH_2-OH$$

Ethyl alcohol

$$CH_3-\underset{}{\overset{\overset{\displaystyle OH}{|}}{CH}}-CH_3$$

Isopropyl alcohol

Acetone or dimethyl ketone is produced in great amounts commercially. Acetone is used as an organic solvent because it dissolves a wide variety of organic substances.

$$CH_3-\overset{\overset{\displaystyle O}{\|}}{C}-CH_3$$

Acetone

Ketones and aldehydes are found in flavorings such as vanilla, cinnamon, and spearmint. When we buy a small bottle of liquid flavorings, the aldehyde or ketone is dissolved in alcohol because the compounds are not very soluble in water. Formaldehyde, HCHO, the simplest aldehyde, is a colorless gas with a pungent odor. Industrially, it is a reactant in the synthesis of polymers used to make fabrics, insulation materials, carpeting, pressed wood products such as plywood, and plastics for kitchen counters. An aqueous solution called formalin, which contains 40% formaldehyde, is used as a germicide and to preserve biological specimens. The aldehyde butyraldehyde adds a buttery taste to foods and margarine.

$$CH_3-CH_2-CH_2-\overset{\overset{\displaystyle O}{\|}}{C}-H$$

Butyraldehyde (butter flavoring)

The sour tastes of vinegar and fruit juices and the pain from ant stings are all due to carboxylic acids. Acetic acid is the carboxylic acid that makes up vinegar. Aspirin also contains a carboxylic acid group. Esters found in fruits produce the pleasant aromas and tastes of bananas, oranges, pears, and pineapples. Esters are also used as solvents in many household cleaners, polishes, and glues.

$$CH_3-\overset{\overset{\displaystyle O}{\|}}{C}-OH$$

Acetic acid (in vinegar)

$$CH_3-\overset{\overset{\displaystyle O}{\|}}{C}-O-CH_2-CH_2-CH_3$$

Propyl acetate (pears)

$$CH_3-NH_2$$

Methylamine

$$CH_3-\overset{\overset{\displaystyle O}{\|}}{C}-O-CH_2-CH_2-CH_2-CH_2-CH_3$$

Pentyl acetate (bananas)

One of the characteristics of fish is their odor, which is due to amines. Amines produced when proteins decay have a particularly pungent and offensive odor.

$$H_2N-CH_2-CH_2-CH_2-CH_2-NH_2$$

Putrescine

$$H_2N-CH_2-CH_2-CH_2-CH_2-CH_2-NH_2$$

Cadaverine

Alkaloids are biologically active amines synthesized by plants to ward off insects and animals. Some typical alkaloids include caffeine, nicotine, histamine, and the decongestant epinephrine. Many are painkillers and hallucinogens such as morphine, LSD, marijuana, and cocaine. Certain parts of our neurons have receptor sites that respond to the various alkaloids. By modifying the structures of certain alkaloids to eliminate side effects, chemists have synthesized painkillers and drugs such as Novocain, codeine, and Valium.

QUESTIONS AND PROBLEMS

Functional Groups

10.25 Identify the class of compounds that contains each of the following functional groups:
- **a.** a hydroxyl group attached to a carbon chain
- **b.** a carbon–carbon double bond
- **c.** a carbonyl group attached to a hydrogen atom
- **d.** a carboxyl group attached to two carbon atoms

10.26 Identify the class of compounds that contains each of the following functional groups:
- **a.** a nitrogen atom attached to one or more carbon atoms
- **b.** a carboxyl group
- **c.** an oxygen atom bonded to two carbon atoms
- **d.** a carbonyl group between two carbon atoms

10.27 Classify the following molecules according to their functional groups. The possibilities are alcohol, ether, ketone, carboxylic acid, or amine.

a. $CH_3—CH_2—O—CH_2—CH_3$

b. $CH_3—CH(OH)—CH_3$

c. $CH_3—C(=O)—CH_2—CH_3$

d. $CH_3—CH_2—CH_2—COOH$

e. $CH_3—CH_2—NH_2$

10.28 Classify the following molecules according to their functional groups. The possibilities are alkene, aldehyde, carboxylic acid, ester, or amide.

a. $CH_3—C(=O)—O—CH_3$

b. $CH_3—C(=O)—NH_2$

c. $CH_3—CH_2—CH_2—C(=O)—H$

d. $CH_3—CH_2—CH_2—CH_2—CH_2—COOH$

e. $CH_3—CH=CH—CH_3$

CONCEPT MAP

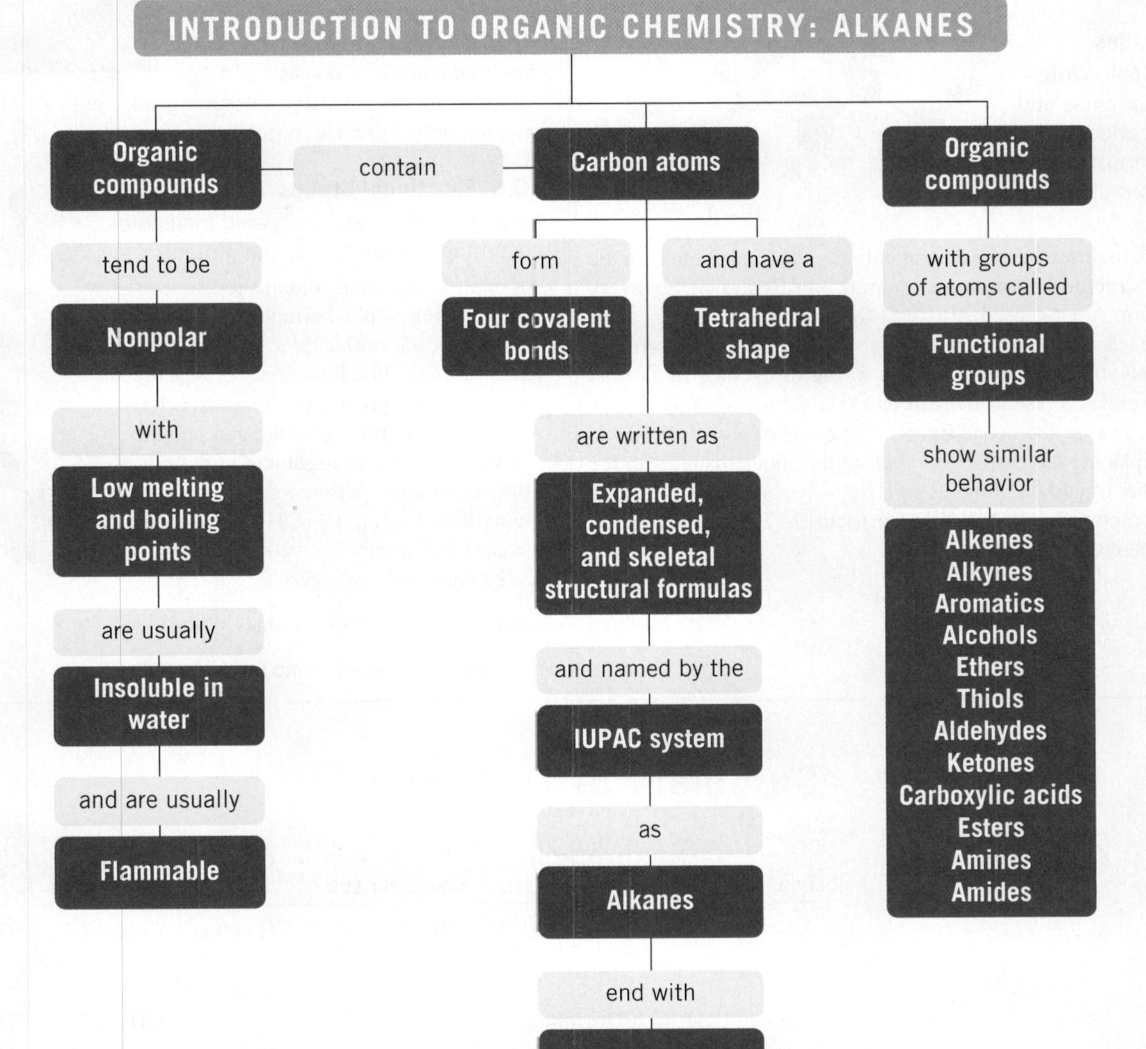

CHAPTER REVIEW

10.1 Organic Compounds

Learning Goal: Identify properties characteristic of organic or inorganic compounds.

Most organic compounds have covalent bonds and form nonpolar molecules. Often they have low melting points and low boiling points, are not very soluble in water, produce molecules in solutions, and burn vigorously in air. In contrast, many inorganic compounds are ionic or contain polar covalent bonds and form polar molecules. Many have high melting and boiling points, are usually soluble in water, produce ions in water, and do not burn in air. Carbon atoms share four valence electrons to form four covalent bonds. In the simplest organic molecule, methane (CH_4), the four bonds that bond hydrogen to the carbon atom are directed to the corners of a tetrahedron with bond angles of 109°.

10.2 Alkanes

Learning Goal: Write the IUPAC names and draw the condensed structural formulas and skeletal formulas for alkanes.

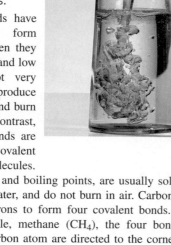

Alkanes are hydrocarbons that have only C—C single bonds. In the expanded structural formula, a separate line is drawn for every bonded atom. A condensed structural formula depicts groups composed of each carbon atom and its attached hydrogen atoms. In cycloalkanes, the carbon atoms form a ring or cyclic structure. A skeletal formula represents the carbon skeleton as ends and corners of a zigzag line or geometric figure. The name of a cycloalkane is written by placing the prefix *cyclo* before the alkane name with the same number of carbon atoms. The IUPAC system is used to name organic compounds in a systematic manner. The IUPAC name indicates the number of carbon atoms.

10.3 Alkanes with Substituents

Learning Goal: Write the IUPAC names for alkanes with substituents.

In an alkane, the carbon atoms are connected in a chain and bonded to hydrogen atoms. Substituents such as alkyl groups can replace hydrogen atoms on an alkane. In the IUPAC system, halogen atoms are named as fluoro, chloro, bromo, or iodo substituents attached to the main chain.

10.4 Properties of Alkanes

Learning Goal: Identify the properties of alkanes and write a balanced equation for combustion.

As nonpolar molecules, alkanes are not soluble in water. They are less dense than water. With only weak attractions, they have low melting and boiling points. Although the C—C bonds in alkanes resist most reactions, alkanes undergo combustion. In combustion, or burning, alkanes react with oxygen to produce carbon dioxide, water, and energy.

10.5 Functional Groups

Learning Goal: Classify organic molecules according to their functional groups.

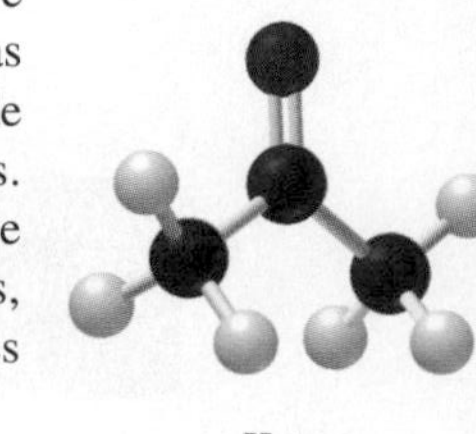

An organic molecule contains a characteristic group of atoms called a functional group that determines the molecule's family name and chemical reactivity. Functional groups are used to classify organic compounds, act as reactive sites in the molecule, and provide a system of naming organic compounds. Some common functional groups include the hydroxyl group (—OH) in alcohols, the carbonyl group (C=O) in aldehydes and ketones, and a nitrogen atom $\left(-\overset{|}{N}-\right)$ in amines, and a nitrogen and carbonyl group in amides.

Summary of Naming

Type	Example	Characteristic	Structure
Alkane	Propane	Single C—C, C—H bonds	$CH_3-CH_2-CH_3$
	2-Methylbutane		$CH_3-CH(CH_3)-CH_2-CH_3$
	1-Chloropropane	Halogen atom	$CH_3-CH_2-CH_2-Cl$
Cycloalkane	Cyclobutane	Carbon ring	□

Summary of Reactions

Combustion

$$\text{Alkane}(g) + O_2(g) \xrightarrow{\Delta} CO_2(g) + H_2O(g) + \text{energy}$$

Key Terms

alcohols A class of organic compounds that contains the hydroxyl (—OH) group bonded to a carbon atom.

aldehydes A class of organic compounds that contains a carbonyl group (C═O) bonded to at least one hydrogen atom.

alkanes Hydrocarbons containing only single bonds between carbon atoms.

alkenes Hydrocarbons that contain carbon–carbon double bonds (C═C).

alkyl group An alkane minus one hydrogen atom. Alkyl groups are named like the alkanes except a *yl* ending replaces *ane*.

alkynes Hydrocarbons that contain carbon–carbon triple bonds (C≡C).

amides A class of organic compounds in which the hydroxyl group of a carboxylic acid is replaced by a nitrogen group.

amines A class of organic compounds that contains a nitrogen atom bonded to one or more carbon atoms.

aromatic compound A compound that contains benzene. Benzene has a six-carbon ring with a hydrogen atom attached to each carbon.

branch A carbon group or halogen bonded to the main carbon chain.

branched alkane A hydrocarbon containing a substituent bonded to the main chain.

carbonyl group A functional group that contains a double bond between a carbon atom and an oxygen atom (C═O).

carboxylic acids A class of organic compounds that contains the functional group —COOH.

condensed structural formula A structural formula that shows the arrangement of the carbon atoms in a molecule but groups each carbon atom with its bonded hydrogen atoms (CH_3—, —CH_2—, or —CH—).

cycloalkane An alkane that is a ring or cyclic structure.

esters A class of organic compounds that contains a —COO— group with an oxygen atom bonded to carbon.

ethers A class of organic compounds that contains an oxygen atom bonded to two carbon atoms.

expanded structural formula A type of structural formula that shows the arrangement of the atoms by drawing each bond in the hydrocarbon as C—H or C—C.

functional group A group of atoms that determine the physical and chemical properties and naming of a class of organic compounds.

hydrocarbons Organic compounds containing only carbon and hydrogen.

hydroxyl group The group of atoms (—OH) characteristic of alcohols.

isomers Organic compounds in which identical molecular formulas have different arrangements of atoms.

IUPAC system A system for naming organic compounds devised by the International Union of Pure and Applied Chemistry.

ketones A class of organic compounds in which a carbonyl group (C═O) is bonded to two carbon atoms.

organic compounds Compounds made of carbon that typically have covalent bonds, nonpolar molecules, low melting and boiling points, are insoluble in water, and are flammable.

substituent Groups of atoms such as an alkyl group or a halogen bonded to the main chain or ring of carbon atoms.

thiol A class of organic molecules that contains the —SH functional group bonded to a carbon atom.

Understanding the Concepts

10.29 Sunscreens contain compounds such as oxybenzone and 2-ethylhexyl *p*-methoxycinnamate that absorb UV light.

Identify the functional groups in each of the following UV-absorbing compounds used in sunscreens:

a. oxybenzone

OH, O, C, CH_3, O

b. 2-ethylhexyl *p*-methoxycinnamate

O, O, CH_3, O

10.30 Oxymetazoline is a vasoconstrictor used in nasal decongestant sprays such as Afrin.

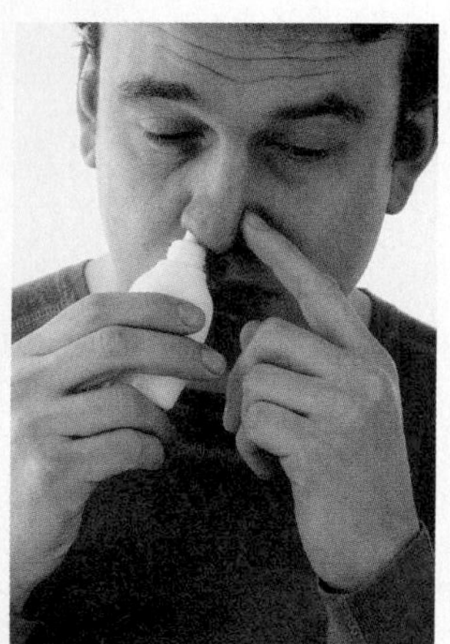

What functional groups are in oxymetazoline?

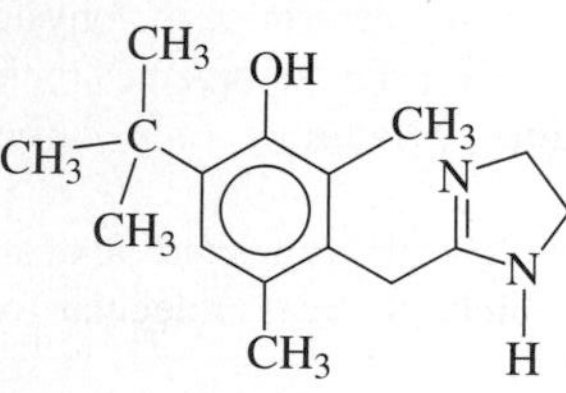

10.31 Decimemide is used as an anticonvulsant.

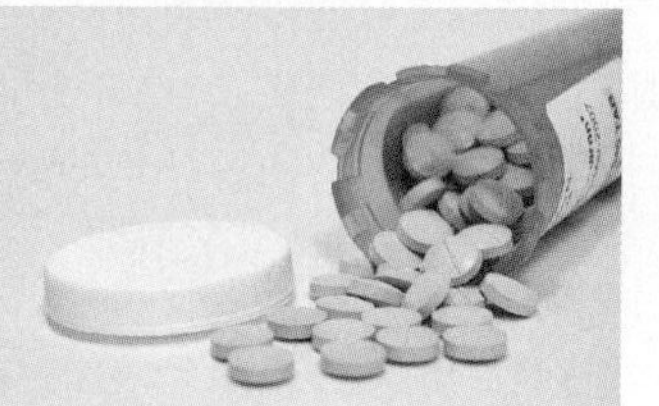

What functional groups are in decimemide?

$$CH_3-O \quad (ring) \quad CH_3-(CH_2)_9-O-C_6H_2(OCH_3)_2-\overset{O}{\overset{\|}{C}}-NH_2$$

10.32 The odor and taste of pineapples is from ethyl butyrate.

What functional group is in ethyl butyrate?

$$CH_3-CH_2-CH_2-\overset{O}{\overset{\|}{C}}-O-CH_2-CH_3$$

Additional Questions and Problems

For instructor-assigned homework, go to **www.masteringchemistry.com**.

10.33 Compare organic and inorganic compounds in terms of:
a. types of bonds **b.** solubility in water
c. melting points **d.** flammability

10.34 Identify each of the following compounds as organic or inorganic:
a. Na_2SO_4 **b.** $CH_2{=}CH_2$
c. Cr_2O_3 **d.** $C_{12}H_{22}O_{11}$

10.35 Match the following physical and chemical properties with potassium chloride, KCl, used in salt substitutes, or butane, C_4H_{10}, used in lighters:

a. melts at −138 °C
b. burns vigorously in air
c. melts at 770 °C
d. contains ionic bonds
e. is a gas at room temperature

10.36 Match the following physical and chemical properties with octane, C_8H_{18}, found in gasoline, or magnesium sulfate, $MgSO_4$, also called Epsom salts:
a. contains only covalent bonds
b. melts at 1124 °C
c. is insoluble in water
d. is a liquid at room temperature
e. is a strong electrolyte

10.37 Identify the functional groups in each of the following flavoring compounds:

a.

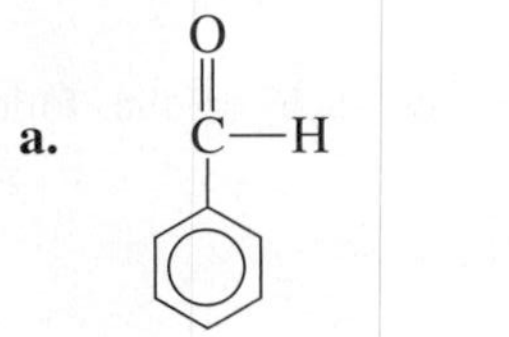

Almonds

b. $C_6H_5-CH{=}CH-\overset{\displaystyle O}{\overset{\|}{C}}-H$

Cinnamon sticks

c. $CH_3-\overset{\displaystyle O}{\overset{\|}{C}}-\overset{\displaystyle O}{\overset{\|}{C}}-CH_3$

Butter

10.38 Identify the functional groups in each of the following compounds:

a. BHA, an antioxidant used as a preservative in foods such as baked goods, butter, meats, and snack foods

(benzene ring with OH, $C(CH_3)_3$, and $O-CH_3$)

b. vanillin, a flavoring, obtained from the seeds of the vanilla bean

(benzene ring with $\overset{\displaystyle O}{\overset{\|}{C}}-H$, $O-CH_3$, and OH)

10.39 The sweetener aspartame is made from two amino acids: aspartic acid and phenylalanine. Identify the functional groups in aspartame.

$$H_2N-\underset{\underset{\underset{\displaystyle \overset{\displaystyle O}{\overset{\|}{C}}-OH}{|}}{CH_2}}{\overset{|}{CH}}-\overset{\displaystyle O}{\overset{\|}{C}}-\overset{\displaystyle H}{\overset{|}{N}}-\overset{\displaystyle C_6H_5-CH_2}{\overset{|}{CH}}-\overset{\displaystyle O}{\overset{\|}{C}}-O-CH_3$$

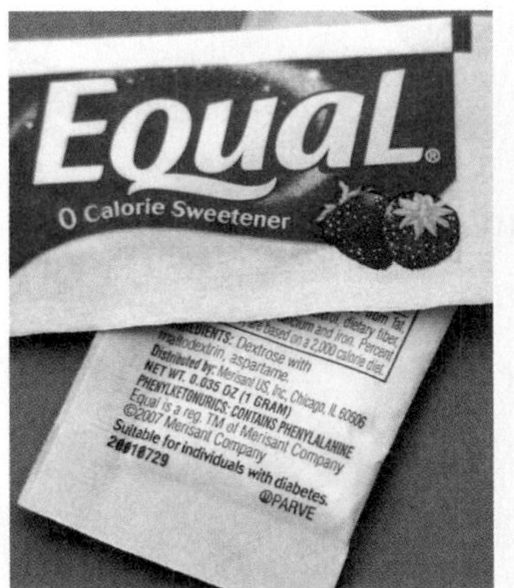

10.40 Some aspirin substitutes contain phenacetin to reduce fever. Identify the functional groups in phenacetin.

$$CH_3-CH_2-O-C_6H_4-NH-\overset{\displaystyle O}{\overset{\|}{C}}-CH_3$$

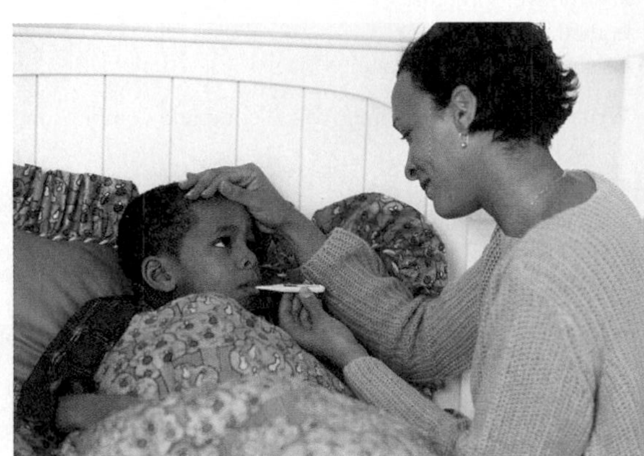

10.41 Give the IUPAC name for each of the following molecules:

a. $CH_3-CH_2-\overset{\displaystyle CH_3}{\overset{|}{\underset{\displaystyle CH_3}{\underset{|}{C}}}}-CH_3$

b. CH_3-CH_2-Cl

c. $CH_3-CH_2-\overset{\displaystyle CH_3-CH_2}{\overset{|}{CH}}-CH_2-\overset{\displaystyle Br}{\overset{|}{CH}}-CH_3$

d. (cyclohexane)

10.42 Give the IUPAC name for each of the following molecules:

a. (cyclohexane with CH_2-CH_3)

b. $Cl-CH_2-\overset{\displaystyle Br}{\overset{|}{CH}}-CH_2-Br$

```
       CH3
       |
c. CH3—CH—CH—CH3
            |
            CH2
            |
            CH2
            |
            CH3
```

```
               Cl
               |
d. CH3—CH2—C—CH2—CH3
               |
               CH2
               |
               CH3
```

10.43 Draw the condensed structural formula for each of the following molecules:
a. 3-ethylhexane
b. 2,3-dimethylpentane
c. 1,3-dichloro-3-methylheptane

10.44 Draw the condensed structural formula for each of the following molecules:
a. ethylcyclopropane
b. methylcyclohexane
c. isopropylcyclopentane

10.45 Write a balanced equation for the complete combustion of each of the following:
a. C_5H_{12}
b. cyclobutane, C_4H_8
c. hexane
d. $CH_3—CH=CH_2$

10.46 Write a balanced equation for the complete combustion of each of the following:
a. heptane
b. $HC\equiv C—CH_2—CH_3$
c. 2-methylpropane
d. $CH_3—CH_2—CH_2—CH_3$

10.47 Identify the functional group in each of the following compounds:
a. $CH_3—NH_2$

```
          OH
          |
b. CH3—C—CH2—CH3
          |
          CH3
```

```
       CH3 O
       |   ||
c. CH3—CH—C—O—CH3
```

```
       CH3
       |
d. CH3—CH—CH=CH2
```

10.48 Identify the functional group in each of the following compounds:
a. $CH_3—C\equiv CH$
b. $CH_3—O—CH_2—CH_3$
c. $CH_3—CH_2—CH_2—SH$

```
              O
              ||
d. CH3—CH2—C—H
```

10.49 Match each description with a term from the following list: alkane, alkene, alkyne, alcohol, ether, aldehyde, ketone, carboxylic acid, ester, amine, amide, functional group, isomers.
a. an organic compound that contains a hydroxyl group bonded to a carbon
b. a hydrocarbon that contains one or more carbon–carbon double bonds
c. an organic compound in which the carbon of a carbonyl group is bonded to a hydrogen
d. a hydrocarbon that contains only carbon–carbon single bonds
e. an organic compound in which the carbon of a carbonyl group is bonded to a hydroxyl group
f. an organic compound that contains a nitrogen atom bonded to one or more carbon atoms

10.50 Match each description with a term from the following list: alkane, alkene, alkyne, alcohol, ether, aldehyde, ketone, carboxylic acid, ester, amine, amide, functional group, isomers.
a. organic compounds with identical molecular formulas that differ in the order the atoms are connected
b. an organic compound in which the hydrogen atom of a carboxyl group is replaced by a carbon atom
c. an organic compound that contains an oxygen atom bonded to two carbon atoms
d. a hydrocarbon that contains a carbon–carbon triple bond
e. a characteristic group of atoms that make compounds behave and react in a particular way
f. an organic compound in which the carbonyl group is bonded to two carbon atoms

10.51 A tank on an outdoor heater contains 5.0 lb of propane.
a. Write the equation for the complete combustion of propane.
b. How many kilograms of CO_2 are produced by the complete combustion of all the propane?

10.52 A butane fireplace lighter contains 56.0 g of butane.
a. Write the equation for the complete combustion of butane.
b. How many grams of oxygen are needed for the complete combustion of the butane in the lighter?

Challenge Questions

10.53 The density of pentane, a component of gasoline, is 0.63 g/mL. The heat of combustion for pentane is 845 kcal/mole.
a. Write the equation for the complete combustion of pentane.
b. What is the molar mass of pentane?
c. How much heat is produced when 1 gallon of pentane is burned (1 gal = 4 qt)?
d. How many liters of CO_2 at STP are produced from the complete combustion of 1 gallon of pentane?

10.54 Draw the condensed structural formulas of two esters and a carboxylic acid that each have molecular formula $C_3H_6O_2$.

10.55 Draw the condensed structural formulas for all the possible alkane isomers that have a total of six carbon atoms and a four-carbon chain.

10.56 Draw the condensed structural formulas for all the possible alkane isomers that have four carbon atoms and a bromine.

10.57 Consider the compound propane.
- **a.** Draw the condensed structural formula.
- **b.** Write the equation for the complete combustion of propane.
- **c.** How many grams of O_2 are needed to react with 12.0 L of propane gas at STP?
- **d.** How many grams of CO_2 would be produced from the reaction in part **c**?

10.58 Consider the compound ethylcyclopentane.
- **a.** Draw the skeletal formula.
- **b.** Write the equation for the complete combustion of ethylcyclopentane.
- **c.** Calculate the grams of O_2 required for the reaction of 25.0 g of ethylcyclopentane.
- **d.** How many liters of CO_2 would be produced at STP from the reaction in part **c**?

10.59 In an automobile engine, "knocking" occurs when the combustion of gasoline occurs too rapidly. The octane number of gasoline represents the ability of a gasoline mixture to reduce knocking. A sample of gasoline is compared with heptane, rated 0 because it reacts with severe knocking, and 2,2,4-trimethylpentane, which has a rating of 100 because of its low knocking.
- **a.** Draw the condensed structural formula.
- **b.** Write the molecular formula.
- **c.** Write the balanced equation for the complete combustion of 2,2,4-trimethylpentane.

Answers

Answers to Study Checks

10.1 $CH_3—CH_2—CH_2—CH_2—CH_2—CH_2—CH_3$, heptane

10.2 cyclobutane

10.3 This condensed structural formula represents another isomer because it has the same molecular formula (C_6H_{14}), but it has a different arrangement of carbon atoms with one $—CH_3$ group attached to the middle (third) carbon of a five-carbon chain.

10.4 1-chloro-2,4-dimethylhexane

10.5 1,2-dibromoethane

10.6 $CH_3—CH(Br)—CH_2—CH(CH_3)—CH_3$ (skeletal formula with Br)

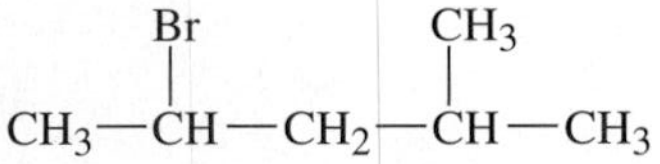

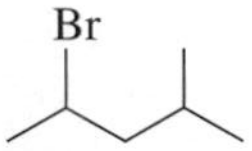

10.7 This represents an ester; the oxygen atom of the carboxyl group is attached to a carbon atom.

Answers to Selected Questions and Problems

10.1 **a.** inorganic **b.** organic **c.** organic **d.** inorganic **e.** inorganic **f.** organic

10.3 **a.** inorganic **b.** organic **c.** organic **d.** inorganic

10.5 **a.** ethane **b.** ethane **c.** NaBr **d.** NaBr

10.7 The four bonds between carbon and hydrogen in CH_4 are as far apart as possible, which means that the hydrogen atoms are at the corners of a tetrahedron.

10.9 **a.** pentane **b.** ethane **c.** hexane

10.11 **a.** CH_4
b. $CH_3—CH_3$
c. $CH_3—CH_2—CH_2—CH_2—CH_3$
d.  (triangle)

10.13 **a.** same molecule **b.** isomers of C_5H_{12} **c.** isomers of C_6H_{14}

10.15 **a.** 2,2-dimethylpropane **b.** 2,3-dimethylpentane **c.** 4-ethyl-2,2-dimethylhexane **d.** methylcyclopentane **e.** 1-fluoropropane **f.** 2-chloropropane

10.17 **a.** 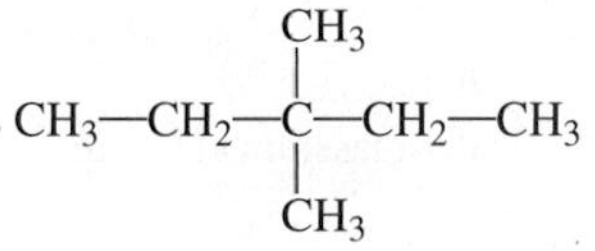 $CH_3—CH_2—C(CH_3)_2—CH_2—CH_3$
b. $CH_3—CH(CH_3)—CH(CH_3)—CH_2—CH(CH_3)—CH_3$
c. $CH_3—CH(CH_3)—CH(CH_2—CH_3)—CH_2—CH(CH_3)—CH_2—CH_2—CH_3$
d. $Br—CH_2—CH_2—Cl$
e. 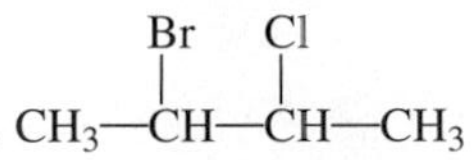 $CH_3—CH(Br)—CH(Cl)—CH_3$

10.19 **a.**  (skeletal formula)
b. 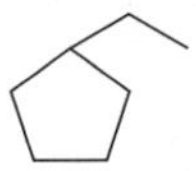 (skeletal formula)
c. 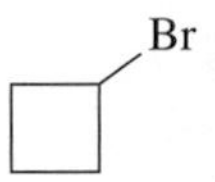 (skeletal formula) Br
d. 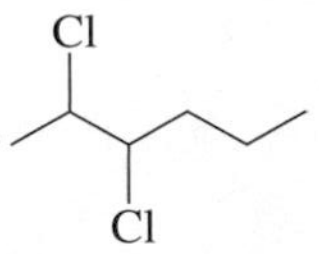 (skeletal formula) Cl Cl

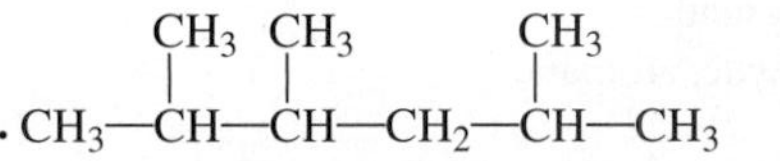

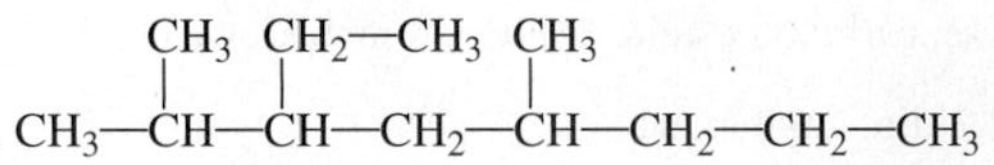

10.21 **a.** $CH_3—CH_2—CH_2—CH_2—CH_2—CH_2—CH_3$ (skeletal formula)
b. liquid
c. no
d. float
e. $C_7H_{16}(g) + 11O_2(g) \xrightarrow{\Delta} 7CO_2(g) + 8H_2O(g) + \text{energy}$

10.23 **a.** $2C_2H_6(g) + 7O_2(g) \xrightarrow{\Delta} 4CO_2(g) + 6H_2O(g) + \text{energy}$
b. $2C_3H_6(g) + 9O_2(g) \xrightarrow{\Delta} 6CO_2(g) + 6H_2O(g) + \text{energy}$
c. $2C_8H_{18}(g) + 25O_2(g) \xrightarrow{\Delta} 16CO_2(g) + 18H_2O(g) + \text{energy}$

10.25 a. alcohol **b.** alkene
c. aldehyde **d.** ester

10.27 a. ether **b.** alcohol
c. ketone **d.** carboxylic acid
e. amine

10.29 a. ether, aromatic, alcohol, ketone
b. ether, aromatic, alkene, ester

10.31 ether, aromatic, amide

10.33 a. Organic compounds have mostly covalent bonds; inorganic compounds have ionic as well as polar covalent bonds, and a few have nonpolar covalent bonds.
b. Most organic compounds are insoluble in water; many inorganic compounds are soluble in water.
c. Most organic compounds have low melting points; inorganic compounds have high melting points.
d. Most organic compounds are flammable; inorganic compounds are not usually flammable.

10.35 a. butane **b.** butane
c. potassium chloride **d.** potassium chloride
e. butane

10.37 a. aldehyde, aromatic
b. alkene, aldehyde, aromatic
c. ketone

10.39 amine, carboxylic acid, amide, aromatic, ester

10.41 a. 2,2-dimethylbutane
b. chloroethane
c. 2-bromo-4-ethylhexane
d. cyclohexane

10.43 a. $CH_3-CH_2-CH(CH_2-CH_3)-CH_2-CH_2-CH_3$

b. $CH_3-CH(CH_3)-CH(CH_3)-CH_2-CH_3$

c. $Cl-CH_2-CH_2-C(Cl)(CH_3)-CH_2-CH_2-CH_2-CH_3$

10.45 a. $C_5H_{12}(g) + 8O_2(g) \xrightarrow{\Delta} 5CO_2(g) + 6H_2O(g) + \text{energy}$

b. $C_4H_8(g) + 6O_2(g) \xrightarrow{\Delta} 4CO_2(g) + 4H_2O(g) + \text{energy}$

c. $2C_6H_{14}(g) + 19O_2(g) \xrightarrow{\Delta} 12CO_2(g) + 14H_2O(g) + \text{energy}$

d. $2C_3H_6(g) + 9O_2(g) \xrightarrow{\Delta} 6CO_2(g) + 6H_2O(g) + \text{energy}$

10.47 a. amine **b.** alcohol
c. ester **d.** alkene

10.49 a. alcohol **b.** alkene
c. aldehyde **d.** alkane
e. carboxylic acid **f.** amine

10.51 a. $C_3H_8(g) + 5O_2(g) \xrightarrow{\Delta} 3CO_2(g) + 4H_2O(g) + \text{energy}$
b. 6.8 kg of CO_2

10.53 a. $C_5H_{12}(g) + 8O_2(g) \xrightarrow{\Delta} 5CO_2(g) + 6H_2O(g) + \text{energy}$
b. 72.1 g/mole
c. 2.8×10^4 kcal
d. 3.7×10^3 L of CO_2 at STP

10.55 $CH_3-CH(CH_3)-CH(CH_3)-CH_3$ $CH_3-C(CH_3)_2-CH_2-CH_3$

10.57 a. $CH_3-CH_2-CH_3$
b. $C_3H_8(g) + 5O_2(g) \xrightarrow{\Delta} 3CO_2(g) + 4H_2O(g) + \text{energy}$
c. 85.7 g of O_2
d. 70.7 g of CO_2

10.59 a. $CH_3-C(CH_3)_2-CH_2-CH(CH_3)-CH_3$

b. C_8H_{18}

c. $2C_8H_{18}(g) + 25O_2(g) \xrightarrow{\Delta} 16CO_2(g) + 18H_2O(g) + \text{energy}$

11 Unsaturated Hydrocarbons

"Dentures replace natural teeth that are extracted due to cavities, bad gums, or trauma," says Dr. Irene Hilton, dentist, La Clinica De La Raza. "I make an impression of teeth using alginate, which is a polysaccharide extracted from seaweed. I mix the compound with water and place the gel-like material in the patient's mouth, where it becomes a hard, cement-like substance. I fill this mold with gypsum ($CaSO_4$) and water, which form a solid to which I add teeth made of plastic or porcelain. When I get a good match to the patient's own teeth, I prepare a preliminary wax denture. This is placed in the patient's mouth to check the bite and adjust the position of the replacement teeth. Then a permanent denture is made using a hard plastic polymer (methyl methacrylate)."

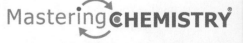

LOOKING AHEAD

In Chapter 10, we looked at alkanes, the hydrocarbons that contain only single bonds. Now we will investigate hydrocarbons that contain double bonds or triple bonds between carbon atoms. When we cook with vegetable oils such as corn oil, safflower oil, or olive oil, we are using organic compounds called lipids, which have one or more double bonds in their long carbon chains. Animal fats also contain long chains of carbon atoms, but with fewer double bonds. If we compare the two types of fats, we find considerable differences in their physical and chemical properties. Vegetable oils are liquid at room temperature, whereas animal fats are solid. Because double bonds are very reactive, they easily add hydrogen atoms (hydrogenation), water (hydration), or halogen atoms (halogenation) to the carbon atoms in the double bond.

LEARNING GOAL

Identify structural formulas as alkenes, cycloalkenes, and alkynes, and write their IUPAC or common names.

11.1 Alkenes and Alkynes

Alkenes and *alkynes* are families of hydrocarbons that contain double and triple bonds, respectively. They are called **unsaturated hydrocarbons** because they do not contain the maximum number of hydrogen atoms as do alkanes. They react with hydrogen gas to increase the number of hydrogen atoms to become alkanes, which are **saturated hydrocarbons**.

Identifying Alkenes and Alkynes

Alkenes contain one or more carbon–carbon double bonds that form when adjacent carbon atoms share two pairs of valence electrons. Recall that a carbon atom always forms four covalent bonds. In the simplest alkene, ethene, C_2H_4, two carbon atoms are connected by a double bond and each is also attached to two H atoms. This gives each carbon atom in the double bond a trigonal planar arrangement with bond angles of 120°. As a result, the ethene molecule is flat because the carbon and hydrogen atoms all lie in the same plane (see Figure 11.1).

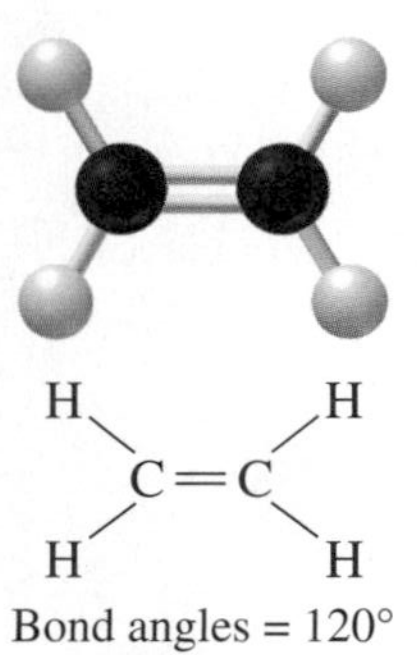

$H_2C{=}CH_2$

Bond angles = 120°

Ethene

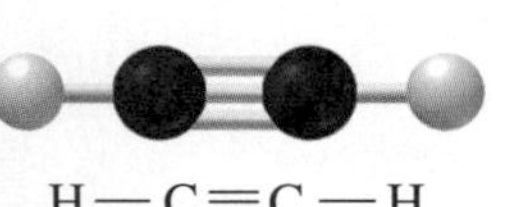

$H{-}C{\equiv}C{-}H$

Bond angles = 180°

Ethyne

FIGURE 11.1 Ball-and-stick models of ethene and ethyne show the functional groups of double or triple bonds and the bond angles.

Q Why are these compounds called unsaturated hydrocarbons?

A mixture of acetylene and oxygen undergoes combustion during the welding of metals.

Ethene, more commonly called ethylene, is an important plant hormone involved in promoting the ripening of fruit. Commercially grown fruit, such as avocados, bananas, and tomatoes, are often picked before they are ripe. Before the fruit is brought to market, it is exposed to ethylene to accelerate the ripening process. Ethylene also accelerates the breakdown of cellulose in plants, which causes flowers to wilt and leaves to fall from trees.

In an **alkyne**, a triple bond forms when two carbon atoms share three pairs of valence electrons. In the simplest alkyne, ethyne (C_2H_2), the two carbon atoms of the triple bond are each attached to one hydrogen atom, which gives a triple bond a linear geometry. Ethyne, commonly called acetylene, is used in welding where it reacts with oxygen to produce flames with temperatures above 3300 °C.

Fruit is ripened with ethene, a plant hormone.

CONCEPT CHECK 11.1

Identifying Unsaturated Compounds

Classify each of the following as an alkane, alkene, or alkyne:

a. $CH_3-C\equiv C-CH_2-CH_3$

b.

c.

ANSWER

a. A condensed structural formula with a carbon–carbon triple bond is an alkyne.
b. A skeletal formula with all single bonds between carbon atoms is an alkane.
c. A skeletal formula with a carbon–carbon double bond is an alkene.

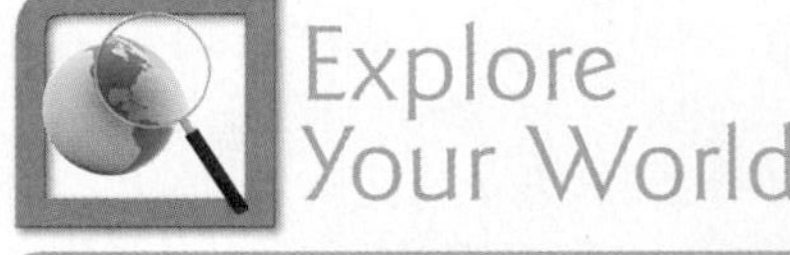

Explore Your World

RIPENING FRUIT

Obtain two unripe green bananas. Place one in a plastic bag and seal the bag. Leave the banana and the bag with a banana on the counter. Check the bananas twice a day to observe any difference in the ripening process.

QUESTIONS

1. What compound helps to ripen the bananas?
2. What are some possible reasons for any difference in the ripening rate?
3. If you wish to ripen an avocado, what procedure might you use?

Naming Alkenes and Alkynes

The IUPAC names for alkenes and alkynes are similar to those of alkanes. Using the alkane name with the same number of carbon atoms, the *ane* ending is replaced with *ene* for an alkene and *yne* for an alkyne (see Table 11.1). Cyclic alkenes are named as *cycloalkenes*.

TABLE 11.1 Comparison of Names for Alkanes, Alkenes, and Alkynes

Alkane	Alkene	Alkyne
CH_3-CH_3 Ethane	$H_2C=CH_2$ Ethene (ethylene)	$HC\equiv CH$ Ethyne (acetylene)
$CH_3-CH_2-CH_3$ Propane	$CH_3-CH=CH_2$ Propene	$CH_3-C\equiv CH$ Propyne

TUTORIAL
Naming Alkenes and Alkynes

Examples of naming an alkene and an alkyne are seen in Sample Problem 11.1.

SAMPLE PROBLEM 11.1

Naming Alkenes and Alkynes

Write the IUPAC name for each of the following:

a. $CH_3-\underset{}{\overset{CH_3}{\overset{|}{CH}}}-CH=CH-CH_3$

b. $CH_3-CH_2-C\equiv C-CH_2-CH_3$

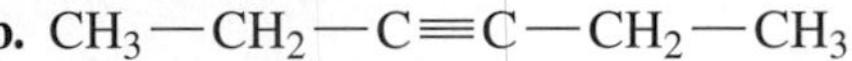

Guide to Naming Alkenes and Alkynes

1. Name the longest carbon chain that contains the double or triple bond.
2. Number the carbon chain starting from the end nearer the double or triple bond.
3. Give the location and name of each substituent (alphabetical order) as a prefix to the alkene or alkyne name.

SOLUTION

a. Step 1 Name the longest carbon chain that contains the double bond. There are five carbon atoms in the longest carbon chain containing the double bond. Replace the *ane* in the corresponding alkane name with *ene* to give pentene.

$$\underset{}{CH_3}-\overset{\overset{\large CH_3}{|}}{CH}-CH=CH-CH_3 \qquad \text{pentene}$$

Step 2 Number the carbon chain starting from the end nearer the double bond. Place the number of the first carbon in the double bond in front of the alkene name.

$$\underset{5}{CH_3}-\underset{4}{\overset{\overset{\large CH_3}{|}}{CH}}-\underset{3}{CH}=\underset{2}{CH}-\underset{1}{CH_3} \qquad \text{2-pentene}$$

Alkenes or alkynes with two or three carbons do not need numbers. For example, the double bond in ethene or propene must be between carbon 1 and carbon 2.

Step 3 Give the location and name of each substituent (alphabetical order) as a prefix to the alkene name. The methyl group is located on carbon 4.

$$\underset{5}{CH_3}-\underset{4}{\overset{\overset{\large CH_3}{|}}{CH}}-\underset{3}{CH}=\underset{2}{CH}-\underset{1}{CH_3} \qquad \text{4-methyl-2-pentene}$$

b. Step 1 Name the longest carbon chain that contains the triple bond. There are six carbon atoms in the longest chain containing the triple bond. Replace the *ane* in the corresponding alkane name with *yne* to give hexyne.

$$CH_3-CH_2-C\equiv C-CH_2-CH_3 \qquad \text{hexyne}$$

Step 2 Number the carbon chain starting from the end nearer the triple bond. Place the number of the first carbon in the triple bond in front of the alkyne name.

$$\underset{1}{CH_3}-\underset{2}{CH_2}-\underset{3}{C}\equiv\underset{4}{C}-\underset{5}{CH_2}-\underset{6}{CH_3} \qquad \text{3-hexyne}$$

Step 3 Give the location and name of each substituent (alphabetical order) as a prefix to the alkyne name. There are no substituents in this formula.

For *cycloalkenes*, the double bond is always between carbons 1 and 2. If there is a substituent, the ring is numbered from carbon 2 in the direction to give the lower number to the substituent.

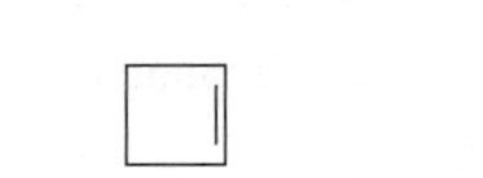

Cyclobutene 3-Methylcyclopentene

STUDY CHECK 11.1

Draw the condensed structural formula for each of the following:

a. 2-pentyne **b.** 2-chloro-1-hexene

TUTORIAL
Drawing Alkenes and Alkynes

Chemistry Link to the Environment

FRAGRANT ALKENES

The odors you associate with lemons, oranges, roses, and lavender are due to volatile compounds that are synthesized by the plants. The pleasant flavors and fragrances of many fruits and flowers are often due to unsaturated compounds. They were some of the first kinds of compounds to be extracted from natural plant material. In ancient times, they were highly valued in their pure forms. Limonene and myrcene give the characteristic odors and flavors to lemons and bay leaves, respectively. Geraniol and citronellal give roses and lemon grass their distinct aromas. In the food and perfume industries, these compounds are extracted or synthesized and used as perfumes and flavorings.

The characteristic odor of a rose is due to geraniol, a 10-carbon alcohol with a double bond.

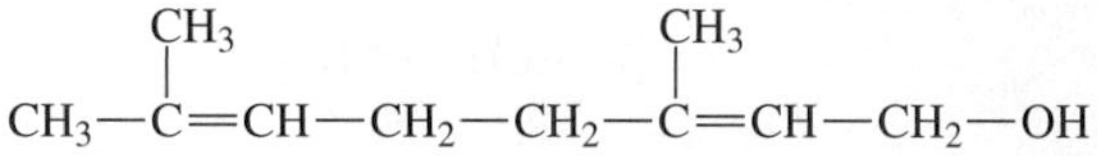

$CH_3-C(CH_3)=CH-CH_2-CH_2-C(CH_3)=CH-CH_2-OH$

Geraniol, roses

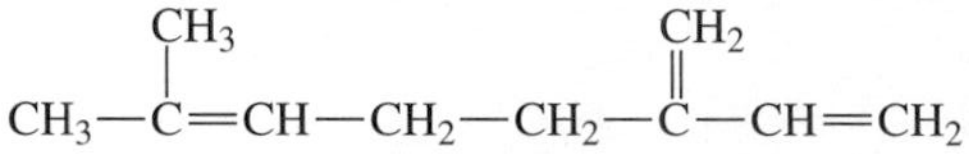

$CH_3-C(CH_3)=CH-CH_2-CH_2-C(=CH_2)-CH=CH_2$

Myrcene, bay leaves

$CH_3-C(CH_3)=CH-CH_2-CH_2-CH(CH_3)-CH_2-CHO$

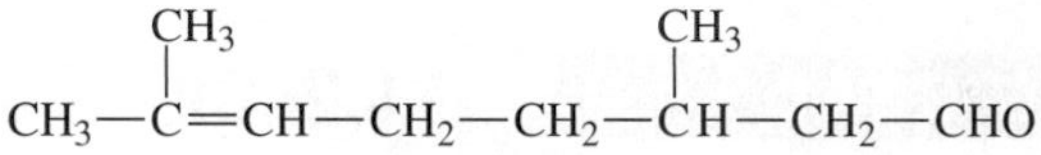

Citronellal, lemon grass

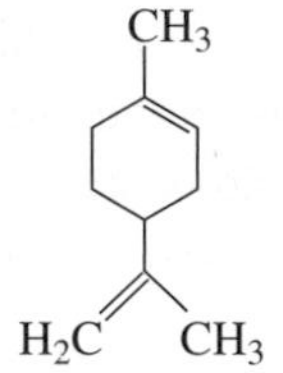

Limonene, lemons

QUESTIONS AND PROBLEMS

Alkenes and Alkynes

11.1 Identify the following as alkanes, alkenes, cycloalkenes, or alkynes:

a. 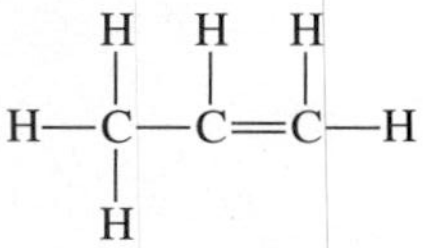 $H-CH_2-CH=CH-H$ (H—C(H)(H)—C(H)=C(H)—H)

b. $CH_3-CH_2-C\equiv C-H$

c. 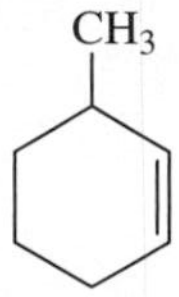 (1-methylcyclohexene structure, CH_3)

d.  (skeletal structure)

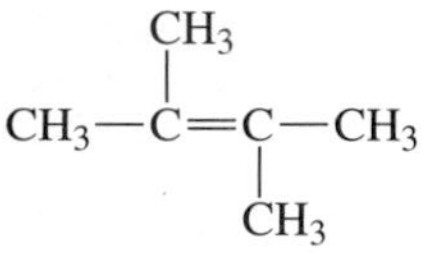

c. $CH_3-CH_2-C\equiv C-CH_2-CH_3$

d. $CH_3-CH_2-CH_2-CH_2-CH_3$

11.2 Identify the following as alkanes, alkenes, cycloalkenes, or alkynes:

a. (methylcyclopropene structure, CH_3)

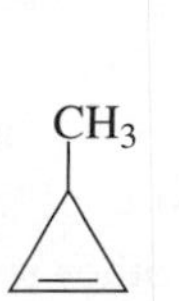

b. $CH_3-C(CH_3)=C(CH_3)-CH_3$

11.3 Compare the condensed structural formulas of propene and propyne.

11.4 Compare the condensed structural formulas of 1-butyne and 2-butyne.

11.5 Give the IUPAC name for each of the following:

a. $H_2C=CH_2$

b. 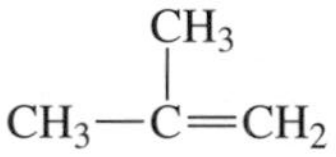 $CH_3-C(CH_3)=CH_2$

c. $CH_3-CH_2-C\equiv C-CH_3$

d.  (cyclobutene structure)

11.6 Give the IUPAC name for each of the following:

a. $H_2C{=}CH-CH_2-CH_3$

b. $CH_3-C{\equiv}C-CH_2-CH_2-\underset{}{CH}(CH_3)-CH_3$

c. 

d. $CH_3-CH_2-CH{=}CH-CH_3$

c. 2-methyl-1-butene

d. cyclohexene

e. 1-butyne

f. 1-bromo-3-hexyne

11.7 Draw the condensed structural formula for each of the following:

a. propene

b. 1-pentene

11.8 Draw the condensed structural formula for each of the following:

a. cyclopentene

b. 3-methyl-1-butyne

c. 3,4-dimethyl-1-pentene

d. cyclobutene

e. propyne

f. 2-methyl-2-hexene

LEARNING GOAL

Draw the condensed structural formulas and give the names for the cis–trans isomers of alkenes.

TUTORIAL
Cis–Trans Isomers

11.2 Cis–Trans Isomers

In any alkene, the double bond is rigid, which means there is no rotation around the double bond (see Explore Your World "Modeling Cis–Trans Isomers"). As a result, the atoms or groups that are attached to the carbon atoms in the double bond remain on one side or the other, which gives two different structures or isomers. For example, 1,2-dichloroethene can be drawn as two isomers when the expanded formulas with bond angles of 120° are drawn to the atoms bonded on the carbons in the double bond.

$$Cl-CH{=}CH-Cl$$

1,2-Dichloroethene

Hydrogen atoms are on the same side of the double bond.

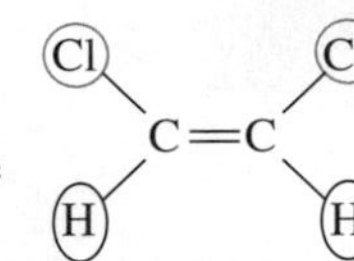

cis-1,2-Dichloroethene

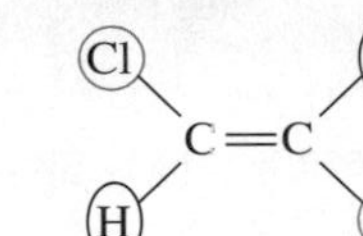

trans-1,2-Dichloroethene

Hydrogen atoms are on the opposite sides of the double bond.

Because there can be two different structures, we need two different names. When the hydrogen atoms are on the same side of the double bond, the structure is the **cis isomer**. When the hydrogen atoms are on the opposite sides of the double bond (across from each other), the structure is the **trans isomer**. Trans means "across" as in transcontinental.

The cis and trans isomers are also possible for alkenes with longer carbon chains. For example, we have drawn the condensed structural formula of 2-butene as

$$CH_3-CH{=}CH-CH_3$$

2-Butene

However, if we look closely at the carbon atoms in the double bond, we find that each one is bonded to a hydrogen atom and a CH_3- group. If we draw the expanded formula using bond angles of 120°, we find that there are two isomers for 2-butene.

cis-2-Butene
(mp –139 °C; bp 3.7 °C)

trans-2-Butene
(mp –106 °C; bp 0.3 °C)

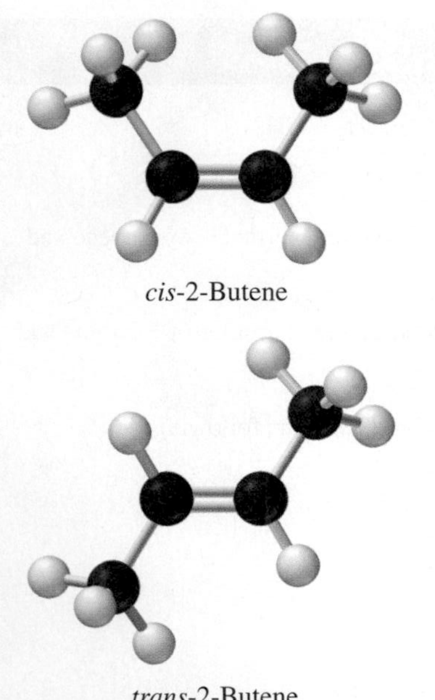

cis-2-Butene

trans-2-Butene

FIGURE 11.2 Ball-and-stick models of the cis and trans isomers of 2-butene.

Q What feature in 2-butene accounts for the cis and trans isomers?

In *cis*-2-butene isomer, the hydrogen atoms appear on the same side of the double bond and the CH_3- groups are attached to the double bond on the other side. In the *trans*-2-butene isomer, the hydrogen atoms in the double bond appear across from each other or on the opposite sides as do the CH_3- groups (see Figure 11.2). Thus molecules of 2-butene have cis and trans isomers, which are named using the prefix *cis* or *trans* to specify each isomer.

In general, trans isomers are more stable than their cis counterparts because the groups that are bigger than hydrogen atoms on the double bond are farther apart. The cis–trans isomers of 2-butene are different compounds with different physical properties, such as melting and boiling points, as well as different chemical properties.

As long as the groups attached to each carbon atom in the double bond are different, an alkene will show cis–trans isomers. It is important to consider the possibility of cis and trans isomers when given the name of an alkene. For example, 3-hexene can be drawn with cis and trans isomers using the H atoms and the $CH_3—CH_2—$ groups attached to the carbon atoms in the double bond.

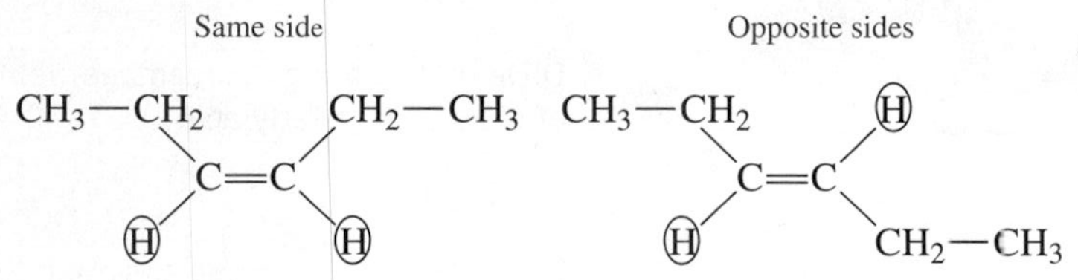

Same side — Opposite sides

$CH_3—CH_2$, $CH_2—CH_3$, C=C, H, H

cis-3-Hexene

$CH_3—CH_2$, H, C=C, H, $CH_2—CH_3$

trans-3-Hexene

CONCEPT CHECK 11.2

Identifying Cis and Trans Isomers

Identify each of the following as a description of a cis isomer or a trans isomer. Give the name of each.

a. In an isomer of 3-hexene, the hydrogen atoms attached to the carbon atoms in the double bond are on the opposite sides of the double bond.

b. In an isomer of 2-pentene, the hydrogen atoms in the double bond are attached on the same side of the double bond.

ANSWER

a. When the hydrogen atoms in the double bond are attached on the opposite sides, the isomer is a trans isomer, which is named *trans*-3-hexene.

b. When the hydrogen atoms in the double bond are attached on the same side, the isomer is a cis isomer, which is named *cis*-2-pentene.

Alkenes that consist of cis and trans isomers must have different groups attached to the carbon atoms in the double bond. If an alkene has identical groups, such as two H atoms on the same carbon atom, that alkene does not produce cis–trans isomers. For example, in 1-butene there are two hydrogen atoms on the first carbon in the double bond. If we switched the two H atoms, the structures stay the same. Thus, 1-butene does not have cis–trans isomers.

Identical atoms — H, H, C=C, $CH_2—CH_3$, H

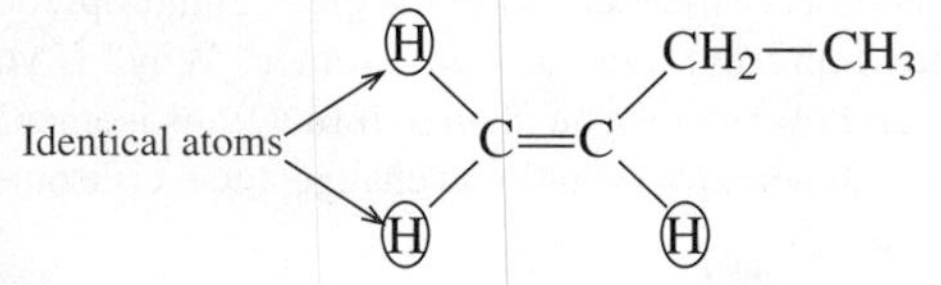

1-Butene

More complex systems of naming alkene isomers are possible but will not be discussed in this text. Alkynes do not have cis–trans isomers because the carbons in the triple bond are each attached to only one group.

CONCEPT CHECK 11.3

Characteristics of Fatty Acids

The formula of oleic acid, found in olive oil, is often written in its skeletal form.

Olive oil has a high percentage of oleic acid, a fatty acid.

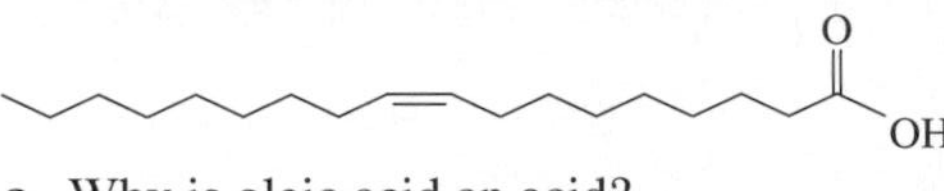

a. Why is oleic acid an acid?
b. Is oleic acid an unsaturated or saturated fatty acid?
c. Is this skeletal formula for oleic acid drawn as the cis isomer or the trans isomer?

ANSWER

a. Oleic acid has a carboxyl functional group, which makes it a carboxylic acid.
b. Oleic acid contains one double bond, which makes it an unsaturated fatty acid.
c. This skeletal formula of oleic acid is drawn as the cis isomer.

Explore Your World

MODELING CIS–TRANS ISOMERS

Because cis–trans isomerism is not easy to visualize, here are some things you can do to understand the difference in rotation around a single bond compared to a double bond and how it affects groups that are attached to the carbon atoms in the double bond.

Put the tips of your index fingers together. This is a model of a single bond. Consider the index fingers as a pair of carbon atoms, and think of your thumbs and other fingers as other parts of a carbon chain. While your index fingers are touching, twist your hands and change the position of the thumbs relative to each other. Notice how the relationship of your other fingers changes.

Now place the tips of your index fingers and middle fingers together in a model of a double bond. As you did before, twist your hands to move the thumbs away from each other. What happens? Can you change the location of your thumbs relative to each other without breaking the double bond? The difficulty of moving your hands with two fingers touching represents the lack of rotation about a double bond. You have made a model of a cis isomer when both thumbs point in the same direction. If you turn one hand over so one thumb points down and the other thumb points up, you have made a model of a trans isomer.

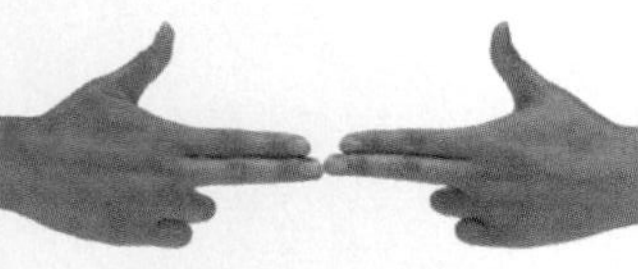

Cis-hands (cis-thumbs/fingers)

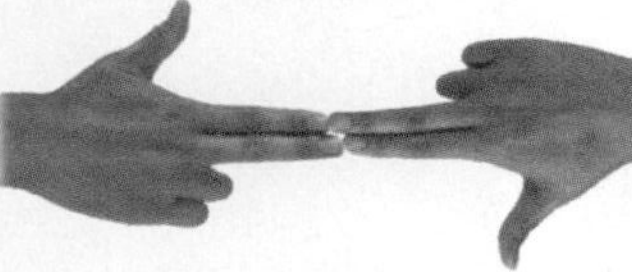

Trans-hands (trans-thumbs/fingers)

Using Gumdrops and Toothpicks to Model Cis–Trans Isomers

Obtain some toothpicks and yellow, green, and black gumdrops. The black gumdrops represent C atoms, the yellow gumdrops represent H atoms, and the green gumdrops represent Cl atoms. Place a toothpick between two black gumdrops. Use six more toothpicks to attach two yellow gumdrops and one green gumdrop to each black gumdrop. Rotate one of the gumdrop carbon atoms to show the conformations of the attached H and Cl atoms.

Remove a toothpick and yellow gumdrop from each black gumdrop. Place a second toothpick between the carbon atoms, which makes a double bond. Try to twist the double bond of toothpicks. Can you do it? When you observe the location of the green gumdrops, does the model you made represent a cis or trans isomer? Why? If your model is a cis isomer, how would you change it to a trans isomer? If your model is a trans isomer, how would you change it to a cis isomer?

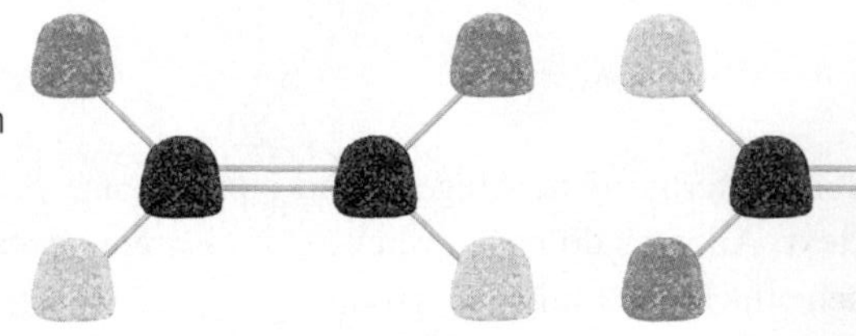

Models can be made from gumdrops to represent cis and trans isomers.

SAMPLE PROBLEM 11.2

Identifying Cis–Trans Isomers

Identify each of the following as the cis or trans isomer and give its name:

a. 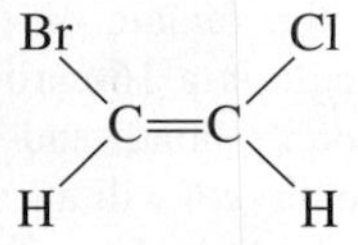 Br and Cl on top; $C{=}C$; H and H on bottom

b. 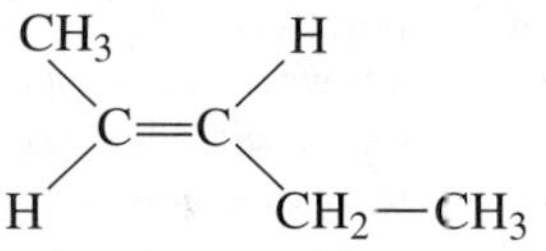 CH_3 and H on top; $C{=}C$; H and $CH_2{-}CH_3$ on bottom

SOLUTION

a. This is a cis isomer because the two H atoms in the double bond are on the same side. The two-carbon alkene with a bromo group on carbon 1 and a chloro group on carbon 2 is named *cis*-1-bromo-2-chloroethene.

b. This is a trans isomer because the two H atoms in the double bond are on the opposite sides of the double bond. The five-carbon alkene is named *trans*-2-pentene.

STUDY CHECK 11.2

Is the following compound *cis*-3-hexene or *trans*-3-hexene?

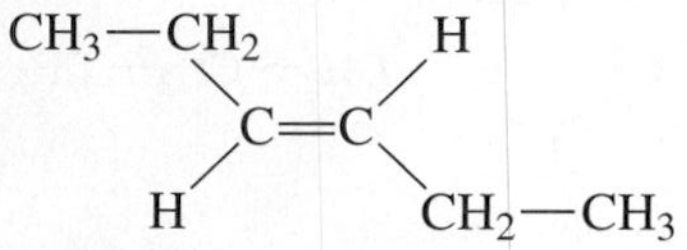

$CH_3{-}CH_2$ and H on top; $C{=}C$; H and $CH_2{-}CH_3$ on bottom

QUESTIONS AND PROBLEMS

Cis–Trans Isomers

11.9 Write the IUPAC name of each of the following, using cis or trans prefixes:

a. 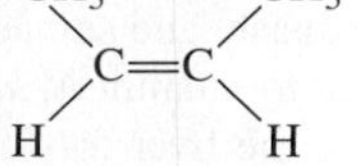 CH_3 and CH_3 on top; $C{=}C$; H and H on bottom

b. $CH_3{-}CH_2$ and H on top; $C{=}C$; H and $CH_2{-}CH_2{-}CH_2{-}CH_3$ on bottom

c. $CH_3{-}CH_2{-}CH_2$ and $CH_2{-}CH_3$ on top; $C{=}C$; H and H on bottom

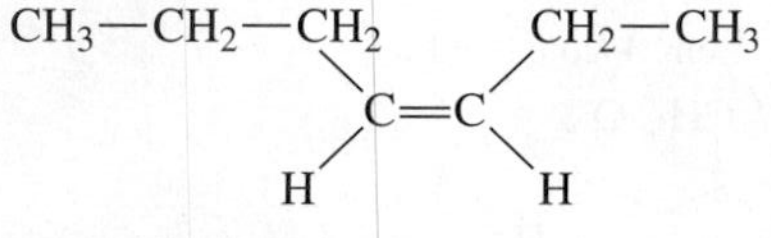

11.10 Write the IUPAC name of each of the following, using cis or trans prefixes:

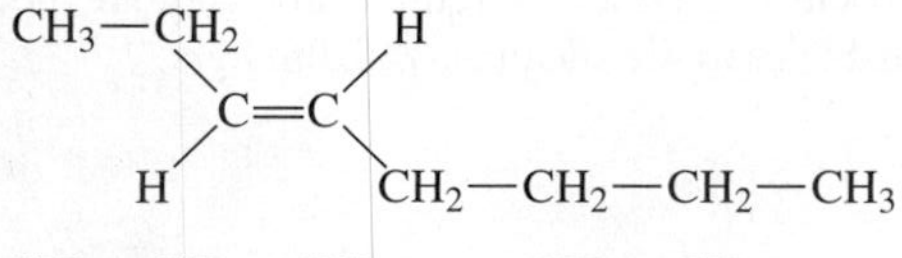

a. 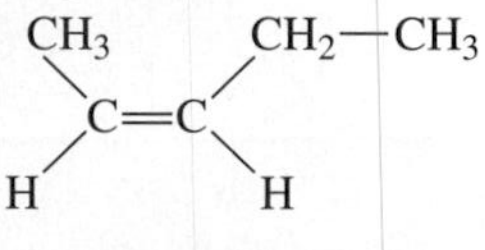 CH_3 and $CH_2{-}CH_3$ on top; $C{=}C$; H and H on bottom

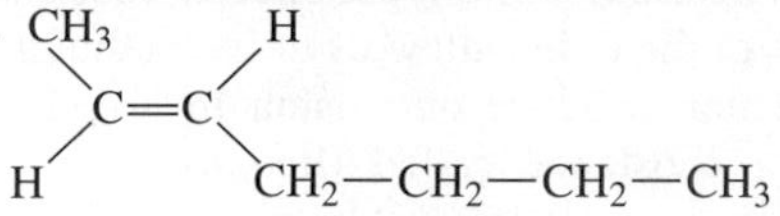

b. CH_3 and H on top; $C{=}C$; H and $CH_2{-}CH_2{-}CH_2{-}CH_3$ on bottom

c. $CH_3{-}CH_2{-}CH_2$ and H on top; $C{=}C$; H and $CH_2{-}CH_3$ on bottom

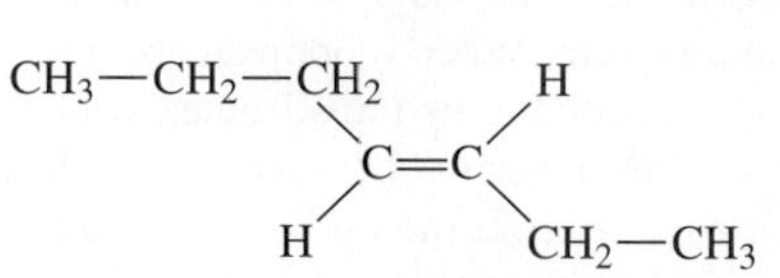

11.11 Draw the condensed structural formula for each of the following:

a. *trans*-1-bromo-2-chloroethene
b. *cis*-2-pentene
c. *trans*-3-heptene

11.12 Draw the condensed structural formula for each of the following:

a. *cis*-1,2-difluoroethene
b. *trans*-2-pentene
c. *cis*-4-octene

Chemistry Link to the Environment

PHEROMONES IN INSECT COMMUNICATION

Insects and many other organisms emit minute quantities of chemicals called pheromones. Insects use pheromones to send messages to individuals of the same species. Some pheromones warn of danger, others call for defense, mark a trail, or attract the opposite sex. In the last 40 years, the structures of many pheromones have been chemically determined. One of the most studied is bombykol, the sex pheromone produced by the female of the silkworm moth species. The bombykol molecule is a 16-carbon chain with one cis double bond, one trans double bond, and an alcohol group. A few molecules of synthetic bombykol will attract male silkworm moths from distances of over 1 kilometer. The effectiveness of many of these pheromones depends on the cis or trans configuration of the double bonds in the molecules. A certain species will respond to one isomer but not the other. Scientists are interested in synthesizing pheromones for use as nontoxic alternatives to pesticides. When used in a trap, bombykol can be used to isolate male silkworm moths. When a synthetic pheromone is released in several areas of a field or crop, the males cannot locate the females, which disrupts the reproductive cycle. This technique has been successful with controlling the oriental fruit moth, the grapevine moth, and the pink bollworm.

Pheromones allow insects to attract mates from a great distance.

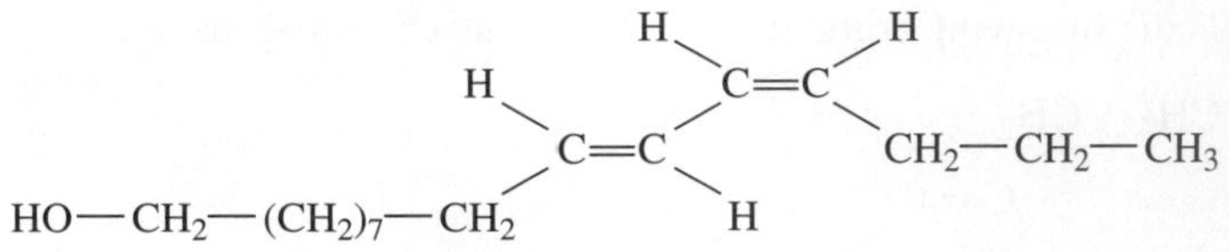

Bombykol, sex attractant for the silkworm moth

Chemistry Link to Health

CIS–TRANS ISOMERS FOR NIGHT VISION

The retinas of the eyes consist of two types of cells: rods and cones. The rods on the edge of the retina allow us to see in dim light, and the cones, in the center, produce our vision in bright light. In the rods, there is a substance called rhodopsin that absorbs light. Rhodopsin is composed of *cis*-11-retinal, an unsaturated compound, attached to a protein. When rhodopsin absorbs light, the *cis*-11-retinal isomer is converted to its trans isomer, which changes its shape. The trans form no longer fits the protein, and it separates from the protein. The change from the cis to trans isomer and its separation from the protein generate an electrical signal that the brain converts into an image.

An enzyme (isomerase) converts the trans isomer back to the *cis*-11-retinal isomer and the rhodopsin re-forms. If there is a deficiency of rhodopsin in the rods of the retina, night blindness may occur. One common cause is a lack of vitamin A in the diet. In our diet, we obtain vitamin A from plant pigments containing β-carotene, which is found in foods such as carrots, squash, and spinach. In the small intestine, the β-carotene is converted to vitamin A, which can be converted to *cis*-11-retinal or stored in the liver for future use. Without a sufficient quantity of retinal, not enough rhodopsin is produced to enable us to see adequately in dim light.

Cis–Trans Isomers of Retinal

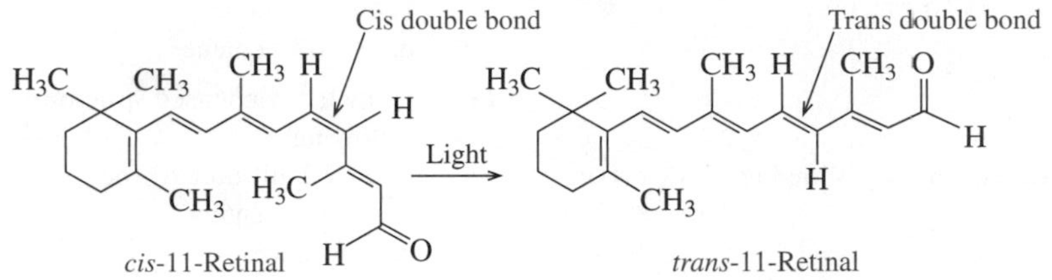

11.3 Addition Reactions

The most characteristic reaction of alkenes and alkynes is the **addition** of atoms or groups of atoms to the carbon atoms in a double or triple bond. Addition occurs because double and triple bonds are easily broken, providing electrons to form new single bonds.

The addition reactions have different names that depend on the type of reactant we add to the alkene or alkyne, as Table 11.2 shows.

LEARNING GOAL

Draw the condensed structural formulas and give the names for the organic products of addition reactions of alkenes and alkynes.

TABLE 11.2 Summary of Addition Reactions

Name of Addition Reaction	Reactants	Catalysts	Product
Hydrogenation	Alkene + H_2	Pt, Ni, or Pd	Alkane
	Alkyne + $2H_2$	Pt, Ni, or Pd	Alkane
Halogenation	Alkene + Br_2 (Cl_2)		Dihaloalkane
Hydration	Alkene + H_2O	H^+ (strong acid)	Alcohol

TUTORIAL
Addition Reactions

TUTORIAL
Hydrogenation and Hydration Reactions

Hydrogenation

In a reaction called **hydrogenation,** H atoms add to each of the carbon atoms in a double bond of an alkene or in the triple bond of an alkyne. During hydrogenation, the double or triple bonds are converted to single bonds in alkanes. A catalyst such as finely divided platinum (Pt), nickel (Ni), or palladium (Pd) is used to speed up the reaction. The general equation for hydrogenation can be written as follows:

$$\mathrm{{>}C{=}C{<}} + \mathrm{H{-}H} \xrightarrow{\text{Pt, Ni, or Pd}} \mathrm{{-}\overset{H}{\underset{|}{\overset{|}{C}}}{-}\overset{H}{\underset{|}{\overset{|}{C}}}{-}}$$

Double bond (alkene) — Single bond (alkane)

Some examples of the hydrogenation of alkenes follow:

$$\underset{\text{2-Butene}}{CH_3{-}CH{=}CH{-}CH_3} + H{-}H \xrightarrow{\text{Pt}} \underset{\text{Butane}}{CH_3{-}\overset{H}{\overset{|}{C}H}{-}\overset{H}{\overset{|}{C}H}{-}CH_3}$$

$$\text{Cyclohexene} + H{-}H \xrightarrow{\text{Ni}} \text{Cyclohexane (with added H, H)}$$

Cyclohexene — Cyclohexane

The hydrogenation of alkynes requires two molecules of hydrogen (H_2) to form the alkane product.

$$\underset{\text{2-Butyne}}{CH_3{-}C{\equiv}C{-}CH_3} + 2\,H{-}H \xrightarrow{\text{Pt}} \underset{\text{Butane}}{CH_3{-}\underset{H}{\overset{H}{C}}{-}\underset{H}{\overset{H}{C}}{-}CH_3}$$

UNSATURATION IN FATS AND OILS

Read the labels on some containers of vegetable oils, margarine, peanut butter, and shortenings.

QUESTIONS

1. What terms on the label tell you that the compounds contain double bonds?
2. A label on a bottle of canola oil lists saturated, polyunsaturated, and monounsaturated fats. What do these terms tell you about the type of bonding in the fats?
3. A peanut butter label states that it contains partially hydrogenated vegetable oils or completely hydrogenated vegetable oils. What does this tell you about the type of reaction that took place in preparing the peanut butter?

SAMPLE PROBLEM 11.3

Writing Equations for Hydrogenation

Draw the condensed structural formula for the product of each of the following hydrogenation reactions:

a. $CH_3—CH═CH_2 + H_2 \xrightarrow{Pt}$

b. (cyclopentene) $+ H_2 \xrightarrow{Pt}$

c. $HC≡CH + 2H_2 \xrightarrow{Ni}$

SOLUTION

In an addition reaction, hydrogen adds to the double or triple bond to give an alkane.

a. $CH_3—CH_2—CH_3$

b. (cyclopentane)

c. $CH_3—CH_3$

STUDY CHECK 11.3

Draw the condensed structural formula of the product of the hydrogenation of 2-methyl-1-butene, using a platinum catalyst.

Chemistry Link to Health

HYDROGENATION OF UNSATURATED FATS

Vegetable oils such as corn oil or safflower oil are unsaturated fats composed of fatty acids that contain double bonds. The process of hydrogenation is used commercially to convert the double bonds in the unsaturated fats in vegetable oils to saturated fats such as margarine, which are more solid. Adjusting the amount of added hydrogen produces partially hydrogenated fats such as soft margarine, solid margarine in sticks, and shortenings, which are used in cooking. For example, oleic acid is a typical unsaturated fatty acid in olive oil and has a cis double bond at carbon 9. When oleic acid is hydrogenated, it is converted to stearic acid, a saturated fatty acid.

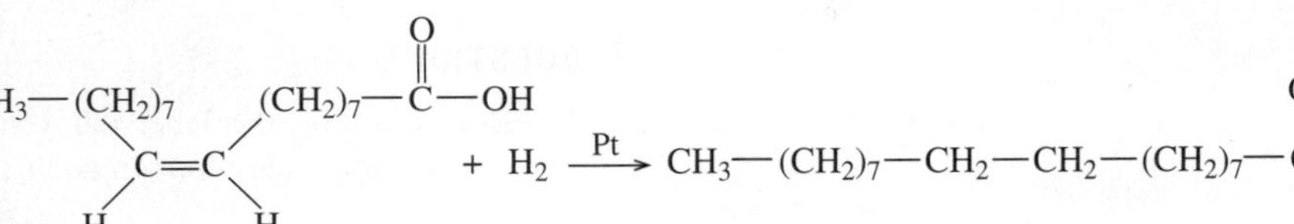

$$CH_3—(CH_2)_7—CH{=}CH—(CH_2)_7—\overset{O}{\overset{\|}{C}}—OH + H_2 \xrightarrow{Pt} CH_3—(CH_2)_7—CH_2—CH_2—(CH_2)_7—\overset{O}{\overset{\|}{C}}—OH$$

Oleic acid (found in olive oil and other unsaturated fats)

Stearic acid (found in saturated fats)

The unsaturated fats in vegetable oils are converted to saturated fats to make a more solid product.

Halogenation

In the **halogenation** reactions of alkenes, halogen atoms such as chlorine or bromine are added to the double bonds. The reaction occurs readily, without the use of a catalyst, and adds halogen atoms to yield a dihaloalkane product. In the general equation for halogenation, the symbol X—X or X_2 is used for Cl_2 or Br_2.

$$>C{=}C< + \mathbf{X—X} \longrightarrow —\overset{\mathbf{X}}{\overset{|}{C}}—\overset{\mathbf{X}}{\overset{|}{C}}—$$

Here are some examples of adding Cl_2 or Br_2 to alkenes:

$$H_2C{=}CH_2 + Cl{-}Cl \longrightarrow \underset{|}{\overset{Cl}{}}CH_2{-}CH_2 \text{ (each C bonded to Cl)}$$

Ethene → 1, 2-Dichloroethane

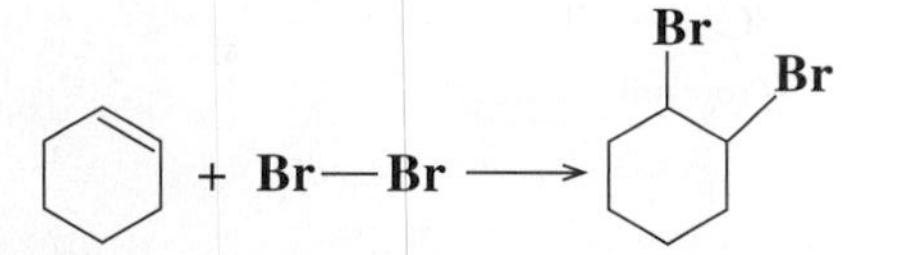

Cyclohexene + Br—Br ⟶ 1,2-Dibromocyclohexane

The addition reaction of bromine is sometimes used to test for the presence of double bonds, as shown in Figure 11.3.

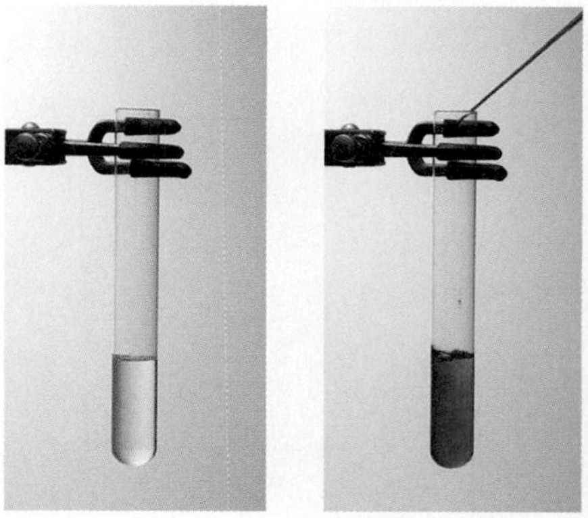

(a)

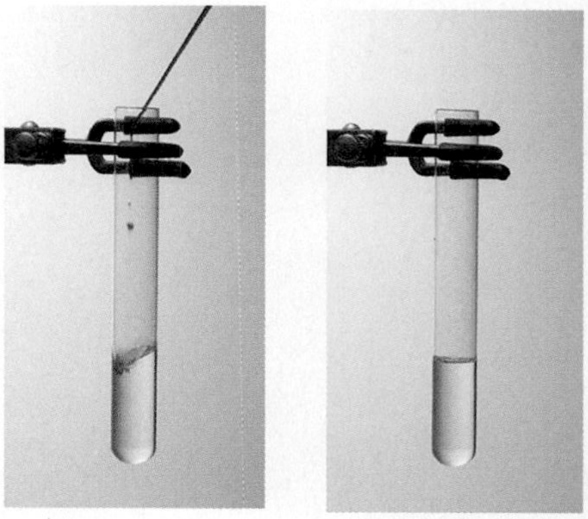

(b)

FIGURE 11.3 **(a)** When bromine is added to an alkane in the first test tube, the orange color of bromine remains because the alkane does not react. **(b)** When bromine is added to an alkene in the second test tube, the orange color immediately disappears as bromine atoms add to the double bond to give colorless products.

Q Why does the orange color disappear when bromine is added to cyclohexene?

SAMPLE PROBLEM 11.4

Halogenation

Draw the condensed structural formula of the product of the following reaction:

$$CH_3{-}\overset{CH_3}{\overset{|}{C}}{=}CH_2 + Br_2 \longrightarrow$$

SOLUTION

The addition of bromine to an alkene places a bromine atom on each of the carbon atoms of the double bond.

$$CH_3{-}\underset{Br}{\underset{|}{\overset{CH_3}{\overset{|}{C}}}}{-}\underset{Br}{\underset{|}{CH_2}}$$

STUDY CHECK 11.4

What is the name of the product formed when chlorine is added to 1-butene?

Hydration

In **hydration**, an alkene reacts with water (H—OH). A hydrogen atom (H) forms a bond with one carbon atom in the double bond, and the oxygen atom in OH forms a bond with the other carbon. The reaction is catalyzed by a strong acid such as H_2SO_4. Hydration is used to prepare alcohols, which have the hydroxyl (—OH) functional group. In the general equation for hydration, the acid catalyst is represented by H^+.

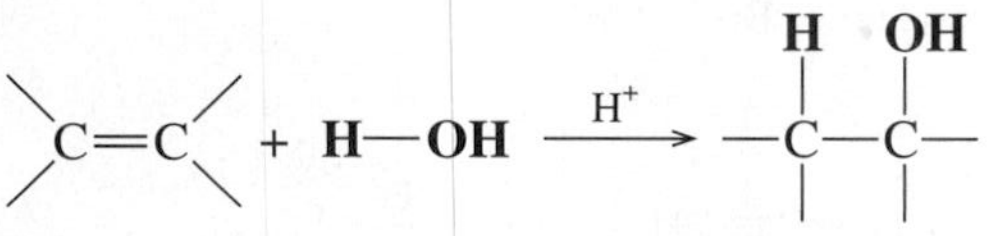

$$\text{>C=C<} + H{-}OH \xrightarrow{H^+} {-}\underset{|}{\overset{H}{\overset{|}{C}}}{-}\underset{|}{\overset{OH}{\overset{|}{C}}}{-}$$

Alkene → Alcohol

$$H_2C{=}CH_2 + H{-}OH \xrightarrow{H^+} \overset{H}{\overset{|}{CH_2}}{-}\overset{OH}{\overset{|}{CH_2}}$$

Ethene → Ethanol (ethyl alcohol)

OH ⟵ Functional group of alcohols

When water adds to a double bond in which the carbon atoms are attached to different numbers of H atoms, the H— from HOH attaches to the carbon that has the greater number of H atoms. In the following example, the H— from HOH attaches to the end carbon of the double bond, which has more hydrogen atoms.

$$\underset{\text{Propene}}{CH_3{-}CH{=}CH_2} + \mathbf{H{-}OH} \xrightarrow{H^+} \underset{\text{2-Propanol}}{CH_3{-}\underset{|}{\overset{\mathbf{OH}}{CH}}{-}\overset{\mathbf{H}}{CH_2}}$$

SAMPLE PROBLEM 11.5

Hydration

Draw the condensed structural formula of the product that forms in the following hydration reaction:

$$CH_3{-}CH_2{-}CH_2{-}CH{=}CH_2 + HOH \xrightarrow{H^+}$$

SOLUTION

The H— and —OH from water (HOH) add to the carbon atoms in the double bond. The H— adds to the carbon with more hydrogens, and the —OH bonds to the carbon with fewer hydrogen atoms.

$$CH_3{-}CH_2{-}CH_2{-}\overset{OH\downarrow}{CH}{=}\overset{H\downarrow}{CH_2} \xrightarrow{H^+} CH_3{-}CH_2{-}CH_2{-}\overset{OH}{\underset{}{CH}}{-}CH_3$$

STUDY CHECK 11.5

Draw the condensed structural formula of the alcohol obtained by the hydration of 2-methyl-2-butene.

QUESTIONS AND PROBLEMS

Addition Reactions

11.13 Draw the condensed structural formula of the product in each of the following reactions:

a. $CH_3{-}CH_2{-}CH_2{-}CH{=}CH_2 + H_2 \xrightarrow{Pt}$

b. $CH_3{-}CH{=}CH{-}CH_3 + HOH \xrightarrow{H^+}$

c. (cyclobutene) + HOH $\xrightarrow{H^+}$

d. cyclopentene + $Cl_2 \longrightarrow$

e. 2-pentyne + $2H_2 \xrightarrow{Pt}$

11.14 Draw the condensed structural formula of the product in each of the following reactions:

a. $CH_3{-}CH_2{-}CH{=}CH_2 + HOH \xrightarrow{H^+}$

b. cyclohexene + $H_2 \xrightarrow{Pt}$

c. *cis*-2-butene + $Br_2 \longrightarrow$

d. $CH_3{-}\overset{CH_3}{\underset{}{CH}}{-}C{\equiv}CH + 2H_2 \xrightarrow{Pt}$

e. $CH_3{-}\overset{CH_3}{C}{=}CH{-}CH_2{-}CH_3 + H_2 \xrightarrow{Pt}$

11.4 Polymers of Alkenes

A **polymer** is a large molecule that consists of small repeating units called **monomers**. In the past hundred years, the plastics industry has made synthetic polymers that are in many of the materials we use every day, such as carpeting, plastic wrap, nonstick pans, plastic cups, and rain gear. In medicine, synthetic polymers are used to replace diseased or damaged body parts such as hip joints, teeth, heart valves, and blood vessels. There are about 100 billion kg of plastics produced every year, which is about 15 kg for every person on Earth.

Many of the synthetic polymers are made by addition reactions of small alkenes. The conditions for many polymerization reactions require high temperatures and very high pressure (over 1000 atm). In an addition reaction, a polymer grows longer as monomers are added to the end of the chain. A polymer may consist of a long carbon chain with as many as 1000 monomers. Polyethylene, a polymer made from ethylene monomers, is used in plastic bottles, film, and plastic dinnerware. More polyethylene is produced worldwide than any other polymer.

$$H_2C{=}CH_2 + H_2C{=}CH_2 + H_2C{=}CH_2 \longrightarrow -CH_2-CH_2-\left[CH_2-CH_2\right]_{many}-CH_2-CH_2-$$

Ethene (ethylene) monomers — Polyethylene section (Monomer unit repeats)

Table 11.3 lists several alkene monomers that are used to produce common synthetic polymers, and Figure 11.4 shows examples of each. The alkane-like nature of these plastic synthetic polymers makes them unreactive. Thus, they do not decompose easily (they are nonbiodegradable) and have become contributors to pollution. Efforts are being made to make them more degradable. It is becoming increasingly important to recycle plastic material, rather than add to our growing landfills.

Polyethylene

Polyvinyl chloride

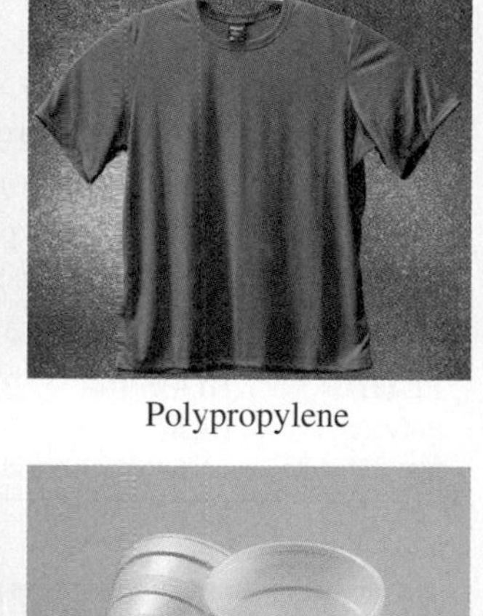
Polypropylene

Polytetrafluoroethylene (Teflon)

Polydichloroethylene (Saran)

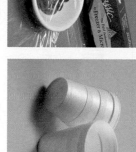

Polystyrene

FIGURE 11.4 Synthetic polymers provide a wide variety of items that we use every day.
Q What are some alkenes used to make the polymers in these plastic items?

LEARNING GOAL

Draw condensed structural formulas of monomers that form a polymer or a three-monomer section of a polymer.

TUTORIAL
Polymers

SELF STUDY ACTIVITY
Introduction to Polymers

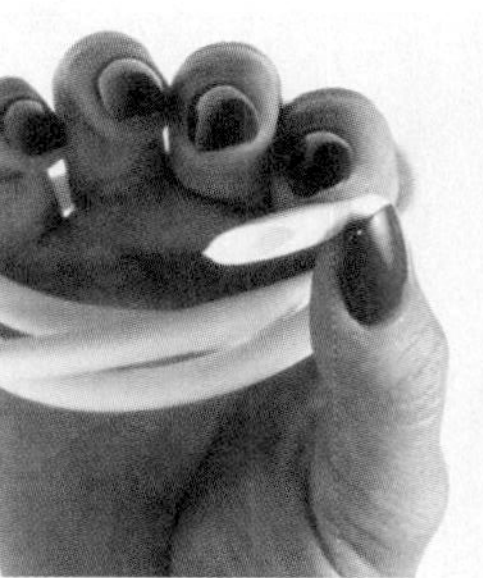
Synthetic polymers are used to replace diseased veins and arteries.

POLYMERS AND RECYCLING PLASTICS

1. Make a list of the items you use or have in your room or home that are made of polymers.
2. Recycling information on the bottom or side of a plastic item includes a triangle with a code number that identifies the type of polymer used to make the plastic. Make a collection of several different kinds of plastic items. Try to find plastic items with each type of polymer.

QUESTIONS

1. What are the most common types of plastics among the plastic items in your collection?
2. What are the monomer units of some of the plastics you looked at?

TABLE 11.3 Some Alkenes and Their Polymers

Monomer	Polymer Section	Common Uses
$H_2C{=}CH_2$ Ethene (ethylene)	$-CH_2-CH_2-CH_2-CH_2-CH_2-CH_2-$ (each C bonded to H above and H below) Polyethylene	Plastic bottles, film, insulation materials
$H_2C{=}CH(Cl)$ Chloroethene (vinyl chloride)	$-CH_2-CHCl-CH_2-CHCl-CH_2-CHCl-$ (above: H, Cl, H, Cl, H, Cl; below: H, H, H, H, H, H) Polyvinyl chloride (PVC)	Plastic pipes and tubing, garden hoses, garbage bags
$H_2C{=}CH(CH_3)$ Propene (propylene)	$-CH_2-CH(CH_3)-CH_2-CH(CH_3)-CH_2-CH(CH_3)-$ (above: H, CH_3, H, CH_3, H, CH_3; below: H, H, H, H, H, H) Polypropylene	Ski and hiking clothing, carpets, artificial joints
$F_2C{=}CF_2$ (F—C=C—F with F on each C) Tetrafluoroethene	$-CF_2-CF_2-CF_2-CF_2-CF_2-CF_2-$ (each C bonded to F above and F below) Polytetrafluoroethylene (Teflon)	Nonstick coatings
$H_2C{=}C(Cl)-Cl$ 1,1-Dichloroethene	$-CH_2-CCl_2-CH_2-CCl_2-CH_2-CCl_2-$ (above: H, Cl, H, Cl, H, Cl; below: H, Cl, H, Cl, H, Cl) Polydichloroethylene (Saran)	Plastic film and wrap
$H_2C{=}CH$ (with a benzene ring on CH) Phenylethene (styrene)	$-CH_2-CH-CH_2-CH-CH_2-CH-$ (a benzene ring on each CH) Polystyrene	Plastic coffee cups and cartons, insulation

The recycling symbol indicates the type of polymer.

Today tables, benches, and trash receptacles are made from recycled plastics.

You can identify the type of polymer used to manufacture a plastic item by looking for the recycling symbol (arrows that form a triangle) found on the label or on the bottom of the plastic container. For example, either the number 5 or the letters PP inside the triangle is a code for a polypropylene plastic.

1	2	3	4	5	6	7
PETE	HDPE	PVC	LDPE	PP	PS	O
Polyethylene terephthalate	High-density polyethylene	Polyvinyl chloride	Low-density polyethylene	Polypropylene	Polystyrene	Other

SAMPLE PROBLEM 11.6

Polymers

Give the name and draw the condensed structural formula of the starting monomers for the following polymers:

a. polypropylene

b. Saran

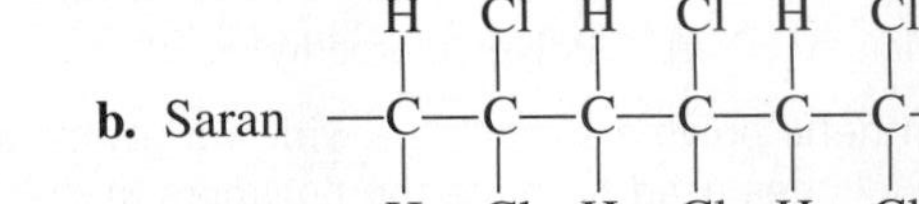
$-CH_2-CCl_2-CH_2-CCl_2-CH_2-CCl_2-$ (above: H, Cl, H, Cl, H, Cl; below: H, Cl, H, Cl, H, Cl)

SOLUTION

a. propene (propylene), $H_2C{=}\overset{\displaystyle CH_3}{\overset{|}{C}}H$ **b.** 1,1-dichloroethene, $H_2C{=}\overset{\displaystyle Cl}{\overset{|}{C}}{-}Cl$

STUDY CHECK 11.6

What is the condensed structural formula of the monomer used in the manufacturing of PVC?

QUESTIONS AND PROBLEMS

Polymers of Alkenes

11.15 What is a polymer?

11.16 What is a monomer?

11.17 Write an equation that represents the formation of a part of the polypropylene polymer from three of the monomer units.

11.18 Write an equation that represents the formation of a part of the polystyrene polymer from three of the monomer units.

11.5 Aromatic Compounds

LEARNING GOAL

Describe the bonding in benzene; name aromatic compounds, and draw their condensed structural formulas.

In 1825, Michael Faraday isolated a hydrocarbon called *benzene*, which had the molecular formula C_6H_6. A molecule of **benzene** consists of a ring of six carbon atoms with one hydrogen atom attached to each carbon. Because many compounds containing benzene had fragrant odors, the family of benzene compounds became known as **aromatic compounds**. In benzene, each carbon atom uses three valence electrons to bond to the hydrogen atom and two adjacent carbons. That leaves one valence electron, which scientists first thought was shared in a double bond with an adjacent carbon. In 1865, August Kekulé proposed that the carbon atoms in benzene were arranged in a flat ring with alternating single and double bonds between the carbon atoms.

However, scientists discovered that benzene did not react like an alkene. Instead, benzene behaved more like an alkane. Today we know that six electrons are shared equally among the six carbon atoms and that all of the carbon–carbon bonds in benzene are identical. This unique feature of benzene makes it especially stable. Today benzene is most often represented as its skeletal formula, which shows a hexagon with a circle in the center. Some of the ways to represent benzene are as follows:

Structures for benzene

Skeletal formula for benzene ring

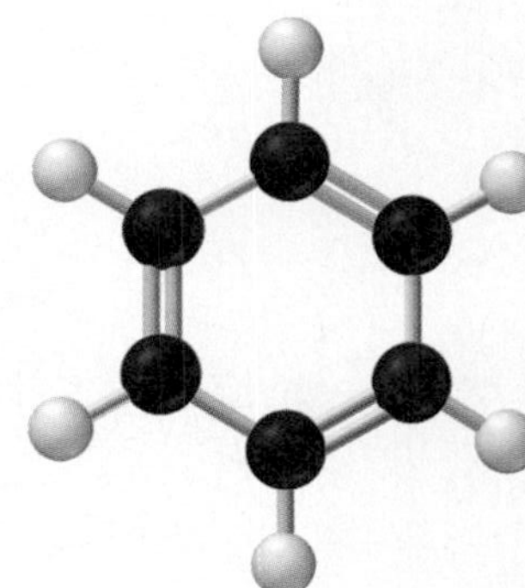

Naming Aromatic Compounds

TUTORIAL
Naming Aromatic Compounds

Aromatic compounds that contain a benzene ring with a single substituent are usually named as benzene derivatives. However, many of these compounds have been important in chemistry for many years and still use their common names. Some widely used names such as toluene, aniline, and phenol are allowed by IUPAC rules.

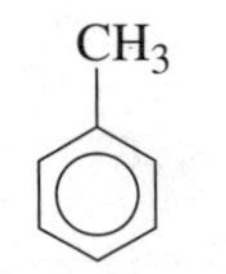

CH_3 NH_2 OH

Toluene (methylbenzene) 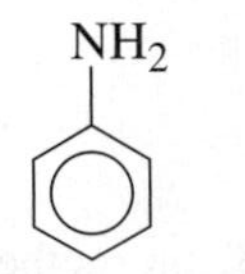 Aniline (aminobenzene) 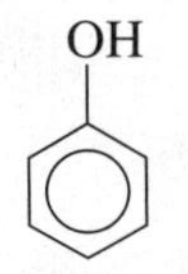 Phenol (hydroxybenzene)

When benzene has only one substituent, the benzene ring is not numbered. When there are two or more substituents, the benzene ring is numbered to give the lower numbers to the substituents.

Cl Cl Cl Cl Cl Cl Cl

Chlorobenzene 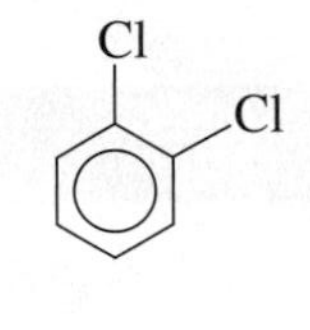 1,2-Dichlorobenzene 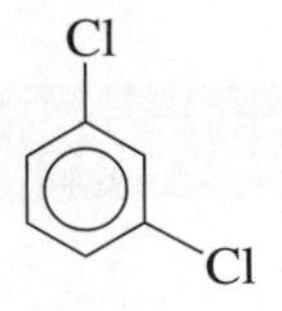 1,3-Dichlorobenzene 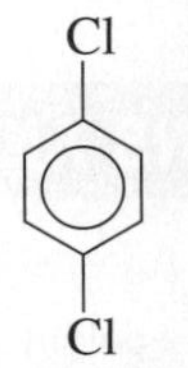 1,4-Dichlorobenzene

When a common name such as toluene, phenol, or aniline can be used, the carbon atom attached to the methyl, hydroxyl, or amine group is numbered as carbon 1. Then the substituents are named alphabetically.

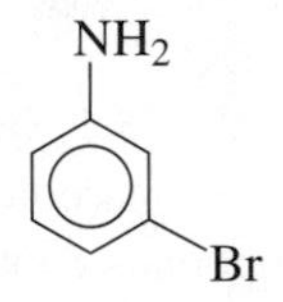

NH_2 Br OH Cl Br CH_3 Br Br Cl

3-Bromoaniline 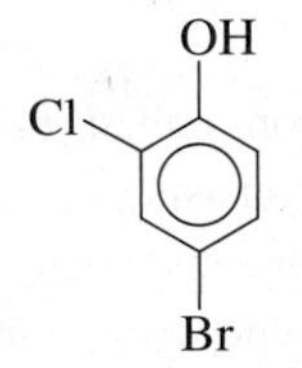 4-Bromo-2-chlorophenol 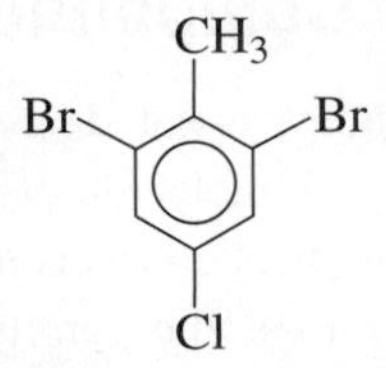 2,6-Dibromo-4-chlorotoluene

SAMPLE PROBLEM 11.7

Naming Aromatic Compounds

Give the IUPAC name for the following:

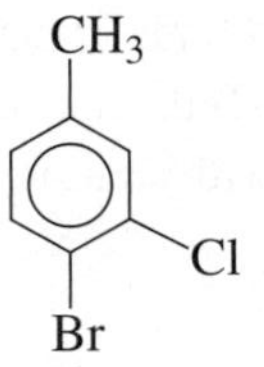

CH_3 Cl Br

SOLUTION

A benzene ring with a methyl substituent is named toluene. The methyl group is attached to carbon 1 and the ring is numbered to give the lower numbers. Naming the substituents in alphabetical order, this aromatic compound is 4-bromo-3-chlorotoluene.

STUDY CHECK 11.7

Name the following compound:

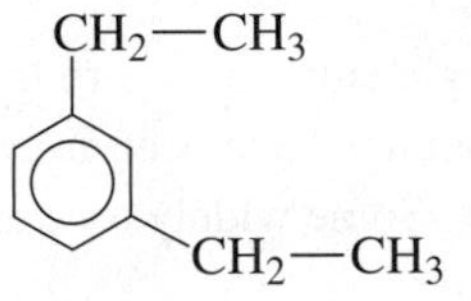

$CH_2—CH_3$ $CH_2—CH_3$

Chemistry Link to Health

SOME COMMON AROMATIC COMPOUNDS

Aromatic compounds are common in nature and in medicine. Toluene is used as a reactant to make drugs, dyes, and explosives such as TNT (trinitrotoluene). The benzene ring is found in some amino acids (the building blocks of proteins); in pain relievers such as aspirin, acetaminophen, and ibuprofen; and in flavorings such as vanillin.

TNT (2,4,6-trinitrotoluene)

Aspirin

Vanillin

Ibuprofen

Acetaminophen

Phenylalanine (amino acid)

QUESTIONS AND PROBLEMS

Aromatic Compounds

11.19 Cyclohexane and benzene each have six carbon atoms. How are they different?

11.20 In the Chemistry Link to Health feature "Some Common Aromatic Compounds," what part of each molecule is the aromatic portion?

11.21 Give the IUPAC name for each of the following:

a. (benzene ring with CH_3 and Cl) **b.** (benzene ring with $CH_2—CH_3$) **c.** (benzene ring with Cl, Cl, Cl)

11.22 Give the IUPAC name for each of the following:

a. (benzene ring with CH_3 and Br) **b.** (benzene ring with OH, Cl, Cl) **c.** (benzene ring with Br, Cl, Cl)

11.23 Draw the condensed structural formula for each of the following compounds:

a. toluene **b.** 1,3-dichlorobenzene
c. 4-ethyltoluene

11.24 Draw the condensed structural formula for each of the following compounds:

a. benzene **b.** 4-bromoaniline
c. 1,2,4-trichlorobenzene

Chemistry Link to Health

POLYCYCLIC AROMATIC HYDROCARBONS (PAHs)

Large aromatic compounds known as polycyclic aromatic hydrocarbons are formed by fusing together two or more benzene rings edge to edge. In a fused-ring compound, neighboring benzene rings share two carbon atoms. Naphthalene, with two benzene rings, is well known for its use in mothballs. Anthracene, with three rings, is used in the manufacture of dyes.

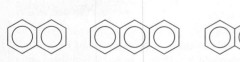

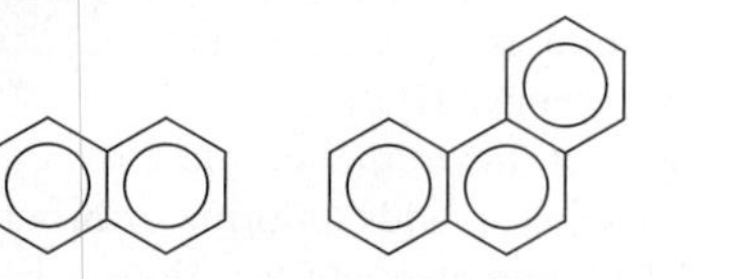

Naphthalene Anthracene Phenanthrene

When a polycyclic compound contains phenanthrene, it may act as a carcinogen, a substance known to cause cancer.

Compounds containing five or more fused benzene rings such as benzo[*a*]pyrene are potent carcinogens. The molecules interact with the DNA in the cells, causing abnormal cell growth and cancer. Increased exposure to carcinogens increases the chance of DNA alterations in the cells. Benzo[*a*]pyrene, a product of combustion, has been identified in coal tar, tobacco smoke, barbecued meats, and automobile exhaust.

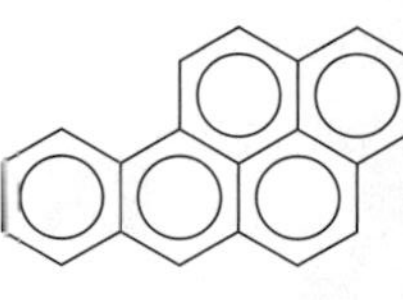

Benzo[*a*]pyrene

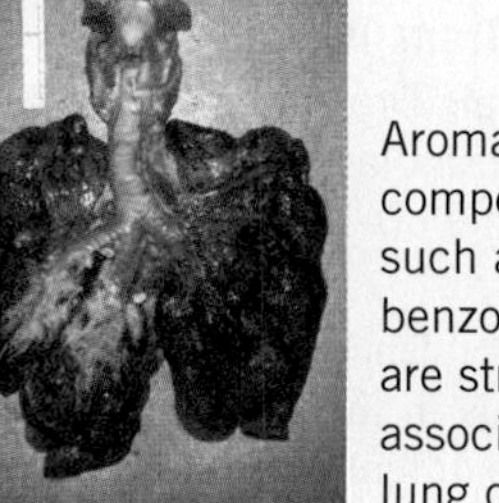

Aromatic compounds such as benzo[*a*]pyrene are strongly associated with lung cancers.

CONCEPT MAP

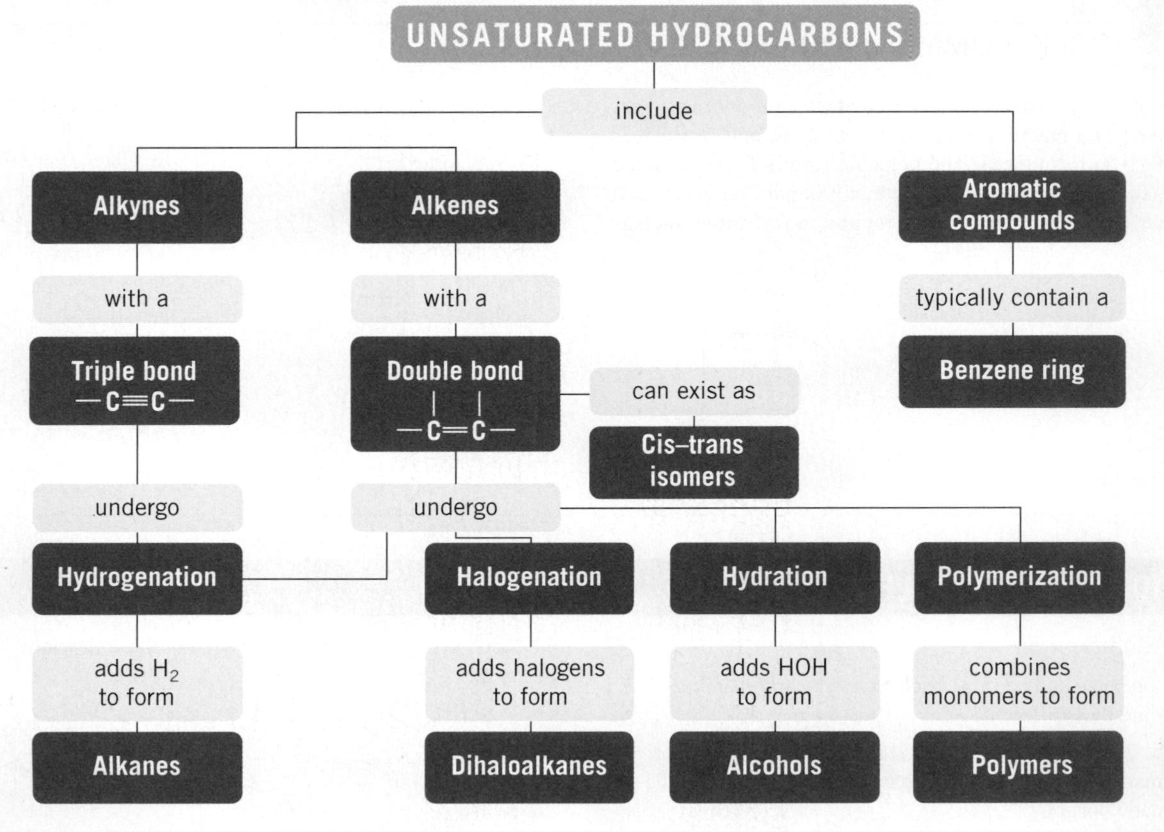

CHAPTER REVIEW

11.1 Alkenes and Alkynes

Learning Goal: Identify structural formulas as alkenes, cycloalkenes, and alkynes, and write their IUPAC or common names.

Alkenes are unsaturated hydrocarbons that contain carbon–carbon double bonds (C═C). Alkynes contain a triple bond (—C≡C—). The IUPAC names of alkenes end with *ene*, while alkyne names end with *yne*. The main chain is numbered from the end nearer the double or triple bond.

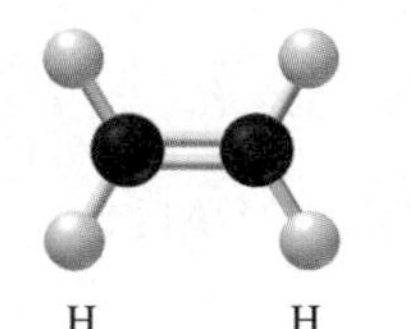

Bond angles = 120°
Ethene

cis-2-Butene

11.2 Cis–Trans Isomers

Learning Goal: Draw the condensed structural formulas and give the names for the cis–trans isomers of alkenes.

Isomers of alkenes occur when the carbon atoms in the double bond are connected to different atoms or groups. In the cis isomer, the similar groups are on the same side of the double bond, whereas in the trans isomer they are connected on opposite sides of the double bond.

11.3 Addition Reactions

Learning Goal: Draw the condensed structural formulas and give the names for the organic products of addition reactions of alkenes and alkynes.

The addition of small molecules to the double bond is a characteristic reaction of alkenes. Hydrogenation adds hydrogen atoms to the double bond of an alkene to yield an alkane. Halogenation adds chlorine or bromine atoms to the double bond. Water can also add to a double bond. When there are a different number of hydrogen atoms attached to the carbons in the double bond, the —H from the HOH adds to the carbon with the greater number of hydrogen atoms, and —OH from the HOH adds to the other carbon.

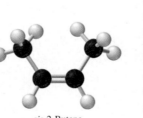

11.4 Polymers of Alkenes

Learning Goal: Draw condensed structural formulas of monomers that form a polymer or a three-monomer section of a polymer.

Polymers are long-chain molecules that consist of many repeating units of smaller carbon molecules called monomers. Many materials that we use every day are synthetic polymers, including carpeting, plastic wrap, and nonstick pans. These synthetic materials are often made by addition reactions in which a catalyst links the carbon atoms from various kinds of alkene molecules.

11.5 Aromatic Compounds

Learning Goal: Describe the bonding in benzene; name aromatic compounds, and draw their condensed structural formulas.

$CH_3-CH(CH_3)-CH_2-C_6H_4-CH(CH_3)-C(=O)-OH$

Ibuprofen

Most aromatic compounds contain benzene, C_6H_6, a cyclic structure drawn as a hexagon with a circle in the center. Aromatic compounds containing benzene are named using the parent name benzene, although common names such as toluene are retained. The benzene ring is numbered, and the substituents are listed in alphabetical order.

Summary of Naming

Type	Example	Characteristic	Structure
Alkene	Propene (propylene)	Double bond	$CH_3-CH=CH_2$
Cycloalkene	Cyclopropene	Double bond in a carbon ring	(triangle with double bond)
Alkyne	Propyne	Triple bond	$CH_3-C\equiv CH$
Aromatic	Benzene	Aromatic ring of six carbons	(benzene ring)
	Methylbenzene, or toluene		CH_3 on benzene ring

Summary of Reactions

Hydrogenation

$$\text{Alkene} + H_2 \xrightarrow{Pt} \text{alkane}$$

$$H_2C=CH-CH_3 + H_2 \xrightarrow{Pt} CH_3-CH_2-CH_3$$

$$\text{Alkyne} + 2H_2 \xrightarrow{Pt} \text{alkane}$$

$$CH_3-C\equiv CH + 2H_2 \xrightarrow{Pt} CH_3-CH_2-CH_3$$

Halogenation

$$\text{Alkene} + X_2 \longrightarrow \text{dihaloalkane}$$

$$H_2C=CH_2 + Cl-Cl \longrightarrow CH_2(Cl)-CH_2(Cl)$$

Hydration

$$\text{Alkene} + H-OH \xrightarrow{H^+} \text{alcohol}$$

$$H_2C=CH-CH_3 + H-OH \xrightarrow{H^+} CH_3-CH(OH)-CH_3$$

Key Terms

addition A reaction in which atoms or groups of atoms bond to a double bond or triple bond. Addition reactions include the addition of hydrogen (hydrogenation), water (hydration), and halogens (halogenation).

alkene An unsaturated hydrocarbon containing a carbon–carbon double bond.

alkyne An unsaturated hydrocarbon containing a carbon–carbon triple bond.

aromatic compounds Compounds that usually have fragrant odors and often contain the ring structure of benzene.

benzene A ring of six carbon atoms, each of which is attached to a hydrogen atom, C_6H_6.

cis isomer An isomer of an alkene in which the hydrogen atoms in the double bond are on the same side.

halogenation An addition reaction in which halogen atoms (chlorine or bromine) are added to a double bond.

hydration An addition reaction in which the components of water, H— and —OH, bond to the carbon–carbon double bond to form an alcohol.

hydrogenation The addition of hydrogen (H_2) to the double bond of alkenes or alkynes to yield alkanes.

monomer The small organic molecule that is repeated many times in a polymer.

polymer A very large molecule that is composed of many small, repeating structural units that are identical.

saturated hydrocarbon A compound of carbon and hydrogen that contains the maximum number of hydrogen atoms.

trans isomer An isomer of an alkene in which the hydrogen atoms in the double bond are on opposite sides.

unsaturated hydrocarbon A compound of carbon and hydrogen in which the carbon chain contains at least one double (alkene) or triple (alkyne) carbon–carbon bond. An unsaturated compound is capable of an addition reaction with hydrogen, which converts the double or triple bonds to single carbon–carbon bonds.

Understanding the Concepts

11.25 Draw a part of the polymer (use three monomers) of Teflon made from 1,1,2,2-tetrafluoroethene.

Teflon is used as a nonstick coating on cooking pans.

11.26 A garden hose is made of polyvinyl chloride (PVC) from chloroethene (vinyl chloride). Draw a part of the polymer (use three monomers) for PVC.

A plastic garden hose is made of PVC.

11.27 Explosives used in mining contain TNT, or trinitrotoluene.

The TNT in explosives is used in mining.

a. If the functional group *nitro* is $—NO_2$, what is the condensed structural formula of 2,4,6-trinitrotoluene, one isomer of TNT?
b. TNT is actually a mixture of isomers of trinitrotoluene. Draw two other possible isomers.

11.28 Margarine is produced from the hydrogenation of vegetable oils, which contain unsaturated fatty acids. How many grams of hydrogen are required to completely saturate 75.0 g of oleic acid, $C_{18}H_{34}O_2$, which has one double bond?

Margarines are produced by hydrogenation of unsaturated fats.

Additional Questions and Problems

For instructor-assigned homework, go to **www.masteringchemistry.com**.

11.29 Compare the formulas and bonding in propane, cyclopropane, propene, and propyne.

11.30 Compare the formulas and bonding in butane, cyclobutane, cyclobutene, and 2-butyne.

11.31 Give the IUPAC name for each of the following compounds:

a. $H_2C{=}C(CH_3){-}CH_2{-}CH_2{-}CH_3$

b. 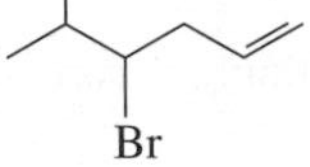 Br

c.

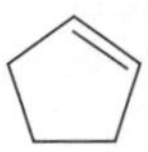

11.32 Give the IUPAC name for each of the following compounds:

a. 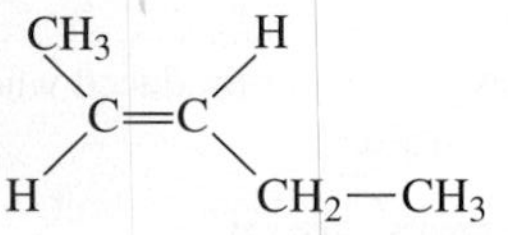 CH_3 and H on one carbon, H and CH_2-CH_3 on the other: $C=C$

b.

c.

11.33 Indicate if the following pairs of structures represent isomers, cis–trans isomers, or the same molecule:

a. 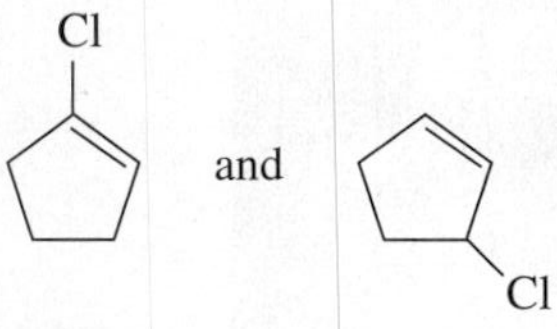 Cl and Cl

b. 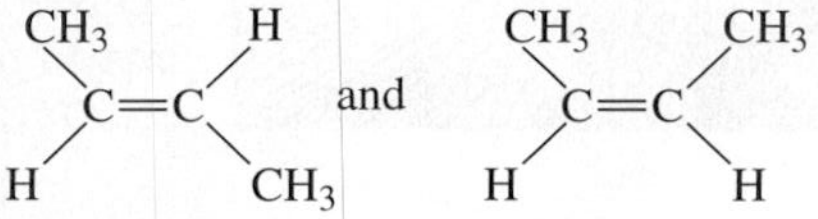 CH_3, H, $C=C$, H, CH_3 and CH_3, CH_3, $C=C$, H, H

11.34 Indicate if the following pairs of structures represent isomers, cis–trans isomers, or the same molecule:

a. 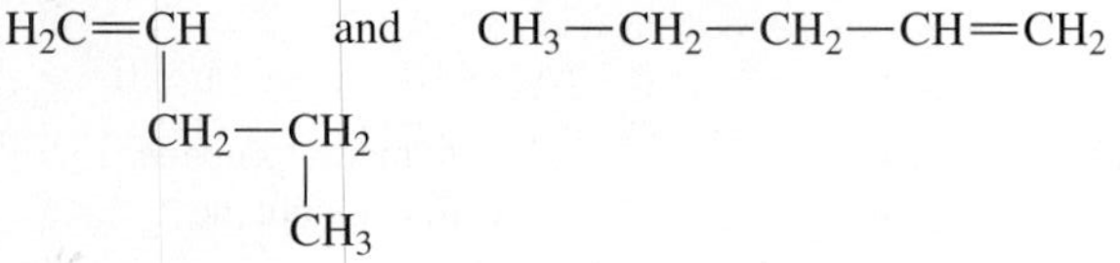 $H_2C=CH-CH_2-CH_2-CH_3$ (drawn as $H_2C=CH$ with $CH_2-CH_2-CH_3$ branching down) and $CH_3-CH_2-CH_2-CH=CH_2$

b.  and

11.35 Draw the condensed structural formula of each of the following compounds:

a. 1,1-dibromo-2-pentyne

b. *cis*-2-heptene

11.36 Draw the condensed structural formula of each of the following compounds:

a. *trans*-2-hexene

b. 2,3-dichloro-1-butene

11.37 Draw the cis and trans isomers for each of the following:

a. 2-pentene

b. 3-hexene

11.38 Draw the cis and trans isomers for each of the following:

a. 2-butene

b. 2-hexene

11.39 Give the name of the product from the hydrogenation of each of the following:

a. 3-methyl-2-pentene

b. cyclohexene

c. 2-pentyne

11.40 Give the name of the product from the hydrogenation of each of the following:

a. 1-hexene

b. 2-methyl-2-butene

c. propyne

11.41 Draw the condensed structural formula (or skeletal formula) of the product for each of the following:

a. $CH_3-CH=CH-CH_3 + Br_2 \longrightarrow$

b. 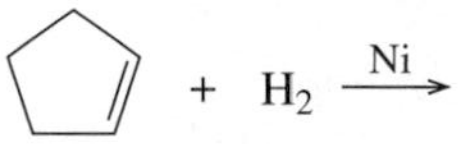 $+ H_2 \xrightarrow{Ni}$

c. 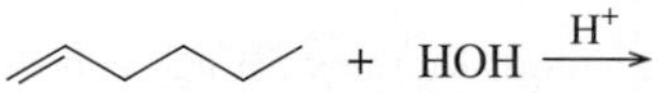 $+ HOH \xrightarrow{H^+}$

11.42 Draw the condensed structural formula (or skeletal formula) of the product for each of the following:

a. 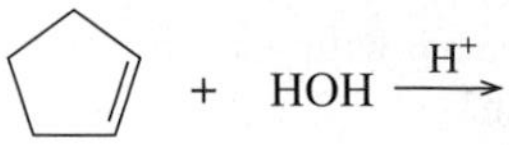 $+ HOH \xrightarrow{H^+}$

b. 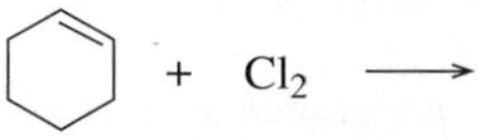 $+ Cl_2 \longrightarrow$

c. $CH_3-CH_2-C\equiv C-CH_3 + 2H_2 \xrightarrow{Pt}$

11.43 A plastic called polyvinylidene difluoride, PVDF, is made from monomers of 1,1-difluoroethene. Draw the condensed structural formula of the polymer formed from the addition of three monomers of 1,1-difluoroethene.

11.44 An alkene called acrylonitrile is the monomer used to form the polymer used in the fabric material called Orlon. Write an equation that represents the formation of a part of the polyacrylonitrile polymer from three of the monomer units. The structure of acrylonitrile is

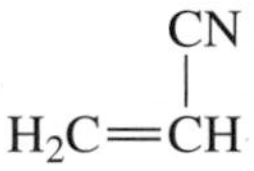

CN

$H_2C=CH$

11.45 Name each of the following aromatic compounds:

a.  CH_3

b. 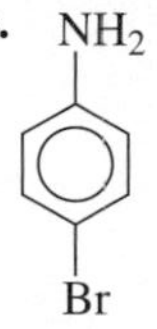 NH_2 Br

c. 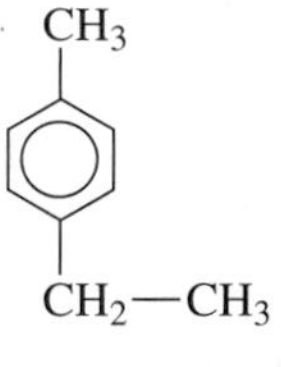 CH_3 CH_2-CH_3

11.46 Draw the condensed structural formula for each of the following:

a. ethylbenzene

b. 2,5-dibromophenol

c. 1,2,4-trimethylbenzene

Challenge Questions

11.47 How many grams of hydrogen are needed to hydrogenate 30.0 g of 2-butene?

11.48 Using each of the following carbon chains for C_5H_{10}, draw the condensed structural formulas and give the names of all the possible alkenes, including those with cis and trans isomers:

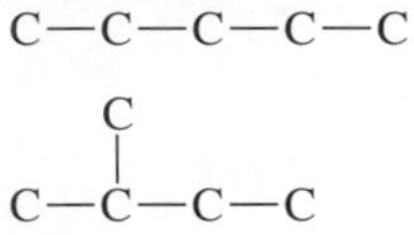

```
C—C—C—C—C

    C
    |
C—C—C—C
```

11.49 If a female silkworm moth secretes 50 ng of bombykol, a sex attractant, how many molecules does she secrete? (See Chemistry Link to the Environment "Pheromones in Insect Communication.")

11.50 Acetylene (ethyne) gas reacts with oxygen and burns at 3300 °C in an acetylene torch.

a. Write the balanced equation for the complete combustion of acetylene.

b. How many grams of oxygen are needed to react with 8.5 L of acetylene gas at STP?

c. How many liters of CO_2 gas at STP are produced when 30.0 g of acetylene undergoes combustion?

Acetylene and oxygen burn at 3300 °C.

Answers

Answers to Study Checks

11.1 a. $CH_3-C\equiv C-CH_2-CH_3$

b. 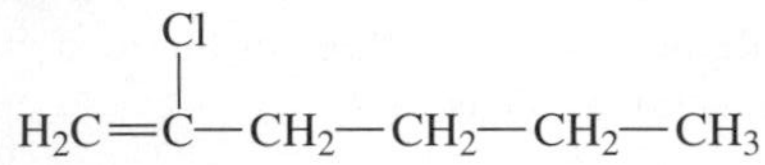
```
       Cl
       |
H2C=C—CH2—CH2—CH2—CH3
```

11.2 *trans*-3-hexene

11.3
```
        CH3
        |
CH3—CH—CH2—CH3
```

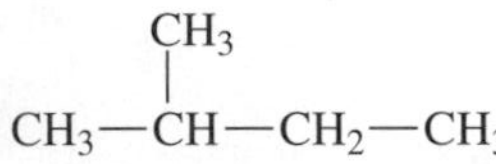

11.4 1,2-dichlorobutane

11.5
```
        OH
        |
CH3—C—CH2—CH3
        |
        CH3
```

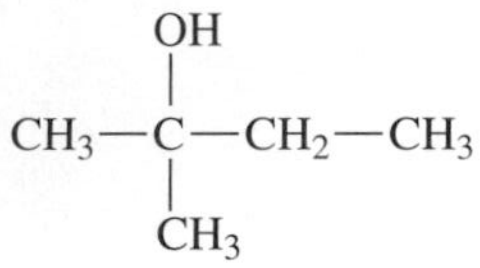

11.6
```
        Cl
        |
H2C=CH
```

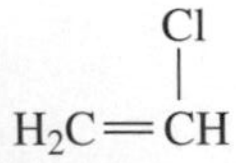

11.7 1,3-diethylbenzene

Answers to Selected Questions and Problems

11.1 a. An alkene has a double bond.
b. An alkyne has a triple bond.
c. A cycloalkene has a double bond in a ring.
d. An alkene has a double bond.

11.3 Propene contains a double bond, and propyne has a triple bond. Propene has six hydrogen atoms, and propyne has only four hydrogen atoms.

11.5 a. ethene **b.** methylpropene
c. 2-pentyne **d.** cyclobutene

11.7 a. $CH_3-CH=CH_2$

b. $H_2C=CH-CH_2-CH_2-CH_3$

c. 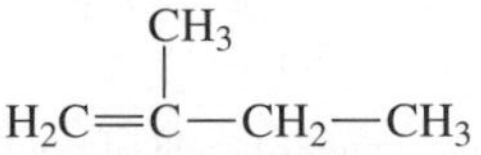
```
        CH3
        |
H2C=C—CH2—CH3
```

d.  (cyclohexene ring)

e. $CH_3-CH_2-C\equiv CH$

f. $Br-CH_2-CH_2-C\equiv C-CH_2-CH_3$

11.9 a. *cis*-2-butene
b. *trans*-3-octene
c. *cis*-3-heptene

11.11 a. 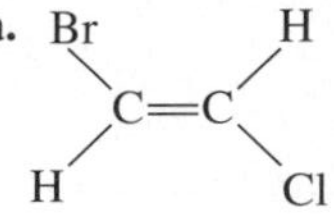
```
Br       H
  \     /
   C=C
  /     \
H        Cl
```

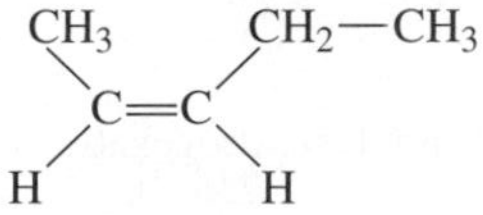

b.
```
CH3       CH2—CH3
   \     /
    C=C
   /     \
H         H
```

c.
```
CH3—CH2        H
       \      /
        C=C
       /      \
     H         CH2—CH2—CH3
```

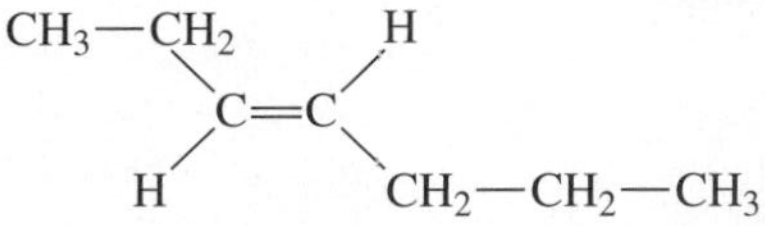

11.13 a. $CH_3-CH_2-CH_2-CH_2-CH_3$

b. 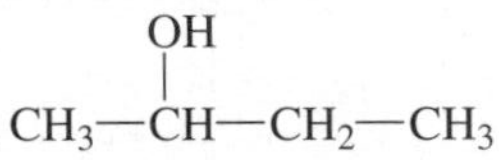
```
          OH
          |
CH3—CH—CH2—CH3
```

c. 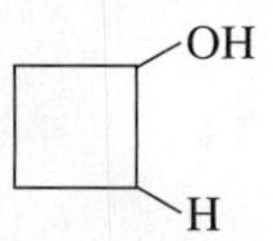 OH, H (cyclobutane structure)

d. 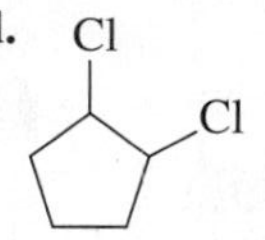 Cl, Cl (cyclopentane structure)

e. $CH_3-CH_2-CH_2-CH_2-CH_3$

11.15 A polymer is a very large molecule composed of small units that are repeated many times.

11.17 $3\ H_2C{=}CH(CH_3) \longrightarrow -C(H)(H)-C(CH_3)(H)-C(H)(H)-C(CH_3)(H)-C(H)(H)-C(CH_3)(H)-$

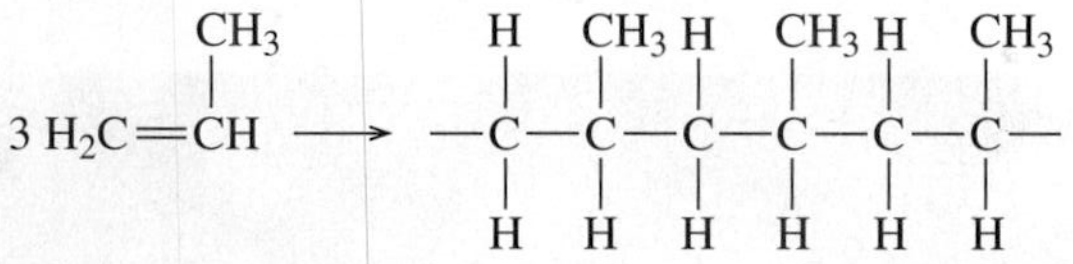

11.19 Cyclohexane, C_6H_{12}, is a cycloalkane with 6 carbon atoms and 12 hydrogen atoms. The carbon atoms are connected in a ring by single bonds. Benzene, C_6H_6, is an aromatic compound with 6 carbon atoms and 6 hydrogen atoms. The carbon atoms are connected in a ring where the electrons are equally shared among the six carbon atoms.

11.21 **a.** 2-chlorotoluene
b. ethylbenzene
c. 1,3,5-trichlorobenzene

11.23 **a.** 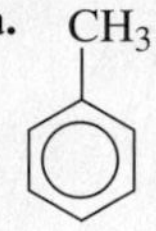 CH_3 (benzene ring) **b.** 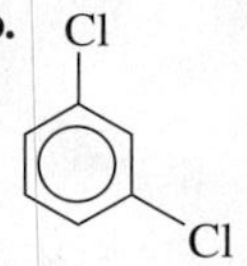 Cl, Cl (benzene ring) **c.** 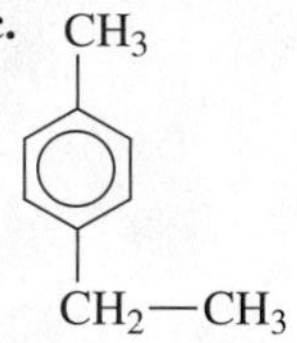 CH_3, CH_2-CH_3 (benzene ring)

11.25 $-CF_2-CF_2-CF_2-CF_2-CF_2-CF_2-$

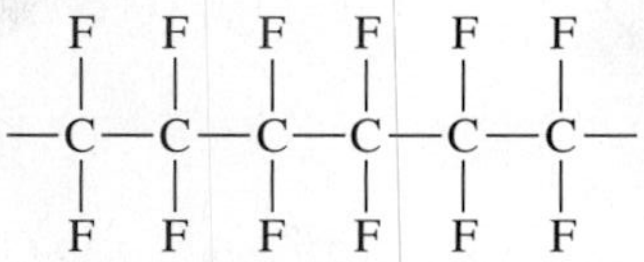

11.27 **a.** 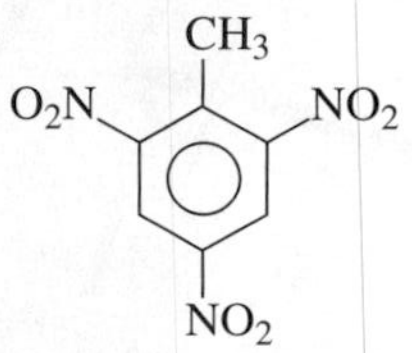 CH_3, O_2N, NO_2, NO_2 (benzene ring)

b. 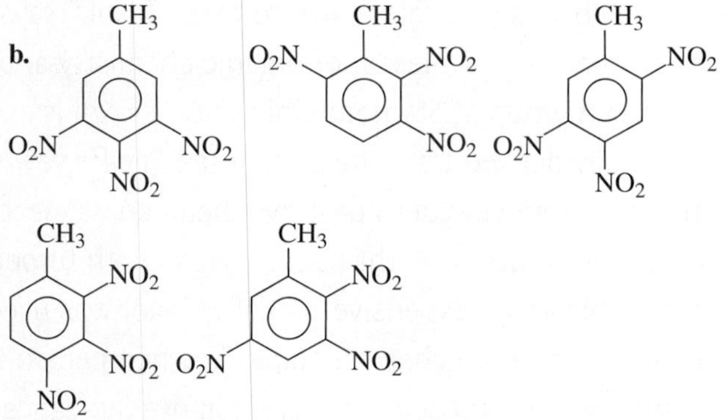 CH_3, O_2N, NO_2, NO_2 (benzene ring); CH_3, O_2N, NO_2, NO_2 (benzene ring); CH_3, NO_2, O_2N, NO_2 (benzene ring); CH_3, NO_2, NO_2, NO_2 (benzene ring); CH_3, NO_2, O_2N, NO_2 (benzene ring)

11.29 All the compounds have three carbon atoms: propane has eight hydrogen atoms, cyclopropane has six hydrogen atoms, propene has six hydrogen atoms, and propyne has four hydrogen atoms. Propane is a saturated alkane, and cyclopropane is a saturated cyclic hydrocarbon. Both propene and propyne are unsaturated hydrocarbons, but propene has a double bond and propyne has a triple bond.

11.31 **a.** 2-methyl-1-pentene
b. 4-bromo-5-methyl-1-hexene
c. cyclopentene

11.33 **a.** isomers **b.** cis–trans isomers

11.35 **a.** 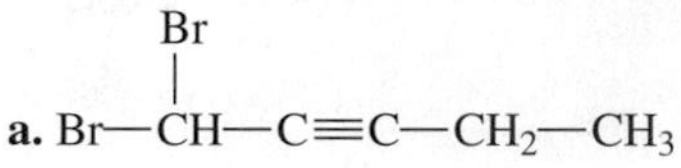 $Br-CH(Br)-C{\equiv}C-CH_2-CH_3$

b. 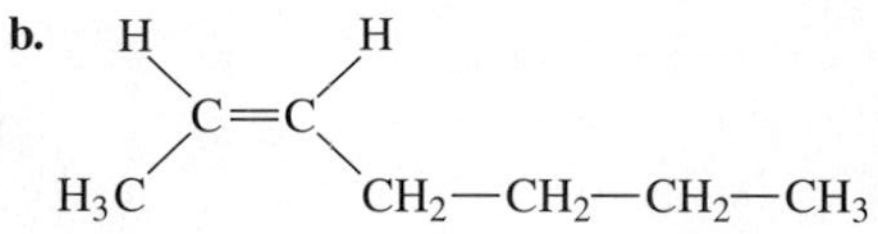 H, H, H_3C, $CH_2-CH_2-CH_2-CH_3$ ($C{=}C$)

11.37 **a.** 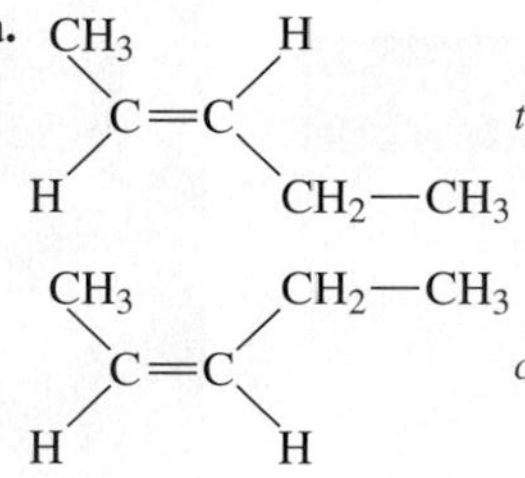 CH_3, H, H, CH_2-CH_3 ($C{=}C$) *trans*-2-Pentene

CH_3, CH_2-CH_3, H, H ($C{=}C$) *cis*-2-Pentene

b. 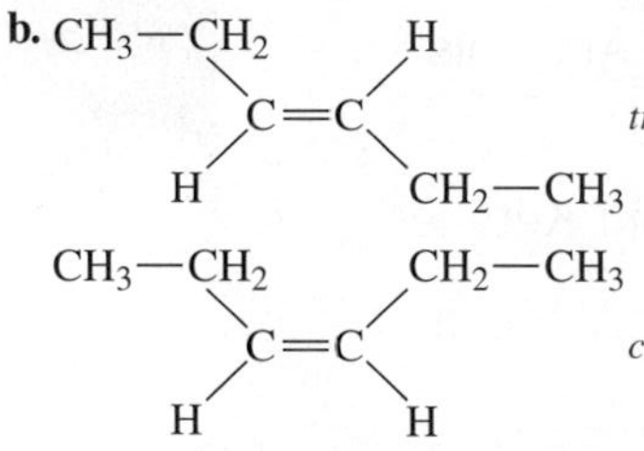 CH_3-CH_2, H, H, CH_2-CH_3 ($C{=}C$) *trans*-3-Hexene

CH_3-CH_2, CH_2-CH_3, H, H ($C{=}C$) *cis*-3-Hexene

11.39 **a.** 3-methylpentane
b. cyclohexane
c. pentane

11.41 **a.** 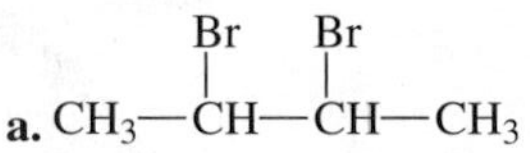 $CH_3-CH(Br)-CH(Br)-CH_3$ **b.** 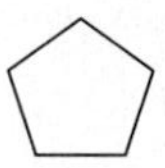 (cyclopentane)

c. 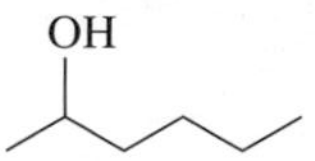 OH (skeletal structure)

11.43 $-CFF-CHH-CFF-CHH-CFF-CHH-$

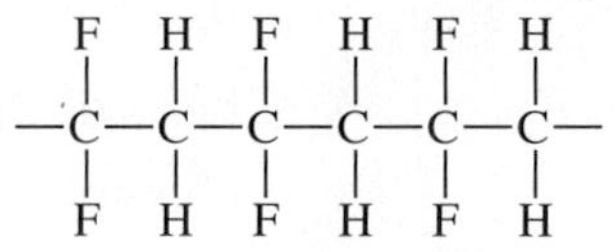

11.45 **a.** toluene
b. 4-bromoaniline
c. 4-ethyltoluene

11.47 1.08 g of H_2

11.49 1×10^{14} molecules

12 Organic Compounds with Oxygen and Sulfur

LOOKING AHEAD

"The purpose of our research was to create a way to make Taxol," says Paul Wender, Francis W. Bergstrom Professor of organic chemistry and head of the Wender research group at Stanford University. "Taxol is a chemotherapy drug originally derived from the bark of the Pacific yew tree. However, removing the bark from yew trees destroys them, so we need a renewable resource. We worked out a synthesis that began with turpentine, which is both renewable and inexpensive. Initially, Taxol was used with patients who did not respond to chemotherapy. The first person to be treated was a woman who was diagnosed with terminal ovarian cancer and given three to six months to live. After a few treatments with Taxol, she was declared 98% disease free. A drug like Taxol can save many lives, which is one reason that a study of organic chemistry is so important."

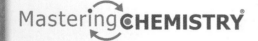

MasteringCHEMISTRY®

Visit **www.masteringchemistry.com** for self-study materials and instructor-assigned homework.

In this chapter, we will look at organic compounds that contain functional groups with oxygen or sulfur atoms. Alcohols, which contain the *hydroxyl group* (—OH), are commonly found in nature and are used in industry and at home. For centuries, grains, vegetables, and fruits have been fermented to produce the ethanol present in alcoholic beverages. The hydroxyl group is important in biomolecules such as sugars and starches as well as in steroids such as cholesterol and estradiol. Menthol is a cyclic alcohol with a minty odor and flavor that is used in cough drops, shaving creams, and ointments. The phenols contain the hydroxyl group (—OH) attached to a benzene ring. Ethers are compounds that contain an oxygen atom connected to two carbon atoms (—O—). Ethers are important solvents in chemistry and medical laboratories. Beginning in 1842, diethyl ether was used for about 100 years as a general anesthetic.

We will also study two other families of organic compounds: aldehydes and ketones. Many of the odors and flavors associated with flavorings and perfumes are due to a carbon–oxygen double bond called a *carbonyl group* (C=O). Aldehydes in foods and perfumes provide the odors and flavors of vanilla, almond, and cinnamon. In biology, you may have seen specimens preserved in a solution of formaldehyde. You probably notice the odor of a ketone when you use paint or nail-polish remover.

12.1 Alcohols, Phenols, Thiols, and Ethers

LEARNING GOAL

Give the IUPAC and common names for alcohols and phenols; identify common names for thiols and ethers.

As we learned in Chapter 10, alcohols and ethers are two classes of organic compounds that contain an oxygen (O) atom, shown in red in the ball-and-stick models. In an **alcohol**, a hydroxyl group (—OH) replaces a hydrogen atom in a hydrocarbon. In a **phenol**, the hydroxyl group replaces a hydrogen atom attached to a benzene ring. A **thiol** contains a sulfur atom, shown in yellow, which makes a thiol similar to an alcohol except that —OH is replaced by —SH. In an **ether**, an oxygen atom is attached to two carbon atoms. Molecules of alcohols, phenols, thiols, and ethers have bent shapes around the oxygen or sulfur atom, similar to water (see Figure 12.1).

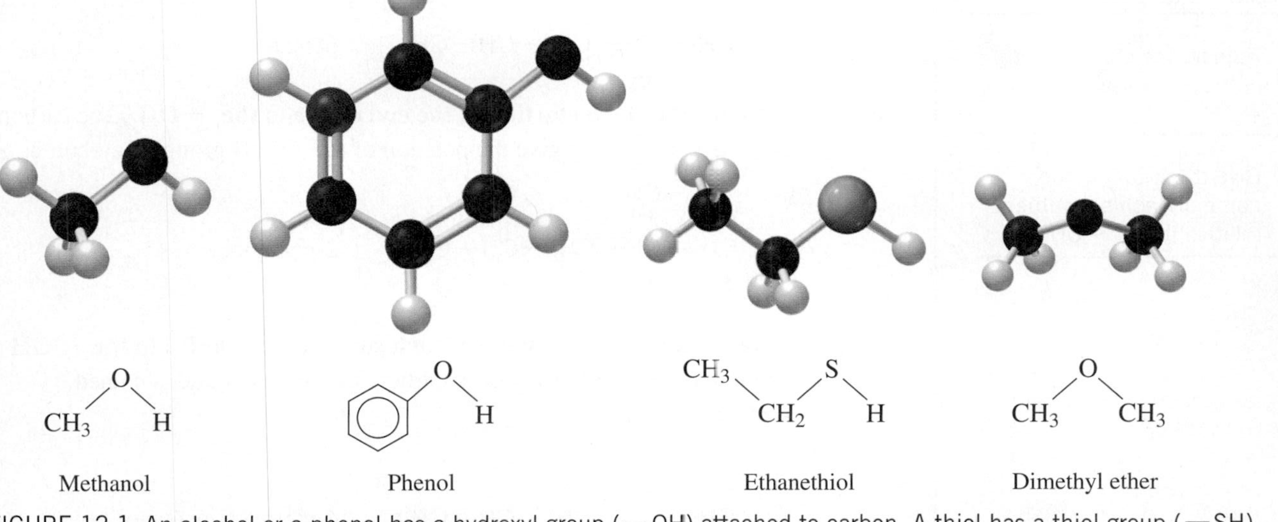

FIGURE 12.1 An alcohol or a phenol has a hydroxyl group (—OH) attached to carbon. A thiol has a thiol group (—SH) attached to carbon. An ether contains an oxygen atom (—O—) bonded to two carbon groups.

Q How is an alcohol different from a thiol?

TUTORIAL
Naming Alcohols, Phenols, and Thiols

TUTORIAL
Drawing Alcohols, Phenols, and Thiols

Naming Alcohols

In the IUPAC system, an alcohol is named by replacing the *e* of the corresponding alkane name with *ol*. The common name of a simple alcohol uses the name of the alkyl group followed by *alcohol*. Alcohols with one or two carbon atoms do not require a number for the hydroxyl group.

$CH_3—H$	$CH_3—OH$	$CH_3—CH_2—H$	$CH_3—CH_2—OH$
Methane	Methanol (methyl alcohol)	Ethane	Ethanol (ethyl alcohol)

When an alcohol consists of a carbon chain with three or more carbon atoms, the chain is numbered to give the position of the —OH and any substituents on the carbon chain. When an alcohol contains two —OH groups, it is named as a *diol*.

$$\underset{3}{CH_3}—\underset{2}{CH_2}—\underset{1}{CH_2}—OH$$

1-Propanol
(propyl alcohol)

$$HO—\underset{1}{CH_2}—\underset{2}{CH_2}—\underset{3}{CH_2}—OH$$

1,3-Propanediol

The condensed structural formula of an alcohol can also be drawn using its skeletal formula as shown for 2-propanol and 2-butanol.

OH

2-Propanol
(isopropyl alcohol)

OH

2-Butanol

SAMPLE PROBLEM 12.1

Naming Alcohols

Give the IUPAC name for each of the following:

a. $CH_3—CH(CH_3)—CH_2—CH(OH)—CH_3$

b. OH, Br (benzene ring)

SOLUTION

Guide to Naming Alcohols

1. Name the longest carbon chain with the —OH group. Name an aromatic alcohol as a *phenol*.
2. Number the chain starting at the end closer to the —OH.
3. Give the location and name of each substituent relative to the —OH group.

a. Step 1 Name the longest carbon chain with the —OH group. Name an aromatic alcohol as a *phenol*. The parent chain with five carbon atoms is pentane; the alcohol is named pentanol.

$$CH_3—\overset{CH_3}{\overset{|}{CH}}—CH_2—\overset{OH}{\overset{|}{CH}}—CH_3 \quad \text{pentanol}$$

Step 2 Number the chain starting at the end closer to the —OH. The carbon chain is numbered to give the position of the —OH group on carbon 2.

$$\underset{5}{CH_3}—\underset{4}{\overset{CH_3}{\overset{|}{CH}}}—\underset{3}{CH_2}—\underset{2}{\overset{OH}{\overset{|}{CH}}}—\underset{1}{CH_3} \quad \text{2-pentanol}$$

Step 3 Give the location and name of each substituent relative to the —OH group. The methyl group is on carbon 4. The compound is named 4-methyl-2-pentanol.

$$\underset{5}{CH_3}—\underset{4}{\overset{CH_3}{\overset{|}{CH}}}—\underset{3}{CH_2}—\underset{2}{\overset{OH}{\overset{|}{CH}}}—\underset{1}{CH_3} \quad \text{4-methyl-2-pentanol}$$

b. Step 1 Name the longest carbon chain with the —OH group. Name an aromatic alcohol as a *phenol*. The compound is a *phenol* because the —OH is attached to a benzene ring.

Br, OH on benzene ring — phenol

Step 2 Number the chain starting at the end closer to the —OH. For a phenol, the carbon atom attached to the —OH is understood to be carbon 1. No number is needed to give the location of the —OH group.

Br (2), OH (1) on benzene ring — phenol

Step 3 Give the location and name of each substituent relative to the —OH group. The phenol is numbered in the direction that gives the lower number to the bromine atom. The compound is named 2-bromophenol.

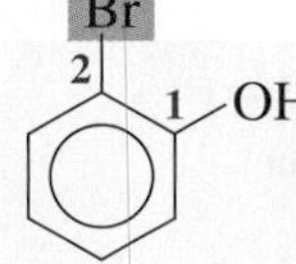

Br (2), OH (1) on benzene ring — 2-bromophenol

STUDY CHECK 12.1

Give the IUPAC name for the following:

Cl, OH (skeletal structure)

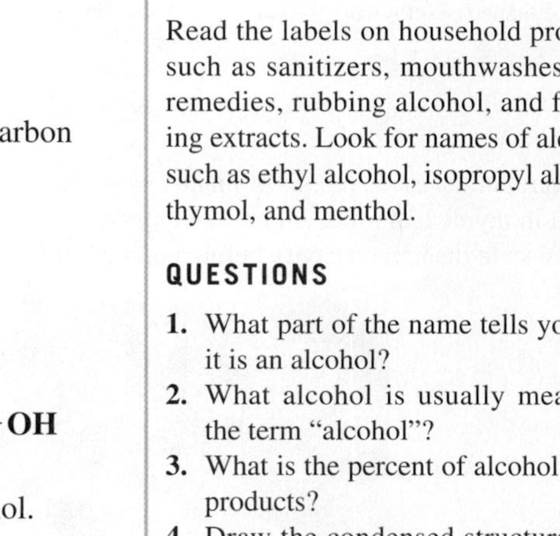

Explore Your World

ALCOHOLS IN HOUSEHOLD PRODUCTS

Read the labels on household products such as sanitizers, mouthwashes, cold remedies, rubbing alcohol, and flavoring extracts. Look for names of alcohols such as ethyl alcohol, isopropyl alcohol, thymol, and menthol.

QUESTIONS

1. What part of the name tells you that it is an alcohol?
2. What alcohol is usually meant by the term "alcohol"?
3. What is the percent of alcohol in the products?
4. Draw the condensed structural formulas of the alcohols you find listed on the labels. You may need to use a reference book for some structures.

Chemistry Link to Health

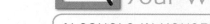

SOME IMPORTANT ALCOHOLS AND PHENOLS

Methanol (methyl alcohol), the simplest alcohol, is found in many solvents and paint removers. If ingested, methanol is oxidized to formaldehyde, which can cause headaches, blindness, and death. Methanol is used to make plastics, medicines, and fuels. In car racing, it is used as a fuel because it is less flammable and has a higher octane rating than does gasoline.

Ethanol (ethyl alcohol) has been known since prehistoric times as an intoxicating product formed by the fermentation of grains and starches.

$$C_6H_{12}O_6 \xrightarrow{\text{Fermentation}} 2CH_3{-}CH_2{-}OH + 2CO_2$$

Today, ethanol for commercial uses is produced by allowing ethene and water to react at high temperatures and pressures. It is used as a solvent for perfumes, varnishes, and some medicines, such as tincture of iodine. "Gasohol" is a mixture of ethanol and gasoline used as a fuel.

$$H_2C{=}CH_2 + H_2O \xrightarrow[\text{Catalyst}]{300\ °C,\ 200\ atm} CH_3{-}CH_2{-}OH$$

1,2,3-Propanetriol (glycerol or glycerin), a trihydroxy alcohol, is a viscous liquid obtained from oils and fats during the production of soaps. The presence of several —OH groups makes it strongly attracted to water, a feature that makes glycerin useful as a skin softener in products such as skin lotions, cosmetics, shaving creams, and liquid soaps.

$$HO{-}CH_2{-}\underset{}{\overset{OH}{\overset{|}{C}H}}{-}CH_2{-}OH$$

1,2,3-Propanetriol (glycerol)

1,2-Ethanediol (ethylene glycol) is used as an antifreeze in heating and cooling systems. It is also a solvent for paints, inks, and plastics and is used in the production of synthetic fibers such as Dacron. If ingested, it is extremely toxic. In the body, it is oxidized to oxalic acid, which forms insoluble salts in the kidneys that cause renal damage, convulsions, and death. Because its sweet taste is attractive to pets and children, ethylene glycol solutions must be carefully stored.

$$\underset{\text{1,2-Ethanediol (ethylene glycol)}}{HO-CH_2-CH_2-OH} \xrightarrow{[O]} \underset{\text{Oxalic acid}}{HO-\overset{\overset{\displaystyle O}{\|}}{C}-\overset{\overset{\displaystyle O}{\|}}{C}-OH}$$

An antifreeze raises the boiling point and decreases the freezing point of water in a radiator.

Several of the essential oils of plants, which produce the odor or flavor of the plant, are derivatives of phenol. Isoeugenol is found in nutmeg, thymol in thyme and mint, eugenol in cloves, and vanillin in vanilla bean. Thymol has a pleasant, minty taste and is used in mouthwashes and by dentists to disinfect a cavity before adding a filling compound.

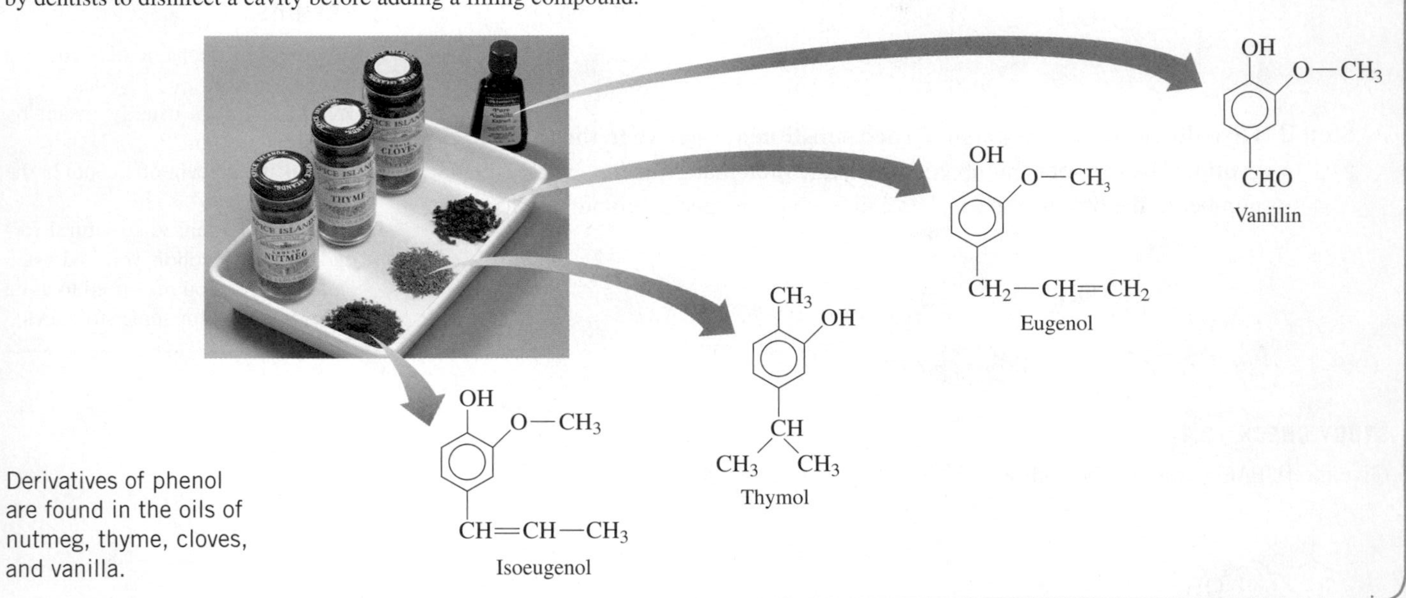

Derivatives of phenol are found in the oils of nutmeg, thyme, cloves, and vanilla.

Thiols

SELF STUDY ACTIVITY
Alcohols and Thiols

Thiols are a family of sulfur-containing organic compounds that have a *thiol* (—SH) group. In the IUPAC system, thiols are named by adding *thiol* to the longest carbon chain connected to the —SH group. An important property of thiols is their strong, sometimes

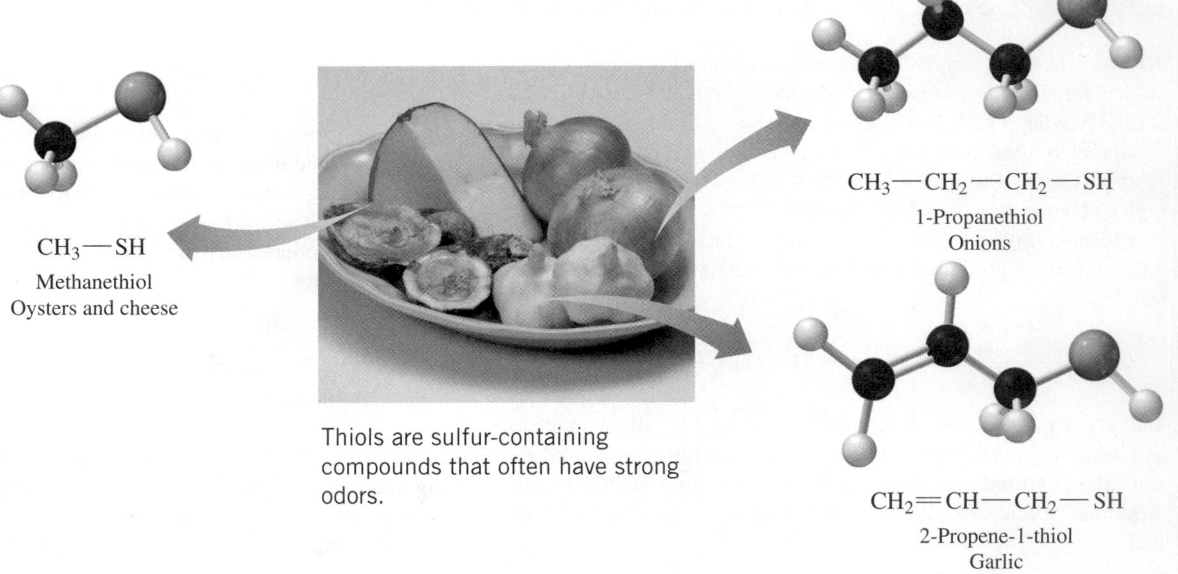

Thiols are sulfur-containing compounds that often have strong odors.

disagreeable, odor, which is characteristic of oysters, cheddar cheese, onions, and garlic. To help us detect natural gas (methane) leaks, a small amount of ethanethiol is added to the gas supply.

Ethers

An ether contains an oxygen atom that is attached by single bonds to two carbon groups that are alkyls or aromatic rings. Ethers have a bent structure like water and alcohols, except both hydrogen atoms are replaced by carbon groups.

Naming Ethers

Most ethers have common names. The name of each alkyl or aromatic group attached to the oxygen atom is written in alphabetical order, followed by the word *ether*. In this text, we will use only the common names of ethers.

MC TUTORIAL Naming Ethers

TUTORIAL Drawing Ethers

Methyl group (↓) Propyl group (↓)

$$\boxed{CH_3}-O-\boxed{CH_2-CH_2-CH_3}$$

Common name: methyl propyl ether

SAMPLE PROBLEM 12.2

Common Names of Ethers

Give the common name for the following:

$$CH_3-O-CH_3$$

SOLUTION

The groups attached to the oxygen atom in the ether are both methyl groups. In the common name, the prefix *di* is used to indicate two of the same group. The name of this ether is dimethyl ether.

STUDY CHECK 12.2

Draw the skeletal formula of ethyl propyl ether.

Isomers of Alcohols and Ethers

Alcohols and ethers can have the same molecular formula. For example, we can draw the condensed structural formulas of the isomers with the molecular formula C_2H_6O as follows:

CH_3-CH_2-OH (Ethyl alcohol) and CH_3-O-CH_3 (Dimethyl ether)

CONCEPT CHECK 12.1

Isomers

Why do the following condensed structural formulas represent a pair of isomers?

$$CH_3-CH_2-CH_2-OH \qquad CH_3-O-CH_2-CH_3$$

ANSWER

Both compounds, one an alcohol and the other an ether, have the same molecular formula, C_3H_8O, but different condensed structural formulas, which makes them isomers.

Chemistry Link to Health

ETHERS AS ANESTHETICS

Anesthesia is the loss of sensation and consciousness. A general anesthetic is a substance that blocks signals to the awareness centers in the brain so the person has a loss of memory, a loss of feeling pain, and an artificial sleep. The term *ether* has been associated with anesthesia because diethyl ether was the most widely used anesthetic for more than a hundred years. Although it is easy to administer, ether is very volatile and highly flammable. A small spark in the operating room could cause an explosion. Since the 1950s, anesthetics such as Forane (isoflurane), Ethrane (enflurane), and Penthrane (methoxyflurane) have been developed that are not as flammable and do not cause nausea. Most of these anesthetics retain the ether group, but the addition of halogen atoms reduces the volatility and flammability of the ethers. More recently, they have been replaced by halothane (1-bromo-1-chloro-2,2,2-trifluoroethane) because of the side effects of the ether-type inhalation anesthetics.

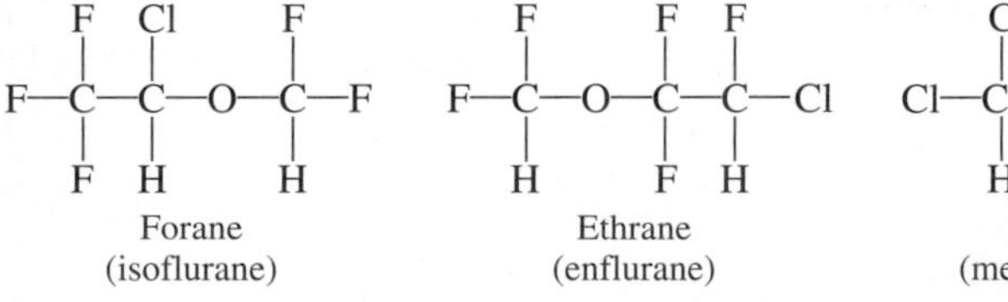

Forane (isoflurane) Ethrane (enflurane) Penthrane (methoxyflurane)

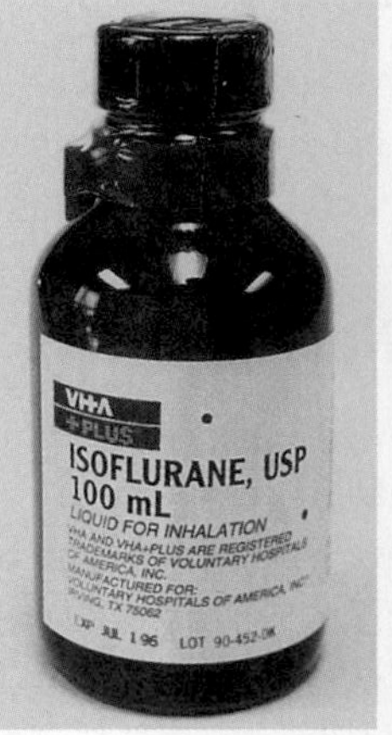

Isoflurane (Forane) is an inhaled anesthetic.

QUESTIONS AND PROBLEMS

Alcohols, Phenols, Thiols, and Ethers

12.1 Give the IUPAC name for each of the following:

a. CH_3-CH_2-OH **b.** $CH_3-CH_2-CH(OH)-CH_3$

c. (skeletal structure with OH) **d.** (benzene ring with OH and Br)

12.2 Give the IUPAC name for each of the following:

a. (benzene ring with OH and CH_2-CH_3)

b. $CH_3-CH_2-CH(CH_3)-CH_2-OH$

c. (skeletal structure with OH)

d. (skeletal structure with OH)

12.3 Draw the condensed structural formula of each of the following:
a. 1-propanol
b. 3-pentanol
c. 2-methyl-2-butanol

12.4 Draw the condensed structural formula of each of the following:
a. ethyl alcohol
b. 3-methyl-1-butanol
c. propyl alcohol

12.5 Give a common name for each of the following:
a. $CH_3-CH_2-CH_2-O-CH_2-CH_2-CH_3$
b. (cyclohexane ring)–O–CH_3

12.6 Give a common name for each of the following:
a. $CH_3-CH_2-O-CH_2-CH_2-CH_3$
b. (cyclopentane ring)–O–CH_3

12.7 Draw the condensed structural formula for each of the following:
a. ethyl methyl ether
b. cyclopropyl ethyl ether

12.8 Draw the condensed structural formula for each of the following:
a. diethyl ether
b. cyclobutyl methyl ether

12.2 Properties of Alcohols and Ethers

LEARNING GOAL

Describe the classification of alcohols; describe the solubility of alcohols and ethers in water.

Alcohols are classified by the number of alkyl groups attached to the carbon atom bonded to the hydroxyl (—OH) group. A **primary (1°) alcohol** has one alkyl group attached to the carbon atom bonded to the —OH, a **secondary (2°) alcohol** has two alkyl groups, and a **tertiary (3°) alcohol** has three alkyl groups.

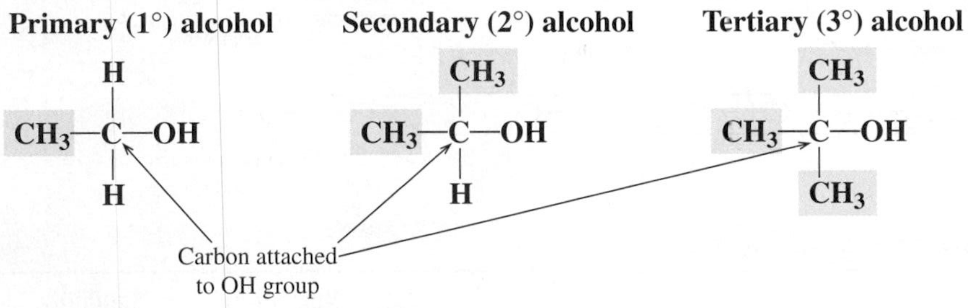

CONCEPT CHECK 12.2

Classifying Alcohols

Classify each of the following alcohols as primary (1°), secondary (2°), or tertiary (3°):

a. $CH_3—CH_2—CH_2—OH$

b. OH

ANSWER

a. The carbon atom bonded to the —OH is attached to one alkyl group, which makes this a primary (1°) alcohol.

b. The carbon atom bonded to the —OH is attached to three alkyl groups, which makes this a tertiary (3°) alcohol.

Solubility of Alcohols and Ethers in Water

In Chapter 10, we learned that hydrocarbons, which are composed of only carbon and hydrogen, are nonpolar and insoluble in water. However, the polar —OH group in an alcohol can form hydrogen bonds with the H and O atoms of water, which makes alcohols more soluble in water. The electronegative O atom in ethers can also form hydrogen bonds with water (see Figure 12.2).

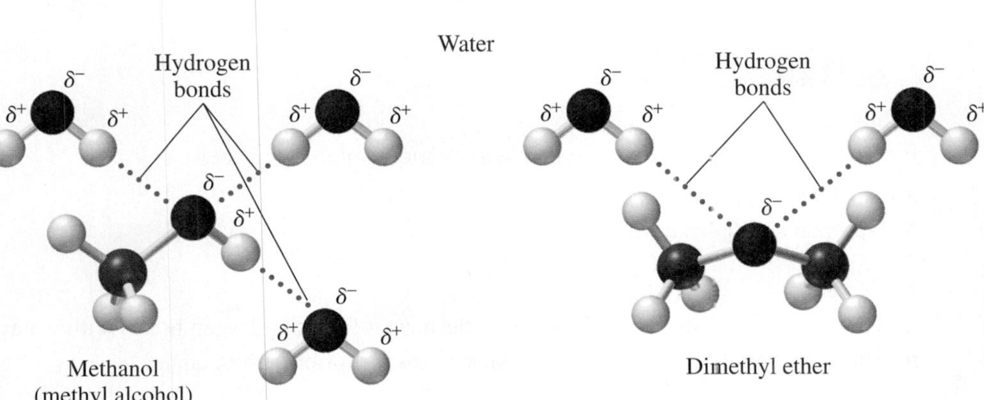

FIGURE 12.2 Methanol and dimethyl ether form hydrogen bonds with water molecules.

Q Why are alcohols more soluble in water than ethers?

Alcohols with one to three carbon atoms are miscible with water, which means any amount of the alcohol is completely soluble in water. However, the solubility provided by the —OH group decreases as the number of carbon atoms increases. Alcohols with four carbon atoms are slightly soluble in water and alcohols with five or more carbon atoms are insoluble. Table 12.1 compares the solubility of some alcohols.

Nonpolar carbon chain → $\boxed{CH_3—CH_2}$—OH

Soluble in water

$\boxed{CH_3—CH_2—CH_2—CH_2—CH_2—CH_2—CH_2—CH_2}$—OH

Insoluble in water

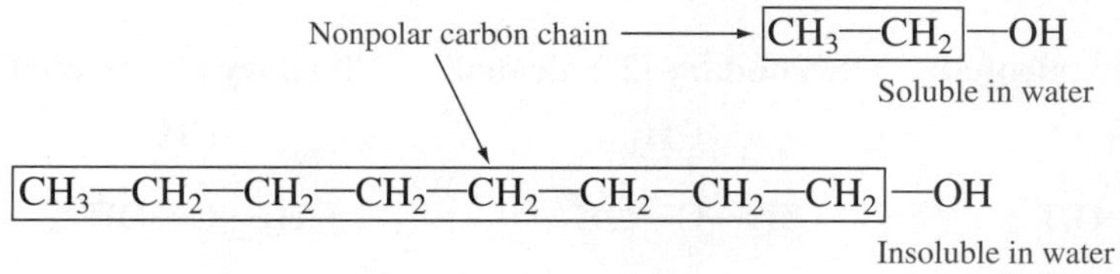

TABLE 12.1 Solubility of Some Alcohols

Compound	Condensed Structural Formula	Number of Carbon Atoms	Solubility in Water
Methanol	$CH_3—OH$	1	Soluble
Ethanol	$CH_3—CH_2—OH$	2	Soluble
1-Propanol	$CH_3—CH_2—CH_2—OH$	3	Soluble
1-Butanol	$CH_3—CH_2—CH_2—CH_2—OH$	4	Slightly soluble
1-Pentanol	$CH_3—CH_2—CH_2—CH_2—CH_2—OH$	5	Insoluble

Although ethers can form hydrogen bonds with water, they do not form as many hydrogen bonds with water as do the alcohols. Thus, ethers with up to four carbon atoms are slightly soluble in water. Ethers are more soluble in water than are alkanes, but not as soluble as alcohols.

Solubility of Phenols

Phenol is slightly soluble in water because the —OH group can form hydrogen bonds with water molecules. In water, the —OH ionizes slightly, which makes phenol a weak acid. In fact, an early name for phenol was *carbolic acid.* A concentrated solution of phenol is very corrosive and highly irritating to the skin; it can cause severe burns and ingestion can be fatal. At one time, dilute solutions of phenol were used as antiseptics, but they have generally been replaced.

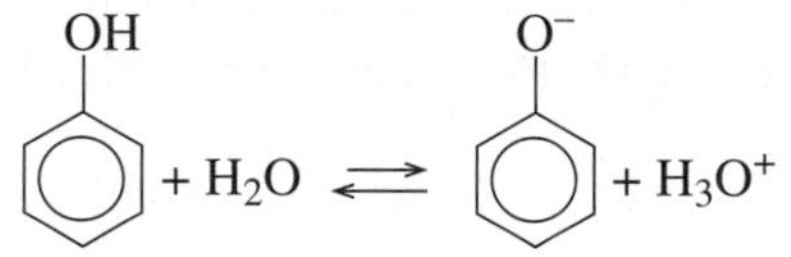

$$C_6H_5OH + H_2O \rightleftarrows C_6H_5O^- + H_3O^+$$

Phenol   Phenoxide ion

CONCEPT CHECK 12.3

Solubility

Predict which compound in each pair will be more soluble in water.

a. propane or ethanol

b. 1-propanol or 1-hexanol

ANSWER

a. The alcohol ethanol is soluble in water because it forms hydrogen bonds with water.

b. The 1-propanol is more soluble because it has a shorter carbon chain.

Chemistry Link to Health

HAND SANITIZERS AND ETHANOL

Hand sanitizers are used as an alternative to washing hands to kill most bacteria and viruses that spread colds and flu. As a gel or liquid solution, many hand sanitizers use ethanol as their active ingredient. While safe for most adults, supervision is recommended when used by children because there is some concern about their risk to the health of children.

The amount of ethanol in an alcohol-containing sanitizer is typically 60% (v/v) but can be as high as 85%. The high volume of ethanol can make hand sanitizers a fire hazard in the home because they are highly flammable. When ethanol undergoes combustion, it produces a transparent blue flame. When using an ethanol-containing sanitizer, it is important to rub hands until they are completely dry. It is also recommended that sanitizers containing ethanol be placed in storage areas that are away from heat sources in the home.

Some sanitizers are alcohol free, but often the active ingredient is triclosan, which contains aromatic, ether, and phenol functional groups. However, the Food and Drug Administration is considering banning triclosan in personal-care products because when mixed with tap water for disposal, the triclosan that accumulates in the environment may promote growth of antibiotic-resistant bacteria.

Hand sanitizers that contain ethanol are used to kill bacteria on the hands.

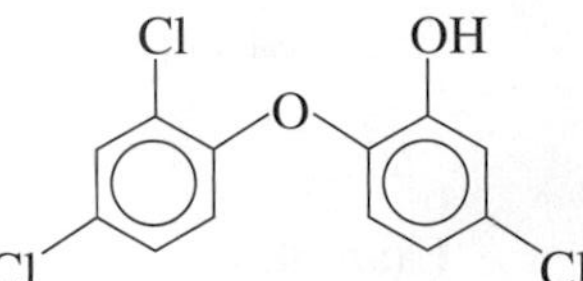

Triclosan is an antibacterial compound used in personal-care products.

QUESTIONS AND PROBLEMS

Properties of Alcohols and Ethers

12.9 Classify each of the following alcohols as primary (1°), secondary (2°), or tertiary (3°):

a. OH

b. $CH_3-CH_2-CH_2-CH_2-OH$

c. $CH_3-C(OH)(CH_3)-CH_2-CH_3$

d. OH, CH_3

12.10 Classify each of the following alcohols as primary (1°), secondary (2°), or tertiary (3°):

a. CH_3, OH

b. OH

c. CH_2-OH

d. $CH_3-CH_2-CH_2-C(CH_3)_2-OH$

12.11 Are each of the following soluble, slightly soluble, or insoluble in water?

a. CH_3-CH_2-OH

b. CH_3-O-CH_3

c. $CH_3-CH_2-CH_2-CH_2-CH_2-CH_2-OH$

12.12 Are each of the following soluble, slightly soluble, or insoluble in water?

a. $CH_3-CH_2-CH_2-OH$

b. $CH_3-CH_2-CH_2-CH_2-CH_3$

c. $CH_3-CH_2-O-CH_2-CH_3$

12.13 Give an explanation for the following observations:

a. Methanol is soluble in water, but ethane is not.

b. 2-Propanol is soluble in water, but 1-butanol is only slightly soluble.

c. 1-Propanol is soluble in water, but ethyl methyl ether is only slightly soluble.

12.14 Give an explanation for the following observations:

a. Ethanol is soluble in water, but propane is not.

b. Dimethyl ether is slightly soluble in water, but pentane is not.

c. 1-Propanol is soluble in water, but 1-hexanol is not.

LEARNING GOAL

Write equations for the combustion, dehydration, and oxidation of alcohols.

FIGURE 12.3 A flaming dessert is prepared using liquor that undergoes combustion.

Q What is the equation for the complete combustion of the ethanol in the liquor?

TUTORIAL
Dehydration and Oxidation of Alcohols

12.3 Reactions of Alcohols and Thiols

In Chapter 10, we learned that hydrocarbons undergo combustion in the presence of oxygen. Alcohols burn with oxygen too. For example, in a restaurant, a flaming dessert may be prepared by pouring liquor on fruit or ice cream and lighting it (see Figure 12.3). The combustion of the ethanol in the liquor proceeds as follows:

$$CH_3{-}CH_2{-}OH(g) + 3O_2(g) \xrightarrow{\Delta} 2CO_2(g) + 3H_2O(g) + \text{energy}$$

Dehydration of Alcohols to Form Alkenes

Earlier, we saw that alkenes can add water to yield alcohols. In a reverse reaction, alcohols lose a water molecule when they are heated with an acid catalyst such as H_2SO_4. During the **dehydration** of an alcohol, H— and —OH are removed from *adjacent carbon atoms of the same alcohol* to produce a water molecule. A double bond forms between the same two carbon atoms to produce an alkene product:

$$\underset{\text{Alcohol}}{-\overset{\text{H}}{\underset{|}{\overset{|}{\text{C}}}}-\overset{\text{OH}}{\underset{|}{\overset{|}{\text{C}}}}-} \xrightarrow[\text{Heat}]{\text{H}^+} \underset{\text{Alkene}}{\text{C}=\text{C}} + \underset{\text{Water}}{\mathbf{H_2O}}$$

Examples

$$\underset{\text{Ethanol}}{\text{H}-\overset{\text{H}}{\underset{\text{H}}{\text{C}}}-\overset{\text{OH}}{\underset{\text{H}}{\text{C}}}-\text{H}} \xrightarrow[\text{Heat}]{\text{H}^+} \underset{\text{Ethene}}{\text{H}_2\text{C}=\text{CH}_2} + \mathbf{H_2O}$$

$$\text{Cyclopentanol} \xrightarrow[\text{Heat}]{\text{H}^+} \text{Cyclopentene} + \mathbf{H_2O}$$

CONCEPT CHECK 12.4

Dehydration of Alcohols

What is the name of the product from the dehydration of 1-pentanol?

ANSWER

In dehydration, the 1-pentanol loses —OH from carbon 1 and —H from carbon 2 to form an alkene and water. The name of the alkene is 1-pentene.

SAMPLE PROBLEM 12.3

Dehydration of Alcohols

Draw the condensed structural formula for the alkene produced by the dehydration of the following:

$$CH_3{-}\overset{\text{OH}}{\overset{|}{\text{CH}}}{-}CH_3 \xrightarrow[\text{Heat}]{\text{H}^+}$$

SOLUTION

In a dehydration, the —OH is removed from carbon 2 and one —H is removed from an adjacent carbon, which forms a double bond.

$CH_3—CH{=}CH_2$

STUDY CHECK 12.3

What is the name of the alkene produced by the dehydration of cyclohexanol?

Oxidation of Alcohols

In Chapter 5, we described oxidation as a loss of hydrogen atoms or the addition of oxygen. We can now apply this definition of oxidation to our study of organic chemistry where **oxidation** occurs when there is an increase in the number of carbon–oxygen bonds. In a *reduction* reaction, the product has fewer bonds between carbon and oxygen.

$$CH_3—CH_3 \underset{\text{Reduction}}{\overset{\text{Oxidation}}{\rightleftharpoons}} CH_3—CH_2—OH \underset{\text{Reduction}}{\overset{\text{Oxidation}}{\rightleftharpoons}} CH_3—C({=}O)—H \underset{\text{Reduction}}{\overset{\text{Oxidation}}{\rightleftharpoons}} CH_3—C({=}O)—OH$$

Alkane; Alcohol (1°) — 1 Bond to O; Aldehyde — 2 Bonds to O; Carboxylic acid — 3 Bonds to O

Oxidation of Primary and Secondary Alcohols

The oxidation of a primary alcohol produces an aldehyde, which contains a double bond between carbon and oxygen. For example, the oxidation of methanol occurs by removing two hydrogen atoms, one from the —OH group and another from the carbon that is bonded to the —OH. To indicate the presence of an oxidizing agent, the symbol [O] is written over the arrow. An oxidation reaction is balanced with H_2O, which assumes an oxygen atom is obtained from the oxidizing agent.

$$H—\underset{\mathbf{H}}{\overset{\mathbf{OH}}{C}}—H \xrightarrow{[O]} H—\overset{O}{\overset{\|}{C}}—H + \mathbf{H_2O}$$

Methanol (methyl alcohol) → Methanal (formaldehyde)

In the oxidation of ethanol, two hydrogen atoms are removed, one from the —OH group and another from the carbon that is bonded to the —OH. The oxidized product contains the same number of carbon atoms (two) as the reactant.

$$CH_3—\overset{OH}{CH_2} \xrightarrow{[O]} CH_3—\overset{O}{\overset{\|}{C}}—H + H_2O$$

Ethanol (ethyl alcohol) → Ethanal (acetaldehyde)

Aldehydes oxidize further by the addition of oxygen to form a carboxylic acid. This step occurs so readily that it is often difficult to isolate the aldehyde product during oxidation. We will learn more about carboxylic acids in Chapter 14.

$$CH_3—\overset{O}{\overset{\|}{C}}—H \xrightarrow{[O]} CH_3—\overset{O}{\overset{\|}{C}}—\mathbf{OH}$$

Ethanal (acetaldehyde) → Ethanoic acid (acetic acid)

Chemistry Link to Health

METHANOL POISONING

Methanol (methyl alcohol), CH_3-OH, is a highly toxic alcohol present in products such as windshield washer fluid, Sterno, and paint strippers. Methanol is rapidly absorbed in the gastrointestinal tract. In the liver, it is oxidized to formaldehyde and then formic acid, a substance that causes nausea, severe abdominal pain, and blurred vision. Blindness can occur because the intermediate products destroy the retina of the eye. As little as 4 mL of methanol can produce blindness. The formic acid, which is not readily eliminated from the body, lowers blood pH so severely that just 30 mL of methanol can lead to coma and death.

The treatment for methanol poisoning involves giving sodium bicarbonate to neutralize the formic acid in the blood. In some cases, ethanol is given intravenously to the patient. The enzymes in the liver pick up ethanol molecules to oxidize instead of methanol molecules. This process gives time for the methanol to be eliminated via the lungs without the formation of its dangerous oxidation products.

In the oxidation of secondary alcohols, the products are ketones. Two hydrogen atoms are removed, one from the —OH group and another from the carbon bonded to the —OH group. The result is a ketone that has the carbon–oxygen double bond attached to alkyl groups on both sides. There is no further oxidation of a ketone because there are no hydrogen atoms attached to the carbonyl group.

$$CH_3-\overset{\mathbf{OH}}{\underset{\mathbf{H}}{\overset{|}{\underset{|}{C}}}}-CH_3 \xrightarrow{[O]} CH_3-\overset{O}{\overset{\|}{C}}-CH_3 + \mathbf{H_2O}$$

2-Propanol (isopropyl alcohol) → Propanone (dimethyl ketone; acetone)

Tertiary alcohols do not oxidize readily because there is no hydrogen atom on the carbon bonded to the —OH group. Because C—C bonds are usually too strong to oxidize, tertiary alcohols resist oxidation.

$$CH_3-\overset{O-\mathbf{H}}{\underset{CH_3}{\overset{|}{\underset{|}{C}}}}-CH_3 \xrightarrow{[O]} \text{No oxidation product readily formed}$$

No double bond forms; No hydrogen on this carbon

3° Alcohol

SAMPLE PROBLEM 12.4

Oxidation of Alcohols

Classify each of the following as primary (1°), secondary (2°), or tertiary (3°) alcohols; draw the condensed structural formula of the aldehyde or ketone formed by its oxidation:

a. $CH_3-CH_2-\overset{OH}{\overset{|}{C}H}-CH_3$

b. $CH_3-CH_2-CH_2-CH_2-OH$

SOLUTION

a. This is a secondary (2°) alcohol, which oxidizes to a ketone.

$$CH_3-CH_2-\overset{O}{\overset{\|}{C}}-CH_3$$

b. This is a primary (1°) alcohol, which can oxidize to an aldehyde.

$$CH_3-CH_2-CH_2-\overset{O}{\overset{\|}{C}}-H$$

STUDY CHECK 12.4

Draw the condensed structural formula of the product of the oxidation of 2-pentanol.

Career Focus

PHARMACIST

"The pharmacy is one of the many factors in the final integration of chemistry and medicine in patient care," says Dorothea Lorimer, pharmacist, Kaiser Hospital. "If someone is allergic to a medication, I have to find out if a new medication has similar structural features. For instance, some people are allergic to sulfur. If there is sulfur in the new medication, there is a chance it will cause a reaction."

A prescription indicates a specific amount of a medication. At the pharmacy, the chemical name, formula, and quantity in milligrams or micrograms are checked. Then the prescribed number of capsules is placed in a container. If it is a liquid medication, a specific volume is measured and poured into a bottle for liquid prescriptions.

Oxidation of Thiols

Thiols also undergo oxidation when there is a loss of hydrogen atoms from each of two —SH groups. The oxidized product contains a disulfide bond —S—S—. Much of the protein in hair is cross-linked by disulfide bonds, which occur mostly between the side chains of the amino acids of cysteine, which contain thiol groups.

$$\text{Protein chain}-CH_2-\mathbf{SH} + \mathbf{HS}-CH_2-\text{Protein chain} \xrightarrow{[O]}$$

Cysteine side groups

$$\text{Protein chain}-CH_2-\mathbf{S}-\mathbf{S}-CH_2-\text{Protein chain} + \mathbf{H_2O}$$

Disulfide bond

When a person is given a "perm," a reducing substance is used to break the disulfide bonds. While the hair is still wrapped around the curlers, an oxidizing substance is applied that causes new disulfide bonds to form between different parts of the protein hair strands, which gives the hair a new shape.

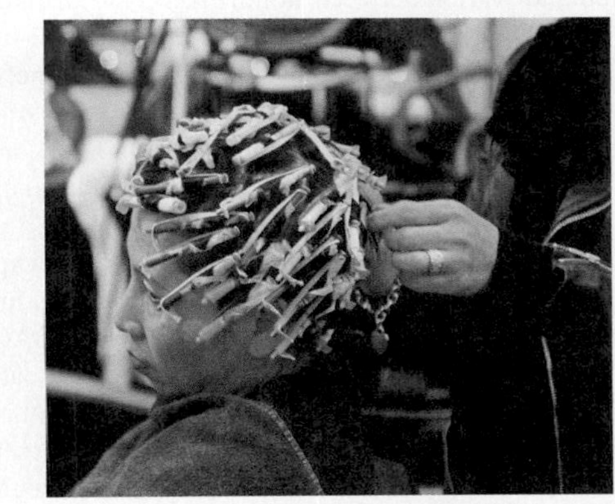

Proteins in the hair take new shapes when disulfide bonds are reduced and oxidized.

CASE STUDY
Alcohol Toxicity

Chemistry Link to Health

OXIDATION OF ALCOHOL IN THE BODY

Ethanol is the most commonly abused drug in the United States. When ingested in small amounts, ethanol may produce a feeling of euphoria in the body although it is a depressant. In the liver, enzymes such as alcohol dehydrogenase oxidize ethanol to acetaldehyde, a substance that impairs mental and physical coordination. If the blood alcohol concentration exceeds 0.4%, coma or death may occur. Table 12.2 gives some of the typical behaviors exhibited at various levels of blood alcohol.

TABLE 12.2 Typical Behaviors Exhibited by a 150-lb Person Consuming Alcohol

Number of Beers (12 oz) or Glasses of Wine (5 oz) in 1 hour	Blood Alcohol Level (% m/v)	Typical Behavior
1	0.025	Slightly dizzy, talkative
2	0.050	Euphoria, loud talking and laughing
4	0.10	Loss of inhibition, loss of coordination, drowsiness, legally drunk in most states
8	0.20	Intoxicated, quick to anger, exaggerated emotions
12	0.30	Unconscious
16–20	0.40–0.50	Coma and death

$$\underset{\substack{\text{Ethanol}\\\text{(ethyl alcohol)}}}{CH_3-CH_2-OH} \xrightarrow{[O]} \underset{\substack{\text{Ethanal}\\\text{(acetaldehyde)}}}{CH_3-\overset{\overset{\large O}{\|}}{C}-H} \xrightarrow{[O]} \underset{\substack{\text{Ethanoic acid}\\\text{(acetic acid)}}}{CH_3-\overset{\overset{\large O}{\|}}{C}-OH} \xrightarrow{[O]} 2CO_2 + H_2O$$

The acetaldehyde produced from ethanol in the liver is further oxidized to acetic acid, which is converted to carbon dioxide and water in the citric acid (Krebs) cycle. Thus, the enzymes in the liver can eventually break down ethanol, but the aldehyde and carboxylic acid intermediates can cause considerable damage while they are present within the cells of the liver.

A person weighing 150 lb requires about one hour to completely metabolize 10 ounces of beer. However, the rate of metabolism of ethanol varies between nondrinkers and drinkers. Typically, nondrinkers and social drinkers can metabolize 12–15 mg of ethanol/dL of blood in one hour, but an alcoholic can metabolize as much as 30 mg of ethanol/dL in one hour. Some effects of alcohol metabolism include an increase in liver lipids (fatty liver), an increase in serum triglycerides, gastritis, pancreatitis, ketoacidosis, alcoholic hepatitis, and psychological disturbances.

When alcohol is present in the blood, it evaporates through the lungs. Thus, the percentage of alcohol in the lungs can be used to calculate the blood alcohol concentration (BAC). Several devices are used to measure the BAC. When a Breathalyzer is used, a suspected drunk driver exhales through a mouthpiece into a solution containing the orange Cr^{6+} ion. Any alcohol present in the exhaled air is oxidized, which reduces the orange Cr^{6+} to a green Cr^{3+}.

$$\underset{\text{Ethanol}}{CH_3-CH_2-OH} + \underset{\text{Orange}}{Cr^{6+}} \xrightarrow{[O]} \underset{\text{Acetic acid}}{CH_3-\overset{\overset{\large O}{\|}}{C}-OH} + \underset{\text{Green}}{Cr^{3+}}$$

The Alcosensor uses an oxidation of alcohol in a fuel cell to generate an electric current that is measured. The Intoxilyzer measures the amount of light absorbed by the alcohol molecules.

Sometimes alcoholics are treated with a drug called Antabuse (disulfiram), which prevents the oxidation of acetaldehyde to acetic acid. As a result, acetaldehyde accumulates in the blood, which causes nausea, profuse sweating, headache, dizziness, vomiting, and respiratory difficulties. Because of these unpleasant side effects, the person is less likely to use alcohol.

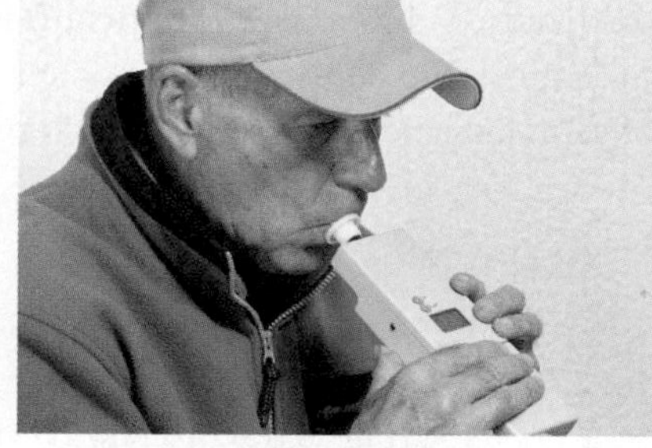

A Breathalyzer is used to determine blood alcohol.

QUESTIONS AND PROBLEMS

Reactions of Alcohols and Thiols

12.15 Draw the condensed structural formula of the alkene produced by each of the following dehydration reactions:

a. $CH_3-CH_2-CH_2-CH_2-OH \xrightarrow[\text{Heat}]{H^+}$

b. 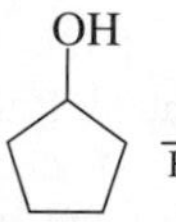 cyclopentanol (OH on cyclopentane ring) $\xrightarrow[\text{Heat}]{H^+}$

c. $CH_3-CH_2-\overset{\overset{\large OH}{|}}{CH}-CH_2-CH_3 \xrightarrow[\text{Heat}]{H^+}$

d. $CH_3-\overset{\overset{\large OH}{|}}{CH}-CH_2-\overset{\overset{\large CH_3}{|}}{CH}-CH_3$

Note: 12.15 c is printed as a structure: cyclohexanol (OH on cyclohexane ring); the condensed formula above labelled c. appears in the left column as c. of 12.15.

12.16 Draw the condensed structural formula of the alkene produced by each of the following dehydration reactions:

a. $CH_3-CH_2-OH \xrightarrow[\text{Heat}]{H^+}$

b. $CH_3-\overset{\overset{\large CH_3}{|}}{CH}-CH_2-OH \xrightarrow[\text{Heat}]{H^+}$

c. 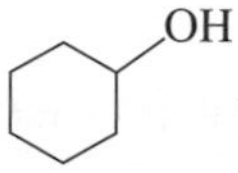 cyclohexanol (OH on cyclohexane ring) $\xrightarrow[\text{Heat}]{H^+}$

12.17 Draw the condensed structural formula of the aldehyde or ketone formed when each of the following alcohols is oxidized [O] (if no reaction, write *none*):

a. $CH_3-CH_2-CH_2-CH_2-CH_2-OH$

b. $CH_3-\overset{\overset{\large OH}{|}}{CH}-CH_2-CH_2-CH_2-CH_3$

c. cyclohexanol (OH on cyclohexane ring)

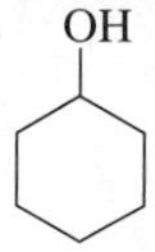

d. $CH_3-\overset{\overset{\large OH}{|}}{CH}-CH_2-\overset{\overset{\large CH_3}{|}}{CH}-CH_3$

12.18 Draw the condensed structural formula of the aldehyde or ketone formed when each of the following alcohols is oxidized [O] (if no reaction, write *none*):

a. $CH_3-\overset{\overset{\large CH_3}{|}}{CH}-CH_2-CH_2-OH$

b. $CH_3-CH_2-\underset{\underset{\large CH_3}{|}}{\overset{\overset{\large OH}{|}}{C}}-CH_3$

c. $CH_3-CH_2-\overset{\overset{\large OH}{|}}{CH}-CH_2-CH_3$

d. 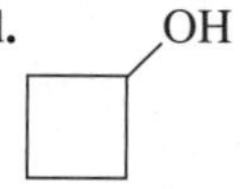 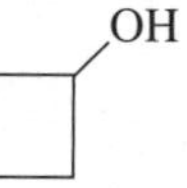 cyclobutanol (OH on cyclobutane ring)

12.4 Aldehydes and Ketones

LEARNING GOAL

Write the IUPAC and common names for aldehydes and ketones; draw the condensed structural formulas when given their names.

MC TUTORIAL
Aldehyde or Ketone?

As we learned in Chapter 10, the **carbonyl group** consists of a carbon–oxygen double bond. The double bond in the carbonyl group is similar to that of alkenes, except the carbonyl group has a dipole. The oxygen atom with two lone pairs of electrons is much more electronegative than the carbon atom. Therefore, the carbonyl group has a strong dipole with a partial negative charge (δ^-) on the oxygen and a partial positive charge (δ^+) on the carbon. The polarity of the carbonyl group strongly influences the physical and chemical properties of aldehydes and ketones.

$O^{\delta-}$
‖
$C^{\delta+}$

In an **aldehyde**, the carbon of the carbonyl group is bonded to at least one hydrogen atom. That carbon may also be bonded to another hydrogen atom, a carbon of an alkyl group, or an aromatic ring (see Figure 12.4).

In Chapter 10, we indicated that the aldehyde group may be written as separate atoms or as —CHO, with the double bond understood. In a **ketone**, the carbonyl group is bonded to two alkyl groups or aromatic rings. As we saw in Chapter 10, the keto group (C═O) can sometimes be written as CO. A skeletal formula may also be used to represent an aldehyde or ketone.

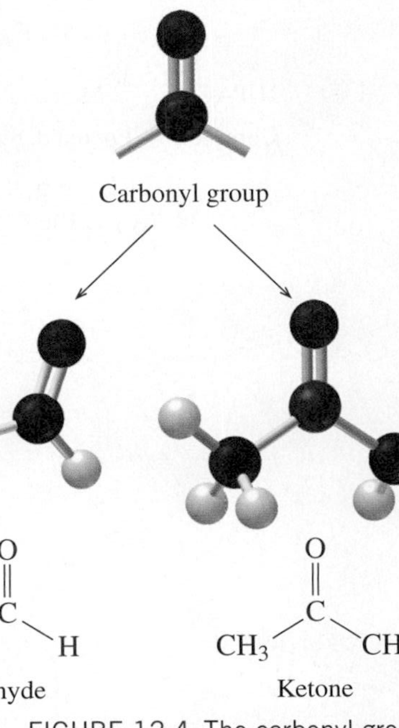

FIGURE 12.4 The carbonyl group is found in aldehydes and ketones.
Q If aldehydes and ketones both contain a carbonyl group, how can you differentiate between compounds from each family?

Formulas for Isomers of C_3H_6O

$CH_3—CH_2—C(=O)—H$ = $CH_3—CH_2—CHO$ = (skeletal formula)
Aldehyde

$CH_3—C(=O)—CH_3$ = $CH_3—CO—CH_3$ = (skeletal formula)
Ketone

CONCEPT CHECK 12.5

Identifying Aldehydes and Ketones

Identify each of the following compounds as an aldehyde or a ketone:

a. (skeletal formula with C=O bonded to H)

b. (cyclopentyl—C(=O)—CH_3)

ANSWER

a. A functional group with a carbonyl (C═O) group bonded to one hydrogen atom makes this an aldehyde.
b. A functional group with a carbonyl (C═O) group bonded to two carbon atoms makes this a ketone.

Naming Aldehydes

MC TUTORIAL
Naming Aldehydes and Ketones

In the IUPAC system, an aldehyde is named by replacing the *e* of the corresponding alkane name with *al*. No number is needed for the aldehyde group because it always appears at the end of the chain. The first four aldehydes are often referred to by their common names, which end in *aldehyde* (see Figure 12.5). The roots (*form*, *acet*, *propion*, and *butyr*) of these common names are derived from Latin or Greek words.

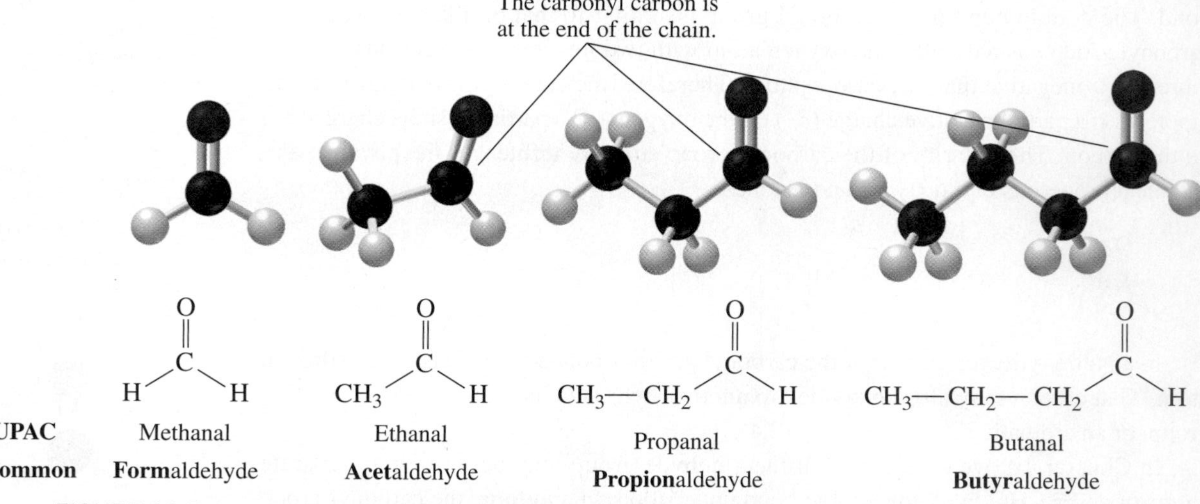

The carbonyl carbon is at the end of the chain.

	$H{-}C({=}O){-}H$	$CH_3{-}C({=}O){-}H$	$CH_3{-}CH_2{-}C({=}O){-}H$	$CH_3{-}CH_2{-}CH_2{-}C({=}O){-}H$
IUPAC	Methanal	Ethanal	Propanal	Butanal
Common	**Form**aldehyde	**Acet**aldehyde	**Propion**aldehyde	**Butyr**aldehyde

FIGURE 12.5 In the structures of aldehydes, the carbonyl group is always the end carbon.
Q Why is the carbon in the carbonyl group in aldehydes always at the end of the chain?

The IUPAC system names the aldehyde of benzene as benzaldehyde.

$C_6H_5{-}C({=}O){-}H$

Benzaldehyde

SAMPLE PROBLEM 12.5

Naming Aldehydes

Give the IUPAC name for each of the following:

a. $CH_3{-}CH(CH_3){-}CH_2{-}C({=}O){-}H$ **b.** $Cl{-}C_6H_4{-}C({=}O){-}H$ (para)

SOLUTION

a. Step 1 **Name the longest carbon chain containing the carbonyl group by replacing the *e* in the alkane name with *al*.** The longest chain has four carbon atoms including the carbon of the carbonyl group, which is named butanal in the IUPAC system.

$CH_3{-}CH(CH_3){-}CH_2{-}C({=}O){-}H$ butanal

Step 2 **Name and number substituents by counting the carbonyl group as carbon 1.**

$\underset{4}{CH_3}{-}\underset{3}{CH}(CH_3){-}\underset{2}{CH_2}{-}\underset{1}{C}({=}O){-}H$ 3-methylbutanal

Guide to Naming Aldehydes

1. Name the longest carbon chain containing the carbonyl group by replacing the *e* in the alkane name with *al*.
2. Name and number substituents by counting the carbonyl group as carbon 1.

b. Step 1 Name the longest carbon chain containing the carbonyl group by replacing the *e* in the alkane name with *al*. The aldehyde of benzene is named benzaldehyde.

Cl—C$_6$H$_4$—C(=O)—H benzaldehyde

Step 2 Name and number substituents by counting the carbonyl group as carbon 1. The benzene ring is numbered from carbon 1 attached to the carbonyl group to carbon 4 attached to the chlorine atom, which gives the name 4-chlorobenzaldehyde.

Cl—[4]C$_6$H$_4$[1]—C(=O)—H 4-chlorobenzaldehyde

STUDY CHECK 12.5

What are the IUPAC and common names of the aldehyde with three carbon atoms?

Chemistry Link to the Environment

VANILLA

Vanilla has been used as a flavoring for over a thousand years. After drinking a beverage made from powdered vanilla and cocoa beans with Emperor Montezuma in Mexico, Cortez took vanilla back to Europe, where it became popular for flavoring and for scenting perfumes and tobacco. Thomas Jefferson introduced vanilla to the United States in the late 1700s. Today, much of the vanilla we use in the world is grown in Mexico, Madagascar, Réunion, Seychelles, Tahiti, Sri Lanka, Java, the Philippines, and Africa.

The vanilla plant is a member of the orchid family. There are many species of *Vanilla*, but *Vanilla planifolia* (or *V. fragrans*) is considered to produce the best flavor. The vanilla plant grows like a vine and can grow to 100 feet in length. Its flowers are hand pollinated to produce a green fruit that is picked in 8 to 9 months. It is sun dried to form a long, dark brown pod, which is called a "vanilla bean" because it looks like a string bean. The flavor and fragrance of the vanilla bean come from the black seeds found inside the dried bean.

The seeds and pod are used to flavor desserts such as custards and ice cream. The extract of vanilla is made by chopping up vanilla beans and mixing them with a 35% alcohol–water mixture. The liquid, which contains the aldehyde vanillin, is drained from the bean residue and used for flavoring.

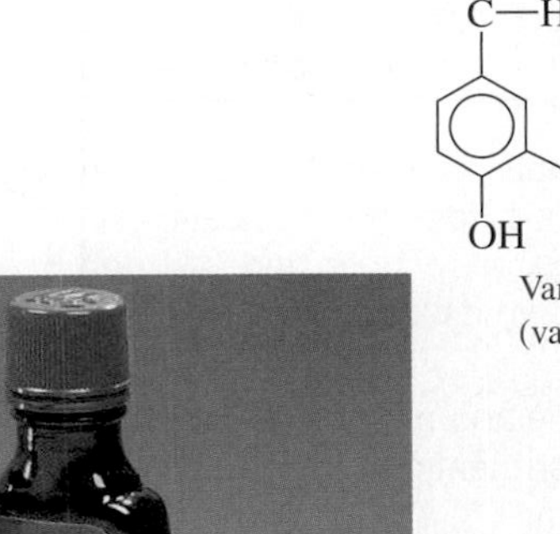

Vanillin (vanilla)

The vanilla bean is the dried fruit of the vanilla plant.

Vanilla flavoring liquid is prepared by soaking vanilla beans in ethanol and water.

Naming Ketones

Aldehydes and ketones are some of the most important classes of organic compounds. Because they have played a major role in organic chemistry for more than a century, the common names for unbranched ketones are still in use. In the common names, the alkyl groups bonded to the carbonyl group are named as substituents and are listed alphabetically, followed by *ketone*. Acetone, which is another name for propanone, has been retained by the IUPAC system.

In the IUPAC system, the name of a ketone is obtained by replacing the *e* in the corresponding alkane name with *one*. Carbon chains with five carbon atoms or more are numbered from the end nearer the carbonyl group.

$$CH_3-\overset{\overset{\displaystyle O}{\|}}{C}-CH_3 \qquad CH_3-CH_2-\overset{\overset{\displaystyle O}{\|}}{C}-CH_3 \qquad CH_3-CH_2-\overset{\overset{\displaystyle O}{\|}}{C}-CH_2-CH_3$$

Propanone (dimethyl ketone; acetone) — Butanone (ethyl methyl ketone) — 3-Pentanone (diethyl ketone)

In a cyclic ketone, the carbonyl carbon is numbered as carbon 1. The ring is numbered in the direction to give substituents the lowest possible numbers.

3-Methylcyclohexanone

Guide to Naming Ketones

1. Name the longest carbon chain that contains the carbonyl group by replacing the *e* in the alkane name with *one.*
2. Number the carbon chain starting from the end nearer the carbonyl group and indicate its location.
3. Name and number any substituents on the carbon chain.

SAMPLE PROBLEM 12.6

Names of Ketones

Give the IUPAC name for the following ketone:

$$CH_3-\overset{\overset{\displaystyle CH_3}{|}}{CH}-CH_2-\overset{\overset{\displaystyle O}{\|}}{C}-CH_3$$

SOLUTION

Step 1 **Name the longest carbon chain that contains the carbonyl group by replacing the *e* in the alkane name with *one*.** The longest chain has five carbon atoms, which is named pentanone.

$$CH_3-\overset{\overset{\displaystyle CH_3}{|}}{CH}-CH_2-\overset{\overset{\displaystyle O}{\|}}{C}-CH_3 \qquad \text{pentanone}$$

Step 2 **Number the carbon chain starting from the end nearer the carbonyl group and indicate its location.** Counting from the right, the carbonyl group is on carbon 2.

$$\underset{5}{CH_3}-\underset{4}{\overset{\overset{\displaystyle CH_3}{|}}{CH}}-\underset{3}{CH_2}-\underset{2}{\overset{\overset{\displaystyle O}{\|}}{C}}-\underset{1}{CH_3} \qquad \text{2-pentanone}$$

Step 3 **Name and number any substituents on the carbon chain.** Counting from the right, the methyl group is on carbon 4. The IUPAC name is 4-methyl-2-pentanone.

$$\underset{5}{CH_3}-\underset{4}{\overset{\overset{\displaystyle CH_3}{|}}{CH}}-\underset{3}{CH_2}-\underset{2}{\overset{\overset{\displaystyle O}{\|}}{C}}-\underset{1}{CH_3} \qquad \text{4-methyl-2-pentanone}$$

STUDY CHECK 12.6

What is the common name of 3-hexanone?

Chemistry Link to Health

SOME IMPORTANT ALDEHYDES AND KETONES

Formaldehyde, the simplest aldehyde, is a colorless gas with a pungent odor. Industrially, it is a reactant in the synthesis of polymers used to make fabrics, insulation materials, carpeting, pressed-wood products such as plywood, and plastics for kitchen counters. An aqueous solution called formalin, which contains 40% formaldehyde, is used as a germicide and to preserve biological specimens. Exposure to formaldehyde fumes can irritate eyes, nose, upper respiratory tract, and can cause skin rashes, headaches, dizziness, and general fatigue. Formaldehyde is classified as a carcinogen.

Acetone, or propanone (dimethyl ketone), which is the simplest ketone, is a colorless liquid with a mild odor that has wide use as a solvent in cleaning fluids, paint and nail-polish removers, and rubber cement (see Figure 12.6). It is extremely flammable and care must be taken when using acetone. In the body, acetone may be produced in uncontrolled diabetes, fasting, and high-protein diets when large amounts of fats are metabolized for energy.

$CH_3-\overset{O}{\overset{\|}{C}}-\overset{O}{\overset{\|}{C}}-CH_3$

Butanedione

The flavor of butter or margarine is from butanedione, muscone is used to make musk perfumes, and oil of spearmint contains carvone.

$$\begin{matrix} CH_2-CH_2-CH_2-CH_2-CH_2-\underset{|}{\overset{CH_3}{CH}}-CH_2 \\ | \qquad\qquad\qquad\qquad\qquad\qquad\qquad \backslash C=O \\ CH_2-CH_2-CH_2-CH_2-CH_2-CH_2-CH_2 \end{matrix}$$

Muscone (musk)

Carvone (spearmint oil)

Several naturally occurring aromatic aldehydes are used to flavor food and as fragrances in perfumes. Benzaldehyde is found in almonds, vanillin in vanilla beans, and cinnamaldehyde in cinnamon.

Benzaldehyde (almond)

Vanillin (vanilla)

Cinnamaldehyde (cinnamon)

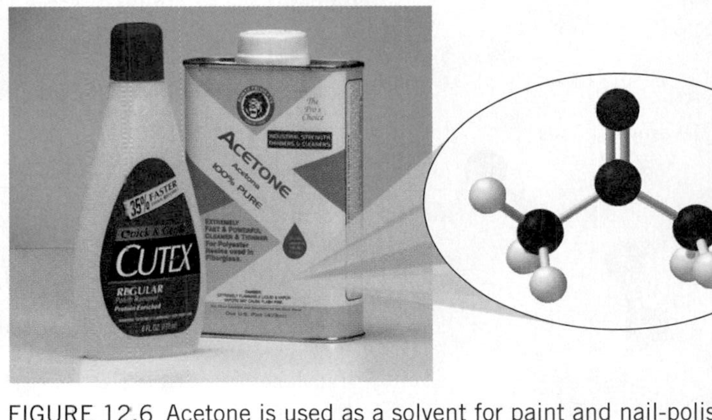

FIGURE 12.6 Acetone is used as a solvent for paint and nail-polish removers.

Q What is the IUPAC name of acetone?

QUESTIONS AND PROBLEMS

Aldehydes and Ketones

12.19 Give the common name for each of the following:

a. $CH_3-\overset{O}{\overset{\|}{C}}-H$

b.

c. $H-\overset{O}{\overset{\|}{C}}-H$

12.20 Give the common name for each of the following:

a. $CH_3-\overset{O}{\overset{\|}{C}}-CH_2-CH_3$

b.

c. $CH_3-CH_2-\overset{O}{\overset{\|}{C}}-H$

12.21 Give the IUPAC name for each of the following:

a. $CH_3-CH_2-\overset{\overset{\displaystyle O}{\|}}{C}-H$ **b.** $CH_3-CH_2-\overset{\overset{\displaystyle O}{\|}}{C}-\overset{\overset{\displaystyle CH_3}{|}}{CH}-CH_3$

c. (cyclohexanone ring with O, CH$_3$ substituent) **d.** (benzene ring with $\overset{\overset{\displaystyle O}{\|}}{C}-H$)

12.22 Give the IUPAC name for each of the following:

a. $CH_3-CH_2-\overset{\overset{\displaystyle CH_3}{|}}{CH}-CH_2-\overset{\overset{\displaystyle O}{\|}}{C}-H$

b. $CH_3-CH_2-CH_2-\overset{\overset{\displaystyle O}{\|}}{C}-CH_3$

c. (cyclopentanone ring with O) **d.** (benzene ring with $\overset{\overset{\displaystyle O}{\|}}{C}-H$ and Cl)

12.23 Draw the condensed structural formula for each of the following:

a. acetaldehyde **b.** 2-pentanone
c. butyl methyl ketone **d.** 3-methylpentanal

12.24 Draw the condensed structural formula for each of the following:

a. propionaldehyde **b.** butanal
c. 4-bromo-2-pentanone **d.** acetone

LEARNING GOAL

Describe the solubility of aldehydes and ketones in water; draw the condensed structural formulas for their oxidation and reduction products.

12.5 Properties of Aldehydes and Ketones

Aldehydes and ketones contain a polar carbonyl group (carbon–oxygen double bond), which has a partially negative oxygen atom and a partially positive carbon atom. Because the electronegative oxygen atom forms hydrogen bonds with water molecules, aldehydes and ketones with one to four carbons are very soluble. However, aldehydes and ketones with five or more carbon atoms are not soluble because longer hydrocarbon chains, which are nonpolar, diminish the solubility effect of the polar carbonyl group. Table 12.3 compares the solubility of some aldehydes and ketones.

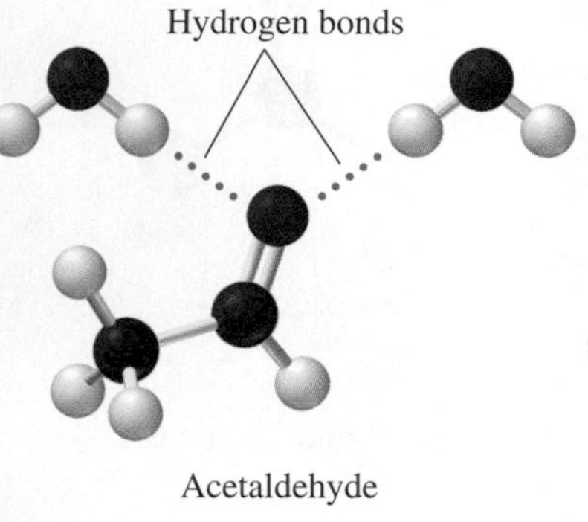

TABLE 12.3 Solubility of Some Aldehydes and Ketones

Compound	Condensed Structural Formula	Number of Carbon Atoms	Solubility in Water
Methanal	$H-CHO$	1	Soluble
Ethanal	CH_3-CHO	2	Soluble
Propanal	CH_3-CH_2-CHO	3	Soluble
Propanone	$CH_3-CO-CH_3$	3	Soluble
Butanal	$CH_3-CH_2-CH_2-CHO$	4	Slightly soluble
Butanone	$CH_3-CO-CH_2-CH_3$	4	Soluble
Pentanal	$CH_3-CH_2-CH_2-CH_2-CHO$	5	Insoluble
2-Pentanone	$CH_3-CO-CH_2-CH_2-CH_3$	5	Insoluble

CONCEPT CHECK 12.6

Solubility of Ketones

Why is acetone soluble in water, but 2-hexanone is not?

ANSWER

Acetone contains a carbonyl group with an electronegative oxygen atom that forms hydrogen bonds with water. 2-Hexanone also contains a carbonyl group, but its long, nonpolar hydrocarbon chain makes it insoluble in water.

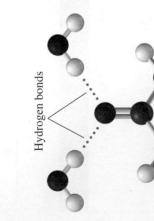

Oxidation of Aldehydes

Earlier in this chapter, we saw that primary alcohols can oxidize to aldehydes. However, aldehydes are easily oxidized further to carboxylic acids. In contrast, ketones produced by the oxidation of secondary alcohols do not undergo further oxidation.

MC TUTORIAL Oxidation-Reduction Reactions of Aldehydes and Ketones

$$\underset{\text{Ethanol (1°)}}{CH_3-CH_2-OH} \xrightarrow{\text{Oxidation}} \underset{\text{Ethanal}}{CH_3-\overset{\overset{\large O}{\|}}{C}-H} \xrightarrow{\text{Further oxidation}} \underset{\text{Ethanoic acid}}{CH_3-\overset{\overset{\large O}{\|}}{C}-OH}$$

$$\underset{\text{2-Propanol (2°)}}{CH_3-\overset{\overset{\large OH}{|}}{CH}-CH_3} \xrightarrow{\text{Oxidation}} \underset{\text{Propanone}}{CH_3-\overset{\overset{\large O}{\|}}{C}-CH_3} \xrightarrow{\text{Further oxidation}} \text{no reaction}$$

CONCEPT CHECK 12.7

Oxidation of Alcohols to Carboxylic Acids

Draw the condensed structural formulas for the aldehyde and carboxylic acid that form in the oxidation of 1-propanol.

ANSWER

The condensed structural formulas of the oxidation products of 1-propanol with three carbon atoms show a change in functional groups on carbon 1 to give a carboxylic acid with three carbon atoms. In the oxidation of 1-propanol, a primary alcohol, the carbon attached to the —OH group is oxidized to the carbonyl group of an aldehyde, which is further oxidized to a carboxylic acid.

$$\underset{\text{Alcohol}}{CH_3-CH_2-CH_2-OH} \xrightarrow{[O]} \underset{\text{Aldehyde}}{CH_3-CH_2-\overset{\overset{\large O}{\|}}{C}-H} \xrightarrow{[O]} \underset{\text{Carboxylic acid}}{CH_3-CH_2-\overset{\overset{\large O}{\|}}{C}-OH}$$

Tollens' Test

Tollens' test, which uses a solution of Ag^+ ($AgNO_3$) and ammonia, oxidizes aldehydes, but not ketones. The silver ion is reduced and forms a "silver mirror" on the inside of the container. Commercially, a similar process is used to make mirrors by applying a solution of $AgNO_3$ and ammonia on glass with a spray gun (see Figure 12.7).

$$\underset{\text{Acetaldehyde}}{CH_3-\overset{\overset{\large O}{\|}}{C}-H} + \underset{\text{Tollens' reagent}}{2Ag^+} \longrightarrow \underset{\text{Silver mirror}}{2Ag(s)} + \underset{\text{Acetic acid}}{CH_3-\overset{\overset{\large O}{\|}}{C}-OH}$$

Another test, called **Benedict's test**, gives a positive result with compounds that have an aldehyde functional group and an adjacent hydroxyl group. When Benedict's solution containing Cu^{2+}($CuSO_4$) ions is added to this type of aldehyde and heated, a brick-red solid of Cu_2O forms (see Figure 12.8). The test is negative with simple aldehydes and ketones.

$$\underset{\text{2-Hydroxypropanal}}{CH_3-\overset{\overset{\large OH}{|}}{CH}-\overset{\overset{\large O}{\|}}{C}-H} + \underset{\text{Benedict's reagent}}{2Cu^{2+}} \longrightarrow \underset{\text{Brick-red solid}}{Cu_2O(s)} + \underset{\text{2-Hydroxypropanoic acid}}{CH_3-\overset{\overset{\large OH}{|}}{CH}-\overset{\overset{\large O}{\|}}{C}-OH}$$

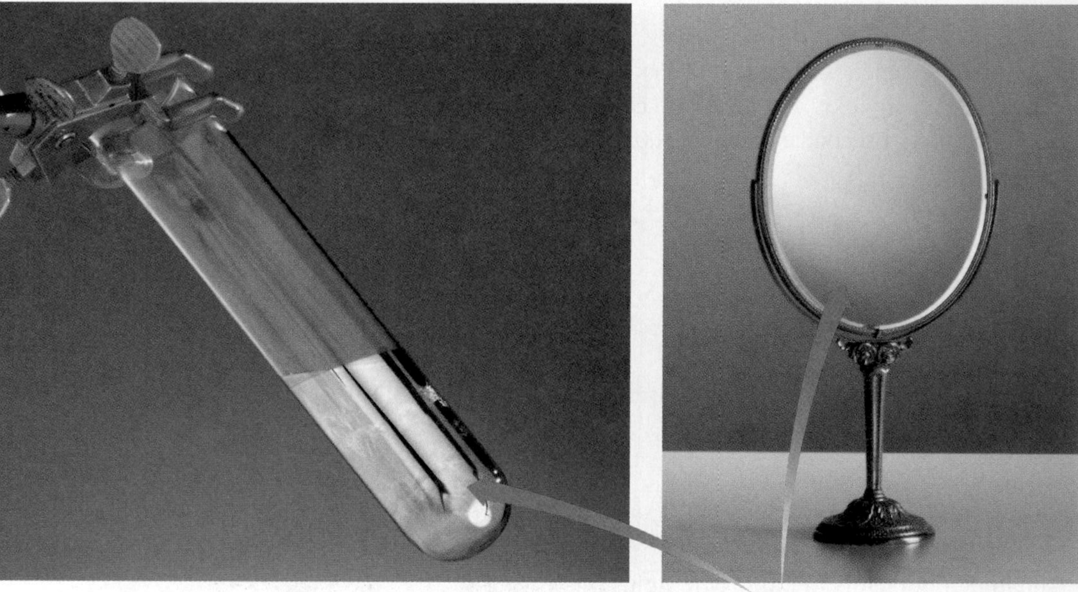

FIGURE 12.7 In Tollens' test, a silver mirror forms when the oxidation of an aldehyde reduces silver ion to metallic silver. The silvery surface of a mirror is formed in a similar way.

Q What is the product of the oxidation of an aldehyde?

FIGURE 12.8 The blue Cu^{2+} in Benedict's solution forms a brick-red solid of Cu_2O in a positive test for many sugars and aldehydes with adjacent hydroxyl groups.

Q Which test tube indicates that glucose is present?

Because many sugars, such as glucose, contain this type of aldehyde grouping, Benedict's reagent can be used to determine the presence of glucose in blood or urine.

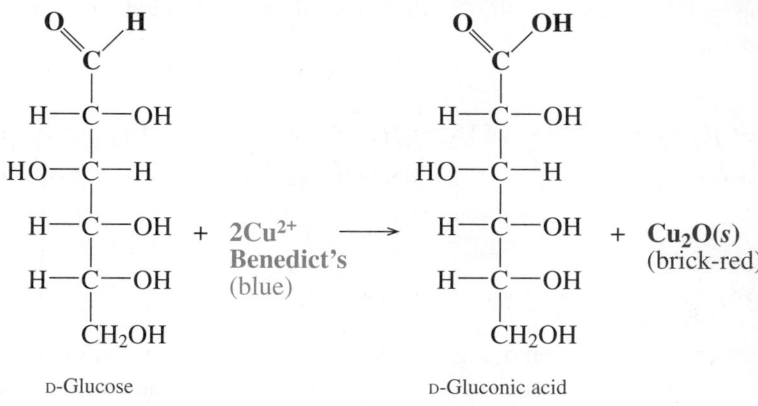

Reduction of Aldehydes and Ketones

Aldehydes and ketones are reduced by sodium borohydride ($NaBH_4$) or hydrogen (H_2). In the **reduction** of organic compounds, there is a decrease in the number of carbon–oxygen bonds by the addition of hydrogen or the loss of oxygen. Aldehydes reduce to primary alcohols, and ketones reduce to secondary alcohols. A catalyst such as nickel, platinum, or palladium is needed for the addition of hydrogen to the carbonyl group.

Aldehydes Reduce to Primary Alcohols

$$CH_3—CH_2—\overset{\overset{\displaystyle O}{\|}}{C}—H + \mathbf{H_2} \xrightarrow{Pt} CH_3—CH_2—\overset{\overset{\displaystyle \mathbf{OH}}{|}}{\underset{\underset{\displaystyle \mathbf{H}}{|}}{C}}—H$$

Propionaldehyde — 1-Propanol (1° alcohol)

Ketones Reduce to Secondary Alcohols

$$CH_3-\overset{\overset{\displaystyle O}{\|}}{C}-CH_3 + \mathbf{H_2} \xrightarrow{Ni} CH_3-\overset{\overset{\displaystyle \mathbf{OH}}{|}}{\underset{\underset{\displaystyle \mathbf{H}}{|}}{C}}-CH_3$$

Dimethyl ketone — 2-Propanol (2° alcohol)

CONCEPT CHECK 12.8

Reduction of Carbonyl Groups

Write an equation for the reduction of cyclopentanone using hydrogen in the presence of a nickel catalyst.

ANSWER

The reacting molecule is a cyclic ketone that has five carbon atoms. During the reduction, hydrogen atoms add to the carbon and oxygen in the carbonyl group, which reduces the ketone to the corresponding secondary alcohol.

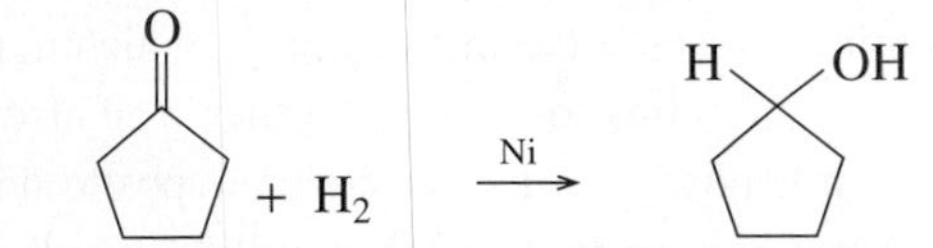

Cyclopentanone + H_2 $\xrightarrow{Ni}$ Cyclopentanol

QUESTIONS AND PROBLEMS

Properties of Aldehydes and Ketones

12.25 Which compound in each of the following pairs would be more soluble in water? Explain.

a. $CH_3-\overset{\overset{\displaystyle O}{\|}}{C}-CH_2-CH_2-CH_3$ or $CH_3-\overset{\overset{\displaystyle O}{\|}}{C}-\overset{\overset{\displaystyle O}{\|}}{C}-CH_2-CH_3$

b. acetone or 2-pentanone

c. propanal or pentanal

12.26 Which compound in each of the following pairs would be more soluble in water? Explain.

a. $CH_3-CH_2-CH_3$ or CH_3-CH_2-CHO

b. propanone or 3-hexanone

c. butane or butanal

12.27 Draw the condensed structural formulas of the aldehyde and carboxylic acid produced when each of the following is oxidized:

a. $CH_3-CH_2-CH_2-CH_2-CH_2-OH$

b. $CH_3-\overset{\overset{\displaystyle CH_3}{|}}{CH}-CH_2-CH_2-OH$

c. 1-butanol

12.28 Draw the condensed structural formula of the aldehyde and carboxylic acid produced when each of the following is oxidized:

a. (cyclobutyl)$-CH_2-OH$

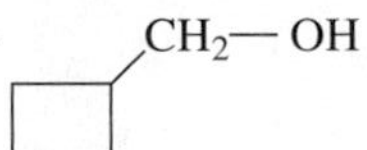

b. CH_3-OH

c. 3-chloro-1-propanol

12.29 Draw the condensed structural formula of the organic product formed when each of the following is reduced by hydrogen in the presence of a nickel catalyst:

a. butyraldehyde

b. acetone

c. hexanal

d. 2-methyl-3-pentanone

12.30 Draw the condensed structural formula of the organic product formed when each of the following is reduced by hydrogen in the presence of a nickel catalyst:

a. ethyl propyl ketone

b. formaldehyde

c. 3-chlorocyclopentanone

d. 2-pentanone

LEARNING GOAL

Identify chiral and achiral carbon atoms in an organic molecule.

12.6 Chiral Molecules

In the preceding chapters, we have looked at isomers. Let's review those now. Molecules are structural isomers when they have the same molecular formula, but different bonding arrangements.

Structural Isomers

C_2H_6O	$CH_3—CH_2—OH$ Ethanol	$CH_3—O—CH_3$ Dimethyl ether
C_3H_6O	$CH_3—CH_2—C(=O)—H$ Propanal	$CH_3—C(=O)—CH_3$ Propanone

Another group of isomers called **stereoisomers** have identical molecular formulas, too, but they are not structural isomers. In stereoisomers, the atoms are bonded in the same sequence but differ in the way they are arranged in space.

Chirality

MC TUTORIAL Chiral Carbon Atoms

When stereoisomers have mirror images that are different, they are said to have "handedness." If you hold your right hand up to a mirror, you see the mirror image, which matches your left hand (see Figure 12.9). If you turn your palms toward each other, you also have mirror images. If you look at the palms of your hands, your thumbs are on opposite sides. If you place your right hand over your left hand, you cannot match up all the parts of the hands: palms, backs, thumbs, and little fingers. The thumbs and little fingers can be matched, but then the palms or backs of your hands are facing each other. Your hands are mirror images that cannot be superimposed on each other. When organic molecules have mirror images that cannot be completely matched, we say they are *nonsuperimposable*.

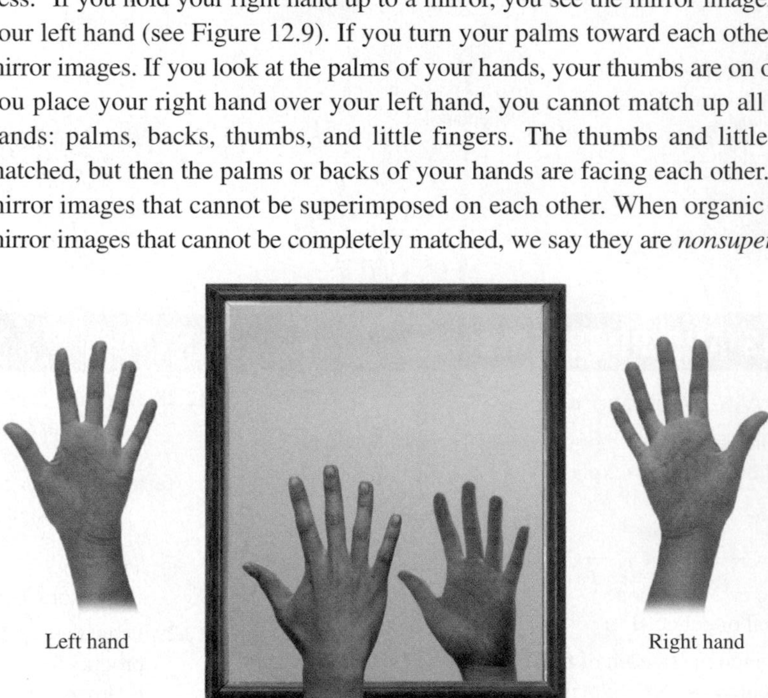

FIGURE 12.9 The left and right hands are chiral because they have mirror images that cannot be superimposed on each other.
Q Why are your shoes chiral objects?

Objects such as hands that have nonsuperimposable mirror images are **chiral** (pronunciation *kai′-ral*). Left and right shoes are chiral; left- and right-handed golf clubs are chiral. If we try to put a left-hand glove on our right hand or a right shoe on our left foot, or use left-handed scissors if we are right-handed, we realize that mirror images that are chiral do not match.

Sometimes a mirror image can be superimposed on the original. For example, all parts of the mirror image of a plain drinking glass can be matched to the original glass. When a mirror image can be superimposed on the other, the object is *achiral* (see Figure 12.10).

Molecules also have mirror images, and often one stereoisomer has a different biological effect than the other one. In some cases, one isomer has a certain odor, and the mirror image has a completely different odor. For example, one stereoisomer of limonene smells like lemons, while its mirror image has the odor of oranges.

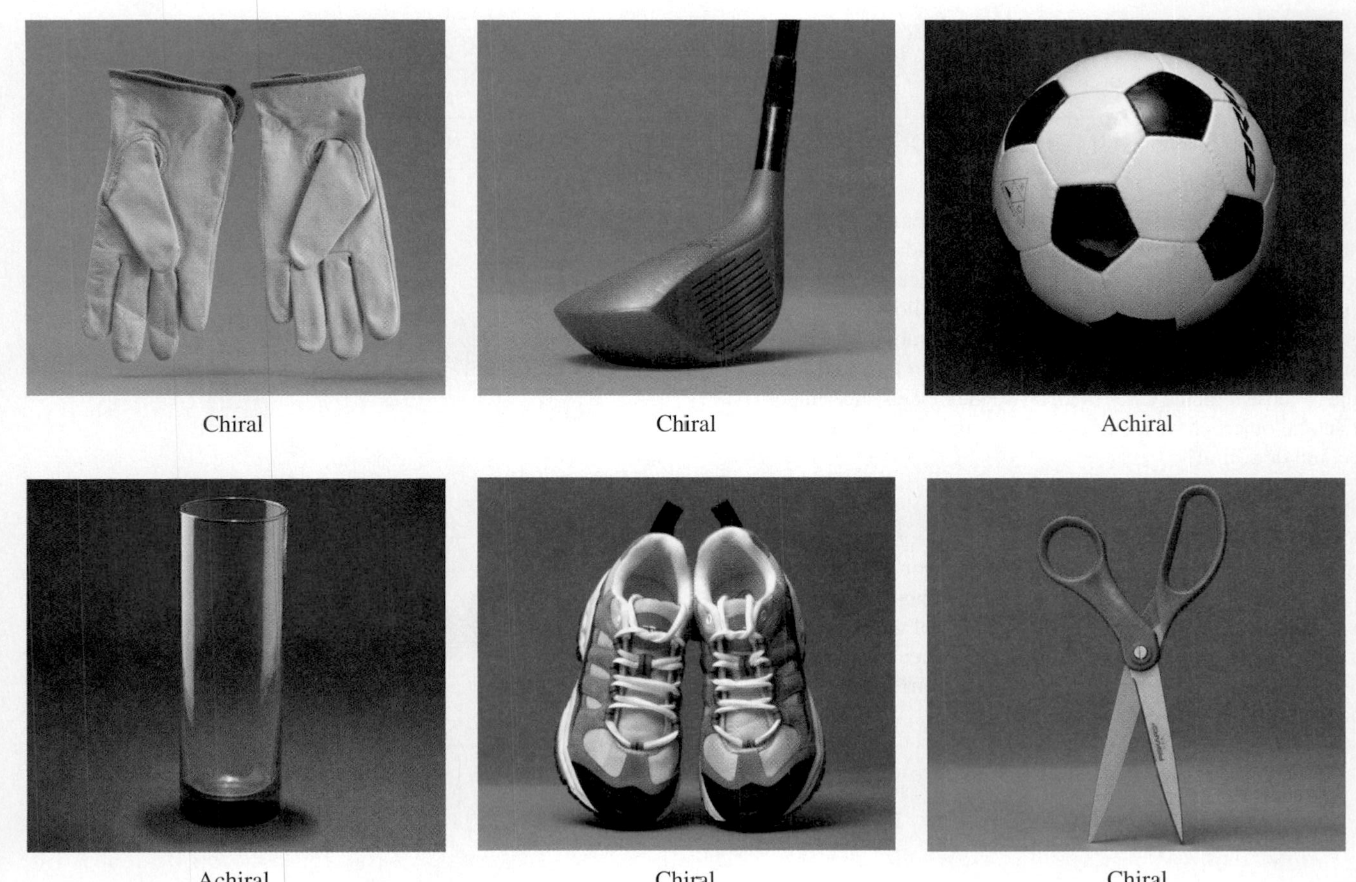

FIGURE 12.10 Everyday objects can be chiral or achiral.
Q Why are some of the objects chiral and others achiral?

CONCEPT CHECK 12.9

Chiral and Achiral Objects

Identify each of the following objects as chiral or achiral:

a. white tennis sock **b.** flip-flop beach shoe **c.** golf ball

ANSWER

a. A white tennis sock is achiral because its mirror image is superimposable.
b. A flip-flop beach shoe is chiral because its mirror image of the left flip-flop shoe is the right flip-flop shoe; the two shoes are not superimposable.
c. A golf ball is achiral because its mirror image is superimposable.

Chiral Carbon Atoms

A carbon compound is chiral if it has at least one carbon atom bonded to four different atoms or groups. This type of carbon atom is called a **chiral carbon** because there are two different ways that it can bond to four atoms or groups of atoms. The resulting structures are nonsuperimposable mirror images. Let's look at the mirror images of a carbon bonded to four different atoms (see Figure 12.11). If we line up the hydrogen and iodine atoms in the mirror images, the bromine and chlorine atoms appear on opposite sides. No matter how we turn the models, we cannot align all four atoms at the same time. When stereoisomers cannot be superimposed, they are called **enantiomers**.

If two or more atoms bonded to a carbon atom are the same, the atoms can be aligned (superimposed), and the mirror images represent the same structure (see Figure 12.12).

Explore Your World

USING GUMDROPS AND TOOTHPICKS TO MODEL CHIRAL OBJECTS

Part 1: Achiral Objects

Obtain some toothpicks and orange, yellow, green, purple, and black gumdrops. Place four toothpicks into the black gumdrop making the ends of toothpicks the corners of a tetrahedron. Attach the gumdrops to the toothpicks: two orange, one green, and one yellow.

Using another black gumdrop, make a second model that is the mirror image of the original model. Now rotate one of the models and try to superimpose it on the other model. Are the models superimposable? If achiral objects have superimposable mirror images, are these models chiral or achiral?

Part 2: Chiral Objects

Using one of the original models, replace one orange gumdrop with a purple gumdrop. Now there are four different colors of gumdrops attached to the black gumdrop. Make its mirror image by replacing one orange gumdrop with purple. Now rotate one of the models and try to superimpose it on the other model. Are the models superimposable? If chiral objects have nonsuperimposable mirror images, are these models chiral or achiral?

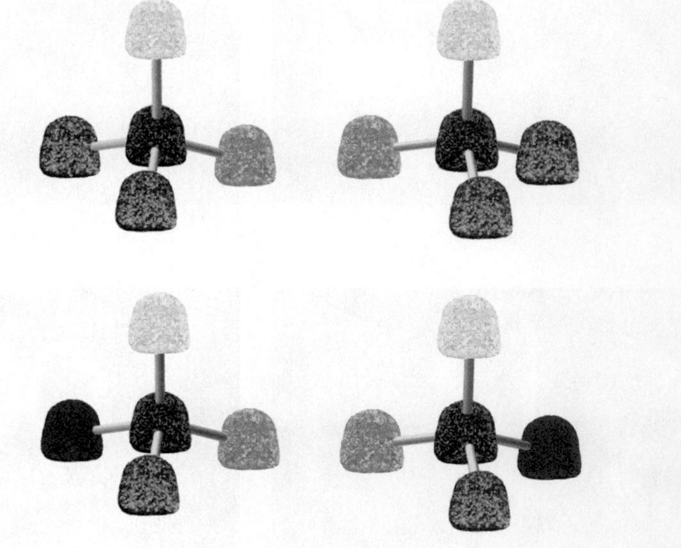

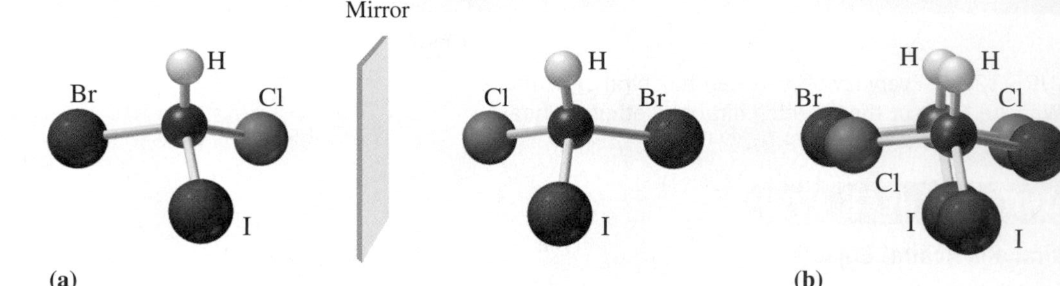

FIGURE 12.11 **(a)** The enantiomers of a chiral molecule are mirror images. **(b)** The enantiomers of a chiral molecule cannot be superimposed on each other.
Q Why is the carbon atom in this compound chiral?

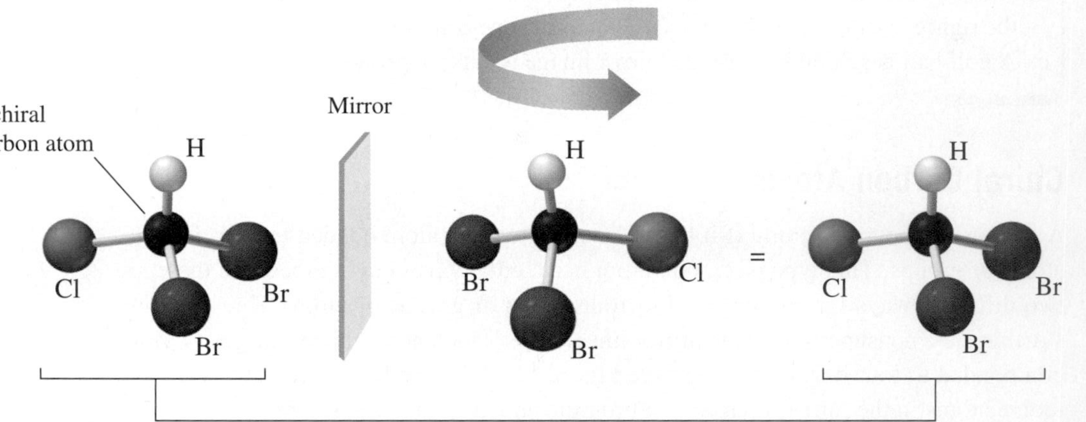

FIGURE 12.12 The mirror images of an achiral compound can be superimposed on each other.
Q Why do the mirror images have the same structure?

SAMPLE PROBLEM 12.7

Chiral Carbons

Indicate whether the carbon in red is chiral or achiral.

a. 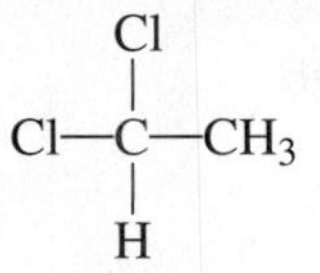 $Cl—C(Cl)(H)—CH_3$

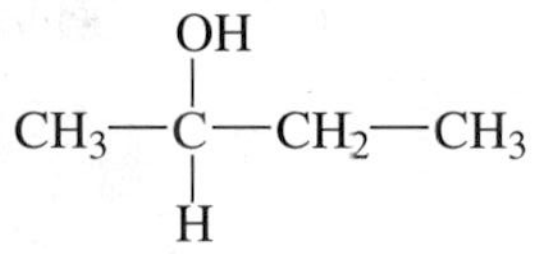

b. $CH_3—C(OH)(H)—CH_2—CH_3$

c. $CH_3—C(=O)—CH_2—CH_3$

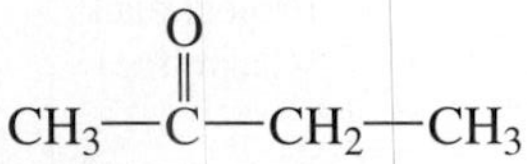

SOLUTION

a. Achiral. Two of the substituents on the carbon in red are the same. A chiral carbon must be bonded to four different groups or atoms.

b. Chiral. The carbon in red is bonded to four different groups: $—OH$, $—CH_3$, $—CH_2—CH_3$, and $—H$.

c. Achiral. The carbon in red is bonded to only three groups, not four.

STUDY CHECK 12.7

Circle the two chiral carbons in the condensed structural formula of the carbohydrate erythrose.

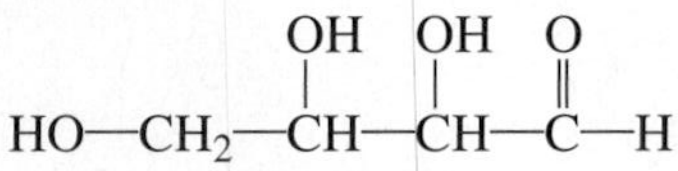

$HO—CH_2—CH(OH)—CH(OH)—C(=O)—H$

Erythrose

Drawing Fischer Projections

Emil Fischer devised a simplified system for drawing stereoisomers that shows the arrangements of the atoms. Fischer received the Nobel Prize in Chemistry in 1902 for his contributions to carbohydrate chemistry. Now we can use his model, called a **Fischer projection**, to represent a three-dimensional structure. Vertical lines represent bonds that project backward from a carbon atom and horizontal lines represent bonds that project forward. The intersection of vertical and horizontal lines represents a carbon atom with the most highly oxidized carbon at the top.

For glyceraldehyde, the carbonyl group, the most highly oxidized group, is written at the top above the chiral carbon. The $—CH_2OH$ group is written below the chiral carbon. The $—H$ and $—OH$ can be written at the ends of a horizontal line, but in two different ways. One stereoisomer has the $—OH$ group on the left, which is designated as the L isomer. When the $—OH$ is written on the right of the chiral carbon, it represents the D isomer (see Figure 12.13). Today, the "D and L" system is used to identify stereoisomers of carbohydrates and amino acids.

Fischer projections can also be drawn for compounds that have two or more chiral carbons. For example, in the mirror images of erythrose, the chiral carbon atom at each intersection is bonded to four different groups. To draw the mirror image of L-erythrose, we reverse $—OH$ groups from left to right on the horizontal lines.

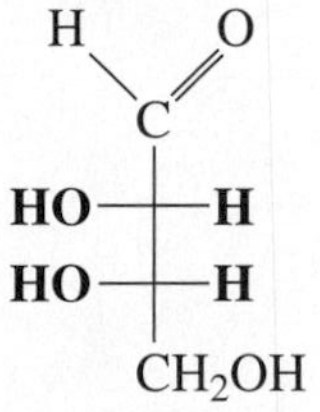

L-Erythrose	D-Erythrose
H–C=O	H–C=O
HO—+—**H**	**H**—+—**OH**
HO—+—**H**	**H**—+—**OH**
CH_2OH	CH_2OH

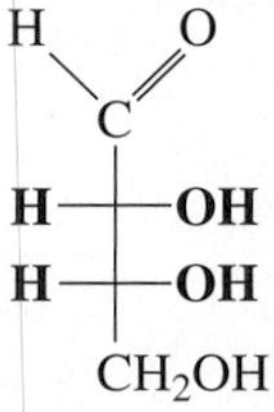

L-Erythrose D-Erythrose

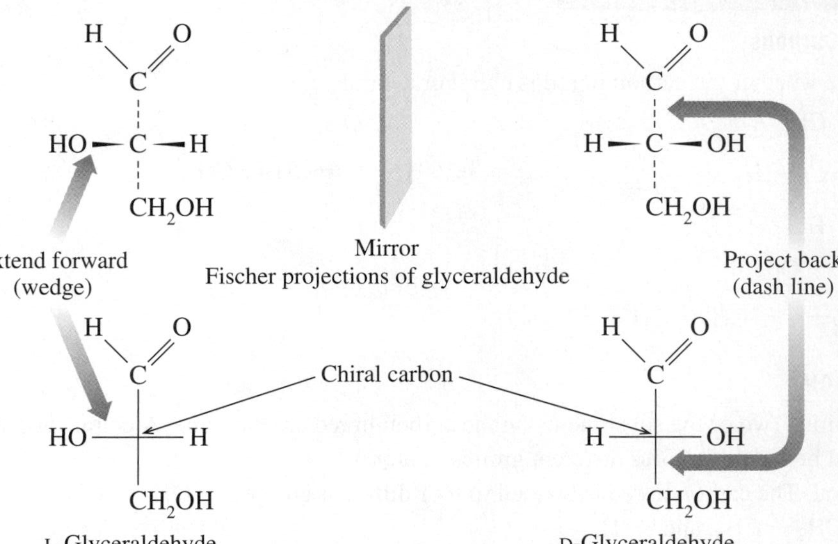

FIGURE 12.13 In a Fischer projection, the chiral carbon atom is at the center with horizontal lines for bonds that extend toward the viewer and vertical lines for bonds that point away.

Q Why does glyceraldehyde have only one chiral carbon atom?

SAMPLE PROBLEM 12.8

Fischer Projections

Determine if each Fischer projection is chiral or achiral. If chiral, identify it as the D or L isomer and draw the mirror image.

a. (top: CH_2OH; left: HO; right: H; bottom: CH_3)

b. (top: CH_3; left: HO; right: H; bottom: CH_3)

SOLUTION

a. Chiral. The carbon at the intersection is attached to four different substituents. It is the L isomer because the —OH is on the left. The mirror image is written by reversing the —H and —OH.

(top: CH_2OH; left: H; right: OH; bottom: CH_3)

b. Achiral. The carbon atom at the intersection is attached to two identical groups ($-CH_3$).

STUDY CHECK 12.8

Does this Fischer projection represent the D or L isomer?

(top: CHO; left: HO; right: H; bottom: CH_3)

QUESTIONS AND PROBLEMS

Chiral Molecules

12.31 Identify each of the following structures as chiral or achiral. If chiral, indicate the chiral carbon.

a. 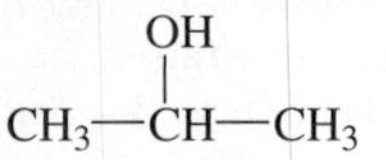 $CH_3—CH(OH)—CH_3$ **b.** 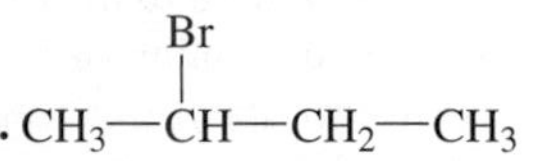 $CH_3—CH(Br)—CH_2—CH_3$

c. 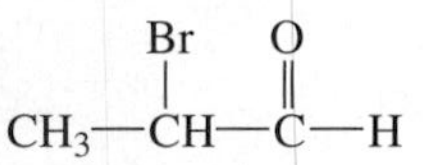 $CH_3—CH(Br)—C(=O)—H$ **d.** 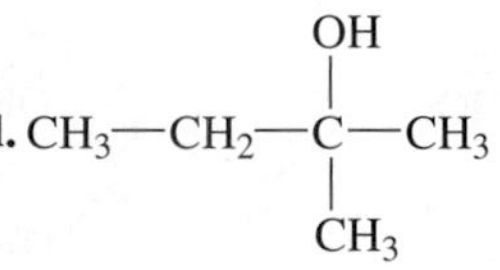 $CH_3—CH_2—C(OH)(CH_3)—CH_3$

12.32 Identify each of the following structures as chiral or achiral. If chiral, indicate the chiral carbon.

a. 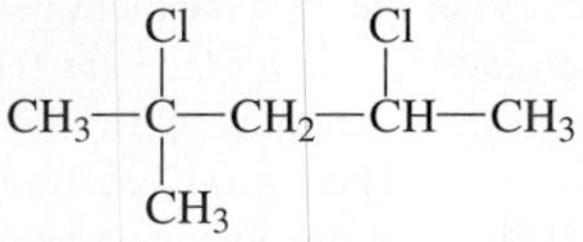 $CH_3—C(Cl)(CH_3)—CH_2—CH(Cl)—CH_3$

b. $CH_3—C(Br)=CH—CH_3$

c. $CH_3—C(OH)(CH_3)—CH(OH)—CH_3$

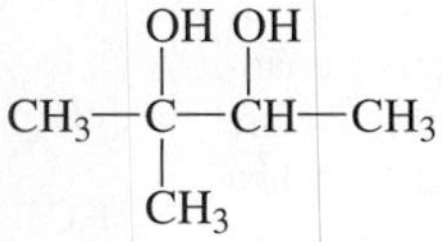

d. 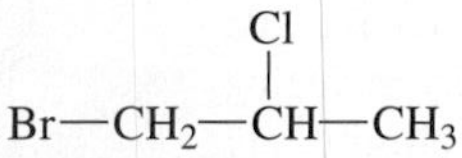 $Br—CH_2—CH(Cl)—CH_3$

12.33 Identify the chiral carbon in each of the following naturally occurring compounds:

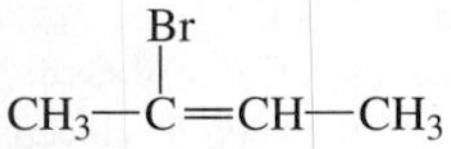

a. citronellol; one enantiomer has the geranium odor

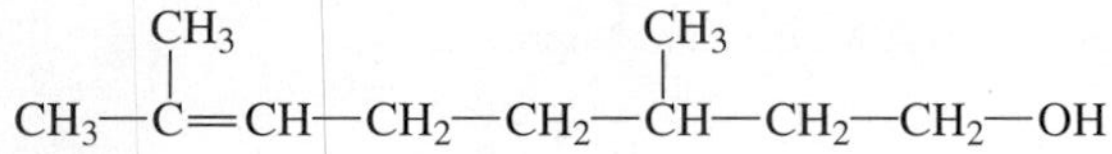

$CH_3—C(CH_3)=CH—CH_2—CH_2—CH(CH_3)—CH_2—CH_2—OH$

b. alanine, an amino acid

$H_2N—CH(CH_3)—C(=O)—OH$

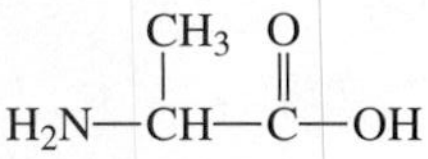

12.34 Identify the chiral carbon in each of the following naturally occurring compounds:

a. amphetamine (Benzedrine), stimulant, used in the treatment of hyperactivity

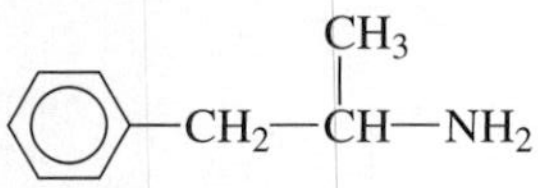

$C_6H_5—CH_2—CH(CH_3)—NH_2$

b. norepinephrine, increases blood pressure and nerve transmission

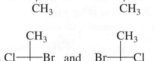

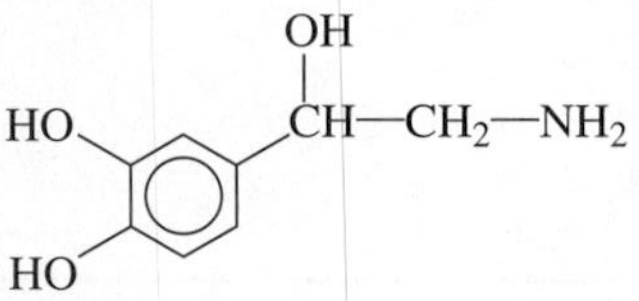

(HO)(HO)$C_6H_3—CH(OH)—CH_2—NH_2$

12.35 Draw Fischer projections for each of the following dash–wedge structures:

a.

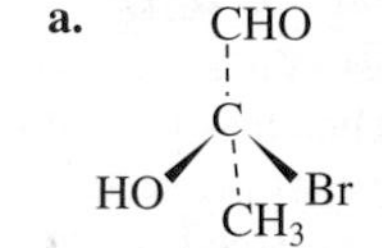

b.

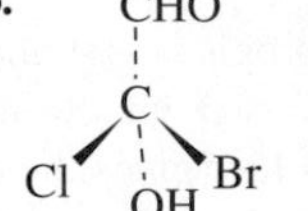

c.

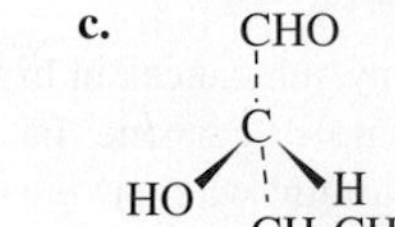

12.36 Draw Fischer projections for each of the following dash–wedge structures:

a. 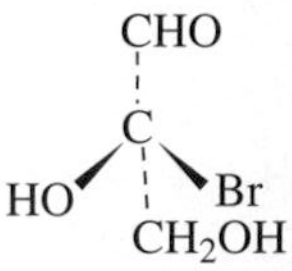 CHO / C / HO, Br / CH_2OH **b.** 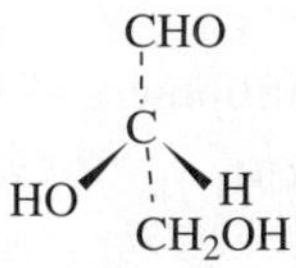 CHO / C / HO, H / CH_2OH **c.**

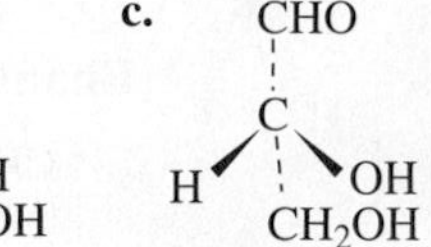

12.37 Indicate whether each pair of Fischer projections represents enantiomers or identical structures.

a. 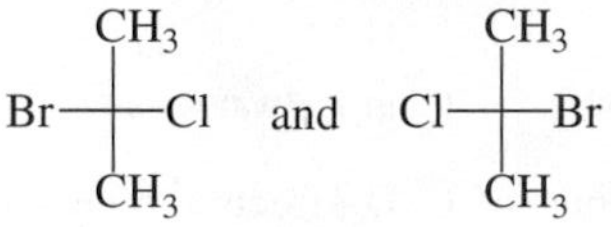 top CH_3, left Br, right Cl, bottom CH_3 and top CH_3, left Cl, right Br, bottom CH_3

b. top CHO, left HO, right H, bottom CH_3 and

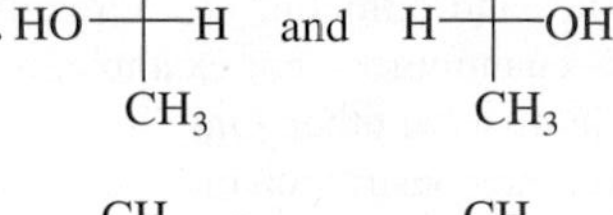

c. top CH_3, left Cl, right Br, bottom H and

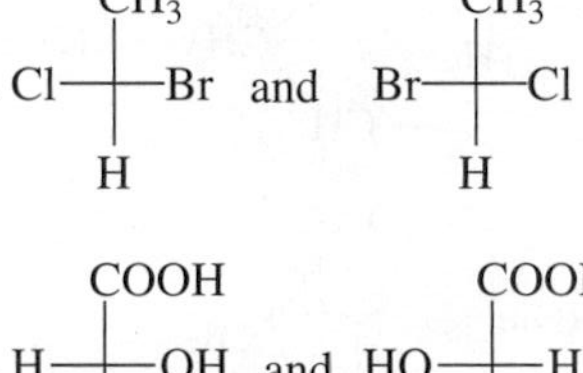

d. 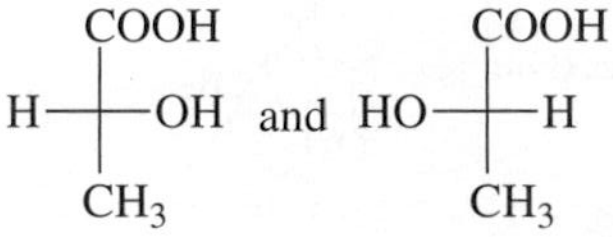 top COOH, left H, right OH, bottom CH_3 and top COOH, left HO, right H, bottom CH_3

12.38 Indicate whether each pair of Fischer projections represents enantiomers or identical structures.

a. 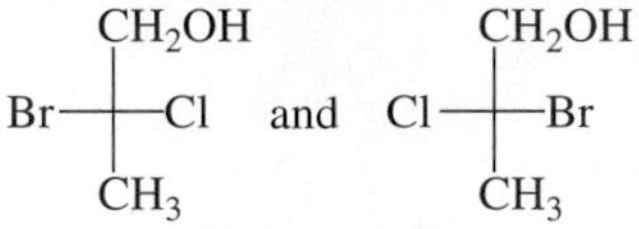 top CH_2OH, left Br, right Cl, bottom CH_3 and top CH_2OH, left Cl, right Br, bottom CH_3

b. 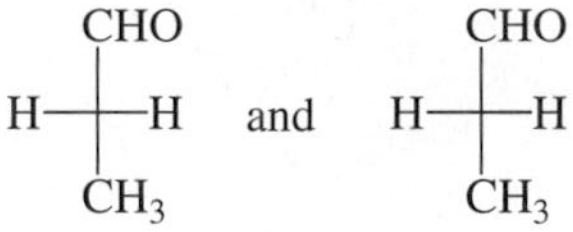 top CHO, left H, right H, bottom CH_3 and top CHO, left H, right H, bottom CH_3

c. 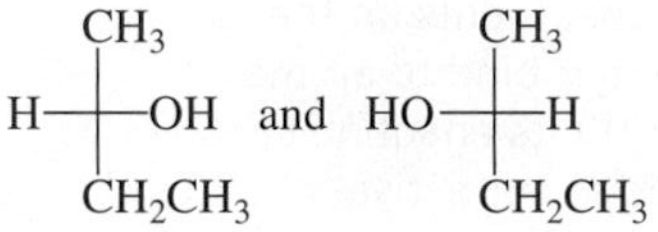 top CH_3, left H, right OH, bottom CH_2CH_3 and top CH_3, left HO, right H, bottom CH_2CH_3

d. 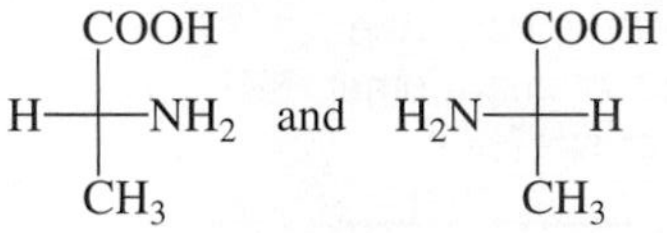 top COOH, left H, right NH_2, bottom CH_3 and top COOH, left H_2N, right H, bottom CH_3

Chemistry Link to Health

ENANTIOMERS IN BIOLOGICAL SYSTEMS

Many substances in biological systems consist of enantiomers. For example, carvone has two enantiomers: One has the odor of spearmint whereas the other gives the odor to caraway seeds. In the olfactory cells in the nasal cavity and the gustatory cells in the taste buds on the tongue, there are receptor sites that fit the shape of only one enantiomer. Thus, our senses of smell and taste are responsive to the chirality of molecules.

Enantiomers of Carvone

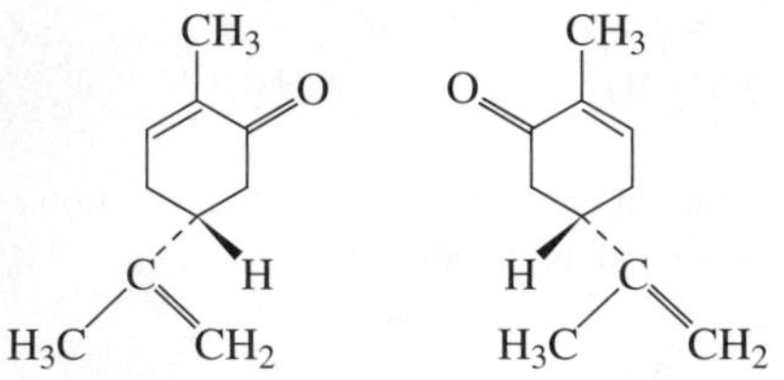

From spearmint oil   From caraway seeds

In the brain, one enantiomer of LSD affects the production of serotonin, which influences sensory perception and may lead to hallucinations. However, its enantiomer produces little effect in the brain. The behavior of nicotine and epinephrine (adrenalin) also depends upon only one of their enantiomers. For example, one enantiomer of nicotine is more toxic than the other. Only one enantiomer of epinephrine is responsible for the constriction of blood vessels.

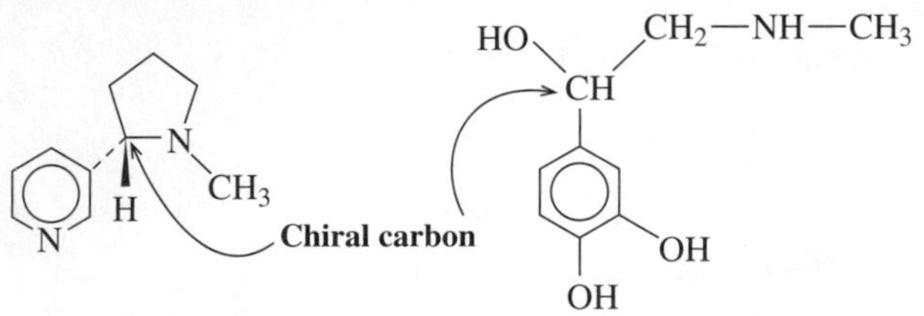

Chiral carbon

Nicotine   Adrenalin (epinephrine)

A substance used to treat Parkinson's disease is L-dopa, which is converted to dopamine in the brain, where it raises the serotonin level. However, the D-dopa enantiomer is not effective for the treatment of Parkinson's disease.

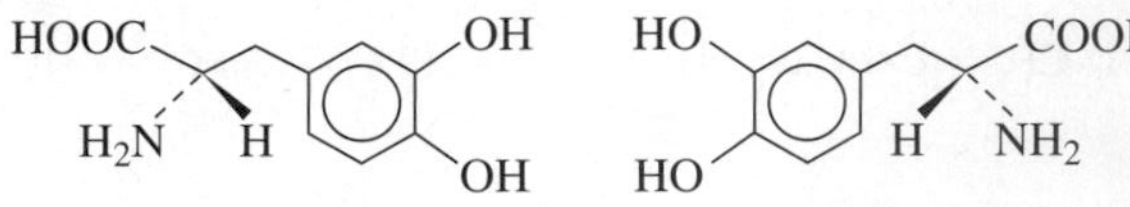

L-Dopa   D-Dopa

Many compounds in biological systems have only one enantiomer that is active. This happens because the enzymes and cell surface receptors on which metabolic reactions take place are themselves chiral. Thus, only one enantiomer interacts with its enzymes or receptors; the other is inactive. The chiral receptor fits the arrangement of the substituents in only one enantiomer; its mirror image does not fit properly (see Figure 12.14).

For many drugs, only one of the enantiomers is biologically active. However, for many years, drugs have been produced that were mixtures of their enantiomers. Today, drug researchers are using *chiral technology* to produce the active enantiomers of chiral drugs. Chiral catalysts are being designed that direct the formation of just one enantiomer rather than both. The benefits of producing only the active enantiomer include using a lower dose, enhancing activity, reducing interactions with other drugs, and eliminating possible harmful side effects from the nonactive enantiomer. Several active enantiomers are now being produced such as L-dopa and the active enantiomer of the popular analgesic ibuprofen used in Advil, Motrin, and Nuprin.

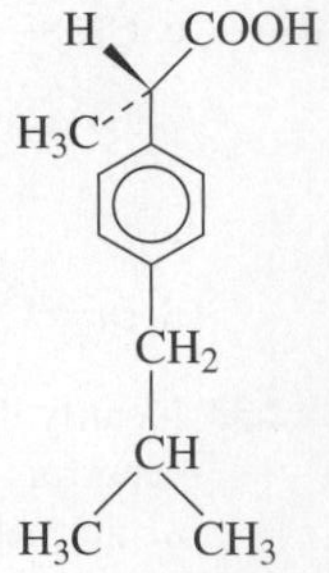

Ibuprofen

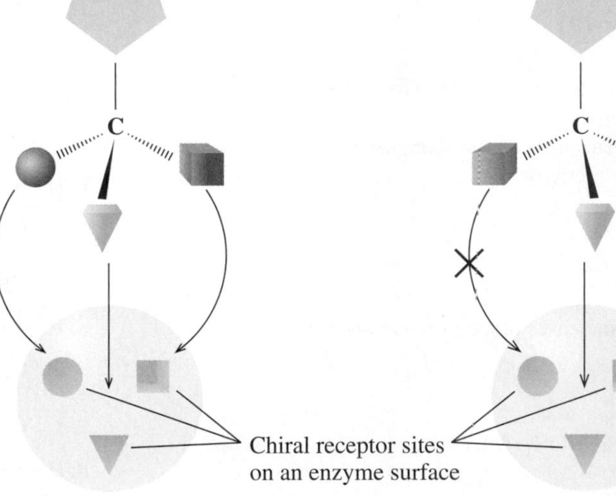

FIGURE 12.14 (**a**) The substituents on the biologically active enantiomer bind to all the sites on a chiral receptor; (**b**) its enantiomer does not bind properly and is not active biologically.

Q Why don't all the substituents of the mirror image of the active enantiomer fit into a chiral receptor site?

CONCEPT MAP

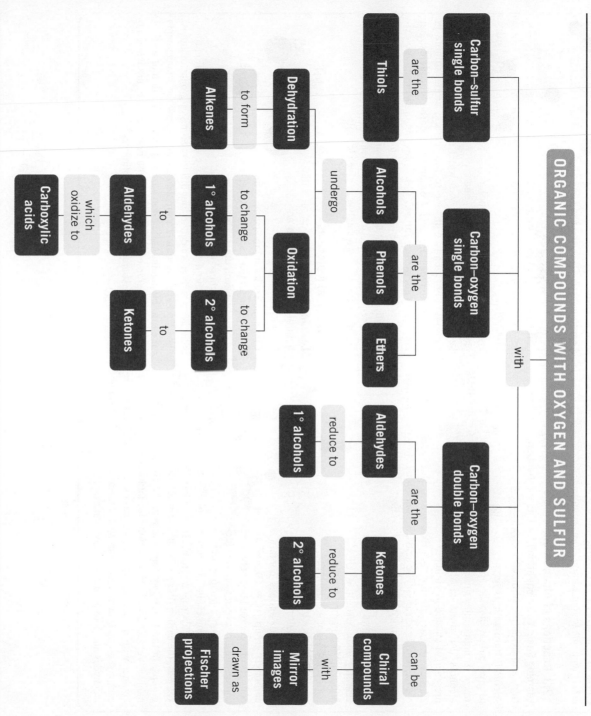

ORGANIC COMPOUNDS WITH OXYGEN AND SULFUR

with

- **Carbon–sulfur single bonds**
  - are the **Thiols**
- **Carbon–oxygen single bonds**
  - are the **Alcohols**, **Phenols**, **Ethers**
    - Alcohols undergo
      - **Dehydration** to form **Alkenes**
      - **Oxidation**
        - to change **1° alcohols** to **Aldehydes** which oxidize to **Carboxylic acids**
        - to change **2° alcohols** to **Ketones**
- **Carbon–oxygen double bonds**
  - are the **Aldehydes**, **Ketones**
    - Aldehydes reduce to **1° alcohols**
    - Ketones reduce to **2° alcohols**
  - can be **Chiral compounds** with **Mirror images** drawn as **Fischer projections**

CHAPTER REVIEW

12.1 Alcohols, Phenols, Thiols, and Ethers

Learning Goal: Give the IUPAC and common names for alcohols and phenols; identify common names for thiols and ethers.

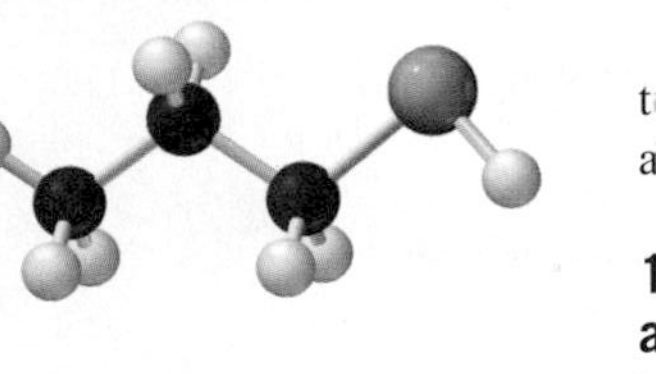

$CH_3—CH_2—CH_2—SH$
1-Propanethiol
Onions

The functional group of an alcohol is the hydroxyl group —OH bonded to a carbon chain. In a phenol, the hydroxyl group is bonded to an aromatic ring. In the IUPAC system, the names of alcohols have *ol* endings, and the location of the —OH group is given by numbering the carbon chain. Simple alcohols are generally named by their common names, with the alkyl name preceding the term *alcohol*. In thiols, the functional group is —SH, which is analogous to the —OH group of alcohols.

In an ether, an oxygen atom is connected by single bonds to two alkyl or aromatic groups. In the common names of ethers, the alkyl groups are listed alphabetically, followed by the name *ether*.

12.2 Properties of Alcohols and Ethers

Learning Goal: Describe the classification of alcohols; describe the solubility of alcohols and ethers in water.

Hydrogen bonds
δ− δ+ δ+ δ− δ+ δ+ δ− δ+ δ− δ+ δ+
Methanol (methyl alcohol)

Alcohols are classified according to the number of alkyl or aromatic groups bonded to the carbon that holds the —OH group. In a primary (1°) alcohol, one alkyl group is attached to the hydroxyl carbon. In a secondary (2°) alcohol, two alkyl groups are attached; and in a tertiary (3°) alcohol, there are three alkyl groups bonded to the hydroxyl carbon.

The polar —OH group in alcohols and the —O— group in ethers form hydrogen bonds with water, which makes alcohols and ethers with one to four carbon atoms soluble in water.

12.3 Reactions of Alcohols and Thiols

Learning Goal: Write equations for the combustion, dehydration, and oxidation of alcohols.

Alcohols burn with oxygen to give carbon dioxide and water. At high temperatures, alcohols dehydrate in the presence of an acid to yield alkenes. Primary alcohols are oxidized to aldehydes, which can oxidize further to carboxylic acids. Secondary alcohols are oxidized to ketones. Tertiary alcohols do not oxidize.

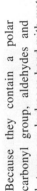

$$\underset{\substack{\text{Ethanol}\\\text{(ethyl alcohol)}}}{CH_3-CH_2-OH} \xrightarrow{[O]} \underset{\substack{\text{Ethanal}\\\text{(acetaldehyde)}}}{CH_3-\overset{O}{\overset{\|}{C}}-H} + H_2O$$

12.4 Aldehydes and Ketones

Learning Goal: Write the IUPAC and common names for aldehydes and ketones; draw the condensed structural formulas when given their names.

Carbonyl group

Aldehyde

Ketone

Aldehydes and ketones contain a carbonyl group (C=O), which consists of a double bond between a carbon and an oxygen atom. In aldehydes, the carbonyl group appears at the end of carbon chains attached to at least one hydrogen atom. In ketones, the carbonyl group occurs between two alkyl or aromatic groups. In the IUPAC system, the *e* in the corresponding alkane name is replaced with *al* for aldehydes and *one* for ketones. For ketones with more than four carbon atoms in the main chain, the carbonyl group is numbered to show its location. Many of the simple aldehydes and ketones use common names.

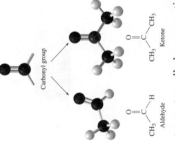

12.5 Properties of Aldehydes and Ketones

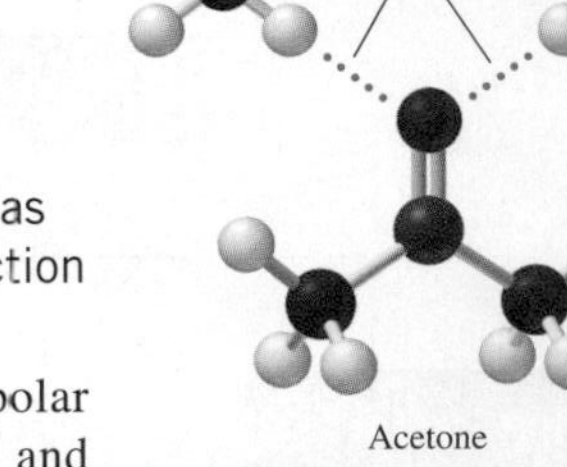

Acetone

Learning Goal: Describe the solubility of aldehydes and ketones in water; draw the condensed structural formulas for their oxidation and reduction products.

Because they contain a polar carbonyl group, aldehydes and ketones can hydrogen bond with water molecules, which makes carbonyl compounds with one to four carbon atoms soluble in water. Aldehydes are easily oxidized to carboxylic acids, but ketones do not oxidize further. Aldehydes, but not ketones, are oxidized by Tollens' reagent to give silver mirrors. In Benedict's test, aldehydes with adjacent hydroxyl groups reduce blue Cu^{2+} to give a brick-red Cu_2O solid. Aldehydes and ketones can be reduced with H_2 in the presence of a catalyst to give alcohols.

12.6 Chiral Molecules

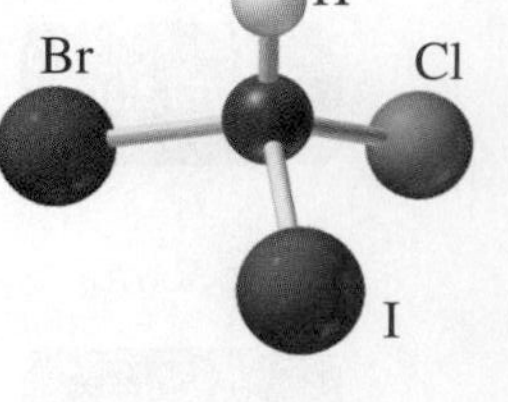

Learning Goal: Identify chiral and achiral carbon atoms in an organic molecule.

Chiral molecules are molecules with mirror images that cannot be superimposed on each other. These types of stereoisomers are called enantiomers. A chiral molecule must have at least one chiral carbon, which is a carbon bonded to four different atoms or groups of atoms. The Fischer projection is a simplified way to draw the arrangements of atoms by placing the carbon atoms at the intersection of vertical and horizontal lines. The names of the mirror images are labeled D or L to differentiate between the enantiomers.

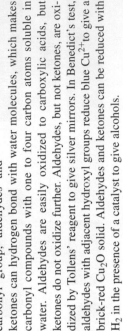

Summary of Naming

Structure	Family	IUPAC Name	Common Name
$CH_3-\mathbf{OH}$	Alcohol	Methanol	Methyl alcohol
$C_6H_5-\mathbf{OH}$ (benzene ring—OH)	Phenol	Phenol	Phenol
$CH_3-\mathbf{SH}$	Thiol	Methanethiol	
$CH_3-\mathbf{O}-CH_3$	Ether		Dimethyl ether
$H-\overset{O}{\overset{\|}{C}}-H$	Aldehyde	Methanal	Formaldehyde
$CH_3-\overset{O}{\overset{\|}{C}}-CH_3$	Ketone	Propanone	Acetone; dimethyl ketone

Summary of Reactions

Combustion of Alcohols

$$\underset{\text{Ethanol}}{CH_3{-}CH_2{-}OH} + \underset{\text{Oxygen}}{3O_2} \xrightarrow{\Delta} \underset{\text{Carbon dioxide}}{2CO_2} + \underset{\text{Water}}{3H_2O} + \text{energy}$$

Dehydration of Alcohols to Form Alkenes

$$\underset{\text{1-Propanol}}{CH_3{-}CH_2{-}CH_2{-}OH} \xrightarrow[\text{Heat}]{H^+} \underset{\text{Propene}}{CH_3{-}CH{=}CH_2} + H_2O$$

Oxidation of Primary Alcohols to Form Aldehydes

$$\underset{\text{Ethanol}}{CH_3{-}\overset{\overset{OH}{|}}{C}H_2} \xrightarrow{[O]} \underset{\text{Acetaldehyde}}{CH_3{-}\overset{\overset{O}{\|}}{C}{-}H} + H_2O$$

Oxidation of Secondary Alcohols to Form Ketones

$$\underset{\text{2-Propanol}}{CH_3{-}\overset{\overset{OH}{|}}{C}H{-}CH_3} \xrightarrow{[O]} \underset{\text{Propanone}}{CH_3{-}\overset{\overset{O}{\|}}{C}{-}CH_3} + H_2O$$

Oxidation of Aldehydes to Form Carboxylic Acids

$$\underset{\text{Acetaldehyde}}{CH_3{-}\overset{\overset{O}{\|}}{C}{-}H} \xrightarrow{[O]} \underset{\text{Acetic acid}}{CH_3{-}\overset{\overset{O}{\|}}{C}{-}OH}$$

Reduction of Aldehydes to Form Primary Alcohols

$$\underset{\text{Acetaldehyde}}{CH_3{-}\overset{\overset{O}{\|}}{C}{-}H} + H_2 \xrightarrow{Pt} \underset{\text{Ethanol}}{CH_3{-}\overset{\overset{OH}{|}}{C}H_2}$$

Reduction of Ketones to Form Secondary Alcohols

$$\underset{\text{Propanone}}{CH_3{-}\overset{\overset{O}{\|}}{C}{-}CH_3} + H_2 \xrightarrow{Ni} \underset{\text{2-Propanol}}{CH_3{-}\overset{\overset{OH}{|}}{C}H{-}CH_3}$$

Key Terms

alcohols Organic compounds that contain the hydroxyl (—OH) functional group attached to a carbon chain.

aldehyde An organic compound with a carbonyl functional group bonded to at least one hydrogen.

$$-\overset{\overset{O}{\|}}{C}{-}H \;=\; -CHO$$

Benedict's test A test for aldehydes with adjacent hydroxyl groups in which Cu^{2+} ($CuSO_4$) ions in Benedict's reagent are reduced to a brick-red solid of Cu_2O.

carbonyl group A functional group that contains a carbon–oxygen double bond (C=O).

chiral Having nonsuperimposable mirror images.

chiral carbon A carbon atom that is bonded to four different atoms or groups.

dehydration A reaction that removes water from an alcohol, in the presence of an acid, to form alkenes.

enantiomers Stereoisomers that are mirror images that cannot be superimposed.

ether An organic compound in which an oxygen atom is bonded to two alkyl or two aromatic groups, or one of each.

Fischer projection A system for drawing stereoisomers with horizontal lines representing bonds that project forward, vertical lines that represent bonds that project backward, and with a carbon atom at each intersection.

ketone An organic compound in which the carbonyl functional group is bonded to two alkyl or aromatic groups.

$$-\overset{\overset{O}{\|}}{C}- \;=\; -CO-$$

oxidation The loss of two hydrogen atoms from a reactant to give a more oxidized compound. For example, primary alcohols oxidize to aldehydes, and secondary alcohols oxidize to ketones. An oxidation can also be the addition of an oxygen atom or an increase in the number of carbon–oxygen bonds.

phenol An organic compound that has a hydroxyl group attached to a benzene ring.

primary (1°) alcohol An alcohol that has one alkyl group bonded to the carbon atom with the —OH group.

reduction A gain of hydrogen when a ketone reduces to a 2° alcohol and an aldehyde reduces to a 1° alcohol.

secondary (2°) alcohol An alcohol that has two alkyl groups bonded to the carbon atom with the —OH group.

stereoisomers Isomers that have atoms bonded in the same order, but with different arrangements in space.

tertiary (3°) alcohol An alcohol that has three alkyl groups bonded to the carbon atom with the —OH group.

thiols Organic compounds that contain a —SH group.

Tollens' test A test for aldehydes in which Ag^+ in Tollens' reagent is reduced to metallic silver forming a "silver mirror" on the walls of the container.

Understanding the Concepts

12.39 A compound called butylated hydroxytoluene, or BHT, has been added to cereal and other foods since 1947 as an antioxidant.

OH

$(CH_3)_3C$ — [benzene ring] — $C(CH_3)_3$

CH_3

BHT is an antioxidant added to preserve foods such as cereal.

a. Identify the functional groups in BHT.

b. What parts of the molecule make up the toluene part of the compound?

12.40 Used in sunless tanning lotions, the compound 1,3-dihydroxy-2-propanone or dihydroxyacetone (DHA) darkens the skin without sun. DHA reacts with amino acids in the outer surface of the skin. Draw the condensed structural formula of DHA.

A sunless tanning lotion contains DHA to darken the skin.

12.41 Identify the functional groups in each of the following:

a. urushiol, a substance in poison ivy and poison oak that causes itching and blistering of the skin

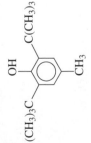

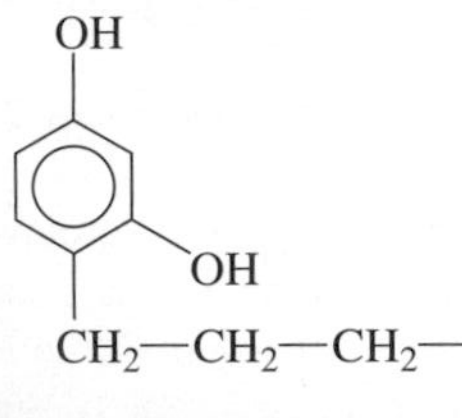

Poison ivy contains urushiol that causes a rash and itching of the skin.

b. menthol, which gives a peppermint taste and odor used in candy and throat lozenges

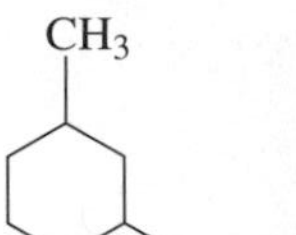

Menthol gives the taste of peppermint in candy.

12.42 Which of the following will give a positive Tollens' test?

a. propanal **b.** ethanol **c.** ethyl methyl ether
d. 1-propanol **e.** 2-propanol **f.** hexanal

12.43 Citronellal, a constituent of oil of citronella as well as lemon and lemon grass, is used in perfumes and as an insect repellent.

An insect-repelling candle contains citronellal, which is found in natural sources.

$$CH_3—\overset{\overset{CH_3}{|}}{C}=CH—CH_2—CH_2—\overset{\overset{CH_3}{|}}{CH}—CH_2—\overset{\overset{O}{\|}}{C}—H$$

a. Complete the IUPAC name for citronellal: __,__ -di______ - __ -octenal

b. What does the *en* in octenal signify?

c. What does the *al* in octenal signify?

d. Write the balanced equation for the combustion of citronellal when burned in a candle to repel insects.

12.44 Draw the condensed structural formula for each of the following:

a. 2-heptanone, an alarm pheromone of bees

b. 2,6-dimethyl-3-heptanone, a communication pheromone of bees

c. 2-nonanol, an alarm pheromone released when a bee stings

Bees emit chemicals called pheromones to communicate.

Additional Questions and Problems

For instructor-assigned homework, go to **www.masteringchemistry.com**.

12.45 Classify each of the following alcohols as primary (1°), secondary (2°), or tertiary (3°):

a. 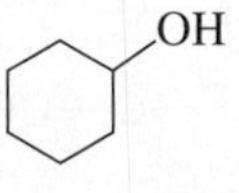 OH

b. 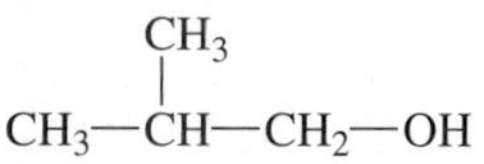 $CH_3-CH(CH_3)-CH_2-OH$

c. 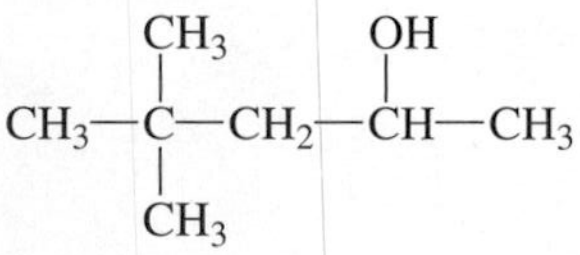 $CH_3-C(CH_3)_2-CH_2-CH(OH)-CH_3$

d. 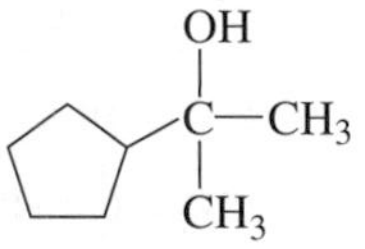 OH, $-C(CH_3)_2$, CH_3

12.46 Classify each of the following alcohols as primary (1°), secondary (2°), or tertiary (3°):

a. 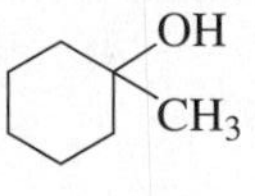 OH, CH_3

b. 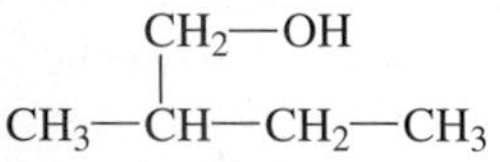 $CH_3-CH(CH_2-OH)-CH_2-CH_3$

c. 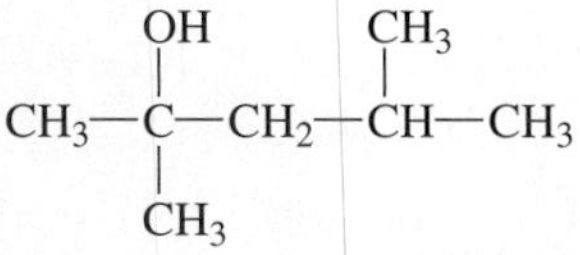 $CH_3-C(OH)(CH_3)-CH_2-CH(CH_3)-CH_3$

d. 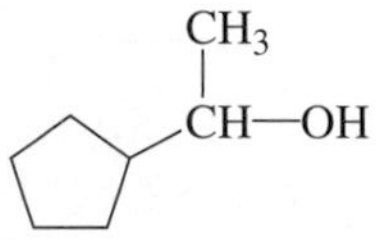 CH_3, $CH-OH$

12.47 Give the IUPAC name of each of the following alcohols and phenols:

a. 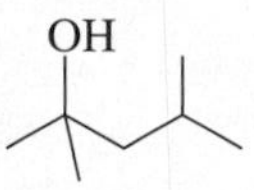 OH

b. 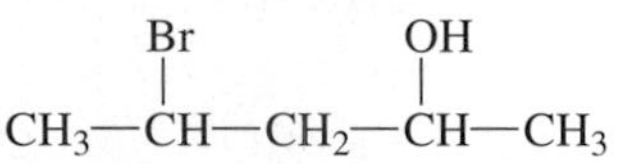 $CH_3-CH(Br)-CH_2-CH(OH)-CH_3$

c. 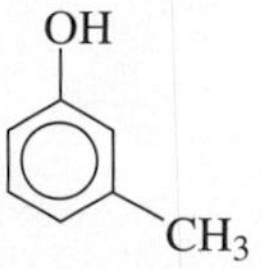 OH, CH_3

12.48 Give the IUPAC name of each of the following alcohols and phenols:

a.  OH, Cl

b. 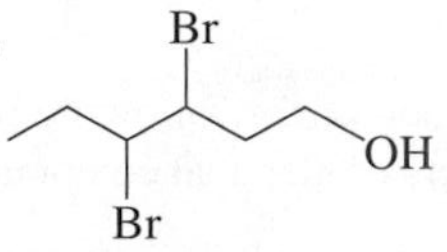 Br, Br, OH

c. 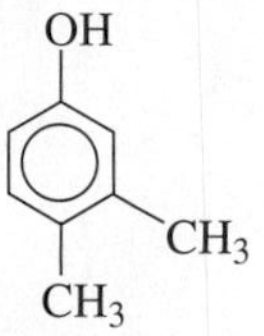 OH, CH_3, CH_3

12.49 Draw the condensed structural formula of each of the following:

a. 4-chlorophenol **b.** 2-methyl-3-pentanol
c. 2-methyl-1-propanol **d.** 2,4-dibromophenol

12.50 Draw the condensed structural formula of each of the following:

a. 2-methyl-1-hexanol **b.** 2-pentanol
c. methyl propyl ether **d.** 3-methyl-2-butanol

12.51 Which compound in each pair would you expect to be more soluble in water? Why?

a. butane or 1-propanol **b.** 1-propanol or diethyl ether
c. ethanol or 1-hexanol

12.52 Which compound in each pair would you expect to be more soluble in water? Why?

a. ethane or ethanol **b.** 2-propanol or 2-pentanol
c. dibutyl ether or 1-butanol

12.53 Explain why each of the following compounds would be soluble or insoluble in water:

a. 2-propanol **b.** 1-hexanol

12.54 Explain why each of the following compounds would be soluble or insoluble in water:

a. butane **b.** 1,3-propanediol

12.55 Draw the condensed structural formula for the alkene, aldehyde, or ketone product of each of the following reactions:

a. $CH_3-CH_2-CH_2-OH \xrightarrow[\text{Heat}]{H^+}$

b. $CH_3-CH_2-CH_2-OH \xrightarrow{[O]}$

c. 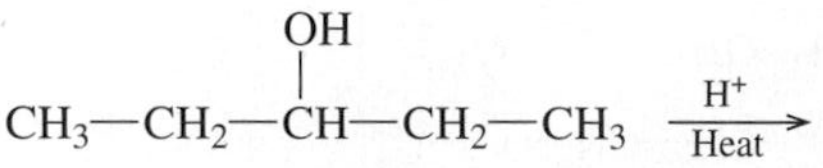 $CH_3-CH_2-CH(OH)-CH_2-CH_3 \xrightarrow[\text{Heat}]{H^+}$

d. 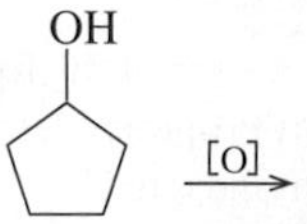 OH $\xrightarrow{[O]}$

12.56 Draw the condensed structural formula for the alkene, aldehyde, or ketone product of each of the following reactions:

a. 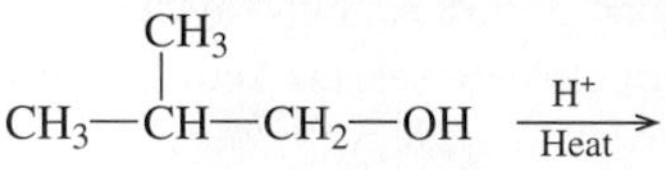 $CH_3-CH(CH_3)-CH_2-OH \xrightarrow[\text{Heat}]{H^+}$

b. 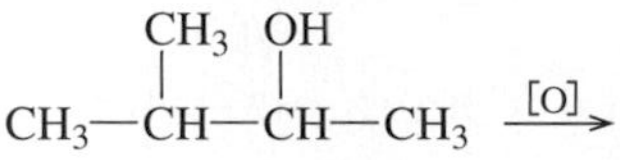 $CH_3-CH(CH_3)-CH(OH)-CH_3 \xrightarrow{[O]}$

c. 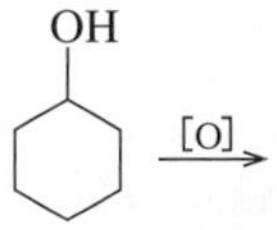 OH $\xrightarrow{[O]}$

d. 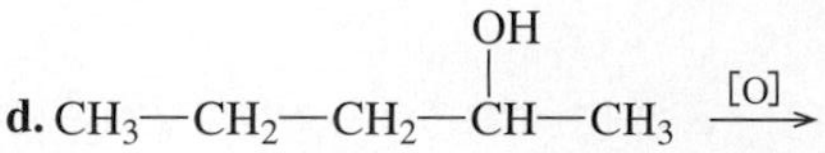 $CH_3-CH_2-CH_2-CH(OH)-CH_3 \xrightarrow{[O]}$

12.57 Draw the condensed structural formula of the organic product when hydrogen and a nickel catalyst reduce each of the following:

a. 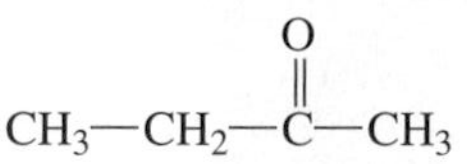 $CH_3-CH_2-C(=O)-CH_3$

b. $CH_2-C(=O)-H$

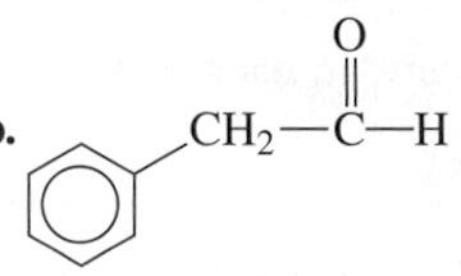

c. O

12.58 Draw the condensed structural formula of the organic product when hydrogen and a nickel catalyst reduce each of the following:

a. 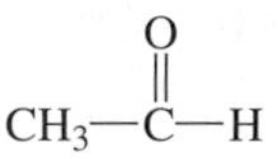 $CH_3-C(=O)-H$

b. O, CH_3

c. $H-C(=O)-H$

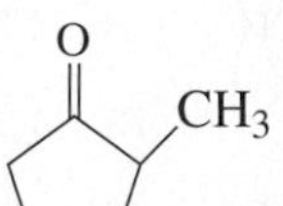

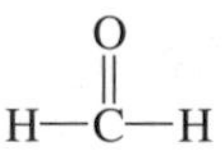

12.59 Give the IUPAC name for each of the following:

a. (benzene ring bearing $-C(=O)-H$, with Br and Cl substituents)

b. $Cl-CH_2-CH_2-C(=O)-H$

c. (skeletal structure: ketone, C=O, with Cl substituent)

12.60 Give the IUPAC name for each of the following:

a. (skeletal structure: ketone, C=O)

b. (benzene ring bearing $-C(=O)H$, with Cl and Cl substituents)

c. $CH_3-CH(CH_3)-CH(OH)-CH_2-C(=O)-H$

12.61 Draw the condensed structural formula of each of the following:

a. 4-chlorobenzaldehyde **b.** 3-chloropropionaldehyde
c. ethyl methyl ketone **d.** 3-methylhexanal

12.62 Draw the condensed structural formula of each of the following:

a. formaldehyde **b.** 2-chlorobutanal
c. 3-methyl-2-hexanone **d.** 3,5-dimethylhexanal

12.63 Which of the following aldehydes and ketones are soluble in water?

a. $CH_3-CH_2-C(=O)-H$ **b.** $CH_3-C(=O)-CH_3$

c. $CH_3-CH_2-C(=O)-CH_2-CH_2-CH_3$

12.64 Which of the following aldehydes and ketones are soluble in water?

a. $CH_3-CH_2-C(=O)-CH_3$ **b.** $CH_3-C(=O)-H$

c. $CH_3-CH_2-CH(CH_3)-CH_2-CH_2-C(=O)-H$

12.65 Identify the chiral carbons, if any, in each of the following compounds:

a. $H-C(Cl)(Cl)-C(Cl)(H)-OH$ **b.** $CH_3-C(H)=C(CH_3)-CH_3$

c. $HO-CH_2-CH(OH)-CH_2-OH$ **d.** $CH_3-CH(NH_2)-C(=O)-H$

e. $CH_3-CH_2-CH(Br)-CH_2-CH_2-CH_3$

12.66 Identify the chiral carbons, if any, in each of the following compounds:

a. $CH_3-CH(O-CH_3)-CH_3$ **b.** $CH_3-CH(OH)-C(=O)-CH_3$

c. $CH_3-C(OH)(OH)-CH_3$ **d.** $CH_3-CH(CH_3)-C(=O)-CH_3$

e. $CH_3-C(Br)(OH)-CH_2-CH_3$

12.67 Identify each of the following pairs of Fischer projections as enantiomers or identical compounds:

a. top CH_2OH; H (left), OH (right); bottom CH_2OH and top CH_2OH; HO (left), H (right); bottom CH_2OH

b. top CHO; H (left), OH (right); bottom CH_2OH and top CHO; HO (left), H (right); bottom CH_2OH

c. top CH_2OH; Cl (left), H (right); bottom CH_3 and top CH_2OH; H (left), Cl (right); bottom CH_3

d. top OH; H (left), OH (right); bottom CH_3 and top OH; HO (left), H (right); bottom CH_3

12.68 Identify each of the following pairs of Fischer projections as enantiomers or identical compounds:

a. top CH_2OH; H (left), Cl (right); bottom CH_2CH_3 and top CH_2OH; Cl (left), H (right); bottom CH_2CH_3

b. top CH_2OH; H (left), OH (right); bottom CH_3 and top CH_2OH; HO (left), H (right); bottom CH_3

c. top CH_2OH; H (left), Cl (right); bottom CH_3 and top CH_2OH; H (left), Cl (right); bottom CH_3

d. top CHO; H (left), OH (right); bottom CH_3 and top CHO; HO (left), H (right); bottom CH_3

12.69 Draw the condensed structural formula of the ketone or carboxylic acid product when each of the following is oxidized:

a. $CH_3—CH_2—CH_2—OH$

b. $CH_3—CH(OH)—CH_2—CH_2—CH_3$

c. $CH_3—CH_2—CH_2—C(=O)—H$

d. cyclohexanol (cyclohexane ring with OH)

12.70 Draw the condensed structural formula of the ketone or carboxylic acid product when each of the following is oxidized:

a. $CH_3—CH_2—CH_2—CH_2—OH$

b. $CH_3—CH_2—CH(OH)—CH_3$

c. $CH_3—CH(CH_3)—CH_2—C(=O)—H$

d. cyclohexyl—$CH(OH)—CH_3$

Challenge Questions

12.71 Draw the condensed structural formulas and give the IUPAC names of all the alcohols that have the formula $C_5H_{12}O$.

12.72 Draw the condensed structural formulas and give the IUPAC names of all the aldehydes and ketones that have the formula $C_5H_{10}O$.

12.73 A compound with the formula C_4H_8O is synthesized from 2-methyl-1-propanol and oxidizes easily to give a carboxylic acid. Draw the condensed structural formula and give the IUPAC name of the compound.

12.74 Methyl *tert*-butyl ether (MTBE) or methyl 2-methyl-2-propyl ether is a fuel additive for gasoline to boost the octane rating. It increases the oxygen content, which reduces CO emissions to an acceptable level determined by the Clean Air Act.

a. If fuel mixtures are required to contain 2.7% oxygen by mass, how many grams of MTBE must be added to each 100. g of gasoline?

b. How many liters of liquid MTBE would be in a liter of fuel if the density of both gasoline and MTBE is 0.740 g/mL?

c. Write the balanced equation for the complete combustion of MTBE.

d. How many liters of air containing 21% (v/v) O_2 are required at STP to completely react (combust) 1.00 L of liquid MTBE?

MTBE was added to gasoline to increase the oxygen content.

12.75 Compound A is 1-propanol. When compound A is heated with strong acid, it dehydrates to form compound B (C_3H_6). When compound A is oxidized, compound C (C_3H_6O) forms. Draw the condensed structural formulas and give the IUPAC names of compounds A, B, and C.

12.76 Compound X is 2-propanol. When compound X is heated with strong acid, it dehydrates to form compound Y (C_3H_6). When compound X is oxidized, compound Z (C_3H_6O) forms, which cannot be oxidized further. Draw the condensed structural formulas and give the IUPAC names of compounds X, Y, and Z.

Answers

Answers to Study Checks

12.1 3-chloro-1-butanol

12.2 (skeletal structure: ethyl propyl ether, O in chain)

12.3 cyclohexene

12.4 2-Pentanol loses two hydrogen atoms to form a ketone.

$CH_3—C(=O)—CH_2—CH_2—CH_3$

12.5 propanal (IUPAC), propionaldehyde (common)

12.6 ethyl propyl ketone

12.7 $HO—CH_2—\text{Ⓒ}H(OH)—\text{Ⓒ}H(OH)—C(=O)—H$

12.8 Because the —OH is on the left, this is the L isomer.

Answers to Selected Questions and Problems

12.1 **a.** ethanol **b.** 2-butanol **c.** 2-pentanol **d.** 3-bromophenol

12.3 **a.** $CH_3—CH_2—CH_2—OH$

b. $CH_3—CH_2—CH(OH)—CH_2—CH_3$

c. $CH_3—C(OH)(CH_3)—CH_2—CH_3$

12.5 **a.** dipropyl ether **b.** cyclohexyl methyl ether

12.7 **a.** $CH_3—CH_2—O—CH_3$

b. $CH_3—CH_2—O—$cyclopropyl

12.9 a. 1° **b.** 1° **c.** 3° **d.** 3°

12.11 a. soluble **b.** slightly soluble **c.** insoluble

12.13 a. Methanol can form hydrogen bonds with water, but ethane cannot.
b. 2-Propanol is more soluble because it has a shorter carbon chain.
c. 1-Propanol is more soluble because it can form more hydrogen bonds.

12.15 a. $CH_3-CH_2-CH{=}CH_2$
b. (cyclopentene) **c.** $CH_3-CH{=}CH-CH_2-CH_3$

12.17 a. $CH_3-CH_2-CH_2-CH_2-C(=O)-H$
b. $CH_3-C(=O)-CH_2-CH_2-CH_2-CH_3$ **c.** (cyclohexanone)
d. $CH_3-C(=O)-CH_2-CH(CH_3)-CH_3$

12.19 a. acetaldehyde **b.** methyl propyl ketone
c. formaldehyde

12.21 a. propanal **b.** 2-methyl-3-pentanone
c. 3-methylcyclohexanone **d.** benzaldehyde

12.23 a. $CH_3-C(=O)-H$ **b.** $CH_3-C(=O)-CH_2-CH_2-CH_3$
c. $CH_3-C(=O)-CH_2-CH_2-CH_2-CH_3$
d. $CH_3-CH_2-CH(CH_3)-CH_2-C(=O)-H$

12.25 a. $CH_3-C(=O)-C(=O)-CH_2-CH_3$; more hydrogen bonding
b. acetone; lower number of carbon atoms
c. propanal; lower number of carbon atoms

12.27 a. $CH_3-CH_2-CH_2-CH_2-C(=O)-H$;
$CH_3-CH_2-CH_2-CH_2-C(=O)-OH$
b. $CH_3-CH(CH_3)-CH_2-C(=O)-H$; $CH_3-CH(CH_3)-CH_2-C(=O)-OH$
c. $CH_3-CH_2-CH_2-C(=O)-H$; $CH_3-CH_2-CH_2-C(=O)-OH$

12.29 a. $CH_3-CH_2-CH_2-CH_2-OH$
b. $CH_3-CH(OH)-CH_3$
c. $CH_3-CH_2-CH_2-CH_2-CH_2-CH_2-OH$
d. $CH_3-CH(CH_3)-CH(OH)-CH_2-CH_3$

12.31 a. achiral
b. chiral $CH_3-CH(Br)-CH_2-CH_3$ (Chiral carbon: the CH bearing Br)
c. chiral $CH_3-CH(Br)-C(=O)-H$ (Chiral carbon: the CH bearing Br)
d. achiral

12.33 a. $CH_3-C(CH_3){=}CH-CH_2-CH_2-CH(CH_3)-CH_2-CH_2-OH$ (Chiral carbon: the CH bearing CH_3)
b. $H_2N-CH(CH_3)-C(=O)-OH$ (Chiral carbon: the CH)

12.35 a. Fischer projection: CHO (top), HO (left), Br (right), CH_3 (bottom)
b. Fischer projection: CHO (top), Cl (left), Br (right), OH (bottom)
c. Fischer projection: CHO (top), HO (left), H (right), CH_2CH_3 (bottom)

12.37 a. identical **b.** enantiomers
c. enantiomers **d.** enantiomers

12.39 a. phenol (aromatic + hydroxyl group)
b. Toluene is the benzene ring attached to the methyl group.

12.41 a. phenol (aromatic + hydroxyl group)
b. alcohol

12.43 a. 3,7-dimethyl-6-octenal
b. The *en* in octenal signifies a double bond. The -6- indicates the double bond is between carbon 6 and carbon 7, counting from the aldehyde as carbon 1.
c. The *al* in octenal signifies that an aldehyde group is carbon 1.
d. $C_{10}H_{18}O(g) + 14O_2(g) \xrightarrow{\Delta} 10CO_2(g) + 9H_2O(g) + \text{energy}$

12.45 a. 2° **b.** 1° **c.** 2° **d.** 3°

12.47 a. 2,4-dimethyl-2-pentanol
b. 4-bromo-2-pentanol
c. 3-methylphenol

12.49 a. (benzene ring with OH and Cl in para positions)
b. $CH_3-CH(CH_3)-CH(OH)-CH_2-CH_3$

c. $CH_3—CH(CH_3)—CH_2—OH$

d. 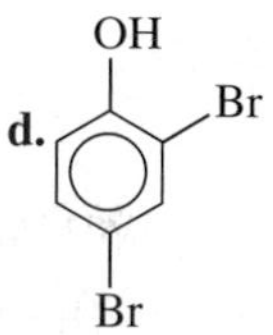 2,4-dibromophenol (benzene ring with OH, Br ortho, Br para)

12.51 a. 1-Propanol can form hydrogen bonds with its polar hydroxyl group.
b. An alcohol forms more hydrogen bonds than an ether.
c. Ethanol has a smaller number of carbon atoms.

12.53 a. soluble; hydrogen bonding
b. insoluble; long carbon chain diminishes the effect of polar —OH on hydrogen bonding

12.55 a. $CH_3—CH=CH_2$ **b.** $CH_3—CH_2—C(=O)—H$

c. $CH_3—CH=CH—CH_2—CH_3$

e. 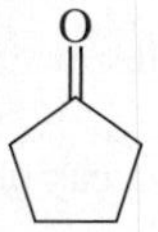 cyclopentanone

12.57 a. $CH_3—CH_2—CH(OH)—CH_3$ **b.** 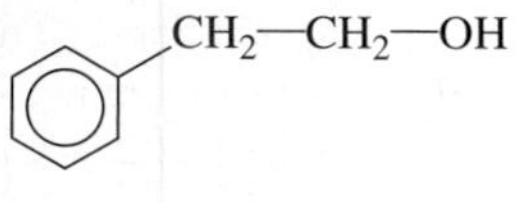 $C_6H_5—CH_2—CH_2—OH$

c. cyclopropanol (cyclopropane ring with OH)

12.59 a. 3-bromo-4-chlorobenzaldehyde
b. 3-chloropropanal
c. 2-chloro-3-pentanone

12.61 a. 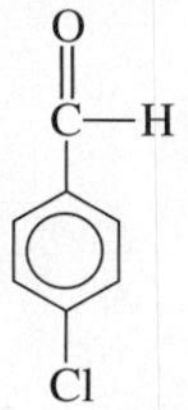 4-chlorobenzaldehyde (benzene ring with —C(=O)—H and Cl para) **b.** $Cl—CH_2—CH_2—C(=O)—H$

c. $CH_3—CH_2—C(=O)—CH_3$

d. $CH_3—CH_2—CH_2—CH(CH_3)—CH_2—C(=O)—H$

12.63 a and **b**

12.65 a. $H—CCl_2—$Ⓒ$(Cl)(H)—OH$ **b.** none

c. none **d.** $CH_3—$Ⓒ$H(NH_2)—C(=O)—H$

e. $CH_3—CH_2—$Ⓒ$H(Br)—CH_2—CH_2—CH_3$

12.67 a. identical **b.** enantiomers
c. enantiomers **d.** identical

12.69 a. $CH_3—CH_2—C(=O)—OH$

b. $CH_3—C(=O)—CH_2—CH_2—CH_3$

c. $CH_3—CH_2—CH_2—C(=O)—OH$ **d.** 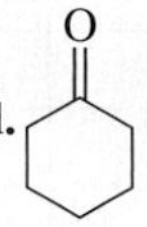 cyclohexanone

12.71

$CH_3—CH_2—CH_2—CH_2—CH_2—OH$	1-Pentanol
$CH_3—CH(OH)—CH_2—CH_2—CH_3$	2-Pentanol
$CH_3—CH_2—CH(OH)—CH_2—CH_3$	3-Pentanol
$HO—CH_2—CH(CH_3)—CH_2—CH_3$	2-Methyl-1-butanol
$HO—CH_2—CH_2—CH(CH_3)—CH_3$	3-Methyl-1-butanol
$CH_3—C(CH_3)(OH)—CH_2—CH_3$	2-Methyl-2-butanol
$CH_3—CH(OH)—CH(CH_3)—CH_3$	3-Methyl-2-butanol
$CH_3—C(CH_3)_2—CH_2—OH$	2,2-Dimethyl-1-propanol

12.73 $CH_3—CH(CH_3)—C(=O)—H$ 2-Methylpropanal

12.75 $CH_3—CH_2—CH_2—OH$ 1-Propanol
A

$CH_3—CH=CH_2$ Propene
B

$CH_3—CH_2—C(=O)—H$ Propanal
C

Combining Ideas from Chapters 9 to 12

CI.19 Some isotopes of silicon are listed in the following table:

Isotope	% Natural Abundance	Atomic Mass	Half-Life	Radiation Emitted
$^{27}_{14}Si$		26.99	4.2 s	Positron
$^{28}_{14}Si$	92.23	27.98	Stable	None
$^{29}_{14}Si$	4.67	28.98	Stable	None
$^{30}_{14}Si$	3.10	29.97	Stable	None
$^{31}_{14}Si$		30.98	2.6 h	Beta

a. In the following table, indicate the number of protons, neutrons, and electrons for each isotope listed:

Isotope	Number of Protons	Number of Neutrons	Number of Electrons
$^{27}_{14}Si$			
$^{28}_{14}Si$			
$^{29}_{14}Si$			
$^{30}_{14}Si$			
$^{31}_{14}Si$			

b. What is the electron arrangement of silicon?
c. Calculate the atomic mass for silicon, using the isotopes that have a natural abundance.
d. Write the nuclear equations for the decay of $^{27}_{14}Si$ and $^{31}_{14}Si$.
e. Draw the electron-dot formula, and predict the shape of $SiCl_4$.
f. How many hours are needed for a sample of $^{31}_{14}Si$ with an activity of 16 μCi to decay to 2.0 μCi?

CI.20 K^+, an electrolyte required by the human body, is found in many foods, and in salt substitutes. One radioisotope of potassium, $^{40}_{19}K$, has a natural abundance of 0.012%, a half-life of 1.30×10^9 y, and an activity of 7.0 μCi/g. The isotope $^{40}_{19}K$ decays to $^{40}_{20}Ca$ or to $^{40}_{18}Ar$.

Potassium chloride is used as a salt substitute.

a. Write a balanced nuclear equation for each type of decay.
b. Identify the particle emitted for each type of decay.
c. How many K^+ ions are in 3.5 oz of KCl?
d. What is the activity of 25 g of KCl, in becquerels?

CI.21 Of much concern to environmentalists is radon-222, which is a radioactive noble gas that can seep from the ground into basements of buildings. Radon-222 is a product of the decay of radium-226 that occurs naturally in rocks and soil in much of the United States. Radon-222, which has a half-life of 3.8 days, decays by emitting an alpha particle. Radon-222 gas can be inhaled into the lungs where it is strongly associated with lung cancer. Environmental agencies have set the maximum level of radon-222 in a home at 4 picocuries per liter (pCi/L) of air.

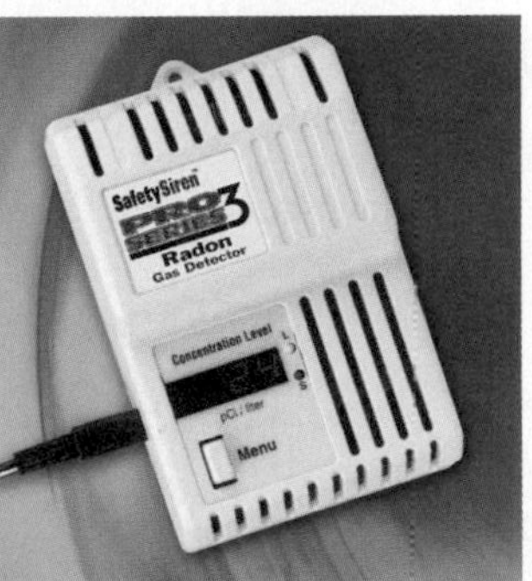

A home detection kit is used to measure the level of radon-222.

a. Write the balanced nuclear equation for the decay of Ra-226.
b. Write the balanced nuclear equation for the decay of Rn-222.
c. If a room contains 24 000 atoms of radon-222, how many atoms of radon-222 remain after 15.2 days?
d. Suppose a room in a home has a volume of 72 000 L. If the radon level is 2.5 pCi/L, how many alpha particles are emitted in one day? (1 Ci $= 3.7 \times 10^{10}$ disintegrations/s)

CI.22 Ionone is a compound that gives violets their aroma. The small, edible, purple flowers of violets are used on salads and to make teas. Liquid ionone has a density of 0.935 g/mL.

The aroma of violets is due to ionone.

Ionone

a. What functional groups are present in ionone?
b. Is the double bond on the side chain cis or trans?
c. What is the molecular formula and molar mass of ionone?
d. How many moles are in 2.00 mL of ionone?

e. When ionone reacts with hydrogen in the presence of a platinum catalyst, hydrogen adds to the double bonds and converts the ketone group to an alcohol. Draw the condensed structural formula and give the molecular formula of the product.

f. How many milliliters of hydrogen gas are needed at STP to completely react 5.0 mL of ionone?

CI.23 Butyraldehyde is a clear liquid solvent with an unpleasant odor. It has a low boiling point and is highly flammable.

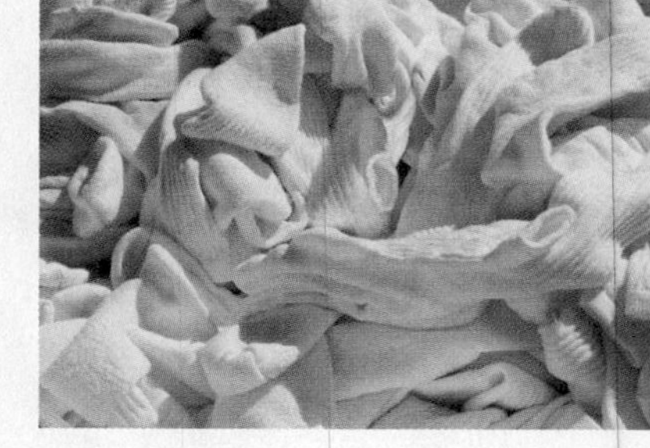

The unpleasant odor of old gym socks is due to butyraldehyde.

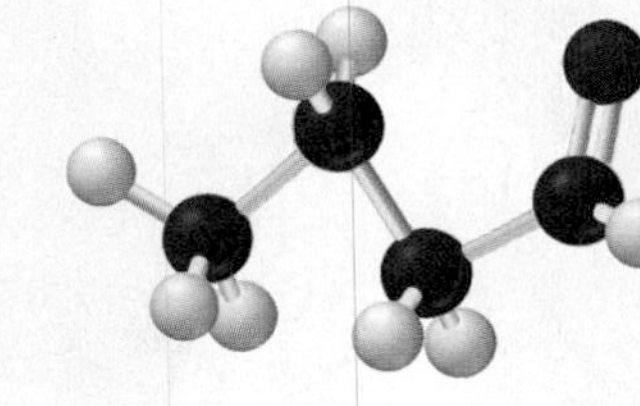

a. Draw the condensed structural formula of butyraldehyde.

b. Draw the skeletal formula of butyraldehyde.

c. What is the IUPAC name of butyraldehyde?

d. Draw the condensed structural formula of the alcohol that is produced when butyraldehyde is reduced.

CI.24 Butyraldehyde has a density of 0.802 g/mL and a heat of combustion of 1520 kJ/mole. Using your solutions for Problem CI.23, answer the following:

a. Write the balanced equation for the complete combustion of butyraldehyde.

b. How many grams of oxygen gas are needed to completely react with 15.0 mL of butyraldehyde?

c. How many liters of carbon dioxide gas are produced at STP in part **b**?

d. Calculate the heat, in kilojoules, released from the combustion of butyraldehyde in part **b**.

Answers

CI.19 a.

Isotope	Number of Protons	Number of Neutrons	Number of Electrons
$^{27}_{14}Si$	14	13	14
$^{28}_{14}Si$	14	14	14
$^{29}_{14}Si$	14	15	14
$^{30}_{14}Si$	14	16	14
$^{31}_{14}Si$	14	17	14

b. 2,8,4

c. Atomic mass calculated from the three isotopes is 28.09 amu.

d. $^{27}_{14}Si \longrightarrow ^{27}_{13}Al + ^{0}_{+1}e$

$^{31}_{14}Si \longrightarrow ^{31}_{15}P + ^{0}_{-1}e$

e.
```
        :C̈l:
         |
   :C̈l—Si—C̈l:      Tetrahedral
         |
        :C̈l:
```

f. 7.8 h (three half-lives)

CI.21 a. $^{226}_{88}Ra \longrightarrow ^{222}_{86}Rn + ^{4}_{2}He$

b. $^{222}_{86}Rn \longrightarrow ^{218}_{84}Po + ^{4}_{2}He$

c. 1500 atoms of radon-222 remain

d. 5.8×10^8 alpha particles/day

CI.23 a. $CH_3—CH_2—CH_2—\overset{\overset{O}{\|}}{C}—H$

b. (skeletal formula: chain with C=O and H at end)

c. butanal

d. $CH_3—CH_2—CH_2—CH_2—OH$

13 Carbohydrates

LOOKING AHEAD

"We measure the sugar content in a small sample of juices from the grapes in different areas of the vineyard," says Leslie Bucher, laboratory director at Bouchaine Winery. "We also measure the alcohol content during fermentation and run tests for sulfur, pH, and total acid."

As grapes ripen, there is an increase in the sugars, which are the monosaccharides fructose and glucose. The sugar content is affected by soil conditions and the amount of sunlight and water. When the grapes are ripe and sugar content is at a desirable level, they are harvested. During fermentation, enzymes from yeast convert the sugar to ethanol, and carbon dioxide. Grapes harvested with 22.5% sugar will ferment to give a wine with 12.5–13.5% alcohol content.

Visit **www.masteringchemistry.com** for self-study materials and instructor-assigned homework.

Carbohydrates are the most abundant of all the organic compounds in nature. In plants, energy from the Sun is used to convert carbon dioxide and water into the carbohydrate glucose. Many of the glucose molecules are made into long-chain polymers of starch that store energy. About 65% of the foods in our diet consist of carbohydrates. Each day we utilize carbohydrates in foods such as bread, pasta, potatoes, and rice. Other carbohydrates called disaccharides include sucrose (table sugar) and lactose in milk. During digestion and cellular metabolism, carbohydrates are converted into glucose, which is oxidized further in our cells to provide our bodies with energy and to provide the cells with carbon atoms for building molecules of protein, lipids, and nucleic acids. In plants, a polymer of glucose called cellulose builds the structural framework. Cellulose has other important uses, too. The wood in our furniture, the pages in this book, and the cotton in our clothing are made of cellulose.

Carbohydrates contained in foods such as pasta and bread provide energy for the body.

LEARNING GOAL

Classify a monosaccharide as an aldose or a ketose, and indicate the number of carbon atoms.

13.1 Carbohydrates

Carbohydrates such as table sugar, lactose in milk, and cellulose are all made of carbon, hydrogen, and oxygen. Simple sugars, which have formulas of $C_n(H_2O)_n$, were once thought to be hydrates of carbon, thus the name *carbohydrate*. In a series of reactions called photosynthesis, energy from the Sun is used to combine the carbon atoms from carbon dioxide (CO_2) and the hydrogen and oxygen atoms of water (H_2O) into the carbohydrate glucose.

$$6CO_2 + 6H_2O + \text{energy} \underset{\text{Respiration}}{\overset{\text{Photosynthesis}}{\rightleftharpoons}} \underset{\text{Glucose}}{C_6H_{12}O_6} + 6C_2$$

In the body, glucose is oxidized in a series of metabolic reactions known as respiration, which releases chemical energy to do work in the cells. Carbon dioxide and water are produced and returned to the atmosphere. The combination of photosynthesis and respiration is called the carbon cycle, in which energy from the Sun is stored in plants by photosynthesis and made available to us when the carbohydrates in our diets are metabolized (see Figure 13.1).

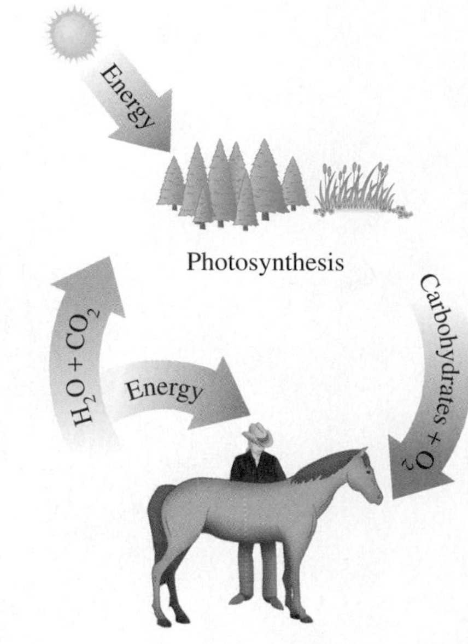

FIGURE 13.1 During photosynthesis, energy from the Sun combines CO_2 and H_2O to form glucose ($C_6H_{12}O_6$) and O_2. During respiration in the body, carbohydrates are oxidized to CO_2 and H_2O, while energy is produced.
Q What are the reactants and products of respiration?

Types of Carbohydrates

The simplest carbohydrates are the **monosaccharides**. A monosaccharide cannot be split or hydrolyzed into smaller carbohydrates. One of the most common carbohydrates, glucose ($C_6H_{12}O_6$), is a monosaccharide. **Disaccharides** consist of two monosaccharide units joined together. Thus, a disaccharide can be split into two monosaccharide units. For example, ordinary table sugar, sucrose ($C_{12}H_{22}O_{11}$), is a disaccharide that can be split by water (hydrolysis) in the presence of an acid or an enzyme to give one molecule of glucose and one molecule of another monosaccharide, fructose.

$$\underset{\text{Sucrose}}{C_{12}H_{22}O_{11}} + H_2O \xrightarrow{H^+ \text{ or enzyme}} \underset{\text{Glucose}}{C_6H_{12}O_6} + \underset{\text{Fructose}}{C_6H_{12}O_6}$$

SELF STUDY ACTIVITY
Carbohydrates

TUTORIAL
Carbohydrates

Polysaccharides are carbohydrates that are naturally occurring polymers containing many monosaccharide units. In the presence of an acid or an enzyme, a polysaccharide can be completely hydrolyzed to yield many molecules of monosaccharides.

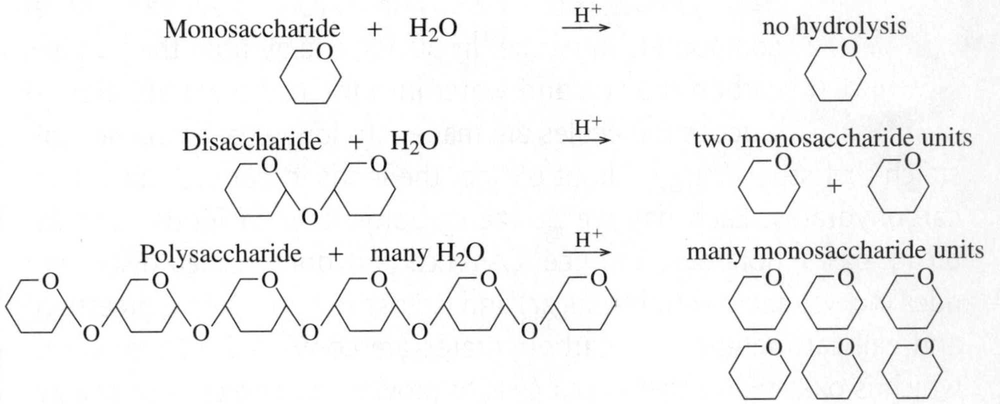

Monosaccharide + H_2O $\xrightarrow{H^+}$ no hydrolysis

Disaccharide + H_2O $\xrightarrow{H^+}$ two monosaccharide units

Polysaccharide + many H_2O $\xrightarrow{H^+}$ many monosaccharide units

Monosaccharides

TUTORIAL
Types of Carbohydrates

TUTORIAL
Carbonyls in Carbohydrates

Monosaccharides are sugars that have a chain of three to eight carbon atoms, one in a carbonyl group and the rest attached to hydroxyl groups. There are two types of monosaccharide structures. In an **aldose**, the carbonyl group is on the first carbon (—CHO), an aldehyde, while a **ketose** contains the carbonyl group on the second carbon atom as a ketone (C═O).

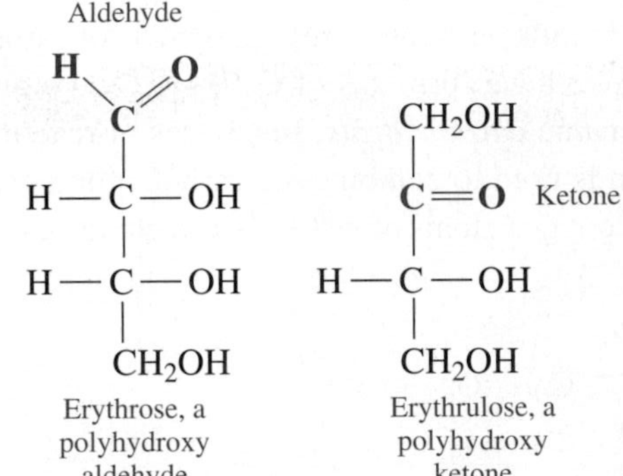

A monosaccharide with three carbon atoms is a *triose*, one with four carbon atoms is a *tetrose*; a *pentose* has five carbons, and a *hexose* contains six carbons. We can use both classification systems to indicate the type of carbonyl group and the number of carbon atoms. An aldopentose is a five-carbon monosaccharide that is an aldehyde; a ketohexose is a six-carbon monosaccharide that is a ketone. Some examples:

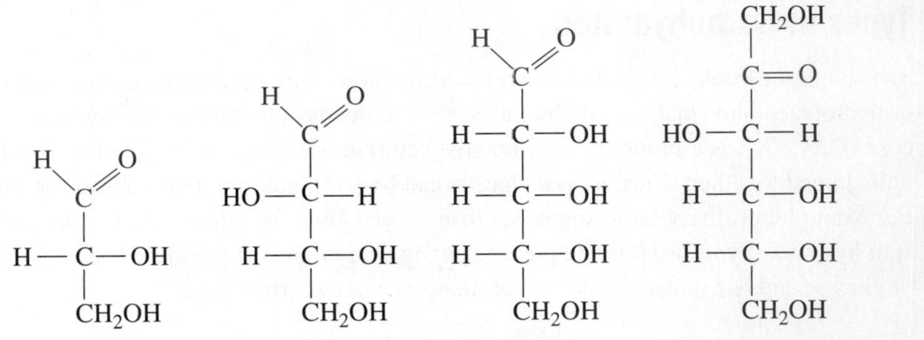

Glyceraldehyde (aldotriose)

Threose (aldotetrose)

Ribose (aldopentose)

Fructose (ketohexose)

CONCEPT CHECK 13.1

Monosaccharides

Classify each of the following monosaccharides as an aldopentose, aldohexose, ketopentose, or ketohexose:

a.
```
    CH2OH
      |
     C=O
      |
  H—C—OH
      |
  H—C—OH
      |
    CH2OH
```
Ribulose

b.
```
   H     O
    \  //
      C
      |
  H—C—OH
      |
 HO—C—H
      |
  H—C—OH
      |
  H—C—OH
      |
    CH2OH
```
Glucose

ANSWER

a. Ribulose has five carbon atoms (pentose) and is a ketone, which makes it a ketopentose.

b. Glucose has six carbon atoms (hexose) and is an aldehyde, which makes it an aldohexose.

QUESTIONS AND PROBLEMS

Carbohydrates

13.1 What reactants are needed for photosynthesis and respiration?

13.2 What is the relationship between photosynthesis and respiration?

13.3 What is a monosaccharide? A disaccharide?

13.4 What is a polysaccharide?

13.5 What functional groups are found in all monosaccharides?

13.6 What is the difference between an aldose and a ketose?

13.7 What are the functional groups and number of carbons in a ketopentose?

13.8 What are the functional groups and number of carbons in an aldohexose?

13.9 Classify each of the following monosaccharides as an aldopentose, ketopentose, aldohexose, or ketohexose:

a.
```
    CH2OH
      |
     C=O
      |
  H—C—OH
      |
  H—C—OH
      |
  H—C—OH
      |
    CH2OH
```
Psicose

b.
```
     CHO
      |
 HO—C—H
      |
 HO—C—H
      |
  H—C—OH
      |
    CH2OH
```
Lyxose

13.10 Classify each of the following monosaccharides as an aldopentose, ketopentose, aldohexose, or ketohexose:

a.
```
     CHO
      |
  H—C—OH
      |
 HO—C—H
      |
  H—C—OH
      |
    CH2OH
```
Xylose

b.
```
    CH2OH
      |
     C=O
      |
 HO—C—H
      |
 HO—C—H
      |
  H—C—OH
      |
    CH2OH
```
Tagatose

LEARNING GOAL

Use Fischer projections to draw the D or L isomers of glucose, galactose, and fructose.

SELF STUDY ACTIVITY
Forms of Carbohydrates

TUTORIAL
Drawing Fischer Projections of Monosaccharides

TUTORIAL
Identifying Chiral Carbons in Monosaccharides

TUTORIAL
Identifying D and L Sugars

13.2 Fischer Projections of Monosaccharides

Fischer Projections

As we discussed in Section 12.6, a Fischer projection is drawn with vertical and horizontal lines with the aldehyde group at the top and —H and —OH groups on the intersecting line. In L-glyceraldehyde, the letter L is assigned to the stereoisomer with the —OH group on the left. In D-glyceraldehyde, the —OH group is on the right. The carbon atom in the $—CH_2OH$ group at the bottom of the Fischer projection is not chiral because it does not have four different groups bonded to it.

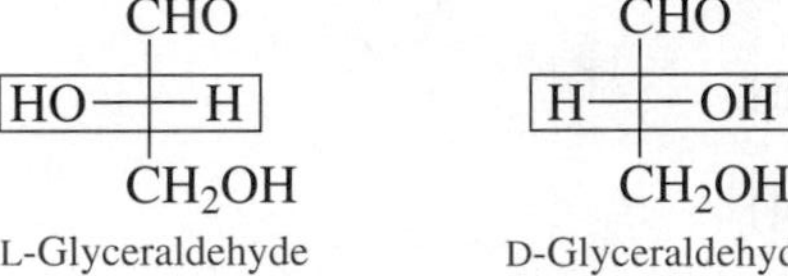

Most of the monosaccharides we will study have five or six carbon atoms with more than one chiral carbon. Then the —OH group on the chiral carbon farthest from the carbonyl group is used to determine the D or L isomer. The following are the Fischer projections for the D and L isomers of ribose, a five-carbon monosaccharide, and the D and L isomers of glucose, a six-carbon monosaccharide. In each of the mirror images, it is important to understand that the —OH groups on *all the chiral carbon atoms* are reversed from one side to the other. For example, in L-ribose, the —OH groups are all written on the left side of the horizontal line. In the mirror image, D-ribose, the —OH groups are all written on the right side of the horizontal line.

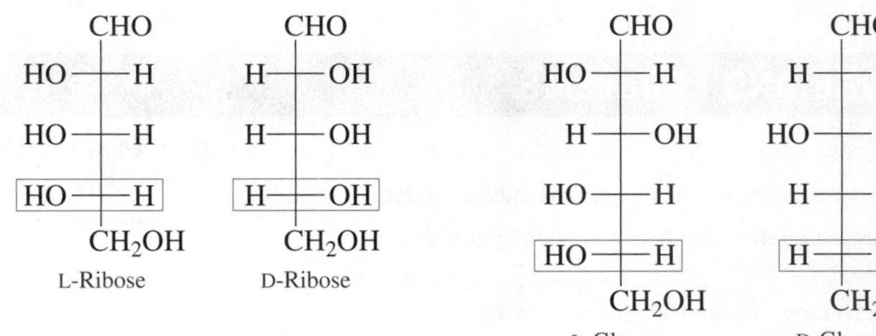

CONCEPT CHECK 13.2

Identifying D and L Isomers of Monosaccharides

Identify the following Fischer projection as D- or L-xylose:

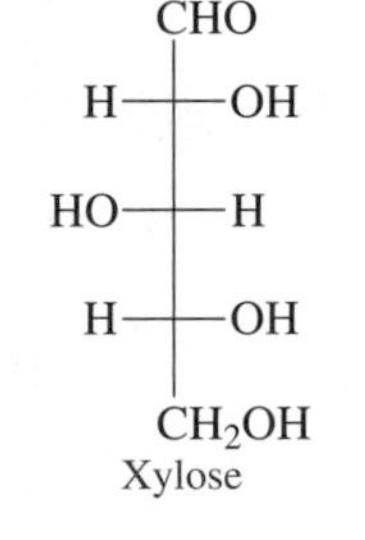

ANSWER

In the Fischer projection, the —OH group on the chiral carbon farthest from the carbonyl group is on the right, which makes this D-xylose.

Structures of Some Important Monosaccharides

The hexoses glucose, galactose, and fructose are the most important monosaccharides. Although we can draw Fischer projections for D and L isomers, the D isomers are more commonly found in nature and used in the cells of the body. The Fischer projections for the D isomers are drawn as follows:

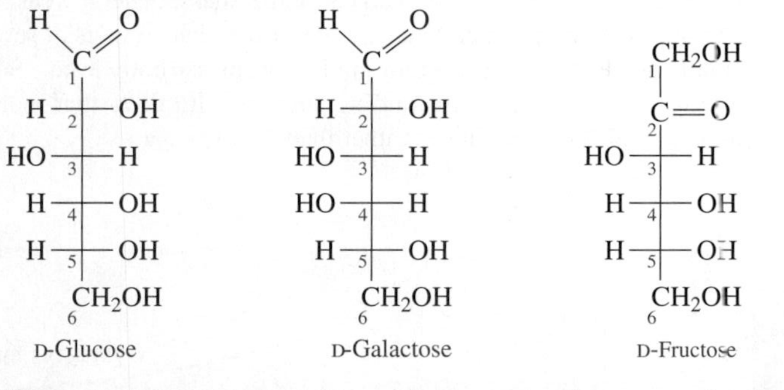

The most common hexose, **D-glucose**, also known as dextrose and blood sugar, is found in fruits, vegetables, corn syrup, and honey (see Figure 13.2). D-glucose is a building block of the disaccharides sucrose, lactose, and maltose, and polysaccharides such as amylose, cellulose, and glycogen.

Galactose is an aldohexose that is obtained from the disaccharide lactose, which is found in milk and milk products. Galactose is important in the cellular membranes of the brain and nervous system. The only difference in the Fischer projections of D-glucose and D-galactose is the arrangement of the —OH group on carbon 4.

In a condition called *galactosemia*, an enzyme needed to convert galactose to glucose is missing. The accumulation of galactose in the blood and tissues can lead to cataracts, mental retardation, and cirrhosis. The treatment for galactosemia is the removal of all galactose-containing foods, mainly milk and milk products, from the diet. If this is done immediately after birth, the damaging effects of galactose accumulation can be avoided.

In contrast to glucose and galactose, **fructose**, also called levulose and fruit sugar, is a ketohexose. The structure of fructose differs at carbons 1 and 2 by the location of the carbonyl group. Fructose is the sweetest of the carbohydrates, twice as sweet as sucrose (table sugar). This makes fructose popular with dieters because less fructose and, therefore, fewer calories are needed to provide a pleasant taste.

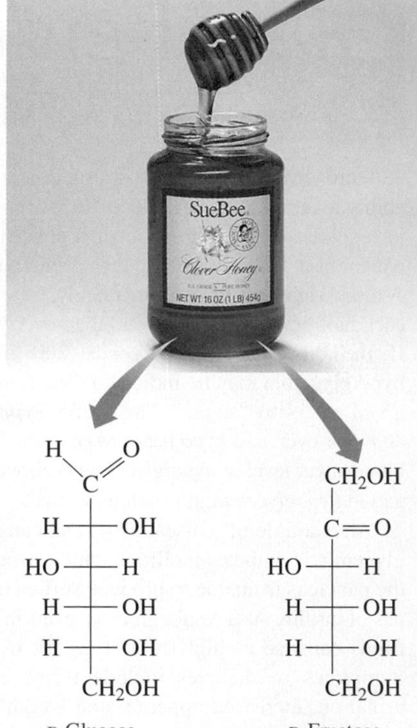

FIGURE 13.2 The sweet taste of honey is due to the monosaccharides D-glucose and D-fructose.

Q What are some differences in the Fischer projections of D-glucose and D-fructose?

CASE STUDY
Diabetes and Blood Glucose

SAMPLE PROBLEM 13.1

Monosaccharides

Ribulose has the following Fischer projection. Identify the compound as D- or L-ribulose.

CH_2OH
|
C=O
H—|—OH
H—|—OH
CH_2OH

SOLUTION

The compound is D-ribulose because the —OH group is on the right side of the chiral carbon farthest from the carbonyl group.

STUDY CHECK 13.1

Draw the Fischer projection for the mirror image of the ribulose in Sample Problem 13.1.

Chemistry Link to Health

HYPERGLYCEMIA AND HYPOGLYCEMIA

A doctor may order a glucose tolerance test to evaluate the body's ability to return to normal blood glucose concentrations (70–90 mg/dL) in response to the ingestion of a specific amount of glucose. The patient fasts for 12 hours and then drinks a solution containing glucose. A blood sample is taken immediately, followed by more blood samples each half hour for 2 hours, and then every hour, for a total of 5 hours. If the blood glucose exceeds 200 mg/dL and remains high, hyperglycemia may be indicated. The term *glyc* or *gluco* refers to "sugar." The prefix *hyper* means above or over, and *hypo* is below or under. Thus the blood sugar level in *hyperglycemia* is above normal, and in *hypoglycemia*, it is below normal.

An example of a disease that can cause hyperglycemia is diabetes mellitus, which occurs when the pancreas is unable to produce sufficient quantities of insulin. As a result, glucose levels in the body fluids can rise as high as 350 mg/dL of plasma. Symptoms of diabetes include thirst, excessive urination, increased appetite, and weight loss. In older persons, diabetes is sometimes a consequence of excessive weight gain.

When a person is hypoglycemic, the blood glucose level rises and then decreases rapidly to levels as low as 40 mg/dL. In some cases, hypoglycemia is caused by overproduction of insulin by the pancreas. Low blood glucose can cause dizziness, general weakness, and muscle tremors. A diet may be prescribed that consists of several small meals high in protein and low in carbohydrate. Some hypoglycemic patients are finding success with diets that include more complex carbohydrates rather than simple sugars.

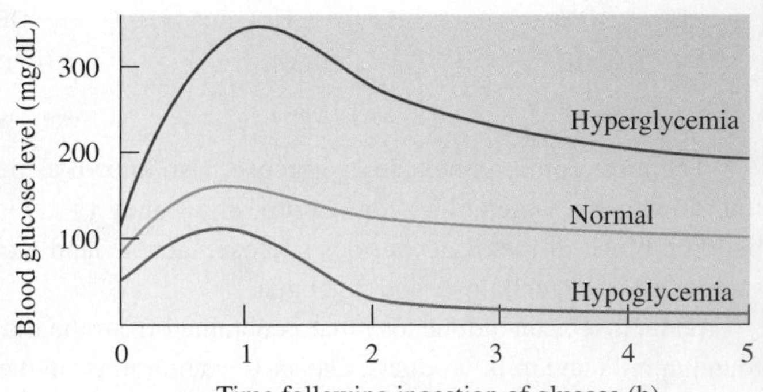

A glucose solution is given to determine blood glucose levels.

QUESTIONS AND PROBLEMS

Fischer Projections of Monosaccharides

13.11 How is a Fischer projection identified as a D or an L isomer?

13.12 Draw the Fischer projection for D-glyceraldehyde and L-glyceraldehyde.

13.13 Identify each of the following as the D or L isomer:

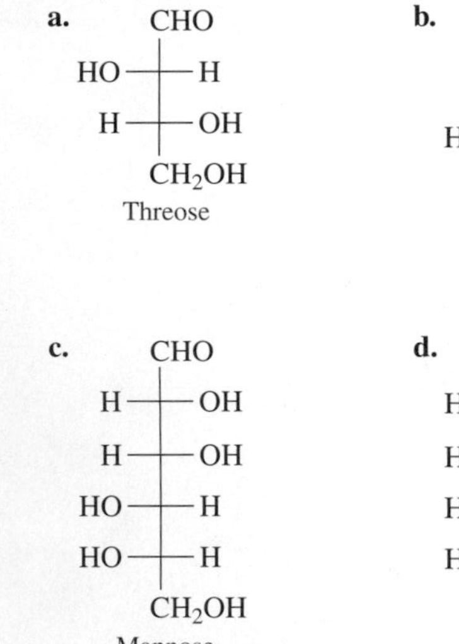

13.14 Identify each of the following as the D or L isomer:

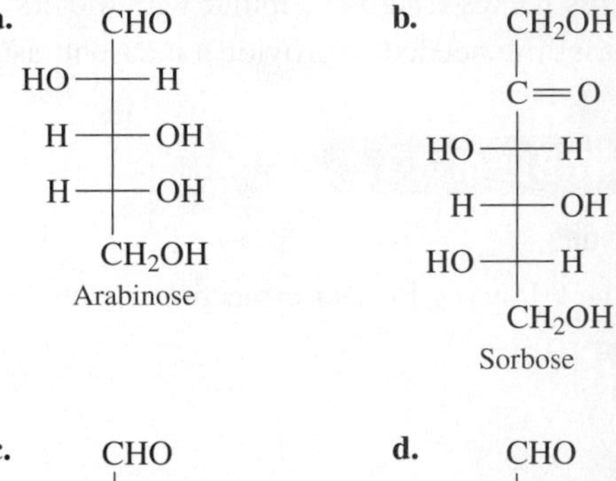

c. CHO; HO—H; HO—H; H—OH; CH_2OH; Lyxose

d. CHO; HO—H; HO—H; HO—H; CH_2OH; Ribose

13.15 Draw the Fischer projections for the mirror images for **a** to **d** in Problem 13.13.

13.16 Draw the Fischer projections for the mirror images for **a** to **d** in Problem 13.14.

13.17 Draw the Fischer projections for D-glucose and L-glucose.

13.18 Draw the Fischer projections for D-fructose and L-fructose.

13.19 How does the Fischer projection of D-galactose differ from D-glucose?

13.20 How does the Fischer projection of D-fructose differ from D-glucose?

13.21 Identify the monosaccharide that fits each of the following descriptions:

a. is also called blood sugar
b. is not metabolized in a condition known as galactosemia
c. is also called fruit sugar

13.22 Identify a monosaccharide that fits each of the following descriptions:

a. found in high blood levels in diabetes
b. obtained as a hydrolysis product of lactose
c. is the sweetest of the monosaccharides

13.3 Haworth Structures of Monosaccharides

LEARNING GOAL

Draw and identify the Haworth structures of monosaccharides.

TUTORIAL
Drawing Cyclic Sugars

Up until now, we have drawn the structures of monosaccharides as open chains. However, molecules of monosaccharides normally exist in a cyclic structure formed when a carbonyl group and a hydroxyl group in the *same* molecule react to give ring structures known as **Haworth structures**. While the carbonyl group in the open chain could react with several of the —OH groups, the most stable form of pentoses and hexoses are five- or six-atom rings. Let's look at how we draw the Haworth structures for some D isomers, starting with the open-chain structure of D-glucose.

Drawing Haworth Structures

Step 1 **Turn the open-chain condensed structural formula clockwise 90°.** This places the —OH groups on the right of carbons 2 and 4 (blue) and carbon 5 (red) of the vertical open chain below the carbon atoms. The —OH group on the left (green) of the open chain is drawn above carbon 3.

D-Glucose (open chain)

Step 2 **Fold the carbon chain into a hexagon and bond the O on carbon 5 to the carbonyl group.** With carbon 2 and 3 as the base of a hexagon, move the remaining carbons upwards. The reacting —OH group (red) on carbon 5 is drawn next to the carbonyl carbon, which moves carbon 6 in $—CH_2OH$ above carbon 5. To complete the Haworth structure, draw a bond from the oxygen of the reacting —OH group (carbon 5) to the carbonyl carbon.

Hydroxyl group on new chiral carbon

Carbon-5 oxygen bonds to carbonyl

Cyclic structure

Step 3 **Draw the new —OH group on carbon 1 down to give the α isomer or up to give β isomer.** In a Haworth structure, the corners of the ring represent carbon atoms. There are two ways to draw the new —OH, either up or down, which gives two possible stereoisomers of D-glucose.

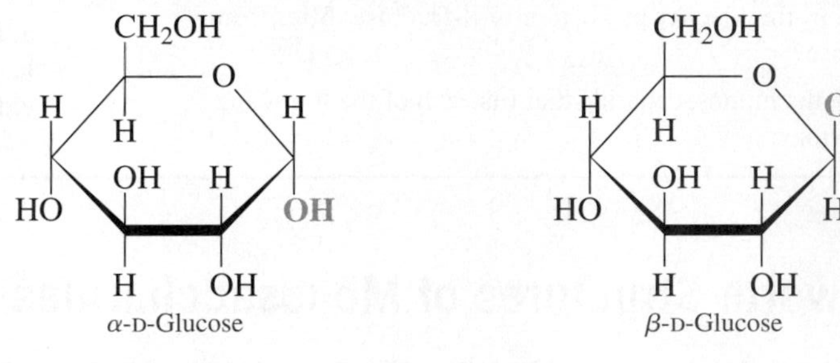

α-D-Glucose

β-D-Glucose

Mutarotation of α- and β-D-Glucose

In an aqueous solution, the Haworth structure of α-D-glucose opens to give the open chain of D-glucose with an aldehyde group. At any given time, there is only a trace amount of the open chain because it closes quickly to form a stable ring structure. However, when the open chain closes again it can form β-D-glucose. In this process called *mutarotation*, each isomer converts to the open chain and back again. As the ring opens and closes, the —OH group on carbon 1 can form either the α or β isomer. An aqueous glucose solution contains a mixture of 36% α-D-glucose and 64% β-D-glucose.

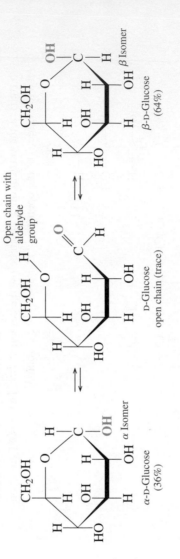

α Isomer

Open chain with aldehyde group

β Isomer

α-D-Glucose (36%)

D-Glucose open chain (trace)

β-D-Glucose (64%)

Haworth Structures of Galactose

Galactose is an aldohexose that differs from glucose only in the arrangement of the —OH group on carbon 4. Thus, its Haworth structure is similar to glucose, except that the —OH on carbon 4 is drawn up. Galactose also exists as α and β isomers.

D-Galactose

α-D-Galactose

β-D-Galactose

Haworth Structures of Fructose

In contrast to glucose and galactose, fructose is a ketohexose. The Haworth structure for fructose is a five-atom ring with carbon 2 at the right corner of the ring. The cyclic structure forms when the hydroxyl group on carbon 5 reacts with carbon 2 in the carbonyl group. The new hydroxyl group on carbon 2 gives the α and β isomers of fructose.

D-Fructose

α-D-Fructose

β-D-Fructose

CONCEPT CHECK 13.3

α and β Isomers

What is the difference between the α and β isomers of D-galactose?

ANSWER

In the cyclic structures of D-galactose, a new —OH group forms on carbon 1. Because there are two ways to draw this —OH group, there are two isomers of D-galactose. In α-D-galactose, the new —OH is drawn down, and in β-D-galactose, the new —OH is drawn up.

SAMPLE PROBLEM 13.2

Drawing Haworth Structures for Sugars

D-Mannose, a carbohydrate found in immunoglobulins, has the following open-chain structure. Draw the Haworth structure for β-D-mannose.

D-Mannose

Guide to Drawing Haworth Structures

1. Turn the open-chain condensed structural formula clockwise 90°.
2. Fold the chain into a hexagon and bond the O on carbon 5 to the carbonyl group.
3. Draw the new —OH group on carbon 1 down to give the α isomer or up to give the β isomer.

SOLUTION

Step 1 Turn the open-chain condensed structural formula clockwise 90°.

Step 2 Fold the chain into a hexagon and bond the O on carbon 5 to the carbonyl group.

Step 3 Draw the new —OH group on carbon 1 up to give the β isomer.

STUDY CHECK 13.2

Draw the Haworth structure for α-D-mannose.

QUESTIONS AND PROBLEMS

Haworth Structures of Monosaccharides

13.23 What are the kind and number of atoms in the ring portion of the Haworth structure of glucose?

13.24 What are the kind and number of atoms in the ring portion of the Haworth structure of fructose?

13.25 Draw the Haworth structures for α- and β-D-glucose.

13.26 Draw the Haworth structures for α- and β-D-fructose.

13.27 Identify each of the following as the α or β isomer:

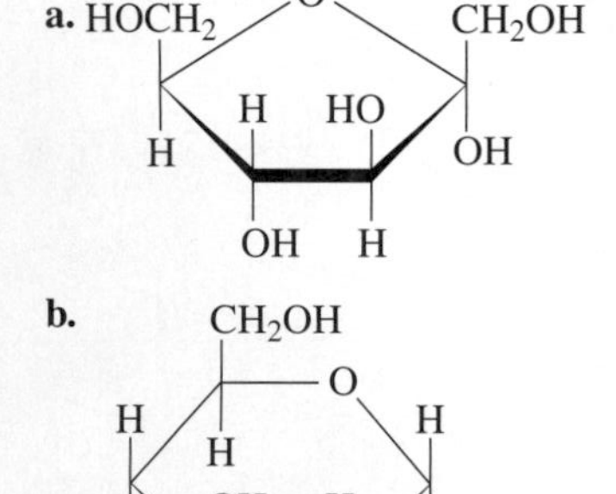

13.28 Identify each of the following as the α or β isomer:

a.

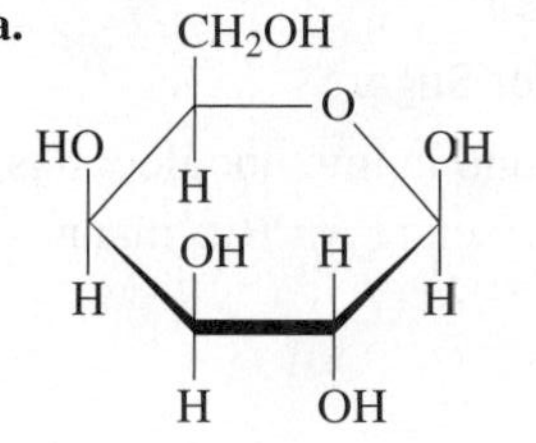

b.

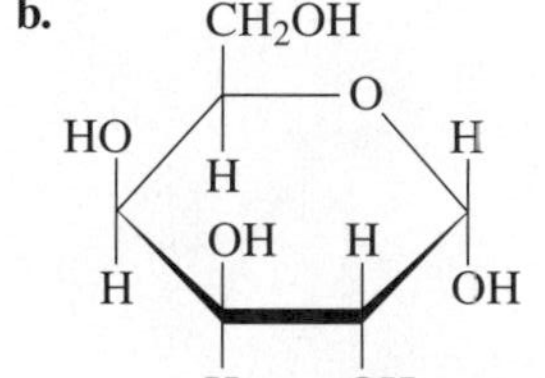

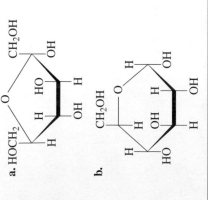

13.4 Chemical Properties of Monosaccharides

LEARNING GOAL

Identify the products of oxidation or reduction of monosaccharides; determine whether a carbohydrate is a reducing sugar.

Monosaccharides contain functional groups that can undergo chemical reactions. In an aldose, the aldehyde group can be oxidized to a carboxylic acid. The carbonyl group in both an aldose and a ketose can be reduced to give a hydroxyl group. The hydroxyl groups can react with other compounds to form a variety of derivatives that are important in biological structures.

TUTORIAL
Reactions of Monosaccharides

Oxidation of Monosaccharides

Although monosaccharides exist mostly in cyclic forms, we have seen that a small amount of the open-chain form is always present, which provides an aldehyde group. As we discussed in Chapter 12, an aldehyde group with an adjacent hydroxyl can be oxidized to a carboxylic acid by an oxidizing agent such as Benedict's reagent. Then the Cu^{2+} is reduced to Cu^{+}, which forms a brick-red precipitate of Cu_2O. A carbohydrate that reduces another substance is called a **reducing sugar**.

```
      O                                    Oxidized
      ‖                                       O
      C—H                                     ‖
      |                                       C—OH
  H—C—OH                                      |
      |                                   H—C—OH
 HO—C—H     + 2Cu²⁺  ———→                    |          Reduced
      |       (blue)                      HO—C—H     + Cu₂O(s)
  H—C—OH                                      |          (brick-red)
      |                                   H—C—OH
  H—C—OH                                      |
      |                                   H—C—OH
    CH₂OH                                     |
                                            CH₂OH
```

Open chain of D-glucose, a reducing sugar → D-Gluconic acid

Fructose, a ketohexose, is also a reducing sugar. Usually a ketone cannot be oxidized. However, in Benedict's solution, which is basic, a rearrangement occurs between the ketone group on carbon 2 and the hydroxyl group on carbon 1. As a result, fructose is converted to glucose, which produces an aldehyde group with an adjacent hydroxyl that can be oxidized.

```
                                  H    O
 ₁CH₂OH                            \  //
   |                                ₁C
   |        Rearrangement           |
 ₂C=O     ⇌                     H—₂C—OH
   |                                |
D-Fructose                      D-Glucose
 (ketose)                        (aldose)
```

Reduction of Monosaccharides

The reduction of the carbonyl group in monosaccharides produces sugar alcohols, which are also called *alditols*. D-Glucose is reduced to D-glucitol, better known as D-sorbitol. Sugar alcohols such as D-sorbitol, D-xylitol from D-xylose, and D-mannitol from D-mannose are used as sweeteners in many sugar-free products such as diet drinks and sugarless gum as well as products for people with diabetes. However, there are some side effects of these sugar substitutes. Some people experience some discomfort such as gas

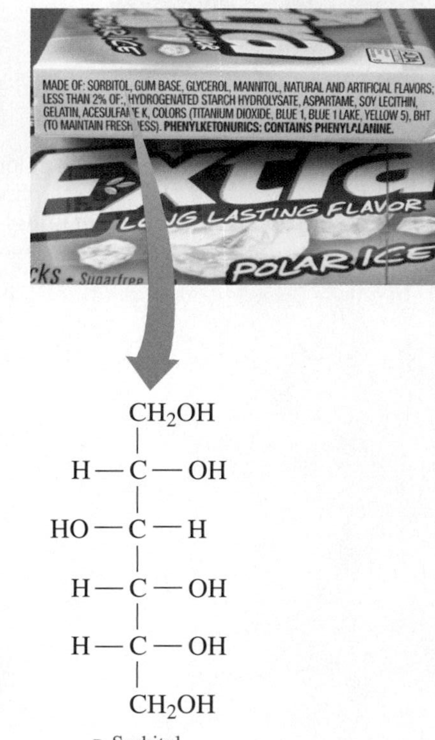


```
    CH₂OH
      |
  H—C—OH
      |
 HO—C—H
      |
  H—C—OH
      |
  H—C—OH
      |
    CH₂OH
```

D-Sorbitol

and diarrhea from the ingestion of sugar alcohols. The development of cataracts in diabetics is attributed to the accumulation of D-sorbitol in the lens of the eye.

```
        O
        ‖
        C—H                         CH₂OH
        |                           |
    H—C—OH                      H—C—OH
        |                           |
   HO—C—H                      HO—C—H
        |          + H₂  —Pt→       |
    H—C—OH                      H—C—OH
        |                           |
    H—C—OH                      H—C—OH
        |                           |
      CH₂OH                       CH₂OH
    D-Glucose               D-Glucitol or D-Sorbitol
```

Chemistry Link to Health

TESTING FOR GLUCOSE IN URINE

Normally, blood glucose flows through the kidneys and is reabsorbed into the bloodstream. However, if the blood level exceeds about 160 mg of glucose/dL of blood, the kidneys cannot reabsorb all of the glucose, and it spills over into the urine, a condition known as glucosuria. A symptom of diabetes mellitus is a high level of glucose in the urine.

Benedict's test can be used to determine the presence of glucose in urine. The amount of copper(I) oxide (Cu_2O) formed is proportional to the amount of reducing sugar present in the urine. Low to moderate levels of reducing sugar turn the solution green; solutions with high glucose levels turn Benedict's reagent yellow or brick-red. Table 13.1 lists some colors associated with the concentration of glucose in the urine.

In another clinical test that is specific for glucose, the enzyme glucose oxidase is used. The oxidase enzyme converts glucose to gluconic acid and oxygen to hydrogen peroxide, H_2O_2. The peroxide produced reacts with a dye in the test strip to give different colors. The level of glucose present in the urine is found by matching the color produced to a color chart on the test strip container.

The color of a test strip determines the glucose level in urine.

TABLE 13.1 Glucose Test Results

	Glucose Present in Urine	
Color of Benedict's test	**% (m/v)**	**mg/dL**
Blue	0	0
Blue-green	0.25	250
Green	0.50	500
Yellow	1.00	1000
Brick-red	2.00	2000

SAMPLE PROBLEM 13.3

Reducing Sugars

Why is D-glucose a reducing sugar?

SOLUTION

The aldehyde group with an adjacent hydroxyl of D-glucose is easily oxidized by Benedict's reagent. A carbohydrate that reduces Cu^{2+} to Cu^{+} is called a reducing sugar.

STUDY CHECK 13.3

A test using Benedict's reagent turns brick-red with a urine sample. According to Table 13.1, what might this result indicate?

QUESTIONS AND PROBLEMS

Chemical Properties of Monosaccharides

13.29 Draw the structure of D-xylitol produced when D-xylose is reduced.

```
        O
        ‖
        C—H
        |
   H—C—OH
        |
  HO—C—H
        |
   H—C—OH
        |
      CH2OH
    D-Xylose
```

13.30 Draw the structure of D-mannitol produced when D-mannose is reduced.

```
        O
        ‖
        C—H
        |
  HO—C—H
        |
  HO—C—H
        |
   H—C—OH
        |
   H—C—OH
        |
      CH2OH
    D-Mannose
```

13.31 Draw the condensed structural formulas of the oxidation and the reduction products of D-arabinose. What is the name of the sugar alcohol produced?

```
        O
        ‖
        C—H
        |
  HO—C—H
        |
   H—C—OH
        |
   H—C—OH
        |
      CH2OH
   D-Arabinose
```

13.32 Draw the condensed structural formulas of the oxidation and the reduction products of D-ribose. What is the name of the sugar alcohol produced?

```
        O
        ‖
        C—H
        |
   H—C—OH
        |
   H—C—OH
        |
   H—C—OH
        |
      CH2OH
     D-Ribose
```

13.5 Disaccharides

LEARNING GOAL

Describe the monosaccharide units and linkages in disaccharides.

A disaccharide is composed of two monosaccharides linked together. The most common disaccharides are maltose, lactose, and sucrose. When they are split by water (hydrolysis) in the presence of an acid or an enzyme, the products are two monosaccharides.

$$\text{Maltose} + H_2O \xrightarrow{H^+ \text{ or maltase}} \text{glucose} + \text{glucose}$$

$$\text{Lactose} + H_2O \xrightarrow{H^+ \text{ or lactase}} \text{glucose} + \text{galactose}$$

$$\text{Sucrose} + H_2O \xrightarrow{H^+ \text{ or sucrase}} \text{glucose} + \text{fructose}$$

Maltose, or malt sugar, is obtained from starch and is found in germinating grains. When maltose in barley and other grains is hydrolyzed by yeast enzymes, glucose is obtained, which can undergo fermentation to give ethanol. Maltose is used in cereals, candies, and the brewing of beverages.

In the Haworth structure of a disaccharide, a **glycosidic bond** is an ether bond that connects two monosaccharides. In maltose, a glycosidic bond forms between the —OH groups of carbons 1 and 4 of two α-D-glucose molecules with a loss of a water molecule. The glycosidic bond in maltose is designated as an α-1,4 linkage to show that an alpha —OH on carbon 1 is joined to carbon 4 of the second glucose molecule. Because the second glucose molecule still has a free —OH group on carbon 1, it can form an open

SUGAR AND SWEETENERS

Add a tablespoon of sugar to a glass of water and stir. Taste. Add more sugar, stir, and taste. If you have other carbohydrates such as fructose, honey, cornstarch, arrowroot, or flour, add some of each to separate glasses of water and stir. If you have some artificial sweeteners, add a few drops of the sweetener or a package, if solid, to a glass of water. Taste each.

QUESTIONS

1. Which substance is the most soluble in water?
2. Place the substances in order from the one that tastes least sweet to the sweetest.
3. How does your list compare to the values in Table 13.2?
4. How does the sweetness of sucrose compare with that of the artificial sweeteners?
5. Check the labels of food products in your kitchen. Look for sugars such as sucrose or fructose, or artificial sweeteners such as aspartame or sucralose on the label. How many grams of sugar are in a serving of the food?

chain, which allows maltose to form both α and β isomers. The open chain provides an aldehyde group that can be oxidized, making maltose a reducing sugar.

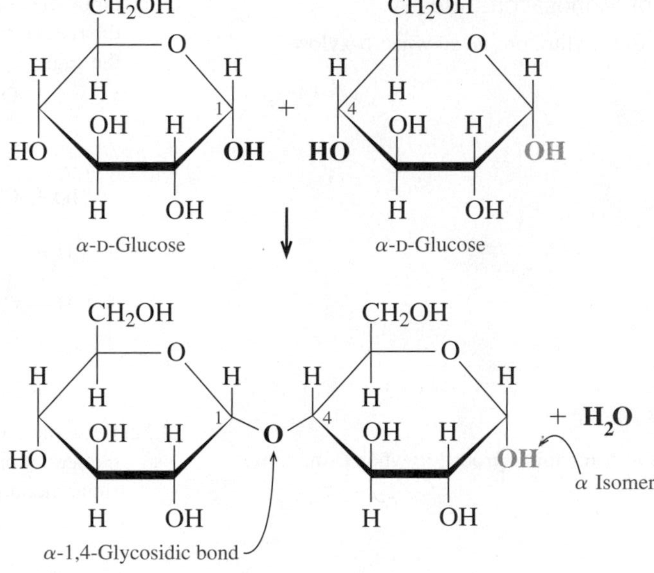

α-Maltose, a disaccharide

Lactose, milk sugar, is a disaccharide found in milk and milk products (see Figure 13.3). The bond in lactose is a β-1,4-glycosidic bond because the —OH on carbon 1 of β-D-galactose forms a glycosidic bond with the —OH group on carbon 4 of a D-glucose molecule. Because D-glucose still has a free —OH group on carbon 1, it can form an open chain, which allows lactose to form both α and β isomers. The open chain provides an aldehyde group that can be oxidized, making lactose a reducing sugar.

Lactose makes up 6–8% of human milk and about 4–5% of cow's milk, and it is used in products that attempt to duplicate mother's milk. Some people do not produce sufficient quantities of the enzyme lactase needed to hydrolyze lactose. Then lactose remains undigested, which can cause abdominal cramps and diarrhea. In some commercial milk products, lactase is added to break down lactose.

Sucrose consists of an α-D-glucose and a β-D-fructose molecule joined by an α,β-1,2-glycosidic bond (see Figure 13.4). Unlike maltose and lactose, the glycosidic bond in sucrose is between carbon 1 of glucose and carbon 2 of fructose. Thus, sucrose

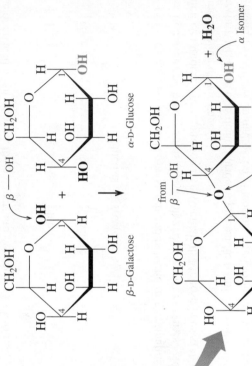

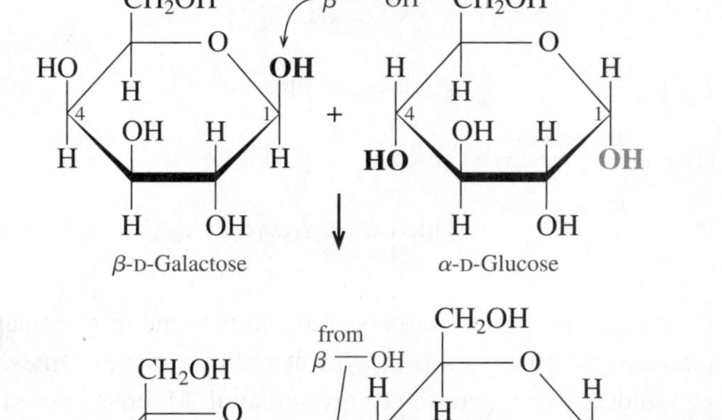

α-Lactose, a disaccharide

FIGURE 13.3 Lactose, a disaccharide found in milk and milk products, contains galactose and glucose.

Q What type of glycosidic bond links galactose and glucose in lactose?

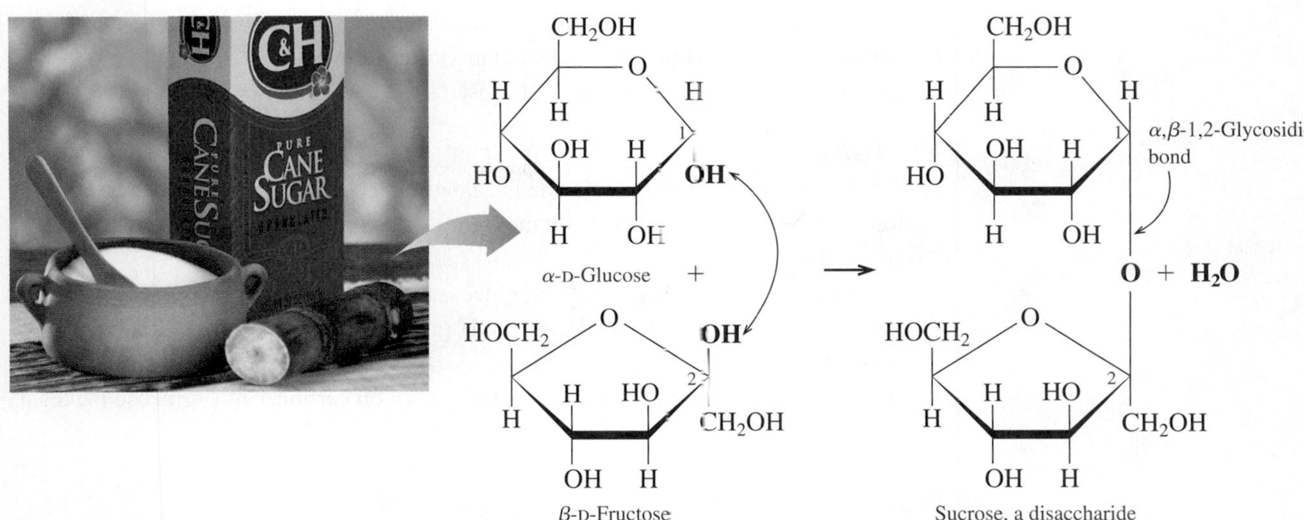

FIGURE 13.4 Sucrose, a disaccharide obtained from sugar beets and sugar cane, contains glucose and fructose.
Q Why is sucrose a nonreducing sugar?

cannot form an open chain and cannot be oxidized. Sucrose cannot react with Benedict's reagent and is not a reducing sugar.

The sugar we use to sweeten our cereal, coffee, or tea is sucrose. Most of the sucrose for table sugar comes from sugar cane (20% by mass) or sugar beets (15% by mass). Both the raw and refined forms of sugar are sucrose. Some estimates in 2003 indicate that each person in the United States consumes an average of 68 kg (150 lb) of sucrose every year, either by itself or in a variety of food products.

CONCEPT CHECK 13.4

Glycosidic Bonds

Why is the glycosidic bond in maltose an α-1,4-glycosidic bond, but in lactose, it is a β-1,4-glycosidic bond?

ANSWER

To produce an α-1,4-glycosidic bond of maltose, the —OH in the bond must come from α-D-glucose that had the —OH on carbon 1 drawn down. To produce a β-1,4-glycosidic bond of lactose, the —OH in the bond comes from β-D-galactose that had the —OH on carbon 1 drawn up.

SAMPLE PROBLEM 13.4

Glycosidic Bonds in Disaccharides

Melibiose is a disaccharide that is 30 times sweeter than sucrose. Melibiose has the following Haworth structure:

Melibiose

Career Focus

PHLEBOTOMIST

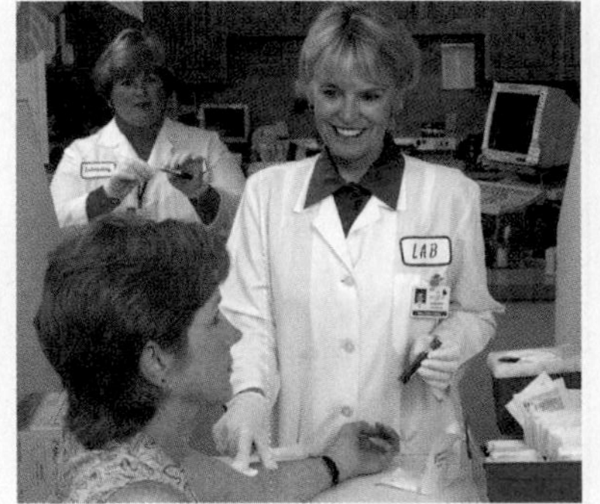

As part of the medical team, phlebotomists collect and process blood for laboratory tests. They work directly with patients, calming them if necessary prior to the collection of blood. Phlebotomists are trained to collect blood in a safe manner and to provide patient care if fainting occurs. Blood is drawn through venipuncture methods such as syringe, vacutainer, blood culture, and finger stick. They also prepare patients for procedures such as glucose tolerance tests. In the preparation of specimens for analysis, a phlebotomist determines media, inoculation method, and reagents for culture setup.

a. What are the monosaccharide units in melibiose?
b. What type of glycosidic bond links the monosaccharides?
c. Identify the structure as α- or β-melibiose.

SOLUTION

a. The monosaccharide on the left side is galactose because the —OH on carbon 4 is drawn up. The monosaccharide on the right is glucose because the —OH on carbon 4 is down.
b. The monosaccharides galactose and glucose are connected by an α-1,6-glycosidic bond because it connects the alpha —OH group on carbon 1 in galactose with carbon 6 in glucose.
c. The downward position of the free —OH group on carbon 1 of D-glucose makes it α-melibiose.

STUDY CHECK 13.4

Cellobiose is a disaccharide composed of two D-glucose molecules linked by a β-1,4-glycosidic linkage. Draw a Haworth structure for β-cellobiose.

QUESTIONS AND PROBLEMS

Disaccharides

13.33 For each of the following give the monosaccharide units produced by hydrolysis, the type of glycosidic bond, and the name of the disaccharide including α or β:

a.

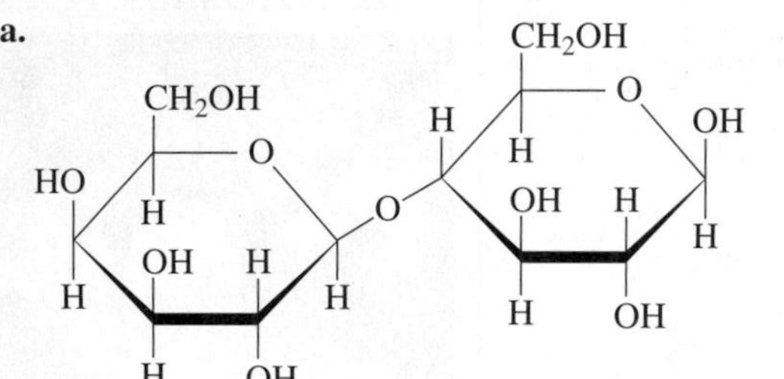

b.

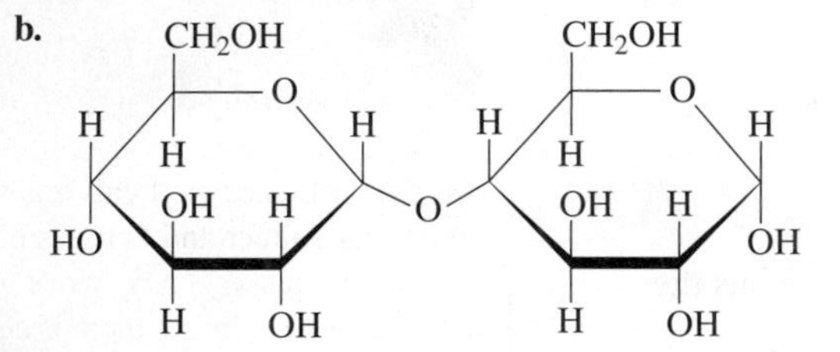

13.34 For each of the following give the monosaccharide units produced by hydrolysis, the type of glycosidic bond, and the name of the disaccharide including α or β:

a.

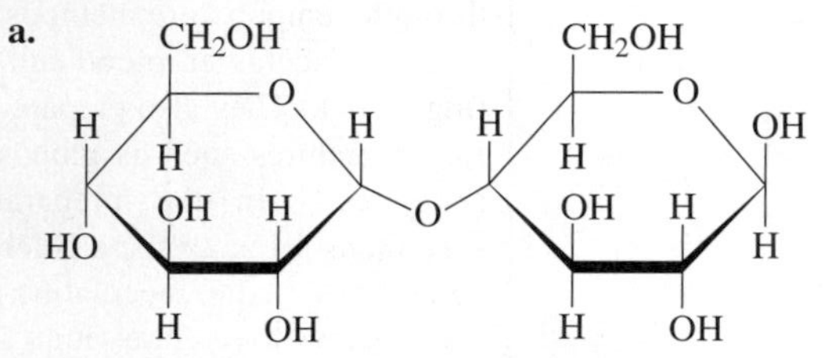

b.

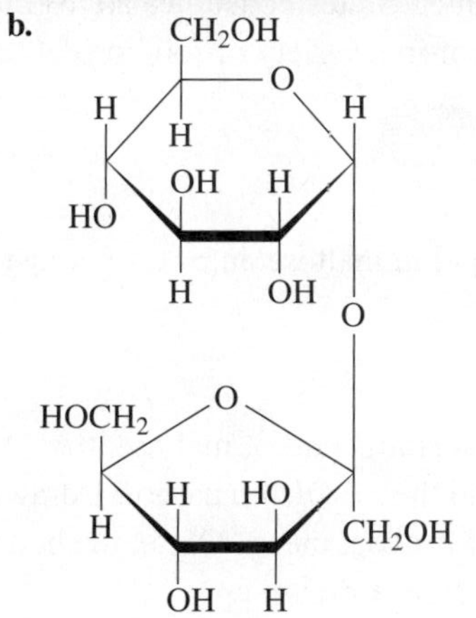

13.35 Indicate whether the disaccharides in Problem 13.33 will undergo oxidation.

13.36 Indicate whether the disaccharides in Problem 13.34 will undergo oxidation.

13.37 Identify the disaccharide that fits each of the following descriptions:
a. ordinary table sugar
b. found in milk and milk products
c. also called malt sugar
d. hydrolysis gives galactose and glucose

13.38 Identify the disaccharide that fits each of the following descriptions:
a. not a reducing sugar
b. composed of two glucose units
c. also called milk sugar
d. hydrolysis gives glucose and fructose

Chemistry Link to Health

HOW SWEET IS MY SWEETENER?

Although many of the monosaccharides and disaccharides taste sweet, they differ considerably in their degree of sweetness. Dietetic foods contain sweeteners that are noncarbohydrate or carbohydrates that are sweeter than sucrose. Some examples of sweeteners compared with sucrose are shown in Table 13.2.

Sucralose is made from sucrose by replacing some of the hydroxyl groups with chlorine atoms.

Sucralose

Aspartame, which is marketed as Nutra-Sweet, is used in a large number of sugar-free products. It is a noncarbohydrate sweetener made of aspartic acid and a methyl ester of phenylalanine. It does have some caloric value, but it is so sweet that a very small quantity is needed. However, phenylalanine, one of the breakdown products, poses a danger to anyone who cannot metabolize it properly, a condition called phenylketonuria (PKU).

Methyl ester

From aspartic acid — From phenylalanine

Aspartame (Nutra-Sweet)

Another artificial sweetener, Neotame, is a modification of the aspartame structure. The addition of a large alkyl group to the amine group prevents enzymes from breaking the amide bond between aspartic acid and phenylalanine. Thus, phenylalanine is not produced when Neotame is used as a sweetener. Very small amounts of Neotame are needed because it is about 10 000 times sweeter than sucrose.

Large alkyl group to modify Aspartame

Neotame

TABLE 13.2 Relative Sweetness of Sugars and Artificial Sweeteners

	Sweetness Relative to Sucrose
Monosaccharides	
Galactose	30
Glucose	75
Fructose	175
Disaccharides	
Lactose	16
Maltose	33
Sucrose	100 ⟵ reference standard
Sugar Alcohols	
Sorbitol	60
Maltitol	80
Xylitol	100
Artificial Sweeteners (Noncarbohydrate)	
Aspartame	18 000
Saccharin	45 000
Sucralose	60 000
Neotame	1 000 000

Saccharin has been used as a noncarbohydrate artificial sweetener for the past 25 years. The use of saccharin has been banned in Canada because studies indicate that it may cause bladder tumors. However, it is still approved for use by the FDA in the United States.

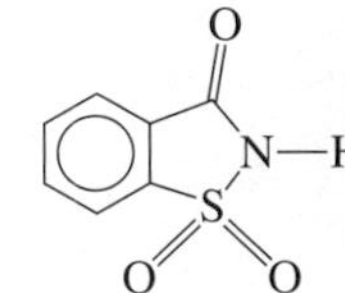

Saccharin

Artificial sweeteners are used as sugar substituents.

Chemistry Link to Health

BLOOD TYPES AND CARBOHYDRATES

Every individual's blood can be typed as one of four blood groups: A, B, AB, or O. Although there is some variation among ethnic groups in the United States, the incidence of blood types in the general population is about 43% O, 40% A, 12% B, and 5% AB.

The blood types A, B, and O are determined by monosaccharides attached to the surface of red blood cells. The blood types, A, B, and O, all contain the monosaccharides *N*-acetylglucosamine, galactose, and fucose. In type A, galactose is bonded to *N*-acetylgalactosamine. In type B, the galactose is bonded to a second galactose. In type AB, the monosaccharides of both blood types A and B are found. The structures of these monosaccharides are as follows:

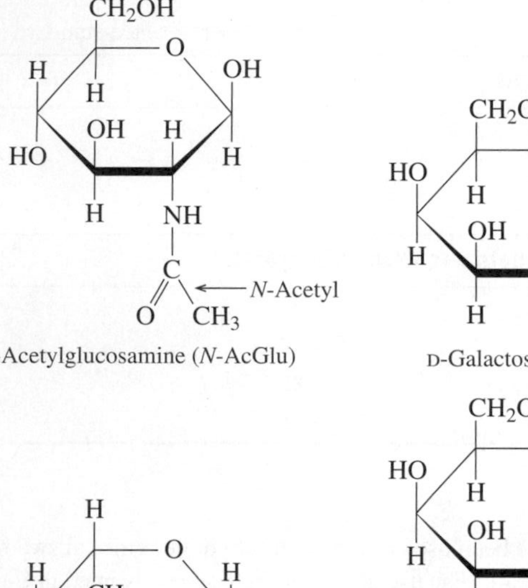

N-Acetylglucosamine (*N*-AcGlu)

L-Fucose (Fuc)

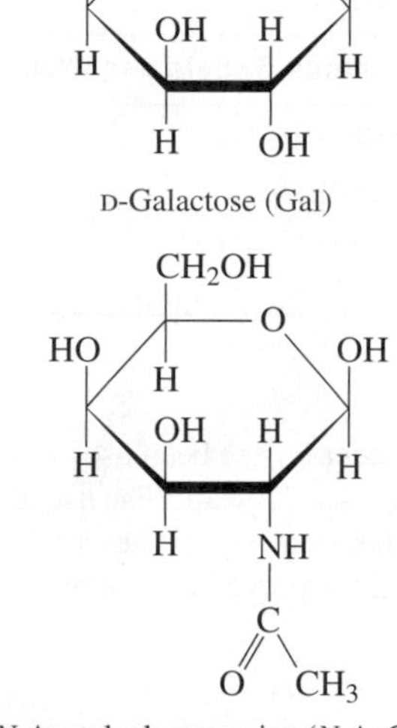

D-Galactose (Gal)

N-Acetylgalactosamine (*N*-AcGal)

Terminal saccharide that determines blood type

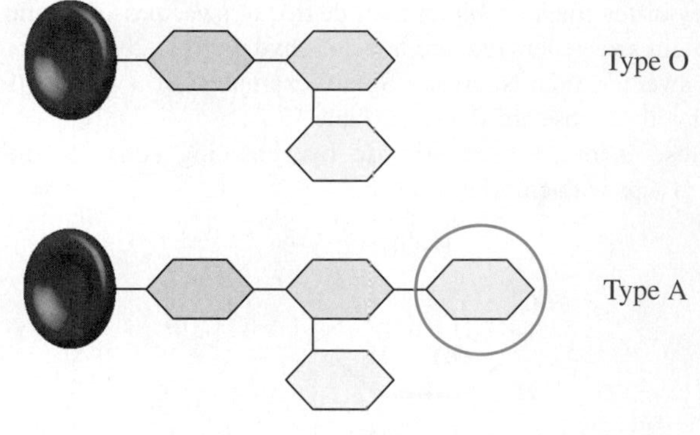

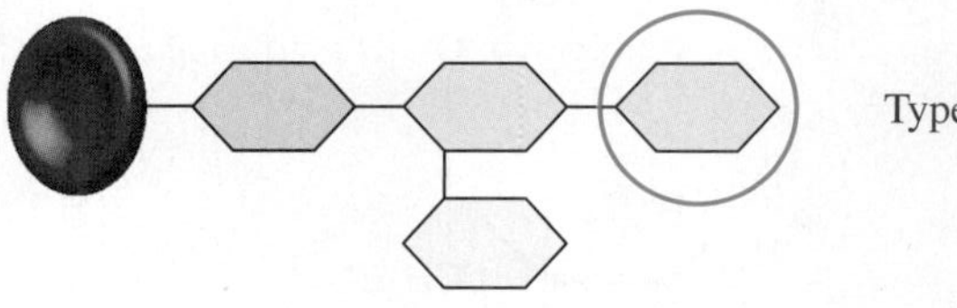

TABLE 13.3 Compatibility of Blood Groups

Blood Type	Can Receive	Produce Antibodies Against
A	A, O	B, AB
B	B, O	A, AB
AB*	A, B, AB, O	None
O**	O	A, B, AB

*AB Universal recipient
**O Universal donor

A person with type A blood produces antibodies against type B, whereas a person with type B blood produces antibodies against type A. A person with type AB blood produces no antibodies, whereas a person with type O blood produces antibodies against both types A and B blood. Thus, if a person with type A blood receives a transfusion of type B blood, factors in the recipient's blood will cause the donor's red blood cells to clump together, or agglutinate.

People with type O blood can donate to individuals with all blood types; they are universal donors. However, a type O person can receive only type O blood. People with type AB blood can receive all blood types because they do not produce antibodies for types A, B, and O; they are universal recipients. Table 13.3 summarizes the compatibility of blood groups for transfusion.

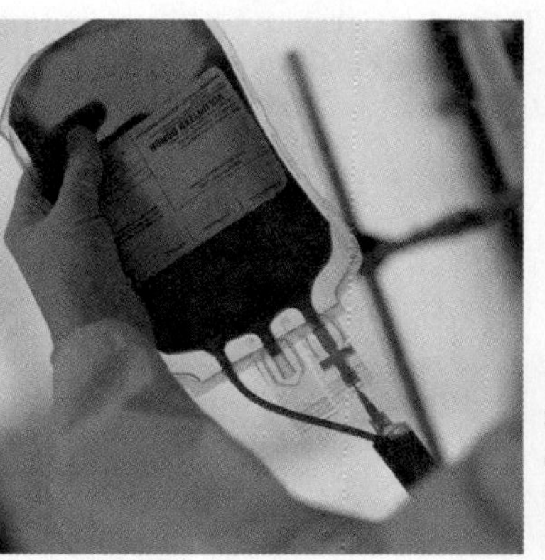

Blood from a donor is screened to make an exact match with the blood type of the recipient.

13.6 Polysaccharides

LEARNING GOAL

Describe the structural features of amylose, amylopectin, glycogen, and cellulose.

A polysaccharide is a polymer of many monosaccharides joined together. Four important polysaccharides—*amylose*, *amylopectin*, *cellulose*, and *glycogen*—are all polymers of D-glucose that differ only in the type of glycosidic bonds and the amount of branching in the molecule.

Starch, a storage form of glucose in plants, is found as insoluble granules in rice, wheat, potatoes, beans, and cereals. Starch is composed of two kinds of polysaccharides, amylose and amylopectin. **Amylose**, which makes up about 20% of starch, consists of 250–4000 α-D-glucose molecules connected by α-1,4-glycosidic bonds in a continuous chain. Sometimes called a straight-chain polymer, polymers of amylose are actually coiled in helical fashion (see Figure 13.5a).

Amylopectin, which makes up as much as 80% of starch, is a branched-chain polysaccharide. Like amylose, the glucose molecules are connected by α-1,4-glycosidic bonds. However, at about every 25 glucose units, there is a branch of glucose molecules attached by an α-1,6-glycosidic bond between carbon 1 of the branch and carbon 6 in the main chain (see Figure 13.5b).

Starches hydrolyze easily in water and acid to give dextrins, which then hydrolyze to maltose and finally glucose. In our bodies, these complex carbohydrates are digested by the enzymes amylase (in saliva) and maltase (in the intestine). The glucose obtained provides about 50% of our nutritional calories.

$$\text{Amylose, amylopectin} \xrightarrow{H^+ \text{ or amylase}} \text{dextrins} \xrightarrow{H^+ \text{ or amylase}} \text{maltose} \xrightarrow{H^+ \text{ or maltase}} \text{many D-glucose units}$$

Glycogen, or animal starch, is a polymer of glucose that is stored in the liver and muscle of animals. It is hydrolyzed in our cells at a rate that maintains the blood level of glucose and provides energy between meals. The structure of glycogen is very similar to that of amylopectin found in plants, except that glycogen is more highly branched. In glycogen, the glucose units are joined by α-1,4-glycosidic bonds, and branches occurring about every 10–15 glucose units are attached by α-1,6-glycosidic bonds.

Cellulose is the major structural material of wood and plants. Cotton is almost pure cellulose. In cellulose, glucose molecules form a long unbranched chain similar to that of amylose. However, the glucose units in cellulose are linked by β-1,4-glycosidic bonds. The cellulose chains do not form coils like amylose but are aligned in parallel rows that are held in place by hydrogen bonds between hydroxyl groups in adjacent chains, making cellulose insoluble in water. This gives a rigid structure to the cell walls in wood and fiber that is more resistant to hydrolysis than are the starches (see Figure 13.6).

Humans have enzymes called α-amylase in saliva and pancreatic juices that hydrolyze the α-1,4-glycosidic bonds of starches, but not the β-1,4-glycosidic bonds of cellulose. Thus, humans cannot digest cellulose. Animals such as horses, cows, and goats can obtain glucose from cellulose because their digestive systems contain bacteria that provide enzymes such as cellulase to hydrolyze β-1,4-glycosidic bonds.

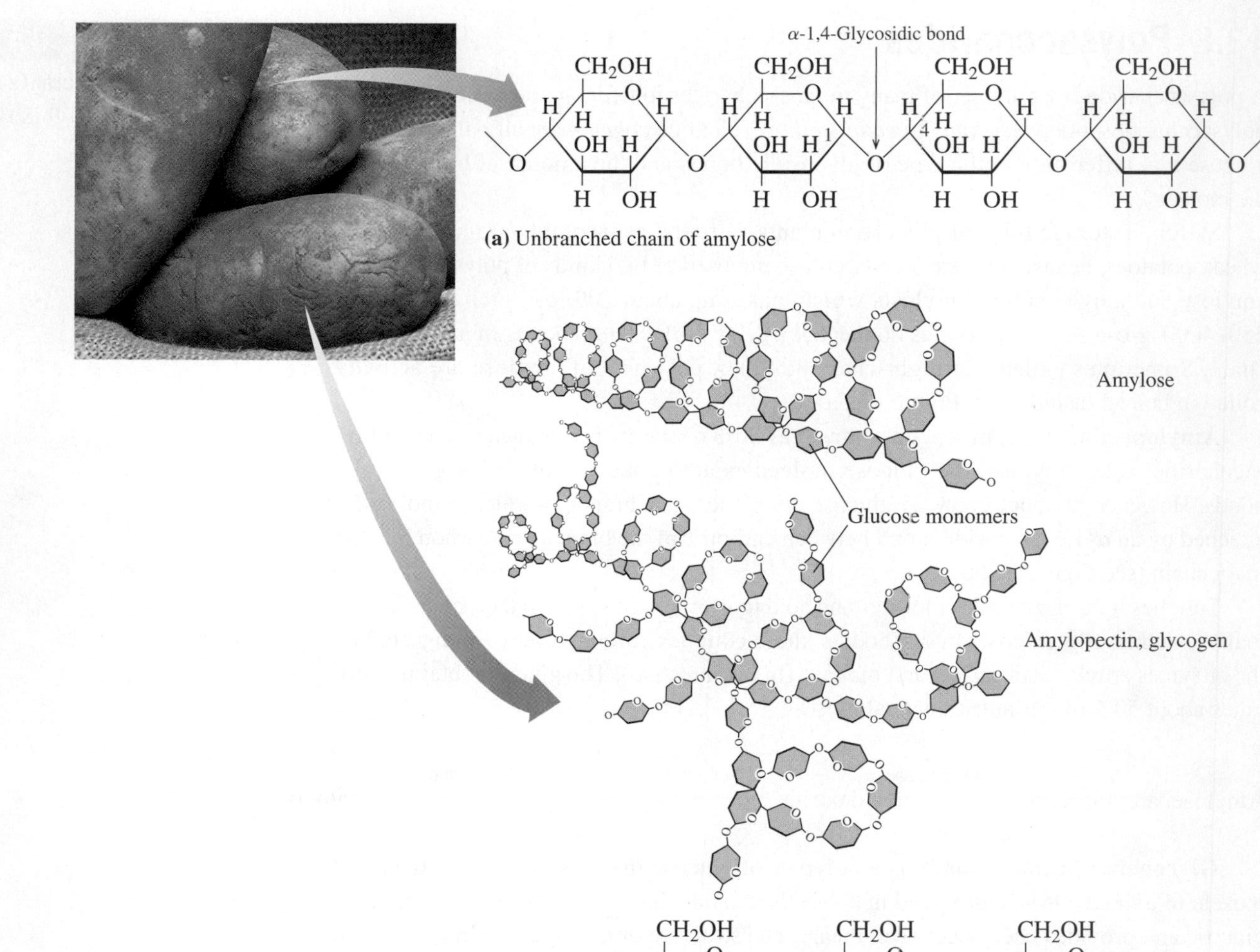

FIGURE 13.5 The structure of **(a)** amylose is a straight-chain polysaccharide of glucose units, and the structure of **(b)** amylopectin is a branched chain of glucose.
Q What are the two types of glycosidic bonds that link glucose molecules in amylopectin?

SAMPLE PROBLEM 13.5

Structures of Polysaccharides

Identify the polysaccharide described by each of the following:

a. a polysaccharide that is stored in the liver and muscle tissues
b. an unbranched polysaccharide containing β-1,4-glycosidic bonds
c. a starch containing α-1,4- and α-1,6-glycosidic bonds

SOLUTION

a. glycogen **b.** cellulose **c.** amylopectin, glycogen

STUDY CHECK 13.5

Cellulose and amylose are both unbranched glucose polymers. How do they differ?

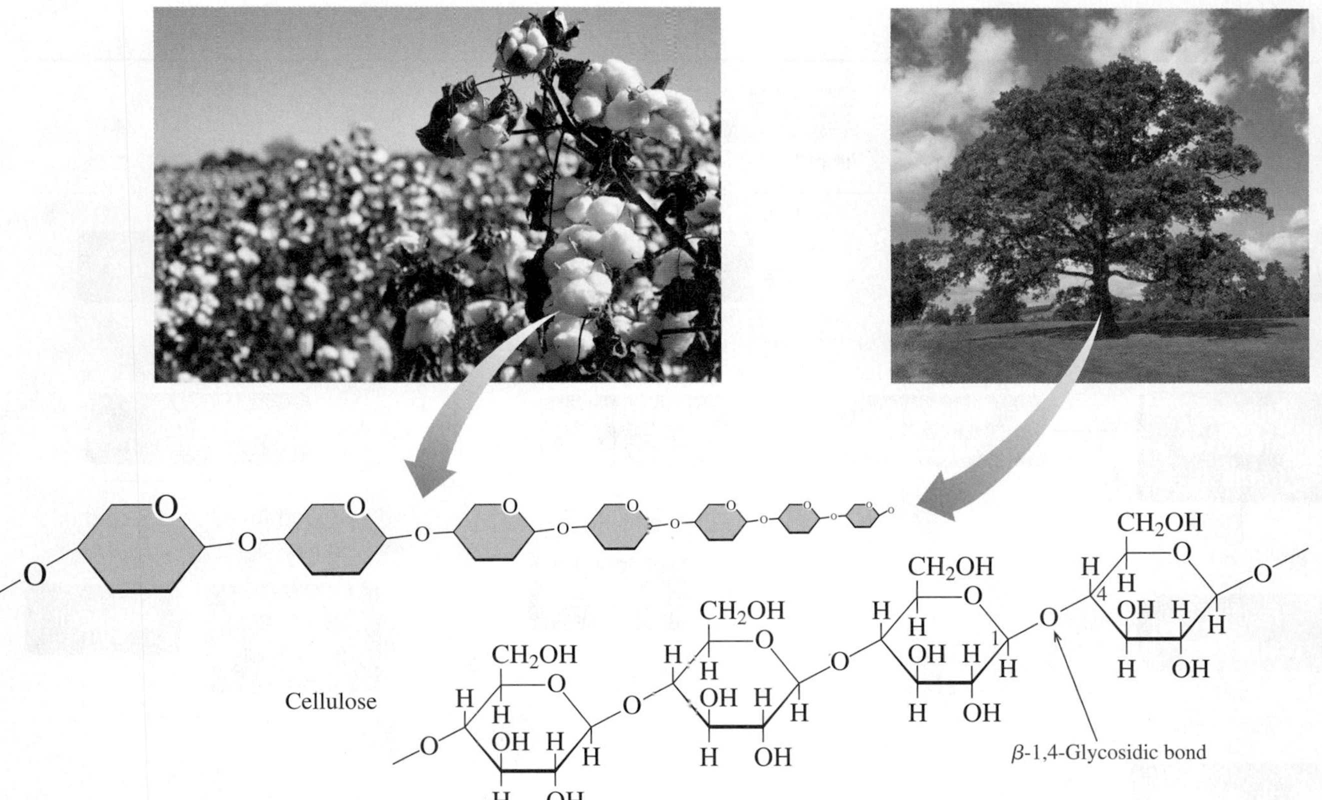

FIGURE 13.6 The polysaccharide cellulose is composed of glucose units connected by β-1,4-glycosidic bonds.

Q Why are humans unable to digest cellulose?

Explore Your World

POLYSACCHARIDES

Read the nutrition label on a box of crackers, cereal, bread, chips, or pasta. The major ingredient in crackers is flour, which contains starch. Chew on a single cracker for 4–5 minutes. Note how the taste changes as you chew the cracker. An enzyme (amylase) in your saliva breaks apart the bonds in starch.

QUESTIONS

1. How are carbohydrates listed on the label?
2. What other carbohydrates are listed?
3. How did the taste of the cracker change during the time that you chewed it?
4. What happens to the starches in the cracker as the amylase enzyme in your saliva reacts with the amylose and amylopectin?

QUESTIONS AND PROBLEMS

Polysaccharides

13.39 Describe the similarities and differences in the following:
a. amylose and amylopectin
b. amylopectin and glycogen

13.40 Describe the similarities and differences in the following:
a. amylose and cellulose **b.** cellulose and glycogen

13.41 Give the name of one or more polysaccharides that matches each of the following descriptions:
a. not digestible by humans
b. the storage form of carbohydrates in plants
c. contains only α-1,4-glycosidic bonds
d. the most highly branched polysaccharide

13.42 Give the name of one or more polysaccharides that matches each of the following descriptions:
a. the storage form of carbohydrates in animals
b. contains only β-1,4-glycosidic bonds
c. contains both α-1,4- and α-1,6-glycosidic bonds
d. produces maltose during digestion

CONCEPT MAP

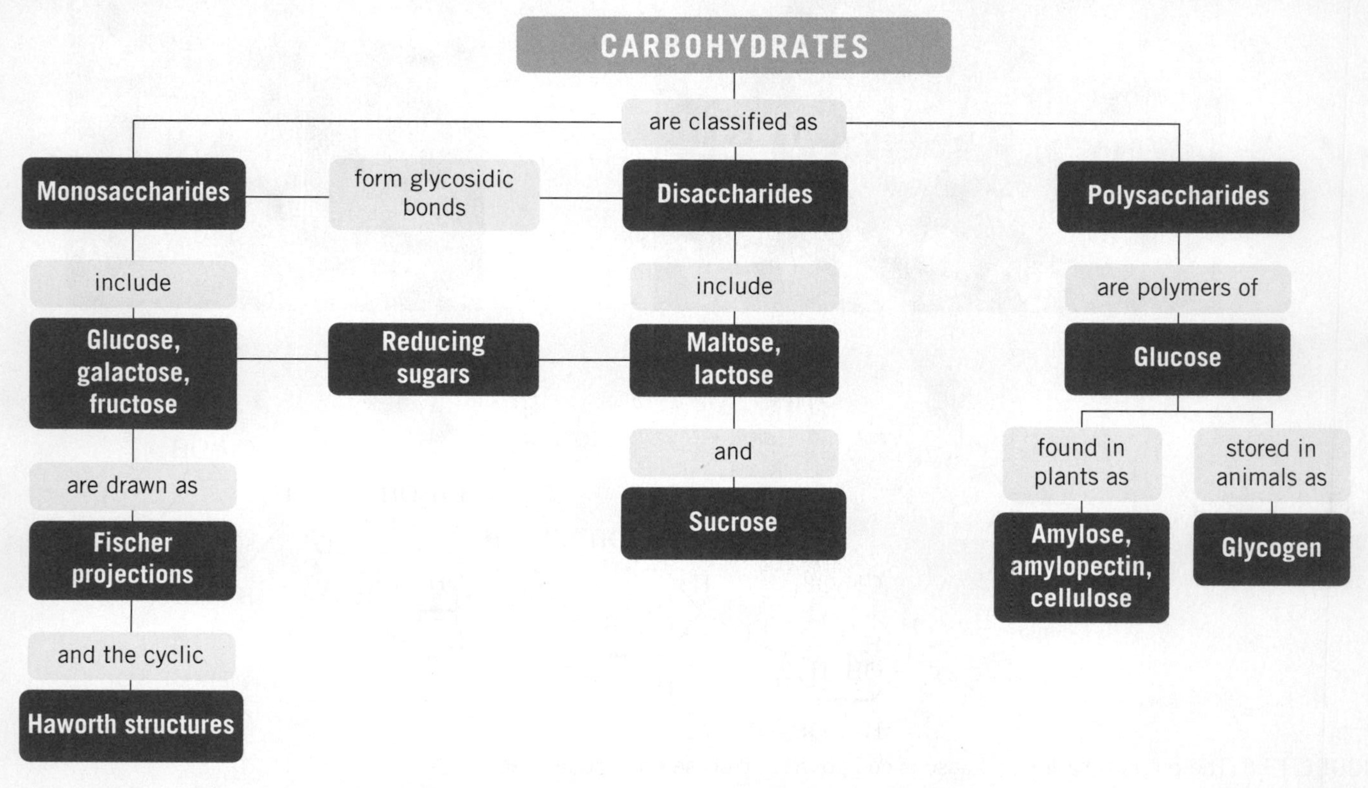

CHAPTER REVIEW

13.1 Carbohydrates

Learning Goal: Classify a monosaccharide as an aldose or a ketose, and indicate the number of carbon atoms.

Carbohydrates are classified as monosaccharides (simple sugars), disaccharides (two monosaccharide units), or polysaccharides (many monosaccharide units). Monosaccharides are polyhydroxy aldehydes (aldoses) or ketones (ketoses). Monosaccharides are also classified by their number of carbon atoms: triose, tetrose, pentose, or hexose.

Aldehyde

Erythrose, a polyhydroxy aldehyde

13.2 Fischer Projections of Monosaccharides

Learning Goal: Use Fischer projections to draw the D or L isomers of glucose, galactose, and fructose.

In a Fischer projection, the prefixes D and L are used to distinguish between mirror images. In a D isomer, the —OH is on the right of the chiral carbon farthest from the carbonyl carbon; it is on the left in the L isomer. Important monosaccharides are the aldohexoses, glucose and galactose, and the ketohexose, fructose.

L-Glucose D-Glucose

13.3 Haworth Structures of Monosaccharides

Learning Goal: Draw and identify the Haworth structures of monosaccharides.

The predominant form of monosaccharides is a ring of five or six atoms. The cyclic structure forms when an —OH (usually the one on carbon 5 in hexoses) reacts with the carbonyl group of the same molecule. The formation of a new hydroxyl group on carbon 1 gives α and β isomers of the cyclic monosaccharide.

α-D-Glucose

13.4 Chemical Properties of Monosaccharides

Learning Goal: Identify the products of oxidation or reduction of monosaccharides; determine whether a carbohydrate is a reducing sugar.

The aldehyde group in an aldose can be oxidized to a carboxylic acid, while the carbonyl group in an aldose or a ketose can be reduced to give a hydroxyl group. Monosaccharides that are reducing sugars have an aldehyde group in the open chain that can be oxidized.

Oxidized

D-Gluconic acid

13.5 Disaccharides

Learning Goal: Describe the monosaccharide units and linkages in disaccharides.

Disaccharides are two monosaccharide units joined together by a glycosidic bond. In the common disaccharides maltose, lactose, and sucrose, there is at least one glucose unit.

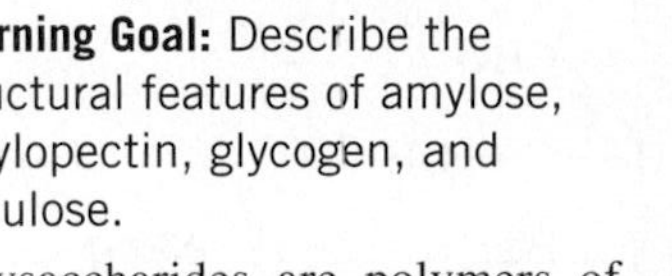

Sucrose, a disaccharide

13.6 Polysaccharides

Learning Goal: Describe the structural features of amylose, amylopectin, glycogen, and cellulose.

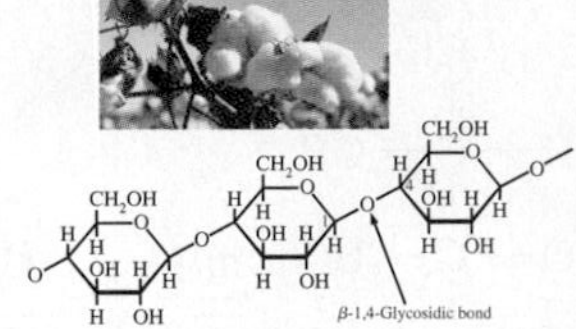

Polysaccharides are polymers of monosaccharide units. Amylose is an unbranched chain of glucose with α-1,4-glycosidic bonds, and amylopectin is a branched polymer of glucose with α-1,4- and α-1,6-glycosidic bonds. Glycogen is similar to amylopectin with more branching. Cellulose is also a polymer of glucose, but in cellulose, the glycosidic bonds are β-1,4- bonds.

Summary of Carbohydrates

Carbohydrate	Food Sources	
Monosaccharides		
Glucose	Fruit juices, honey, corn syrup	
Galactose	Lactose hydrolysis	
Fructose	Fruit juices, honey, sucrose hydrolysis	
Disaccharides		**Monosaccharides**
Maltose	Germinating grains, starch hydrolysis	Glucose + glucose
Lactose	Milk, yogurt, ice cream	Glucose + galactose
Sucrose	Sugar cane, sugar beets	Glucose + fructose
Polysaccharides		
Amylose	Rice, wheat, grains, cereals	Unbranched polymer of glucose joined by α-1,4-glycosidic bonds
Amylopectin	Rice, wheat, grains, cereals	Branched polymer of glucose joined by α-1,4- and α-1,6-glycosidic bonds
Glycogen	Liver, muscles	Highly branched polymer of glucose joined by α-1,4- and α-1,6-glycosidic bonds
Cellulose	Plant fiber, bran, beans, celery	Unbranched polymer of glucose joined by β-1,4-glycosidic bonds

Summary of Reactions

Formation of Disaccharides

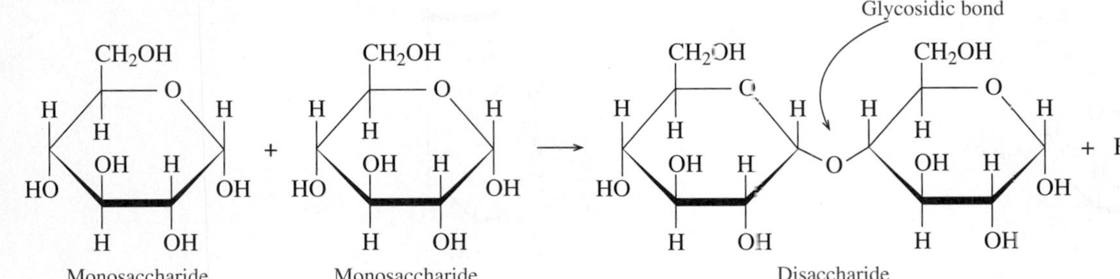

Oxidation and Reduction of Monosaccharides

CH_2OH / H—C—OH / HO—C—H / H—C—OH / H—C—OH / CH_2OH
D-Glucitol

$\xleftarrow{\text{Reduction}}$

CHO / H—C—OH / HO—C—H / H—C—OH / H—C—OH / CH_2OH
D-Glucose

$\xrightarrow{\text{Oxidation}}$

COOH / H—C—OH / HO—C—H / H—C—OH / H—C—OH / CH_2OH
D-Gluconic acid

Hydrolysis of Disaccharides

$$\text{Sucrose} + H_2O \xrightarrow[\text{sucrase}]{H^+ \text{ or}} \text{glucose} + \text{fructose}$$

$$\text{Lactose} + H_2O \xrightarrow[\text{lactase}]{H^+ \text{ or}} \text{glucose} + \text{galactose}$$

$$\text{Maltose} + H_2O \xrightarrow[\text{maltase}]{H^+ \text{ or}} \text{glucose} + \text{glucose}$$

Hydrolysis of Polysaccharides

$$\text{Amylose, amylopectin} \xrightarrow[\text{enzymes}]{H^+ \text{ or}} \text{many D-glucose units}$$

Key Terms

aldose A monosaccharide that contains an aldehyde group.

amylopectin A branched-chain polymer of starch composed of glucose units joined by α-1,4- and α-1,6-glycosidic bonds.

amylose An unbranched polymer of starch composed of glucose units joined by α-1,4-glycosidic bonds.

carbohydrate A simple or complex sugar composed of carbon, hydrogen, and oxygen.

cellulose An unbranched polysaccharide composed of glucose units linked by β-1,4-glycosidic bonds that cannot be hydrolyzed by the human digestive system.

disaccharides Carbohydrates composed of two monosaccharides joined by a glycosidic bond.

fructose A monosaccharide that is also called levulose and fruit sugar and is found in honey and fruit juices; it is combined with glucose in sucrose.

galactose A monosaccharide that occurs combined with glucose in lactose.

glucose An aldohexose found in fruits, vegetables, corn syrup, and honey that is also known as blood sugar and dextrose. The most prevalent monosaccharide in the diet. Most polysaccharides are polymers of glucose.

glycogen A polysaccharide formed in the liver and muscles for the storage of glucose as an energy reserve. It is composed of glucose in a highly branched polymer joined by α-1,4- and α-1,6-glycosidic bonds.

glycosidic bond The bond that forms when the hydroxyl group of one monosaccharide reacts with the hydroxyl group of another monosaccharide. It is the type of bond that links monosaccharide units in di- or polysaccharides.

Haworth structure The ring structure of a monosaccharide.

ketose A monosaccharide that contains a ketone group.

lactose A disaccharide consisting of glucose and galactose found in milk and milk products.

maltose A disaccharide consisting of two glucose units; it is obtained from the hydrolysis of starch and is found in germinating grains.

monosaccharide A polyhydroxy compound that contains an aldehyde or ketone group.

polysaccharides Polymers of many monosaccharide units, usually glucose. Polysaccharides differ in the types of glycosidic bonds and the amount of branching in the polymer.

reducing sugar A carbohydrate with an aldehyde group capable of reducing the Cu^{2+} in Benedict's reagent.

sucrose A disaccharide composed of glucose and fructose; a nonreducing sugar, commonly called table sugar or sugar.

Understanding the Concepts

13.43 Isomaltose, obtained from the breakdown of starch, has the following Haworth structure:

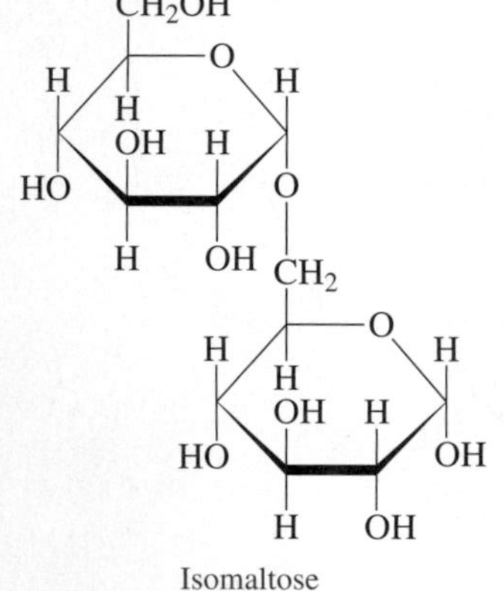

Isomaltose

a. Is isomaltose a mono-, di-, or polysaccharide?
b. What are the monosaccharides in isomaltose?
c. What is the glycosidic link in isomaltose?
d. Is this the α or β isomer of isomaltose?
e. Is isomaltose a reducing sugar?

13.44 Sophorose, a carbohydrate found in certain types of beans, has the following Haworth structure:

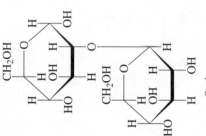

Sophorose

a. Is sophorose a mono-, di-, or polysaccharide?
b. What are the monosaccharides in sophorose?
c. What is the glycosidic link in sophorose?
d. Is this the α or β isomer of sophorose?
e. Is sophorose a reducing sugar?

13.45 Melezitose, a substance secreted by insects, has the following Haworth structure:

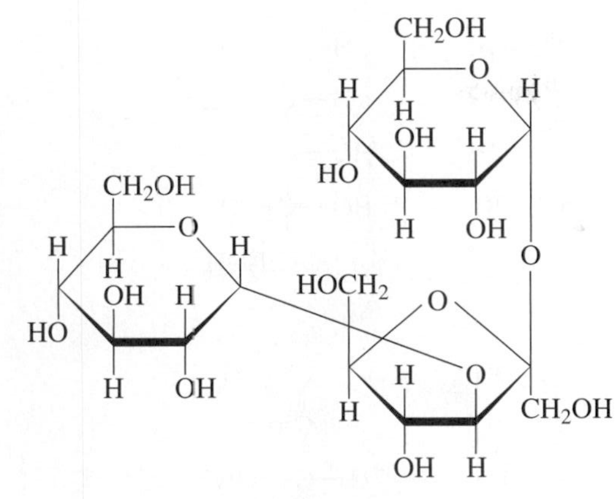

Melezitose

a. Is melezitose a mono-, di-, or trisaccharide?
b. What ketohexose and aldohexose are present in melezitose?
c. Is melezitose a reducing sugar?

13.46 What are the disaccharides and polysaccharides present in each of the following?

(a) (b)

(c)

(d)

Additional Questions and Problems

For instructor-assigned homework, go to **www.masteringchemistry.com**.

13.47 What are the differences in the Fischer projections of D-fructose and D-galactose?

13.48 What are the differences in the Fischer projections of D-glucose and D-fructose?

13.49 What are the differences in the Fischer projections of D-galactose and L-galactose?

13.50 What are the differences in the Haworth structures of α-D-glucose and β-D-glucose?

13.51 The sugar D-gulose is a sweet-tasting syrup.

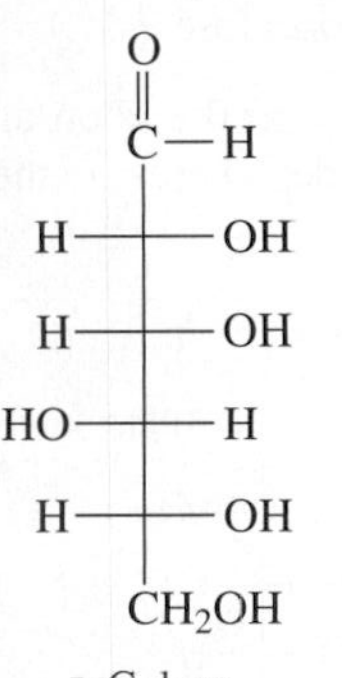

D-Gulose

a. Draw the Fisher projection for L-gulose.
b. Draw the Haworth structures for α- and β-D-gulose.

13.52 Consider the open-chain structure for D-gulose in Question 13.51.
a. Draw the Fischer projection and name the product formed by the reduction of D-gulose.
b. Draw the Fischer projection and name the product formed by the oxidation of D-gulose.

13.53 D-Sorbitol, a sweetener found in seaweed and berries, contains only hydroxyl functional groups. When D-sorbitol is oxidized, it forms D-glucose. Draw the Fischer projection of D-sorbitol.

13.54 Raffinose is a trisaccharide found in Australian manna and in cottonseed meal. It is composed of three different monosaccharides. Identify the monosaccharides in raffinose.

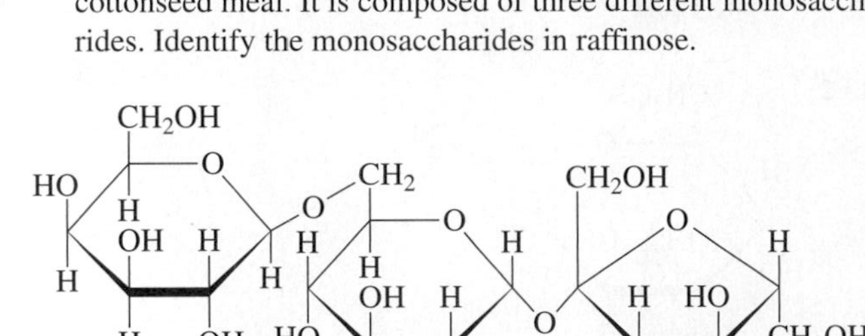
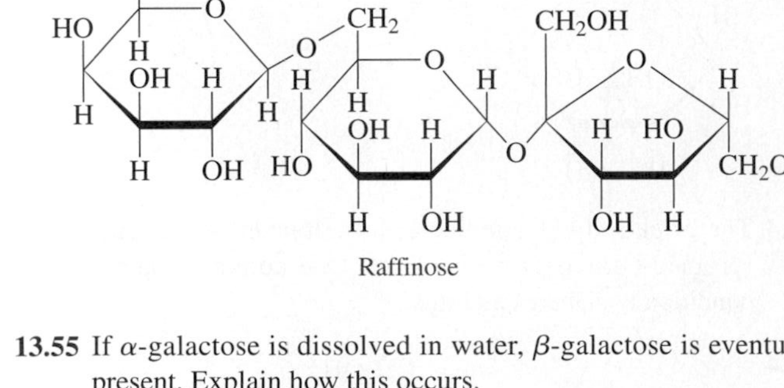

Raffinose

13.55 If α-galactose is dissolved in water, β-galactose is eventually present. Explain how this occurs.

13.56 Why are lactose and maltose reducing sugars, but sucrose is not?

Challenge Questions

13.57 α-Cellobiose is a disaccharide obtained from the hydrolysis of cellulose. It is quite similar to maltose except it has a β-1,4-glycosidic bond. Draw the Haworth structure of α-cellobiose.

13.58 The disaccharide trehalose found in mushrooms is composed of two α-D-glucose molecules joined by an α-1,1-glycosidic bond. Draw the Haworth structure of trehalose.

13.59 Gentiobiose is found in saffron.

a. Gentiobiose contains two glucose molecules linked by a β-1,6-glycosidic bond. Draw the Haworth structure of β-gentiobiose.

b. Is gentiobiose a reducing sugar? Explain.

13.60 Identify the open-chain formula **A–D** that matches each of the following:

a. the L isomer of mannose **b.** a ketopentose

c. an aldopentose **d.** a ketohexose

A: CHO; H—C—OH; HO—C—H; H—C—OH; CH_2OH

B: CHO; H—C—OH; H—C—OH; HO—C—H; HO—C—H; CH_2OH

C: CH_2OH; C=O; HO—C—H; H—C—OH; CH_2OH

D: CH_2OH; C=O; H—C—OH; HO—C—H; HO—C—H; CH_2OH

Answers

Answers to Study Checks

13.1 To draw the mirror image, all the —OH groups on the chiral carbon atoms are written on the opposite side. L-Ribulose has the following Fischer projection:

CH_2OH; C=O; HO—H; HO—H; CH_2OH

13.2 [Haworth structure: CH_2OH, O, H, H, OH, OH, HO, OH, H, H]

13.3 The brick-red color indicates a high level of reducing sugar (probably glucose) in the urine. One common cause of this condition is diabetes mellitus.

13.4

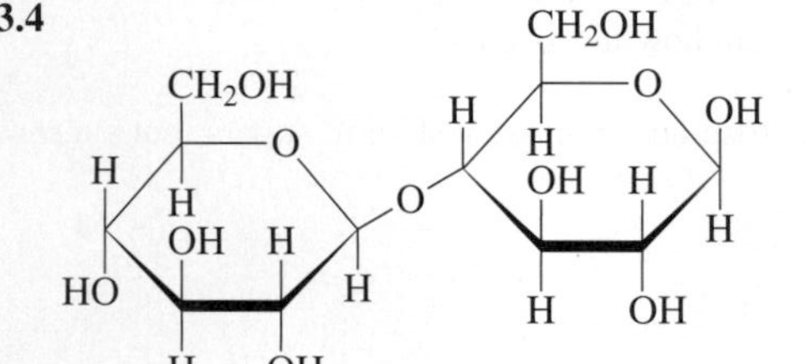

13.5 Cellulose contains glucose units connected by β-1,4-glycosidic bonds, whereas the glucose units in amylose are connected by α-1,4-glycosidic bonds.

Answers to Selected Questions and Problems

13.1 Photosynthesis requires CO_2, H_2O, and the energy from the Sun. Respiration requires O_2 from the air and glucose from our foods.

13.3 Monosaccharides are a chain of three to eight carbon atoms, one in a carbonyl group as an aldehyde or ketone, and the rest attached to hydroxyl groups. A monosaccharide cannot be split or hydrolyzed into smaller carbohydrates. A disaccharide consists of two monosaccharide units joined together that can be split.

13.5 Hydroxyl groups are found in all monosaccharides along with a carbonyl on the first or second carbon.

13.7 A ketopentose contains hydroxyl and ketone functional groups and has five carbon atoms.

13.9 **a.** ketohexose **b.** aldopentose

13.11 In the D isomer, the —OH on the chiral carbon atom at the bottom of the chain is on the right side, whereas in the L isomer, the —OH appears on the left side.

13.13 **a.** D **b.** D **c.** L **d.** D

13.15 **a.** CHO; H—OH; HO—H; CH_2OH

b. CH_2OH; C=O; H—OH; HO—H; CH_2OH

c. CHO; HO—H; HO—H; H—OH; H—OH; CH_2OH

d. CHO; HO—H; HO—H; HO—H; HO—H; CH_2OH

13.17

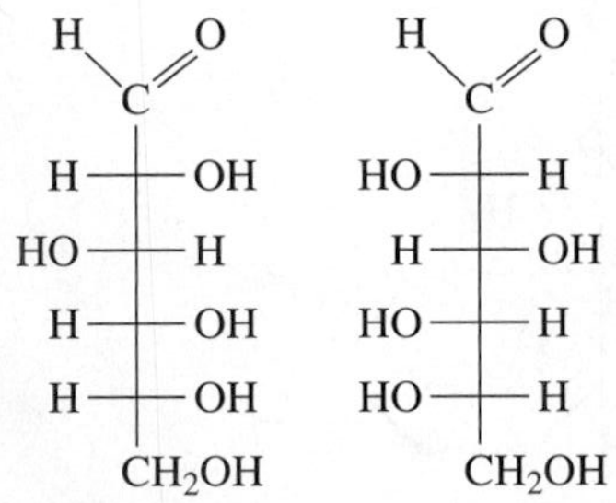

H—C=O
H—OH
HO—H
H—OH
H—OH
CH_2OH
D-Glucose

H—C=O
HO—H
H—OH
HO—H
HO—H
CH_2OH
L-Glucose

13.19 In D-galactose, the hydroxyl on carbon 4 extends to the left. In D-glucose, this hydroxyl goes to the right.

13.21 **a.** glucose **b.** galactose **c.** fructose

13.23 In the cyclic structure of glucose, there are five carbon atoms and an oxygen atom.

13.25

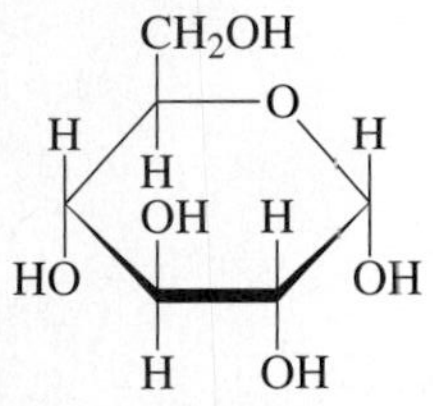

CH_2OH, O, H, H, OH, H, HO, OH, H, OH
α-D-Glucose

CH_2OH, O, H, OH, H, OH, H, HO, H, H, OH
β-D-Glucose

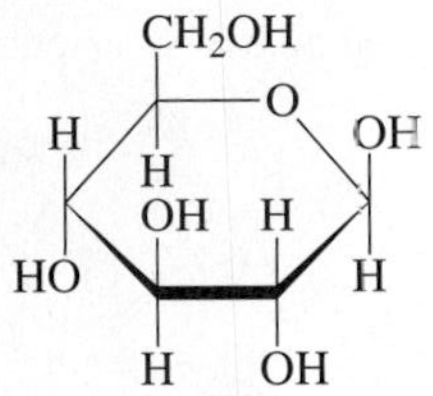

13.27 **a.** α isomer
b. α isomer

13.29

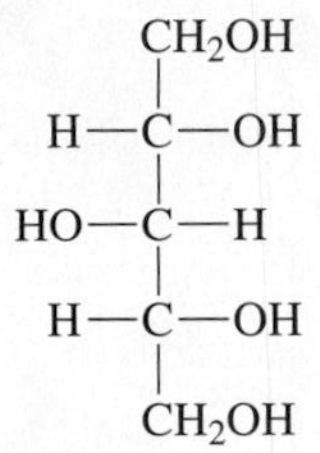

CH_2OH
H—C—OH
HO—C—H
H—C—OH
CH_2OH
D-Xylitol

13.31 Oxidation product:

O
‖
C—OH
HO—C—H
H—C—OH
H—C—OH
CH_2OH

Reduction product (sugar alcohol):

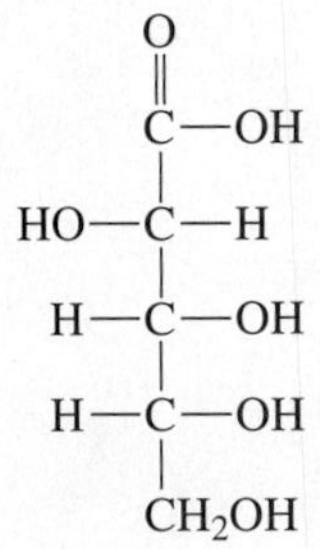

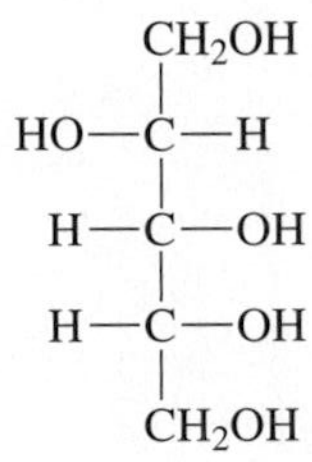

CH_2OH
HO—C—H
H—C—OH
H—C—OH
CH_2OH
D-Arabitol

13.33 **a.** galactose and glucose; β-1,4 bond; β-lactose
b. glucose and glucose; α-1,4 bond; α-maltose

13.35 **a.** can be oxidized
b. can be oxidized

13.37 **a.** sucrose
b. lactose
c. maltose
d. lactose

13.39 **a.** Amylose is an unbranched polymer of glucose units joined by α-1,4 bonds; amylopectin is a branched polymer of glucose joined by α-1,4 and α-1,6 bonds.
b. Amylopectin, which is produced in plants, is a branched polymer of glucose, joined by α-1,4 and α-1,6 bonds. Glycogen, which is produced in animals, is a highly branched polymer of glucose, joined by α-1,4 and α-1,6 bonds.

13.41 **a.** cellulose
b. amylose, amylopectin
c. amylose
d. glycogen

13.43 **a.** disaccharide
b. α-glucose
c. α-1,6
d. α
e. Yes.

13.45 **a.** trisaccharide
b. two glucose and one fructose
c. Melezitose has no free —OH groups on the glucose or fructose molecules; like sucrose it is not a reducing sugar.

13.47 D-Fructose is a ketohexose, whereas D-galactose is an aldohexose. In galactose, the —OH on carbon 4 is on the left; in fructose, the —OH is on the right.

13.49 L-Galactose is the mirror image of D-galactose. In the Fischer projection of D-galactose, the —OH on carbon 2 and carbon 5 is on the right side, but on the left for carbon 3 and carbon 4. In L-galactose, the —OH groups are reversed; carbons 2 and 5 have —OH on the left, and carbons 3 and 4 have —OH on the right.

13.51 **a.**

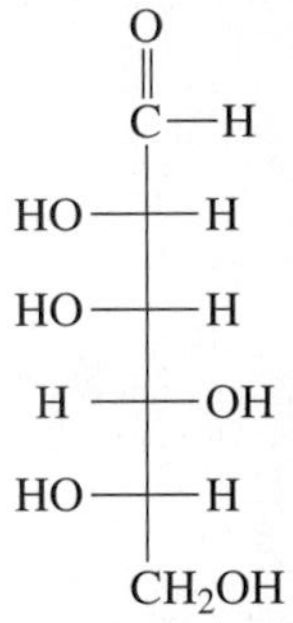

O
‖
C—H
HO—H
HO—H
H—OH
HO—H
CH_2OH
L-Gulose

b.

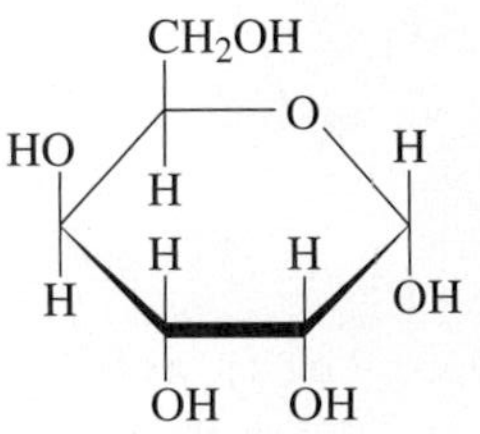

CH_2OH, O, HO, H, H, H, H, OH, H, OH, OH
a-D-Gulose

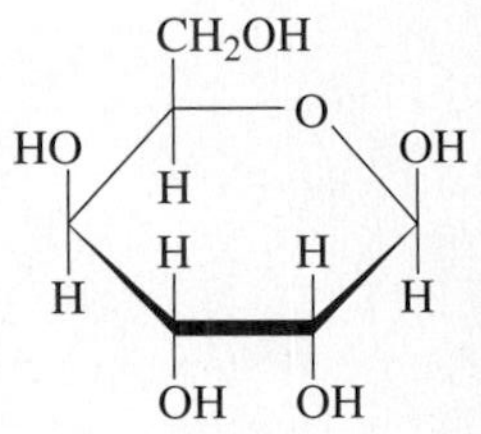

CH_2OH, O, HO, OH, H, H, H, H, H, OH, OH
b-D-Gulose

13.53

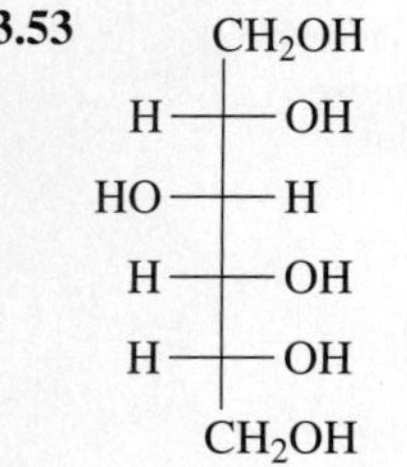

13.55 When α-galactose forms an open-chain structure, it can close to form either α- or β-galactose.

13.57

13.59 a.

b. Yes. Gentiobiose is a reducing sugar. The ring on the right can open up to form an aldehyde that can be oxidized.

14 Carboxylic Acids, Esters, Amines, and Amides

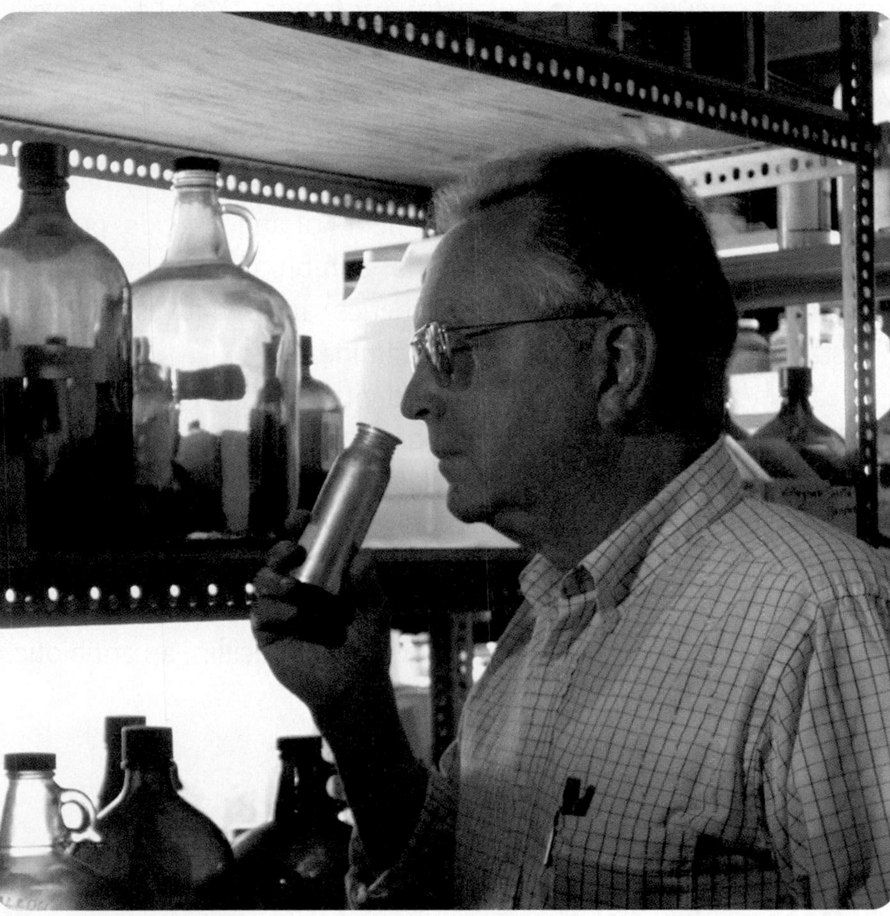

"There are many carboxylic acids, including the alpha hydroxy acids, found today in skin products," says Dr. Ken Peterson, pharmacist and cosmetic chemist in Oakland, California. "When you take a carboxylic acid called a fatty acid and react it with a strong base, you get a soap. Soap has a high pH because the weak fatty acid and the strong base won't have a neutral pH of 7. If you take soap and drop its pH down to 7, you will convert the soap to the fatty acid. When I create fragrances, I use my nose and my chemistry background to identify the reactions that produce good scents. Many fragrances are esters, which form when an alcohol reacts with a carboxylic acid. For example, the ester that smells like pineapple is made from ethanol and butyric acid."

LOOKING AHEAD

Visit **www.masteringchemistry.com** for self-study materials and instructor-assigned homework.

Carboxylic acids are weak acids, which we studied in Chapter 8. They have a sour or tart taste, produce hydronium ions in water, and neutralize bases. You encounter carboxylic acids when you use a salad dressing containing vinegar, which is a solution of acetic acid and water, or experience the sour taste of citric acid in a grapefruit or lemon. When a carboxylic acid reacts with an alcohol, an ester and water are produced. Fats and oils are esters of glycerol and fatty acids, which are long-chain carboxylic acids. Esters produce the pleasant aromas and flavors of many fruits, such as bananas, strawberries, and oranges.

Amines and amides are organic compounds that contain nitrogen. Many nitrogen-containing compounds are important to life as components of amino acids, proteins, and nucleic acids (DNA and RNA). Many amines that exhibit strong physiological activity are used in medicine as decongestants, anesthetics, and sedatives. Examples include dopamine, histamine, epinephrine, and amphetamine.

Alkaloids such as caffeine, nicotine, cocaine, and digitalis, which have powerful physiological activity, are naturally occurring amines obtained from plants. In an amide, the functional group consists of a carboxyl group attached to an amine. Amides, which are derived from carboxylic acids, are important in biology in proteins. In biochemistry, the amide bond that links amino acids in a protein is called a peptide bond. Some medically important amides include acetaminophen (Tylenol) used to reduce fever; phenobarbital, a sedative and anticonvulsant medication; and penicillin, an antibiotic.

LEARNING GOAL

Give the IUPAC names and common names of carboxylic acids; draw their condensed structural formulas.

SELF STUDY ACTIVITY
Carboxylic Acids

14.1 Carboxylic Acids

In Chapters 10 and 12, we described the carbonyl group (C=O) as the functional group in aldehydes and ketones. We also described the oxidation of an aldehyde to produce a carboxylic acid. In a **carboxylic acid**, a hydroxyl group is attached to the carbonyl group, forming a **carboxyl group**. The carboxyl functional group may be attached to an alkyl group or an aromatic group.

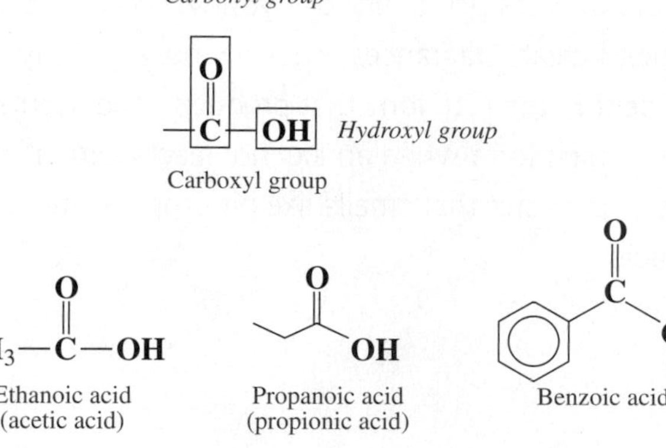

IUPAC Names of Carboxylic Acids

The IUPAC names of carboxylic acids replace the *e* of the corresponding alkane name with *oic acid*.

TUTORIAL
Naming and Drawing Carboxylic Acids

$$\underset{\text{Methanoic acid}}{\mathrm{H{-}\overset{O}{\overset{\|}{C}}{-}OH}} \qquad \underset{\text{Propanoic acid}}{\mathrm{CH_3{-}CH_2{-}\overset{O}{\overset{\|}{C}}{-}OH}}$$

Common Names of Carboxylic Acids

Many carboxylic acids still use common names. In Chapter 12, we looked at the common names of aldehydes, which have the same prefixes as the common names of carboxylic acids. These prefixes are related to the natural sources of the simple carboxylic acids. For example, formic acid is injected under the skin from bee or red ant stings and other insect bites. Acetic acid is the oxidation product of the ethanol in wines and apple cider. A solution of acetic acid and water is known as vinegar. Butyric acid gives the foul odor to rancid butter (see Table 14.1).

A red ant sting contains formic acid that irritates the skin.

TABLE 14.1 Names and Condensed Structural Formulas of Some Carboxylic Acids

Condensed Structural Formula	IUPAC Name	Common Name	Ball-and-Stick Model
$\mathrm{H{-}\overset{O}{\overset{\|}{C}}{-}OH}$	Methanoic acid	Formic acid	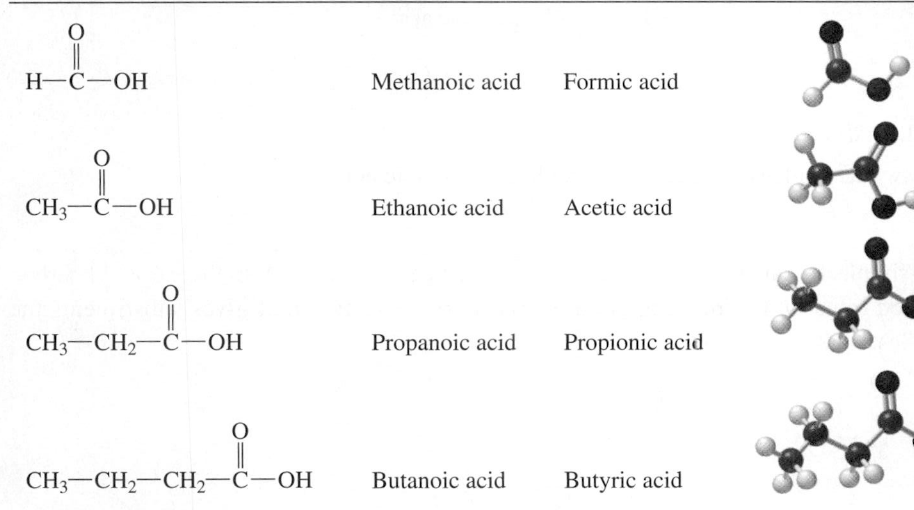
$\mathrm{CH_3{-}\overset{O}{\overset{\|}{C}}{-}OH}$	Ethanoic acid	Acetic acid	
$\mathrm{CH_3{-}CH_2{-}\overset{O}{\overset{\|}{C}}{-}OH}$	Propanoic acid	Propionic acid	
$\mathrm{CH_3{-}CH_2{-}CH_2{-}\overset{O}{\overset{\|}{C}}{-}OH}$	Butanoic acid	Butyric acid	

The sour taste of vinegar is due to ethanoic acid (acetic acid).

CONCEPT CHECK 14.1

Carboxylic Acids

Why can the carboxylic acid with three carbon atoms be named as propanoic acid or propionic acid?

ANSWER

The longest carbon chain containing the carboxyl group has three carbon atoms. In the IUPAC system, the *e* in propane is replaced by *oic acid*, to give the name propanoic acid. It also has a common name, which is propionic acid.

Guide to Naming Carboxylic Acids

1. Identify the carbon chain containing the carboxyl group and replace the *e* in the alkane name with *oic acid*.
2. Give the location and name of each substituent on the main chain by counting the carboxyl carbon as 1.

SAMPLE PROBLEM 14.1

Naming Carboxylic Acids

Give the IUPAC name for the following:

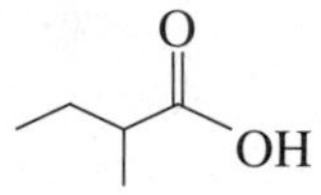

O

OH

SOLUTION

Step 1 **Identify the carbon chain containing the carboxyl group and replace the *e* in the alkane name with *oic acid*.** The IUPAC name of a carboxylic acid with four carbon atoms is butanoic acid.

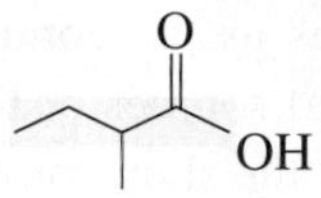

O

OH butanoic acid

Step 2 **Give the location and name of each substituent on the main chain by counting the carboxyl carbon as 1.** This carboxylic acid has a methyl group on carbon 2 adjacent to the carboxyl group, which is carbon 1. The IUPAC name is 2-methylbutanoic acid.

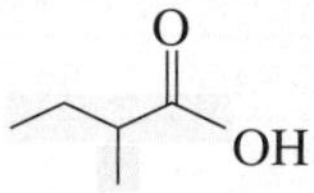

O

OH 2-methylbutanoic acid

4 3 2 1

STUDY CHECK 14.1

Draw the condensed structural formula of pentanoic acid.

The simplest aromatic carboxylic acid is named benzoic acid. With the carboxyl carbon bonded to carbon 1, the ring is numbered in the direction that gives substituents the smallest possible numbers.

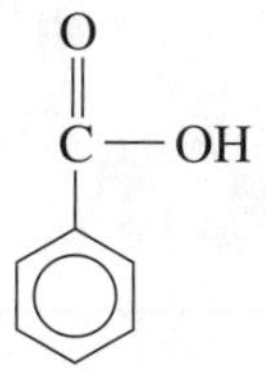

O ‖ C—OH

O ‖ C—OH 1 3 4 Cl Cl

Benzoic acid 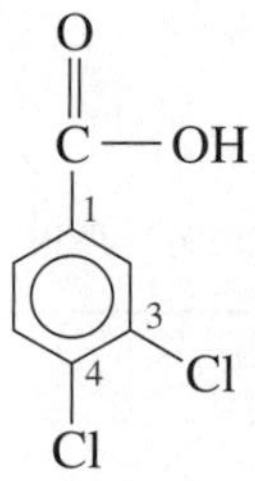 3,4-Dichlorobenzoic acid

Chemistry Link to Health

ALPHA HYDROXY ACIDS (AHAs)

Alpha hydroxy acids (AHAs) are naturally occurring carboxylic acids with a hydroxyl group (—OH) on the carbon adjacent to the carboxyl group. Cleopatra, queen of Egypt, reportedly bathed in sour milk to smooth her skin. Dermatologists have been using products with a high concentration (20–70%) of AHAs to remove acne scars and in skin peels to reduce irregular pigmentation and age spots. Lower concentrations (8–10%) of AHAs are added to skin care products for the purpose of smoothing fine lines, improving skin texture, and cleansing pores. Glycolic acid and lactic acid are most frequently used.

Recent studies indicate that products with AHAs increase sensitivity of the skin to sun and UV radiation. It is recommended that a sunscreen with a sun protection factor (SPF) of at least 15 be used when treating the skin with products that include AHAs. Products containing AHAs at concentrations under 10% and pH values greater than 3.5 are generally considered safe. However, the Food and Drug Administration (FDA) has reports of AHAs causing skin irritation including blisters, rashes, and discoloration of the skin. The FDA does not require product safety reports from cosmetic manufacturers, although they are responsible for marketing safe products. The FDA advises that you test any product containing AHAs on a small area of skin before you use it on a large area.

Alpha hydroxy carboxylic acids are used in many skin care products.

Alpha Hydroxy Acid (Source)	Condensed Structural Formula
Glycolic acid (sugar cane, sugar beet)	$HO-CH_2-C(=O)-OH$
Lactic acid (sour milk)	$CH_3-CH(OH)-C(=O)-OH$
Tartaric acid (grapes)	$HO-C(=O)-CH(OH)-CH(OH)-C(=O)-OH$
Citric acid (lemons, oranges, grapefruit)	$HO-C(CH_2-COOH)(CH_2-COOH)-COOH$
Malic acid (apples, grapes)	$HO-C(=O)-CH_2-CH(OH)-C(=O)-OH$

QUESTIONS AND PROBLEMS

Carboxylic Acids

14.1 What carboxylic acid is responsible for the pain of an ant sting?

14.2 What carboxylic acid is found in vinegar?

14.3 Explain the differences in the condensed structural formulas of propanal and propanoic acid.

14.4 Explain the differences in the condensed structural formulas of benzaldehyde and benzoic acid.

14.5 Give the IUPAC and common names (if any) for the following carboxylic acids:

a. $CH_3-C(=O)-OH$

b. [benzene ring bearing $-C(=O)-OH$, with Br adjacent to it and Br para to it]

c. [skeletal structure: branched chain ending in $C(=O)OH$]

14.6 Give the IUPAC and common names (if any) for the following carboxylic acids:

a. $H-C(=O)-OH$

b. [benzene ring bearing $-C(=O)-OH$, with Cl para to it]

c. [skeletal structure: branched chain ending in $C(=O)OH$]

14.7 Draw the condensed structural formula of each of the following:

a. benzoic acid
b. chloroethanoic acid
c. butyric acid
d. heptanoic acid

14.8 Draw the condensed structural formula of each of the following:

a. 3-ethylbenzoic acid
b. 2,4-dibromobutanoic acid
c. 2,3-dimethylpentanoic acid
d. hexanoic acid

LEARNING GOAL

Describe the solubility and ionization of carboxylic acids in water.

MasteringC

TUTORIAL
Properties of Carboxylic Acids

14.2 Properties of Carboxylic Acids

Carboxylic acids are among the most polar organic compounds because their functional group consists of two polar groups: a hydroxyl (—OH) group and a carbonyl (C═O) group. The —OH group is similar to the functional group in alcohols, and the C═O double bond is similar to that of aldehydes and ketones.

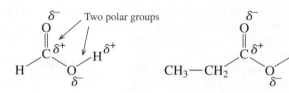

Two polar groups

$H-C^{\delta+}(=O^{\delta-})-O^{\delta-}-H^{\delta+}$ $\quad$ $CH_3-CH_2-C^{\delta+}(=O^{\delta-})-O^{\delta-}-H^{\delta+}$

Solubility in Water

Carboxylic acids with one to four carbons are very soluble in water because the carboxyl group forms hydrogen bonds with several water molecules (see Figure 14.1). However, as the length of the hydrocarbon chain increases, the nonpolar portion reduces the solubility of carboxylic acids in water. Carboxylic acids having five or more carbons are not very soluble in water. Table 14.2 lists the solubility for some selected carboxylic acids.

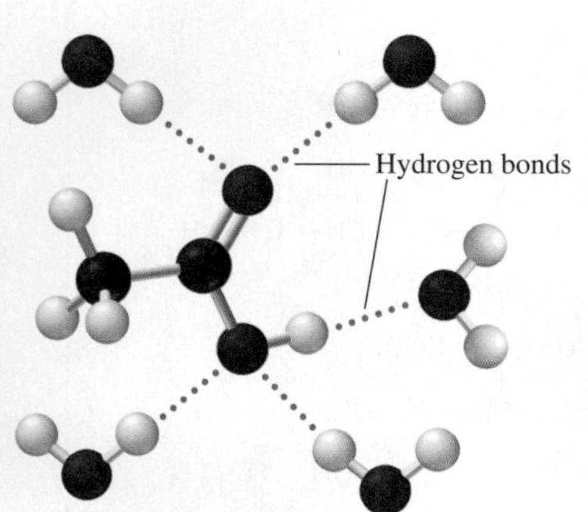

FIGURE 14.1 Acetic acid forms hydrogen bonds with water molecules.
Q Why do the atoms in the carboxyl group hydrogen bond with water molecules?

TABLE 14.2 **Solubility of Selected Carboxylic Acids**

IUPAC Name	Condensed Structural Formula	Solubility in Water
Methanoic acid	$HCOOH$	Soluble
Ethanoic acid	CH_3-COOH	Soluble
Propanoic acid	CH_3-CH_2-COOH	Soluble
Butanoic acid	$CH_3-CH_2-CH_2-COOH$	Soluble
Pentanoic acid	$CH_3-CH_2-CH_2-CH_2-COOH$	Slightly soluble
Hexanoic acid	$CH_3-CH_2-CH_2-CH_2-CH_2-COOH$	Slightly soluble

CONCEPT CHECK 14.2

Solubility of Carboxylic Acids

Why is butanoic acid completely soluble in water, but 1-butanol is only slightly soluble in water?

ANSWER

Butanoic acid contains the carboxyl group, which contains a polar carbonyl group (C═O) as well as polar hydroxyl group (—OH). Two polar groups make butanoic acid completely soluble in water. The 1-butanol is only slightly soluble in water because it has a polar hydroxyl group (—OH). With a hydrocarbon chain of four carbon atoms, the solubility of 1-butanol in water is less.

Acidity of Carboxylic Acids

An important property of carboxylic acids is their ionization in water. When a carboxylic acid ionizes in water, a proton is transferred to a water molecule to give a negatively charged **carboxylate ion** and a hydronium ion. Carboxylic acids are more acidic than most other organic compounds including phenols. However, they are weak acids because only a few carboxylic acid molecules ionize in water. Carboxylic acids can lose protons because the negative charge of the carboxylate anion is stabilized by the two oxygen atoms. By comparison, alcohols do not lose the H in their —OH group because they cannot stabilize the negative charge that would result on a single oxygen atom.

$$\underset{\substack{\text{Ethanoic acid}\\ \text{(acetic acid)}}}{CH_3-\overset{\displaystyle O}{\overset{\|}{C}}-OH} + H_2O \rightleftharpoons \underset{\substack{\text{Ethanoate ion}\\ \text{(acetate ion)}}}{CH_3-\overset{\displaystyle O}{\overset{\|}{C}}-O^-} + \underset{\substack{\text{Hydronium}\\ \text{ion}}}{H_3O^+}$$

SAMPLE PROBLEM 14.2

Ionization of Carboxylic Acids in Water

Write the balanced equation for the ionization of propionic acid in water.

SOLUTION

The ionization of propionic acid produces a carboxylate ion and a hydronium ion.

$$CH_3-CH_2-\overset{\displaystyle O}{\overset{\|}{C}}-OH + H_2O \rightleftharpoons CH_3-CH_2-\overset{\displaystyle O}{\overset{\|}{C}}-O^- + H_3O^+$$

STUDY CHECK 14.2

Write a balanced equation for the ionization of formic acid in water.

Neutralization of Carboxylic Acids

Because carboxylic acids are weak acids, they are completely neutralized by strong bases such as NaOH and KOH. The products are a **carboxylate salt** and water. The carboxylate ion is named by replacing the *ic acid* ending of the acid name with *ate*.

$$\underset{\text{Formic acid}}{H-\overset{\displaystyle O}{\overset{\|}{C}}-OH} + NaOH \longrightarrow \underset{\text{Sodium formate}}{H-\overset{\displaystyle O}{\overset{\|}{C}}-O^-Na^+} + H_2O$$

$$\underset{\text{Benzoic acid}}{C_6H_5-\overset{\displaystyle O}{\overset{\|}{C}}-OH} + KOH \longrightarrow \underset{\text{Potassium benzoate}}{C_6H_5-\overset{\displaystyle O}{\overset{\|}{C}}-O^-K^+} + H_2O$$

Preservatives and flavor enhancers in soups and seasonings are often carboxylic acids or their salts.

Sodium propionate, a preservative, is added to cheeses, bread, and other bakery items to inhibit the spoilage of the food by microorganisms. Sodium benzoate, an inhibitor of mold and bacteria, is added to juices, margarine, relishes, salads, and jams. Monosodium glutamate (MSG) is added to meats, fish, vegetables, and bakery items to enhance flavor, although it causes headaches in some people.

$$\underset{\text{Sodium propionate}}{CH_3-CH_2-\overset{\displaystyle O}{\overset{\|}{C}}-O^-Na^+} \qquad \underset{\text{Sodium benzoate}}{C_6H_5-\overset{\displaystyle O}{\overset{\|}{C}}-O^-Na^+} \qquad \underset{\text{Monosodium glutamate}}{HO-\overset{\displaystyle O}{\overset{\|}{C}}-\overset{\displaystyle NH_2}{\overset{|}{C}H}-CH_2-CH_2-\overset{\displaystyle O}{\overset{\|}{C}}-O^-Na^+}$$

Carboxylate salts are ionic compounds with strong attractions between ions of metals such as Li^+, Na^+, and K^+ and the negatively charged carboxylate ion. Like most salts, the carboxylate salts are solids at room temperature, have high melting points, and are usually soluble in water.

SAMPLE PROBLEM 14.3

Neutralization of a Carboxylic Acid

Write the balanced chemical equation for the neutralization of propionic acid with sodium hydroxide.

SOLUTION

The neutralization of an acid with a base produces the salt of the acid and water.

$$\underset{\text{Propionic acid}}{CH_3{-}CH_2{-}\overset{\overset{\displaystyle O}{\|}}{C}{-}OH} + NaOH \longrightarrow \underset{\text{Sodium propionate}}{CH_3{-}CH_2{-}\overset{\overset{\displaystyle O}{\|}}{C}{-}O^-Na^+} + H_2O$$

STUDY CHECK 14.3

What carboxylic acid will produce potassium butanoate (potassium butyrate) when it is neutralized by KOH?

QUESTIONS AND PROBLEMS

Properties of Carboxylic Acids

14.9 Identify the compound in each group that is most soluble in water. Explain.
a. propanoic acid, hexanoic acid, benzoic acid
b. pentane, 1-hexanol, propanoic acid

14.10 Identify the compound in each group that is most soluble in water. Explain.
a. butanone, butanoic acid, butane
b. ethanoic acid (acetic acid), hexanoic acid, octanoic acid

14.11 Write balanced equations for the ionization of each of the following carboxylic acids in water:
a. pentanoic acid
b. acetic acid

14.12 Write balanced equations for the ionization of each of the following carboxylic acids in water:
a. $CH_3{-}\overset{\overset{\displaystyle CH_3}{|}}{CH}{-}\overset{\overset{\displaystyle O}{\|}}{C}{-}OH$
b. butanoic acid

14.13 Write balanced equations for the reaction of each of the following carboxylic acids with NaOH:
a. formic acid
b. 3-chloropropanoic acid
c. benzoic acid

14.14 Write balanced equations for the reaction of each of the following carboxylic acids with KOH:
a. acetic acid
b. 2-methylbutanoic acid
c. 4-chlorobenzoic acid

LEARNING GOAL

Name an ester; write balanced equations for the formation and hydrolysis of an ester.

14.3 Esters

A carboxylic acid reacts with an alcohol to form an **ester** and water. In an ester, the —H of the carboxylic acid is replaced by an alkyl group. Fats and oils in our diets contain esters of glycerol and fatty acids, which are long-chain carboxylic acids. The aromas and flavors of many fruits, including bananas, oranges, and strawberries, are due to esters.

Carboxylic acid	**Ester**
$CH_3{-}\overset{\overset{\displaystyle O}{\|}}{C}{-}O{-}H$	$CH_3{-}\overset{\overset{\displaystyle O}{\|}}{C}{-}O{-}CH_3$
Ethanoic acid (acetic acid)	Methyl ethanoate (methyl acetate)

Esterification

In a reaction called **esterification**, an ester is produced when a carboxylic acid and an alcohol react in the presence of an acid catalyst (usually H_2SO_4). In esterification, the —OH is removed from the carboxylic acid along with the —H from the alcohol, which combine to form water. An excess of the alcohol is used to favor the formation of the ester product.

MC TUTORIAL
Writing Esterification Equations

TUTORIAL
Formation of Esters from Carboxylic Acids

$$\underset{\substack{\text{Ethanoic acid}\\\text{(acetic acid)}}}{CH_3-\overset{\overset{\displaystyle O}{\|}}{C}-O-H} + \underset{\substack{\text{Methanol}\\\text{(methyl alcohol)}}}{H-O-CH_3} \underset{}{\overset{H^+,\ \text{heat}}{\rightleftharpoons}} \underset{\substack{\text{Methyl ethanoate}\\\text{(methyl acetate)}}}{CH_3-\overset{\overset{\displaystyle O}{\|}}{C}-O-CH_3} + H-O-H$$

For example, the ester responsible for the flavor and odor of pears can be prepared using ethanoic acid and 1-propanol. The equation for the esterification is written as

$$\underset{\substack{\text{Ethanoic acid}\\\text{(acetic acid)}}}{CH_3-\overset{\overset{\displaystyle O}{\|}}{C}-OH} + \underset{\substack{\text{1-Propanol}\\\text{(propyl alcohol)}}}{H-O-CH_2-CH_2-CH_3} \overset{H^+,\ \text{heat}}{\rightleftharpoons} \underset{\substack{\text{Propyl ethanoate}\\\text{(propyl acetate)}}}{CH_3-\overset{\overset{\displaystyle O}{\|}}{C}-O-CH_2-CH_2-CH_3} + H_2O$$

SAMPLE PROBLEM 14.4

Writing Esterification Equations

The ester that gives the flavor and odor of apples can be synthesized from butyric acid and methyl alcohol. What is the equation for the formation of the ester in apples?

SOLUTION

$$\underset{\substack{\text{Butanoic acid}\\\text{(butyric acid)}}}{CH_3-CH_2-CH_2-\overset{\overset{\displaystyle O}{\|}}{C}-OH} + \underset{\substack{\text{Methanol}\\\text{(methyl alcohol)}}}{H-O-CH_3} \overset{H^+,\ \text{heat}}{\rightleftharpoons} \underset{\substack{\text{Methyl butanoate}\\\text{(methyl butyrate)}}}{CH_3-CH_2-CH_2-\overset{\overset{\displaystyle O}{\|}}{C}-O-CH_3} + H_2O$$

STUDY CHECK 14.4

What are the IUPAC names of the carboxylic acid and alcohol needed to form the following ester, which gives the flavor and odor to apricots? (*Hint*: Separate the O and C=O of the ester group and add —H and —OH to give the original alcohol and carboxylic acid.)

$$CH_3-CH_2-\overset{\overset{\displaystyle O}{\|}}{C}-O-CH_2-CH_2-CH_2-CH_2-CH_3$$

MC TUTORIAL
Naming Esters

Naming Esters

The name of an ester consists of two words, which are derived from the names of the alcohol and the acid in that ester. The first word indicates the *alkyl* part of the alcohol. The second word is the *carboxylate* name of the carboxylic acid. The IUPAC names of esters use the IUPAC names for the acid, while the common names of esters use the common

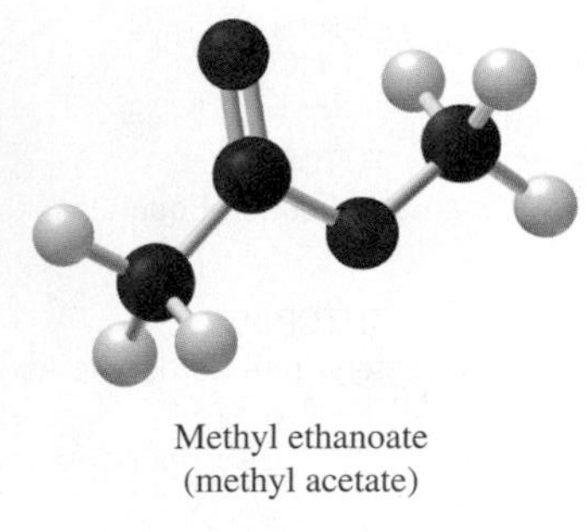
Methyl ethanoate (methyl acetate)

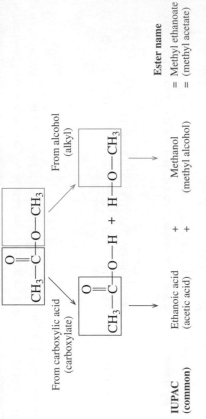

names for the acid. Let's take a look at the following ester, which has a pleasant, fruity odor. We start by separating the ester bond to identify the alkyl part from the alcohol and the carboxylate part from the acid. Then we name the ester as an alkyl carboxylate.

$$CH_3-\overset{\overset{\Large O}{\|}}{C}-O-CH_3$$

From carboxylic acid (carboxylate) | From alcohol (alkyl)

$$CH_3-\overset{\overset{\Large O}{\|}}{C}-O-H \;+\; H-O-CH_3$$

					Ester name
IUPAC	Ethanoic acid	+	Methanol	=	Methyl ethanoate
(common)	(acetic acid)	+	(methyl alcohol)	=	(methyl acetate)

The following examples of some typical esters show the IUPAC, as well as the common names, of esters.

$$CH_3-CH_2-O-\overset{\overset{\Large O}{\|}}{C}-CH_3$$
Ethyl ethanoate (ethyl acetate)

$$CH_3-O-\overset{\overset{\Large O}{\|}}{C}-CH_2-CH_3$$
Methyl propanoate (methyl propionate)

$$CH_3-CH_2-O-\overset{\overset{\Large O}{\|}}{C}-C_6H_5$$
Ethyl benzoate

Chemistry Link to Health

SALICYLIC ACID AND ASPIRIN

Chewing on a piece of willow bark was used as a way of relieving pain for many centuries. By the 1800s, chemists discovered that salicylic acid was the agent in the bark responsible for the relief of pain. However, salicylic acid, which has both a carboxylic group and a hydroxyl group, irritates the stomach lining. A less irritating ester of salicylic acid and acetic acid, called acetylsalicylic acid or "aspirin," was prepared in 1899 by the Bayer chemical company in Germany. In some aspirin preparations, a buffer is added to neutralize the carboxylic acid group and lessen its irritation of the stomach. Aspirin is used as an analgesic (pain reliever), antipyretic (fever reducer), and anti-inflammatory agent. Many people take a daily low-dose aspirin, which has been found to lower the risk of heart attack and stroke.

$$\text{Salicylic acid } (HO-C_6H_4-COOH) + HO-\overset{\overset{\Large O}{\|}}{C}-CH_3 \text{ (Acetic acid)} \xrightleftharpoons{H^+,\text{ heat}} \text{Acetylsalicylic acid (aspirin)} + H_2O$$

Salicylic acid — Acetic acid — Acetylsalicylic acid (aspirin)

Oil of wintergreen, or methyl salicylate, has a spearmint odor and flavor. Because it can pass through the skin, methyl salicylate is used in skin ointments where it acts as a counterirritant, producing heat to soothe sore muscles.

$$\text{Salicylic acid} + HO-CH_3 \text{ (Methyl alcohol)} \xrightleftharpoons{H^+,\text{ heat}} \text{Methyl salicylate (oil of wintergreen)} + H_2O$$

Salicylic acid — Methyl alcohol — Methyl salicylate (oil of wintergreen)

Ointments containing methyl salicylate are used to soothe sore muscles.

SAMPLE PROBLEM 14.5

Naming Esters

What are the IUPAC and common names of the following ester?

$$CH_3-CH_2-\overset{\overset{\displaystyle O}{\|}}{C}-O-CH_2-CH_2-CH_3$$

SOLUTION

Step 1 **Write the name of the carbon chain from the alcohol as an *alkyl* group.** The alcohol part of the ester is from propanol or propyl alcohol. The alkyl group is propyl.

$$CH_3-CH_2-\overset{\overset{\displaystyle O}{\|}}{C}-O-CH_2-CH_2-CH_3 \qquad \text{propyl}$$

Step 2 **Change the *ic acid* of the acid name to *ate*.** The carboxylic acid part of the ester is from propanoic (propionic) acid, which becomes propanoate (propylpropionate). The ester is named propyl propanoate (propyl propionate).

$$CH_3-CH_2-\overset{\overset{\displaystyle O}{\|}}{C}-O-CH_2-CH_2-CH_3 \qquad \text{propyl propanoate (propyl propionoate)}$$

STUDY CHECK 14.5

Draw the condensed structural formula of pentyl ethanoate (pentyl acetate), which gives the odor and flavor to bananas.

Guide to Naming Esters

1. Write the name of the carbon chain from the alcohol as an *alkyl* group.
2. Change the *ic acid* of the acid name to *ate*.

Chemistry Link to the Environment

PLASTICS

Terephthalic acid (an aromatic acid with two carboxyl groups) is produced in large quantities for the manufacture of polyesters such as Dacron. When terephthalic acid reacts with ethylene glycol, ester bonds form on both ends of the molecules, allowing many molecules to combine into a long polyester polymer.

Dacron polyester is used to make permanent press fabrics, carpets, and clothes. In medicine, artificial blood vessels and valves are made of Dacron, which is biologically inert and does not clot the blood. The polyester can also be made as a film called Mylar and as a plastic known as PETE (**poly**ethylene**te**rephthalate). PETE is used for plastic soft drink bottles as well as for containers of salad dressing, shampoos, and dishwashing liquids.

Today PETE (recycling symbol "1") is the most widely recycled of all the plastics. After it is separated from other plastics, PETE can be changed into other useful items, including polyester fabric for T-shirts and coats, fill for sleeping bags, doormats, and containers for tennis balls.

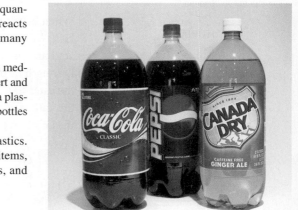

$$HOOC-C_6H_4-COOH + HO-CH_2-CH_2-OH \xrightarrow{H^+,\ heat}$$

Terephthalic acid — Ethylene glycol

$$-O-\overset{\overset{\displaystyle O}{\|}}{C}-C_6H_4-\overset{\overset{\displaystyle O}{\|}}{C}-O-CH_2-CH_2-O-\overset{\overset{\displaystyle O}{\|}}{C}-C_6H_4-\overset{\overset{\displaystyle O}{\|}}{C}-O-CH_2-CH_2-O-$$

Ester bonds

A section of the polyester Dacron

Polyester, in the form of the plastic PETE, is used to make soft drink bottles.

Esters in Plants

Many of the fragrances of perfumes and flowers and the flavors of fruits are due to esters. Small esters are volatile, so we can smell them, and they are soluble in water, so we can taste them. Several of these, with their flavor and odor, are listed in Table 14.3.

TABLE 14.3 Some Esters in Fruits and Flavorings

Condensed Structural Formula and Name	Flavor/Odor
$CH_3-\overset{\overset{\displaystyle O}{\|\|}}{C}-O-CH_2-CH_2-CH_3$ Propyl ethanoate (propyl acetate)	Pears
$CH_3-\overset{\overset{\displaystyle O}{\|\|}}{C}-O-CH_2-CH_2-CH_2-CH_2-CH_3$ Pentyl ethanoate (pentyl acetate)	Bananas
$CH_3-\overset{\overset{\displaystyle O}{\|\|}}{C}-O-CH_2-CH_2-CH_2-CH_2-CH_2-CH_2-CH_2-CH_3$ Octyl ethanoate (octyl acetate)	Oranges
$CH_3-CH_2-CH_2-\overset{\overset{\displaystyle O}{\|\|}}{C}-O-CH_2-CH_3$ Ethyl butanoate (ethyl butyrate)	Pineapples
$CH_3-CH_2-CH_2-\overset{\overset{\displaystyle O}{\|\|}}{C}-O-CH_2-CH_2-CH_2-CH_2-CH_3$ Pentyl butanoate (pentyl butyrate)	Apricots

Esters such as ethyl butanoate provide the odor and flavor of many fruits such as pineapples.

TUTORIAL
Hydrolysis of Esters

Acid Hydrolysis of Esters

When esters are heated with water in the presence of a strong acid, usually H_2SO_4 or HCl, *hydrolysis* occurs. In **hydrolysis**, water reacts with the ester to form a carboxylic acid and an alcohol. Therefore, hydrolysis is the reverse of the esterification reaction. During hydrolysis, a water molecule provides —OH to convert the carbonyl group of the ester to a carboxyl group. A large quantity of water is used to favor the formation of the carboxylic acid and alcohol products. When hydrolysis of biological esters occurs in the cells, an enzyme replaces the acid as the catalyst.

$$\underset{\text{Methyl ethanoate (methyl acetate)}}{CH_3-\overset{\overset{\displaystyle O}{\|}}{C}-O-CH_3} + \underset{\text{Water}}{H-OH} \overset{H^+}{\rightleftharpoons} \underset{\text{Ethanoic acid (acetic acid)}}{CH_3-\overset{\overset{\displaystyle O}{\|}}{C}-O-H} + \underset{\text{Methanol (methyl alcohol)}}{CH_3-OH}$$

CONCEPT CHECK 14.3

Acid Hydrolysis of Esters

Aspirin that has been stored for a long time may undergo hydrolysis in the presence of water and heat. What are the hydrolysis products of aspirin? Why does a bottle of old aspirin smell like vinegar?

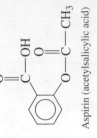

Aspirin (acetylsalicylic acid)

ANSWER

To write the hydrolysis products, separate the compound at the ester bond. Complete the formula of the carboxylic acid by adding —OH (from water) to the carbonyl group and a —H to complete the alcohol. The acetic acid in the products gives the odor of vinegar to a sample of aspirin that has hydrolyzed.

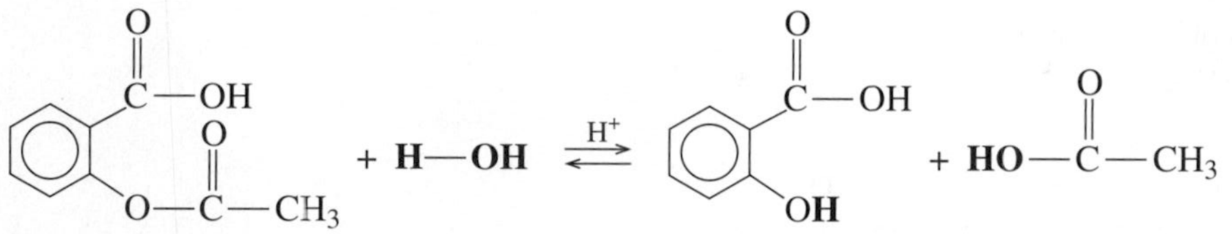

$$\text{Aspirin} + \mathbf{H{-}OH} \underset{}{\overset{H^+}{\rightleftharpoons}} \text{Salicylic acid} + \mathbf{HO}{-}\overset{\overset{\displaystyle O}{\|}}{C}{-}CH_3 \text{ (Acetic acid)}$$

Aspirin   Salicylic acid   Acetic acid

Base Hydrolysis of Esters

When an ester undergoes hydrolysis with a strong base such as NaOH or KOH, the products are the carboxylate salt and the corresponding alcohol. This hydrolysis in a base is also called **saponification**, which refers to the reaction of a long-chain fatty acid with NaOH to make soap. Thus a carboxylic acid, which is produced in acid hydrolysis, is converted to its carboxylate salt when hydrolyzed by a strong base.

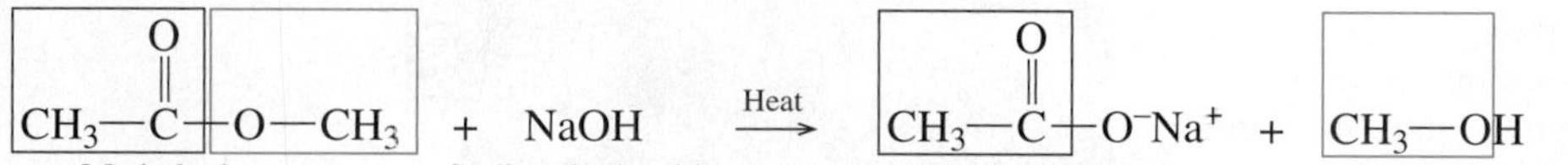

$$CH_3{-}\overset{\overset{\displaystyle O}{\|}}{C}{-}O{-}CH_3 + NaOH \xrightarrow{\text{Heat}} CH_3{-}\overset{\overset{\displaystyle O}{\|}}{C}{-}O^-Na^+ + CH_3{-}OH$$

Methyl ethanoate (methyl acetate)   Sodium hydroxide   Sodium ethanoate (sodium acetate)   Methanol (methyl alcohol)

SAMPLE PROBLEM 14.6

Base Hydrolysis of Esters

Ethyl acetate is a solvent widely used for fingernail polish, plastics, and lacquers. Write the balanced equation of the hydrolysis of ethyl acetate by NaOH.

SOLUTION

The hydrolysis of ethyl acetate by NaOH gives the carboxylate salt, sodium acetate, and ethyl alcohol.

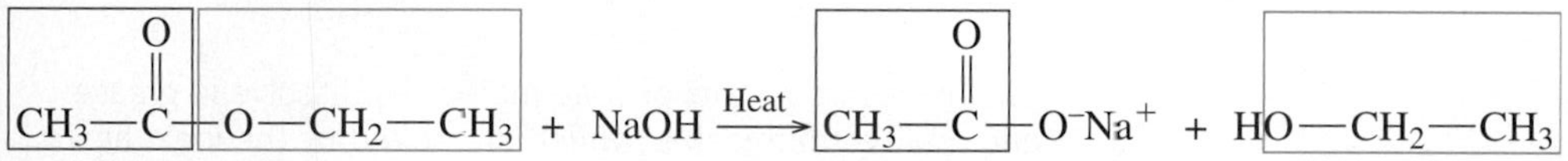

$$CH_3{-}\overset{\overset{\displaystyle O}{\|}}{C}{-}O{-}CH_2{-}CH_3 + NaOH \xrightarrow{\text{Heat}} CH_3{-}\overset{\overset{\displaystyle O}{\|}}{C}{-}O^-Na^+ + HO{-}CH_2{-}CH_3$$

Ethyl acetate   Sodium acetate   Ethyl alcohol

STUDY CHECK 14.6

Draw the condensed structural formulas of the products from the hydrolysis of methyl benzoate by KOH.

Chemistry Link to the Environment

CLEANING ACTION OF SOAPS

For many centuries, soaps were made by heating a mixture of animal fats (tallow) with lye, a basic solution obtained from wood ashes. In the soap-making process, fats, which are esters of long-chain fatty acids, undergo saponification with the strong base in lye.

Carboxylate salt ("soap")

$$\boxed{CH_3CH_2CH_2CH_2CH_2CH_2CH_2CH_2CH_2CH_2CH_2CH_2CH_2CH_2CH_2CH_2}-\overset{O}{\overset{\|}{C}}-O^-Na^+$$

Nonpolar tail (hydrophobic) — Polar head (hydrophilic)

A "soap" is the salt of a long-chain carboxylic acid.

Soaps are typically prepared from fats such as coconut oil. Perfumes are added to give a pleasant-smelling product. We say that soap is the salt of a long-chain fatty acid because soap has a long hydrocarbon-like chain with a carboxylate negative end combined with a positive sodium or potassium ion. Thus, the two ends of a soap molecule have different polarities. The sodium or potassium carboxylate end is ionic and very soluble in water ("hydrophilic") but not in oils or grease. However, the long hydrocarbon end is not soluble in water ("hydrophobic") but does dissolve in nonpolar substances such as oil or grease.

When soap is used, the hydrocarbon tails of the soap molecules dissolve in the nonpolar fats and oils that accompany dirt. The soap molecules coat the oil or grease, forming clusters called *micelles*, in which the water-loving salt ends of the soap molecules extend outside where they can dissolve in water. As a result, small globules of oil and fat coated with soap molecules are pulled into the water and rinsed away. One of the problems of using soaps is that the carboxylate end reacts with ions in hard water, such as Ca^{2+} and Mg^{2+}, and forms insoluble substances.

$$\underset{\text{Stearate ion}}{2CH_3(CH_2)_{16}COO^-} + \underset{\text{Magnesium ion}}{Mg^{2+}} \longrightarrow \underset{\substack{\text{Magnesium stearate}\\ \text{(insoluble)}}}{[CH_3(CH_2)_{16}COO^-]_2Mg^{2+}}$$

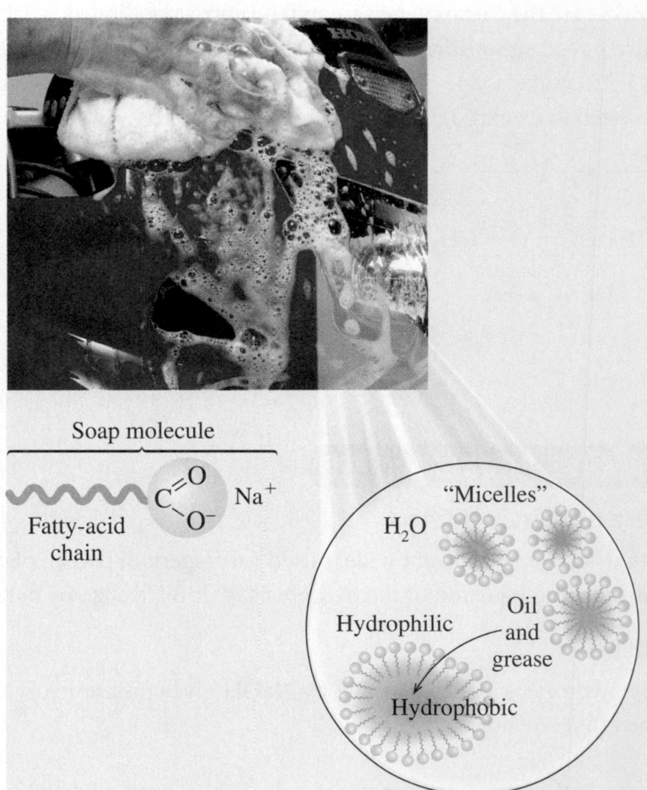

The hydrocarbon tails of soap molecules dissolve in grease and oil to form micelles, which are pulled by the ionic heads into rinse water.

QUESTIONS AND PROBLEMS

Esters

14.15 Draw the condensed structural formula of the ester formed when each of the following reacts with methyl alcohol:
a. acetic acid **b.** butyric acid
c. benzoic acid

14.16 Draw the condensed structural formula of the ester formed when each of the following reacts with ethyl alcohol:
a. formic acid **b.** propionic acid
c. 2-methylpentanoic acid

14.17 Draw the condensed structural formula of the ester formed in each of the following reactions:

a. $CH_3—CH_2—CH_2—CH_2—\overset{O}{\overset{\|}{C}}—OH + HO—\underset{}{\overset{CH_3}{\overset{|}{CH}}}—CH_3 \underset{Heat}{\overset{H^+}{\rightleftarrows}}$

b. $CH_3—CH_2—\overset{O}{\overset{\|}{C}}—OH + HO—CH_2—CH_2—CH_3 \underset{Heat}{\overset{H^+}{\rightleftarrows}}$

14.18 Draw the condensed structural formula of the ester formed in each of the following reactions:

a. $CH_3—CH_2—\overset{O}{\overset{\|}{C}}—OH + HO—CH_3 \underset{Heat}{\overset{H^+}{\rightleftarrows}}$

b. 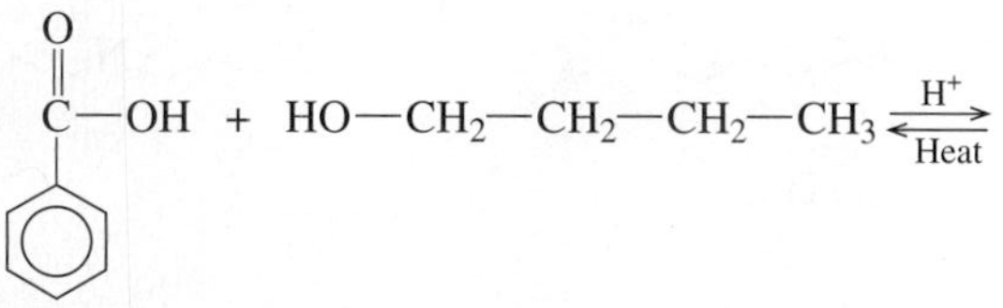 $C_6H_5—\overset{O}{\overset{\|}{C}}—OH + HO—CH_2—CH_2—CH_2—CH_3 \underset{Heat}{\overset{H^+}{\rightleftarrows}}$

14.19 Give the IUPAC and common names, if any, of each of the following:

a. $H—\overset{O}{\overset{\|}{C}}—O—CH_3$ **b.** $CH_3—\overset{O}{\overset{\|}{C}}—O—CH_3$

c.

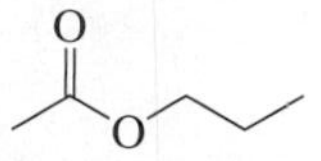

14.20 Give the IUPAC and common names, if any, of each of the following:

a. 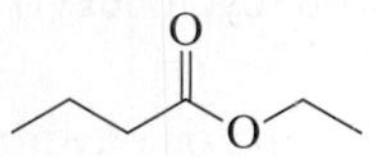 **b.**

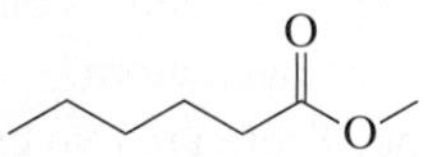

c. $CH_3—CH_2—\overset{O}{\overset{\|}{C}}—O—CH_2—CH_2—CH_2—CH_3$

14.21 Draw the condensed structural formula of each of the following:
a. propyl butyrate **b.** butyl formate
c. ethyl pentanoate

14.22 Draw the condensed structural formula of each of the following:
a. hexyl acetate **b.** propyl methanoate
c. propyl benzoate

14.23 What is the ester responsible for the flavor and odor of the following fruit?
a. banana **b.** orange **c.** pear

14.24 What flavor would you notice if you smelled or tasted the following?
a. ethyl butanoate **b.** propyl acetate
c. pentyl acetate

14.25 What are the products of the acid hydrolysis of an ester?

14.26 What are the products of the base hydrolysis of an ester?

14.27 Draw the condensed structural formulas of the products from the acid- or base-catalyzed hydrolysis of each of the following:

a. $CH_3—CH_2—\overset{O}{\overset{\|}{C}}—O—CH_3 + NaOH \xrightarrow{Heat}$

b. $CH_3—\overset{O}{\overset{\|}{C}}—O—CH_2—CH_2—CH_3 + H_2O \overset{H^+}{\rightleftarrows}$

c. $CH_3—CH_2—CH_2—\overset{O}{\overset{\|}{C}}—O—CH_2—CH_3 + H_2O \overset{H^+}{\rightleftarrows}$

d. 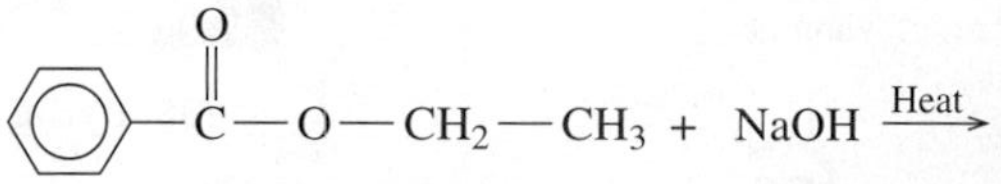 $C_6H_5—\overset{O}{\overset{\|}{C}}—O—CH_2—CH_3 + NaOH \xrightarrow{Heat}$

14.28 Draw the condensed structural formulas of the products from the acid- or base-catalyzed hydrolysis of each of the following:

a. $CH_3—CH_2—\overset{O}{\overset{\|}{C}}—O—CH_2—CH_2—CH_2—CH_3 + H_2O \overset{H^+}{\rightleftarrows}$

b. $H—\overset{O}{\overset{\|}{C}}—O—CH_2—CH_3 + NaOH \xrightarrow{Heat}$

c. $CH_3—CH_2—\overset{O}{\overset{\|}{C}}—O—CH_3 + H_2O \overset{H^+}{\rightleftarrows}$

d. 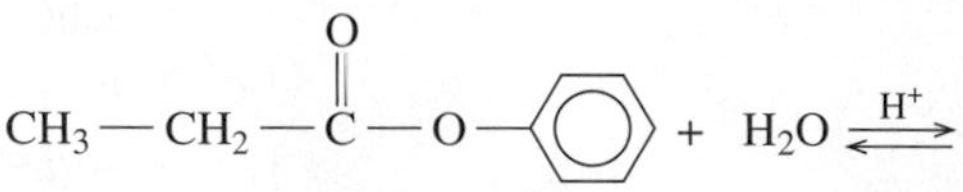 $CH_3—CH_2—\overset{O}{\overset{\|}{C}}—O—C_6H_5 + H_2O \overset{H^+}{\rightleftarrows}$

14.4 Amines

LEARNING GOAL

Classify amines as primary (1°), secondary (2°), or tertiary (3°). Name amines using common names; draw the condensed structural formulas given the names. Describe the solubility, ionization, and neutralization of amines.

Amines are considered derivatives of ammonia (NH_3), in which one or more hydrogen atoms attached to the nitrogen atom are replaced with alkyl or aromatic groups. In methylamine, a methyl group replaces one hydrogen atom in ammonia. The bonding of two methyl groups gives dimethylamine, and the three methyl groups in trimethylamine replace all the hydrogen atoms in ammonia.

Ammonia

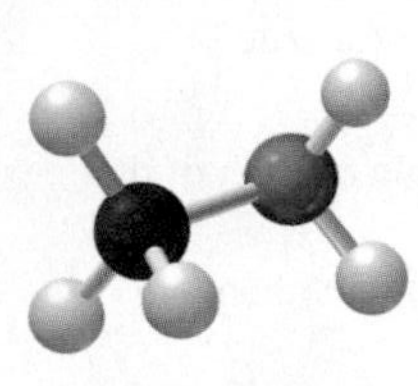

Methylamine

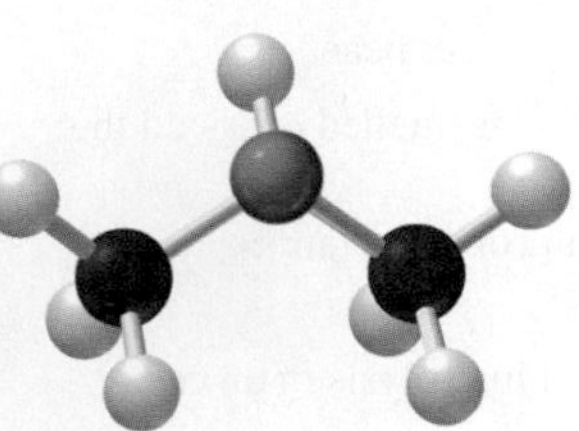

Dimethylamine

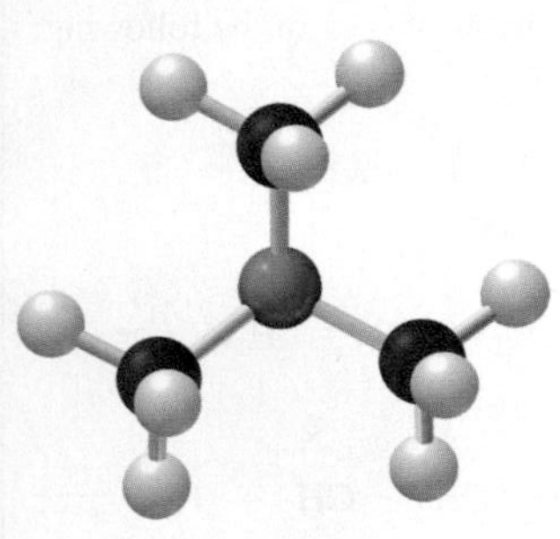

Trimethylamine

SELF STUDY ACTIVITY
Amine and Amide Functional Groups

Classification of Amines

In the classification of amines, we determine if the nitrogen atom (shown in blue) is bonded to one, two, or three alkyl or aromatic groups. In a *primary (1°) amine*, a nitrogen atom is bonded to one alkyl group. In a *secondary (2°) amine*, a nitrogen atom is bonded to two alkyl groups, and in a *tertiary (3°) amine*, a nitrogen atom is bonded to three alkyl groups. Thus, the nitrogen atom in a primary amine and a secondary amine are attached to two hydrogen atoms and one hydrogen atom, respectively. The nitrogen atom in a tertiary amine does not have any hydrogen atoms attached.

Ammonia	**Primary (1°) amine**	**Secondary (2°) amine**	**Tertiary (3°) amine**
$H-\underset{\underset{H}{\vert}}{N}-H$	$CH_3-\underset{\underset{H}{\vert}}{N}-H$	$CH_3-\underset{\underset{H}{\vert}}{N}-CH_3$	$CH_3-\underset{\underset{CH_3}{\vert}}{N}-CH_3$

Skeletal Formulas for Amines

We can draw the skeletal formulas for amines just as we did for other organic compounds. For example, we can draw the following and classify each of the amines:

Primary (1°) amine **Secondary (2°) amine** **Tertiary (3°) amine**

NH_2

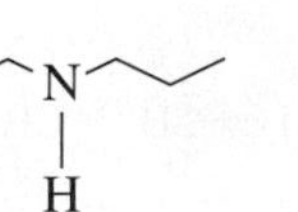

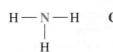

CONCEPT CHECK 14.4

Classifying Amines

Classify the following amines as primary (1°), secondary (2°), or tertiary (3°):

a. NH_2 **b.** 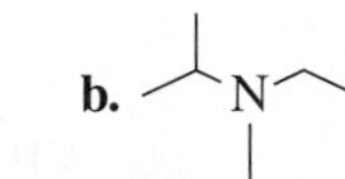 **c.** $N-CH_3$, H

ANSWER

a. This is a primary (1°) amine because there is one alkyl group (cyclohexyl) and two hydrogen atoms attached to a nitrogen atom.

b. This is a secondary (2°) amine. There are two carbon chains and one hydrogen attached to the nitrogen atom.

c. This is a secondary (2°) amine with an alkyl group, an aromatic group, and one hydrogen atom bonded to the nitrogen atom.

MC TUTORIAL
Name that Amine

TUTORIAL
Drawing Amines

Naming Amines

For simple amines, the common names are often used. When there are two or more alkyl groups bonded to the nitrogen atom, they are listed in alphabetical order. The prefixes *di* and *tri* are used to indicate two and three identical substituents.

CH_3-NH_2

Methylamine

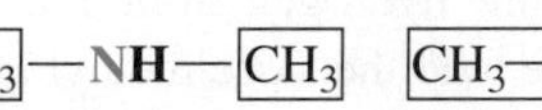

Dimethylamine

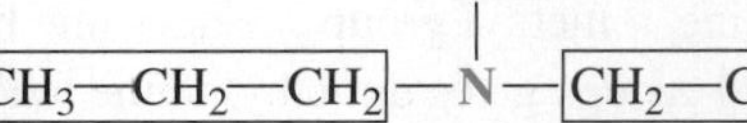

Ethylmethylpropylamine

SAMPLE PROBLEM 14.7

Naming Amines

Give the common name for each of the following amines:

a. $CH_3—CH_2—NH_2$

b. $CH_3—\underset{}{N}(CH_3)—CH_2—CH_3$ (the CH_3 group is attached to N)

SOLUTION

a. This amine has one ethyl group attached to the nitrogen atom; its name is ethylamine.

b. This amine has two methyl groups and one ethyl group attached to the nitrogen atom; its name is ethyldimethylamine.

STUDY CHECK 14.7

Draw the condensed structural formula of ethylpropylamine.

Chemistry Link to Health

AMINES IN HEALTH AND MEDICINE

In response to allergic reactions or injury to cells, the body increases the production of histamine, which causes blood vessels to dilate and increases the permeability of the cells. Redness and swelling occur in the area. Administering an antihistamine such as diphenhydramine helps block the effects of histamine.

Histamine: imidazole ring (N—H) with $—CH_2—CH_2—NH_2$

Diphenhydramine: $(C_6H_5)_2CH—O—CH_2—CH_2—N(CH_3)—CH_3$

In the body, hormones called biogenic amines carry messages between the central nervous system and nerve cells. Epinephrine (adrenaline) and norepinephrine (noradrenaline) are released by the adrenal medulla in "fight or flight" situations to raise the blood glucose level and move the blood to the muscles. Used in remedies for colds, hay fever, and asthma, norepinephrine contracts the capillaries in the mucous membranes of the respiratory passages. The prefix *nor* in a drug name means there is one less $CH_3—$ group on the nitrogen atom. Parkinson's disease is a result of a deficiency in another biogenic amine called dopamine.

Epinephrine (adrenaline): $(HO)_2C_6H_3—CH(OH)—CH_2—NH—CH_3$

Norepinephrine (noradrenaline): $(HO)_2C_6H_3—CH(OH)—CH_2—NH_2$

Dopamine: $(HO)_2C_6H_3—CH_2—CH_2—NH_2$

Produced synthetically, amphetamines (known as "uppers") are stimulants of the central nervous system much like epinephrine, but they also increase cardiovascular activity and depress the appetite. They are sometimes used to bring about weight loss, but they can cause chemical dependency. Benzedrine and Neo-Synephrine (phenylephrine) are used in medications to reduce respiratory congestion from colds, hay fever, and asthma. Sometimes, benzedrine is taken to combat the desire to sleep, but it has side effects. Methedrine is used to treat depression and in the illegal form is known as "speed" or "crank." The prefix *meth* means that there is one more $CH_3—$ group on the nitrogen atom.

Benzedrine (amphetamine): $C_6H_5—CH_2—CH(CH_3)—NH_2$

Neo-Synephrine (phenylephrine): $HO—C_6H_4—CH(OH)—CH_2—NH—CH_3$

Methamphetamine (methedrine): $C_6H_5—CH_2—CH(CH_3)—NH—CH_3$

Aromatic Amines

The aromatic amines use the name *aniline*. Alkyl groups attached to the nitrogen of aniline are named with the prefix *N*- followed by the alkyl name.

Aniline ($C_6H_5-NH_2$) 4-Bromoaniline ($Br-C_6H_4-NH_2$) *N*-Methylaniline ($C_6H_5-NH-CH_3$)

Properties of Amines

Because amines contain a polar N—H bond, they form hydrogen bonds with water. In primary (1°) amines, $-NH_2$ can form more hydrogen bonds than the secondary (2°) amines. A tertiary (3°) amine, which has no hydrogen on the nitrogen atom, can form only hydrogen bonds with water from the N atom in the amine to the H of a water molecule.

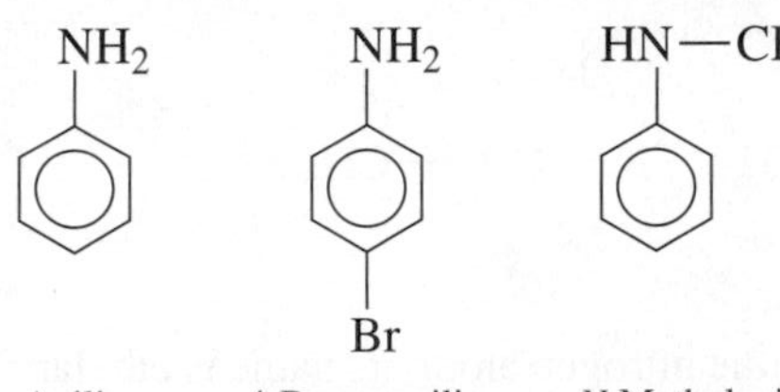

Solubility in Water

Like alcohols, the smaller amines, including tertiary ones, are soluble because they form hydrogen bonds with water. However, in amines with more than six carbon atoms, the effect of hydrogen bonding is diminished. Then the nonpolar hydrocarbon chains of the amine decreases its solubility in water.

Amines React as Bases

In Chapter 8, we saw that ammonia (NH_3) acts as a Brønsted–Lowry base because it accepts a proton (H^+) from water to produce an ammonium ion (NH_4^+) and a hydroxide ion (OH^-). Let's review that equation:

$$\ddot{N}H_3 + H_2O \rightleftharpoons NH_4^+ + OH^-$$

Ammonia ⇌ Ammonium ion + Hydroxide ion

TUTORIAL
Reactions of Amines

In water, amines act as Brønsted–Lowry bases because the lone electron pair on the nitrogen atom accepts a proton from water and produces alkylammonium and hydroxide ions.

$$CH_3-\ddot{N}H_2 + H_2O \rightleftharpoons CH_3-\overset{+}{N}H_3 + OH^-$$

Methylamine ⇌ Methylammonium ion + Hydroxide ion

$$CH_3-\ddot{N}H(CH_3) + H_2O \rightleftharpoons CH_3-\overset{+}{N}H_2(CH_3) + OH^-$$

Dimethylamine ⇌ Dimethylammonium ion + Hydroxide ion

The amines in fish react with the acid in lemon juice to remove the "fishy" odor.

Amine Salts

When you squeeze lemon juice on fish, the "fishy" odor of the amines is removed by converting them to *amine salts*. In a neutralization reaction, an amine acts as a base and reacts with an acid to form an **amine salt**. The lone pair of electrons on the nitrogen atom accepts a proton H^+ from an acid to give an amine salt; no water is formed. An amine salt is named by replacing the *amine* part of the name with *ammonium*, followed by the name of the negative ion.

Neutralization of an Amine

$$CH_3{-}\ddot{N}H_2 + HCl \longrightarrow CH_3{-}\overset{+}{N}H_3\,Cl^-$$

Methyl**amine** → Methyl**ammonium chloride**

$$CH_3{-}\ddot{N}H(CH_3) + HCl \longrightarrow CH_3{-}\overset{+}{N}H_2(CH_3)\,Cl^-$$

Dimethyl**amine** → Dimethyl**ammonium chloride**

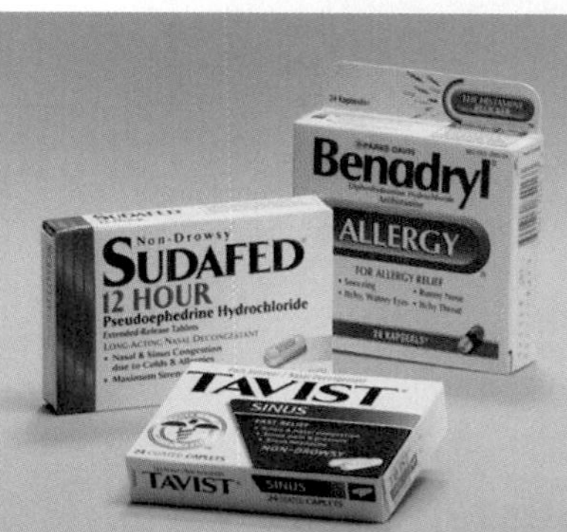

FIGURE 14.2 Decongestants and products that relieve itch and skin irritations often contain ammonium salts.

Q Why are the ammonium salts used rather than the biologically active amines?

Properties of Amine Salts

Amine salts are ionic compounds with strong attractions between the positively charged ammonium ion and an anion, usually chloride. Like most other salts, amine salts are solids at room temperature, odorless, and soluble in water and body fluids. For this reason, amines used as drugs are usually converted to their amine salts. The amine salt of ephedrine is used as a bronchodilator and in decongestant products such as Sudafed. The amine salt of diphenhydramine is used in products such as Benadryl for relief of itching and pain from skin irritations and rashes (see Figure 14.2). In pharmaceuticals, the naming of the amine salt follows an older method of giving the amine name followed by the name of the acid.

$$HO{-}CH(C_6H_5){-}CH(CH_3){-}\overset{+}{N}H_2{-}CH_3\,Cl^-$$

Ephedrine hydrochloride
Ephedrine HCl
Sudafed

$$(C_6H_5)_2CH{-}O{-}CH_2{-}CH_2{-}\overset{+}{N}(CH_3)_2{-}H\,Cl^-$$

Diphenhydramine hydrochloride
Diphenhydramine HCl
Benadryl

When an amine salt reacts with a strong base such as NaOH, it is converted back to the amine, which is also called the free amine or free base.

$$CH_3{-}NH_3^+Cl^- + NaOH \longrightarrow CH_3{-}NH_2 + NaCl + H_2O$$

Cocaine is typically extracted from coca leaves, using an acidic HCl solution to give a white, solid amine salt, cocaine hydrochloride. This is the form in which cocaine is smuggled and sold illegally on the street to be snorted or injected. "Crack cocaine" is the free amine or free base of the amine obtained by treating the cocaine hydrochloride with NaOH and ether, a process known as "free-basing." The solid product is known as "crack cocaine" because it makes a crackling noise when heated. The free amine is rapidly absorbed when smoked and gives stronger highs than the cocaine hydrochloride. Unfortunately, these effects of crack cocaine have caused a rise in addiction to cocaine.

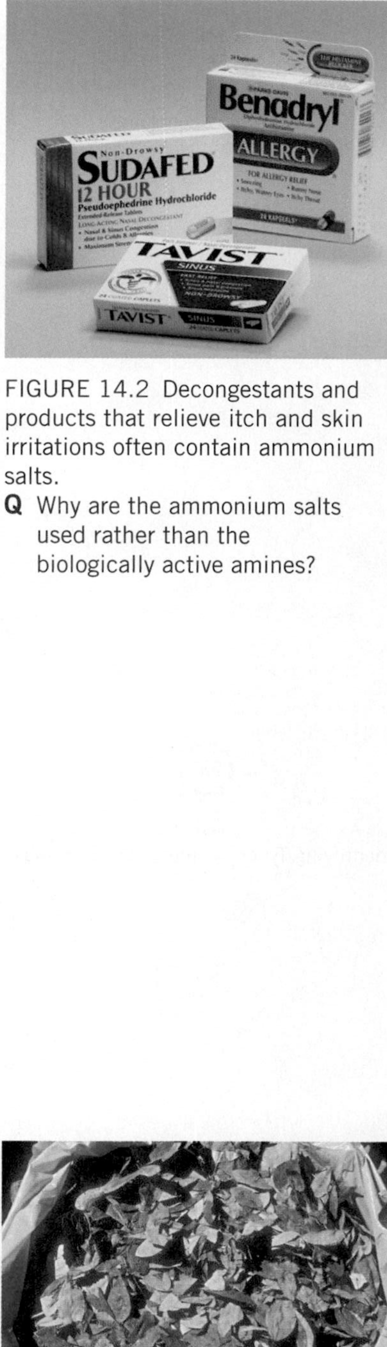

Coca leaves are a source of cocaine.

Cocaine hydrochloride + NaOH → Cocaine (free base) + NaCl + H_2O

Cocaine hydrochloride

Cocaine (free base)

CASE STUDY
Death by Chocolate?

SAMPLE PROBLEM 14.8

Reactions of Amines

Write a balanced equation that shows ethylamine

a. ionizing as a weak base in water

b. neutralized by HCl

SOLUTION

a. In water, ethylamine acts as a weak base by accepting a proton from water to produce ethylammonium and hydroxide ions.

$$CH_3{-}CH_2{-}NH_2 + H{-}OH \rightleftharpoons CH_3{-}CH_2{-}NH_3^+ + OH^-$$

b. In a reaction with an acid, ethylamine acts as a weak base by accepting a proton from HCl to produce ethylammonium chloride.

$$CH_3{-}CH_2{-}NH_2 + HCl \longrightarrow CH_3{-}CH_2{-}NH_3^+Cl^-$$

STUDY CHECK 14.8

What is the condensed structural formula of the salt formed by the reaction of trimethylamine and HCl?

TUTORIAL
Identifying Types of Heterocyclic Amines

Heterocyclic Amines

A *heterocyclic amine* is a cyclic compound that contains one or more nitrogen atoms in the ring. The heterocyclic amine rings typically consist of five or six atoms and one or more nitrogen atoms. Of the five-atom rings, the simplest one is pyrrolidine, which is a ring of four carbon atoms and a nitrogen atom, all with single bonds. Pyrrole is a five-atom ring with one nitrogen atom and two double bonds. Imidazole is a five-atom ring that contains two nitrogen atoms.

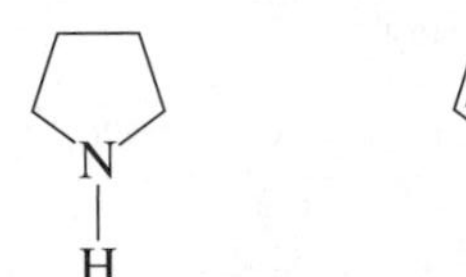
Pyrrolidine

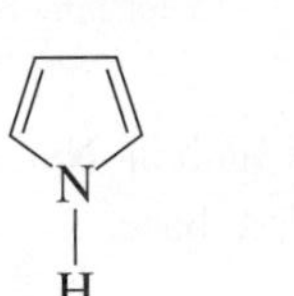
Pyrrole

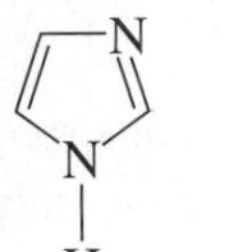
Imidazole

The aroma of pepper is due to piperidine, a heterocyclic amine.

Some of the pungent aroma and taste we associate with black pepper is due to a compound called *piperidine*, which is a six-atom heterocyclic ring with a nitrogen atom. The fruit from the black pepper plant is dried and ground to give the black pepper we use to season our foods.

Many of the other six-atom heterocyclic amines are aromatic. Pyridine is similar to benzene, except it has a nitrogen atom in place of a carbon atom. Pyrimidine, which is found in DNA and RNA, has two nitrogen atoms. Purine, which has two heterocyclic rings fused together, is also found in DNA and RNA.

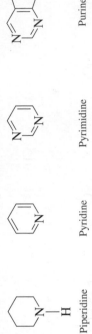

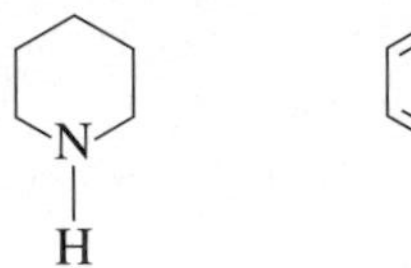
Piperidine Pyridine

Pyrimidine

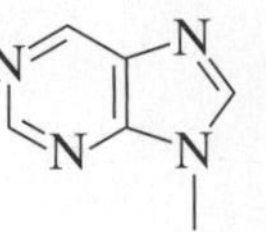
Purine

Alkaloids: Amines in Plants

Alkaloids are physiologically active nitrogen-containing compounds produced by plants. The term *alkaloid* refers to the alkali-like or basic characteristics we have seen for amines. Certain alkaloids are used in anesthetics, in antidepressants, and as stimulants, although many are habit forming.

As a stimulant, nicotine increases the level of adrenaline in the blood, which increases the heart rate and blood pressure. It is well known that smoking cigarettes can damage the lungs and that exposure to tars and other carcinogens in cigarette smoke can lead to lung cancer. However, nicotine is responsible for the addiction of smoking. Coniine, which is obtained from hemlock, is an extremely toxic alkaloid.

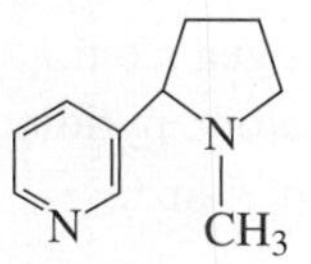

Nicotine

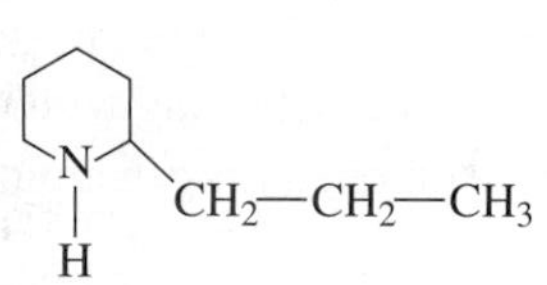

$CH_2—CH_2—CH_3$

Coniine

Caffeine is a stimulant of the central nervous system. Present in coffee, tea, soft drinks, energy drinks, chocolate, and cocoa, caffeine increases alertness but may cause nervousness and insomnia. Caffeine is also used in certain pain relievers to counteract the drowsiness caused by an antihistamine.

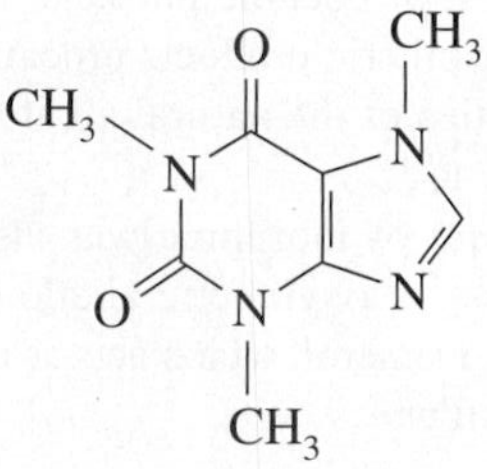

Caffeine

Caffeine is a stimulant found in coffee, tea, energy drinks, and chocolate.

Several alkaloids are used in medicine. Quinine obtained from the bark of the cinchona tree has been used in the treatment of malaria since the 1600s. Atropine from belladonna is used in low concentrations to accelerate slow heart rates and as an anesthetic for eye examinations.

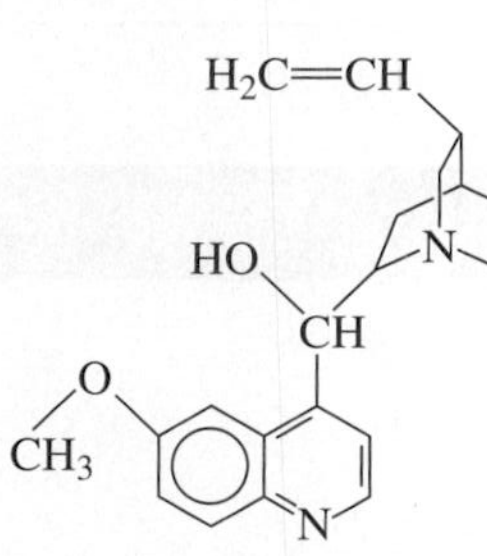

Quinine

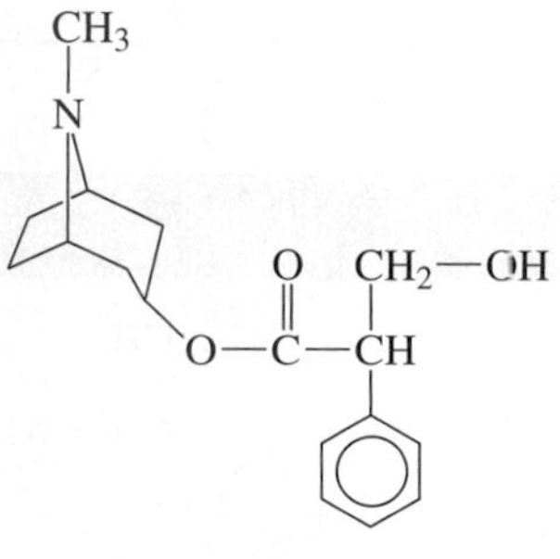

Atropine

For many centuries, morphine and codeine, alkaloids found in the oriental poppy plant, have been used as effective painkillers. Codeine, which is structurally similar to morphine, is used in some prescription painkillers and cough syrups. Heroin, obtained by a chemical modification of morphine, is strongly addicting and is no longer used medically.

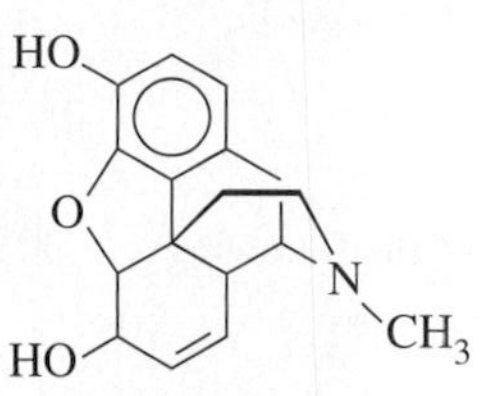

The sap of the opium poppy contains the alkaloids morphine and codeine.

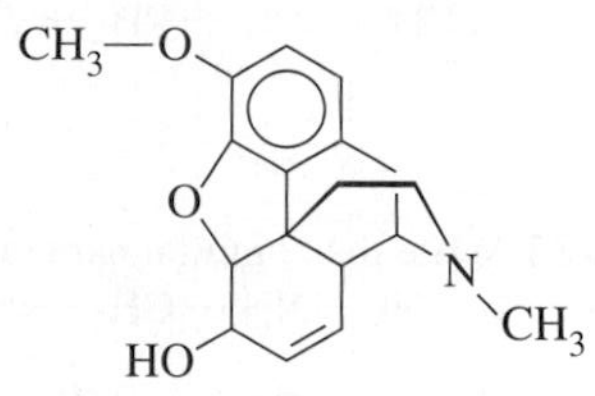

Morphine

Codeine

CONCEPT CHECK 14.5

Heterocyclic Amines

Identify the heterocyclic amines in the alkaloids nicotine and coniine.

ANSWER

In nicotine, the heterocyclic amines are a five-atom ring with one nitrogen atom, which is pyrrolidine, and a six-atom aromatic ring with one nitrogen atom, pyridine. Coniine contains a six-atom ring with one nitrogen atom, which is piperidine.

Chemistry Link to Health

SYNTHESIZING DRUGS

One area of research in pharmacology is the synthesis of compounds that retain the anesthetic characteristic of naturally occurring alkaloids such as cocaine and morphine without the addictive side effects. For example, cocaine is an effective anesthetic, but it is addictive.

Research chemists modified the structure of cocaine but kept the benzene group and nitrogen atom. The synthetic products procaine and lidocaine retain the anesthetic qualities of the natural alkaloid without the addictive side effects.

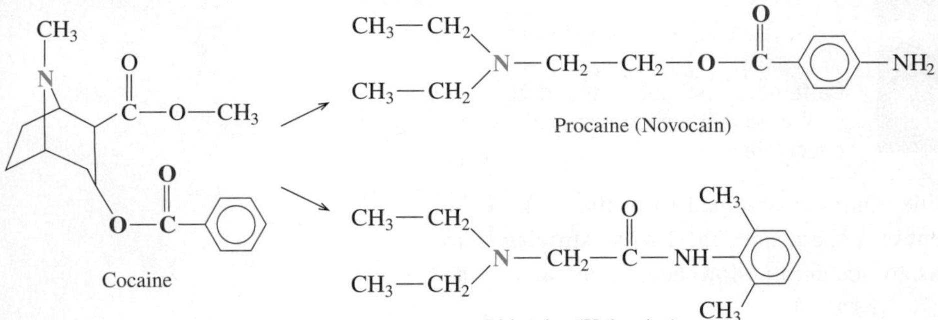

The structure of morphine was also modified to make a synthetic alkaloid, meperidine, or Demerol, which acts as an effective painkiller.

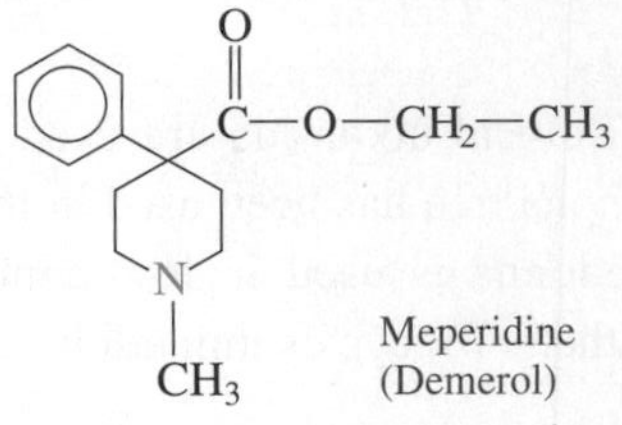

Cocaine

$(CH_3-CH_2)_2N-CH_2-CH_2-O-C(=O)-C_6H_4-NH_2$

Procaine (Novocain)

$(CH_3-CH_2)_2N-CH_2-C(=O)-NH-C_6H_3(CH_3)_2$

Lidocaine (Xylocaine)

Meperidine (Demerol)

QUESTIONS AND PROBLEMS

Amines

14.29 Classify each of the following amines as primary (1°), secondary (2°), or tertiary (3°):

a. $CH_3-CH_2-CH_2-NH_2$

b. 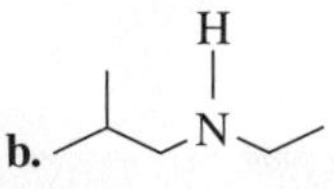 (skeletal structure with N–H)

c. 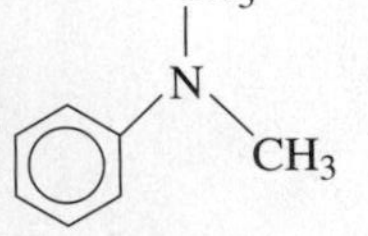 (benzene ring bonded to N with two CH_3 groups)

d. 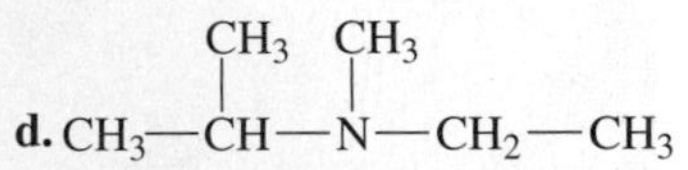 $CH_3-CH(CH_3)-N(CH_3)-CH_2-CH_3$

14.30 Classify each of the following amines as primary (1°), secondary (2°), or tertiary (3°):

a. 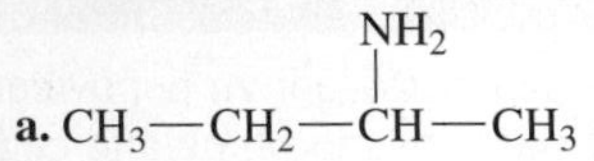 $CH_3-CH_2-CH(NH_2)-CH_3$

b. 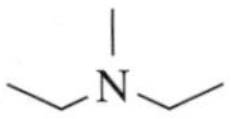 (skeletal structure with N)

c. 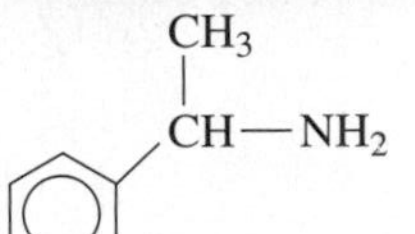 (benzene ring bonded to $CH(CH_3)-NH_2$)

d. 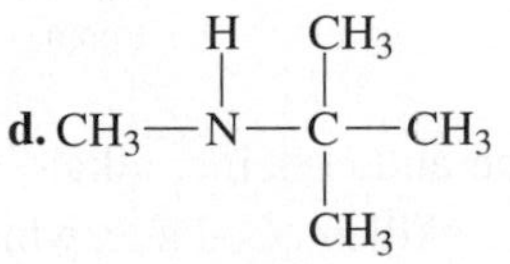 $CH_3-NH-C(CH_3)_2-CH_3$

14.31 Write the common name for each of the following:

a. $CH_3-CH_2-CH_2-NH_2$

b. $CH_3-NH-CH_2-CH_2-CH_3$

c. 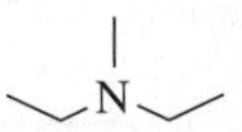 (skeletal structure with N)

14.32 Write the common name for each of the following:

a. $CH_3-NH-CH_2-CH_3$

b. 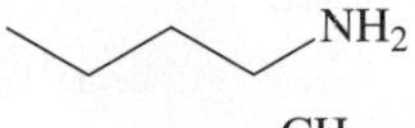 (skeletal chain with NH_2)

c. 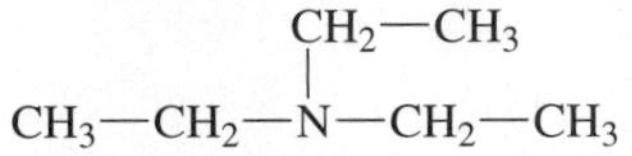 $CH_3-CH_2-N(CH_2-CH_3)-CH_2-CH_3$

14.33 Draw the condensed structural formula for each of the following amines:
a. ethylamine **b.** *N*-methylaniline
c. butylpropylamine

14.34 Draw the condensed structural formula for each of the following amines:
a. diethylamine **b.** 4-chloroaniline
c. *N,N*-diethylaniline

14.35 Indicate if each of the following is soluble in water. Explain.
a. $CH_3-CH_2-NH_2$ **b.** $CH_3-NH-CH_3$
c. 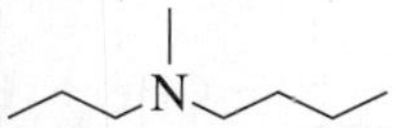 (skeletal structure with N)
d. $CH_3-CH(NH_2)-CH_2-CH_3$

14.36 Indicate if each of the following is soluble in water. Explain.
a. $CH_3-CH_2-CH_2-NH_2$
b. $CH_3-CH_2-CH_2-NH-CH_2-CH_3$
c. $CH_3-N(CH_3)-CH_3$
d. 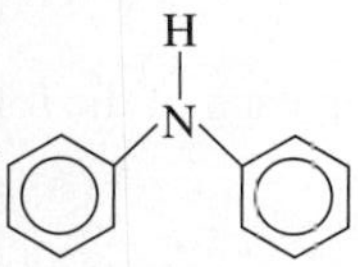 (diphenylamine structure, N—H)

14.37 Write a balanced equation for the ionization of each of the following amines in water:
a. methylamine **b.** dimethylamine **c.** aniline

14.38 Write a balanced equation for the ionization of each of the following amines in water:
a. ethylmethylamine **b.** propylamine **c.** *N*-methylaniline

14.39 Identify the heterocyclic amines in each of the following:
a. 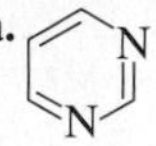 (ring structure with two N) **b.** 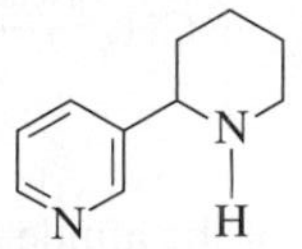 (structure)

Anabasine from tree tobacco

14.40 Identify the heterocyclic amines in each of the following:
a. 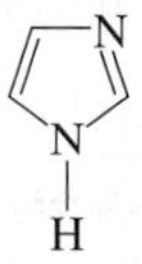 (ring structure with N and N—H)

b. 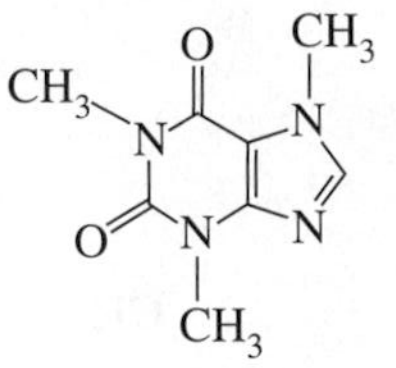 (structure)

Caffeine

14.41 Low levels of serotonin in the brain appear to be associated with depressed states. What type of heterocyclic amine is serotonin?

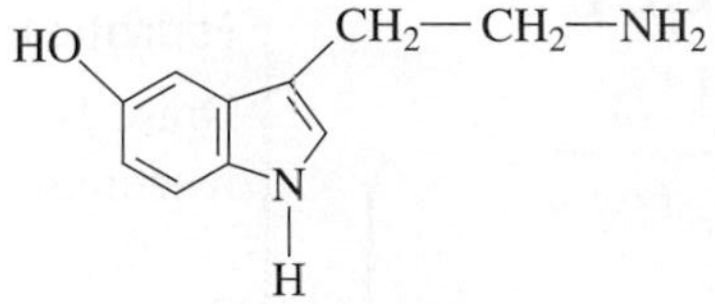

Serotonin

14.42 Bufotenin is a poisonous alkaloid found in the skin of some toads and *amanita* mushrooms. What is the type of heterocyclic amine in bufotenin?

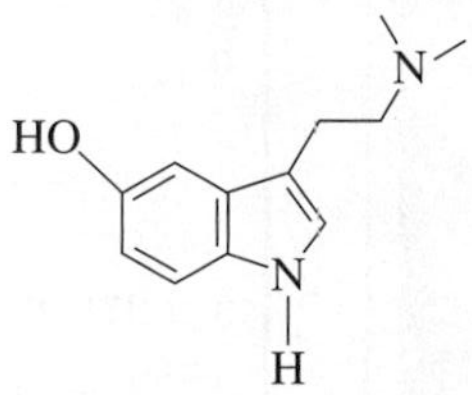

Bufotenin

14.5 Amides

LEARNING GOAL
Write the common and IUPAC names of amides and the products of formation and hydrolysis.

The **amides** are derivatives of carboxylic acids in which a nitrogen group replaces the hydroxyl group.

Amidation

An amide is produced in a reaction called *amidation*, in which a carboxylic acid reacts with ammonia or an amine. An —OH is removed from the carboxylic acid and an —H is removed from ammonia or an amine to form water. The remaining part of the carboxylic acid and ammonia or amine molecule join to form an amide, much like the

TUTORIAL
Amidation Reactions

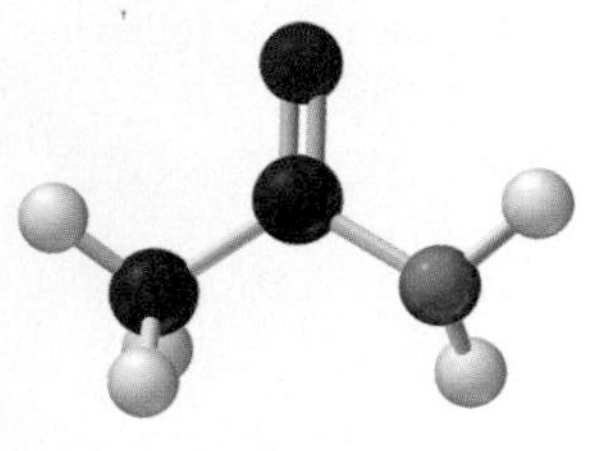
Ethanamide (acetamide)

formation of an ester. Because a hydrogen atom must be removed from ammonia or an amine, only primary and secondary amines undergo amidation.

$$\underset{\text{Ethanoic acid (acetic acid)}}{CH_3-\overset{O}{\overset{\|}{C}}-\mathbf{OH}} + \underset{\text{Ammonia}}{\mathbf{H}-\overset{H}{\overset{|}{N}}-H} \xrightarrow{\text{Heat}} \underset{\text{Ethanamide (acetamide)}}{CH_3-\boxed{\overset{O}{\overset{\|}{C}}-\overset{H}{\overset{|}{N}}}-H} + \mathbf{H_2O}$$

$$\underset{\text{Ethanoic acid (acetic acid)}}{CH_3-\overset{O}{\overset{\|}{C}}-\mathbf{OH}} + \underset{\text{Methylamine}}{\mathbf{H}-\overset{H}{\overset{|}{N}}-CH_3} \xrightarrow{\text{Heat}} \underset{\text{N-Methylethanamide (N-methylacetamide)}}{CH_3-\boxed{\overset{O}{\overset{\|}{C}}-\overset{H}{\overset{|}{N}}}-CH_3} + \mathbf{H_2O}$$

SAMPLE PROBLEM 14.9

Formation of Amides

Draw the condensed structural formula of the amide product in each of the following reactions:

a. $C_6H_5-\overset{O}{\overset{\|}{C}}-OH + NH_3 \xrightarrow{\text{Heat}}$

b. $CH_3-\overset{O}{\overset{\|}{C}}-OH + NH_2-CH_2-CH_3 \xrightarrow{\text{Heat}}$

SOLUTION

a. The condensed structural formula of the amide product can be written by attaching the carbonyl group from the acid to the nitrogen atom of the amine. —OH is removed from the acid and —H from the amine to form water.

$$C_6H_5-\overset{O}{\overset{\|}{C}}-NH_2$$

b. $CH_3-\overset{O}{\overset{\|}{C}}-\overset{H}{\overset{|}{N}}-CH_2-CH_3$

STUDY CHECK 14.9

What are the condensed structural formulas of the carboxylic acid and amine needed to prepare the following amide? (*Hint*: Separate the N and C═O of the amide group, and add —H and —OH to give the original amine and carboxylic acid.)

$$H-\overset{O}{\overset{\|}{C}}-\overset{CH_3}{\overset{|}{N}}-CH_3$$

Career Focus

CLINICAL LABORATORY TECHNOLOGIST

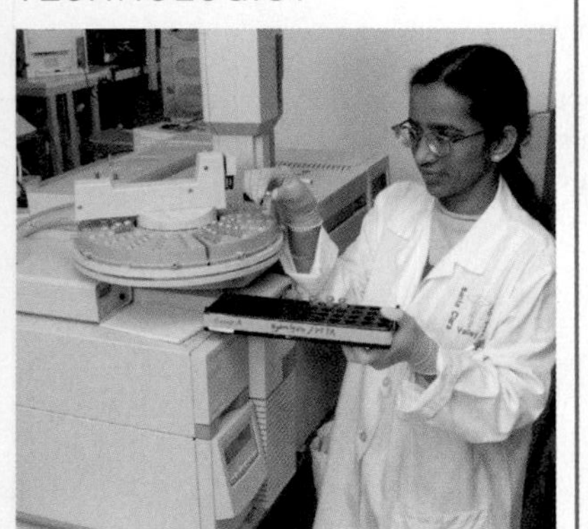

"We use mass spectrometry to analyze and confirm the presence of drugs," says Valli Vairavan, clinical lab technologist—Mass Spectrometry, Santa Clara Valley Medical Center. "A mass spectrometer separates and identifies compounds, including drugs, by mass. When we screen a urine sample, we look for metabolites, which are the products of drugs that have metabolized in the body. If the presence of one or more drugs such as heroin and cocaine is indicated, we confirm it by using mass spectrometry."

Drugs or their metabolites are detected in urine 24–48 hours after use. Cocaine metabolizes to benzoylecgonine and hydroxycocaine, morphine to morphine-3-glucuronide, and heroin to acetylmorphine. Amphetamines and methamphetamines are detected unchanged.

Naming Simple Amides

In both the common and IUPAC names, simple amides are named by dropping the *oic acid* or *ic acid* from the carboxylic acid names and adding the suffix *amide*. An alkyl group attached to the nitrogen of an amide is named with the prefix *N*-, followed by the alkyl name.

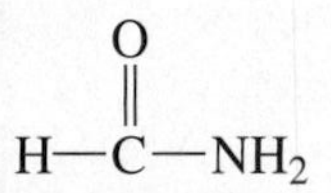

$$\mathrm{H{-}\overset{\overset{\displaystyle O}{\|}}{C}{-}NH_2} \qquad \mathrm{CH_3{-}\overset{\overset{\displaystyle O}{\|}}{C}{-}NH_2} \qquad \mathrm{CH_3{-}CH_2{-}\overset{\overset{\displaystyle O}{\|}}{C}{-}\overset{\overset{\displaystyle H}{|}}{N}{-}CH_3}$$

Methanamide (formamide) 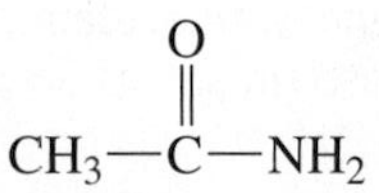 — Ethanamide (acetamide) 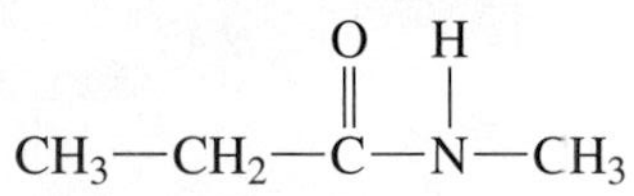 — *N*-Methylpropanamide (*N*-methylpropionamide) 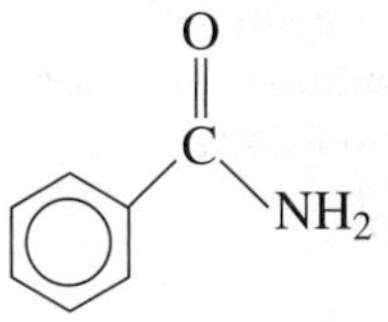 — Benzamide

SAMPLE PROBLEM 14.10

Naming Amides

Give the IUPAC and common names for

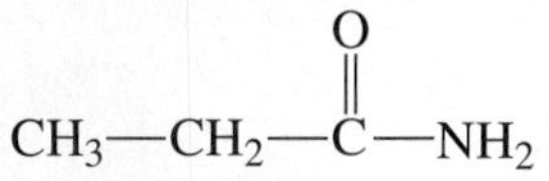

$$\mathrm{CH_3{-}CH_2{-}\overset{\overset{\displaystyle O}{\|}}{C}{-}NH_2}$$

SOLUTION

The IUPAC name of the carboxylic acid is propanoic acid; the common name is propionic acid. Replacing the *oic acid* or *ic acid* ending with *amide* gives the IUPAC name of propanamide and the common name of propionamide.

STUDY CHECK 14.10

Draw the condensed structural formula of benzamide.

Solubility of Amides

The amides do not have the properties of bases that we saw for the amines. The amides with one to five carbon atoms are soluble in water because they can hydrogen bond with water molecules. However, in amides with more than five carbon atoms, the effect of hydrogen bonding is diminished as the longer carbon chain decreases the solubility of an amide in water.

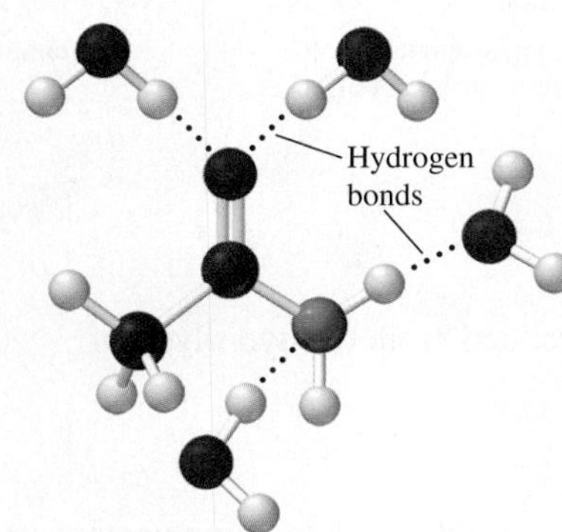

Chemistry Link to Health

AMIDES IN HEALTH AND MEDICINE

The simplest natural amide is urea, an end product of protein metabolism in the body. The kidneys remove urea from the blood and provide for its excretion in urine. If the kidneys malfunction, urea is not removed and builds to a toxic level, a condition called uremia. Urea is also used as a component of fertilizer, to increase nitrogen in the soil.

$$NH_2—\overset{\large O}{\overset{\|}{C}}—NH_2$$ Urea

Many barbiturates are cyclic amides of barbituric acid that act as sedatives in small dosages or sleep inducers in larger dosages. They are often habit forming. Barbiturate drugs include phenobarbital (Luminal) and pentobarbital (Nembutal).

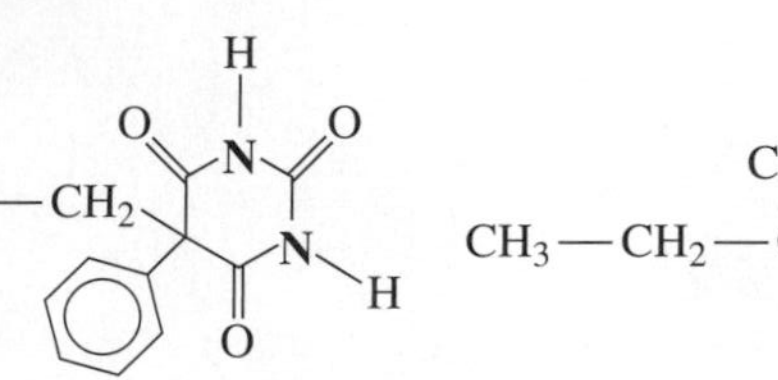

Luminal (phenobarbital)

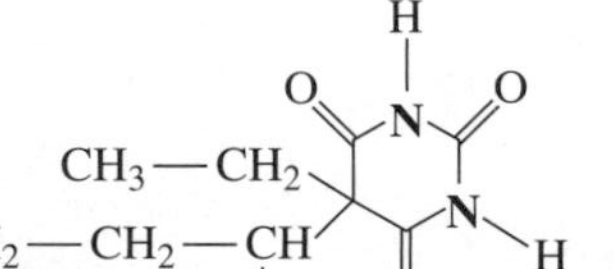

Nembutal (pentobarbital)

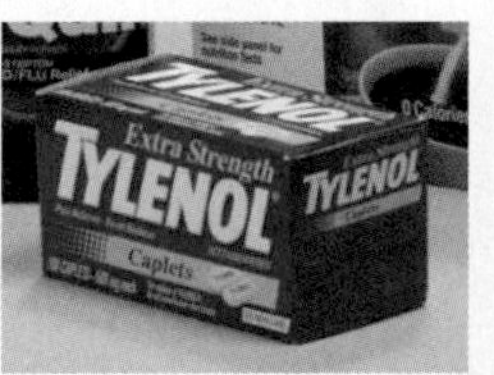

Tylenol with acetaminophen is an aspirin substitute.

Aspirin substitutes contain phenacetin or acetaminophen, which is used in Tylenol. Like aspirin, acetaminophen reduces fever and pain, but it has little anti-inflammatory effect.

$$CH_3—CH_2—O—C_6H_4—\underset{\large H}{N}—\overset{\large O}{\overset{\|}{C}}—CH_3$$

Phenacetin

$$HO—C_6H_4—\underset{\large H}{N}—\overset{\large O}{\overset{\|}{C}}—CH_3$$

Acetaminophen

Hydrolysis of Amides

TUTORIAL
Hydrolysis of Amides

Amides undergo hydrolysis when water is added to the amide bond to split the molecule. When an acid is used, the hydrolysis products of an amide are the carboxylic acid and the ammonium salt. In base hydrolysis, the amide produces the salt of the carboxylic acid and ammonia or an amine.

Acid Hydrolysis of Amides

$$CH_3—\overset{\large O}{\overset{\|}{C}}—NH_2 + HOH + HCl \rightleftarrows CH_3—\overset{\large O}{\overset{\|}{C}}—OH + NH_4^+Cl^-$$

Ethanamide (acetamide); Ethanoic acid (acetic acid); Ammonium chloride

Base Hydrolysis of Amides

$$CH_3—CH_2—\overset{\large O}{\overset{\|}{C}}—NH—CH_3 + NaOH \longrightarrow CH_3—CH_2—\overset{\large O}{\overset{\|}{C}}—O^-Na^+ + NH_2—CH_3$$

N-Methylpropanamide (*N*-methylpropionamide); Sodium propanoate, a salt (sodium propionate); Methylamine

SAMPLE PROBLEM 14.11

Hydrolysis of Amides

Draw the condensed structural formulas of the products from the hydrolysis of *N*-methylpentanamide with NaOH.

SOLUTION

Hydrolysis of the amide with a base produces a carboxylate salt (sodium pentanoate) and the corresponding amine (methylamine).

$$CH_3-CH_2-CH_2-CH_2-\overset{\overset{\displaystyle O}{\|}}{C}-O^-Na^+ + NH_2-CH_3$$

STUDY CHECK 14.11

What are the condensed structural formulas of the products from the hydrolysis of *N*-methylbutyramide with HBr?

QUESTIONS AND PROBLEMS

Amides

14.43 Draw the condensed structural formula of the amide formed in each of the following reactions:

a. $CH_3-\overset{\overset{\displaystyle O}{\|}}{C}-OH + NH_3 \xrightarrow{\text{Heat}}$

b. $CH_3-\overset{\overset{\displaystyle O}{\|}}{C}-OH + NH_2-CH_2-CH_3 \xrightarrow{\text{Heat}}$

c. $C_6H_5-\overset{\overset{\displaystyle O}{\|}}{C}-OH + NH_2-CH_2-CH_2-CH_3 \xrightarrow{\text{Heat}}$

14.44 Draw the condensed structural formula of the amide formed in each of the following reactions:

a. $CH_3-CH_2-CH_2-CH_2-\overset{\overset{\displaystyle O}{\|}}{C}-OH + NH_3 \xrightarrow{\text{Heat}}$

b. $CH_3-\overset{\overset{\displaystyle CH_3}{|}}{CH}-CH_2-\overset{\overset{\displaystyle O}{\|}}{C}-OH + NH_2-CH_2-CH_2-CH_3 \xrightarrow{\text{Heat}}$

c. $CH_3-CH_2-\overset{\overset{\displaystyle O}{\|}}{C}-OH + C_6H_5-NH_2 \xrightarrow{\text{Heat}}$

14.45 Give the IUPAC and common names (if any) for each of the following amides:

a. $CH_3-\overset{\overset{\displaystyle O}{\|}}{C}-NH_2$ **b.** $CH_3-CH_2-CH_2-\overset{\overset{\displaystyle O}{\|}}{C}-NH_2$

c. $H-\overset{\overset{\displaystyle O}{\|}}{C}-NH_2$

14.46 Give the IUPAC and common names (if any) for each of the following amides:

a. $CH_3-CH_2-CH_2-CH_2-\overset{\overset{\displaystyle O}{\|}}{C}-NH_2$

b. $C_6H_5-\overset{\overset{\displaystyle O}{\|}}{C}-NH_2$

c. $CH_3-CH_2-CH_2-CH_2-CH_2-\overset{\overset{\displaystyle O}{\|}}{C}-NH_2$

14.47 Draw the condensed structural formula for each of the following amides:
a. propionamide
b. 2-methylpentanamide
c. methanamide

14.48 Draw the condensed structural formula for each of the following amides:
a. formamide **b.** 3-chlorobenzamide
c. 3-methylbutyramide

14.49 Draw the condensed structural formulas for the products of the hydrolysis of each of the following amides with HCl:

a. $CH_3-\overset{\overset{\displaystyle O}{\|}}{C}-NH_2$ **b.** $CH_3-CH_2-\overset{\overset{\displaystyle O}{\|}}{C}-NH_2$

c. $CH_3-CH_2-CH_2-\overset{\overset{\displaystyle O}{\|}}{C}-NH-CH_3$

d. $C_6H_5-\overset{\overset{\displaystyle O}{\|}}{C}-NH_2$

e. *N*-ethylpentanamide

14.50 Draw the condensed structural formulas for the products of the hydrolysis of each of the following amides with NaOH:

a. $CH_3-CH_2-\overset{\overset{\displaystyle CH_3}{|}}{CH}-\overset{\overset{\displaystyle O}{\|}}{C}-NH_2$

b. $CH_3-CH_2-CH_2-\overset{\overset{\displaystyle O}{\|}}{C}-\overset{\overset{\displaystyle CH_2-CH_3}{|}}{N}-CH_2-CH_3$

c. $C_6H_5-\overset{\overset{\displaystyle O}{\|}}{C}-\overset{\overset{\displaystyle CH_3}{|}}{N}-CH_2-CH_2-CH_2-CH_3$

d. $CH_3-\overset{\overset{\displaystyle Cl}{|}}{CH}-\overset{\overset{\displaystyle O}{\|}}{C}-\overset{\overset{\displaystyle CH_3}{|}}{N}-CH_2-CH_3$

e. $C_6H_5-\overset{\overset{\displaystyle O}{\|}}{C}-\overset{\overset{\displaystyle H}{|}}{N}-CH_2-CH_2-CH_3$

CONCEPT MAP

CARBOXYLIC ACIDS, ESTERS, AMINES, AND AMIDES

- **Carboxylic acids**
  - have a → **Carboxyl group**
  - are → **Polar** and **Soluble in water up to 5 C atoms**
  - ionize as → **Weak acids**
  - form → **Amides** (which hydrolyze → Amines)
  - react with → **Alcohols** to form → **Esters**; Esters hydrolyze → Alcohols
- **Amines**
  - have a → **Nitrogen atom**
  - are → **Soluble in water up to 6 C atoms**
  - ionize as → **Weak bases**

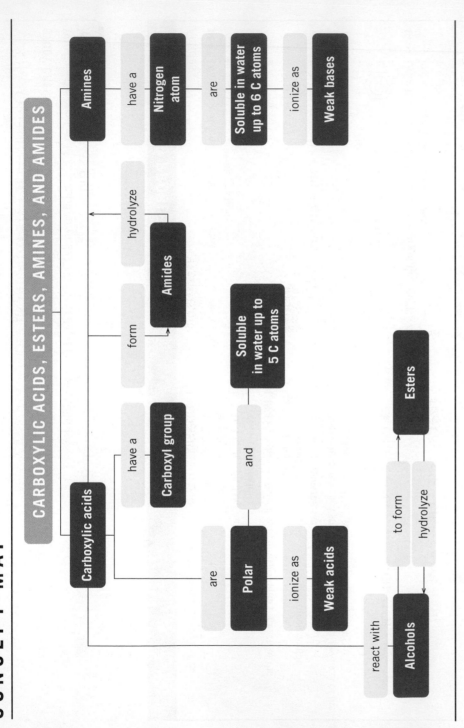

CHAPTER REVIEW

14.1 Carboxylic Acids

Learning Goal: Give the IUPAC names and common names of carboxylic acids; draw their condensed structural formulas.

A carboxylic acid contains the carboxyl functional group, which is a hydroxyl group connected to the carbonyl group. In the IUPAC system, carboxylic acids are named by replacing the *e* in the name of the corresponding alkane with *oic acid*.

14.2 Properties of Carboxylic Acids

Learning Goal: Describe the solubility and ionization of carboxylic acids in water.

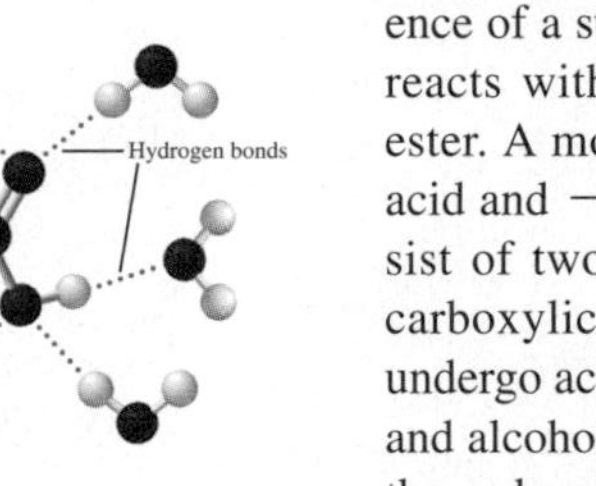

The carboxyl group contains polar bonds of O—H and C=O, which makes a carboxylic acid with one to four carbon atoms very soluble in water. As weak acids, carboxylic acids ionize slightly by donating a proton to water to form carboxylate and hydronium ions. Carboxylic acids are neutralized by base, producing the carboxylate salt and water.

14.3 Esters

Learning Goal: Name an ester; write balanced equations for the formation and hydrolysis of an ester.

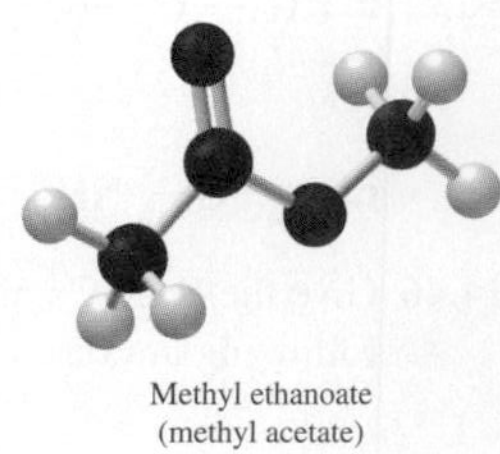
Methyl ethanoate (methyl acetate)

In an ester, an alkyl or aromatic group has replaced the H of the hydroxyl group of a carboxylic acid. In the presence of a strong acid, a carboxylic acid reacts with an alcohol to produce an ester. A molecule of water is removed: —OH from the carboxylic acid and —H from the alcohol molecule. The names of esters consist of two words, one from the alcohol and the other from the carboxylic acid with the *ic acid* ending replaced by *ate*. Esters undergo acid hydrolysis by adding water to yield the carboxylic acid and alcohol. Base hydrolysis, or saponification, of an ester produces the carboxylate salt and an alcohol.

14.4 Amines

Learning Goal: Classify amines as primary (1°), secondary (2°), or tertiary (3°). Name amines using common names; draw the condensed structural formulas given the names. Describe the solubility, ionization, and neutralization of amines.

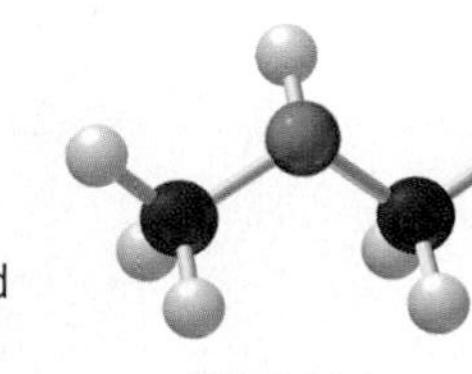
Dimethylamine

A nitrogen atom attached to one, two, or three alkyl or aromatic groups forms a primary, secondary, or tertiary amine. Many amines, synthetic or naturally occurring, have physiological activity. Heterocyclic amines contain a nitrogen atom in a ring. In the common names of simple amines, the alkyl groups are listed alphabetically followed by the suffix *amine*. Small amines are soluble in water. In water, amines act as weak bases to produce ammonium and hydroxide ions. When amines react with acids, they form amine salts. As ionic compounds, amine salts are solids, soluble in water, and odorless compared to the amines.

14.5 Amides

Learning Goal: Write the common and IUPAC names of amides and the products of formation and hydrolysis.

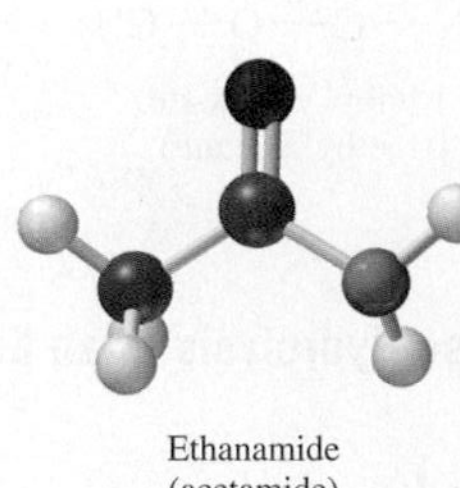
Ethanamide (acetamide)

Amides are derivatives of carboxylic acids in which the hydroxyl group is replaced by a nitrogen group. Amides are named by replacing the *ic acid* or *oic acid* with *amide*. Hydrolysis of an amide by an acid produces an amine salt and a carboxylic acid. Hydrolysis by a base produces the salt of the carboxylic acid and an amine.

Summary of Naming

Family	Condensed Structural Formula	IUPAC Name	Common Name
Carboxylic acid	$CH_3-\overset{O}{\overset{\|\|}{C}}-OH$	Ethanoic acid	Acetic acid
Ester	$CH_3-\overset{O}{\overset{\|\|}{C}}-O-CH_3$	Methyl ethanoate	Methyl acetate
Amine	$CH_3-CH_2-NH_2$		Ethylamine
Amide	$CH_3-\overset{O}{\overset{\|\|}{C}}-NH_2$	Ethanamide	Acetamide

Summary of Reactions

Ionization of a Carboxylic Acid in Water

$$\underset{\text{Ethanoic acid (acetic acid)}}{CH_3-\overset{O}{\overset{\|}{C}}-OH} + H_2O \rightleftarrows \underset{\text{Ethanoate ion (acetate ion)}}{CH_3-\overset{O}{\overset{\|}{C}}-O^-} + \underset{\text{Hydronium ion}}{H_3O^+}$$

Neutralization of a Carboxylic Acid

$$\underset{\text{Propanoic acid (propionic acid)}}{CH_3-CH_2-\overset{O}{\overset{\|}{C}}-OH} + \underset{\text{Sodium hydroxide}}{NaOH} \longrightarrow \underset{\text{Sodium propanoate (sodium propionate)}}{CH_3-CH_2-\overset{O}{\overset{\|}{C}}-O^-Na^+} + H_2O$$

Esterification: Carboxylic Acid and an Alcohol

$$\underset{\text{Ethanoic acid (acetic acid)}}{CH_3-\overset{O}{\overset{\|}{C}}-OH} + \underset{\text{Methanol (methyl alcohol)}}{HO-CH_3} \underset{\text{Heat}}{\overset{H^+}{\rightleftarrows}} \underset{\text{Methyl ethanoate (methyl acetate)}}{CH_3-\overset{O}{\overset{\|}{C}}-O-CH_3} + H_2O$$

Acid Hydrolysis of an Ester

$$CH_3-\overset{\overset{O}{\|}}{C}-O-CH_3 + H-OH \overset{H^+}{\rightleftharpoons} CH_3-\overset{\overset{O}{\|}}{C}-OH + HO-CH_3$$

Methyl ethanoate (methyl acetate); Ethanoic acid (acetic acid); Methanol (methyl alcohol)

Base Hydrolysis of an Ester: Saponification

$$CH_3-CH_2-\overset{\overset{O}{\|}}{C}-O-CH_3 + NaOH \xrightarrow{Heat} CH_3-CH_2-\overset{\overset{O}{\|}}{C}-O^-Na^+ + HO-CH_3$$

Methyl propanoate (methyl propionate); Sodium hydroxide; Sodium propanoate (sodium propionate); Methanol (methyl alcohol)

Ionization of an Amine in Water

$$CH_3-\underset{\underset{H}{|}}{\overset{\overset{H}{|}}{N}} + HOH \rightleftharpoons CH_3-\underset{\underset{H}{|}}{\overset{\overset{H}{|}}{N^+}}-H + OH^-$$

Methylamine; Methylammonium ion; Hydroxide ion

Formation of an Amine Salt

$$CH_3-\underset{\underset{H}{|}}{\overset{\overset{H}{|}}{N}} + HCl \longrightarrow CH_3-\underset{\underset{H}{|}}{\overset{\overset{H}{|}}{N^+}}-H\ Cl^-$$

Methylamine; Methylammonium chloride

Amidation: Carboxylic Acid and an Amine

$$CH_3-CH_2-\overset{\overset{O}{\|}}{C}-OH + H-\overset{\overset{H}{|}}{N}-H \xrightarrow{Heat} CH_3-CH_2-\overset{\overset{O}{\|}}{C}-\overset{\overset{H}{|}}{N}-H + H_2O$$

Propanoic acid (propionic acid); Ammonia; Propanamide (propionamide)

Acid Hydrolysis of an Amide

$$CH_3-\overset{\overset{O}{\|}}{C}-NH_2 + HOH + HCl \rightleftharpoons CH_3-\overset{\overset{O}{\|}}{C}-OH + NH_4^+Cl^-$$

Ethanamide (acetamide); Ethanoic acid (acetic acid); Ammonium chloride

Base Hydrolysis of an Amide

$$CH_3-CH_2-\overset{\overset{O}{\|}}{C}-NH-CH_3 + NaOH \longrightarrow CH_3-CH_2-\overset{\overset{O}{\|}}{C}-O^-Na^+ + NH_2-CH_3$$

N-methylpropanamide (*N*-methylpropionamide); Sodium propanoate (sodium propionate); Methylamine

Key Terms

alkaloids Physiologically active amines that are produced in plants.

amides Organic compounds containing the carbonyl group attached to an amino group or a substituted nitrogen atom.

$$-\overset{\overset{\displaystyle O}{\|}}{C}-NH_2 \qquad -\overset{\overset{\displaystyle O}{\|}}{C}-\overset{|}{N}-$$

Amide

amines Organic compounds containing a nitrogen atom attached to one, two, or three hydrocarbon groups.

amine salt An ionic compound produced from an amine and an acid.

carboxyl group A functional group found in carboxylic acids composed of carbonyl and hydroxyl groups.

$$-\overset{\overset{\displaystyle O}{\|}}{C}-OH$$

Carboxyl group

carboxylate ion The anion produced when a carboxylic acid donates a proton to water.

carboxylic acids A family of organic compounds containing the carboxyl group.

$$-\overset{\overset{\displaystyle O}{\|}}{C}-OH$$

Carboxylic acid

carboxylate salt A carboxylate ion and the metal ion from the base that is the product of neutralization of a carboxylic acid.

esterification The formation of an ester from a carboxylic acid and an alcohol, with the elimination of a molecule of water in the presence of an acid catalyst.

esters A family of organic compounds in which an alkyl group replaces the hydrogen atom in a carboxylic acid.

$$-\overset{\overset{\displaystyle O}{\|}}{C}-O-\underset{|}{\overset{|}{C}}-$$

Ester

hydrolysis The splitting of a molecule by the addition of water. Esters hydrolyze to produce a carboxylic acid and an alcohol. Amides yield the corresponding carboxylic acid and amine or their salts.

saponification The hydrolysis of an ester with a strong base to produce a salt of the carboxylic acid and an alcohol.

Understanding the Concepts

14.51 There are four amine isomers with the molecular formula C_3H_9N. Draw their condensed structural formulas. Name and classify each as a primary (1°), secondary (2°), or tertiary (3°) amine.

14.52 There are four amide isomers with the molecular formula C_3H_7NO. Draw their condensed structural formulas.

14.53 Neo-Synephrine is the active ingredient in some nose sprays used to reduce swelling of nasal membranes. What functional groups are in the structure of Neo-Synephrine?

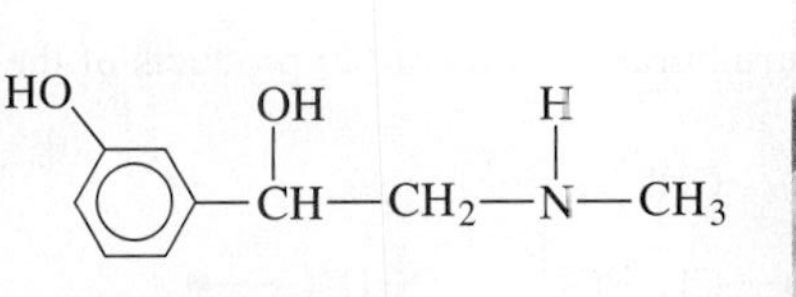

Neo-Synephrine

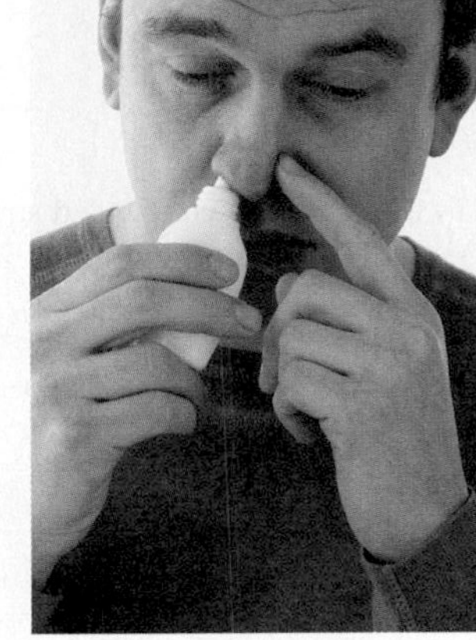

14.54 Melatonin is a naturally occurring compound in plants and animals, where it regulates the biological time clock. Melatonin is sometimes used to counteract jet lag. What functional groups are in the structure of melatonin?

O—CH₃ / HN / HN / CH₃ / O

Melatonin

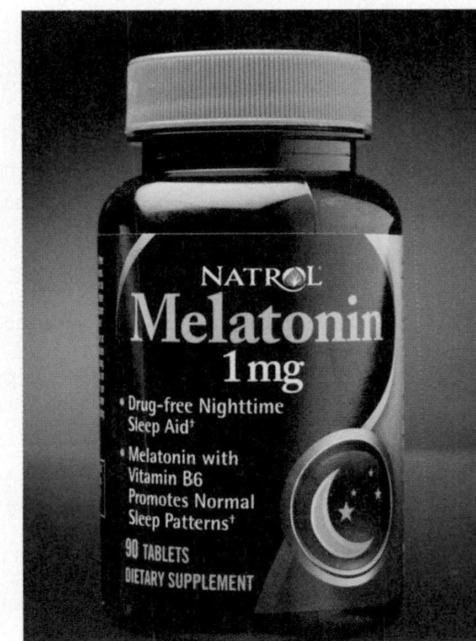

Additional Questions and Problems

For instructor-assigned homework go to **www.masteringchemistry.com**.

14.55 Give the IUPAC and common names (if any) for each of the following compounds:

a. 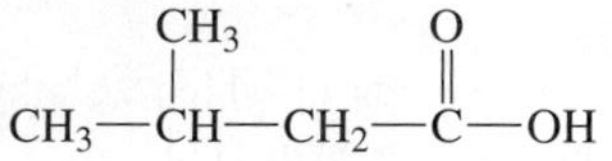 $CH_3—\overset{CH_3}{\overset{|}{CH}}—CH_2—\overset{O}{\overset{\|}{C}}—OH$

b. $\overset{O}{\overset{\|}{C}}—O—CH_2—CH_3$

c. $CH_3—CH_2—O—\overset{O}{\overset{\|}{C}}—CH_2—CH_3$

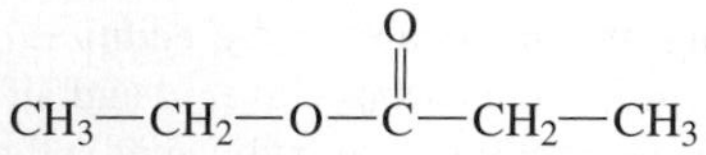

d. 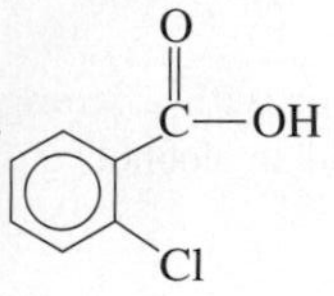 $\overset{O}{\overset{\|}{C}}—OH$, Cl

e. 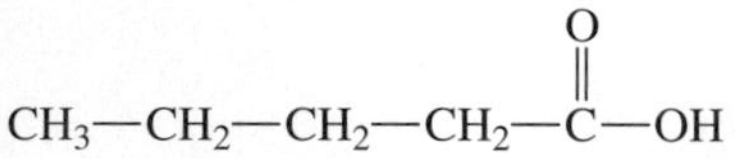 $CH_3—CH_2—CH_2—CH_2—\overset{O}{\overset{\|}{C}}—OH$

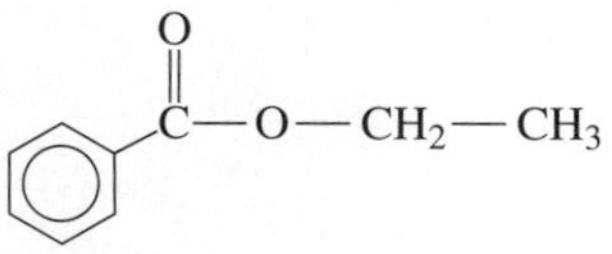

14.56 Give the IUPAC and common names (if any) for each of the following compounds:

a. 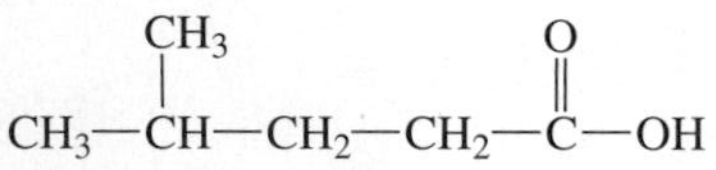 $CH_3—\overset{CH_3}{\overset{|}{CH}}—CH_2—CH_2—\overset{O}{\overset{\|}{C}}—OH$

b. $\overset{O}{\overset{\|}{C}}—OH$, Cl, Cl

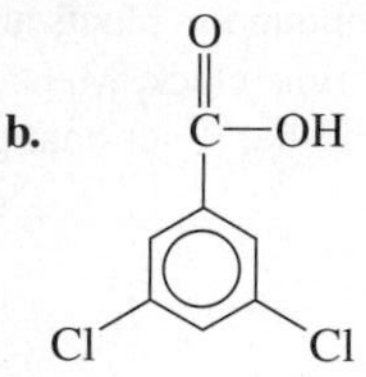

c. $\overset{O}{\overset{\|}{C}}—O—CH_3$

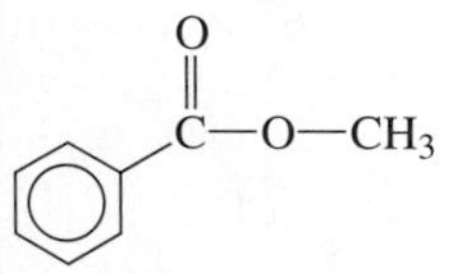

d. 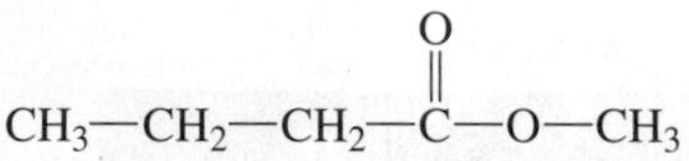 $CH_3—CH_2—CH_2—\overset{O}{\overset{\|}{C}}—O—CH_3$

e. 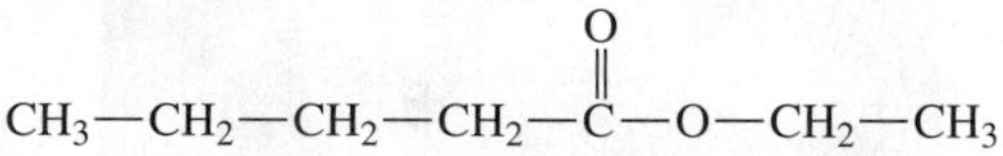 $CH_3—CH_2—CH_2—CH_2—\overset{O}{\overset{\|}{C}}—O—CH_2—CH_3$

14.57 Draw the condensed structural formula of each of the following:

a. methyl acetate
b. ethyl butanoate
c. 3-methylpentanoic acid
d. ethyl benzoate

14.58 Draw the condensed structural formula of each of the following:

a. pentyl formate
b. 2,3-dichlorobutanoic acid
c. 3,5-dimethylhexanoic acid
d. butyl propanoate

14.59 Draw the condensed structural formula of the products of the following reactions:

a. 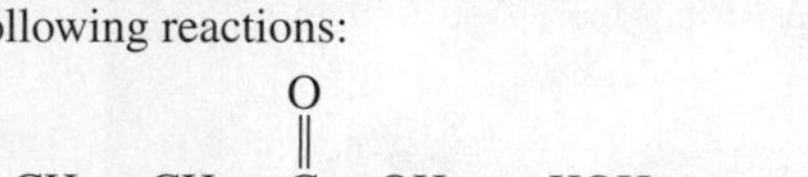 $CH_3—CH_2—\overset{O}{\overset{\|}{C}}—OH + KOH \longrightarrow$

b. $CH_3—CH_2—\overset{O}{\overset{\|}{C}}—OH + HO—CH_3 \overset{H^+}{\rightleftarrows}$

c. $\overset{O}{\overset{\|}{C}}—OH + HO—CH_2—CH_3 \overset{H^+}{\rightleftarrows}$

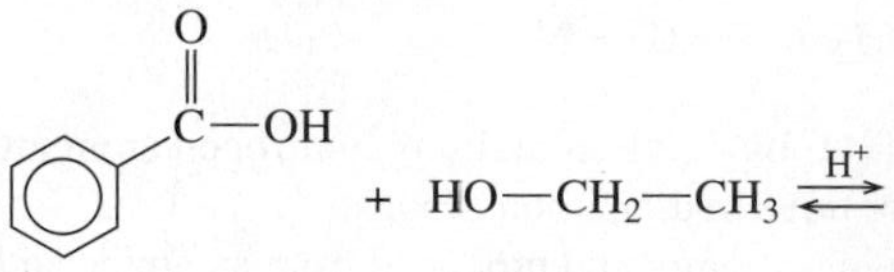

d. 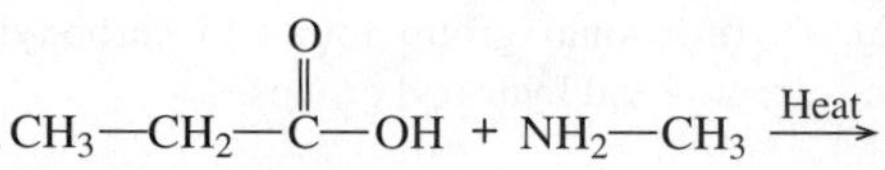 $CH_3—CH_2—\overset{O}{\overset{\|}{C}}—OH + NH_2—CH_3 \xrightarrow{\text{Heat}}$

14.60 Draw the condensed structural formula of the products of the following reactions:

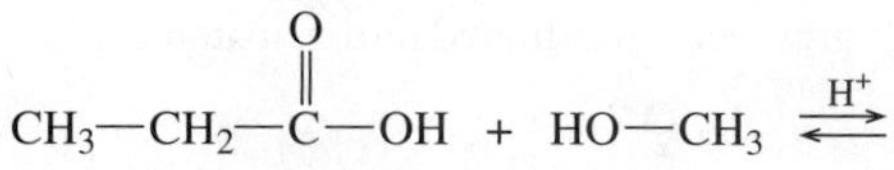

a. 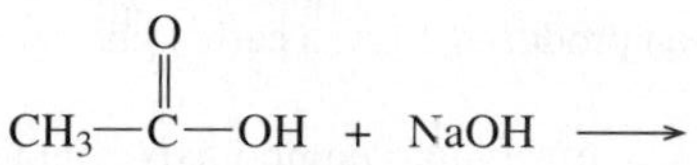 $CH_3—\overset{O}{\overset{\|}{C}}—OH + NaOH \longrightarrow$

b. $CH_3—\overset{CH_3}{\overset{|}{CH}}—\overset{O}{\overset{\|}{C}}—OH + KOH \longrightarrow$

c. $CH_3—\overset{CH_3}{\overset{|}{CH}}—\overset{O}{\overset{\|}{C}}—OH + HO—CH_3 \overset{H^+}{\rightleftarrows}$

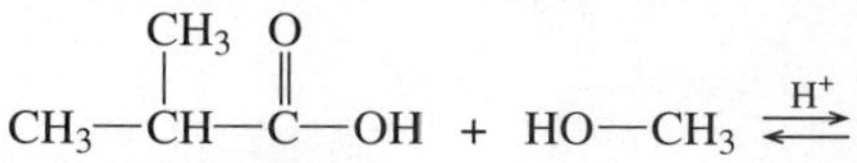

d. 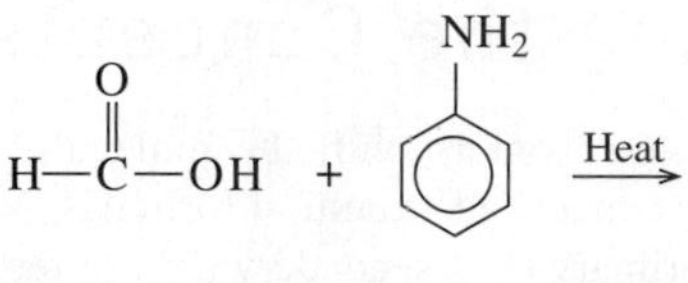 $H—\overset{O}{\overset{\|}{C}}—OH$ + NH_2 $\xrightarrow{\text{Heat}}$

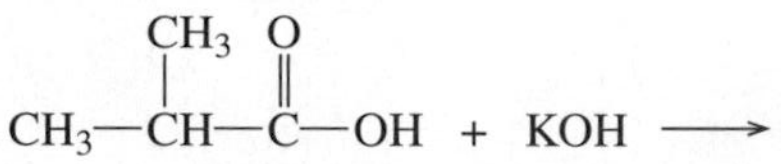

14.61 Draw the condensed structural formula of the products of the following reactions:

a. 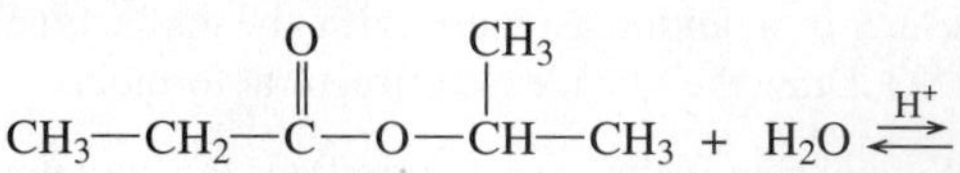 $CH_3—CH_2—\overset{O}{\overset{\|}{C}}—O—\overset{CH_3}{\overset{|}{CH}}—CH_3 + H_2O \overset{H^+}{\rightleftarrows}$

b. $CH_3—\overset{CH_3}{\overset{|}{CH}}—\overset{O}{\overset{\|}{C}}—O—CH_2—CH_2—CH_3 + H_2O \overset{H^+}{\rightleftarrows}$

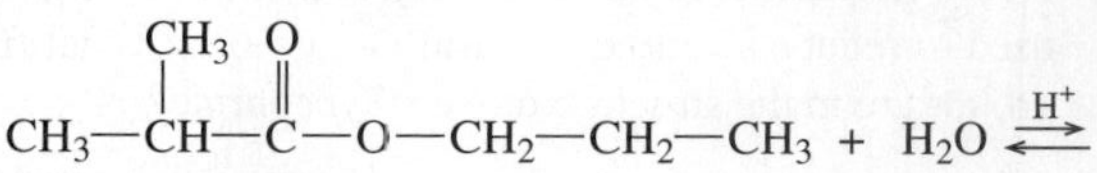

c. NH_2 + HCl $\longrightarrow$

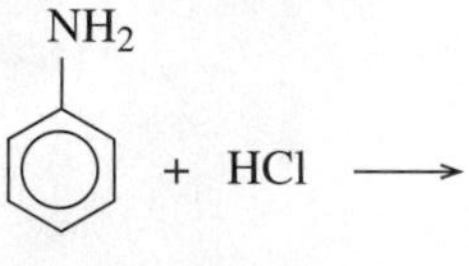

14.62 Draw the condensed structural formula of the products of the following reactions:

a. 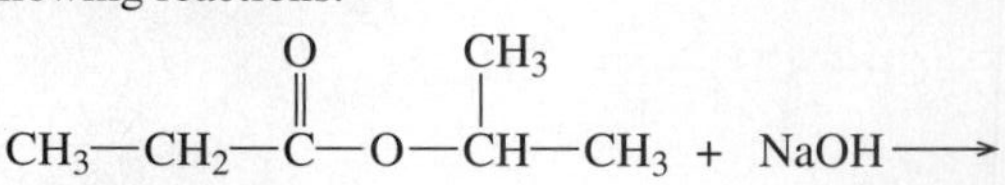 $CH_3—CH_2—\overset{O}{\overset{\|}{C}}—O—\overset{CH_3}{\overset{|}{CH}}—CH_3 + NaOH \longrightarrow$

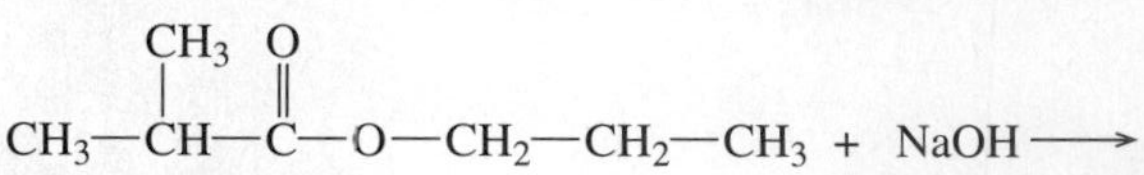

b. $CH_3—\overset{CH_3}{\overset{|}{CH}}—\overset{O}{\overset{\|}{C}}—O—CH_2—CH_2—CH_3 + NaOH \longrightarrow$

c. $CH_3—CH_2—\overset{O}{\overset{\|}{C}}—NH_2 + NaOH \longrightarrow$

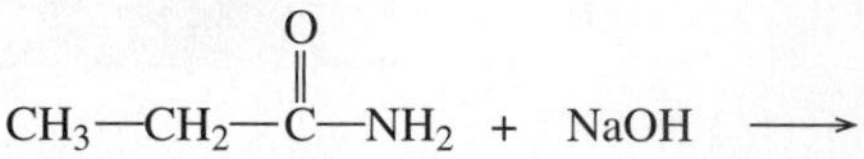

14.63 Draw the condensed structural formula of each of the following compounds:

a. dimethylamine
b. cyclohexylamine
c. dimethylammonium chloride
d. triethylamine
e. *N*-ethylaniline

14.64 Give the IUPAC name for each of the following amides:

a. 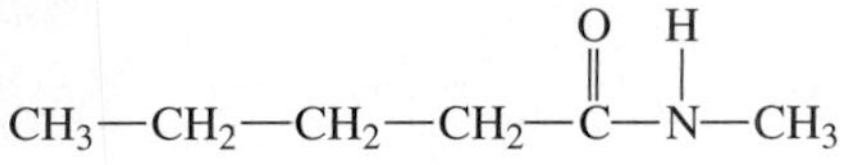 $CH_3-CH_2-CH_2-CH_2-\overset{O}{\overset{\|}{C}}-\overset{H}{\overset{|}{N}}-CH_3$

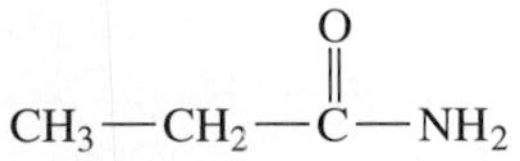

b. $CH_3-CH_2-\overset{O}{\overset{\|}{C}}-NH_2$

c. $CH_3-CH_2-CH_2-\overset{Cl}{\overset{|}{C}H}-CH_2-\overset{O}{\overset{\|}{C}}-NH_2$

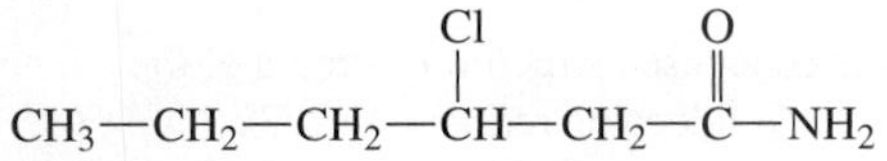

d. 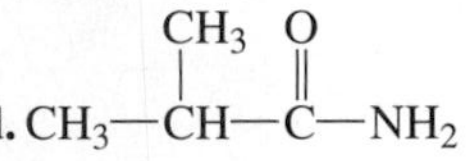 $CH_3-\overset{CH_3}{\overset{|}{C}H}-\overset{O}{\overset{\|}{C}}-NH_2$

14.65 Draw the condensed structural formula of the products of the following reactions:

a. $CH_3-CH_2-NH_2 + H_2O \rightleftharpoons$

b. $CH_3-CH_2-NH_2 + HCl \longrightarrow$

c. $CH_3-CH_2-NH-CH_3 + H_2O \rightleftharpoons$

d. $CH_3-CH_2-NH-CH_3 + HCl \longrightarrow$

14.66 Toradol is used in dentistry to relieve pain. Name the functional groups in this molecule.

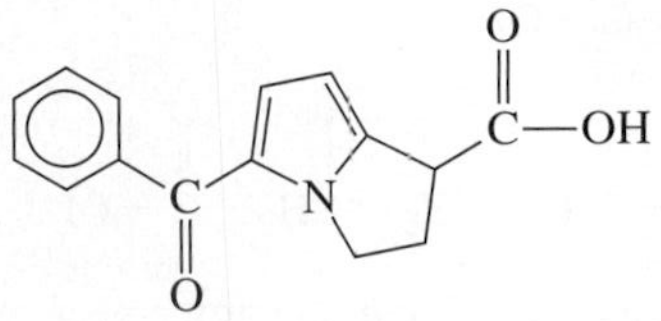

Toradol

14.67 Voltaren is indicated for acute and chronic treatment of the symptoms of rheumatoid arthritis. Name the functional groups in this molecule.

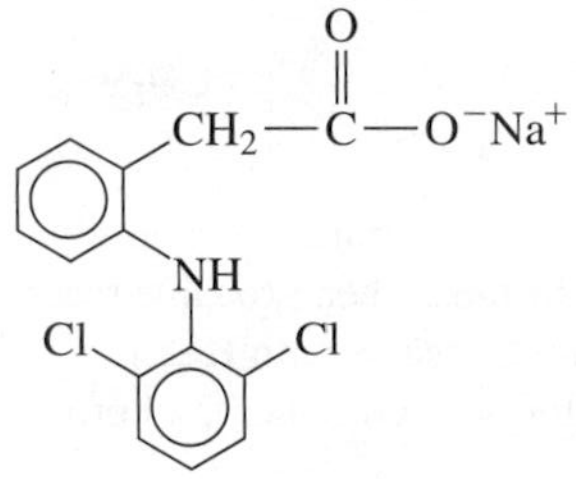

Voltaren

14.68 Using a reference book such as *The Merck Index* or *Physicians' Desk Reference* or the Internet, look up the condensed structural formula of the following medicinal drugs, and list the functional groups in the compounds.

a. Keflex, an antibiotic

b. Inderal, a beta-blocker used to treat heart irregularities

c. ibuprofen, an anti-inflammatory agent

d. Aldomet (methyldopa)

e. Percodan, a narcotic pain reliever

f. triamterene, a diuretic

Challenge Questions

14.69 Propyl acetate is the ester that gives the odor and smell of pears.

a. Draw the condensed structural formula of propyl acetate.

b. Write a balanced equation for the formation of propyl acetate.

c. Write a balanced equation for the acid hydrolysis of propyl acetate.

d. Write a balanced equation for the base hydrolysis of propyl acetate with NaOH.

e. How many milliliters of a 0.208 M NaOH solution is needed to completely hydrolyze (saponify) 1.58 g of propyl acetate?

14.70 Ethyl octanoate is a flavor component of mangoes.

a. Draw the condensed structural formula of ethyl octanoate.

b. Write a balanced equation for the formation of ethyl octanoate.

c. Write a balanced equation for the acid hydrolysis of ethyl octanoate.

d. Write a balanced equation for the base hydrolysis of ethyl octanoate with NaOH.

e. How many milliliters of a 0.315 M NaOH solution is needed to completely hydrolyze (saponify) 2.84 g of ethyl octanoate?

14.71 Novocain, a local anesthetic, is the amine salt of procaine.

$$H_2N-C_6H_4-C(=O)-O-CH_2-CH_2-N(CH_2-CH_3)_2$$

Procaine

a. Draw the condensed structural formula of the amine salt (procaine hydrochloride) formed when procaine reacts with HCl. (*Hint*: The tertiary amine reacts with HCl.)
b. Why is procaine hydrochloride used rather than procaine?

14.72 Lidocaine (xylocaine) is used as a local anesthetic and cardiac depressant.

$$(CH_3)_2C_6H_3-NH-C(=O)-CH_2-N(CH_2-CH_3)_2$$

Lidocaine (xylocaine)

a. Draw the condensed structural formula of the amine salt formed when lidocaine reacts with HCl.
b. Why is the amine salt of lidocaine used rather than the amine?

Answers

Answers to Study Checks

14.1 $CH_3-CH_2-CH_2-CH_2-C(=O)-OH$

14.2 $H-C(=O)-OH + H_2O \rightleftharpoons H-C(=O)-O^- + H_3O^+$

14.3 butanoic acid (butyric acid)

14.4 propanoic acid (propionic acid) and 1-pentanol

14.5 $CH_3-C(=O)-O-CH_2-CH_2-CH_2-CH_2-CH_3$

14.6 $C_6H_5-C(=O)-O^-K^+ + CH_3-OH$

14.7 $CH_3-CH_2-NH-CH_2-CH_2-CH_3$

14.8 $CH_3-N^+(CH_3)_2-H\ Cl^-$

14.9 $H-C(=O)-OH$ and $H-N(CH_3)-CH_3$

14.10 $C_6H_5-C(=O)-NH_2$

14.11 $CH_3-CH_2-CH_2-C(=O)-OH + CH_3-NH_3{}^+Br^-$

Answers to Selected Questions and Problems

14.1 methanoic acid (formic acid)

14.3 Each compound contains three carbon atoms. They differ because propanal, an aldehyde, contains a carbonyl group bonded to a hydrogen; in propanoic acid, the carbonyl group connects to a hydroxyl group, forming a carboxyl group.

14.5 **a.** ethanoic acid (acetic acid) **b.** 2,4-dibromobenzoic acid
c. 4-methylpentanoic acid

14.7 **a.** $C_6H_5-C(=O)-OH$ **b.** $Cl-CH_2-C(=O)-OH$

c. $CH_3-CH_2-CH_2-C(=O)-OH$

d. $CH_3-CH_2-CH_2-CH_2-CH_2-CH_2-C(=O)-OH$

14.9 **a.** Propanoic acid is the most soluble of the group because it has the fewest number of carbon atoms in its hydrocarbon chain. Solubility of carboxylic acids decreases as the number of carbon atoms in the hydrocarbon chain increases.
b. Propanoic acid is more soluble than 1-hexanol because it has fewer carbon atoms in its hydrocarbon chain. Propanoic acid is also more soluble because the carboxyl group forms more hydrogen bonds with water than does the hydroxyl group of an alcohol. An alkane is not soluble in water.

14.11 **a.** $CH_3-CH_2-CH_2-CH_2-C(=O)-OH + H_2O \rightleftharpoons CH_3-CH_2-CH_2-CH_2-C(=O)-O^- + H_3O^+$

b. $CH_3-C(=O)-OH + H_2O \rightleftharpoons CH_3-C(=O)-O^- + H_3O^+$

14.13 **a.** $H-C(=O)-OH + NaOH \longrightarrow H-C(=O)-O^-Na^+ + H_2O$

b. $Cl-CH_2-CH_2-C(=O)-OH + NaOH \longrightarrow Cl-CH_2-CH_2-C(=O)-O^-Na^+ + H_2O$

c. $C_6H_5-C(=O)-OH + NaOH \longrightarrow C_6H_5-C(=O)-O^-Na^+ + H_2O$

14.15 **a.** $CH_3-\overset{O}{\overset{\|}{C}}-O-CH_3$

b. $CH_3-CH_2-CH_2-\overset{O}{\overset{\|}{C}}-O-CH_3$ **c.** 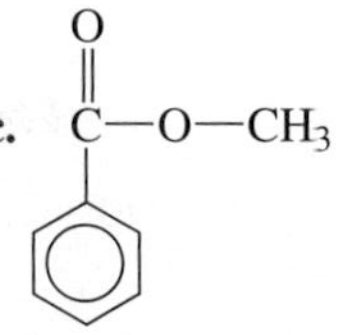 $C_6H_5-\overset{O}{\overset{\|}{C}}-O-CH_3$

14.17 **a.** $CH_3-CH_2-CH_2-CH_2-\overset{O}{\overset{\|}{C}}-O-\underset{}{\overset{CH_3}{\overset{|}{CH}}}-CH_3$

b. $CH_3-CH_2-\overset{O}{\overset{\|}{C}}-O-CH_2-CH_2-CH_3$

14.19 **a.** methyl methanoate (methyl formate)
b. methyl ethanoate (methyl acetate)
c. propyl ethanoate (propyl acetate)

14.21 **a.** $CH_3-CH_2-CH_2-\overset{O}{\overset{\|}{C}}-O-CH_2-CH_2-CH_3$

b. $H-\overset{O}{\overset{\|}{C}}-O-CH_2-CH_2-CH_2-CH_3$

c. $CH_3-CH_2-CH_2-CH_2-\overset{O}{\overset{\|}{C}}-O-CH_2-CH_3$

14.23 **a.** pentyl ethanoate (pentyl acetate)
b. octyl ethanoate (octyl acetate)
c. propyl ethanoate (propyl acetate)

14.25 The products of the acid hydrolysis of an ester are an alcohol and a carboxylic acid.

14.27 **a.** $CH_3-CH_2-\overset{O}{\overset{\|}{C}}-O^-Na^+ \; + \; CH_3-OH$

b. $CH_3-\overset{O}{\overset{\|}{C}}-OH \; + \; CH_3-CH_2-CH_2-OH$

c. $CH_3-CH_2-CH_2-\overset{O}{\overset{\|}{C}}-OH \; + \; CH_3-CH_2-OH$

d. 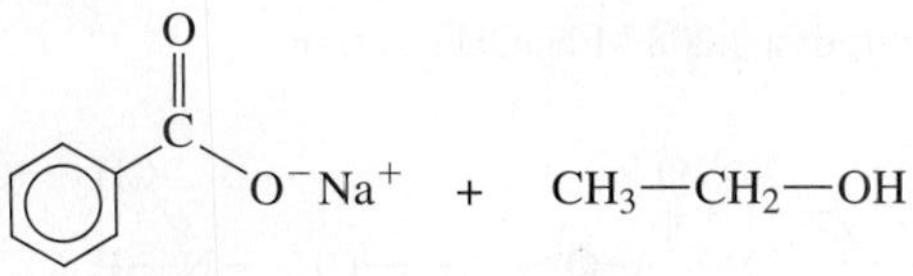 $C_6H_5-\overset{O}{\overset{\|}{C}}-O^-Na^+ \; + \; CH_3-CH_2-OH$

14.29 **a.** primary **b.** secondary
c. tertiary **d.** tertiary

14.31 **a.** propylamine **b.** methylpropylamine
c. diethylmethylamine

14.33 **a.** $CH_3-CH_2-NH_2$ **b.** 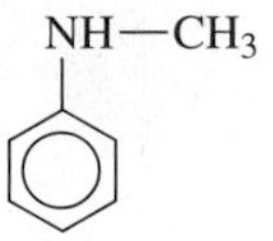 $C_6H_5-NH-CH_3$

c. $CH_3-CH_2-CH_2-CH_2-\overset{H}{\overset{|}{N}}-CH_2-CH_2-CH_3$

14.35 **a.** Yes; amines with fewer than six carbon atoms hydrogen bond with water molecules and are soluble in water.
b. Yes; amines with fewer than six carbon atoms hydrogen bond with water molecules and are soluble in water.
c. No; an amine with eight carbon atoms is insoluble in water.
d. Yes; amines with fewer than six carbon atoms hydrogen bond with water molecules and are soluble in water.

14.37 **a.** $CH_3-NH_2 + H_2O \rightleftharpoons CH_3-NH_3^+ + OH^-$

b. $CH_3-NH-CH_3 + H_2O \rightleftharpoons CH_3-\overset{+}{N}H_2-CH_3 + OH^-$

c. 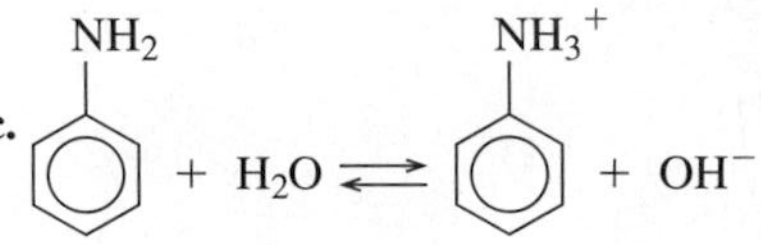 $C_6H_5-NH_2 + H_2O \rightleftharpoons C_6H_5-NH_3^+ + OH^-$

14.39 **a.** pyrimidine
b. piperidine, pyridine

14.41 pyrrole

14.43 **a.** $CH_3-\overset{O}{\overset{\|}{C}}-NH_2$ **b.** $CH_3-\overset{O}{\overset{\|}{C}}-\overset{H}{\overset{|}{N}}-CH_2-CH_3$

c. 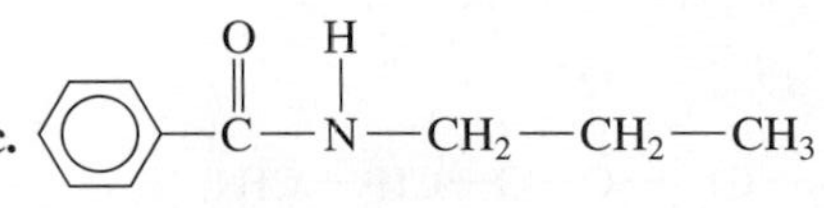 $C_6H_5-\overset{O}{\overset{\|}{C}}-\overset{H}{\overset{|}{N}}-CH_2-CH_2-CH_3$

14.45 **a.** ethanamide (acetamide)
b. butanamide (butyramide)
c. methanamide (formamide)

14.47 **a.** $CH_3-CH_2-\overset{O}{\overset{\|}{C}}-NH_2$

b. $CH_3-CH_2-CH_2-\overset{CH_3}{\overset{|}{CH}}-\overset{O}{\overset{\|}{C}}-NH_2$

c. $H-\overset{O}{\overset{\|}{C}}-NH_2$

14.49 **a.** $CH_3-\overset{O}{\overset{\|}{C}}-OH \; + \; NH_4^+Cl^-$

b. $CH_3-CH_2-\overset{O}{\overset{\|}{C}}-OH \; + \; NH_4^+Cl^-$

c. $CH_3-CH_2-CH_2-\overset{O}{\overset{\|}{C}}-OH \; + \; CH_3-NH_3^+Cl^-$

d. 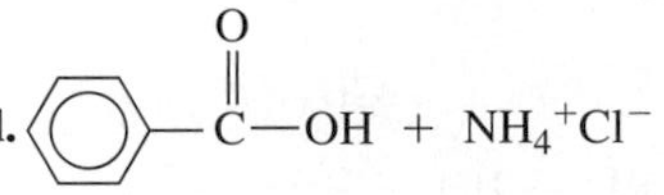 $C_6H_5-\overset{O}{\overset{\|}{C}}-OH \; + \; NH_4^+Cl^-$

e. $CH_3-CH_2-CH_2-CH_2-\overset{O}{\overset{\|}{C}}-OH$

$+ \; CH_3-CH_2-NH_3^+Cl^-$

14.51 $CH_3—CH_2—CH_2—NH_2$
Propylamine (1°)

$CH_3—CH_2—NH—CH_3$
Ethylmethylamine (2°)

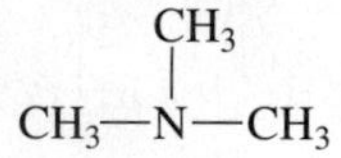

$CH_3—N(CH_3)—CH_3$
Trimethylamine (3°)

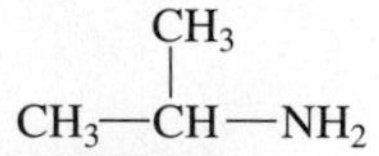

$CH_3—CH(CH_3)—NH_2$
Isopropylamine (1°)

14.53 aromatic, alcohol, amine

14.55 **a.** 3-methylbutanoic acid **b.** ethyl benzoate
c. ethyl propanoate; ethyl propionate
d. 2-chlorobenzoic acid **e.** pentanoic acid

14.57 **a.** 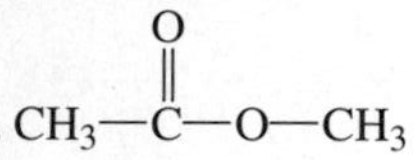 $CH_3—C(=O)—O—CH_3$

b. $CH_3—CH_2—CH_2—C(=O)—O—CH_2—CH_3$

c. $CH_3—CH_2—CH(CH_3)—CH_2—C(=O)—OH$

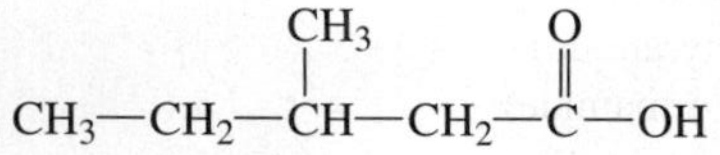

d. 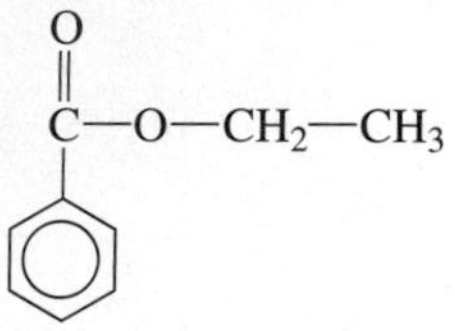 $C_6H_5—C(=O)—O—CH_2—CH_3$

14.59 **a.** 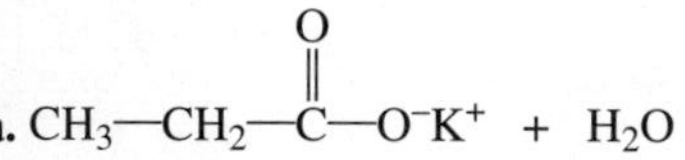 $CH_3—CH_2—C(=O)—O^-K^+ + H_2O$

b. $CH_3—CH_2—C(=O)—O—CH_3 + H_2O$

c. $C_6H_5—C(=O)—O—CH_2—CH_3 + H_2O$

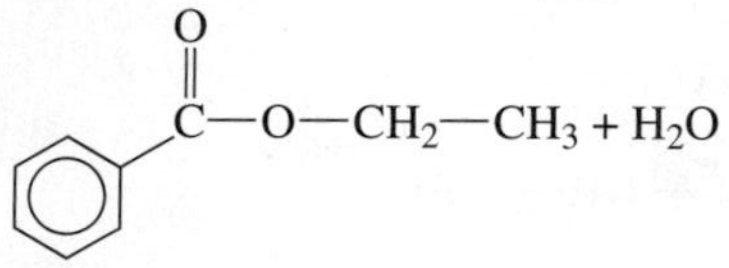

d. 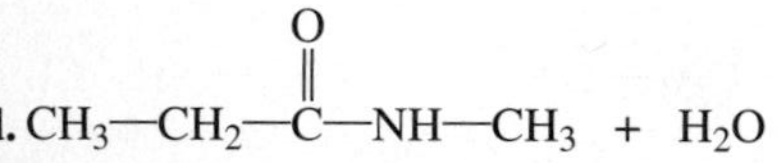 $CH_3—CH_2—C(=O)—NH—CH_3 + H_2O$

14.61 **a.** 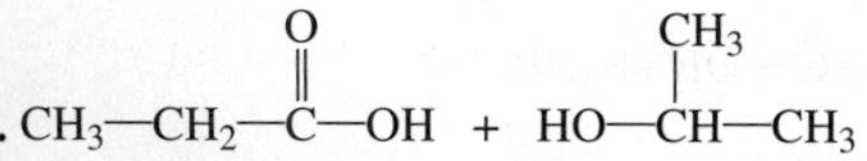 $CH_3—CH_2—C(=O)—OH + HO—CH(CH_3)—CH_3$

b. $CH_3—CH(CH_3)—C(=O)—OH + HO—CH_2—CH_2—CH_3$

c. $C_6H_5—\overset{+}{N}H_3Cl^-$

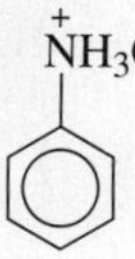

14.63 **a.** 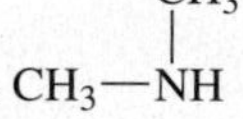 $CH_3—NH—CH_3$ **b.** 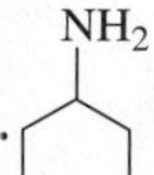 cyclohexyl—NH_2 **c.** 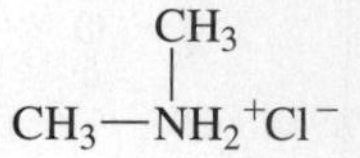 $CH_3—NH_2^+(CH_3)Cl^-$

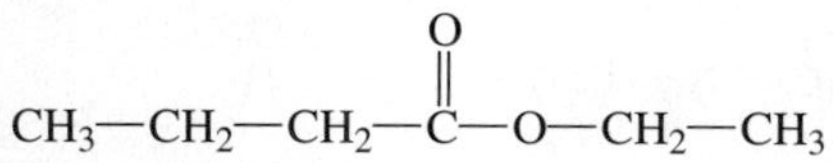

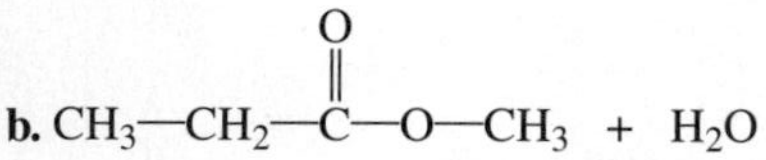

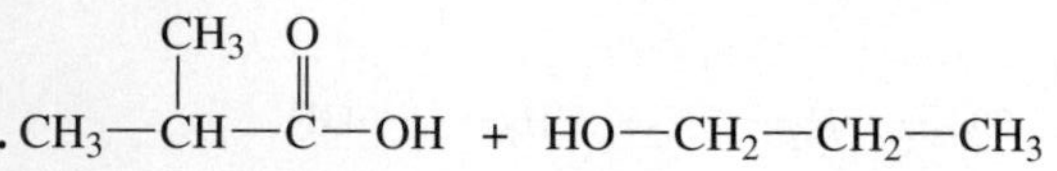

d. 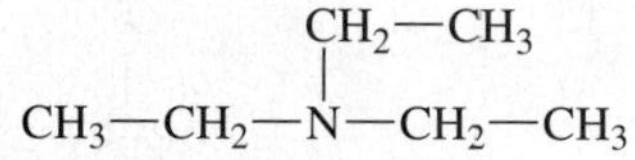 $CH_3—CH_2—N(CH_2—CH_3)—CH_2—CH_3$ **e.** 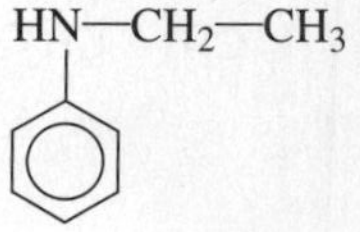 $C_6H_5—HN—CH_2—CH_3$

14.65 **a.** $CH_3—CH_2—\overset{+}{N}H_3 + OH^-$

b. $CH_3—CH_2—\overset{+}{N}H_3Cl^-$

c. $CH_3—CH_2—\overset{+}{N}H_2—CH_3 + OH^-$

d. $CH_3—CH_2—\overset{+}{N}H_2—CH_3Cl^-$

14.67 carboxylate salt, aromatic, amine

14.69 **a.** 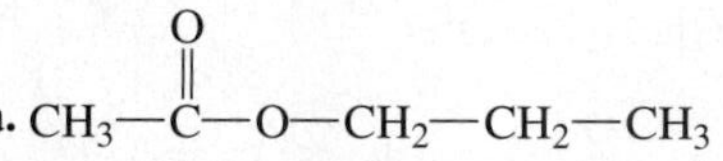 $CH_3—C(=O)—O—CH_2—CH_2—CH_3$

b. $CH_3—C(=O)—OH + HO—CH_2—CH_2—CH_3 \xrightleftharpoons[\text{Heat}]{H^+}$

$CH_3—C(=O)—O—CH_2—CH_2—CH_3 + H_2O$

c. $CH_3—C(=O)—O—CH_2—CH_2—CH_3 + H_2O \xrightleftharpoons{H^+}$

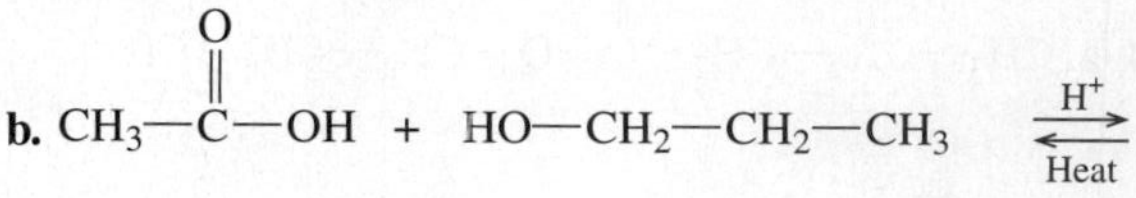

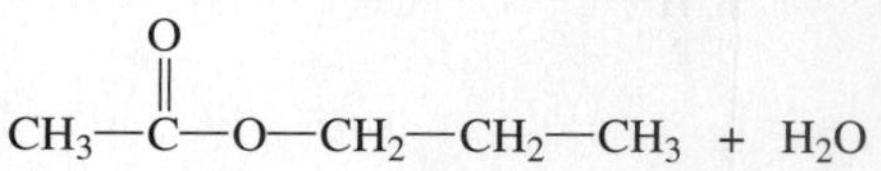

$CH_3—C(=O)—OH + HO—CH_2—CH_2—CH_3$

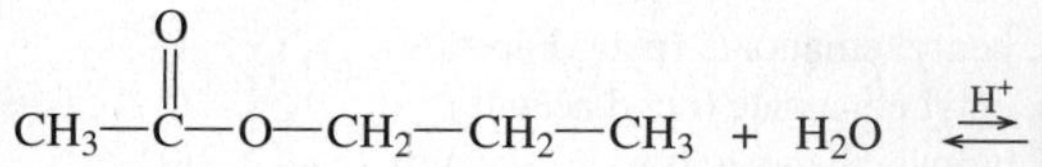

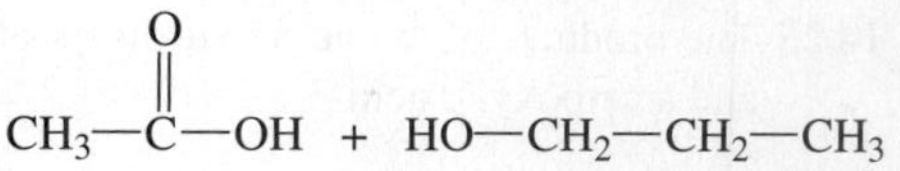

d. 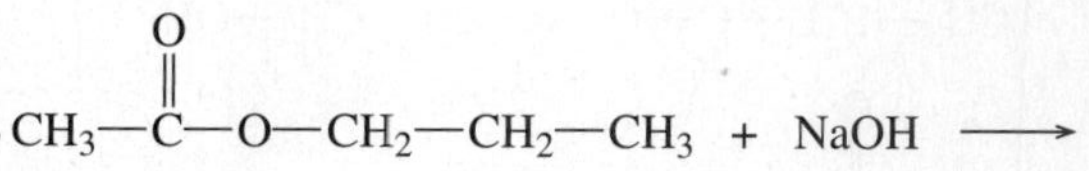 $CH_3—C(=O)—O—CH_2—CH_2—CH_3 + NaOH \longrightarrow$

$CH_3—C(=O)—O^-Na^+ + HO—CH_2—CH_2—CH_3$

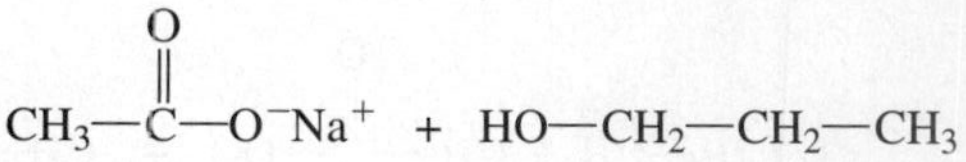

e. 74.4 mL of a 0.208 M NaOH solution

14.71 **a.** 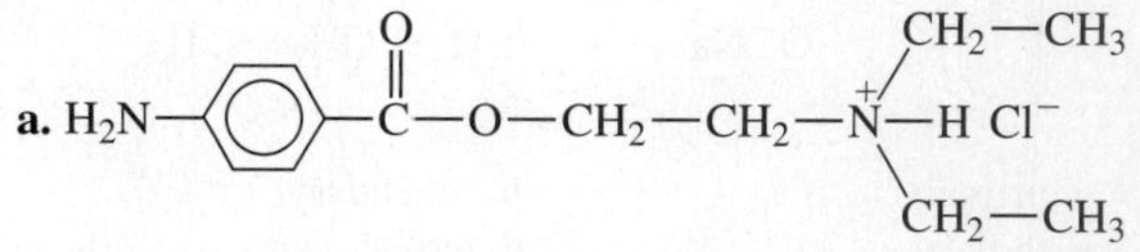 $H_2N—C_6H_4—C(=O)—O—CH_2—CH_2—\overset{+}{N}H(CH_2—CH_3)_2\ Cl^-$

b. The amine salt (Novocain) is more soluble in body fluids than procaine.

Lipids

LOOKING AHEAD

"In our toxicology lab, we measure the drugs in samples of urine or blood," says Penny Peng, assistant supervisor of chemistry, Toxicology Lab, Santa Clara Valley Medical Center. "But first we extract the drugs from the fluid and concentrate them so they can be detected in the machine we use. We extract the drugs by using different organic solvents such as methanol, ethyl acetate, or methylene chloride, and by changing the pH. We evaporate most of the organic solvent to concentrate any drugs it may contain. A small sample of the concentrate is placed into a machine called a gas chromatograph. As the gas moves over a column, the drugs in it are separated. From the results, we can identify as many as 10 to 15 different drugs from one urine sample."

Visit **www.masteringchemistry.com** for self-study materials and instructor-assigned homework.

When we talk of fats and oils, waxes, steroids, cholesterol, and fat-soluble vitamins, we are discussing lipids. All the lipids are naturally occurring compounds that vary considerably in structure but share a common feature of being soluble in nonpolar solvents, but not in water. Fats, which are one family of lipids, have many functions in the body; they store energy and protect and insulate internal organs. Other types of lipids are found in nerve fibers and in hormones, which act as chemical messengers. Because lipids are not soluble in water, one of their major functions as components of cell membranes is to separate the internal contents of cells from the external environment.

Many people are concerned about the amounts of saturated fats and cholesterol in our diets. Researchers suggest that saturated fats and cholesterol are associated with diseases such as diabetes; cancers of the breast, pancreas, and colon; and atherosclerosis, a condition in which deposits of lipid materials called *plaque* accumulate in the coronary blood vessels. In atherosclerosis, plaque restricts the flow of blood to the tissue, causing necrosis (death). An accumulation of plaque in the heart could result in a *myocardial infarction* (heart attack).

The American Institute for Cancer Research has recommended that we increase the fiber and starch content of our diets by adding more vegetables, fruits, and whole grains with moderate amounts of foods with low levels of fat and cholesterol such as fish, poultry, lean meats, and low-fat dairy products. The AICR has also suggested that we limit our intake of foods high in fat and cholesterol such as eggs, nuts, fatty or organ meats, cheeses, butter, and coconut and palm oil.

LEARNING GOAL

Describe the classes of lipids.

TUTORIAL
Classes of Lipids

15.1 Lipids

Lipids are a family of biomolecules that have the common property of being soluble in organic solvents but not in water. The word "lipid" comes from the Greek word *lipos*, meaning "fat" or "lard." Typically, the lipid content of a cell can be extracted using a nonpolar solvent such as ether or chloroform. Lipids are an important feature in cell membranes, fat-soluble vitamins, and steroid hormones.

Types of Lipids

Within the lipid family, there are distinct structures that distinguish the different types of lipids. Lipids such as waxes, fats, oils, and glycerophospholipids are esters that can be hydrolyzed to give fatty acids along with other products including an alcohol. Steroids are characterized by the steroid nucleus of four fused carbon rings. They do not contain fatty acids and cannot be hydrolyzed. Figure 15.1 illustrates the types and general structure of lipids we will discuss in this chapter.

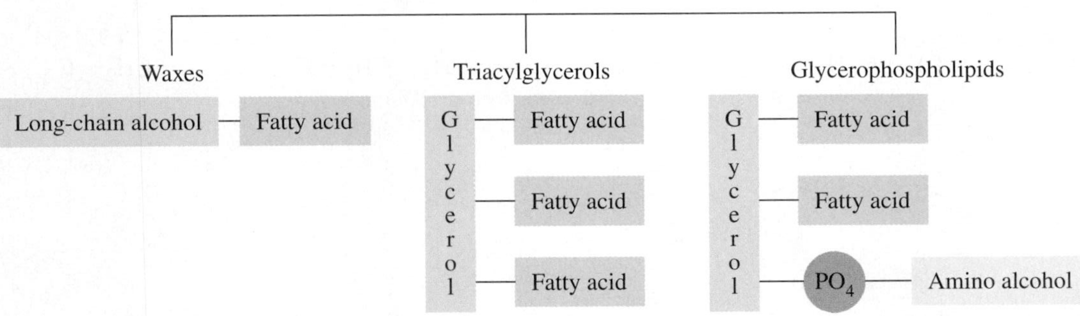

FIGURE 15.1 Lipids are naturally occurring compounds that are soluble in organic solvents but not in water.

Q What chemical property do waxes, triacylglycerols, and steroids have in common?

CONCEPT CHECK 15.1

Classes of Lipids

What type of lipid does not contain fatty acids?

ANSWER

The steroids are a group of lipids with no fatty acids.

QUESTIONS AND PROBLEMS

Lipids

15.1 What are some functions of lipids in the body?

15.2 What are some of the different kinds of lipids?

15.3 Lipids are not soluble in water. Are lipids polar or nonpolar molecules?

15.4 Which of the following solvents might be used to dissolve an oil stain?
a. water
b. CCl_4
c. diethyl ether
d. benzene
e. NaCl solution

Explore Your World

SOLUBILITY OF FATS AND OILS

Place some water in a small bowl. Add a drop of a vegetable oil. Then add a few more drops of the oil. Now add a few drops of liquid soap and mix. Record your observations.

Place a small amount of fat such as margarine, butter, shortening, or vegetable oil on a dish or plate. Run water over it. Record your observations. Mix some soap with the fat substance and run water over it again. Record your observations.

QUESTIONS

1. Do the drops of oil in the water separate or do they come together? Explain.
2. How does the soap affect the oil layer?
3. Why don't the fats on the dish or plate wash off with water?
4. In general, what is the solubility of lipids in water?
5. Why does soap help to wash the fats off the plate?

LEARNING GOAL

Draw the condensed structural formula of a fatty acid and identify it as saturated or unsaturated.

TUTORIAL
Structures and Properties of Fatty Acids

TUTORIAL
Lipids and Fatty Acids

15.2 Fatty Acids

The fatty acids are the simplest type of lipids and are found as components in more complex lipids. A **fatty acid** contains a long, unbranched carbon chain attached to a carboxylic acid group at one end. Although the carboxylic acid part is hydrophilic, the long hydrophobic carbon chain makes long-chain fatty acids insoluble in water.

Naturally occurring fatty acids have an even number of carbon atoms, usually between 10 and 20. An example of a fatty acid is lauric acid, a 12-carbon acid found in coconut oil. In a skeletal formula of a fatty acid, the ends and bends of the line are the carbon atoms. The structural formula of lauric acid can be drawn in several forms.

Drawing Formulas for Lauric Acid

$$CH_3{-}(CH_2)_{10}{-}\overset{\overset{\large O}{\|}}{C}{-}OH \qquad CH_3{-}(CH_2)_{10}{-}COOH$$

$$CH_3{-}CH_2{-}CH_2{-}CH_2{-}CH_2{-}CH_2{-}CH_2{-}CH_2{-}CH_2{-}CH_2{-}CH_2{-}C(=O){-}OH$$

Condensed structural formula

O, OH

Skeletal formula

In a **saturated fatty acid**, the long carbon chain is like an alkane because there are only single carbon–carbon bonds. In a **monounsaturated fatty acid**, the long carbon chain has one double bond, which makes its properties similar to those of an alkene. In a **polyunsaturated fatty acid**, there are at least two carbon–carbon double bonds. Table 15.1 lists some of the typical fatty acids in lipids.

Cis and Trans Isomers of Unsaturated Fatty Acids

Unsaturated fatty acids can be drawn as cis and trans isomers in the same way as the cis and trans alkene structures we looked at in Chapter 11. For example, oleic acid, a monounsaturated fatty acid found in olives and corn, has one double bond at carbon 9. We can show its cis and trans structural formulas using the line-bond notation. The cis structure is the most prevalent isomer found in naturally occurring unsaturated fatty acids. In the cis isomer, the carbon chain has a "kink" at the double bond site. As we will see, the cis bond has a major impact on the physical properties and uses of unsaturated fatty acids.

H, H, C=C, O, C, OH

cis-Oleic acid
cis double bond

H, C, C, H, O, C, OH

trans-Oleic acid
trans double bond

The cis structures of fatty acids are more prevalent in nature than the trans structures.

TABLE 15.1 Structures and Melting Points of Common Fatty Acids

Name	Carbon Atoms	Source	Melting Point (°C)	Structures
Saturated				
Lauric acid	12	Coconut	43	$CH_3{-}(CH_2)_{10}{-}COOH$
Myristic acid	14	Nutmeg	54	$CH_3{-}(CH_2)_{12}{-}COOH$
Palmitic acid	16	Palm	62	$CH_3{-}(CH_2)_{14}{-}COOH$
Stearic acid	18	Animal fat	69	$CH_3{-}(CH_2)_{16}{-}COOH$
Monounsaturated				
Palmitoleic acid	16	Butter	0	$CH_3{-}(CH_2)_5{-}CH{=}CH{-}(CH_2)_7{-}COOH$
Oleic acid	18	Olives, corn	13	$CH_3{-}(CH_2)_7{-}CH{=}CH{-}(CH_2)_7{-}COOH$
Polyunsaturated				
Linoleic acid	18	Soybean, safflower, sunflower	−9	$CH_3{-}(CH_2)_4{-}CH{=}CH{-}CH_2{-}CH{=}CH{-}(CH_2)_7{-}COOH$
Linolenic acid	18	Corn	−17	$CH_3{-}CH_2{-}CH{=}CH{-}CH_2{-}CH{=}CH{-}CH_2{-}CH{=}CH{-}(CH_2)_7{-}COOH$
Arachidonic acid	20	Meat, eggs, fish	−50	$CH_3{-}(CH_2)_3{-}(CH_2{-}CH{=}CH)_4{-}(CH_2)_3{-}COOH$

The human body is capable of synthesizing most fatty acids from carbohydrates or other fatty acids. However, humans do not synthesize sufficient amounts of polyunsaturated fatty acids such as linoleic acid, linolenic acid, and arachidonic acid. Because they must be obtained from the diet, they are known as *essential fatty acids*. In infants, a deficiency of essential fatty acids can cause skin dermatitis. However, the role of fatty acids in adult nutrition is not well understood. Adults do not usually have a deficiency of essential fatty acids.

Physical Properties of Fatty Acids

The saturated fatty acids fit closely together in a regular pattern, which allows strong attractions to occur between the carbon chains. As a result, a significant amount of energy and high temperatures are required to separate the fatty acids and melt the fat. As the length of the carbon chain increases, more interactions occur between the carbon chains, requiring higher melting points. Saturated fatty acids are usually solids at room temperature.

In unsaturated fatty acids, the cis double bonds cause the carbon chain to bend or kink, giving the molecules an irregular shape. As a result, fewer interactions occur between carbon chains. Consequently, less energy is required to separate the molecules, making the melting points of unsaturated fats lower than those of saturated fats (see Figure 15.2). Most unsaturated fats are liquid oils at room temperature.

We might think of saturated fatty acids as chips with matching shapes that fit closely together in a can. Then irregularly shaped chips would be like unsaturated fatty acids that do not fit closely together.

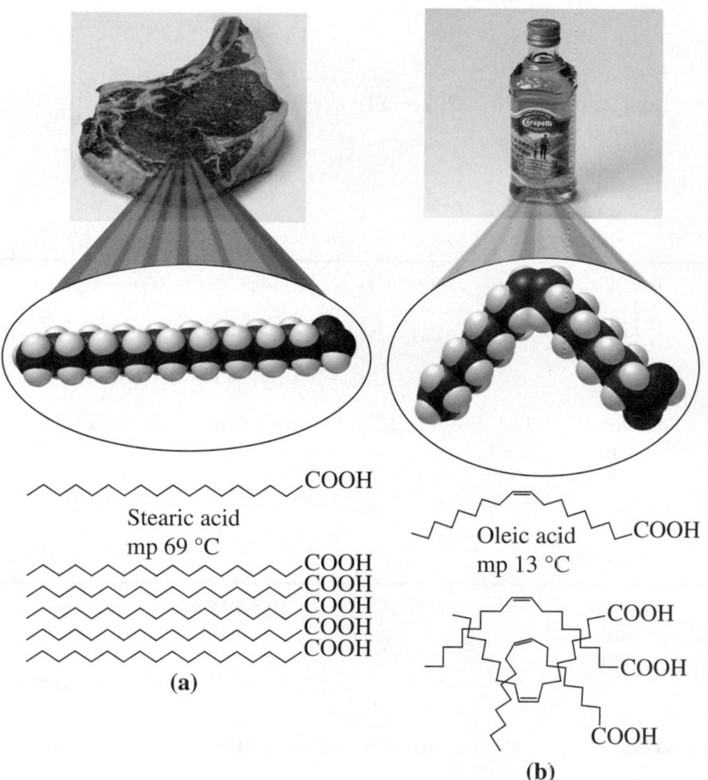

FIGURE 15.2 **(a)** In saturated fatty acids, the molecules fit closely together to give high melting points. **(b)** In unsaturated fatty acids, molecules cannot fit closely together, resulting in lower melting points.
Q Why does the cis double bond affect the melting points of unsaturated fatty acids?

SAMPLE PROBLEM 15.1

Structures and Properties of Fatty Acids

Consider the condensed structural formula of oleic acid.

$$CH_3—(CH_2)_7—CH{=}CH—(CH_2)_7—\overset{\displaystyle O}{\overset{\|}{C}}—OH$$

a. Why is the substance an acid?
b. How many carbon atoms are in oleic acid?

c. Is it a saturated or unsaturated fatty acid?
d. Is it most likely to be solid or liquid at room temperature?
e. Would it be soluble in water?

SOLUTION

a. Oleic acid contains a carboxylic acid group.
b. It contains 18 carbon atoms.
c. It is a monounsaturated fatty acid.
d. It is liquid at room temperature.
e. No, its long hydrocarbon chain makes it insoluble in water.

STUDY CHECK 15.1

Palmitoleic acid is a fatty acid with the following condensed structural formula:

$$CH_3-(CH_2)_5-CH=CH-(CH_2)_7-\overset{\overset{\large O}{\|}}{C}-OH$$

a. How many carbon atoms are in palmitoleic acid?
b. Is it a saturated or unsaturated fatty acid?
c. Is it most likely to be solid or liquid at room temperature?

Prostaglandins

Prostaglandins are hormone-like substances produced in small amounts in most cells of the body. The prostaglandins, also known as *eicosanoids*, are formed from arachidonic acid, the polyunsaturated fatty acid with 20 carbon atoms (*eicos* is the Greek word for "20"). Swedish chemists first discovered prostaglandins and named them "prostaglandin E" (soluble in ether) and "prostaglandin F" (soluble in phosphate buffer or *fosfat* in Swedish). The various kinds of prostaglandins differ by the substituents attached to the five-carbon ring. Prostaglandin E (PGE) has a ketone group on carbon 9, whereas prostaglandin F (PGF) has a hydroxyl group. The number of double bonds is shown as a subscript 1 or 2.

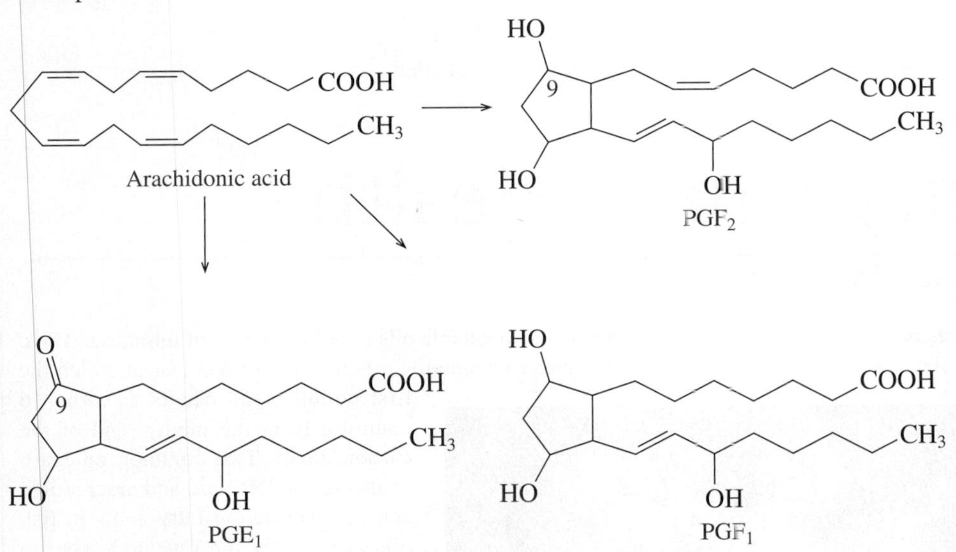

COOH
CH_3
Arachidonic acid

HO
9
COOH
CH_3
HO
OH
PGF_2

O
9
COOH
CH_3
HO
OH
PGE_1

HO
COOH
CH_3
HO
OH
PGF_1

Although prostaglandins are broken down quickly, they have potent physiological effects. Some prostaglandins increase blood pressure, and others lower blood pressure. Other prostaglandins stimulate contraction and relaxation in the smooth muscle of the uterus. When tissues are injured, arachidonic acid is converted to prostaglandins such as PGE and PGF that produce inflammation and pain in the area.

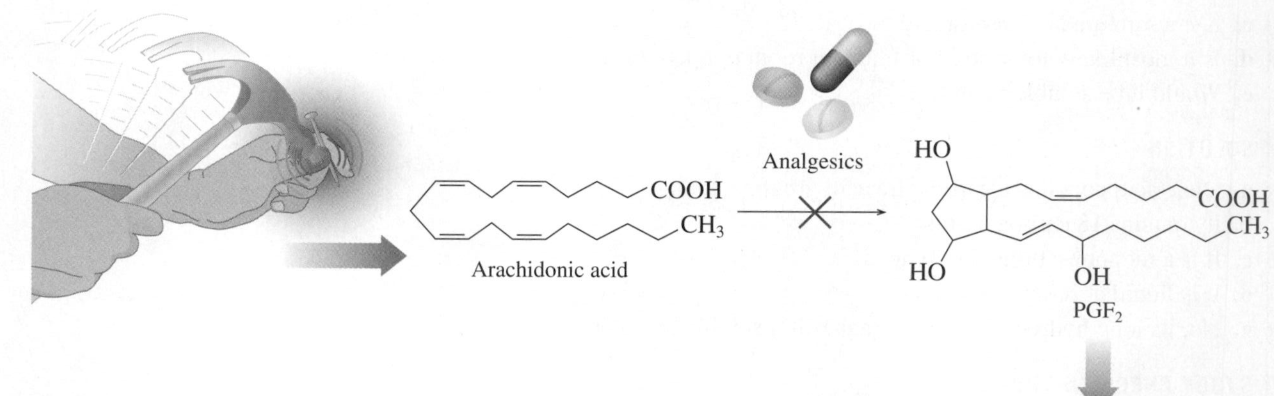

When tissues are injured, prostaglandins are produced, which cause pain and inflammation.

The treatment of pain, fever, and inflammation is based on inhibiting the enzymes that convert arachidonic acid to prostaglandins. Several nonsteroidal anti-inflammatory drugs (NSAIDs), such as aspirin, block the production of prostaglandins and in doing so decrease pain and inflammation and reduce fever (antipyretics). Ibuprofen has similar anti-inflammatory and analgesic effects. Other NSAIDs include naproxen (Aleve and Naprosyn), ketoprofen (Actron), and nabumetone (Relafen). Long-term use of such products can result in liver, kidney, and gastrointestinal damage. Some forms of PGE are being tested as inhibitors of gastric secretion for use in the treatment of stomach ulcers.

Aspirin (acetylsalicylic acid) Ibuprofen (Advil, Motrin) Naproxen (Aleve, Naprosyn)

Chemistry Link to Health

OMEGA-3 FATTY ACIDS IN FISH OILS

Because unsaturated fats are now recognized as being more beneficial to health than saturated fats, American diets have changed to include more unsaturated fats and less saturated fats. This change is a response to research that indicates that atherosclerosis and heart disease are associated with high levels of fats in the diet. However, the Inuit people of Alaska have a diet with high levels of unsaturated fats as well as high levels of blood cholesterol, but a very low occurrence of atherosclerosis and heart disease. The fats in the Inuit diet are primarily unsaturated fats from fish rather than from land animals.

Cold-water fish are a source of omega-3 fatty acids.

Both fish and vegetable oils have high levels of unsaturated fats. The fatty acids in vegetable oils are omega-6 acids, in which the first double bond occurs at carbon 6 counting from the methyl end of the carbon chain. Two common omega-6 acids are linoleic acid and arachidonic acid. However, the fatty acids in fish oils are mostly the omega-3 type, in which the first double bond occurs at the third carbon counting from the methyl group. Three common omega-3 fatty acids in fish are linolenic acid, eicosapentaenoic acid, and docosahexaenoic acid. The omega-6 fatty acids are mostly found in grains, oils from plants, and eggs, and the omega-3 fatty acids are mostly found in cold-water fish such as tuna and salmon.

In atherosclerosis and heart disease, cholesterol forms plaques that adhere to the walls of the blood vessels. Blood pressure rises as blood has to squeeze through a smaller opening in the blood vessel. As more plaque forms, there is also a possibility of blood clots blocking the blood vessels and causing a heart attack. Omega-3 fatty acids lower the tendency of blood platelets to stick together, thereby reducing the possibility of blood clots. However, high levels of omega-3 fatty acids can increase bleeding if the ability of the platelets to form blood clots is reduced too much. It does seem that a diet that includes fish such as salmon, tuna, and herring can provide higher amounts of the omega-3 fatty acids, which help lessen the possibility of developing heart disease.

Omega-6 Fatty Acids

Linoleic acid (LA) $\underset{1}{CH_3}-(CH_2)_4-\underset{6}{CH}=CH-CH_2-CH=CH-(CH_2)_7-COOH$

Arachidonic acid (AA) $\underset{1}{CH_3}-(CH_2)_4-(\underset{6}{CH}=CH-CH_2)_4-(CH_2)_2-COOH$

Omega-3 Fatty Acids

Linolenic acid (ALA) $\underset{1}{CH_3}-CH_2-(\underset{3}{CH}=CH-CH_2)_3-(CH_2)_6-COOH$

Eicosapentaenoic acid (EPA) $\underset{1}{CH_3}-CH_2-(\underset{3}{CH}=CH-CH_2)_5-(CH_2)_2-COOH$

Docosahexaenoic acid (DHA) $\underset{1}{CH_3}-CH_2-(\underset{3}{CH}=CH-CH_2)_6-CH_2-COOH$

QUESTIONS AND PROBLEMS

Fatty Acids

15.5 Describe some similarities and differences in the structures of a saturated fatty acid and an unsaturated fatty acid.

15.6 Stearic acid and linoleic acid both have 18 carbon atoms. Why does stearic acid melt at 69 °C but linoleic acid melts at −9 °C?

15.7 Draw the skeletal formula of the following fatty acids:
a. palmitic acid **b.** oleic acid

15.8 Draw the skeletal formula of the following fatty acids:
a. stearic acid **b.** linoleic acid

15.9 Which of the following fatty acids are saturated and which are unsaturated?
a. lauric acid **b.** linolenic acid
c. palmitoleic acid **d.** stearic acid

15.10 Which of the following fatty acids are saturated and which are unsaturated?
a. linoleic acid **b.** palmitic acid
c. myristic acid **d.** oleic acid

15.11 How does the structure of a fatty acid with a cis double bond differ from the structure of a fatty acid with a trans double bond?

15.12 In each pair, identify the fatty acid with the lower melting point. Explain.
a. myristic acid and stearic acid
b. stearic acid and linoleic acid
c. oleic acid and linolenic acid

15.13 What is the difference in the location of the first double bond in an omega-3 and an omega-6 fatty acid (see Chemistry Link to Health "Omega-3 Fatty Acids in Fish Oils")?

15.14 a. What are some sources of omega-3 and omega-6 fatty acids (see Chemistry Link to Health "Omega-3 Fatty Acids in Fish Oils")?
b. How may omega-3 fatty acids help in lowering the risk of heart disease?

15.15 Compare the structures and functional groups of arachidonic acid and prostaglandin PGE_1.

15.16 Compare the structures and functional groups of PGF_1 and PGF_2.

15.17 What are some effects of prostaglandins in the body?

15.18 How does an anti-inflammatory drug reduce inflammation?

15.3 Waxes, Fats, and Oils

LEARNING GOAL

Draw the condensed structural formula of a wax, fat, or oil produced by the reaction of a fatty acid and an alcohol or glycerol.

Waxes are found in many plants and animals. Coatings of carnauba wax on fruits, leaves, and stems of plants help to prevent loss of water and damage from pests. Waxes on the skin, fur, and feathers of animals provide a waterproof coating. A **wax** is an ester of a saturated fatty acid and a long-chain alcohol, each containing from 14 to 30 carbon atoms.

The formulas of some common waxes are given in Table 15.2. Beeswax obtained from honeycombs and carnauba wax from palm trees are used to give a protective coating to furniture, cars, and floors. Jojoba wax is used in making candles and cosmetics such as lipstick. Lanolin, a mixture of waxes obtained from wool, is used in hand and facial lotions to aid retention of water, softening the skin.

Ester bond

$$\text{Fatty acid}-\overset{\overset{\large O}{\|}}{C}-O-\text{Long-chain alcohol}$$

Wax

Waxes are esters of long-chain alcohols and fatty acids.

TABLE 15.2 Some Typical Waxes

Type	Condensed Structural Formula	Source	Uses
Beeswax	$CH_3-(CH_2)_{14}-\overset{O}{\overset{\|}{C}}-O-(CH_2)_{29}-CH_3$	Honeycomb	Candles, shoe polish, wax paper
Carnauba wax	$CH_3-(CH_2)_{24}-\overset{O}{\overset{\|}{C}}-O-(CH_2)_{29}-CH_3$	Brazilian palm tree	Waxes for furniture, cars, floors, shoes
Jojoba wax	$CH_3-(CH_2)_{18}-\overset{O}{\overset{\|}{C}}-O-(CH_2)_{19}-CH_3$	Jojoba bush	Candles, soaps, cosmetics

SELF STUDY ACTIVITY Triacylglycerols

SELF STUDY ACTIVITY Fats

Fats and Oils: Triacylglycerols

In the body, fatty acids are stored as fats and oils known as **triacylglycerols**. These substances, also called *triglycerides*, are triesters of glycerol (a trihydroxy alcohol) and fatty acids. The general formula of a triacylglycerol follows:

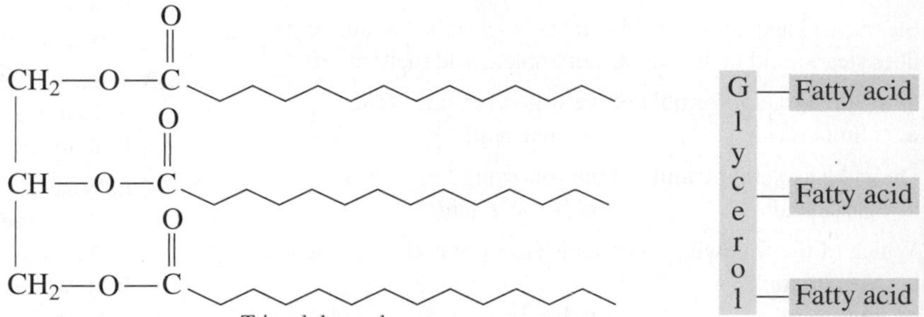

$$\begin{array}{l} CH_2-O-\overset{O}{\overset{\|}{C}}\sim\sim\sim\sim\sim\sim \\ | \\ CH-O-\overset{O}{\overset{\|}{C}}\sim\sim\sim\sim\sim\sim \\ | \\ CH_2-O-\overset{O}{\overset{\|}{C}}\sim\sim\sim\sim\sim\sim \end{array}$$

Triacylglycerol

Glycerol	
G	— Fatty acid
l y c e r	— Fatty acid
o l	— Fatty acid

In Chapter 13, we saw that esters are produced from the esterification reaction between a carboxylic acid and an alcohol. In a triacylglycerol, three hydroxyl groups on glycerol form ester bonds with the carboxyl groups of three fatty acids. For example, glycerol and three molecules of stearic acid form glyceryl tristearate which is commonly named tristearin.

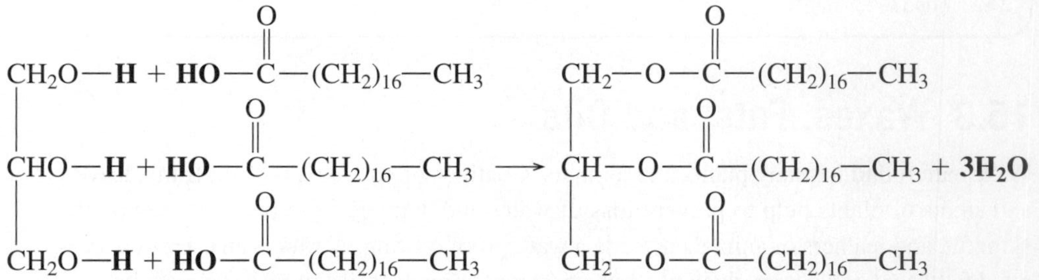

$$\begin{array}{l} CH_2O-\mathbf{H} + \mathbf{HO}-\overset{O}{\overset{\|}{C}}-(CH_2)_{16}-CH_3 \\ | \\ CHO-\mathbf{H} + \mathbf{HO}-\overset{O}{\overset{\|}{C}}-(CH_2)_{16}-CH_3 \\ | \\ CH_2O-\mathbf{H} + \mathbf{HO}-\overset{O}{\overset{\|}{C}}-(CH_2)_{16}-CH_3 \end{array} \longrightarrow \begin{array}{l} CH_2-O-\overset{O}{\overset{\|}{C}}-(CH_2)_{16}-CH_3 \\ | \\ CH-O-\overset{O}{\overset{\|}{C}}-(CH_2)_{16}-CH_3 + \mathbf{3H_2O} \\ | \\ CH_2-O-\overset{O}{\overset{\|}{C}}-(CH_2)_{16}-CH_3 \end{array}$$

Glycerol — 3 Stearic acid molecules — Glyceryl tristearate (tristearin, a fat)

Most fats and oils are mixed triacylglycerols that contain two or three different fatty acids. For example, a mixed triacylglycerol might be made from lauric acid, myristic acid, and palmitic acid. One possible structure for the mixed triacylglycerol follows:

$$CH_2-O-\overset{\overset{\large O}{\|}}{C}-(CH_2)_{10}-CH_3 \quad \text{Lauric acid}$$
$$CH-O-\overset{\overset{\large O}{\|}}{C}-(CH_2)_{12}-CH_3 \quad \text{Myristic acid}$$
$$CH_2-O-\overset{\overset{\large O}{\|}}{C}-(CH_2)_{14}-CH_3 \quad \text{Palmitic acid}$$

A mixed triacylglycerol

Prior to hibernation, a polar bear eats food with a high content of fats and oils.

Triacylglycerols are the major form of energy storage for animals. Animals that hibernate eat large quantities of plants, seeds, and nuts that contain high levels of fats and oils. Prior to hibernation, these animals, such as polar bears, gain as much as 14 kilograms a week. As the external temperature drops, the animal goes into hibernation. The body temperature drops to nearly freezing, and cellular activity, respiration, and heart rate are drastically reduced. Animals that live in extremely cold climates hibernate for 4–7 months. During this time, stored fat is the only source of energy.

CONCEPT CHECK 15.2

Triacylglycerols

Name the triacylglycerol that is formed by the esterification of glycerol with each of the following fatty acids:

a. $CH_3-(CH_2)_{12}-COOH$ **b.** $CH_3-(CH_2)_5-CH=CH-(CH_2)_7-COOH$

ANSWER

a. The saturated fatty acid with 14 carbon atoms is myristic acid. The triacylglycerol formed from glycerol and myristic acid is named glyceryl trimyristate or trimyristin (common name).

b. The monounsaturated fatty acid with 16 carbon atoms is palmitoleic acid. The triacylglycerol formed from glycerol and palmitoleic acid is named glyceryl tripalmitoleate or tripalmitolein (common name).

SAMPLE PROBLEM 15.2

Drawing the Structure of a Triacylglycerol

Draw the condensed structural formula of glyceryl trioleate (triolein).

SOLUTION

Glyceryl trioleate (triolein) is the triacylglycerol that contains ester bonds between glycerol and three oleic acid molecules.

$$CH_2-O-\overset{\overset{\large O}{\|}}{C}-(CH_2)_7-CH=CH-(CH_2)_7-CH_3$$
$$CH-O-\overset{\overset{\large O}{\|}}{C}-(CH_2)_7-CH=CH-(CH_2)_7-CH_3$$
$$CH_2-O-\overset{\overset{\large O}{\|}}{C}-(CH_2)_7-CH=CH-(CH_2)_7-CH_3$$

Glyceryl trioleate (triolein)

STUDY CHECK 15.2

Draw the condensed structural formula of the triacylglycerol containing three molecules of myristic acid.

Melting Points of Fats and Oils

A **fat** is a triacylglycerol that is solid at room temperature, and it usually comes from animal sources such as meat, whole milk, butter, and cheese.

An **oil** is a triacylglycerol that is usually a liquid at room temperature and is obtained from a plant source. Olive oil and peanut oil are monounsaturated because they contain large amounts of oleic acid. Oils from corn, cottonseed, safflower, and sunflower are polyunsaturated because they contain large amounts of fatty acids with two or more double bonds (see Figure 15.3). Palm oil and coconut oil are solids at room temperature because they consist mostly of saturated fatty acids. Although coconut oil is 92% saturated fats, about half is lauric acid, which contains 12 carbon atoms rather than 18 carbon atoms found in the stearic acid of animal sources. Thus coconut oil has a melting point that is higher than typical vegetable oils, but not as high as fats from animal sources that contain stearic acid. The amounts of saturated, monounsaturated, and polyunsaturated fatty acids in some typical fats and oils are listed in Figure 15.4.

Saturated fatty acids have higher melting points than unsaturated fatty acids because they pack together more tightly. Animal fats usually contain more saturated fatty acids than do vegetable oils. Therefore the melting points of animal fats are higher than those of vegetable oils.

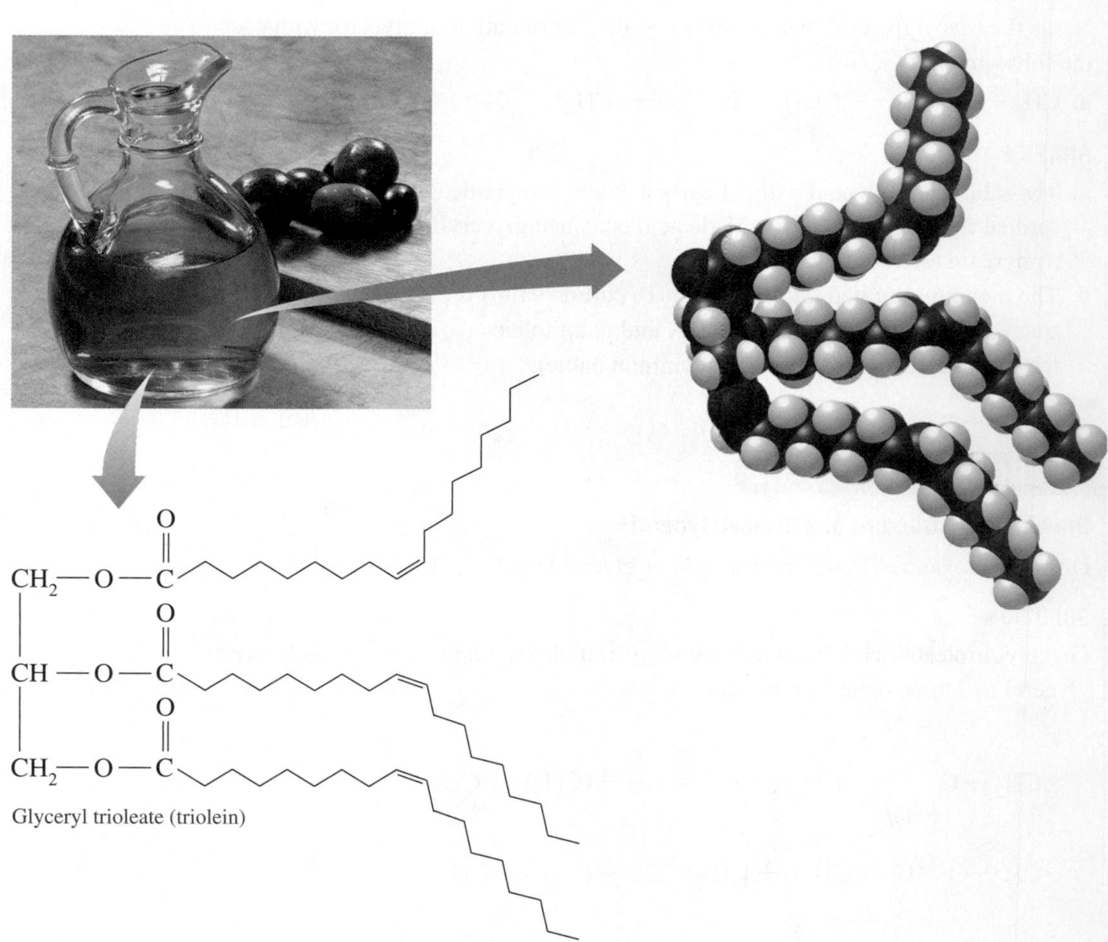

FIGURE 15.3 Vegetable oils such as olive oil, corn oil, and safflower oil contain unsaturated fats.
Q Why is olive oil a liquid at room temperature?

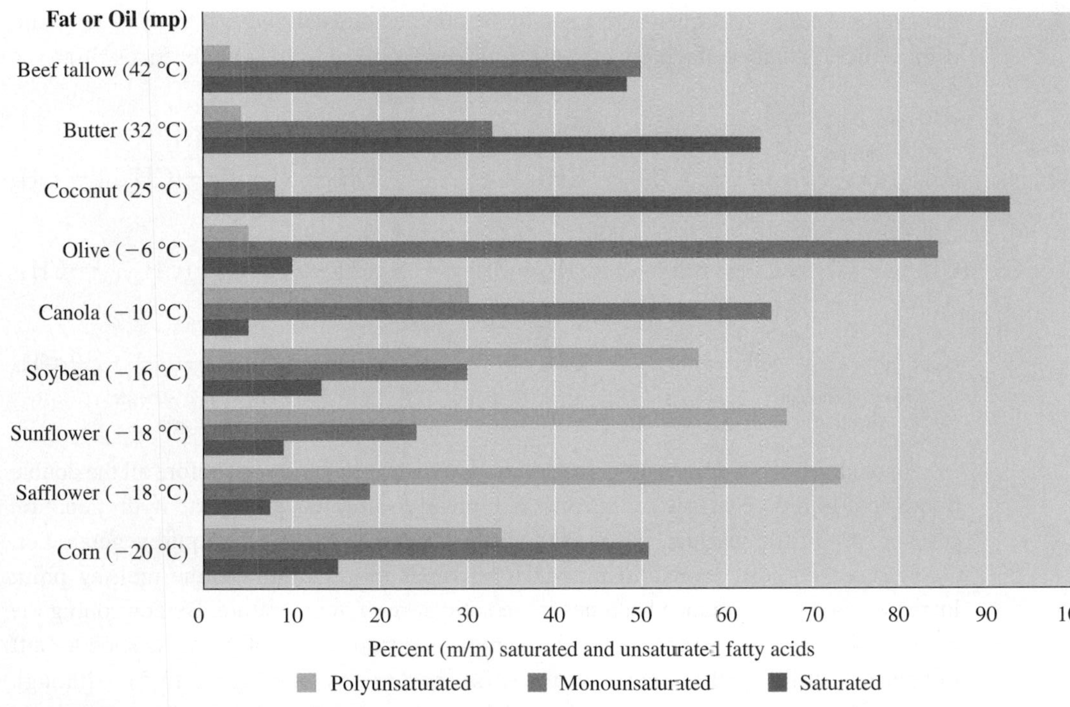

FIGURE 15.4 Vegetable oils have low melting points because they have a higher percentage of unsaturated fatty acids than do animal fats.

Q Why is the melting point of butter higher than olive oil or canola oil?

QUESTIONS AND PROBLEMS

Waxes, Fats, and Oils

15.19 Draw the condensed structural formula of an ester in beeswax formed from myricyl alcohol, $CH_3-(CH_2)_{29}-OH$, and palmitic acid.

15.20 Draw the condensed structural formula of an ester in jojoba wax formed from arachidic acid, a 20-carbon saturated fatty acid, and 1-docosanol, $CH_3-(CH_2)_{21}-OH$.

15.21 Draw the condensed structural formula of a triacylglycerol that contains stearic acid and glycerol.

15.22 A mixed triacylglycerol contains two palmitic acid molecules and one oleic acid molecule. Draw two possible condensed structural formulas (isomers) for the compound.

15.23 Capric acid is a 10-carbon saturated fatty acid. Draw the condensed structural formula of glyceryl tricaprate (tricaprin).

15.24 Draw the condensed structural formula of glyceryl trilinoleate (trilinolein).

15.25 Safflower oil is polyunsaturated, whereas olive oil is monounsaturated. Explain.

15.26 Why does olive oil have a lower melting point than butter fat?

15.27 Why does coconut oil, a vegetable oil, have a melting point similar to that of fats from animal sources?

15.28 A label on a bottle of 100% sunflower seed oil states that it is lower in saturated fats than all the leading oils.

a. How does the percentage of saturated fats in sunflower seed oil compare to that of safflower, corn, and canola oils (see Figure 15.4)?

b. Is the claim valid?

15.4 Chemical Properties of Triacylglycerols

LEARNING GOAL

Draw the condensed structural formula of the product from the reaction of a triacylglycerol with hydrogen, or an acid or base.

The chemical reactions of the triacylglycerols (fats and oils) are the same as those we discussed for the hydrogenation of alkenes (Chapter 11) and the saponification of esters (Chapter 13).

Hydrogenation

In the **hydrogenation** of an unsaturated fat, hydrogen is added to one or more carbon–carbon double bonds to form carbon–carbon single bonds. The hydrogen gas is bubbled through the heated oil typically in the presence of a nickel catalyst.

MC TUTORIAL Hydrolysis and Hydrogenation of Triacylglycerols

$$-CH=CH- + \mathbf{H_2} \xrightarrow{Ni} -\overset{\displaystyle H}{\underset{\displaystyle H}{\overset{|}{\underset{|}{C}}}}-\overset{\displaystyle H}{\underset{\displaystyle H}{\overset{|}{\underset{|}{C}}}}-$$

For example, when hydrogen adds to all of the double bonds of glyceryl trioleate (triolein) using a nickel catalyst, the product is the saturated fat glyceryl tristearate (tristearin).

$$\begin{array}{l} CH_2-O-\overset{\overset{\displaystyle O}{\|}}{C}-(CH_2)_7-CH=CH-(CH_2)_7-CH_3 \\ | \\ CH-O-\overset{\overset{\displaystyle O}{\|}}{C}-(CH_2)_7-CH=CH-(CH_2)_7-CH_3 \\ | \\ CH_2-O-\overset{\overset{\displaystyle O}{\|}}{C}-(CH_2)_7-CH=CH-(CH_2)_7-CH_3 \end{array} + 3H_2 \xrightarrow{Ni} \begin{array}{l} CH_2-O-\overset{\overset{\displaystyle O}{\|}}{C}-(CH_2)_{16}-CH_3 \\ | \\ CH-O-\overset{\overset{\displaystyle O}{\|}}{C}-(CH_2)_{16}-CH_3 \\ | \\ CH_2-O-\overset{\overset{\displaystyle O}{\|}}{C}-(CH_2)_{16}-CH_3 \end{array}$$

Glyceryl trioleate (triolein) → Glyceryl tristearate (tristearin)

In commercial hydrogenation, the addition of hydrogen is stopped before all the double bonds in a liquid vegetable oil become completely saturated. Complete hydrogenation gives a very brittle product, whereas the partial hydrogenation of a liquid vegetable oil changes it to a soft, semisolid fat. As it becomes more saturated, the melting point increases, and the substance becomes more solid at room temperature. By controlling the amount of hydrogen, manufacturers can produce various types of products such as soft margarines, solid stick margarines, and solid shortenings (see Figure 15.5). Although these products now contain more saturated fatty acids than the original oils, they contain no cholesterol, unlike similar products from animal sources, such as butter and lard.

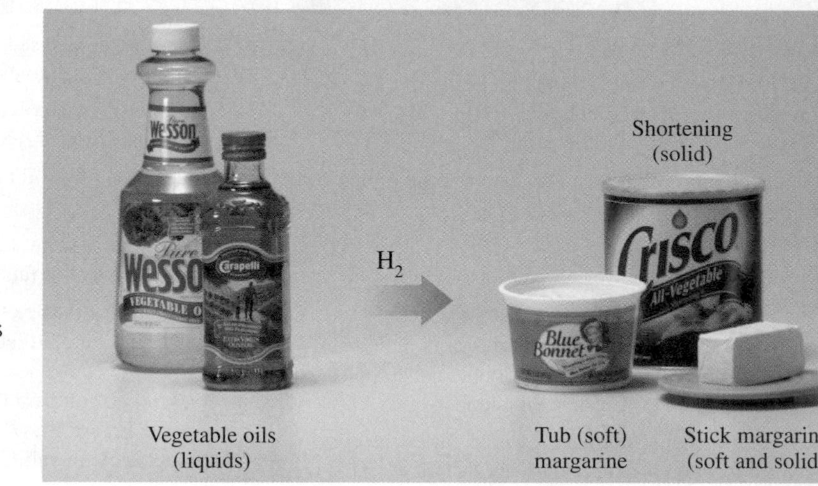

FIGURE 15.5 Many soft margarines, stick margarines, and solid shortenings are produced by the partial hydrogenation of vegetable oils.
Q How does hydrogenation change the structure of the fatty acids in the vegetable oils?

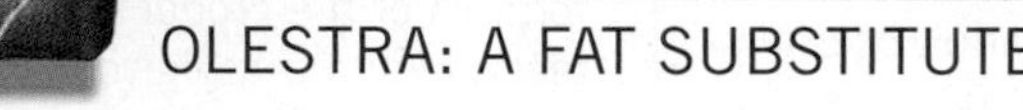

Chemistry Link to Health

OLESTRA: A FAT SUBSTITUTE

In 1968, food scientists designed an artificial fat called *olestra* as a source of nutrition for premature babies. However, olestra could not be digested and was never used for that purpose. Then scientists realized that olestra had the flavor and texture of a fat without the calories.

Olestra is manufactured by obtaining the fatty acids from the fats in cottonseed or soybean oils and bonding the fatty acids with the hydroxyl groups on sucrose. Chemically, olestra is composed of six to eight long-chain fatty acids attached by ester links to a sucrose molecule rather than to a glycerol molecule found in fats. This structure makes olestra a very large molecule that cannot be absorbed through the intestinal walls. The enzymes and bacteria in the intestinal tract are unable to break down the olestra molecule, and it travels through the intestinal tract undigested.

The large molecule of olestra also combines with fat-soluble vitamins (A, D, E, and K) as well as the carotenoids from the foods we eat before they can be absorbed through the intestinal wall. Carotenoids are plant pigments in fruits and vegetables that protect against cancer, heart disease, and macular degeneration, a form of blindness in the elderly. The FDA now requires manufacturers to add the four vitamins, but not the carotenoids, to olestra products. There have been reports of some adverse reactions including diarrhea and abdominal cramps indicating that olestra may act as a laxative in some people. However, the manufacturers contend there is no direct proof that olestra is the cause of those effects.

Snack foods with olestra such as potato chips, tortilla chips, crackers, and fried snacks are now in supermarkets nationwide, but it remains to be seen whether olestra will have any significant effect on reducing the problem of obesity.

Olestra

Fatty acids
$CH_3-(CH_2)_6-COOH$
$CH_3-(CH_2)_8-COOH$

Chemistry Link to Health

TRANS FATTY ACIDS AND HYDROGENATION

In the early 1900s, margarine became a popular replacement for the highly saturated fats such as butter and lard. Margarine is produced by partially hydrogenating the unsaturated fats in vegetable oils such as safflower oil, corn oil, canola oil, cottonseed oil, and sunflower oil. Fats that are more saturated are more resistant to oxidation.

In vegetable oils, the unsaturated fats usually contain cis double bonds. As hydrogenation occurs, double bonds are converted to single bonds. However, some of the cis double bonds are converted to trans double bonds, which causes a change in the overall structure of the fatty acids. If the label on a product states that the oils have been "partially" or "fully hydrogenated," that product will also contain trans fatty acids. In the United States, it is estimated that 2–4% of our total calories come from trans fatty acids.

The concern about trans fatty acids is that their altered structure may make them behave like saturated fatty acids in the body. In the 1980s, research indicated that trans fatty acids have an effect on blood cholesterol similar to that of saturated fats, although study results vary. Several studies reported that trans fatty acids raise the levels of LDL-cholesterol, low-density lipoproteins containing cholesterol that can accumulate in the arteries. (LDLs and HDLs are described in the section on lipoproteins later in this chapter.) Some studies also report that trans fatty acids lower HDL-cholesterol, high-density lipoproteins that carry cholesterol to the liver to be excreted. But other studies did not report any decrease in HDL-cholesterol. In some American and European studies, an increased risk of breast cancer was associated with increased intake of trans fatty acids. However, these studies are not conclusive and not all studies have supported such findings. Current evidence does not yet indicate that the intake of trans fatty acids is a significant risk factor for heart disease. The trans fatty acids controversy will continue to be debated as more research is done.

Foods containing trans fatty acids include milk, bread, fried foods, ground beef, baked goods, stick margarine, butter, soft margarine, cookies, crackers, and vegetable shortening. The American Heart Association recommends that margarine should have no more than 2 grams of saturated fat per tablespoon and a liquid vegetable oil should be the first ingredient. They recommend the use of soft margarine, which is lower in trans fatty acids because soft margarine is only slightly hydrogenated, and diet margarine also, because it has less fat and therefore fewer trans fatty acids.

Many health organizations agree that fat should account for less than 30% of daily calories (the current average for Americans is 34%) and saturated fat should be less than 10% of total calories. Lowering the overall fat intake would also decrease the amount of trans fatty acids. The FDA and the U.S. Department of Agriculture are encouraging the use of new food labels to inform consumers of the fat content of food. The best advice may be to reduce total fat in the diet by using fats and oils sparingly, cooking with little or no fat, substituting olive oil or canola oil for other oils, and limiting the use of coconut oil and palm oil, which are high in saturated fatty acids.

There are several products including peanut butter and butter-like spreads on the market that have 0% trans fatty acids, as indicated on the labels. However, in the list of natural vegetable oils, such as soy and canola oil, there is also palm oil. Because palm oil has a melting point of 30 °C, it increases the overall melting point of the spread and gives a product that is solid at room temperature. However, palm oil contains high amounts of saturated fatty acids and has a similar effect in the body as do stearic acid (18 carbons) and fats derived from animal sources. Health experts recommend that we limit the amount of saturated fats, including palm oil, in our diets.

cis-Oleic acid

H_2/Ni

Ni catalyst

H_2 Isomerization

Addition of H_2

Undesired side product (*trans*-oleic acid)

Desired saturated product (stearic acid)

Hydrolysis

Triacylglycerols are hydrolyzed (split by water) in the presence of strong acids or digestive enzymes called *lipases*. The products of hydrolysis of the ester bonds are glycerol and three fatty acids. The polar glycerol is soluble in water, but the fatty acids with their long hydrocarbon chains are not.

Water adds to ester bonds

$$\begin{array}{l} CH_2-O-\overset{O}{\overset{\|}{C}}-(CH_2)_{14}-CH_3 \\ | \\ CH-O-\overset{O}{\overset{\|}{C}}-(CH_2)_{14}-CH_3 + \mathbf{3H_2O} \\ | \\ CH_2-O-\overset{O}{\overset{\|}{C}}-(CH_2)_{14}-CH_3 \end{array} \xrightarrow[\text{lipase}]{H^+ \text{ or}} \begin{array}{l} CH_2-\mathbf{OH} \\ | \\ CH-\mathbf{OH} + \mathbf{3HO}-\overset{O}{\overset{\|}{C}}-(CH_2)_{14}-CH_3 \\ | \\ CH_2-\mathbf{OH} \end{array}$$

Glyceryl tripalmitate (tripalmitin) — Glycerol — 3 Palmitic acid molecules

Saponification

Saponification occurs when a fat is heated with a strong base such as sodium hydroxide to give glycerol and the sodium salts of the fatty acids, which are soaps. When NaOH is used, a solid soap is produced that can be molded into a desired shape; KOH produces a softer, liquid soap. Oils that are polyunsaturated produce softer soaps. Names like "coconut" or "avocado shampoo" tell you the sources of the oil used in the reaction.

$$\text{Fat or oil} + \text{strong base} \longrightarrow \text{glycerol} + \text{salts of fatty acids (soaps)}$$

$$\begin{array}{l} CH_2-O-\overset{O}{\overset{\|}{C}}-(CH_2)_{14}-CH_3 \\ | \\ CH-O-\overset{O}{\overset{\|}{C}}-(CH_2)_{14}-CH_3 + \mathbf{3NaOH} \\ | \\ CH_2-O-\overset{O}{\overset{\|}{C}}-(CH_2)_{14}-CH_3 \end{array} \longrightarrow \begin{array}{l} CH_2-\mathbf{OH} \\ | \\ CH-\mathbf{OH} + \mathbf{3Na^{+}\,{}^{-}O}-\overset{O}{\overset{\|}{C}}-(CH_2)_{14}-CH_3 \\ | \\ CH_2-\mathbf{OH} \end{array}$$

Glyceryl tripalmitate (tripalmitin) — Glycerol — 3 Sodium palmitate (soap)

The reactions we have discussed for fatty acids and triacylglycerols are similar to the reactions of hydrogenation of alkenes, esterification, hydrolysis, and saponification as summarized in Table 15.3.

TABLE 15.3 Summary of Organic and Lipid Reactions

Reaction	Organic Reactants and Products	Lipid Reactants and Products
Esterification	Carboxylic acid + alcohol $\xrightarrow{H^+, \text{Heat}}$ ester + water	3 Fatty acids + glycerol $\xrightarrow{\text{Enzyme}}$ triacylglycerol (fat) + 3 water
Hydrogenation	Alkene (double bond) + hydrogen $\xrightarrow{Pt}$ alkane (single bonds)	Unsaturated fat (double bonds) + hydrogen $\xrightarrow{Ni}$ saturated fat (single bonds)
Hydrolysis	Ester + water $\xrightarrow{H^+, \text{Heat}}$ carboxylic acid + alcohol	Triacylglycerol (fat) + 3 water $\xrightarrow{\text{Enzyme}}$ 3 fatty acids + glycerol
Saponification	Ester + sodium hydroxide $\longrightarrow$ sodium salt of carboxylic acid + alcohol	Triacylglycerol (fat) + 3 sodium hydroxide $\longrightarrow$ 3 sodium salts of fatty acid (soaps) + glycerol

Explore Your World

TYPES OF FATS

Read the labels on food products that contain fats such as butter, margarine, vegetable oils, peanut butter, and potato chips. Look for terms such as saturated, monounsaturated, polyunsaturated, and partially or fully hydrogenated.

QUESTIONS

1. What type(s) of fats or oils are in the product?
2. How many grams of saturated, monounsaturated, and polyunsaturated fat are in one serving of the product?
3. What percent of the total fat is saturated fat? Unsaturated fat?
4. The label on a container of peanut butter states that the cottonseed and canola oils used to make the peanut butter have been fully hydrogenated. What are the typical products that would form when hydrogen is added?
5. For each packaged food, determine the following:
   a. How many grams of fat are in one serving of the food?
   b. Using the caloric value for fat (9 kcal/g), how many Calories (kilocalories) come from the fat in one serving?
   c. What is the percentage of Cal (kcal) from fat in one serving?

Chemistry Link to the Environment

BIODIESEL AS AN ALTERNATIVE FUEL

Biodiesel is a name of a nonpetroleum fuel that can be used in place of diesel fuel. Biodiesel is produced from renewable biological resources such as vegetable oils (primarily soybean), waste vegetable oils from restaurants, and some animal fats. Biodiesel is nontoxic and biodegradable.

Biodiesel is prepared from triacylglycerols and alcohols (usually ethanol) to form ethyl esters and glycerol. The glycerol that separates from the fat is used in soaps and other products. The reaction of triacylglycerols is catalyzed by a base such as NaOH or KOH at low temperatures to produce the biodiesel fuel.

$$\text{Triacylglycerol} + 3\ \text{ethanol} \longrightarrow 3\ \text{ethyl esters (biodiesel)} + \text{glycerol}$$

Compared to diesel fuel from petroleum, biodiesel burns in an engine to produce much lower levels of carbon dioxide emissions, particulates, unburned hydrocarbons, and aromatic hydrocarbons that cause lung cancer. Because biodiesel has extremely low sulfur content, it does not contribute to the formation of the sulfur oxides that produce acid rain. The energy output from the combustion of biodiesel is almost the same as energy produced by petroleum diesel.

Biodiesel 20 contains 20% ethyl esters and 80% standard diesel fuel.

$$\begin{array}{l} CH_2-O-C(=O)-(CH_2)_{12}-CH_3 \\ H-C-O-C(=O)-(CH_2)_7-CH{=}CH-(CH_2)_7-CH_3 \\ CH_2-O-C(=O)-(CH_2)_{16}-CH_3 \end{array} + 3CH_3-CH_2-OH$$

Triacylglycerol from vegetable oil; Ethanol

$$\xrightarrow{\text{NaOH catalyst}} \begin{array}{l} CH_2-OH \\ H-C-OH \\ CH_2-OH \end{array} + \begin{array}{l} CH_3-CH_2-O-C(=O)-(CH_2)_{12}-CH_3 \\ CH_3-CH_2-O-C(=O)-(CH_2)_7-CH{=}CH-(CH_2)_7-CH_3 \\ CH_3-CH_2-O-C(=O)-(CH_2)_{16}-CH_3 \end{array}$$

Glycerol; Ethyl esters used for biodiesel

CONCEPT CHECK 15.3

Hydrogenation, Hydrolysis, and Saponification

Identify each of the following as hydrogenation, hydrolysis, or saponification and state the products:

a. the reaction of palm oil with KOH
b. the reaction of glyceryl trilinoleate from safflower oil with water and HCl
c. the reaction of corn oil and hydrogen (H_2) using a nickel catalyst

ANSWER

a. The reaction of palm oil with KOH is saponification, and the products are glycerol and the potassium salts of the fatty acids, which are soaps.
b. The reaction of glyceryl trilinoleate from safflower oil with water and HCl is a hydrolysis, which splits the ester bonds to produce glycerol and three molecules of linoleic acid.
c. The reaction of corn oil and hydrogen (H_2) using a nickel catalyst is hydrogenation, which adds H_2 to the double bonds in the corn oil to produce a more saturated triacylglycerol (fat) that is more solid.

SAMPLE PROBLEM 15.3

Reactions of Lipids

Write the equation for the reaction catalyzed by the enzyme lipase that hydrolyzes glyceryl trilaurate (trilaurin) during the digestion process.

SOLUTION

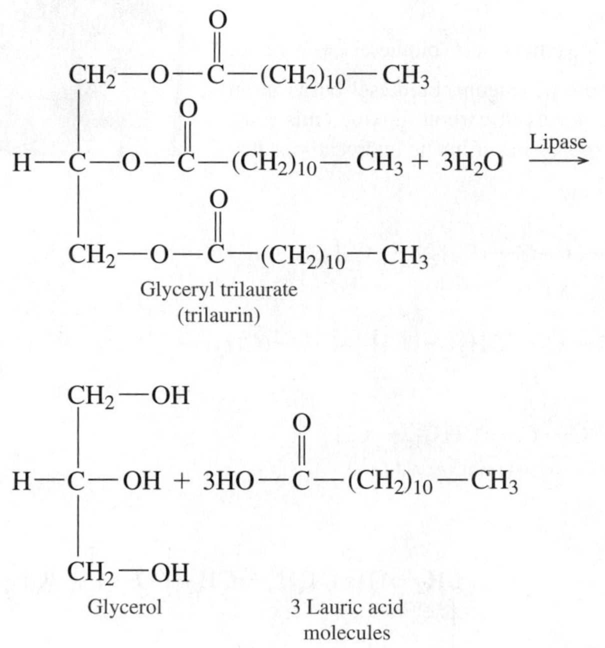

$$\begin{array}{l} CH_2-O-\overset{\overset{O}{\|}}{C}-(CH_2)_{10}-CH_3 \\ | \\ H-C-O-\overset{\overset{O}{\|}}{C}-(CH_2)_{10}-CH_3 + 3H_2O \xrightarrow{\text{Lipase}} \\ | \\ CH_2-O-\overset{\overset{O}{\|}}{C}-(CH_2)_{10}-CH_3 \end{array}$$

Glyceryl trilaurate (trilaurin)

$$\begin{array}{l} CH_2-OH \\ | \\ H-C-OH + 3HO-\overset{\overset{O}{\|}}{C}-(CH_2)_{10}-CH_3 \\ | \\ CH_2-OH \end{array}$$

Glycerol; 3 Lauric acid molecules

STUDY CHECK 15.3

What is the name of the product formed when a triacylglycerol containing oleic acid and linoleic acid is completely hydrogenated?

QUESTIONS AND PROBLEMS

Chemical Properties of Triacylglycerols

15.29 Write an equation for the hydrogenation of glyceryl trioleate, a fat containing glycerol and three oleic acid units.

15.30 Write an equation for the hydrogenation of glyceryl trilinolenate, a fat containing glycerol and three linolenic acid units.

15.31 **a.** Write an equation for the acid hydrolysis of glyceryl trimyristate (trimyristin).
b. Write an equation for the NaOH saponification of glyceryl trimyristate (trimyristin).

15.32 **a.** Write an equation for the acid hydrolysis of glyceryl trioleate (triolein).
b. Write an equation for the NaOH saponification of glyceryl trioleate (triolein).

15.33 Compare the structure of a triacylglycerol to the structure of olestra (see Chemistry Link to Health "Olestra: A Fat Substitute").

15.34 A vegetable oil is partially hydrogenated.
a. Are all or just some of the double bonds converted to single bonds?
b. What happens to many of the cis double bonds during hydrogenation?
c. How can you reduce the amount of trans fatty acids in your diet?

15.35 Draw the condensed structural formula of the product of the hydrogenation of the following triacylglycerol:

$$\begin{array}{l} CH_2-O-\overset{\large O}{\overset{\|}{C}}-(CH_2)_{16}-CH_3 \\ | \\ CH-O-\overset{\large O}{\overset{\|}{C}}-(CH_2)_7-CH=CH-(CH_2)_7-CH_3 \\ | \\ CH_2-O-\overset{\large O}{\overset{\|}{C}}-(CH_2)_{16}-CH_3 \end{array}$$

15.36 Draw the condensed structural formulas of all the products that would be obtained when the triacylglycerol in Problem 15.35 undergoes complete hydrolysis.

15.5 Glycerophospholipids

LEARNING GOAL
Describe the characteristics of glycerophospholipids.

The **glycerophospholipids** are a family of lipids similar to triacylglycerols except that one hydroxyl group of glycerol is replaced by the ester of phosphoric acid and an amino alcohol, bonded through a phosphodiester bond. We can compare the general structures of a triacylglycerol and a glycerophospholipid as follows:

Glycerol	
G	—Fatty acid
	—Fatty acid
	—Fatty acid

Triacylglycerol (triglyceride)

Glycerol		
G	—Fatty acid	
	—Fatty acid	
	—Phosphate	—Amino alcohol

Glycerophospholipid

TUTORIAL
The Split Personality of Glycerophospholipids

Three amino alcohols found in glycerophospholipids are choline, serine, and ethanolamine. In the body, at physiological pH of 7.4, they are ionized.

$$HO-CH_2-CH_2-\overset{CH_3}{\underset{CH_3}{\overset{|}{\underset{|}{N^+}}}}-CH_3$$

Choline

$$HO-CH_2-\overset{\overset{+}{N}H_3}{\overset{|}{CH}}-COO^-$$

Serine

$$HO-CH_2-CH_2-\overset{+}{N}H_3$$

Ethanolamine

Lecithins and **cephalins** are two types of glycerophospholipids that are particularly abundant in brain and nerve tissues as well as in egg yolks, wheat germ, and yeast. Lecithins contain choline, and cephalins contain ethanolamine and sometimes serine. In the following structural formulas, palmitic acid is used as an example of a fatty acid.

$$\begin{array}{l} CH_2-O-\overset{O}{\overset{\|}{C}}-(CH_2)_{14}-CH_3 \\ CH-O-\overset{O}{\overset{\|}{C}}-(CH_2)_{14}-CH_3 \\ CH_2-O-\underset{O^-}{\underset{|}{\overset{O}{\overset{\|}{P}}}}-O-CH_2-CH_2-\overset{+}{N}-(CH_3)_3 \end{array}$$

Nonpolar fatty acids

Polar

Choline

A lecithin

$$\begin{array}{l} CH_2-O-\overset{O}{\overset{\|}{C}}-(CH_2)_{14}-CH_3 \\ CH-O-\overset{O}{\overset{\|}{C}}-(CH_2)_{14}-CH_3 \\ CH_2-O-\underset{O^-}{\underset{|}{\overset{O}{\overset{\|}{P}}}}-O-CH_2-CH_2-\overset{+}{N}H_3 \end{array}$$

Ethanolamine

A cephalin

Glycerophospholipids contain both polar and nonpolar regions, which allow them to interact with both polar and nonpolar substances. The ionized amino alcohol and phosphate portion called the head is polar and strongly attracted to water (see Figure 15.6). The two fatty acids connected to the glycerol molecule represent the nonpolar tails of the glycerophospholipid. The hydrocarbon chains that make up the tails are only soluble in other nonpolar substances, mostly lipids.

Glycerophospholipids are the most abundant lipids in cell membranes, where they play an important role in cellular permeability. They make up much of the myelin sheath that protects nerve cells. In the body fluids, they combine with the less polar triglycerides and cholesterol to make them more soluble as they are transported in the body.

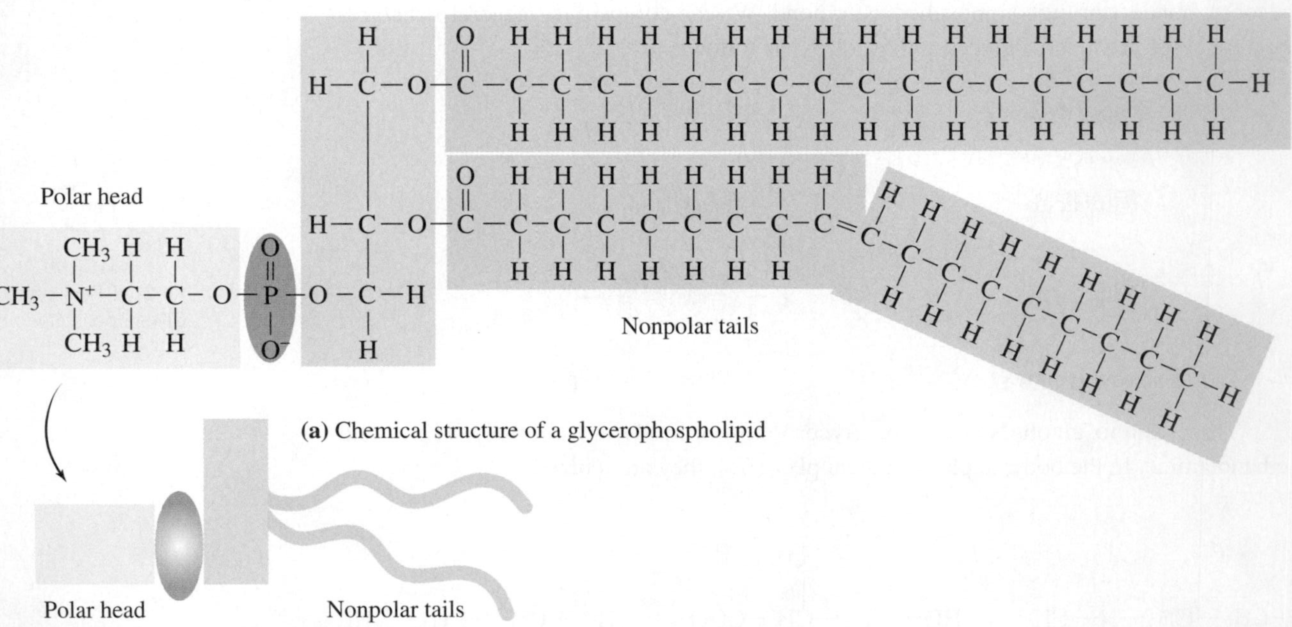

FIGURE 15.6 (**a**) In a glycerophospholipid, a polar head contains the ionized amino alcohol and phosphate groups, while the two fatty acids make up the nonpolar tails. (**b**) A simplified drawing indicates the polar region and the nonpolar region.
Q Why are glycerophospholipids polar?

CONCEPT CHECK 15.4

Glycerophospholipid Structure

Identify each part of the following glycerophospholipid as (A) glycerol, (B) fatty acid, (C) phosphate, or (D) amino alcohol:

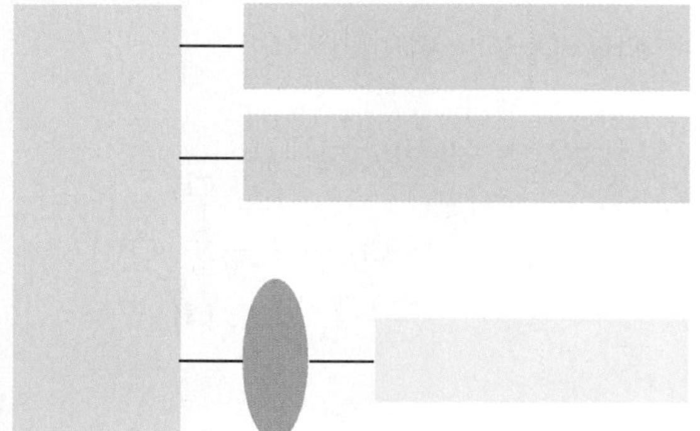

ANSWER

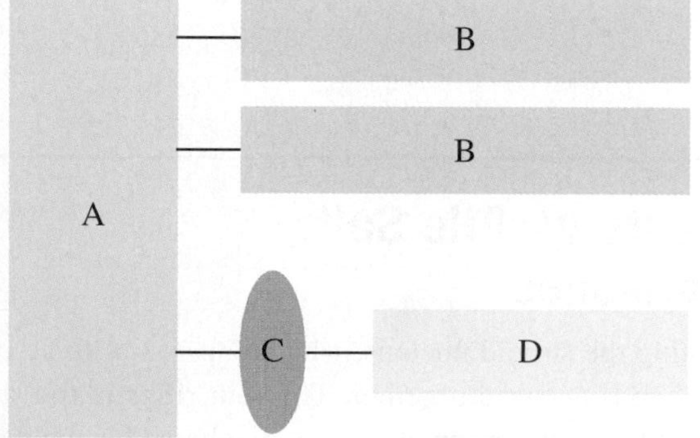

SAMPLE PROBLEM 15.4

Drawing Glycerophospholipid Structures

Draw the condensed structural formula of a cephalin that contains stearic acid and serine. Describe each component in the glycerophospholipid.

SOLUTION

In general, glycerophospholipids are composed of a glycerol molecule in which two carbon atoms are attached to fatty acids such as stearic acid. The third carbon atom is attached by an ester bond to phosphate linked to an amino alcohol. In this example, the amino alcohol is serine.

$$
\begin{array}{l}
CH_2-O-\overset{\overset{\displaystyle O}{\|}}{C}-(CH_2)_{16}-CH_3 \\
| \\
CH-O-\overset{\overset{\displaystyle O}{\|}}{C}-(CH_2)_{16}-CH_3 \\
| \\
CH_2-O-\underset{\underset{\displaystyle O^-}{|}}{\overset{\overset{\displaystyle O}{\|}}{P}}-O-CH_2-\overset{\overset{\displaystyle \overset{+}{N}H_3}{|}}{CH}-COO^-
\end{array}
\quad
\begin{array}{l}
\text{Stearic acids} \\
\\
\text{Serine}
\end{array}
$$

STUDY CHECK 15.4

Draw the structure of a lecithin, using myristic acid for the fatty acids and choline as the amino alcohol.

Career Focus

PHYSICAL THERAPIST

"We do all kinds of activities that are typical of the things kids would do," says Helen Tong, physical therapist. "For example, we play with toys to help improve motor control. When a child can't do something, we use adaptive equipment to help him or her do that activity in a different way, which still allows participation. In school, we learn how the body works, and why it does not work. Then we can figure out what to do to help a child learn new skills. For example, this child with Rett syndrome has motor difficulties. Although she has difficulty talking, she does amazing work at a computer. There has been a real growth in assisted technology for children, and it has changed our work tremendously."

QUESTIONS AND PROBLEMS

Glycerophospholipids

15.37 Describe the differences between triacylglycerols and glycerophospholipids.

15.38 Describe the differences between lecithins and cephalins.

15.39 Draw the structure of a glycerophospholipid that contains palmitic acid and ethanolamine. What is another name for this type of glycerophospholipid?

15.40 Draw the structure of a glycerophospholipid that contains choline and palmitic acid.

15.41 Identify the type of the following glycerophospholipid, and list its components:

$$\begin{array}{l} CH_2{-}O{-}\overset{O}{\overset{\|}{C}}{-}(CH_2)_7{-}CH{=}CH{-}(CH_2)_7{-}CH_3 \\ \ | \\ CH{-}O{-}\overset{O}{\overset{\|}{C}}{-}(CH_2)_{16}{-}CH_3 \\ \ | \\ CH_2{-}O{-}\overset{O}{\overset{\|}{P}}{-}O{-}CH_2{-}CH_2{-}\overset{+}{N}H_3 \\ \qquad\quad\ \ | \\ \qquad\quad\ O^- \end{array}$$

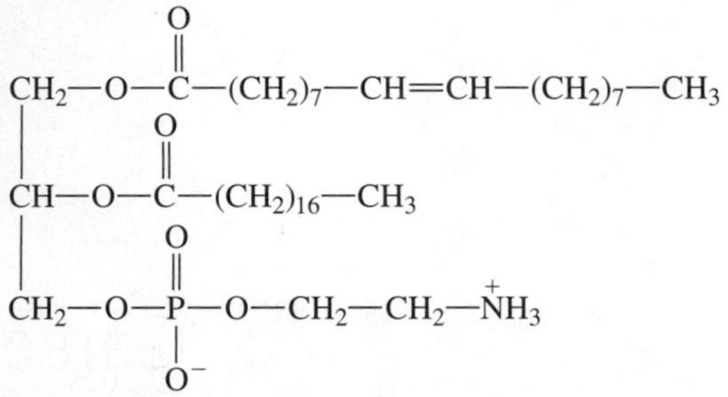

15.42 Identify the type of the following glycerophospholipid, and list its components:

$$\begin{array}{l} CH_2{-}O{-}\overset{O}{\overset{\|}{C}}{-}(CH_2)_{14}{-}CH_3 \\ \ | \\ CH{-}O{-}\overset{O}{\overset{\|}{C}}{-}(CH_2)_{16}{-}CH_3 \\ \ | \\ CH_2{-}O{-}\overset{O}{\overset{\|}{P}}{-}O{-}CH_2{-}CH_2{-}\overset{CH_3}{\overset{|}{\overset{+}{N}}}{-}CH_3 \\ \qquad\quad\ \ | \qquad\qquad\qquad\quad\ \ | \\ \qquad\quad\ O^- \qquad\qquad\qquad\ CH_3 \end{array}$$

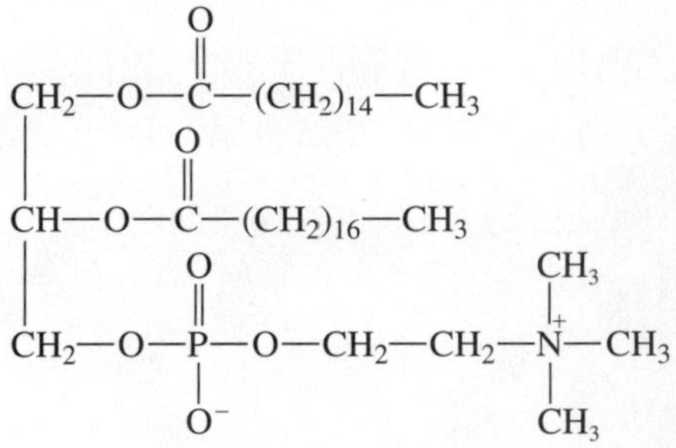

LEARNING GOAL

Describe the structures of steroids.

15.6 Steroids: Cholesterol, Bile Salts, and Steroid Hormones

Steroids are compounds containing the steroid nucleus, which consists of three cyclohexane rings and one cyclopentane ring fused together. The four rings in the steroid nucleus are designated A, B, C, and D. The carbon atoms are numbered beginning with the carbons in ring A, and in steroids like cholesterol, ending with two methyl groups.

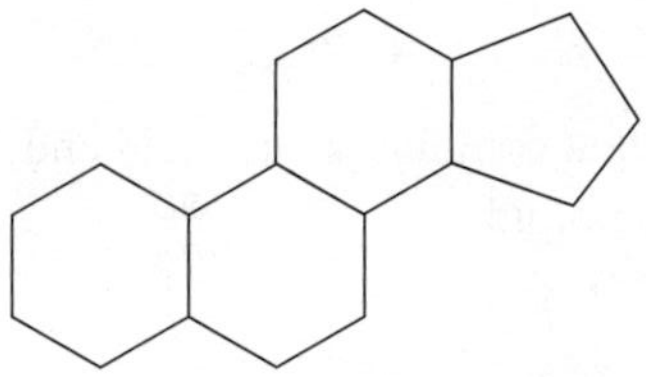

Steroid nucleus

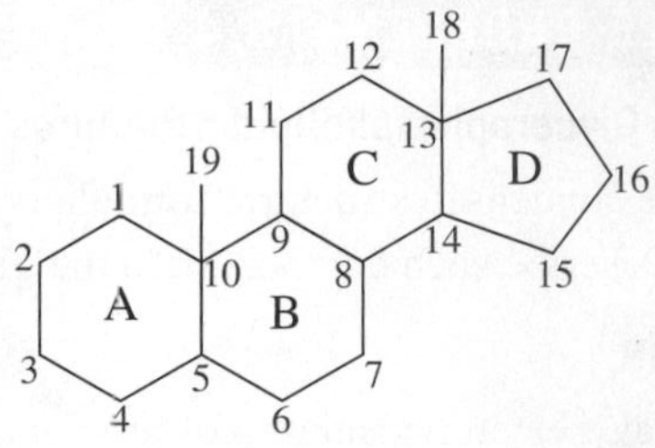

Steroid numbering system

Cholesterol

MC TUTORIAL Cholesterol

Attaching other atoms and groups of atoms to the steroid structure forms a wide variety of steroid compounds. **Cholesterol**, which is one of the most important and abundant steroids in the body, is a *sterol* because it contains an oxygen atom as a hydroxyl (—OH) group on carbon 3. Like many steroids, cholesterol has methyl groups at carbons 10 and 13, a carbon chain at carbon 17, and a double bond between carbons 5 and 6. In other steroids, the oxygen atom forms a carbonyl (C=O) group at carbon 3.

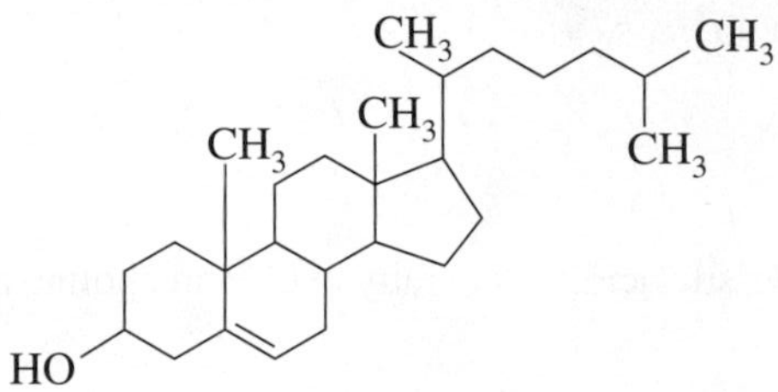

Cholesterol

Cholesterol in the Body

Cholesterol is a component of cellular membranes, myelin sheath, and brain and nerve tissue. It is also found in the liver, bile salts, and skin, where it forms vitamin D. In the adrenal gland, it is used to synthesize steroid hormones. Cholesterol in the body is obtained from eating meats, milk, and eggs, and it is also synthesized by the liver from fats, carbohydrates, and proteins. There is no cholesterol in vegetable and plant products.

If a diet is high in cholesterol, the liver produces less. A typical daily American diet includes 400–500 mg of cholesterol, one of the highest in the world. The American Heart Association has recommended that we consume no more than 300 mg of cholesterol a day. The cholesterol contents of some typical foods are listed in Table 15.4.

TABLE 15.4 Cholesterol Content of Some Foods

Food	Serving Size	Cholesterol (mg)
Liver (beef)	3 oz	370
Large egg	1	200
Lobster	3 oz	175
Fried chicken	$3\frac{1}{2}$ oz	130
Hamburger	3 oz	85
Chicken (no skin)	3 oz	75
Fish (salmon)	3 oz	40
Whole milk	1 cup	35
Butter	1 tablespoon	30
Skim milk	1 cup	5
Margarine	1 tablespoon	0

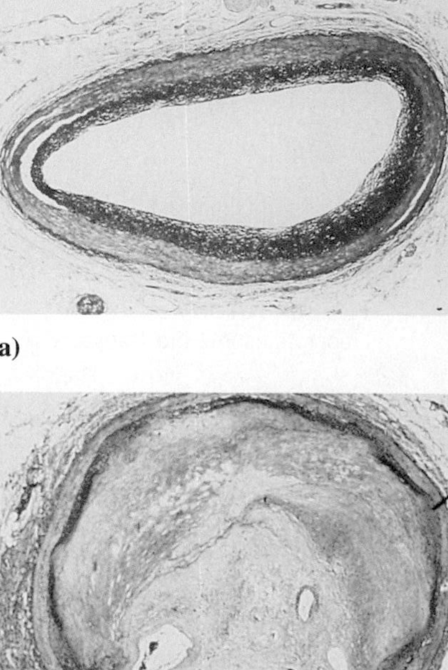

(a)

(b)

FIGURE 15.7 Excess cholesterol forms plaque that can block an artery, resulting in a heart attack. **(a)** A normal, open artery shows no buildup of plaque. **(b)** An artery that is almost completely clogged by atherosclerotic plaque.

Q What property of cholesterol would cause it to form deposits along the coronary arteries?

High levels of cholesterol are associated with the accumulation of lipid deposits (plaque) that line and narrow the coronary arteries (see Figure 15.7). Clinically, cholesterol levels are considered elevated if the total plasma cholesterol level exceeds 200 mg/dL.

Saturated fats in the diet may stimulate the production of cholesterol by the liver. A diet that is low in foods containing cholesterol and saturated fats appears to be helpful in reducing the serum cholesterol level. Other factors that may also increase the risk of heart disease are family history, lack of exercise, smoking, obesity, diabetes, gender, and age.

SAMPLE PROBLEM 15.5

Cholesterol

Refer to the structure of cholesterol for each of the following questions:

a. What part of cholesterol is the steroid nucleus?
b. What features have been added to the steroid nucleus in cholesterol?
c. What classifies cholesterol as a sterol?

SOLUTION

a. The four fused rings form the steroid nucleus.
b. The cholesterol molecule contains hydroxyl group ($-OH$) on the first ring, one double bond in the second ring, the methyl groups ($-CH_3$) at carbons 10 and 13, and a branched carbon chain.
c. The hydroxyl group determines the sterol classification.

STUDY CHECK 15.5

Why is cholesterol in the lipid family?

Bile Salts

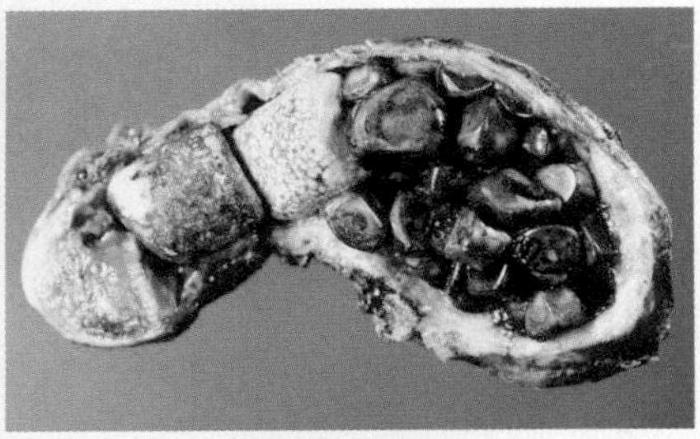

FIGURE 15.8 Gallstones form in the gallbladder when cholesterol levels are high.

Q What type of steroid is stored in the gallbladder?

The *bile salts* are synthesized in the liver from cholesterol and stored in the gallbladder. When bile is secreted into the small intestine, the bile salts mix with the water-insoluble fats and oils in our diets. The bile salts with their nonpolar and polar regions act much like soaps, breaking apart and emulsifying large globules of fat. The emulsions that form have a larger surface area for the lipases, enzymes that digest fat. The bile salts also help in the absorption of cholesterol into the intestinal mucosa.

When large amounts of cholesterol accumulate in the gallbladder, cholesterol can precipitate out and form gallstones (see Figure 15.8). If a gallstone passes into the bile duct, the pain can be severe. If the gallstone obstructs the duct, bile cannot be excreted. Then bile pigments known as bilirubin enter the blood where they cause jaundice, which gives a yellow color to the skin and eyes.

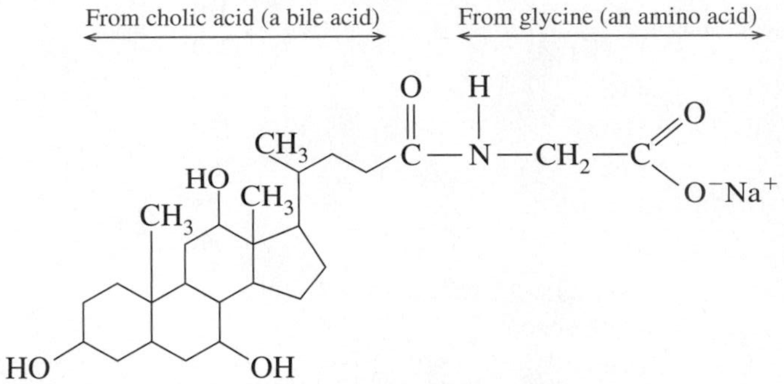

Sodium glycocholate (a bile salt)

Lipoproteins: Transport of Lipids

TUTORIAL
Lipoproteins in Lipid Transport

In the body, lipids must be transported through the bloodstream to tissues where they are stored, used for energy, or transformed to hormones. However, lipids are nonpolar and insoluble in the aqueous environment of blood. They are made more soluble by combining them with glycerophospholipids and proteins to form water-soluble complexes called **lipoproteins**. In general, lipoproteins are spherical particles with an outer surface of polar proteins and glycerophospholipids that surround hundreds of nonpolar molecules of triacylglycerols and cholesteryl esters (see Figure 15.9). Cholesteryl esters are the prevalent form of cholesterol in the blood. They are formed by the esterification of the hydroxyl group in cholesterol with a fatty acid.

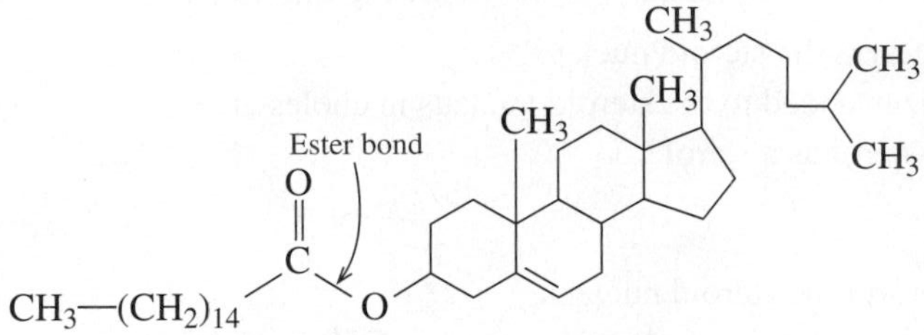

Cholesteryl ester

Types of lipoproteins differ in density, lipid composition, and function. They include chylomicrons, very-low-density lipoprotein (VLDL), low-density lipoprotein (LDL), and high-density lipoprotein (HDL). The LDLs form when the triacylglycerol portion is removed from VLDLs. The density of the lipoproteins increases as the percentage of protein in each type also increases (see Table 15.5).

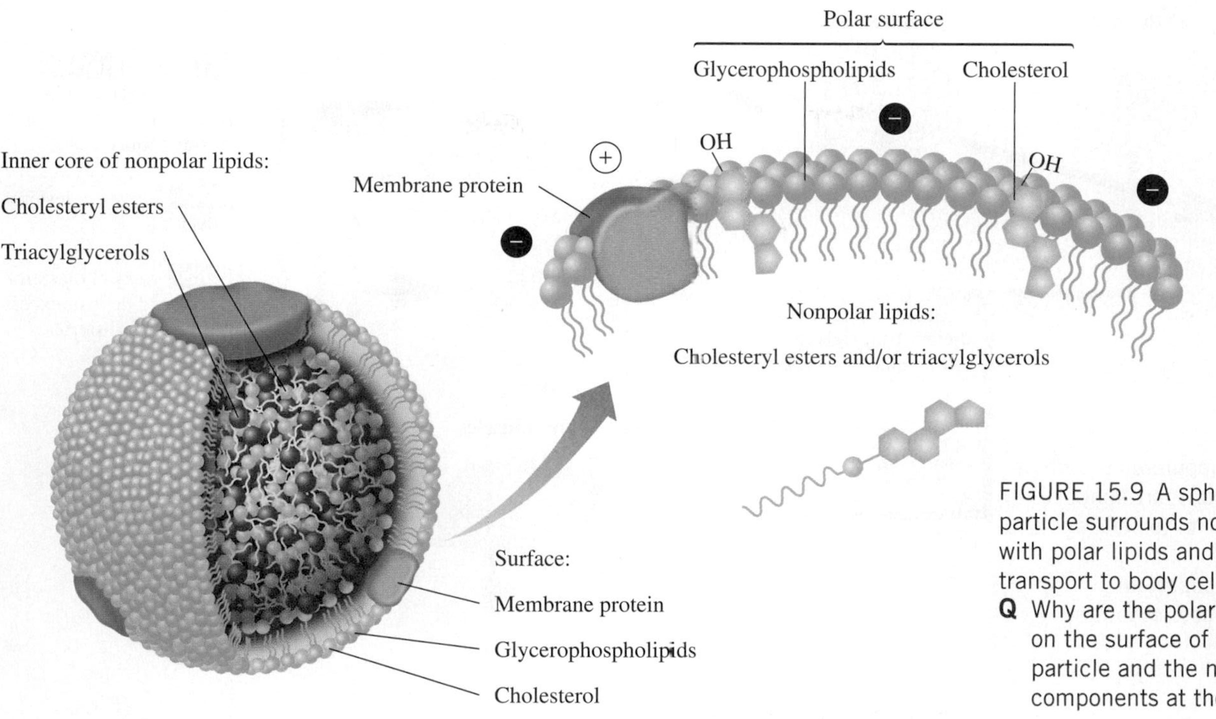

FIGURE 15.9 A spherical lipoprotein particle surrounds nonpolar lipids with polar lipids and protein for transport to body cells.

Q Why are the polar components on the surface of a lipoprotein particle and the nonpolar components at the center?

TABLE 15.5 Composition and Properties of Plasma Lipoproteins

	Chylomicron	VLDL	LDL	HDL
Density (g/mL)	<0.95	0.950–1.006	1.006–1.063	1.063–1.210
Composition (% by mass)				
Triacylglycerol	86	55	6	4
Phospholipids	7	18	22	24
Cholesterol	2	7	8	2
Cholesteryl esters	3	12	42	15
Protein	2	8	22	55

The chylomicrons and the VLDLs transport triacylglycerols, glycerophospholipids, and cholesterol to the tissues for storage or to the muscles for energy (see Figure 15.10). The LDLs transport cholesterol to tissues where it is used for the synthesis of cell membranes, steroid hormones, and bile salts. When the level of LDL exceeds the amount of cholesterol needed by the tissues, the LDLs deposit cholesterol in the arteries, restricting blood flow and increasing the risk of developing heart disease and/or myocardial infarctions (heart attacks). This is why LDL cholesterol is called "bad" cholesterol.

The HDLs remove excess cholesterol from the tissues and carry it to the liver, where it is converted to bile salts and eliminated. When HDL levels are high, cholesterol that is not needed by the tissues is carried to the liver for elimination rather than deposited in the arteries, which gives the HDLs the name of "good" cholesterol. Most of the cholesterol in the body is synthesized in the liver, although some comes from the diet. However, a person on a high-fat diet reabsorbs cholesterol from the bile salts, causing less cholesterol to be eliminated. In addition, higher levels of saturated fats stimulate the synthesis of cholesterol by the liver.

Because high cholesterol levels are associated with the onset of atherosclerosis and heart disease, the serum levels of LDL and HDL are generally determined as part of a medical examination. For adults, recommended levels for total cholesterol are less than 200 mg/dL, with LDL less than 130 mg/dL and HDL higher than 40 mg/dL. A lower level of serum cholesterol decreases the risk of heart disease. Increased HDL levels are found in people who exercise regularly and eat less saturated fat.

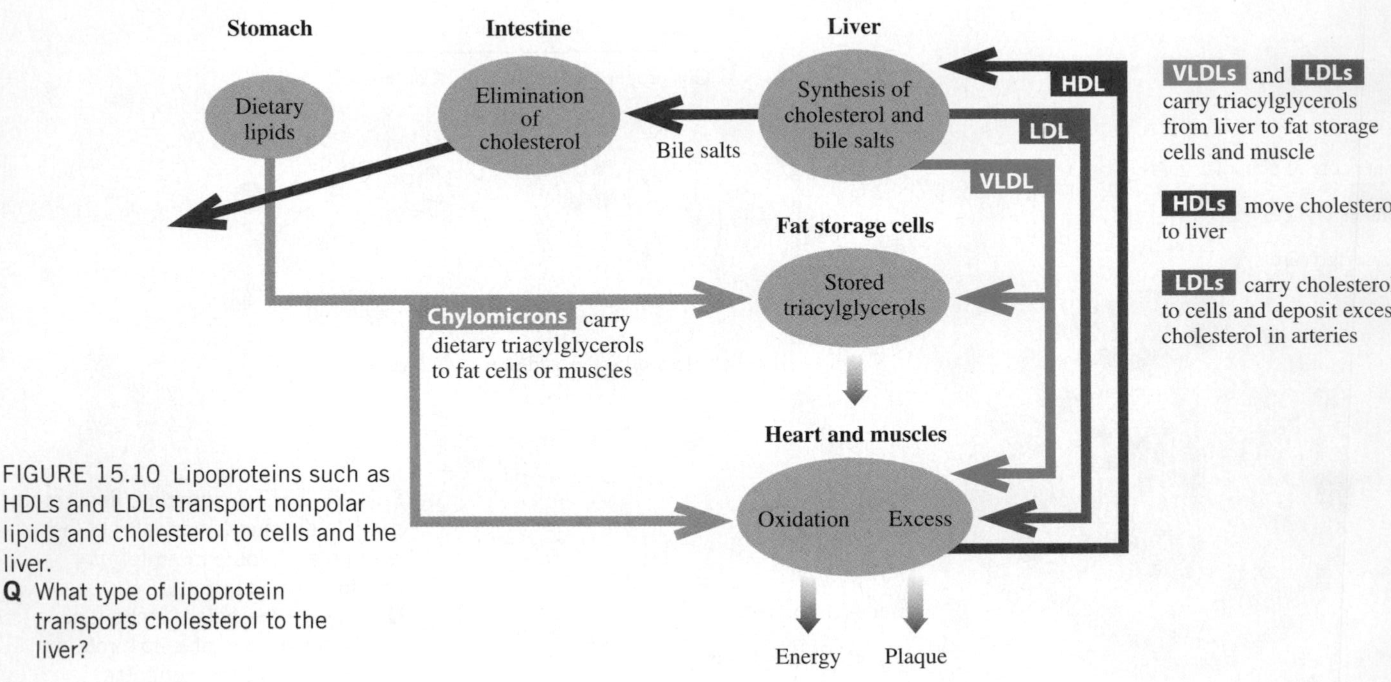

FIGURE 15.10 Lipoproteins such as HDLs and LDLs transport nonpolar lipids and cholesterol to cells and the liver.

Q What type of lipoprotein transports cholesterol to the liver?

Steroid Hormones

The word *hormone* comes from the Greek "to arouse" or "to excite." Hormones are chemical messengers that serve as a kind of communication system from one part of the body to another. The *steroid* hormones, which include the sex hormones and the adreno-cortical hormones, are closely related in structure to cholesterol and depend on cholesterol for their synthesis.

Two of the male sex hormones, *testosterone* and *androsterone*, promote the growth of muscle and facial hair and the maturation of the male sex organs and of sperm.

The *estrogens*, a group of female sex hormones, direct the development of female sexual characteristics: The uterus increases in size, fat is deposited in the breasts, and the pelvis broadens. *Progesterone* prepares the uterus for the implantation of a fertilized egg. If an egg is not fertilized, the levels of progesterone and estrogen drop sharply, and menstruation follows. Synthetic forms of the female sex hormones are used in birth-control pills. As with other kinds of steroids, side effects can include weight gain and a greater risk of forming blood clots. The structures of some steroid hormones follow:

Testosterone (androgen)
(produced in testes)

Estradiol (estrogen)
(produced in ovaries)

Progesterone
(produced in ovaries)

Norethindrone
(synthetic progestin)

Chemistry Link to Health

ANABOLIC STEROIDS

Some of the physiological effects of testosterone are to increase muscle mass and decrease body fat. Derivatives of testosterone called *anabolic steroids* that enhance these effects have been synthesized. Although they have some medical uses, anabolic steroids have been used in high dosages by some athletes in an effort to increase muscle mass. Such use is illegal.

Use of anabolic steroids in attempting to improve athletic strength can cause side effects including hypertension, fluid retention, increased hair growth, sleep disturbances, and acne. Over a long period of time, their use may cause irreversible liver damage and decreased sperm production.

Some Anabolic Steroids

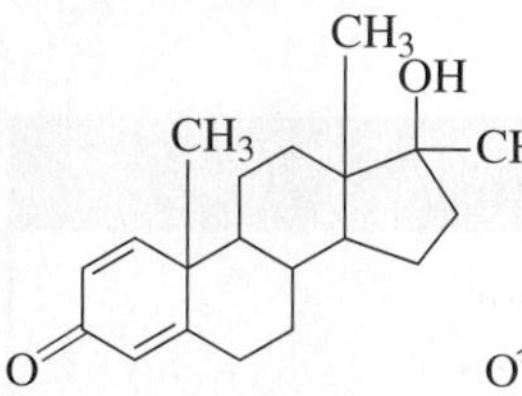

Methandienone 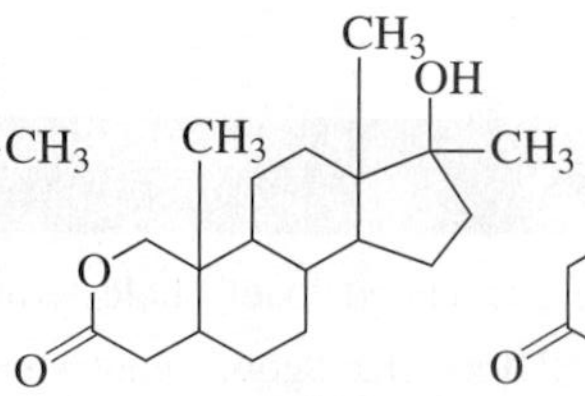   Oxandrolone 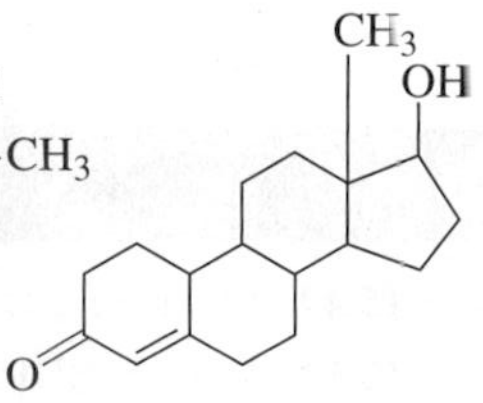   Nandrolone 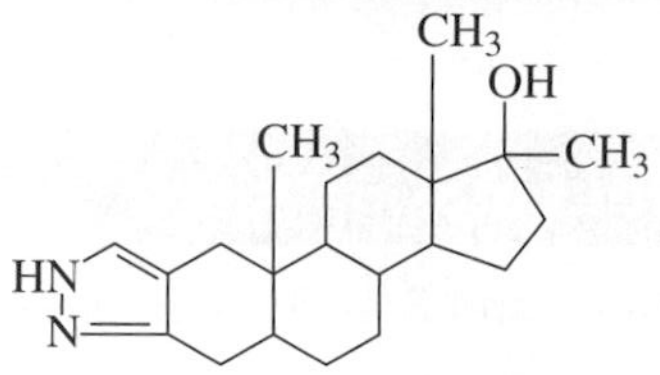   Stanozolol

Adrenal Corticosteroids

The adrenal glands, located on the top of each kidney, produce the corticosteroids. *Aldosterone*, a mineralocorticoid, is responsible for electrolyte and water balance by the kidneys. *Cortisone*, a glucocorticoid, increases the blood glucose level and stimulates the synthesis of glycogen from amino acids in the liver. Synthetic corticoids such as *prednisone* are derived from cortisone and used medically for reducing inflammation and treating asthma and rheumatoid arthritis, although precautions are given for long-term use.

Corticosteroids

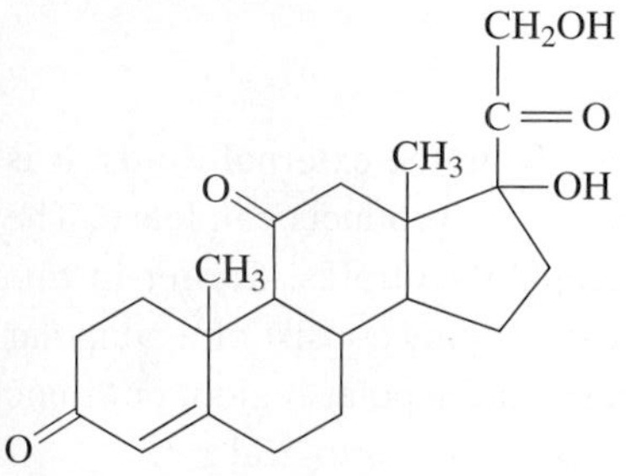

Cortisone (produced in adrenal gland)	Aldosterone (mineralocorticoid) (produced in adrenal gland)	Prednisone (synthetic corticoid)
Biological Effects		
Increases the blood glucose and glycogen levels from fatty acids and amino acids	Increases the reabsorption of Na^+ in kidneys; retention of water	Reduces inflammation; treatment of asthma and rheumatoid arthritis

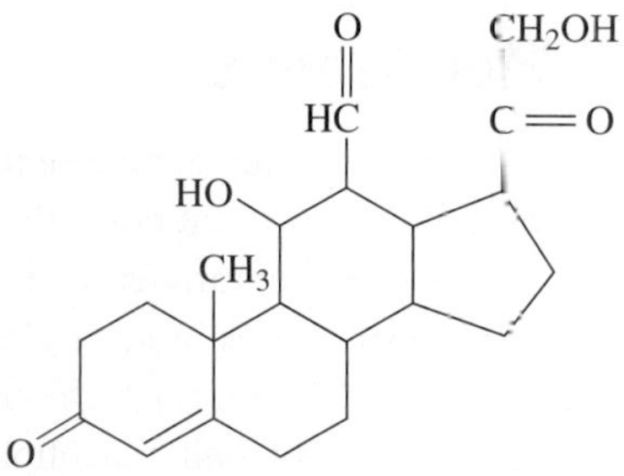

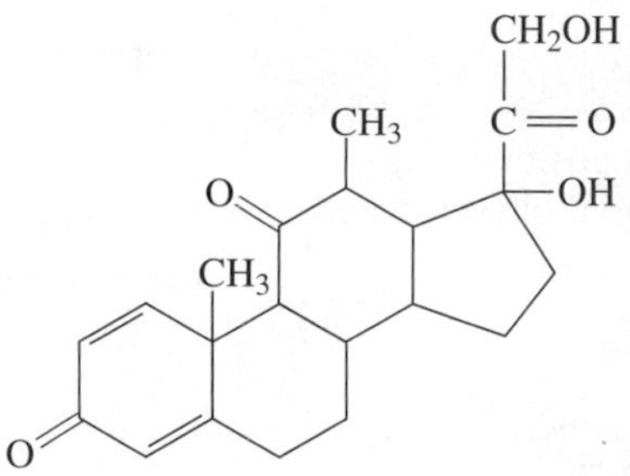

CONCEPT CHECK 15.5

Steroid Hormones

a. What are the substituents on the steroid nucleus in the hormones estradiol and testosterone?

b. What are the similarities and differences in the structures of testosterone and the anabolic steroid nandrolone? (See Chemistry Link to Health "Anabolic Steroids.")

ANSWER

a. Estradiol contains an aromatic ring, one methyl group, and two hydroxyl groups. Testosterone contains a ketone group, a double bond, two methyl groups, and a hydroxyl group.

b. Testosterone and nandrolone both contain a steroid nucleus with one double bond and a ketone group in the first ring, and a methyl and a hydroxyl group on the five-carbon ring. They differ because nandrolone does not have a methyl group where the first and second rings join.

QUESTIONS AND PROBLEMS

Steroids: Cholesterol, Bile Salts, and Steroid Hormones

15.43 Draw the structure for the steroid nucleus.

15.44 Which of the following compounds are derived from cholesterol?
a. glyceryl tristearate **b.** cortisone
c. testosterone **d.** estradiol
e. bile salts

15.45 What is the function of bile salts in digestion?

15.46 Why are lipoproteins needed to transport lipids in the bloodstream?

15.47 How do chylomicrons differ from VLDLs (very-low-density lipoproteins)?

15.48 How do LDLs differ from HDLs?

15.49 Why are LDLs called "bad" cholesterol?

15.50 Why are HDLs called "good" cholesterol?

15.51 What are the similarities and differences between the sex hormones estradiol and testosterone?

15.52 What are the similarities and differences between the adrenal hormone cortisone and the synthetic corticoid prednisone?

15.53 Which of the following are male sex hormones?
a. cholesterol **b.** aldosterone
c. estrogen **d.** testosterone
e. choline

15.54 Which of the following are adrenal steroids?
a. cholesterol **b.** aldosterone
c. estrogen **d.** testosterone
e. choline

LEARNING GOAL

Describe the composition and function of the lipid bilayer in cell membranes.

TUTORIAL
Phospholipids and the Cell Membrane

15.7 Cell Membranes

The membrane of a cell separates the contents of a cell from the external fluids. It is semipermeable so that nutrients can enter the cell and waste products can leave. The main components of a cell membrane are the glycerophospholipids. Earlier in this chapter, we saw that the structures of such glycerophospholipids consist of a nonpolar region, or "hydrocarbon tail," with long-chain fatty acids and a polar region, or "ionic head," from phosphoric acid and amino alcohols that ionize at physiological pH.

In a cell (plasma) membrane, two layers of glycerophospholipids are arranged with their hydrophobic tails in the center and their hydrophilic heads at the outer and inner layers of the membrane. This arrangement of glycerophospholipids is called a **lipid bilayer** (see Figure 15.11). The outer layer of glycerophospholipids is in contact with the external fluids, and the inner layer is in contact with the internal contents of the cell.

Most of the glycerophospholipids in the lipid bilayer contain unsaturated fatty acids. Because of the kinks in the carbon chains at the cis double bonds, the glycerophospholipids

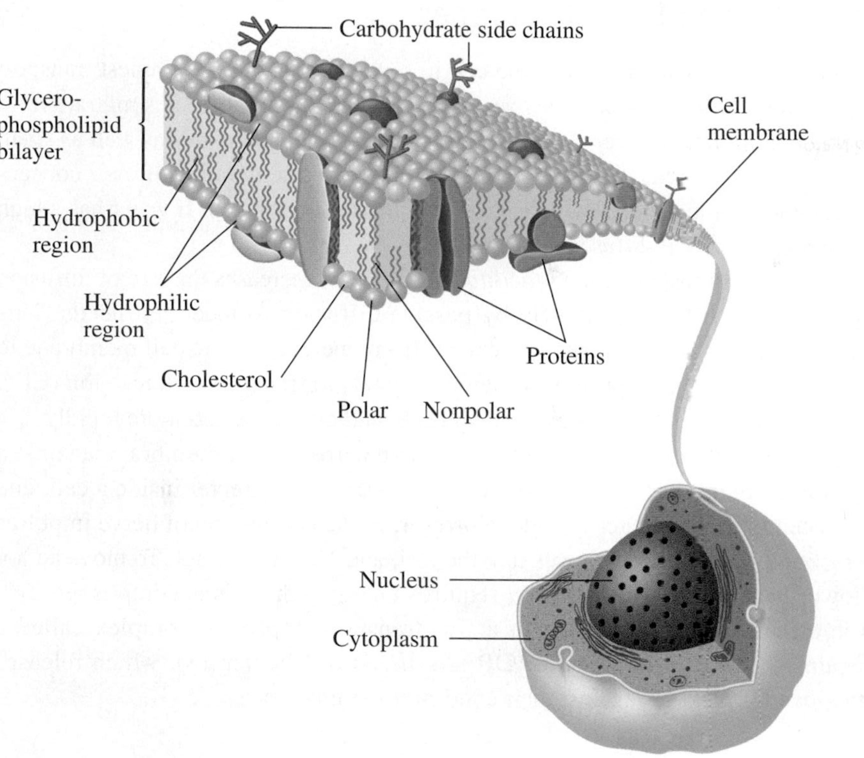

FIGURE 15.11 The fluid mosaic model of a cell membrane. Proteins and cholesterol are embedded in a lipid bilayer of glycerophospholipids. The bilayer forms a membrane-type barrier with polar heads at the membrane surfaces and the nonpolar tails in the center away from the water.

Q What types of fatty acids are found in the glycerophospholipids of the lipid bilayer?

do not fit closely together. As a result, the lipid bilayer is not a rigid, fixed structure, but one that is dynamic and fluid-like. In this liquid-like bilayer, there are also proteins, carbohydrates, and cholesterol molecules. For this reason, the model of biological membranes is referred to as the **fluid mosaic model** of membranes

In the fluid mosaic model, proteins known as *peripheral proteins* emerge on just one of the surfaces, outer or inner. The *integral proteins* extend through the entire lipid bilayer and appear on both surfaces of the membrane. Some proteins and lipids on the outer surface of the cell membrane are attached to carbohydrates. These carbohydrate chains project into the surrounding fluid environment where they are responsible for cell recognition and communication with chemical messengers such as hormones and neurotransmitters. In animals, cholesterol molecules embedded among the glycerophospholipids make up 20–25% of the lipid bilayer. Because cholesterol molecules are large and rigid, they reduce the flexibility of the lipid bilayer and add strength to the cell membrane.

SAMPLE PROBLEM 15.6

Lipid Bilayer in Cell Membranes

Describe the role of glycerophospholipids in the lipid bilayer.

SOLUTION

Glycerophospholipids consist of polar and nonpolar parts. In a cell membrane, an alignment of the nonpolar sections toward the center with the polar sections on the outside produces a barrier that prevents the contents of a cell from mixing with the fluids on the outside of the cell.

STUDY CHECK 15.6

What is the function of cholesterol in the cell membrane?

SELF STUDY ACTIVITY
Active Transport

Transport through Cell Membranes

Ions and molecules flow in and out of the cell in several ways. In the simplest transport mechanism called *diffusion* or *passive transport*, ions and small molecules migrate from a higher concentration to a lower concentration. For example, some ions as well as small molecules such as O_2, urea, and water diffuse through cell membranes. If their concentration is greater outside the cell than inside, they diffuse into the cell. If water has a high concentration in the cell, it diffuses out of the cell.

Another type of transport, called *facilitated transport*, increases the rate of diffusion for substances that diffuse too slowly by passive diffusion to meet cell needs. This process utilizes the integral proteins that extend from one edge of the cell membrane to the other. Groups of integral proteins provide channels to transport chloride ion (Cl^-), bicarbonate ion (HCO_3^-), and glucose molecules in and out of the cell more rapidly.

Certain ions such as K^+, Na^+, and Ca^{2+} move across a cell membrane against a concentration gradient. For example, the K^+ concentration is greater inside a cell, and the Na^+ concentration is greater outside. However, in the conduction of nerve impulses and contraction of muscles, K^+ moves into the cell, and Na^+ moves out. To move an ion from a lower to a higher concentration requires energy, which is accomplished by a process known as *active transport*. In active transport, a protein complex called a Na^+/K^+ pump breaks down ATP to ADP (discussed in Chapter 18), which releases energy to move Na^+ and K^+ against their concentration gradients.

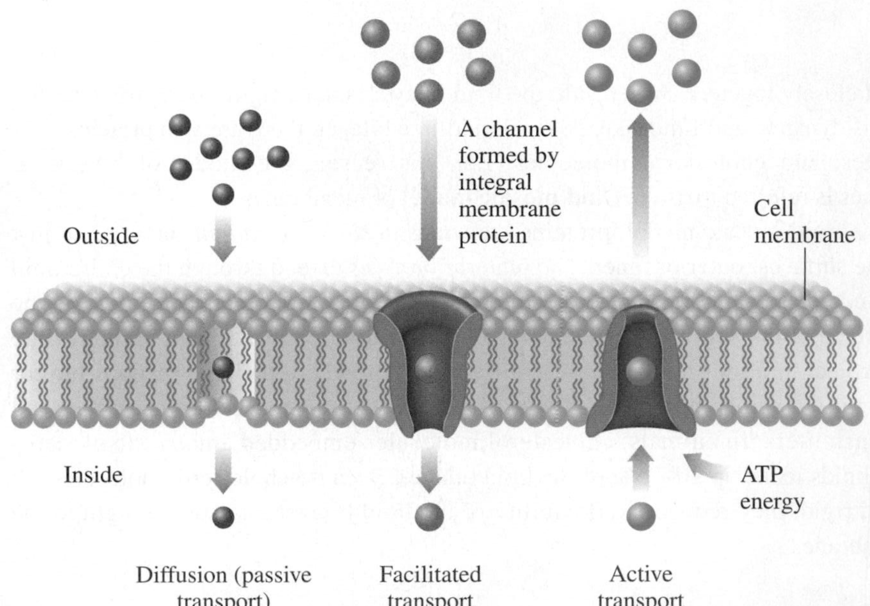

Substances are transported across a cell membrane by either diffusion, facilitated transport, or active transport.

CONCEPT CHECK 15.6

Transport through Cell Membranes

Why are protein channels needed in the lipid bilayer?

ANSWER

Protein channels in the lipid bilayer allow ions and polar molecules to flow into and out of the cell.

QUESTIONS AND PROBLEMS

Cell Membranes

15.55 What types of lipids are found in cell membranes?

15.56 Describe the structure of a lipid bilayer.

15.57 What is the function of the lipid bilayer in a cell membrane?

15.58 How do the unsaturated fatty acids in the glycerophospholipids affect the structure of cell membranes?

15.59 Where are proteins located in cell membranes?

15.60 What is the difference between peripheral and integral proteins?

15.61 What is the function of the carbohydrates on a cell membrane surface?

15.62 Why is a cell membrane semipermeable?

CONCEPT MAP

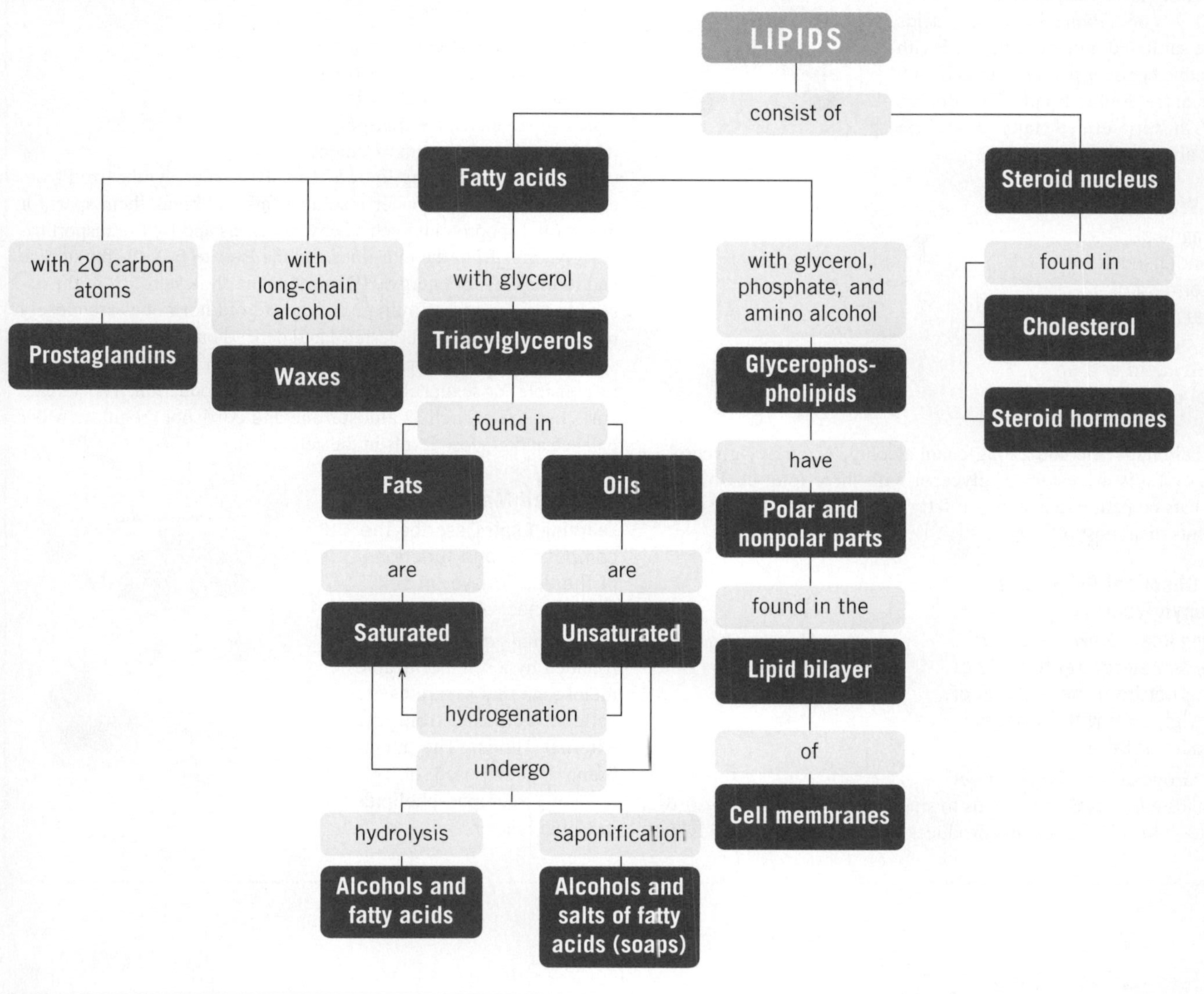

CHAPTER REVIEW

15.1 Lipids

Learning Goal: Describe the classes of lipids.

Lipids are nonpolar compounds that are not soluble in water. Classes of lipids include fatty acids, prostaglandins, waxes, triacylglycerols, glycerophospholipids, and steroids.

Glycerophospholipids

G l y c e r o l — Fatty acid — Fatty acid — PO_4 — Amino alcohol

15.2 Fatty Acids

Learning Goal: Draw the condensed structural formula of a fatty acid and identify it as saturated or unsaturated.

Fatty acids are unbranched carboxylic acids that typically contain an even number (10–20) of carbon atoms. Fatty acids may be saturated, monounsaturated with one double bond, or polyunsaturated with two or more double bonds. The double bonds in unsaturated fatty acids are almost always cis.

15.3 Waxes, Fats, and Oils

Learning Goal: Draw the condensed structural formula of a wax, fat, or oil produced by the reaction of a fatty acid and an alcohol or glycerol.

$$\begin{array}{l} CH_2-O-\overset{O}{\overset{\|}{C}}\sim\sim\sim \\ CH-O-\overset{O}{\overset{\|}{C}}\sim\sim\sim \\ CH_2-O-\overset{O}{\overset{\|}{C}}\sim\sim\sim \end{array}$$

Triacylglycerol

A wax is an ester of a long-chain fatty acid and a long-chain alcohol. The triacylglycerols of fats and oils are esters of glycerol with three long-chain fatty acids. Fats contain more saturated fatty acids and have higher melting points than most oils.

15.4 Chemical Properties of Triacylglycerols

Learning Goal: Draw the condensed structural formula of the product from the reaction of a triacylglycerol with hydrogen, or an acid or base.

$$3Na^{+}\,{}^{-}O-\overset{O}{\overset{\|}{C}}-(CH_2)_{14}-CH_3$$

3 Sodium palmitate (soap)

The hydrogenation of unsaturated fatty acids converts double bonds to single bonds. The hydrolysis of the ester bonds in fats or oils produces glycerol and fatty acids. In saponification, a fat heated with a strong base produces glycerol and the salts of the fatty acids or soaps.

15.5 Glycerophospholipids

Learning Goal: Describe the characteristics of glycerophospholipids.

Glycerophospholipids are esters of glycerol with two fatty acids and a phosphate group attached to an amino alcohol.

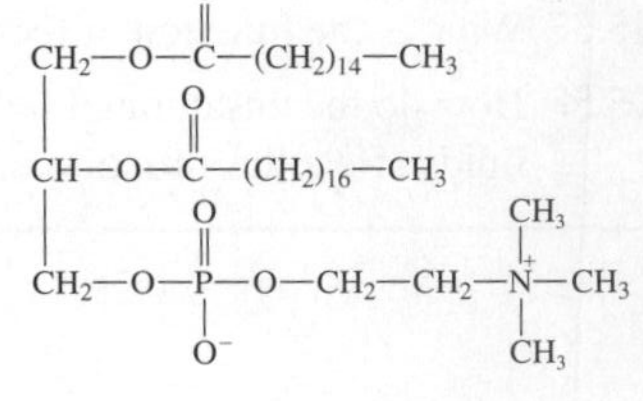

15.6 Steroids: Cholesterol, Bile Salts, and Steroid Hormones

Learning Goal: Describe the structures of steroids.

Steroid nucleus

Steroids are lipids containing the steroid nucleus, which is a fused structure of four rings. Lipids, which are nonpolar, are transported through the aqueous environment of the blood by forming lipoproteins. Bile salts, synthesized from cholesterol, mix with water-insoluble fats and break them apart for digestion. Lipoproteins such as chylomicrons and LDL transport triacylglycerols from the intestines and the liver to fat cells for storage and to muscles for energy. HDLs transport cholesterol from the tissues to the liver for elimination. The steroid hormones are closely related in structure to cholesterol and depend on cholesterol for their synthesis. The sex hormones, such as estrogen and testosterone, are responsible for sexual characteristics and reproduction. The adrenal corticosteroids, such as aldosterone and cortisone, regulate water balance and glucose levels in the cells.

15.7 Cell Membranes

Learning Goal: Describe the composition and function of the lipid bilayer in cell membranes.

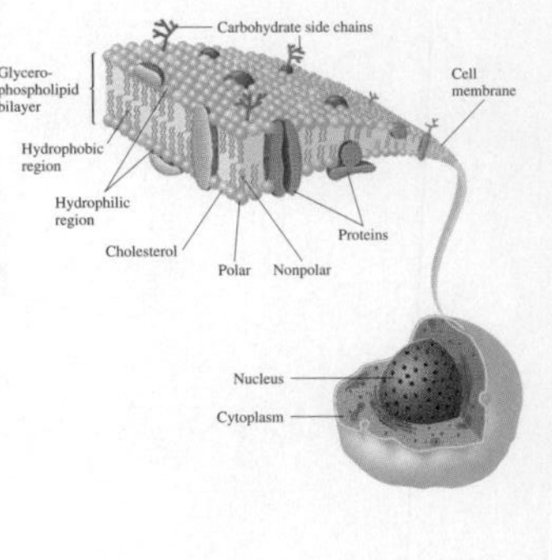

All animal cells are surrounded by a semipermeable membrane that separates the cellular contents from the external fluids. The membrane is composed of two rows of glycerophospholipids in a lipid bilayer.

Summary of Reactions

Esterification

$$\text{Glycerol} + 3\text{ fatty acid molecules} \xrightarrow{\text{Enzyme}} \text{triacylglycerol} + 3H_2O$$

Hydrogenation of Triacylglycerols

$$\text{Triacylglycerol (unsaturated)} + H_2 \xrightarrow{\text{Ni}} \text{triacylglycerol (saturated)}$$

Hydrolysis of Triacylglycerols

$$\text{Triacylglycerol} + 3H_2O \xrightarrow{H^+ \text{ or lipase}} \text{glycerol} + 3 \text{ fatty acid molecules}$$

Saponification of Triacylglycerols

$$\text{Triacylglycerol} + 3NaOH \longrightarrow \text{glycerol} + 3 \text{ sodium salts of fatty acids (soaps)}$$

Key Terms

cephalins Glycerophospholipids found in brain and nerve tissues that incorporate the amino alcohol serine or ethanolamine.

cholesterol The most prevalent of the steroid compounds; needed for cellular membranes, the synthesis of vitamin D, hormones, and bile salts.

fat A triacylglycerol obtained from an animal source; usually a solid at room temperature.

fatty acids Long-chain carboxylic acids found in many lipids.

fluid mosaic model The concept that cell membranes are lipid bilayer structures that contain an assortment of polar lipids and proteins in a dynamic, fluid arrangement.

glycerophospholipids Polar lipids of glycerol attached to two fatty acids and a phosphate group connected to an amino alcohol such as choline, serine, or ethanolamine.

hydrogenation The addition of hydrogen to unsaturated fats.

lecithins Glycerophospholipids containing choline as the amino alcohol.

lipid bilayer A model of a cell membrane in which glycerophospholipids are arranged in two rows.

lipids A family of compounds that is nonpolar in nature and not soluble in water; includes fatty acids, prostaglandins, waxes, triacylglycerols, glycerophospholipids, and steroids.

lipoprotein A combination of nonpolar lipids with glycerophospholipids and proteins to form a polar complex that can be transported through body fluids.

monounsaturated fatty acids Fatty acids with one double bond.

oil A triacylglycerol obtained from a plant source; usually liquid at room temperature.

polyunsaturated fatty acids Fatty acids that contain two or more double bonds.

prostaglandins A number of compounds derived from arachidonic acid that regulate several physiological processes.

saturated fatty acids Fatty acids that have no double bonds, they have higher melting points than unsaturated lipids, and are usually solid at room temperature.

steroids Types of lipid composed of a multicyclic ring system.

triacylglycerols A family of lipids composed of three fatty acids bonded through ester bonds to glycerol, a trihydroxy alcohol.

wax The ester of a long-chain alcohol and a long-chain saturated fatty acid.

Understanding the Concepts

15.63 Palmitic acid is obtained from palm oil as glyceryl tripalmitate (tripalmitin). Draw the condensed structural formula of glyceryl tripalmitate.

Palm fruit from palm trees are a source of palm oil.

15.64 Jojoba wax in candles consists of a 20-carbon saturated fatty acid and a 20-carbon saturated alcohol. Draw the condensed structural formula of jojoba wax.

Candles contain jojoba wax.

15.65 Sunflower oil can be used to make margarine. A triacylglycerol in sunflower oil contains 2 linoleic acids and 1 oleic acid.
a. Draw the condensed structural formulas of two isomers for the triacylglycerol in sunflower oil.
b. Using one of the isomers, write the reaction that would be used when sunflower oil is used to make solid margarine.

Sunflower oil is obtained from the seeds of the sunflower.

15.66 Identify each of the following as a saturated, monounsaturated, polyunsaturated, omega-3, or omega-6 fatty acid:
a. $CH_3-(CH_2)_4-CH=CH-CH_2-CH=CH-(CH_2)_7-COOH$
b. linolenic acid
c. $CH_3-(CH_2)_{14}-COOH$
d. $CH_3-(CH_2)_7-CH=CH-(CH_2)_7-COOH$

Salmon is a good source of omega-3 unsaturated fatty acids.

Additional Questions and Problems

For instructor-assigned homework, go to **www.masteringchemistry.com**.

15.67 Among the ingredients in lipstick are beeswax, carnauba wax, hydrogenated vegetable oils, and glyceryl tricaprate (tricaprin). What types of lipids have been used? Draw the condensed structural formula of glyceryl tricaprate (tricaprin). Capric acid is a saturated 10-carbon fatty acid.

15.68 Because peanut oil floats on the top of peanut butter, many brands of peanut butter are hydrogenated. A solid product then forms that is mixed into the peanut butter and does not separate. If a triacylglycerol in peanut oil that contains one palmitic acid, one oleic acid, and one linoleic acid is completely hydrogenated, what is the product?

15.69 Trans fats are produced during the hydrogenation of polyunsaturated oils.
a. What is the typical configuration of the double bond in a monounsaturated fatty acid?
b. How does a trans fatty acid differ from a cis fatty acid?
c. Draw the condensed structural formula of *trans*-oleic acid.

15.70 One mole of glyceryl trioleate (triolein) is completely hydrogenated.
a. Draw the condensed structural formula of the product.
b. How many moles of hydrogen are required?
c. How many liters of hydrogen are needed if the reaction is run at STP?

15.71 On the list of ingredients of a cosmetic product are glyceryl tristearate and a lecithin. Draw the structure of glyceryl tristearate and a lecithin with palmitic acids and choline.

15.72 Some typical meals at fast-food restaurants are listed here. Calculate the number of kilocalories from fat and the percentage of total kilocalories that are due to fat (1 gram of fat = 9 kcal). Round off the kilocalories to the tens place. Would you expect the fats to be mostly saturated or unsaturated? Why?
a. a chicken dinner, 830 kcal, 46 g of fat
b. a quarter-pound cheeseburger, 520 kcal, 29 g of fat
c. pepperoni pizza (three slices), 560 kcal, 18 g of fat
d. beef burrito, 470 kcal, 21 g of fat
e. deep-fried fish (three pieces), 480 kcal, 28 g of fat

15.73 Identify the following as fatty acids, soaps, triacylglycerols, waxes, glycerophospholipids, or steroids:
a. beeswax
b. cholesterol
c. lecithin
d. glyceryl tripalmitate (tripalmitin)
e. sodium stearate
f. safflower oil
g. whale blubber
h. adipose tissue
i. progesterone
j. cortisone
k. stearic acid

15.74 Why would an animal that lives in a cold climate have more unsaturated triacylglycerols in its body fat than an animal that lives in a warm climate?

15.75 Identify the components (**1–5**) contained in each of the following types of lipids (**a–e**):
1. glycerol
2. fatty acid
3. phosphate
4. amino alcohol
5. steroid nucleus

a. estrogen
b. cephalin
c. wax
d. triacylglycerol
e. glycerophospholipid

15.76 Which of the following are found in cell membranes?
a. cholesterol
b. triacylglycerols
c. carbohydrates
d. proteins
e. waxes
f. glycerophospholipids

Challenge Questions

15.77 Identify the lipoprotein (**1–4**) in each of the following descriptions (**a–h**):

1. chylomicrons **2.** VLDL
3. LDL **4.** HDL

a. "good" cholesterol
b. transports most of the cholesterol to the cells
c. carries triacylglycerols from the intestine to the fat cells
d. transports cholesterol to the liver
e. has the greatest abundance of protein
f. "bad" cholesterol
g. carries triacylglycerols synthesized in the liver to the muscles
h. has the lowest density

15.78 **a.** Which of the following fatty acids has the lowest melting point? Explain.
b. Which of the following fatty acids has the highest melting point? Explain.

$CH_3-(CH_2)_{16}-COOH$
Stearic acid

$CH_3-(CH_2)_4-CH{=}CH-CH_2-CH{=}CH-(CH_2)_7-COOH$
Linoleic acid

$CH_3-(CH_2)_7-CH{=}CH-(CH_2)_7-COOH$
Oleic acid

15.79 Draw the condensed structural formula of a glycerophospholipid that contains two stearic acids and a phosphate bonded to ethanolamine.

15.80 One of the triglycerols in olive oil is glyceryl tripalmitoleate (tripalmitolein).
a. Draw the condensed structural formula for glyceryl tripalmitoleate (tripalmitolein).
b. How many liters of H_2 gas at STP are needed to completely saturate 100. g of glyceryl tripalmitoleate (tripalmitolein)?

Olive oil from olives contains glyceryl tripalmitoleate (tripalmitolein).

c. How many milliliters of a 0.250 M NaOH solution are needed to completely saponify 100. g of glyceryl tripalmitoleate (tripalmitolein)?

15.81 A sink drain can become clogged with solid fat such as glyceryl tristearate (tristearin).
a. How would adding lye (NaOH) to the sink drain remove the blockage?
b. Write an equation for the reaction that occurs.

A sink drain can become clogged with saturated fats.

Answers

Answers to Study Checks

15.1 **a.** 16
b. unsaturated
c. liquid

15.2

```
            O
            ‖
CH2—O—C—(CH2)12—CH3
|           O
|           ‖
CH—O—C—(CH2)12—CH3
|           O
|           ‖
CH2—O—C—(CH2)12—CH3
```

15.3 glyceryl tristearate (tristearin)

15.4

```
            O
            ‖
CH2—O—C—(CH2)12—CH3
|           O
|           ‖
CH—O—C—(CH2)12—CH3
|           O                            CH3
|           ‖                            |
CH2—O—P—O—CH2—CH2—N+—CH3
            |                            |
            O−                           CH3
```

15.5 Cholesterol is not soluble in water; it is part of the lipid family.

15.6 Cholesterol adds strength and rigidity to the cell membrane.

Answers to Selected Questions and Problems

15.1 Lipids provide energy and protection and insulation for the organs in the body.

15.3 Because lipids are not soluble in water, a polar solvent, they are nonpolar molecules.

15.5 All fatty acids contain a long chain of carbon atoms with a carboxylic acid group. Saturated fatty acids contain only carbon–carbon single bonds; unsaturated fatty acids contain one or more double bonds.

15.7 **a.** palmitic acid

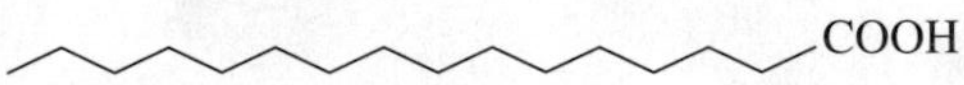

COOH

b. oleic acid

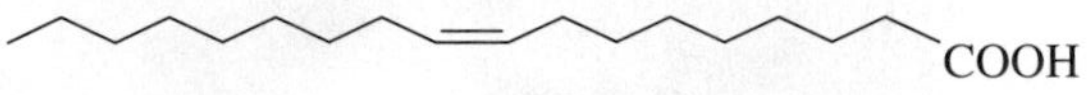

COOH

15.9 **a.** saturated **b.** unsaturated
c. unsaturated **d.** saturated

15.11 In a cis fatty acid, the hydrogen atoms are on the same side of the double bond, which produces a kink in the carbon chain. In a trans fatty acid, the hydrogen atoms are on opposite sides of the double bond, which gives a carbon chain without any kink.

15.13 In an omega-3 fatty acid, there is a double bond on carbon 3, counting from the methyl group, whereas in an omega-6 fatty acid, there is a double bond beginning at carbon 6, counting from the methyl group.

15.15 Arachidonic acid contains four double bonds and no side groups. In PGE_1, a part of the chain forms cyclopentane and there are hydroxyl and ketone functional groups.

15.17 Prostaglandins raise blood pressure, stimulate contraction and relaxation of smooth muscle, and may cause inflammation and pain.

15.19 $CH_3{-}(CH_2)_{14}{-}\overset{O}{\overset{\|}{C}}{-}O{-}(CH_2)_{29}{-}CH_3$

15.21

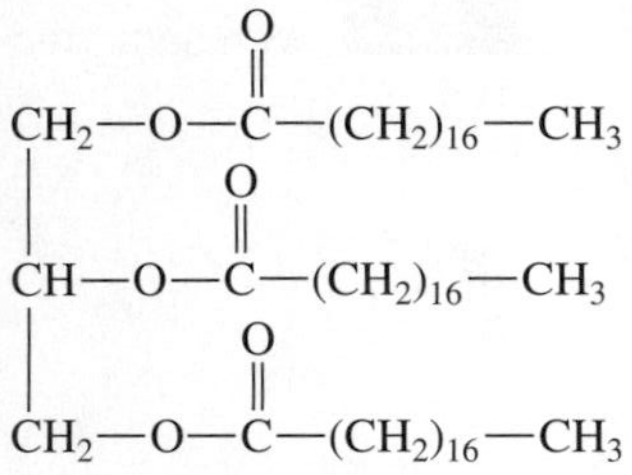

$$\begin{array}{l} CH_2{-}O{-}\overset{O}{\overset{\|}{C}}{-}(CH_2)_{16}{-}CH_3 \\ CH{-}O{-}\overset{O}{\overset{\|}{C}}{-}(CH_2)_{16}{-}CH_3 \\ CH_2{-}O{-}\overset{O}{\overset{\|}{C}}{-}(CH_2)_{16}{-}CH_3 \end{array}$$

15.23

$$\begin{array}{l} CH_2{-}O{-}\overset{O}{\overset{\|}{C}}{-}(CH_2)_8{-}CH_3 \\ CH{-}O{-}\overset{O}{\overset{\|}{C}}{-}(CH_2)_8{-}CH_3 \\ CH_2{-}O{-}\overset{O}{\overset{\|}{C}}{-}(CH_2)_8{-}CH_3 \end{array}$$

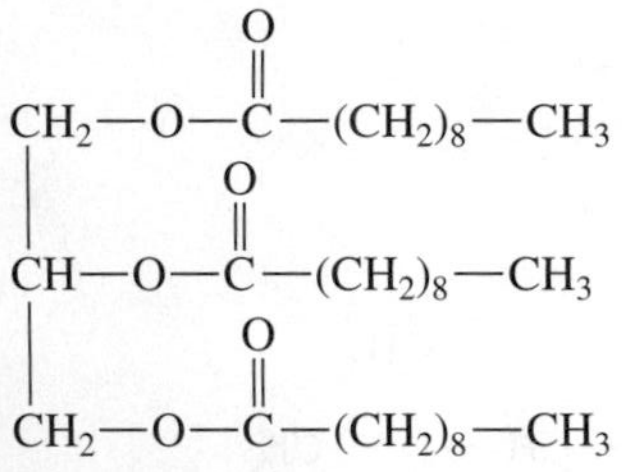

15.25 Safflower oil contains fatty acids with two or three double bonds; olive oil contains a large amount of oleic acid, which has only one (monounsaturated) double bond.

15.27 Although coconut oil comes from a plant source, it has large amounts of saturated fatty acids and small amounts of unsaturated fatty acids.

15.29

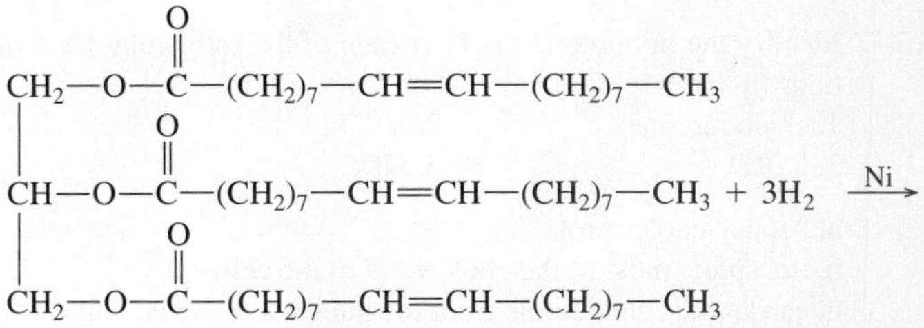

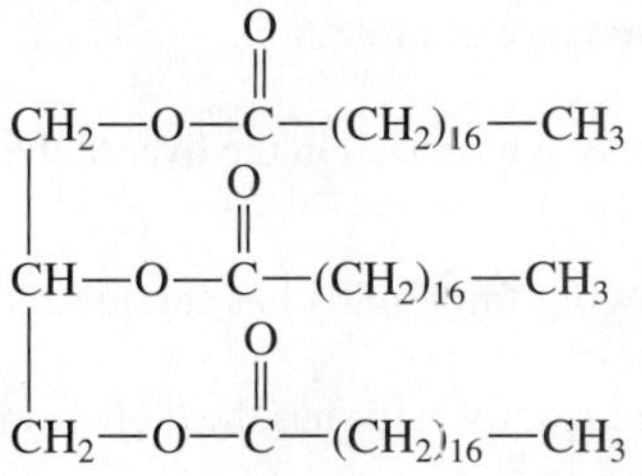

$$\begin{array}{l} CH_2{-}O{-}\overset{O}{\overset{\|}{C}}{-}(CH_2)_7{-}CH{=}CH{-}(CH_2)_7{-}CH_3 \\ CH{-}O{-}\overset{O}{\overset{\|}{C}}{-}(CH_2)_7{-}CH{=}CH{-}(CH_2)_7{-}CH_3 + 3H_2 \xrightarrow{Ni} \\ CH_2{-}O{-}\overset{O}{\overset{\|}{C}}{-}(CH_2)_7{-}CH{=}CH{-}(CH_2)_7{-}CH_3 \end{array}$$

$$\begin{array}{l} CH_2{-}O{-}\overset{O}{\overset{\|}{C}}{-}(CH_2)_{16}{-}CH_3 \\ CH{-}O{-}\overset{O}{\overset{\|}{C}}{-}(CH_2)_{16}{-}CH_3 \\ CH_2{-}O{-}\overset{O}{\overset{\|}{C}}{-}(CH_2)_{16}{-}CH_3 \end{array}$$

15.31 **a.**

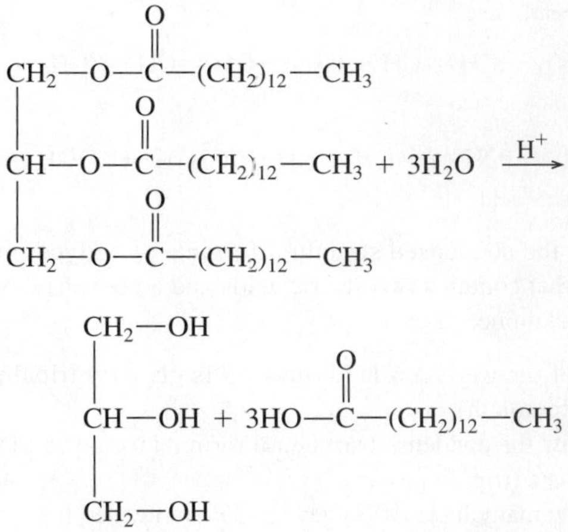

$$\begin{array}{l} CH_2{-}O{-}\overset{O}{\overset{\|}{C}}{-}(CH_2)_{12}{-}CH_3 \\ CH{-}O{-}\overset{O}{\overset{\|}{C}}{-}(CH_2)_{12}{-}CH_3 + 3H_2O \xrightarrow{H^+} \\ CH_2{-}O{-}\overset{O}{\overset{\|}{C}}{-}(CH_2)_{12}{-}CH_3 \end{array}$$

$$\begin{array}{l} CH_2{-}OH \\ CH{-}OH + 3HO{-}\overset{O}{\overset{\|}{C}}{-}(CH_2)_{12}{-}CH_3 \\ CH_2{-}OH \end{array}$$

b.

$$\begin{array}{l} CH_2{-}O{-}\overset{O}{\overset{\|}{C}}{-}(CH_2)_{12}{-}CH_3 \\ CH{-}O{-}\overset{O}{\overset{\|}{C}}{-}(CH_2)_{12}{-}CH_3 + 3NaOH \longrightarrow \\ CH_2{-}O{-}\overset{O}{\overset{\|}{C}}{-}(CH_2)_{12}{-}CH_3 \end{array}$$

$$\begin{array}{l} CH_2{-}OH \\ CH{-}OH + 3Na^{+}\,{}^{-}O{-}\overset{O}{\overset{\|}{C}}{-}(CH_2)_{12}{-}CH_3 \\ CH_2{-}OH \end{array}$$

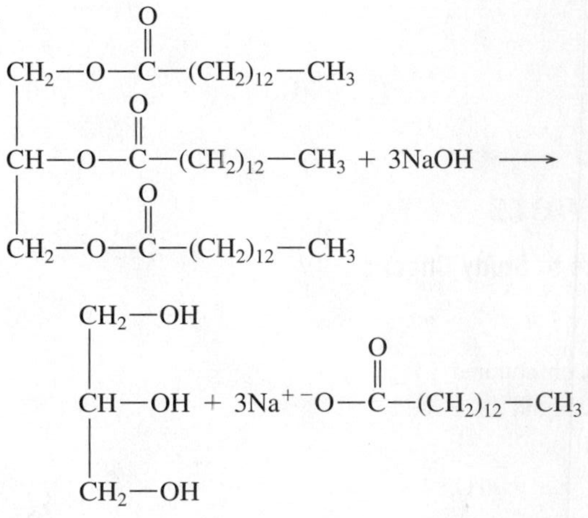

15.33 A triacylglycerol is composed of glycerol with three hydroxyl groups that form ester links with three long-chain fatty acids. In olestra, six to eight long-chain fatty acids form ester links with the hydroxyl groups on sucrose, a sugar. The olestra cannot be digested because our enzymes cannot break down the large olestra molecule.

15.35

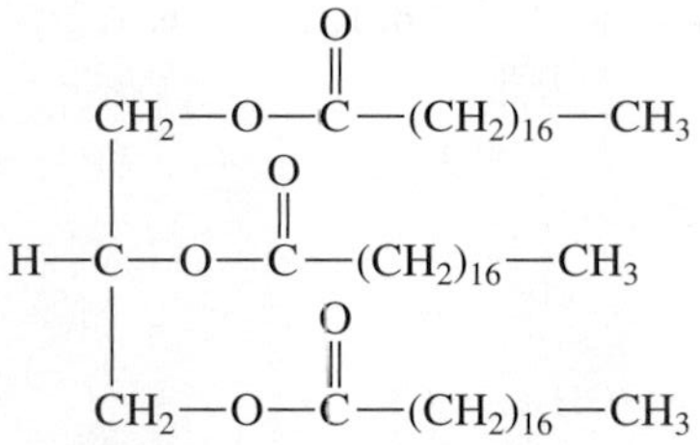

$$\begin{array}{l}\mathrm{CH_2{-}O{-}\overset{O}{\overset{\|}{C}}{-}(CH_2)_{16}{-}CH_3}\\ \mathrm{H{-}C{-}O{-}\overset{O}{\overset{\|}{C}}{-}(CH_2)_{16}{-}CH_3}\\ \mathrm{CH_2{-}O{-}\overset{O}{\overset{\|}{C}}{-}(CH_2)_{16}{-}CH_3}\end{array}$$

15.37 A triacylglycerol consists of glycerol and three fatty acids. A glycerophospholipid consists of glycerol, two fatty acids, a phosphate group, and an amino alcohol.

15.39

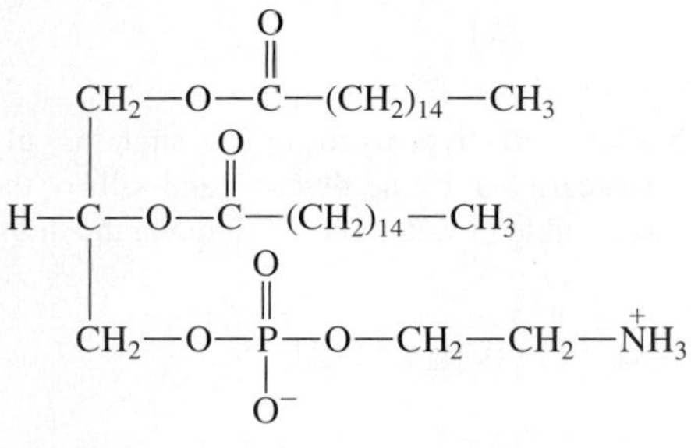

$$\begin{array}{l}\mathrm{CH_2{-}O{-}\overset{O}{\overset{\|}{C}}{-}(CH_2)_{14}{-}CH_3}\\ \mathrm{H{-}C{-}O{-}\overset{O}{\overset{\|}{C}}{-}(CH_2)_{14}{-}CH_3}\\ \mathrm{CH_2{-}O{-}\underset{O^-}{\underset{|}{\overset{O}{\overset{\|}{P}}}}{-}O{-}CH_2{-}CH_2{-}\overset{+}{N}H_3}\end{array}$$

This is a cephalin.

15.41 This glycerophospholipid is a cephalin. It contains glycerol, oleic acid, stearic acid, a phosphate, and ethanolamine.

15.43

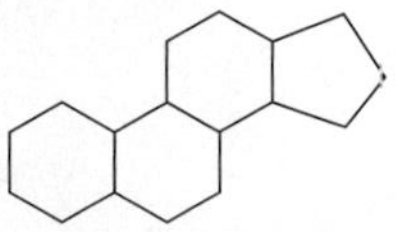

15.45 Bile salts act to emulsify fat globules, allowing the fat to be more easily digested.

15.47 Chylomicrons have a lower density than VLDLs. They pick up triacylglycerols from the intestine, whereas VLDLs transport triacylglycerols synthesized in the liver.

15.49 "Bad" cholesterol is the cholesterol carried by LDLs that can form deposits in the arteries called plaque, which narrows the arteries.

15.51 Both estradiol and testosterone contain the steroid nucleus and a hydroxyl group. Testosterone has a ketone group, a double bond, and two methyl groups. Estradiol has an aromatic ring, a hydroxyl group in place of the ketone, and a methyl group.

15.53 **d.** Testosterone is a male sex hormone.

15.55 The lipids in a cell membrane are glycerophospholipids with smaller amounts of cholesterol.

15.57 The lipid bilayer in a cell membrane surrounds the cell and separates the contents of the cell from the external fluids.

15.59 The peripheral proteins in the membrane emerge on the inner or outer surface only, whereas the integral proteins extend through the membrane to both surfaces.

15.61 The carbohydrates attached to proteins and lipids on the surface of cells act as receptors for cell recognition and chemical messengers such as neurotransmitters.

15.63

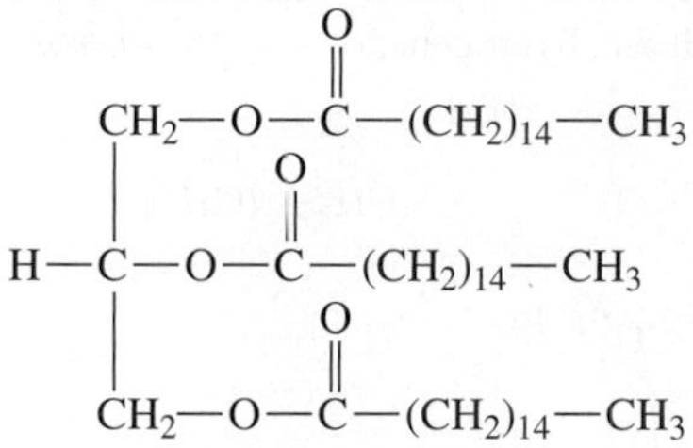

$$\begin{array}{l}\mathrm{CH_2{-}O{-}\overset{O}{\overset{\|}{C}}{-}(CH_2)_{14}{-}CH_3}\\ \mathrm{H{-}C{-}O{-}\overset{O}{\overset{\|}{C}}{-}(CH_2)_{14}{-}CH_3}\\ \mathrm{CH_2{-}O{-}\overset{O}{\overset{\|}{C}}{-}(CH_2)_{14}{-}CH_3}\end{array}$$

15.65

a.

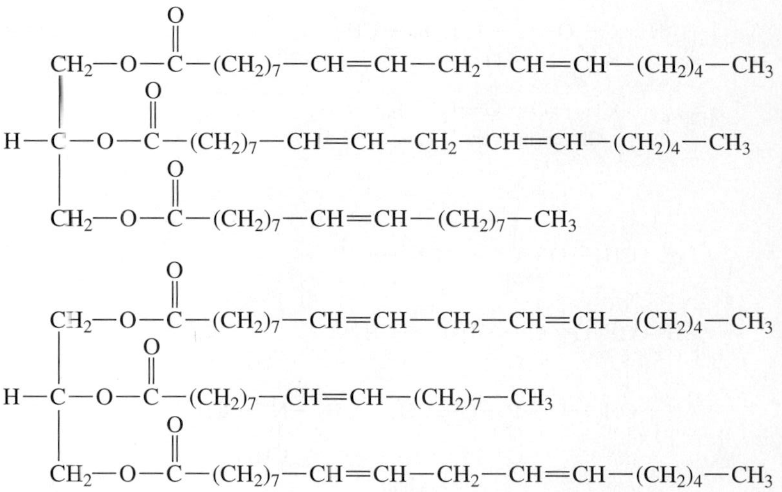

$$\begin{array}{l}\mathrm{CH_2{-}O{-}\overset{O}{\overset{\|}{C}}{-}(CH_2)_7{-}CH{=}CH{-}CH_2{-}CH{=}CH{-}(CH_2)_4{-}CH_3}\\ \mathrm{H{-}C{-}O{-}\overset{O}{\overset{\|}{C}}{-}(CH_2)_7{-}CH{=}CH{-}CH_2{-}CH{=}CH{-}(CH_2)_4{-}CH_3}\\ \mathrm{CH_2{-}O{-}\overset{O}{\overset{\|}{C}}{-}(CH_2)_7{-}CH{=}CH{-}(CH_2)_7{-}CH_3}\end{array}$$

$$\begin{array}{l}\mathrm{CH_2{-}O{-}\overset{O}{\overset{\|}{C}}{-}(CH_2)_7{-}CH{=}CH{-}CH_2{-}CH{=}CH{-}(CH_2)_4{-}CH_3}\\ \mathrm{H{-}C{-}O{-}\overset{O}{\overset{\|}{C}}{-}(CH_2)_7{-}CH{=}CH{-}(CH_2)_7{-}CH_3}\\ \mathrm{CH_2{-}O{-}\overset{O}{\overset{\|}{C}}{-}(CH_2)_7{-}CH{=}CH{-}CH_2{-}CH{=}CH{-}(CH_2)_4{-}CH_3}\end{array}$$

b.

$$\begin{array}{l}\mathrm{CH_2{-}O{-}\overset{O}{\overset{\|}{C}}{-}(CH_2)_7{-}CH{=}CH{-}CH_2{-}CH{=}CH{-}(CH_2)_4{-}CH_3}\\ \mathrm{H{-}C{-}O{-}\overset{O}{\overset{\|}{C}}{-}(CH_2)_7{-}CH{=}CH{-}(CH_2)_7{-}CH_3} + 5H_2 \xrightarrow{\mathrm{Ni}}\\ \mathrm{CH_2{-}O{-}\overset{O}{\overset{\|}{C}}{-}(CH_2)_7{-}CH{=}CH{-}CH_2{-}CH{=}CH{-}(CH_2)_4{-}CH_3}\end{array}$$

$$\begin{array}{l}\mathrm{CH_2{-}O{-}\overset{O}{\overset{\|}{C}}{-}(CH_2)_{16}{-}CH_3}\\ \mathrm{H{-}C{-}O{-}\overset{O}{\overset{\|}{C}}{-}(CH_2)_{16}{-}CH_3}\\ \mathrm{CH_2{-}O{-}\overset{O}{\overset{\|}{C}}{-}(CH_2)_{16}{-}CH_3}\end{array}$$

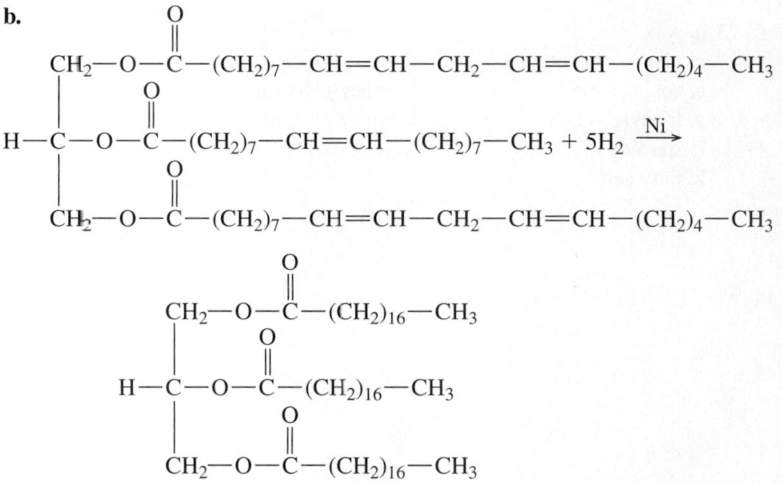

15.67 Beeswax and carnauba are waxes. Vegetable oil and glyceryl tricaprate (tricaprin) are triacylglycerols.

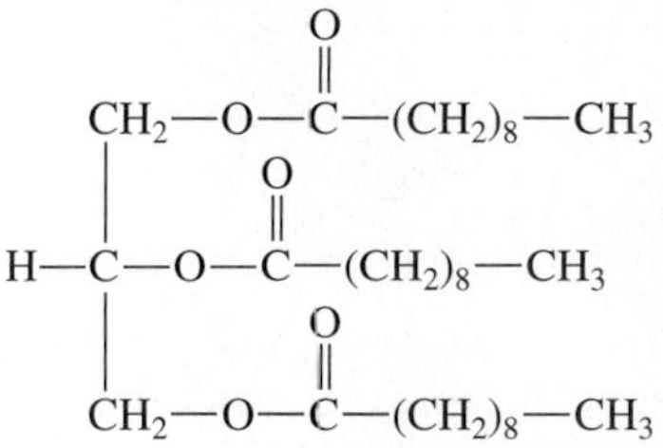

$$\begin{array}{l}\mathrm{CH_2{-}O{-}\overset{O}{\overset{\|}{C}}{-}(CH_2)_8{-}CH_3}\\ \mathrm{H{-}C{-}O{-}\overset{O}{\overset{\|}{C}}{-}(CH_2)_8{-}CH_3}\\ \mathrm{CH_2{-}O{-}\overset{O}{\overset{\|}{C}}{-}(CH_2)_8{-}CH_3}\end{array}$$

Glyceryl tricaprate (tricaprin)

15.69 **a.** A typical unsaturated fatty acid has a cis double bond.
b. A cis unsaturated fatty acid contains hydrogen atoms on the same side of each double bond. A trans unsaturated fatty acid has hydrogen atoms on the opposite sides of each double bond that forms during hydrogenation.
c.

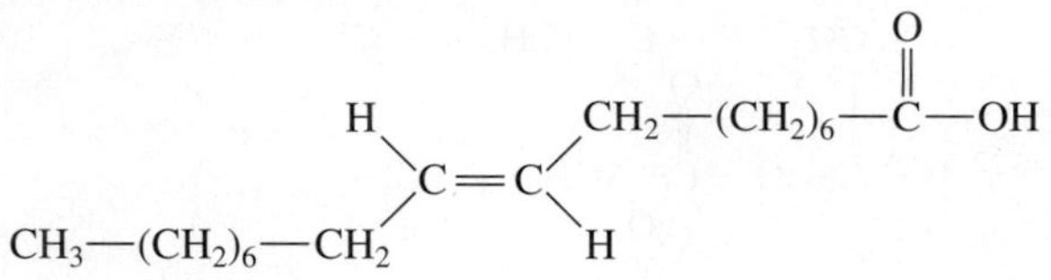

$$\mathrm{CH_3{-}(CH_2)_6{-}CH_2{-}CH{=}CH{-}CH_2{-}(CH_2)_6{-}\overset{O}{\overset{\|}{C}}{-}OH}\ \text{(trans: H atoms on opposite sides of C=C)}$$

15.71

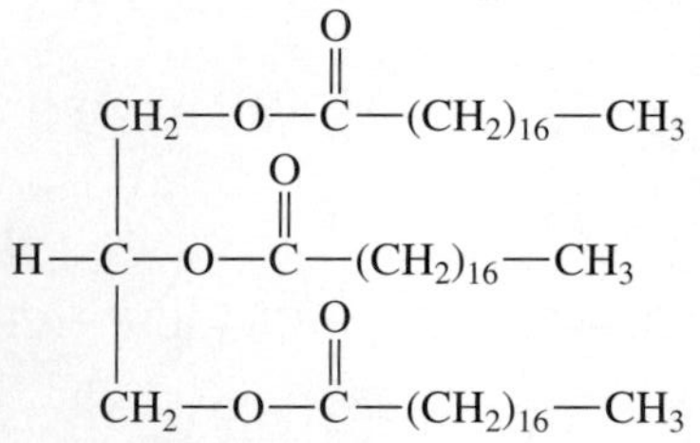

$$\begin{array}{l} \mathrm{CH_2{-}O{-}\overset{O}{\overset{\|}{C}}{-}(CH_2)_{16}{-}CH_3} \\ \mathrm{H{-}C{-}O{-}\overset{O}{\overset{\|}{C}}{-}(CH_2)_{16}{-}CH_3} \\ \mathrm{CH_2{-}O{-}\overset{O}{\overset{\|}{C}}{-}(CH_2)_{16}{-}CH_3} \end{array}$$

Glyceryl tristearate (tristearin)

$$\begin{array}{l} \mathrm{CH_2{-}O{-}\overset{O}{\overset{\|}{C}}{-}(CH_2)_{14}{-}CH_3} \\ \mathrm{H{-}C{-}O{-}\overset{O}{\overset{\|}{C}}{-}(CH_2)_{14}{-}CH_3} \\ \mathrm{CH_2{-}O{-}\overset{O}{\overset{\|}{P}}(O^-){-}O{-}CH_2{-}CH_2{-}N^+(CH_3)_3} \end{array}$$

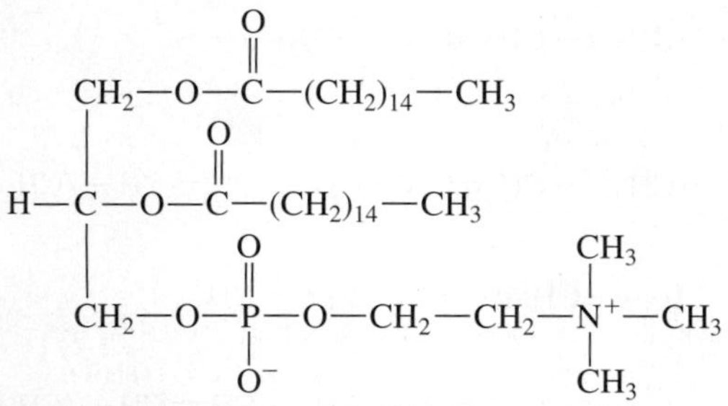

Lecithin

15.73 **a.** wax **b.** steroid
c. glycerophospholipid **d.** triacylglycerol
e. soap **f.** triacylglycerol
g. triacylglycerol **h.** triacylglycerol
i. steroid **j.** steroid
k. fatty acid

15.75 **a.** 5 **b.** 1, 2, 3, 4 **c.** 2 **d.** 1, 2 **e.** 1, 2, 3, 4

15.77 **a.** 4 **b.** 3 **c.** 1
d. 4 **e.** 4 **f.** 3
g. 2 **h.** 1

15.79

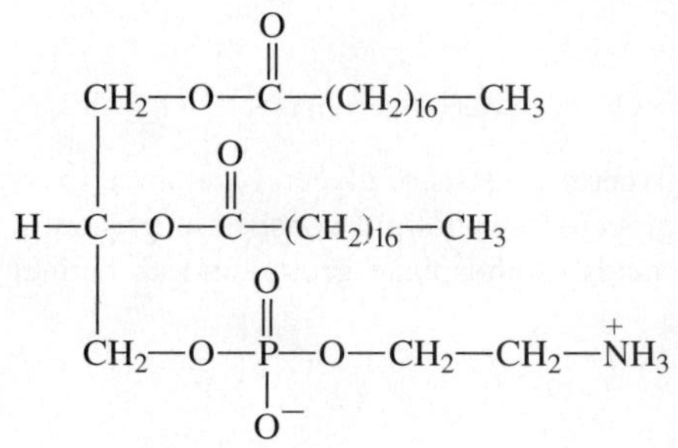

$$\begin{array}{l} \mathrm{CH_2{-}O{-}\overset{O}{\overset{\|}{C}}{-}(CH_2)_{16}{-}CH_3} \\ \mathrm{H{-}C{-}O{-}\overset{O}{\overset{\|}{C}}{-}(CH_2)_{16}{-}CH_3} \\ \mathrm{CH_2{-}O{-}\overset{O}{\overset{\|}{P}}(O^-){-}O{-}CH_2{-}CH_2{-}\overset{+}{N}H_3} \end{array}$$

15.81 **a.** Adding NaOH will hydrolyze lipids such as glyceryl tristearate (tristearin), forming glycerol and salts of the fatty acids that are soluble in water and wash down the drain.
b.

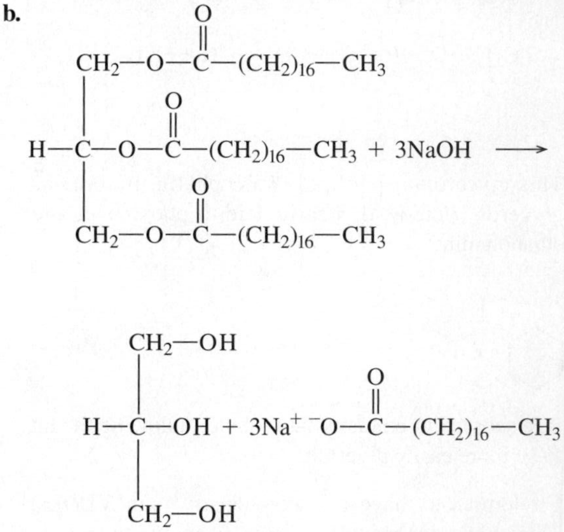

$$\begin{array}{l} \mathrm{CH_2{-}O{-}\overset{O}{\overset{\|}{C}}{-}(CH_2)_{16}{-}CH_3} \\ \mathrm{H{-}C{-}O{-}\overset{O}{\overset{\|}{C}}{-}(CH_2)_{16}{-}CH_3 + 3NaOH \longrightarrow} \\ \mathrm{CH_2{-}O{-}\overset{O}{\overset{\|}{C}}{-}(CH_2)_{16}{-}CH_3} \end{array}$$

$$\begin{array}{l} \mathrm{CH_2{-}OH} \\ \mathrm{H{-}C{-}OH + 3Na^+\ {}^-O{-}\overset{O}{\overset{\|}{C}}{-}(CH_2)_{16}{-}CH_3} \\ \mathrm{CH_2{-}OH} \end{array}$$

Glycerol Sodium stearate

Combining Ideas from Chapters 13 to 15

CI.25 The plastic known as PETE (**p**oly**e**thylene**te**rephthalate) is used to make plastic soft drink bottles and containers for salad dressing, shampoos, and dishwashing liquids. Today, PETE is the most widely recycled of all the plastics. PETE is a polymer of terephthalic acid and ethylene glycol. In one year, 1.7 billion pounds of PETE are recycled. After it is separated from other plastics, PETE can be used in polyester fabric, door mats, tennis ball containers, and fill for sleeping bags. The density of PETE is 1.38 g/mL.

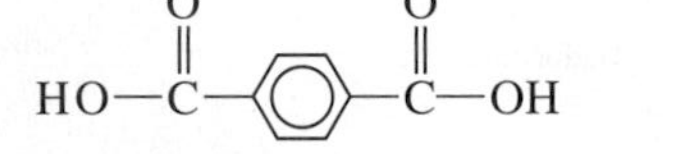

Plastic bottles made of PETE are ready to be recycled.

a. Draw the condensed structural formula of the ester formed from one molecule of terephthalic acid and one molecule of ethylene glycol.
b. Draw the condensed structural formula of the product formed when a second molecule of ethylene glycol reacts with the ester you drew in part **a**.
c. How many kilograms of PETE are recycled in one year?
d. What volume, in liters, of PETE is recycled in one year?
e. Suppose a landfill with an area of a football field and a depth of 5.0 m holds 2.7×10^7 L of recycled PETE. If all the PETE that is recycled in a year were placed instead in landfills, how many would it fill?

CI.26 Using the Internet or a reference book such as *The Merck Index* or *Physicians' Desk Reference*, look up the condensed structural formulas of the following medicinal drugs and list the functional groups in the compounds. You may need to refer to the cross-index of names at the back of the reference book.
a. baclofen, a muscle relaxant
b. anethole, a licorice flavoring agent in anise and fennel
c. alibendol, antispasmodic drug
d. pargyline, antihypertensive drug
e. naproxen, non-steroidal anti-inflammatory drug

CI.27 The insect repellent DEET is an amide that can be made from 3-methylbenzoic acid and diethylamine. A 6.0-fl oz can of DEET repellent contains 25% DEET (m/v) (1 qt = 32 fluid ounces).
a. Draw the condensed structural formula of DEET.
b. Give the molecular formula of DEET.
c. What is the molar mass, to three significant figures, of DEET?
d. How many grams of DEET are in one spray can?
e. How many molecules of DEET are in one spray can?

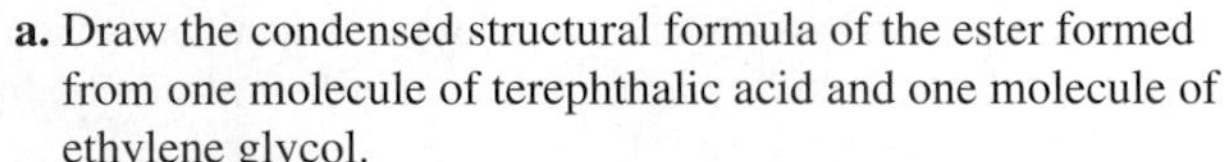

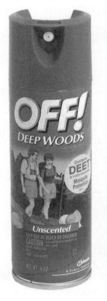

DEET is used in insect repellent.

CI.28 Glyceryl trimyristate (trimyristin) is found in the seeds of the nutmeg (*Myristica fragrans*). The oil known as nutmeg butter contains 75% glyceryl trimyristate (trimyristin). Ground nutmeg, which is sweet, is used to flavor many foods. It is used as a lubricant and fragrance in soaps and shaving creams. Isopropyl myristate is used to increase absorption of skin creams. Draw the condensed structural formula for each of the following:

Nutmeg contains high levels of glyceryl trimyristate.

a. myristic acid
b. glyceryl trimyristate (trimyristin)
c. isopropyl myristate
d. products of the hydrolysis of glyceryl trimyristate with an acid catalyst
e. products of the saponification of glyceryl trimyristate with KOH

CI.29 Panose is a trisaccharide that is being considered as a possible sweetener by the food industry.

A B C

a. What are the monosaccharide units A, B, and C in panose?
b. What type of glycosidic bond connects monosaccharides A and B?
c. What type of glycosidic bond connects monosaccharides B and C?
d. Is the structure drawn as α- or β-panose?
e. Why would panose be a reducing sugar?

CI.30 Hyaluronic acid (HA), a polymer of about 25 000 disaccharide units, is a natural component of eye and joint fluid as well as skin and cartilage. Because of the ability of HA to absorb water, it is used in skin-care products and injections to smooth wrinkles and for treatment of arthritis. The repeating disaccharide units in HA consist of D-gluconic acid and N-acetyl-D-glucosamine. N-Acetyl-D-glucosamine is an amide derived from acetic acid and D-glucosamine, in which an amine group ($—NH_2$) replaces the hydroxyl on carbon 2 of D-glucose. Another natural polymer called chitin is found in the shells of lobsters and crabs. Chitin is made of repeating units of N-acetyl-D-glucosamine connected by β-1,4-glycosidic bonds.

Hyaluronic acid

The shells of crabs and lobsters contain chitin.

a. Draw the Haworth structures for the oxidation reaction of the hydroxyl group on carbon 6 in β-D-glucose to form β-D-gluconic acid.
b. Draw the Haworth structure of D-glucosamine.
c. Draw the Haworth structure for the amide of D-glucosamine and acetic acid.
d. What are the two types of glycosidic bonds that link the monosaccharides in hyaluronic acid?
e. Draw the structure of a section of chitin with three N-acetyl-D-glucosamine units linked by β-1,4-glycosidic bonds.

Answers

CI.25 a. $HO—C(=O)—C_6H_4—C(=O)—O—CH_2—CH_2—OH$

b. $HO—CH_2—CH_2—O—C(=O)—C_6H_4—C(=O)—O—CH_2—CH_2—OH$

c. 7.7×10^8 kg of PETE
d. 5.6×10^8 L of PETE
e. 21 landfills

CI.27 a. $CH_3—C_6H_4—C(=O)—N(CH_2—CH_3)—CH_2—CH_3$ **b.** $C_{12}H_{17}NO$ (DEET)

c. 191 g/mole (DEET) **d.** 44 g of DEET
e. 1.4×10^{23} molecules

CI.29 a. A, B, and C are all glucose.
b. An α-1,6-glycosidic bond links A and B.
c. An α-1,4-glycosidic bond links B and C.
d. β-panose
e. Panose is a reducing sugar because it has a free hydroxyl group on carbon 1 of structure C, which allows glucose (C) to form an aldehyde.

Amino Acids, Proteins, and Enzymes

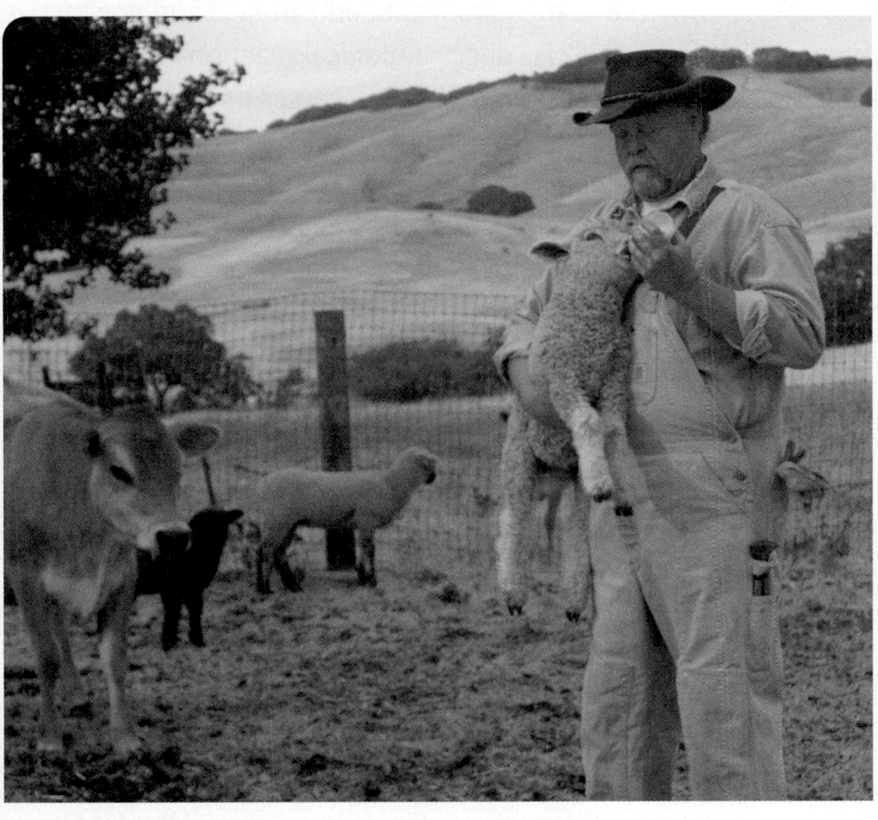

LOOKING AHEAD

"This lamb is fed with Lamb Lac, which is a chemically formulated replacement for ewe's milk," says part-time farmer Dennis Samuelson. "Its mother had triplets and didn't have enough milk to feed them all, so they weren't thriving the way the other lambs were. The Lamb Lac includes dried skim milk, dried whey, milk proteins, egg albumin, the amino acids methionine and lysine, vitamins, and minerals."

A veterinary technician diagnoses and treats diseases of animals, takes blood and tissue samples, and administers drugs and vaccines. Agricultural technologists assist in the study of farm crops to increase productivity and ensure a safe food supply. They look for ways to improve crop yields, develop safer methods of weed and pest control, and design methods to conserve soil and water.

Visit **www.masteringchemistry.com** for self-study materials and instructor-assigned homework.

The word "protein" is derived from the Greek word *proteios*, meaning "first." All proteins in humans are polymers made up from 20 different amino acids. Each kind of protein is composed of amino acids arranged in a specific order that determines the characteristics of the protein and its biological action. Proteins provide structure in membranes, build cartilage and connective tissue, transport oxygen in blood and muscle, direct biological reactions as enzymes, defend the body against infection, and control metabolic processes as hormones. They can even be a source of energy.

The different functions of proteins depend on the structures and chemical behavior of amino acids, the building blocks of proteins. We will see how peptide bonds link amino acids and how the order of the amino acids in these protein polymers directs the formation of unique three-dimensional structures.

Every second, thousands of chemical reactions occur in the cells of the human body at rates that meet our physiological and metabolic needs. To make this happen, enzymes catalyze the chemical reactions in our cells, with a different enzyme for every reaction. Digestive enzymes in the mouth, stomach, and small intestine catalyze the hydrolysis of carbohydrates, fats, and proteins. Enzymes in the mitochondria extract energy from biomolecules to give us energy.

LEARNING GOAL

Classify proteins by their functions. Give the name and abbreviation of an amino acid and draw its ionized structure.

SELF STUDY ACTIVITY
Functions of Proteins

16.1 Proteins and Amino Acids

The many kinds of proteins perform different functions. There are proteins that form structural components such as cartilage, muscles, hair, and nails. Wool, silk, feathers, and horns in animals are made of proteins (see Figure 16.1). Proteins that function as enzymes regulate biological reactions such as digestion and cellular metabolism. Other proteins, such as hemoglobin and myoglobin, transport oxygen in the blood and muscle. Table 16.1 gives examples of proteins that are classified by their functions.

TABLE 16.1 Classification of Some Proteins and Their Functions

Class of Protein	Function	Examples
Structural	Provide structural components	*Collagen* is in tendons and cartilage. *Keratin* is in hair, skin, wool, and nails.
Contractile	Make muscles move	*Myosin* and *actin* contract muscle fibers.
Transport	Carry essential substances throughout the body	*Hemoglobin* transports oxygen. *Lipoproteins* transport lipids.
Storage	Store nutrients	*Casein* stores protein in milk. *Ferritin* stores iron in the spleen and liver.
Hormone	Regulate body metabolism and nervous system	*Insulin* regulates blood glucose level. *Growth hormone* regulates body growth.
Enzyme	Catalyze biochemical reactions in the cells	*Sucrase* catalyzes the hydrolysis of sucrose. *Trypsin* catalyzes the hydrolysis of proteins.
Protection	Recognize and destroy foreign substances	*Immunoglobulins* stimulate immune responses.

FIGURE 16.1 The horns of animals are made of proteins.
Q What class of protein would be in horns?

TUTORIAL
Proteins "R" Us

TUTORIAL
Protein Building Blocks

TUTORIAL
Amino Acids

CONCEPT CHECK 16.1

Classifying Proteins by Function

Give the class of protein that would perform each of the following functions:

a. catalyzes metabolic reactions of lipids
b. carries oxygen in the bloodstream
c. stores amino acids in milk

ANSWER

a. Enzymes catalyze metabolic reactions.
b. Transport proteins carry substances such as oxygen through the bloodstream.
c. Storage proteins store nutrients such as amino acids in milk.

Amino Acids

Proteins are composed of molecular building blocks called amino acids. However, there are only 20 different amino acids present in human proteins. Every **amino acid** consists of a central carbon atom called the α-carbon bonded to two functional groups: an *amino* group ($-NH_2$) and a *carboxylic acid* group ($-COOH$). The α-carbon is also bonded to a hydrogen atom and an R group. It is the R group, which differs in each of the 20 amino acids, that provides unique characteristics to each type of amino acid. For example, alanine has a methyl, $-CH_3$, as its R group.

Structures of an α-Amino Acid

The side chain (R)
Carboxylic acid group
α-Carbon
Amino group

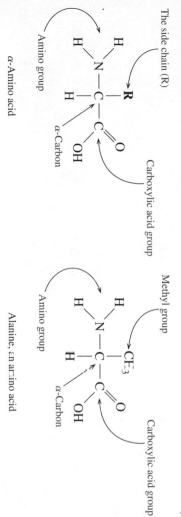

α-Amino acid

Methyl group
Carboxylic acid group
α-Carbon
Amino group

Alanine, an amino acid

Ionization of Amino Acids

Although we have drawn an amino acid with uncharged amino ($-NH_2$) and carboxylic acid ($-COOH$) groups, these groups are ionized for amino acids in most body fluids. At physiological pH, the $-NH_2$ group gains H^+ to give its ionized form $-NH_3^+$ and the $-COOH$ group loses H^+ to give its ionized form $-COO^-$. An ionized amino acid,

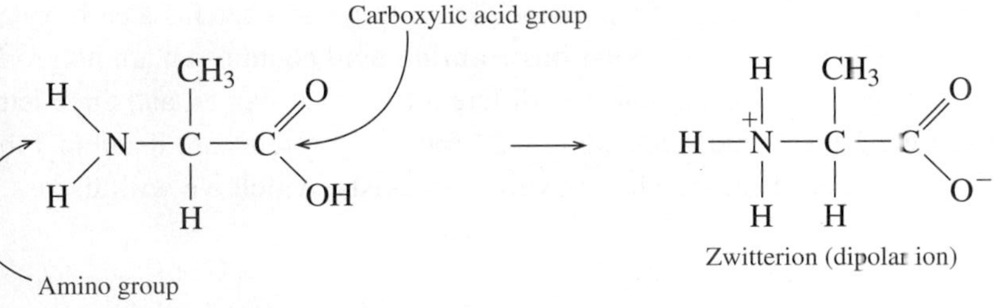

Alanine, an amino acid, forms a zwitterion in body fluids.

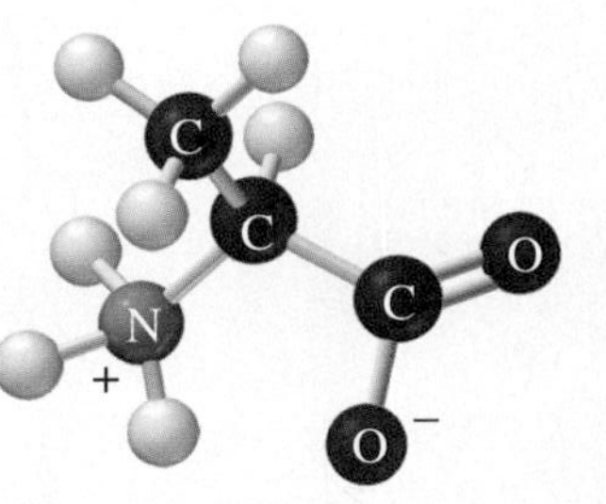

Ball-and-stick model of the zwitterion of alanine

which has both a positive charge and a negative charge, is a dipolar ion called a **zwitterion**. In the zwitterion, the ionized regions have charge balance, which means that the ionized amino acid has an overall zero charge. As zwitterions, amino acids are similar to salts. Thus, amino acids have high melting points and are soluble in water, but not in organic solvents.

Amino Acid Stereoisomers

All the α-amino acids except for glycine are chiral because the α-carbon is attached to four different atoms. Thus amino acids can exist as D and L isomers. We can draw Fischer projections for α-amino acids as we did in Chapter 12 for aldehydes by placing the carboxylate group at the top and the R group at the bottom. In the L isomer, the $-NH_3^+$ is on the left, and in the D isomer, it is on the right. In biological systems, the only amino acids incorporated into proteins are the L isomers. There are D amino acids found in nature, but not in proteins. Let's take a look at the isomers for L- and D-glyceraldehyde and the isomers of L- and D-alanine and L- and D-cysteine.

CHO / HO—C—H / CH_2OH
L-Glyceraldehyde

CHO / H—C—OH / CH_2OH
D-Glyceraldehyde

COO^- / $H_3\overset{+}{N}$—C—H / CH_3
L-Alanine

COO^- / H—C—$\overset{+}{N}H_3$ / CH_3
D-Alanine

COO^- / $H_3\overset{+}{N}$—C—H / CH_2SH
L-Cysteine

COO^- / H—C—$\overset{+}{N}H_3$ / CH_2SH
D-Cysteine

Classification of Amino Acids

We can now classify amino acids using their specific R groups, which determine their characteristics in aqueous solution. The **nonpolar amino acids** have hydrogen, alkyl, or aromatic R groups, which make them *hydrophobic* ("water fearing"). The **polar amino acids** have R groups that interact with water, which makes them *hydrophilic* ("water loving"). The *polar neutral* amino acids contain hydroxyl ($-OH$), thiol ($-SH$), or amide ($-CONH_2$) groups. The R group of a polar **acidic amino acid** contains a carboxylate group ($-COO^-$). The R group of a polar **basic amino acid** contains an amino group, which ionizes to give an ammonium ion. The different R groups, names, and three-letter abbreviations of the 20 α-amino acids commonly found in proteins are listed in Table 16.2. The pI value is the pH at which the zwitterion exists, which we will discuss in Section 16.2.

TABLE 16.2 The 20 Amino Acids (Ionized) in Proteins

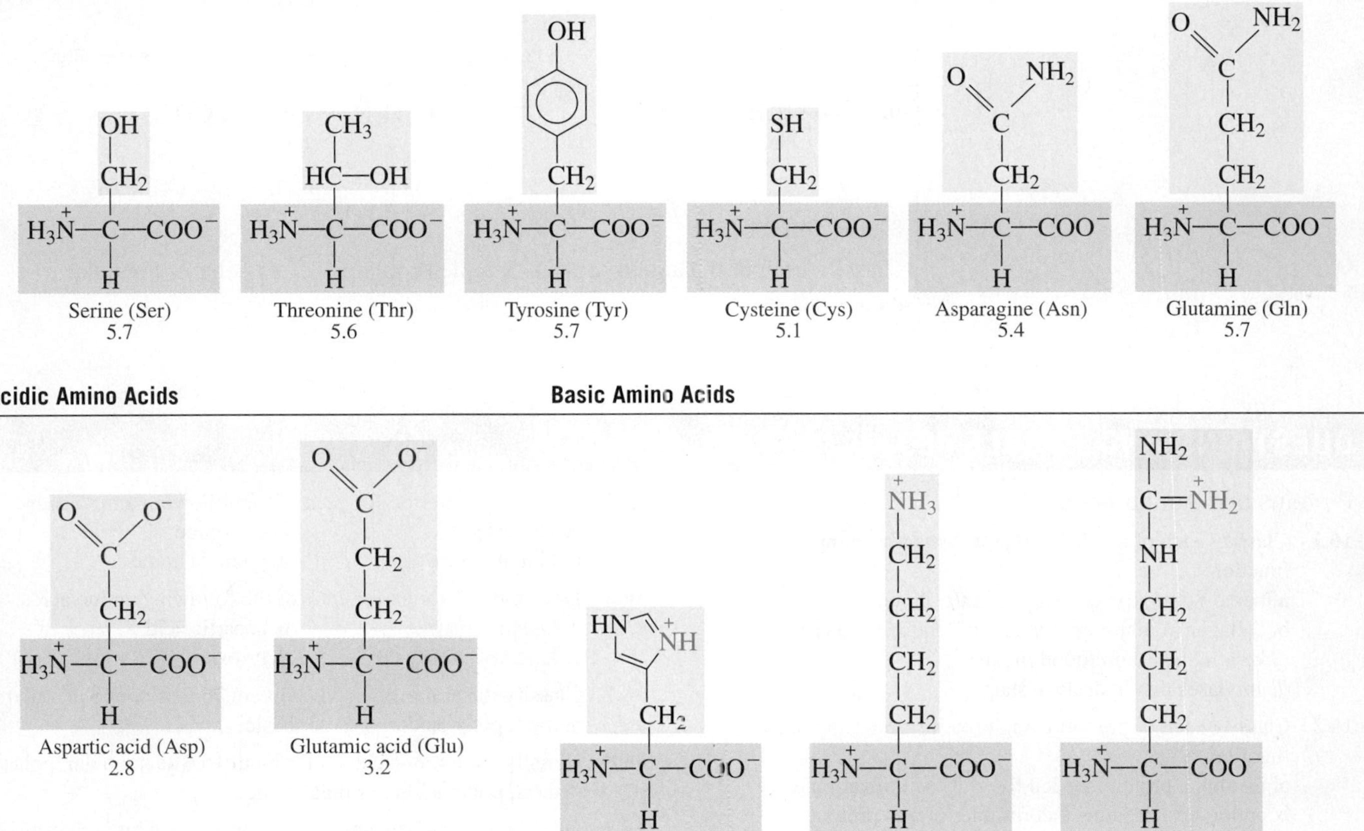

*Isoelectric points (pI)

CONCEPT CHECK 16.2

R Groups and Polarity of Amino Acids

Compare the types of atoms and the polarity of the R groups in each of the following:

a. valine **b.** threonine **c.** lysine

ANSWER

a. The R group in valine contains only C and H atoms, which makes valine a nonpolar amino acid.

b. The R group in threonine contains C atoms, H atoms, and an $—OH$ group, which makes threonine a polar neutral amino acid.

c. The R group in lysine contains C atoms, H atoms, and an $—NH_3^+$ group, which makes lysine a polar basic amino acid.

SAMPLE PROBLEM 16.1

Structural Formulas of Amino Acids

Draw the condensed structural formula for the zwitterion for each of the following amino acids:

a. valine **b.** serine

SOLUTION

a. $H_3\overset{+}{N}—CH(—CH(CH_3)_2)—COO^-$ (R group: $CH(CH_3)_2$)
valine

b. $H_3\overset{+}{N}—CH(—CH_2—OH)—COO^-$ (R group: $CH_2—OH$)
serine

STUDY CHECK 16.1

Classify each of the amino acids in Sample Problem 16.1 as polar or nonpolar.

QUESTIONS AND PROBLEMS

Proteins and Amino Acids

16.1 Classify each of the following proteins according to its function:
a. hemoglobin, oxygen carrier in the blood
b. collagen, a major component of tendons and cartilage
c. keratin, a protein found in hair
d. amylases that hydrolyze starch

16.2 Classify each of the following proteins according to its function:
a. insulin, a protein needed for glucose utilization
b. antibodies, proteins that disable foreign proteins
c. casein, milk protein
d. lipases that hydrolyze lipids

16.3 What functional groups are found in all α-amino acids?

16.4 How does the polarity of the R group in leucine compare to the R group in serine?

16.5 Draw the zwitterion for each of the following amino acids:
a. glycine **b.** threonine
c. glutamic acid **d.** phenylalanine

16.6 Draw the zwitterion for each of the following amino acids:
a. lysine **b.** aspartic acid
c. leucine **d.** tyrosine

16.7 Classify the amino acids in Problem 16.5 as nonpolar, polar neutral, polar acidic, or polar basic.

16.8 Classify the amino acids in Problem 16.6 as nonpolar, polar neutral, polar acidic, or polar basic.

16.9 Give the name of the amino acid represented by each of the following three-letter abbreviations:
a. Ala **b.** Val **c.** Lys **d.** Cys

16.10 Give the name of the amino acid represented by each of the following three-letter abbreviations:
a. Trp **b.** Met **c.** Pro **d.** Gly

Chemistry Link to Health

ESSENTIAL AMINO ACIDS

Of the 20 amino acids used to build the proteins in the body, only 10 can be synthesized in the body. The other 10 amino acids, listed in Table 16.3, are *essential amino acids* that cannot be synthesized and must be obtained from the proteins in the diet.

Complete proteins, which contain all of the essential amino acids, are found in most animal products such as eggs, milk, meat, fish, and poultry. However, gelatin and plant proteins such as grains, beans, and nuts are *incomplete proteins* because they are deficient in one or more of the essential amino acids. Diets that rely on plant foods for protein must contain a variety of protein sources to obtain all the essential amino acids. For example, a diet of rice and beans contains all the essential amino acids because they are complementary protein sources. Rice contains the methionine and tryptophan that are deficient in beans, while beans contain the lysine that is lacking in rice (see Table 16.4).

TABLE 16.3 Essential Amino Acids

Arginine (Arg)*	Methionine (Met)
Histidine (His)*	Phenylalanine (Phe)
Isoleucine (Ile)	Threonine (Thr)
Leucine (Leu)	Tryptophan (Trp)
Lysine (Lys)	Valine (Val)

*Required in diets of children, not adults.

TABLE 16.4 Amino Acid Deficiency in Selected Vegetables and Grains

Food Source	Amino Acids Missing
Eggs, milk, meat, fish, poultry	None
Wheat, rice, oats	Lysine
Corn	Lysine, tryptophan
Beans	Methionine, tryptophan
Peas	Methionine
Almonds, walnuts	Lysine, tryptophan
Soy	Low in methionine

16.2 Amino Acids as Acids and Bases

LEARNING GOAL

Draw the ionized condensed structural formula of an amino acid at pH values above or below its pI.

As we discuss in Section 16.1, amino acids typically exist as zwitterions due to the ionization of both the amino ($—NH_2$) and carboxylic acid ($—COOH$) groups. A zwitterion with positive and negative charges and thus an overall neutral charge forms only at a certain pH called the **isoelectric point (pI)**. However, an amino acid can exist as a positive ion if a solution is more acidic (has a lower pH) than its pI or as a negative ion if a solution is more basic (has a higher pH) than its pI.

Ionized Forms of Nonpolar and Polar Neutral Amino Acids

TUTORIAL
pH, pI, and Amino Acid Ionization

The pI values for nonpolar and polar neutral amino acids are from pH 5.1 to 6.3. Let's take a look at the different ionic forms of alanine. Alanine forms its zwitterion in a solution with a pH of 6.0, which is also its pI value. In the zwitterion form, alanine contains a carboxylate anion ($—COO^-$) and an ammonium cation ($—NH_3^+$), which give an overall charge of zero. In a solution with a pH lower than its pI (pH < 6.0), the $—COO^-$ group of alanine gains H^+ to form its carboxylic acid ($—COOH$). However, the ammonium cation ($—NH_3^+$) gives alanine an overall positive charge (1−).

$$H_3\overset{+}{N}—\underset{\underset{CH_3}{|}}{CH}—\overset{\overset{O}{\|}}{C}—O^- \xrightleftharpoons{H^+} H_3\overset{+}{N}—\underset{\underset{CH_3}{|}}{CH}—\overset{\overset{O}{\|}}{C}—OH$$

Zwitterion of alanine pH = 6.0 (charge = 0)

Alanine at a pH < 6.0 (charge = 1+)

When alanine is in a solution with a pH greater than its pI (pH > 6.0), the $—NH_3^+$ loses H^+ to form an amino group ($—NH_2$). Because the $—COO^-$ remains ionized, alanine has an overall negative charge (1−). The changes in the ionized groups with pH above, below, or equal to the pI are summarized in Table 16.5.

$$\underset{\substack{\text{Zwitterion of alanine}\\ \text{pH} = 6.0\\ (\text{charge} = 0)}}{H_3\overset{+}{N}—\underset{\substack{|\\ CH_3}}{CH}—\overset{\substack{O\\ \|}}{C}—O^-} \xrightleftharpoons{OH^-} \underset{\substack{\text{Alanine at a pH} > 6.0\\ (\text{charge} = 1-)}}{H_2N—\underset{\substack{|\\ CH_3}}{CH}—\overset{\substack{O\\ \|}}{C}—O^-}$$

TABLE 16.5 Ionized Forms of Nonpolar and Polar Neutral Amino Acids

Condition	pH < pI	pH = pI	pH > pI
Change in H^+	$[H^+]$ increases	none	$[H^+]$ decreases
Change in ionized groups	$—COOH$	$—COO^-$	$—COO^-$
	$—\overset{+}{N}H_3$	$—\overset{+}{N}H_3$	$—NH_2$
Charge of amino acid	1+	0	1−

Ionized Forms of Polar Acidic and Polar Basic Amino Acids

The pI values of the polar acidic amino acids (aspartic acid, glutamic acid) are about pH 3. At pH values of 3, the carboxylic acid group in the R groups of their zwitterions is not ionized. However, at physiological pH values, which are greater than 3, the carboxylic acid in the R group loses H^+ to form a negatively charged $—COO^-$.

The pI values of basic amino acids are typically higher than physiological pH value, ranging from pH 7.6 to 10.8. Thus at physiological pH values, the amines in the R groups of the basic amino acids (lysine, arginine, and histidine) gain H^+ to form an overall positive charge $—NH_3^+$. The pI values are included below the names of the amino acids in Table 16.2.

CONCEPT CHECK 16.3

Ionized Forms of Amino Acids

Consider the amino acid cysteine.

a. What is the pI of cysteine and what does it mean?
b. At a pH of 2.0, how does the zwitterion of cysteine change?
c. At a pH of 8.0, how does the zwitterion of cysteine change?

ANSWER

a. From Table 16.2, the pI of cysteine is 5.1, which means that at a pH of 5.1, cysteine exists as a zwitterion with an overall charge of zero.
b. Because a pH of 2.0 is more acidic and below the pI of cysteine, the $—COO^-$ gains H^+ to give $—COOH$. The remaining $—NH_3^+$ gives cysteine an overall positive charge (1+).
c. Because a pH of 8.0 is more basic and above the pI of cysteine, the $—NH_3^+$ loses H^+ to give $—NH_2$. The remaining $—COO^-$ gives cysteine an overall negative charge (1−).

Career Focus

REHABILITATION SPECIALIST

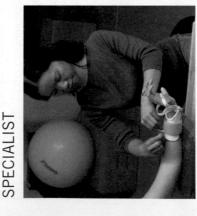

"I am interested in the biomechanical part of rehabilitation, which involves strengthening activities that help people return to the activities of daily living," says Minna Robles, rehabilitation specialist. "Here I am fitting a patient with a wrist extension splint that allows her to lift her hand. This exercise will also help the muscles and soft tissues in that wrist area to heal. An understanding of the chemicals of the body, how they interact, and how we can affect the body on a chemical level is important in understanding our work. One technique we use is called myofacial release. We apply pressure to a part of the body, which helps to increase circulation. By increasing circulation, we can move the soft tissues better, which improves movement and range of motion."

SAMPLE PROBLEM 16.2

Amino Acids in Acid or Base

The pI of glycine is 6.0. Draw the condensed structural formulas for glycine at pH 6.0 and at pH 8.0. State the overall charge for each.

SOLUTION

At its pI of 6.0, glycine exists as a zwitterion with both the amino and carboxylic acid groups ionized. The overall charge on the zwitterion is zero (0).

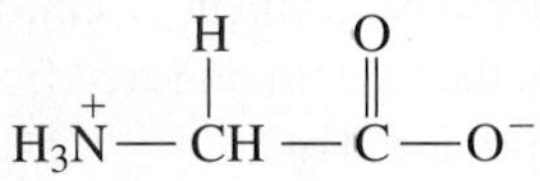

$$H_3\overset{+}{N}-\underset{}{\overset{H}{\overset{|}{C}H}}-\overset{O}{\overset{\|}{C}}-O^-$$ Zwitterion of glycine

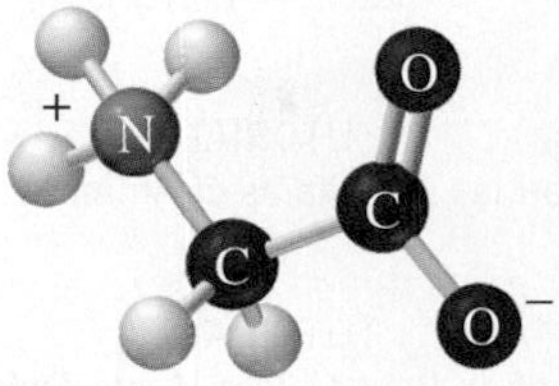

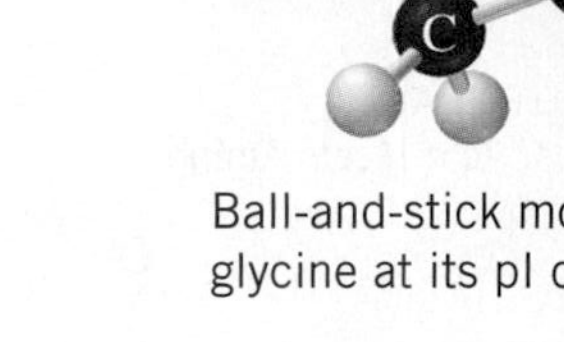

Ball-and-stick model of glycine at its pI of 6.0.

At a pH of 8.0 (basic solution), the $-NH_3^+$ loses H^+. Because the carboxyl group remains ionized, the overall charge is negative (1−).

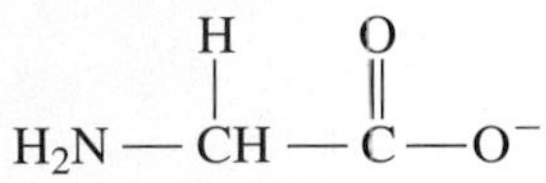

$$H_2N-\overset{H}{\overset{|}{C}H}-\overset{O}{\overset{\|}{C}}-O^-$$ Charge 1− at pH 8.0

STUDY CHECK 16.2

Draw the condensed structural formula of serine at pH 3.0 and give the overall charge.

QUESTIONS AND PROBLEMS

Amino Acids as Acids and Bases

16.11 What is the overall charge of a nonpolar amino acid at a pH below its pI?

16.12 What is the overall charge of a polar neutral amino acid at a pH above its pI?

16.13 Draw the condensed structural formula for each of the following amino acids at pH below 1.0:

a. glycine **b.** cysteine
c. serine **d.** threonine

16.14 Draw the condensed structural formula for each of the following amino acids at pH above 12.0:

a. glutamine **b.** methionine
c. isoleucine **d.** valine

16.15 Would each of the following ions of valine exist at a pH above, below, or at pI?

a. 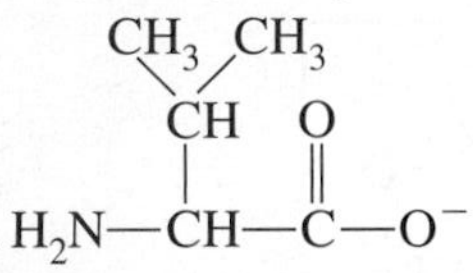 $H_2N-\overset{CH_3\ CH_3\ \diagdown\ \diagup\ CH}{\overset{|}{C}H}-\overset{O}{\overset{\|}{C}}-O^-$ **b.** $H_3\overset{+}{N}-\overset{CH_3\ CH_3\ \diagdown\ \diagup\ CH}{\overset{|}{C}H}-\overset{O}{\overset{\|}{C}}-OH$

c. $H_3\overset{+}{N}-\overset{CH_3\ CH_3\ \diagdown\ \diagup\ CH}{\overset{|}{C}H}-\overset{O}{\overset{\|}{C}}-O^-$

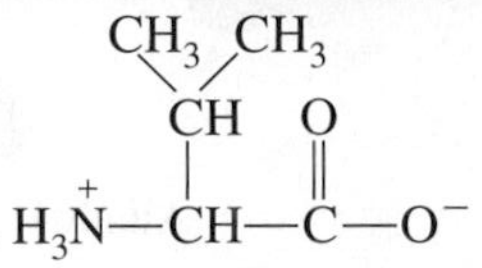

16.16 Would each of the following ions of serine exist at a pH above, below, or at pI?

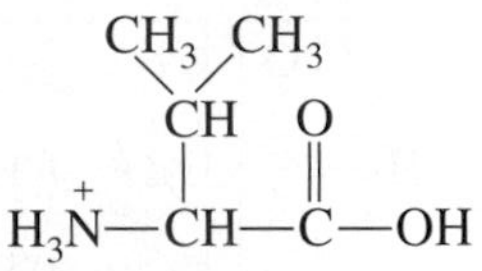

a. 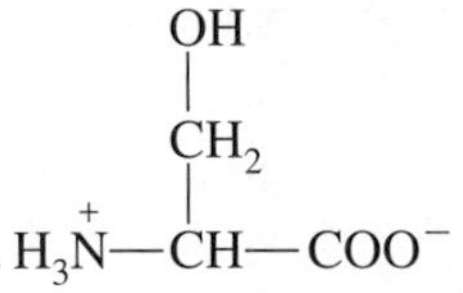 $H_3\overset{+}{N}-\overset{OH-CH_2}{\overset{|}{C}H}-COO^-$ **b.** $H_3\overset{+}{N}-\overset{OH-CH_2}{\overset{|}{C}H}-COOH$

c. $H_2N-\overset{OH-CH_2}{\overset{|}{C}H}-COO^-$

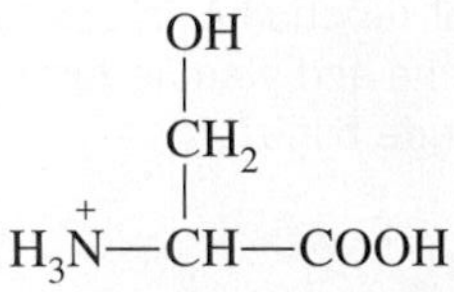

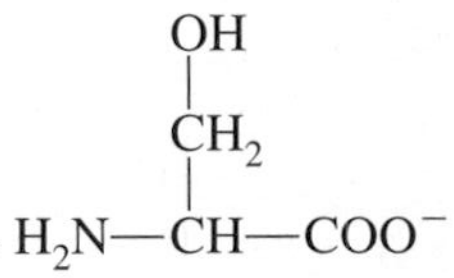

LEARNING GOAL

Draw the condensed structural formula of a dipeptide and give its name.

SELF STUDY ACTIVITY
Structure of Proteins

TUTORIAL
Peptides Are Chains of Amino Acids

TUTORIAL
Peptide Bonds: Acid Meets Amino

16.3 Formation of Peptides

A **peptide bond** is an amide bond that forms when the $-COO^-$ of one amino acid reacts with the $-NH_3^+$ of the next amino acid. The linking of two or more amino acids by peptide bonds forms a **peptide**. Two amino acids form a *dipeptide*, three amino acids form a *tripeptide*, and four amino acids form a *tetrapeptide*. A chain of five amino acids is a *pentapeptide*, and long chains of amino acids are called *polypeptides*.

We can look now at the amidation reaction between the zwitterions of glycine and alanine (see Figure 16.2). The amino acid glycine on the left with a free $-NH_3^+$ is called the **N terminal** amino acid. The amino acid alanine on the right with a free $-COO^-$ is called the **C terminal** amino acid. The dipeptide forms when the carbonyl group in glycine bonds to the N atom in $-NH_3^+$ of alanine. During the amidation, the O atom removed from the carboxylate combines with two H atoms from $-NH_3^+$ to produce H_2O.

$$\underset{\text{Glycine}}{H_3\overset{+}{N}-CH_2-\overset{\overset{O}{\|}}{C}-O^-} + \underset{\text{Alanine}}{H_3\overset{+}{N}-\underset{\underset{CH_3}{|}}{CH}-\overset{\overset{O}{\|}}{C}-O^-} \longrightarrow \underset{\text{Glycylalanine (Gly–Ala)}}{\underset{\text{N terminal}}{H_3\overset{+}{N}}-CH_2-\boxed{\overset{\overset{O}{\|}}{C}-\overset{\overset{H}{|}}{N}}_{\text{Peptide bond}}-\underset{\underset{CH_3}{|}}{CH}-\underset{\text{C terminal}}{\overset{\overset{O}{\|}}{C}-O^-}} + H_2O$$

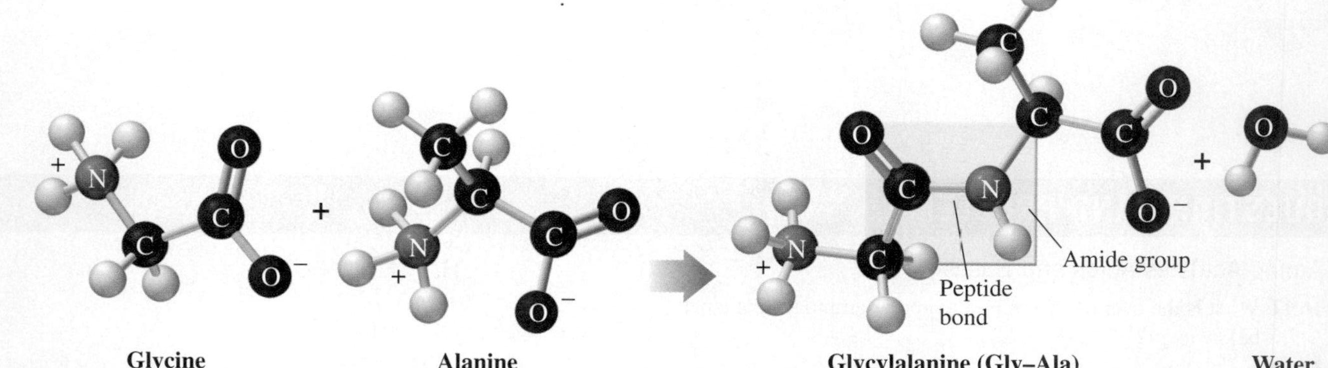

FIGURE 16.2 A peptide bond between zwitterions of glycine and alanine forms the dipeptide glycylalanine.
Q What functional groups in glycine and alanine form the peptide bond?

Naming Peptides

With the exception of the C terminal amino acid, the names of all the other amino acids in a peptide end with *yl*. For example, a tripeptide consisting of alanine at the N terminal, glycine, and serine at the C terminal is named as one word: alan**yl**glyc**yl**serine. For convenience, the order of amino acids in the peptide is often written as the sequence of three-letter abbreviations.

N Terminal — C Terminal

$$\underset{\substack{\text{From alanine}\\\text{Alanyl}}}{\boxed{H_3\overset{+}{N}-\underset{\underset{CH_3}{|}}{CH}-\overset{\overset{O}{\|}}{C}}}\underset{\substack{\text{From glycine}\\\text{Glycyl}}}{\boxed{NH-CH_2-\overset{\overset{O}{\|}}{C}}}\underset{\substack{\text{From serine}\\\text{Serine}}}{\boxed{NH-\underset{\underset{CH_2-OH}{|}}{CH}-\overset{\overset{O}{\|}}{C}-O^-}}$$

Alanylglycylserine
(Ala–Gly–Ser)

CONCEPT CHECK 16.4

Structure and Names of Peptides

Consider the dipeptide Val–Pro:

a. What amino acid is the N terminal amino acid?
b. What amino acid is the C terminal amino acid?
c. How are the amino acids connected?
d. Give the name of the dipeptide.
e. Give the name of the dipeptide in which the amino acid order is reversed as Pro–Val.

ANSWER

a. Valine, the first amino acid in the peptide, has a free $-NH_3^+$, which makes it the N terminal amino acid.
b. Proline, the last amino acid in the peptide, has a free $-COO^-$, which makes it the C terminal amino acid.
c. The amino acids valine and proline are connected by a peptide bond, which is an amide bond between the C=O of valine and the N—H of proline.
d. In the dipeptide Val–Pro, the N terminal amino acid, valine, is named valyl and the C terminal amino acid, proline, retains its name. The dipeptide is named valylproline.
e. In the dipeptide Pro–Val, the N terminal amino acid, proline, is named prolyl and the C terminal amino acid, valine, retains its name. The dipeptide is named prolylvaline.

SAMPLE PROBLEM 16.3

Identifying a Tripeptide

Consider the following tripeptide:

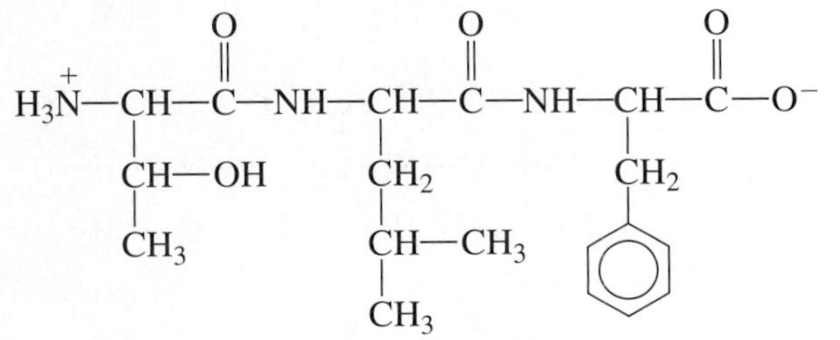

$$\overset{+}{H_3N}-\underset{\underset{CH_3}{|}}{\underset{CH-OH}{|}}{CH}-\overset{O}{\overset{\|}{C}}-NH-\underset{\underset{\underset{CH_3}{|}}{CH-CH_3}}{\underset{CH_2}{|}}{CH}-\overset{O}{\overset{\|}{C}}-NH-\underset{\underset{C_6H_5}{|}}{\underset{CH_2}{|}}{CH}-\overset{O}{\overset{\|}{C}}-O^-$$

a. What is the amino acid at the N terminal? What is the amino acid at the C terminal?
b. Use the three-letter abbreviation of amino acids to give their order in the tripeptide.

SOLUTION

a. Threonine is the N terminal amino acid; phenylalanine is the C terminal amino acid.
b. Thr–Leu–Phe

STUDY CHECK 16.3

What is the full name of the tripeptide in Sample Problem 16.3?

QUESTIONS AND PROBLEMS

Formation of Peptides

16.17 Draw the condensed structural formula of each of the following peptides, and give its abbreviation:
a. alanylcysteine
b. serylphenylalanine
c. glycylalanylvaline
d. valylisoleucyltryptophan

16.18 Draw the condensed structural formula of each of the following peptides, and give its abbreviation:
a. methionylaspartic acid
b. threonyltryptophan
c. methionylglutaminyllysine
d. histidylglycylglutamylisoleucine

LEARNING GOAL

Identify the structural levels of a protein.

16.4 Levels of Protein Structure

When there are more than 50 amino acids in a chain, the polypeptide is usually called a **protein**. Each protein in our cells has a unique sequence of amino acids that determines its biological function.

TUTORIAL
Levels of Structure in Proteins

SELF STUDY ACTIVITY
Primary and Secondary Structure

Primary Structure

The **primary structure** of a protein is the particular sequence of amino acids held together by peptide bonds. For example, a hormone that stimulates the thyroid to release thyroxin is a tripeptide with the amino acid sequence Glu–His–Pro. Although other amino acid sequences are possible, such as His–Pro–Glu or Pro–His–Glu, they do not produce hormonal activity. Thus the biological function of peptides and proteins depends on the sequence of the amino acids.

Glu His Pro

Chemistry Link to Health

POLYPEPTIDES IN THE BODY

Enkephalins and endorphins are natural painkillers produced in the body. They are polypeptides that bind to receptors in the brain to give relief from pain. This effect appears to be responsible for the runner's high, for the temporary loss of pain when severe injury occurs, and for the analgesic effects of acupuncture.

The *enkephalins*, found in the thalamus and the spinal cord, are pentapeptides, the smallest molecules with opiate activity. The amino acid sequence of met-enkephalin is found in the longer amino acid sequence of the endorphins.

Four groups of *endorphins* have been identified: α-endorphin contains 16 amino acids, β-endorphin contains 31 amino acids, γ-endorphin has 17 amino acids, and δ-endorphin has 27 amino acids. Endorphins may produce their sedating effects by preventing the release of substance P, a polypeptide with 11 amino acids, which has been found to transmit pain impulses to the brain.

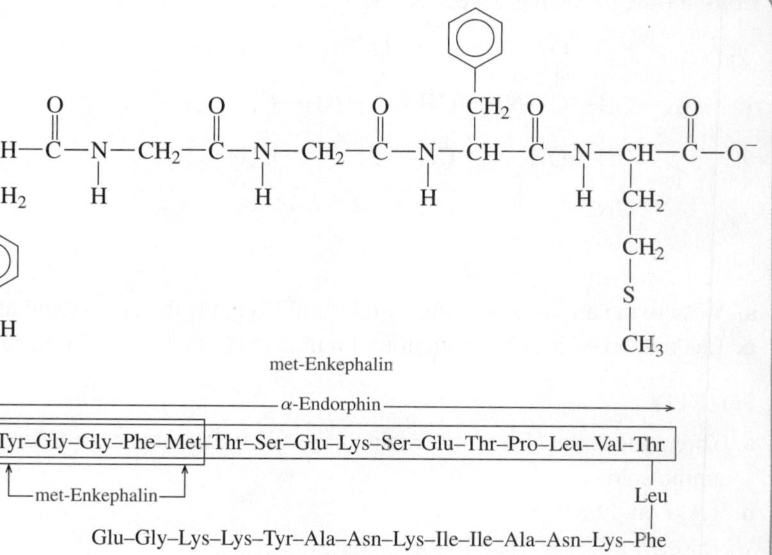

met-Enkephalin

α-Endorphin

Tyr–Gly–Gly–Phe–Met–Thr–Ser–Glu–Lys–Ser–Glu–Thr–Pro–Leu–Val–Thr

met-Enkephalin

Leu

Glu–Gly–Lys–Lys–Tyr–Ala–Asn–Lys–Ile–Ile–Ala–Asn–Lys–Phe

β-Endorphin

Oxytocin
Vasopressin

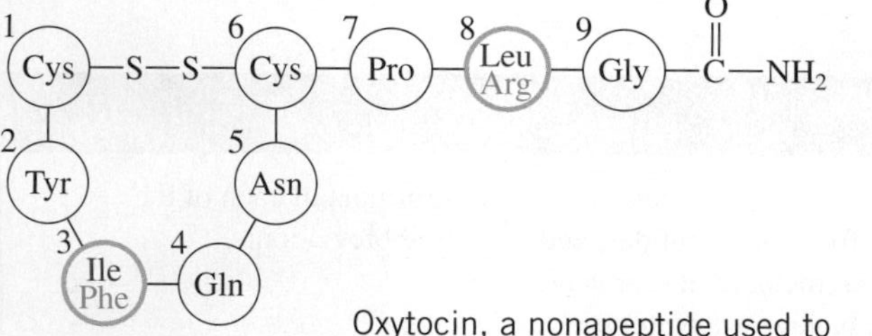

Oxytocin, a nonapeptide used to initiate labor, was the first hormone to be synthesized in the laboratory.

Two hormones produced by the pituitary gland are the nonapeptides (nine-amino-acid peptides) oxytocin and vasopressin. Oxytocin stimulates uterine contractions in labor and vasopressin is an antidiuretic hormone that regulates blood pressure by adjusting the amount of water reabsorbed by the kidneys. The structures of these nonapeptides are very similar. Only the amino acids in positions 3 and 8 are different. However, the difference of two amino acids greatly affects how the two hormones function in the body.

The first protein to have its primary structure determined was insulin, which is a hormone that regulates the glucose level in the blood. In the primary structure of human insulin, there are two polypeptide chains. In chain A, there are 21 amino acids, and chain B has 30 amino acids. The polypeptide chains are held together by disulfide bonds formed by the thiol groups in the cysteine amino acids in each of the chains (see Figure 16.3). Today, human insulin produced through genetic engineering is used in the treatment of diabetes.

The shape of an alpha helix is similar to that of a spiral staircase.

CONCEPT CHECK 16.5

Primary Structure

What are the abbreviations of the possible tetrapeptides containing two valines, one proline, and one histidine if the C terminal is proline?

ANSWER

The C terminal of proline in the possible tetrapeptides would be preceded by three different sequences of two valines and one histidine: Val–Val–His–Pro, Val–His–Val–Pro, and His–Val–Val–Pro.

TUTORIAL
The Shapes of Protein Chains: Helices and Sheets

Secondary Structures

The **secondary structure** of a protein describes the type of structure that forms when amino acids form hydrogen bonds within a polypeptide or between polypeptides. The three most common types of secondary structure are the *alpha helix*, the *beta-pleated sheet*, and the *triple helix*.

In an **alpha helix (α helix)**, hydrogen bonds form between the oxygen of the C═O groups and the hydrogen of N—H groups of the amide bonds in the next turn of the α helix (see Figure 16.4). Because there are many hydrogen bonds along the polypeptide, it has the helical shape of a spiral staircase. The R groups of the different α-amino acids extend to the outside of the helix.

Another type of secondary structure is the **beta-pleated sheet (β-pleated sheet)**. In a β-pleated sheet, hydrogen bonds form between the oxygen atoms in the carbonyl groups of one polypeptide chain and the hydrogen atoms in the N—H groups of the amide bonds in adjacent polypeptide chains. As a result, several polypeptide chains are held together side by side like folded or pleated sheets. In silk fibers, the predominance of glycine, alanine, and serine with small R groups allows the β-pleated sheets to stack closely together. This compact arrangement gives strength, flexibility, and durability to proteins such as silk (see Figure 16.5).

The secondary structure in silk is a beta-pleated sheet.

Collagen, which is the most abundant protein in the body, makes up from 25% to 35% of all protein in vertebrates. It is found in connective tissue, blood vessels, skin, tendons, ligaments, the cornea of the eye, and cartilage. The strong structure of collagen

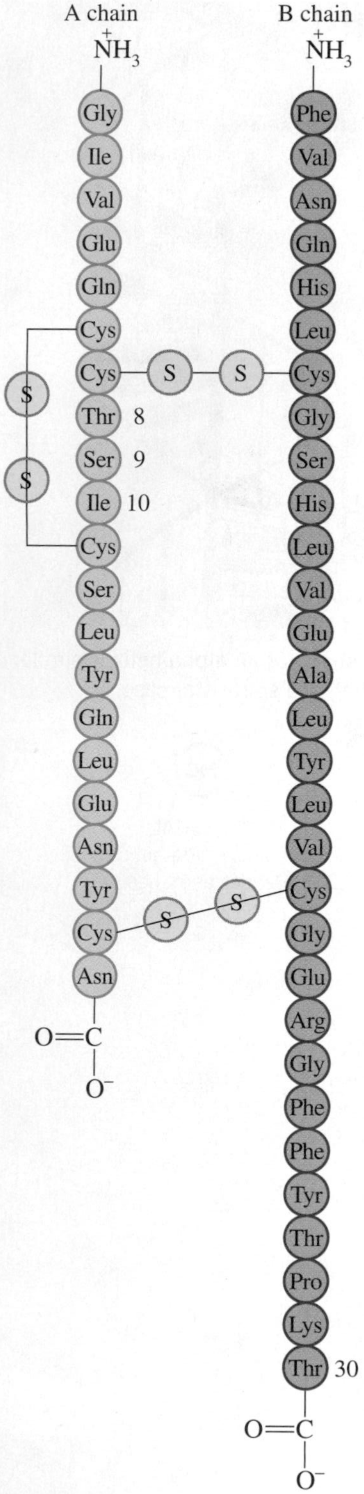

A chain: $\overset{+}{N}H_3$–Gly–Ile–Val–Glu–Gln–Cys–Cys–Thr 8–Ser 9–Ile 10–Cys–Ser–Leu–Tyr–Gln–Leu–Glu–Asn–Tyr–Cys–Asn–COO^-

B chain: $\overset{+}{N}H_3$–Phe–Val–Asn–Gln–His–Leu–Cys–Gly–Ser–His–Leu–Val–Glu–Ala–Leu–Tyr–Leu–Val–Cys–Gly–Glu–Arg–Gly–Phe–Phe–Tyr–Thr–Pro–Lys–Thr 30–COO^-

(S–S disulfide bridges: A chain Cys–Cys within the A chain; A chain Cys to B chain Cys, two S–S links)

FIGURE 16.3 The sequence of amino acids in human insulin is its primary structure.
Q What kinds of bonds occur in the primary structure of a protein?

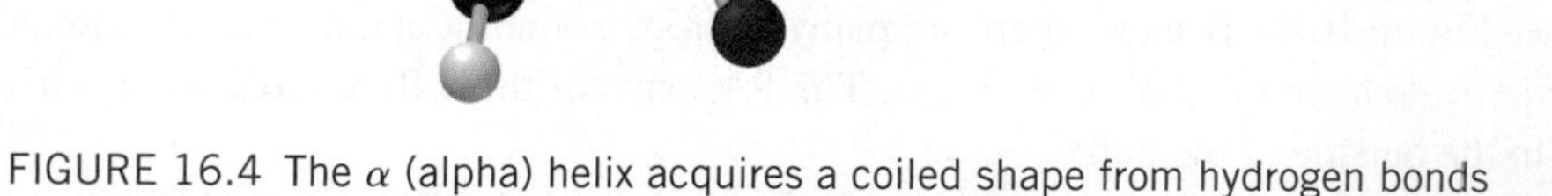

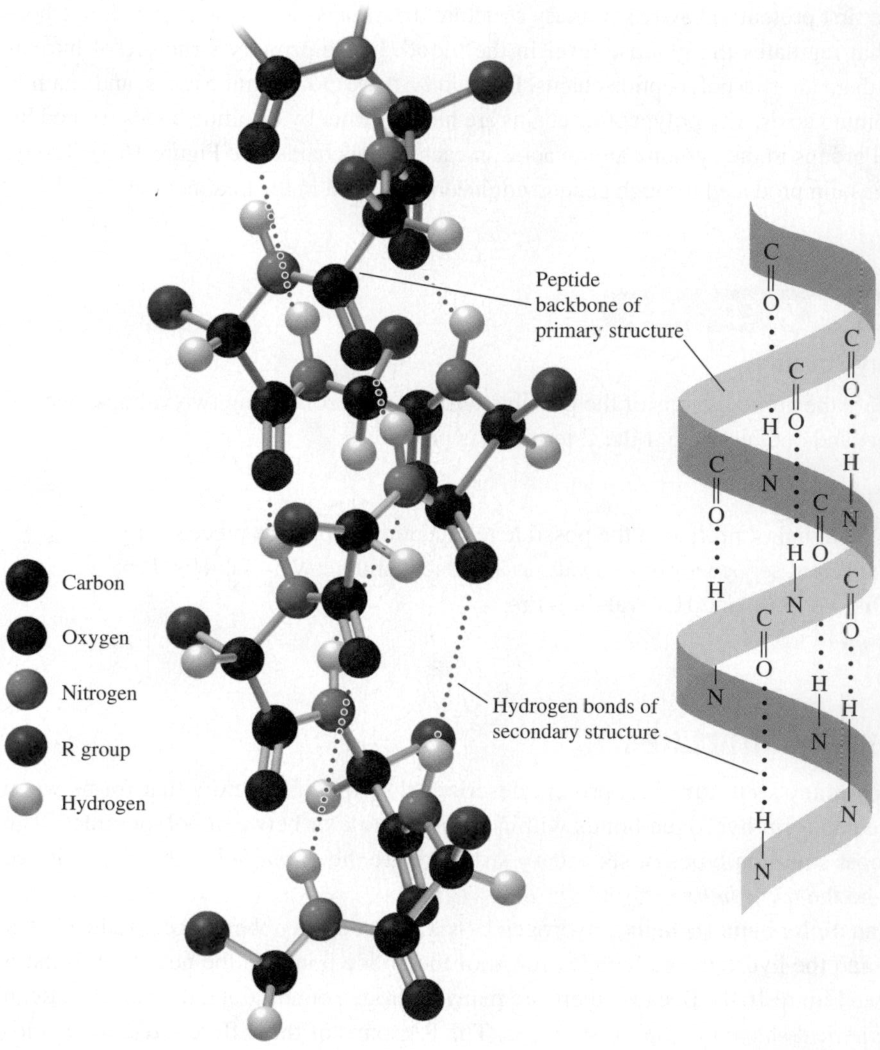

FIGURE 16.4 The α (alpha) helix acquires a coiled shape from hydrogen bonds between the N—H of the peptide bond in one loop and the C═O of the peptide bond in the next loop.
Q What are the partial charges of the H in N—H and the O in C═O that permit hydrogen bonds to form?

is a result of three α helical polypeptides woven together like a braid to form a **triple helix** (see Figure 16.6). Collagen has a high content of glycine (33%), proline (22%), alanine (12%), and smaller amounts of hydroxyproline and hydroxylysine. The hydroxy forms of proline and lysine contain —OH groups that form additional hydrogen bonds across the peptide chains to give strength to the collagen triple helix.

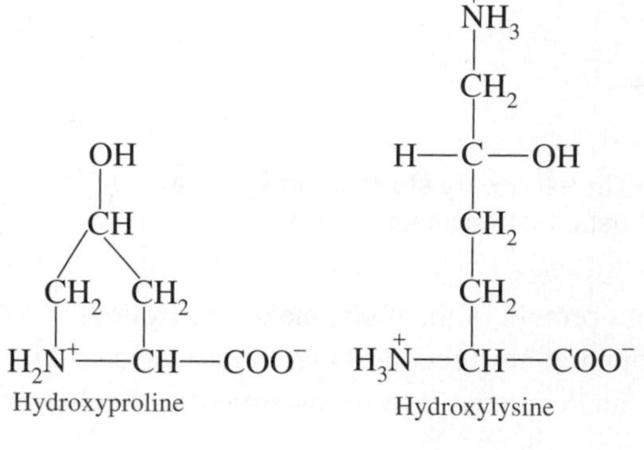

Hydroxyproline: $H_2\overset{+}{N}$—CH—COO^- (ring: CH_2–CH(OH)–CH_2)

Hydroxylysine: $H_3\overset{+}{N}$—CH—COO^- with side chain —CH_2—CH_2—CH(OH)—CH_2—$\overset{+}{N}H_3$

Hydroxyproline and hydroxylysine provide additional hydrogen bonds in the triple helices of collagen.

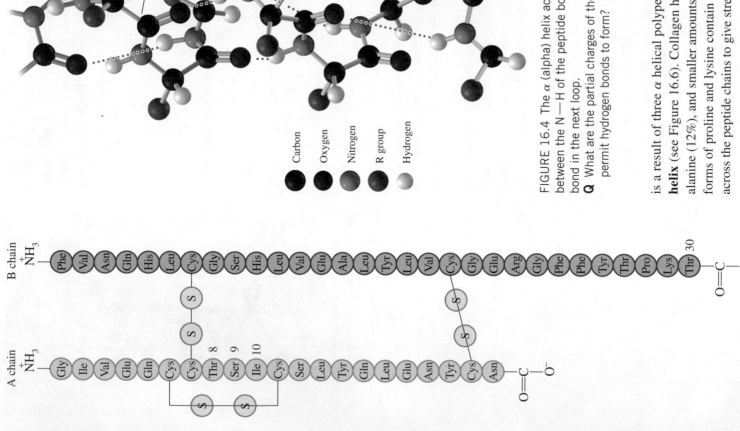

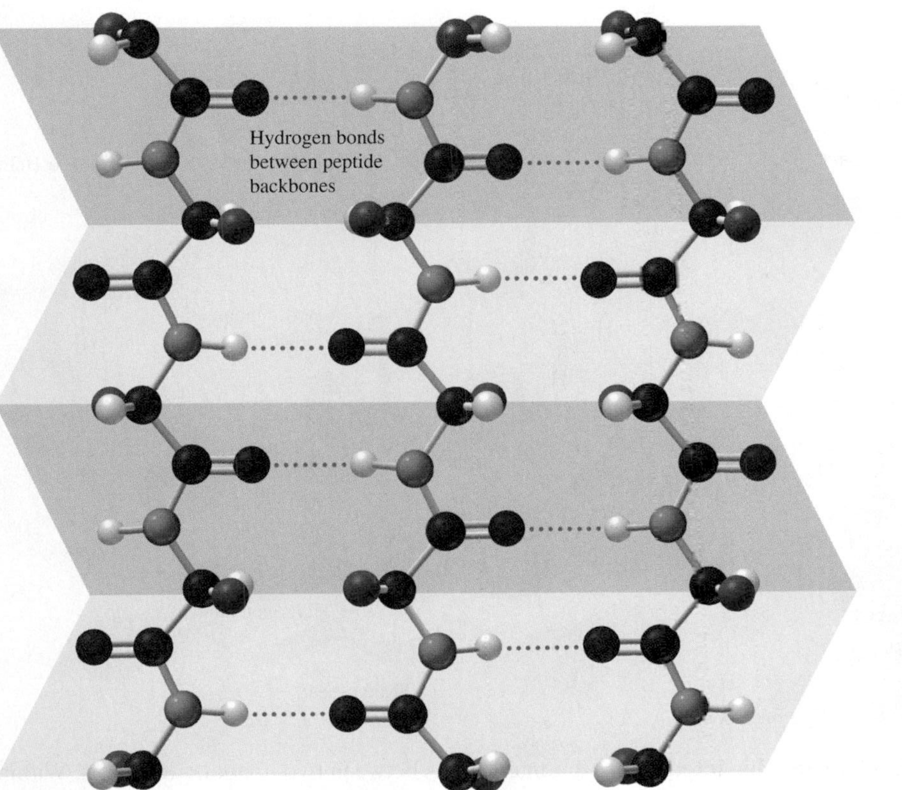

FIGURE 16.5 In a β (beta)-pleated sheet secondary structure, hydrogen bonds form between adjacent peptide chains.

Q How do the hydrogen bonds in a β-pleated sheet differ from the way hydrogen bonds form in an α helix?

When a diet is deficient in vitamin C, collagen fibrils are weakened because the enzymes needed to form hydroxyproline and hydroxylysine require vitamin C. Because there are fewer —OH groups, there is less hydrogen bonding between collagen fibrils. Collagen becomes less elastic as a person ages because additional cross-links form between the fibrils. Bones, cartilage, and tendons become more brittle, and wrinkles are seen as the skin loses elasticity.

FIGURE 16.6 Collagen fibers are triple helices of polypeptide chains held together by hydrogen bonds.

Q What are some of the amino acids in collagen that form hydrogen bonds between the polypeptide chains?

CONCEPT CHECK 16.6

Identifying Secondary Structures

Indicate the secondary structure (α helix, β-pleated sheet, or triple helix) described in each of the following statements:

a. a structure that has hydrogen bonds between adjacent polypeptide chains
b. three α helical polypeptides woven together
c. a coiled peptide chain held in place by hydrogen bonding between peptide bonds in the same polypeptide chain

ANSWER

a. β-pleated sheet **b.** triple helix **c.** α helix

Tertiary Structure

The **tertiary structure** of a protein involves attractions and repulsions between the R groups of the amino acids in the polypeptide chain. As interactions occur between different parts of the peptide chain, segments of the chain twist and bend until the protein acquires a specific three-dimensional shape. The tertiary structure of a protein is stabilized by interactions between the R groups of the amino acids in one region of the polypeptide chain and the R groups of amino acids in other regions of the protein (see Figure 16.7).

SELF STUDY ACTIVITY
Tertiary and Quaternary Structure

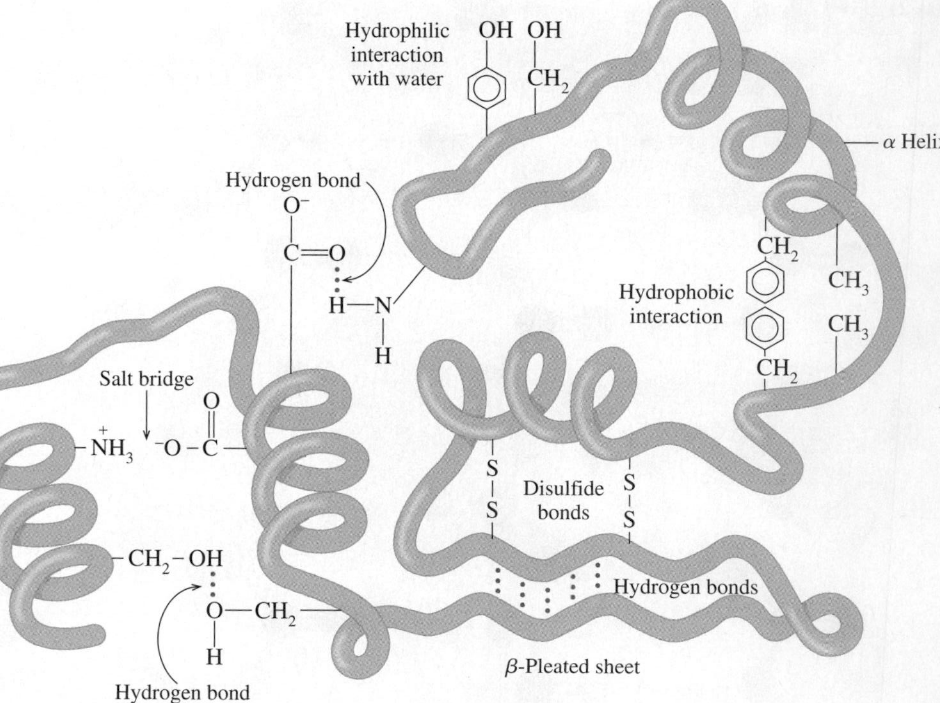

FIGURE 16.7 Interactions between amino acid R groups fold a protein into a specific three-dimensional shape called its tertiary structure.
Q Why would one section of a protein move to the center while another section remains on the surface of the tertiary structure?

1. **Hydrophobic interactions** are interactions between two nonpolar R groups. Within a protein, the amino acids with nonpolar R groups move away from the aqueous environment to form a hydrophobic center at the interior of the protein molecule.
2. **Hydrophilic interactions** are attractions between the external aqueous environment and the R groups of polar amino acids moving the polar amino acids toward the outer surface of globular proteins where they form hydrogen bonds with water.
3. **Salt bridges** are ionic bonds between ionized R groups of basic and acidic amino acids. For example, the ionized R group of arginine, which has a positive charge, can form a salt bridge (ionic bond) with the R group in aspartic acid, which has a negative charge.
4. **Hydrogen bonds** form between H of a polar R group and the O or N of another amino acid. For example, a hydrogen bond can form between the —OH groups of two serines or between the —OH of serine and the $-NH_2$ in the R group of glutamine.
5. **Disulfide bonds** (—S—S—) are covalent bonds that form between the —SH groups of cysteines in a polypeptide chain.

SAMPLE PROBLEM 16.4

Cross-Links in Tertiary Structures

What type of attraction would you expect between the R groups of each of the following?

a. cysteine and cysteine **b.** glutamic acid and lysine

SOLUTION

a. Because the R group of cysteine contains —SH, a disulfide bond could form.
b. A salt bridge can form by the attraction of the $-COO^-$ in the R group of glutamic acid and the $-NH_3^+$ in the R group of lysine.

STUDY CHECK 16.4

Would you expect to find valine and leucine on the outside or the inside of the tertiary structure? Why?

Globular and Fibrous Proteins

A group of proteins known as **globular proteins** have compact, spherical shapes because sections of the polypeptide chain fold over on top of each other due to the various interactions between R groups. It is the globular proteins that carry out the work of the cells: functions such as synthesis, transport, and metabolism.

Myoglobin is a globular protein that stores oxygen in skeletal muscle. High concentrations of myoglobin are found in the muscles of sea mammals, such as seals and whales, allowing them to stay under the water for long periods. Myoglobin contains 153 amino acids in a single polypeptide chain with about three-fourths of the chain in the α-helix secondary structure. The polypeptide chain, including its helical regions, forms a compact tertiary structure by folding upon itself (see Figure 16.8). Within the tertiary structure, a pocket of amino acids and a heme group binds and stores oxygen (O_2).

The **fibrous proteins** are proteins that consist of long, thin, fiber-like shapes. They are typically involved in the structure of cells and tissues. Two types of fibrous protein are the α- and β-keratins. The α-keratins are the proteins that make up hair, wool, skin, and nails. In hair, three α-helices coil together like a braid to form a fibril. Within the fibril, the α helices are held together by disulfide (—S—S—) linkages between the thiol groups of the many cysteine amino acids in hair. Several fibrils bind together to form a strand of hair (see Figure 16.9). The β-keratins are the type of proteins found in the feathers of birds and scales of reptiles. In β-keratins, the proteins consist of large amounts of β-pleated sheet structure.

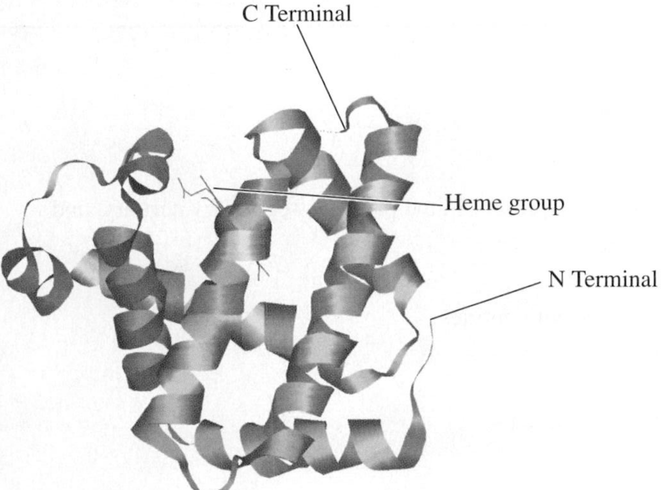

FIGURE 16.8 The ribbon structure represents the tertiary structure of the polypeptide chain of myoglobin. The globular protein contains a heme group that binds oxygen to be carried to the tissues.

Q Would hydrophilic amino acids be found on the outside or inside of the myoglobin structure?

FIGURE 16.9 The fibrous proteins of α-keratin wrap together to form fibrils of hair and wool.

Q Why does hair have a large amount of cysteine?

Quaternary Structure: Hemoglobin

When a biologically active protein consists of two or more polypeptide chains or subunits, the structural level is referred to as a **quaternary structure**. Hemoglobin, a globular protein that transports oxygen in blood, consists of four polypeptide chains or

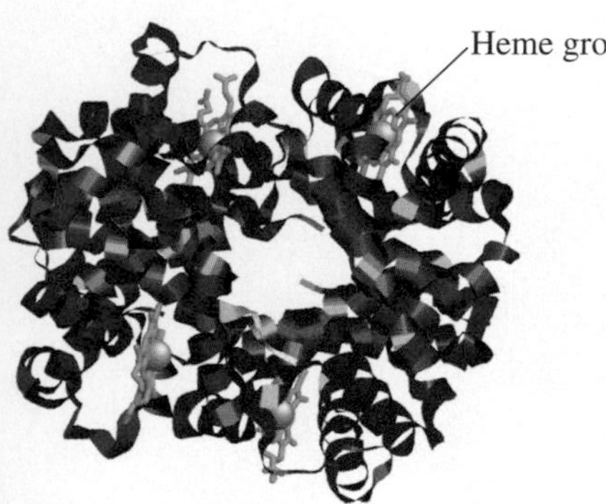

FIGURE 16.10 In the ribbon structure of hemoglobin, the quaternary structure is made up of four polypeptide subunits, two (red) are α chains and two (blue) are β chains. The heme groups (green) in the four subunits bind oxygen.

Q What is the difference between a tertiary structure and a quaternary structure?

subunits: two α chains, and two β chains (see Figure 16.10). In the quaternary structure, the subunits are held together by the same interactions that stabilize tertiary structures, such as hydrogen bonds, salt bridges, disulfide links, and hydrophobic interactions between R groups. Each subunit of the hemoglobin contains a heme group that binds oxygen. In the adult hemoglobin molecule, all four subunits ($\alpha_2\beta_2$) *must* be combined for hemoglobin to properly function as an oxygen carrier. Therefore, the complete quaternary structure of hemoglobin can bind and transport four molecules of oxygen.

Hemoglobin and myoglobin have similar biological functions. Hemoglobin carries oxygen in the blood, whereas myoglobin carries oxygen in muscle. Myoglobin, a single polypeptide chain with a molar mass of 17 000 g/mole, has about one-fourth the molar mass of hemoglobin (67 000 g/mole). The tertiary structure of the single polypeptide myoglobin is almost identical to the tertiary structure of each of the subunits of hemoglobin. Myoglobin stores just one molecule of oxygen, just as each subunit of hemoglobin carries one oxygen molecule. The similarity in tertiary structures allows each protein to bind and release oxygen in a similar manner. Table 16.6 and Figure 16.11 summarize the structural levels of proteins.

TABLE 16.6 Summary of Structural Levels in Proteins

Structural Level	Characteristics
Primary	The sequence of amino acids
Secondary	The coiled α helix, β-pleated sheet, or a triple helix form by hydrogen bonding between peptide bonds along the chain
Tertiary	A protein folds into a compact, three-dimensional shape stabilized by interactions between R groups of amino acids
Quaternary	Two or more protein subunits combine to form a biologically active protein

TUTORIAL
Levels of Structure in Proteins

CONCEPT CHECK 16.7

Structures of Proteins

Indicate which of the following are present in the primary, secondary, tertiary, and quaternary structures of proteins:

a. peptide bonds
b. hydrogen bonds between adjacent peptides
c. hydrogen bonds within a single peptide
d. hydrophobic interactions
e. association of four polypeptide chains

ANSWER

a. Peptide bonds are present at all levels of protein structures.
b. Hydrogen bonds between adjacent peptides occur in the secondary structures such as β-pleated sheets, triple helices, and in quaternary structures.
c. Hydrogen bonds within a single peptide is a characteristic of the secondary structure of an α helix as well as in tertiary structures.
d. Hydrophobic interactions between two nonpolar R groups occur in the tertiary and quaternary structures of proteins.
e. The association of four polypeptide chains occurs in the quaternary structures of proteins.

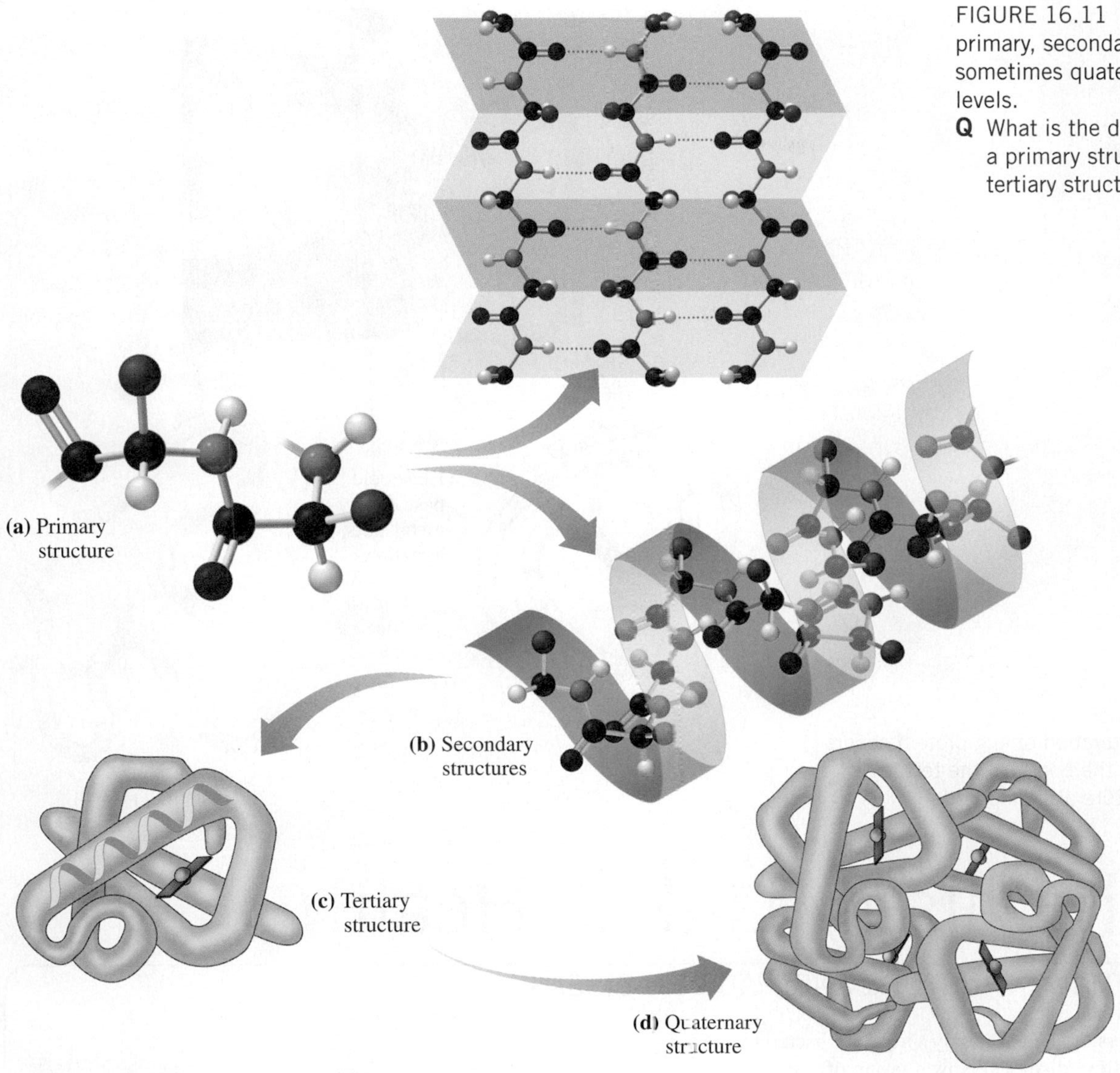

FIGURE 16.11 Proteins consist of primary, secondary, tertiary, and sometimes quaternary structural levels.

Q What is the difference between a primary structure and a tertiary structure?

SAMPLE PROBLEM 16.5

Identifying Protein Structure

Indicate whether the following conditions are responsible for primary, secondary, tertiary, or quaternary protein structures:

a. Disulfide bonds form between portions of a protein chain.

b. Peptide bonds form a chain of amino acids.

SOLUTION

a. Disulfide bonds help to stabilize the tertiary structure of a protein.

b. The sequence of amino acids in a polypeptide is a primary structure.

STUDY CHECK 16.5

What structural level is represented by the interactions of the two subunits in insulin?

TUTORIAL
Protein Demolition

TUTORIAL
Understanding Protein Degradation

Denaturation of Proteins

Denaturation of a protein occurs when there is a change that disrupts the interactions between R groups that stabilize the secondary, tertiary, or quaternary structure. However, the covalent amide bonds of the primary structure are not affected.

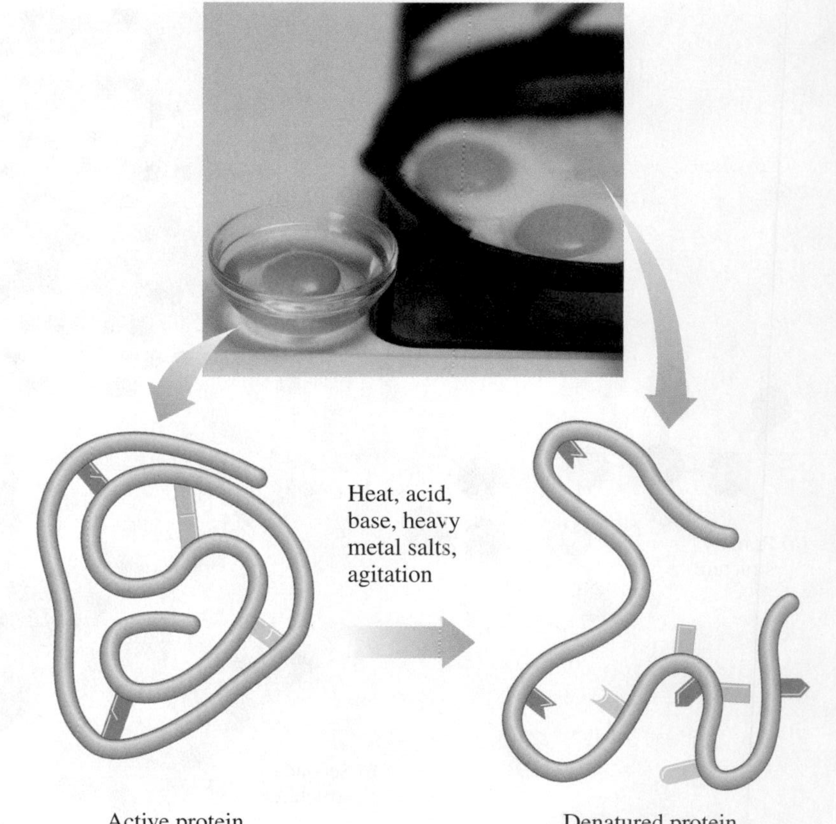

Denaturation of egg protein occurs when the bonds of the tertiary structure are disrupted.

Chemistry Link to Health

PROTEIN STRUCTURE AND MAD COW DISEASE

Until recently, researchers thought that only viruses or bacteria were responsible for transmitting diseases. Now a group of diseases has been found in which the infectious agents are proteins called *prions*. Bovine spongiform encephalopathy (BSE), or "mad cow disease," is a fatal brain disease of cattle in which the brain fills with cavities resembling a sponge. In the noninfectious form of the prion PrP^{c}, the N-terminal portion is a random coil. Although the noninfectious form may be ingested from meat products, its structure can change to what is known as PrP^{sc} or *prion-related protein scrapie*. In this infectious form, the end of the peptide chain folds into a β-pleated sheet, which has disastrous effects on the brain and spinal cord. The conditions that cause this structural change are not yet known.

BSE was diagnosed in Great Britain in 1986. The protein is present in nerve tissue of cattle, but it is not found in their meat. Control measures that exclude brain and spinal cord from animal feed are now in place to reduce the incidence of BSE.

The human variant of this disease is called Creutzfeldt–Jakob (CJD) disease. Around 1955, Dr. Carleton Gajdusek was studying the Fore tribe in Papua New Guinea, where many tribe members were dying of neurological disease known as "kuru." Among the Fore, it was a custom to cannibalize members of the tribe upon their death. Gajdusek eventually determined that this practice was responsible for transmitting the infectious agent from one tribe member to another.

After Gajdusek identified the infection agent in kuru as similar to the prions that cause BSE, he received the Nobel Prize in Physiology or Medicine in 1976.

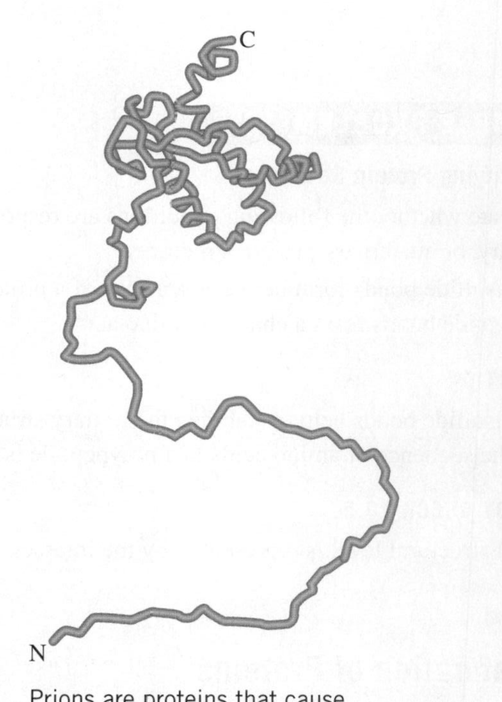

Prions are proteins that cause infections such as "mad cow disease."

TABLE 16.7 Examples of Protein Denaturation

Denaturing Agent	Bonds Disrupted	Examples
Heat above 50 °C	Hydrogen bonds; hydrophobic interactions between nonpolar R groups	Cooking food and autoclaving surgical items
Acids and bases	Hydrogen bonds between polar R groups; salt bridges	Lactic acid from bacteria, which denatures milk protein in the preparation of yogurt and cheese
Organic compounds	Hydrophobic interactions	Ethanol and isopropyl alcohol, which disinfect wounds and prepare the skin for injections
Heavy metal ions Ag^+, Pb^{2+}, and Hg^{2+}	Disulfide bonds in proteins by forming ionic bonds	Mercury and lead poisoning
Agitation	Hydrogen bonds and hydrophobic interactions by stretching polypeptide chains and disrupting stabilizing interactions	Whipped cream, meringue made from egg whites

The loss of secondary and tertiary structures occurs when conditions change, such as increasing the temperature or making the pH very acidic or basic. If the pH changes, the basic and acidic R groups lose their ionic charges and cannot form salt bridges, which causes a change in the shape of the protein. Denaturation can also occur by adding certain organic compounds or heavy metal ions or by mechanical agitation (see Table 16.7). When the interactions between the R groups are disrupted, a globular protein unfolds like a loose piece of spaghetti. With the loss of its overall shape (tertiary structure), the protein is no longer biologically active.

CONCEPT CHECK 16.8

Denaturation of Proteins

Describe the denaturation process in each of the following:

a. An appetizer known as ceviche is prepared without heat by placing slices of raw fish in a solution of lemon or lime juice. After 3 or 4 hours, the fish appears to be "cooked."

b. In baking sliced potatoes and milk to prepare scalloped potatoes, the milk curdles (forms solids).

ANSWER

a. The acids in lemon or lime juice break down the hydrogen bonds between polar R groups and disrupt salt bridges, which denature the proteins of the fish.

b. The heat during baking breaks apart hydrogen bonds and hydrophobic bonds between nonpolar R groups. When the milk denatures, the proteins become insoluble and form solids called curds.

Explore Your World

DENATURATION OF MILK PROTEIN

Place some milk in five glasses. Add the following to the milk samples in glasses 1 to 4. The fifth glass of milk is a reference sample.

1. Vinegar, drop by drop. Stir.
2. One-half teaspoon of meat tenderizer. Stir.
3. One teaspoon of fresh pineapple juice. (Canned juice has been heated and cannot be used.)
4. One teaspoon of fresh pineapple juice after the juice is heated to boiling.

QUESTIONS

1. How did the appearance of the milk change in each of the samples?
2. What enzyme is listed on the package label of the tenderizer?
3. How does the effect of the heated pineapple juice compare with that of the fresh juice? Explain.
4. Why is cooked pineapple used when making gelatin (a protein) desserts?

QUESTIONS AND PROBLEMS

Levels of Protein Structure

16.19 Three peptides each contain one molecule of valine and two molecules of serine. Use their three-letter abbreviations to write the possible tripeptides they can form.

16.20 How can two proteins with exactly the same number and type of amino acids have different primary structures?

16.21 What happens to the primary structure of a protein when a protein forms a secondary structure?

16.22 What are three different types of secondary protein structure?

16.23 What is the difference in bonding between an α helix and a β-pleated sheet?

16.24 In an α helix, how does bonding occur between the amino acids in the polypeptide chain?

16.25 What type of interaction would you expect between the R groups of the following amino acids in a tertiary structure?

a. cysteine and cysteine **b.** aspartic acid and lysine
c. serine and aspartic acid **d.** leucine and leucine

16.26 What type of interaction would you expect between the R groups of the following amino acids in a quaternary structure?
a. phenylalanine and isoleucine
b. aspartic acid and histidine
c. asparagine and tyrosine
d. alanine and proline

16.27 A portion of a polypeptide chain contains the following sequence of amino acids:

—Leu—Val—Cys—Asp—

a. Which amino acids can form a disulfide cross-link?
b. Which amino acids are likely to be found on the inside of the protein structure? Why?
c. Which amino acids would be found on the outside of the protein? Why?
d. How does the primary structure of a protein affect its tertiary structure?

16.28 In myoglobin, about one-half of the 153 amino acids have nonpolar R groups.
a. Where would you expect those amino acids to be located in the tertiary structure?
b. Where would you expect the polar R groups to be in the tertiary structure?
c. Why is myoglobin more soluble in water than silk or wool?

16.29 State whether the following statements describe primary, secondary, tertiary, or quaternary protein structure:
a. R groups interact to form disulfide bonds or ionic bonds.
b. Peptide bonds join amino acids in a polypeptide chain.
c. Adjacent chains of polypeptides are held together by hydrogen bonds between the O of the carbonyl group of one chain and the H of an amide bond in another chain.
d. Hydrogen bonding between amino acids in the same polypeptide gives a coiled shape to the protein.

16.30 State whether the following statements describe primary, secondary, tertiary, or quaternary protein structure:
a. Hydrophobic R groups seeking a nonpolar environment move toward the inside of the folded protein.
b. Protein chains of collagen form a triple helix.
c. An active protein contains four tertiary subunits.
d. In sickle-cell anemia, valine replaces glutamic acid in the β chain.

16.31 Indicate the changes in protein structure for each of the following:
a. An egg placed in water at 100 °C is soft boiled in about 3 minutes.
b. Prior to giving an injection, the skin is wiped with an alcohol swab.
c. Surgical instruments are placed in a 120 °C autoclave.
d. During surgery, a wound is closed by cauterization (heat).

16.32 Indicate the changes in protein structure for each of the following:
a. Tannic acid is placed on a burn.
b. Milk is heated to 60 °C to make yogurt.
c. To avoid spoilage, seeds are treated with a solution of $HgCl_2$.
d. Hamburger is cooked at high temperatures to destroy *E. coli* bacteria that may cause intestinal illness.

Chemistry Link to Health

SICKLE-CELL ANEMIA

Sickle-cell anemia is a disease caused by an abnormality in the shape of one of the subunits of the hemoglobin protein. In the β chain, the sixth amino acid, glutamic acid, which is polar acidic, is replaced by valine, a nonpolar amino acid.

Because valine has a nonpolar R group, it is attracted to the nonpolar regions within the beta hemoglobin chains. The affected red blood cells change from a rounded shape to a crescent shape, like a sickle, which interferes with their ability to transport adequate quantities of oxygen. Hydrophobic interactions also cause sickle-cell hemoglobin molecules to stick together. They form insoluble fibers of sickle-cell hemoglobin that clog capillaries, where they cause inflammation, pain, and organ damage. Critically low oxygen levels may occur in the affected tissues.

Polar acidic amino acid

Normal β chain: Val–His–Leu–Thr–Pro–[Glu]–Glu–Lys–

Sickled β chain: Val–His–Leu–Thr–Pro–[Val]–Glu–Lys–

Nonpolar amino acid

In sickle-cell anemia, both genes for the altered hemoglobin must be inherited. However, a few sickled cells are found in persons who carry one gene for sickle-cell hemoglobin, a condition that is also known to provide protection from malaria.

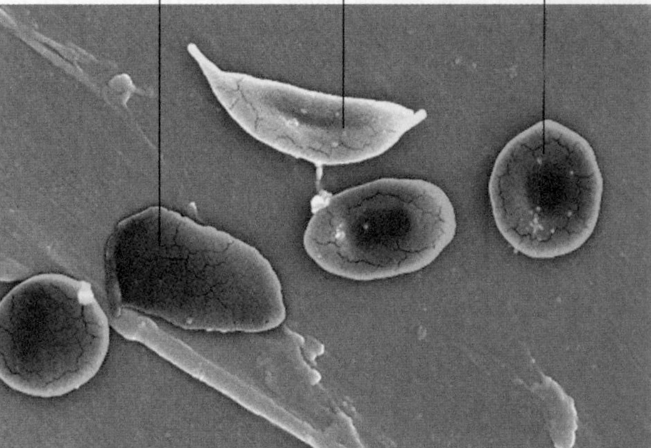

16.5 Enzymes

LEARNING GOAL

Describe how enzymes function as catalysts and give their names.

TUTORIAL
Enzymes and Activation Energy

Biological catalysts known as **enzymes** are needed for most chemical reactions that take place in the body. As we discussed in Chapter 5, a catalyst increases the reaction rate by changing the way a reaction takes place but is itself not changed at the end of the reaction. An uncatalyzed reaction in a cell may take place eventually, but not at a rate fast enough for survival. For example, the hydrolysis of proteins in our diet would eventually occur without a catalyst, but not fast enough to meet the body's requirements for amino acids. The chemical reactions in our cells must occur at incredibly fast rates under the mild conditions of pH 7.4 and a body temperature of 37 °C.

As catalysts, enzymes lower the activation energy for a chemical reaction (see Figure 16.12). Less energy is required to convert reactant molecules to products, which increases the rate of a biochemical reaction compared to the rate of the uncatalyzed reaction. For example, an enzyme in the blood called carbonic anhydrase converts carbon dioxide and water to carbonic acid. In 1 minute, one molecule of carbonic anhydrase catalyzes the reaction of about 1 million molecules of carbon dioxide.

$$CO_2 + H_2O \xrightarrow{\text{Carbonic anhydrase}} H_2CO_3$$

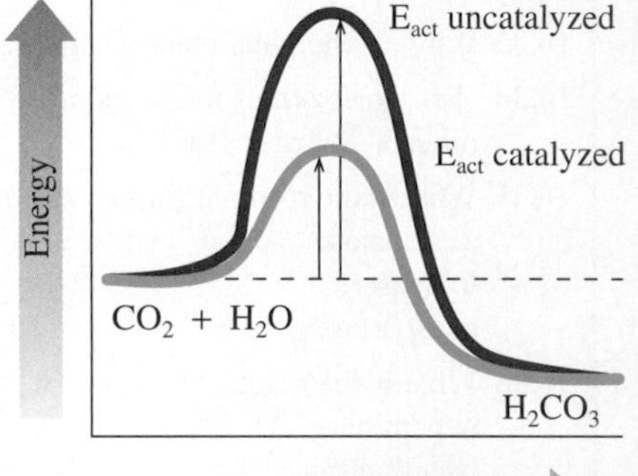

FIGURE 16.12 The enzyme carbonic anhydrase lowers the activation energy needed for the reaction of CO_2 and H_2O.

Q Why are enzymes used in biological reactions?

Names and Classification of Enzymes

The names of enzymes describe the compound or the reaction that is catalyzed. The actual names of enzymes are derived by replacing the end of the name of the reaction or reacting compound with the suffix *ase*. For example, an *oxidase* catalyzes an oxidation reaction, and a *dehydrogenase* removes hydrogen atoms. The compound sucrose is hydrolyzed by the enzyme *sucrase*, and a lipid is hydrolyzed by a *lipase*. Some early known enzymes use names that end in the suffix *in*, such as *papain* found in papaya, *rennin* found in milk, and *pepsin* and *trypsin*, enzymes that catalyze the hydrolysis of proteins.

More recently, a systematic method of classifying and naming enzymes has been established. The name and class of each indicates the type of reaction it catalyzes. There are six main classes of enzymes, as described in Table 16.8.

TABLE 16.8 Classes of Enzymes

Class	Reaction Catalyzed	Examples
Oxidoreductases	Oxidation–reduction reactions	*Oxidases* oxidize a substance. *Reductases* reduce a substance. *Dehydrogenases* remove 2H atoms to form a double bond.
Transferases	Transfer a group between two compounds	*Transaminases* transfer amino groups. *Kinases* transfer phosphate groups.
Hydrolases	Hydrolysis reactions	*Proteases* hydrolyze peptide bonds in proteins. *Lipases* hydrolyze ester bonds in lipids. *Carbohydrases* hydrolyze glycosidic bonds in carbohydrates. *Phosphatases* hydrolyze phosphate-ester bonds. *Nucleases* hydrolyze nucleic acids.
Lyases	Add or remove groups involving a double bond without hydrolysis	*Carboxylases* add CO_2. *Deaminases* remove NH_3.
Isomerases	Rearrange atoms in a molecule to form an isomer	*Isomerases* convert cis to trans isomers or trans to cis isomers. *Epimerases* convert D to L isomers or L to D.
Ligases	Form bonds between molecules using ATP energy	*Synthetases* combine two molecules.

SAMPLE PROBLEM 16.6

Naming Enzymes

What chemical reaction would each of the following enzymes catalyze?

a. aminotransferase

b. lactate dehydrogenase

SOLUTION

a. the transfer of an amino group

b. the removal of hydrogen from lactate

STUDY CHECK 16.6

What is the name of the enzyme that catalyzes the hydrolysis of lipids?

QUESTIONS AND PROBLEMS

Enzymes

16.33 Why do chemical reactions in the body require enzymes?

16.34 How do enzymes make chemical reactions in the body proceed at faster rates?

16.35 What is the reactant for each of the following enzymes?
- **a.** galactase
- **b.** lipase
- **c.** aspartase

16.36 What is the reactant for each of the following enzymes?
- **a.** peptidase
- **b.** cellulase
- **c.** lactase

16.37 What is the name of the class of enzymes that would catalyze each of the following reactions?
- **a.** hydrolysis of sucrose
- **b.** addition of oxygen
- **c.** converting glucose ($C_6H_{12}O_6$) to fructose ($C_6H_{12}O_6$)
- **d.** moving an amino group from one molecule to another

16.38 What is the name of the class of enzymes that would catalyze each of the following reactions?
- **a.** addition of water to a double bond
- **b.** removing hydrogen atoms
- **c.** splitting peptide bonds in proteins
- **d.** removing CO_2 from pyruvate

LEARNING GOAL

Describe the role of an enzyme in an enzyme-catalyzed reaction.

16.6 Enzyme Action

Nearly all enzymes are globular proteins. Each has a unique three-dimensional shape that recognizes and binds a small group of reacting molecules, which are called **substrates**. The tertiary structure of an enzyme plays an important role in how that enzyme catalyzes reactions.

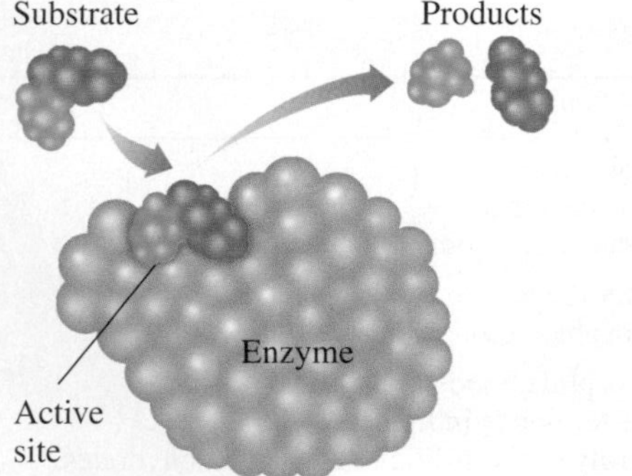

FIGURE 16.13 On the surface of an enzyme, a small region called an active site binds a substrate and catalyzes a reaction of that substrate.
Q Why does an enzyme catalyze a reaction of only certain substrates?

Active Site

In a catalyzed reaction, an enzyme must first bind to a substrate in a way that favors catalysis. A typical enzyme is much larger than its substrate. However, within an enzyme's large tertiary structure is a region called the **active site**, where the enzyme binds one or more substrates and catalyzes the reaction. This active site is often a small pocket that closely fits the shape of the substrate (see Figure 16.13).

Within the active site, the R groups of amino acids interact with the functional groups of the substrate to form hydrogen bonds, salt bridges, or hydrophobic interactions. The active site of a particular enzyme fits the shape of only a few types of substrates, which makes an enzyme very specific about the type of substrate it binds.

Enzyme-Catalyzed Reaction

The proper alignment of a substrate within the active site forms an **enzyme–substrate (ES) complex**. This combination of enzyme and a substrate provides an alternative pathway for the reaction that has a lower activation energy. Within the active site, the amino acid R groups take part in catalyzing the chemical reaction. For example, acidic and basic

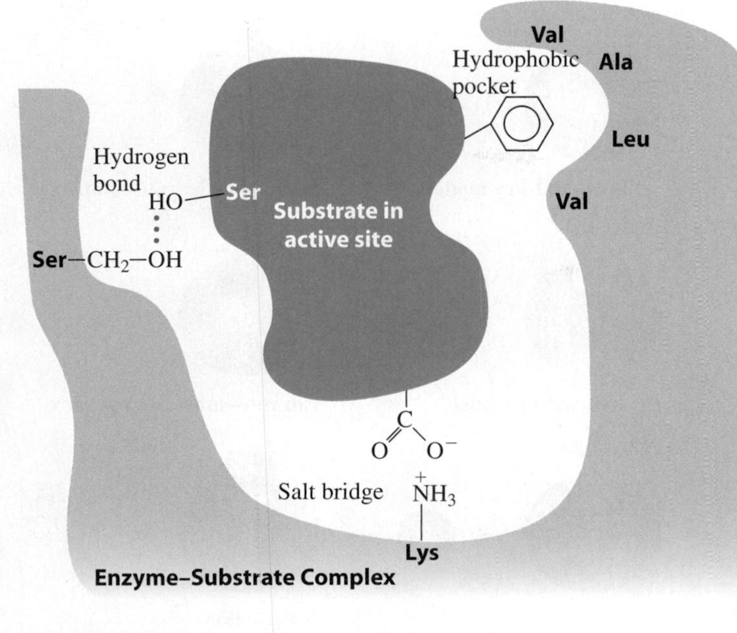

Binding of a substrate occurs when it interacts with the amino acids within the active site.

R groups remove protons from or provide protons for the substrate. As soon as the catalyzed reaction is complete, the products are released from the enzyme so it can bind to another substrate molecule.

We can write the catalyzed reaction of an enzyme (E) with a substrate (S) to form product (P) as follows:

$$\textbf{Step 1}\quad E + S \rightleftharpoons ES$$
$$\textbf{Step 2}\quad ES \longrightarrow E + P$$
$$\underset{\text{Enzyme + substrate}}{E + S} \rightleftharpoons \underset{\text{ES complex}}{ES} \longrightarrow \underset{\text{Enzyme + product}}{E + P}$$

Let's consider the hydrolysis of sucrose by sucrase. When a molecule of sucrose binds to the active site of sucrase, its glycosidic bond is placed in a position favorable for reaction. The amino acid R groups catalyze the hydrolysis of sucrose with water to give the products glucose and fructose. The products in the active site are released, and the sucrase binds another sucrose substrate (see Figure 16.14).

$$\underset{\text{Sucrase + sucrose}}{E + S} \rightleftharpoons \underset{\text{sucrase–sucrose complex}}{\text{ES complex}} \longrightarrow \underset{\text{sucrase + glucose + fructose}}{E + P_1 + P_2}$$

Lock-and-Key and Induced-Fit Models

An early theory of enzyme action, called the **lock-and-key model**, describes the active site as having a rigid, nonflexible shape. Thus, only those substrates with shapes that fit exactly into the active site are able to bind with that enzyme. The shape of the active site is analogous to a lock, and the proper substrate is the key that fits into the lock (see Figure 16.15a).

While the lock-and-key model explains the binding of substrates for many enzymes, certain enzymes have a broader range of activity than the lock-and-key model allows. In the **induced-fit model**, the active site adjusts to fit the shape of the substrate more closely (see Figure 16.15b). At the same time, the substrate adjusts its shape to better adapt to the geometry of the active site. As a result, the reacting section of the substrate becomes properly aligned with the groups in the active site that catalyze the reaction.

In the induced-fit model, substrate and enzymes work together to acquire a geometrical arrangement that lowers the activation energy. A different substrate could not induce these changes, and no catalysis would occur (see Figure 16.15c).

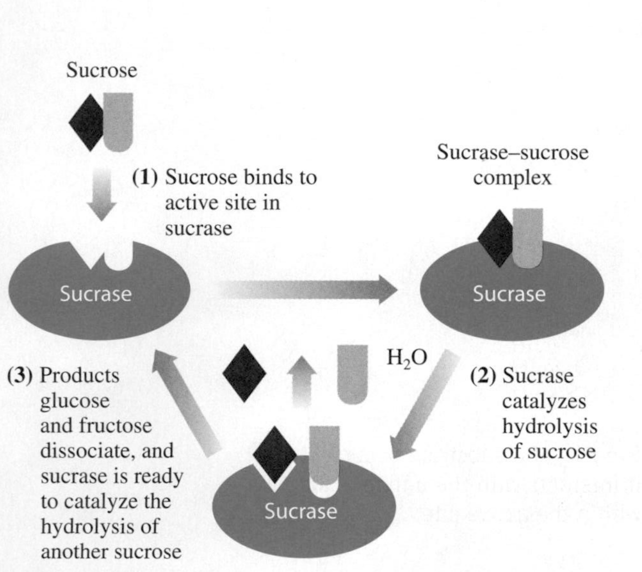

FIGURE 16.14 At the active site, sucrose is aligned for the hydrolysis reaction. The monosaccharides produced dissociate from the active site, and the enzyme is ready to bind to another sucrose molecule.

Q Why does the enzyme-catalyzed hydrolysis of sucrose go faster than the hydrolysis of sucrose in the chemistry laboratory?

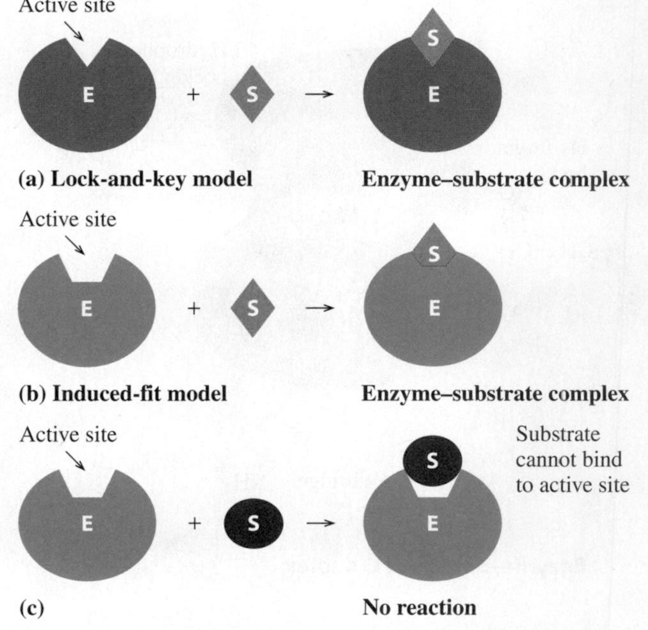

FIGURE 16.15 **(a)** In the lock-and-key model, a substrate fits the shape of the active site and forms an enzyme–substrate complex. **(b)** In the induced-fit model, a flexible active site and substrate adjust shape to provide the best fit for the reaction. **(c)** A substrate that does not fit or induce a fit in the active site cannot undergo catalysis by the enzyme.

Q How does the induced-fit model differ from the lock-and-key model?

CONCEPT CHECK 16.9

Enzyme Action

Compare the lock-and-key and induced-fit models in how they are similar and how they differ in their description of the enzyme action.

ANSWER

In both models, an enzyme catalyzes a reaction by binding a substrate to the R groups within the active site. However, in the lock-and-key model, the shape of a substrate must fit the shape of the active site exactly. In the induced-fit model, the substrate and the active site both adjust their shapes to provide the best fit.

QUESTIONS AND PROBLEMS

Enzyme Action

16.39 Match the terms, (1) enzyme–substrate complex, (2) enzyme, and (3) substrate, with each of the following:
a. has a tertiary structure that recognizes the substrate
b. the combination of an enzyme with the substrate
c. has a structure that fits the active site of an enzyme

16.40 Match the terms, (1) active site, (2) lock-and-key model, and (3) induced-fit model, with each of the following:
a. the portion of an enzyme where catalytic activity occurs
b. an active site that adapts to the shape of a substrate
c. an active site that has a rigid shape

16.41 a. Write an equation that represents an enzyme-catalyzed reaction.
b. How is the active site different from the whole enzyme structure?

16.42 a. How does an enzyme speed up the reaction of a substrate?
b. After the products have formed, what happens to the enzyme?

16.43 What are isoenzymes?

16.44 How is the LDH isoenzyme in the heart different from LDH isoenzyme in the liver? (Refer to the Chemistry Link to Health "Isoenzymes as Diagnostic Tools.")

16.45 A patient arrives in an emergency department complaining of chest pains. What enzymes would you test for in the blood serum? (Refer to the Chemistry Link to Health "Isoenzymes as Diagnostic Tools.")

16.46 A patient has elevated blood serum levels of LDH and AST. What condition might be indicated? (Refer to the Chemistry Link to Health "Isoenzymes as Diagnostic Tools.")

Chemistry Link to Health

ISOENZYMES AS DIAGNOSTIC TOOLS

Isoenzymes are different forms of an enzyme that catalyze the same reaction in different cells or tissues of the body. They consist of quaternary structures with slight variations in the amino acids in the polypeptide subunits. For example, there are five isoenzymes of *lactate dehydrogenase* (*LDH*) that catalyze the conversion between lactate and pyruvate.

$$\underset{\text{Lactate}}{CH_3-\overset{\displaystyle OH}{\overset{|}{C}H}-COO^-} \underset{}{\overset{\text{Lactate dehydrogenase}}{\rightleftharpoons}} \underset{\text{Pyruvate}}{CH_3-\overset{\displaystyle O}{\overset{\|}{C}}-COO^-} + 2H$$

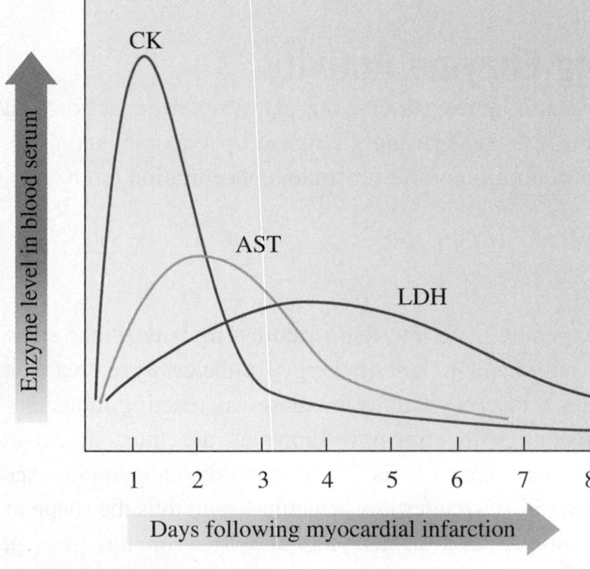

A myocardial infarction may be indicated by an increase in the levels of creatine kinase (CK), aspartate transaminase (AST), and lactate dehydrogenase (LDH).

Each LDH isoenzyme contains a mix of polypeptide subunits, M and H. In the liver and muscle, lactate is converted to pyruvate by the LDH_5 isoenzyme with four M subunits designated M_4. In the heart, the same reaction is catalyzed by the LDH_1 isoenzyme (H_4) containing four H subunits. Different combinations of the M and H subunits are found in the LDH isoenzymes of the brain, red blood cells, kidneys, and white blood cells.

The different forms of an enzyme allow a medical diagnosis of damage or disease to a particular organ or tissue. In healthy tissues, isoenzymes function within the cells. However, when a disease damages a particular organ, cells die, which releases their contents including the isoenzymes into the blood. Measurements of the elevated levels of specific isoenzymes in the blood serum help to identify the disease and its location in the body. For example, an elevation in serum LDH_5, which is the M_4 isoenzyme of lactate dehydrogenase, indicates liver damage or disease. When a myocardial infarction (MI) or heart attack damages heart muscle, an increase in the level of LDH_1 (H_4) isoenzyme is detected in the blood serum (see Table 16.9).

Isoenzymes of lactate dehydrogenase	**Highest levels found in**
H M	
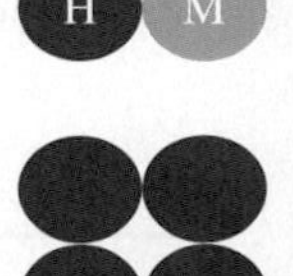 H_4 (LDH_1)	Heart, kidneys
 H_3M (LDH_2)	Red blood cells, heart, kidney, brain
 H_2M_2 (LDH_3)	Brain, lung, white blood cells
 HM_3 (LDH_4)	Lung, skeletal muscle
 M_4 (LDH_5)	Skeletal muscle, liver

The different isoenzymes of lactate dehydrogenase (LDH) indicate damage to different organs in the body.

TABLE 16.9 Isoenzymes of Lactate Dehydrogenase and Creatine Kinase

Isoenzyme	Abundant in	Subunits
Lactate Dehydrogenase (LDH)		
LDH_1	Heart, kidneys	H_4
LDH_2	Red blood cells, heart, kidney, brain	H_3M
LDH_3	Brain, lung, white blood cells	H_2M_2
LDH_4	Lung, skeletal muscle	HM_3
LDH_5	Skeletal muscle, liver	M_4
Creatine Kinase (CK)		
CK_1	Brain, lung	BB
CK_2	Heart muscle	MB
CK_3	Skeletal muscle, red blood cells	MM

Another isoenzyme used diagnostically is creatine kinase (CK), which consists of two types of polypeptide subunits. Subunit B is prevalent in the brain, and subunit M predominates in muscle. Normally CK_3 (subunits MM) is present at low levels in the blood serum. However, in a patient who has suffered a myocardial infarction, the level of CK_2 (subunits MB) is elevated within 4 to 6 h and reaches a peak in about 24 h. Table 16.10 lists some enzymes used to diagnose tissue damage and diseases of certain organs.

TABLE 16.10 Serum Enzymes Used in Diagnosis of Tissue Damage

Condition	Diagnostic Enzymes Elevated
Heart attack or liver disease (cirrhosis, hepatitis)	Lactate dehydrogenase (LDH) Aspartate transaminase (AST)
Heart attack	Creatine kinase (CK)
Hepatitis	Alanine transaminase (ALT)
Liver (carcinoma) or bone disease (rickets)	Alkaline phosphatase (ALP)
Pancreatic disease	Pancreatic amylase (PA), cholinesterase (CE), lipase (LPS)
Prostate carcinoma	Acid phosphatase (ACP) Prostate specific antigen (PSA)

LEARNING GOAL

Describe the effect of temperature, pH, concentration of substrate, and inhibitors on enzyme activity.

16.7 Factors Affecting Enzyme Activity

The **activity** of an enzyme describes how fast an enzyme catalyzes the reaction that converts a substrate to product. This activity is strongly affected by reaction conditions, which include temperature, pH, concentration of the substrate, concentration of enzyme, and the presence of inhibitors.

Temperature

Thermophiles survive in the high temperatures in a hot spring.

Enzymes are very sensitive to temperature. At low temperatures, most enzymes show little activity because there is not a sufficient amount of energy for the catalyzed reaction to take place. At higher temperatures, enzyme activity increases as reacting molecules move faster to cause more collisions with enzymes. Enzymes are most active at **optimum temperature**, which for most enzymes is 37 °C or body temperature (see Figure 16.16). At temperatures above 50 °C, the tertiary structure—and thus the shape of most proteins—is destroyed, causing a loss in enzyme activity. For this reason, equipment in hospitals and laboratories is sterilized in autoclaves, where the high temperatures denature the enzymes in harmful bacteria.

Certain organisms, known as thermophiles, live in environments where temperatures range from 50 °C to 80 °C. In order to survive in these extreme conditions, thermophiles must have enzymes with tertiary structures that are not destroyed by such high temperatures. Some research shows that their enzymes are very similar to ordinary enzymes except they contain more arginine and tyrosine. These slight changes allow the enzymes in thermophiles to form more hydrogen bonds and salt bridges that stabilize the tertiary structures at high temperatures and resist unfolding and the loss of enzymatic activity.

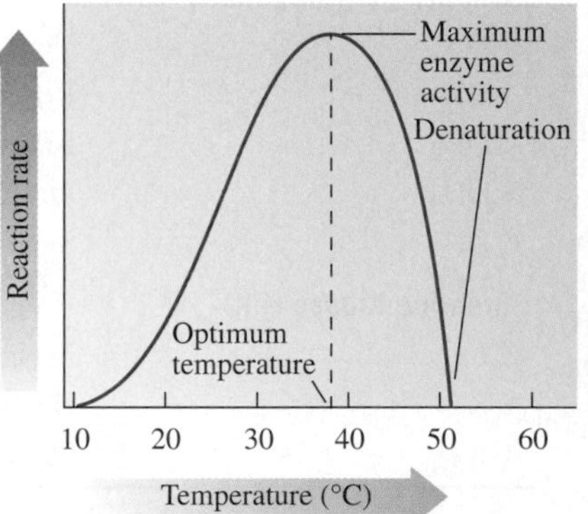

FIGURE 16.16 An enzyme attains maximum activity at its optimum temperature, usually 37 °C. Lower temperatures slow the rate of reaction, and temperatures above 50 °C denature most enzymes with a loss of catalytic activity.

Q Why is 37 °C the optimum temperature for many enzymes?

TUTORIAL
Enzyme and Substrate Concentrations

pH

Enzymes are most active at their **optimum pH**, the pH that maintains the proper tertiary structure of the protein (see Figure 16.17). If a pH value is above or below the optimum pH, the R group interactions are disrupted, which destroys the tertiary structure and the active site. As a result, the enzyme no longer binds substrate, and no catalysis occurs. Small changes in pH are reversible, which allows an enzyme to regain its structure and activity. However, large variations from optimum pH permanently destroy the structure of the enzyme.

Enzymes in most cells have optimum pH values around 7.4. However, enzymes in the stomach have a low optimum pH because they hydrolyze proteins at the acidic pH in the stomach. For example, pepsin, a digestive enzyme in the stomach, has an optimum pH of 1.5–2.0. Between meals, the pH in the stomach is 4 or 5 and pepsin shows little or no digestive activity. When food enters the stomach, the secretion of HCl lowers the pH to about 2, which activates pepsin. Table 16.11 lists the optimum pH values for selected enzymes.

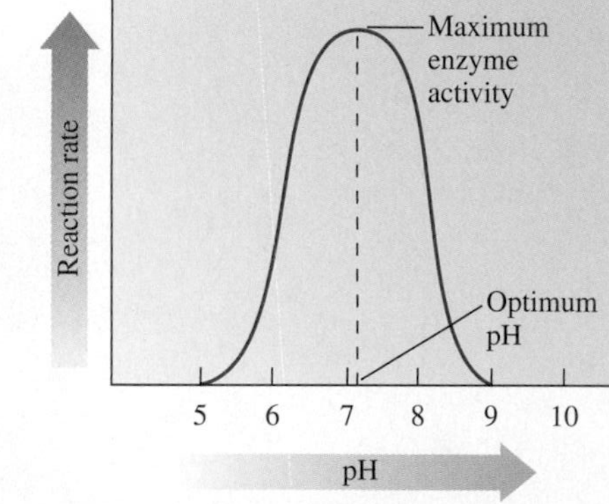

FIGURE 16.17 Enzymes are most active at their optimum pH. At a higher or lower pH, denaturation of the enzyme causes a loss of catalytic activity.

Q Why does the digestive enzyme pepsin have an optimum pH of 2?

TABLE 16.11 Optimum pH for Selected Enzymes

Enzyme	Location	Substrate	Optimum pH
Pepsin	Stomach	Peptide bonds	1.5–2.0
Sucrase	Small intestine	Sucrose	6.2
Pancreatic amylase	Pancreas	Amylose	6.7–7.0
Urease	Liver	Urea	7.0
Trypsin	Small intestine	Peptide bonds	7.7–8.0
Lipase	Pancreas	Lipid (ester bonds)	8.0
Arginase	Liver	Arginine	9.7

Explore Your World

ENZYME ACTIVITY

The enzymes on the surface of a freshly cut apple, avocado, or banana react with oxygen in the air to turn the surface brown. An antioxidant, such as vitamin C in lemon juice, prevents the oxidation reaction. Cut an apple, an avocado, or a banana into several slices. Place one slice in a plastic zipper lock bag, squeeze out all the air, and close the zipper lock. Dip another slice in lemon juice and place it on a plate. Sprinkle another slice with a crushed vitamin C tablet. Leave another slice alone as a control. Observe the surface of each of your samples. Record your observations immediately, then every hour for 6 hours or longer.

QUESTIONS

1. Which slice(s) shows the most oxidation (turns brown)?
2. Which slice(s) shows little or no oxidation?
3. How was the oxidation reaction on each slice affected by treatment with an antioxidant?

Enzyme and Substrate Concentration

In any catalyzed reaction, the substrate must first bind with the enzyme to form the enzyme–substrate complex. For a particular substrate concentration, an increase in enzyme concentration increases the rate of the catalyzed reaction. At higher enzyme concentrations, more molecules are available to bind and catalyze the reaction. As long as the substrate concentration is greater than the enzyme concentration, there is a direct relationship between the enzyme concentration and enzyme activity (see Figure 16.18a). In most enzyme-catalyzed reactions, the concentration of the substrate is much greater than the concentration of the enzyme.

When enzyme concentration is kept constant, increasing the substrate concentration increases the rate of the catalyzed reaction until the substrate saturates the enzyme. With all the available enzyme molecules bonded to substrate, the rate of the catalyzed reaction reaches its maximum. Adding more substrate molecules cannot increase the rate further (see Figure 16.18b).

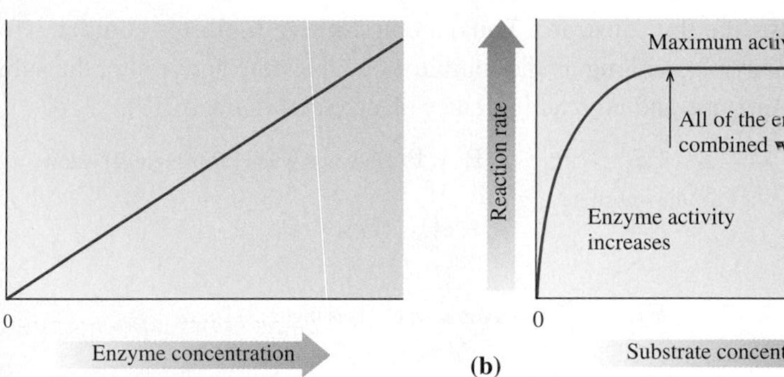

FIGURE 16.18 **(a)** Increasing enzyme concentration increases the rate of reaction. **(b)** Increasing substrate concentration increases the rate of reaction until the enzyme molecules are saturated.

Q What happens to the rate of reaction when the substrate saturates the enzyme?

CONCEPT CHECK 16.10

Enzyme Activity

Describe how each of the following affects the activity of an enzyme:

a. decreasing the pH from the optimum pH
b. increasing the temperature from the optimum temperature
c. increasing the substrate concentration at constant temperature and pH

ANSWER

a. A more acidic environment disrupts the hydrogen bonds and salt bridges of the tertiary structure, which causes a loss of enzymatic activity.
b. When the temperature is increased above the optimum temperature, the tertiary structure breaks down (denaturation), the shape of the active site deteriorates, and enzyme activity is lost.
c. An increase in a substrate concentration increases the rate of reaction until all the enzyme has combined with substrate. Then the reaction rate of an enzyme continues at a constant rate.

TUTORIAL
Enzyme Inhibition

TUTORIAL
Enzyme and Substrate Concentrations

SAMPLE PROBLEM 16.7

Factors Affecting Enzymatic Activity

Describe what effect the changes in parts **a** and **b** would have on the rate of the reaction catalyzed by urease.

$$H_2N-\overset{\overset{\displaystyle O}{\|}}{C}-NH_2 + H_2O \xrightarrow{\text{Urease}} 2NH_3 + CO_2$$

Urea

a. increasing the urea concentration
b. lowering the temperature to 10 °C

SOLUTION

a. An increase in urea concentration will increase the rate of reaction until all the enzyme molecules are bound to the urea substrate.
b. Because 10 °C is lower than the optimum temperature of 37 °C, the lower temperature will decrease the rate of the reaction.

STUDY CHECK 16.7

If urease has an optimum pH of 7.0, what is the effect of lowering the pH to 3?

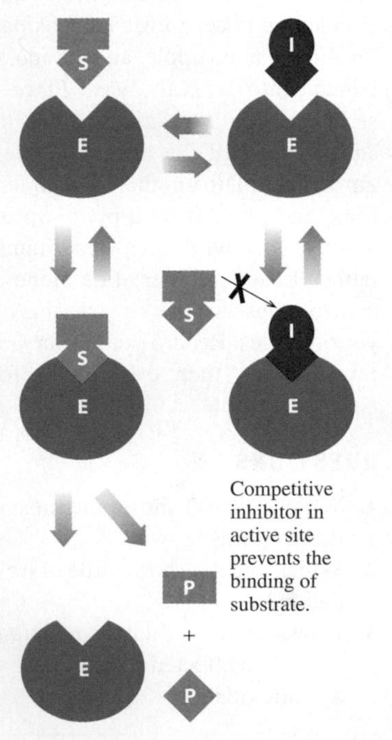

FIGURE 16.19 With a structure similar to the substrate for an enzyme, a competitive inhibitor also fits the active site and competes with the substrate when both are present.
Q Why does increasing the substrate concentration reverse the inhibition by a competitive inhibitor?

Enzyme Inhibition

Many kinds of molecules called **inhibitors** cause enzymes to lose catalytic activity. Although inhibitors act differently, they all prevent the active site from binding with a substrate. An enzyme with a reversible inhibitor can regain enzymatic activity, but an enzyme attached to an irreversible inhibitor loses enzymatic activity permanently.

A **competitive inhibitor** has a structure that is so similar to the substrate it can bond to the enzymes just like the substrate. Thus, a competitive inhibitor competes for the active site on the enzyme. As long as the inhibitor occupies the active site, the substrate cannot bind to the enzyme and no reaction takes place (see Figure 16.19).

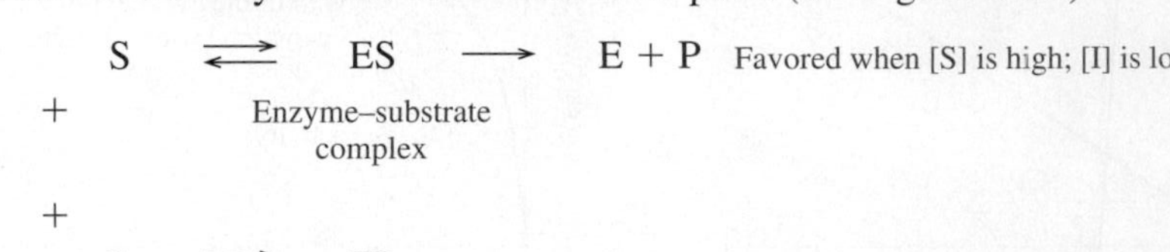

$$E + S \rightleftarrows \underset{\text{Enzyme–substrate complex}}{ES} \longrightarrow E + P \quad \text{Favored when [S] is high; [I] is low}$$

$$E + I \rightleftarrows \underset{\text{Enzyme–inhibitor complex}}{EI} \quad \text{Favored when [I] is high; [S] is low}$$

As long as the concentration of the inhibitor is substantial, there is a loss of enzyme activity. However, adding more substrate displaces the competitive inhibitor. As more enzyme molecules bind to substrate (ES), enzyme activity is regained.

The structure of a **noncompetitive inhibitor** does not resemble the substrate and does not compete for the active site. Instead, a noncompetitive inhibitor binds to a site on the enzyme that is not the active site. When the noncompetitive inhibitor is bonded to the enzyme, the shape of the enzyme is distorted. Inhibition occurs because the substrate cannot fit in the active site or it does not fit properly. Without the proper alignment of substrate with the amino acid side groups, no catalysis can take place (see Figure 16.20).

Because a noncompetitive inhibitor is not competing for the active site, the addition of more substrate does not reverse this type of inhibition. Examples of noncompetitive inhibitors are the heavy metal ions Pb^{2+}, Ag^{+}, and Hg^{2+} that bond with amino acid side groups such as $—COO^-$ or —OH. Catalytic activity is restored when chemical reagents remove the inhibitors.

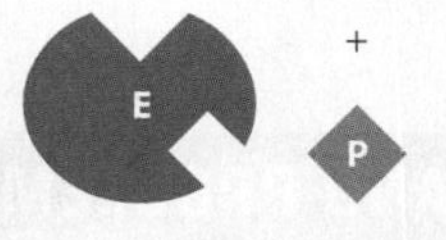

FIGURE 16.20 A noncompetitive inhibitor binds to an enzyme at a site other than the active site, which distorts the enzyme and prevents the proper binding and catalysis of the substrate at the active site.

Q Will an increase in substrate concentration reverse the inhibition by a noncompetitive inhibitor?

Irreversible Inhibition

In irreversible inhibition, a molecule causes an enzyme to lose all enzymatic activity. Most irreversible inhibitors are toxic substances that destroy enzymes. Usually an irreversible inhibitor forms a covalent bond with an amino acid side group within the active site, which prevents the substrate from binding to the active site or prevents catalytic activity.

Insecticides and nerve gases act as irreversible inhibitors of acetylcholinesterase, an enzyme needed for nerve conduction. The compound DFP (diisopropyl fluorophosphate), an organophosphate insecticide, forms a covalent bond with the side chain $—CH_2OH$ of serine in the active site. When acetylcholinesterase is inhibited, the transmission of nerve impulses is blocked, and paralysis occurs.

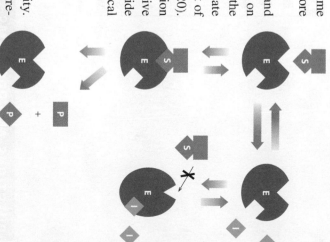

$$\text{(E)—}CH_2\text{—OH} + \text{F—P(=O)(O—CH(CH}_3)_2)_2 \longrightarrow \text{(E)—}CH_2\text{—O—P(=O)(O—CH(CH}_3)_2)_2 + \text{HF}$$

Enzyme—Serine; DFP (diisopropyl fluorophosphate); Serine covalently bonded to DFP

Antibiotics produced by bacteria, mold, or yeast are inhibitors used to stop bacterial growth. For example, penicillin inhibits an enzyme needed for the formation of cell walls in bacteria, but not human cell membranes. With an incomplete cell wall, bacteria cannot survive and the infection is stopped. However, some bacteria are resistant to penicillin because they produce penicillinase, an enzyme that breaks down penicillin. Over the years, derivatives of penicillin to which bacteria have not yet become resistant have been produced.

R Groups for Penicillin Derivatives

Penicillin (R—C(=O)—NH— attached to the β-lactam ring fused to the thiazolidine ring with S, two CH_3, and COO^-)

Penicillin G	Penicillin V	Ampicillin	Amoxicillin
$C_6H_5—CH_2—$	$C_6H_5—O—CH_2—$	$C_6H_5—CH(NH_2)—$	$HO—C_6H_4—CH(NH_2)—$

SAMPLE PROBLEM 16.8

Enzyme Inhibition

State the type of inhibition in the following:

a. The inhibitor has a structure that is similar to the substrate.

b. This inhibitor binds to the surface of the enzyme, changing its shape in such a way that it cannot bind to substrate.

SOLUTION

a. competitive inhibition

b. noncompetitive inhibition

STUDY CHECK 16.8

What type of enzyme inhibition can be reversed by adding more substrate?

QUESTIONS AND PROBLEMS

Factors Affecting Enzyme Activity

16.47 Trypsin, a peptidase that hydrolyzes polypeptides, functions in the small intestine at an optimum pH of 7.7–8.0. How is the rate of a trypsin-catalyzed reaction affected by each of the following conditions?

a. lowering the concentration of polypeptides
b. changing the pH to 3.0
c. running the reaction at 75 °C
d. adding more trypsin

16.48 Pepsin, a peptidase that hydrolyzes proteins, functions in the stomach at an optimum pH of 1.5–2.0. How is the rate of a pepsin-catalyzed reaction affected by each of the following conditions?

a. increasing the concentration of proteins
b. changing the pH to 5.0
c. running the reaction at 0 °C
d. using less pepsin

16.49 The following graph shows the curves for pepsin, sucrase, and trypsin. Estimate the optimum pH for each.

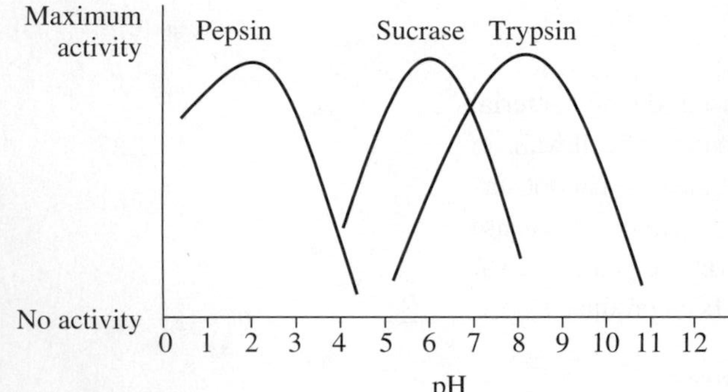

16.50 Refer to the graph in Problem 16.49 to determine if the reaction rate in each condition will be at the optimum rate or not.

a. trypsin, pH 5.0
b. sucrase, pH 5.0
c. pepsin, pH 4.0
d. trypsin, pH 8.0
e. pepsin, pH 2.0

16.51 Indicate whether each of the following describes a competitive or a noncompetitive enzyme inhibitor:

a. The inhibitor has a structure similar to the substrate.
b. The effect of the inhibitor cannot be reversed by adding more substrate.
c. The inhibitor competes with the substrate for the active site.
d. The structure of the inhibitor is not similar to the substrate.
e. The addition of more substrate reverses the inhibition.

16.52 Oxaloacetate is an inhibitor of succinate dehydrogenase.

Succinate	Oxaloacetate
COO^-	COO^-
CH_2	CH_2
CH_2	$C{=}O$
COO^-	COO^-

a. Would you expect oxaloacetate to be a competitive or a noncompetitive inhibitor? Why?
b. Would oxaloacetate bind to the active site or elsewhere on the enzyme?
c. How would you reverse the effect of the inhibitor?

16.53 Methanol and ethanol are oxidized by alcohol dehydrogenase. In methanol poisoning, ethanol is given intravenously to prevent the formation of formaldehyde that has toxic effects.

a. Draw the condensed structural formulas of methanol and ethanol.
b. Would ethanol compete for the active site or bind to a different site?
c. Would ethanol be a competitive or noncompetitive inhibitor of methanol oxidation?

16.54 In humans, the antibiotic amoxicillin (a type of penicillin) is used to treat certain bacterial infections.

a. Does the antibiotic inhibit enzymes in humans?
b. Why does the antibiotic kill bacteria, but not humans?

16.8 Enzyme Cofactors

LEARNING GOAL

Describe the types of cofactors found in enzymes.

TUTORIAL
Enzyme Cofactors and Vitamins

Enzymes known as **simple enzymes** consist only of proteins. However, many enzymes require small molecules or metal ions called **cofactors** to catalyze reactions properly. When the cofactor is a small organic molecule, it is known as a **coenzyme**. If an enzyme requires a cofactor, neither the protein structure nor the cofactor alone has catalytic activity.

Forms of Active Enzymes

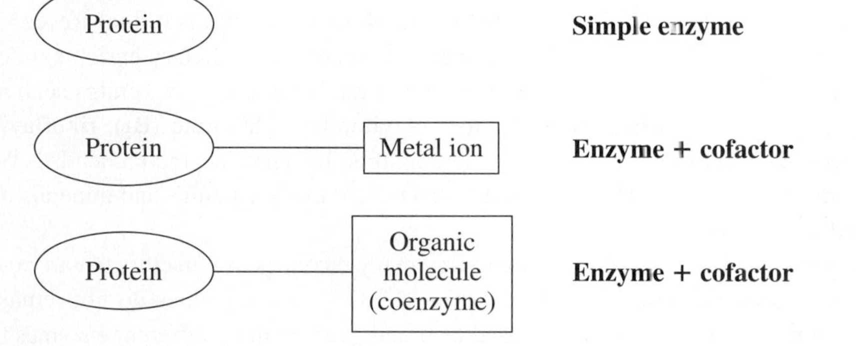

Metal Ions

Many enzymes must contain a metal ion to carry out their catalytic activity. The metal ions are bonded to one or more of the amino acid R groups. The metal ions from the minerals that we obtain from foods in our diet have various functions in catalysis. Ions such as Fe^{2+} and Cu^{2+} are used by oxidases because they lose or gain electrons in oxidation and reduction reactions. Other metals ions such as Zn^{2+} stabilize the amino acid R groups during hydrolysis reactions. Some metal cofactors required by enzymes are listed in Table 16.12.

TABLE 16.12 Enzymes and the Metal Ions Required as Cofactors

Metal Ion Cofactor	Function	Enzyme
Cu^{+}/Cu^{2+}	Oxidation–reduction	Cytochrome oxidase
Fe^{2+}/Fe^{3+}	Oxidation–reduction	Catalase Cytochrome oxidase
Zn^{2+}	Used with NAD^{+}	Alcohol dehydrogenase Carbonic anhydrase Carboxypeptidase A
Mg^{2+}	Hydrolyzes phosphate esters	Glucose-6-phosphatase
Mn^{2+}	Oxidation	Arginase
Ni^{2+}	Hydrolyzes amides	Urease

Vitamins and Coenzymes

Vitamins are organic molecules that are essential for normal health and growth. They are required in trace amounts and must be obtained from the diet because they are not synthesized in the body. Before vitamins were discovered, it was known that lime juice prevented the disease scurvy in sailors and that cod liver oil could prevent rickets. In 1912, scientists found that, in addition to carbohydrates, fats, and proteins, certain other factors called vitamins must be obtained from the diet.

Vitamins are classified into two groups by solubility: water-soluble and fat-soluble. **Water-soluble vitamins** have polar groups such as —OH and —COOH, which make them soluble in the aqueous environment of the cells. The **fat-soluble vitamins** are nonpolar compounds that are soluble in the fat (lipid) components of the body such as fat deposits and cell membranes.

FIGURE 16.21 Oranges, lemons, peppers, and tomatoes contain vitamin C, or ascorbic acid.
Q What happens to the excess vitamin C that may be consumed in a day?

Most water-soluble enzymes are not stored in the body, and excess amounts are eliminated in the urine each day. Therefore, the water-soluble vitamins must be in the foods of our daily diets (see Figure 16.21). Because many water-soluble vitamins are easily destroyed by heat, oxygen, and ultraviolet light, care must be taken in food preparation, processing, and storage. In the 1940s, a Committee on Food and Nutrition of the National Research Council began to recommend dietary enrichment of cereal grains. It was known that refining grains such as wheat caused a loss of vitamins. Thiamine (B_1), riboflavin (B_2), niacin (B_3), and iron were in the first group of added nutrients recommended. We now see the Recommended Daily Allowance (RDA) for many vitamins and minerals on most food product labels.

The water-soluble vitamins are required by many enzymes as cofactors to carry out certain aspects of catalytic action (see Table 16.13). The coenzymes do not remain bonded to a particular enzyme but are used over and over again by different enzymes to facilitate an enzyme-catalyzed reaction (see Figure 16.22). Thus, only small amounts of coenzymes are required in the cells.

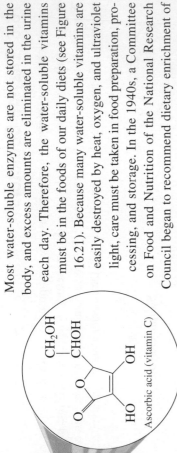

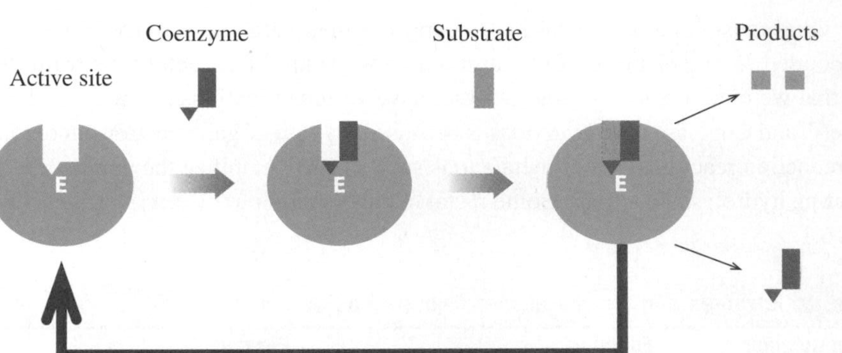

FIGURE 16.22 The active forms of many enzymes require the combination of the protein with a coenzyme.
Q What is the function of water-soluble vitamins in enzymes?

TABLE 16.13 Vitamins and Function

Water-Soluble Vitamins	Coenzyme	Function
Thiamine (vitamin B_1)	Thiamine pyrophosphate	Decarboxylation
Riboflavin (vitamin B_2)	Flavin adenine dinucleotide (FAD); flavin mononucleotide (FMN)	Electron transfer
Niacin (vitamin B_3)	Nicotinamide adenine dinucleotide (NAD^+); nicotinamide adenine dinucleotide phosphate ($NADP^+$)	Oxidation–reduction
Pantothenic acid (vitamin B_5)	Coenzyme A	Acetyl group transfer
Pyridoxine (vitamin B_6)	Pyridoxal phosphate	Transamination
Cobalamin (vitamin B_{12})	Methylcobalamin	Methyl group transfer
Ascorbic acid (vitamin C)	Vitamin C	Collagen synthesis, healing of wounds
Biotin	Biocytin	Carboxylation
Folic acid	Tetrahydrofolate	Methyl group transfer
Fat-Soluble Vitamins		
Vitamin A		Formation of visual pigments; development of epithelial cells
Vitamin D		Absorption of calcium and phosphate; deposition of calcium and phosphate in bone
Vitamin E		Antioxidant; prevents oxidation of vitamin A and unsaturated fatty acids
Vitamin K		Synthesis of prothrombin for blood clotting

The fat-soluble vitamins A, D, E, and K are not involved as coenzymes, but they are important in processes such as vision, formation of bone, protection from oxidation, and proper blood clotting. Because the fat-soluble vitamins are stored in the body and not eliminated, it is possible to take too much, which could be toxic (see Figure 16.23).

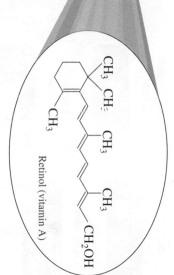

FIGURE 16.23 The orange pigment in carrots is used to synthesize vitamin A in the body.

Q Why is vitamin A a fat-soluble vitamin?

SAMPLE PROBLEM 16.9

Cofactors

Indicate whether each of the following enzymes is active as a simple enzyme or requires a cofactor:

a. a polypeptide that needs Mg^{2+} for catalytic activity
b. an active enzyme composed only of a polypeptide chain
c. an enzyme that consists of a quaternary structure attached to vitamin B_6

SOLUTION

a. The enzyme requires a cofactor.
b. An active enzyme that consists of only a polypeptide chain is a simple enzyme.
c. The enzyme requires a cofactor.

STUDY CHECK 16.9

Which of the nonprotein portions of the enzymes in Sample Problem 16.9 is a coenzyme?

QUESTIONS AND PROBLEMS

Enzyme Cofactors

16.55 Is the enzyme described in each of the following statements a simple enzyme or one that requires a cofactor?
a. requires vitamin B_1 (thiamine)
b. needs Zn^{2+} for catalytic activity
c. its active form consists of two polypeptide chains

16.56 Is the enzyme described in each of the following statements a simple enzyme or one that requires a cofactor?
a. requires vitamin B_2 (riboflavin)
b. its active form is composed of 155 amino acids
c. uses Cu^{2+} during catalysis

16.57 Identify a vitamin that is a component of each of the following coenzymes:
a. coenzyme A **b.** tetrahydrofolate (THF) **c.** NAD^+

16.58 Identify a vitamin that is a component of each of the following coenzymes:
a. thiamine pyrophosphate **b.** FAD
c. pyridoxal phosphate

CONCEPT MAP

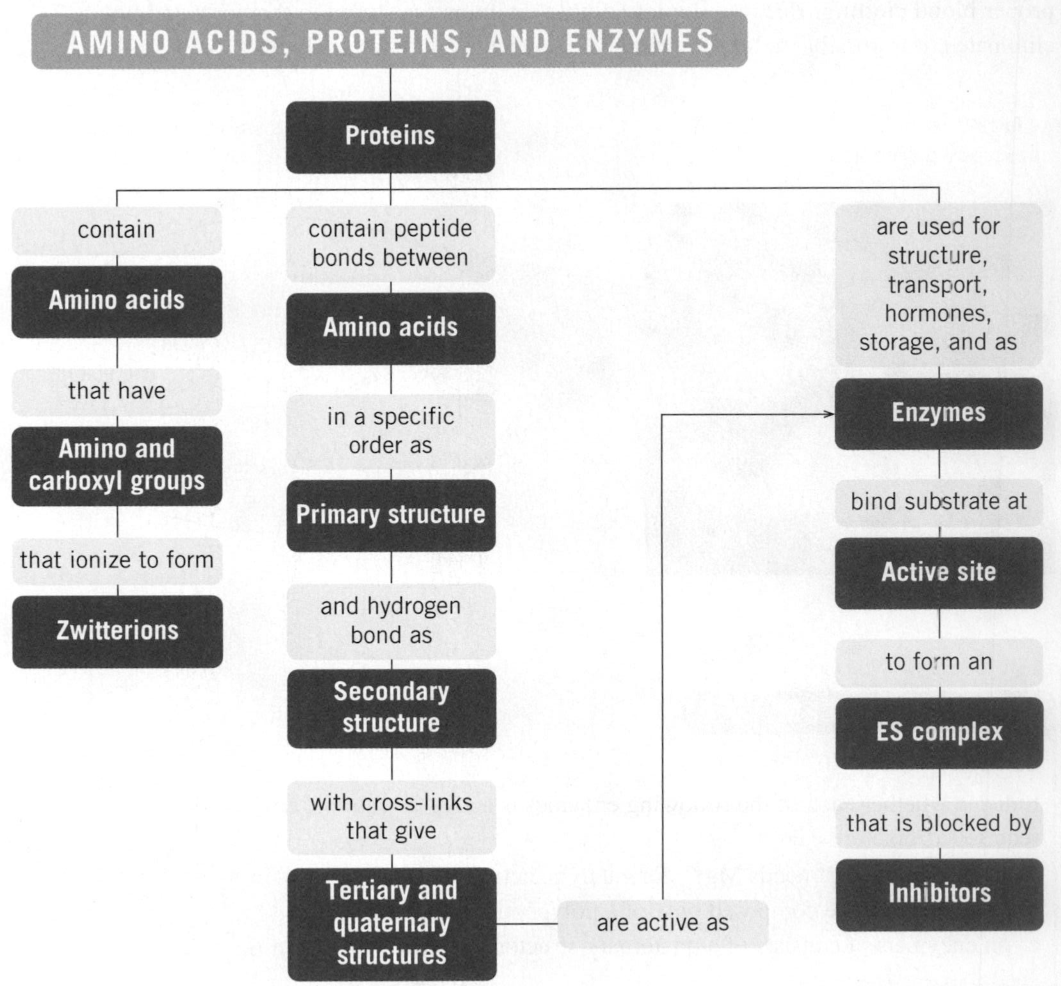

CHAPTER REVIEW

16.1 Proteins and Amino Acids

Learning Goal: Classify proteins by their functions. Give the name and abbreviation of an amino acid and draw its ionized structure.

Some proteins are enzymes or hormones, whereas others are important in structure, transport, protection, storage, and muscle contraction. A group of 20 amino acids provides the molecular building blocks of proteins. Attached to the central α(alpha)-carbon of each amino acid are an amino group, a carboxyl group, and a unique R group. The R group gives an amino acid the property of being nonpolar, polar, acidic, or basic. At physiological pH, nonpolar and polar neutral amino acids exist as dipolar ions called zwitterions.

16.2 Amino Acids as Acids and Bases

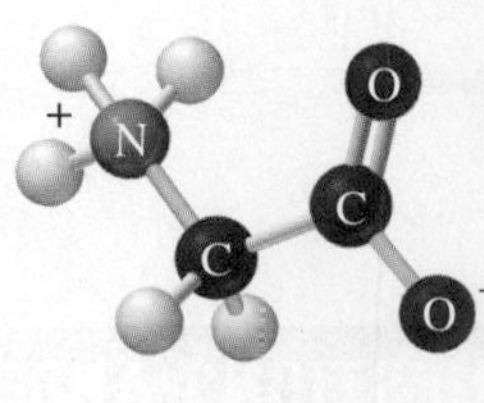

Learning Goal: Draw the ionized condensed structural formula of an amino acid at pH values above or below its pI.

Amino acids exist as positive ions at pH values below their pI values, and as negative ions at pH values above their pI values. At their isoelectric points, zwitterions have an overall charge of zero.

16.3 Formation of Peptides

Learning Goal: Draw the condensed structural formula of a dipeptide and give its name.

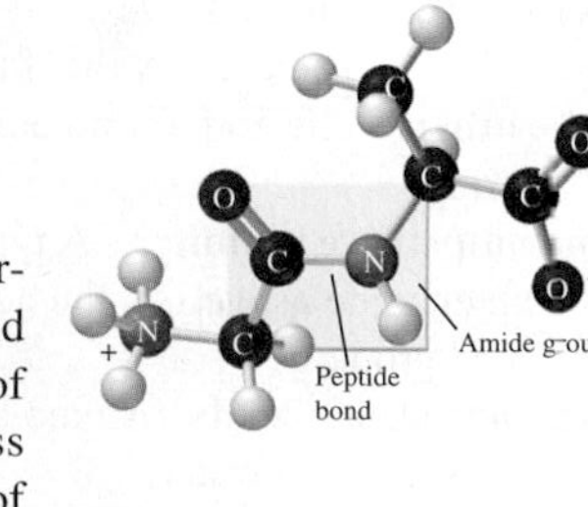

A peptide bond forms when the carboxylate group of one amino acid attaches to the ammonium group of a second amino acid with the loss of a water molecule. Long chains of amino acids are called proteins. Peptides are named from the N terminal by replacing the *ine* or *ic acid* of each amino acid name with *yl* followed by the amino acid name of the C terminal.

16.4 Levels of Protein Structure

Learning Goal: Identify the structural levels of a protein.

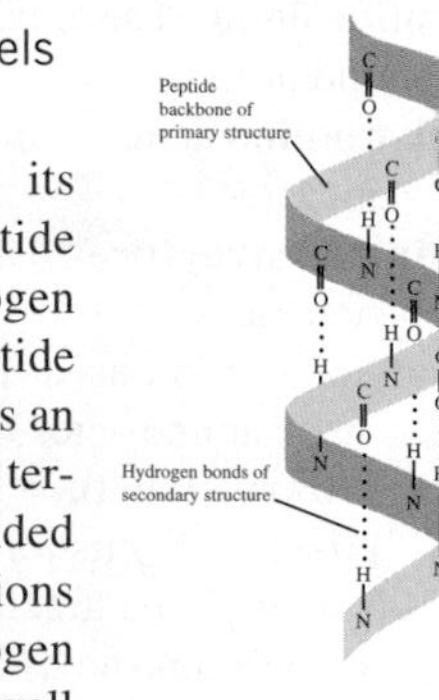

The primary structure of a protein is its sequence of amino acids joined by peptide bonds. In the secondary structure, hydrogen bonds that form between atoms in the peptide bonds produce a characteristic shape such as an α helix, β-pleated sheet, or a triple helix. A tertiary structure that is globular is a folded polypeptide chain stabilized by interactions between R groups, which form hydrogen bonds, disulfide bonds, and salt bridges as well as interactions that move hydrophobic R groups to the inside and hydrophilic R groups to the outside surface. In a quaternary structure, two or more tertiary subunits needed for biological activity are held together by the same interactions found in tertiary structures.

Denaturation of a protein occurs when high temperatures, acids or bases, organic compounds, metal ions, or agitation destroy the secondary, tertiary, or quaternary structures of a protein with a loss of biological activity.

16.5 Enzymes

Learning Goal: Describe how enzymes function as catalysts and give their names.

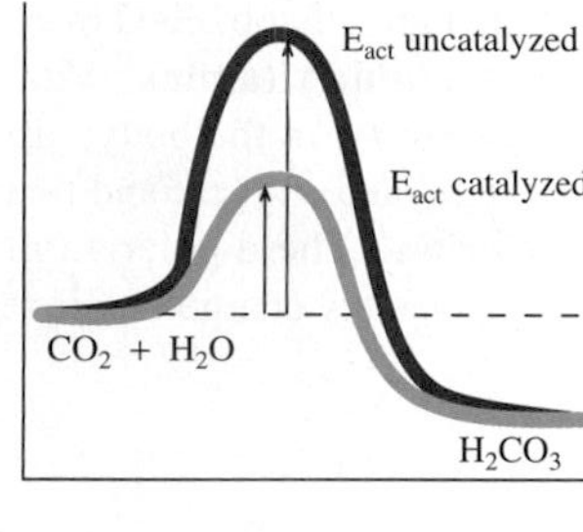

Enzymes act as biological catalysts by lowering activation energy and accelerating the rate of cellular reactions. The names of most enzymes ending in *ase* describe the compound or reaction catalyzed by the enzyme. Enzymes are classified by the main type of reaction they catalyze, such as oxidoreductase, transferase, or isomerase.

16.6 Enzyme Action

Learning Goal: Describe the role of an enzyme in an enzyme-catalyzed reaction.

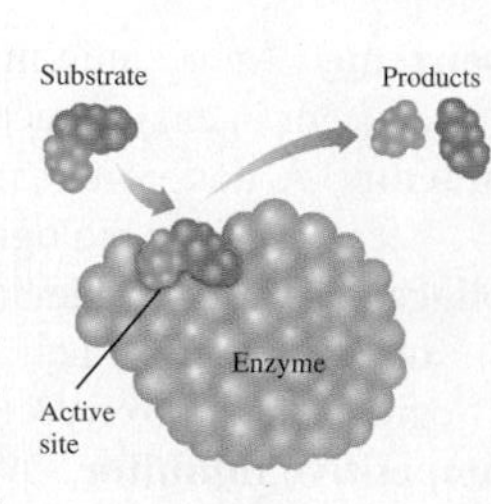

Within the tertiary structure of an enzyme, a small pocket called the active site binds the substrates. In the lock-and-key model, a substrate precisely fits the shape of the active site. In the induced-fit model, substrates induce the active site to change structure to give an optimal fit by the substrate. In the enzyme–substrate complex, catalysis takes place when amino acid R groups react with a substrate. The products are released and the enzyme is available to bind another substrate molecule.

16.7 Factors Affecting Enzyme Activity

Learning Goal: Describe the effect of temperature, pH, concentration of substrate, and inhibitors on enzyme activity.

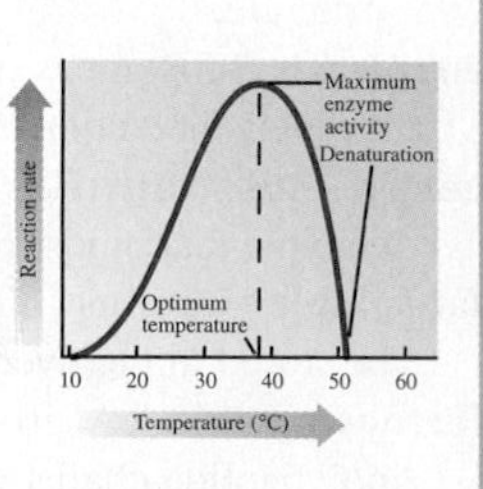

Enzymes are most effective at optimum temperature, usually 37 °C, and optimum pH, usually 7.4. The rate of an enzyme-catalyzed reaction decreases as temperature and pH values go above or below the optimum. An increase in substrate concentration increases the reaction rate of an enzyme-catalyzed reaction. If an enzyme is saturated, adding more substrate will not increase the rate further.

An inhibitor reduces the activity of an enzyme or makes it inactive. A competitive inhibitor has a structure similar to the substrate and competes for the active site. When the active site is occupied, the enzyme cannot catalyze the reaction of the substrate. A noncompetitive inhibitor attaches elsewhere on the enzyme, changing the shape of both the enzyme and its active site. As long as the noncompetitive inhibitor is attached, the altered active site cannot bind with substrate.

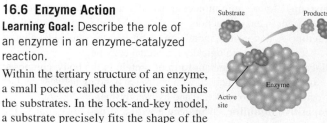

16.8 Enzyme Cofactors

Learning Goal: Describe the types of cofactors found in enzymes.

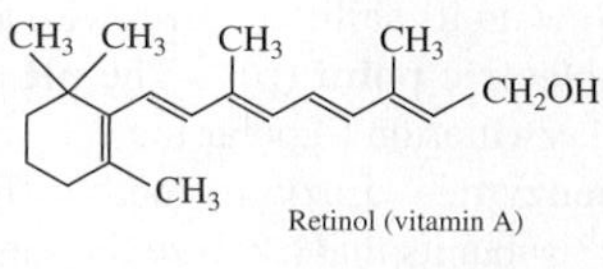

Simple enzymes are biologically active as a protein only, whereas other enzymes require small organic molecules or metal ions called cofactors. A cofactor may be a metal ion such as Cu^{2+} or Fe^{2+} or an organic molecule called a coenzyme. A vitamin is a small organic molecule needed for health and normal growth. Vitamins are obtained in small amounts through foods in the diet. The water-soluble vitamins B and C function as coenzymes.

Key Terms

acidic amino acid An amino acid that has a carboxylic acid R group (—COOH), which ionizes as a weak acid.

active site A pocket in a part of the tertiary enzyme structure that binds substrate and catalyzes a reaction.

activity The rate at which an enzyme catalyzes the reaction that converts substrate to product.

α (alpha) helix A secondary level of protein structure, in which hydrogen bonds connect the N—H of one peptide bond with the C=O of a peptide bond later in the chain to form a coiled or corkscrew structure.

amino acid The building block of proteins, consisting of an amino group, a carboxylic acid group, and a unique side group attached to the alpha carbon.

basic amino acid An amino acid that contains an amino ($—NH_2$) R group that can ionize as a weak base.

β (beta)-pleated sheet A secondary level of protein structure that consists of hydrogen bonds between peptide links in parallel polypeptide chains.

C terminal The end amino acid in a peptide chain with a free $—COO^-$ group.

coenzyme An organic molecule, usually a vitamin, required as a cofactor in enzyme action.

cofactor A nonprotein metal ion or an organic molecule that is necessary for a biologically functional enzyme.

collagen The most abundant form of protein in the body, composed of fibrils of triple helices with hydrogen bonding between —OH groups of hydroxyproline and hydroxylysine.

competitive inhibitor A molecule with a structure similar to the substrate that inhibits enzyme action by competing for the active site.

denaturation The loss of secondary and tertiary protein structure, caused by heat, acids, bases, organic compounds, heavy metals, and/or agitation.

disulfide bonds Covalent —S—S— bonds that form between the —SH groups of cysteines in a protein to stabilize the tertiary structure.

enzymes Substances including globular proteins that catalyze biological reactions.

enzyme–substrate (ES) complex An intermediate consisting of an enzyme that binds to a substrate in an enzyme-catalyzed reaction.

fat-soluble vitamins Vitamins that are not soluble in water and can be stored in the liver and body fat.

fibrous protein A protein that is insoluble in water; consists of polypeptide chains with α helices or β-pleated sheets that make up the fibers of hair, wool, skin, nails, and silk.

globular proteins Proteins that acquire a compact shape from attractions between R groups of the amino acids in the protein.

hydrogen bonds Attractions between polar R groups such as —OH, $—NH_2$, and —COOH of amino acids in a polypeptide chain.

hydrophilic interactions The attractions between water and polar R groups on the outside of the protein.

hydrophobic interactions The attractions between nonpolar R groups on the inside of a globular protein.

induced-fit model A model of enzyme action in which the shapes of the substrate and its active site are modified to give an optimal fit.

inhibitors Substances that make an enzyme inactive by interfering with its ability to react with a substrate.

isoelectric point (pI) The pH at which an amino acid exists as a zwitterion (dipolar ion).

isoenzymes Enzymes with different combinations of polypeptide subunits that catalyze the same reaction in different tissues of the body.

lock-and-key model A model of enzyme action in which the substrate is like a key that fits the specific shape of the active site.

N terminal The end amino acid in a peptide with a free $—NH_3^+$ group.

noncompetitive inhibitor A type of inhibitor that alters the shape of an enzyme as well as the active site so that the substrate cannot bind properly.

nonpolar amino acids Amino acids with nonpolar R groups that are not soluble in water.

optimum pH The pH at which an enzyme is most active.

optimum temperature The temperature at which an enzyme is most active.

peptide The combination of two or more amino acids joined by peptide bonds; dipeptide, tripeptide, and so on.

peptide bond The amide bond that joins amino acids in polypeptides and proteins.

polar amino acids Amino acids with polar R groups that are soluble in water.

primary structure The specific sequence of the amino acids in a protein.

protein A term used for biologically active polypeptides that have many amino acids linked together by peptide bonds.

quaternary structure A protein structure in which two or more protein subunits form an active protein.

salt bridge The attraction between ionized side groups of basic and acidic amino acids in the tertiary structure of a protein.

secondary structure The formation of an α helix, a β-pleated sheet, or a triple helix.

simple enzyme An enzyme that is active as a polypeptide only.

substrate The molecule that reacts in the active site in an enzyme-catalyzed reaction.

tertiary structure The folding of the secondary structure of a protein into a compact structure that is stabilized by the interactions of R groups.

triple helix The protein structure found in collagen, consisting of three polypeptide chains woven together like a braid.

vitamins Organic molecules, which are essential for normal health and growth, obtained in small amounts from the diet.

water-soluble vitamins Vitamins that are soluble in water; cannot be stored in the body; are easily destroyed by heat, ultraviolet light, and oxygen; and function as coenzymes.

zwitterion The dipolar form of an amino acid consisting of two oppositely charged ionic regions, $—NH_3^+$ and $—COO^-$.

Understanding the Concepts

16.59 Ethylene glycol ($HO—CH_2—CH_2—OH$) is a major component of antifreeze. In the body, it is first converted to HOOC—CHO (oxoethanoic acid) and then to HOOC—COOH (oxalic acid), which is toxic.

Ethylene glycol is added to a radiator to prevent freezing and boiling.

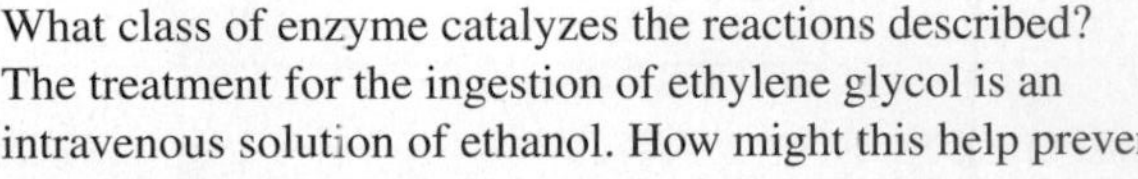

a. What class of enzyme catalyzes the reactions described?

b. The treatment for the ingestion of ethylene glycol is an intravenous solution of ethanol. How might this help prevent toxic levels of oxalic acid in the body?

16.60 Adults who are lactose intolerant cannot break down the disaccharide in milk products. To help digest dairy food, a product known as Lactaid can be added to milk and the milk then refrigerated for 24 hours.

The disaccharide lactose is present in milk products.

a. What enzyme is present in Lactaid, and what is the major class?
b. What might happen to the enzyme if the Lactaid were stored at 55 °C?

16.61 Aspartame, which is used in artificial sweeteners, contains the following dipeptide:

$$H_3\overset{+}{N}-\underset{\underset{\underset{O^{\nearrow}\!\!\!\!\!\!\!\!\!\!\quad C\;\;O^-}{|}}{CH_2}}{CH}-\overset{O}{\overset{\|}{C}}-NH-\underset{\underset{C_6H_5}{|}}{\overset{}{CH}}\!\!\!\!\quad(CH_2)-\overset{O}{\overset{\|}{C}}-O^-$$

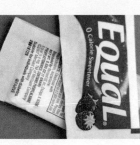

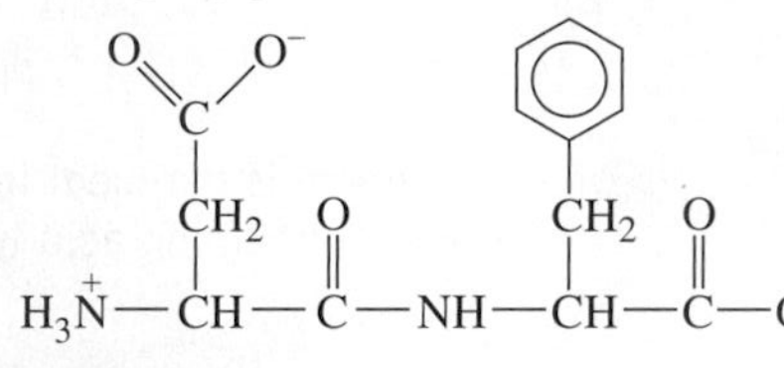

Some artificial sweeteners contain aspartame, which is an ester of a dipeptide.

a. What are the amino acids in aspartame?
b. How would you name the dipeptide in aspartame?

16.62 Fresh pineapple contains the enzyme bromelain that hydrolyzes peptide bonds in proteins.

Pineapple contains the enzyme bromelain.

a. The directions for a gelatin dessert say not to add fresh pineapple. However, canned pineapple where pineapple is heated to high temperatures can be added. Why?
b. Fresh pineapple is used in a marinade to tenderize tough meat. Why?
c. What structural level of a protein does the bromelain enzyme destroy?

16.63 Cysteine, an amino acid prevalent in hair, has a pI of 5.1.

The proteins in hair contain many cysteine R groups that form disulfide bonds.

Consider the following condensed structural formulas of cysteine **1–4** to answer Problems 16.63 and 16.64:

(1) $H_2N-CH(CH_2-SH)-COO^-$

(2) $H_3\overset{+}{N}-CH(CH_2-SH)-COOH$

(3) $H_3\overset{+}{N}-CH(CH_2-SH)-COO^-$

(4) $H_2N-CH(CH_2-SH)-COOH$

Which condensed structural formula would it have in solutions with each of the following pH values?
a. 10.5 **b.** 5.1 **c.** 1.8

16.64 For the cysteine discussed in Problem 16.63, which condensed structural formula would it have in solutions with each of the following pH values?
a. 2.0 **b.** 3.5 **c.** 9.1

16.65 Identify the amino acids and type of cross-link that occurs between the following side groups in tertiary protein structures:

a. $-CH_2-\overset{O}{\overset{\|}{C}}-NH_2$ and $HO-CH_2-$

b. $-CH_2-\overset{O}{\overset{\|}{C}}-O^-$ and $H_3\overset{+}{N}-(CH_2)_4-$

c. $-CH_2-SH$ and $HS-CH_2-$

d. $-CH_2-\overset{CH_3}{\overset{|}{CH}}-CH_3$ and CH_3-

16.66 What type of interaction would you expect between the following in a tertiary structure?
a. threonine and glutamine
b. valine and alanine
c. arginine and glutamic acid

16.67 **a.** Draw the condensed structural formula of Ser–Lys–Asp.
b. Would you expect to find this segment at the center or at the surface of a globular protein? Why?

16.68 **a.** Draw the condensed structural formula of Val–Ala–Leu.
b. Would you expect to find this segment at the center or at the surface of a globular protein? Why?

16.69 Seeds and vegetables are often deficient in one or more essential amino acids. Using the following table, state whether each combination provides all of the essential amino acids:

Source	Lysine	Tryptophan	Methionine
Oatmeal	No	Yes	Yes
Rice	No	Yes	Yes
Garbanzo beans	Yes	No	Yes
Lima beans	Yes	No	No
Cornmeal	No	No	Yes

a. rice and garbanzo beans
b. lima beans and cornmeal
c. a salad of garbanzo beans and lima beans

16.70 Seeds and vegetables are often deficient in one or more essential amino acids. Using the table in Problem 16.69, state whether each combination provides all of the essential amino acids.
a. rice and lima beans
b. rice and oatmeal
c. oatmeal and lima beans

Oatmeal is deficient in the essential amino acid lysine.

Additional Questions and Problems

For instructor-assigned homework, go to **www.masteringchemistry.com**.

16.71 What are some differences between the following pairs?
a. secondary and tertiary protein structures
b. essential and nonessential amino acids
c. polar and nonpolar amino acids
d. dipeptides and tripeptides

16.72 What are some differences between the following pairs?
a. an ionic bond (salt bridge) and a disulfide bond
b. fibrous and globular proteins
c. α helix and β-pleated sheet
d. tertiary and quaternary structures of proteins

16.73 If glutamic acid were replaced by proline in a protein, how would the tertiary structure be affected?

16.74 If glycine were replaced by alanine in a protein, how would the tertiary structure be affected?

16.75 How do enzymes differ from catalysts used in chemical laboratories?

16.76 Why do enzymes function only under mild conditions?

16.77 Lactase is an enzyme that hydrolyzes lactose to glucose and galactose.
a. What are the reactants and products of the reaction?
b. Draw an energy diagram for the reaction with and without lactase.
c. How does lactase make the reaction go faster?

16.78 Maltase is an enzyme that hydrolyzes maltose to two glucose molecules.
a. What are the reactants and products of the reaction?
b. Draw an energy diagram for the reaction with and without maltase.
c. How does maltase make the reaction go faster?

16.79 Indicate whether each of the following would be a substrate (S) or an enzyme (E):
a. lactose
b. lipase
c. urease
d. trypsin
e. pyruvate
f. transaminase

16.80 Indicate whether each of the following would be a substrate (S) or an enzyme (E):
a. glucose
b. hydrolase
c. maleate isomerase
d. alanine
e. amylose
f. lactase

16.81 Give the substrate of each of the following enzymes:
a. urease
b. succinate dehydrogenase
c. aspartate transaminase
d. tyrosinase

16.82 Give the substrate of each of the following enzymes:
a. maltase
b. fructose oxidase
c. phenolase
d. sucrase

16.83 How would the lock-and-key model explain that sucrase hydrolyzes sucrose, but not lactose?

16.84 How does the induced-fit model of enzyme action allow an enzyme to catalyze a reaction of a group of substrates?

16.85 If a blood test indicates a high level of LDH and CK, what could be the cause?

16.86 If a blood test indicates a high level of ALT, what could be the cause?

16.87 Indicate whether an enzyme is saturated or not saturated in each of the following conditions:
a. Adding more substrate does not increase the rate of reaction.
b. Doubling the substrate concentration doubles the rate of reaction.

16.88 Indicate whether each of the following enzymes would be functional:
a. pepsin, a digestive enzyme, at pH 2
b. an enzyme at 37 °C, if the enzyme is from a type of thermophilic bacteria that has an optimum temperature of 100 °C

16.89 Does each of the following statements describe a simple enzyme or an enzyme that requires a cofactor?
a. contains Mg^{2+} in the active site
b. has catalytic activity as a tertiary protein structure
c. requires folic acid for catalytic activity

16.90 Does each of the following statements describe a simple enzyme or an enzyme that requires a cofactor?
a. contains only amino acids
b. has four subunits of polypeptide chains
c. requires Fe^{3+} in the active site for catalytic activity

Challenge Questions

16.91 Indicate the overall charge of each of the following amino acids at the following pH values as 0, 1+, or 1−:
a. serine at pH 5.7
b. threonine at pH 2.0
c. isoleucine at pH 3.0
d. leucine at pH 9.0

16.92 Indicate the overall charge of each of the following amino acids at the following pH values as 0, 1+, or 1−:
a. tyrosine at pH 3.0
b. glycine at pH 10.0
c. phenylalanine at pH 8.5
d. methionine at pH 5.7

16.93 Consider the amino acids lysine, valine, and aspartic acid in an enzyme. Which of these amino acids have R groups that would
a. be found in hydrophobic regions
b. be found in hydrophilic regions
c. form hydrogen bonds
d. form salt bridges

16.94 Consider the amino acids histidine, phenylalanine, and serine in an enzyme. Which of these amino acids have R groups that would
a. be found in hydrophobic regions
b. be found in hydrophilic regions
c. form hydrogen bonds
d. form salt bridges

Answers

Answers to Study Checks

16.1 a. nonpolar
b. polar (neutral)

16.2 $H_3\overset{+}{N}-CH(CH_2-OH)-C(=O)-OH$

16.3 threonylleucylphenylalanine

16.4 Both are nonpolar and would be found on the inside of the tertiary structure.

16.5 quaternary

16.6 lipase

16.7 At a pH lower than the optimum pH, denaturation of the tertiary structure will decrease the activity of urease.

16.8 competitive inhibition

16.9 vitamin B_6

Answers to Selected Questions and Problems

16.1 a. transport **b.** structural
c. structural **d.** enzyme

16.3 All amino acids contain a carboxylic acid group and an amine group on the α-carbon.

16.5 a. $H_3\overset{+}{N}-CH(H)-C(=O)-O^-$ **b.** $H_3\overset{+}{N}-CH(H-C(OH)-CH_3)-C(=O)-O^-$

c. $H_3\overset{+}{N}-CH(CH_2-CH_2-C(=O)O^-)-C(=O)-O^-$ **d.** $H_3\overset{+}{N}-CH(CH_2-C_6H_5)-C(=O)-O^-$

16.7 a. nonpolar **b.** polar neutral
c. polar acidic **d.** nonpolar

16.9 a. alanine **b.** valine
c. lysine **d.** cysteine

16.11 positive (1+)

16.13 a. $H_3\overset{+}{N}-CH(H)-C(=O)-OH$ **b.** $H_3\overset{+}{N}-CH(CH_2-SH)-C(=O)-OH$

c. $H_3\overset{+}{N}-CH(CH_2-OH)-C(=O)-OH$ **d.** $H_3\overset{+}{N}-CH(HO-CH-CH_3)-C(=O)-OH$

16.15 a. above pI **b.** below pI **c.** at pI

16.17 a. $H_3\overset{+}{N}-CH(CH_3)-C(=O)-NH-CH(CH_2-SH)-C(=O)-O^-$

Ala–Cys

b. $H_3\overset{+}{N}-CH(CH_2-OH)-C(=O)-NH-CH(CH_2-C_6H_5)-C(=O)-O^-$

Ser–Phe

c. $H_3\overset{+}{N}-CH(H)-C(=O)-NH-CH(CH_3)-C(=O)-NH-CH(CH(CH_3)_2)-C(=O)-O^-$

Gly–Ala–Val

d. $H_3\overset{+}{N}-CH(CH(CH_3)_2)-C(=O)-NH-CH(CH(CH_3)-CH_2-CH_3)-C(=O)-NH-CH(CH_2\text{-indolyl})-C(=O)-O^-$

Val–Ile–Trp

16.19 Val–Ser–Ser, Ser–Val–Ser, or Ser–Ser–Val

16.21 The primary structure remains unchanged and intact as hydrogen bonds form between carbonyl oxygen atoms and hydrogen atoms of amide groups in the polypeptide chains.

16.23 In the α helix, hydrogen bonds form between the carbonyl oxygen atom and the hydrogen atom of an amide group in the next turn of the helical chain. In the β-pleated sheet, hydrogen bonds occur between parallel peptide chains.

16.25 **a.** disulfide bond
b. salt bridge
c. hydrogen bond
d. hydrophobic interaction

16.27 **a.** cysteine
b. Leucine and valine will be found on the inside of the protein because they have R groups that are hydrophobic.
c. The cysteine and aspartic acid would be on the outside of the protein because they have R groups that are polar.
d. The order of the amino acids (the primary structure) provides R groups that interact to determine the tertiary structure of the protein.

16.29 **a.** tertiary, quaternary
b. primary
c. secondary
d. secondary

16.31 **a.** Placing an egg in boiling water disrupts hydrogen bonds and hydrophobic interactions, which changes secondary and tertiary structures and causes loss of overall shape.
b. Using an alcohol swab disrupts hydrophobic interactions and changes the tertiary structure.
c. The heat from an autoclave will disrupt hydrogen bonds and hydrophobic interactions, which changes secondary and tertiary structures of the protein in any bacteria present.
d. Heat will cause changes in the secondary and tertiary structure of surrounding protein, which results in the formation of solid protein that helps to close the wound.

16.33 The chemical reactions can occur without enzymes, but the rates are too slow. Catalyzed reactions, which are many times faster, provide the amounts of products needed by the cell at a particular time.

16.35 **a.** galactose
b. lipid
c. aspartic acid

16.37 **a.** hydrolase
b. oxidoreductase
c. isomerase
d. transferase

16.39 **a.** enzyme
b. enzyme–substrate complex
c. substrate

16.41 **a.** $E + S \rightleftarrows ES \longrightarrow E + P$
b. The active site is a region or pocket within the tertiary structure of an enzyme that accepts the substrate, aligns the substrate for reaction, and catalyzes the reaction.

16.43 Isoenzymes are slightly different forms of an enzyme that catalyze the same reaction in different organs and tissues of the body.

16.45 A doctor might run tests for the enzymes CK, LDH, and AST to determine if the patient had a heart attack.

16.47 **a.** The rate would decrease.
b. The rate would decrease.
c. The rate would decrease.
d. The rate would increase.

16.49 pepsin, pH 2; sucrase, pH 6; trypsin, pH 8

16.51 **a.** competitive
b. noncompetitive
c. competitive
d. noncompetitive
e. competitive

16.53 **a.** methanol, CH_3-OH; ethanol, CH_3-CH_2-OH
b. Ethanol has a structure similar to methanol and could compete for the active site.
c. Ethanol is a competitive inhibitor of methanol oxidation.

16.55 **a.** an enzyme that requires a cofactor
b. an enzyme that requires a cofactor
c. a simple enzyme

16.57 **a.** pantothenic acid (vitamin B_5)
b. folic acid
c. niacin (vitamin B_3)

16.59 **a.** oxidoreductase
b. At high concentration, ethanol, which acts as a competitive inhibitor of ethylene glycol, would saturate the alcohol dehydrogenase enzyme to allow ethylene glycol to be removed from the body without producing oxalic acid.

16.61 **a.** aspartic acid and phenylalanine
b. aspartylphenylalanine

16.63 **a.** (1) **b.** (3) **c.** (2)

16.65 **a.** asparagine and serine; hydrogen bond
b. aspartic acid and lysine; salt bridge
c. cysteine and cysteine; disulfide bond
d. leucine and alanine; hydrophobic interaction

16.67 **a.** $H_3\overset{+}{N}-CH(CH_2OH)-C(=O)-N(H)-CH((CH_2)_4\overset{+}{N}H_3)-C(=O)-N(H)-CH(CH_2COO^-)-C(=O)-O^-$

b. This segment contains polar R groups, which would be found on the surface of a globular protein where they hydrogen bond with water.

16.69 **a.** yes **b.** no **c.** no

16.71 **a.** The secondary structure of a protein depends on hydrogen bonds to form a helix or a pleated sheet; the tertiary structure is determined by the interactions of R groups such as disulfide bonds and salt bridges.
b. Nonessential amino acids can be synthesized by the body; essential amino acids must be supplied in the diet.
c. Polar amino acids have hydrophilic side groups, whereas nonpolar amino acids have hydrophobic side groups.
d. Dipeptides contain two amino acids, whereas tripeptides contain three.

16.73 Glutamic acid is a polar acidic amino acid, whereas proline is nonpolar. The polar acidic R group of glutamic acid would be located on the outside surface of the protein where it would form hydrophilic interactions with water. However, the nonpolar proline would move to the hydrophobic center of the protein.

16.75 In chemical laboratories, reactions are often run at high temperatures using catalysts that are strong acids or bases. Enzymes, which function at physiological temperatures and pH, are denatured rapidly if high temperatures or acids or bases are used.

16.77 **a.** The reactant is lactose, and the products are glucose and galactose.
b.

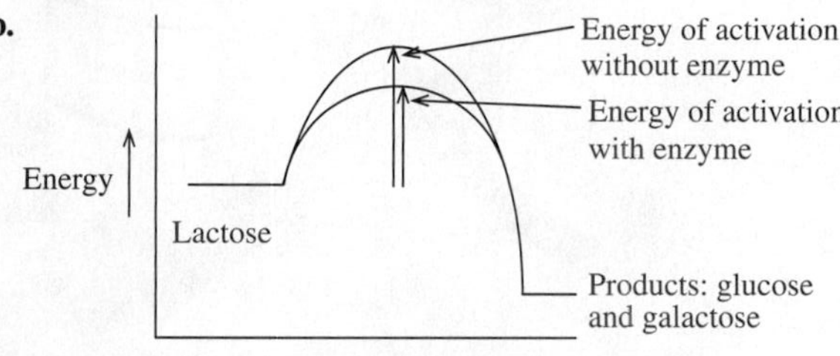

c. By lowering the energy of activation, the enzyme furnishes a lower energy pathway by which the reaction can take place.

16.79 **a.** S **b.** E **c.** E
d. E **e.** S **f.** E

16.81 **a.** urea
b. succinate
c. aspartate
d. tyrosine

16.83 Sucrose fits the shape of the active site in sucrase, but lactose does not.

16.85 A heart attack may be the cause.

16.87 **a.** saturated
b. unsaturated

16.89 **a.** an enzyme that requires a cofactor
b. a simple enzyme
c. an enzyme that requires a cofactor (coenzyme)

16.91 **a.** 0
b. 1+
c. 1+
d. 1−

16.93 **a.** valine
b. lysine, aspartic acid
c. lysine, aspartic acid
d. lysine, aspartic acid

17 Nucleic Acids and Protein Synthesis

LOOKING AHEAD

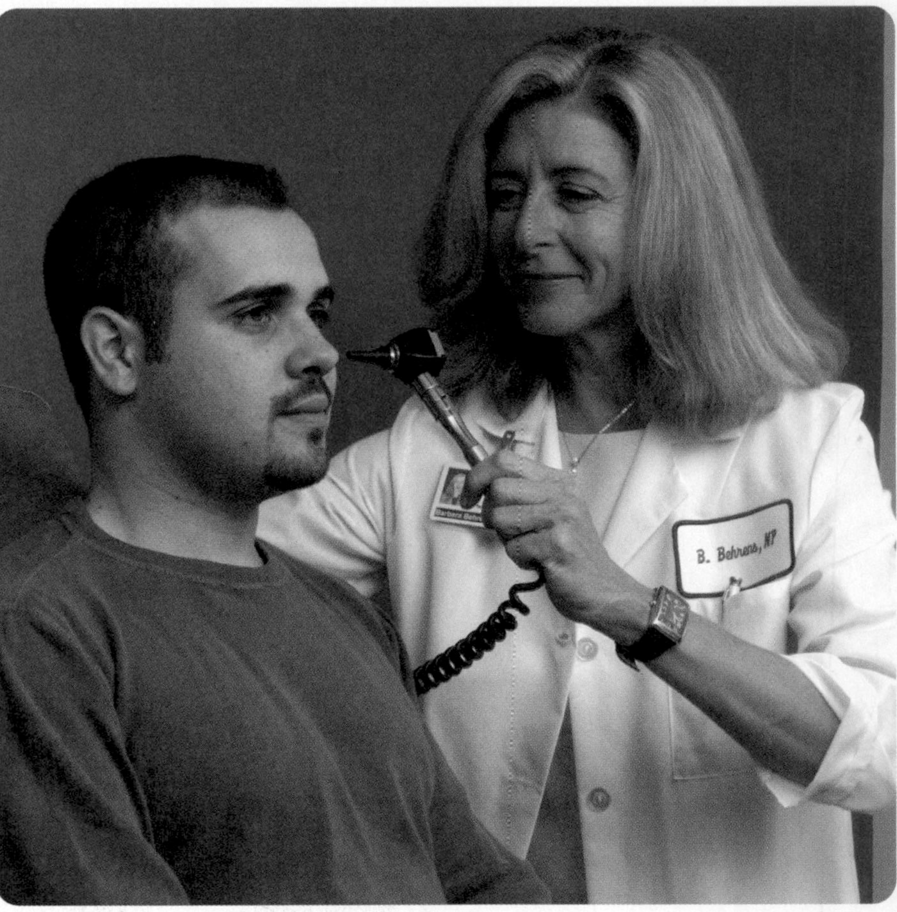

"I run the Hepatitis C Clinic, where patients are often anxious when diagnosed," says Barbara Behrens, nurse practitioner, Hepatitis C Clinic, Kaiser Hospital. "The treatment for hepatitis C can produce significant reactions such as a radical drop in blood count. When this happens, I get help to them within 24 hours. I monitor our patients very closely, and many call me whenever they need to."

Hepatitis C is a RNA virus, or retrovirus, that causes liver inflammation, often resulting in chronic liver disease. Unlike many other viruses to which we eventually develop immunity, the hepatitis C virus undergoes mutations so rapidly that scientists have not been able to produce vaccines. People who carry the virus are contagious throughout their lives and are able to pass the virus to other people.

The nucleic acids are large molecules found in the nuclei of the cells in our bodies that store information and direct activities for cellular growth and reproduction. Deoxyribonucleic acid (DNA), the genetic material in the nucleus of a cell, contains all the information needed for the development of a complete living system. The way you grow, your hair, your eyes, your physical appearance, the activities of the cells in your body are all determined by a set of directions contained in the DNA of your cells.

All of the genetic information in the cell is called the *genome*. Every time a cell divides, the information in the genome is copied and passed on to the new cells. This replication process must duplicate the genetic instructions exactly. Some sections of DNA called *genes* contain the information to make a particular protein.

As a cell requires protein, another type of nucleic acid, RNA, interprets the genetic information in DNA and carries that information to the ribosomes, where the synthesis of protein takes place. However, mistakes sometimes lead to mutations that affect the synthesis of a certain protein.

17.1 Components of Nucleic Acids

LEARNING GOAL

Describe the bases and ribose sugars that make up the nucleic acids DNA and RNA.

There are two closely related types of nucleic acids: *deoxyribonucleic acid* (**DNA**), and *ribonucleic acid* (**RNA**). Both are unbranched polymers of repeating monomer units known as *nucleotides*. A DNA molecule may contain several million nucleotides; smaller RNA molecules may contain up to several thousand. Each nucleotide has three components: a base, a five-carbon sugar, and a phosphate group (see Figure 17.1).

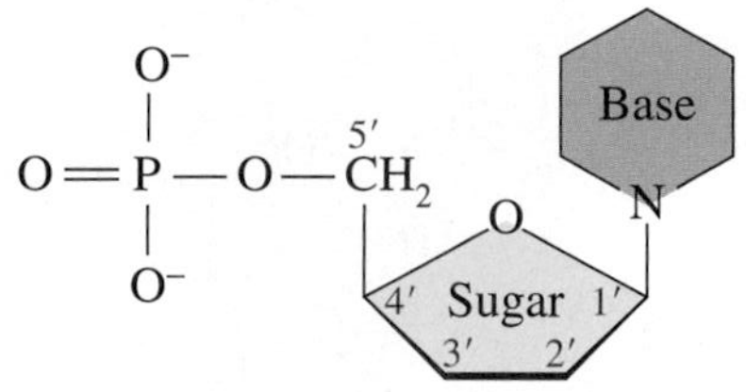

FIGURE 17.1 The general structure of a nucleotide includes a nitrogen-containing base, a sugar, and a phosphate group.

Q In a nucleotide, what types of groups are bonded to a five-carbon sugar?

Bases

The nitrogen-containing **bases** in nucleic acids are derivatives of *pyrimidine* or *purine*.

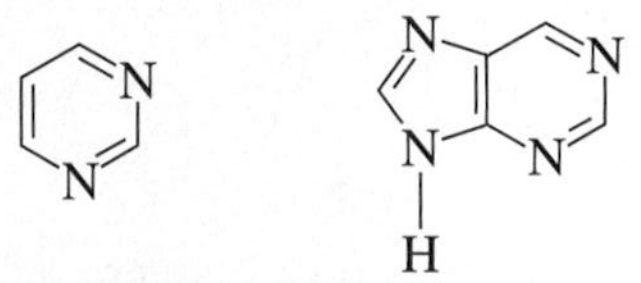

Pyrimidine Purine

TUTORIAL
Nucleic Acid Building Blocks

In DNA, the purine bases with double rings are adenine (A) and guanine (G); and the pyrimidine bases with single rings are cytosine (C) and thymine (T). RNA contains the same bases, except thymine (5-methyluracil) is replaced by uracil (U) (see Figure 17.2).

Pyrimidines

Cytosine (C)
(DNA and RNA)

Thymine (T)
(DNA only)

Uracil (U)
(RNA only)

Purines

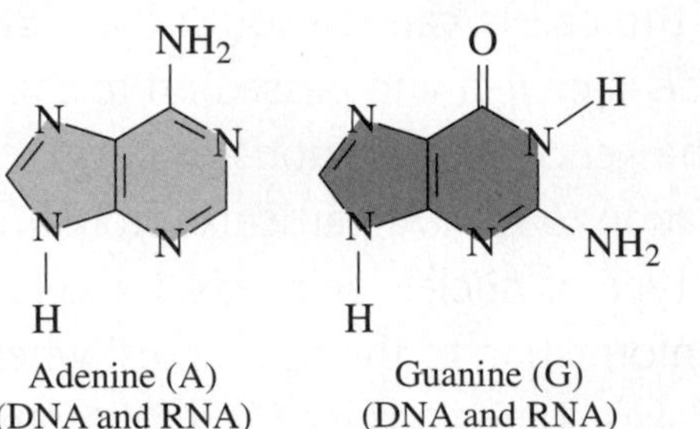

FIGURE 17.2 DNA contains the bases A, G, C, and T; RNA contains A, G, C, and U.
Q Which bases are found in DNA?

Pentose sugars in RNA and DNA

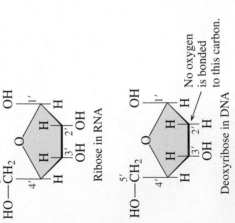

Ribose in RNA

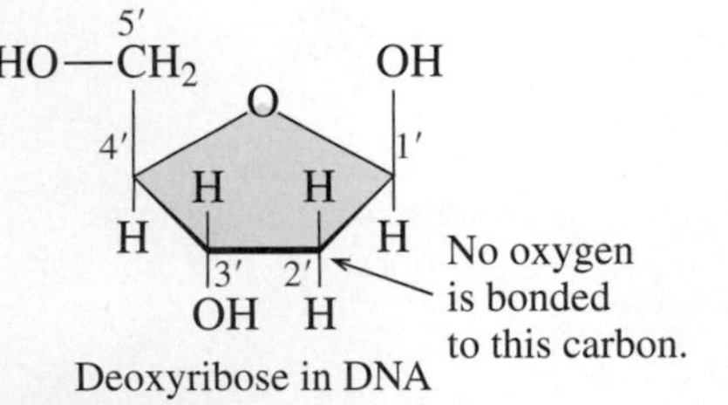

FIGURE 17.3 The five-carbon pentose sugar found in RNA is ribose and in DNA, deoxyribose.
Q What is the difference between ribose and deoxyribose?

CONCEPT CHECK 17.1

Components of Nucleic Acids

Identify each of the following bases as a purine or a pyrimidine. Indicate if each base is found in RNA, DNA, or both.

a.

b.

ANSWER

a. Guanine is a purine found in both RNA and DNA.
b. Uracil is a pyrimidine found only in RNA.

Pentose Sugars

In RNA, the five-carbon sugar is *ribose*, which gives the letter R in the abbreviation RNA. The atoms in the pentose sugars are numbered with primes (1′, 2′, 3′, 4′, and 5′) to differentiate them from the atoms in the bases (see Figure 17.3). In DNA, the five-carbon sugar is *deoxyribose*, which is similar to ribose except that there is no hydroxyl group (—OH) on C2′. The *deoxy* prefix means "without oxygen" and provides the D in DNA.

Nucleosides and Nucleotides

Nucleosides, which are a combination of a sugar and a base, are produced when the nitrogen atom in a pyrimidine or a purine base forms an *N*-glycosidic bond to carbon 1 (C1′) of a ribose or deoxyribose sugar. For example, adenine and ribose form a nucleoside called adenosine.

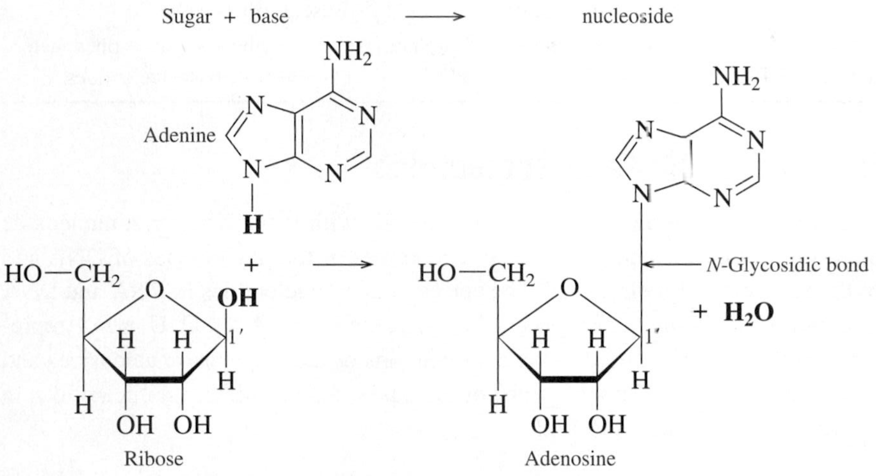

Sugar + base ⟶ nucleoside

Adenine + Ribose ⟶ Adenosine + H_2O (*N*-Glycosidic bond)

A base forms an *N*-glycosidic bond with a ribose or deoxyribose sugar to form a nucleoside.

Nucleotides are nucleosides in which a phosphate group bonds to —OH on carbon 5 (C5′) of a ribose or deoxyribose sugar. The product is a phosphate ester. Other hydroxyl groups on ribose can form phosphate esters too, but only the 5′-monophosphate nucleotides are found in RNA and DNA. All the nucleotides in RNA and DNA are shown in Figure 17.4.

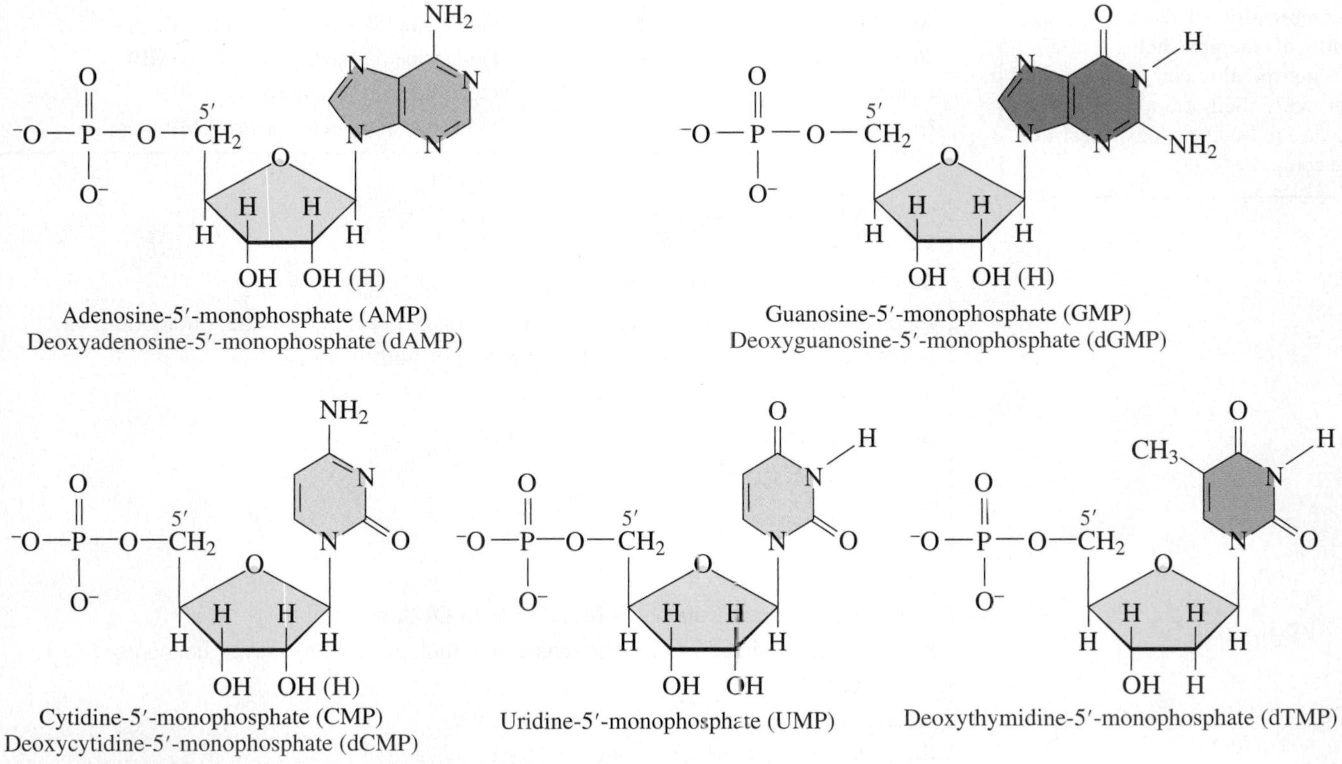

Adenosine-5′-monophosphate (AMP)
Deoxyadenosine-5′-monophosphate (dAMP)

Guanosine-5′-monophosphate (GMP)
Deoxyguanosine-5′-monophosphate (dGMP)

Cytidine-5′-monophosphate (CMP)
Deoxycytidine-5′-monophosphate (dCMP)

Uridine-5′-monophosphate (UMP)

Deoxythymidine-5′-monophosphate (dTMP)

FIGURE 17.4 The nucleotides of RNA are identical to those of DNA, except in DNA the sugar is deoxyribose and deoxythymidine replaces uridine.

Q What are two differences in the nucleotides of RNA and DNA?

Career Focus

OCCUPATIONAL THERAPIST

"Occupational therapists teach children and adults the skills they need for the job of living," says occupational therapist Leslie Wakasa. "When working with the pediatric population, we are crucial in educating disabled children, their families, caregivers, and school staff in ways to help them be as independent as they can be in all aspects of their daily lives. It's rewarding when you can show children how to feed themselves, which is a huge self-esteem issue for them. The opportunity to help people become more independent is very rewarding."

A combination of technology and occupational therapy helps children who are nonverbal to communicate and interact with their environment. By leaning on a red switch, Alex is learning to use a computer.

Table 17.1 summarizes the components in DNA and RNA.

TABLE 17.1 Components in DNA and RNA

Component	DNA	RNA
Bases	A, G, C, and T	A, G, C, and U
Sugar	Deoxyribose	Ribose
Nucleoside	Base + deoxyribose sugar	Base + ribose sugar
Nucleotide	Base + deoxyribose sugar + phosphate	Base + ribose sugar + phosphate
Nucleic acid	Polymer of deoxyribose nucleotides	Polymer of ribose nucleotides

Naming Nucleosides and Nucleotides

The name of a nucleoside that contains a purine ends with *osine* whereas a nucleoside that contains a pyrimidine ends with *idine*. The names of the nucleosides of DNA add *deoxy* to the beginning of their names. The corresponding nucleotides in RNA and DNA are named by adding *-5′-monophosphate*. Although the letters A, G, C, U, and T represent the bases, they are often used in the abbreviations of the respective nucleosides and nucleotides. The names and abbreviations of the bases, nucleosides, and nucleotides in DNA and RNA are listed in Table 17.2.

TABLE 17.2 Nucleosides and Nucleotides in DNA and RNA

Base	Nucleosides	Nucleotides
DNA		
Adenine (A)	Deoxyadenosine (A)	Deoxyadenosine-5′-monophosphate (dAMP)
Guanine (G)	Deoxyguanosine (G)	Deoxyguanosine-5′-monophosphate (dGMP)
Cytosine (C)	Deoxycytidine (C)	Deoxycytidine-5′-monophosphate (dCMP)
Thymine (T)	Deoxythymidine (T)	Deoxythymidine-5′-monophosphate (dTMP)
RNA		
Adenine (A)	Adenosine (A)	Adenosine-5′-monophosphate (AMP)
Guanine (G)	Guanosine (G)	Guanosine-5′-monophosphate (GMP)
Cytosine (C)	Cytidine (C)	Cytidine-5′-monophosphate (CMP)
Uracil (U)	Uridine (U)	Uridine-5′-monophosphate (UMP)

CONCEPT CHECK 17.2

Components of Nucleic Acid

Identify each of the following as a pentose sugar, base, nucleoside, nucleotide, or nucleic acid and if it is found in DNA, RNA, or both:

a. guanine
b. deoxyadenosine-5′-monophosphate (dAMP)
c. ribose
d. cytidine

ANSWER

a. Guanine is one of the bases found in both DNA and RNA.
b. Deoxyadenosine-5′-monophosphate is a nucleotide (*tides* have phosphate) that is found in DNA.
c. Ribose is the pentose sugar found in RNA.
d. Cytidine is a nucleoside found in RNA.

SAMPLE PROBLEM 17.1

Nucleotides

For each of the following nucleotides, identify the components and whether the nucleotide is found in DNA, RNA, or both:

a. deoxyguanosine-5′-monophosphate (dGMP)
b. adenosine-5′-monophosphate (AMP)

SOLUTION

a. This DNA nucleotide consists of deoxyribose, guanine, and phosphate.
b. This RNA nucleotide contains ribose, adenine, and phosphate.

STUDY CHECK 17.1

What are the name and abbreviation of the DNA nucleotide of cytosine?

QUESTIONS AND PROBLEMS

Components of Nucleic Acids

17.1 Identify each of the following bases as a purine or a pyrimidine:
a. thymine
b. (structure: pyrimidine ring with NH_2, N, O, N–H)

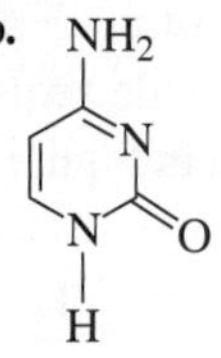

17.2 Identify each of the following bases as a purine or a pyrimidine:
a. guanine
b. (structure: fused ring with NH_2, N, N, N, N–H)

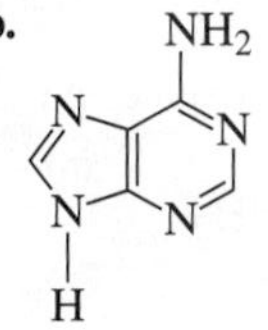

17.3 Identify each of the bases in Problem 17.1 as a component of RNA, DNA, or both.

17.4 Identify each of the bases in Problem 17.2 as a component of RNA, DNA, or both.

17.5 What are the names and abbreviations of the four nucleotides in DNA?

17.6 What are the names and abbreviations of the four nucleotides in RNA?

17.7 Identify each of the following as a nucleoside or nucleotide:
a. adenosine
b. deoxycytidine
c. uridine
d. cytidine-5′-monophosphate

17.8 Identify each of the following as a nucleoside or nucleotide:
a. deoxythymidine
b. guanosine
c. deoxyadenosine-5′-monophosphate
d. uridine-5′-monophosphate

17.9 Draw the structure of deoxyadenosine-5′-monophosphate (dAMP).

17.10 Draw the structure of uridine-5′-monophosphate (UMP).

17.2 Primary Structure of Nucleic Acids

LEARNING GOAL

Describe the primary structures of RNA and DNA.

SELF STUDY ACTIVITY
DNA and RNA Structure

The **nucleic acids** are polymers of many nucleotides in which the 3′—OH group of the sugar in one nucleotide bonds to the phosphate group on the 5′-carbon atom in the sugar of the next nucleotide. This phosphate link between the sugars in adjacent nucleotides is referred to as a **phosphodiester bond**. As more nucleotides are added using phosphodiester bonds, a backbone forms that consists of alternating sugar and phosphate groups. The bases, which are attached to each sugar, extend out from the sugar-phosphate backbone.

Formation of an ester bond

3′–5′ Phosphodiester bond

Free 5′-phosphate

Free 3′-hydroxyl

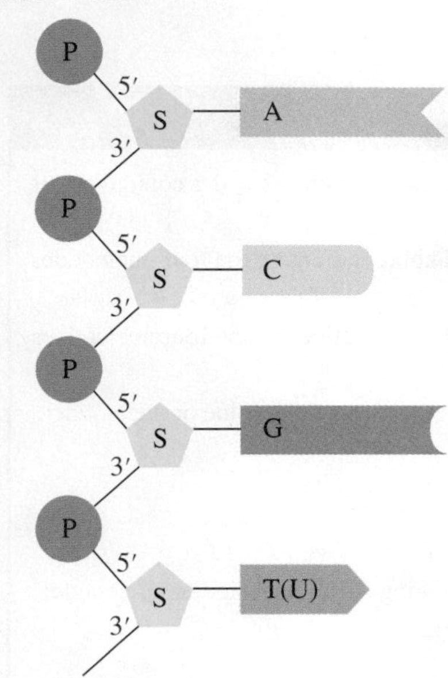

In the primary structure of nucleic acids, each sugar in a sugar-phosphate backbone is attached to a base.

Each nucleic acid has its own unique sequence of bases, which is known as its **primary structure**. It is this sequence of bases that carries the genetic information from one cell to the next. Along a DNA or RNA chain, the bases attached to each of the sugars extend out from the nucleic acid backbone. In any nucleic acid, the sugar at the one end has an unreacted or free 5′-phosphate terminal end, and the sugar at the other end has a free 3′-hydroxyl group.

A nucleic acid sequence is read from the sugar with free 5′-phosphate to the sugar with the free 3′-hydroxyl group. The order of nucleotides is often written using only the letters of the bases. For example, the nucleotide sequence starting with adenine (free 5′-phosphate end) in the section of RNA shown in Figure 17.5 is 5′—A—C—G—U—3′.

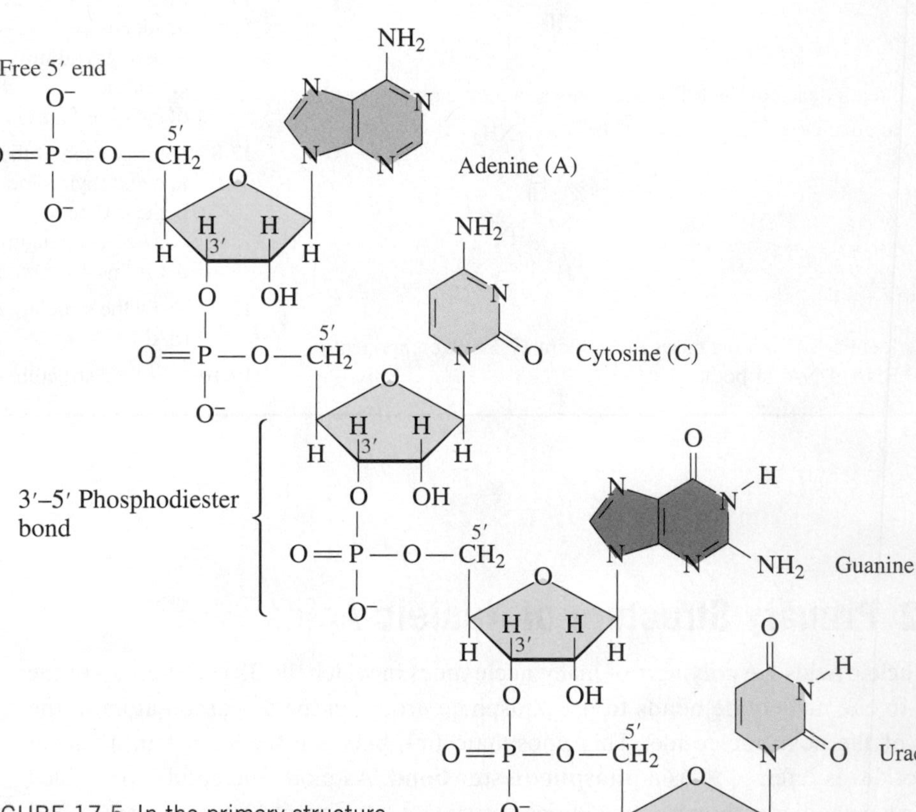

FIGURE 17.5 In the primary structure of RNA, A, C, G, and U are linked by 3′–5′-phosphodiester bonds.

Q Where are the free 5′-phosphate and 3′-hydroxyl groups?

SAMPLE PROBLEM 17.2

Bonding of Nucleotides

Draw the structure of the RNA dinucleotide formed by two cytidine-5′-monophosphates.

SOLUTION

The dinucleotide is drawn by connecting the 3′—OH on the first cytidine-5′-monophosphate with the 5′-phosphate group on the second.

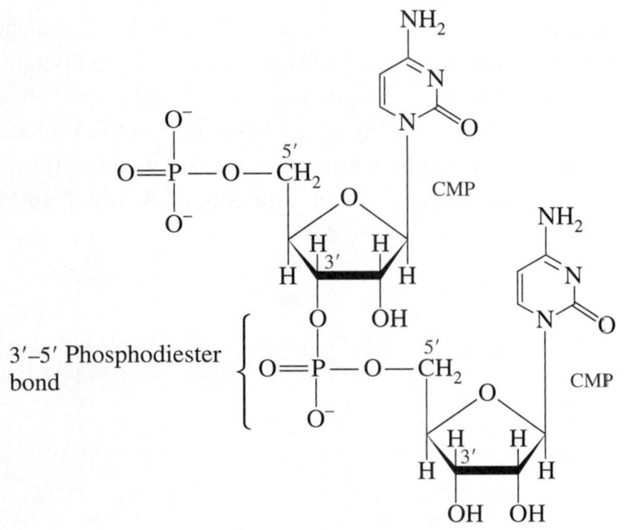

3′–5′ Phosphodiester bond

CMP

CMP

STUDY CHECK 17.2

In the dinucleotide of cytidine shown in the solution to Sample Problem 17.2, identify the free 5′-phosphate group and the free 3′-hydroxyl (—OH) group.

QUESTIONS AND PROBLEMS

Primary Structure of Nucleic Acids

17.11 How are the nucleotides held together in a nucleic acid polymer?

17.12 How do the ends of a nucleic acid polymer differ?

17.13 Draw the structure of the dinucleotide G—C in RNA.

17.14 Draw the structure of the dinucleotide A—T in DNA.

17.3 DNA Double Helix

LEARNING GOAL

Describe the double helix of DNA.

During the 1940s, biologists determined the bases in DNA from a variety of organisms had a specific relationship: the amount of adenine (A) was equal to the amount of thymine (T), and the amount of guanine (G) was equal to the amount of cytosine (C). Eventually, biologists determined that adenine is always paired (1:1) with thymine, and guanine is always paired (1:1) with cytosine.

Number of purine molecules = Number of pyrimidine molecules

A = T

G = C

TUTORIAL
The Double Helix

FIGURE 17.6 A model of a DNA molecule shows the double helix as the characteristic shape of DNA molecules.
Q What is meant by the term *double helix?*

In 1953, James Watson and Francis Crick proposed that DNA was a **double helix** that consists of two polynucleotide strands winding about each other like a spiral staircase (see Figure 17.6). The sugar-phosphate backbones are analogous to the outside railings with the bases arranged like steps along the inside. The two strands run in opposite directions. One strand goes from the 5′ to 3′ direction, and the other strand goes in the 3′ to 5′ direction.

Complementary Base Pairs

Each of the bases along one polynucleotide strand forms hydrogen bonds to a specific base on the opposite DNA strand. Adenine forms hydrogen bonds only to thymine, and guanine bonds only to cytosine (see Figure 17.7). The pairs A—T and G—C are called **complementary base pairs**. The specific pairing of the bases occurs because adenine and thymine form only two hydrogen bonds, while cytosine and guanine form three hydrogen bonds. This explains why DNA has equal amounts of A and T and equal amounts of G and C.

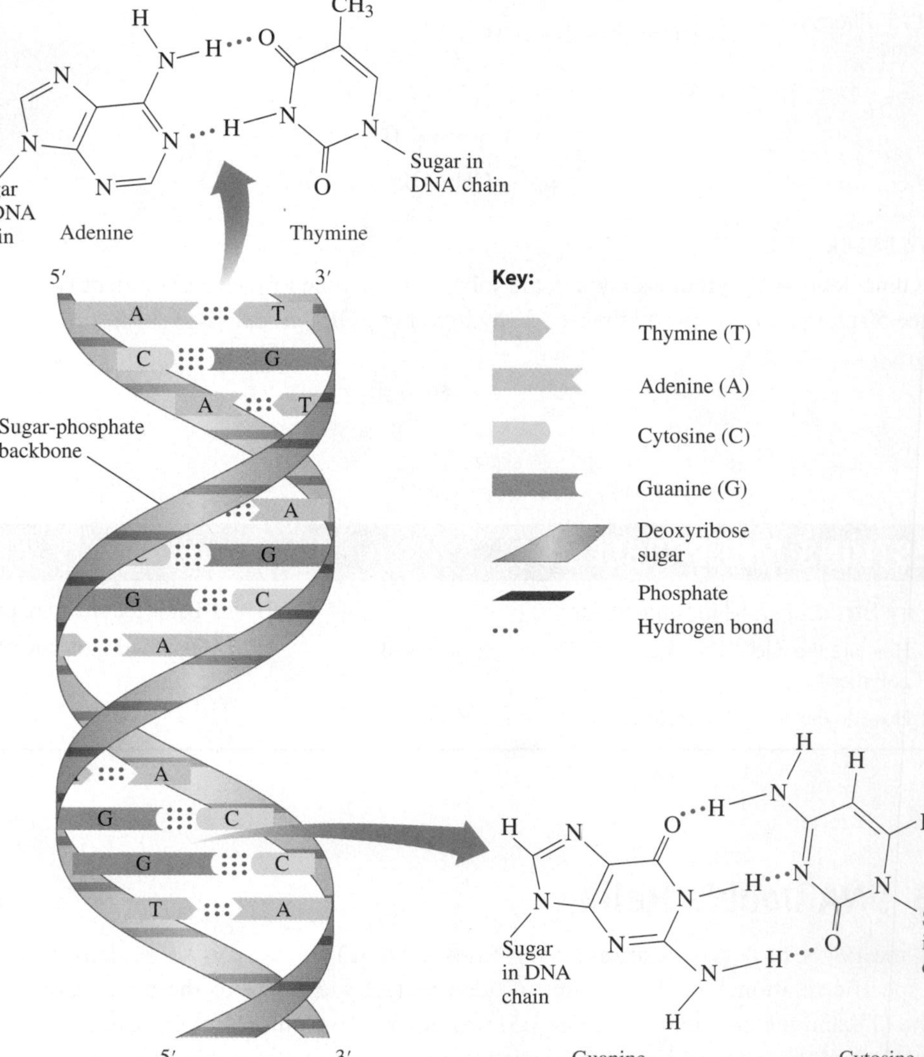

FIGURE 17.7 Hydrogen bonds between complementary base pairs hold the polynucleotide strands in the double helix of DNA.
Q Why are G—C base pairs more stable than A—T base pairs?

SAMPLE PROBLEM 17.3

Complementary Base Pairs

Write the complementary base sequence for the following DNA segment:

—A—C—G—A—T—C—T—

SOLUTION

In the complementary segment of DNA, A must be paired with T, and G must be paired with C:

Given segment of DNA: —A—C—G—A—T—C—T—
: : : : : : :
Complementary segment: —T—G—C—T—A—G—A—

STUDY CHECK 17.3

What sequence of bases is complementary to a DNA segment with the base sequence of —G—G—T—T—A—A—C—C—?

DNA Replication

TUTORIAL
DNA Replication

SELF STUDY ACTIVITY
DNA Replication

The function of DNA in cells of animals and plants as well as in bacteria is to preserve genetic information. As cells divide, copies of DNA are produced that transfer genetic information to the new cells.

In DNA **replication**, the strands in the parent DNA separate, which allows the synthesis of complementary DNA strands. The process begins when an enzyme called *helicase* catalyzes the unwinding of a portion of the double helix by breaking the hydrogen bonds between the complementary bases. These single strands now act as templates for the synthesis of new complementary DNA strands (see Figure 17.8).

As the complementary base pairs form, *DNA polymerase* catalyzes the formation of phosphodiester bonds between the nucleotides. Eventually the entire double helix of the parent DNA is copied. In each new DNA molecule, one strand of the double helix is from the original DNA and one is a newly synthesized strand. This process produces two new DNAs called *daughter DNAs* that are identical to each other and exact copies of the original parent DNA. In the process of DNA replication, complementary base pairing ensures the correct placement of bases in the new DNA strands.

CONCEPT CHECK 17.3

DNA Replication

In an original DNA strand, a segment has the base sequence —G—C—A—A—T—C—.
What is the sequence of nucleotides in the DNA daughter strands that is complementary to this sequence?

ANSWER

Only one possible nucleotide can pair with each base in the original sequence. Thymine pairs only with adenine and cytosine pairs only with guanine to give the complementary base sequence —C—G—T—T—A—G—.

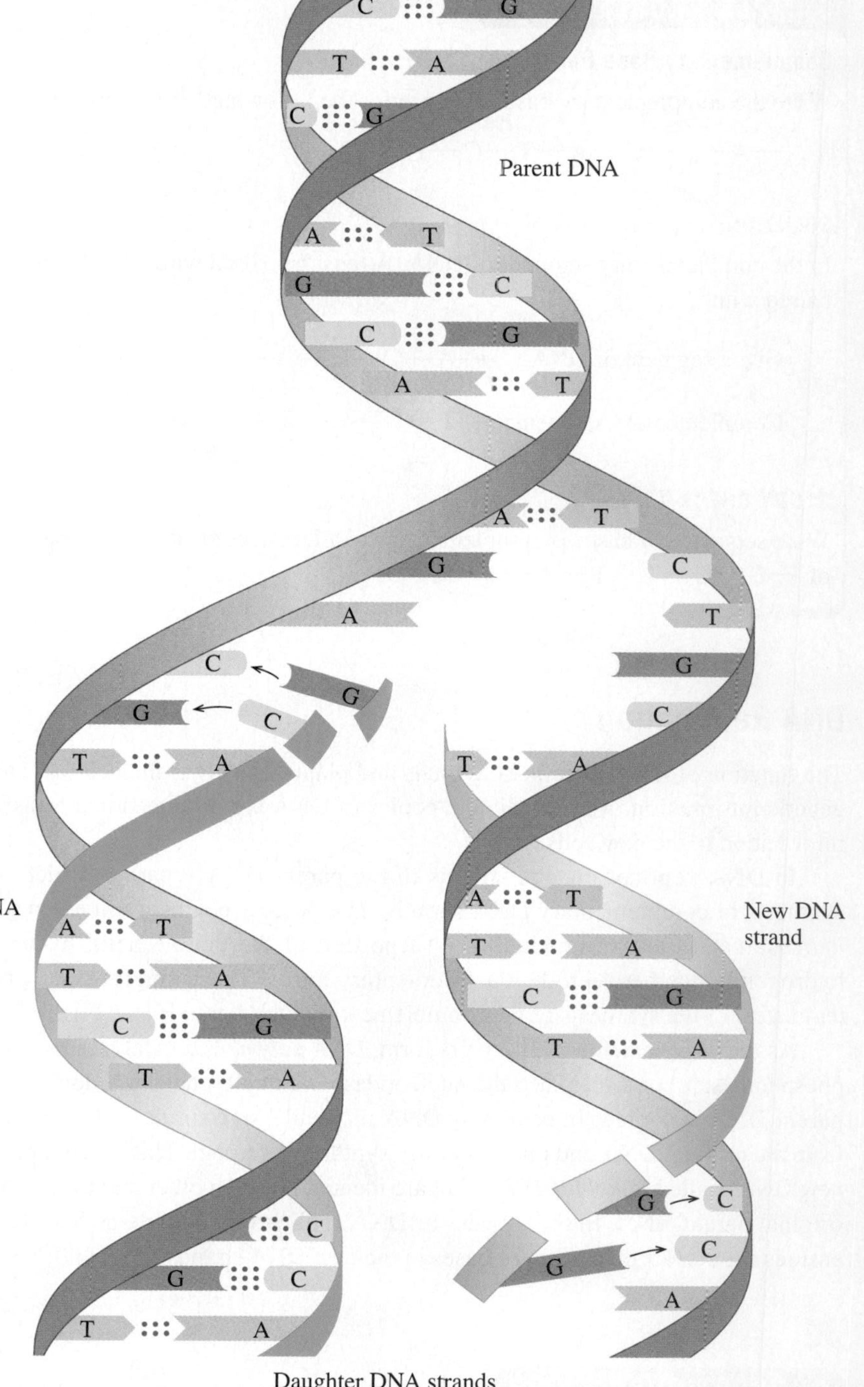

FIGURE 17.8 In DNA replication, the separate strands of the parent DNA are the templates for the synthesis of complementary strands, which produces two exact copies (daughter DNAs).

Q How many strands of the parent DNA are in each of the new double-stranded copies of DNA?

QUESTIONS AND PROBLEMS

DNA Double Helix

17.15 How are the two strands of nucleic acid in DNA held together?

17.16 What is meant by complementary base pairing?

17.17 Complete the base sequence in a complementary DNA segment if a portion of the parent strand has each of the following base sequences:

a. —A—A—A—A—A—A—
b. —G—G—G—G—G—G—
c. —A—G—T—C—C—A—G—G—T—
d. —C—T—G—T—A—T—A—C—G—T—T—A—

17.18 Complete the base sequence in a complementary DNA segment if a portion of the parent strand has each of the following base sequences:

a. —T—T—T—T—T—T—
b. —C—C—C—C—C—C—C—C—C—
c. —A—T—G—G—C—A—
d.
—A—T—A—T—G—C—G—C—T—A—A—A—

Chemistry Link to Health

DNA FINGERPRINTING

In a process called DNA fingerprinting, enzymes are used to cut DNA into smaller sections. The resulting DNA fragments are then separated by size and treated with a radioactive isotope that adheres to specific base sequences in the fragments. A piece of X-ray film that is placed over the gel is exposed by the radiation. The pattern of dark and light bands on the film is known as a DNA fingerprint. It has been estimated that the odds of two persons who are not identical twins producing the same DNA fingerprint is less than one in a billion.

One application of DNA fingerprinting is in forensic science, where DNA from samples such as blood, hair, or semen is used to connect a suspect with a crime. Recently, it has also been used to gain the release of persons who were wrongly convicted. Other applications of DNA fingerprinting are to determine the biological parents of a child, to establish the identity of a deceased person, and to match recipients with organ donors.

Human Genome Project

During the 1970s, scientists began to map the location of *genes* within the DNA of the *genome*, which contains the hereditary information of an organism. By 1987, the genome of *E. coli* was determined. More recently, these techniques combined with new computer programs have compiled the map of the human genome, which contains about 30 000 genes.

Scientists think that most of the genome is not functional and has perhaps been carried from generation to generation for millions of years. Large blocks of genes are copied from one human chromosome to another even though they no longer code for needed proteins. Thus, the coding portions of the genes seem to make up only about 1% of the total genome. The results of the genome project will help us identify defective genes that lead to genetic disease. Today DNA fingerprinting is used to screen for genes responsible for genetic diseases such as sickle-cell anemia, cystic fibrosis, breast cancer, colon cancer, Huntington's disease, and Lou Gehrig's disease.

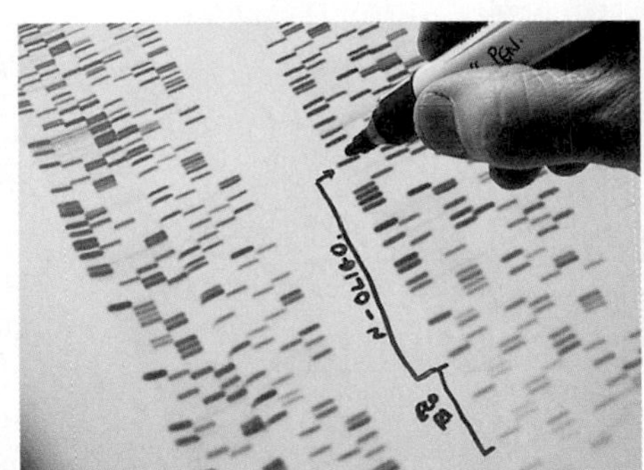

Dark and light bands on a film indicate the nucleotide sequence in DNA from a human gene.

17.4 RNA and the Genetic Code

LEARNING GOAL

Identify the different RNAs; describe the synthesis of mRNA.

Ribonucleic acid, RNA, which makes up most of the nucleic acid found in the cell, is involved with transmitting the genetic information needed to operate the cell. Similar to DNA, RNA molecules are polymers of nucleotides. However, RNA differs from DNA in several important ways:

1. The sugar in RNA is ribose rather than the deoxyribose found in DNA.
2. The base uracil replaces thymine.
3. RNA molecules are single stranded, not double stranded.
4. RNA molecules are much smaller than DNA molecules.

Types of RNA

TUTORIAL
Types of RNA

There are three major types of RNA in the cells: *messenger RNA*, *ribosomal RNA*, and *transfer RNA*. Ribosomal RNA (**rRNA**), the most abundant type of RNA, is combined with proteins in the ribosomes. Ribosomes, which are the sites for protein synthesis, consist of two subunits: a large subunit and a small subunit (see Figure 17.9). Cells that synthesize large numbers of proteins have thousands of ribosomes.

Messenger RNA (**mRNA**) carries genetic information from the DNA, located in the nucleus of the cell, to the ribosomes located in the cytoplasm. Each gene segment of DNA produces a separate mRNA for a particular protein that is needed in the cell. The size of mRNA depends on the number of nucleotides in that gene.

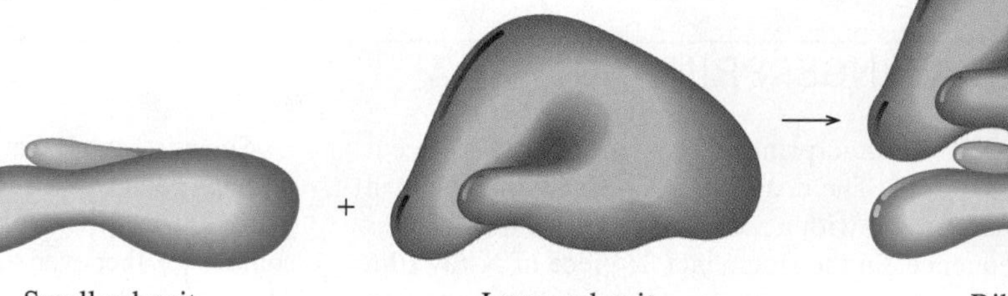

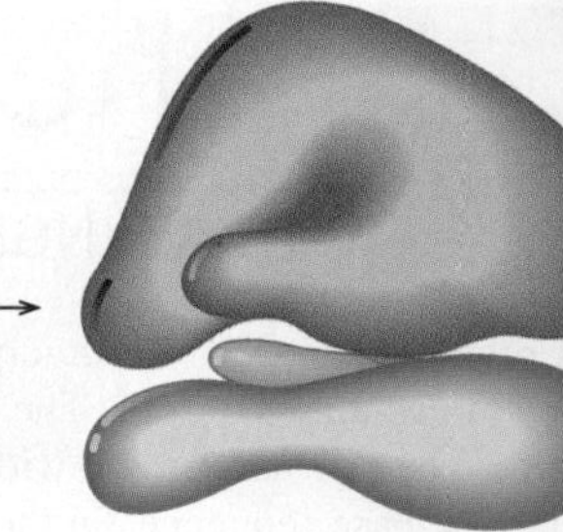

FIGURE 17.9 A typical ribosome consists of a small subunit and a large subunit.
Q Why would there be many thousands of ribosomes in a cell?

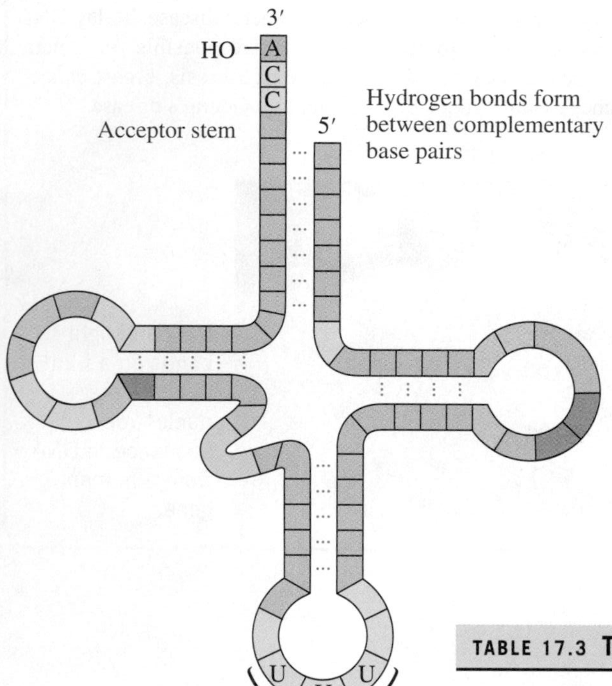

FIGURE 17.10 A tRNA has a cloverleaf shape when hydrogen bonds form between complementary bases within the tRNA. The acceptor stem attaches to an amino acid and its anticodon bonds with a codon on mRNA.
Q Why will different tRNAs have different bases in the anticodon loop?

Transfer RNA (**tRNA**), the smallest of the RNA molecules, interprets the genetic information in mRNA and brings specific amino acids to the ribosome for protein synthesis. Only tRNA can translate the genetic information into amino acids for proteins. There can be more than one tRNA for each of the 20 amino acids. The structures of the transfer RNAs are similar, consisting of 70 to 90 nucleotides. Hydrogen bonds between some of the complementary bases in the chain produce loops that give some double-stranded regions.

Although the structure of tRNA is complex, we draw tRNA as a cloverleaf to illustrate its features. All tRNA molecules have a 3′ end with the nucleotide sequence ACC, which is known as the *acceptor stem.* An enzyme attaches an amino acid by forming an ester bond with the free —OH at the end of the acceptor stem. Each tRNA contains an *anticodon,* which is a series of three bases that complements three bases on mRNA (see Figure 17.10). Table 17.3 summarizes the three types of RNA.

TABLE 17.3 Types of RNA Molecules in Humans

Type	Abbreviation	Percentage of Total RNA	Function in the Cell
Ribosomal RNA	rRNA	80	Major component of the ribosomes; protein synthesis
Messenger RNA	mRNA	5	Carries information for protein synthesis from the DNA in the nucleus to the ribosomes
Transfer RNA	tRNA	15	Brings amino acids to the ribosomes for protein synthesis

CONCEPT CHECK 17.4

Types of RNA

a. What is the function of mRNA in a cell?
b. What is the function of tRNA in a cell?

ANSWER

a. mRNA carries the instructions for the synthesis of a protein from the DNA in the nucleus to the ribosomes in the cytoplasm.
b. Each type of tRNA brings specific amino acids to the ribosome for protein synthesis.

We now look at the overall process involved in transferring genetic information encoded in the DNA to the production of proteins. In the nucleus, genetic information for the synthesis of a protein is copied from a gene in DNA to make messenger

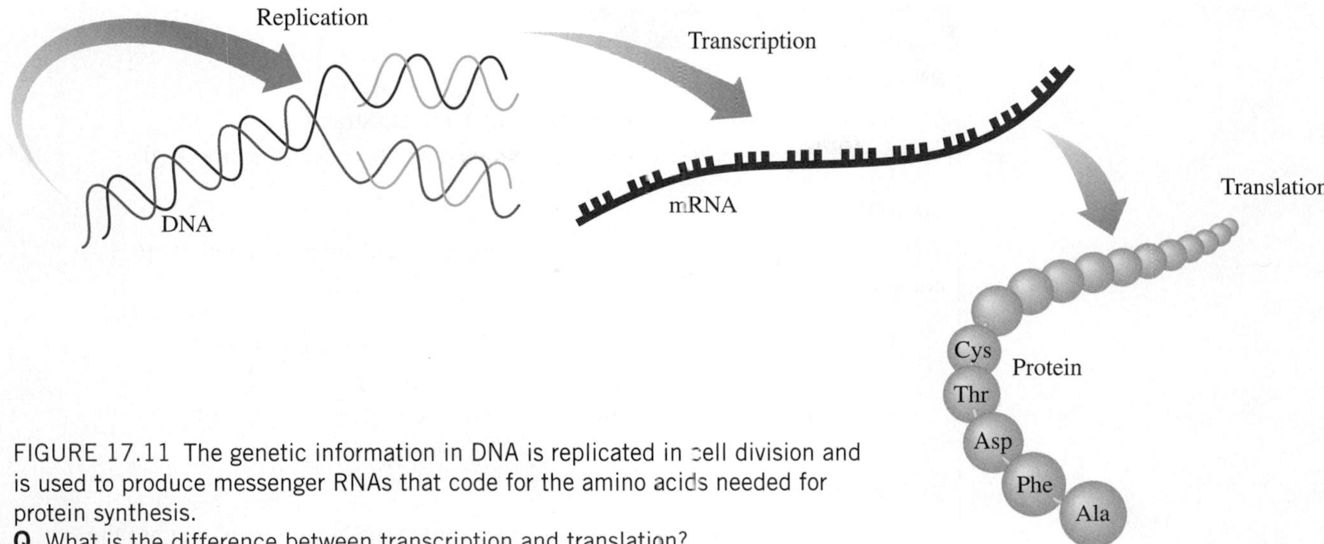

FIGURE 17.11 The genetic information in DNA is replicated in cell division and is used to produce messenger RNAs that code for the amino acids needed for protein synthesis.
Q What is the difference between transcription and translation?

RNA (mRNA), a process called **transcription**. The mRNA molecules move out of the nucleus into the cytoplasm where they combine with ribosomes. Then in a process called **translation**, tRNA molecules convert the information in the mRNA into amino acids, which are placed in the proper sequence to synthesize a protein (see Figure 17.11).

Protein Synthesis: Transcription

Transcription: Synthesis of mRNA

Transcription begins when the section of a DNA that contains the gene to be copied unwinds. Within the unwound DNA, RNA polymerase enzyme uses one of the strands as a template to synthesize mRNA. Just as in DNA synthesis, C pairs with G, T pairs with A, but in mRNA, U (not T) pairs with A. The RNA polymerase moves along the DNA template strand, forming bonds between the bases. When the RNA polymerase reaches the termination site, transcription ends and the new mRNA is released. The unwound section of the DNA returns to its double helix structure (see Figure 17.12).

SELF STUDY ACTIVITY
Transcription

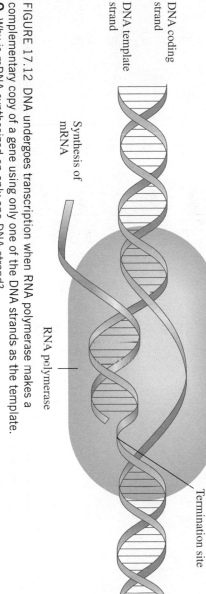

RNA polymerase →

Sequence of bases on DNA template:	—G—A—A—C—T—
Transcription	↓ ↓ ↓ ↓ ↓
Complementary base sequence in mRNA:	—C—U—U—G—A—

FIGURE 17.12 DNA undergoes transcription when RNA polymerase makes a complementary copy of a gene using only one of the DNA strands as the template.
Q Why is mRNA synthesized on only one DNA strand?

SAMPLE PROBLEM 17.4

RNA Synthesis

The sequence of bases in a part of the DNA template strand is —C—G—A—T—C—A—. What corresponding mRNA is produced?

SOLUTION

To form mRNA, the bases in the DNA template strand are paired with their complementary bases: G with C, C with G, T with A, and A with U.

DNA template —C—G—A—T—C—A—

↓ ↓ ↓ ↓ ↓ ↓

Complementary bases in mRNA —G—C—U—A—G—U—

STUDY CHECK 17.4

What is the DNA template strand that codes for the mRNA segment that has the sequence —G—G—G—U—U—U—A—A—A—?

TUTORIAL
Genetic Code

TUTORIAL
Following the Instructions in DNA

The Genetic Code

The **genetic code** consists of a series of three nucleotides (triplet) in mRNA called **codons** that specify the amino acids and their sequence in the protein. Early work on protein synthesis showed that repeating triplets of uracil (UUU) in mRNA produced a polypeptide that contained only phenylalanine. Therefore, a sequence of —UUU—UUU—UUU— codes for three phenylalanines.

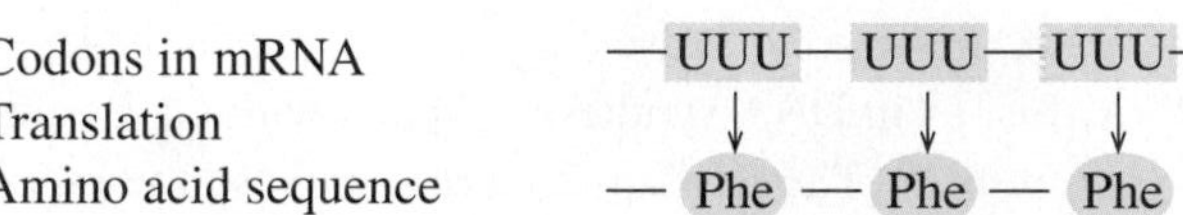

TABLE 17.4 Codons in mRNA: The Genetic Code for Amino Acids

First Letter	Second Letter: U	Second Letter: C	Second Letter: A	Second Letter: G	Third Letter
U	UUU Phe	UCU Ser	UAU Tyr	UGU Cys	U
U	UUC Phe	UCC Ser	UAC Tyr	UGC Cys	C
U	UUA Leu	UCA Ser	**UAA Stop	**UGA Stop	A
U	UUG Leu	UCG Ser	**UAG Stop	UGG Trp	G
C	CUU Leu	CCU Pro	CAU His	CGU Arg	U
C	CUC Leu	CCC Pro	CAC His	CGC Arg	C
C	CUA Leu	CCA Pro	CAA Gln	CGA Arg	A
C	CUG Leu	CCG Pro	CAG Gln	CGG Arg	G
A	AUU Ile	ACU Thr	AAU Asn	AGU Ser	U
A	AUC Ile	ACC Thr	AAC Asn	AGC Ser	C
A	AUA Ile	ACA Thr	AAA Lys	AGA Arg	A
A	*AUG Met/Start	ACG Thr	AAG Lys	AGG Arg	G
G	GUU Val	GCU Ala	GAU Asp	GGU Gly	U
G	GUC Val	GCC Ala	GAC Asp	GGC Gly	C
G	GUA Val	GCA Ala	GAA Glu	GGA Gly	A
G	GUG Val	GCG Ala	GAG Glu	GGG Gly	G

*Codon that signals the start of a peptide chain.

**Stop codons signal the end of a peptide chain.

Codons have been determined for all 20 amino acids. A total of 64 codons are possible from the triplet combinations of A, G, C, and U. Three of these, UGA, UAA, and UAG, are stop signals that code for the termination of protein synthesis. All the other three-base codons shown in Table 17.4 specify amino acids. Thus one amino acid can have several codons. For example, glycine has four codons: GGU, GGC, GGA, and GGG. The triplet AUG has two roles in protein synthesis. At the beginning of an mRNA, the codon AUG signals the start of protein synthesis. In the middle of a series of codons, the AUG codon specifies the amino acid methionine.

CONCEPT CHECK 17.5

The Genetic Code

Indicate the nucleotides in mRNA that code for the following:

a. the amino acid phenylalanine
b. the amino acid proline
c. the start of polypeptide synthesis

ANSWER

a. In mRNA, the codons for the amino acid phenylalanine are UUU and UUC.
b. In mRNA, the codons for the amino acid proline are CCU, CCC, CCA, and CCG.
c. In mRNA, the codon AUG signals the start of polypeptide synthesis.

SAMPLE PROBLEM 17.5

Codons

What is the sequence of amino acids specified by the following codons in mRNA?

—GUC—AGC—CCA—

SOLUTION

According to Table 17.4, GUC codes for valine, AGC for serine, and CCA for proline. The sequence of amino acids is Val–Ser–Pro.

STUDY CHECK 17.5

The codon UGA does not code for an amino acid. What is its function in mRNA?

QUESTIONS AND PROBLEMS

RNA and the Genetic Code

17.19 What are the three different types of RNA?

17.20 What are the functions of each type of RNA?

17.21 What is meant by the term "transcription"?

17.22 What bases in mRNA are used to complement the bases A, T, G, and C in DNA?

17.23 Write the segment of mRNA produced from the following section of a DNA template strand:

—C—C—G—A—A—G—G—T—T—C—A—C—

17.24 Write the segment of mRNA produced from the following section of a DNA template strand:

—T—A—C—G—G—C—A—A—G—C—T—A—

17.25 What is a codon?

17.26 What is the genetic code?

17.27 What amino acid is coded for by each codon?
a. CUU **b.** UCA
c. GGU **d.** AGG

17.28 What amino acid is coded for by each codon?
a. AAA **b.** UUC
c. CGG **d.** GCA

17.29 When does the codon AUG signal the start of a protein, and when does it code for the amino acid methionine?

17.30 The codons UGA, UAA, and UAG do not code for amino acids. What is their role as codons in mRNA?

17.5 Protein Synthesis

LEARNING GOAL

Describe the process of protein synthesis from mRNA.

Once the mRNA is synthesized, it migrates out of the nucleus into the cytoplasm to the ribosomes. In the *translation* process, tRNA molecules, amino acids, and enzymes convert the codons on mRNA into a protein.

SELF STUDY ACTIVITY
Overview of Protein Synthesis

SELF STUDY ACTIVITY
Translation

Protein Synthesis: Translation

Activation of tRNA

Each tRNA molecule contains a loop called an **anticodon**, which is a triplet of bases that complements a codon in mRNA (see Figure 17.13). A tRNA is activated for protein synthesis by aminoacyl tRNA synthetase, which attaches the correct amino acid to the acceptor stem.

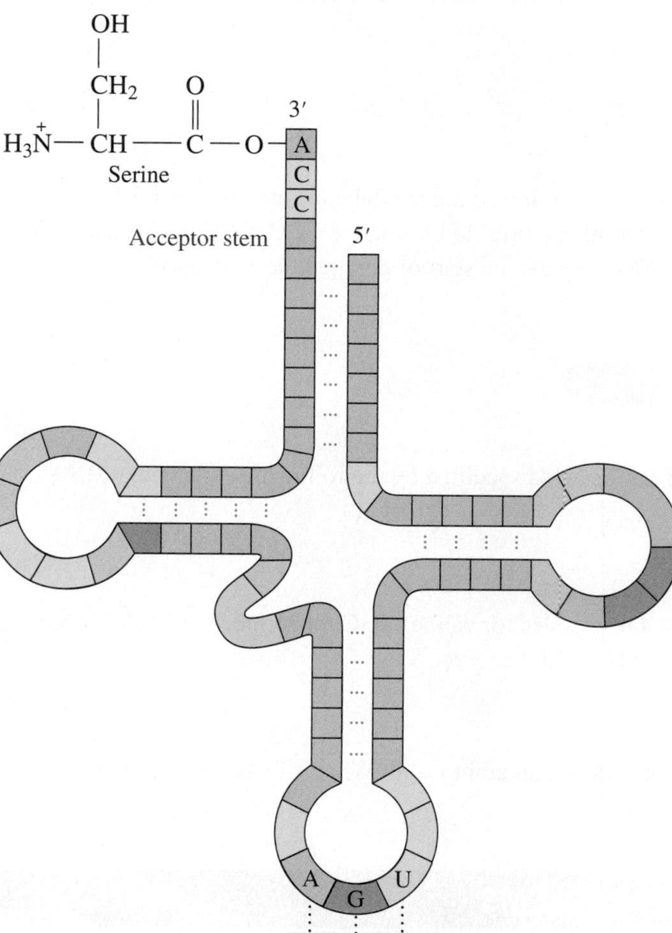

FIGURE 17.13 An activated tRNA with anticodon AGU bonds to serine at the acceptor stem.
Q What is the codon for serine for this tRNA?

Initiation and Chain Elongation

Protein synthesis begins when mRNA combines with a ribosome. The first codon in mRNA is a *start* codon, AUG, which forms hydrogen bonds with methionine-tRNA. Another tRNA hydrogen bonds to the next codon placing a second amino acid adjacent to methionine. Then a peptide bond forms between the C terminal of methionine and the N terminal of the second amino acid (see Figure 17.14). Once the peptide bond is formed, the initial tRNA detaches from the ribosome, which shifts to the next available codon, a process called *translocation*. During *chain elongation*, the ribosome moves along the mRNA from codon to codon, attaching new amino acids to the growing polypeptide chain. Sometimes several ribosomes, called a polysome, translate the same strand of mRNA to produce several copies of the peptide chain at the same time.

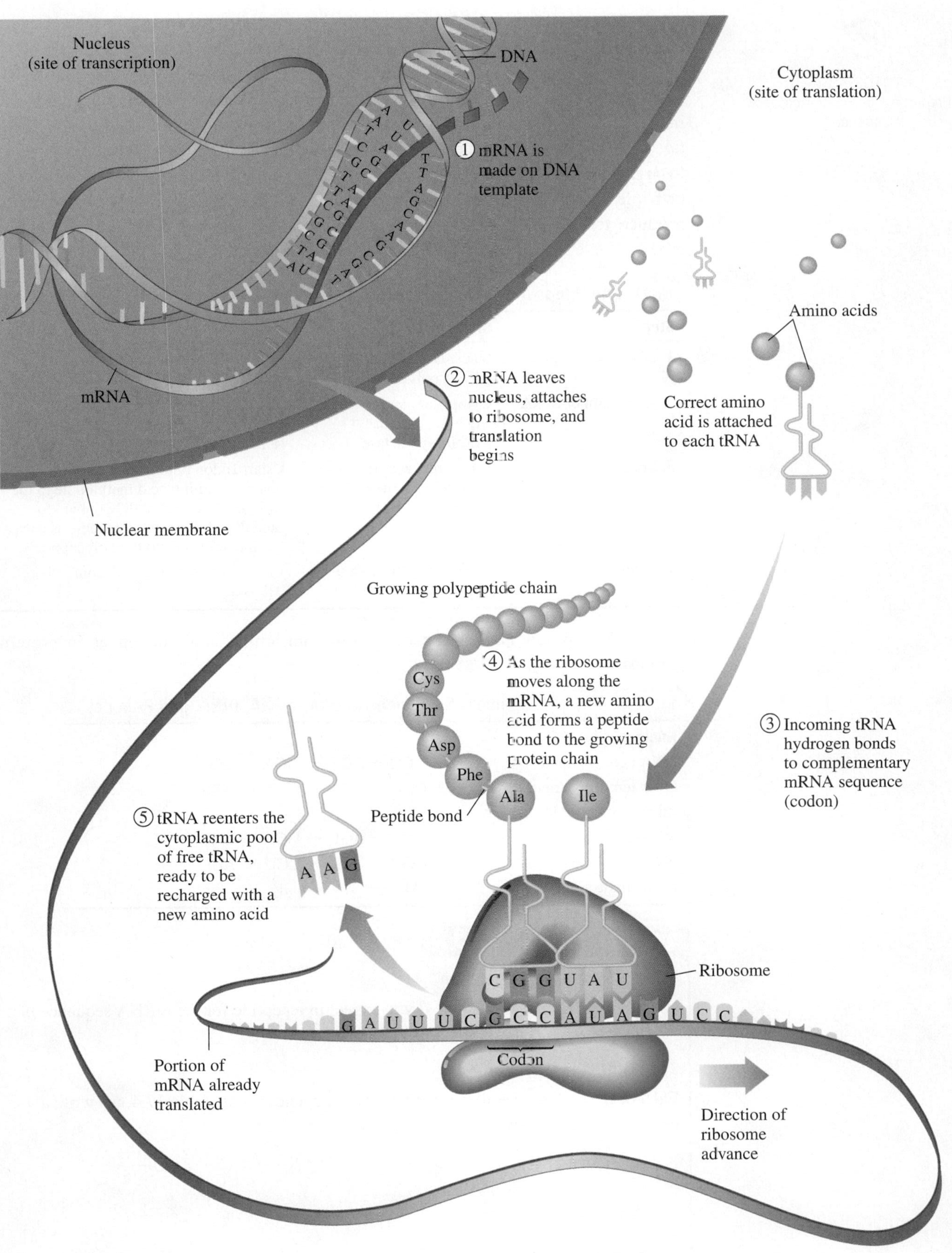
Nucleus
(site of transcription)
DNA
Cytoplasm
(site of translation)
① mRNA is made on DNA template
mRNA
Amino acids
② mRNA leaves nucleus, attaches to ribosome, and translation begins
Correct amino acid is attached to each tRNA
Nuclear membrane
Growing polypeptide chain
④ As the ribosome moves along the mRNA, a new amino acid forms a peptide bond to the growing protein chain
Cys
Thr
Asp
Phe
Ala
Ile
③ Incoming tRNA hydrogen bonds to complementary mRNA sequence (codon)
Peptide bond
⑤ tRNA reenters the cytoplasmic pool of free tRNA, ready to be recharged with a new amino acid
A A G
Ribosome
C G G U A U
G A U U U C G C C A U A G U C C
Codon
Portion of mRNA already translated
Direction of ribosome advance

FIGURE 17.14 In the translation process, the mRNA synthesized by transcription attaches to a ribosome and tRNAs pick up their amino acids and place them in a growing peptide chain.
Q How is the correct amino acid placed in the peptide chain?

TUTORIAL
Predicting an Amino Acid Sequence

Chain Termination

Eventually, the ribosome encounters one of the stop codons—UAA, UGA, or UAG—that have no corresponding tRNAs. A stop codon signals the termination of polypeptide synthesis and the polypeptide chain is released from the ribosome. The initial amino acid methionine is usually removed from the beginning of the polypeptide chain. Once released, the R groups of the amino acids in the new polypeptide can form hydrogen bonds to give the secondary structures of α helices, β-pleated sheets, or triple helices and form cross-links such as salt bridges and disulfide bonds to produce tertiary and quaternary structures, which makes it a biologically active protein. Table 17.5 summarizes the steps in protein synthesis.

TABLE 17.5 Steps in Protein Synthesis

Step	Site: Materials	Process
1. DNA Transcription	Nucleus: nucleotides, RNA polymerase	A DNA template is used to produce mRNA.
2. Activation of tRNA	Cytoplasm: amino acids, tRNAs, aminoacyl tRNA synthetase	Molecules of tRNA pick up specific amino acids according to their anticodons.
3. Initiation and Chain Elongation	Ribosome: Met-tRNA, mRNA, amino acyl tRNAs	A start codon binds the first tRNA carrying amino acid methionine to the mRNA. Successive tRNAs bind to and detach from the ribosome as each amino acid adds to the polypeptide.
4. Chain Termination	Ribosome: stop codon on mRNA	A polypeptide is released from ribosome.

Table 17.6 summarizes the nucleotide and amino acid sequences in protein synthesis.

TABLE 17.6 Complementary Sequences in DNA, mRNA, tRNA, and Peptides

Nucleus	
DNA coding strand	—GCG—AGT—GGA—TAC—
DNA template strand	—CGC—TCA—CCT—ATG—
Ribosome (cytoplasm)	
mRNA	—GCG—AGU—GGA—UAC—
tRNA anticodons	—CGC—UCA—CCU—AUG—
Polypeptide amino acids	—Ala—Ser—Gly—Tyr—

SAMPLE PROBLEM 17.6

Protein Synthesis: Translation

What order of amino acids would you expect in a peptide for the mRNA sequence of —UCA—AAA—GCC—CUU—?

SOLUTION

Each of the codons specifies a particular amino acid. Using Table 17.4, we write a peptide with the following amino acid sequence:

mRNA codons:	—UCA—AAA—GCC—CUU—
	↓ ↓ ↓ ↓
Amino acid sequence:	—Ser—Lys—Ala—Leu—

STUDY CHECK 17.6

Where would protein synthesis stop in the following series of bases in mRNA?

—GGG—AGC—AGU—UAG—GUU—

Chemistry Link to Health

MANY ANTIBIOTICS INHIBIT PROTEIN SYNTHESIS

Several antibiotics stop bacterial infections by interfering with the synthesis of proteins needed by the bacteria. Some antibiotics act only on bacterial cells by binding to the ribosomes in bacteria but do not act on human cells. A description of some of these antibiotics is given in Table 17.7.

TABLE 17.7 Antibiotics That Inhibit Protein Synthesis in Bacterial Cells

Antibiotic	Effect on Ribosomes to Inhibit Protein Synthesis
Chloramphenicol	Inhibits peptide bond formation and prevents the binding of tRNAs
Erythromycin	Inhibits peptide chain growth by preventing the translocation of the ribosome along the mRNA
Puromycin	Causes release of an incomplete protein by ending the growth of the polypeptide early
Streptomycin	Prevents the proper attachment of tRNAs
Tetracycline	Prevents the binding of tRNAs

QUESTIONS AND PROBLEMS

Protein Synthesis

17.31 What is the difference between a *codon* and an *anticodon*?

17.32 Why are there at least 20 different tRNAs?

17.33 What amino acid sequence would you expect from each of the following mRNA segments?

a. —ACC—ACA—ACU—
b. —UUC—CCG—UUC—CCA—
c. —UAC—GGG—AGA—UGU—

17.34 What amino acid sequence would you expect from each of the following mRNA segments?

a. —AAU—CCC—UUG—GCU—
b. —CCU—CGC—AGC—CCA—UGA—
c. —AUG—CAC—AAG—GAA—GUA—CUG—

17.35 How is a peptide chain extended?

17.36 What is meant by "translocation"?

17.37 The following portion of DNA is in the DNA template strand:

—GCT—TTT—CAA—AAA—

a. What is the corresponding mRNA section?
b. What are the anticodons of the tRNAs?
c. What amino acids will be placed in the peptide chain?

17.38 The following portion of DNA is in the DNA template strand:

—TGT—GGG—GTT—ATT—

a. What is the corresponding mRNA section?
b. What are the anticodons of the tRNAs?
c. What amino acids will be placed in the peptide chain?

17.6 Genetic Mutations

LEARNING GOAL

Describe some ways in which DNA is altered to cause mutations.

A **mutation** is a change in the nucleotide sequence of DNA. Such a change may alter the sequence of amino acids, affecting the structure and function of a protein in a cell. Mutations may result from X-rays, overexposure to sun (ultraviolet, or UV, light), chemicals called mutagens, and possibly some viruses. If a mutation occurs in a somatic cell (a cell other than a reproductive cell), the altered DNA will be limited to that cell and its daughter cells. If there is uncontrolled growth, the mutation could lead to cancer. If a mutation occurs in a germ cell (egg or sperm), then the DNA produced in a new individual will contain the same genetic change. When a mutation severely alters the function of structural proteins or enzymes, the new cells may not survive or the person may exhibit a genetic disease.

TUTORIAL
Genetic Mutations

Types of Mutations

Consider a triplet of bases CCG in the template strand of DNA, which produces the codon GGC in mRNA. At the ribosome, tRNA would place the amino acid glycine in the peptide chain (see Figure 17.15a). Now, suppose that T replaces the first C in the DNA triplet, which gives TCG as the triplet. Then the codon produced in the mRNA is AGC, which brings the tRNA with the amino acid serine to add to the peptide chain. The replacement of one base in the template strand of DNA with another is called a **substitution** or a *point* mutation. When there is a change of nucleotides in the codon, a different amino acid may be inserted into the polypeptide. However, if a substitution gives a codon for the same amino acid, there is no effect on the amino acid sequence in the protein. This type of substitution is a *silent mutation*. Substitution is the most common way in which mutations occur (see Figure 17.15b).

In a **frameshift mutation**, a base is inserted into or deleted from the normal order of bases in the template strand of DNA. Suppose that an A is deleted from the triplet AAA, giving a new triplet of AAC (see Figure 17.15c). The next triplet becomes CGA rather than CCG and so on. All the triplets shift by one base, which changes all the codons that follow and leads to a different sequence of amino acids.

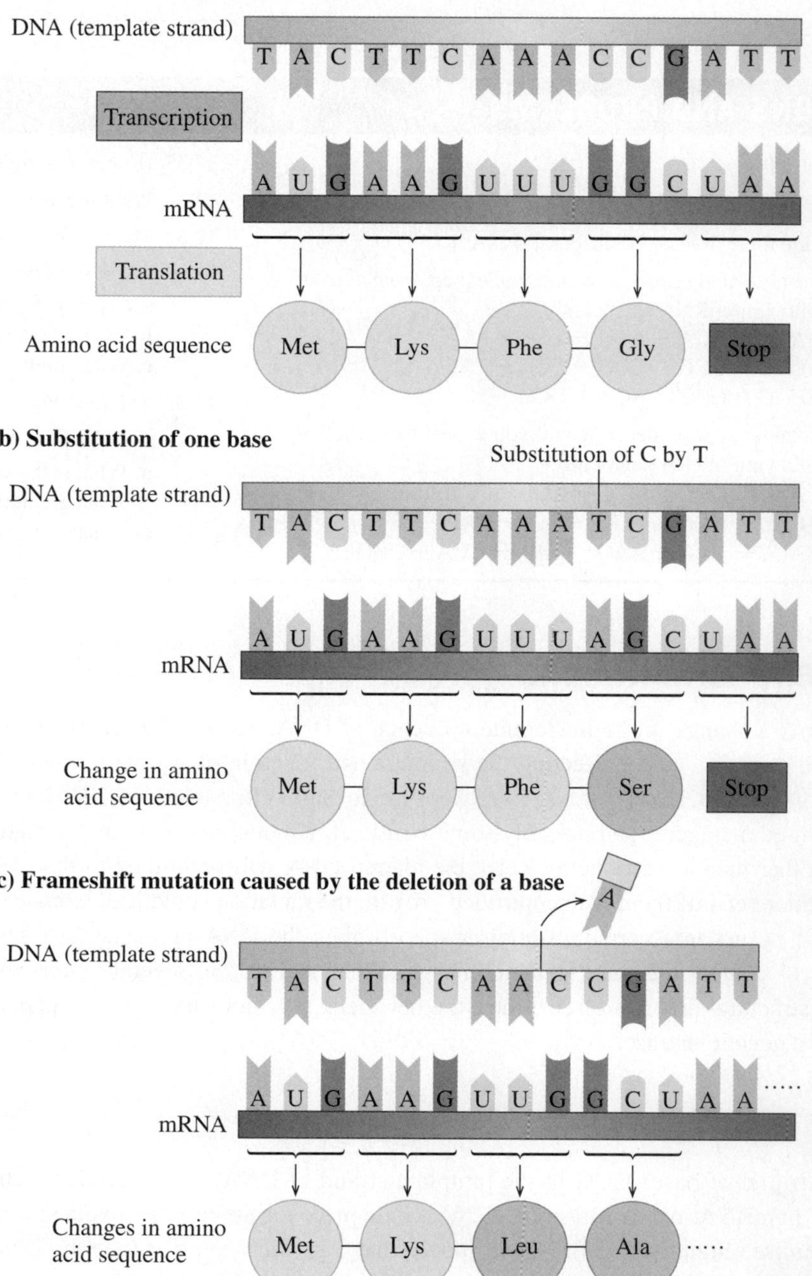

FIGURE 17.15 An alteration in the DNA template strand produces a change in the sequence of amino acids in the protein, which may lead to a mutation. **(a)** A normal DNA leads to the correct amino acid order in a protein. **(b)** The substitution of a base in DNA leads to a change in the mRNA codon and a change in the amino acid. **(c)** The deletion of a base causes a frameshift mutation, which changes the subsequent amino acid order.

Q When would a substitution mutation cause protein synthesis to stop?

Effect of Mutations

When a mutation causes a change in the amino acid sequence, the structure of the resulting protein can be severely altered and it may lose biological activity. If the protein is an enzyme, it may no longer bind to its substrate or react with the substrate at the active site. When an altered enzyme cannot catalyze a reaction, certain substances may accumulate until they act as poisons in the cell or substances vital to survival may not be synthesized. If a defective enzyme occurs in a major metabolic pathway or in the building of a cell membrane, the mutation can be lethal. When a protein deficiency is hereditary, the condition is called a **genetic disease.**

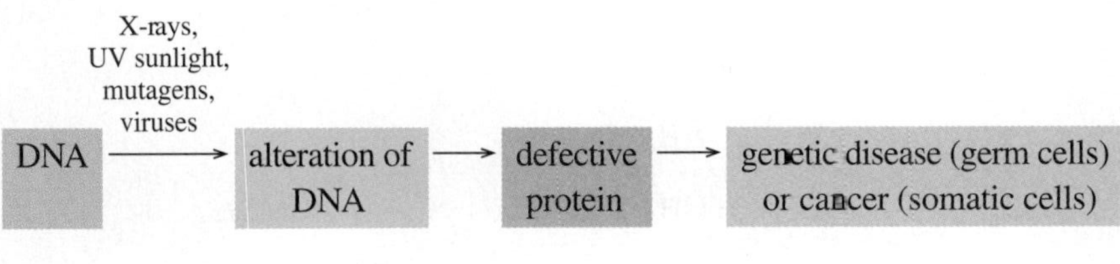

SAMPLE PROBLEM 17.7

Mutations

An mRNA has the sequence of codons —CCC—AGA—GCC—. If a base substitution in the DNA changes the mRNA codon of AGA to GGA, how is the amino acid sequence affected in the resulting protein?

SOLUTION

The initial mRNA sequence of —CCC—AGA—GCC— codes for the following amino acids: proline, arginine, and alanine. When the mutation occurs, the new sequence of mRNA codons is —CCC—GGA—GCC—, which codes for proline, glycine, and alanine. The basic amino acid arginine is replaced by nonpolar glycine.

	Normal	After Mutation
mRNA codons	—CCC—AGA—GCC—	—CCC—GGA—GCC—
Amino acids	— Pro — Arg — Ala —	— Pro — Gly — Ala —

STUDY CHECK 17.7

How might the protein made from this mRNA in Sample Problem 17.7 be affected by this mutation?

Genetic Diseases

A genetic disease is the result of a defective enzyme caused by a mutation in its genetic code. For example, phenylketonuria, PKU, results when DNA cannot direct the synthesis of the enzyme phenylalanine hydroxylase, required for the conversion of phenylalanine to tyrosine. In an attempt to break down the phenylalanine, other enzymes in the cells convert it to phenylpyruvate. The accumulation of phenylalanine and phenylpyruvate in the blood can lead to severe brain damage and mental retardation. If PKU is detected in a newborn baby, a diet is prescribed that eliminates all the foods that contain phenylalanine. Preventing the buildup of the phenylpyruvate ensures normal growth and development.

The amino acid tyrosine is needed in the formation of melanin, the pigment that gives the color to our skin and hair. If the enzyme that converts tyrosine to melanin is defective, no melanin is produced and a genetic disease known as albinism results.

Explore Your World

A MODEL FOR DNA REPLICATION AND MUTATION

1. Cut out 16 rectangular pieces of paper. Using 8 rectangular pieces for strand 1, label two each with each of the following nucleotide symbols: A═, T═, G≡, and C≡.
2. Using the other 8 rectangular pieces for strand 2, label two each with the following nucleotide symbols: ═A, ═T, ≡G, and ≡C.
3. Place the pieces for strand 1 in random order.
4. Using the DNA segment you made in part 3, select the correct bases to build the complementary segment of DNA strand 2.
5. Use the rectangular pieces for nucleotides to make a DNA segment of —A—T—T—G—C—C—.
6. In the DNA segment of part **5**, change the G to an A. What is the mRNA from this segment of DNA? What is the dipeptide that forms? What type of mutation is this?

FIGURE 17.16 This peacock with albinism does not produce the melanin needed to make the bright colors of its feathers.

Q Why are traits such as albinism related to the gene?

TUTORIAL
Mutations and Genetic Diseases

Persons and animals with no melanin have no skin or hair pigment (see Figure 17.16). Table 17.8 lists some other common genetic diseases and the type of metabolism or area affected.

$$C_6H_5{-}CH_2{-}\overset{\overset{+}{N}H_3}{\overset{|}{C}H}{-}COO^- \longrightarrow C_6H_5{-}CH_2{-}\overset{O}{\overset{\|}{C}}{-}COO^-$$

Phenylalanine → Phenylpyruvate → Phenylketonuria (PKU)

Phenylalanine hydroxylase (blocked)

$$HO{-}C_6H_4{-}CH_2{-}\overset{\overset{+}{N}H_3}{\overset{|}{C}H}{-}COO^- \not\longrightarrow \text{Melanin (pigments)}$$

Tyrosine → Albinism

TABLE 17.8 Some Genetic Diseases

Genetic Disease	Result
Galactosemia	In galactosemia, the transferase enzyme required for the metabolism of galactose-1-phosphate is absent, resulting in the accumulation of Gal-1-P, which leads to cataracts and mental retardation. Galactosemia occurs in about 1 in every 50 000 births.
Cystic fibrosis	Cystic fibrosis (CF) is caused by a mutation in the gene for the protein that regulates the production of stomach fluids and mucus. CF is one of the most common inherited diseases in children, in which thick mucus secretions make breathing difficult and block pancreatic function.
Down syndrome	Down syndrome is the leading cause of mental retardation, occurring in about 1 of every 800 live births, although mother's age strongly influences its occurrence. Mental and physical problems include heart and eye defects, which are the result of the formation of three chromosomes (trisomy), usually number 21.
Familial hyper-cholesterolemia	Familial hypercholesterolemia occurs when there is a mutation of a gene on chromosome 19, which produces high cholesterol levels that lead to early coronary heart disease.
Muscular dystrophy (Duchenne)	Muscular dystrophy, Duchenne form, is caused by a mutation in the X chromosome that destroys muscles at about age 5, with death by age 20. It occurs in about 1 of 10 000 males.
Huntington's disease (HD)	Huntington's disease affects the nervous system leading to total physical impairment. It is the result of a mutation in a gene on chromosome 4, which can now be mapped to test people in families with HD. There are about 30 000 people with Huntington's disease in the United States.
Sickle-cell anemia	Sickle-cell anemia is caused by a defective form of hemoglobin, resulting from a mutation in a gene on chromosome 11. It decreases the oxygen-carrying ability of red blood cells, which take on a sickled shape, causing anemia and plugged capillaries from red blood cell aggregation. In the United States, about 72 000 people are affected by sickle-cell anemia.
Hemophilia	Hemophilia is the result of one or more defective blood-clotting factors that lead to poor coagulation, excessive bleeding, and internal hemorrhages. There are about 20 000 hemophilia patients in the United States.
Tay-Sachs disease	Tay-Sachs disease is the result of a defective hexosaminidase A, which causes an accumulation of gangliosides and leads to mental retardation, loss of motor control, and early death.

QUESTIONS AND PROBLEMS

Genetic Mutations

17.39 What is a substitution mutation?

17.40 How does a substitution mutation in the genetic code for an enzyme affect the order of amino acids in that protein?

17.41 What is the effect of a frameshift mutation on the amino acid sequence in the polypeptide?

17.42 How can a mutation decrease the activity of a protein?

17.43 How is protein synthesis affected if the normal base sequence TTT in the DNA template strand is changed to TTC?

17.44 How is protein synthesis affected if the normal base sequence CCC in the DNA template strand is changed to ACC?

17.45 Consider the following segment in mRNA, which is produced by the normal order of DNA nucleotides:

—ACA—UCA—CGG—GUA—

a. What is the amino acid order produced from this mRNA?
b. What is the amino acid order if a mutation changes UCA to ACA?
c. What is the amino acid order if a mutation changes CGG to GGG?
d. What happens to protein synthesis if a mutation changes UCA to UAA?
e. What happens if a G is added to the beginning of the mRNA segment?
f. What happens if the A is removed from the beginning of the mRNA segment?

17.46 Consider the following segment in mRNA produced by the normal order of DNA nucleotides:

—CUU—AAA—CGA—GUU—

a. What is the amino acid order produced from this mRNA?
b. What is the amino acid order if a mutation changes CUU to CCU?
c. What is the amino acid order if a mutation changes CGA to AGA?
d. What happens to protein synthesis if a mutation changes AAA to UAA?
e. What happens if a G is added to the beginning of the mRNA segment?
f. What happens if the C is removed from the beginning of the mRNA segment?

17.47 **a.** A base substitution changes a codon for an enzyme from GCC to GCA. Why is there no change in the amino acid order in the protein?
b. In sickle-cell anemia, mutation causes valine to replace glutamic acid in the resulting hemoglobin. Why does the replacement of one amino acid cause such a drastic change in biological function?

17.48 **a.** A mutation causes alanine to replace leucine in the resulting enzyme. Why does this change in amino acids have little effect on the biological activity of the enzyme?
b. A base substitution replaces cytosine in the codon UCA with adenine. How would this substitution affect the amino acids in the protein?

17.7 Viruses

LEARNING GOAL

Describe the methods by which a virus infects a cell.

Viruses are small particles of 3 to 200 genes that cannot replicate without a host cell. A typical virus contains a nucleic acid, DNA or RNA, but not both, inside a protein coat. It does not have the necessary material such as nucleotides and enzymes to make proteins and grow. The only way a virus can replicate is to invade a host cell and take over the materials necessary for protein synthesis and growth. Some infections caused by viruses invading human cells are listed in Table 17.9. There are also viruses that attack bacteria, plants, and animals.

A viral infection begins when an enzyme in the protein coat makes a hole in the host cell, allowing the viral nucleic acids to enter and mix with the materials in the host cell (see Figure 17.17). If the virus contains DNA, the host cell begins to replicate the viral DNA in the same way it would replicate normal DNA. Viral DNA produces viral RNA, and a protease produces a protein coat to form a viral particle that leaves the cell (see Figure 17.18). The cell synthesizes so many virus particles that it eventually bursts and releases new viruses to infect more cells.

Vaccines are inactive forms of viruses that boost the immune response by causing the body to produce antibodies to the virus. Several childhood diseases such as polio, mumps, chicken pox, and measles can be prevented through the use of vaccines.

TABLE 17.9 Some Diseases Caused by Viral Infection

Disease	Virus
Common cold	Coronavirus (over 100 types)
Influenza	Orthomyxovirus
Warts	Papovavirus
Herpes	Herpesvirus
HPV	Human papillomavirus
Leukemia, cancers, AIDS	Retrovirus
Hepatitis	Hepatitis A virus (HAV), hepatitis B virus (HBV), hepatitis C virus (HCV)
Mumps	Paramyxovirus
Epstein–Barr	Epstein–Barr virus (EBV)
Chicken pox (shingles)	Varicella zoster virus (VZV)

Reverse Transcription

A virus that contains RNA as its genetic material is a **retrovirus**. Once inside the host cell, it must first make viral DNA using a process known as *reverse transcription*. A retrovirus contains a polymerase enzyme called *reverse transcriptase* that uses the viral

RNA template to synthesize complementary strands of DNA. Once produced, the DNA strands form double-stranded DNA using the nucleotides present in the host cell. This newly formed viral DNA, called a *provirus*, integrates with the DNA of the host cell.

Head
DNA
Tail
Tail fiber
Infected cell
Viral DNA replicates
Cellular enzymes → Viral protein + Viral DNA
Cellular enzymes make viral proteins and viral DNA, which assemble into viruses
Cell bursts
New viruses released
Uninfected cell
Uninfected cell

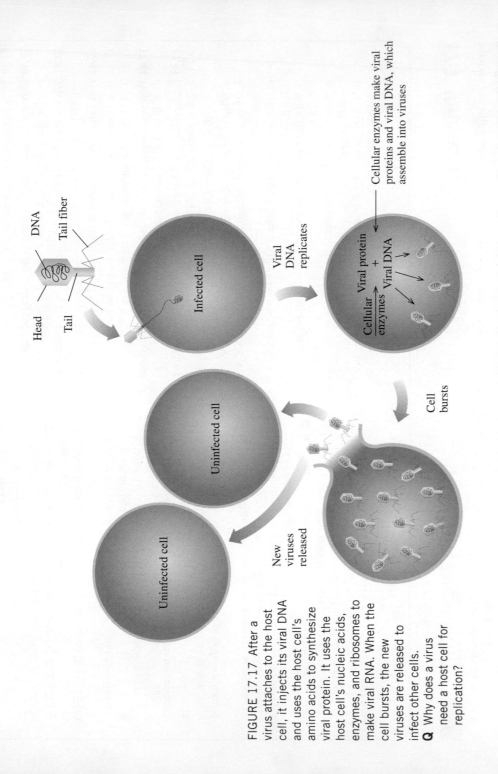

FIGURE 17.17 After a virus attaches to the host cell, it injects its viral DNA and uses the host cell's amino acids to synthesize viral protein. It uses the host cell's nucleic acids, enzymes, and ribosomes to make viral RNA. When the cell bursts, the new viruses are released to infect other cells.
Q Why does a virus need a host cell for replication?

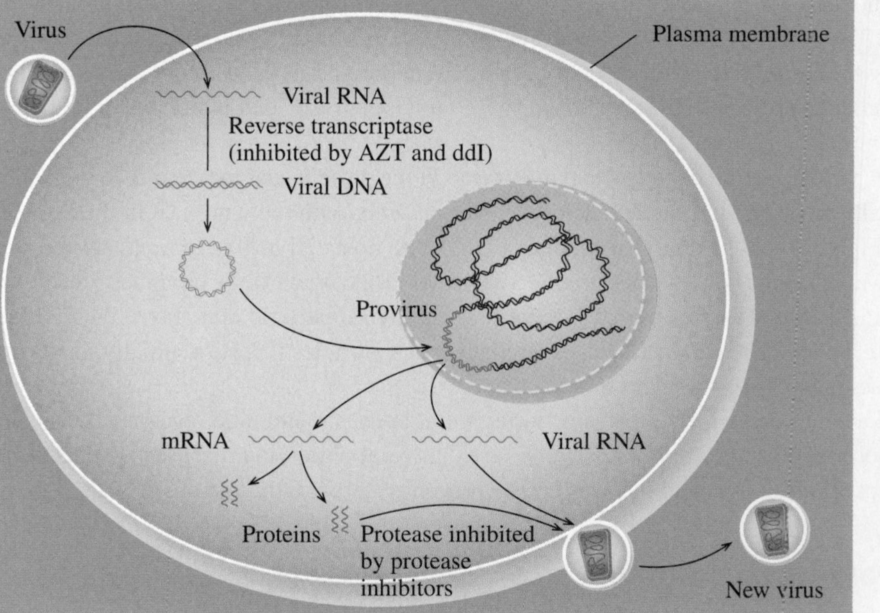

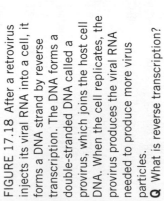

FIGURE 17.18 After a retrovirus injects its viral RNA into a cell, it forms a DNA strand by reverse transcription. The DNA forms a double-stranded DNA called a provirus, which joins the host cell DNA. When the cell replicates, the provirus produces the viral RNA needed to produce more virus particles.
Q What is reverse transcription?

AIDS

In the early 1980s, a disease called AIDS (acquired immune deficiency syndrome) began to claim an alarming number of lives. An HIV-1 virus (human immunodeficiency virus type 1) is now known to be the AIDS-causing agent (see Figure 17.19). HIV is a retrovirus that infects and destroys T4 lymphocyte cells, which are involved in the immune response. After the HIV-1 virus binds to receptors on the surface of a T4 cell, the virus injects viral RNA into the host cell. As a retrovirus, the genes of the viral RNA direct the formation of viral DNA. The gradual depletion of T4 cells reduces the ability of the immune system to destroy harmful organisms. The AIDS syndrome is characterized by opportunistic infections such as *Pneumocystis carinii*, which causes pneumonia, and *Kaposi's sarcoma*, a skin cancer.

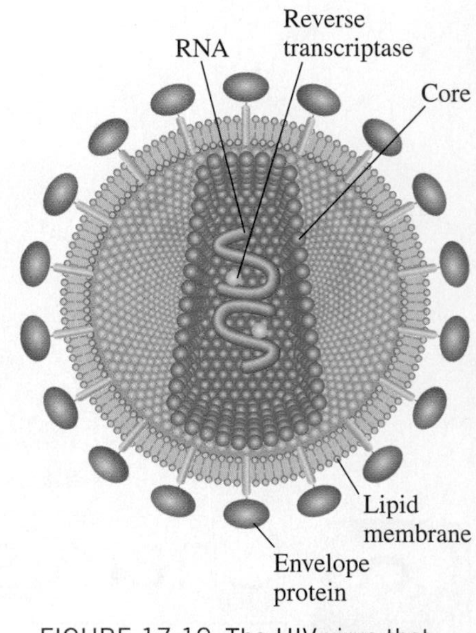

FIGURE 17.19 The HIV virus that causes AIDS destroys the immune system in the body.

Q Does the HIV virus contain DNA or RNA?

SELF STUDY ACTIVITY
HIV Reproductive Cycle

Treatment for AIDS is based on attacking the HIV-1 at different points in its life cycle, including reverse transcription and protein synthesis. Nucleoside analogs mimic the structures of the nucleosides used for DNA synthesis, which inhibit the reverse transcriptase enzyme. For example, the drug AZT (3′-azido-3′-deoxythymidine) is similar to thymidine, and ddI (2′,3′-dideoxyinosine) is similar to guanosine. Two other drugs are ddC (2′,3′-dideoxycytidine) and d4T (2′,3′-didehydro-2′,3′-dideoxythymidine). Such compounds are found in the "cocktails" that are providing extended remission of HIV infections. When a nucleoside analog is incorporated into viral DNA, the lack of a hydroxyl group on the 3′-carbon in the sugar prevents the formation of the sugar-phosphate bonds and stops the replication of the virus.

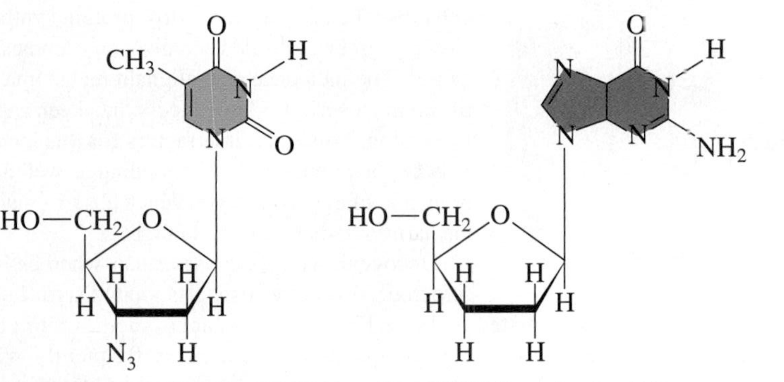

3′-Azido-3′-deoxythymidine (AZT)

2′,3′-Dideoxyinosine (ddI)

2′,3′-Dideoxycytidine (ddC)

2′,3′-Didehydro-2′,3′-dideoxythymidine (d4T)

Treatment of AIDS often combines reverse transcriptase inhibitors with protease inhibitors such as saquinavir (Invirase), indinavir, lexiva (fosamprenavir), nelfinavir, and ritonavir. The inhibition of a protease enzyme prevents the synthesis of proteins used by viruses to make more copies.

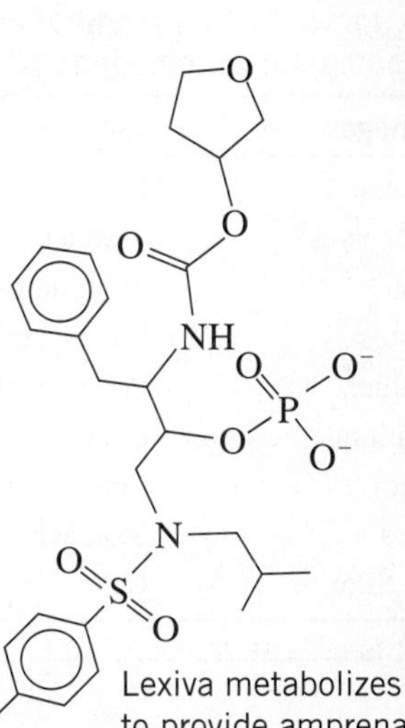

Lexiva metabolizes slowly to provide amprenavir, an HIV-protease inhibitor.

CONCEPT CHECK 17.6

Viruses

Why are viruses unable to replicate on their own?

ANSWER

Viruses contain only packets of DNA or RNA, but not the necessary replication machinery that includes enzymes and nucleosides.

Chemistry Link to Health

CANCER

In an adult body, many cells do not continue to reproduce. When cells in the body begin to grow and multiply without control, they invade neighboring cells and appear as a tumor. If these tumors are limited, they are benign. When they invade other tissues and interfere with normal functions of the body, the tumors are cancerous. Cancer can be caused by chemical and environmental substances, by radiation, or by oncogenic viruses, which are viruses associated with human cancers.

Some reports estimate that 70–80% of all human cancers are initiated by chemical and environmental substances. A carcinogen is any substance that increases the chance of inducing a tumor. Known carcinogens include aniline dyes, cigarette smoke, and asbestos. More than 90% of all persons with lung cancer are smokers. A carcinogen causes cancer by reacting with molecules in a cell, probably DNA, and altering the growth of that cell. Some known carcinogens are listed in Table 17.10.

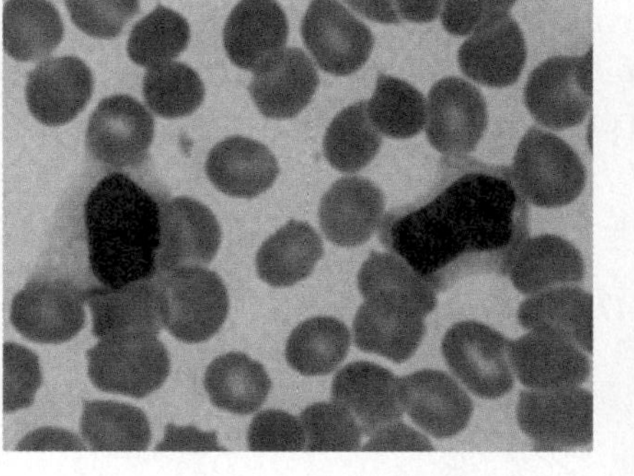

Epstein–Barr virus (EBV), herpesvirus 4, causes cancer in humans.

Radiant energy from sunlight or medical radiation is another type of environmental factor. Skin cancer has become one of the most prevalent forms of cancer. It appears that DNA damage in the exposed areas of the skin causes mutations. The cells lose their ability to control protein synthesis. This type of uncontrolled cell division becomes skin cancer. The incidence of malignant melanoma, one of the most serious skin cancers, has been rapidly increasing. Some possible factors for this increase may be the popularity of sun tanning as well as the reduction of the ozone layer, which absorbs much of the harmful radiation from sunlight.

Oncogenic viruses cause cancer when cells are infected. Several viruses associated with human cancers are listed in Table 17.11. Some cancers such as retinoblastoma and breast cancer appear to occur more frequently within families. There is some indication that a missing or defective gene may be responsible.

TABLE 17.10 Some Chemical and Environmental Carcinogens

Carcinogen	Tumor Site
Aflatoxin	Liver
Aniline dyes	Bladder
Arsenic	Skin, lungs
Asbestos	Lungs, respiratory tract
Cadmium	Prostate, kidneys
Chromium	Lungs
Nickel	Lungs, sinuses
Nitrites	Stomach
Vinyl chloride	Liver

TABLE 17.11 Human Cancers Caused by Oncogenic Viruses

Virus	Disease
RNA viruses	
Human T-cell lymphotropic virus-type 1 (HTLV-1)	Leukemia
DNA viruses	
Epstein–Barr virus (EBV)	Epstein–Barr
	Burkitt's lymphoma (cancer of white blood B cells)
	Nasopharyngeal carcinoma
	Hodgkin's disease
Hepatitis B virus (HBV)	Liver cancer
Herpes simplex virus (type 2)	Cervical and uterine cancer
Papilloma virus	Cervical and colon cancer, genital warts

QUESTIONS AND PROBLEMS

Viruses

17.49 What type of genetic information is found in a virus?

17.50 Why do viruses need to invade a host cell?

17.51 A specific virus contains RNA as its genetic material.
a. Why would reverse transcription be used in the life cycle of this type of virus?
b. What is the name of this type of virus?

17.52 What is the purpose of a vaccine?

17.53 How do nucleoside analogs disrupt the life cycle of the HIV-1 virus?

17.54 How do protease inhibitors disrupt the life cycle of the HIV-1 virus?

CONCEPT MAP

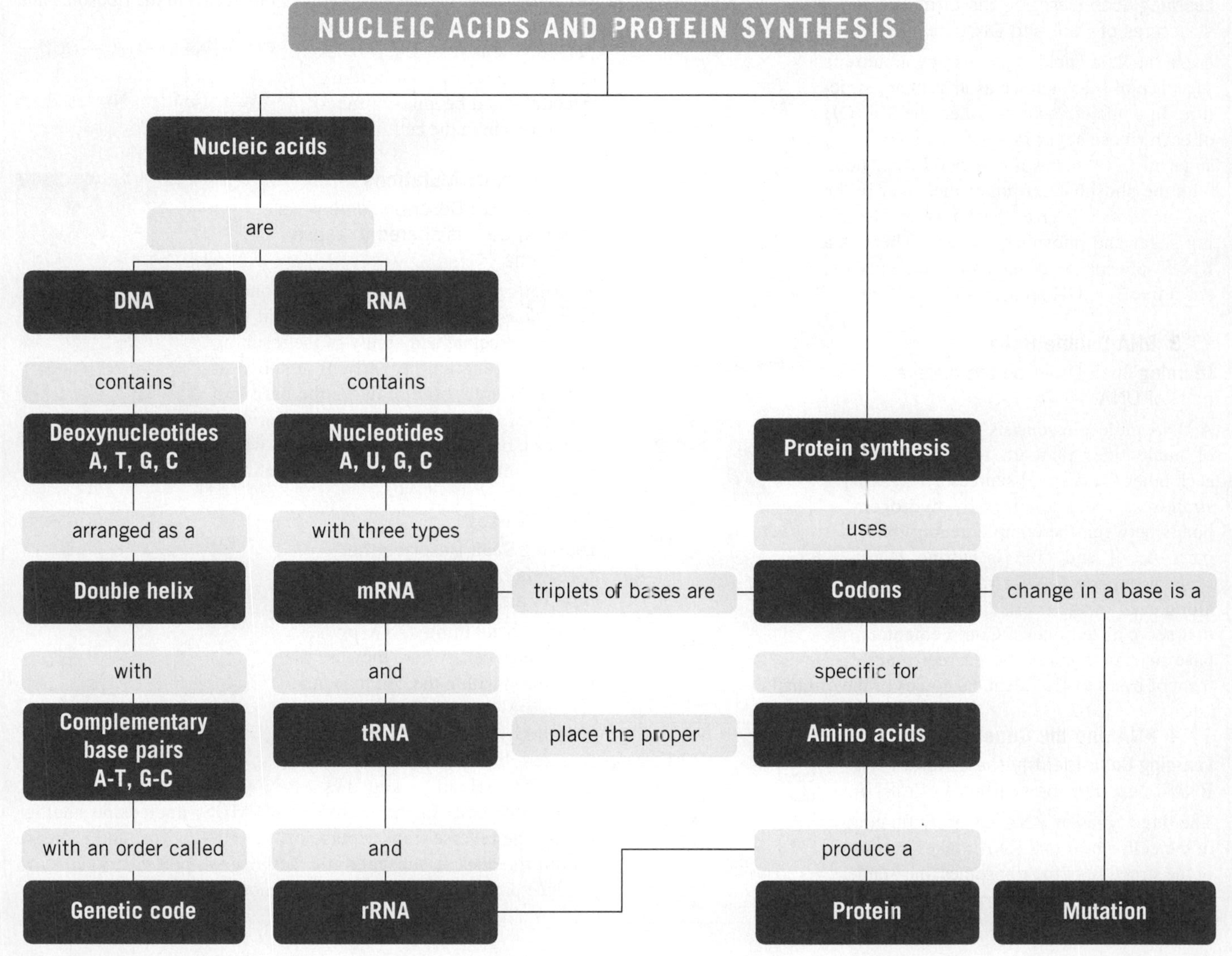

CHAPTER REVIEW

17.1 Components of Nucleic Acids

Learning Goal: Describe the bases and ribose sugars that make up the nucleic acids DNA and RNA.

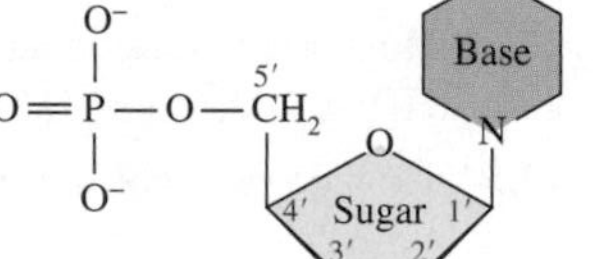

A nucleoside is a combination of a pentose sugar and a base. A nucleotide is composed of three parts: a base, a sugar, and a phosphate group. Nucleic acids, deoxyribonucleic acid (DNA) and ribonucleic acid (RNA), are polymers of nucleotides. In DNA, the sugar is deoxyribose and the base can be adenine, thymine, guanine, or cytosine. In RNA, the sugar is ribose and uracil replaces thymine.

17.2 Primary Structure of Nucleic Acids

Learning Goal: Describe the primary structures of RNA and DNA.

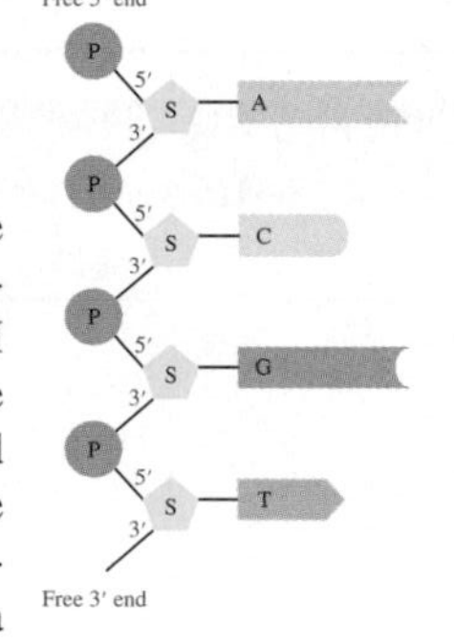

Each nucleic acid has its own unique sequence of bases known as its primary structure. In a nucleic acid polymer, the 3′—OH of each ribose sugar in RNA or deoxyribose sugar in DNA forms a phosphodiester bond with the phosphate group of the sugar in the next nucleotide to give a backbone of alternating sugar and phosphate groups. There is a free 5′-phosphate at one end of the polymer and a free 3′—OH group at the other end.

17.3 DNA Double Helix

Learning Goal: Describe the double helix of DNA.

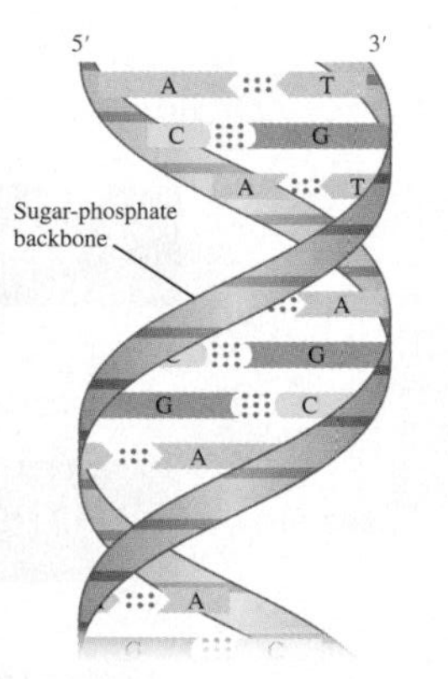

A DNA molecule consists of two strands of nucleotides that are wound around each other like a spiral staircase. The two strands are held together by hydrogen bonds between the complementary base pairs A—T and G—C. During DNA replication, new DNA strands are made along each of the original DNA strands that serve as templates. Complementary base pairing ensures the correct placement of bases to give identical copies of the original DNA.

17.4 RNA and the Genetic Code

Learning Goal: Identify the different RNAs; describe the synthesis of mRNA.

U		C	
UUU, UUC	Phe	UCU, UCC, UCA, UCG	Ser
UUA, UUG	Leu		
CUU, CUC, CUA, CUG	Leu	CCU, CCC, CCA, CCG	Pro

The three types of RNA differ by function in the cell: ribosomal RNA makes up most of the structure of the ribosomes, messenger RNA carries genetic information from the DNA to the ribosomes, and transfer RNA places the correct amino acids in the protein. Transcription is the process by which RNA polymerase produces mRNA from one strand of DNA. The bases in the mRNA are complementary to the DNA, except that A in DNA is paired with U in RNA. The production of mRNA occurs when certain proteins are needed in the cell. The genetic code consists of a sequence of three bases (triplet) that specifies the order for the amino acids in a protein. The codon AUG signals the start of transcription and codons UAG, UGA, and UAA signal it to stop.

17.5 Protein Synthesis

Learning Goal: Describe the process of protein synthesis from mRNA.

Proteins are synthesized at the ribosomes in a translation process that includes initiation, translocation, and termination. During translation, tRNAs bring the appropriate amino acids to the ribosome and peptide bonds form. When the polypeptide is released, it takes on its secondary and tertiary structures and becomes a functional protein in the cell.

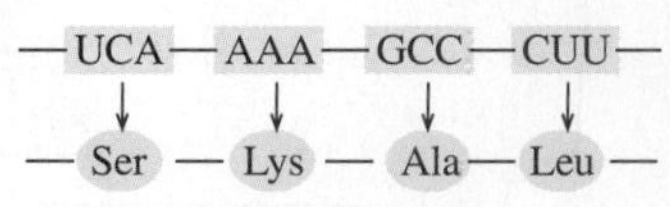

17.6 Genetic Mutations

Learning Goal: Describe some ways in which DNA is altered to cause mutations.

A genetic mutation is a change of one or more bases in the DNA sequence that may alter the structure and ability of the resulting protein to function properly. In a substitution, a different base replaces the base in a codon. A frameshift mutation inserts or deletes a base in the DNA sequence, which alters the mRNA codons after the mutation.

17.7 Viruses

Learning Goal: Describe the methods by which a virus infects a cell.

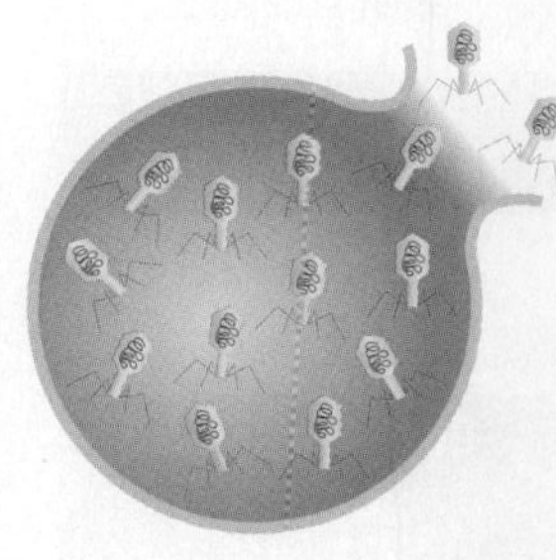

Viruses containing DNA or RNA invade host cells where they use the machinery within the cell to synthesize more viruses. For a retrovirus containing RNA, viral DNA is synthesized by reverse transcription using the nucleotides and enzymes in the host cell. In the treatment of AIDS, nucleoside analogs inhibit the reverse transcriptase of the HIV-1 virus, and protease inhibitors disrupt the catalytic activity of protease needed to produce proteins for the synthesis of more viruses.

Key Terms

anticodon The triplet of bases in the center loop of tRNA that is complementary to a codon on mRNA.

base Nitrogen-containing compounds of purine and pyrimidine found in DNA and RNA: adenine (A), thymine (T), cytosine (C), guanine (G), and uracil (U).

codon A sequence of three bases in mRNA that specifies a certain amino acid to be placed in a protein. A few codons signal the start or stop of protein synthesis.

complementary base pairs In DNA, adenine is always paired with thymine (A—T or T—A), and guanine is always paired with cytosine (G—C or C—G) . In forming RNA, adenine is paired with uracil (A—U).

DNA Deoxyribonucleic acid, the genetic material of cells, contains deoxyribose sugar, phosphate, and bases adenine, thymine, guanine, and cytosine.

double helix The helical shape of the double chain of DNA that is like a spiral staircase with a sugar-phosphate backbone on the outside and base pairs like stair steps on the inside.

frameshift mutation A mutation that inserts or deletes a base in a DNA sequence.

genetic code The information in DNA that is transferred to mRNA as a sequence of codons for the synthesis of protein.

genetic disease A physical malformation or metabolic dysfunction caused by a mutation in the base sequence of DNA.

mRNA Messenger RNA; produced in the nucleus by DNA to carry the genetic information to the ribosomes for the construction of a protein.

mutation A change in the DNA base sequence that alters the formation of a protein in the cell.

nucleic acids Large molecules composed of nucleotides, found as a double helix in DNA and as the single strands of RNA.

nucleoside The combination of a pentose sugar and a base.

nucleotides Building blocks of a nucleic acid, consisting of a base, a pentose sugar (ribose or deoxyribose), and a phosphate group.

phosphodiester bond The phosphate link that joins the 3′-hydroxyl group in one nucleotide to the phosphate group on the 5′-carbon atom in the next nucleotide.

primary structure The sequences of nucleotides in nucleic acids.

replication The process of duplicating DNA by pairing the bases on each parent strand with their complementary base.

retrovirus A virus that contains RNA as its genetic material and synthesizes a complementary DNA strand inside a cell.

RNA Ribonucleic acid is a single strand of nucleotides containing adenine, cytosine, guanine, and uracil.

rRNA Ribosomal RNA, the most prevalent type of RNA, is a major component of the ribosomes.

substitution A mutation that replaces one base in a DNA with a different base.

transcription The transfer of genetic information from DNA by the formation of mRNA.

translation The interpretation of the codons in mRNA as amino acids in a peptide.

tRNA Transfer RNA places a specific amino acid into a peptide chain at the ribosome using an anticodon. There is one or more tRNA for each of the 20 different amino acids.

virus Small particles containing DNA or RNA in a protein coat that require a host cell for replication.

Understanding the Concepts

17.55 Answer the following questions for the given section of DNA:

a. Complete the bases for the parent and new strands.

A		G	T				C	T	Parent strand
	C			G	C	G			New strand

b. Using the new strand as a template, write the mRNA sequence.

c. Write the three-letter symbols of the amino acids that would form the tripeptide from the mRNA you wrote in part **b**.

17.56 Suppose a mutation occurs in the DNA section in Problem 17.55 and the first base in the parent chain, adenine, is replaced by guanine.

a. What type of mutation has occurred?

b. Using the new strand that results from this mutation as a template, write the order of bases in the altered mRNA.

c. Write the three-letter symbols of the amino acids that would form the tripeptide from the mRNA you wrote in part **b**.

d. What effect, if any, might this mutation have on the structure and/or function of the resulting protein?

Additional Questions and Problems

For instructor-assigned homework, go to **www.masteringchemistry.com**.

17.57 Identify each of the following bases as a pyrimidine or a purine:
a. cytosine **b.** adenine
c. uracil **d.** thymine
e. guanine

17.58 Indicate if each of the bases in Problem 17.57 is found in DNA, RNA, or both.

17.59 Identify the base and sugar in each of the following nucleosides:
a. deoxythymidine **b.** adenosine
c. cytidine **d.** deoxyguanosine

17.60 Identify the base and sugar in each of the following nucleotides:
a. CMP
b. dAMP
c. dTMP
d. UMP

17.61 How do the bases thymine and uracil differ?

17.62 How do the bases cytosine and uracil differ?

17.63 What is similar about the primary structure of RNA and DNA?

17.64 What is different about the primary structure of RNA and DNA?

17.65 Write the complementary base sequence for each of the following DNA segments:
a. —G—A—C—T—T—A—G—G—C—
b. —T—G—C—A—A—A—C—T—A—G—C—T—
c. —A—T—C—G—A—T—C—G—A—T—C—G—

17.66 Write the complementary base sequence for each of the following DNA segments:
a. —T—T—A—C—G—G—A—C—C—G—C—
b. —A—T—A—G—C—C—C—T—T—A—C—T—G—G—
c. —G—G—C—C—T—A—C—C—T—T—A—A—C—G—

17.67 Match the following statements with rRNA, mRNA, or tRNA:
a. is the smallest type of RNA
b. makes up the highest percentage of RNA in the cell
c. carries genetic information from the nucleus to the ribosomes

17.68 Match the following statements with rRNA, mRNA, or tRNA:
a. combines with proteins to form ribosomes
b. brings amino acids to the ribosomes for protein synthesis
c. acts as a template for protein synthesis

17.69 What are the possible codons for each of the following amino acids?
a. threonine **b.** serine **c.** cysteine

17.70 What are the possible codons for each of the following amino acids?
a. valine **b.** arginine **c.** histidine

17.71 What is the amino acid for each of the following codons?
a. AAG **b.** AUU **c.** CGA

17.72 What is the amino acid for each of the following codons?
a. CAA **b.** GGC **c.** AAC

17.73 Endorphins are polypeptides that reduce pain. What is the amino acid order for the following mRNA that codes for a pentapeptide that is an endorphin called leucine enkephalin?
—AUG—UAC—GGU—GGA—UUU—CUA—UAA—

17.74 Endorphins are polypeptides that reduce pain. What is the amino acid order for the following mRNA that codes for a pentapeptide that is an endorphin called methionine enkephalin?
—AUG—UAC—GGU—GGA—UUU—AUG—UAA—

17.75 What is the anticodon on tRNA for each of the following codons in mRNA?
a. AGC **b.** UAU **c.** CCA

17.76 What is the anticodon on tRNA for each of the following codons in mRNA?
a. GUG **b.** CCC **c.** GAA

Challenge Questions

17.77 Oxytocin is a nonapeptide. How many nucleotides would be found in the mRNA for this protein?

17.78 A protein contains 36 amino acids. How many nucleotides would be needed to form the mRNA for this protein?

17.79 **a.** If the DNA double helix in salmon contains 28% adenine, what is the percent of thymine, guanine, and cytosine?
b. If the DNA double helix in humans contains 20% cytosine, what is the percent of guanine, adenine, and thymine?

17.80 Why are there no base pairs in DNA between adenine and guanine, or thymine and cytosine?

17.81 What is the difference between a DNA virus and a retrovirus?

17.82 Consider the following segment of a template strand of DNA:
—ATA—AGC—TTC—GAC—
a. What is the mRNA produced for the segment?
b. What is the amino acid sequence from the mRNA in part **a**?
c. What is the mRNA if a mutation changes AGC to AAC?
d. What is the amino acid sequence from the mRNA in part **c**?
e. What is the mRNA produced if G is inserted at the beginning of the DNA segment?
f. What is the amino acid sequence from the mRNA in part **e**?

Answers

Answers to Study Checks

17.1 deoxycytidine-5′-monophosphate (dCMP)

17.2

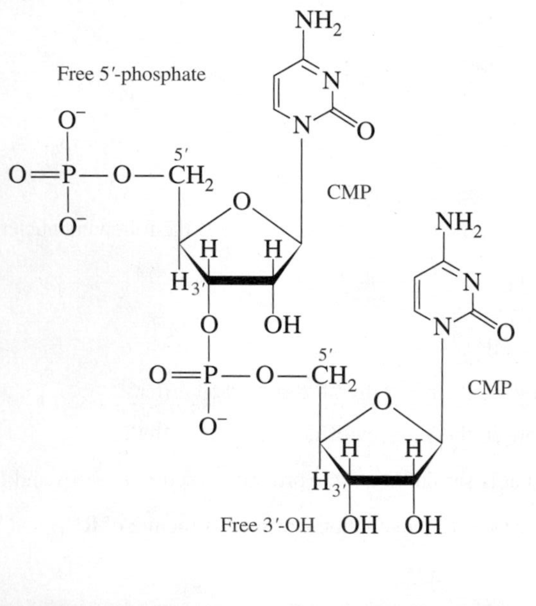

Free 5′-phosphate
CMP
CMP
Free 3′-OH

17.3 —C—C—A—A—T—T—G—G—

17.4 —C—C—C—A—A—A—T—T—T—

17.5 UGA is a stop codon that signals the termination of translation.

17.6 at UAG

17.7 If the substitution of an amino acid in the polypeptide affects an interaction essential to functional structure or the binding of a substrate, the resulting protein could be less effective or nonfunctional. In this example, glycine, a nonpolar amino acid, replaces arginine, a polar basic amino acid. The shape of this protein is likely to change because there will be disruptions of cross-links such as salt bridges and hydrophilic attractions.

Answers to Selected Questions and Problems

17.1 **a.** pyrimidine
b. pyrimidine

17.3 **a.** DNA
b. both DNA and RNA

17.5 deoxyadenosine-5′-monophosphate (dAMP), deoxythymidine-5′-monophosphate (dTMP), deoxycytidine-5′-monophosphate (dCMP), and deoxyguanosine-5′-monophosphate (dGMP)

17.7 **a.** nucleoside **b.** nucleoside
c. nucleoside **d.** nucleotide

17.9

17.11 The nucleotides in nucleic acids are held together by phosphodiester bonds between the 3′-OH of a sugar (ribose or deoxyribose) and a phosphate group on the 5′-carbon of another sugar.

17.13

Guanosine (G)

Cytidine (C)

17.15 The two DNA strands are held together by hydrogen bonds between the complementary bases in each strand.

17.17 **a.** —T—T—T—T—T—T—
b. —C—C—C—C—C—C—
c. —T—C—A—G—G—T—C—C—A—
d. —G—A—C—A—T—A—T—G—C—A—A—T—

17.19 ribosomal RNA, messenger RNA, and transfer RNA

17.21 In transcription, the sequence of nucleotides on a DNA template strand is used to produce the base sequences of a messenger RNA.

17.23 —G—G—C—U—U—C—C—A—A—G—U—G—

17.25 A codon is a three-base sequence in mRNA that codes for a specific amino acid in a protein.

17.27 **a.** leucine **b.** serine
c. glycine **d.** arginine

17.29 When AUG is the first codon, it signals the start of protein synthesis. Thereafter, AUG codes for methionine.

17.31 A codon is a base triplet in the mRNA. An anticodon is the complementary triplet on a tRNA for a specific amino acid.

17.33 **a.** —Thr—Thr—Thr—
b. —Phe—Pro—Phe—Pro—
c. —Tyr—Gly—Arg—Cys—

17.35 The new amino acid is joined by a peptide bond to the peptide chain. The ribosome moves to the next codon, which attaches to a tRNA carrying the next amino acid.

17.37 **a.** —CGA—AAA—GUU—UUU—
b. GCU, UUU, CAA, AAA
c. Using codons in mRNA: —Arg—Lys—Val—Phe—

17.39 In a substitution mutation, a base in DNA is replaced by a different base.

17.41 In a frameshift mutation, a base is lost or gained, which changes the remaining sequence of nucleotides in the codons and therefore the amino acids in the remaining polypeptide chain.

17.43 The normal triplet TTT forms a codon AAA, which codes for lysine. The mutation TTC forms a codon AAG, which also codes for lysine. There is no effect on the amino acid sequence.

17.45 **a.** —Thr—Ser—Arg—Val—
b. —Thr—Thr—Arg—Val—
c. —Thr—Ser—Gly—Val—
d. —Thr—STOP Protein synthesis would terminate early. If this occurs early in the formation of the polypeptide, the resulting protein will probably be nonfunctional.
e. The new protein will contain the sequence —Asp—Ile—Thr—Gly—.
f. The new protein will contain the sequence —His—His—Gly—.

17.47 **a.** GCC and GCA both code for alanine.
b. A cross-link in the tertiary structure of hemoglobin cannot be formed when the polar glutamic acid is replaced by nonpolar valine.

17.49 A virus contains DNA or RNA, but not both.

17.51 **a.** Viral RNA is used to synthesize viral DNA, which produces the mRNA to make the protein coat that allows the virus to replicate and leave the cell.
b. retrovirus

17.53 Nucleoside analogs such as AZT and ddI are similar to the nucleosides required to make viral DNA in reverse transcription. When they are incorporated into viral DNA, the lack of a hydroxyl group on the 3′-carbon in the sugar prevents the formation of the sugar-phosphate bonds and stops the replication of the virus.

17.55 **a.** **b.** **c.**

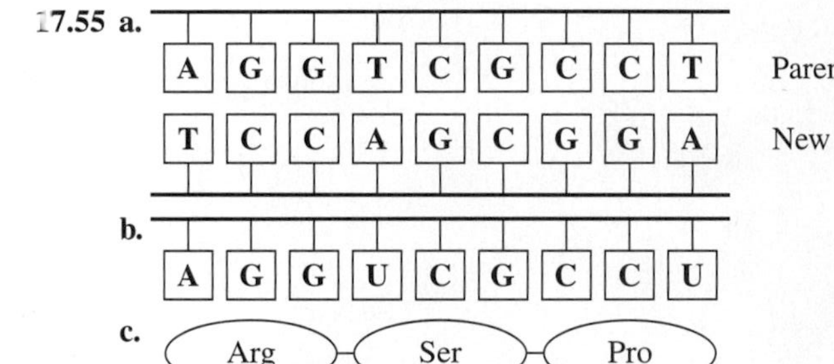

17.57 **a.** pyrimidine
b. purine
c. pyrimidine
d. pyrimidine
e. purine

17.59 **a.** thymine and deoxyribose
b. adenine and ribose
c. cytosine and ribose
d. guanine and deoxyribose

17.61 They are both pyrimidines, but thymine has a methyl group.

17.63 They are both polymers of nucleotides connected through phosphodiester bonds between alternating sugar and phosphate groups with bases extending out from each sugar.

17.65 **a.** —C—T—G—A—A—T—C—C—G—
b. —A—C—G—T—T—T—G—A—T—C—G—A—
c. —T—A—G—C—T—A—G—C—T—A—G—C—

17.67 **a.** tRNA **b.** rRNA **c.** mRNA

17.69 **a.** ACU, ACC, ACA, and ACG
b. UCU, UCC, UCA, UCG, AGU, and AGC
c. UGU and UGC

17.71 **a.** lysine
b. isoleucine
c. arginine

17.73 START—Tyr—Gly—Gly—Phe—Leu—STOP

17.75 **a.** UCG
b. AUA
c. GGU

17.77 Three nucleotides are needed for each amino acid, plus a start and stop triplet, making a minimum total of 33 nucleotides.

17.79 **a.** Because A bonds with T, T is also 28%. Thus the sum of A + T is 56%, which leaves 44% divided equally between G and C, 22% G and 22% C.
b. Because C bonds with G, G is also 20%. Thus the sum of C + G is 40%, which leaves 60% divided equally between A and T, 30% A and 30% T.

17.81 A DNA virus attaches to a cell and injects viral DNA that uses the host cell to produce copies of DNA to make viral RNA. A retrovirus injects viral RNA from which complementary DNA is produced by reverse transcription.

Metabolic Pathways and Energy Production

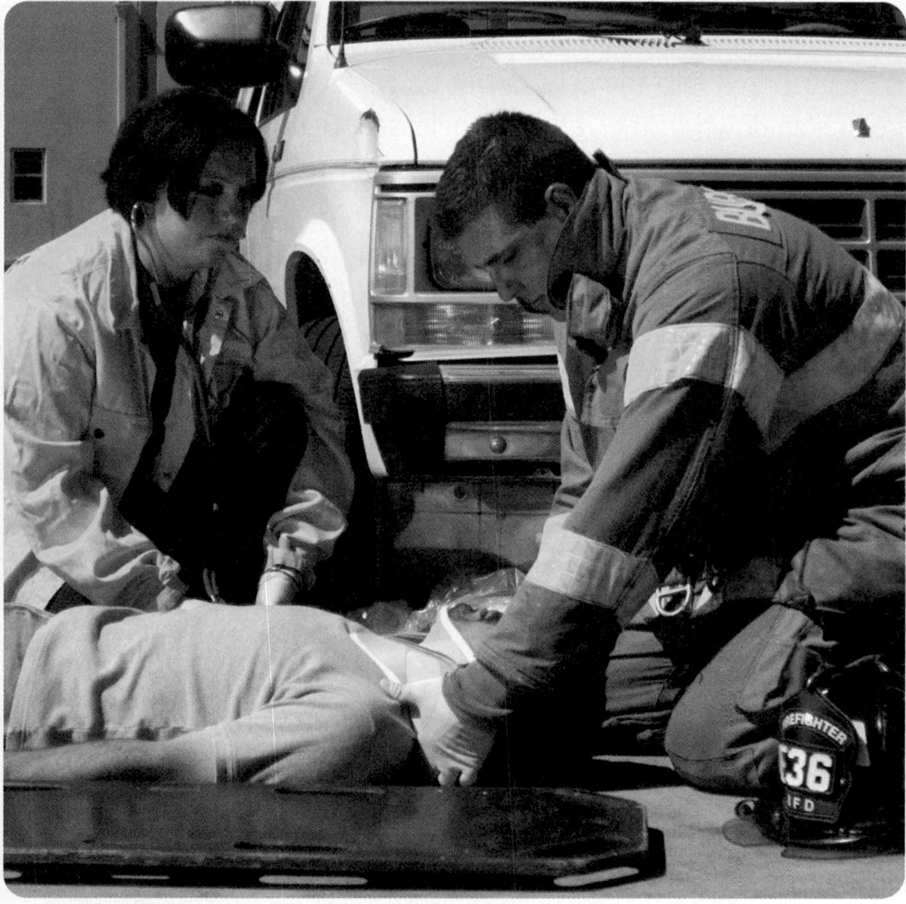

LOOKING AHEAD

"I am trained in basic life support. I work with the ER staff to assist in patient care," says Mandy Dornell, emergency medical technician at Seaton Medical Center. "In the ER, I take vital signs, do patient assessment, and do CPR. If someone has a motor vehicle accident, I may suspect a neck or back injury. Then I may use a backboard or a cervical collar, which prevents the patient from moving and causing further damage. When people have difficulty breathing, I insert an airway—nasal or oral—to assist ventilation. I also set up and monitor IVs, and I am trained in childbirth."

When someone is critically ill or injured, the quick reactions of emergency medical technicians (EMTs) and paramedics provide immediate medical care and transport to an ER or trauma center.

Visit **www.masteringchemistry.com** for self-study material and instructor-assigned homework.

All the chemical reactions that take place in cells to break down or build molecules are known as *metabolism*. A *metabolic pathway* is a series of linked reactions, each catalyzed by a specific enzyme. In this chapter, we will look at these pathways and the way they produce energy and cellular compounds.

When we eat food, the polysaccharides, lipids, and proteins are digested to smaller molecules that can be absorbed into the cells of our body. As the glucose, fatty acids, and amino acids are broken down further, energy is released. Because we do not use all the energy from our foods at one time, we store energy in the cells as high-energy adenosine triphosphate, ATP. Our cells can then break down ATP and obtain energy to do work in our bodies: contracting muscles, synthesizing large molecules, sending nerve impulses, and moving substances across cell membranes.

In the *citric acid cycle*, a series of metabolic reactions in the mitochondria oxidizes the two carbon atoms in the acetyl component of acetyl CoA to two molecules of CO_2. The reduced coenzymes NADH and $FADH_2$ enter *electron transport* where they provide hydrogen ions and electrons that combine with oxygen (O_2) to form H_2O. The energy released during electron transport is used to synthesize ATP from ADP and P_i.

LEARNING GOAL

Describe the three stages of metabolism and the role of ATP.

18.1 Metabolism and ATP Energy

The term **metabolism** refers to all the chemical reactions that provide energy and the substances required for continued cell growth. There are two types of metabolic reactions: catabolic and anabolic. In **catabolic reactions**, complex molecules are broken down to simpler ones with an accompanying release of energy. **Anabolic reactions** utilize energy available in the cell to build large molecules from simple ones.

We can think of the catabolic processes in metabolism as consisting of three stages (see Figure 18.1).

Stage 1 Catabolism begins with the processes of **digestion** in which enzymes in the digestive tract break down large molecules into smaller ones. The polysaccharides break down to monosaccharides, fats break down to glycerol and fatty acids, and the proteins yield amino acids. These digestion products diffuse into the bloodstream for transport to cells.

Stage 2 Within the cells, catabolic reactions continue as the digestion products are broken down further to yield two- and three-carbon compounds such as pyruvate and acetyl CoA.

Stage 3 The major production of energy takes place in the mitochondria, as the two-carbon acetyl group is oxidized in the citric acid cycle, which produces reduced coenzymes NADH and $FADH_2$. As long as the cells have oxygen, the hydrogen ions and electrons from the reduced coenzymes are transferred to electron transport to synthesize ATP.

Stages of Metabolism

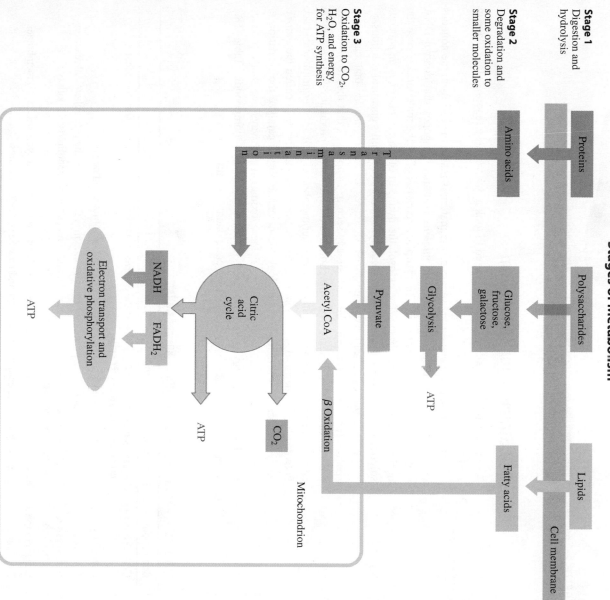

FIGURE 18.1 In the three stages of catabolism, large molecules from foods are digested and degraded to give smaller molecules that can be oxidized to produce energy.

Q Where is most of the ATP energy produced in the cells?

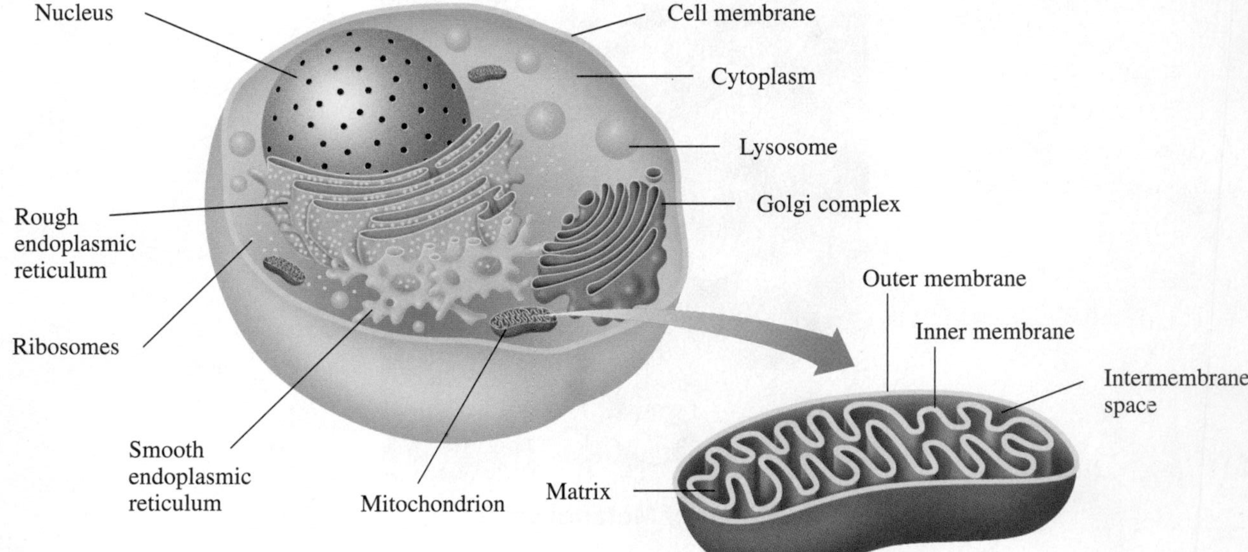

FIGURE 18.2 The diagram illustrates the major components of a typical animal cell.
Q What is the function of the mitochondria in a cell?

TUTORIAL
Metabolism and Cell Structure

Cell Structure for Metabolism

To understand metabolic reactions, we need to look at where these reactions take place in the cell (see Figure 18.2).

In animals, a *cell membrane* separates the materials inside the cell from the aqueous environment surrounding the cell. The *nucleus* contains the genes that control DNA replication and protein synthesis. The **cytoplasm** consists of all the materials between the nucleus and the cell membrane. The **cytosol**, the fluid part of the cytoplasm, is an aqueous solution of electrolytes and enzymes that catalyze many of the cell's chemical reactions.

Within the cytoplasm are specialized structures called *organelles* that carry out specific functions in the cell. We have already seen in Chapter 17 that the *ribosomes* are the sites of protein synthesis. The **mitochondria** are the energy-producing factories of the cells. A mitochondrion has an outer and an inner membrane, with an intermembrane space between them. The fluid section surrounded by the inner membrane is called the *matrix.* Enzymes located in the matrix and along the inner membrane catalyze the oxidation of carbohydrates, fats, and amino acids. All these oxidation pathways eventually produce CO_2, H_2O, and energy, which is used to form energy-rich compounds. Table 18.1 summarizes some of the functions of the components in animal cells.

TABLE 18.1 Functions of Components in Animal Cells

Component	Description and Function
Cell membrane	Separates the contents of a cell from the external environment and contains structures that communicate with other cells
Cytoplasm	Consists of the cellular contents between the cell membrane and nucleus
Cytosol	Is the fluid part of the cytoplasm that contains enzymes for many of the cell's chemical reactions
Endoplasmic reticulum	Rough type processes proteins for secretion and synthesizes phospholipids; smooth type synthesizes fats and steroids
Golgi complex	Modifies and secretes proteins from the endoplasmic reticulum and synthesizes cell membranes
Lysosome	Contains hydrolytic enzymes that digest and recycle old cell structures
Mitochondrion	Contains the structures for the synthesis of ATP from energy-producing reactions
Nucleus	Contains genetic information for the replication of DNA and the synthesis of protein
Ribosome	Is the site of protein synthesis using mRNA templates

CONCEPT CHECK 18.1

Metabolism and Cell Structure

Identify each of the following as a catabolic or an anabolic reaction:

a. digestion of polysaccharides **b.** synthesis of proteins

c. oxidation of glucose to CO_2 and H_2O

ANSWER

a. The breakdown of large molecules is a catabolic reaction.

b. The synthesis of large molecules requires energy and involves anabolic reactions.

c. The breakdown of glucose to smaller molecules is a catabolic reaction.

ATP and Energy

SELF STUDY ACTIVITY
ATP

In our cells, the energy released from the oxidation of the food we eat is stored in the form of a "high-energy" compound called *adenosine triphosphate* (ATP). The **ATP** molecule is composed of the base adenine, a ribose sugar, and three phosphate groups.

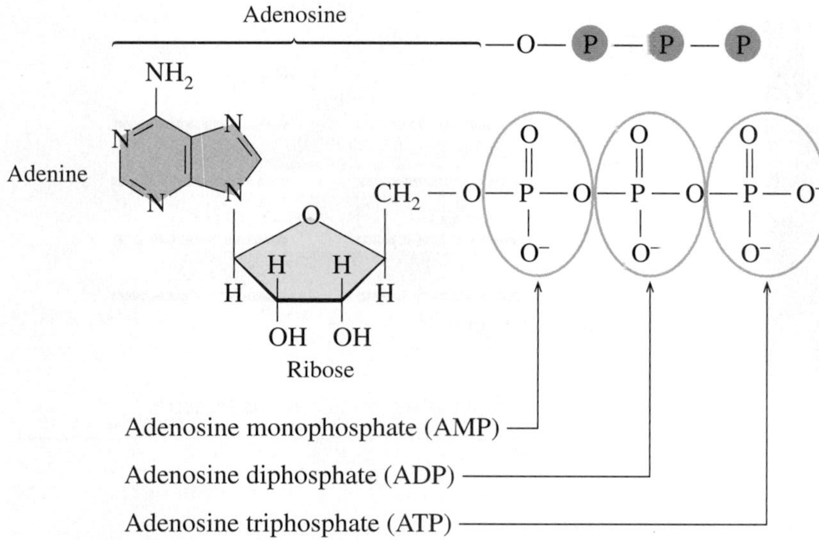

ATP is the energy-storage molecule in the body.

TUTORIAL
ATP: Energy Storage

When ATP undergoes hydrolysis, the products are adenosine diphosphate (ADP), a phosphate group abbreviated as P_i, and energy of 7.3 kcal per mole of ATP. We can write this reaction as

$$ATP + H_2O \longrightarrow ADP + P_i + 7.3\text{ kcal/mole}$$

The ADP can also hydrolyze to form adenosine monophosphate (AMP) and an inorganic phosphate (P_i).

$$ADP + H_2O \longrightarrow AMP + P_i + 7.3\text{ kcal/mole}$$

Every time we contract muscles, move substances across cellular membranes, send nerve signals, or synthesize an enzyme, we use energy from ATP hydrolysis. In a cell that is doing work (anabolic processes), 1–2 million ATP molecules may be hydrolyzed in one second. The amount of ATP hydrolyzed in one day can be as much as our body mass, even though only about 1 gram of ATP is present in all our cells at any given time.

When we take in food, the resulting catabolic reactions provide energy to regenerate ATP in our cells. Then 7.3 kcal/mole is used to make ATP from ADP and P_i.

$$ADP + P_i + 7.3\text{ kcal/mole} \longrightarrow ATP + H_2O$$

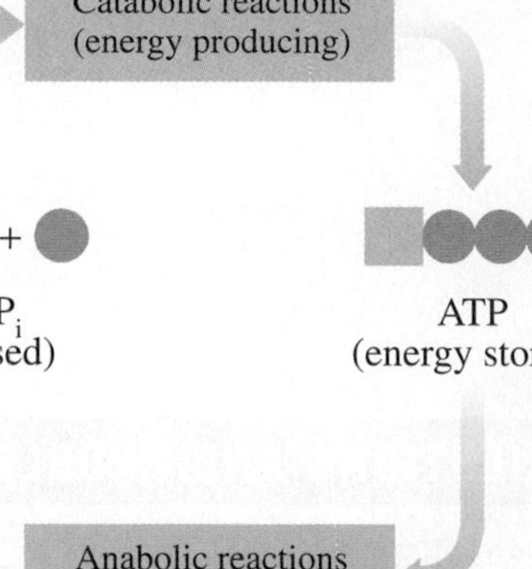

ATP, the energy-storage molecule, links energy-producing reactions with energy-requiring reactions in the cells.

Chemistry Link to Health

ATP ENERGY AND Ca^{2+} NEEDED TO CONTRACT MUSCLES

Our muscles consist of thousands of parallel fibers. Within these muscle fibers are fibrils composed of two kinds of proteins, called filaments. Arranged in alternating rows, the thick filaments of myosin overlap the thin filaments containing actin. During a muscle contraction, the thin filaments slide inward over the thick filaments, causing a shortening of the muscle fibers.

Calcium ion, Ca^{2+}, and ATP play an important role in muscle contraction. An increase in the Ca^{2+} concentration in the muscle fibers causes the filaments to slide, while a decrease stops the process. In a relaxed muscle, the Ca^{2+} concentration is low. However, when a nerve impulse reaches the muscle, calcium channels in the membrane open, and Ca^{2+} flows into the fluid surrounding the filaments. The muscle contracts as myosin binds to actin and pulls the filaments inward. The energy for the contraction is provided by splitting ATP to ADP + P_i. Muscle contraction continues as long as both ATP and Ca^{2+} levels are high around the filaments. When the nerve impulse ends, the calcium channels close. The Ca^{2+} concentration decreases as energy from ATP pumps Ca^{2+} out of the filaments, which causes the muscle to relax. In rigor mortis, Ca^{2+} concentration remains high within the muscle fibers causing a continued state of rigidity. After approximately 24 hours, Ca^{2+} decreases because of cellular deterioration and muscles relax.

Muscle contraction uses the energy from the breakdown of ATP.

Actin / Myosin / Actin

Relaxed muscle

Ca^{2+} — ATP → ADP + P_i

Contracted muscle

Muscles contract when myosin binds to actin.

CONCEPT CHECK 18.2

Components of ATP

Describe the components of ADP and ATP.

ANSWER

ADP and ATP both contain adenosine, which is the nitrogen-containing base adenine and pentose sugar ribose. In ADP, adenosine is attached to two phosphate groups, but in ATP, it is attached to three phosphate groups.

QUESTIONS AND PROBLEMS

Metabolism and ATP Energy

18.1 What stage of metabolism involves the digestion of polysaccharides?

18.2 What stage of metabolism involves the conversion of small molecules to CO_2, H_2O, and energy?

18.3 What is meant by a catabolic reaction in metabolism?

18.4 What is meant by an anabolic reaction in metabolism?

18.5 Why is ATP considered an energy-rich compound?

18.6 How much energy is obtained from the hydrolysis of ATP?

18.2 Digestion of Foods

In stage 1 of catabolism, foods undergo digestion, a process that converts large molecules to smaller ones that can be absorbed by the body.

LEARNING GOAL

Give the sites and products of digestion for carbohydrates, triacylglycerols, and proteins.

TUTORIAL
Breakdown of Carbohydrates

Digestion of Carbohydrates

We begin the digestion of carbohydrates as soon as we chew food. Enzymes produced in the salivary glands hydrolyze some of the α-glycosidic bonds in amylose and amylopectin, producing maltose, glucose, and smaller polysaccharides called dextrins, which contain three to eight glucose units. After swallowing, the partially digested starches enter the acidic environment of the stomach, where the low pH stops carbohydrate digestion (see Figure 18.3).

In the small intestine, which has a pH of about 8, enzymes produced in the pancreas hydrolyze the remaining dextrins to maltose and glucose. Then enzymes produced in the mucosal cells that line the small intestine hydrolyze maltose as well as lactose and sucrose. The monosaccharides are absorbed through the intestinal wall into the bloodstream, which carries them to the liver, where the hexoses fructose and galactose are converted to glucose.

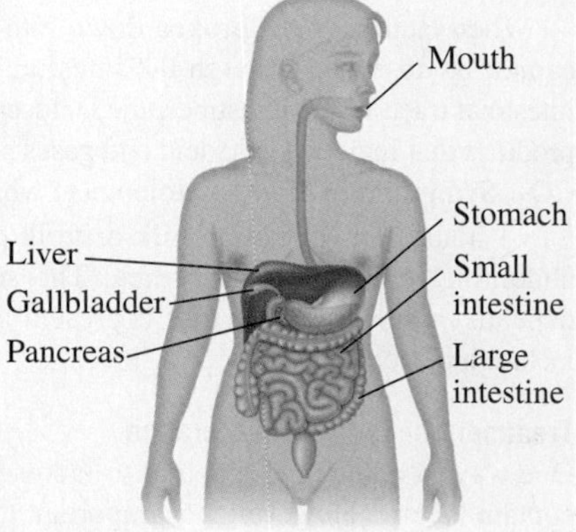

Carbohydrates begin digestion in the mouth, lipids in the small intestine, and proteins in the stomach.

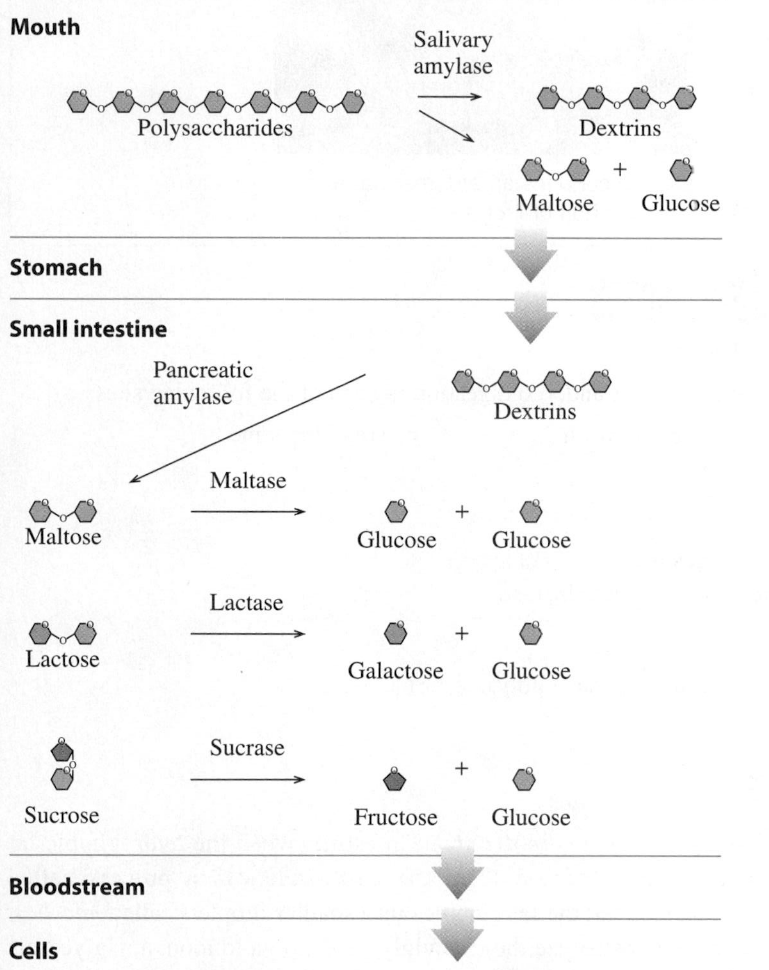

FIGURE 18.3 In stage 1 of metabolism, the digestion of carbohydrates begins in the mouth and is completed in the small intestine.

Q Why is there little or no digestion of carbohydrates in the stomach?

CARBOHYDRATE DIGESTION

1. Obtain a cracker or small piece of bread and chew it for 4–5 minutes. During that time, observe any change in the taste.
2. Some milk products contain Lactaid, which is the lactase that digests lactose. Look for the brands of milk and ice cream that contain Lactaid or lactase enzyme.

QUESTIONS

1a. How does the taste of the cracker or bread change after you have chewed it for 4–5 minutes? What could be an explanation?

b. What part of carbohydrate digestion occurs in the mouth?

2a. Write an equation for the hydrolysis of lactose.

b. Where does lactose undergo digestion?

Chemistry Link to Health

LACTOSE INTOLERANCE

The disaccharide in milk is lactose, which is broken down by *lactase* in the intestinal tract to monosaccharides that are a source of energy. Infants and small children produce lactase to break down the lactose in milk. It is rare for an infant to lack the ability to produce lactase. However, the production of lactase decreases as many people age, which causes lactose intolerance. This condition affects approximately 25% of the people in the United States. A deficiency of lactase occurs in adults throughout the world, but in the United States it is prevalent among the African American, Hispanic, and Asian populations.

When lactose is not broken down into glucose and galactose, it cannot be absorbed through the intestinal wall and remains in the intestinal tract. In the intestines, the lactose undergoes fermentation to products that include lactic acid and gases such as methane (CH_4) and CO_2. Symptoms of lactose intolerance, which appear approximately $\frac{1}{2}$ to 1 hour after ingesting milk or milk products, include nausea, abdominal cramps, and diarrhea. The severity of the symptoms depends on how much lactose is present in the food and how much lactase a person produces.

Treatment of Lactose Intolerance

One way to reduce the reaction to lactose is to avoid products that contain lactose. However, it is important to consume foods that provide the body with calcium. Many people with lactose intolerance seem to tolerate yogurt, which is a good source of calcium. Although there is lactose in yogurt, the bacteria in yogurt may produce some lactase, which helps to digest the lactose. A person who is lactose intolerant should also know that some foods that may not seem to be dairy products contain lactose. For example, baked goods, cereals, breakfast drinks, salad dressings, and even lunchmeat can contain lactose in their ingredients. Labels must be read carefully to see if the ingredients include "milk" or "lactose."

The enzyme lactase is now available in many forms such as tablets that are taken with meals, drops that are added to milk, or as additives in many dairy products such as milk. When lactase is added to milk that is left in the refrigerator for 24 hours, the lactose level is reduced by 70–90%. Lactase pills or chewable tablets are taken when a person begins to eat a meal that contains dairy foods. If taken too far ahead of the meal, too much of the lactase will be degraded by stomach acid. If taken following a meal, the lactose will have entered the lower intestine.

Lactaid contains an enzyme that aids the digestion of lactose.

SAMPLE PROBLEM 18.1

Digestion of Carbohydrates

Indicate the carbohydrates that undergo digestion in each of the following sites:

a. mouth **b.** stomach **c.** small intestine

SOLUTION

a. polysaccharides
b. no carbohydrate digestion
c. dextrins, maltose, sucrose, and lactose

STUDY CHECK 18.1

Describe the digestion of amylose, a polysaccharide.

TUTORIAL
Digestion of Triacylglycerols

Digestion of Fats

The digestion of dietary fats begins in the small intestine, when the hydrophobic fat globules mix with bile salts released from the gallbladder. In a process called *emulsification*, the bile salts break the fat globules into smaller droplets called *micelles*. Enzymes from the pancreas hydrolyze the triacylglycerols to yield monoacylglycerols and fatty acids, which are absorbed into the intestinal lining where they recombine to form triacylglycerols. These nonpolar compounds are then coated with proteins to form *chylomicrons*, which are more polar and soluble in the aqueous environment of the lymph and bloodstream (see Figure 18.4).

Small intestine

$$\begin{array}{l} CH_2 - \text{Fatty acid} \\ | \\ HC - \text{Fatty acid} + 2H_2O \\ | \\ CH_2 - \text{Fatty acid} \end{array} \xrightarrow{\text{Pancreatic lipase}} \begin{array}{l} CH_2 - OH \\ | \\ HC - \text{Fatty acid} + 2 \text{ Fatty acids} \\ | \\ CH_2 - OH \end{array}$$

Triacylglycerol → Monoacylglycerol

Intestinal wall

Triacylglycerols + Protein

Chylomicrons

Lymphatic system

Bloodstream

Cells

Glycerol + Fatty acids

FIGURE 18.4 The triacylglycerols are hydrolyzed in the small intestine and re-formed in the intestinal wall where they bind to proteins for transport through the lymphatic system and bloodstream to the cells.

Q Why do chylomicrons form in the intestinal wall?

The chylomicrons transport the triacylglycerols to the cells of the heart, muscle, and adipose tissues. When energy is needed in the cells, enzymes hydrolyze the triacylglycerols to yield glycerol and fatty acids.

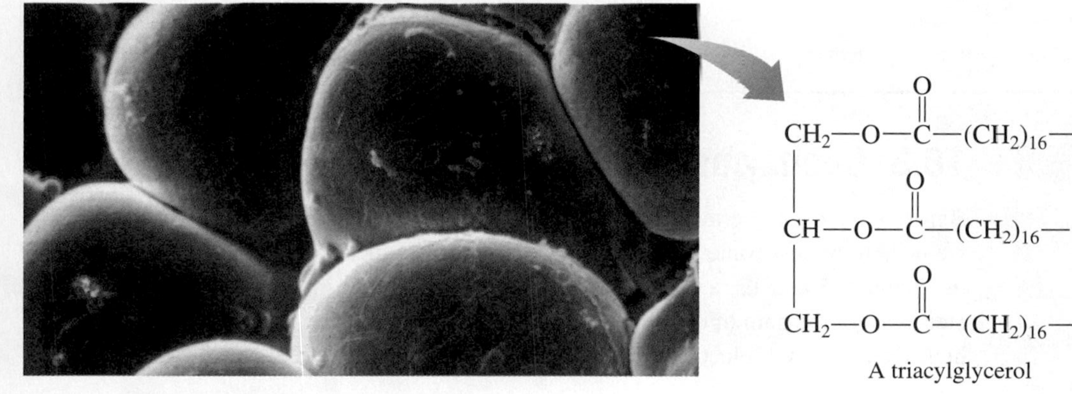

$$\begin{array}{l} CH_2-O-\overset{O}{\overset{\|}{C}}-(CH_2)_{16}-CH_3 \\ | \\ CH-O-\overset{O}{\overset{\|}{C}}-(CH_2)_{16}-CH_3 \\ | \\ CH_2-O-\overset{O}{\overset{\|}{C}}-(CH_2)_{16}-CH_3 \end{array}$$

A triacylglycerol

The fat cells, which make up adipose tissue, store unlimited quantities of triacylglycerols.

CONCEPT CHECK 18.3

Fats and Digestion

a. What are the sites and products for the digestion of triacylglycerols?

b. How are the products from the hydrolysis of triacylglycerols transported to the cells?

ANSWER

a. Triacylglycerols are hydrolyzed in the small intestine to yield monoacylglycerols and fatty acids.

b. In the membrane of the small intestine, monoacylglycerols and fatty acids recombine to form new triacylglycerols that are coated with proteins to form chylomicrons, which are more soluble in water for transport through the lymphatic system and the bloodstream to the cells.

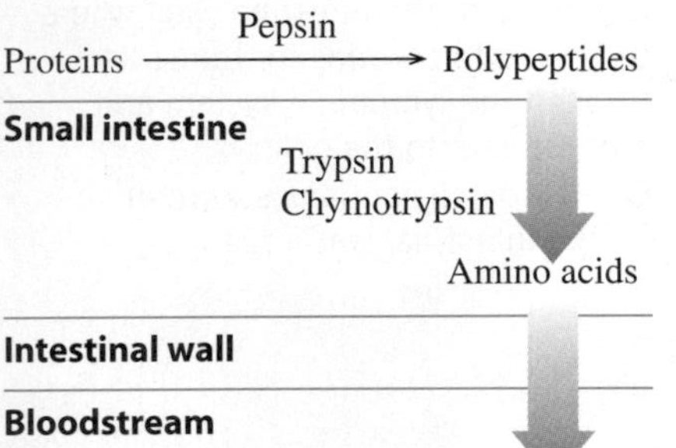

FIGURE 18.5 Proteins are hydrolyzed in the stomach and the small intestine.

Q Where in digestion are peptides hydrolyzed to amino acids?

TUTORIAL
Protein Digestion

Digestion of Proteins

The major role of proteins is to provide amino acids for the synthesis of new proteins for the body and nitrogen for the synthesis of compounds such as nucleotides. In stage 1, the digestion of proteins begins in the stomach, where hydrochloric acid (HCl) at pH 2 denatures the proteins and activates enzymes that begin to hydrolyze peptide bonds. Smaller peptides from the stomach move into the small intestine, where they are completely hydrolyzed to amino acids. The amino acids are absorbed through the intestinal walls into the bloodstream for transport to the cells (see Figure 18.5).

SAMPLE PROBLEM 18.2

Digestion of Proteins

What are the sites and end products for the digestion of proteins?

SOLUTION

Proteins begin digestion in the stomach and complete digestion in the small intestine to yield amino acids.

STUDY CHECK 18.2

What is the function of HCl in the stomach?

QUESTIONS AND PROBLEMS

Digestion of Foods

18.7 What is the general type of reaction that occurs during the digestion of carbohydrates?

18.8 What are the end products of the digestion of proteins?

18.9 What is the role of bile salts in lipid digestion?

18.10 How are insoluble triacylglycerols transported to the cells?

18.11 Where do dietary proteins undergo digestion in the body?

18.12 What is the purpose of digestion in stage 1?

LEARNING GOAL

Describe the components and functions of the coenzymes NAD^+, FAD, and coenzyme A.

18.3 Coenzymes in Metabolic Pathways

Before we look at the metabolic reactions that extract energy from the food we digest, we need to review some ideas about oxidation and reduction reactions. As we discussed in Chapters 5 and 12, an *oxidation* reaction involves the loss of hydrogen, the loss of electrons, or the gain of oxygen by a substance. In a *reduction* reaction, there is a gain of hydrogen, a gain of electrons, or a loss of oxygen (see Table 18.2).

TABLE 18.2 Characteristics of Oxidation and Reduction in Metabolic Pathways

Oxidation	Reduction
Loss of electrons	Gain of electrons
Loss of hydrogen	Gain of hydrogen
Gain of oxygen	Loss of oxygen

NAD^+

NAD^+ (nicotinamide adenine dinucleotide) is an important coenzyme in which the B_3 vitamin *niacin* provides the *nicotinamide* group, which is bonded to adenosine diphosphate (ADP) (see Figure 18.6). The oxidized form of NAD^+ undergoes reduction when a carbon in the nicotinamide ring reacts with one hydrogen ion and two electrons, leaving one H^+.

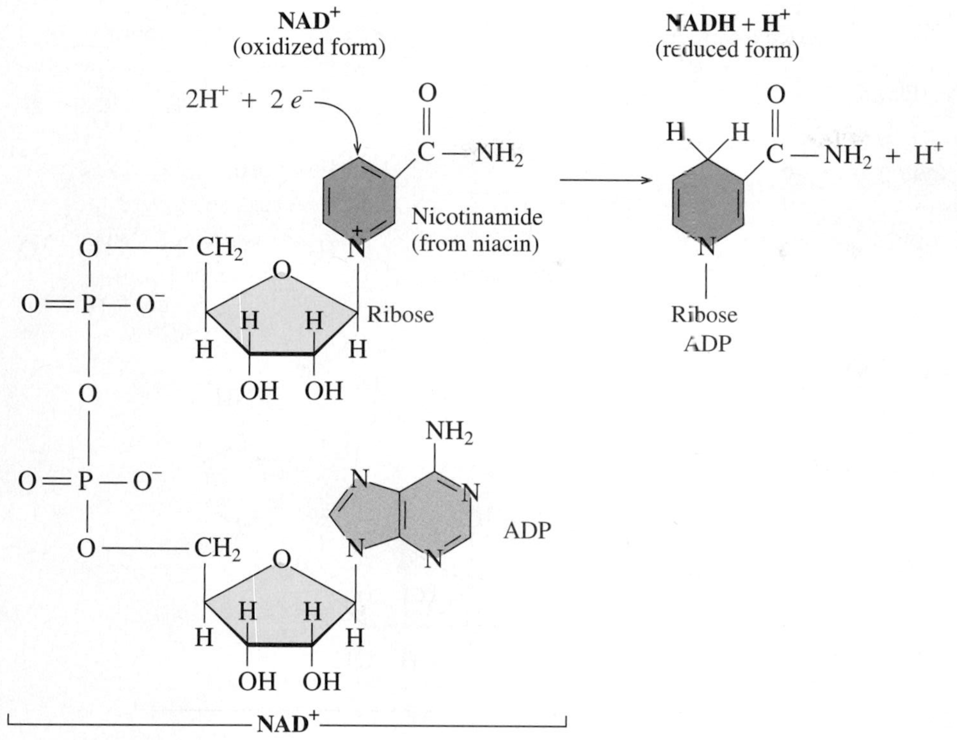

NAD$^+$ (oxidized form) $2H^+ + 2e^-$ → NADH + H$^+$ (reduced form)

Nicotinamide (from niacin), Ribose, ADP

FIGURE 18.6 The coenzyme NAD$^+$ (nicotinamide adenine dinucleotide), which consists of a nicotinamide portion from the vitamin niacin, ribose, and adenosine diphosphate, is reduced to NADH + H$^+$.

Q Why is the conversion of NAD$^+$ to NADH and H$^+$ called a reduction?

The NAD$^+$ coenzyme is required for metabolic reactions that produce carbon–oxygen (C=O) double bonds, such as in the oxidation of alcohols to aldehydes and ketones. An example of an oxidation–reduction reaction that utilizes NAD$^+$ is the oxidation of ethanol in the liver to ethanal and NADH.

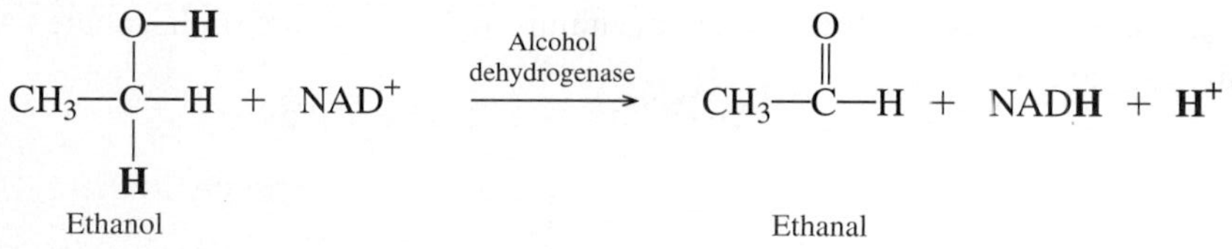

$$CH_3-CH(OH)-H + NAD^+ \xrightarrow{\text{Alcohol dehydrogenase}} CH_3-C(=O)-H + NADH + H^+$$

Ethanol → Ethanal

FAD

FAD (flavin adenine dinucleotide) is a coenzyme that contains adenosine diphosphate (ADP) and riboflavin. Riboflavin, also known as vitamin B_2, consists of ribitol (a sugar alcohol) and flavin. The oxidized form of FAD undergoes reduction when the two nitrogen atoms in the flavin part of the FAD coenzyme react with two hydrogen atoms reducing FAD to $FADH_2$ (see Figure 18.7).

FAD is used as a coenzyme when a dehydrogenation reaction converts a carbon–carbon single bond to a carbon–carbon (C=C) double bond. An example of a reaction in the citric acid cycle that utilizes FAD is the conversion of the carbon–carbon single bond in succinate to a double bond in fumarate and $FADH_2$.

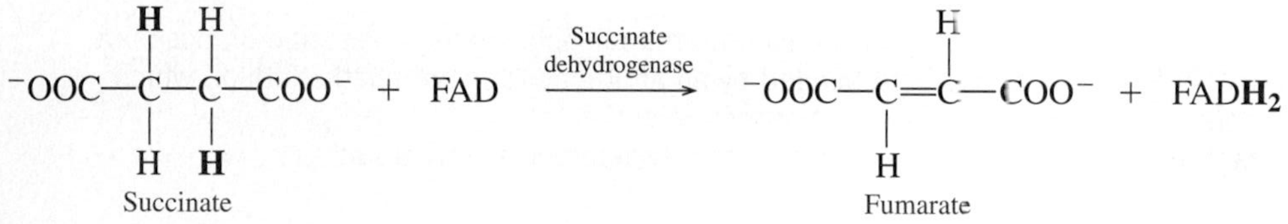

$$^-OOC-CH_2-CH_2-COO^- + FAD \xrightarrow{\text{Succinate dehydrogenase}} {}^-OOC-CH=CH-COO^- + FADH_2$$

Succinate → Fumarate

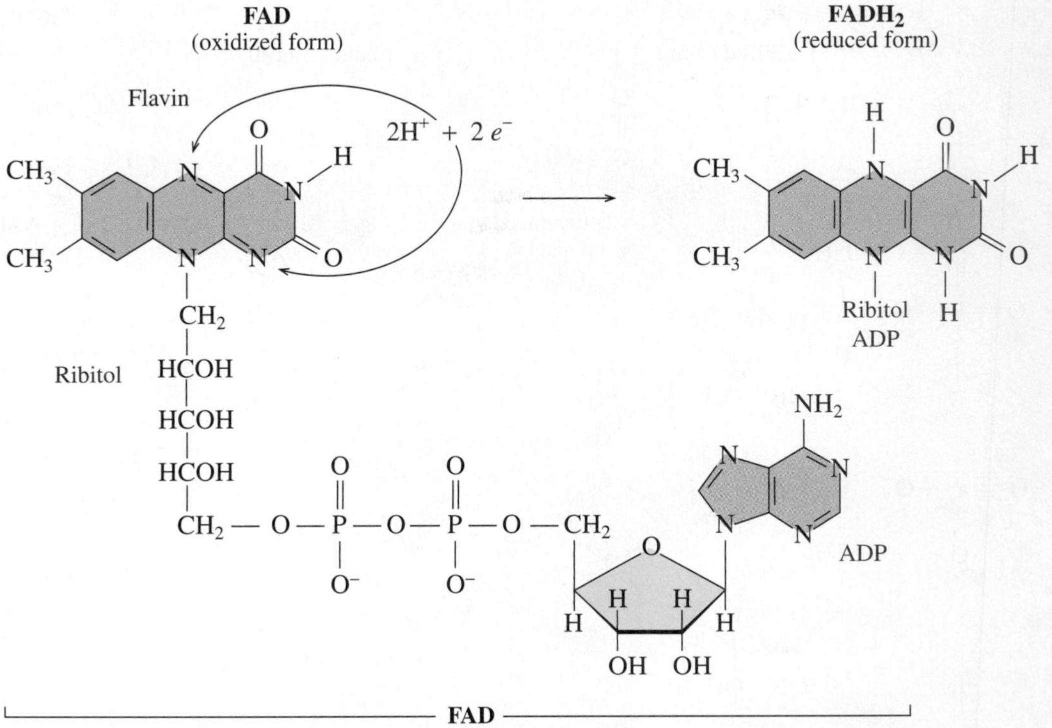

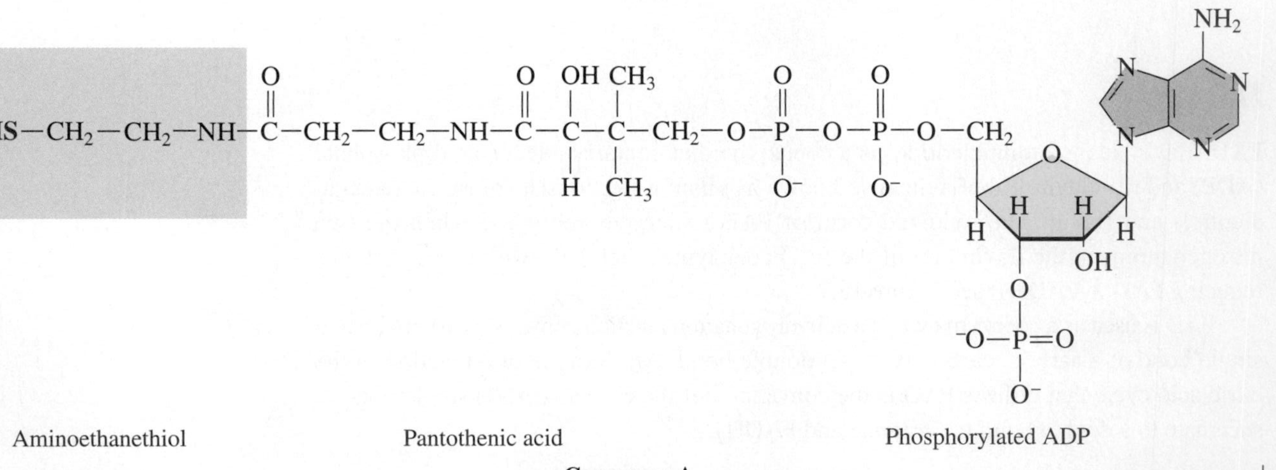

FIGURE 18.7 The coenzyme FAD (flavin adenine dinucleotide), made from riboflavin (vitamin B_2) and adenosine diphosphate, is reduced to $FADH_2$.
Q What is the type of reaction in which FAD accepts hydrogen?

Coenzyme A

Coenzyme A (CoA), which is not involved in oxidation–reduction reactions, is made up of several components: pantothenic acid (vitamin B_5), adenosine diphosphate (ADP), and aminoethanethiol (see Figure 18.8).

$$HS{-}CH_2{-}CH_2{-}NH{-}\overset{O}{\overset{\|}{C}}{-}CH_2{-}CH_2{-}NH{-}\overset{O}{\overset{\|}{C}}{-}\underset{H}{\overset{OH}{C}}{-}\underset{CH_3}{\overset{CH_3}{C}}{-}CH_2{-}O{-}\underset{O^-}{\overset{O}{P}}{-}O{-}\underset{O^-}{\overset{O}{P}}{-}O{-}CH_2\text{–(phosphorylated adenosine)}$$

Aminoethanethiol | Pantothenic acid | Phosphorylated ADP

Coenzyme A

FIGURE 18.8 Coenzyme A is derived from a phosphorylated adenosine diphosphate (ADP) and pantothenic acid bonded by an amide bond to aminoethanethiol, which contains the —SH reactive part of the molecule.
Q What part of coenzyme A reacts with a two-carbon acetyl group?

The important feature of coenzyme A (abbreviated HS–CoA) is the thiol group, which bonds to two-carbon acetyl groups to give the energy-rich thioester **acetyl CoA**.

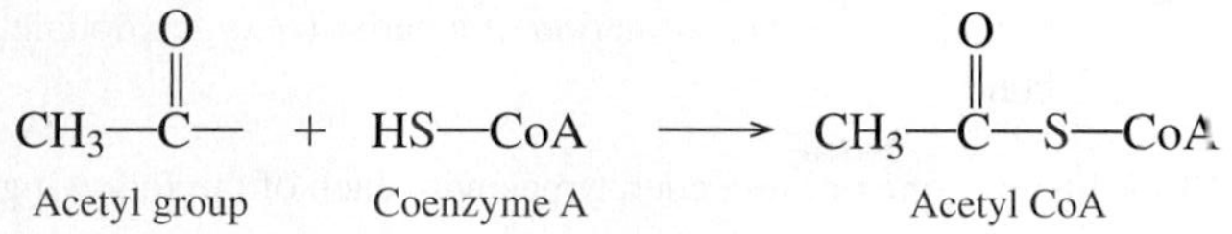

$$\underset{\text{Acetyl group}}{CH_3\text{—}\overset{\overset{\displaystyle O}{\|}}{C}\text{—}} + \underset{\text{Coenzyme A}}{HS\text{—}CoA} \longrightarrow \underset{\text{Acetyl CoA}}{CH_3\text{—}\overset{\overset{\displaystyle O}{\|}}{C}\text{—}S\text{—}CoA}$$

CONCEPT CHECK 18.4

Coenzymes

Describe the reactive part of each of the following coenzymes and the way each participates in metabolic pathways:

a. FAD

b. NAD^+

c. coenzyme A

ANSWER

a. When two nitrogen atoms in the flavin accept $2H^+$ and $2\,e^-$, FAD is reduced to $FADH_2$. FAD is the coenzyme in dehydrogenation (oxidation with a loss of 2H) reactions that produce a carbon–carbon (C=C) double bond.

b. When a carbon atom in the pyridine ring of nicotinamide accepts H^+ and $2\,e^-$, NAD^+ is reduced to NADH. The NAD^+ coenzyme participates in oxidation reactions that produce a carbon–oxygen (C=O) double bond.

a. The thiol (—SH) in coenzyme A combines with an acetyl group to form acetyl coenzyme A, which participates in the transfer of acetyl groups.

Types of Metabolic Reactions

Many reactions within the cells are similar to the types of reactions we looked at in organic chemistry such as hydration, dehydration, hydrogenation, oxidation, and reduction. Organic reactions typically require strong acids (low pH), high temperatures, and/or metallic catalysts. However, metabolic reactions take place at body temperature and physiological pH, which requires enzymes and often their coenzymes. We discussed the classes and names of enzymes in Chapter 16 and can associate the type of enzymes required for the metabolic reactions (see Table 18.3).

TABLE 18.3 Metabolic Reactions and Enzymes

Reaction	Enzyme	Coenzyme
Oxidation	Dehydrogenase	NAD^-, FAD
Reduction	Dehydrogenase	$NADH + H^+$, $FADH_2$
Hydration	Hydrase	
Dehydration	Dehydrase	
Rearrangement	Isomerase	
Transfer of phosphate	Transferase, kinase	ATP, GDP, ADP
Transfer of acetyl group	Acetyl CoA transferase	CoA
Decarboxylation	Decarboxylase	
Hydrolysis	Hydrolase, protease, lipase	

QUESTIONS AND PROBLEMS

Coenzymes in Metabolic Pathways

18.13 Give the abbreviation for each of the following coenzymes:
a. reduced form of NAD^+
b. oxidized form of $FADH_2$
c. participates in the formation of a carbon–carbon double bond

18.14 Give the abbreviation for each of the following coenzymes:
a. reduced form of FAD
b. oxidized form of NADH
c. participates in the formation of a carbon–oxygen double bond

18.15 Identify one or more coenzymes with each of the following components:
a. pantothenic acid **b.** niacin **c.** ribitol

18.16 Identify one or more coenzymes with each of the following components:
a. riboflavin **b.** adenine **c.** aminoethanethiol

LEARNING GOAL

Describe the conversion of glucose to pyruvate in glycolysis and the subsequent conversion of pyruvate to acetyl CoA or lactate.

SELF STUDY ACTIVITY
Glycolysis

TUTORIAL
The Glycolysis Pathway

TUTORIAL
Energy Use and Capture by Glycolysis

18.4 Glycolysis: Oxidation of Glucose

The major source of energy for the body is the glucose produced when we digest the carbohydrates in our food, or from glycogen, a polysaccharide stored in the liver and skeletal muscle. In stage 2 of metabolism, glucose in the bloodstream enters our cells for further degradation in a pathway called *glycolysis*. Early organisms used glycolysis to produce energy from simple nutrients long before there was any oxygen in Earth's atmosphere. Glycolysis is an **anaerobic** process; no oxygen is required.

In **glycolysis**, a six-carbon glucose molecule is broken down to two molecules of three-carbon pyruvate (see Figure 18.9). All the reactions in glycolysis take place in the cytoplasm of the cell. In the first five reactions (1–5), the energy of two ATPs is required to form sugar phosphates. In reactions 4 and 5, the six-carbon sugar phosphate is split to yield two molecules of three-carbon sugar phosphate. In the last five reactions (6–10), energy is obtained from the hydrolysis of the energy-rich phosphate compounds to form four ATPs.

Summary of Glycolysis

In the glycolysis pathway, a six-carbon glucose molecule is converted to two three-carbon pyruvates. Initially, two ATPs are required to form fructose-1,6-bisphosphate. In later reactions, phosphate transfers produce a total of four ATPs. Overall, glycolysis yields two ATPs and two NADHs when a glucose molecule is converted to two pyruvates.

$$\underset{\text{Glucose}}{C_6H_{12}O_6} + 2NAD^+ \xrightarrow{2ADP + 2P_i \;\curvearrowright\; 2ATP} \underset{\text{Pyruvate}}{2CH_3{-}\overset{\overset{\displaystyle O}{\|}}{C}{-}COO^-} + 2NADH + 4H^+$$

CONCEPT CHECK 18.5

Reactions in Glycolysis

Identify each of the statements as one of the following reactions:

(1) isomerization, (2) phosphorylation, (3) dehydration, (4) cleavage

a. a phosphate is transferred to ADP to form ATP
b. 3-phosphoglycerate is converted to 2-phosphoglycerate
c. water is lost from 2-phosphoglycerate
d. fructose-1,6-bisphosphate splits to form two three-carbon compounds

ANSWER

a. Phosphorylation (2) involves the transfer of phosphate to ADP to form ATP.
b. The change in location of a phosphate on a carbon chain is isomerization (1).
c. The loss of water is dehydration (3).
d. The splitting of fructose-1,6-bisphosphate is cleavage (4).

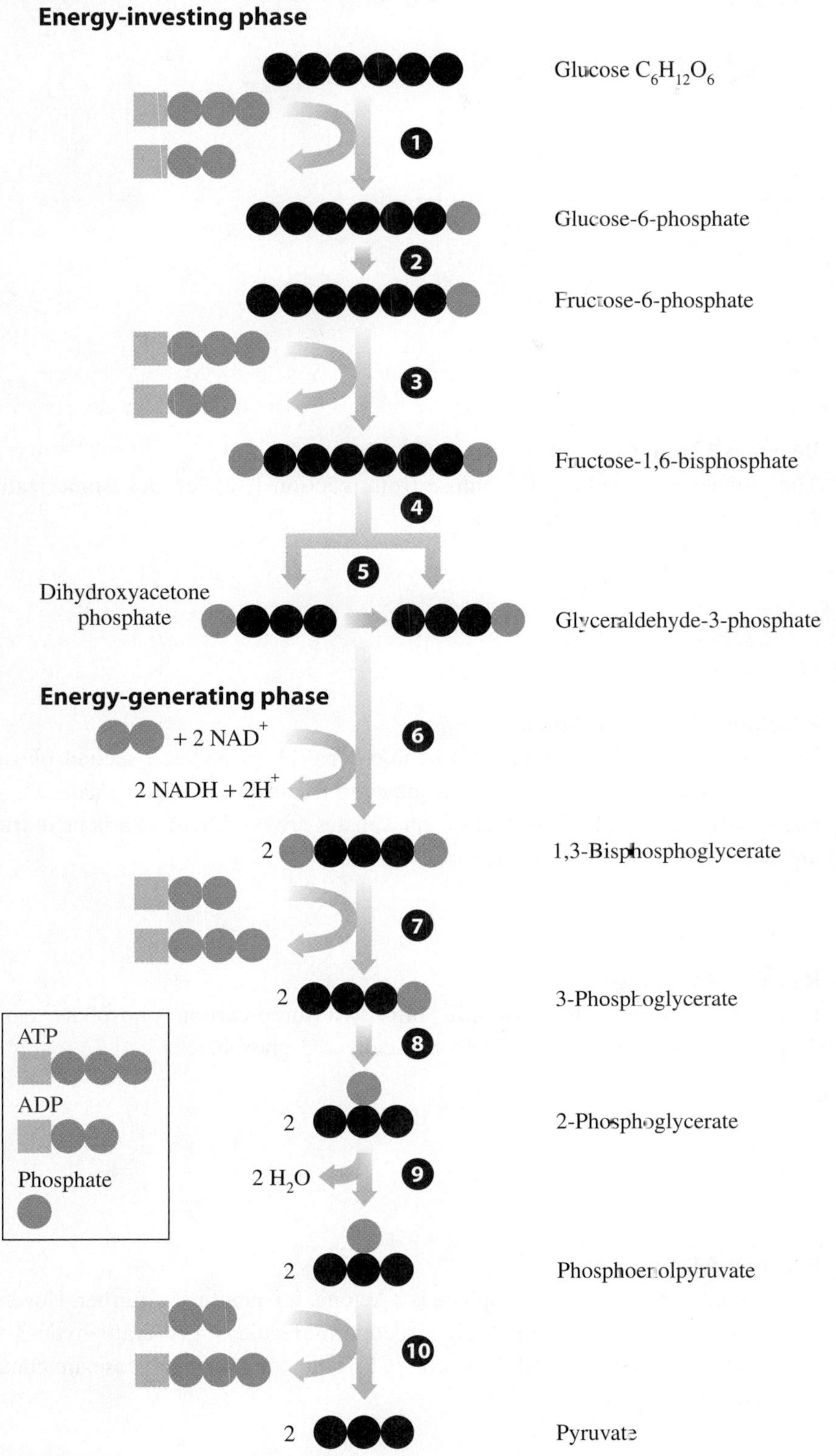

FIGURE 18.9 In glycolysis, the six-carbon glucose molecule is degraded to yield two three-carbon pyruvate molecules. A net of two ATPs is produced along with two NADH.

Q Where in the glycolysis pathway is glucose cleaved to yield two three-carbon compounds?

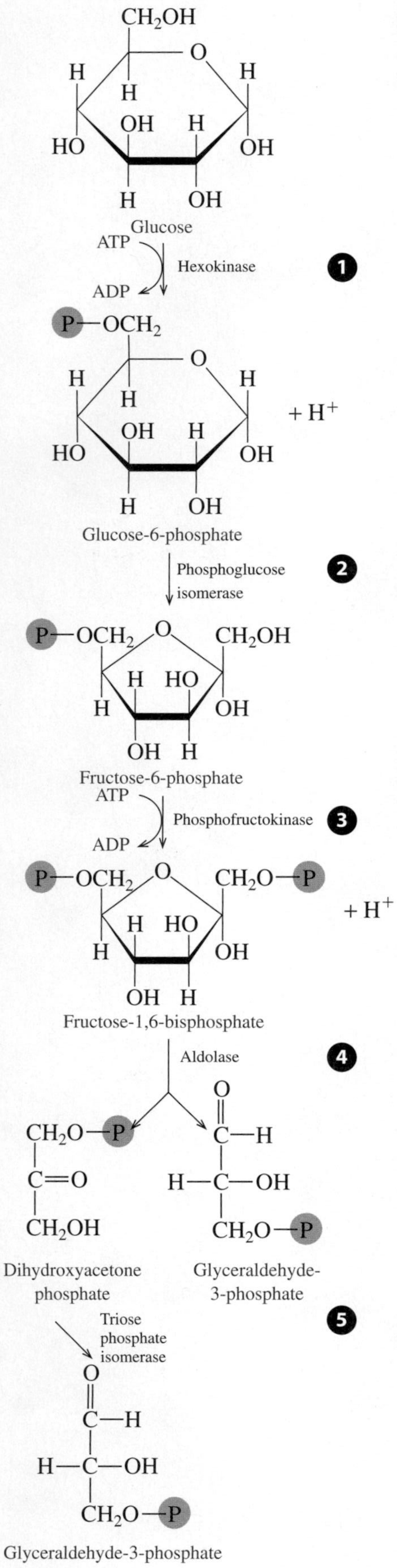

Glucose

ATP → ADP (Hexokinase) **1**

Glucose-6-phosphate + H^+

Phosphoglucose isomerase **2**

Fructose-6-phosphate

ATP → ADP (Phosphofructokinase) **3**

Fructose-1,6-bisphosphate + H^+

Aldolase **4**

Dihydroxyacetone phosphate

Glyceraldehyde-3-phosphate

Triose phosphate isomerase **5**

Glyceraldehyde-3-phosphate

Energy-Investing Reactions 1–5

Reaction 1 Phosphorylation

In the initial reaction, a phosphate from ATP is added to glucose to form glucose-6-phosphate and ADP.

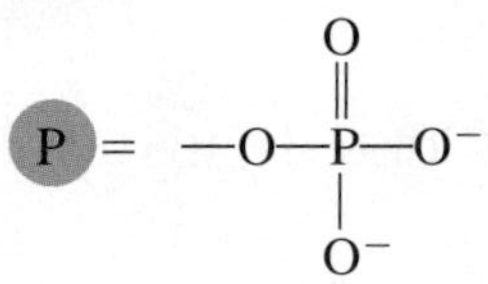

$$\text{P} = -O-\overset{\overset{\displaystyle O}{\|}}{\underset{\underset{\displaystyle O^-}{|}}{P}}-O^-$$

Reaction 2 Isomerization

The glucose-6-phosphate, the aldose from reaction 1, undergoes isomerization to fructose-6-phosphate, which is a ketose.

Reaction 3 Phosphorylation

The hydrolysis (energy-requiring) of another ATP provides a second phosphate group, which converts fructose-6-phosphate to fructose-1,6-bisphosphate. The word *bisphosphate* is used to show that the phosphates are on different carbons in fructose and not connected to each other.

Reaction 4 Cleavage

Fructose-1,6-bisphosphate is split into two three-carbon phosphate isomers: dihydroxyacetone phosphate and glyceraldehyde-3-phosphate.

Reaction 5 Isomerization

Because dihydroxyacetone phosphate is a ketone, it cannot react further. However, it undergoes isomerization to provide a second molecule of glyceraldehyde-3-phosphate, which can be oxidized. Now all six carbon atoms from glucose are contained in two identical triose phosphates.

Energy-Generating Reactions 6–10

Reaction 6 Oxidation and Phosphorylation

The aldehyde group of each glyceraldehyde-3-phosphate is oxidized to a carboxyl group by the coenzyme NAD^+, which is reduced to NADH and H^+. A phosphate adds to the new carboxyl groups to form two molecules of the high-energy compound, 1,3-bisphosphoglycerate.

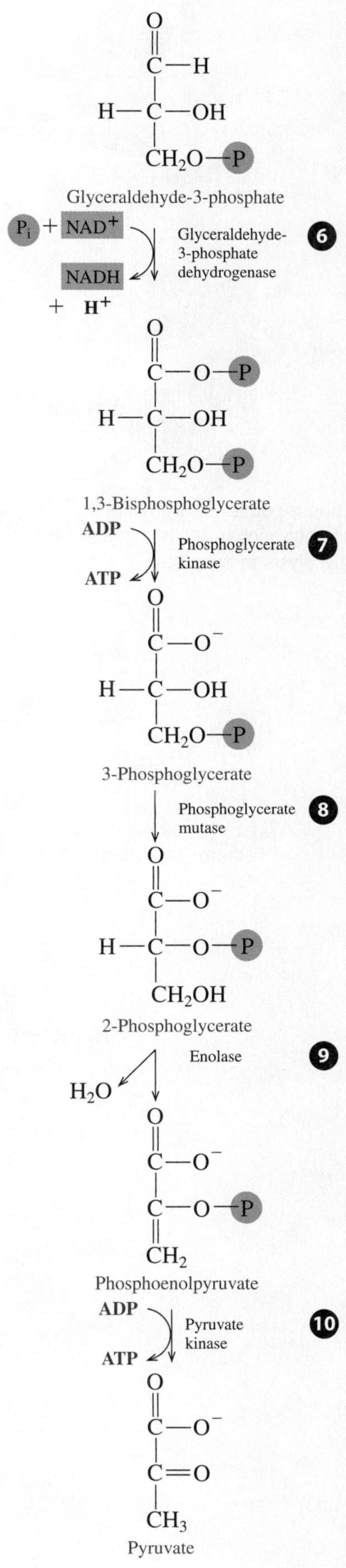

Reaction 7 Phosphate Transfer

A phosphorylation transfers a phosphate group from each 1,3-bisphosphoglycerate to ADP to produce two molecules of ATP. At this point in glycolysis, two ATPs are produced, which balance the two ATPs consumed in reactions 1 and 3.

Reaction 8 Isomerization

Two 3-phosphoglycerate molecules undergo isomerization, which moves the phosphate group from carbon 3 to carbon 2 yielding two molecules of 2-phosphoglycerate.

Reaction 9 Dehydration

Each of the phosphoglycerate molecules undergoes dehydration (loss of water) to give two high-energy molecules of phosphoenolpyruvate.

Reaction 10 Phosphate Transfer

In a second direct phosphate transfer, phosphate groups from two phosphoenolpyruvate are transferred to two ADPs to form two pyruvates and two ATPs.

Glyceraldehyde-3-phosphate

6 P_i + NAD^+ → NADH + H^+ (Glyceraldehyde-3-phosphate dehydrogenase)

1,3-Bisphosphoglycerate

7 ADP → ATP (Phosphoglycerate kinase)

3-Phosphoglycerate

8 Phosphoglycerate mutase

2-Phosphoglycerate

9 Enolase → H_2O

Phosphoenolpyruvate

10 ADP → ATP (Pyruvate kinase)

Pyruvate

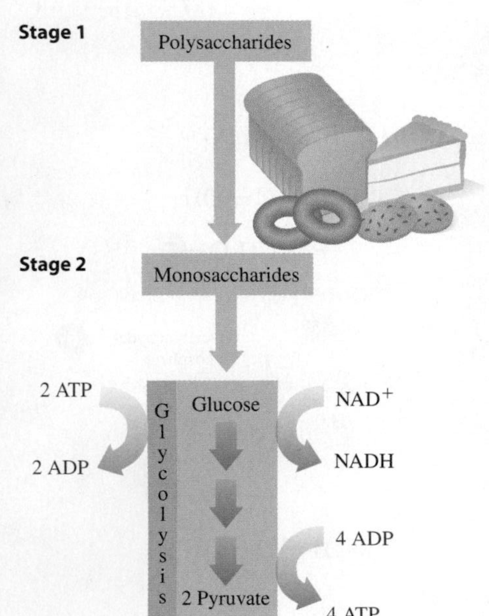

Glucose obtained from the digestion of polysaccharides is degraded in glycolysis to pyruvate.

SAMPLE PROBLEM 18.3

Glycolysis

What are the reactions in glycolysis that generate ATP?

SOLUTION

ATP is produced when phosphate groups are transferred directly to ADP from 1,3-bisphosphoglycerate (reaction 7) and from phosphoenolpyruvate (reaction 10).

STUDY CHECK 18.3

If four ATP molecules are produced in glycolysis, why is there a net yield of only two ATPs?

Pathways for Pyruvate

The pyruvate produced from glucose can now enter pathways that continue to extract energy. The available pathway depends on whether there is sufficient oxygen in the cell. During **aerobic** conditions, oxygen is available to convert pyruvate to acetyl coenzyme A (CoA). When oxygen levels are low, pyruvate is reduced to lactate.

TUTORIAL
Pathways for Pyruvate

Aerobic Conditions

In glycolysis, two ATP molecules were generated when one glucose molecule was converted to two pyruvates. However, much more energy is obtained from glucose when oxygen levels are high in the cells. Under aerobic conditions, pyruvate moves from the cytoplasm into the mitochondria to be oxidized further. In a complex reaction, pyruvate is oxidized, and a carbon atom is removed as CO_2. The coenzyme NAD^+ is reduced during the oxidation. The resulting two-carbon acetyl compound is attached to CoA, producing acetyl CoA, an important intermediate in many metabolic pathways (see Figure 18.10).

$$\underset{\text{Pyruvate}}{CH_3-\overset{O}{\overset{\|}{C}}-\overset{O}{\overset{\|}{C}}-O^-} + HS-CoA + NAD^+ \xrightarrow[\text{dehydrogenase}]{\text{Pyruvate}} \underset{\text{Acetyl CoA}}{CH_3-\overset{O}{\overset{\|}{C}}-S-CoA} + CO_2 + NADH$$

Anaerobic Conditions

When we engage in strenuous exercise, the oxygen stored in our muscle cells is quickly depleted. Under anaerobic conditions, pyruvate is reduced to lactate. NAD^+ is produced and is used to oxidize more glyceraldehyde-3-phosphate in the glycolysis pathway, which produces a small but needed amount of ATP.

$$\underset{\substack{\text{Pyruvate}\\\text{(oxidized)}}}{CH_3-\overset{O}{\overset{\|}{C}}-\overset{O}{\overset{\|}{C}}-O^-} \xrightleftharpoons[\text{Lactate dehydrogenase}]{NAD\mathbf{H} + \mathbf{H}^+ \;\to\; NAD^+} \underset{\substack{\text{Lactate}\\\text{(reduced)}}}{CH_3-\overset{\mathbf{OH}}{\overset{|}{\underset{\mathbf{H}}{\underset{|}{C}}}}-\overset{O}{\overset{\|}{C}}-O^-}$$

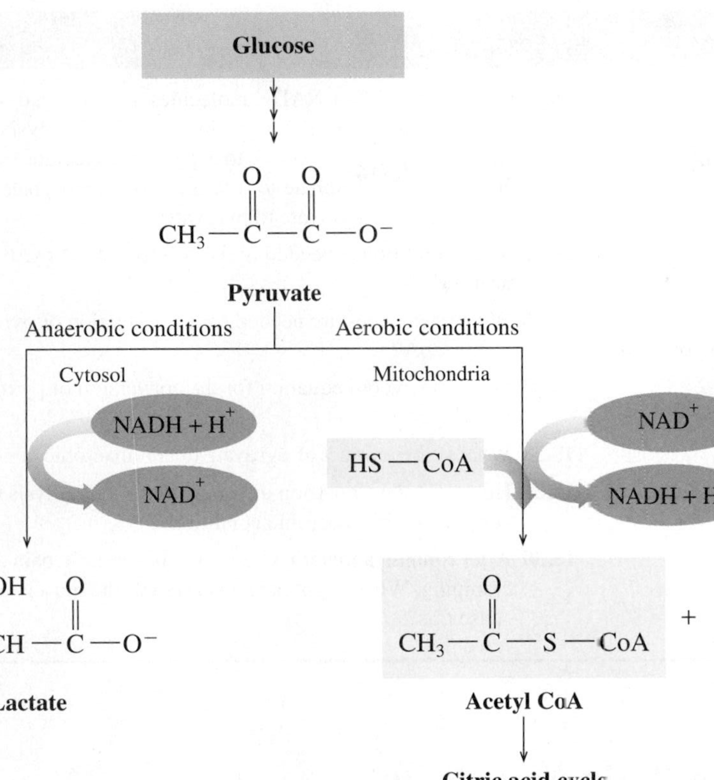

FIGURE 18.10 Pyruvate is converted to acetyl CoA under aerobic conditions and to lactate under anaerobic conditions.

Q During vigorous exercise, why does lactate accumulate in the muscles?

The accumulation of lactate causes the muscles to tire and become sore. After exercise, a person continues to breathe rapidly to repay the oxygen debt incurred during exercise. Most of the lactate is transported to the liver, where it is converted back into pyruvate. Under anaerobic conditions, the only ATP production in glycolysis occurs during the steps that phosphorylate ADP, giving a net gain of only two ATP molecules.

$$\underset{\text{Glucose}}{C_6H_{12}O_6} + 2\,ADP + 2\,P_i \longrightarrow 2\,\underset{\text{Lactate}}{CH_3{-}CH(OH){-}COO^-} + 2\,ATP$$

Bacteria also convert pyruvate to lactate under anaerobic conditions. In the preparation of kimchee and sauerkraut, cabbage is covered with salt brine. The glucose from the starches in the cabbage is converted to lactate. This acid environment acts as a preservative that prevents the growth of other bacteria. The pickling of olives and cucumbers gives similar products. When cultures of bacteria that produce lactate are added to milk, the acid denatures the milk proteins to give sour cream and yogurt.

After vigorous exercise, rapid breathing helps to repay the oxygen debt.

CONCEPT CHECK 18.6

Fates of Pyruvate

When is pyruvate converted to each of the following?

a. acetyl CoA **b.** lactate

ANSWER

a. Pyruvate is converted to acetyl CoA and NADH under aerobic conditions when oxygen is plentiful. The NADH is oxidized back to NAD^+ to allow glycolysis to continue.

b. Pyruvate is converted to lactate and NAD^+ under anaerobic conditions, which provides NAD^+ for glycolysis.

QUESTIONS AND PROBLEMS

Glycolysis: Oxidation of Glucose

18.17 What is the starting compound of glycolysis?

18.18 What is the end product of glycolysis?

18.19 How is ATP used in the initial steps of glycolysis?

18.20 How many ATP molecules are used in the initial steps of glycolysis?

18.21 How does phosphorylation account for the production of ATP in glycolysis?

18.22 Why are there two ATP molecules formed for one molecule of glucose?

18.23 How many ATP or NADH molecules are produced (or required) in each of the following steps in glycolysis?
a. glucose to glucose-6-phosphate
b. glyceraldehyde-3-phosphate to 1,3-bisphosphoglycerate
c. glucose to pyruvate

18.24 How many ATP or NADH molecules are produced (or required) in each of the following steps in glycolysis?
a. 1,3-bisphosphoglycerate to 3-phosphoglycerate
b. fructose-6-phosphate to fructose-1,6-bisphosphate
c. phosphoenolpyruvate to pyruvate

18.25 What condition is needed in the cell to convert pyruvate to acetyl CoA?

18.26 What coenzymes are needed for the oxidation of pyruvate to acetyl CoA?

18.27 Write the overall equation for the conversion of pyruvate to acetyl CoA.

18.28 What is the product of pyruvate under anaerobic conditions?

18.29 How does the formation of lactate permit glycolysis to continue under anaerobic conditions?

18.30 After running a marathon, a runner has muscle pain and cramping. What might have occurred in the muscle cells to cause this?

LEARNING GOAL

Describe the oxidation of acetyl CoA in the citric acid cycle.

18.5 The Citric Acid Cycle

The citric acid cycle is a series of reactions that connects the intermediate acetyl CoA from the metabolic pathways in stages 1 and 2 with electron transport and the synthesis of ATP in stage 3. As a central pathway in metabolism, the **citric acid cycle** uses the two-carbon acetyl group in acetyl CoA to produce CO_2, NADH + H^+, and $FADH_2$. The citric acid cycle is named for the citrate ion from citric acid ($C_6H_8O_7$), a tricarboxylic acid, which forms in the first reaction.

TUTORIAL
Citric Acid Cycle

Overview of the Citric Acid Cycle

There are a total of eight reactions and eight enzymes in the citric acid cycle, which we can separate into two parts. In part 1, an acetyl group (2C) in acetyl CoA bonds with oxaloacetate (4C) to yield citrate (6C) (see Figure 18.11). Then two decarboxylation reactions remove two carbon atoms as CO_2 molecules to give succinyl CoA (4C). In part 2, succinyl CoA is converted to a series of four-carbon compounds and eventually to oxaloacetate. The cycle starts all over as oxaloacetate combines with another acetyl CoA. In one turn of the citric acid cycle, four oxidation reactions provide hydrogen ions and electrons, which are used to reduce FAD and NAD^+ coenzymes.

SELF STUDY ACTIVITY
Krebs Cycle

Part 1 Decarboxylation Removes Two Carbon Atoms

Reaction 1 Formation of Citrate

In the first reaction of the citric acid cycle, the acetyl group (2C) from acetyl CoA bonds with oxaloacetate (4C) to yield citrate (6C) (see Figure 18.12).

Reaction 2 Isomerization

The citrate produced in reaction 1 contains a tertiary alcohol group that cannot be oxidized further. In reaction 2, citrate undergoes isomerization to yield its isomer isocitrate, which provides a secondary alcohol group that can be oxidized in the next reaction.

Reaction 3 Oxidation and Decarboxylation

In reaction 3, an oxidation and a *decarboxylation* occur together. The secondary alcohol group in isocitrate is oxidized to a ketone. A **decarboxylation** converts a carboxylate group ($—COO^-$) to a CO_2 molecule producing α-ketoglutarate, which has five carbon atoms. The oxidation reaction also produces hydrogen ions and electrons that reduce NAD^+ to NADH and H^+.

Reaction 4 Oxidation and Decarboxylation

In reaction 4, α-ketoglutarate undergoes oxidation and decarboxylation to produce a four-carbon group that combines with CoA to form succinyl CoA (4C). As in reaction 3, this oxidation reaction also produces hydrogen ions and electrons that reduce NAD^+ to NADH and H^+.

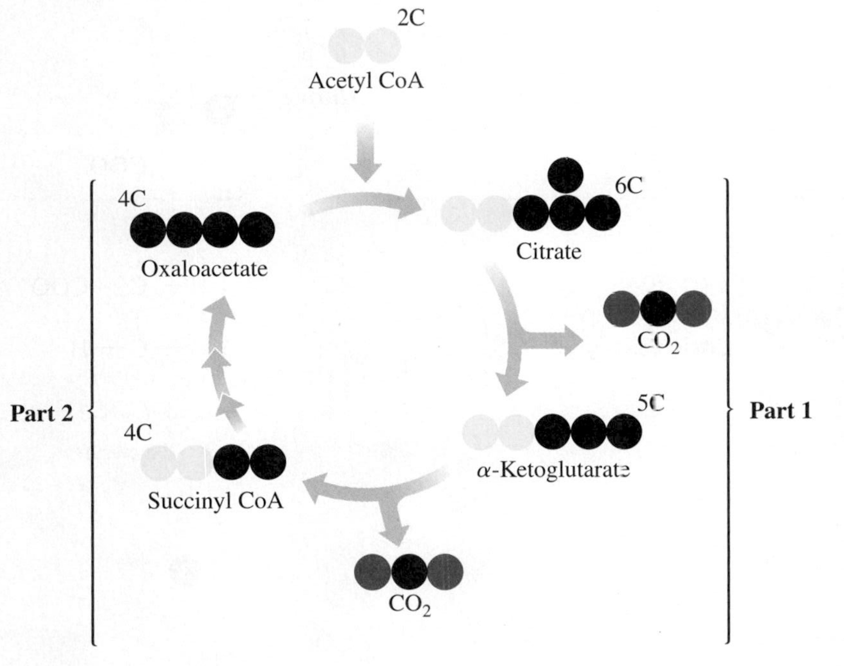

FIGURE 18.11 In Part 1 of the citric acid cycle, two carbon atoms are removed as CO_2 from six-carbon citrate to give four-carbon succinyl CoA, which is converted in part 2 to yield four-carbon oxaloacetate.

Q What is the difference in part 1 and part 2 of the citric acid cycle?

CONCEPT CHECK 18.7

Reactions of Part 1 of the Citric Acid Cycle

a. What is the function of reaction 1 in the citric acid cycle?
b. Why does citrate undergo isomerization in reaction 2?

ANSWER

a. Reaction 1 in the citric acid cycle combines an acetyl group (2C) with oxaloacetate (4C) to produce citrate (6C).
b. In reaction 2, citrate, a tertiary alcohol, isomerizes to isocitrate, a secondary alcohol that can be oxidized.

Part 2 Converting Four Carbon Compounds to Oxaloacetate

Reaction 5 Hydrolysis

In reaction 5, succinyl CoA undergoes hydrolysis to succinate and CoA. The energy released is used to add a phosphate group (P_i) to GDP (guanosine diphosphate), which yields GTP (guanosine triphosphate), a high-energy compound similar to ATP.

TUTORIAL
Citric Acid Cycle

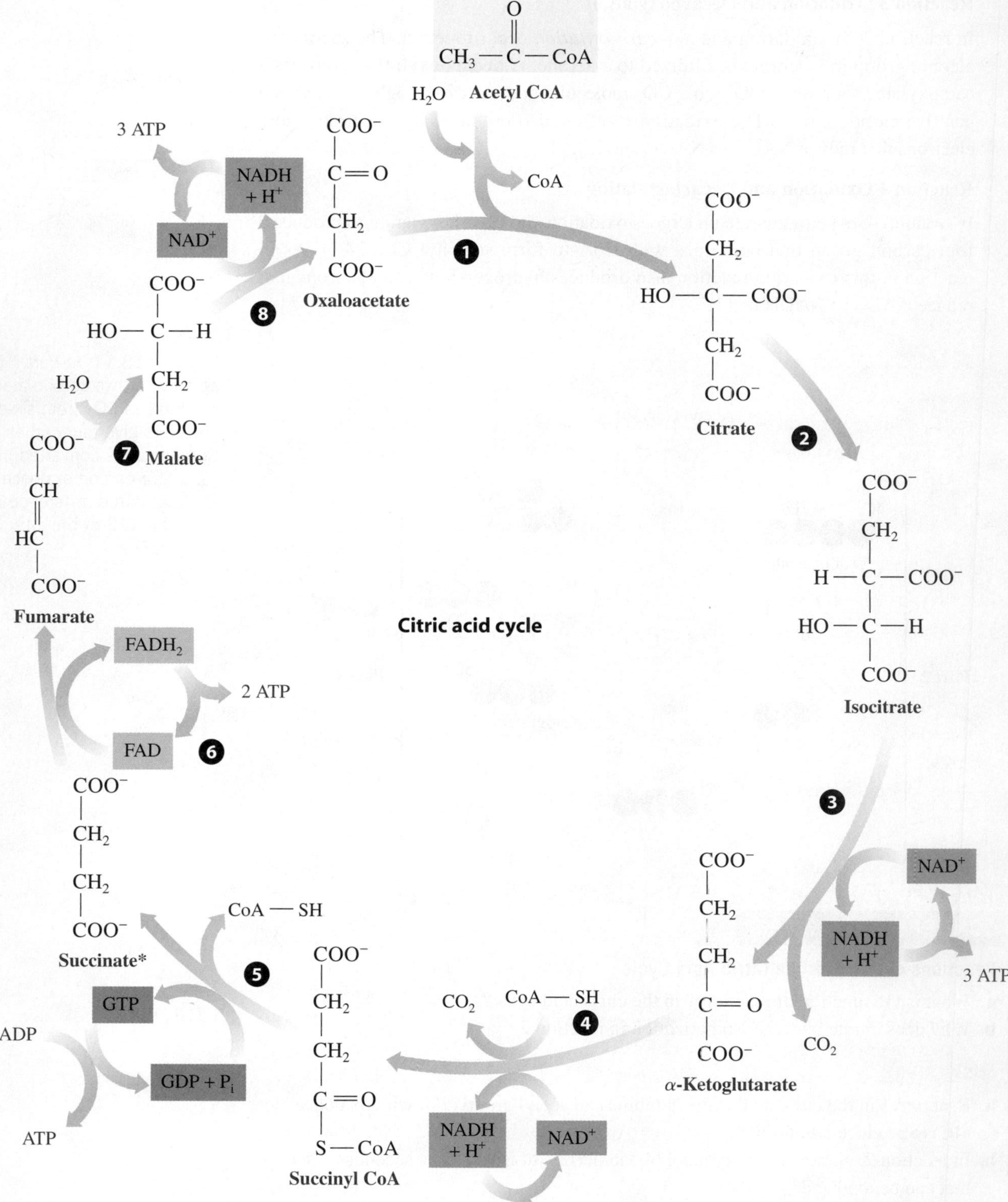

* Succinate is a symmetrical compound.

FIGURE 18.12 In the citric acid cycle, oxidation reactions produce two CO_2 and reduced coenzymes NADH and $FADH_2$, and it also regenerates oxaloacetate.

Q How many reactions in the citric acid cycle produce a reduced coenzyme?

Eventually, the GTP undergoes hydrolysis with a release of energy that is used to add a phosphate group to ADP to form ATP. This is the only time in the citric acid cycle that ATP is produced by a direct transfer of phosphate.

$$GTP + ADP \longrightarrow GDP + ATP$$

Reaction 6 Dehydrogenation

In reaction 6, hydrogen is removed from each of two carbon atoms in succinate, which produces fumarate, a compound with a trans double bond. The hydrogens lost are used to reduce the coenzyme FAD to $FADH_2$.

Reaction 7 Hydration

In reaction 7, a hydration adds water to the double bond of fumarate to yield malate, which is a secondary alcohol.

Reaction 8 Oxidation

In reaction 8, the last step of the citric acid cycle, the secondary alcohol group in malate is oxidized to oxaloacetate, which has a ketone group. The oxidation reaction also produces hydrogen ions and electrons that reduce the coenzyme NAD^+ to NADH and H^+.

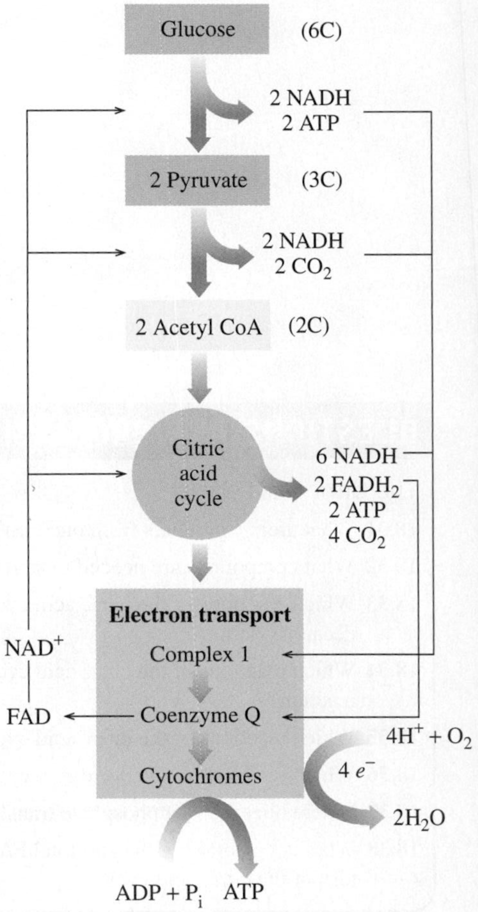

Glucose (6C)
2 NADH
2 ATP
2 Pyruvate (3C)
2 NADH
$2\ CO_2$
2 Acetyl CoA (2C)
Citric acid cycle
6 NADH
$2\ FADH_2$
2 ATP
$4\ CO_2$
Electron transport
Complex 1
Coenzyme Q
Cytochromes
NAD^+
FAD
$4H^+ + O_2$
$4\ e^-$
$2H_2O$
ADP + P_i ATP

The reduced coenzymes NADH and $FADH_2$ that provide hydrogens and electrons to produce ATP in electron transport are regenerated as NAD^+ and FAD.

CONCEPT CHECK 18.8

Reactions of Part 2 of the Citric Acid Cycle

a. Where is ATP produced by a direct phosphate transfer in the citric acid cycle?
b. Where is oxaloacetate generated in the citric acid cycle?

ANSWER

a. Reaction 5 in the citric acid cycle produces GTP, which is used to add a phosphate group to ADP to give ATP.
b. In reaction 8, malate is oxidized to generate oxaloacetate.

Summary of Products from the Citric Acid Cycle

We have seen that the citric acid cycle begins when a two-carbon acetyl group from acetyl CoA combines with oxaloacetate to form citrate. In part 1 of the cycle, two carbon atoms are removed from citrate to yield two CO_2 and a four-carbon compound.

In part 2, the four-carbon compounds eventually regenerate oxaloacetate. In the four oxidation reactions of one turn of the citric acid cycle, three NAD^+s and one FAD are reduced to three NADHs and one $FADH_2$. One GDP is converted to GTP, which is used to convert one ADP to ATP.

We can write an overall chemical equation for one complete turn of the citric acid cycle as follows:

$$\text{Acetyl CoA} + 3NAD^+ + FAD + GDP + P_i + 2H_2O \longrightarrow 2CO_2 + 3NADH + 3H^+ + FADH_2 + CoA + GTP$$

Products from One Turn of Citric Acid Cycle

$2\ CO_2$
3 NADH
$1\ FADH_2$
1 GTP (1 ATP)
1 CoA
$3\ H^+$

SAMPLE PROBLEM 18.4

Citric Acid Cycle

When one acetyl CoA completes the citric acid cycle, how many of each of the following is produced?

a. NADH **b.** ketone group **c.** CO_2

SOLUTION

a. One turn of the citric acid cycle produces three molecules of NADH.

b. Two ketone groups form when the secondary alcohol groups in isocitrate and malate are oxidized by NAD^+.

c. Two molecules of CO_2 are produced by the decarboxylation of isocitrate and α-ketoglutarate.

STUDY CHECK 18.4

What is the substrate in the first reaction of the citric acid cycle, as well as a product in the last reaction?

QUESTIONS AND PROBLEMS

The Citric Acid Cycle

18.31 What are the products from one turn of the citric acid cycle?

18.32 What compounds are needed to start the citric acid cycle?

18.33 Which reactions of the citric acid cycle involve oxidation and decarboxylation?

18.34 Which reactions of the citric acid cycle involve a hydration reaction?

18.35 Which reactions of the citric acid cycle reduce NAD^+?

18.36 Which reactions of the citric acid cycle reduce FAD?

18.37 Where does a direct phosphate transfer occur?

18.38 What is the total NADH and total $FADH_2$ produced in one turn of the citric acid cycle?

18.39 Refer to the diagram of the citric acid cycle in Figure 18.12 to answer each of the following:
a. What are the six-carbon compounds?
b. How is the number of carbon atoms decreased?
c. What are the five-carbon compounds?
d. Which reactions are oxidation reactions?
e. In which reactions are secondary alcohols oxidized?

18.40 Refer to the diagram of the citric acid cycle in Figure 18.12 to answer each of the following:
a. What is the yield of CO_2 molecules?
b. What are the four-carbon compounds?
c. What is the yield of GTP molecules?
d. What are the decarboxylation reactions?
e. Where does a hydration occur?

LEARNING GOAL

Describe electron transport and the process of oxidative phosphorylation; calculate the ATP from the complete oxidation of glucose.

18.6 Electron Transport and Oxidative Phosphorylation

At this point in stage 3, each glucose molecule that completes glycolysis and the citric acid cycle produces four ATP along with ten NADH and two $FADH_2$.

From One Glucose	ATP	Reduced Coenzymes
Glycolysis	2	2 NADH
Oxidation of 2 pyruvate		2 NADH
Citric acid cycle with 2 acetyl CoA	2	6 NADH 2 $FADH_2$
Total for one glucose	**4**	**10 NADH 2 $FADH_2$**

MC

TUTORIAL
Electron Transport

SELF STUDY ACTIVITY
Electron Transport

Electron Transport

In **electron transport**, hydrogen ions and electrons from NADH and $FADH_2$ are passed from one electron carrier to the next until they combine with oxygen to form H_2O. The energy released during electron transport is used to synthesize ATP from ADP and P_i.

A mitochondrion consists of an outer membrane, an intermembrane space, and an inner membrane that surrounds the matrix. Along the highly folded inner membrane are the enzymes and electron carriers required for electron transport.

The electron transport system consists of a series of electron carriers embedded in four enzyme complexes: I, II, III, and IV. Electrons and hydrogen ions from the oxidation of NADH and $FADH_2$ flow between the electron carriers in each complex. Two electron carriers, coenzyme Q (Q) and cytochrome *c* (cyt *c*) are not firmly attached to the membrane. They function as mobile carriers by shuttling electrons between the protein complexes (see Figure 18.13).

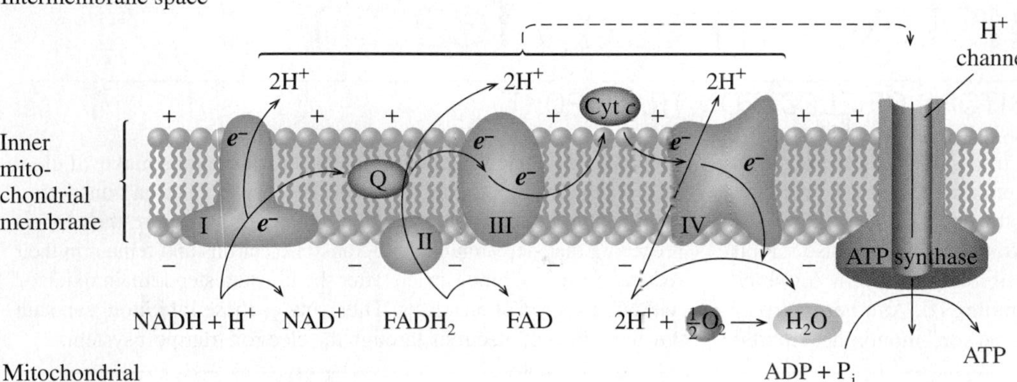

FIGURE 18.13 In electron transport, coenzymes NADH and $FADH_2$ are oxidized in enzyme complexes providing electrons and hydrogen ions for ATP synthesis.
Q What is the major source of NADH for electron transport?

Complex I

Electron transport begins when NADH transfers hydrogen ions and electrons to the enzymes and electron carriers of complex I and forms the oxidized coenzyme NAD^+. The hydrogen ions and electrons are transferred to the mobile electron carrier coenzyme Q (CoQ), which carries electrons to complex II. The overall reaction in complex I is written as follows:

$$\text{NAD}\mathbf{H} + \mathbf{H^+} + \text{Q} \longrightarrow \text{NAD}^+ + \text{Q}\mathbf{H_2}$$

Complex II

Complex II is used when $FADH_2$ is generated in the citric acid cycle from the conversion of succinate to fumarate. The hydrogen ions and electrons from $FADH_2$ are transferred to coenzyme Q to yield QH_2 and the oxidized coenzyme FAD.

Because complex II is at a lower energy level than complex I, the electrons from $FADH_2$ enter electron transport at a lower energy level than those from NADH. The overall reaction in complex II is written as follows:

$$\text{FA}\mathbf{DH_2} + \text{Q} \longrightarrow \text{FAD} + \text{Q}\mathbf{H_2}$$

Complex III

In complex III, electrons from QH_2 are transferred to cytochrome *c*. Cytochrome *c*, which contains Fe^{3+}/Fe^{2+}, is reduced when it gains an electron and oxidized when an electron is lost. As a mobile carrier, cytochrome *c* carries electrons to complex IV. The overall reaction in complex III is written as follows:

$$\text{Q}\mathbf{H_2} + 2\,\text{cyt}\,c\,(\text{oxidized}) \longrightarrow \text{Q} + \mathbf{2H^+} + 2\,\text{cyt}\,c\,(\text{reduced})$$

Complex IV

At complex IV, electrons from cytochrome *c* are passed to other electron carriers until the electrons combine with hydrogen ions and oxygen (O_2) to form water. The overall reaction in complex IV is written as follows:

$$4\,e^- + 4H^+ + O_2 \longrightarrow 2H_2O$$

This equation may also be written as:

$$2\,e^- + 2H^+ + \tfrac{1}{2}O_2 \longrightarrow H_2O$$

Chemistry Link to Health

TOXINS: INHIBITORS OF ELECTRON TRANSPORT

Several substances can inhibit the electron carriers in the electron transport system. Rotenone, a product from a plant root used in South America to poison fish, blocks electron transport between complex I and coenzyme Q. The barbiturates amytal and Demerol also inhibit complex I. Another inhibitor is the antibiotic antimycin A, which blocks the flow of electrons within complex III. Another group of compounds, including cyanide and carbon monoxide, inhibit cytochrome *c*. The toxic nature of these compounds makes it clear that an organism relies heavily on the process of electron transport.

When an inhibitor blocks a step in electron transport, the carriers preceding that step are unable to transfer electrons and remain in their reduced forms. All the carriers after the blocked step remain oxidized without a source of electrons. Thus, any of these inhibitors can shut down the flow of electrons through the electron transport system.

TUTORIAL
Power from Protons: ATP Synthase

TUTORIAL
The Chemiosmotic Model

Oxidative Phosphorylation

We have seen how hydrogen ions and electrons from NADH and $FADH_2$ are carried through electron transport to combine with oxygen and form water. Now we will look at how H^+ ions are involved in providing energy needed to produce ATP, a process called *oxidative phosphorylation*.

In 1978, Peter Mitchell received the Nobel Prize in Chemistry for his theory called the **chemiosmotic model**, which links the energy from electron transport to a proton gradient that drives the synthesis of ATP. In this model, three of the complexes (I, III, and IV) extend through the inner mitochondrial membrane with one end of each complex in the matrix and the other end in the intermembrane space. Each complex acts as a **proton pump** by pushing H^+ ions generated from the oxidation of NADH and $FADH_2$ out of the matrix and into the intermembrane space. The increase in H^+ concentration in the intermembrane space lowers the pH and creates a proton gradient. Because protons are positively charged, the lower pH and the electrical charge of the proton gradient make it an electrochemical gradient (see Figure 18.13).

To equalize the pH between the intermembrane space and the matrix, there is a tendency for the protons to return to the matrix. However, protons cannot diffuse through the inner membrane. The only way protons can return to the matrix is to pass through a protein complex called **ATP synthase**. We might think of the flow of protons as a river that turns a waterwheel. As the protons flow through ATP synthase, energy is generated that is used to convert ADP to ATP. Thus the process of **oxidative phosphorylation** couples the energy from electron transport to the synthesis of ATP from ADP and P_i.

$$\text{ADP} + P_i + \text{energy} \xrightarrow{\text{ATP synthase}} \text{ATP}$$

CONCEPT CHECK 18.9

Chemiosmotic Model

Consider the process of proton pumping in the chemiosmotic model.

a. What changes in pH take place in the mitochondrial matrix and in the intermembrane space?
b. How do the protons return to the matrix?
c. How is energy obtained for the synthesis of ATP?

ANSWER

a. The process of proton pumping "pushes" protons (H^+) out of the mitochondrial matrix and into the intermembrane space. The loss of H^+ increases the pH in the matrix whereas the increase of protons (H^+) in the intermembrane space decreases its pH.
b. Protons return to the matrix by passing through ATP synthase.
c. The energy from the proton flow through ATP synthase is used to synthesize ATP.

ATP Synthesis

When NADH enters electron transport at complex I, the energy released from its oxidation is used to synthesize three ATP molecules. $FADH_2$, which enters electron transport at complex II, provides energy to produce 2 ATPs. However, recent measurements indicate that the oxidation of NADH has an energy yield closer to 2.5 ATPs and that one $FADH_2$ has an energy yield closer to 1.5 ATPs. Because there are still disagreements about the actual values for ATP yield, we will use the traditional values of 3 ATPs for NADH and 2 ATPs for $FADH_2$. The overall equation for the oxidation of NADH and $FADH_2$ can be written as follows:

$$\textbf{NADH + H}^{+} + \tfrac{1}{2}O_2 + 3ADP + 3P_i \longrightarrow NAD^{+} + H_2O + \textbf{3ATP}$$

$$\textbf{FADH}_2 + \tfrac{1}{2}O_2 + 2ADP + 2P_i \longrightarrow FAD + H_2O + \textbf{2ATP}$$

SAMPLE PROBLEM 18.5

ATP Synthesis

Why does the oxidation of NADH provide energy for the formation of three ATP molecules, whereas $FADH_2$ produces two ATPs?

SOLUTION

Electrons from the oxidation of NADH enter electron transport at complex I. They pass through three complexes (I, III, and IV) pumping protons into the intermembrane space. Electrons from $FADH_2$ enter electron transport at complex II, pumping protons through only two complexes (III and IV) that pump protons.

STUDY CHECK 18.5

Which complexes in electron transport act as proton pumps?

ATP from Glycolysis

In glycolysis, the oxidation of glucose stores energy in two NADH molecules as well as two ATP molecules from direct phosphate transfer. However, glycolysis occurs in the cytoplasm, and the NADH produced cannot pass through the mitochondrial membrane.

Therefore the hydrogen ions and electrons from NADH in the cytoplasm are transferred to compounds that can enter the mitochondria. In this shuttle system, $FADH_2$ is produced. The overall reaction for the shuttle is

$$\underset{\text{Cytoplasm}}{NADH} + H^{+} + FAD \longrightarrow NAD^{+} + \underset{\text{Mitochondria}}{FADH_2}$$

Therefore the transfer of electrons from NADH in the cytoplasm to $FADH_2$ produces two ATPs, rather than three. In glycolysis, glucose yields a total of six ATPs: four ATPs from two NADHs, and two ATPs from phosphorylation.

$$\text{Glucose} \longrightarrow 2\text{ pyruvate} + 2ATP + 2NADH\ (\longrightarrow 2FADH_2)$$

$$\text{Glucose} \longrightarrow 2\text{ pyruvate} + 6ATP$$

ATP from the Oxidation of Two Pyruvates

Under aerobic conditions, pyruvate enters the mitochondria, where it is oxidized to give acetyl CoA, CO_2, and NADH. Because glucose yields two pyruvates, two NADHs enter electron transport. The oxidation of two pyruvates leads to the production of six ATPs.

$$2 \text{ Pyruvate} \longrightarrow 2 \text{ acetyl CoA} + 6\text{ATP}$$

ATP from the Citric Acid Cycle

One turn of the citric acid cycle produces two CO_2, three NADHs, one $FADH_2$, and one ATP molecule by direct phosphate transfer.

$$3 \text{ NADH} \times 3 \text{ ATP/NADH} = 9 \text{ ATP}$$
$$1 \text{ FADH}_2 \times 2 \text{ ATP/FADH}_2 = 2 \text{ ATP}$$
$$1 \text{ GTP} \times 1 \text{ ATP/1 GTP} = 1 \text{ ATP}$$
$$\text{Total (one turn)} = 12 \text{ ATP}$$

Because one glucose produces two acetyl CoAs, two turns of the citric acid cycle produce a total of 24 ATPs.

$$\text{Acetyl CoA} \longrightarrow 2CO_2 + 12\text{ATP (one turn of citric acid cycle)}$$
$$2 \text{ Acetyl CoA} \longrightarrow 4CO_2 + 24\text{ATP (two turns of citric acid cycle)}$$

Chemistry Link to Health

ATP SYNTHASE AND HEATING THE BODY

Some types of compounds, called *uncouplers*, separate the electron transport system from ATP synthase. They do this by disrupting the proton gradient needed for the synthesis of ATP. The electrons are transported to O_2 in electron transport, but ATP is not formed by ATP synthase.

Some uncouplers transport the protons through the inner membrane, which is normally impermeable to protons; others block the channel in ATP synthase. Compounds such as dicumarol and 2,4-dinitrophenol (DNP) are hydrophobic and bind with protons to carry them across the inner membrane. An antibiotic, oligomycin, blocks the channel, which does not allow any protons to return to the matrix. By removing protons or blocking the channel, there is no proton flow to generate energy for ATP synthesis.

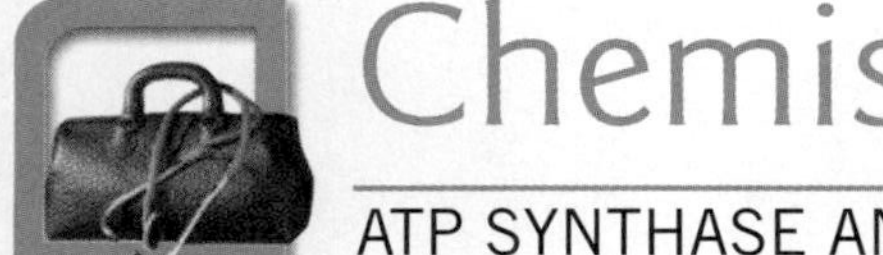

Dicumarol　　2,4-Dinitrophenol (DNP)

When there is no mechanism for ATP synthesis, the energy of electron transport is released as heat. Certain animals that are adapted to cold climates have developed their own uncoupling system, which allows them to use electron transport energy for heat production. These animals have large amounts of a tissue called brown fat, which contains a high concentration of mitochondria. This tissue is brown because of the iron in the cytochromes of the mitochondria. The proton pumps still operate in brown fat, but a protein embedded in the inner mitochondrial membrane allows the protons to bypass ATP synthase. The energy that would be used to synthesize ATP is released as heat. In newborn babies, brown fat is used to generate heat because newborns have a small mass but a large surface area and need to produce more heat than adults. The brown fat deposits are located near major blood vessels, which carry the warmed blood to the body. Most adults have little or no brown fat, although someone who works outdoors in a cold climate will develop some brown fat deposits.

Plants also use uncouplers. In early spring, heat is used to warm early shoots of plants under the snow, which helps them melt the snow around the plants. Some plants, such as skunk cabbage, use uncoupling agents to volatilize fragrant compounds that attract insects to pollinate the plants.

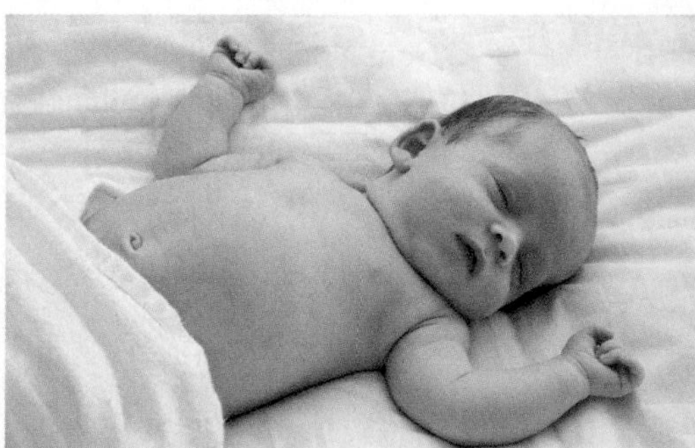

Brown fat helps babies to keep warm.

FIGURE 18.14 The complete oxidation of glucose to CO_2 and H_2O yields a total of 36 ATPs.

Q What metabolic pathway produces most of the ATP from the oxidation of glucose?

ATP from the Complete Oxidation of Glucose

The total ATP for the complete oxidation of glucose is calculated by combining the ATP produced from glycolysis, the oxidation of pyruvate, and the citric acid cycle (see Figure 18.14). The ATP produced for these reactions is given in Table 18.4.

TUTORIAL
ATP Energy from Glucose

SAMPLE PROBLEM 18.6

ATP Production

Indicate the amount of ATP produced by each of the following oxidation reactions:

a. pyruvate to acetyl CoA

b. glucose to acetyl CoA

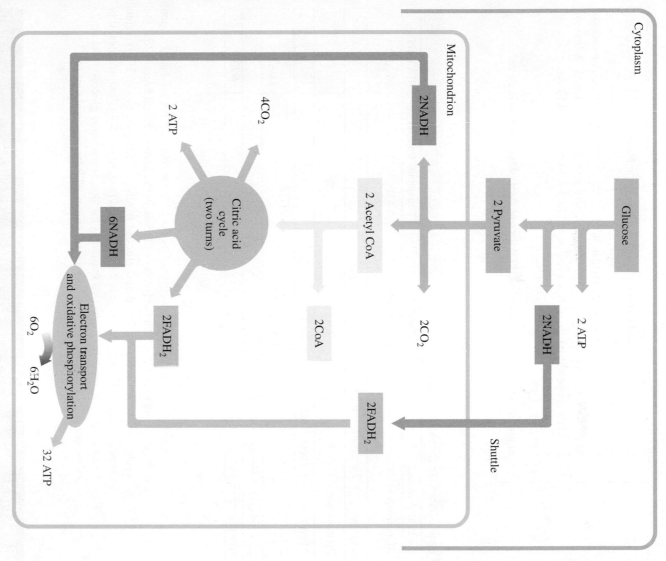

SOLUTION

a. The oxidation of pyruvate to acetyl CoA produces one NADH, which yields three ATPs.

b. Six ATPs are produced from the oxidation of glucose to two pyruvates. Another six ATPs are produced when two pyruvates oxidize to give two acetyl CoAs. A total of 12 ATPs are produced from the oxidation of glucose to two acetyl CoAs.

STUDY CHECK 18.6

What are the sources and amounts of ATP for one turn of the citric acid cycle?

TABLE 18.4 ATP Produced by the Complete Oxidation of Glucose

Pathway	Substrate/Product	Coenzymes	Phosphate Transfer	ATP Yield
Glycolysis	Glucose/pyruvate	2 NADH (shuttle 2 $FADH_2$)	2 ATP	6 ATP
	Glucose $\longrightarrow$ 2 pyruvate + 6 ATP			
Oxidation and Decarboxylation	Pyruvate/acetyl CoA + 2 CO_2	2 NADH		6 ATP
	2 Pyruvate $\longrightarrow$ 2 acetyl CoA + 2 CO_2 + 6 ATP			
Citric Acid Cycle (two turns)	2 Acetyl CoA/ 4 CO_2	6 NADH 2 $FADH_2$	2 GTP	24 ATP
	2 Acetyl CoA $\longrightarrow$ 4 CO_2 + 24 ATP			
Total Yield	Glucose $\longrightarrow$ 6 CO_2 + 36 ATP			36 ATP

When glucose is not immediately used by the cells for energy, it is stored as glycogen in the liver and muscles. When the levels of glucose in the brain or blood become low, the glycogen reserves are hydrolyzed and glucose is released into the blood. If glycogen stores are depleted, some glucose can be synthesized from noncarbohydrate sources. It is the balance of all these reactions that maintains the necessary blood glucose level available to our cells and provides the necessary amount of ATP for our energy needs.

QUESTIONS AND PROBLEMS

Electron Transport and Oxidative Phosphorylation

18.41 What reduced coenzymes provide the electrons for electron transport?

18.42 What happens to the energy level as electrons are passed along in electron transport?

18.43 How are electrons carried from complex I to complex III?

18.44 How are electrons carried from complex III to complex IV?

18.45 How is NADH oxidized in electron transport?

18.46 How is $FADH_2$ oxidized in electron transport?

18.47 What is meant by oxidative phosphorylation?

18.48 How is the proton gradient established?

18.49 According to the chemiosmotic theory, how does the proton gradient provide energy to synthesize ATP?

18.50 How does the phosphorylation of ADP occur?

18.51 How are glycolysis and the citric acid cycle linked to the production of ATP by electron transport?

18.52 Why does $FADH_2$ yield two ATPs, using the electron transport system, but NADH yields three ATPs?

18.53 What is the energy yield in ATP molecules associated with each of the following?
a. NADH $\longrightarrow$ NAD^+
b. glucose $\longrightarrow$ 2 pyruvate
c. 2 pyruvate $\longrightarrow$ 2 acetyl CoA + $2CO_2$
d. acetyl CoA $\longrightarrow$ $2CO_2$

18.54 What is the energy yield in ATP molecules associated with each of the following?
a. $FADH_2$ $\longrightarrow$ FAD
b. glucose + $6O_2$ $\longrightarrow$ $6CO_2$ + $6H_2O$
c. glucose $\longrightarrow$ 2 lactate
d. pyruvate $\longrightarrow$ lactate

18.7 Oxidation of Fatty Acids

LEARNING GOAL

Describe the metabolic pathway of β oxidation; calculate the ATP from the complete oxidation of a fatty acid.

A large amount of energy is obtained when fatty acids undergo oxidation in the mitochondria to yield acetyl CoA. In stage 2 of fat metabolism, fatty acids undergo **beta oxidation (β oxidation)**, which removes two-carbon segments from the carboxyl end.

β oxidation occurs here

$$CH_3-(CH_2)_{14}-\underset{\beta}{CH_2}-\underset{\alpha}{CH_2}-\overset{\overset{O}{\|}}{C}-OH$$

Stearic acid

TUTORIAL
Oxidation of Fatty Acids

Each cycle in β oxidation produces acetyl CoA and a fatty acid that is shorter by two carbons. The cycle repeats until the original fatty acid is completely degraded to two-carbon acetyl CoA units. Each acetyl CoA can then enter the citric acid cycle in the same way as the acetyl CoA units derived from glucose.

Fatty Acid Activation

Fatty acids are prepared for transport through the inner membrane of the mitochondria by adding a CoA group to the fatty acid in the cytosol. This process, called *activation*, combines a fatty acid with coenzyme A to yield fatty acyl CoA. The energy for the activation is obtained from the hydrolysis of ATP to give AMP (adenosine monophosphate) and two inorganic phosphates ($2P_i$). This is equivalent to the energy released from the hydrolysis of two ATPs to two ADPs.

$$\underset{\text{Fatty acid}}{R-CH_2-CH_2-\overset{\overset{O}{\|}}{C}}-\mathbf{OH}+ATP+\mathbf{HS-CoA}\xrightarrow[\text{synthetase}]{\text{Acyl CoA}}\underset{\text{Fatty acyl CoA}}{R-CH_2-CH_2-\overset{\overset{O}{\|}}{C}}-\mathbf{S-CoA}+AMP+2P_i+\mathbf{H_2O}$$

Reactions of the β Oxidation Cycle

In the matrix, fatty acyl CoA molecules undergo β oxidation, which is a cycle of four reactions that convert the $-CH_2-$ of the β carbon to a β-keto group. Once the β-keto group is formed, a two-carbon acetyl group can be split from the carbon chain, which shortens the fatty acyl chain. The reaction for one cycle of β oxidation is written as follows:

$$\underset{\text{Fatty acyl CoA}}{R-\mathbf{CH_2}-\mathbf{CH_2}-\overset{\overset{\mathbf{O}}{\|}}{C}-S-\mathbf{CoA}}+NAD^++FAD+H_2O+HS-CoA\longrightarrow$$

$$\underset{\text{New fatty acyl (–2C) CoA}}{R-\overset{\overset{\mathbf{O}}{\|}}{\mathbf{C}}-S-CoA}+\underset{\text{Acetyl CoA}}{\mathbf{CH_3}-\overset{\overset{O}{\|}}{C}-\mathbf{S-CoA}}+NADH+H^++FADH_2$$

The specific reactions in β oxidation are described next for capric acid (10C) (see Figure 18.15).

Reaction 1 Dehydrogenation

Hydrogen atoms removed by FAD from the α and β carbons form a carbon–carbon double bond and $FADH_2$.

Reaction 2 Hydration

Water adds to the double bond forming a secondary $-OH$ on the β carbon.

Reaction 3 Oxidation

The —OH group on the β carbon is oxidized by NAD^+ to form a β ketone and $NADH + H^+$.

Reaction 4 Cleavage

The $C_\alpha—C_\beta$ bond is cleaved to yield acetyl CoA and a shorter (8C) fatty acyl CoA molecule. The 8C fatty acyl CoA repeats the β oxidation cycle where it is cleaved to a 6C fatty acyl CoA, then to a 4C fatty acyl CoA, which splits to give two acetyl CoA molecules. All the acetyl CoA molecules produced from the fatty acid can enter the citric acid cycle to produce energy.

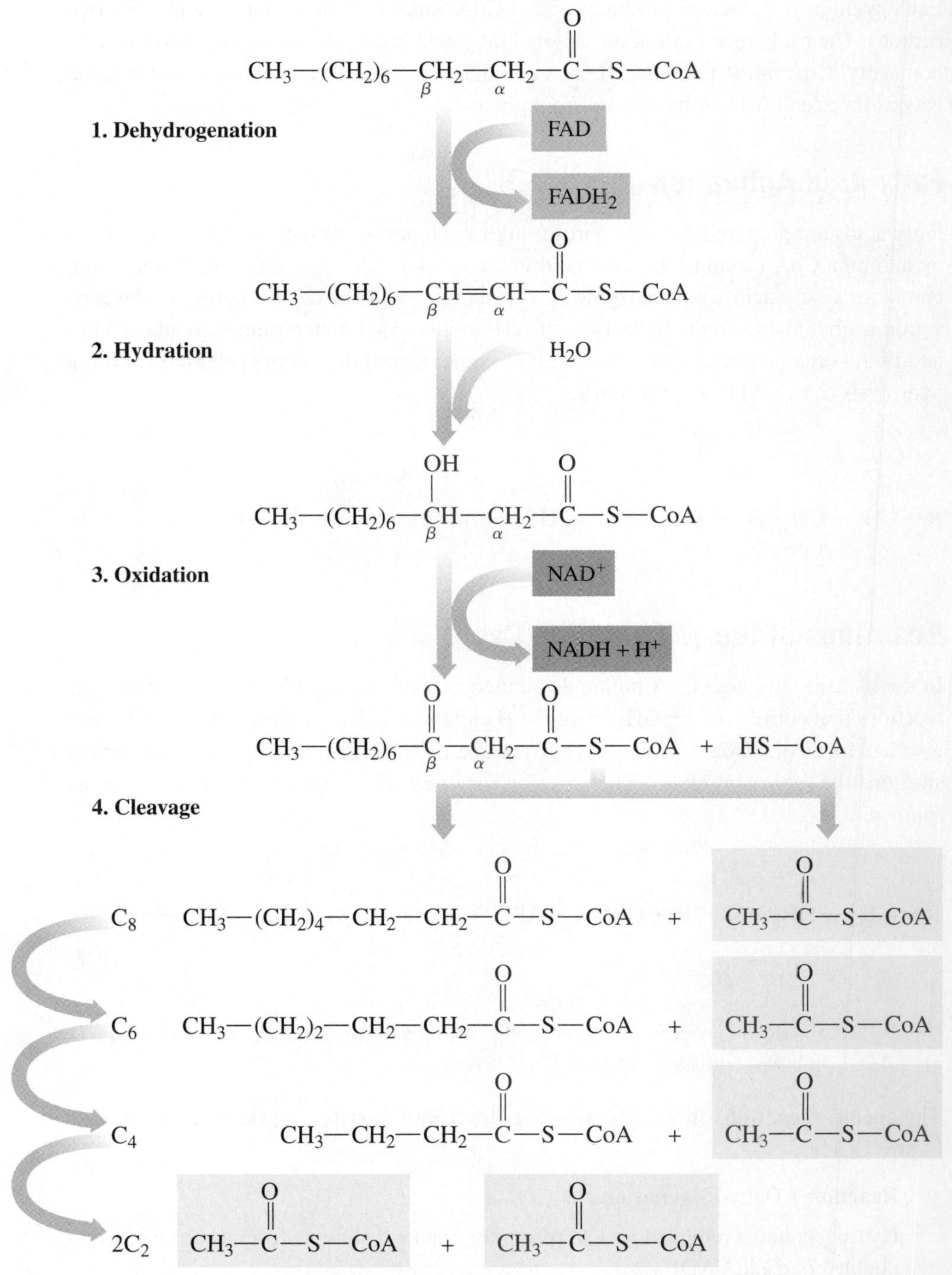

$CH_3—(CH_2)_6—\underset{\beta}{CH_2}—\underset{\alpha}{CH_2}—\overset{O}{\overset{\|}{C}}—S—CoA$

1. Dehydrogenation FAD → $FADH_2$

$CH_3—(CH_2)_6—\underset{\beta}{CH}=\underset{\alpha}{CH}—\overset{O}{\overset{\|}{C}}—S—CoA$

2. Hydration H_2O

$CH_3—(CH_2)_6—\underset{\beta}{\overset{OH}{\overset{|}{CH}}}—\underset{\alpha}{CH_2}—\overset{O}{\overset{\|}{C}}—S—CoA$

3. Oxidation NAD^+ → $NADH + H^+$

$CH_3—(CH_2)_6—\underset{\beta}{\overset{O}{\overset{\|}{C}}}—\underset{\alpha}{CH_2}—\overset{O}{\overset{\|}{C}}—S—CoA + HS—CoA$

4. Cleavage

C_8 $CH_3—(CH_2)_4—CH_2—CH_2—\overset{O}{\overset{\|}{C}}—S—CoA + CH_3—\overset{O}{\overset{\|}{C}}—S—CoA$

C_6 $CH_3—(CH_2)_2—CH_2—CH_2—\overset{O}{\overset{\|}{C}}—S—CoA + CH_3—\overset{O}{\overset{\|}{C}}—S—CoA$

C_4 $CH_3—CH_2—CH_2—\overset{O}{\overset{\|}{C}}—S—CoA + CH_3—\overset{O}{\overset{\|}{C}}—S—CoA$

$2C_2$ $CH_3—\overset{O}{\overset{\|}{C}}—S—CoA + CH_3—\overset{O}{\overset{\|}{C}}—S—CoA$

FIGURE 18.15 Capric acid (10C) undergoes four β oxidation cycles that repeat reactions 1 to 4 to yield 5 acetyl CoA, 4 NADH, and 4 $FADH_2$.

Q How many NADH and $FADH_2$ are produced in one cycle of β oxidation?

SAMPLE PROBLEM 18.7

β Oxidation

Match each of the following with reactions in the β-oxidation cycle:

(1) dehydrogenation **(2)** hydration
(3) oxidation **(4)** cleavage

a. Water is added to a double bond. **b.** An acetyl CoA is removed.
c. FAD is reduced to $FADH_2$. **d.** NAD^+ is reduced to NADH.

SOLUTION

a. (2) hydration **b.** (4) cleavage **c.** (1) dehydrogenation **d.** (3) oxidation

STUDY CHECK 18.7

Which coenzyme is needed in reaction 3 when a β-hydroxyl group is converted to a β-keto group?

Fatty Acid Length Determines Cycle Repeats

The number of carbon atoms in a fatty acid determines the number of times the cycle repeats and the number of acetyl CoA units it produces. For example, the complete β oxidation of myristic acid (C_{14}) produces seven acetyl CoA groups, which is equal to one-half the number of carbon atoms in the fatty acid. Because the final turn of the cycle produces two acetyl CoA groups, the total number of times the cycle repeats is one less than the total number of acetyl groups it produces. Therefore, the C_{14} fatty acid goes through the cycle six times.

Fatty Acid	Number of Carbon Atoms	Number of Acetyl CoA	Number of β-Oxidation Cycles
Myristic acid	14	7	6
Palmitic acid	16	8	7
Stearic acid	18	9	8

ATP from Fatty Acid Oxidation

We can now determine the total energy yield from the oxidation of a particular fatty acid. In each β-oxidation cycle, one NADH, one $FADH_2$, and one acetyl CoA are produced. Each NADH generates sufficient energy to synthesize three ATPs, whereas $FADH_2$ leads to the synthesis of two ATPs. However, the greatest amount of energy produced from a fatty acid is generated by the production of the acetyl CoA units that enter the citric acid cycle. Earlier, we saw that one acetyl CoA leads to the synthesis of a total of 12 ATPs.

So far we know that the C_{14} acid goes through six turns of the cycle and produces seven acetyl CoA units. We also need to remember that activation of the myristic acid requires the equivalent of two ATPs. We can set up the calculation as follows:

ATP Production from β Oxidation for Myristic Acid (14C)	
Activation	−2 ATP
7 acetyl CoA	
(14 C atoms × 1 acetyl CoA/2 C atoms)	
7 acetyl CoA × 12 ATP/acetyl CoA	84 ATP
6 β-oxidation cycles (coenzymes)	
6 $FADH_2$ × 2 ATP/$FADH_2$	12 ATP
6 NADH × 3 ATP/NADH	18 ATP
Total	112 ATP

FAT STORAGE AND BLUBBER

Obtain two medium-sized plastic baggies and a can of Crisco or other type of shortening used for cooking. You will also need a bucket or large container with water and ice cubes. Place several tablespoons of the shortening in one of the baggies. Place the second baggie inside and tape the outside edges of the bag. With your hand inside the inner baggie, move the shortening around to cover your hand. With one hand inside the double baggie, submerge both your hands in the container of ice water. Measure the time it takes for one hand to feel uncomfortably cold. Experiment with different amounts of shortening.

QUESTIONS

1. How effective is the bag with "blubber" in protecting your hand from the cold?
2. How does blubber help an animal survive starvation?
3. How would twice the amount of shortening affect your results?
4. Why would animals in warm climates, such as camels and migratory birds, need to store fat?
5. If you placed 300 g of shortening in the baggie, how many moles of ATP could it provide if used for energy production (assume it produces the same ATP as myristic acid)?

TUTORIAL
Energy from Fatty Acid Oxidation

Chemistry Link to Health

STORED FAT AND OBESITY

The storage of fat is an important survival feature in the lives of many animals. In hibernating animals, large amounts of stored fat provide the energy for the entire hibernation period, which could be several months. In camels, large amounts of food are stored in the camel's hump, which is actually a huge fat deposit. When food resources are low, the camel can survive months without food or water by utilizing the fat reserves in the hump. Migratory birds preparing to fly long distances also store large amounts of fat. Whales are kept warm by a layer of body fat called "blubber" under their skin, which can be as thick as 2 feet. Blubber also provides energy when whales must survive long periods of starvation. Penguins also have blubber, which protects them from the cold and provides energy when they are sitting on a nest of eggs.

Humans also have the capability to store large amounts of fat, although they do not hibernate or usually have to survive for long periods of time without food. When early humans survived on sparse diets that were mostly vegetarian, about 20% of the calories came from fats. Today, a typical diet includes more dairy products and foods with high fat levels, which increases the calories supplied by fats to as much as 60%. The U.S. Public Health Service now estimates that in the United States, more than one-third of adults are obese. Obesity is defined as a body weight that is more than 20 percent over an ideal weight. Obesity is a major factor in health problems such as diabetes, heart disease, high blood pressure, stroke, and gallstones as well as some cancers and some forms of arthritis.

At one time, we thought that obesity was simply a problem of eating too much. However, research now indicates that certain pathways in lipid and carbohydrate metabolism may cause excessive weight gain in some people. In 1995, scientists discovered a hormone called *leptin* that is produced in fat cells. When fat cells are full, high levels of leptin signal the brain to limit the intake of food. When fat stores are low, leptin production decreases, which signals the brain to increase food intake. Some obese persons have high levels of leptin, which means that leptin does not cause them to decrease how much they eat.

Research on obesity has become a major research field. Scientists are studying differences in the rate of leptin production, degrees of resistance to leptin, and possible combinations of these factors. After a person has dieted and lost weight, the leptin level drops. This decrease in leptin may cause slow metabolism, increased hunger, and increased food intake, which starts the weight-gain cycle all over again. Currently, studies are being made to assess the safety of leptin therapy following weight loss.

Marine mammals have thick layers of blubber that serve as insulation as well as energy storage.

Camels store large amounts of fat in their hump.

SAMPLE PROBLEM 18.8

ATP Production from β Oxidation

How much ATP will be produced from the β oxidation of palmitic acid, a C_{16} saturated fatty acid?

SOLUTION

A 16-carbon fatty acid will produce 8 acetyl CoA units and go through 7 β-oxidation cycles. Each acetyl CoA can produce 12 ATPs, by way of the citric acid cycle. In electron transport, each $FADH_2$ produces 2 ATPs, and each NADH produces 3 ATPs. For activation, 2 ATPs are required.

ATP Production from β Oxidation for Palmitic Acid (16C)	
Activation	-2 ATP
8 Acetyl CoA	
8 acetyl CoA $\times$ 12 ATP/acetyl CoA	96 ATP
7 β-oxidation cycles (coenzymes)	
7 $FADH_2$ $\times$ 2 ATP/$FADH_2$	14 ATP
7 NADH $\times$ 3 ATP/NADH	21 ATP
Total	129 ATP

STUDY CHECK 18.8

In β oxidation, why is more ATP produced from acetyl CoA than from reduced coenzymes?

Oxidation of Unsaturated Fatty Acids

Many of the fats in our diets, especially the oils, contain unsaturated fatty acids which have one or more double bonds. Thus, in β oxidation, the fatty acid is ready for hydration. Since no $FADH_2$ is formed in this step, the energy from the β oxidation of unsaturated fatty acids is slightly less than the energy from saturated fatty acids. However, for simplicity, we will assume that the total ATP production is the same for both saturated and unsaturated fatty acids.

Ketone Bodies

When carbohydrates are not available to meet energy needs, the body breaks down body fat. However, the oxidation of large amounts of fatty acids can cause acetyl CoA molecules to accumulate in the liver. Then acetyl CoA molecules combine to form keto compounds called **ketone bodies** (see Figure 18.16).

TUTORIAL
Ketogenesis and Ketone Bodies

CASE STUDY
Diabetes and Blood Glucose

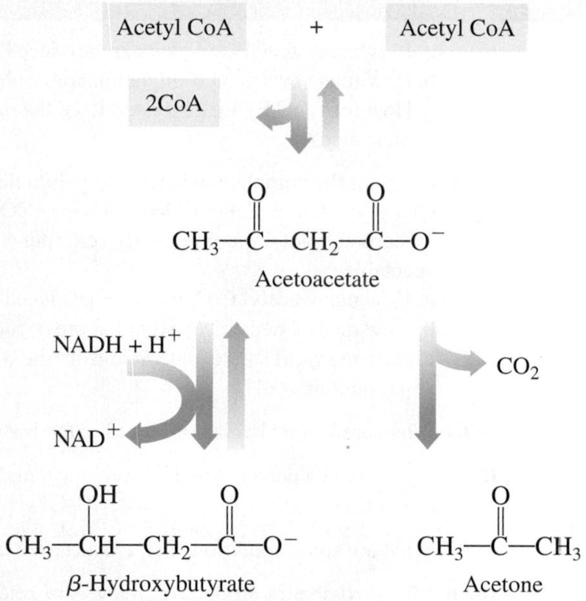

Acetyl CoA + Acetyl CoA

2CoA

$CH_3-C(=O)-CH_2-C(=O)-O^-$

Acetoacetate

$NADH + H^+$

NAD^+

CO_2

$CH_3-CH(OH)-CH_2-C(=O)-O^-$

β-Hydroxybutyrate

$CH_3-C(=O)-CH_3$

Acetone

FIGURE 18.16 In ketogenesis, acetyl CoA molecules combine to produce ketone bodies: acetoacetate, β-hydroxybutyrate, and acetone.

Q What condition in the body leads to the formation of ketone bodies?

Ketone bodies are produced mostly in the liver and transported to cells in the heart, brain, and skeletal muscle, where small amounts of energy can be obtained by converting acetoacetate or β-hydroxybutyrate back to acetyl CoA.

$$\beta\text{-Hydroxybutyrate} \longrightarrow \text{acetoacetate} + 2\text{ CoA} \longrightarrow 2\text{ acetyl CoA}$$

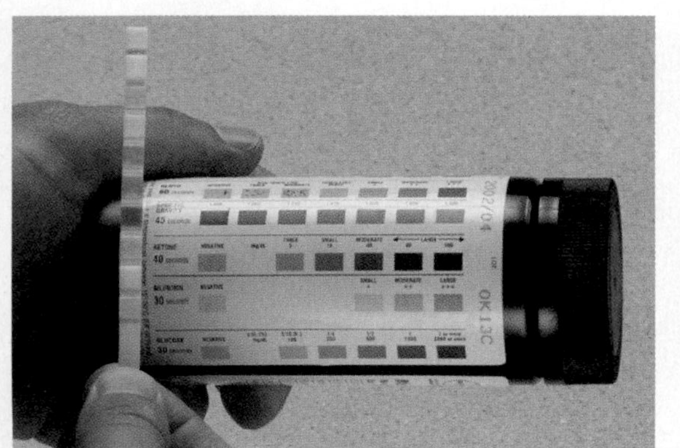

A test strip determines ketone bodies in a urine sample.

Ketosis

When ketone bodies accumulate, they may be incompletely metabolized by the body. This may lead to a condition called **ketosis**, which is found in severe diabetes, diets high in fat and low in carbohydrates, and starvation. Because two of the ketone bodies are acids, they can lower the blood pH below 7.4, which is **acidosis**, a condition that often accompanies ketosis. A drop in blood pH can interfere with the ability of the blood to carry oxygen and cause breathing difficulties.

Fatty Acid Synthesis from Carbohydrates

When the body has met all its energy needs and the glycogen stores are full, excess acetyl CoA from the breakdown of carbohydrates is used to form new fatty acids. Two-carbon acetyl units are linked together to give fatty acids such as palmitic and stearic acid. Several of the reactions in the synthesis of fatty acids are the reverse of the reactions we described for the oxidation of fatty acids. However, fatty acid oxidation occurs in the mitochondria, whereas fatty acid synthesis occurs in the cytosol. The new fatty acids are attached to glycerol to make triacylglycerols, which can be stored in unlimited quantities as body fat.

QUESTIONS AND PROBLEMS

Oxidation of Fatty Acids

18.55 Where in the cell is a fatty acid activated?

18.56 What coenzymes are required for β oxidation?

18.57 Consider the complete oxidation of caprylic acid $CH_3{-}(CH_2)_6{-}COOH$, a C_8 fatty acid.
a. How many acetyl CoA units are produced?
b. How many cycles of β oxidation are needed?
c. How many ATPs are generated from the oxidation of caprylic acid?

18.58 Consider the complete oxidation of arachidic acid, $CH_3{-}(CH_2)_{18}{-}COOH$, a C_{20} fatty acid.
a. How many acetyl CoA units are produced?
b. How many cycles of β oxidation are needed?
c. How many ATPs are generated from the oxidation of arachidic acid?

18.59 Consider the complete oxidation of oleic acid, $CH_3{-}(CH_2)_7{-}CH{=}CH{-}(CH_2)_7{-}COOH$, which is an 18-carbon monounsaturated fatty acid.
a. How many acetyl CoA units are produced?
b. How many cycles of β oxidation are needed?
c. How many ATPs are generated from the oxidation of oleic acid?

18.60 Consider the complete oxidation of palmitoleic acid, $CH_3{-}(CH_2)_5{-}CH{=}CH{-}(CH_2)_7{-}COOH$, which is a 16-carbon monounsaturated fatty acid found in animal and vegetable oils.
a. How many acetyl CoA units are produced?
b. How many cycles of β oxidation are needed?
c. How many ATPs are generated from the oxidation of palmitoleic acid?

18.61 When are ketone bodies produced in the body?

18.62 Why would a person who is fasting have high levels of acetyl CoA?

18.63 What are some conditions that characterize ketosis?

18.64 Why do diabetics produce high levels of ketone bodies?

Chemistry Link to Health

KETONE BODIES AND DIABETES

Blood glucose is elevated within 30 minutes following a meal containing carbohydrates. The elevated level of glucose stimulates the secretion of the hormone insulin from the pancreas, which increases the flow of glucose into muscle and adipose tissue for the synthesis of glycogen. As blood glucose levels drop, the secretion of insulin decreases. When blood glucose is low, another hormone, glucagon, is secreted by the pancreas, which stimulates the breakdown of glycogen in the liver to yield glucose.

In *diabetes mellitus*, glucose cannot be utilized or stored as glycogen because insulin is not secreted or does not function properly. In type 1, *insulin-dependent diabetes*, which often occurs in childhood, the pancreas produces inadequate levels of insulin. This type of diabetes can result from damage to the pancreas by viral infections or from genetic mutations. In type 2, *insulin-resistant diabetes*, which usually occurs in adults, insulin is produced, but insulin receptors are not responsive. Thus a person with type 2 diabetes does not respond to insulin therapy. *Gestational diabetes* can occur during pregnancy, but blood glucose levels usually return to normal after the baby is born. Mothers with diabetes tend to gain weight and have large babies.

In all types of diabetes, insufficient amounts of glucose are available in the muscle, liver, and adipose tissue. As a result, liver cells synthesize glucose from noncarbohydrate sources (gluconeogenesis) and break down fat, elevating the acetyl CoA level. Excess acetyl CoA undergoes ketogenesis, and ketone bodies accumulate in the blood. As the level of acetone increases, its odor can be detected on the breath of a person with uncontrolled diabetes who is in ketosis.

In uncontrolled diabetes, the concentration of blood glucose exceeds the ability of the kidney to reabsorb glucose, and glucose appears in the urine. High levels of glucose increase the osmotic pressure in the blood, which leads to an increase in urine output. Symptoms of diabetes include frequent urination and excessive thirst. Treatment for diabetes includes a change to a diet limiting carbohydrate intake and may require medication such as a daily injection of insulin or pills taken by mouth.

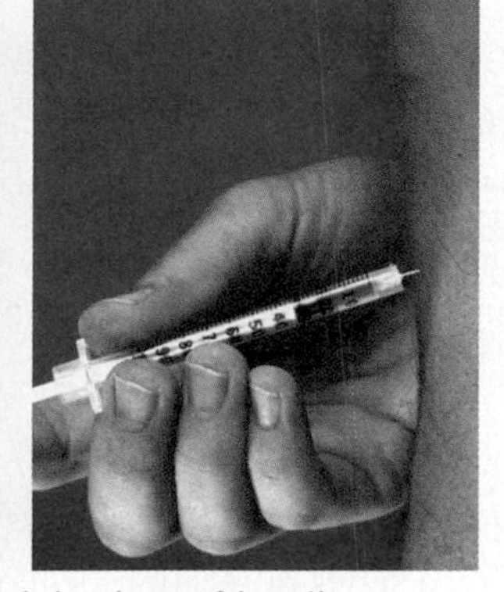

Diabetes can be treated with injections of insulin.

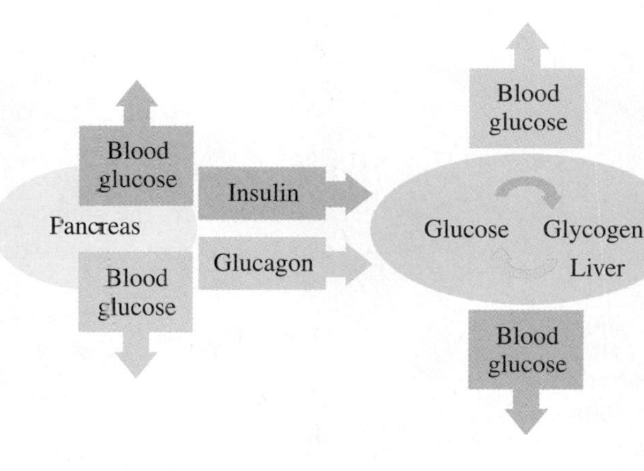

18.8 Degradation of Amino Acids

LEARNING GOAL

Describe the reactions of transamination, oxidative deamination, and the entry of amino acid carbons into the citric acid cycle.

When dietary protein exceeds the nitrogen needed by the body, the excess amino acids are degraded. The α-amino group is removed to yield a keto acid, which can be converted to an intermediate of other metabolic pathways. The carbon atoms from amino acids are used in the citric acid cycle as well as for the synthesis of fatty acids, ketone bodies, and glucose. Most of the amino groups are converted to urea.

Transamination

TUTORIAL
Transamination and Deamination

The degradation of amino acids occurs primarily in the liver. In a **transamination** reaction, an α-amino group is transferred from an amino acid to an α-keto acid, usually α-ketoglutarate. A new amino acid and a new α-keto acid are produced. We can write an equation to show the transfer of the amino group from alanine to α-ketoglutarate to yield glutamate, the new amino acid, and the α-keto acid pyruvate. Pyruvate can now enter the citric acid cycle for the production of energy.

$$\underset{\text{Alanine}}{CH_3-\overset{\overset{+}{N}H_3}{\overset{|}{C}H}-COO^-} + \underset{\alpha\text{-Ketoglutarate}}{^-OOC-\overset{O}{\overset{\|}{C}}-CH_2-CH_2-COO^-} \xrightarrow{\text{Alanine aminotransferase}}$$

$$\underset{\text{Pyruvate}}{CH_3-\overset{O}{\overset{\|}{C}}-COO^-} + \underset{\text{Glutamate}}{^-OOC-\overset{\overset{+}{N}H_3}{\overset{|}{C}H}-CH_2-CH_2-COO^-}$$

Oxidative Deamination

In a process called **oxidative deamination**, the ammonium group ($-NH_3^+$) in glutamate is removed as an ammonium ion, NH_4^+. The reaction regenerates α-ketoglutarate, which can enter transamination with an amino acid. The oxidation provides hydrogens for the NAD^+ coenzyme.

$$\underset{\text{Glutamate}}{^-OOC-\overset{\overset{+}{N}H_3}{\overset{|}{C}H}-CH_2-CH_2-COO^-} + H_2O + NAD^+ \xrightarrow{\text{Glutamate dehydrogenase}}$$

$$\underset{\alpha\text{-Ketoglutarate}}{^-OOC-\overset{O}{\overset{\|}{C}}-CH_2-CH_2-COO^-} + NH_4^+ + NADH + H^+$$

Therefore, the amino group from any amino acid can be used to form glutamate, which undergoes oxidative deamination, converting the amino group to an ammonium ion.

Urea Cycle

TUTORIAL
Detoxifying Ammonia in the Body

The ammonium ion, which is the end product of amino acid degradation, is toxic if it is allowed to accumulate. Therefore, a series of reactions, called the **urea cycle**, detoxifies ammonium ion (NH_4^+) by forming urea, which is excreted in the urine.

$$2NH_4^+ + CO_2 \longrightarrow \underset{\text{Urea}}{H_2N-\overset{O}{\overset{\|}{C}}-NH_2} + 2H^+ + H_2O$$

In one day, a typical adult may excrete about 25–30 g of urea in the urine. This amount increases when a diet is high in protein. If urea is not properly excreted, it builds up quickly to a toxic level. To detect renal disease, the blood urea nitrogen (BUN) level is measured. If the BUN is high, protein intake must be reduced, and hemodialysis may be needed to remove toxic nitrogen waste from the blood.

CONCEPT CHECK 18.10

Degradation of Amino Acids

Describe the reactions and functions of transamination and oxidative deamination.

ANSWER

Transamination involves the transfer of an amino group from an amino acid to an α-keto acid such as α-ketoglutarate to form a new amino acid (glutamate) and an α-keto acid. This new α-keto acid provides a carbon skeleton that can enter the citric acid cycle for energy production.

Oxidative deamination involves the removal of an amino group from an amino acid as an ammonium ion, and the oxidation of that carbon to regenerate an α-keto acid. The α-keto acid is usually α-ketoglutarate, which can undergo transamination reactions with other amino acids.

SAMPLE PROBLEM 18.9

Transamination and Oxidative Deamination

Indicate whether each of the following represents a transamination or an oxidative deamination:

a. Glutamate is converted to α-ketoglutarate and NH_4^+.
b. Alanine and α-ketoglutarate react to form pyruvate and glutamate.

SOLUTION

a. oxidative deamination **b.** transamination

STUDY CHECK 18.9

What is the source of the nitrogen atoms used to produce urea?

ATP Energy from Amino Acids

TUTORIAL
The Fate of Carbon Atoms in Amino Acids

Normally, only a small amount (about 10%) of our energy needs is supplied by amino acids. However, more energy is extracted from amino acids in conditions such as fasting or starvation, when carbohydrate and fat stores are exhausted. If amino acids remain the only source of energy for a long period of time, the breakdown of body proteins eventually leads to a destruction of essential body tissues.

Once the carbon skeletons of amino acids are obtained from transamination, they can be used as intermediates of the citric acid cycle to provide reduced coenzymes for electron transport. Some amino acids can enter the citric acid cycle at more than one place (see Figure 18.17).

SAMPLE PROBLEM 18.10

Carbon Atoms from Amino Acids

At what point do the carbon atoms of each of the following amino acids enter metabolic pathways for the production of energy?

a. valine **b.** tyrosine **c.** glycine

SOLUTION

a. Carbon atoms from valine enter the citric acid cycle as succinyl CoA.
b. Carbon atoms from tyrosine enter the citric acid cycle as fumarate.
c. Carbon atoms from glycine enter metabolic pathways as pyruvate.

STUDY CHECK 18.10

Which amino acids enter the citric acid cycle as α-ketoglutarate?

Overview of Metabolism

In this chapter, we have seen that catabolic pathways degrade large molecules to small molecules that are used for energy production, using the citric acid cycle and electron transport. We have also indicated that anabolic pathways lead to the synthesis of larger molecules in the cell. In the overall view of metabolism, there are several branch points

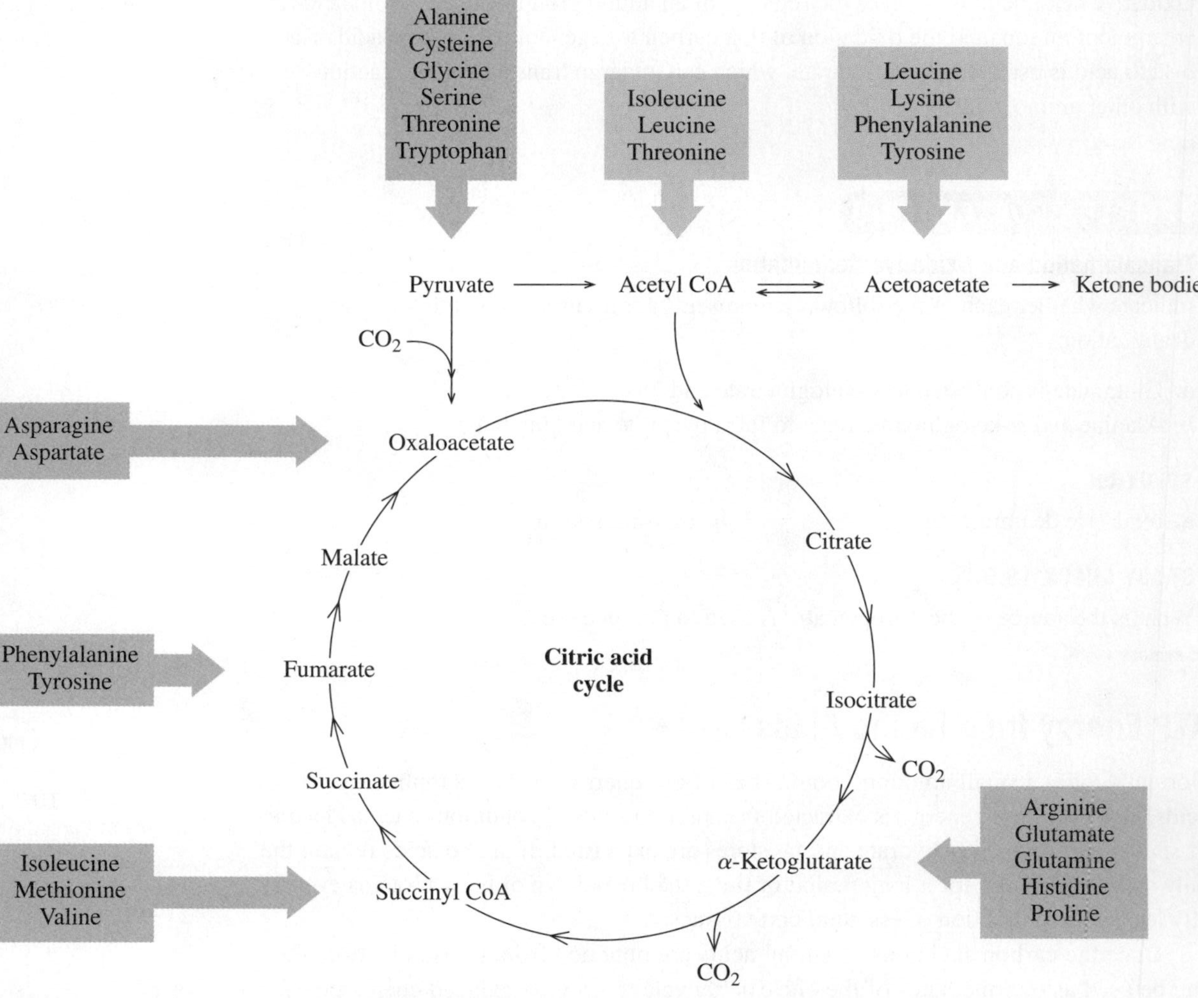

FIGURE 18.17 Carbon atoms from amino acids can be converted to the intermediates of the citric acid cycle. Carbon atoms from amino acids in green boxes can also produce ketone bodies.

Q What compound in the citric acid cycle is obtained from the carbon atoms of alanine and glycine?

from which compounds may be degraded for energy or used to synthesize larger molecules. For example, glucose can be degraded to acetyl CoA for the citric acid cycle to produce energy or converted to glycogen for storage. When glycogen stores are depleted, fatty acids are degraded for energy. Amino acids normally used to synthesize nitrogen-containing compounds in the cells can also be used for energy after they are degraded to intermediates of the citric acid cycle. In the synthesis of certain amino acids, α-keto acids of the citric acid cycle enter a variety of reactions that convert them to amino acids through transamination by glutamate (see Figure 18.18).

QUESTIONS AND PROBLEMS

Degradation of Amino Acids

18.65 Draw the condensed structural formula of the α-keto acid produced from each of the following in transamination:

a. $H{-}CH(\overset{+}{N}H_3){-}COO^-$ Glycine

b. $CH_3{-}CH(\overset{+}{N}H_3){-}COO^-$ Alanine

18.66 Draw the condensed structural formula of the α-keto acid produced from each of the following in transamination:

a. $^-OOC{-}CH_2{-}CH(\overset{+}{N}H_3){-}COO^-$ Aspartate

b. $CH_3{-}CH_2{-}CH(CH_3){-}CH(\overset{+}{N}H_3){-}COO^-$ Isoleucine

18.67 Why does the body convert NH_4^+ to urea?

18.68 What is the structure of urea?

18.69 What metabolic substrate(s) are produced from the carbon atoms of each of the following amino acids?
a. alanine
b. aspartate
c. tyrosine
d. glutamine

18.70 What metabolic substrate(s) are produced from the carbon atoms of each of the following amino acids?
a. leucine
b. asparagine
c. cysteine
d. arginine

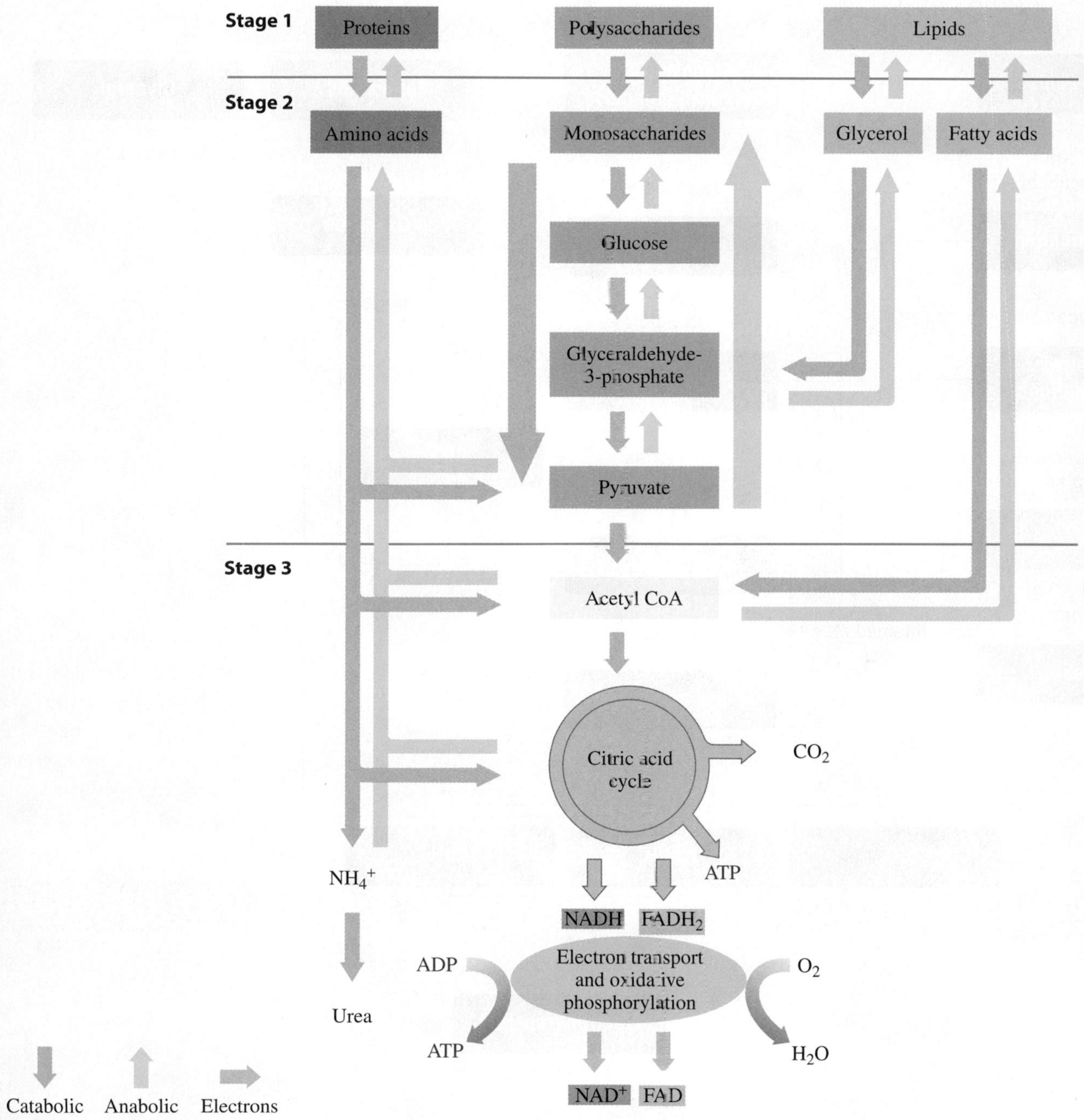

FIGURE 18.18 Catabolic and anabolic pathways in the cells provide the energy and necessary compounds for the cells.
Q Under what conditions in the cell are amino acids degraded for energy?

CONCEPT MAP

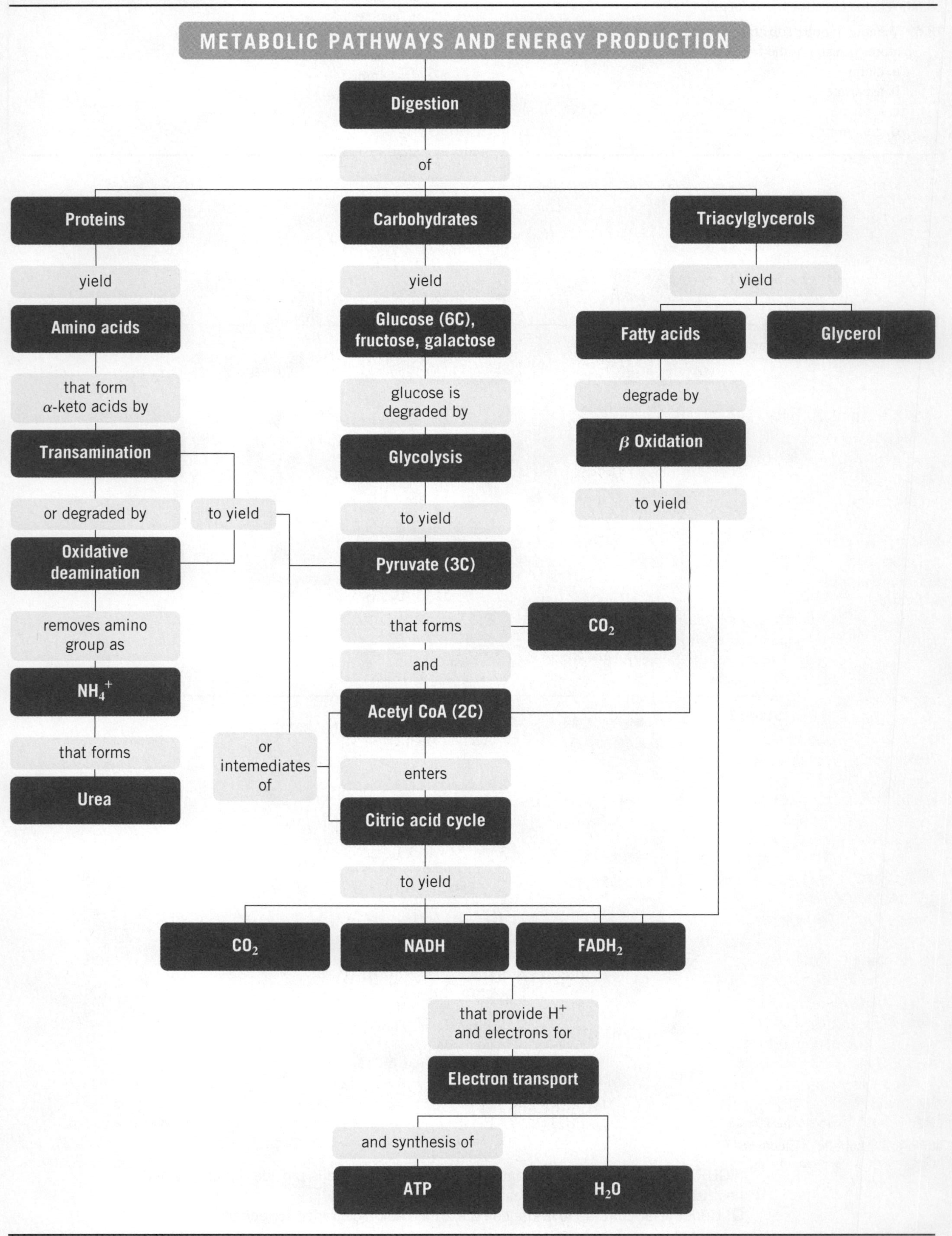
METABOLIC PATHWAYS AND ENERGY PRODUCTION
Digestion
of
Proteins
Carbohydrates
Triacylglycerols
yield
Amino acids
that form
α-keto acids by
Transamination
or degraded by
Oxidative deamination
to yield
removes amino group as
NH4+
that forms
Urea
yield
Glucose (6C), fructose, galactose
glucose is degraded by
Glycolysis
to yield
Pyruvate (3C)
that forms
CO2
and
Acetyl CoA (2C)
or intemediates of
enters
Citric acid cycle
to yield
yield
Fatty acids
Glycerol
degrade by
β Oxidation
to yield
CO2
NADH
FADH2
that provide H+ and electrons for
Electron transport
and synthesis of
ATP
H2O

CHAPTER REVIEW

18.1 Metabolism and ATP Energy

Learning Goal: Describe the three stages of metabolism and the role of ATP.

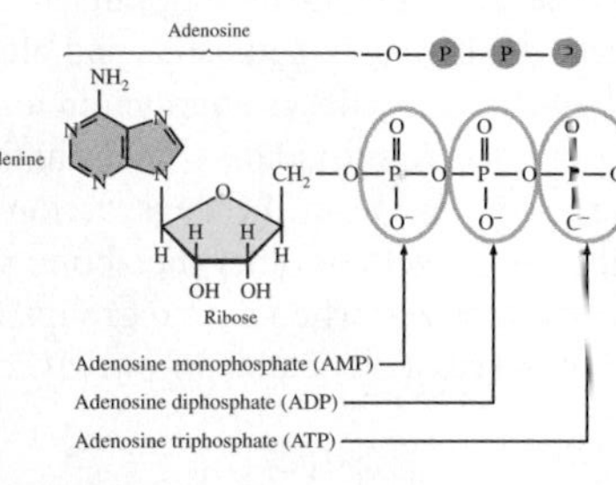

Metabolism includes all the catabolic and anabolic reactions that occur in the cells. Catabolic reactions degrade large molecules into smaller ones with an accompanying release of energy. Anabolic reactions require energy to synthesize larger molecules from smaller ones. The three stages of metabolism are digestion of food, degradation of monomers such as glucose to pyruvate, and the extraction of energy from the two- and three-carbon compounds. Energy obtained from catabolic reactions is stored in adenosine triphosphate (ATP), a high-energy compound that is hydrolyzed when energy is required by anabolic reactions.

18.2 Digestion of Foods

Learning Goal: Give the sites and products of digestion of carbohydrates, triacylglycerols, and proteins.

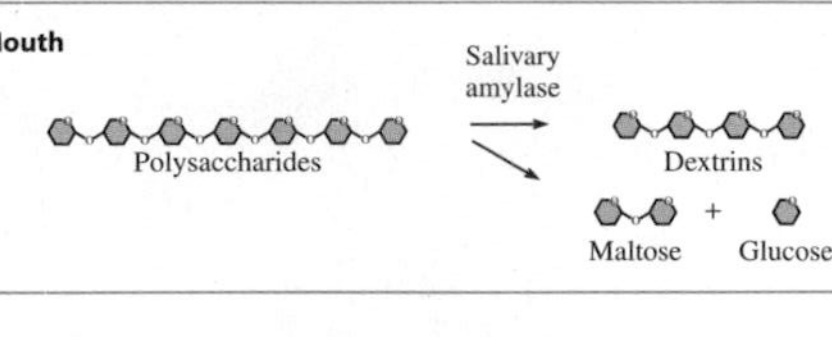

The digestion of carbohydrates is a series of reactions that breaks down polysaccharides into smaller polysaccharides (dextrins), and eventually to monosaccharides glucose, galactose, and fructose. These monomers can be absorbed through the intestinal wall into the bloodstream to be carried to cells where they provide energy and carbon atoms for synthesis of new molecules. Triacylglycerols are hydrolyzed in the small intestine to monoacylglycerol and fatty acids, which enter the intestinal wall and form new triacylglycerols. They bind with proteins to form chylomicrons, which transport them through the lymphatic system and bloodstream to the tissues. The digestion of proteins, which begins in the stomach and continues in the small intestine, involves the hydrolysis of peptide bonds to yield amino acids that are absorbed through the intestinal wall and transported to the cells.

18.3 Coenzymes in Metabolic Pathways

Learning Goal: Describe the components and functions of the coenzymes NAD^+, FAD, and coenzyme A.

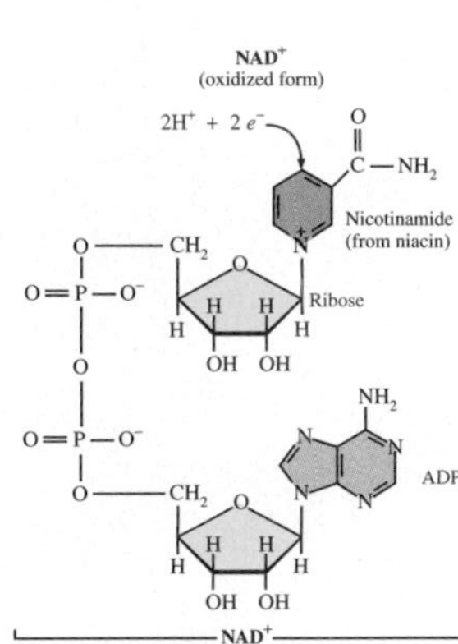

FAD and NAD^+ are the oxidized forms of coenzymes that participate in oxidation–reduction reactions. When they pick up hydrogen ions and electrons, they are reduced to $FADH_2$ and $NADH + H^+$. Coenzyme A contains a thiol group that usually bonds with a two-carbon acetyl group (acetyl CoA).

18.4 Glycolysis: Oxidation of Glucose

Learning Goal: Describe the conversion of glucose to pyruvate in glycolysis and the subsequent conversion of pyruvate to acetyl CoA or lactate.

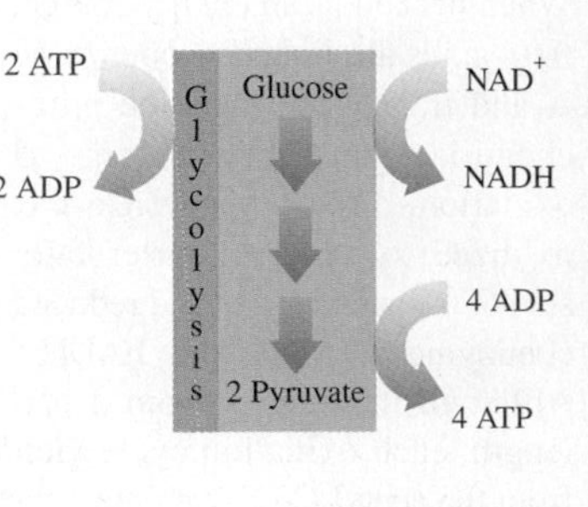

Glycolysis, which occurs in the cytosol, consists of 10 reactions that degrade glucose (6C) to two pyruvates (3C). The overall series of reactions yields two molecules of the reduced coenzyme NADH and two ATPs. Under aerobic conditions, pyruvate is oxidized in the mitochondria to acetyl CoA. In the absence of oxygen, pyruvate is reduced to lactate and NAD^+ is regenerated for the continuation of glycolysis.

18.5 The Citric Acid Cycle

Learning Goal: Describe the oxidation of acetyl CoA in the citric acid cycle.

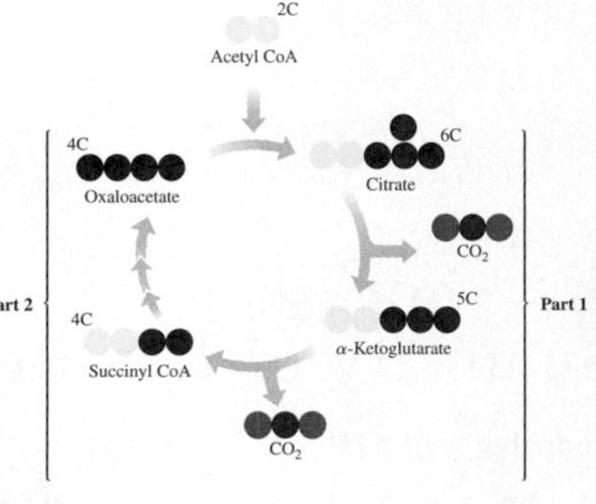

In a sequence of reactions called the citric acid cycle, an acetyl group is combined with oxaloacetate to yield citrate. Citrate undergoes oxidation and decarboxylation to yield two CO_2, GTP, three NADHs, and $FADH_2$, with the regeneration of oxaloacetate. The direct phosphorylation of ADP by GTP yields ATP.

18.6 Electron Transport and Oxidative Phosphorylation

Learning Goal: Describe electron transport and the process of oxidative phosphorylation; calculate the ATP from the complete oxidation of glucose.

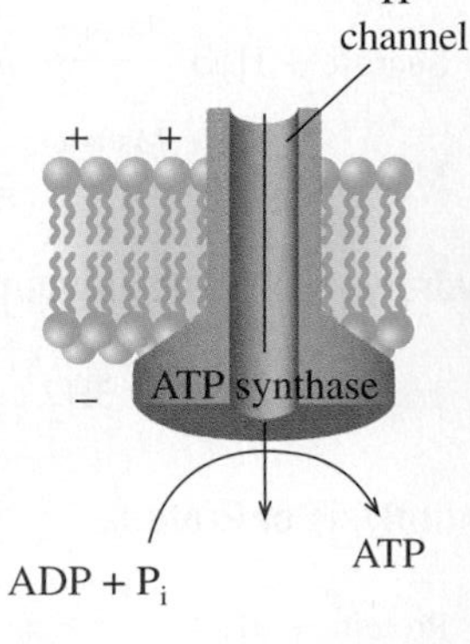

The reduced coenzymes NADH and $FADH_2$ from various metabolic pathways are oxidized to NAD^+ and FAD when their hydrogens and electrons enter the electron transport system. The final acceptor, O_2, combines with protons and electrons to yield H_2O.

The protein complexes in electron transport act as proton pumps to move hydrogen ions into the intermembrane space, which produces a proton gradient. As the protons return to the matrix by way of ATP synthase, energy is generated. This energy is used to drive the synthesis of ATP in a process known as oxidative phosphorylation. The oxidation of NADH yields three ATP molecules, and $FADH_2$ yields two ATPs. Under aerobic conditions, the complete oxidation of glucose yields a total of 36 ATPs.

18.7 Oxidation of Fatty Acids

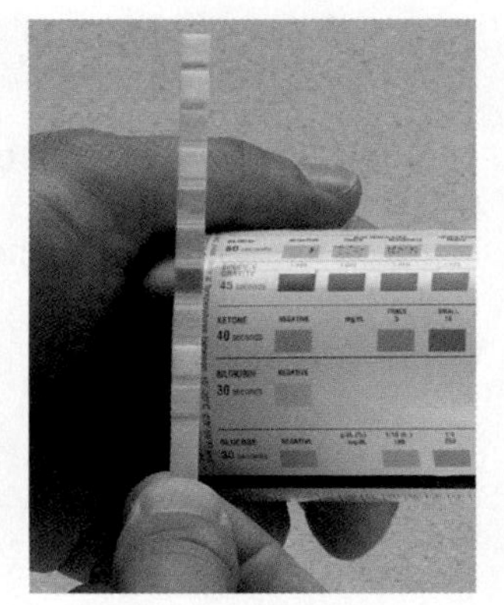

Learning Goal: Describe the metabolic pathway of β oxidation; calculate the ATP from the complete oxidation of a fatty acid.

When needed as an energy source, fatty acids are linked to coenzyme A and transported into the mitochondria where they undergo β oxidation. The fatty acyl CoA is oxidized to yield a shorter fatty acyl CoA, acetyl CoA, and reduced coenzymes NADH and $FADH_2$.

Although the energy from a particular fatty acid depends on its length, each oxidation cycle yields 5 ATPs with another 12 ATPs from the acetyl CoA that enters the citric acid cycle. When high levels of acetyl CoA are present in the cell, they enter the ketogenesis pathway, forming ketone bodies such as acetoacetate, which cause ketosis and acidosis.

18.8 Degradation of Amino Acids

Learning Goal: Describe the reactions of transamination, oxidative deamination, and the entry of amino acid carbons into the citric acid cycle.

When the amount of amino acids in the cells exceeds that needed for synthesis of nitrogen compounds, the process of transamination converts them to α-keto acids and glutamate. Oxidative deamination of glutamate produces ammonium ions and α-ketoglutarate. Ammonium ions from oxidative deamination are converted to urea. The carbon atoms from the degradation of amino acids can enter the citric acid cycle or other metabolic pathways. Certain amino acids are synthesized when amino groups from glutamate are transferred to an α-keto acid obtained from glycolysis or the citric acid cycle.

$$\underset{\text{Alanine}}{CH_3-CH(\overset{+}{N}H_3)-COO^-} + \underset{\alpha\text{-Ketoglutarate}}{^-OOC-C(=O)-CH_2-CH_2-COO^-} \xrightarrow{\text{Alanine aminotransferase}} \underset{\text{Pyruvate}}{CH_3-C(=O)-COO^-} + \underset{\text{Glutamate}}{^-OOC-CH(\overset{+}{N}H_3)-CH_2-CH_2-COO^-}$$

Summary of Key Reactions

Hydrolysis of ATP

$$ATP + H_2O \longrightarrow ADP + P_i + 7.3\text{ kcal/mole}$$

Formation of ATP

$$ADP + P_i + 7.3\text{ kcal/mole} \longrightarrow ATP + H_2O$$

Hydrolysis of Disaccharides

$$\text{Lactose} + H_2O \xrightarrow{\text{Lactase}} \text{galactose} + \text{glucose}$$

$$\text{Sucrose} + H_2O \xrightarrow{\text{Sucrase}} \text{fructose} + \text{glucose}$$

$$\text{Maltose} + H_2O \xrightarrow{\text{Maltase}} \text{glucose} + \text{glucose}$$

Hydrolysis of Triacylglycerols

$$\text{Triacylglycerols} + 3H_2O \xrightarrow{\text{Lipases}} \text{glycerol} + 3\text{ fatty acids}$$

Hydrolysis of Proteins

$$\text{Protein} + H_2O \xrightarrow{H^+} \text{amino acids}$$

Reduction of NAD^+ and FAD

$$NAD^+ + 2H^+ + 2e^- \longrightarrow NADH + H^+$$

$$FAD + 2H^+ + 2e^- \longrightarrow FADH_2$$

Glycolysis

$$\underset{\text{Glucose}}{C_6H_{12}O_6} + 2ADP + 2P_i + 2NAD^+ \longrightarrow 2\underset{\text{Pyruvate}}{CH_3{-}C({=}O){-}COO^-} + 2ATP + 2NADH + 4H^+$$

Reduction of Pyruvate to Lactate (Anaerobic)

$$\underset{\text{Pyruvate}}{CH_3{-}C({=}O){-}COO^-} + NADH + H^+ \longrightarrow \underset{\text{Lactate}}{CH_3{-}CH(OH){-}COO^-} + NAD^+$$

Oxidation of Pyruvate to Acetyl CoA (Aerobic)

$$\underset{\text{Pyruvate}}{CH_3{-}C({=}O){-}COO^-} + NAD^+ + HS{-}CoA \longrightarrow \underset{\text{Acetyl CoA}}{CH_3{-}C({=}O){-}S{-}CoA} + NADH + CO_2$$

Citric Acid Cycle

$$\text{Acetyl CoA} + 3NAD^+ + FAD + GDP + P_i + 2H_2O \longrightarrow 2CO_2 + 3NADH + 3H^+ + FADH_2 + CoA + GTP$$

Synthesis of ATP (ATP Synthase)

$$NADH + H^+ + 3ADP + 3P_i + \tfrac{1}{2}O_2 \longrightarrow NAD^+ + 3ATP + H_2O$$

$$FADH_2 + 2ADP + 2P_i + \tfrac{1}{2}O_2 \longrightarrow FAD + 2ATP + H_2O$$

Oxidation of Glucose (Complete)

$$C_6H_{12}O_6 + 6O_2 + 36ADP + 36P_i \longrightarrow 6CO_2 + 6H_2O + 36ATP$$

β Oxidation of Fatty Acids

$$\underset{\text{Fatty acyl CoA}}{R{-}CH_2{-}CH_2{-}C({=}O){-}S{-}CoA} + NAD^+ + FAD + H_2O + HS{-}CoA \longrightarrow$$

$$\underset{\text{New fatty acyl (–2C) CoA}}{R{-}C({=}O){-}S{-}CoA} + \underset{\text{Acetyl CoA}}{CH_3{-}C({=}O){-}S{-}CoA} + NADH + H^+ + FADH_2$$

Transamination

$$\underset{\text{Alanine}}{CH_3{-}CH(\overset{+}{N}H_3){-}COO^-} + \underset{\alpha\text{-Ketoglutarate}}{{}^-OOC{-}C({=}O){-}CH_2{-}CH_2{-}COO^-} \xrightarrow{\text{Alanine aminotransferase}}$$

$$\underset{\text{Pyruvate}}{CH_3{-}C({=}O){-}COO^-} + \underset{\text{Glutamate}}{{}^-OOC{-}CH(\overset{+}{N}H_3){-}CH_2{-}CH_2{-}COO^-}$$

Oxidative Deamination

$$\underset{\text{Glutamate}}{{}^-OOC{-}CH(\overset{+}{N}H_3){-}CH_2{-}CH_2{-}COO^-} + H_2O + NAD^+ \xrightarrow{\text{Glutamate dehydrogenase}}$$

$$\underset{\alpha\text{-Ketoglutarate}}{{}^-OOC{-}C({=}O){-}CH_2{-}CH_2{-}COO^-} + NH_4^+ + NADH + H^+$$

Urea Cycle

$$2NH_4^+ + CO_2 \longrightarrow \underset{\text{Urea}}{H_2N{-}\overset{\displaystyle O}{\overset{\|}{C}}{-}NH_2} + 2H^+ + H_2O$$

Key Terms

acetyl CoA A two-carbon acetyl unit from oxidation of pyruvate that bonds to coenzyme A.

acidosis Low blood pH resulting from the formation of acidic ketone bodies.

aerobic An oxygen-containing environment in the cells.

anabolic reaction A metabolic reaction that requires energy.

anaerobic A condition in cells when there is no oxygen.

ATP Adenosine triphosphate, a high-energy compound that stores energy in the cells and that consists of adenine, a ribose sugar, and three phosphate groups.

ATP synthase An enzyme complex that uses the energy released by protons returning to the matrix to synthesize ATP from ADP and P_i.

beta (β) oxidation The degradation of fatty acids that removes two-carbon segments from the fatty acid at the oxidized β carbon.

catabolic reaction A metabolic reaction that produces energy for the cell by the degradation and oxidation of glucose and other molecules.

chemiosmotic model The conservation of energy from electron transport by pumping protons into the intermembrane space to produce a proton gradient that provides the energy to synthesize ATP.

citric acid cycle A series of oxidation reactions in the mitochondria that convert acetyl CoA to CO_2 and yield NADH and $FADH_2$. It is also called the Krebs cycle.

coenzyme A (CoA) A coenzyme that transports acyl and acetyl groups.

cytoplasm The material in eukaryotic cells between the nucleus and the plasma membrane.

cytosol The fluid of the cytoplasm, which is an aqueous solution of electrolytes and enzymes.

decarboxylation The loss of a carbon atom in the form of CO_2.

digestion The processes in the gastrointestinal tract that break down large food molecules to smaller ones that pass through the intestinal membrane into the bloodstream.

electron transport A series of reactions in the mitochondria that transfer electrons from NADH and $FADH_2$ to electron carriers and finally to O_2, which produces H_2O. The energy released is used to synthesize ATP from ADP and P_i.

FAD A coenzyme (flavin adenine dinucleotide) for dehydrogenase enzymes that form carbon–carbon double bonds.

glycolysis The ten oxidation reactions of glucose that yield two pyruvate molecules.

ketone bodies The products of ketogenesis: acetoacetate, β-hydroxybutyrate, and acetone.

ketosis A condition in which high levels of ketone bodies cannot be metabolized, leading to lower blood pH.

metabolism All the chemical reactions in living cells that carry out molecular and energy transformations.

mitochondria The organelles of the cells where energy-producing reactions take place.

NAD^+ The hydrogen acceptor (nicotinamide adenine dinucleotide) used in oxidation reactions that form carbon–oxygen double bonds.

oxidative deamination The loss of ammonium ion when glutamate is degraded to α-ketoglutarate.

oxidative phosphorylation The synthesis of ATP from ADP and P_i, using energy generated by the oxidation reactions during electron transport.

proton pumps The enzyme complexes I, III, and IV that move protons from the matrix into the intermembrane space, creating a proton gradient.

transamination The transfer of an amino group from an amino acid to an α-keto acid.

urea cycle The process in which ammonium ions from the degradation of amino acids are converted to urea.

Understanding the Concepts

18.71 Lauric acid, $CH_3{-}(CH_2)_{10}{-}COOH$, which is found in coconut oil, is a saturated fatty acid.

Coconut oil is high in lauric acid.

a. Draw the condensed structural formula of lauric acid acyl CoA.

b. Indicate the α and β carbon atoms in the fatty acyl molecule.

c. How many acetyl CoA units are produced?

d. How many cycles of β oxidation are needed?

e. Account for the total ATP yield from β oxidation of lauric acid by completing the following calculation:

activation	⟶	−2 ATP
______ acetyl CoA	⟶	______ ATP
______ $FADH_2$	⟶	______ ATP
______ NADH	⟶	______ ATP
Total		______ **ATP**

18.72 A hiker expends 450 kcal. How many moles of ATP does this require?

ATP is consumed when a person exercises.

18.73 Identify the type of food as carbohydrate, fat, or protein that gives each of the following digestion products:

a. glucose **b.** fatty acid
c. maltose **d.** glycerol
e. amino acids **f.** dextrins

Digestion is the first stage of metabolism.

18.74 Identify each of the following as a six-carbon or a three-carbon compound and arrange them in the order in which they occur in glycolysis:

a. 3-phosphoglycerate **b.** pyruvate
c. glucose-6-phosphate **d.** glucose
e. fructose-1,6-bisphosphate

Additional Questions and Problems

For instructor-assigned homework, go to **www.masteringchemistry.com.**

18.75 Write an equation for the hydrolysis of ATP to ADP.

18.76 At the gym, you expend 310 kcal riding the stationary bicycle for 1 h. How many moles of ATP will this require?

18.77 How and where does lactose undergo digestion in the body? What are the products?

18.78 How and where does sucrose undergo digestion in the body? What are the products?

18.79 What are the reactant and product of glycolysis?

18.80 What is the general type of reaction that takes place in the digestion of carbohydrates?

18.81 When is pyruvate converted to lactate in the body?

18.82 When pyruvate is used to form acetyl CoA, the product has only two carbon atoms. What happened to the third carbon?

18.83 What is the main function of the citric acid cycle in energy production?

18.84 Most metabolic pathways are not considered cycles. Why is the citric acid cycle considered to be a metabolic cycle?

18.85 If there are no reactions in the citric acid cycle that use oxygen, O_2, why does the cycle operate only in aerobic conditions?

18.86 What products of the citric acid cycle are needed for electron transport?

18.87 In the chemiosmotic model, how is energy provided to synthesize ATP?

18.88 What is the effect of proton accumulation in the intermembrane space?

18.89 How many ATPs are produced when glucose is oxidized to pyruvate, compared to when glucose is oxidized to CO_2 and H_2O?

18.90 What metabolic substrate(s) can be produced from the carbon atoms of each of the following amino acids?

a. histidine **b.** isoleucine **c.** serine **d.** phenylalanine

Challenge Questions

18.91 One cell at work may break down 2 million (2 000 000) ATP molecules in one second. Some researchers estimate that the human body has about 10^{13} cells.

a. How much energy, in kcal, would be used by the cells in the body in one day?
b. If ATP has a molar mass of 507 g/mole, how many grams of ATP are hydrolyzed?

18.92 State if each of the following processes produce or consume ATP:

a. citric acid cycle
b. glucose forms two pyruvates
c. pyruvate forms acetyl CoA
d. glucose forms glucose-6-phosphate
e. oxidation of α-ketoglutarate
f. transport of NADH across the mitochondrial membrane
g. activation of a fatty acid

18.93 Match the following ATP yields to reactions **a–g**:

2 ATP	3 ATP	6 ATP	12 ATP
18 ATP	36 ATP	44 ATP	

a. Glucose forms two pyruvates.
b. Pyruvate forms acetyl CoA.
c. Glucose forms two acetyl CoAs.
d. Acetyl CoA goes through one turn of the citric acid cycle.
e. Caproic acid (C_6) is completely oxidized.
f. NADH + H^+ is oxidized to NAD^+.
g. $FADH_2$ is oxidized to FAD.

18.94 Identify each of the following reactions **a** to **e** in the β oxidation of palmitic acid, a C_{16} fatty acid, as
(1) activation
(2) dehydrogenation
(3) hydration
(4) oxidation
(5) cleavage

a. Palmityl CoA and FAD form α,β-unsaturated palmityl CoA and $FADH_2$.
b. β-Keto palmityl CoA forms myristyl CoA and acetyl CoA.
c. Palmitic acid, CoA, and ATP form palmityl CoA.
d. α,β-Unsaturated palmityl CoA and H_2O form β-hydroxy palmityl CoA.
e. β-Hydroxy palmityl CoA and NAD^+ form β-keto palmityl CoA and NADH + H^+.

18.95 Which of the following molecules will produce the most ATP per mole?
a. glucose or maltose
b. myristic acid, $CH_3-(CH_2)_{12}-COOH$, or stearic acid, $CH_3-(CH_2)_{16}-COOH$
c. glucose or two acetyl CoAs
d. glucose or caprylic acid (C_8)
e. citrate or succinate in one turn of the citric acid cycle

18.96 Which of the following molecules will produce the most ATP per mole?
a. glucose or stearic acid (C_{18})
b. glucose or two pyruvate
c. two acetyl CoAs or one palmitic acid (C_{16})
d. lauric acid (C_{12}) or palmitic acid(C_{16})
e. α-ketoglutarate or fumarate in one turn of the citric acid cycle

Answers

Answers to Study Checks

18.1 The digestion of amylose begins in the mouth where enzymes hydrolyze some of the glycosidic bonds. In the small intestine, enzymes hydrolyze more glycosidic bonds, and finally maltose is hydrolyzed to yield glucose.

18.2 HCl denatures proteins and activates enzymes in the stomach that digest proteins.

18.3 In the initial reactions of glycolysis, energy in the form of two ATPs is invested to convert glucose to fructose-1,6-bisphosphate.

18.4 oxaloacetate

18.5 The protein complexes I, III, and IV pump protons from the matrix to the intermembrane space.

18.6 One turn of the citric acid cycle produces a total of 12 ATPs: 9 ATPs from 3 NADHs, 2 ATPs from $FADH_2$, and 1 ATP from a phosphorylation.

18.7 NAD^+

18.8 In β oxidation, each acetyl CoA enters the citric acid cycle to produce 3 NADHs, $FADH_2$, and GTP for a total of 12 ATPs. The reduced coenzymes NADH and $FADH_2$ enter electron transport to give 5 ATPs.

18.9 The process of oxidative deamination involves the removal of the amino group of glutamate as ammonium.

18.10 Arginine, glutamate, glutamine, histidine, and proline provide carbon atoms for α-ketoglutarate.

Answers to Selected Questions and Problems

18.1 The digestion of polysaccharides takes place in stage 1.

18.3 In metabolism, a catabolic reaction breaks apart large molecules, releasing energy.

18.5 The hydrolysis of the phosphodiester bond (P—O—P) in ATP releases energy that is sufficient for energy-requiring processes in the cell.

18.7 Hydrolysis is the main reaction involved in the digestion of carbohydrates.

18.9 The bile salts emulsify fat to give small fat globules for lipase hydrolysis.

18.11 The digestion of proteins begins in the stomach and is completed in the small intestine.

18.13 **a.** NADH
b. FAD
c. FAD

18.15 **a.** coenzyme A
b. NAD^+
c. FAD

18.17 glucose

18.19 ATP is required in phosphorylation reactions.

18.21 ATP is produced in glycolysis by transferring a phosphate from 1,3-bisphosphoglycerate and from phosphoenolpyruvate directly to ADP.

18.23 **a.** 1 ATP is required.
b. 1 NADH is produced for each triose.
c. 2 ATPs and 2 NADHs are produced.

18.25 Aerobic (oxygen) conditions are needed.

18.27 The oxidation of pyruvate converts NAD^+ to NADH and produces acetyl CoA and CO_2.

$$\text{Pyruvate} + NAD^+ + \text{CoA} \longrightarrow \text{acetyl CoA} + CO_2 + \text{NADH} + H^+$$

18.29 When pyruvate is reduced to lactate, the NAD^+ is used to oxidize glyceraldehyde-3-phosphate, which recycles NADH.

18.31 $2CO_2$, 3NADH, $FADH_2$, GTP (ATP), CoA, and $3H^+$.

18.33 Two reactions, 3 and 4, involve oxidation and decarboxylation.

18.35 NAD^+ is reduced in reactions 3, 4, and 8 of the citric acid cycle.

18.37 In reaction 5, GDP undergoes a direct phosphate transfer.

18.39 **a.** citrate and isocitrate
b. In decarboxylation, a carbon atom is lost as CO_2.
c. α-ketoglutarate
d. isocitrate $\longrightarrow$ α-ketoglutarate; α-ketoglutarate $\longrightarrow$ succinyl CoA; succinate $\longrightarrow$ fumarate; malate $\longrightarrow$ oxaloacetate
e. reactions 3 and 8

18.41 NADH and $FADH_2$

18.43 Coenzyme Q transfers electrons from complex I to III.

18.45 NADH transfers electrons to complex I to give NAD^+.

18.47 In oxidative phosphorylation, the energy from the oxidation reactions in electron transport is used to drive ATP synthesis.

18.49 Protons return to a lower energy in the matrix by passing through ATP synthase, which releases energy to drive the synthesis of ATP.

18.51 The reduced coenzymes NADH and $FADH_2$ from glycolysis and the citric acid cycle transfer electrons to the electron transport system, which generates energy to drive the synthesis of ATP.

18.53 **a.** 3 ATP **b.** 6 ATP
c. 6 ATP **d.** 12 ATP

18.55 In the cytosol at the outer mitochondrial membrane

18.57 **a.** 4 acetyl CoA units
b. 3 cycles of β oxidation
c. 48 ATP from 4 acetyl CoA (citric acid cycle) + 9 ATP from 3 NADH + 6 ATP from 3 $FADH_2$ − 2 ATP (activation) = 63 − 2 = 61 ATP

18.59 **a.** 9 acetyl CoA units
b. 8 cycles of β oxidation
c. 108 ATP from 9 acetyl CoA (citric acid cycle) + 24 ATP from 8 NADH + 16 ATP from 8 $FADH_2$ − 2 ATP (activation) = 148 − 2 = 146 ATP

18.61 Ketone bodies form in the body when excess acetyl CoA results from the breakdown of large amounts of fat.

18.63 High levels of ketone bodies lead to ketosis, a condition characterized by acidosis (a drop in blood pH values), excessive urination, and strong thirst.

18.65 **a.** $H{-}\overset{\overset{\large O}{\|}}{C}{-}COO^-$ **b.** $CH_3{-}\overset{\overset{\large O}{\|}}{C}{-}COO^-$

18.67 NH_4^+ is toxic if allowed to accumulate in the body.

18.69 **a.** pyruvate **b.** oxaloacetate
c. fumarate and acetoacetate **d.** α-ketoglutarate

18.71 **a.** and **b.**

$$\underset{}{CH_3{-}(CH_2)_8{-}}\underset{\beta}{CH_2}{-}\underset{\alpha}{CH_2}{-}\overset{\overset{O}{\|}}{C}{-}S{-}CoA$$

c. Six acetyl CoA units are produced.
d. Five cycles of β oxidation are needed.
e.

activation		⟶	−2 ATP
6 acetyl CoA	× 12 ATP/acetyl CoA	⟶	72 ATP
5 $FADH_2$	× 2 ATP/$FADH_2$	⟶	10 ATP
5 NADH	× 3 ATP/NADH	⟶	15 ATP
		Total	95 ATP

18.73 **a.** carbohydrate **b.** fat
c. carbohydrate **d.** fat
e. protein **f.** carbohydrate

18.75 $ATP + H_2O \longrightarrow ADP + P_i + 7.3\ kcal/mole$

18.77 Lactose undergoes digestion in the mucosal cells of the small intestine to yield galactose and glucose.

18.79 Glucose is the reactant, and pyruvate is the product of glycolysis.

18.81 Pyruvate is converted to lactate when oxygen is not present in the cell (anaerobic) to regenerate NAD^+ for glycolysis.

18.83 The oxidation reactions of the citric acid cycle produce a source of reduced coenzymes for electron transport and ATP synthesis.

18.85 The oxidized coenzymes NAD^+ and FAD needed for the citric acid cycle are regenerated by the electron transport system.

18.87 Energy released as protons flow through ATP synthase and then back to the matrix is utilized for the synthesis of ATP.

18.89 The oxidation of glucose to pyruvate produces 6 ATPs whereas the oxidation of glucose to CO_2 and H_2O produces 36 ATPs.

18.91 **a.** 21 kcal **b.** 1500 g of ATP

18.93 **a.** 6 ATP/glucose **b.** 3 ATP/pyruvate
c. 12 ATP/glucose **d.** 12 ATP/acetyl CoA
e. 44 ATP/C_6 acid **f.** 3 ATP/NADH
g. 2 ATP/$FADH_2$

18.95 **a.** maltose **b.** stearic acid
c. glucose **d.** caprylic acid
e. citrate

Combining Ideas from Chapters 16 to 18

CI.31 Beano contains an enzyme that breaks down polysaccharides into mono- and disaccharides that are more digestible. It is used to diminish gas formation that can occur after eating foods such as beans or cruciferous vegetables like cabbage, Brussels sprouts, and broccoli.

Beano is a dietary supplement that is used to reduce gas.

a. The label on beano says "contains alpha-galactosidase." What class of enzyme is present in beano?
b. What is the substrate for the enzyme?
c. The directions indicate you should not heat or cook beano. Why?

CI.32 Kevlar is a lightweight polymer used in tires and bulletproof vests. Part of the strength of Kevlar is due to hydrogen bonds between polymer chains.

Kevlar is used to make bulletproof vests.

a. Draw the condensed structural formula of the carboxylic acid and amine that are polymerized to make Kevlar.
b. What feature of Kevlar will produce hydrogen bonds between the polymer chains?
c. What type of secondary protein structure has bonds that are similar to those in Kevlar?

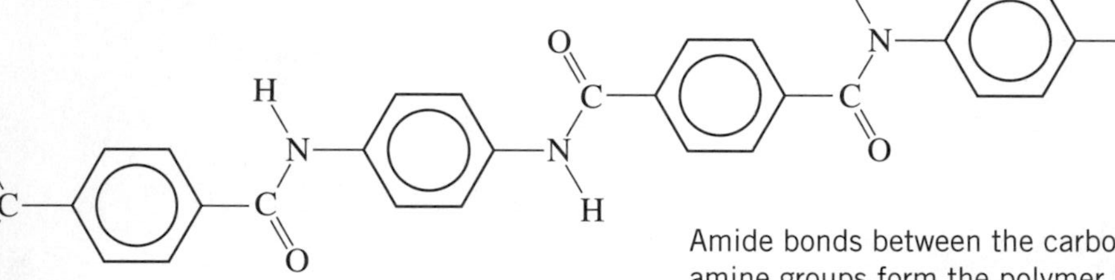

Amide bonds between the carboxyl and amine groups form the polymer Kevlar.

CI.33 Identify each of the following as a substance that is part of the citric acid cycle, electron transport, or both:
a. succinate
b. QH_2
c. FAD
d. cyt *c*
e. H_2O
f. malate
g. NAD^+

CI.34 Use the energy value of 7.3 kcal per mole of ATP and determine the total energy, in kilocalories, stored as ATP from each of the following:
a. 1 mole of glucose in glycolysis
b. the oxidation of 2 moles of pyruvate to 2 moles of acetyl CoA
c. complete oxidation of 1 mole of glucose to CO_2 and H_2O
d. 1 mole of lauric acid, a C_{12} fatty acid, in β oxidation
e. 1 mole of glutamate (from protein) in the citric acid cycle

CI.35 The formula of acetyl coenzyme A is $C_{23}H_{38}N_7O_{17}P_3S$.
a. What are the components of acetyl coenzyme A?
b. What is the function of acetyl coenzyme A?
c. Where does the acetyl group attach to CoA?
d. What is the molar mass, to 3 significant figures, for acetyl CoA?

CI.36 Behenic acid is a saturated 22-carbon fatty acid found in peanut and canola oils.

Behenic acid is one of the fatty acids found in peanut oil.

a. Draw the condensed structural formula of behenic acid acyl CoA.
b. Indicate the α and β carbon atoms in the fatty acyl condensed structural formula in **a.**
c. How many acetyl CoA units are produced from the beta oxidation of behenic acid?
d. How many cycles of β oxidation are needed?
e. What is the total ATP yield from the complete β oxidation of behenic acid?

activation		$\longrightarrow$	−2	ATP
______	acetyl CoA	$\longrightarrow$	______	ATP
______	$FADH_2$	$\longrightarrow$	______	ATP
______	NADH	$\longrightarrow$	______	ATP
Total		$\longrightarrow$	______	**ATP**

CI.37 Butter is a fat that contains 80.% triacylglycerols; the rest is water. Assume the triacylglycerol in butter is glyceryl tripalmitate.

Butter is high in triacylglycerols.

a. Write an equation for the hydrolysis of glyceryl tripalmitate.
b. What is the molar mass of glyceryl tripalmitate, $C_{51}H_{98}O_6$?
c. Calculate the ATP yield from the complete oxidation of palmitic acid.
d. How many kilocalories are released from the palmitic acid in a 0.50-oz pat of butter?
e. If running for 1 h uses 750 kcal, how many pats of butter would provide the energy (kcal) for a 45-min run?

CI.38 Thalassemia is an inherited genetic mutation that limits the production of the beta chain needed for hemoglobin formation. With low levels of beta hemoglobin, there is a shortage of red blood cells (anemia); as a result, the body does not have sufficient amounts of oxygen. In one form of thalassemia, a single nucleotide is deleted in the section of DNA that codes for the beta chain. The mutation involves the deletion of thymine (T) from section 91 in the coding strand of DNA.

89	90	91	92	93	94	95
—AGT—	GAG—	CTG—	CAC—	TGT—	GAC—	A. . . .

a. Write the template strand for this DNA section.
b. Write the mRNA sequence from DNA using the template strand in part **a.**
c. What amino acids are placed in the beta chain by this portion of mRNA?
d. What is the order of nucleotides in the coding strand of the mutant DNA?
e. Write the template strand for the mutant DNA section.
f. Write the mRNA sequence of mutant DNA, using the template strand in part **e.**
g. What amino acids are placed in the beta chain by the mutation?
h. What type of mutation occurs in this form of thalassemia?
i. How might the properties of this section of the beta chain be different from the properties of the normal protein?
j. How might the level of structure in hemoglobin be affected if beta chains are not produced?

CI.39 In response to signals from the nervous system, the hypothalamus secretes a polypeptide hormone known as gonadotropin-releasing factor (GnRF), which stimulates the pituitary gland to release other hormones into the bloodstream. Two of these hormones are luteinizing hormone (LH) in males, and follicle-stimulating hormone (FSH) in females. GnRF is a decapeptide with the following primary structure:

Glu–His–Tyr–Ser–Tyr–Gly–Leu–Arg–Pro–Gly

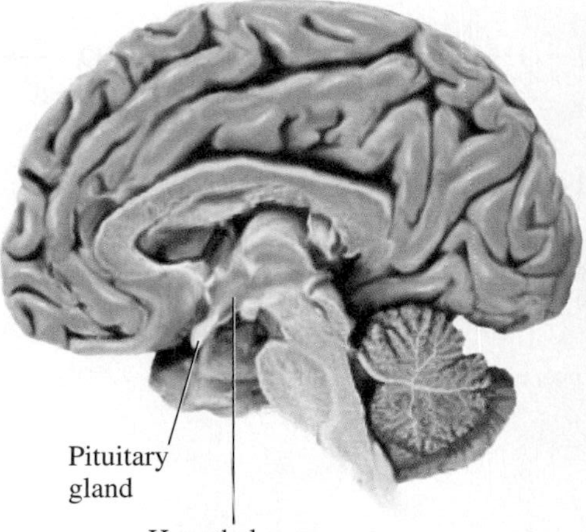

The hypothalamus secretes GnRF.

a. What is the N terminal amino acid?
b. What is the C terminal amino acid?
c. Which amino acids are nonpolar or polar neutral?
d. Draw the condensed structural formulas of the acidic or basic amino acids at physiological pH.

Answers

CI.31 a. An alpha-galactosidase is a hydrolase.
b. The substrate is the α-1,4-glycosidic bond of galactose.
c. High temperatures will denature the hydrolase enzyme so it no longer functions.

CI.33 a. citric acid cycle
b. electron transport
c. both
d. electron transport
e. both
f. citric acid cycle
g. both

CI.35 a. aminoethanethiol, pantothenic acid, phosphorylated adenosine diphosphate
b. Coenzyme A carries an acetyl group to the citric acid cycle for oxidation.
c. The acetyl group links to the S atom in the aminoethanethiol part of CoA.
d. 809 g/mole

CI.37 a.

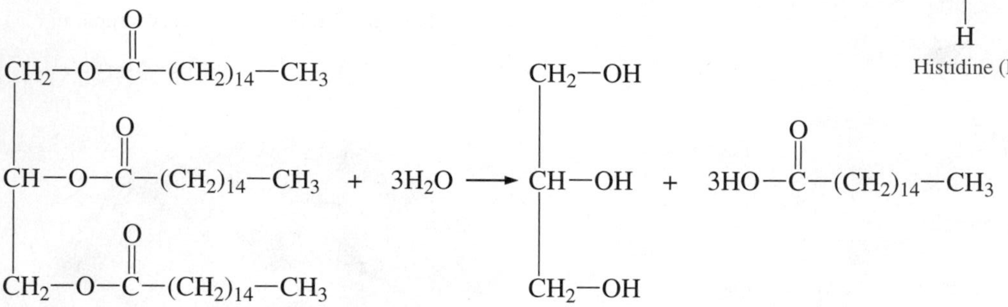

$$\begin{array}{l} CH_2-O-\overset{O}{\overset{\|}{C}}-(CH_2)_{14}-CH_3 \\ CH-O-\overset{O}{\overset{\|}{C}}-(CH_2)_{14}-CH_3 \\ CH_2-O-\overset{O}{\overset{\|}{C}}-(CH_2)_{14}-CH_3 \end{array} + 3H_2O \longrightarrow \begin{array}{l} CH_2-OH \\ CH-OH \\ CH_2-OH \end{array} + 3HO-\overset{O}{\overset{\|}{C}}-(CH_2)_{14}-CH_3$$

b. glyceryl tripalmitate, 807 g/mole
c. 129 ATP
d. 40. kcal
e. 14 pats of butter

CI.39 a. glutamic acid (Glu)
b. glycine (Gly)
c. The nonpolar amino acids are glycine (Gly), leucine (Leu), and proline (Pro). The polar neutral amino acids are serine (Ser) and tyrosine (Tyr).
d. The acidic amino acid is glutamic acid (Glu) and the basic amino acids are histidine (His) and arginine (Arg). At the pH of the body (about 7.4), glutamic acid will have a negative charge while histidine and arginine will be positively charged.

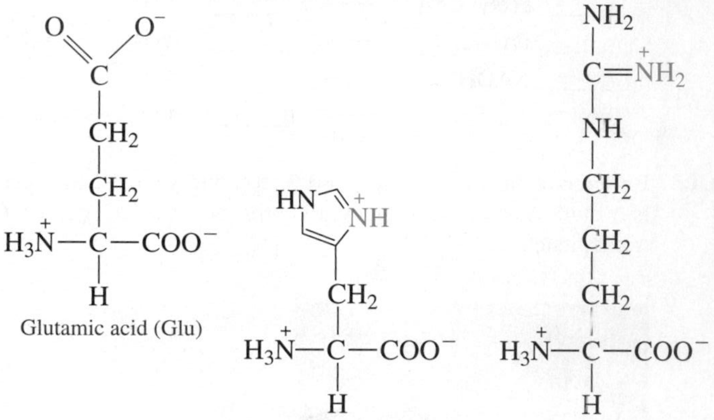

$$H_3\overset{+}{N}-\underset{H}{\underset{|}{C}}(CH_2-CH_2-COO^-)-COO^-$$

Glutamic acid (Glu)

$$H_3\overset{+}{N}-\underset{H}{\underset{|}{C}}(CH_2-\text{imidazolium: }HN\text{—}\overset{+}{N}H)-COO^-$$

Histidine (His)

$$H_3\overset{+}{N}-\underset{H}{\underset{|}{C}}(CH_2-CH_2-CH_2-NH-\overset{NH_2}{\overset{|}{C}}=\overset{+}{N}H_2)-COO^-$$

Arginine (Arg)

CREDITS

Unless otherwise acknowledged, all photographs are the property of Benjamin Cummings Publishers, Pearson Education.

Chapter 1

p. 2 *top:* Patrick Clark/Getty Images, Inc.
p. 2 *bottom:* Shutterstock.
p. 3 *middle:* iStockphoto.
p. 3 *left:* The Alchemist's Workshop, 1570, Jan van der Straet (Jannes Stradanus), Palazzo Vecchio, Florence, Italy/Bridgeman Art Library.
p. 3 *right:* Erich Lessing/Art Resource.
p. 4 *top:* Eric Schrader–Pearson Science.
p. 4 *bottom:* Photos.com.
p. 5 iStockphoto.
p. 6 Design Pics Inc./Alamy.
p. 8 iStockphoto.
p. 9 National Institute of Standards and Technology.
p. 10 *top:* Richard Megna–Fundamental Photographs.
p. 11 Photolibrary.com.
p. 11 Anatomy, University La Sapienza, Rome/Science Photo Library.
p. 12 Science Photo Library/Alamy.
p. 15 iStockphoto.
p. 17 Shutterstock.
p. 18 Shutterstock.
p. 22 *top:* Shutterstock.
p. 22 *bottom:* Shutterstock.
p. 23 *middle:* iStockphoto.
p. 26 Eric Schrader–Pearson Science.
p. 27 *top:* Eric Schrader–Pearson Science.
p. 28 *top:* Eric Schrader–Pearson Science.
p. 28 *bottom:* Carlos Alvarez/iStockphoto.
p. 32 Shutterstock.
p. 35 *bottom left:* Prof. P. La Motta, Dept. Anatomy, University La Sapienza, Rome/Photo Researchers, Inc.
p. 35 *bottom middle:* Prof. La Motta, Dep. Anatomy, University La Sapienza, Rome/Photo Researchers, Inc.
p. 35 *bottom right:* Phanie/Photo Researchers, Inc.
p. 36 iStockphoto.
p. 37 *top:* Michael Wright.
p. 37 *bottom:* Shutterstock.
p. 43 Cristian Ciureanu/Alamy.

Chapter 2

p. 46 *right:* iStockphoto.
p. 46 *left:* Shutterstock.
p. 47 *left:* Richard Megna–Fundamental Photographs.
p. 47 *top:* Eric Schrader–Pearson Science.
p. 48 *left:* iStockphoto.
p. 48 *middle right:* Shutterstock.
p. 48 *middle:* Dorling Kindersley.
p. 48 *bottom:* Dorling Kindersley.
p. 49 *top:* Richard Megna–Fundamental Photographs.
p. 49 *right:* Eric Schrader–Pearson Science.
p. 49 *bottom:* Stephen Frink Collection/Alamy.
p. 50 *middle:* Siede Preis.
p. 50 *left:* Raphel Auvray/SuperStock, Inc.
p. 50 *right:* iStockphoto.
p. 51 *top:* Shutterstock.
p. 51 *left:* Getty Images. - PhotoDisc.
p. 51 *middle:* Getty Images. - PhotoDisc.
p. 51 *right:* Mark Downey/Getty Images-Photodisc.
p. 52 *top:* Westend61 GmbH/Alamy.
p. 52 *bottom:* Shutterstock.
p. 54 *top:* iStockphoto.
p. 54 *bottom:* iStockphoto.
p. 55 iStockphoto.
p. 57 Digital Vision/Alamy.
p. 63 iStockphoto.
p. 56 NASA.
p. 66 Getty Images - Photodisc.
p. 69 *bottom:* Helene Canada/iStockphoto.
p. 70 *top:* iStockphoto.
p. 71 Comstock Complete.
p. 77 *middle:* Masterfile Royalty Free Division.
p. 77 *right:* Spencer Jones/Getty Images.
p. 78 *top left:* iStockphoto.
p. 78 *bottom left:* iStockphoto.
p. 78 *top right:* Jonathan Wood/Getty Images.
p. 78 *bottom right:* Mark Dahners/AP.
p. 80 Shutterstock.
p. 83 *top:* Photolibrary.
p. 83 *middle:* Plush Studios/Getty Images.
p. 83 *bottom:* Eric Schrader–Pearson Science.
p. 84 *left:* Photolibrary
p. 84 *right:* Masterfile.

Chapter 3

p. 88 Mary Ann Sullivan.
p. 89 *left:* iStockphoto.
p. 89 *right:* Shutterstock.
p. 92 *right:* Shutterstock.
p. 95 Getty Images/Digital Vision.
p. 96 *top:* Russ Lappa - Prentice Hall School Division.
p. 96 *bottom:* IBM Research, Almaden Research Center.
p. 100 *left:* OSF/Photolibrary.
p. 107 *top:* iStockphoto.
p. 107 *bottom:* Shutterstock.
p. 111 Public domain, US Government.
p. 121 iStockphoto.

Chapter 4

p. 126 iStockphoto.
p. 127 Shutterstock.
p. 132 *left:* Richard Megna–Fundamental Photographs.
p. 134 Eric Schrader–Pearson Science.
p. 136 Shutterstock.
p. 137 Fotolia.
p. 139 *top:* AP Photo/Eckehard Schulz.
p. 139 *bottom:* Norma Zuniga/Getty Images.
p. 140 Shutterstock.

Chapter 5

p. 169 Gary Rutherford/Photo Researchers, Inc.
p. 170 Charles D. Winters/Photo Researchers, Inc.
p. 171 *bottom:* Michael Hill/iStockphoto.
p. 173 *bottom:* Shutterstock.
p. 175 *top:* Eric Schrader–Pearson Science.
p. 175 *bottom:* Shutterstock.
p. 176 Tamara Kulikova/iStockphoto.
p. 182 Richard Megna–Fundamental Photographs.
p. 185 *top:* Tom Bochsler–Pearson Education.
p. 187 *top:* Richard Megna–Fundamental Photographs.
p. 188 Patrick Clark/Getty Images.
p. 189 Getty Images.
p. 190 *top:* stinkyt/iStockphoto.
p. 191 Chris Price/iStockphoto.
p. 193 NASA.
p. 196 Richard Megna–Fundamental Photographs.
p. 198 Comstock.
p. 200 *top:* Richard Megna–Fundamental Photographs.
p. 205 *top:* Creative Digital Vision CDV LLC.
p. 208 *top:* iStockphoto.
p. 211 *left:* Eric Schrader–Pearson Science.
p. 211 *right:* iStockphoto.
p. 211 *bottom:* Alamy Images.
p. 212 *left:* Photolibrary.
p. 212 *bottom:* Moriori via Wikipedia.
p. 212 *right:* iStockphoto.

Chapter 6

p. 216 NASA.
p. 218 Shutterstock.
p. 224 Getty Images, Inc.-Photodisc./Royalty Free.
p. 232 Getty Images - Photodisc.
p. 234 iStockphoto.
p. 238 Yuri Bathan/iStockphoto.
p. 239 Phanie/Photo Researchers, Inc.
p. 243 Library of Congress.

Chapter 7

p. 248 *bottom:* Jupiter Images Royalty Free.
p. 256 *left:* Custom Medical Stock Photo, Inc.
p. 256 *right:* CMSP/Newscom.
p. 259 CNRI/Photo Researchers, Inc.

p. 271 iStockphoto.
p. 272 *top:* Shutterstock.
p. 272 *bottom:* Florida Images/Alamy.
p. 275 *left:* Sam Singer.
p. 275 *middle:* Sam Singer.
p. 275 *right:* Sam Singer.
p. 277 Hans Eick/DocStock/Photolibrary.
p. 280 *bottom:* Corbis Royalty Free.
p. 280 *top:* Jupiter Images Royalty Free.

Chapter 8

p. 286 iStockphoto.
p. 303 *bottom:* Michael Fernahl/iStockphoto.
p. 303 *top:* Kristen Brochmann/Fundamental Photographs.
p. 307 Creative Digital Vision, Pearson Education.
p. 313 iStockphoto.
p. 315 Peter Arnold/Photolibrary.
p. 317 *top left:* Center for Liquefied Natural Gas.
p. 317 *bottom left:* Shutterstock.
p. 317 *top right:* iStockphoto.
p. 317 *bottom right:* Eric Schrader–Pearson Science.
p. 318 *top left:* iStockphoto.
p. 318 *top right:* Alamy.
p. 318 *bottom:* Charles D. Winters/Photo Researchers, Inc.
p. 318 *middle:* Eric Schrader–Pearson Science.

Chapter 9

p. 323 *top:* iStockphoto.
p. 323 *bottom:* Health Protection Agengy/Photo Researchers, Inc.
p. 325 iStockphoto.
p. 330 Australian Nuclear Science and Technology Organisation.
p. 332 *top:* AP Photo/James MacPherson.
p. 332 *bottom:* Don Farral/Getty Images.
p. 333 *top:* Stanford Dosimetry.
p. 338 Corbis.
p. 337 JIHAD SIQLAWI/AFP/Getty Images.
p. 339 *left:* Phannie/Photo Researchers, Inc.
p. 339 *right:* Pasieka/Science Photo Library/Photo Researchers, Inc.
p. 340 *top:* Getty Images.
p. 340 *bottom:* Getty Images.
p. 341 GJLP/Photo Researchers, Inc.
p. 342 *left:* Cytyc Hologic Corporation.
p. 342 *right:* Cytyc Hologic Corporation.
p. 345 iStockphoto.
p. 350 AP Photo.

Chapter 10

p. 355 *left:* lillisphotography/iStockphoto.
p. 367 *top:* iStockphoto.
p. 367 *bottom:* AP Photo/Marc Dahners.
p. 368 *bottom:* AP Photo/Kamran Jebreili.
p. 370 Getty Images.
p. 379 S. Pearce/PhotoLink/Getty Images.
p. 380 *top left:* Image100/Photolibrary.
p. 380 *top right:* iStockphoto.com.
p. 380 *middle:* Dorling Kindersley.
p. 380 *bottom right:* iStockphoto.
p. 381 *top left:* iStockphoto.
p. 381 *top middle:* C Squared Studios/Getty Images.
p. 381 *middle:* Shutterstock.com.
p. 381 *middle left:* iStockphoto.
p. 381 *top right:* Eric Schrader–Pearson Science.
p. 381 *middle right:* Ariel Skelley/CORBIS.

Chapter 11

p. 386 Shutterstock.com.
p. 387 Getty Images.
p. 389 Shutterstock.
p. 394 Alastair Shay/Papilio/CORBIS.
p. 399 *top:* Southern Illinois University/ Photo Researchers, Inc.
p. 400 *bottom:* iStockphoto.
p. 403 Medical-on-Line/Alamy.
p. 406 *top left:* iStockphoto.
p. 406 *bottom left:* Getty Images.
p. 406 *top right:* Jupiter Images.
p. 406 *bottom right:* Shutterstock.
p. 408 Comstock Images.

Chapter 12

p. 416 Koihealth.org.
p. 419 iStockphoto.
p. 420 Al Assid/CORBIS.
p. 423 *bottom:* dbimages/Alamy.
p. 424 Marjan Laznik/iStockphoto.
p. 444 *top left:* iStockphoto.
p. 444 *middle left:* Eric Schrader–Pearson Science.
p. 444 *bottom left:* iStockphoto.
p. 444 *top right:* Getty Images.
p. 444 *bottom right:* Dorling Kindersley.
p. 447 Shutterstock.
p. 450 *right:* Shutterstock.
p. 450 *top:* Shutterstock.
p. 450 *left:* AlsoSalt.
p. 451 KAKIMAGE/Alamy.

Chapter 13

p. 463 Eric Schrader–Pearson Science.
p. 464 Alamy.
p. 470 Getty Images - PhotoDisc.
p. 473 *left:* Shutterstock.
p. 473 *right:* Getty Images - PhotoDisc.
p. 477 *top right:* Shutterstock.

Chapter 14

p. 483 *top:* iStockphoto.
p. 499 *bottom:* iStockphoto.
p. 501 *top:* Pichugin Dmitry/Shutterstock.
p. 501 *bottom:* Michael S. Yamashita/CORBIS.
p. 511 *left:* Photolibrary.
p. 511 *right:* Eric Schrader–Pearson Science.
p. 513 *left:* Michael Hill/iStockphoto.
p. 513 *right:* iStockphoto.

Chapter 15

p. 524 Photos.com.
p. 526 Tischenko Irina/Shutterstock.
p. 527 John Pitcher/iStockphoto.
p. 533 AP Photo/L. G. Patterson.
p. 539 *top:* National Heart, Lung, and Blood Institute.
p. 539 *bottom:* National Heart, Lung, and Blood Institute.
p. 540 Emmanuelm, via Wikipedia, used under a Creative Commons Licence, *http://creativecommons.org/licenses/by/3.0/*
p. 543: Getty Images.
p. 549 *left:* iStockphoto.
p. 549 *right:* iStockphoto.
p. 550 *left:* iStockphoto.
p. 551 *bottom:* Lauree Feldman/Photolibrary.com.
p. 555 *left:* David Zaitz/Alamy.
p. 555 *bottom:* Marcelo Gabriel Domenichelli/Shutterstock.
p. 556 Shutterstock.

Chapter 16

p. 558 iStockphoto.
p. 568 APP Pharmaceuticals.
p. 569 *bottom:* iStockphoto.com.
p. 569 *top:* Shutterstock.
p. 573 iStockphoto.
p. 578 Centers for Disease Control.
p. 584 iStockphoto.
p. 591 iStockphoto.
p. 594 *left:* iStockphoto.
p. 595 *top:* Eric Schrader–Pearson Science.
p. 595 *left:* Dorling Kindersley.
p. 595 *right:* Dorling Kindersley.

Chapter 17

p. 608 M. Freeman/PhotoLink/Getty Images - Photodisc.
p. 611 Library of Congress.
p. 622 iStockphoto.
p. 626 Ed Uthman via Flickr. Used under a Creative Commons License, *http://creativecommons.org/licenses/by/2.0/*

Chapter 18

p. 635 iStockphoto.
p. 638 iStockphoto.
p. 641 Custom Medical Stock Photography.
p. 651 Joel W. Rogers/CORBIS.
p. 660 Shutterstock.
p. 666 *left:* iStockphoto.
p. 666 *right:* iStockphoto.
p. 669 *right:* Custom Medical Stock Photography.
p. 669 *left:* Custom Medical Stock Photography.
p. 678 iStockphoto.
p. 679 *right:* iStockphoto.
p. 682 *top:* GlaxoSmithKline.
p. 682 *left:* Shutterstock.
p. 682 *right:* Eric Schrader–Pearson Science.

INDEX

A

C

D

E

F

G

H

J

K

N

P

METRIC AND SI UNITS AND SOME USEFUL CONVERSION FACTORS

Length SI unit meter (m)	Volume SI unit cubic meter (m^3)	Mass SI unit kilogram (kg)
1 meter (m) = 100 centimeters (cm)	1 liter (L) = 1000 milliliters (mL)	1 kilogram (kg) = 1000 grams (g)
1 meter (m) = 1000 millimeters (mm)	1 mL = 1 cm^3	1 g = 1000 milligrams (mg)
1 cm = 10 mm	1 L = 1.06 quart (qt)	1 kg = 2.20 lb
1 kilometer (km) = 0.621 mile (mi)	1 qt = 946 mL	1 lb = 454 g
1 inch (in.) = 2.54 cm (exact)		1 mole = 6.02×10^{23} particles
		Water
		density = 1.00 g/mL (at 4 °C)

Temperature SI unit kelvin (K)	Pressure SI unit pascal (Pa)	Energy SI unit joule (J)
$°F = 1.8(°C) + 32$	1 atm = 760 mmHg	1 calorie (cal) = 4.184 J
$°C = \frac{(°F - 32)}{1.8}$	1 atm = 101.325 kPa	1 kcal = 1000 cal
	1 atm = 760 torr	**Water**
$K = °C + 273$	1 mole of gas (STP) = 22.4 L	Heat of fusion = 334 J/g; 80. cal/g
		Heat of vaporization = 2260 J/g; 540 cal/g
		Specific heat (*SH*) = 4.184 J/g °C; 1.00 cal/g °C

PREFIXES FOR METRIC (SI) UNITS

Prefix	Symbol	Power of Ten
Values greater than 1		
tera	T	10^{12}
giga	G	10^{9}
mega	M	10^{6}
kilo	k	10^{3}
Values less than 1		
deci	d	10^{-1}
centi	c	10^{-2}
milli	m	10^{-3}
micro	μ	10^{-6}
nano	n	10^{-9}
pico	p	10^{-12}

FORMULAS AND MOLAR MASSES OF SOME TYPICAL COMPOUNDS

Name	Formula	Molar Mass (g/mole)	Name	Formula	Molar Mass (g/mole)
Ammonia	NH_3	17.0	Hydrogen chloride	HCl	36.5
Ammonium chloride	NH_4Cl	53.5	Iron(III) oxide	Fe_2O_3	159.8
Ammonium sulfate	$(NH_4)_2SO_4$	132.2	Magnesium oxide	MgO	40.3
Bromine	Br_2	159.8	Methane	CH_4	16.0
Butane	C_4H_{10}	58.1	Nitrogen	N_2	28.0
Calcium carbonate	$CaCO_3$	100.1	Oxygen	O_2	32.0
Calcium chloride	$CaCl_2$	111.0	Potassium carbonate	K_2CO_3	138.2
Calcium hydroxide	$Ca(OH)_2$	74.1	Potassium nitrate	KNO_3	101.1
Calcium oxide	CaO	56.1	Propane	C_3H_8	44.1
Carbon dioxide	CO_2	44.0	Sodium chloride	$NaCl$	58.5
Chlorine	Cl_2	71.0	Sodium hydroxide	$NaOH$	40.0
Copper(II) sulfide	CuS	95.7	Sulfur trioxide	SO_3	80.1
Hydrogen	H_2	2.02	Water	H_2O	18.0